Formulas of Acoustics

Springer
Berlin
Heidelberg
New York
Barcelona
Hong Kong
London
Milan
Paris
Tokyo

F. P. Mechel (Ed.)

Formulas of Acoustics

With contributions by
M. L. Munjal · M. Vorländer · P. Költzsch · M. Ochmann
A. Cummings · W. Maysenhölder · W. Arnold · O. V. Rudenko

With approximate 620 Figures and 70 Tables

Springer

Editor:

Prof. Dr. Fridolin P. Mechel

Landhausstrasse 12
71120 Grafenau
Germany

ISBN 3-540-42548-9 Springer-Verlag Berlin Heidelberg New York

Library of Congress Cataloging-in-Publication Data
Formulas of acoustics / F.P. Mechel ; with contributions by W. Arnold ... [et al.]
p. cm. Includes bibliographical references and index.
ISBN 3540425489
1. Sound--Mathematics--Handbooks, manuals, etc. I. Mechel, Fridolin P.
QC228.8 .F67 2002 534'.01'51--dc21 2002017904

Springer-Verlag Berlin Heidelberg New York
a member of BertelsmannSpringer Science+Business Media GmbH

http://www.springer.de

Typesetting: Camera ready by author
Cover-design: de'blik, Berlin
Printed on acid-free paper SPIN: 10831453 62 / 3020 Hu - 5 4 3 2 1 0 -

A
Editorial

A.1 Preface :

Modern acoustics is more and more based on computations, and computations are based on formulas. Such work needs previous and contemporary results. It consumes much time and effort to search needed formulas during the actual work. Therefore, fundamentals and results of acoustics that can be expressed as formulas will be collected in this book.

The author has compiled a private formula collection during all his professional life. This personal collection gave the stimulus to expand it and to make it available to colleagues.

The present formula collection is subdivided into fields of acoustics. For some fields, in which this author is not expert enough, he invited co-authors to contribute. Most colleagues contacted for possible contributions were convinced of the project and agreed spontaneously.

The material within a field of acoustics is subdivided in sections which deal with a defined task. Some overlap of sections has to be tolerated; but the subdivision into well-defined sections will be helpful to the reader to find a particular topic of interest.

The present formula collection should not be considered a textbook in a condensed form. Derivations of a presented result will be described only as far as they are helpful in understanding the problem; the more interested reader is referred to the "source" of the result. Useful principles and computational procedures will also be included, even if they need more description. Symbols and quantities will be well defined, and wherever useful a sketch will help to explain the object and the task.

One of the advantages of a formula collection is seen in uniform definitions, notations and symbols for quantities. A strict uniformity in the form of a central list of symbols used never works, according to this author's observation. Therefore, only commonly used symbols (such as medium density, speed of sound, circular frequency, etc.) are collected in a central list of symbols; other symbols are defined in the relevant chapter.

Many sections contain, below their title or in the text, a reference to the literature. It cannot be the task and intention of this book either to indicate time priorities of publications concerning a topic or to give a survey of the existing literature. The reference quoted is the source of more information, which the author has used.

Computation and formulas are a form of mathematics. Higher transcendental functions are used in the mathematics of acoustics. Functions will be explained by reference to mathematical literature, if necessary. If functions are used with different definitions in the literature, the definition used will be presented. For more details of higher transcendental functions the reader is referred to mathematical formula collections and handbooks.

To the author's knowledge, this is the first formula collection existing in acoustics, and, like every first realisation of an idea, it is an experiment. Also for the publisher, who has limited the first edition to about 1200 pages. This frame is one of the reasons why not all important results from the literature are contained in the book; another reason is the fact that the author(s) had no knowledge of such results. If this book will be accepted by the "acoustics community" further editions surely will follow without page number limitation. Colleagues who wish to find formulas in a future edition which not yet are included are kindly invited to give suitable indications to the author.

Nevertheless the authors think that the book in its present form contains most of traditional and modern results of both fundamental and special character so that the book can be helpful to researchers and engineers in the fields of physical acoustics, noise control, and room acoustics.

The manuscript was written in a camera ready form (in order to avoid proof reading). So printing errors are the responsibility of the author. He would be grateful for indications of such errors.

The author gratefully acknowledges the support given to the project by the co-authors and by the publisher.

Grafenau, October 2001

A.2 Contents

A. Editorials

B. General Linear Fluid Acoustics

E. Scattering of Sound

G. Porous Absorbers

H. Compound Absorbers

I. Sound Transmission

J. Duct Acoustics

K. Acoustic Mufflers

M. Room Acoustics

N. Flow Acoustics

O. Analytical and Numerical Methods in Acoustics

P. Variational Principles in Acoustics

Q. Elasto-Acoustics

R. Ultrasound Absorption in Solids

S. Nonlinear Acoustics

Index

A.3
Conventions

The following conventions will be valid in the book. Exceptions will be clearly noted.

Time factor

- The time factor for harmonic oscillations and waves is $e^{j\omega t}$.

This choice implies that the imaginary part of impedances with mass reaction are positive, with spring reaction negative, and the imaginary part of admittances with mass reaction are negative, with spring reaction positive.

- If not stated differently, the time function is assumed to be $e^{j\omega t}$; the time factor then is dropped, mostly.

Impedance and admittance

- "Impedance" is used for the ratio of sound pressure p to the vector component in some direction of particle velocity v : $Z=p/v$.
- "Mechanical impedance" is used for the ratio of vector components of force F to particle velocity v : $Z_m=F/v$.
- "Flow impedance" is used for the ratio of sound pressure p to volume flow $q=S\cdot v$ through a surface S , with v the velocity component normal to S .
- "Admittance" is the ratio of the vector $\vec{v}$ of particle velocity to sound pressure p . It is a true vector.
- "Mechanical admittance" is the reciprocal of mechanical impedance.
- "Flow admittance" is the ratio of the flow vector $\vec{q}$ to sound pressure; it is a true vector.

Sound intensity and power

Sound intensity is the vector $\vec{I} = p \cdot \vec{v}\,^*$ (where the asterisk indicates the complex conjugate); $\vec{I}$ stands for the momentary sound power in the direction of $\vec{v}$ through a unit surface. The effective intensity is the real part thereof in the time average. (The formally possible definition $\vec{I} = p^* \cdot \vec{v} = \vec{I}\,^*$ would produce conflicts at sound sources.)

Sound power is the integral of the scalar product of sound intensity with the surface element vector $d\vec{s}$ over a surface S: $\Pi = \int_S \vec{I} \cdot d\vec{s}$.

Dimensions

Where necessary, the dimension of a quantity is indicated in brackets [...] .

Complex quantities

Field quantities, such as sound pressure p, particle velocity v, oscillating parts of density ρ, and of temperature T, etc. are mostly complex. If you record such a quantity in an oscillogram, you may take either the real or the imaginary part of a complex expression, after multiplication with the dropped time factor. If you record the amplitude, this corresponds to taking the magnitude of the complex quantity.

Symbol "decorations"

Unnecessary symbol decorations, such as hats for amplitudes, underbars for complex quantities, etc. are avoided. If necessary in the local context, an arrow indicates a vector $\vec{v}$, a star is used for the complex conjugate p^* , primes are used either for the derivative of functions, $f'(x), f''(x)$, or (where no ambiguity is possible) for the real and imaginary parts of complex quantities.

Commonly used symbols

These symbols are commonly used in most sections of the book. If a section uses the same symbol with a different definition, it will be noted.

c_0 adiabatic sound speed in the medium (e.g. in air) [m/s];

f frequency [Hz];

j $=\sqrt{(-1)}$ imaginary unit;

k_0 $=\omega/c_0$ free field wave number of a plane wave [1/m];

p sound pressure [Pa/m^2];

q volume flow [1/(m·s)];

t time [sc];

v velocity [m/s];

Z_0 $=\rho_0 c_0$ wave impedance of free plane wave [Pa·s/m];

α sound absorption coefficient;

κ adiabatic exponent of the medium;

λ wavelength [m];

λ_0 wavelength of free plane wave;

ρ mass density [kg/m^3];

ρ_0 mass density of the medium;

ω $=2\pi/f$ angular (or circular) frequency [1/s];

ϑ polar angle of cylindrical and spherical co-ordinates;

φ azimuthal angle of spherical co-ordinates;

Π power;

Ξ $\Delta p/(d\cdot v)$ flow resistivity of porous material [Pa·s/m^2]
(flow resistance per unit thickness d);

Numbering of equations

Equations are numbered, beginning with number (1) in each Section. Reference to an equation is made as, e.g., "eq.(x)" to the equation with number (x) in the same Section, or as, e.g., "equ.(K.y.x)" to the equation with number (x) in the Section with number y of the Chapter K .

B
General Linear Fluid Acoustics

The medium does not support shear stresses, except viscous shear. The medium parameters are constant in time; stationary flow does not exist, or its velocity is low enough, to be neglected in its influence on the sound field; see the chapter "N. Flow Acoustics" for sound fields in flows.

B.1 Fundamental differential equations

No viscous and/or caloric losses:

Conservation of mass: $$\frac{\partial \rho}{\partial t} + \rho_0 \operatorname{div} \vec{v} = \rho_0 q \cdot \delta(r - r_q) \quad (1)$$

Conservation of impulse: $$\rho_0 \frac{\partial v}{\partial t} = -\operatorname{grad} p \quad (2)$$

Equation of state: $$p = c_0^2 \cdot \rho \quad (3)$$

Relation between pressure and particle velocity: $$\vec{v} = \frac{j}{k_0 Z_0} \operatorname{grad} p \quad (4)$$

Homogeneous wave equation for a harmonic wave: $$\Delta p + k_0^2 \cdot p = 0 \quad (5)$$

Wave equation for harmonic wave with monopole source at r_q: $$\left(\Delta + k_0^2\right) p = -j k_0 Z_0 \cdot q \cdot \delta(r - r_q) \quad (6)$$

Adiabatic sound velocity: $$c_0^2 = \frac{\kappa P_0}{\rho_0} \quad (7)$$

p=	sound pressure;
$\vec{v}$	particle velocity;
ρ=	density;
Δ=	Laplace operator;
P_0=	atmospheric pressure
q=	volume flow density of monopole source;
r=	space co-ordinate;
r_q=	source position;
δ=	Dirac delta function

for other symbols, see "Conventions"

Boundary conditions: (on both sides of the boundary) matching of sound pressures, and matching of normal particle velocities;

or : matching of phase velocities parallel to boundary, and of normal field admittances on both sides of boundary.

Medium with viscous and caloric losses :

Ref.: Mechel, [B.1]

Field quantities, pressure p , density ρ and (absolute) temperature T, are composed of stationary parts (with index $_0$) and oscillating parts (with index $_1$). Velocities v are oscillating particle velocities. The sound field is composed of three coupled waves: the density wave (index ρ), the viscous shear wave (index ν), the heat wave (index α).

Impulse equation:
$$\frac{\partial v}{\partial t}+\frac{1}{\rho_0}\,\mathrm{grad}\,p_1-\nu\,\Delta\vec{v}-\frac{1}{3}\nu\,\mathrm{grad\,div}\,\vec{v}=0 \tag{8}$$

Heat balance:
$$\frac{\partial T_1}{\partial t}+(\kappa-1)T_0\,\mathrm{div}\,\vec{v}-\alpha\,\Delta T_1=0 \tag{9}$$

Conservation of mass:
$$\frac{\partial \rho_1}{\partial t}+\rho_0\,\mathrm{div}\,\vec{v}=0$$

Equation of state:
$$\frac{p_1}{p_0}-\frac{\rho_1}{\rho_0}-\frac{T_1}{T_0}=0 \tag{10}$$

Heat conduction inside a bounding medium (index i):
$$\frac{\partial T_{i1}}{\partial t}-\alpha_i\,\Delta T_{i1}=0 \tag{11}$$

p_0= atmospheric pressure;
ρ_0= stationary density;
T_0= absolute temperature;
c_0= adiabatic speed of sound;
κ= adiabatic exponent;
ν= kinematic viscosity;
α= temperature conductivity; $=\Lambda/(\rho_0 c_p)$;
Λ= heat conductivity;
c_p= specific heat at constant pressure

Field composition with potentials (according to Rayleigh)
$$\vec{v}=-\mathrm{grad}\,\Phi+\mathrm{rot}\,\vec{\Psi} \tag{12}$$

Φ is a scalar potential ;
$\vec{\Psi}$ is a vector potential with
$$\mathrm{rot}(\mathrm{grad}\,\Phi)\equiv 0 \quad ; \quad \mathrm{div}\,\vec{\Psi}\equiv 0 \tag{13}$$

With vector identity
$$\Delta=\mathrm{grad\,div}-\mathrm{rot\,rot} \tag{14}$$

one gets:
$$-\mathrm{grad}[j\omega\Phi-\frac{p_1}{\rho_0}-\frac{4}{3}\nu\Delta\Phi]+\mathrm{rot}[j\omega\vec{\Psi}-\nu\Delta\vec{\Psi}]\equiv 0 \tag{15}$$

Both terms vanish individually (according to Rayleigh)
$$j\omega\Phi-\frac{p_1}{\rho_0}-\frac{4}{3}\nu\Delta\Phi=0 \quad ; \quad j\omega\vec{\Psi}-\nu\Delta\vec{\Psi}=0 \tag{16}$$

Equivalent to wave equations:
$$(\Delta+k_\nu^2)\vec{\Psi}=0 \quad ; \quad (\Delta+k_\rho^2)(\Delta+k_\alpha^2)\Phi=0 \tag{17}$$

Characteristic wave numbers :

for viscous wave :
$$k_\nu^2=-j\frac{\omega}{\nu} \tag{18}$$

for density wave k_ρ and thermal wave k_α :

$$\left.\begin{matrix} k_\rho^2 \\ k_\alpha^2 \end{matrix}\right\} = j\omega \frac{-[\frac{c_0^2}{\omega} + j(\alpha + \frac{4}{3}\nu)] \pm \sqrt{[\frac{c_0^2}{\omega} + j(\alpha + \frac{4}{3}\nu)]^2 - 4j\alpha(\frac{c_0^2}{\kappa\omega} + \frac{4}{3}j\nu)}}{2\alpha(\frac{c_0^2}{\kappa\omega} + \frac{4}{3}j\nu)} \tag{19}$$

Approximations to wave numbers: $$\left.\begin{matrix} k_\rho^2 \\ k_\alpha^2 \end{matrix}\right\} \approx j\frac{\kappa\omega}{2\alpha}\left(-1 \pm \sqrt{1 - 4j\frac{\alpha\omega}{\kappa c_0^2}}\right) \tag{20}$$

or with lower degree of precision: $k_\rho^2 \approx (\omega / c_0)^2 \quad ; \quad k_\alpha^2 \approx -j\kappa\omega / \alpha$

Decomposition of scalar potential for density wave Φ_ρ and thermal wave Φ_α :

$$\Phi = \Phi_\rho + \Phi_\alpha$$

with wave equations: $$(\Delta + k_\rho^2)\Phi_\rho = 0 \quad ; \quad (\Delta + k_\alpha^2)\Phi_\alpha = 0 \tag{21}$$

Relative variation of density: $$\frac{\rho_1}{\rho_0} = \frac{j}{\omega}\left[k_\rho^2\Phi_\rho + k_\alpha^2\Phi_\alpha\right] \tag{22}$$

Relative variation of pressure: $$\frac{p_1}{p_0} = \Pi_\rho\Phi_\rho + \Pi_\alpha\Phi_\alpha$$

with sound pressure coefficients: $$\Pi_{\rho,\alpha} = \frac{\kappa}{c_0^2}\left(j\omega + \frac{4}{3}\nu k_{\rho,\alpha}^2\right) = \frac{jk_{\rho,\alpha}^2}{\omega}\frac{\kappa\omega - j\alpha k_{\rho,\alpha}^2}{\omega - j\alpha k_{\rho,\alpha}^2} \tag{23}$$

Relative variation of temperature: $$\frac{T_1}{T_0} = \Theta_\rho\Phi_\rho + \Theta_\alpha\Phi_\alpha$$

with temperature coefficients: $$\Theta_{\rho,\alpha} = \frac{4}{3}\frac{\kappa\nu}{c_0^2}k_{\rho,\alpha}^2 + j\left(\omega\frac{\kappa}{c_0^2} - \frac{k_{\rho,\alpha}^2}{\omega}\right) = \frac{j(\kappa - 1)k_{\rho,\alpha}^2}{\omega - j\alpha k_{\rho,\alpha}^2} \tag{24}$$

Approximations to wave numbers and coefficients:

with wave number definitions: $$k_0^2 = (\frac{\omega}{c_0})^2 \quad ; \quad k_\nu^2 = -j\frac{\omega}{\nu} \quad ; \quad k_{\alpha 0}^2 = -j\frac{\kappa\omega}{\alpha} \tag{25}$$

$$\left.\begin{matrix} k_\rho^2 \\ k_\alpha^2 \end{matrix}\right\} = \frac{[\frac{1}{k_0^2} + \frac{4}{3k_\nu^2} + \frac{\kappa}{k_{\alpha 0}^2}] \mp \sqrt{[\frac{1}{k_0^2} + \frac{4}{3k_\nu^2} + \frac{\kappa}{k_{\alpha 0}^2}]^2 - \frac{4\kappa}{k_{\alpha 0}^2}(\frac{1}{\kappa k_0^2} + \frac{4}{3k_\nu^2})}}{\frac{2\kappa}{k_{\alpha 0}^2}(\frac{1}{\kappa k_0^2} + \frac{4}{3k_\nu^2})}$$

$$\approx \tfrac{1}{2}k_{\alpha 0}^2 \cdot \left(1 \mp \sqrt{1 - 4\frac{k_0^2}{k_{\alpha 0}^2}}\right) \approx \begin{cases} k_0^2 \\ k_{\alpha 0}^2 \cdot (1 - k_0^2 / k_{\alpha 0}^2) \end{cases} \approx \begin{cases} k_0^2 \\ k_{\alpha 0}^2 \end{cases} \tag{26}$$

$$k_\alpha^2 \approx k_{\alpha 0}^2 \frac{1+\kappa(1+\frac{4\,\mathrm{Pr}}{3}-\frac{1}{\kappa})\frac{k_0^2}{k_{\alpha 0}^2}}{1+\frac{4\kappa^2\,\mathrm{Pr}}{3}\frac{k_0^2}{k_{\alpha 0}^2}} = k_{\alpha 0}^2 \frac{1+1{,}2165\frac{k_0^2}{k_{\alpha 0}^2}}{1+1{,}8259\frac{k_0^2}{k_{\alpha 0}^2}} \tag{27}$$

$$\frac{\Theta_\rho}{\Theta_\alpha} \approx -(\kappa-1)\frac{k_0^2}{k_{\alpha 0}^2} \tag{28}$$

Pr=ν/α Prandtl number

$$\Pi_\rho \approx \kappa\, j\frac{k_\rho^2}{\omega} \approx \kappa\, j\frac{k_0^2}{\omega}$$

$$\Pi_\alpha \approx \frac{jk_{\alpha 0}^2}{\omega}\,\frac{\kappa\,(1-\frac{4\kappa\,\mathrm{Pr}}{3})\frac{k_0^2}{k_{\alpha 0}^2}}{1+\kappa\frac{k_0^2}{k_{\alpha 0}^2}} \approx \frac{jk_0^2}{\omega}\,\kappa\,(1-\frac{4\kappa\,\mathrm{Pr}}{3}) \tag{29}$$

$$\frac{\Pi_\alpha}{\Pi_\rho} \approx 1-\frac{4\kappa\,\mathrm{Pr}}{3} = -0{,}3033 \tag{30}$$

Boundary conditions with v_t= tangential velocity, v_n= normal velocity, T_i= temperature behind the boundary; Λ= heat conductivity of the medium with the sound wave, Λ_i= heat conductivity of the medium behind the boundary:

$$v_t = 0 \quad ; \quad v_n = v_{n,i}$$
$$T_l = T_{l,i} \quad ; \quad \Lambda\frac{\partial}{\partial n}T_l = \Lambda_i\frac{\partial}{\partial n}T_{l,i} \tag{31}$$

Isothermal boundary condition: $T_1 = 0$ (32)

Adiabatic boundary condition: $\frac{\partial}{\partial n}T_l = 0$ (33)

B.2 Material constants of air

Ref.: Mechel, [B.1] ; VDI-Wärmeatlas, [B.2]

For definitions of symbols see Section B.1 of this Chapter and Table B.2.1. Regressions using measured data are given in the form $f(T) = \sum_{i=-4}^{4} a_i \cdot T^{i/2}$ for the material constants of dry air as functions of (absolute) temperature (in Kelvin degrees K) .

The atmospheric pressure is assumed to be P_0=1 [bar]=10^5 [Pa] . The range of application of the regressions is 100 K $\le T \le$ 1500 K .

V=	volume;
P=	static pressure;
ß=	coefficient of thermal volume expansion;

Interrelations are :

Prandtl number: $\mathrm{Pr} = \nu / \alpha$ (1)

Specific heat at constant volume: $c_v = c_p - \dfrac{\beta^2 T}{K\rho_0}$ (2)

Isothermal compressibility: $K = -\dfrac{1}{V}\left(\dfrac{\partial V}{\partial P}\right)_T = \dfrac{1}{\rho}\left(\dfrac{\partial \rho}{\partial P}\right)_T$ (3)

Quantity	Symbol	Value	Dimension	Remark
Molekular weight	M	28.96	kg/kmol	Dry air
Gas constant	R	287.10	J/kgK	Ideal gas
Density	ρ_0	1.1886	kg/m^3	
Sound velocity	c_0	343.30	m/s	$c_0^2=\kappa P_0/\rho_0$
Dynamical viscosity	η	$17.99 \cdot 10^{-6}$	N·s/m^2	
Kinematic viscosity	ν	$15.13 \cdot 10^{-6}$	m^2/s	$\nu=\eta/\rho_0$
Adiabatic exponent	κ	1.401	–	$\kappa=c_p/c_v$
Specific heat	c_p	$1.007 \cdot 10^{3}$	J/kgK	P const.
Temperature expansion	ß	$3.421 \cdot 10^{-3}$	1/K	
Heat cconductivity	Λ	0.02603	W/mK	
Temperature conductivity	α	$21.74 \cdot 10^{-6}$	m^2/s	$\alpha=\Lambda/(\rho_0 \cdot c_p)$
Prandtl number	Pr	0.6977	–	$\mathrm{Pr}=\nu/\alpha$

Table B.2.1 :
Material constants of air at standard conditions (20° C ; 1 bar)

Temperature dependence of Prandtl number:

$$\mathrm{Pr} = 0.66000 + 6.5853 \cdot 10^{-6} \cdot (T-700) + 3.97457 \cdot 10^{-7} \cdot (T-700)^2 - \\ -1.43416 \cdot 10^{-12} \cdot (T-700)^4 + 3.05114 \cdot 10^{-18} \cdot (T-700)^6 \quad (4)$$

Quantity:	a_o	$a_{\pm 1}$	$a_{\pm 2}$	$a_{\pm 3}$	$a_{\pm 4}$
ρ_o kg/m^3	–29.2987	1.38519 363.205	–0.0384181 $-2.08219\cdot 10^3$	$5.78952\cdot 10^{-4}$ $6.48716\cdot 10^3$	$-3.65858\cdot 10^{-6}$ $3.25451\cdot 10^3$
η Ns/m^2	$-3.30199\ 10^{-4}$	$1.39487\cdot 10^{-5}$ $4.35462\cdot 10^{-3}$	$-2.29854\cdot 10^{-7}$ –0.0294172	$1.43167\cdot 10^{-9}$ 0.0740619	$4.55963\cdot 10^{-12}$ 0.03768996
ν m^2/s	$1.04734\ 10^{-4}$	$-1.00547\cdot 10^{-5}$ $-2.16340\cdot 10^{-4}$	$4.03090\cdot 10^{-7}$ $-3.69703\cdot 10^{-3}$	$-3.87707\cdot 10^{-9}$ 0.0183863	$6.20832\cdot 10^{-11}$ $9.00314\cdot 10^{-3}$
κ –	25.9651	1.08207 –313.593	0.0273543 $2.04477\cdot 10^3$	$-3.79526\cdot 10^{-4}$ $-4.89956\cdot 10^3$	$2.26518\cdot 10^{-6}$ $-2.50299\cdot 10^3$
c_p J/kgK	$1.66918\ 10^3$	–983.174 $-1.29648\cdot 10^5$	32.7843 $4.55412\cdot 10^5$	–0.540032 $-1.55411\cdot 10^5$	$3.48332\cdot 10^{-3}$ $-1.17469\cdot 10^5$
Λ W/mK	7.13849	–0.400186 –72.5736	0.0129070 386.078	$-2.17156\cdot 10^{-4}$ –778.310	$1.4793566\cdot 10^{-6}$ –403.616
α m^2/s	0.0128841	$-7.09636\cdot 10^{-4}$ –0.133306	$2.21478\cdot 10^{-5}$ 0.723019	$-3.54709\cdot 10^{-7}$ –1.49085	$2.33895\cdot 10^{-9}$ –0.771509
ß 1/K	0.0762123	$4.36358\cdot 10^{-3}$ –0.695016	$1.37872\cdot 10^{-4}$ 3.55119	$-2.28121\cdot 10^{-6}$ 0.516673	$1.54530\cdot 10^{-8}$ –0.098786

Table B.2.2 : Regression coefficients for material data as functions of (absolute) Temperature T .

Example for measured data (points) and regression (curve) :

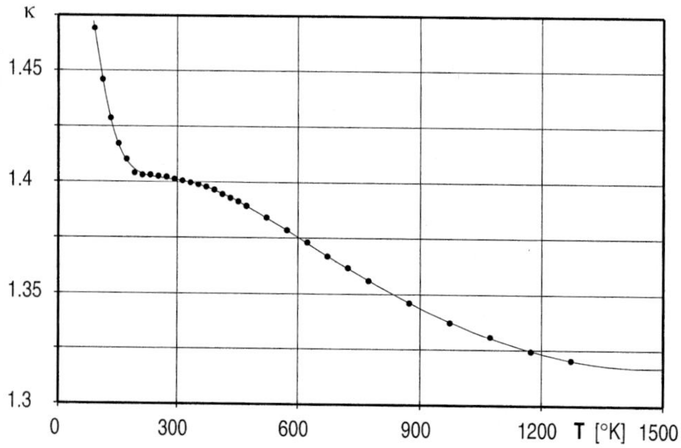

Fig. B.2.1 : Adiabatic exponent κ as function of absolute temperature T . Points: measured; curve: regression

Sound velocity:

$$c_0 = \sqrt{\frac{\kappa P_0}{\rho_0}} = \sqrt{\frac{\tilde{p}}{\tilde{\rho}}} = \frac{1}{\sqrt{K_0\rho_0}} \approx \sqrt{\frac{\langle v^2\rangle}{3}} = \sqrt{\frac{\kappa RT}{M}}$$

$$= 108.28\sqrt{\frac{T}{M}} \approx 333_{m/s} + 0.6\cdot\Theta_{°C} \approx 20{,}05\sqrt{T_{°K}} \tag{5}$$

Sound velocity of a mixture of two gas components:
(x is the concentration of the component with primes)

$$c_x^2 = \frac{RT}{x\cdot M+(1-x)\cdot M'}\,\frac{x\cdot c_p+(1-x)\cdot c_p'}{x\cdot c_p/\kappa+(1-x)\cdot c_p'/\kappa'} \tag{6}$$

κ= adiabatic exponent;
P_0= atmospheric pressure;
ρ_0= atmospheric density;
$\tilde{p}$= sound pressure;
$\tilde{\rho}$= oscillating density;
K_0=compressibility;
$<v^2>$= average square of molecular velocities;
R= universal gas constant;
M= molecular weight,
Θ= temperature in Celsius;
c_p= specific heat at constant pressure

B.3 General relation for field admittance and intensity

Ref.: Mechel, [B.3]

The vector component G_n in a direction n of the field admittance G is defined by :

$$G_n = \frac{v_n}{p} = \frac{j}{k_0 Z_0}\frac{\partial p/\partial n}{p} \tag{1}$$

If the sound pressure is described by magnitude and phase

$$p(r) = |\,p(r)\,|\cdot e^{j\varphi(r)} \tag{2}$$

the field admittance is given by:

$$G_n(r) = \frac{1}{k_0 Z_0}\left[-\frac{\partial}{\partial n}\varphi(r) + j\cdot\frac{\partial}{\partial n}\ln(|\,p(r)\,|)\right] \tag{3}$$

Near an absorbing wall the reactance of the wall admittance determines the slope of $\ln(|p(r)|)$.

The time averaged intensity of a harmonic wave is :

$$I_n = \tfrac{1}{2}\,p\cdot v_n^* = \tfrac{1}{2}\,G_n^*\cdot|\,p\,|^2 \tag{4}$$

With the admittance relation:

$$I_n = -\frac{|\,p(r)\,|^2}{2k_0 Z_0}\cdot\left[\frac{\partial\varphi(r)}{\partial n} + j\cdot\frac{\partial}{\partial n}\ln(|\,p(r)\,|)\right] \tag{5}$$

The real part thereof is the effective intensity, and the imaginary part the reactive intensity.

In vector notation:

$$\vec{G}(r) = \frac{1}{k_0 Z_0}\left[-\mathrm{grad}\,\varphi(r) + j\cdot\mathrm{grad}(|\,p(r)\,|)\right]$$

$$\vec{I} = -\frac{|\,p(r)\,|^2}{2k_0 Z_0}\cdot\left[\mathrm{grad}\,\varphi(r) + j\cdot\mathrm{grad}(|\,p(r)\,|)\right] \tag{6}$$

System of two coupled differential equations for magnitude and phase of sound pressure (with the relations from above)

$$\Delta|p(r)| + k_0^2\left(1 - Z_0^2 \cdot \mathrm{Re}\{G(r)\}\right) \cdot |p(r)| = 0$$

$$\Delta\varphi(r) - 2k_0^2 Z_0^2 \cdot \mathrm{Re}\{G(r)\} \cdot \mathrm{Im}\{G(r)\} = 0$$

If a sound field has no sources or sinks : $\mathrm{div}\,\mathrm{Re}\{I\}=0$. (7)

The effective intensity $I_{eff}=\mathrm{Re}\{I\}$ has the rotation :

$$\mathrm{rot}\, I_{eff} = \frac{-1}{k_0 Z_0} |p(r)| \cdot \mathrm{grad}\,\varphi(r) \times \mathrm{grad}(|p(r)|)$$

(with $\times$ for the cross product of vectors). It follows that $\mathrm{rot}\, I_{eff}=0$, if phase $\varphi(r)$ and magnitude $|p(r)|$ have parallel gradients (like in a plane wave).

B.4 Integral relations

Ref.: A.D.Pierce, [B.4], and other

Consider two different sound fields p_1,p_2 in a volume V with a bounding surface S (with outwards directed surface element $d\vec{s}$). Green's integral is then:

$$\iiint_V (p_1 \cdot \Delta p_2 - p_2 \cdot \Delta p_1)\, d\mathbf{r} = \oiint_S \left(p_1 \cdot \vec{\nabla} p_2 - p_2 \cdot \vec{\nabla} p_1\right) \cdot d\vec{s} \qquad (1)$$

The fields may differ either by different source strengths and/or locations, and/or by different boundary conditions on S , and/or are different forms (modes) for the same sources and boundaries. The surface S is either soft $(p(S)=0)$ or hard $(\partial p/\partial n=0)$ on parts S_0 , or locally reacting on parts S_a with surface admittance G , or parts S_∞ are at infinity, where the fields obey Sommerfeld's condition.

With the fundamental relations of Section B.1 it follows that:

$$\oiint_S p_1 \cdot \vec{v}_2 \cdot d\vec{s} - \oiint_S p_2 \cdot \vec{v}_1 \cdot d\vec{s} = \iiint_V p_1 \cdot q_2 \cdot \delta(r - r_2)\, dr - \iiint_V p_2 \cdot q_1 \cdot \delta(r - r_1)\, dr \qquad (2)$$

if the field p_1 has a source with volume flow q_1 at r_1 and the field p_2 has a source with volume flow q_2 at r_2 . Integration over the Dirac delta functions gives:

$$\oiint_S p_1 \cdot \vec{v}_2 \cdot d\vec{s} - \oiint_S p_2 \cdot \vec{v}_1 \cdot d\vec{s} = p_1(r_2) \cdot q_2 - p_2(r_1) \cdot q_1 \qquad (3)$$

The *reciprocity principle* follows, if both p_1, p_2 everywhere satisfy the same boundary conditions:

$$p_1(r_2) \cdot q_2 = p_2(r_1) \cdot q_1 \qquad (4)$$

If both fields are source-free:
$$\oint\!\!\oint_S p_1 \cdot \vec{v}_2 \cdot d\vec{s} - \oint\!\!\oint_S p_2 \cdot \vec{v}_1 \cdot d\vec{s} = 0 \qquad (5)$$

If, additionally, they satisfy the same boundary condition on a part, e.g. S_0, of the surface S :
$$\iint_{S_a+S_\infty} p_1 \cdot \vec{v}_2 \cdot d\vec{s} - \iint_{S_a+S_\infty} p_2 \cdot \vec{v}_1 \cdot d\vec{s} = 0 \qquad (6)$$

If S_a is hard for p_1 and/or soft for p_2 :
$$\iint_{S_a+S_\infty} p_1 \cdot \vec{v}_2 \cdot d\vec{s} - \iint_{S_\infty} p_2 \cdot \vec{v}_1 \cdot d\vec{s} = 0 \qquad (7)$$

and if they obey the same far field conditions :
$$\iint_{S_a} p_1 \cdot \vec{v}_2 \cdot d\vec{s} = 0 \qquad (8)$$

If one or both fields have sources, the relevant source terms appear on the right-hand sides.

B.5 Green's functions and formalism

Ref: Skudrzyk, [B.9]

In a loss-free medium it is convenient to formulate the wave equation for the sound pressure field p. The particle velocity is then
$$\vec{v} = \frac{j}{k_0 Z_0} \operatorname{grad} p \qquad (1)$$

Let r be a general co-ordinate.
The homogeneous wave equation is:
$$\Delta p(r) + k_0^2 \cdot p(r) = 0 \qquad (2)$$

The inhomogeneous wave equation with a source $q(r)$ of volume flow is:
$$\Delta p(r) + k_0^2 \cdot p(r) = -j k_0 Z_0 \, q(r) \qquad (3)$$

Here $q(r)$ is the rate of volume generation per unit volume and unit time.

The Green's formalism uses a potential function g for the field (instead of the sound pressure function)

i.e.
$$\vec{v} = -\operatorname{grad} g \quad ; \quad p = \rho_0 \frac{\partial g}{\partial t} \qquad (4)$$

The Green's function $g(r \mid r_q, \omega)$ is the solution of the inhomogeneous wave equation for a time harmonic
$$\Delta g(r \mid r_q, \omega) + k_0^2 \, g(r \mid r_q, \omega) = -\delta(r - r_q) \qquad (5)$$

excitation by a point source in r_q of unit strength, which satisfies specified boundary conditions.

With the Dirac delta function :
$$\delta(r - r_q) = \delta(x - x_q) \cdot \delta(y - y_q) \cdot \delta(z - z_q)$$

Any solution h(r) of the homogeneous wave equation, satisfying the boundary conditions, can be added to give a solution $G(r|r_q)$.
$$G(r \mid r_q) = g(r \mid r_q) + h(r) \tag{6}$$

The Green's function of a point source in free space is :

$$g(r \mid r_q, \omega) = \frac{e^{-jk_0R}}{4\pi R} \quad ; \quad R = \sqrt{(x - x_q)^2 + (y - y_q)^2 + (z - z_q)^2} \tag{7}$$

The volume flow of the source is given by :
$$\lim_{R \to 0} \left(-4\pi R^2 \frac{\partial g}{\partial R} \right) \tag{8}$$

From that it follows in three dimensions:
$$\lim_{r \to r_q} \left(g(r \mid r_q, \omega) \right) = \frac{1}{4\pi R} \tag{9a}$$

in two dimensions :
$$\lim_{r \to r_q} \left(g(r \mid r_q, \omega) \right) = \frac{1}{2\pi} \ln |\, r - r_q \,| \tag{9b}$$

in one dimension :
$$\left(\frac{\partial g}{\partial x} \right)_{x_q + \varepsilon} - \left(\frac{\partial g}{\partial x} \right)_{x_q - \varepsilon} = -1 \tag{9c}$$

i.e., the one-dimensional Green's function has a discontinuity in slope at $x = x_q$.

Green's functions are reciprocal :
$$g(r \mid r_q, \omega) = g(r_q \mid r, \omega) \tag{10}$$

The sound pressure field p(r,ω) in a finite space with given boundary conditions and a volume source distribution f(r,ω) has to be a solution of the wave equation :

$$\Delta p(r, \omega) + k_0^2 p(r, \omega) = -f(r, \omega) \tag{11}$$

The solution can be expressed with Green's functions of the infinite space as :

$$p(r, \omega) =$$
$$\iiint f(r_q, \omega) \cdot g(r \mid r_q, \omega)\, dV_q + \iint \left[g(r \mid r_q, \omega) \frac{\partial}{\partial n_q} p(r_q, \omega) - p(r_q, \omega) \frac{\partial}{\partial n_q} g(r \mid r_q, \omega) \right] dS_q \tag{12}$$

where the subscript q indicates the variable for differentiation and integration. This is the *Helmholtz Huygens equation.* The surface integral simplifies if either $g(r \mid r_q, \omega)$ or its normal derivative vanishes.

Green's functions are also defined for non-harmonic sources but with time functions with unit spectral density, i.e., for time the function $\delta(t-t_0)$:

$$g(r\,|\,r_q,t-t_0):=g(r,t\,|\,r_q,t_0):=\int_0^\infty g(r\,|\,r_q,\omega)\cdot e^{j\omega(t-t_0)}\,\frac{d\omega}{2\pi} \tag{13}$$

Then :

$$p(r,t)=\int_{-\infty}^{t} f(r_q,t_0)\cdot g(r\,|\,r_q,t-t_0)\,dt_0 \tag{14}$$

where $f(r_q,t_0)$ is the time function of the source having the spectrum $f(r,\omega)$.

The Green's function

$$g(r\,|\,r_q,\omega)=\frac{e^{-jk_0r+j\omega t}}{4\pi R} \tag{15a}$$

belongs to the time function

$$\frac{1}{4\pi R}\delta\left(t-r/c_0\right) \tag{15b}$$

Green's functions in closed spaces can be expanded in *modes* Ψ_n ; these are solutions of the homogeneous wave equation

$$\Delta\Psi_n+k_n^2\Psi_n=0 \tag{16}$$

satisfying the boundary conditions (and Sommerfeld's far field condition, if the space is infinite in one dimension). The wave number k_n (instead of k_0) recalls that in a finite size space harmonic solutions exist only if the frequency is a resonant frequency. The modes are orthogonal, and may be made orthonormal, i.e.

$$\int\Psi_n\cdot\Psi_m\,dV=\delta_{nm}=\begin{cases}0 & \text{if}\quad n\neq m\\ 1 & \text{if}\quad n=m\end{cases} \tag{17}$$

supposed the boundary conditions have one of the forms :

$$\Psi_n=0\quad\text{or}\quad\partial\Psi_n/\partial n=0\quad\text{or}\quad\Psi_n=-\alpha\cdot\partial\Psi_n/\partial n \tag{18}$$

Then Green's function can be expanded :

$$g(r\,|\,r_q,\omega)=\sum_n A_n\Psi_n=\sum_n\frac{\Psi_n(r)\cdot\Psi_n(r_q)}{k_n^2-k_0^2}\quad;\quad A_n=\frac{\Psi_n(r_q)}{k_n^2-k_0^2} \tag{19}$$

The residues at the poles $k_0=\pm k_n$ are

$$\pm\frac{1}{2k_n}\Psi_n(r)\cdot\Psi_n(r_q) \tag{20}$$

If the space is infinite, complex modes are convenient. The orthogonality integral then should be (instead of (17)):

$$\int\Psi_n\cdot\Psi_m^*\,dV=\delta_{nm}=\begin{cases}0 & \text{if}\quad n\neq m\\ 1 & \text{if}\quad n=m\end{cases} \tag{21}$$

(the asterisk indicates the complex conjugate). The Green's function is then :

$$g(r\,|\,r_q,\omega) = \sum_n \frac{\Psi_n(r)\cdot\Psi_n^*(r_q)}{k_n^2 - k_0^2} \tag{22}$$

If one sets the condition that the physical solution should be real, the relations follow :

$$\begin{aligned} &\Psi_n(r) = \Psi_{-n}^*(r) \\ &\Psi_n(r) + \Psi_{-n}(r) = \Psi_n(r) + \Psi_n^*(r) = 2\,\mathrm{Re}\{\Psi_n(r)\} \end{aligned} \tag{23}$$

The eigenvalues k_n need not be a discrete set of values, but may be continuous. Then the Green's function is :

$$g(r\,|\,r_q,\omega) = \frac{1}{2}\int_{-\infty}^{+\infty} \frac{\Psi(r)\cdot\Psi^*(r_q)}{k_n^2 - k_0^2}\left(\frac{\partial k_n}{\partial n}\right)^{-1} dk_n \tag{24}$$

The form of Green's functions for continuous eigenvalues is similar to *integral transforms* :

$$\begin{aligned} &S_F(x) = \int F(z)\cdot\Psi^*(x,z)\cdot w(z)\,dz \\ &F(z) = \int S_F(x)\cdot\Psi^*(x,z)\cdot w(x)\,dx \end{aligned} \tag{25}$$

The weight function $w(z)$ is often introduced by the co-ordinate system; generally it represents the density of eigenvalues in z space. The following orthogonality and normalizing relations are used :

$$\begin{aligned} &w(k)\int \Psi(k,z)\cdot\Psi^*(x,z)\cdot w(z)\,dz = \delta(k-x) \\ &w(k)\int \Psi(k,z)\cdot\Psi^*(x,\zeta)\cdot w(k)\,dk = \delta(z-\zeta) \end{aligned} \tag{26}$$

Especially, the Dirac delta function is represented by :

$$\delta(r-r_q) = \int_{-\infty}^{+\infty} \Psi(k,r)\cdot\Psi^*(x,r_q)\cdot w(k)\cdot w(r_q)\,dk \tag{27}$$

thus :

$$S_F(x) = \frac{-w(r_q)\cdot\Psi^*(x,r_q)}{k^2 - x^2} \tag{28}$$

and the Green's function becomes:

$$w(r_q)\cdot g(r,r_q\,|\,\omega)=F(r)=\int\limits_{-\infty}^{+\infty}\frac{\Psi(k,r)\cdot\Psi^*(x,r_q)\cdot w(k)\cdot w(r_q)}{k^2-x^2}\,dx \tag{29}$$

Some examples of Green's functions are given below.

A set of plane waves :

Substitute above $x\rightarrow\vec{\kappa}$ indicating a wave number vector; denote with $\vec{r}$ the co-ordinate vector of a point. A set of plane waves is represented by :

$$\Psi(\vec{\kappa},\vec{r})=A(\vec{\kappa})\cdot e^{-j\vec{\kappa}\cdot\vec{r}} \tag{30}$$

(with the scalar product $\vec{\kappa}\cdot\vec{r}$ in the exponent). The density $w(\kappa)$ is unity. The amplitudes are $A(\vec{\kappa})=1/(2\pi)$ (for normalisation). In a two-dimensional space (x,y) with $\vec{\kappa}\cdot\vec{r}=\kappa_x x+\kappa_y y$ the Green's function becomes :

$$g(r,r_q\,|\,\omega)=\int\limits_{-\infty}^{+\infty}\int\limits_{-\infty}^{+\infty}\frac{e^{-j\vec{\kappa}\cdot(\vec{r}-\vec{r}_q)}}{\kappa^2-k_0^2}\,\frac{d\kappa_x\,d\kappa_y}{4\pi^2}\quad;\quad \kappa^2=\kappa_x^2+\kappa_y^2 \tag{31}$$

It can be shown that the integral goes over to the Hankel function of the 2nd kind :

$$g(r,r_q\,|\,\omega)=\frac{-j}{4}H_0^{(2)}(k_0R)\quad;\quad R=|r-r_q| \tag{32}$$

Cylindrical waves :

A set of eigenfunctions of the Bessel differential equation is :

$$\begin{aligned}&\psi_n(r)=J_m(\alpha_n r/a)\quad;\quad J_m(\alpha_n)=0\\&k_n=(\alpha_n/a)\quad;\quad n=0,1,2,\ldots\end{aligned} \tag{33}$$

where α_n are zeros of the Bessel function of order m. The orthogonality relation is :

$$\int\limits_0^a J_m(\alpha_n r/a)\cdot J_m(\alpha_\ell r/a)\,r\,dr=\begin{cases}0 & ;\ \ell\neq n\\ -\dfrac{a^2}{2}J_{m+1}(\alpha_n)\cdot J_{m-1}(\alpha_n) & ;\ \ell=n\end{cases} \tag{34}$$

If $a\rightarrow\infty$ the eigenvalues become continuous. One sets

$$\Psi(k,z)=A\cdot J_m(kz)\quad;\quad A=1 \tag{35}$$

where A=1 follows from the normalisation.

Two-dimensional infinite space in polar co-ordinates :

Two-dimensional eigenfunctions of the wave equation satisfying the normalisation conditions (26) in polar co-ordinates (r,φ) are :

$$\sqrt{w(\kappa)\,w(r)}\,\Psi(\kappa,z)=\sqrt{\frac{\kappa r}{2\pi}}\,J_m(\kappa r)\cdot e^{-jm\varphi} \tag{36}$$

The Green's function becomes :

$$g(\vec{r},\vec{r}_q\,|\,\omega)=\frac{1}{8\pi^2}\sum_{m=0}^{\infty}\delta_m\cos(m(\varphi-\varphi_q))\int_{-\infty}^{+\infty}\frac{J_m(\kappa r)\cdot J_m(\kappa r_q)}{\kappa^2-k_0^2}\,\kappa\,d\kappa \tag{37}$$

One gets after evaluation of the integral :

$$g(\vec{r},\vec{r}_q\,|\,\omega)=\frac{-j}{4}\sum_{m=0}^{\infty}\delta_m\cos(m(\varphi-\varphi_q))\begin{cases}J_m(k_0 r)\cdot H_m^{(2)}(k_0 r_q) & ;\; r\le r_q\\ J_m(k_0 r_q)\cdot H_m^{(2)}(k_0 r) & ;\; r\ge r_q\end{cases} \tag{38}$$

Three-dimensional infinite space :

The Green's function is ;

$$g(\vec{r},\vec{r}_q\,|\,\omega)=\frac{e^{-jk_0R}}{4\pi R}\quad ;\quad R=|\vec{r}-\vec{r}_q| \tag{39}$$

Green's function in spherical harmonics :

In the spherical co-odinates $r,\ \varphi,\ \vartheta$ the Green's function is :

$$\begin{aligned} g(\vec{r},\vec{r}_q\,|\,\omega)&=\frac{e^{-jk_0R}}{4\pi R}=-\frac{jk_0}{4\pi}h_0^{(2)}(k_0R)\\ &=-\frac{jk_0}{4\pi}\sum_{n=0}^{\infty}(2n+1)\begin{cases}j_n(k_0 r)\cdot h_n^{(2)}(k_0 r_q) & ;\; r<r_q\\ j_n(k_0 r_q)\cdot h_n^{(2)}(k_0 r) & ;\; r>r_q\end{cases}\;.\\ &\quad\cdot\sum_{m=0}^{n}\delta_m\frac{(n-m)!}{(n+m)!}\cos(m(\varphi-\varphi_q))\cdot P_n^m(\cos\vartheta_q)\cdot P_n^m(\cos\vartheta)\end{aligned} \tag{40}$$

where $j_m(z),\ h_m^{(2)}(z)$ are spherical Bessel and Hankel functions, and $P_n^m(x)$ are associated Legendre functions.

If the source distance r_q goes to infinity one gets for the plane wave from the spherical directions φ_q,ϑ_q (in the spherical angles φ,ϑ)

$$e^{-j\vec{k}_0\cdot\vec{r}} = \sum_{n=0}^{\infty}(-j)^n(2n+1)\cdot j_n(k_0 r)\sum_{m=0}^{n}\delta_m\frac{(n-m)!}{(n+m)!}\cos(m(\varphi-\varphi_q))\cdot P_n^m(\cos\vartheta_q)\cdot P_n^m(\cos\vartheta) \tag{41}$$

Green's function in cylindrical co-ordinates :

In the cylindrical co-odinates r,ϑ,z the Green's function is :

$$g(\vec{r},\vec{r}_q\,|\,\omega) = \frac{-1}{8\pi^3}\int_0^{2\pi}d\alpha\int_{-\infty}^{+\infty}\int_{-\infty}^{+\infty}\frac{e^{+j\kappa_r r\cos(\alpha-\vartheta)-j\kappa_r r_q\cos(\alpha-\vartheta_q)-j\kappa_z(z-z_q)}}{k_0^2-\kappa_r^2-\kappa_z^2}\,\kappa_r\,d\kappa_r\,d\kappa_z \tag{42a}$$

Performing the integration over κ_z with $\kappa_z = \pm\sqrt{k_0^2-\kappa_r^2} = \pm\sigma$:

$$g(\vec{r},\vec{r}_q\,|\,\omega) = \frac{j}{8\pi^2}\int_0^{2\pi}d\alpha\int_0^{\infty}\frac{\kappa_r}{\sigma}e^{-j\kappa_r r\cos(\alpha-\vartheta)+j\kappa_r r_q\cos(\alpha-\vartheta_q)-j\kappa_z(z-z_0)}\cdot e^{-j\sigma|z-z_q|}\,d\kappa_r \tag{42b}$$

With the exponentials expressed by Bessel functions, one gets :

$$g(\vec{r},\vec{r}_q\,|\,\omega) = \frac{-j}{4\pi}\sum_{m\ge 0}\delta_m\cos(m(\vartheta-\vartheta_q))\int_0^{\infty}\frac{\kappa_r}{\sigma}J_m(\kappa_r r)\cdot J_m(\kappa_r r_q)\cdot e^{-j\sigma|z-z_q|}\,d\kappa_r \tag{43}$$

with :

$$\sigma = \begin{cases}\sqrt{k_0^2-\kappa_r^2} & \text{if } 0<\kappa_r<k_0\\ -j\sqrt{\kappa_r^2-k_0^2} & \text{if } 0<k_0<\kappa_r\end{cases} \tag{44}$$

For $r_q=0$ eq.(43) reduces to the term with $m=0$.

Point source above hard or soft plane :

The Green's function for a hard plane is :

$$g(\vec{r},\vec{r}_q\,|\,\omega) = \frac{e^{-jk_0 r}}{4\pi r} + \frac{e^{-jk_0 r'}}{4\pi r'} \tag{45}$$

where r is the distance from the source to the field point, and r' is the distance from the image source (in a mirror-reflected position relative to the plane) to the field point.

If the plane is soft :

$$g(\vec{r},\vec{r}_q\,|\,\omega)=\frac{e^{-jk_0r}}{4\pi r}-\frac{e^{-jk_0r'}}{4\pi r'} \tag{46}$$

Point source above a locally reacting plane :

The plane is at $x=0$; the source at x_q, y_q, z_q ; the image source at $-x_q, y_q, z_q$. Let $\vec{n}\cdot\vec{r}=n_x x+n_y y+n_z z$ be the scalar product of the wave direction vector $\vec{n}$ with the co-ordinate vector $\vec{r}$. A plane wave, reflected at the plane can be represented as :

$$p_r=R\cdot e^{-jk_0(n_x x+n_y y+n_z z)} \quad ; \quad R=\frac{\zeta n_x-1}{\zeta n_x+1} \tag{47}$$

where ζ is the normalised surface impedance of the plane.

The Green's function in Cartesian co-ordinates is :

$$g(\vec{r},\vec{r}_q\,|\,\omega)=\frac{1}{8\pi^3}\int_{-\infty}^{+\infty}\frac{e^{j[\kappa_x x+\kappa_y(y-y_q)+\kappa_z(z-z_q)]}}{\kappa^2-k_0^2}\left(e^{-j\kappa_x x_q}+R\,e^{+j\kappa_x x_q}\right)d^3\kappa$$

$$\kappa^2=\kappa_x^2+\kappa_y^2+\kappa_z^2 \quad ; \quad d^3\kappa=d\kappa_x\cdot d\kappa_y\cdot d\kappa_z \tag{48}$$

and in cylindrical co-ordinates :

$$g(\vec{r},\vec{r}_q\,|\,\omega)=$$

$$\frac{-j}{4\pi}\sum_{m\geq 0}\delta_m\cos(m(\vartheta-\vartheta_q))\int_0^\infty\frac{\kappa}{\kappa_x}J_m(\kappa' r)\cdot J_m(\kappa' r_q)\left(e^{-j\kappa_x|x-x_q|}+R\,e^{-j\kappa_x|x+x_q|}\right)d\kappa' \tag{49}$$

where $\kappa'^2=k_0^2-\kappa_x^2$.

An approximate expression (if field point and/or source are distant to the plane) is :

$$g(\vec{r},\vec{r}_q\,|\,\omega)=\frac{1}{4\pi}\left[\frac{e^{-jk_0|\vec{r}-\vec{r}_q|}}{|\vec{r}-\vec{r}_q|}+R\frac{e^{+jk_0|\vec{r}-\vec{r}_q'|}}{|\vec{r}-\vec{r}_q'|}\right] \tag{50}$$

where $\vec{r}$ is the vector of the field point, $\vec{r}_q$ the vector to the original source and $\vec{r}_q'$ the vector to the mirror source.

B.6 Orthogonality of modes in a duct with locally reacting walls

Ref.: Mechel, [B.6]

Consider a duct, the interior contour of which follows a co-ordinate surface of a separable system of co-ordinates, and whose contour surface is either totally or in parts locally reacting with an admittance G (the other parts are either hard or soft). Let the cross-section normal to the axial co-ordinate x be A ; r be the one- or two-dimensional co-ordinate normal to x .

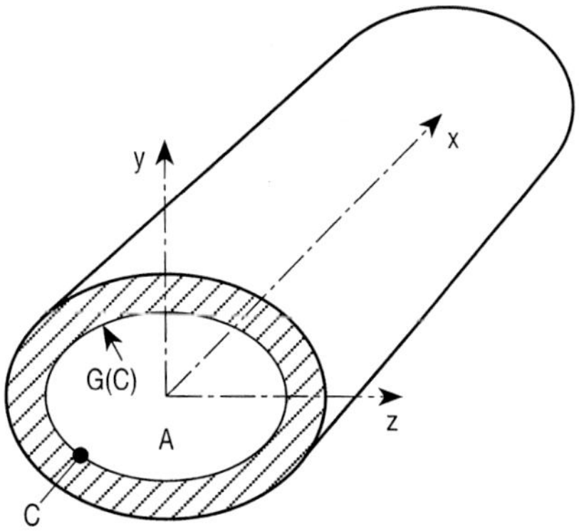

Let $p_m(x,r) = T_m(r) \cdot R_m(x)$
be a mode in the duct, i.e. a field which satisfies the homogeneous wave equation and the boundary conditions, with the transversal function $T_m(r)$ and the axial function $R_m(x)$.

Such modes are orthogonal over the cross section A , i.e. :

$$\int_A T_m(r) \cdot T_n(r) \cdot g(r)\, dr = \delta_{m,n} \cdot N_m \qquad (1)$$

g(r) is weight function which is induced by some co-ordinate systems; it is independent of the mode order m ; $\delta_{m,n}$ is the Kronecker symbol; N_m is the *norm* of the mode.

Orthogonality of modes, under the conditions mentioned, holds whatever the value of G is, and also if the medium in the duct has losses (i.e. k_0, Z_0 complex). They form a complete set of solutions (see MORSE/FESHBACH, part I, section 6.3, p.738 ff) if the defining boundaries normal to r are either hard or soft or locally reacting, and if in this case the derivative $\partial p/\partial r$ does not appear in the separated wave equation of the co-ordinate r . Modes may be one, two, or three dimensional to the number of pairs of walls that define the boundary conditions.

B.7 Orthogonality of modes in a duct with bulk reacting walls

Ref.: Mechel, Cummings [B.7]

Assume a duct like that in Section B.6 , but whose duct lining is laterally (bulk) reacting, and whose outer wall (behind the lining) is hard. The field in the interior volume of the duct, with cross-section A_1 , is marked with an index i=(1) , the field in the lining with an index i=(2), and its cross-section is A_2 . Let the characteristic wave number and wave impedance in A_1 be k_0, Z_0 ; the characteristic propagation constant and wave impedance of the lining material in A_2 be Γ_a, Z_a . The transversal functions of a mode $T_m^{(i)}(r)$ are different in the two sections; its axial function $R_m(x)$ is the same.

The modes are orthogonal over the cross-section A_1+A_2 with the mode norm in the case of a single homogeneous layer of the lining in a cylindrical duct:

$$N_m = \frac{1}{jk_0Z_0} \iint_{A_1} \left(T_m^{(1)}(r)\right)^2 dr + \frac{1}{\Gamma_a Z_a} \iint_{A_2} \left(T_m^{(2)}(r)\right)^2 dr \qquad (1)$$

In the case of multiple layers, an integral must be added for each layer.

B.8 Source conditions :

See also sections B.1, B.4, B.5 .

A special form of the boundary conditions, the source condition, must be satisfied, if the sound field p(r) is excited by a sound source. Commonly used are volume flow sources $q(r_q)$ either as a point source in 3-dimensional fields, or as a line source in 2-dimensional fields, or, more general, as a source distribution on a surface S_q , or as a source distribution in a volume V_q . In the case of distributed sources, $q(r_q)$ is the spatial density of emanating volume flow.

The source condition requires that the integral of the outward normal velocity over a small spherical surface around a point source, or over a narrow cylindrical surface around a line source, or on S_q around distributed sources, equals the given source strength q . This form of the source condition is sometimes difficult to evaluate. A form more suitable to evaluation shall be given.

First consider a *point source or a line source*. This case is illustrated with a line source (for simplicity); a point source is treated similarly. Let the source be located at (r_q,ϑ_q) in a cylindrical co-ordinate system (r,ϑ). In general ϑ stands for a co-ordinate over which orthogonal modes exist (i.e. the modes satisfy the homogeneous wave equation, Sommerfeld's condition, and the boundary conditions at the surfaces normal to ϑ). The line source is located at Q .

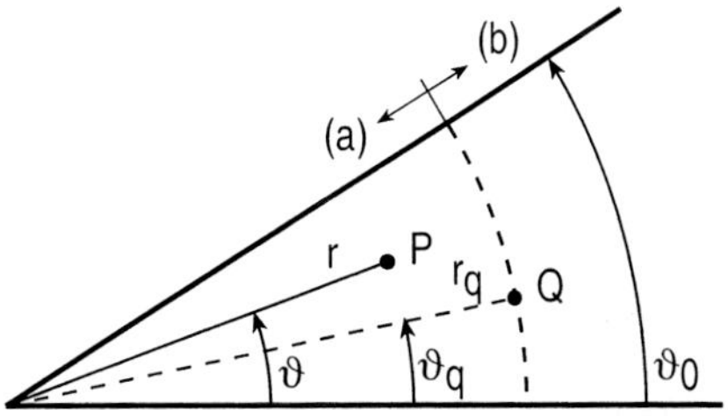

This defines two zones: zone (a) with $0\le r\le r_q$, and zone (b) with $r_q\le r<\infty$.

The modes have the form
$$p_m(r,\vartheta)=T(\eta_m\vartheta)\cdot R_m(r) \tag{1}$$
They are orthogonal over $(0,\vartheta_0)$ with norms N_m .

The radial functions $R_m(r)$ are formulated so that they are continuous at $r=r_q$, but discontinuous in their radial derivatives.

The source condition can be written in the form :
$$Z_0\left[v_r(r_q+0)-v_r(r_q-0)\right]\stackrel{!}{=}\frac{k_0Z_0\,q}{k_0r_q}\cdot\delta(\vartheta-\vartheta_q) \tag{2}$$

The Dirac delta function can be expanded in modes:
$$\delta(\vartheta-\vartheta_q)=\sum_{m\ge 0}b_m\cdot\cos(\eta_m\vartheta) \tag{3}$$

By application of
$$\int_0^{\vartheta_0}\ldots\cdot\cos(\eta_m\vartheta)\,d\vartheta \tag{4}$$

it follows that:
$$\delta(\vartheta-\vartheta_q)=\frac{1}{(2)\vartheta_0}\sum_{m\ge 0}\frac{\cos(\eta_m\vartheta_q)}{N_m}\cdot\cos(\eta_m\vartheta) \tag{5}$$

The factor (2) is applied, if the source is on a boundary, else $(2)\to 1$.

If the sources $q(\vartheta)$ are *distributed* over the surface at $r=r_q$, this distribution is synthesised with the mode norms.

B.9 Sommerfeld's condition

Ref.: Skudrzyk, [B.9]

If the field extends to infinity, it must approach zero there, unless it is a plane wave in a loss-free medium. A sufficient condition is a medium with losses.

Otherwise :
$$\lim_{r\to\infty}\left[r\left(\frac{\partial p}{\partial r}+j\,k_0 p\right)\right]=0 \qquad (1)$$

A weaker but simpler condition is, with A an arbitrary constant :
$$\lim_{r\to\infty}|r\,p|<A$$

B.10 Principles of superposition

Ref.: Ochmann/Donner; Mechel, [B.10]

Some principles of superposition may help to reduce more general problems to a repetition of simpler standard tasks.

First principle of superposition:

Two opposite walls, which are normal to the same co-ordinate, are locally reacting with different admittances G_1, G_2. The sound fields at the walls have the corresponding indices 1,2 .

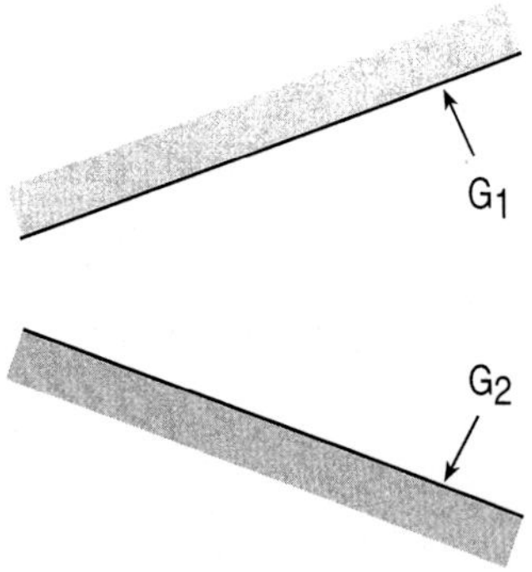

The boundary conditions at these surfaces are (with normal particle velocity components):

$$v_1 = G_1 \cdot p_1$$
$$v_2 = G_2 \cdot p_2$$

Set (with G_1, G_2 selected so that $\mathrm{Re}\{G_a\}\geq 0$) : G_s is the symmetrical, and G_a the antisymmetrical part of the boundary conditions.

$$G_s = \frac{1}{2}(G_1 + G_2)$$
$$G_a = \frac{1}{2}(G_1 - G_2) \qquad (1)$$

Suppose the sound fields p_s, p_a are known for symmetrical linings G_s, G_a, respectively, on each side;

$$v_s = G_s \cdot p_s = \frac{1}{2}(G_1 + G_2)\cdot p_s$$
$$v_a = G_a \cdot p_a = \frac{1}{2}(G_1 - G_2)\cdot p_a \qquad (2)$$

i.e. with the boundary conditions at both flanks : p_s is the symmetrical solution to G_s, p_a the antisymmetrical solution to G_a ; both types of solutions exist for symmetrical boundary conditions.

It follows immediately that:

$$\begin{aligned} v_{s1,2} + v_{a1,2} &= G_1 \cdot (p_{s1,2} + p_{a1,2}) \\ v_{s1,2} - v_{a1,2} &= G_2 \cdot (p_{s1,2} - p_{a1,2}) \end{aligned} \qquad (3)$$

If one compares this with the boundary conditions of the original task, one sees the correspondence :

$$p_1 = p_{s1,2} + p_{a1,2}$$
$$p_2 = p_{s1,2} - p_{a1,2}$$

The desired solution is evidently $p=p_s+p_a$, because both lines formally merge at the walls.

Second principle of superposition:

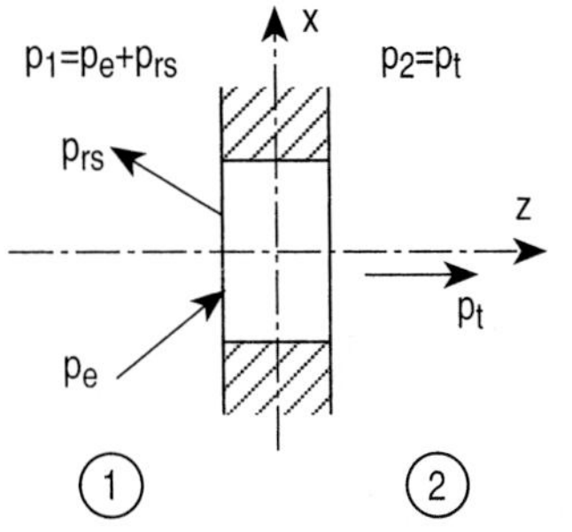

Suppose the object has a plane of symmetry. The medium is steady across the plane of symmetry, and no sound transmissive foil or sheet is in this plane. Let a co-ordinate z be normal to the plane of symmetry, directed from the side of incidence to the side of transmission, with $z=0$ on the plane of symmetry. Co-ordinates transversal to z are represented by x .

An index 1 marks the half-space with the incident wave p_e and a reflected and/or backscattered wave p_{rs} ; an index 2 marks the half-space with the transmitted wave p_t .
The fields in the two half-spaces are $p_1(x,z)=p_e(x,z)+p_{rs}(x,z)$, $p_2(x,z)=p_t(x,z)$.

Replace the original task by two sub-tasks; in the first one the sound transmissive parts of the plane of symmetry are assumed to be hard, in the second one they are assumed to be soft. Both conditions are marked by upper indices (h),(s), respectively. Solve the problems of reflection and/or backscattering for the two sub-tasks.

The sound field components of the original task are :

$$\begin{aligned} p_{rs}(x,z) &= \frac{1}{2}\left(p_{rs}^{(h)}(x,z) + p_{rs}^{(s)}(x,z)\right) \quad ; \quad z \le 0 \\ p_t(x,z) &= \frac{1}{2}\left(p_{rs}^{(h)}(x,-z) - p_{rs}^{(s)}(x,-z)\right) \quad ; \quad z \ge 0 \end{aligned} \qquad (4)$$

Third principle of superposition :

The task:

Find the sound field p_a with (part of) the boundaries absorbent with local reaction, described by a wall admittance G.

Suppose the solutions are known for the same source and geometry, but all walls ideally reflecting, i.e. either hard or soft or mixed.

The third principle of superposition composes p_a with such solutions.

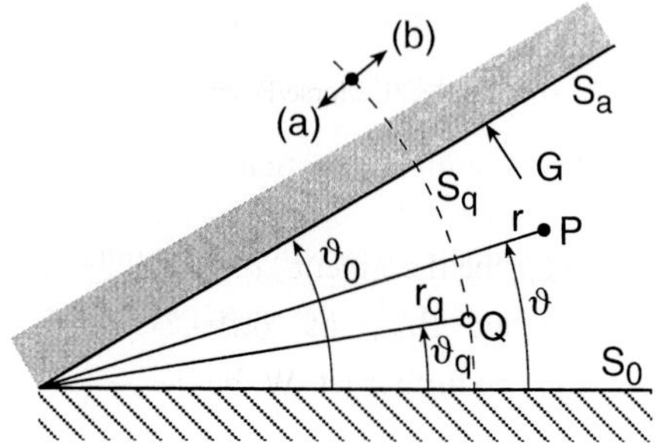

The example assumes a line source at Q in a wedge-shaped space with one hard flank at $\vartheta=0$, and one locally absorbing flank at $\vartheta=\vartheta_0$. The standard situation with a soft flank at $\vartheta=0$ is treated similarly; other situations are treated after application of the first and second principles of superposition.

It is supposed, that the field p_h is known, for which the flank at $\vartheta=\vartheta_0$ is hard, as well as p_s with a soft flank at $\vartheta=\vartheta_0$. Both fields further shall satisfy the source condition at Q individually (see Section B.8).

The desired field p_a is:

$$p_a(r,\vartheta)=\frac{1}{1+G\cdot X(r)}\left[p_h(r,\vartheta)+G\cdot X(r)\cdot p_s(r,\vartheta)\right]$$

with the "cross impedance"

$$X(r)=\frac{p_h(r,\vartheta_0)}{v_{sn}(r,\vartheta_0)}=-jk_0Z_0\frac{p_h(r,\vartheta_0)}{\mathrm{grad}_n p_s(r,\vartheta_0)} \qquad (6)$$

The index n indicates the vector component normal to the absorbing wall and directed into it. $X(r)$ is an impedance, formed with the sound pressure if the flank at $\vartheta=\vartheta_0$ is hard divided by the normal particle velocity if the flank is soft.

The third principle of superposition returns an exact solution, if X is constant with respect to the co-ordinate on the absorbing wall (r in the example); otherwise an approximation to p_a is obtained.

B.11 Hamilton's principle

Ref.: Cremer/Heckl; Morse/Feshbach, [B.11]

Let E_{kin} be the kinetic (effective) energy of a vibrating system, associated with oscillating masses, and E_{pot} its (effective) potential energy, associated with displacements against stresses; be further let W be the (effective) work done by external forces on the system.

Lagrange function: $L = E_{kin} - E_{pot}$

Hamilton's principle: If the system starts to oscillate from reasonable initial conditions, the form of oscillation which it assumes is such that the time average of its Lagrange function is an extreme, if the form of the oscillation is varied (δ stands for such variations) :

$$\delta \int_{t_1}^{t_2} L\,dt + \int_{t_1}^{t_2} \delta W\,dt = 0 \tag{1}$$

If the work W of external forces is constant over time intervals, and time average values of L and W are used : $\delta\langle L\rangle + \delta\langle W\rangle = 0$. If the system is adiabatic, i.e. W=0 , Hamilton's principle requires $\delta\langle L\rangle = 0$. The form of the system's oscillation is governed by amplitudes either of system elements or of field components, such as modes. The variation is applied to these amplitudes a_m . On the other hand, many systems have to obey boundary conditions, which are constraints in terms of variational methods. These boundary conditions are formulated as equations $g_k(a_m)=0$, and they are introduced into Hamilton's principle using the *Lagrange multipliers* λ_k (see MORSE/FESHBACH, part I, section 3.1) , leading to the form of Hamilton's principle suited for application to mechanical systems :

$$\frac{1}{T}\int_0^T \left(E_{kin} - E_{pot}\right)dt + \sum_k \left(\lambda_k^* \cdot g_k + \lambda_k \cdot g_k^*\right) = \min \tag{2}$$

The λ_k are treated in the application of the principle like the amplitudes a_m , i.e. they are parts of the variation.

This expression is formulated as a function $f(a_m,\lambda_k)$. The energies will be sums with products $a_m \cdot a_n^*$ as factors. The minimum is found where the equations hold :

$$\frac{\partial f}{\partial a_n^*} = 0 \quad ; \quad \frac{\partial f}{\partial \lambda_k^*} = 0 \tag{3}$$

This gives a set of linear equations for the a_m,λ_k .

In distributed systems and/or wave fields, the integration is not only over time, but also over space. If the sound field is described by a velocity potential function $\psi(r)$, the *Lagrange density* Λ is :

$$L = \iiint_V \Lambda(r)\,dr$$

$$\Lambda(r) = E_{kin}(r) - E_{pot}(r) = \frac{\rho_0}{2}\left[|\operatorname{grad}\psi|^2 - \frac{1}{c_0^2}\left(\frac{\partial\psi}{\partial t}\right)^2\right] = \frac{\rho_0}{2}\left[|\operatorname{grad}\psi|^2 + k_0^2\,\psi^2\right] \quad (4)$$

B.12 Adjoined wave equation

L and Λ here must not be confused with these symbols in Section B.11.

The wave equation is a second order linear differential equation, with p,q possibly functions of r .

$$L(f(r)) = f''(r) + p\cdot f'(r) + q\cdot f(r) = 0 \quad (1)$$

The *adjoined wave* equation is :

$$\Lambda(g(r)) = g''(r) - p\cdot g'(r) + (q - p')\cdot g(r) = 0 \quad (2)$$

Both satisfy the identity :

$$g\cdot L(f) - f\cdot\Lambda(g) = \frac{d\,P(g,f)}{dr} \quad (3)$$

P(g,f) is the *bilinear concomitant*.
If $\Lambda(g)=0$ can be solved,
then solutions of $L(f)=0$ are :

$$f_1(r) = g(r)\cdot e^{-\int p\,dr}$$

$$f_2(r) = f_1(r)\cdot\int \frac{e^{-\int p(s)\,ds}}{g^2}\,dr \quad (4)$$

The general solution is $f(r)=a\cdot f_1(r)+b\cdot f_2(r)$.

In the special case $q(r)=dp(r)/dr$:

$$g(r) = \int e^{-\int p(s)\,ds}\,dr$$

$$f_1(r) = \frac{g(r)}{g'(r)} \;;\quad f_2(r) = \frac{1}{g'(r)} \quad (5)$$

B.13 Vector and tensor formulation of fundamentals

Co-ordinate systems :

Let (x_1,x_2,x_3) be a rectilinear, orthogonal co-ordinate system. The vector components of a point P are given by $\vec{R} = \vec{OP} = [x_1, x_2, x_3]$

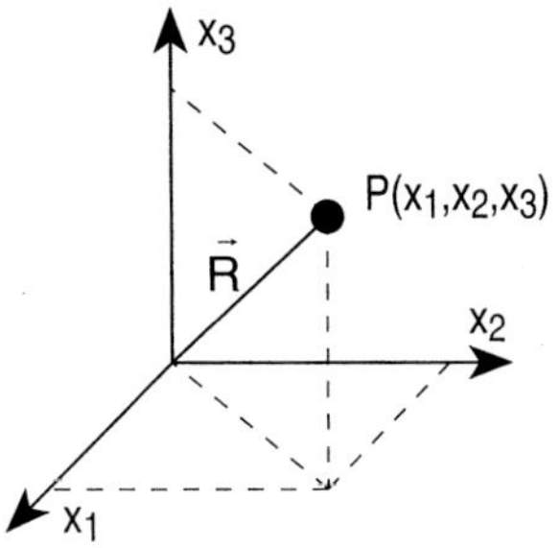

Let (u_1,u_2,u_3) be a curvilinear, orthogonal co-ordinate system. The co-ordinate surfaces are given by :

$$u_1(x_1,x_2,x_3) = \text{const}$$
$$u_2(x_1,x_2,x_3) = \text{const}$$
$$u_3(x_1,x_2,x_3) = \text{const}$$

The intersection of two co-ordinate surfaces is a co-ordinate line.

Tangent vectors at co-ordinate lines :

$$\vec{R}_1 = \frac{\partial \vec{R}}{\partial u_1} = \left[\frac{\partial x_1}{\partial u_1}, \frac{\partial x_2}{\partial u_1}, \frac{\partial x_3}{\partial u_1}\right]$$
$$\vec{R}_2 = \frac{\partial \vec{R}}{\partial u_2} = \left[\frac{\partial x_1}{\partial u_2}, \frac{\partial x_2}{\partial u_2}, \frac{\partial x_3}{\partial u_2}\right] \qquad (1)$$
$$\vec{R}_3 = \frac{\partial \vec{R}}{\partial u_3} = \left[\frac{\partial x_1}{\partial u_3}, \frac{\partial x_2}{\partial u_3}, \frac{\partial x_3}{\partial u_3}\right]$$

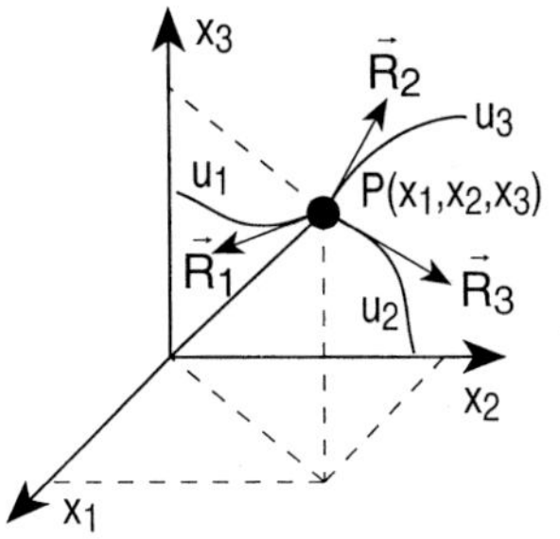

Normal vectors on co-ordinate surfaces :

$$\vec{N}^1 = \text{grad}\, u_1 = \left[\frac{\partial u_1}{\partial x_1}, \frac{\partial u_1}{\partial x_2}, \frac{\partial u_1}{\partial x_3}\right]$$
$$\vec{N}^2 = \text{grad}\, u_2 = \left[\frac{\partial u_2}{\partial x_1}, \frac{\partial u_2}{\partial x_2}, \frac{\partial u_2}{\partial x_3}\right] \qquad (2)$$
$$\vec{N}^3 = \text{grad}\, u_3 = \left[\frac{\partial u_3}{\partial x_1}, \frac{\partial u_3}{\partial x_2}, \frac{\partial u_3}{\partial x_3}\right]$$

If the $\vec{R}_i$ form the basis vectors of a system of co-ordinates, then the $\vec{N}^i$ are the basis of the "reciprocal" system, with :

$$\vec{R}_i \bullet \vec{N}^k = \delta_{i,k} \quad ; \quad \delta_{i,k} = \begin{cases} 1\,; i = k \\ 0\,; i \neq k \end{cases}$$

with the "dot product" or "scalar product".

Unitary tensors:

$$g_{ik} = \vec{R}_i \bullet \vec{R}_k = g_{ik} \quad \text{covariant co-ordinates}$$
$$g_i^k = \vec{R}_i \bullet \vec{N}^k \quad \text{mixed co-ordinates} \qquad (3)$$
$$g^{ik} = \vec{N}^i \bullet \vec{N}^k = g^{ki} \quad \text{contravariant co-ordinates}$$

with $g^{ij} \bullet g_{jk} = g^{i1} \cdot g_{1k} + g^{i2} \cdot g_{2k} + g^{i3} \cdot g_{3k} = \delta_{i,k}$. The determinant of g_{ik} is the square of the scalar triple product $g = \det(g_{ik}) = \vec{R}_1 \vec{R}_2 \vec{R}_3$.

Vector components of a vector $\vec{a}$:

covariant components : $a_i = \vec{a} \bullet \vec{R}_i$

contravariant components: $a^i = \vec{a} \bullet \vec{R}^i$ (4)

Vector representation in a covariant basis : $\vec{a} = a^1 \cdot \vec{R}_1 + a^2 \cdot \vec{R}_2 + a^3 \cdot \vec{R}_3 = \vec{a}^i \bullet \vec{R}_i$

Vector representation in a contravariant basis : $\vec{a} = a_1 \cdot \vec{R}^1 + a_2 \cdot \vec{R}^2 + a_3 \cdot \vec{R}^3 = \vec{a}_i \bullet \vec{R}^i$

It follows that :

$$a^i = g^{i1} a_1 + g^{i2} a_2 + g^{i3} a_3 = g^{ij} a_j$$
$$a_i = g_{i1} a^1 + g_{i2} a^2 + g_{i3} a^3 = g_{ij} a^j \qquad (5)$$

where the last notations use the "summation rule" (summation over multiple indices).

Thus : $\vec{N}^i = g^{ij} \vec{R}_j$; $\vec{R}_i = g_{ij} \vec{N}^j$ (6)

Transformation between systems of co-ordinates $U(u^1, u^2, u^3) \to V(v^1, v^2, v^3)$:

With definitions :

$$\Delta = \frac{\partial\left(v^1, v^2, v^3\right)}{\partial\left(u^1, u^2, u^3\right)} = \begin{vmatrix} \frac{\partial v^1}{\partial u^1} & \frac{\partial v^1}{\partial u^2} & \frac{\partial v^1}{\partial u^3} \\ \frac{\partial v^2}{\partial u^1} & \frac{\partial v^2}{\partial u^2} & \frac{\partial v^2}{\partial u^3} \\ \frac{\partial v^3}{\partial u^1} & \frac{\partial v^3}{\partial u^2} & \frac{\partial v^3}{\partial u^3} \end{vmatrix} = \det\left(A_k^i\right) \neq 0 \qquad (7)$$

and

$$A_k^i = \frac{\partial v^i}{\partial u^k} \quad ; \quad B_i^k = \frac{\partial u^k}{\partial v^i} \qquad (8)$$

it follows that :

$$\Delta^{-1} = \det\left(B_i^k\right)$$
$$\sum_{j=1}^{3} A_j^i \cdot B_k^j = \sum_{j=1}^{3} B_j^i \cdot A_k^j = \delta_{i,k} \qquad (9)$$

$$\vec{R}_i = \sum_k B_i^k \vec{R}_i = \sum_k A_i^k \vec{R}_k \quad ; \quad \vec{N}^i = \sum_k A_k^i \vec{N}^k = \sum_k B_k^i \vec{N}^k$$
$$A_i^k = \vec{R}_i \bullet \vec{N}^k \quad ; \quad B_i^k = \vec{R}_k \bullet \vec{N}^i \tag{10}$$

and :

$$a^i = \sum_k A_k^i a^k = \sum_k B_k^i a^k \quad ; \quad a_i = \sum_k B_i^k a_k = \sum_k A_i^k a_k \tag{11}$$

Vector algebra :

Consider the vectors :

$$\vec{a} = \sum_i a^i \cdot \vec{R}_i = \sum_i a_i \cdot \vec{R}^i \quad ; \quad \vec{b} = \sum_i b^i \cdot \vec{R}_i = \sum_i b_i \cdot \vec{R}^i \quad ; \quad \vec{c} = \sum_i c^i \cdot \vec{R}_i = \sum_i c_i \cdot \vec{R}^i \tag{12}$$

Scalar product :

$$\vec{a} \bullet \vec{b} = \sum_i a^i b_i = \sum_i a_i b^i = \sum_{i,k} g_{ik} a^i b^k = \sum_{i,k} g^{ik} a_i b_k = \frac{-1}{g} \begin{vmatrix} g_{11} & g_{12} & g_{13} & b_1 \\ g_{21} & g_{22} & g_{23} & b_2 \\ g_{31} & g_{32} & g_{33} & b_3 \\ a_1 & a_2 & a_3 & 0 \end{vmatrix} \tag{13}$$

Length of a vector :

$$a = |\vec{a}| = \sqrt{\vec{a} \bullet \vec{a}} = \sum_{i,k} \sqrt{g_{ik} a^i a^k} = \sum_{i,k} \sqrt{a^i a_k} = \sum_{i,k} \sqrt{g^{ik} a_i a_k} \tag{14}$$

Cosine of the *angle between two vectors* :

$$\cos(\vec{a}, \vec{b}) = \frac{\vec{a} \bullet \vec{b}}{ab} = \frac{\sum_{i,k} g_{ik} a^i b^k}{\sum_{i,k} \sqrt{g_{ik} a^i a^k} \sum_{i,k} \sqrt{g_{ik} b^i b^k}} \tag{15}$$

Vector (cross) product :

$$\vec{a} \times \vec{b} = \sqrt{g} \begin{vmatrix} \vec{R}^1 & \vec{R}^2 & \vec{R}^3 \\ a^1 & a^2 & a^3 \\ b^1 & b^2 & b^3 \end{vmatrix} = \frac{1}{\sqrt{g}} \begin{vmatrix} \vec{R}_1 & \vec{R}_2 & \vec{R}_3 \\ a_1 & a_2 & a_3 \\ b_1 & b_2 & b_3 \end{vmatrix} \tag{16}$$

Vector triple product :

$$\vec{a}\vec{b}\vec{c} = \sqrt{g} \begin{vmatrix} a^1 & a^2 & a^3 \\ b^1 & b^2 & b^3 \\ c^1 & c^2 & c^3 \end{vmatrix} = \frac{1}{\sqrt{g}} \begin{vmatrix} a_1 & a_2 & a_3 \\ b_1 & b_2 & b_3 \\ c_1 & c_2 & c_3 \end{vmatrix} \tag{17}$$

Derivatives of basis vectors :

Notation : $\underset{ik}{\vec{R}} = \underset{ki}{\vec{R}} = \dfrac{\partial^2 \vec{R}}{\partial u_i\, \partial u_k}$; $\underset{mn}{\vec{R}} = \underset{nm}{\vec{R}} = \dfrac{\partial^2 \vec{R}}{\partial v_m\, \partial v_n}$ (18)

Transformation:

$$\underset{mn}{\vec{R}} = \sum_{i,k} B^i_m B^k_n \underset{ik}{\vec{R}} + \sum_s \frac{\partial B^s_n}{\partial u_m} \vec{R}_s \qquad (19)$$

Christoffel symbols of 2nd kind : $\left\{ \begin{matrix} j \\ ik \end{matrix} \right\}$:

$$\underset{ik}{\vec{R}} = \sum_j \left\{ \begin{matrix} j \\ ik \end{matrix} \right\} \vec{R}_j \quad \text{or} \quad \left\{ \begin{matrix} j \\ ik \end{matrix} \right\} = \underset{ik}{\vec{R}} \bullet \vec{R}^j \qquad (20)$$

Transformation :

$$\left\{ \begin{matrix} r \\ mn \end{matrix} \right\} = \sum_{i,j,k} B^i_m B^k_n B^j_r \left\{ \begin{matrix} j \\ ik \end{matrix} \right\} + \sum_s \frac{\partial B^s_n}{\partial u_m} \vec{R}_s \qquad (21)$$

Christoffel symbols of 1st kind : $\{ikj\}$:

$$\underset{ik}{\vec{R}} = \sum_j \{ikj\} \vec{R}^j \quad \text{or} \quad \{ikj\} = \underset{ik}{\vec{R}} \bullet \vec{R}_k \qquad (22)$$

Transformation :

$$\{mnr\} = \sum_{i,j,k} B^i_m B^k_n B^j_r \{ikj\} + \sum_{s,t} g_{st} \frac{\partial B^s_n}{\partial u_m} B^t_r \qquad (23)$$

Relations with unitary tensors of the co-ordinate systems :

$$\left\{ \begin{matrix} j \\ ik \end{matrix} \right\} = \sum_s g^{sj} \{iks\} \quad ; \quad \{ikj\} = \sum_s g_{sj} \left\{ \begin{matrix} s \\ ik \end{matrix} \right\} \qquad (24)$$

$$\left\{ \begin{matrix} j \\ ik \end{matrix} \right\} = \frac{1}{2} \sum_s g^{sj} \left(\frac{\partial g_{is}}{\partial u_i} + \frac{\partial g_{is}}{\partial u_k} - \frac{\partial g_{ik}}{\partial u_s} \right) \qquad (25)$$

Derivative of a vector along a curve :

Let a curve be defined by the equations: $u^i = u^i(\tau)$; $i = 1,2,3$
with the parameter τ varying along the curve.

Let further be a vector $\vec{a} = \sum_i a^i \vec{R}_i = \sum_i a^i \vec{R}^i$

with functions $a^i = a^i\big(u_1(\tau), u_2(\tau), u_2(\tau)\big)$; $a_i = a_i\big(u_1(\tau), u_2(\tau), u_2(\tau)\big)$

The complete derivative of the vector components is :

$$\frac{Da^i}{d\tau} = \sum_{j,k} \left(\frac{\partial a^i}{\partial u_k} + \begin{Bmatrix} i \\ jk \end{Bmatrix} a^j \right) \frac{du_k}{d\tau} = \sum_k \frac{du_k}{d\tau} \nabla_k a^i$$

$$\frac{Da_i}{d\tau} = \sum_{j,k} \left(\frac{\partial a_i}{\partial u_k} - \begin{Bmatrix} j \\ ik \end{Bmatrix} a_j \right) \frac{du_k}{d\tau} = \sum_k \frac{du_k}{d\tau} \nabla_k a_i \qquad (26)$$

with notation : $\nabla_k a^i = \frac{\partial a^i}{\partial u_k} + \sum_j \begin{Bmatrix} i \\ jk \end{Bmatrix} a^j \quad ; \quad \nabla_k a_i = \frac{\partial a_i}{\partial u_k} - \sum_j \begin{Bmatrix} j \\ ik \end{Bmatrix} a_j$ (27)

Derivative of a tensor :

$$\nabla_j a^{ik} = \frac{\partial a^{ik}}{\partial u_j} + \sum_s \begin{Bmatrix} i \\ js \end{Bmatrix} a^{sk} + \begin{Bmatrix} k \\ js \end{Bmatrix} a^{is}$$

$$\nabla_j a_i^k = \frac{\partial a_i^k}{\partial u_j} + \sum_s \begin{Bmatrix} k \\ js \end{Bmatrix} a_i^s - \begin{Bmatrix} s \\ ji \end{Bmatrix} a_s^k \qquad (28)$$

$$\nabla_j a_{ik} = \frac{\partial a_{ik}}{\partial u_j} - \sum_s \begin{Bmatrix} s \\ ji \end{Bmatrix} a_{sk} - \begin{Bmatrix} s \\ jk \end{Bmatrix} a_{is}$$

It holds that :

$$\nabla_j\left(a^i b_k\right) = \nabla_j\left(a^i\right) b_k + a^i \nabla_j\left(b_k\right)$$

$$\nabla_j(\delta_{ik}) = \nabla_j g_{ik} = \nabla_j g_i^k = \nabla_j g^{ik} = 0 \qquad (29)$$

Orthonormal basis vectors :

Orthonormal basis vectors named $\vec{e}_i$; i=1,2,3 .

The basis vector components are : $\vec{R}_i = H_i \vec{e}_i \quad ; \quad \vec{R}^i = h_i \vec{e}_i$

with : (30)

$$H_i = |\vec{R}_i| = \left(\sum_k \left(\frac{\partial x_k}{\partial u_i} \right)^2 \right)^{1/2} \quad ; \quad h_i = |\vec{R}^i| = \left(\sum_k \left(\frac{\partial u_k}{\partial x_i} \right)^2 \right)^{1/2} = 1/H_i \quad ; \quad \vec{e}_i \bullet \vec{e}_j = \delta_{i,j}$$

$$g_{ij} = H_i H_j \delta_{i,j} \quad ; \quad g^{ij} = \frac{1}{H_i H_j} \delta_{i,j}$$

$$\left\{g_{ij}\right\} = \begin{pmatrix} H_1^2 & 0 & 0 \\ 0 & H_2^2 & 0 \\ 0 & 0 & H_3^2 \end{pmatrix} \quad ; \quad g = \det(g_{ij}) = H_1 H_2 H_3 \qquad (31)$$

Vector components : $\vec{a} = \sum_i a_i^* \vec{e}_i \quad ; \quad a_i^* = \vec{a} \bullet \vec{e}_i = \frac{a_i}{H_i} = H_i a^i$ (32)

Scalar product : $$\vec{a}\bullet\vec{b}=\sum_i \frac{a_i b_i}{H_i^2}=\sum_i \overset{*}{a}_i \overset{*}{b}_i \tag{33}$$

Vector (cross) product :

$$\vec{a}\times\vec{b}=\frac{1}{H_1H_2H_3}\begin{vmatrix} H_1\vec{e}_1 & H_2\vec{e}_2 & H_3\vec{e}_3 \\ a_1 & a_2 & a_3 \\ b_1 & b_2 & b_3 \end{vmatrix}=\begin{vmatrix} \vec{e}_1 & \vec{e}_2 & \vec{e}_3 \\ \overset{*}{a}_1 & \overset{*}{a}_2 & \overset{*}{a}_3 \\ \overset{*}{b}_1 & \overset{*}{b}_2 & \overset{*}{b}_3 \end{vmatrix} \tag{34}$$

Vector triple product :

$$\vec{a}\vec{b}\vec{c}=\frac{1}{H_1H_2H_3}\begin{vmatrix} a_1 & a_2 & a_3 \\ b_1 & b_2 & b_3 \\ c_1 & c_2 & c_3 \end{vmatrix}=\begin{vmatrix} \overset{*}{a}_1 & \overset{*}{a}_2 & \overset{*}{a}_2 \\ \overset{*}{b}_1 & \overset{*}{b}_2 & \overset{*}{b}_3 \\ \overset{*}{c}_1 & \overset{*}{c}_2 & \overset{*}{c}_3 \end{vmatrix} \tag{35}$$

Differential operators :

The *gradient* of a scalar function is a vector :

$$\operatorname{grad}\varphi=\vec{\nabla}\varphi=\sum_i \frac{1}{H_i}\frac{\partial\varphi}{\partial u_i}\vec{e}_i=\sum_i \operatorname{grad}_i \varphi\ \vec{e}_i \tag{36}$$

Nabla operator (a vector): $$\vec{\nabla}=\left[\frac{1}{H_1}\frac{\partial}{\partial u_1},\frac{1}{H_2}\frac{\partial}{\partial u_2},\frac{1}{H_3}\frac{\partial}{\partial u_3}\right]$$

The *divergence* of a vector is a scalar :

$$\operatorname{div}\vec{a}=\vec{\nabla}\bullet\vec{a}=\frac{1}{H_1H_2H_3}\sum_i \frac{\partial}{\partial u_i}\left(H_1H_2H_3\cdot a^i\right)=\frac{1}{H_1H_2H_3}\sum_i \frac{\partial}{\partial u_i}\left(\frac{H_1H_2H_3}{H_i}\cdot \overset{*}{a}_i\right) \tag{37}$$

The *rotation* of a vector is a vector : (38)

$$\operatorname{rot}\vec{a}=\vec{\nabla}\times\vec{a}=\frac{1}{H_1H_2H_3}\begin{pmatrix} H_1\vec{e}_1 & H_2\vec{e}_2 & H_3\vec{e}_3 \\ \partial/\partial u_1 & \partial/\partial u_2 & \partial/\partial u_3 \\ a_1 & a_2 & a_3 \end{pmatrix}=\frac{1}{H_1H_2H_3}\begin{pmatrix} H_1\vec{e}_1 & H_2\vec{e}_2 & H_3\vec{e}_3 \\ \partial/\partial u_1 & \partial/\partial u_2 & \partial/\partial u_3 \\ H_1\overset{*}{a}_1 & H_2\overset{*}{a}_2 & H_3\overset{*}{a}_3 \end{pmatrix}$$

The *Laplacian* of a scalar function :

$$\Delta\varphi=(\vec{\nabla}\bullet\vec{\nabla})\varphi=\frac{1}{H_1H_2H_3}\sum_i \frac{\partial}{\partial u_i}\left(\frac{H_1H_2H_3}{H_i^2}\frac{\partial\varphi}{\partial u_i}\right) \tag{39}$$

The Laplacian of a vector is a vector : $$\Delta\vec{a}=\operatorname{grad}(\operatorname{div}\vec{a})-\operatorname{rot}(\operatorname{rot}\vec{a}) \tag{40}$$

Identities :

$$\begin{aligned}
&\operatorname{grad}(U_1U_2) = U_1 \cdot \operatorname{grad} U_2 + U_2 \cdot \operatorname{grad} U_1 \\
&\operatorname{grad}(\vec{V}_1 \bullet \vec{V}_2) = (\vec{V}_1 \bullet \operatorname{grad})\vec{V}_2 + (\vec{V}_2 \bullet \operatorname{grad})\vec{V}_1 + \vec{V}_1 \times \operatorname{rot}\vec{V}_2 + \vec{V}_2 \times \operatorname{rot}\vec{V}_1 \\
&\operatorname{div}(U \cdot \vec{V}) = U \cdot \operatorname{div}\vec{V} + \vec{V} \cdot \operatorname{grad} U \\
&\operatorname{div}(\vec{V}_1 \times \vec{V}_2) = \vec{V}_2 \bullet \operatorname{rot}\vec{V}_1 - \vec{V}_1 \bullet \operatorname{rot}\vec{V}_2 \\
&\operatorname{rot}(U \cdot \vec{V}) = U \cdot \operatorname{rot}\vec{V} + \operatorname{grad} U \times \vec{V} \\
&\operatorname{rot}(\vec{V}_1 \times \vec{V}_2) = (\vec{V}_2 \bullet \operatorname{grad})\vec{V}_1 - (\vec{V}_1 \bullet \operatorname{grad})\vec{V}_2 + \vec{V}_1 \operatorname{div}\vec{V}_2 - \vec{V}_2 \operatorname{div}\vec{V}_1
\end{aligned} \qquad (41)$$

$$\vec{\nabla} \bullet (\vec{\nabla} \times \vec{V}) = \operatorname{div}\ \operatorname{rot}\vec{V} = 0$$

$$\vec{\nabla} \times (\vec{\nabla} U) = \operatorname{rot}\ \operatorname{grad} U = 0$$

$$\vec{\nabla} \bullet (\vec{\nabla} U) = \operatorname{div}\ \operatorname{grad} U = \Delta U$$

Some co-ordinate systems (see [B.11], [B.12] for more systems) :

A vector: $\vec{V}$; a scalar function: U

Cartesian co-ordinates : [x,y,z]

Line, surface , and volume elements :

$$\begin{aligned}
&(ds)^2 = dx^2 + dy^2 + dz^2 \\
&dF_x = dy \cdot dz\ ;\ dF_y = dz \cdot dx\ ;\ dF_z = dx \cdot dy \\
&dV = dx \cdot dy \cdot dz \\
&a_i = a^i = a_i^*
\end{aligned} \qquad (42)$$

Differential operators:

$$\begin{aligned}
&\operatorname{grad} U = \frac{\partial U}{\partial x}\vec{e}_x + \frac{\partial U}{\partial y}\vec{e}_y + \frac{\partial U}{\partial z}\vec{e}_z \\
&\operatorname{div}\vec{V} = \frac{\partial V_x}{\partial x} + \frac{\partial V_y}{\partial y} + \frac{\partial V_z}{\partial z}
\end{aligned} \qquad (43)$$

$$\operatorname{rot}\vec{V} = \begin{vmatrix} \vec{e}_x & \vec{e}_y & \vec{e}_z \\ \partial/\partial x & \partial/\partial y & \partial/\partial z \\ V_x & V_y & V_z \end{vmatrix} \qquad (44)$$

$$\Delta U = \frac{\partial^2 U}{\partial x^2} + \frac{\partial^2 U}{\partial y^2} + \frac{\partial^2 U}{\partial z^2} \qquad (45)$$

Circular cylindrical co-ordinates : [r,φ,z]

A vector: $\vec{V} = [V_r, V_\varphi, V_z]$; a scalar function U

Transformation :

$$x = r\cdot\cos\varphi \qquad r=\sqrt{x^2+y^2}$$
$$y = r\cdot\sin\varphi \qquad \varphi=\arctan(y/x)$$
$$z=z \qquad z=z$$

Line, surface, and volume elements:

$$(ds)^2 = dr^2 + r^2 d\varphi^2 + dz^2$$
$$dF_r = r\,d\varphi\cdot dz\,;\; dF_\varphi = dr\cdot dz\,;\; dF_z = r\,dr\,d\varphi$$
$$dV = r\,dr\cdot d\varphi\cdot dz \tag{46}$$
$$H_r = 1 \;;\quad H_\varphi = r \;;\quad H_z = 1$$

Differential operators:

$$\operatorname{grad} U = \frac{\partial U}{\partial x}\vec{e}_r + \frac{1}{r}\frac{\partial U}{\partial \varphi}\vec{e}_\varphi + \frac{\partial U}{\partial z}\vec{e}_z$$

$$\operatorname{div}\vec{V} = \frac{1}{r}\left(\frac{\partial(rV_x)}{\partial r} + \frac{\partial V_\varphi}{\partial \varphi}\right) + \frac{\partial V_z}{\partial z}$$

$$\operatorname{rot}\vec{V} = \left(\frac{1}{r}\frac{\partial V_z}{\partial \varphi} - \frac{\partial V_\varphi}{\partial z}\right)\vec{e}_r + \left(\frac{\partial V_r}{\partial z} - \frac{\partial V_z}{\partial r}\right)\vec{e}_\varphi + \frac{1}{r}\left(\frac{\partial(rV_\varphi)}{\partial r} - \frac{\partial V_r}{\partial \varphi}\right)\vec{e}_z \tag{47}$$

$$\Delta U = \frac{1}{r}\frac{\partial}{\partial r}\left(r\frac{\partial U}{\partial r}\right) + \frac{1}{r^2}\frac{\partial^2 U}{\partial \varphi^2} + \frac{\partial^2 U}{\partial z^2}$$

Spherical co-ordinates: $[r,\varphi,\vartheta]$

A vector: $\vec{V} = [V_r, V_\varphi, V_\vartheta]$; a scalar function U

Transformation :

$$x = r\cdot\sin\vartheta\cdot\cos\varphi \qquad r = \sqrt{x^2+y^2+z^2}$$
$$y = r\cdot\sin\vartheta\cdot\sin\varphi \qquad \vartheta = \arctan(\sqrt{x^2+y^2}/z) \tag{48}$$
$$z = r\cdot\cos\vartheta \qquad \varphi = \arctan(y/x)$$

Line, surface, and volume elements:

$$(ds)^2 = dr^2 + r^2 d\vartheta^2 + r^2\sin^2\vartheta\, d\varphi^2$$
$$dF_r = r^2\sin\vartheta\, d\varphi\cdot d\vartheta\,;\; dF_\varphi = r\,dr\cdot d\vartheta\,;\; dF_\vartheta = r\,dr\,d\varphi \tag{49}$$
$$dV = r^2\sin\vartheta\cdot dr\cdot d\varphi\cdot d\vartheta$$
$$H_r = 1 \;;\quad H_\varphi = r\sin\vartheta \;;\quad H_\vartheta = r$$

Differential operators:

$$\operatorname{grad} U = \frac{\partial U}{\partial r}\vec{e}_r + \frac{1}{r\sin\vartheta}\frac{\partial U}{\partial \varphi}\vec{e}_\varphi + \frac{1}{r}\frac{\partial U}{\partial \vartheta}\vec{e}_\vartheta \tag{50}$$

$$\operatorname{div}\vec{V} = \frac{1}{r^2}\frac{\partial(r^2V_r)}{\partial r} + \frac{1}{r\sin\vartheta}\frac{\partial(\sin\vartheta\, V_\vartheta)}{\partial \vartheta} + \frac{1}{r\sin\vartheta}\frac{\partial V_\varphi}{\partial \varphi}$$

$$\mathrm{rot}\,\vec{V}=\frac{1}{r\sin\vartheta}\left(\frac{\partial(\sin\vartheta\,V_\varphi)}{\partial\vartheta}-\frac{\partial V_\vartheta}{\partial\varphi}\right)\vec{e}_r+\left(\frac{1}{r\sin\vartheta}\frac{\partial V_r}{\partial\varphi}-\frac{1}{r}\frac{\partial(rV_\varphi)}{\partial r}\right)\vec{e}_\vartheta+\frac{1}{r}\left(\frac{\partial(rV_\vartheta)}{\partial r}-\frac{\partial V_r}{\partial\vartheta}\right)\vec{e}_\varphi$$

$$\Delta U=\frac{1}{r^2}\frac{\partial}{\partial r}\left(r^2\frac{\partial U}{\partial r}\right)+\frac{1}{r^2\sin\vartheta}\frac{\partial}{\partial\vartheta}\left(\sin\vartheta\frac{\partial U}{\partial\vartheta}\right)+\frac{1}{r^2\sin^2\vartheta}\frac{\partial^2 U}{\partial\varphi^2}$$

Differential relations of acoustics :

Field variables (overbar: total quantity; index 0 : stationary value)

density:	$\overline{\rho}=\rho_0+\rho$
pressure:	$\overline{p}=p_0+p$
temperature:	$\overline{T}=T_0+T$
entropy:	$\overline{S}=S_0+S$
velocity:	$\overline{v}=v_0+v$

η= dynamic viscosity;
μ= volume viscosity;
Λ= heat conductivity;

Total time derivative: $$\frac{D\ldots}{Dt}=\frac{\partial\ldots}{\partial t}+(\overline{v}\bullet\mathrm{grad})\ldots \quad (51)$$

Equation of continuity : $$\frac{D\overline{\rho}}{Dt}+\overline{\rho}\cdot\mathrm{div}\,\overline{v}=0 \quad (52)$$

linearised and v_0=0 : $$\frac{\partial\rho}{\partial t}+\rho_0\,\mathrm{div}\,v=0 \quad (53)$$

Navier-Stokes equation: $$\overline{\rho}\frac{D\overline{v}}{Dt}=-\mathrm{grad}\left(\overline{p}+\overline{\rho}\Phi-(\mu+\tfrac{4}{3}\eta)\mathrm{div}\,\overline{v}\right)-\eta\,\mathrm{rot}(\mathrm{rot}\,\overline{v}) \quad (54)$$

with: Φ= potential of an external force per unit mass (e.g. gravity);

linearised and v_0=0 ; Φ=0: $$\rho_0\frac{\partial v}{\partial t}=-\mathrm{grad}\,p+(\mu+\tfrac{4}{3}\eta)\,\mathrm{grad}(\mathrm{div}\,v)-\eta\,\mathrm{rot}(\mathrm{rot}\,v) \quad (55)$$

Apply $$[\mathrm{grad}(\mathrm{div}\,v)]_x=\frac{\partial}{\partial x}\left(\frac{\partial u_x}{\partial x}+\frac{\partial u_y}{\partial y}+\frac{\partial u_z}{\partial z}\right)=\Delta v_x+[\mathrm{rot}(\mathrm{rot}\,v)]_x \quad (56)$$

and compose: $$v=v_\ell+v_t \quad \text{with} \quad \begin{cases}\mathrm{rot}\,v_\ell=0\\ \mathrm{div}\,v_t=0\end{cases} \quad (57)$$

This leads to the following two differential equations:

$$\rho_0\frac{\partial v_\ell}{\partial t}=-\mathrm{grad}\,p+(\mu+\tfrac{4}{3}\eta)\cdot\Delta v_\ell$$

$$\rho_0\frac{\partial v_t}{\partial t}=-\eta\cdot\mathrm{rot}(\mathrm{rot}\,v_t)$$

(58)

ℓ= molecular mean free path length;
κ= adiabatic exponent;
C_v=specific heat at constant volume;
C_p=specific heat at constant pressure;
η= dynamic viscosity;

Energy equations :

A) *Heat conduction* : $\vec{J}_h = -\Lambda \cdot \operatorname{grad} T$

with Λ= heat conductibility.

From molecular gas dynamics: $$\Lambda = 1.6 \frac{\ell \rho_0 c_0 C_v}{\sqrt{\kappa}} \approx \tfrac{5}{3} \eta C_v \tag{59}$$

Energy balance with heat conduction : $$\frac{\partial T}{\partial t} = \frac{\Lambda}{\rho_0 C_p} \operatorname{div} \operatorname{grad} T \tag{60}$$

B) *Viscous energy loss*

per unit volume : $$D = \sum_{i,k} \frac{\partial v_i}{\partial x_k} D_{ik} \tag{61}$$

Shear stresses by viscosity
$$D_{ii} = -\left(\mu - \tfrac{2}{3}\eta\right) \operatorname{div} v - 2\eta \frac{\partial v_i}{\partial x_i}$$
$$D_{ik} = -\eta \left(\frac{\partial v_i}{\partial x_k} + \frac{\partial v_k}{\partial x_i} \right) \tag{62}$$

C) *Balance of internal energy* E per unit mass :

$$\frac{dE}{dt} = \frac{1}{\rho_0} \frac{dU}{dt} = \frac{1}{\rho_0} \left[\operatorname{div}(\Lambda \operatorname{grad} T) + D - P \cdot \operatorname{div} u \right] \tag{63}$$

with: E= internal energy per unit mass;
U= internal energy per unit volume;
D= viscous energy loss;
P= static pressure.

D) *Balance of entropy* S per unit mass :

$$\rho \frac{dS}{dt} = \frac{D}{T} + \frac{1}{T} \operatorname{div}(\Lambda \operatorname{grad} T) = \frac{D}{T} + \frac{\Lambda}{T^2} |\operatorname{grad} T|^2 + \operatorname{div}\left(\frac{\Lambda}{T} \operatorname{grad} T \right) \tag{64}$$

E) *Balance of heat* Q per unit volume :

$$\frac{dQ}{dt} = \rho_0 T \frac{dS}{dt} = D + \operatorname{div}(\Lambda \operatorname{grad} T) = D + \frac{\Lambda}{T} |\operatorname{grad} T|^2 + T \cdot \operatorname{div}\left(\frac{\Lambda}{T} \operatorname{grad} T \right) \tag{65}$$

Equation of state :

For an ideal gas : $$\bar{p} = \bar{\rho} \cdot R_0 \cdot \bar{T} \tag{66}$$

with R_0= gas constant.

Equation of state for the mass density variation $\tilde{\rho}$ in a sound wave
(sound field quantities with a $\tilde{}$, stationary quantities with a $\bar{}$, atmospheric values with $_0$):

$$\tilde{\rho} = \left(\frac{\partial \overline{\rho}}{\partial \overline{p}}\right)_T \cdot \tilde{p} + \left(\frac{\partial \overline{\rho}}{\partial \overline{T}}\right)_p \cdot \tilde{T} = \kappa \rho_0 K_S (\tilde{p} - \alpha \tilde{T}) \xrightarrow[\text{ideal gas}]{} \frac{\rho_0}{P_0}\tilde{p} - \frac{\rho_0}{T_0}\tilde{T} \approx \frac{\rho_0}{P_0}\tilde{p} \qquad (67)$$

Equation of state for the entropy variation $\tilde{S}$ in a sound wave

$$\tilde{S} = \left(\frac{\partial \overline{S}}{\partial \overline{p}}\right)_T \cdot \tilde{p} + \left(\frac{\partial \overline{S}}{\partial \overline{T}}\right)_p \cdot \tilde{T} = \frac{C_p}{T_0}\left(\tilde{T} - \frac{\kappa - 1}{\alpha \kappa}\tilde{p}\right) \xrightarrow[\text{ideal gas}]{} C_p\left(\frac{\tilde{T}}{T_0} - \frac{\kappa - 1}{\kappa}\frac{\tilde{p}}{P_0}\right) \qquad (68)$$

Thermodynamic relations :

$K_T = -\frac{1}{V}\left(\frac{\partial V}{\partial P}\right)_T$ isothermal compressibility; (69)

$K_S = -\frac{1}{V}\left(\frac{\partial V}{\partial p}\right)_S$ isotrope compressibility; (70)

$ß = \frac{1}{V}\left(\frac{\partial V}{\partial T}\right)_p$ thermal expansion coefficient; (71)

$\alpha = \left(\frac{\partial p}{\partial T}\right)_V = \frac{ß}{K_T}$ thermal pressure coefficient; (72)

$\kappa = \frac{C_p}{C_v}$ adiabatic exponent; (73)

$$K_T = \kappa K_s \qquad (74)$$

$$\kappa - 1 = \frac{T ß^2}{K_s \rho C_p} \qquad (75)$$

$c_0^2 = \frac{1}{\rho_0 K_s} = \frac{1}{\kappa \rho_0 K_T}$ adiabatic sound velocity c_0 ; (76)

$$K_s = \frac{1}{\kappa \rho_0}\left(\frac{\partial \rho}{\partial P}\right)_T = \frac{1}{\rho_0 c_0^2} \xrightarrow[\text{ideal gas}]{} \frac{1}{\kappa P_0} \qquad (77)$$

$$ß = -\frac{1}{\rho_0}\left(\frac{\partial \rho}{\partial T}\right)_P \xrightarrow[\text{ideal gas}]{} -\frac{1}{T_0} \qquad (78)$$

$$\alpha = \left(\frac{\partial p}{\partial T}\right)_V = \frac{ß}{K_t} = \frac{ß \rho_0 c_0^2}{\kappa} \xrightarrow[\text{ideal gas}]{} \frac{P_0}{T_0} \qquad (79)$$

$\ell_h = \frac{\Lambda}{\rho_0 c_0 C_p} \approx 1.6 \frac{\ell}{\sqrt{\kappa}}$ characteristic mean free molecular path length for heat conduction effects; ℓ mean free path length; (80)

$\ell_v = \dfrac{\eta}{\rho_0 c_0} \approx \dfrac{\ell}{\sqrt{\kappa}}$ characteristic mean free molecular path length for shear viscosity effects; (81)

$\ell'_v = \dfrac{\mu + 4/3\,\eta}{\rho_0 c_0} = \left(\dfrac{4}{3} + \dfrac{\mu}{\eta}\right)\ell_v$ characteristic mean free molecular path length for shear and bulk viscosity; (82)

Linearised fundamental equations for a density wave (time factor $e^{+j\omega t}$):

$$\text{with: } k_\rho^2 \approx \left(\frac{\omega}{c_0}\right)^2 \left[1 + \frac{\omega}{c_0}\ell'_v - j(\kappa-1)\frac{\omega}{c_0}\ell_h\right];\quad \Delta\tilde{p} = -k_\rho^2\tilde{p};\quad \frac{\partial\tilde{p}}{\partial t} = j\omega\cdot\tilde{p};\quad c_0^2 = \frac{1}{\rho_0 K_s} \tag{83}$$

temperature variation:
$$\tilde{T} = \frac{\kappa-1}{\alpha\kappa}\left(1 + j\frac{\omega}{c_0}\ell_h\right)\cdot\tilde{p} \tag{84}$$

density variation:
$$\tilde{\rho} = \frac{1}{c_0^2}\left(1 - j(\kappa-1)\frac{\omega}{c_0}\ell_h\right)\cdot\tilde{p} \tag{85}$$

entropy variation:
$$\tilde{S} = j\frac{C_p}{T_0}\frac{\kappa-1}{\alpha\kappa}\frac{\omega}{c_0}\ell_h\cdot\tilde{p} \tag{86}$$

longitudinal particle velocity:
$$v_\ell = \left(\frac{j}{\omega\rho_0} - \frac{\ell'_v}{\rho_0 c_0}\right)\cdot \operatorname{grad}\tilde{p} \tag{87}$$

Linearised fundamental equations for a temperature wave :

$$\text{with: } k_T^2 \approx \frac{-j\omega}{c_0\ell_h};\quad \Delta\tilde{T} = -k_T^2\tilde{T};\quad \frac{\partial\tilde{T}}{\partial t} = j\omega\cdot\tilde{T} \tag{88}$$

pressure variation:
$$\tilde{p} = \frac{j\kappa\alpha\omega}{c_0}\left(\ell'_v - \ell_h\right)\cdot\tilde{T} \tag{89}$$

density variation:
$$\tilde{\rho} = \frac{-\alpha\kappa}{c_0^2}\left(1 + \frac{j\kappa\omega}{c_0}\left(\ell_h - \ell'_v\right)\right)\cdot\tilde{T} \tag{90}$$

entropy variation:
$$\tilde{S} = \frac{C_p}{T_0}\left(1 + j(\kappa-1)\frac{\omega}{c_0}\left(\ell_h - \ell'_v\right)\right)\cdot\tilde{T} \tag{91}$$

B.14 Boundary condition at a moving boundary

Ref.: Kleinstein / Gunzburger, [B.13]

A boundary separates two media with density and sound velocity ρ_1, c_1 and ρ_2, c_2, respectively (other quantities are distinguished with the same indices). The boundary moves with a velocity U_0 normal to its surface. A wave is incident from the side with ρ_1, c_1. The co-ordinate normal to the surface is x, directed $1 \to 2$.

One-dimensional wave equation (in fixed co-ordinates) for density ρ_i on both sides i=1,2 :

$$\left(\partial/\partial t + U_0\, \partial/\partial x\right)^2 \rho = c_i^2\, \partial^2\rho/\partial x^2 \tag{1}$$

with general solutions :

$$\rho_1 = F_1\left(\omega_1 t - \frac{\omega_1 x}{c_1(1+M_1)}\right) + G_1\left(\overline{\omega}_1 t + \frac{\overline{\omega}_1 x}{c_1(1-M_1)}\right)$$
$$\rho_1 = F_2\left(\omega_2 t - \frac{\omega_2 x}{c_2(1+M_2)}\right) \tag{2}$$

$M_1 = U_0/c_1$;
$M_2 = U_0/c_2$;
$F_1 =$ incident wave;
$G_1 =$ reflected wave;
$F_2 =$ transmitted wave;

Boundary conditions:

$$\delta v = 0 \to \left(\partial/\partial t + U_0\, \partial/\partial x\right)\delta v = 0$$
$$\delta p = 0 \to \left(\partial/\partial t + U_0\, \partial/\partial x\right)\delta p = 0 \tag{3}$$

This leads to the Doppler shifted frequencies :

$$\frac{\omega_1}{1+M_1} = \frac{\overline{\omega}_1}{1-M_1} = \frac{\omega_2}{1+M_2} \tag{4}$$

Wave numbers :

$$k_1 = \omega_1/(c_1 + U_0)$$
$$\overline{k}_1 = -\overline{\omega}_1/(c_1 - U_0) \tag{5}$$
$$k_2 = \omega_2/(c_2 + U_0)$$

Rule of conservation of wave numbers:

$$\partial k/\partial t + \partial\omega/\partial x = 0 \quad \text{or} \quad -U_0 \cdot \delta k + \delta\omega = 0 \quad \text{or} \quad U_0 = \frac{\delta\omega}{\delta k} \tag{6}$$

Applications :

A) The boundary is a shock front in the undisturbed medium: i.e. $\omega_2 = k_2 = 0$;
it follows that: $U_0 = \omega_1 / k_1 =$ phase velocity in the medium i=1

B) A shock front with jump in the state of the medium :
Shock front equation of gas dynamics: $$U_0^2 = \delta\left(p + \rho v^2\right) / \delta\rho \tag{7}$$

- in the limit of small amplitudes, i.e. : $$\rho_2 \approx \rho_1 \; ; \; p_2 \approx p_1 \; ; \; v_1 \approx v_2 \tag{8}$$

it follows that: $$U_0 = \frac{\delta\omega}{\delta k} \to \frac{d\omega}{dk} = \text{group velocity} \tag{9}$$

C) Stationary shock front, i.e. $U_0 = 0$,
follows with $\omega_1 = \omega_2$ $$k_2 = k_1 \frac{c_1}{c_2} \frac{1 + M_1}{1 + M_2} \tag{10}$$

D) Shock front with velocity U_0 and $U_2 =$ flow velocity behind shock :

$$\omega_1 = k_1 c_1 \quad ; \quad \omega_2 = \left(c_2 + U_2\right) k_2$$

$$\frac{\omega_2}{\omega_1} = \left(1 + \frac{U_2}{c_2}\right) \frac{1 + M_1}{1 + M_2} \quad ; \quad \frac{k_2}{k_1} = \frac{c_1}{c_2} \frac{1 + M_1}{1 + M_2} \tag{11}$$

B.15 Boundary conditions at liquids and solids

Ref.: Gottlieb, [B.14]

Let a plane pressure front be parallel to a plane boundary. Let the density and sound velocities in a solid on both sides be ρ_i , c_i , respectively, and tensions and velocities σ_i , u_i ; i=1,2; i=1 input side. In a liquid let $p_i = -\sigma_i$ and v_i be the sound pressure and particle velocity, respectively.

A) In a homogeneous solid with ρ_1, c_1 : $\sigma_1 - \sigma_2 = \rho_1 c_1 (u_2 - u_1)$

B) In a homogeneous liquid with ρ_1, c_1 : $p_1 - p_2 = \rho_1 c_1 (v_1 - v_2)$

C) At a solid-liquid interface

$$u_3 = \frac{-\sigma_1 - p_2 + \rho_1 c_1 \cdot u_1 + \rho_2 c_2 \cdot v_2}{\rho_1 c_1 + \rho_2 c_2}$$

$$v_4 = u_3$$

$$\sigma_3 = \frac{\rho_2 c_2 \cdot \sigma_1 - \rho_1 c_1 \cdot p_2 - \rho_1 c_1 \cdot \rho_2 c_2 \cdot (u_1 - v_2)}{\rho_1 c_1 + \rho_2 c_2} \tag{1}$$

$$p_4 = -\sigma_3$$

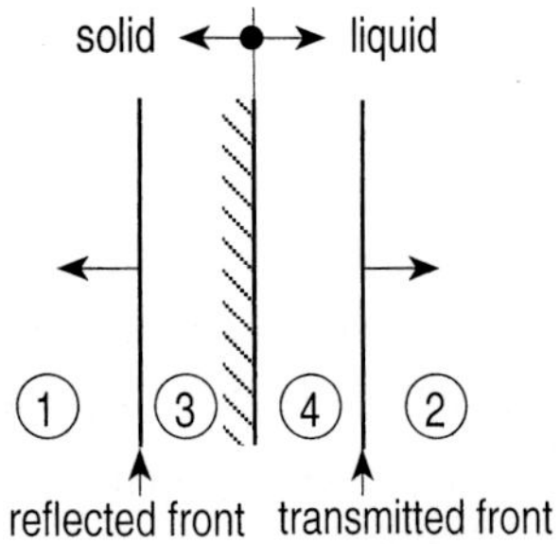

B.16 Corner conditions

Ref.: FELSEN / MARCUVITZ, [B.15]

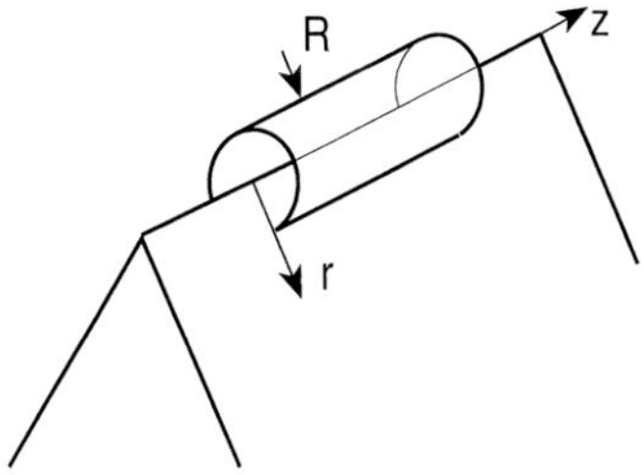

Two-dimensional corner :

Consider a field $f(r)\cdot g(\varphi,z)$.

The condition in the corner at $r=0$ is :

$$\int_0^R |f(r)|^2 \cdot r \cdot dr = \text{finite} \qquad (1)$$

from it which follows that:

$$|f(r)|^2 \le \frac{1}{r^{2(1-\alpha)}} \quad ; \quad \text{for } r \to 0 \text{ ; } \alpha \text{ small, positive}$$

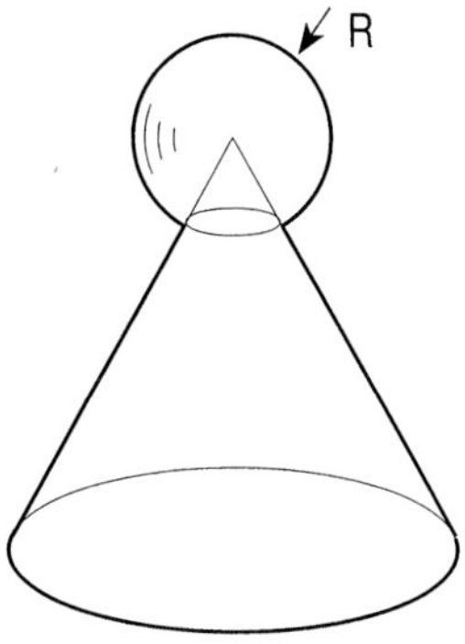

Three-dimensional corner :

Consider a field $f(r)$.

The condition in the corner at $r=0$ is :

$$\int_0^R |f(r)|^2 \cdot r^2 \cdot dr = \text{finite} \qquad (2)$$

from which it follows that:

$$|f(r)|^2 \le \frac{1}{r^{3(1-2\alpha/3)}} \quad ; \quad \text{for } r \to 0 \text{ ; } \alpha \text{ small, positive}$$

B.17 Surface wave at locally reacting plane

Ref.: Mechel, [B.16]

Surface waves are well known in elastic bodies (e.g. as Rayleigh waves). Here surface waves in the fluid are considered, but not those which, as a consequence of a surface wave in an elastic boundary are produced in the fluid. Synonyms are “guided wave”, because surface waves may follow curved boundaries, “creeping wave” in the scattering at cylinders and spheres as slow waves propagation around the scattering objects.

Consider a plane boundary in the x,z-plane, with air in the half space $y \geq 0$. Let the surface be characterised either by a surface impedance Z or by surface admittance $G = 1/Z$.

A wave of the form $$p(x,y) = P_0 \cdot e^{-\Gamma_x x} \cdot e^{-\Gamma_y y} \cdot e^{j\omega t} \tag{1}$$

satisfies the wave equation if $$\Gamma_x^2 + \Gamma_y^2 = -k_0^2 \tag{2}$$

the radiation condition if $$\mathrm{Re}\{\Gamma_x\} \geq 0 \; ; \; \mathrm{Re}\{\Gamma_y\} \geq 0 \tag{3}$$

and the boundary condition if $$Z_0 G \overset{!}{=} Z_0 G_y := -\left.\frac{Z_0 v_y}{p}\right|_{y=0} = \frac{j\Gamma_y}{k_0} = \frac{-\Gamma_y'' + j\Gamma_y'}{k_0} \tag{4}$$

If one compares the last relation with the general admittance relation of section B.3 , in which the sound field is written as $p(x,y) = |p(x,y)| \cdot e^{j\varphi(x,y)}$

which is : $$Z_0 G = \frac{1}{k_0}\left[-\frac{\partial}{\partial n}\varphi(x,y) + j \cdot \frac{\partial}{\partial n}\ln(|\,p(x,y)\,|)\right] \tag{5}$$

one gets : $$\Gamma_y'' = \frac{\partial}{\partial y}\varphi(x,y) \quad ; \quad \Gamma_y' = \frac{\partial}{\partial y}\ln|\,p(x,y)\,| \tag{6}$$

These are just the definitions of Γ_y'', Γ_y' as phase and level measures. So a surface wave is a wave type which satisfies the fundamental equations and the boundary condition "by definition".

The graph shows curves $\mathrm{Re}\{\Gamma_x/k_0\}$= const (about horizontal lines) and curves $\mathrm{Im}\{\Gamma_x/k_0\}$= const (about vertical lines) in the complex plane of $Z_0G = Z_0G' + j \cdot Z_0G''$. The parameter steps $\Delta\mathrm{Re}\{\Gamma_x/k_0\}$, $\Delta\mathrm{Im}\{\Gamma_x/k_0\}$ are 0.2 over the values 0,…,3 . The curve $\mathrm{Im}\{\Gamma_x/k_0\}$= 1 is thick. Values $\mathrm{Im}\{\Gamma_x/k_0\} < 1$ are on the left of the curve for $\mathrm{Im}\{\Gamma_x/k_0\}$=1 The waves there are "fast"; the waves on the right of that curve are "slow". Because $\mathrm{Re}\{\Gamma_x/k_0\} > 0$, the waves are attenuated along the surface.

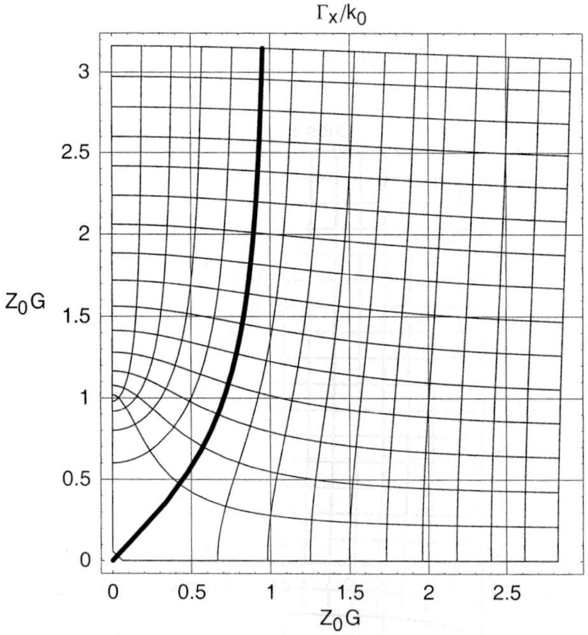

B.18 Surface wave along a locally reacting cylinder

Ref.: Mechel, [B.16]

The topic here is a surface wave *along* a cylinder, not *around* a cylinder. The cylinder has the diameter 2a and is locally reacting at its surface with the normalised radial impedance $W = Z/Z_0 = 1/(Z_0G)$ (G= normalised admittance). The wave is supposed to have an axial symmetry. It is formulated as :

$$p(r,z) = P_0 \cdot K_0(\Gamma_r r) \cdot e^{-\Gamma_z z} \quad ; \quad \Gamma_r^2 + \Gamma_z^2 = -k_0^2$$
$$Z_0 v_r(r,z) = \frac{j}{k_0} \mathrm{grad}_r p = \frac{-j\Gamma_r}{k_0} P_0 \cdot K_1(\Gamma_r r) \cdot e^{-\Gamma_z z} \tag{1}$$

with the modified Bessel function $K_0(z)$ of the second kind with zero order. The boundary condition at r=a leads to the characteristic equation for $\Gamma_r a$:

$$\Gamma_r a \cdot \frac{K_1(\Gamma_r a)}{K_0(\Gamma_r a)} = -j k_0 a \cdot Z_0 G = -j \cdot U \tag{2}$$

Start values for the numerical solution are $\Gamma_r a / k_0 a \approx -j Z_0 G$.

With its solution the axial propagation constant Γ_z is evaluated from :

$$\frac{\Gamma_z a}{k_0 a} = j\sqrt{1 + (\Gamma_r a / k_0 a)^2} \tag{3}$$

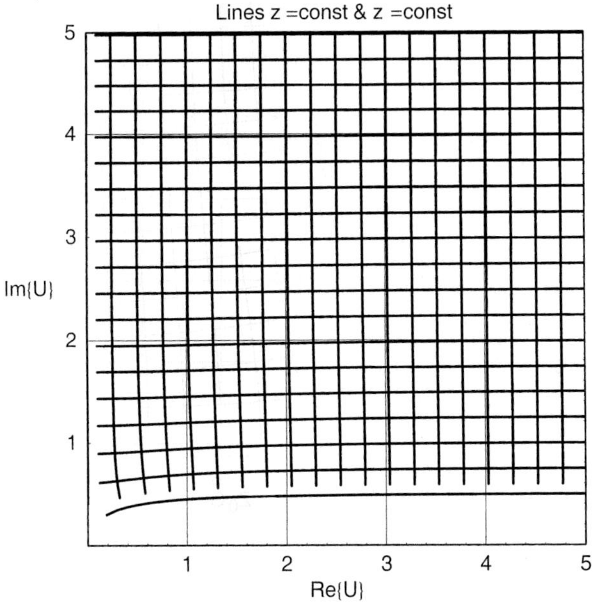

Lines for constant real or imaginary parts of $z = \Gamma_r a = z' + j \cdot z''$ in the plane of $U = k_0 a \cdot Z_0 G$; for z′, z″= 0 to 5; Dz=0.2
(The parameter values at the lines are approximately equal to the co-ordinate values of U)

B.19 Periodic structures, admittance grid

Ref.: MECHEL, [B.17]

An object with a periodic surface is a special case of an object with an inhomogeneous surface (other inhomogeneous surfaces which are amenable to analysis are those in which either the scale of the inhomogeneities and their distances is small compared to λ_0 , then the average admittance is relevant, or the inhomogeneities are at large distances from each other, then scatter matrices can be set up). The method to be applied with periodic structures will be displayed in this and the next sections with some typical examples.

In principle, the quantities thet describe the periodic surface, such as its surface admittance, or the sound field at the surface, are synthesised with a Fourier series. The Fourier terms are waves which have different names in the literature: "spatial harmonics" (used here), "Hartree harmonics" (often used in microwave technology), "Bloch waves" (used in solid state physics). The most important quality of these waves is their orthogonality over a period, which makes them suited for the synthesis of field quantities.

Plane surface with periodic admittance function $G(z)$ *and incident plane wave* :

Consider a plane with a periodic surface admittance $G(z)$ and a plane wave p_e incident with a polar angle ϑ (the wave vector in the x,z plane).

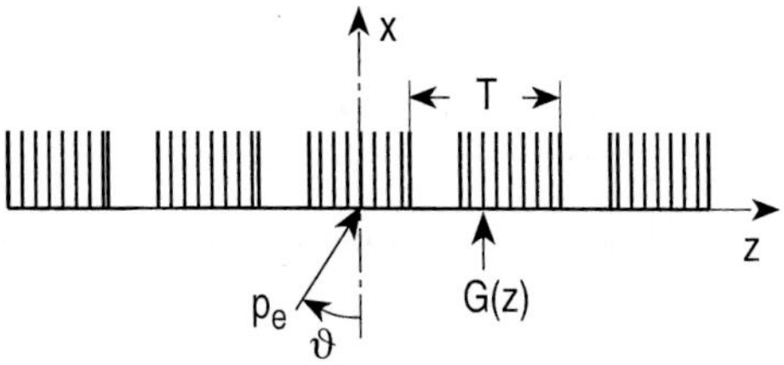

The plane wave p_e is :

$$p_e(x,z) = P_e \cdot e^{-j(k_x x + k_z z)}$$
$$k_x = k_0 \cos\vartheta \quad ; \quad k_z = k_0 \sin\vartheta \qquad (1)$$

The field in the half space $x \le 0$ is written as $p(x,z) = p_e(x,z) + p_s(x,z)$ with the scattered wave p_s formulated as :

$$p_s(x,z) = \sum_{n=-\infty}^{+\infty} A_n \cdot e^{\gamma_n x} \cdot e^{-j\beta_n z} \quad ; \quad \gamma_n^2 = \beta_n^2 - k_0^2 \quad ; \quad \mathrm{Re}\{\gamma_n\} \ge 0 \qquad (2)$$

The relation for γ_n ensures the (term-wise) satisfaction of the wave equation, and the condition for γ_n the satisfaction of Sommerfeld's far field condition. The scattered field p_s can be written as a product $p_s(x,z) = e^{-j\beta_0 z}\cdot S(x,z)$ of a propagation factor $e^{-j\beta_0 z}$ and a factor S(x,z) which must be periodic in z : $S(x,z) = S(x,z+T)$.

This gives for the wave numbers in the z direction:

$$\beta_n = \beta_0 + n\frac{2\pi}{T} \quad ; \quad n = 0, \pm 1, \pm 2, \ldots \tag{3}$$

The spatial harmonic with the order n=0 evidently must agree in its z pattern with the trace of the incident wave at the surface : $\beta_0 = k_z = k_0 \sin\vartheta$.

$$\text{Thus :} \quad \beta_n = k_0\left(\sin\vartheta + n\,\lambda_0/T\right) \quad ; \quad \gamma_n^2 = k_0^2\left[\left(\sin\vartheta + n\,\lambda_0/T\right)^2 - 1\right] \tag{4}$$

and the sound field in $x \le 0$:

$$p(x,z) = \left[P_e \cdot e^{-jk_0 x\cdot\cos\vartheta} + A_0 \cdot e^{+jk_0 x\cdot\cos\vartheta} + \right.$$
$$\left. + \sum_{n\neq 0} A_n \cdot e^{k_0 x\sqrt{(\sin\vartheta + n\,\lambda_0/T)^2 - 1}} \cdot e^{-j(2n\pi/T)z}\right] \cdot e^{-jk_0 z\sin\vartheta} \tag{5}$$

The 2nd term in the brackets is a homogeneously reflected plane wave; the terms in the sum are higher scattered waves. The exponent of the exponential factor with x under the sum must be zero or imaginary, if the spatial harmonic should extend to infinity, i.e. the harmonic is "radiating". The condition for radiating harmonics (order n_s) is :

$$-\frac{T}{\lambda_0}\left(1 + \sin\vartheta\right) \le n_s \le \frac{T}{\lambda_0}\left(1 - \sin\vartheta\right) \tag{6}$$

At (and near) the lower limit the harmonic propagates in the opposite z direction of the incident wave; at (and near) the upper limit the harmonic propagates in the same z direction as the incident wave (if the limits are reached exactly, the harmonic propagates as a plane wave parallel to the surface). The lower limit is attained (or surpassed) the first time for $n_s < 0$ with $1/2 \le T/\lambda_0 \le 1$; the upper limit for $n_s > 0$ with $1 \le T/\lambda_0 < \infty$. A radiating harmonic does not exist for $T/\lambda_0 < 1/2$. The non-radiating harmonics shape the near field at the surface.

The amplitudes A_n are determined from the boundary condition $Z_0 v_x(0,z) \stackrel{!}{=} G(z)\cdot p(0,z)$ at the surface. One expands :

$$G(z) = \sum_{n=-\infty}^{+\infty} g_n \cdot e^{-j(2n\pi/T)z} \quad ; \quad g_n = \frac{1}{T}\int_{-T/2}^{+T/2} G(z)\cdot e^{+j(2n\pi/T)z}\,dz \tag{7}$$

alternatively :

$$G(z) = \frac{a_0}{2} + \sum_{n=1}^{\infty}\left[a_n \cdot \cos(2n\pi z/T) + j \cdot b_n \cdot \sin(2n\pi z/T)\right]$$

$$a_n = \frac{2}{T}\int_{-T/2}^{+T/2} G(z) \cdot \cos(2n\pi z/T)\,dz \quad ; \quad b_n = \frac{2}{T}\int_{-T/2}^{+T/2} G(z) \cdot \sin(2n\pi z/T)\,dz \tag{8}$$

$$g_n = \frac{1}{2}\left[a_n + j \cdot b_n\right] \quad ; \quad n = \pm 1, \pm 2, \ldots \quad ; \quad g_0 = \frac{a_0}{2}$$

The boundary condition gives for $m=0$:

$$A_0(g_0 + \cos\vartheta) + \sum_{n \neq 0} A_n \cdot g_{-n} = P_e \cdot (\cos\vartheta - g_0) \tag{9}$$

and for $m \neq 0$:

$$A_0\, g_m + \sum_{n \neq 0} A_n \cdot \left[g_{m-n} - j\delta_{m,n}\sqrt{(\sin\vartheta + m\,\lambda_0/T)^2 - 1}\,\right] = -P_e \cdot g_m \tag{10}$$

with the Kronecker symbol $\delta_{m,n}$. This is a linear, inhomogeneous system of equations for the amplitudes A_n .

The special case $G(z)=$ const leads to $A_{n \neq 0} = 0$ and the known reflection factor :

$$\frac{A_0}{P_e} = r_0 = \frac{\cos\vartheta - g_0}{\cos\vartheta - g_0} \tag{11}$$

The absorbed effective power (on a period length) is :

$$\Pi' = \frac{T}{2Z_0}\left[\left(|P_e|^2 - |A_0|^2\right)\cos\vartheta - \sum_{n_s \neq 0} |A_{n_s}|^2 \sqrt{1 - (\sin\vartheta + n_s\,\lambda_0/T)^2}\,\right] \tag{12}$$

Referring this to the incident effective power : $\quad \Pi'_e = \dfrac{T}{2Z_0}|P_e|^2 \cdot \cos\vartheta \qquad (13)$

gives the absorption coefficient :

$$\alpha(\vartheta) = \frac{\Pi'}{\Pi'_e} = 1 - \left|\frac{A_0}{P_e}\right|^2 - \frac{1}{\cos\vartheta}\sum_{n_s \neq 0}\left|\frac{A_{n_s}}{P_e}\right|^2 \sqrt{1 - (\sin\vartheta + n_s\,\lambda_0/T)^2} \tag{14}$$

The last term is a correction for the absorption of a homogeneous surface (represented by the first two terms) due to the structured surface; only radiating spatial harmonics enter in this correction. This is plausible, because only the radiating harmonics transport energy into the far field.

Grooved wall with narrow, absorber-filled grooves :

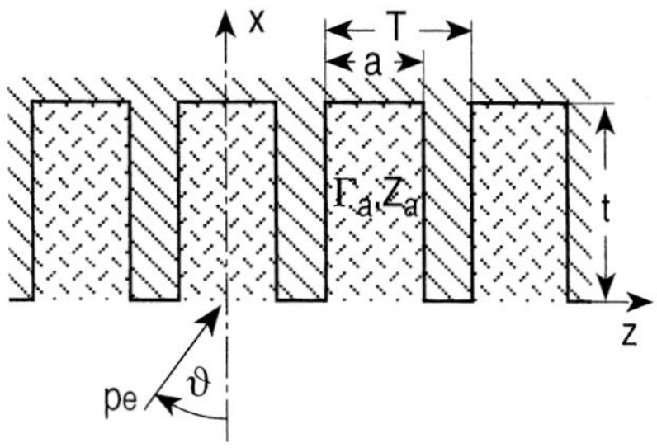

Consider, as a simple example, a plane wall with rectangular grooves, width a , distance T , depth t , the grooves being filled with a porous material with characteristic values Γ_a, Z_a .

A plane wave p_e is incident under a polar angle ϑ (the wave vector in the x,z plane).

The grooves are narrow $(a< \lambda_0/4)$ so that only a plane wave can be assumed to exist in the grooves. Then the grooves can be characterised by an admittance G_s in the groove orifice, and the admittance of the arrangement is $G(z)=0$ in front of the ribs between the grooves.

$$G_s = \frac{1}{Z_{an}} \tanh(k_0 t \cdot \Gamma_{an}) \quad ; \quad \Gamma_{an} = \Gamma_a / k_0 \quad ; \quad Z_{an} = Z_a / Z_0 \tag{15}$$

The Fourier coefficients of G(z) are :

$$g_0 = \frac{a}{T} G_s \quad ; \quad g_m = g_{-m} = \frac{a}{T} G_s \frac{\sin(m\pi a/T)}{m\pi a/T} \quad ; \quad m = 1,2,3,\ldots \tag{16}$$

The system of equations for the amplitudes A_n of the spatial harmonics becomes (with $P_e=1$):

<u>m = 0:</u>

$$A_0 \cdot \left(s(0) + \frac{T}{a} Z_s \cos(\vartheta) \right) + \sum_{n\geq 1} (A_n + A_{-n}) \cdot s(n) = \frac{T}{a} Z_s \cos(\vartheta) - s(0)$$

<u>m = ±1, ± 2,…:</u>

$$A_0 \cdot s(m) + \sum_{n=\pm 1, \pm 2, \ldots} A_n \left[s(m-n) + \delta_{m,n} \frac{T}{a} Z_s \begin{cases} \sqrt{1-(\sin\vartheta + m_s\lambda_0/T)^2} \\ -j\sqrt{(\sin\vartheta + m_s\lambda_0/T)^2 - 1} \end{cases} \right] = -s(m) \tag{17}$$

with $Z_s = 1/G_s$ and the abbreviations :

$$s(m) = \frac{\sin(m\pi a/T)}{m\pi a/T} \quad ; \quad s(0) = 1 \quad ; \quad s(-m) = s(m) \tag{18}$$

The upper form after the brace holds if $m \leq m_s$, with m_s the limit of orders of radiating harmonics, otherwise the lower form holds.

For $a/T \to 1$ it follows that $A_{n\neq 0}=0$ and $A_0 = (Z_s \cos\vartheta - 1)/(Z_s \cos\vartheta + 1)$. This is the analytic justification for making a homogeneous (bulk reacting) absorber layer locally reacting by thin partition walls with small distances.

The examples shown below use the parameters $F= f\cdot t/c_0= t/\lambda_0$; $R= \Xi\cdot t/Z_0$ with the flow resistivity Ξ of the porous material (glass fibres) in the grooves; a/T ; T/t . The 1st graph shows the magnitude of the reflection factor $|r|$ for a homogeneous surface (dashed) and a periodic surface (full) (correspondence: "theta" $\to \vartheta$).

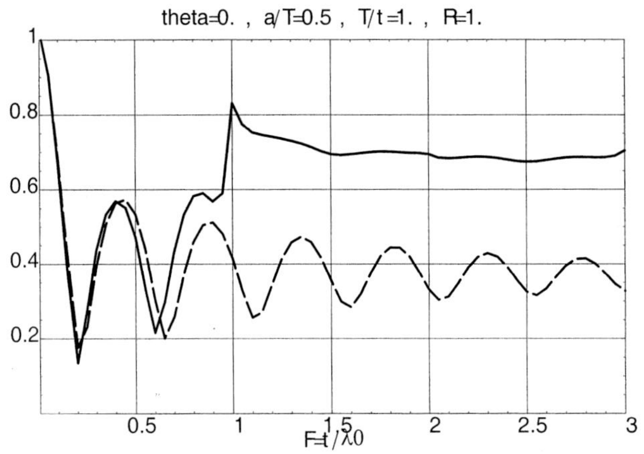

Magnitude of the reflection factor |r| of wall with grooves for normal incidence.
Full line: periodic surface; dashed: homogeneous.

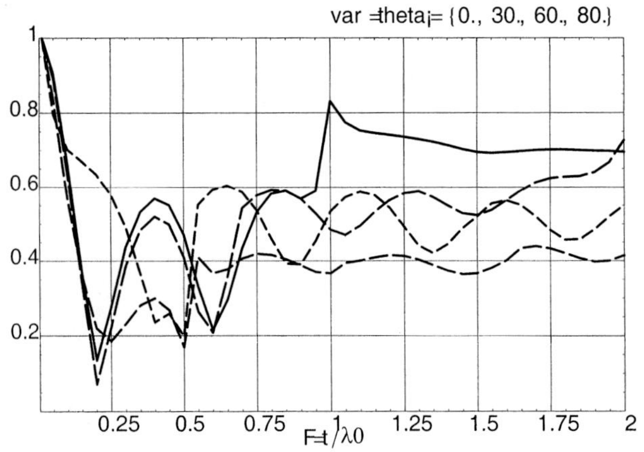

Magnitude of the reflection factor |r| of wall with grooves for normal incidence, as a periodic surface for a list of ϑ values (dashes become shorter for increasing list place). $a/T= 0.5$; $T/t= 1$; $R= 1$.

B.20 Plane wall with wide grooves

Ref.: Mechel, [B.17]

In contrast to the previous section B.19 the grooves are no longer narrow; higher modes may exist in them.

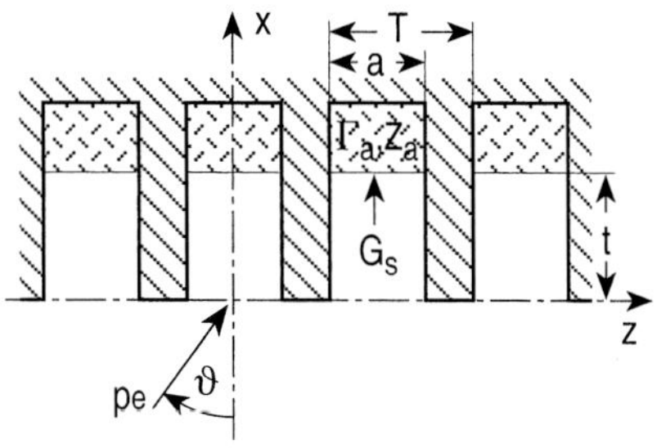

The ground of the grooves is supposed to be terminated with an admittance G_s (e.g. produced by a porous layer there).

Where possible, the relations are taken from the previous section B.19 .

The grooves are numbered $\nu = 0, \pm 1, \pm 2, \ldots$ and a co-ordinate $z_\nu = z - \nu \cdot T$ is used in the ν-th groove with $-a/2 \le z_\nu \le +a/2$. The field in the groove is formulated as :

$$p_k(x, z_\nu) = e^{-j\beta_0 \cdot \nu T} \sum_{m \ge 0} \left[B_m \cdot e^{-j\kappa_m x} + C_m \cdot e^{+j\kappa_m x} \right] \cdot \cos\left(m\pi \left(\frac{z_\nu}{a} - \frac{1}{2} \right) \right) \tag{1}$$

with :

$$\kappa_m = \begin{cases} \sqrt{k_0^2 - (m\pi/a)^2} \ge 0 & ; \; k_0 \ge m\pi/a \\ -j\sqrt{(m\pi/a)^2 - k_0^2} \ge 0 & ; \; k_0 < m\pi/a \end{cases} \tag{2}$$

The amplitudes C_m of the groove modes reflected at the ground are :

$$C_m = -B_m \cdot \frac{G_s - \kappa_m / k_0}{G_s + \kappa_m / k_0} \cdot e^{-2j\kappa_m t} := B_m \cdot R_m \tag{3}$$

The R_m are modal reflection factors "measured" in the groove orifice.

The boundary conditions in the plane $x=0$ lead to the inhomogeneous linear system of equations $(m = 0, \pm 1, \pm 2, \ldots)$ for the amplitudes A_n of the spatial harmonics, which exist in the half space $x < 0$ (with $\delta_{n.m}$= Kronecker symbol; $\delta_0 = 1$; $\delta_{m>0} = 2$):

$$\sum_{n=-\infty}^{+\infty} A_n \left[\frac{a}{2T} \sum_{\mu=0}^{\infty} \frac{\delta_\mu}{2} \frac{\kappa_\mu}{k_0} \frac{1 - R_\mu}{1 + R_\mu} (-1)^\mu s_{-\mu,n} \cdot s_{\mu,m} - \delta_{n,m} \frac{j\gamma_m}{k_0} \right] = P_e \cdot \left[\delta_{0,m} \cdot \cos\vartheta - \frac{a}{2T} \sum_{\mu=0}^{\infty} \frac{\delta_\mu}{2} \frac{\kappa_\mu}{k_0} \frac{1 - R_\mu}{1 + R_\mu} (-1)^\mu s_{-\mu,0} \cdot s_{\mu,m} \right] \tag{4}$$

with the abbreviation :

$$s_{m,n} = e^{j\,m\pi/2}\left[\frac{\sin(m\pi/2 - ß_n a/2)}{m\pi/2 - ß_n a/2} + (-1)^m \frac{\sin(m\pi/2 + ß_n a/2)}{m\pi/2 + ß_n a/2}\right] \tag{5}$$

The amplitudes B_m follow with the A_n from :

$$B_m = \frac{(-1)^m \delta_m}{2(1+R_m)}\left[P_e \cdot s_{-m,0} + \sum_{n=0,\pm1,\pm2\ldots} A_n \cdot s_{-m,n}\right] \tag{6}$$

The reflection factor r of the arrangement follows with the A_n as in the previous section B.19.

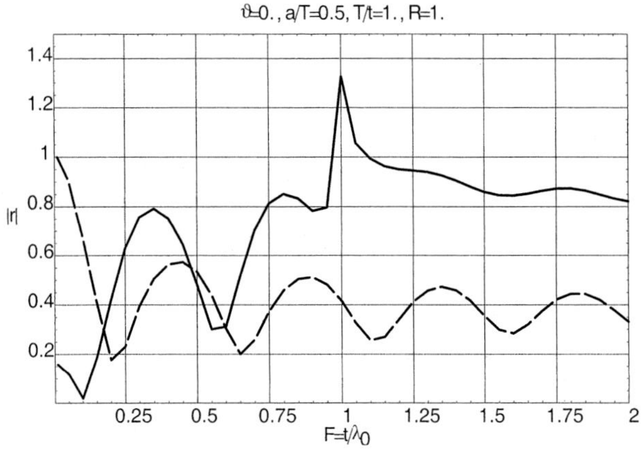

Magnitude of the reflection factor |r| of a wall with wide grooves. The grooves are completely filled with glass fibre material;
t= groove depth;
$R=\Xi\cdot t/Z_0$.
Full: with spatial harmonics
dashed: homogeneous

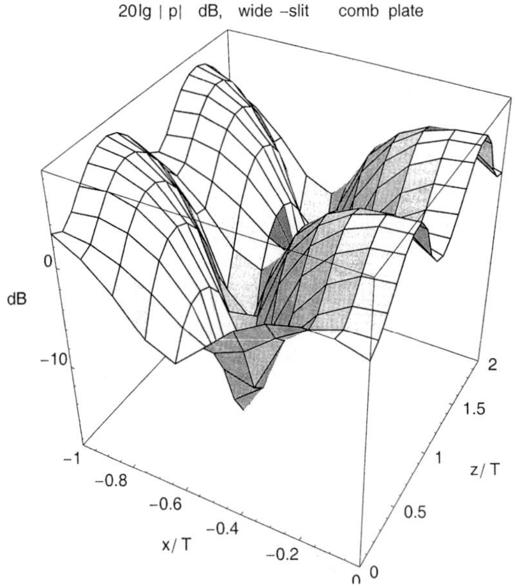

Sound pressure level in front of a wall with wide grooves, completely filled with glass fibre material.
$F= T/\lambda_0= 0.75$; $\vartheta= 45°$;
$a/T= 0.5$; $T/t= 0.25$;
$R= \Xi\cdot t/Z_0= 1$.
One spatial harmonic is radiating, therefore the periodicity extends to far distances.

B.21 Thin grid on half-infinite porous layer

Ref.: Mechel, [B.17]

A thin grid with slits of width a at mutual distance T covers a half space of porous absorber material with flow resistivity Ξ and characteristic values Γ_a, Z_a (or in a normalised form $\Gamma_{an}=\Gamma_a/k_0$, $Z_{an}=Z_a/Z_0$).

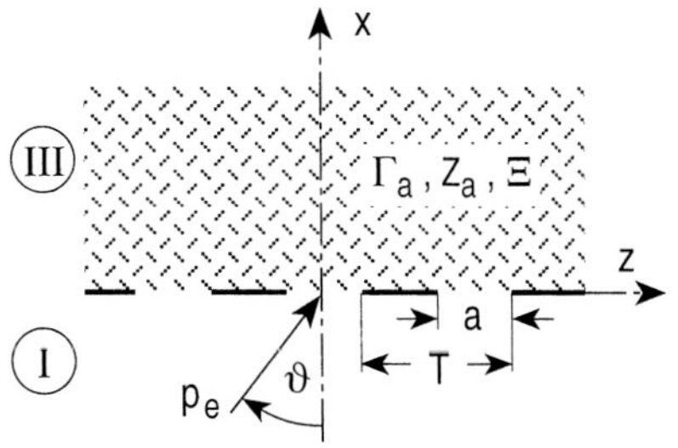

A plane wave p_e is incident at a polar angle ϑ (wave vector in the x,z-plane).

Field formulation in the zone I ($x \le 0$) :

$$p_I(x,z) = p_e(x,z) + p_s(x,z) = P_e \cdot e^{-j(k_x x + k_z z)} + \sum_{n=-\infty}^{+\infty} A_n \cdot e^{\gamma_n x} \cdot e^{-j\beta_n z} \tag{1}$$

$$\begin{gathered} k_x = k_0 \cos\vartheta \quad ; \quad k_z = k_0 \sin\vartheta \quad ; \quad k_x^2 + k_z^2 = k_0^2 \\ \beta_0 = k_z = k_0 \sin\vartheta \quad ; \quad \beta_n = \beta_0 + 2n\pi/T = k_0(\sin\vartheta + n\lambda_0/T) \\ \gamma_n^2 = \beta_n^2 - k_0^2 \quad ; \quad \gamma_0 = jk_0\cos\vartheta \quad ; \quad \gamma_n = k_0\sqrt{(\sin\vartheta + n\lambda_0/T)^2 - 1} \end{gathered} \tag{2}$$

Radiating spatial harmonics with order n_s in the limits :

$$-\frac{T}{\lambda_0}(1+\sin\vartheta) \le n_s \le \frac{T}{\lambda_0}(1-\sin\vartheta) \tag{3}$$

Field formulation in the zone III ($x \ge 0$) :

$$\begin{gathered} p_{III}(x,z) = \sum_{n=-\infty}^{+\infty} D_n \cdot e^{-\varepsilon_n x} \cdot e^{-j\beta_n z} = e^{-j\beta_0 z} \sum_{n=-\infty}^{+\infty} D_n \cdot e^{-\varepsilon_n x} \cdot e^{-j(2n\pi/T)z} \\ v_{IIIx}(x,z) = \frac{k_0}{\Gamma_a Z_a} e^{-j\beta_0 z} \sum_{n=-\infty}^{+\infty} D_n \frac{\varepsilon_n}{k_0} \cdot e^{-\varepsilon_n x} \cdot e^{-j(2n\pi/T)z} \end{gathered} \tag{4}$$

$$\varepsilon_n^2 = \beta_n^2 + \Gamma_a^2 \quad ; \quad \frac{\varepsilon_n}{k_0} = \sqrt{(\sin\vartheta + n\lambda_0/T)^2 + \Gamma_{an}^2} \tag{5}$$

The boundary conditions on the front and back side of the grid, together with the orthogonality of the spatial harmonics, leads to the linear inhomogeneous system of equations (m=0,±1,±2,...):

m=0 :
$$A_0\left(\frac{T}{a}\Gamma_{an}Z_{an}\cos\vartheta+\frac{\varepsilon_0}{k_0}\right)+\sum_{n\neq 0}A_n\frac{\varepsilon_n}{k_0}s(n)=P_e\left(\frac{T}{a}\Gamma_{an}Z_{an}\cos\vartheta-\frac{\varepsilon_0}{k_0}\right) \tag{6}$$

m≠0 :
$$A_0\cdot\frac{\varepsilon_0}{k_0}s(m)+\sum_{n\neq 0}A_n\left[\frac{\varepsilon_n}{k_0}s(m-n)-\delta_{m,n}\cdot\frac{T}{a}\Gamma_{an}Z_{an}\left(j\frac{\gamma_m}{k_0}\right)\right]=-P_e\cdot\frac{\varepsilon_0}{k_0}s(m) \tag{7}$$

with the Kronecker symbol $\delta_{m,n}$ and the abbreviation :

$$s(n):=\frac{\sin(n\pi a/T)}{n\pi a/T} \tag{8}$$

The amplitudes D_n follow from $D_0=P_e+A_0$; $D_{n\neq 0}=A_n$.

Special case $a/T\to 0$: (i.e. hard plane) it follows that $A_0=P_e$; $A_{n\neq 0}=0$;

special case $a/T\to 1$: (i.e. open absorber) it follows that $A_{n\neq 0}=0$ and

$$\frac{A_0}{P_e}=\left(\Gamma_{an}Z_{an}\frac{\cos\vartheta}{\sqrt{\sin^2\vartheta+\Gamma_{an}^2}}-1\right)\Bigg/\left(\Gamma_{an}Z_{an}\frac{\cos\vartheta}{\sqrt{\sin^2\vartheta+\Gamma_{an}^2}}+1\right) \tag{9}$$

which is just the reflection factor at a semi-infinite absorber layer.

Special case : ignore all higher spatial harmonics, i.e. $A_{n\neq 0}=0$:

$$\frac{A_0}{P_e}=\left(\frac{T}{a}\Gamma_{an}Z_{an}\frac{\cos\vartheta}{\sqrt{\sin^2\vartheta+\Gamma_{an}^2}}-1\right)\Bigg/\left(\frac{T}{a}\Gamma_{an}Z_{an}\frac{\cos\vartheta}{\sqrt{\sin^2\vartheta+\Gamma_{an}^2}}+1\right) \tag{10}$$

Special case : the material in the zone III is air : i.e $\Gamma_{an}=j$; $Z_{an}=1$; $\varepsilon_n/k_0\to\gamma_n/k_0$:

$$\sum_n A_n\cdot j\frac{\gamma_n}{k_0}\left[s(m-n)+\delta_{m,n}\cdot\frac{T}{a}\right]=P_e\cos\vartheta\cdot\left[s(m)+\delta_{0,m}\cdot\frac{T}{a}\right]\quad;\quad m=0,\pm 1,\pm 2,\ldots \tag{11}$$

The reflection coefficient $|r|^2$ is evaluated by :

$$|r|^2=\left|\frac{A_0}{P_e}\right|^2+\frac{1}{\cos\vartheta}\sum_{n=n_s\neq 0}\left|\frac{A_n}{P_e}\right|^2\sqrt{1-(\sin\vartheta+n\lambda_0/T)^2} \tag{12}$$

Parameters in the examples shown below are ϑ , a/T , $R=\Xi\cdot T/Z_0$, $F=T/\lambda_0$.

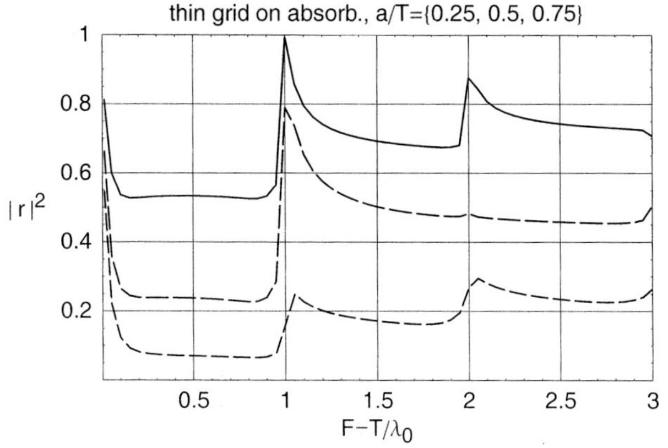

Reflection coefficient $|r|^2$ for a thin grid on an infinite glass fibre layer for different ratios a/T (dashes are shorter with increasing values).
$\vartheta=0$; $R=\Xi T/Z_0=1$.

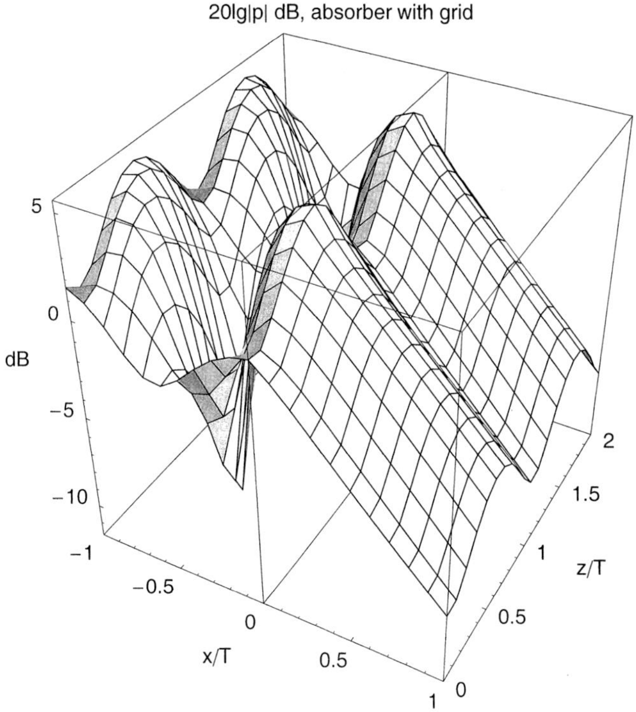

Sound pressure level in front of (x/T<0) and in (x/T>0) the absorber, covered with a thin grid.
$\vartheta=45°$; $T/\lambda_0=0.75$; $a/T=0.5$; $R=\Xi T/Z_0=1$.

B.22 Grid of finite thickness with narrow slits on half-infinite porous layer

Ref.: Mechel, [B.17]

In contrast to the object in the previous section B.21 the grid now has a finite thickness d, and the slits of the grid are assumed to be narrow so that only a plane wave must be assumed in the slits. The slits form the new zone II. The slit and grid period at z=0 can be taken as the representatives for the other slits.

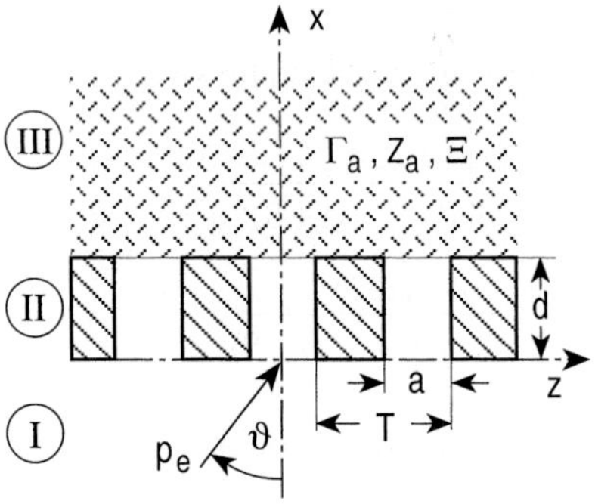

The field formulations in the zones I, III remain as in section B.21.

The field in the ν-th slit, with $z=\nu\cdot T+z_\nu$; $\nu=0,\pm1,\pm2,\ldots$, is formulated as :

$$p_{II}(x,z_\nu)=e^{-j\beta_0 z_\nu}\left[B\cdot e^{-jk_0x}+C\cdot e^{+jk_0x}\right]$$
$$Z_0 v_{xII}(x,z_\nu)=e^{-j\beta_0 z_\nu}\left[B\cdot e^{-jk_0x}-C\cdot e^{+jk_0x}\right] \qquad (1)$$

The boundary conditions lead to :

$$\frac{B}{P_e}=\frac{\sin(\beta_0 a/2)}{\beta_0 a/2}\,\frac{(1+S_a)e^{+jk_0d}}{(S+S_a)\cos(k_0d)+j(1+SS_a)\sin(k_0d)}$$
$$\frac{C}{P_e}=-\frac{\sin(\beta_0 a/2)}{\beta_0 a/2}\,\frac{(1-S_a)e^{-jk_0d}}{(S+S_a)\cos(k_0d)+j(1+SS_a)\sin(k_0d)} \qquad (2)$$

with the abbreviations :

$$S:=\frac{a}{T}\sum_{n=-\infty}^{+\infty} j\frac{k_0}{\gamma_n}\left(\frac{\sin(\beta_n a/2)}{\beta_n a/2}\right)^2 \quad ; \quad S_a:=\frac{a}{T}\frac{\Gamma_a Z_a}{k_0 Z_0}\sum_{n=-\infty}^{+\infty}\frac{k_0}{\varepsilon_n}\left(\frac{\sin(\beta_n a/2)}{\beta_n a/2}\right)^2 \qquad (3)$$

With B, C it follows that :

$$A_0=P_e-(B-C)\frac{a}{T\cos\vartheta}\,\frac{\sin(\beta_0 a/2)}{\beta_0 a/2}$$
$$A_m=-j(B-C)\frac{a}{T}\frac{k_0}{\gamma_m}\,\frac{\sin(\beta_m a/2)}{\beta_m a/2} \quad ; \quad m=\pm1,\pm2,\ldots \qquad (4)$$

$$D_m = \frac{a}{T}\frac{\Gamma_a Z_a}{k_0 Z_0}\left[B\cdot e^{-jk_0 d} - C\cdot e^{+jk_0 d}\right]\frac{k_0}{\varepsilon_m}\frac{\sin(\beta_m a/2)}{\beta_m a/2} \quad ; \quad m = 0, \pm 1, \pm 2, \ldots \tag{5}$$

The reflection coefficient $|r|^2$ follows from :

$$|r|^2 = \left|1 - \frac{a/T}{\cos\vartheta}\frac{\sin(\beta_0 a/2)}{\beta_0 a/2}(B - C)\right|^2 + \frac{a/T}{\cos\vartheta}|B - C|^2\left[\mathrm{Re}\{S\} - \frac{a/T}{\cos\vartheta}\left(\frac{\sin(\beta_0 a/2)}{\beta_0 a/2}\right)^2\right] \tag{6}$$

In the special case of air, instead of a porous material, behind the grid :

$$\frac{B}{P_e} = 2\frac{\sin(\beta_0 a/2)}{\beta_0 a/2}\frac{(1+S)e^{+jk_0 d}}{(1+S)^2 e^{+jk_0 d} - (1-S)^2 e^{-jk_0 d}}$$
$$\frac{C}{P_e} = -2\frac{\sin(\beta_0 a/2)}{\beta_0 a/2}\frac{(1-S)e^{-jk_0 d}}{(1+S)^2 e^{+jk_0 d} - (1-S)^2 e^{-jk_0 d}} \tag{7}$$

The parameters in the following examples are ϑ , a/T , d/T , $R = \Xi\cdot T/Z_0$, $F = T/\lambda_0$.

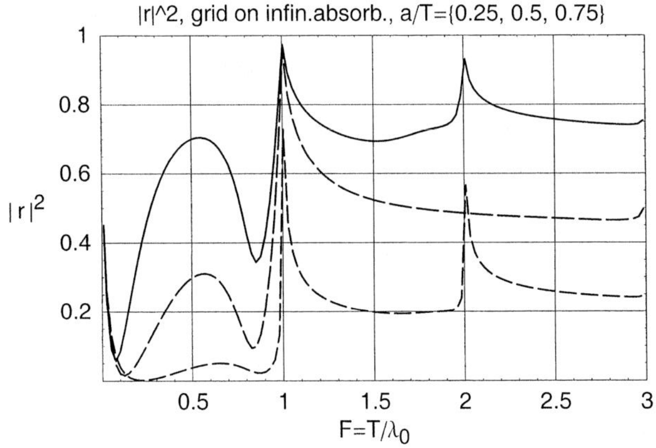

Reflection coefficient $|r|^2$ of a porous half-space covered with a grid with finite thickness for some ratios a/T (the dash becomes shorter for larger a/T).
$\vartheta = 0$; $d/T = 0.25$; $R = \Xi T/Z_0 = 1$

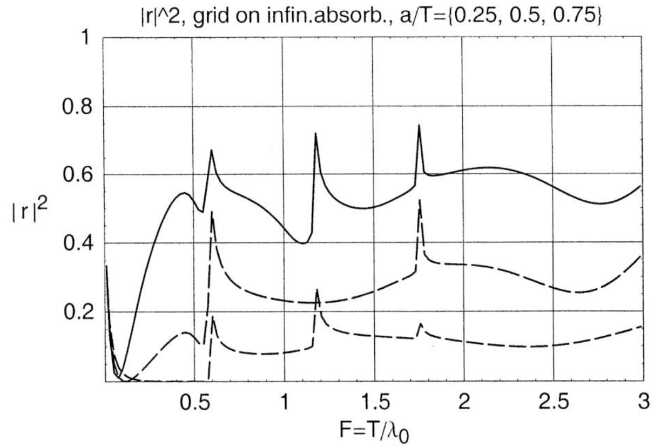

As above, but for $\vartheta = 45°$.

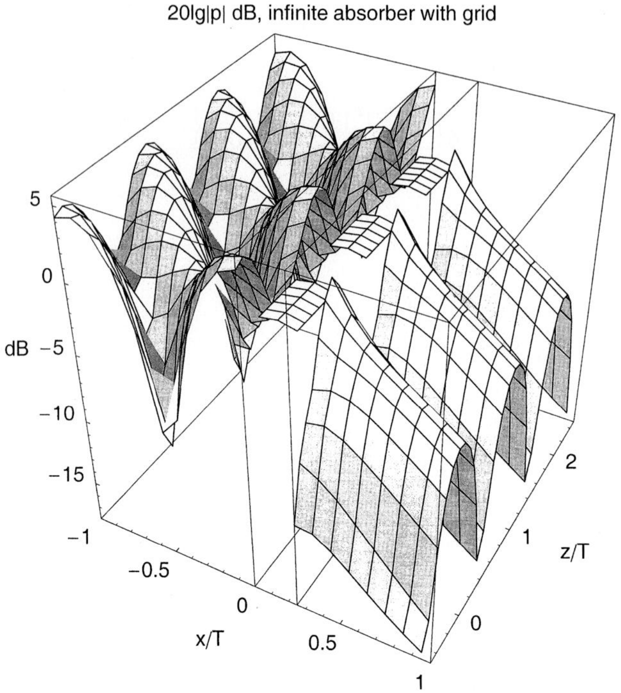

Sound pressure level in front of, in, and behind the grid. $\vartheta = 45°$, $F = T/\lambda_0 = 0.75$, $a/T = 0.5$, $d/T = 0.25$, $R = 1$. The assumption of only a plane wave in the slit channel produces only a least square error matching at the orifices.

B.23 Grid of finite thickness with wide slits on half-infinite porous layer

Ref.: Mechel, [B.17]

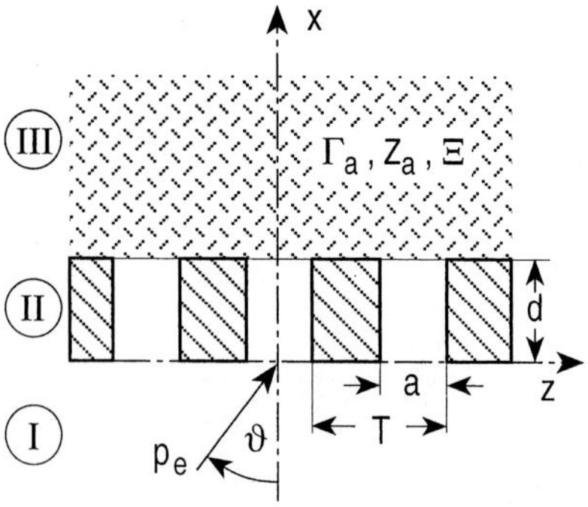

The object is the same as in the previous section B.22 , but the slit channels are no longer assumed to be narrow, i.e. higher modes are assumed in the slits.

The field formulations remain the same as in section B.21 .

The field in the ν-th slit, $\nu = 0, \pm 1, \pm 2, \ldots$, with $z_\nu = z - \nu \cdot T$, is formulated as :

$$p_{II}(x,z_\nu) = e^{-j\beta_0 \nu T} \sum_{m\ge 0} \left[B_m \cdot e^{-j\kappa_m x} + C_m \cdot e^{+j\kappa_m x} \right] \cdot \cos\left(m\pi(z_\nu/a - 1/2)\right)$$

$$\kappa_m = \begin{cases} \sqrt{k_0^2 - (m\pi/a)^2} & ;\ m \le m_g \\ -j\sqrt{(m\pi/a)^2 - k_0^2} & ;\ m > m_g \end{cases} \quad ; \quad m_g = \mathrm{Int}(k_0 a/\pi) \tag{1}$$

For the auxiliary amplitudes X_m, Y_m $(m\ge 0)$:

$$X_m := B_m - C_m \quad ; \quad Y_m := B_m \cdot e^{-j\kappa_m d} - C_m \cdot e^{+j\kappa_m d}$$

$$B_m = \frac{X_m \cdot e^{+j\kappa_m d} - Y_m}{e^{+j\kappa_m d} - e^{-j\kappa_m d}} \quad ; \quad C_m = \frac{X_m \cdot e^{-j\kappa_m d} - Y_m}{e^{+j\kappa_m d} - e^{-j\kappa_m d}} \tag{2}$$

a combined system of equations (m=0,1,2...) is derived from the boundary conditions :

$$\sum_{n\ge 0} X_n \left[\frac{\kappa_n}{k_0} \sum_{\nu=0,\pm1,\dots} \left(j\frac{k_0}{\gamma_\nu} \right) \cdot s_{-m,\nu} \cdot s_{n,\nu} + (-1)^m \frac{\delta_{m,n}}{\delta_m} \frac{4T}{a} \frac{1+e^{-2j\kappa_m d}}{1-e^{-2j\kappa_m d}} \right] = \frac{4T}{a} \left[P_e \cdot s_{-m,0} + \frac{2(-1)^m}{\delta_m} \frac{e^{-j\kappa_m d}}{1-e^{-2j\kappa_m d}} \cdot Y_m \right]$$

$$\sum_{n\ge 0} Y_n \left[\frac{\kappa_n}{k_0} \sum_{\nu=0,\pm1,\dots} \frac{k_0}{\varepsilon_\nu} \cdot s_{-m,\nu} \cdot s_{n,\nu} + (-1)^m \frac{\delta_{m,n}}{\delta_m} \frac{4T}{a} \frac{k_0 Z_0}{\Gamma_a Z_a} \frac{1+e^{-2j\kappa_m d}}{1-e^{-2j\kappa_m d}} \right] = \frac{8T}{a} \frac{k_0 Z_0}{\Gamma_a Z_a} \frac{(-1)^m}{\delta_m} \frac{e^{-j\kappa_m d}}{1-e^{-2j\kappa_m d}} \cdot X_m \tag{3}$$

with $\delta_{m,n}$ the Kronecker symbol ; $\delta_0=1$, $\delta_{n>0}=2$; and the abbreviation :

$$s_{m,n} := e^{j m\pi/2} \left[\frac{\sin(m\pi/2 - \beta_n a/2)}{m\pi/2 - \beta_n a/2} + (-1)^m \frac{\sin(m\pi/2 + \beta_n a/2)}{m\pi/2 + \beta_n a/2} \right] \tag{4}$$

with $s_{m,n} = s_{-m,n}$ and :

$$s_{m,n} = 2\frac{\beta_n a/2}{(\beta_n a/2)^2 - (m\pi/2)^2} \cdot \begin{cases} \sin(\beta_n a/2) & ;\ m = \text{even} \\ -j\cos(\beta_n a/2) & ;\ m = \text{odd} \end{cases} \tag{5}$$

The amplitudes A_n, D_n $(n=0,\pm1,\pm2,\dots)$ follow from :

$$A_n = j\frac{k_0}{\gamma_n} \left[\delta_{0,\nu} \cdot P_e \cdot \cos\vartheta - \frac{a}{2T} \sum_{m\ge 0} (B_m - C_m) \frac{\kappa_m}{k_0} \cdot s_{m,n} \right]$$

$$D_n = \frac{a}{2T} \frac{\Gamma_a Z_a}{k_0 Z_0} \frac{k_0}{\varepsilon_n} e^{+\varepsilon_n d} \sum_{m\ge 0} \left(B_m e^{-j\kappa_m d} - C_m e^{+j\kappa_m d} \right) \frac{\kappa_m}{k_0} \cdot s_{m,n} \tag{6}$$

The reflection coefficient $|r|^2$ is evaluated as in the previous section B.22 .

The parameters in the following examples are ϑ , a/T , d/T , R= $\Xi \cdot T/Z_0$, F= T/λ_0 .

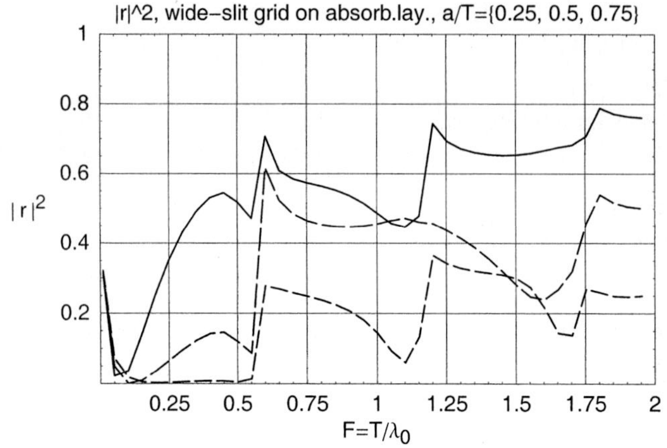

Reflection coefficient of a thick layer of glass fibres, covered with a grid with wide slits for some ratios a/T (the dashes are shorter for higher a/T). ϑ= 45° ; d/T= 0.25 ; R= $\Xi \cdot T/Z_0$= 1 .

References to part B :

General Linear Fluid acoustics

[B.1] MECHEL, F.P.
"Schallabsorber", Vol. II, Chapter 3 :
"Field equations for viscous and heat conducting media"
S.Hirzel Verlag, Stuttgart, 1995

[B.2] VDI-Waermeatlas, 4th edition, 1984
VDI Verlag, Duesseldorf

[B.3] MECHEL, F.P.
"Schallabsorber", Vol. I, Chapter 3
"Sound fields ; Fundamentals"
S.Hirzel Verlag, Stuttgart, 1989

[B.4] PIERCE, A.D.
"Acoustics", Chapter 4:
Acoust.Soc.America

[B.6] MECHEL, F.P.
"Schallabsorber", Vol. III, Chapter 26:
"Rectangular duct with local lining"
S.Hirzel Verlag, Stuttgart, 1998

[B.7] MECHEL, F.P.
"Schallabsorber", Vol. III, Chapter 27:
"Rectangular duct with lateral lining"
S.Hirzel Verlag, Stuttgart, 1998

[B.7] CUMMINGS, A, Proc.Inst.of Acoust. 11 (1989) part 5, 643-650
"Sound Generation in a Duct with a Bulk-Reacting Liner"

[B.9] SKUDRZYK, E.
The Foundations of Acoustics",
Springer, N.Y., 1971

[B.10] OCHMANN, M.; DONNER, U.1994 Acta Acustica **2**, 247-255
„Investigation of silencers with asymmetrical lining; I: Theory"

[B.10] MECHEL, F.P.
„Schallabsorber", Vol..III, Chapter 18:
"Multi-layer finite walls"
S.Hirzel Verlag, Stuttgart, 1998

[B.10] MECHEL, F.P. Acta Acustica, 2000
"A Principle of Superposition"

[B.11] CREMER, L., HECKL, M.
"Koerperschall", 2nd edit.
Springer, Berlin, 1995

[B.11] MORSE, P.M., FESHBACH, H.
"Metheods of Theoretical Physics", part I
McGraw-Hill, N.Y., 1953

[B.12] MOON, P.; SPENCER, D.E.
"Field Theory Handbook", 2nd Edition
Springer, Heidelberg, 1971

[B.13] KLEINSTEIN, GUNZBURGER J. Sound Vibr. 48 (1976) 169-178

[B.14] GOTTLIEB J.Sound Vibr. 40 (1975) 521-533

[B.15] FELSEN, L.B., MARCUVITZ, N.
Radiation and Scattering of Waves", p.89
Prentice Hall, London, 1973

[B.16] MECHEL, F.P.
"Schallabsorber", Vol. I, Chapter 11
"Surface Waves"
S.Hirzel Verlag, Stuttgart, 1989

[B.17] MECHEL, F.P.
"Schallabsorber", Vol. I, Chapter 12
"Periodic Structures"
S.Hirzel Verlag, Stuttgart, 1989

C
Equivalent Networks

The application of equivalent networks is a useful method for the solution of many tasks in acoustics. The method is applicable, if the sound field at any value of the co-ordinate x in the "direction of propagation" has the same lateral distribution. Plane waves are just a special case.

The conception of *end corrections* or, equivalently, *oscillating mass* extends the range of application even to a space with contractions. The method of equivalent networks is based on the analogies (*electro-acoustic analogies*) with electrical circuits.

C.1 Fundamentals of equivalent networks

Ref.: Mechel, [C.1]

Electromagnetic quantities and their relations (A is a cross-sectional area, or Ampere) :

Quantities					
electric			**magnetic**		
Quantity	**Relation**	**Dimension**	**Quantity**	**Relation**	**Dimension**
Voltage	U	Volt, V	Current	I	Ampere, A
El. field strength	E	V/m	Magn. field strength	H	A/m
El. induction	D=q/A	A·sec/m	Magn. induction	$B=\Phi_m/A$	V·sec/m
Charge, flow	$q=\int I\,dt = D\cdot A = \Phi_e$	A·sec	Flow	$\Phi_m=\int U\,dt = B\cdot A$	V·sec
Voltage	$U=\int E\,ds$	V	Current	$I=\oint H\cdot ds$	A
Current	I= dq/dt	A	Voltage	$U=d\Phi_m/dt$	V
Capacity	$C= q/U =\Phi_e/U$	A·sec/V	Inductivity	$L= U/(dI/dt) =\Phi_m/(dI/dt)$	V·sec/A

Table C.1 : Electromagnetic quantities and their relations.

Table C. 2 & 3 : Passive electrical and mechanical circuit components .

Element	Quantity	Symbol	Letter	Definition
Resistor	Resistance		R	$R = \frac{\Delta U}{I}$
Capacitor	Capacity	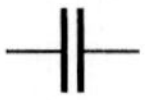	C	$C = \frac{\int I \cdot dt}{\Delta U}$
Coil	Inductivity	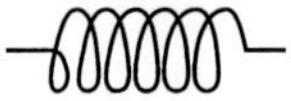	L	$L = \frac{\Delta U}{dI/dt}$
Complex Impedance			Z	$Z = \frac{\Delta U}{I}$
Complex Admittance			G	$G = \frac{I}{\Delta U} = \frac{1}{Z}$
Connection				$\Delta U = 0$; $\Delta I = 0$
Transformer		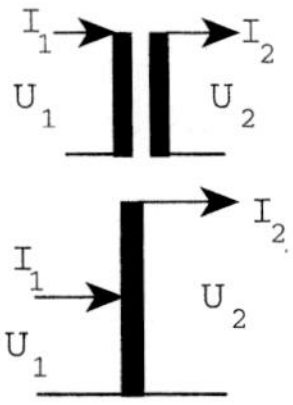	$u = \frac{w_2}{w_1}$	$u = \frac{U_2}{U_1} = \frac{I_1}{I_2}$

Element	Quantity	Symbol	Letter	Definition
Resistor	Friction		R	$R = \frac{\Delta F}{v}$
Spring	Compliance		C	$C = \frac{\int \Delta v\, dt}{\Delta F}$
Mass	Inertance		M	$M = \frac{\Delta F}{dv/dt}$
Complex Impedance			Z	$Z = \frac{\Delta F}{v}$
Complex Admittance			G	$G = \frac{v}{\Delta F} = \frac{1}{Z}$
Rigid connection				$\Delta F = 0$; $\Delta v = 0$
Lever		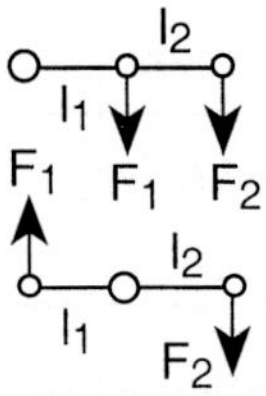	$u = \frac{l_2}{l_1}$	$u = \frac{F_2}{F_1} = \frac{v_1}{v_2}$

Table C.4 : Defining relations for passive mechanical circuit elements.

Element	Co-ordinates	Relations
Friction	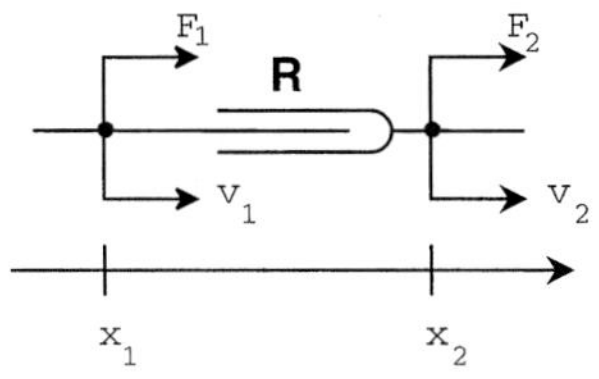	$F = F_1 - F_2$ $v = v_1 - v_2$ $F = R \cdot v$
Spring	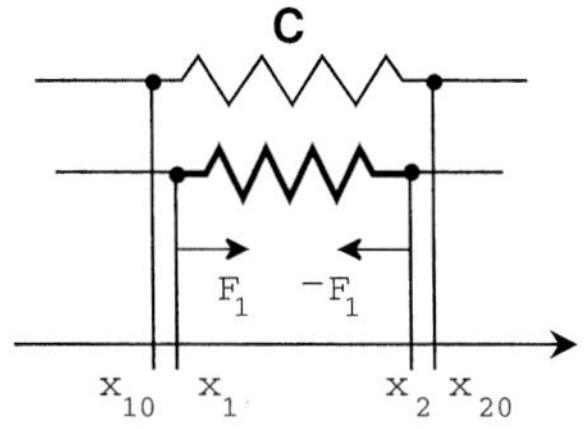	$\xi_1 = (x_1 - x_{10})$ $\xi_2 = (x_2 - x_{20})$ $\xi = \xi_1 - \xi_2$ $F = F_1 = \frac{1}{j\omega C} \cdot v$
Mass	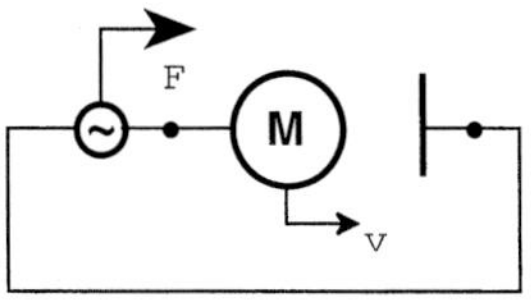	$F = j\omega M \cdot v$
Lever	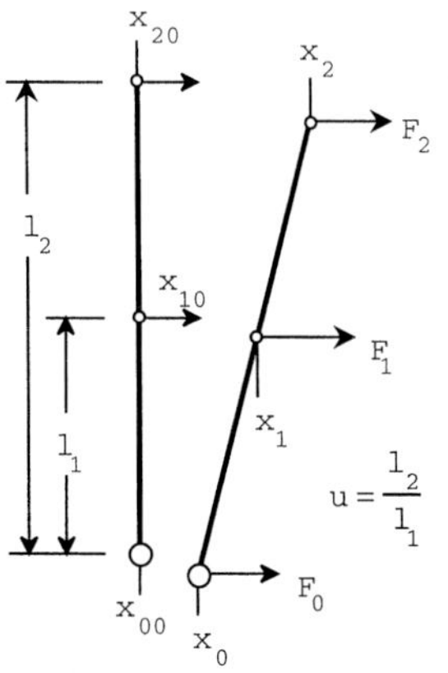	$\xi_{00} = x_0 - x_{00}$ $\xi_{10} = x_1 - x_{10}$ $\xi_{20} = x_2 - x_{20}$ $\xi_1 = \xi_{10} - \xi_{00}$ $\xi_2 = \xi_{20} - \xi_{00}$ $F_0 = -(F_1 + F_2)$ $F_2 = (l_1/l_2) \cdot F_1$ $\xi_2 = (l_2/l_1) \cdot \xi_1$

F= force;
v= velocity;
x= position;
ξ= deformation;
R= friction factor;
M= mass;
C= compliance;
l= length;

The velocity of a resistance is the relative velocity of both ends of the resistance.
The velocity of a spring is the relative velocity of both ends of the spring.
The velocity of a mass is its velocity relative to the point on which the force source is supported.
The force acting on a resistance or a spring is the force difference at both ends of the element.

Boundary conditions :

Node theorem : The sum of all forces acting on an immaterial node point is zero.

Mesh theorem: The sum of the velocities in a closed mesh is zero.

A spring is supposed to have no mass; a mass is supposed to be incompressible.

A hard (or rigid) termination with $v=0$ corresponds in electrical circuits

- to an open termination in the UK-analogy (see below),
- to a short-circuited termination in the Uv-analogy;

a soft (or pressure release) termination with $F=0$ (or $p=0$) corresponds in electrical circuits

- to a short-circuited termination in the UK-analogy,
- to an open termination in the Uv-analogy.

Rules :

- There is no force difference across a spring.
- There is no velocity difference across a mass.
- The second pole of a mass is at the point on which the driving force is supported.

The relations for levers with a leverage u: $\begin{pmatrix} v_2 \\ F_2 \end{pmatrix} = \begin{pmatrix} u & 0 \\ 0 & \frac{1}{u} \end{pmatrix} \cdot \begin{pmatrix} v_1 \\ F_1 \end{pmatrix}$

Sources :

Helmholtz theorem :

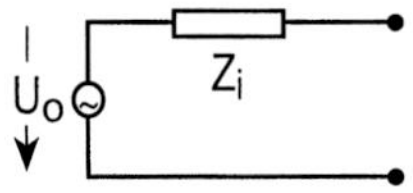

$U_o = Z_i \cdot I_s$

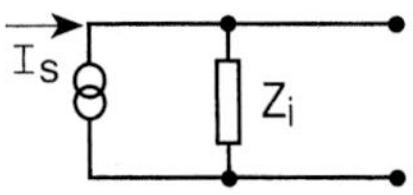

Z_i is the internal source impedance;

U_o is the open-circuit source voltage,

I_s is the short-circuit source current.

Reciprocal networks :

A reciprocal network is composed of elements Z_r which follow from the elements Z of the original network by the rule $Z \rightarrow Z_d = r^2/Z$ with the *reciprocal invariant* r . With suitably normalised impedances, one can take $r=1$. Voltage sources change to current sources, and inversely.

In both networks voltage transfer ratios $\leftrightarrow$ current transfer ratios correspond to each other, with same frequency response curves.

An advantage of the reciprocal network possibly is its easier conception and realisation.

The shape of the reciprocal network changes : a mesh changes to a node; a node changes to a mesh (see below for a more precise rule).

Table C.5 : Reciprocal electrical elements.

← reciprocal →			
Resistance	R 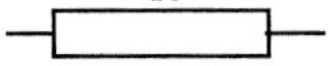	r^2/R 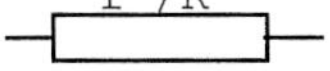	reciprocal resistance
Inductivity	L 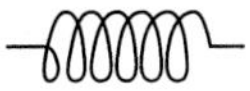	L/r^2 	Capacity
Capacity	C 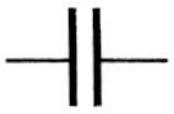	r^2C 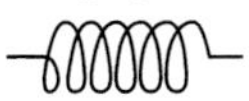	Inductivity
Impedance	Z 	r^2/Z 	Admittance
Admittance	G 	r^2G 	Impedance
Voltage source	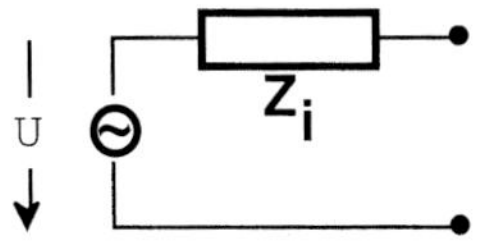	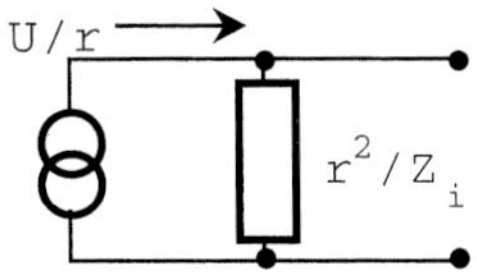	Current source
Current source	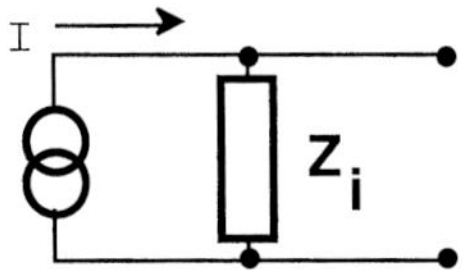	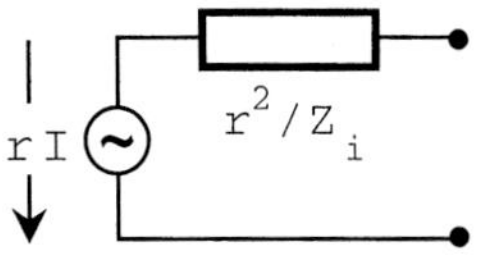	Voltage source

Rule for the construction of the reciprocal network of networks, which can be drawn in one plane:

- Draw a point into every mesh of the original network, and one point outside the network.
- Connect all pairs of points with each other by lines, which cross circuit elements.
- Replace the crossed elements with their reciprocal elements.
- If necessary, redraw the reciprocal network in a better form.

Below is an example for the application of this rule.

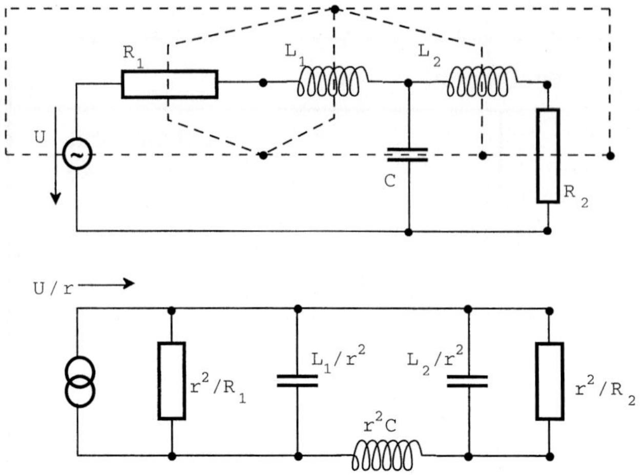

Fig. C.1 :
Example of reciprocal networks.

Electro-acoustic UK-analogy :

Table C.6 :   Corresponding elements in the UK-analogy.

	electrical				mechanical			
Voltage	U				Force	F		
Current	I				Velocity	v		
Resistance	R_e	—▭—	$U = R_e \cdot I$		Resistance	R_m	—▭—	$F = R_m \cdot v$
Coil	L	—⏛—	$U = j\omega L \cdot I$		Mass	M	—◯—	$F = j\omega M \cdot v$
Condensator	C_e	—∥—	$U = \dfrac{1}{j\omega C_e} \cdot I$		Spring	C_m	—◊◊◊—	$F = \dfrac{1}{j\omega C_m} \cdot v$
Impedance	Z_e	—▭—	$U = Z_e \cdot I$		Impedance	Z_m	—▭—	$F = Z_m \cdot v$
Admittance	G_e	—▭—	$U = \dfrac{1}{G_e} \cdot I$		Admittance	G_m	—▭—	$F = \dfrac{1}{G_m} \cdot v$
Voltage source	$U_0 \ominus \quad Z_i$				Force source	$F_0 \ominus \quad Z_i$		
Current source	$I_s \quad Z_i$				Velocity source	$v_s \quad Z_i$		
Node	✳	$\sum I = 0$			Mesh	◇	$\sum v = 0$	
Mesh	◇	$\sum U = 0$			Node	✳	$\sum F = 0$	

Electro-acoustic Uv-analogy :

Table C.7 : Corresponding elements in the Uv-analogy.

elektrical				mechanical			
Voltage	**U**			Velocity	**v**		
Current	**I**			Force	**F**		
Resistance	R_e		$U=R_e\cdot I$	Resistance	$1/R_m$		$v=\frac{1}{R_m}\cdot F$
Coil	**L**		$U=j\omega L\ I$	Spring	C_m		$v=j\omega C_m\cdot F$
Condensator	C_e		$U=\frac{1}{j\omega C_e}\cdot I$	Mass	**M**		$v=\frac{1}{j\omega M}\cdot F$
Impedance	Z_e		$U=Z_e\cdot I$	Admittance	G_m		$v=G_m\cdot F$
Admittance	G_e		$U=\frac{1}{G_e}\cdot I$	Impedance	Z_m		$v=\frac{1}{Z_m}\cdot F$
Voltage source	U_o, Z_i			Velocity source	v_o, Z_i		
Current source	I_s, Z_i			Force source	K_s, Z_i		
Node			$\sum I=0$	Node			$\sum F=0$
Mesh			$\sum U=0$	Mesh			$\sum v=0$

Networks in the UK and Uv analogy, respectively, are reciprocal to each other.

C.2 Distributed network elements

Ref.: Mechel, [C.1]

One distinguishes between "lumped" elements, as in Section C.1, with no sound propagation within one element, and "distributed" elements with internal sound propagation. Distributed elements are homogeneous, i.e. without change of the cross-section and/or material. They are introduced into the network analysis as four-poles, whereas lumped elements are two-poles.

Four-poles themselves can be represented as equivalent networks, either as T-networks, or as Π-networks. The four-pole representation is used for duct sections and/or layers with internal axial sound propagation. In the following formulas t is the duct section length or layer thickness t .

The axial propagation constant Γ_a is either the characteristic propagation constant of the medium in the duct or layer for a plane wave propagating in the axial direction, or the axial component for oblique propagation. Correspondingly, Z_a is the characteristic impedance of the medium, or the axial component, respectively. If the medium in the duct or layer is air, then $\Gamma_a = j\cdot k_0$; $Z_a = Z_0$.

Four-pole equations:

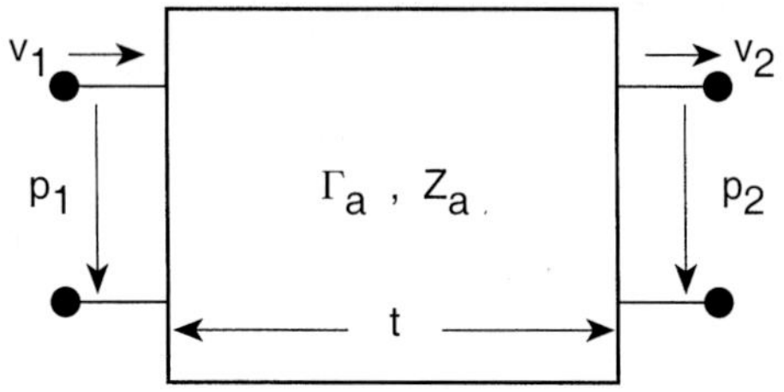

$$p_1 = \cosh(\Gamma_a t)\cdot p_2 + Z_a \sinh(\Gamma_a t)\cdot v_2 \qquad (1)$$

$$Z_a \cdot v_1 = \sinh(\Gamma_a t)\cdot p_2 + Z_a \cosh(\Gamma_a t)\cdot v_2$$

Equivalent T-circuit impedances :

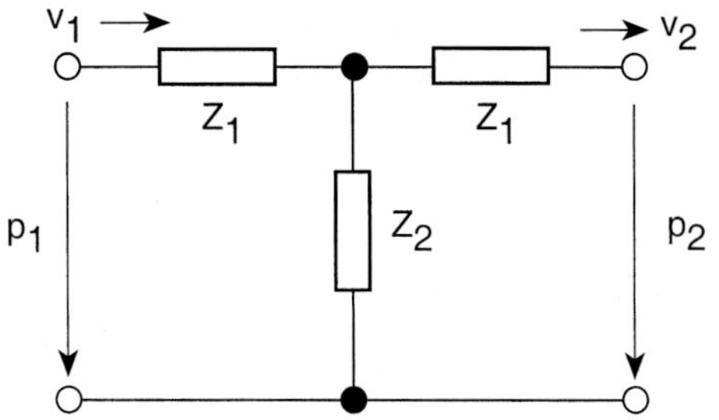

$$Z_1 = Z_a \cdot \coth(\Gamma_a t) - Z_2 = Z_a \frac{\cosh(\Gamma_a t) - 1}{\sinh(\Gamma_a t)} \qquad (2)$$

$$Z_2 = \frac{Z_a}{\sinh(\Gamma_a t)} \qquad (3)$$

Equivalent Π-circuit impedances :

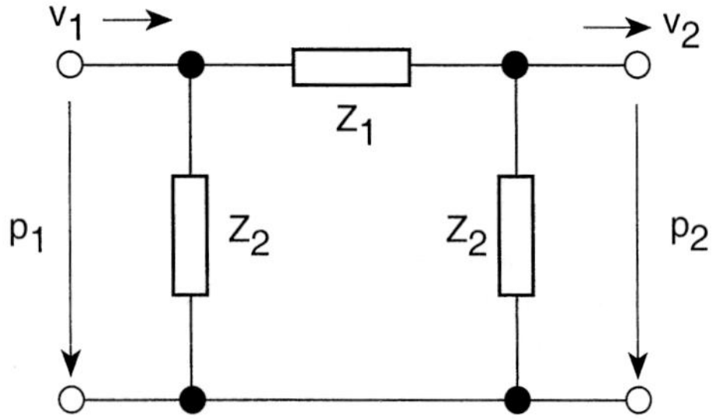

$$Z_1 = Z_a \cdot \sinh(\Gamma_a t) \qquad (4)$$

$$Z_2 = \frac{Z_a \sinh(\Gamma_a t)}{\cosh(\Gamma_a t) - 1}$$

$$\frac{1}{Z_2} = \frac{1}{Z_a \tanh(\Gamma_a t)} - \frac{1}{Z_1} \qquad (5)$$

Some simple systems with distributed network elements :

Tube section with hard termination :

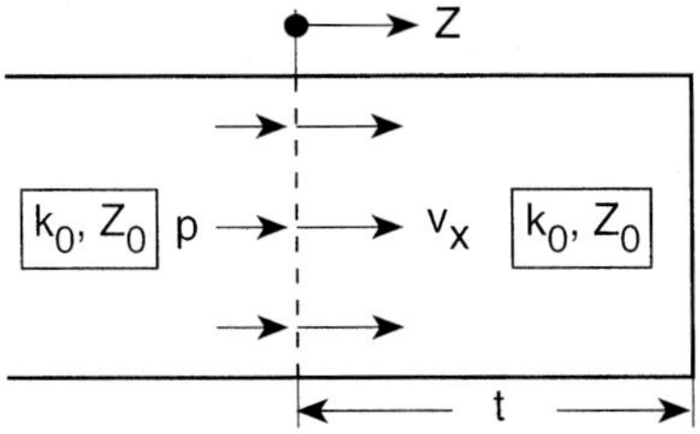

$$\frac{Z}{Z_0} = \frac{p}{v_x} = -j\cot(k_0 t)$$

$$Z \xrightarrow[t \ll \lambda_0]{} \frac{1}{j\omega}\frac{\rho_0 c_0^2}{t} = \frac{1}{j\omega C} \tag{6}$$

$$C = \frac{t}{\rho_0 c_0^2}$$

Remarks : • for $t \ll \lambda_0$ a spring-type reactance;

- 1st resonance at $k_0 t = \pi/2$; $t = \lambda_0/4$;
- for $\pi/2 < k_0 t < \pi$ a mass-type reactance;

Tube section with hard termination, filled with porous material :

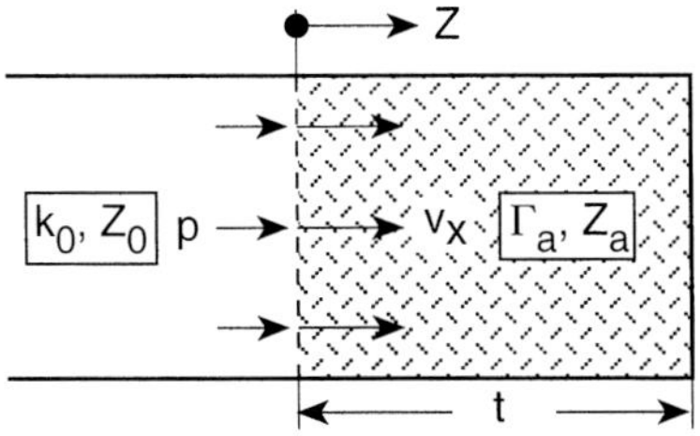

$$\frac{Z}{Z_0} = \frac{p}{v_x} = \frac{Z_a}{Z_0}\coth(\Gamma_a t)$$

$$Z \xrightarrow[t \ll \lambda_a]{} \frac{Z_a}{\Gamma_a t} \approx \frac{1}{j\omega}\frac{\rho_0 c_0^2}{\kappa\sigma t} \tag{7}$$

Remarks: • κ= adiabatic exponent of air ;

- σ= porosity of porous material;

Tube section with open termination :

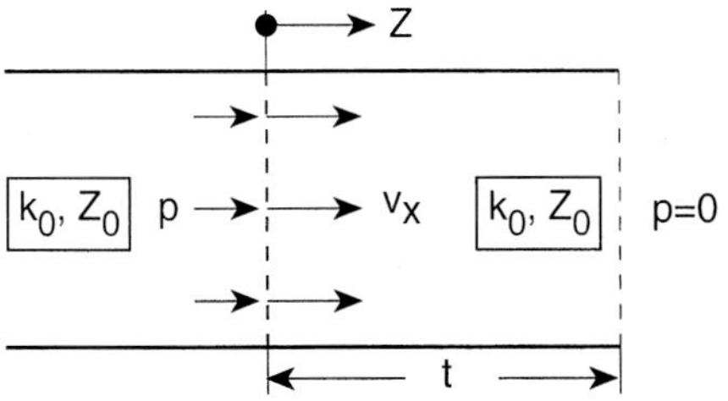

$$\frac{Z}{Z_0} = \frac{p}{v_x} = j\tan(k_0 t)$$

$$Z \xrightarrow[t \ll \lambda_0]{} j\omega\rho_0 t = j\omega M \tag{8}$$

$$M = \rho_0 t$$

Remarks: • the assumption p=0 at the orifice is an approximation for narrow tubes;

- without load of radiation impedance ! ;

- $t \ll \lambda_0$ mass-type reactance;
- $\pi/2 < k_0 t < \pi$ spring-type reactance ;

Tube section with open termination , filled with porous material:

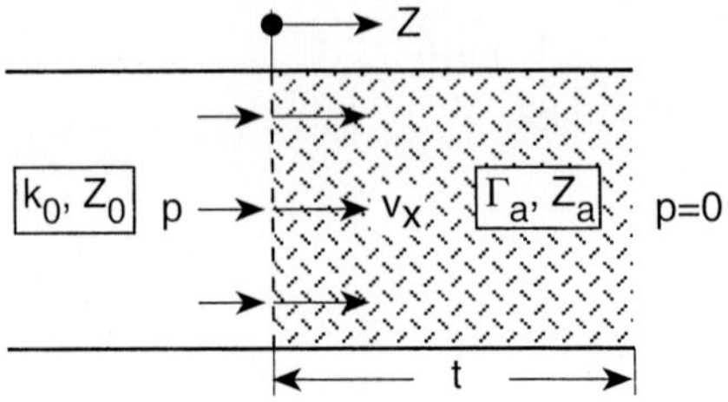

$$\frac{Z}{Z_0} = \frac{p}{v_x} = \frac{Z_a}{Z_0}\tanh(\Gamma_a t)$$

$$Z \xrightarrow[t\ll\lambda_a]{} \Gamma_a Z_a t \approx \frac{j}{\kappa\sigma} Z_0 k_0 t \cdot (\Gamma_a / k_0)^2 \quad (9)$$

$$\xrightarrow[E\ll 1]{} \frac{\Xi t}{\sigma} \; ; \quad \xrightarrow[E\gg 1]{} \frac{j\omega\rho_0 t}{\sigma}$$

Remarks:
- the assumption $p=0$ at the orifice is an approximation for narrow tubes;
- without load of radiation impedance ! ;
- $E=\rho_0 f/\Xi$ absorber parameter;
- Ξ flow resistivity of porous material;
- σ porosity of porous material;
- λ_a wavelength in absorber material ;
- $t \ll \lambda_a$ and $E \ll 1$: $Z \approx$ resistance,
  and $E > 1$: $Z \approx$ mass reactance.

Tube terminated with Helmholtz resonator with thin resonator plate :

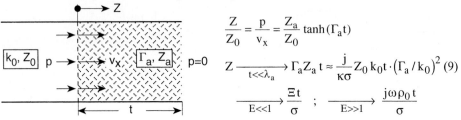

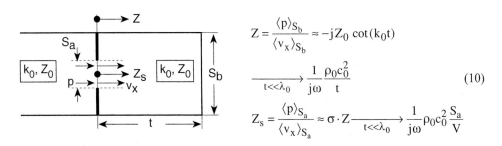

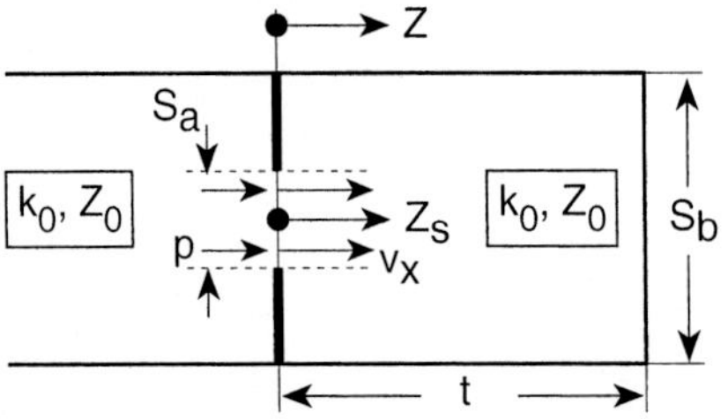

$$Z = \frac{\langle p \rangle_{S_b}}{\langle v_x \rangle_{S_b}} \approx -jZ_0 \cot(k_0 t)$$

$$\xrightarrow[t\ll\lambda_0]{} \frac{1}{j\omega}\frac{\rho_0 c_0^2}{t} \quad (10)$$

$$Z_s = \frac{\langle p \rangle_{S_a}}{\langle v_x \rangle_{S_a}} \approx \sigma \cdot Z \xrightarrow[t\ll\lambda_0]{} \frac{1}{j\omega}\rho_0 c_0^2 \frac{S_a}{V}$$

Remarks:
- end corrections of orifices neglected ! ;
- $S_a \cdot S_b \ll \lambda_0^2$;
- Z= "homogeneous" impedance;
- Z_s= interior orifice impedance;
- $\sigma = S_a/S_b$= resonator plate porosity;
- $V = t \cdot S_b$= resonator volume.

Perforated plate in a tube :

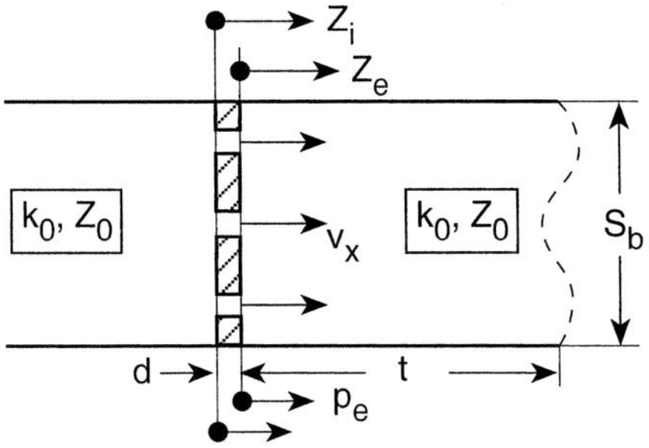

$$Z_i = Z_t + Z_e = \frac{1}{\sigma} Z_s + Z_h \qquad Z_t = \frac{1}{\sigma} Z_s \tag{11}$$

Remarks: • $d \ll \lambda_0$;
• Z_e= "homogeneous" load impedance;
• Z_i= "homogeneous" input impedance;
• Z_t= "homogeneous" partition impedance of plate;
• Z_s= partition impedance of perforations.

A layer of air (transformation of impedances by a layer):

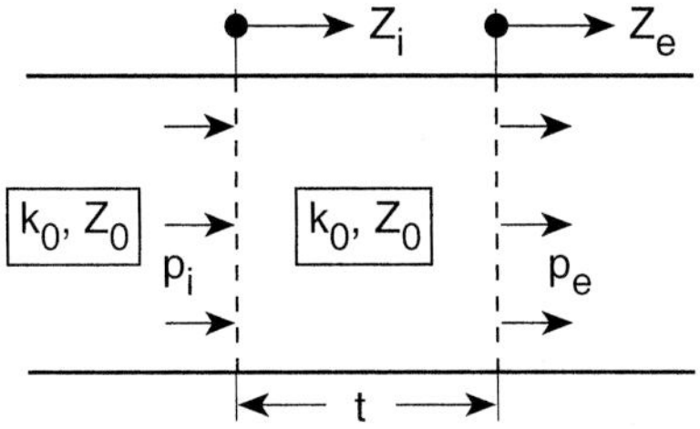

$$\frac{Z_i}{Z_0} = \frac{j\tan(k_0 t) + Z_e/Z_0}{1 + j\, Z_e/Z_0 \cdot \tan(k_0 t)} \qquad Z_i \xrightarrow[t \ll \lambda_0]{} \frac{j\omega\rho_0 + Z_e}{1 + Z_e \cdot j\dfrac{t}{\rho_0 c_0^2}} \tag{12}$$

Remarks: • Z_i= "homogeneous" input impedance;
• Z_e= "homogeneous" load impedance.

A layer of porous material (transformation of impedances by a layer):

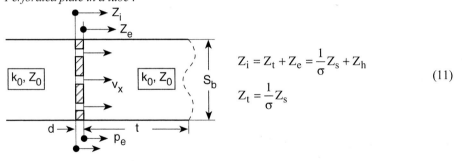

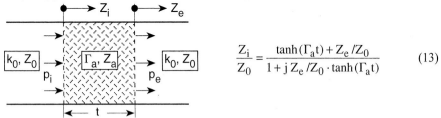

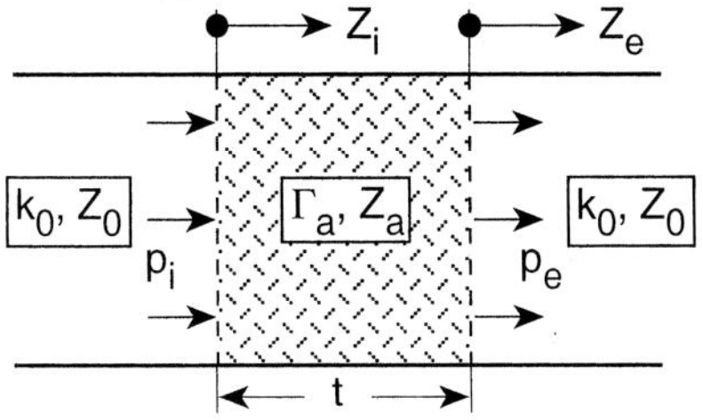

$$\frac{Z_i}{Z_0} = \frac{\tanh(\Gamma_a t) + Z_e/Z_0}{1 + j\, Z_e/Z_0 \cdot \tanh(\Gamma_a t)} \tag{13}$$

Remarks: • Z_i= "homogeneous" input impedance;
• Z_e= "homogeneous" load impedance.

C.3 Elements with constrictions

Ref.: Mechel, [C.1]

Equivalent networks can be used for sound in multi-layer absorbers or in ducts without constrictions, because the lateral sound distribution functions can be divided out. Nevertheless, constrictions can also be represented with equivalent networks, if their lateral dimensions and the lateral dimensions in front of and behind the constriction are small compared to the wavelength, or more precisely, if no higher modes can propagate in the wide cross-sections. Then the constrictions produce only near fields. These can be represented by *equivalent oscillating masses* M_i at the orifice on the side of sound incidence, and M_e at the orifice of sound exit.

The equivalent oscillating mass is proportional to the *end correction* .

The following examples show two Helmholtz resonators, one excited by an incident wave p_i on the resonator, the other excited by a sound pressure p_i at the back side of the resonator volume.

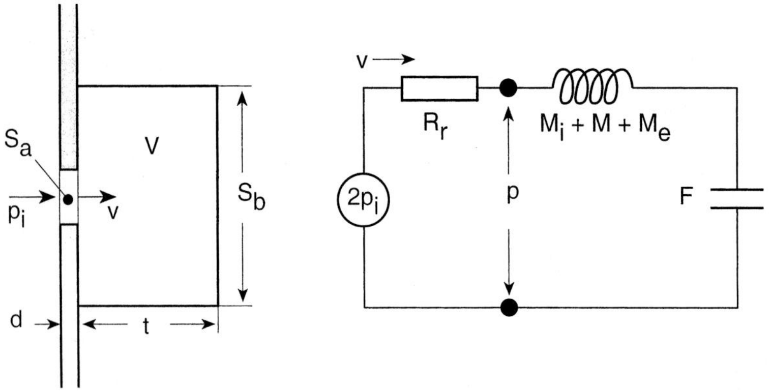

R_r represents the radiation resistance of the orifice towards the side of incidence;
M_i is the equivalent oscillating mass on the side of incidence (outer oscillating mass);
M_e is the equivalent oscillating mass on the side of exit (interior oscillating mass).

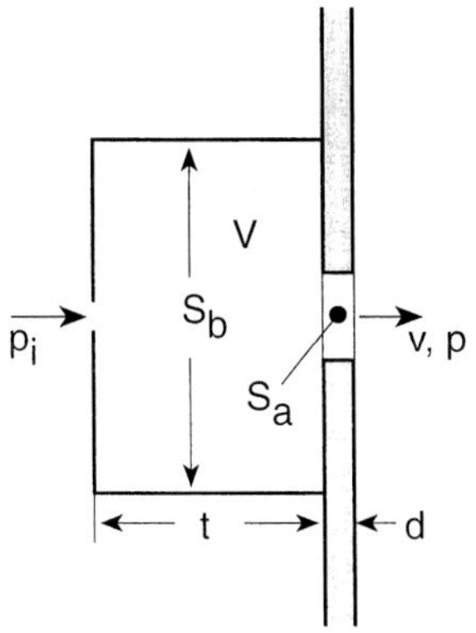

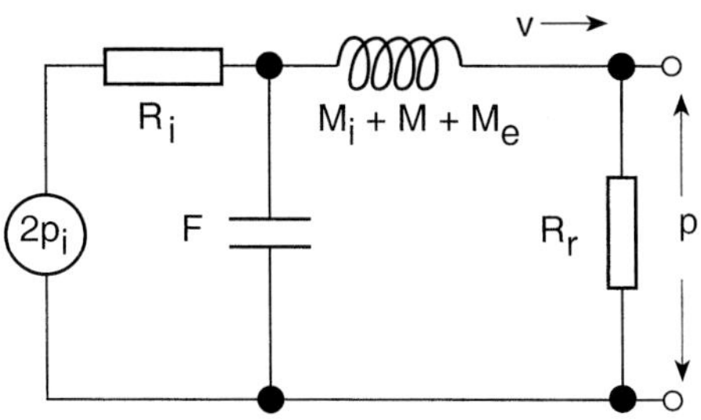

R_i represents the interior source resistance;
M_i is the equivalent oscillating mass on the side of incidence (outer oscillating mass);
M_e is the equivalent oscillating mass on the side of exit (interior oscillating mass).

C.4 Superposition of multiple sources in a network

Helmholtz's theorem of superposition for multiple sources :

If a network is excited by more than one voltage sources (current sources) with the same frequency, the state of the network with common excitation is a superposition of states, in which only one source is active, and the network terminals at the other sources are short-circuited (open).

C.5 Chain circuit

Ref.: Mechel, [C.1]

A chain circuit is a useful representation of multi-layer absorbers (see Section D.4).

A chain network consists of longitudinal impedances Z_n and lateral admittance G_n . Its sound pressures p_n in the nodes and velocities v_{ln} in the longitudinal elements, as well as v_{qn} in the transversal elements can be evaluated by iteration.

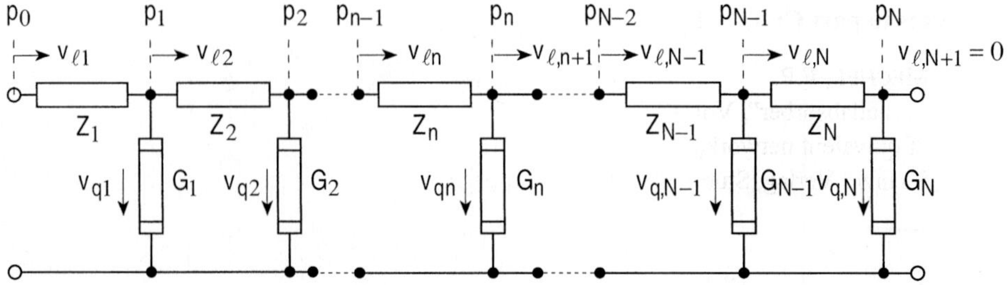

If the network is open (as shown), i.e. $v_{l,N+1}=0$, one begins with an assumed value $p_N=1$.

The backward recursion is :

$$\begin{aligned} v_{q,n} &= p_n \cdot G_n \\ v_{\ell,n} &= v_{q,n} + v_{\ell,n+1} \\ p_{n-1} &= p_n + v_{\ell,n} \cdot Z_n \end{aligned} \qquad (1)$$

One iterates over $n=N, N-1,\ldots,1$. The last result is p_0. All field quantities are proportional to p_N. To replace this by p_0 as the reference pressure, divide all (saved) quantities by the value of p_0. If parameters Z_n, G_n are used which are normalised with Z_0, the velocities are returned as $Z_0 v_n$.

If the real network is terminated with a load impedance Z_{load}, add $1/Z_{load}$ to G_N, so the network to be evaluated is open again.

The input impedance of the network is

$$Z = \frac{p_0}{v_{\ell,1}} = Z_1 + \frac{p_1}{v_{\ell,1}} \qquad (2)$$

The impedance of the part of the network behind the node n is

$$Y_n = \frac{p_n}{v_{\ell,n+1}} \quad ; \quad n = 1,2,\ldots,N-1 \qquad (3)$$

The load impedance at the node n is

$$X_n = \frac{p_n}{v_{\ell,n}} = \frac{p_n}{v_{q,n} + v_{\ell,n+1}} \quad ; \quad n = 1,2,\ldots,N \qquad (4)$$

Suitable representations of a layer of material with thickness t, propagation constant Γ_a and wave impedance Z_a of the material are :

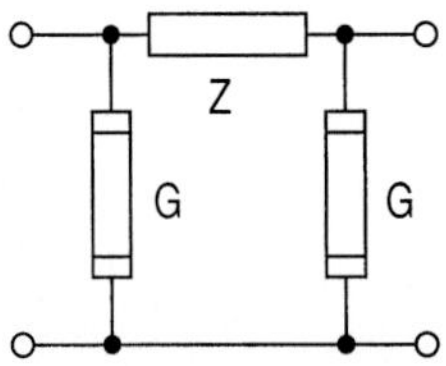

$$Z = Z_a \cdot \sinh(\Gamma_a t)$$
$$G = \frac{\cosh(\Gamma_a t) - 1}{Z_a \cdot \sinh(\Gamma_a t)}$$

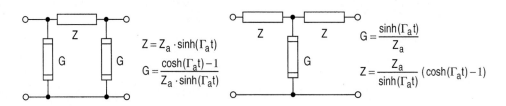

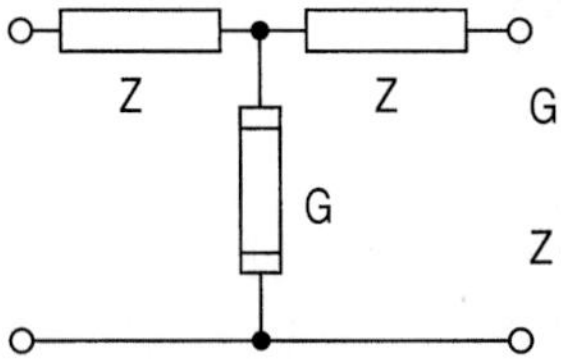

$$G = \frac{\sinh(\Gamma_a t)}{Z_a}$$
$$Z = \frac{Z_a}{\sinh(\Gamma_a t)}(\cosh(\Gamma_a t) - 1)$$

References to part C: Equivalent Network

[C.1] MECHEL, F.P.
"Schallabsorber", Vol. II, Chapter 2:
"Equivalent networks"
S.Hirzel Verlag, Stuttgart, 1995

D
Reflection of Sound

The limit between *reflection* and *scattering* of sound is not sharp. A generally applicable distinction could be: reflection sends sound back only into the half-space of incidence, scattering sends sound also in the forward direction. This is the guideline for placing topics either in this chapter about reflection of sound or in the later chapter about scattering of sound.

The general reference in this chapter is Mechel, [D.1]. Formulas for the input admittance and/or absorption coefficient can also be found in the chapter "Compound Absorbers".

D.1 Plane wave reflection at a locally reacting plane

An absorber is said to be *locally reacting*, if there is no sound propagation inside the absorber parallel to the absorber surface.

A plane wave with amplitude A is incident in the plane (x,y) on an absorbent plane; the y axis is in the absorber surface; the x axis is normal to the absorber and directed into the absorber; the wave vector of the incident wave $p_i(x,y)$ forms a polar angle Θ with the normal to the surface.

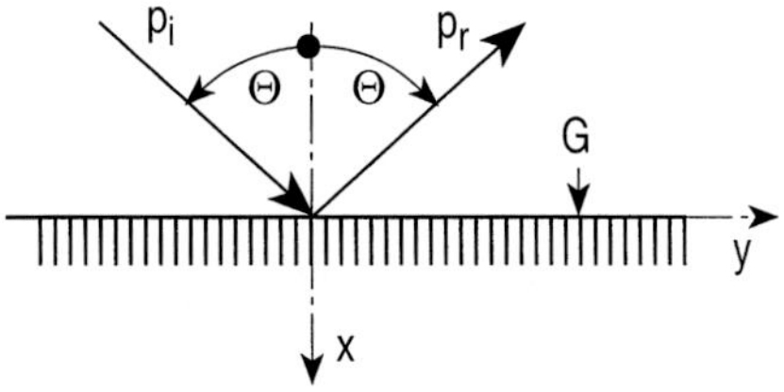

The acoustic quality of the absorber is defined by the wall admittance :

$$G = \frac{v_x(0,y)}{p(0,y)} = \frac{j}{k_0 Z_0}\,\frac{\partial p(0,y)/\partial x}{p(0,y)} \qquad (1)$$

Field formulation :

$$p(x,y) = p_i(x,y) + p_r(x,y)$$
$$p_i(x,y) = A\cdot e^{-jk_0(x\cdot\cos\Theta + y\cdot\sin\Theta)} \qquad (2)$$
$$p_r(x,y) = r\cdot A\cdot e^{-jk_0(-x\cdot\cos\Theta + y\cdot\sin\Theta)}$$

Reflection factor :

$$r = \frac{p_r(0,y)}{p_i(0,y)} = \frac{\cos\Theta - G}{\cos\Theta + G} \qquad (3)$$
$$\xrightarrow[G\to 0]{} 1 \quad ; \quad \xrightarrow[|G|\to\infty]{} -1$$

Absorption coefficient with the normalised absorber: input impedance $Z=Z'+j\cdot Z''$.

$$\alpha(\Theta) = 1-|\,r(\Theta)\,|^2 = \frac{4Z'\cos\Theta}{(1+Z'\cos\Theta)^2+(Z''\cos\Theta)^2} \tag{4}$$

Sometimes the derivatives $r^{(n)}(\Theta)=\partial^n r(\Theta)/\partial\Theta^n$ are needed : (5)

$$r'(\Theta) = \frac{-2G\sin\Theta}{(G+\cos\Theta)^2}$$

$$r''(\Theta) = \frac{-G\left(3+2G\cos\Theta-\cos(2\Theta)\right)}{(G+\cos\Theta)^3}$$

$$r^{(3)}(\Theta) = \frac{-G\left(11-2G^2+8G\cos\Theta-\sin\Theta\cdot\cos(2\Theta)\right)}{(G+\cos\Theta)^4}$$

$$r^{(4)}(\Theta) =$$

$$\frac{-G\left(115-20G^2+2G\,(47-4G^2)\cos\Theta-4\,(19-11G^2)\cos(2\Theta)-22G\cos(3\Theta)+\cos(4\Theta)\right)}{4(G+\cos\Theta)^5}$$

D.2 Plane wave reflection at an infinitely thick porous layer

"Porous layer" here stands for any homogeneous, isotropic material with characteristic propagation constant Γ_a and wave impedance Z_a. If the material is air: $\Gamma_a \to jk_0$; $Z_a \to Z_0$.

Sound incidence as in section D.1 .

Sound field above absorber : $p_1=p_i+p_r$ (as in section D.1)

sound field in absorber : $$p_2(x,y) = p_t(x,y) = B\cdot e^{-\Gamma_a\left(x\cos\Theta_a+y\sin\Theta_a\right)} \tag{1}$$

The boundary conditions are :

- equal propagation constant in y direction on both sides;
- equal normal admittance component on both sides;

(is equivalent to matching sound pressure and normal particle velocity).

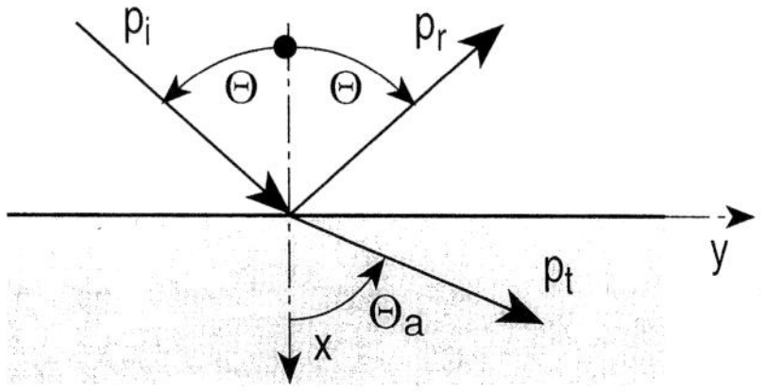

Refracted angle Θ_a (complex !): $$\frac{\sin\Theta_a}{\sin\Theta} = \frac{jk_0}{\Gamma_a} \tag{2}$$

Reflection factor r :
(in the 2nd form Z_{an}, Z_{0n} indicate normal components of the impedances)

$$r = \frac{Z_a/\cos\Theta_a - Z_0/\cos\Theta}{Z_a/\cos\Theta_a + Z_0/\cos\Theta} = \frac{Z_{an} - Z_{0n}}{Z_{an} + Z_{0n}} \qquad (3)$$

Absorption coefficient again is $\alpha = 1 - |r|^2$.

D.3 Plane wave reflection at a porous layer of finite thickness :

The absorber layer of thickness d is backed by a rigid wall.

The input impedance of the layer is :

$$Z_2 = \frac{Z_a}{\cos\Theta_a} \cdot \coth(\Gamma_a d \cdot \cos\Theta_a) \qquad (1)$$

Reflection factor :

$$r = \frac{Z_2/\cos\Theta_a - Z_0/\cos\Theta}{Z_2/\cos\Theta_a + Z_0/\cos\Theta} \qquad (2)$$

With normalised characteristic values $\Gamma_{an} = \Gamma_a/k_0$, $Z_{an} = Z_a/Z_0$ it is convenient to evaluate :

$$\Gamma_{an} \cdot \cos\Theta_a = \sqrt{\Gamma_{an}^2 + \sin^2\Theta_a} = \sqrt{1 + \Gamma_{an}^2 + \cos^2\Theta_a} \qquad (3)$$

Θ_a follows from the law of refraction $\cos\Theta_a = \sqrt{1 + (\sin\Theta/\Gamma_{an})^2}$ in section D.2 . (4)

Limit of layer thickness d above which the layer effectively behaves like an infinitely thick layer for normal sound incidence :

The layer is *locally reacting* (either due to large R or by internal partitions):

The limit follows from one of the relations for locally reacting layers:

$$R \ge 5.158/F^{0.5886}$$

$$F \ge 16.233/R^{1.699} \quad ; \quad F \ge 2.81 \cdot E^{0.629} \qquad (5)$$

$$f_{[Hz]} \cdot d_{[m]}^{2.699} \cdot \Xi_{[Pa\cdot s/m^2]}^{1.699} \ge 2.274 \cdot 10^6$$

The layer is *bulk reacting* :

The limit follows from one of the relations for bulk reacting layers:

$$R \ge 3.209/F^{0.7245}$$

$$F \ge 5.00/R^{1.380} \quad ; \quad F \ge 1.966 \cdot E^{0.580} \qquad (6)$$

$$f_{[Hz]} \cdot d_{[m]}^{2.380} \cdot \Xi_{[Pa\cdot s/m^2]}^{1.380} \ge 0.70 \cdot 10^6$$

with the non-dimensional quantities :

$$R = \Xi \cdot d/Z_0 \quad ; \quad F = f \cdot d/c_0 = d/\lambda_0 \qquad (7)$$

$$E = \rho_0 f/\Xi$$

Contour diagrams of $\alpha(\Theta)$ of a porous absorber layer with hard back.

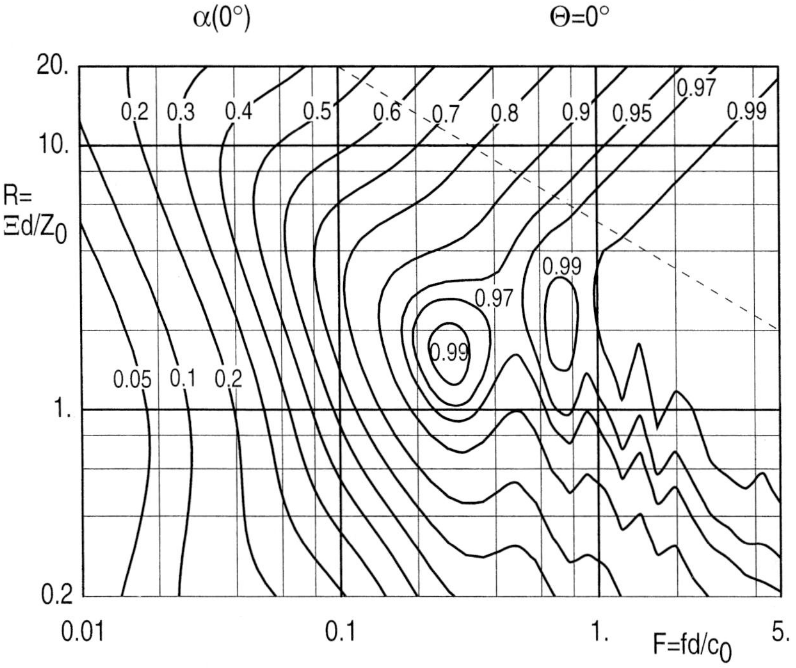

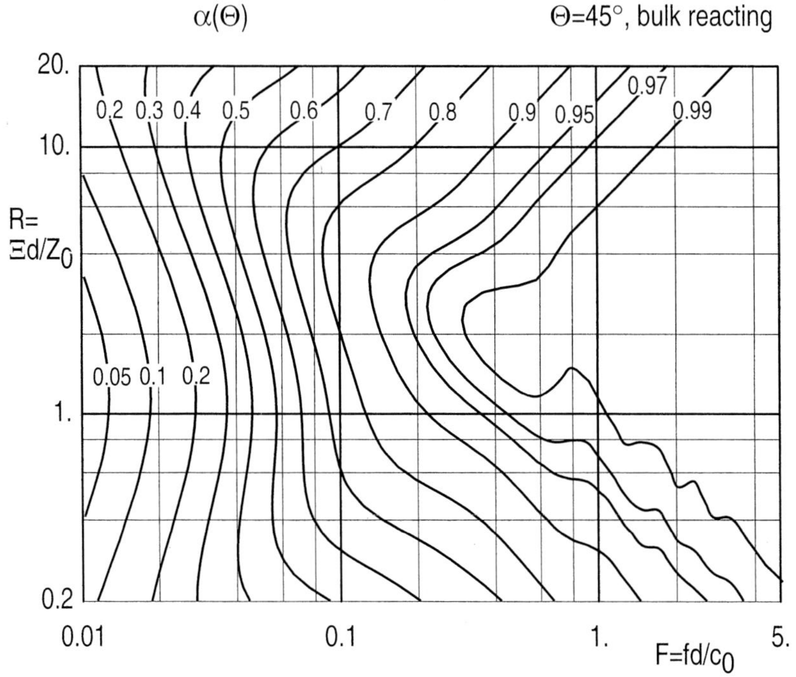

D.4 Plane wave reflection at a multiple layer absorber

The absorber consists of M layers of homogeneous porous material (or air); the layers are numbered with m=1,2,...,M. The space in front of the layer (with characteristic values k_0,Z_0) talkes the index m=0 .

Layer thicknesses : d_m ; m=1,2,...,M

characteristic values : Γ_{am},Z_{am} ; m=0,1,...,M ; (with Γ_{a0}=j , Z_{a0}=1)

incidence and refracted angles : Θ_m ; m=0,1,...,M;

reflection factors r_m ; m=0,1,...,M

(r_0= reflection factor of the arrangement)

layer input admittances : G_m ; m=0,1,...,M

(G_0= input admittance of the arrangement).

acoustic layer thicknesses : D_m; m=1,2,...,M

One can apply the chain circuit algorithm of section C.5 for the evaluation of the input admittance and therewith of the reflection factor, using the equivalent four-poles of section C.2 .

Here will be given a more explicit scheme of iteration with the iteration of the reflection factors (r_m is the reflection factor at the back side of layer m=0,1,2,...,M) :

$$r_{m-1} = \frac{\frac{W_m}{W_{m-1}}\left(1 + r_m \cdot e^{-2D_m}\right) - \left(1 - r_m \cdot e^{-2D_m}\right)}{\frac{W_m}{W_{m-1}}\left(1 + r_m \cdot e^{-2D_m}\right) + \left(1 - r_m \cdot e^{-2D_m}\right)} \tag{1}$$

Auxiliary quantities:

$$W_m = Z_{am}/\cos\Theta_m \quad ; \quad W_0 = Z_0/\cos\Theta_0$$

$$\cos\Theta_m = \sqrt{1 + (k_0/\Gamma_{am})^2 \cdot (1 - \cos^2\Theta_0)} \tag{2}$$

$$D_m = \Gamma_{am} d_m \cdot \cos\Theta_m = k_0 d_m \sqrt{1 + (\Gamma_{am}/k_0)^2 - \cos^2\Theta_0}$$

If the arrangement has a rigid backing, start the iteration with r_M=1 . If the back side of the arrangement is in contact with free space (without a back cover of the last layer), start with :

$$r_M = \frac{Z_0/\cos\Theta_0 - Z_{aM}/\cos\Theta_M}{Z_0/\cos\Theta_0 + Z_{aM}/\cos\Theta_M} \tag{3}$$

Input admittance G_m of the m-th layer (m=1,...,M) :

$$G_m = \frac{1}{W_m}\frac{1 - r_m}{1 + r_m} \tag{4}$$

D.5 Diffuse sound reflection at a locally reacting plane

Generally, the absorption coefficient α_{dif} follows from the absorption coefficient $\alpha(\Theta)$ for oblique incidence by integration over the polar angle Θ. The integrals are :

in 2-dimensional space :
$$\alpha_{2-dif} = \int_0^{\pi/2} \alpha(\Theta)\cdot\cos\Theta\, d\Theta \tag{1}$$

in 3-dimensional space :
$$\alpha_{3-dif} = 2\int_0^{\pi/2} \alpha(\Theta)\cdot\cos\Theta\cdot\sin\Theta\, d\Theta \tag{2}$$

The integral in 3 dimensions has an analytical solution for a locally reacting plane with normalised input admittance $Z_0G=g'+j\cdot g''$:

$$\alpha_{3-dif} = 8g'\left[1+\frac{g'^2-g''^2}{g''}\cdot\arctan\frac{g''}{g'+g'^2+g''^2} - g'\cdot\ln\left(1+\frac{1+2g'}{g'^2+g''^2}\right)\right]$$
$$\xrightarrow[g'\neq 0;\,g''=0]{} 8g'\left[1+\frac{g'^2}{g'+g'^2} - g'\cdot\ln\left(1+\frac{1+2g'}{g'^2}\right)\right] \tag{3}$$
$$\xrightarrow[g'=0;\,g''\neq 0]{} 0$$

or with the normalised input impedance $Z/Z_0=z'+j\cdot z''$:

$$\alpha_{3-dif} = 8\frac{z'}{z'^2+z''^2}\left[1+\frac{1}{z''}\frac{z'^2-z''^2}{z'^2+z''^2}\cdot\arctan\frac{z''}{1+z'} - \frac{z'}{z'^2+z''^2}\cdot\ln\left(1+2z'+z'^2+z''^2\right)\right] \tag{4}$$

Under condition $z''^2 \ll (1+z')^2$, the absorption coefficient α_{3-dif} can be evaluated from the (measured or computed) absorption coefficient α_0 for normal sound incidence :

$$\alpha_{3-dif} = 8\left[\frac{2}{1-\sqrt{1-\alpha_0}} - \frac{1-\sqrt{1-\alpha_0}}{2} + 2\ln\frac{1-\sqrt{1-\alpha_0}}{2}\right]\cdot\left[\frac{1-\sqrt{1-\alpha_0}}{1+\sqrt{1-\alpha_0}}\right]^2 \tag{5}$$

The maximum possible value of α_{3-dif} for locally reacting absorbent planes is $\alpha_{3-dif}=0.951$

The analytical solution for the integral of α_{2-dif} follows from :

$$\alpha_{2-dif} = \int_0^{\pi/2} \alpha(\Theta)\cdot\cos\Theta\, d\Theta = \int_0^{\pi/2} \frac{4z'\cos^2\Theta}{(1+z'\cos\Theta)^2+(z''\cos\Theta)^2}\, d\Theta$$
$$= \int_0^1 \frac{4z'x^2}{(1+z'x)^2+z''^2x^2}\frac{dx}{\sqrt{1-x^2}} = \int_0^1 \frac{4Z'x^2}{1+2z'x+(z'^2+z''^2)x^2}\frac{dx}{\sqrt{1-x^2}} \tag{6}$$

and is :
$$\alpha_{2-dif} = 2z'\left[\frac{\pi}{z'^2 + z''^2} + 2\,\mathrm{Im}\left\{\frac{\ln\left(z + \sqrt{z^2 - 1}\right)}{z'' \cdot z\sqrt{z^2 - 1}}\right\}\right] \tag{7}$$

The difference $\alpha_{2\text{-dif}} - \alpha_{3\text{-dif}}$ is small, in general, so that one absorption coefficient can be approximated by the other. This can be seen from the following diagram showing the difference over the complex plane of the normalised input impedance Z of a locally reacting plane.

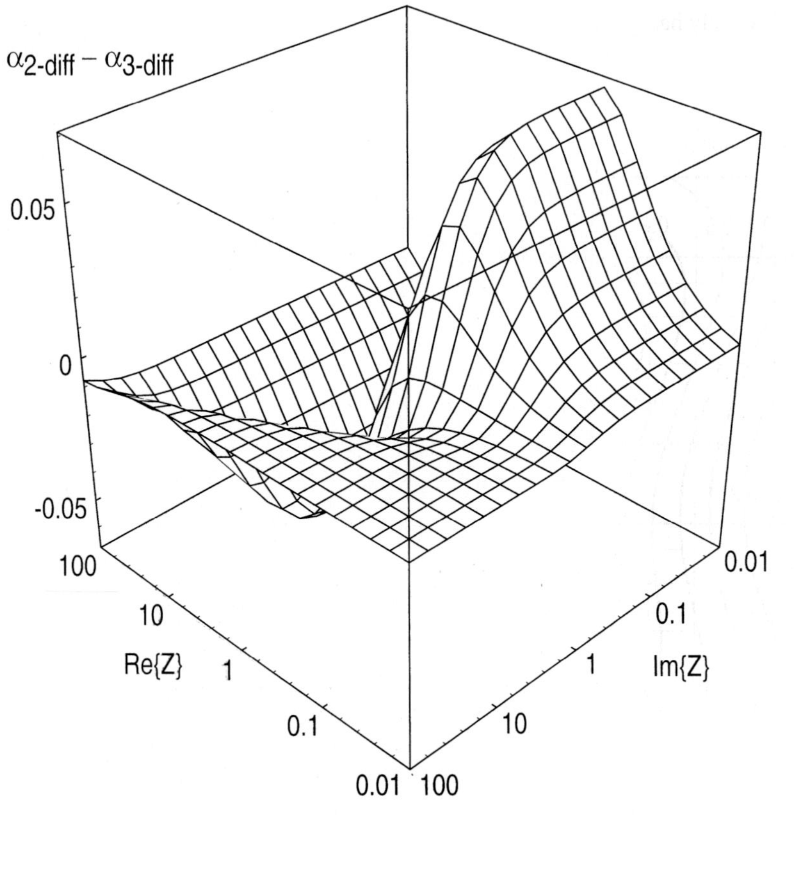

D.6 Diffuse sound reflection at a bulk reacting porous layer

The integral for α_{dif} from section D.5 of bulk reacting absorbers generally must be evaluated numerically. The diagram below shows a contour plot of α_{dif} for a porous layer of thickness d with hard back over the non-dimensional parameters $F=f\cdot d/c_0$ and $R=\Xi\cdot d/Z_0$. In the parameter range above and on the left-hand side of the dash-dotted curve, the absorption coefficients of a locally reacting and a bulk reacting porous layer agree with each other. In the range above and to the right-hand side of the dashed straight line, the bulk reacting layer effectively has an infinite thickness.

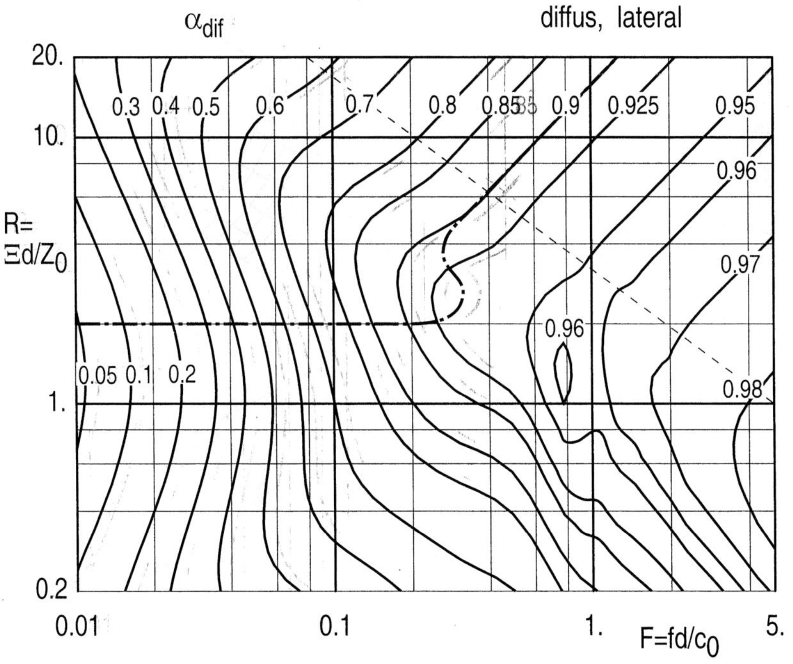

D.7 Sound reflection and scattering at finite-size local absorbers

The wording "local" in this heading (and at other places in this book) is a shorter form of "locally reacting"; the corresponding abbreviation for "bulk reacting" will be "lateral".

If the side dimensions of the plane absorber are finite, scattering takes place at the borders between the absorber and the baffle wall. In fact, some theories determine the sound absorption of finite-size absorbers from the solution of the scattering problem.

Let the absorber with area A be in the plane (x,y); the z co-ordinate shows into the space above the absorber. The sketch shows co-ordinates and angles used, as well as the incident plane wave p_i and the specularly reflected wave p_r . The field point is in P . Let s be a general co-ordinate.

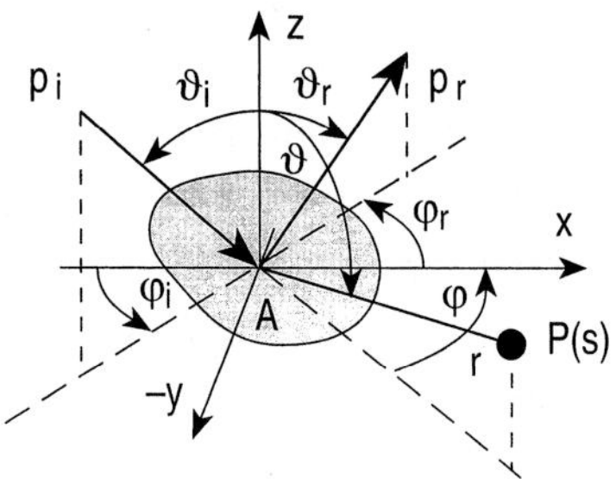

Field composition : $p(s) = p_i(s) + p_r(s) + p_s(s)$

with : $p_i(s)$= incident plane wave,

$p_r(s)$= specularly reflected plane wave,

$p_s(s)$= scattered wave ,

with r_u= reflection factor of the baffle wall,

F_0= normalised admittance of the baffle wall.

$$p_i(s) = P_i \cdot e^{-j(k_x x + k_y y - k_z z)}$$
$$p_r(s) = r_u \cdot P_i \cdot e^{-j(k_x x + k_y y + k_z z)} \tag{1}$$

Wave number components:

$$k_x = k_0 \cdot \sin\vartheta_i \cdot \cos\varphi_i \quad ; \quad k_y = k_0 \cdot \sin\vartheta_i \cdot \sin\varphi_i \quad ; \quad k_z = k_0 \cdot \cos\vartheta_i$$
$$k_x^2 + k_y^2 + k_z^2 = k_0^2 \tag{2}$$

Scattered wave :
$$p_s(s) = \iint_A \left[G(s\,|\,s_0) \cdot \frac{\partial}{\partial n_0} p(s_0) - p(s_0) \cdot \frac{\partial}{\partial n_0} G(s\,|\,s_0) \right] ds_0 \tag{3}$$

with Green's function (in which s is a radius, see sketch) for field points at a large distance:

$$G(s\,|\,s_0) = \frac{e^{-jk_0 s}}{4\pi s} \left[e^{jk_0 z_0 \cos\vartheta} + r_u(\vartheta) \cdot e^{-jk_0 z_0 \cos\vartheta} \right] \cdot e^{jk_0 \sin\vartheta \cdot (x_0 \cos\varphi + y_0 \sin\varphi)} \tag{4}$$

The Green's function corresponds to a superposition of the fields of point sources at Q and at the mirror-reflected point Q′ . It satisfies the boundary condition at the baffle wall.

Source point Q , mirror source point Q′ and field point P in the construction of Green's function :

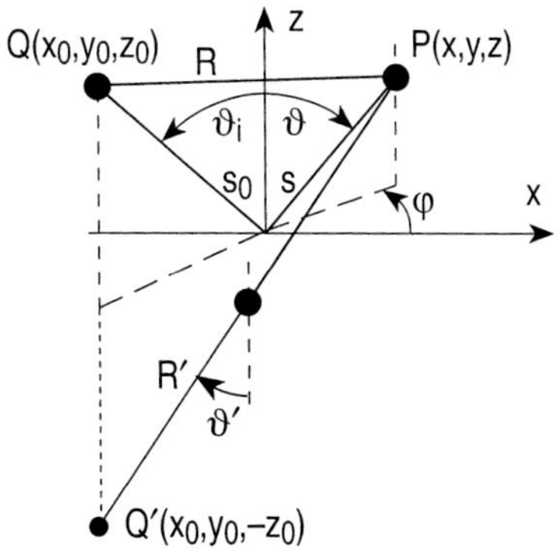

The integral equation above for p_s holds also on the absorber surface A , if the integral is multiplied with 1/2 .

It may be solved by iteration n=0,1,... for p_{sn}

- start with a suitable p_{so} on A (e.g. with the value of p_i+p_r on A);
- insert $p=p_i+p_r+p_s$ in the integrand;
- evaluate the first approximation p_{s1} , and so on. The n-th iteration gives

$$p_n(s)=p_i(s)+p_r(s)+p_{s1}(s)+\ldots+p_{sn}(s) \tag{5}$$

and for $p_s(s)$:

$$p_s(s)= p_{s1}(s)+\ldots+p_{sn}(s)$$

Scattering cross section of A :

$$Q_s=\frac{\Pi_s}{I_i}=\oiint \mathrm{Re}\left\{\frac{p_s^*}{P_i^*}\cdot\frac{v_{ns}}{P_i/Z_0}\right\}dS \tag{6}$$

Absorption cross section of A :

$$Q_a=\frac{\Pi_a}{I_i}=-\oiint \mathrm{Re}\left\{\frac{p^*}{P_i^*}\cdot\frac{v_n}{P_i/Z_0}\right\}dS \tag{7}$$

Extinction cross section of A :

$$Q_e=Q_a+Q_s \tag{8}$$

with Π_s= scattered effective power,
Π_a= absorbed effective power,
I_i= incident effective intensity,
and the integrals over large hemispheres surrounding A .

The absorption coefficient is :

$$\alpha(\vartheta_i,\varphi_i)=\frac{\Pi_a}{\Pi_i}=\frac{\Pi_a}{I_i\cdot A\cdot\cos\vartheta_i}=\frac{Q_a(\vartheta_i,\varphi_i)}{A\cdot\cos\vartheta_i} \tag{9}$$

Q_a can be expressed with the far field angular distribution of $p_s(s)$ with the help of the *extinction theorem* :

The far field p_s can be separated into an angular and a radial function :

$$p_s(s)\xrightarrow[s\to\infty]{} P_i\cdot\Phi_s(\vartheta_i,\varphi_i\,|\,\vartheta,\varphi)\cdot\frac{e^{-jk_0s}}{s} \tag{10}$$

The extinction theorem :

$$Q_e=-\frac{4\pi}{k_0}\cdot \mathrm{Im}\{\Phi_s(\vartheta_i,\varphi_i\,|\,\vartheta_r,\varphi_r)\} \tag{11}$$

(with ϑ_r, φ_r in the direction of the mirror-reflected wave, i.e. in our case $\vartheta_r=\vartheta_i$, $\varphi_r=\varphi_i$).

Finally with $Q_a=Q_e-Q_s$:

$$Q_a = -\frac{4\pi}{k_0}\cdot \mathrm{Im}\{\Phi_s(\vartheta_i,\varphi_i\,|\,\vartheta_r,\varphi_r)\} \int_0^{2\pi} d\varphi \int_0^{\pi/2} |\Phi_s(\vartheta_i,\varphi_i\,|\,\vartheta,\varphi)|^2 \cdot \sin\vartheta\, d\vartheta \tag{12}$$

So one needs the angular distribution of the scattered far field.

Examples :

(1) :

The absorber area has the normalised admittance G, which possibly is a function of surface co-ordinates, $G(x_0,y_0)$; then $\langle G\rangle$ is its average over A . The baffle wall has the constant normalised admittance F_0 . An approximation to the angular far field distribution of the scattered field is :

$$\begin{aligned} \Phi_s &= \frac{-jk_0\cos\vartheta\cdot\cos\vartheta_i}{2\pi(\cos\vartheta+F_0)(\cos\vartheta_i+\langle G\rangle)}\iint_A (G-F_0)e^{-j(\mu_x x_0+\mu_y y_0)}\,dx_0\,dy_0 \\ \mu_x &= k_0(\sin\vartheta_i\cdot\cos\varphi_i-\sin\vartheta\cdot\cos\varphi) \\ \mu_y &= k_0(\sin\vartheta_i\cdot\sin\varphi_i-\sin\vartheta\cdot\sin\varphi) \end{aligned} \tag{13}$$

Because $G-F_0=0$ outside A , the integral can be extended over the whole plane $z=0$; then it just represents the 2-dimensional Fourier integral of the admittance difference.

In the special case $F_0=0$, i.e. a hard baffle wall :

$$\Phi_s = \frac{-jk_0\cos\vartheta_i}{2\pi\,(\cos\vartheta_i+\langle G\rangle)}\iint_A G e^{-j(\mu_x x_0+\mu_y y_0)}\,dx_0\,dy_0 \tag{14}$$

(2) :

The absorber surface $A=a\cdot b$ is a rectangle, centred at the origin, with side length a in the x direction, side length b in the y direction, and G=const. The Fourier transform gives :

$$\Phi_s = \frac{-jk_0ab(G-F_0)\cos\vartheta_i\cdot\cos\vartheta}{2\pi\,(\cos\vartheta+F_0)(\cos\vartheta_i+G)}\;\frac{\sin(a\mu_x/2)}{a\mu_x/2}\;\frac{\sin(b\mu_y/2)}{b\mu_y/2} \tag{15}$$

and for $F_0=0$:

$$\Phi_s = \frac{-jk_0ab\cdot G\cos\vartheta_i}{2\pi\,(\cos\vartheta_i+G)}\;\frac{\sin(a\mu_x/2)}{a\mu_x/2}\;\frac{\sin(b\mu_y/2)}{b\mu_y/2} \tag{16}$$

(3) :

A circular absorber with radius a , centred at the origin, with a constant normalised admittance G (with $J_1(z)$ the Bessel function of 1st order) :

$$\Phi_s = \frac{-j k_0 \pi a^2 (G - F_0)\cos\vartheta_i \cdot \cos\vartheta}{2\pi\,(\cos\vartheta + F_0)(\cos\vartheta_i + G)} \, \frac{2 J_1(\gamma k_0 a)}{\gamma k_0 a} \qquad (17)$$

$$(\gamma k_0)^2 = \mu_x^2 + \mu_y^2$$

The diagram shows a directivity diagram of Φ_{sn} over φ,ϑ of a square with $k_0a=8$; G=1 ; $\varphi_i=\vartheta_i=45°$. Contour lines of Φ_{sn} are displayed ; the thick lines separate ranges with different signs of Φ_{sn} .

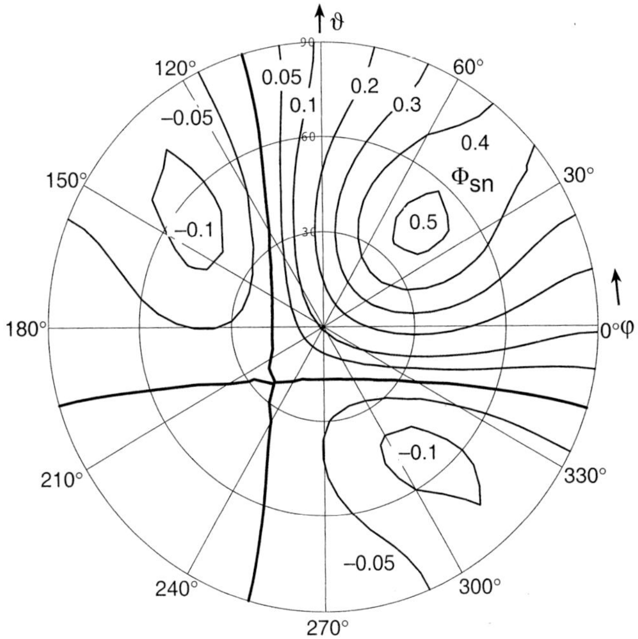

The method of this section can also be applied for diffuse sound incidence. The scattered sound field for diffuse incidence is :

$$p_{s,dif}(s,\vartheta,\varphi) = P_i \frac{e^{-jk_0 s}}{s} \int_0^{2\pi} d\varphi_i \int_0^{\pi/2} \Phi_s(\vartheta_i,\varphi_i \,|\, \vartheta,\varphi)\cdot \sin\vartheta_i \, d\vartheta_i \qquad (18)$$

D.8 Uneven, local absorber surface

This section uses the method described in the previous section D.7 . The "unevenness" may be modelled either with a variation of the normalised absorber admittance $G(x,y)$ or with a variation of the co-ordinates of the surface. The 1st method can be applied to grooves and narrow valleys, for example; the 2nd method is applicable for slow or random variations.

If the absorber surface can be represented by a reference plane with a variable admittance $G(x,y)$, then the admittance first is described by its Fourier series. The following example assumes in A a 1-dimensional variation of the admittance F_0 of the surrounding baffle wall in the form of a cosine modulation $G(x) = F_0 + B\cdot\cos(2\pi x/\ell)$ (1)

with the period length ℓ , and the average admittance over the absorber side length a in the x direction $\langle G\rangle = F_0 + B\cdot \mathrm{si}(\pi a/\ell)$ (2)

of a rectangular absorber with $A = a\cdot b$ (3)

with the function $\mathrm{si}(z) = \sin(z)/z$ (4)

The angular far field function of the scattered field is :

$$\Phi_s = \frac{-jk_0 ab\cdot B\cdot\cos\vartheta_i\cos\vartheta}{4\pi(\cos\vartheta+F_0)(\cos\vartheta_i+\langle G\rangle)}\left[\mathrm{si}\left(a\mu_x/2-\pi a/\ell\right)+\mathrm{si}\left(a\mu_x/2+\pi a/\ell\right)\right]\cdot\mathrm{si}\left(b\mu_y/2\right) \quad (5)$$

$$\begin{aligned}\mu_x &= k_0\left(\sin\vartheta_i\cdot\cos\varphi_i-\sin\vartheta\cdot\cos\varphi\right)\\ \mu_y &= k_0\left(\sin\vartheta_i\cdot\sin\varphi_i-\sin\vartheta\cdot\sin\varphi\right)\end{aligned} \quad (6)$$

Next, the geometrical profile of the absorber surface can be represented by $z=\zeta(x,y)$ and the normalised admittance $G(x,y)$ is given at this surface. The co-ordinate s_0 of the absorber surface in section D.7 now has a non-zero z component $s_0=(x_0,y_0,\zeta(x_0,y_0))$. The derivative normal to the surface becomes :

$$\frac{\partial}{\partial n_0} = -\frac{\partial}{\partial z_0}+\frac{\partial\zeta}{\partial x_0}\cdot\frac{\partial}{\partial x_0}+\frac{\partial\zeta}{\partial y_0}\cdot\frac{\partial}{\partial y_0}$$

If the variation of height is smaller than about half a wavelength, the angular far field distribution of the scattered field is : (7)

$$\Phi_s =$$

$$\frac{-j\cos\vartheta\cos\vartheta_i}{2\pi(\cos\vartheta+F_0)(\cos\vartheta_i+\langle G\rangle)}\iint_A\left(k_0(G-F_0)+\mu_x\frac{\partial\zeta}{\partial x_0}+\mu_y\frac{\partial\zeta}{\partial y_0}\right)\cdot e^{-j(\mu_x x_0+\mu_y y_0)}\,dx_0\,dy_0$$

In a special case (often met with in reverberant room measurements) the profile $\zeta(x,y)$ is a constant height h of A over the surrounding baffle wall, with

$$\begin{aligned}\frac{\partial\zeta}{\partial x_0} &= h\cdot\left[\delta(x_0+a/2)-\delta(x_0-a/2)\right] \quad \text{for} \quad -b/2<y_0<b/2 \\ \frac{\partial\zeta}{\partial y_0} &= h\cdot\left[\delta(y_0+b/2)-\delta(y_0-b/2)\right] \quad \text{for} \quad -a/2<x_0<a/2\end{aligned} \tag{8}$$

with the Dirac delta function $\delta(z)$. The contribution of the height step h to the far field angular distribution of the scattered field is :

$$\Phi_{s\zeta} = \frac{-jk_0ab\cdot k_0h\cdot\cos\vartheta_i\cos\vartheta}{\pi(\cos\vartheta+F_0)(\cos\vartheta_i+\langle G\rangle)}\gamma^2\cdot \mathrm{si}\left(a\mu_x/2\right)\mathrm{si}\left(b\mu_y/2\right) \tag{9}$$

with $(\gamma k_0)^2=\mu_x^2+\mu_y^2$. The ratio of the contribution $\Phi_{s\zeta}$ to the contribution Φ_{sG} which describes the difference of the absorber admittance G from the baffle wall admittance F_0 is :

$$\frac{\Phi_{s\zeta}}{\Phi_{sG}} = 2\frac{k_0h}{G-F_0}\gamma^2 \tag{10}$$

Next, the normalised absorber admittance $G(s_0)$ and/or the absorber surface contour $\zeta(s_0)$ have random variations with correlation distances d_G, d_ζ , respectively, and correlation functions :

$$\begin{aligned}K_G(d) &= \langle(G-\langle G\rangle)^2\rangle_A\cdot e^{-(d/d_G)^2/2} \\ K_\zeta(d) &= \langle\zeta^2\rangle_A\cdot e^{-(d/d_\zeta)^2/2}\end{aligned} \tag{11}$$

With $G_t(k)$ and $\zeta_t(k)$ the Fourier transforms of $G(s_0)-\langle G\rangle$ and $\zeta(s_0)$, respectively, and using the relation between the far field effective intensity I_s and angular distribution Φ_s of the scattered field $I_s=|p_s|^2/(2Z_0)=I_i\cdot|\Phi_s|^2/s^2$ one gets for the far field contribution of the variations in G and/or ζ to the effective intensity :

$$\begin{aligned}I_{s,G,\zeta} &= 4\pi^2\cdot I_i\cdot\frac{A}{s^2}\left|\frac{\cos\vartheta_i\cos\vartheta}{(\cos\vartheta+F_0)(\cos\vartheta_i+\langle G\rangle)}\right|^2\cdot\left(k_0^2\cdot|G_t(k_0\gamma)|^2+k_0^4\gamma^4\cdot|\zeta_t(k_0\gamma)|^2\right) \\ &= I_i\cdot\frac{A}{2\pi s^2}\left|\frac{\cos\vartheta_i\cos\vartheta}{(\cos\vartheta+F_0)(\cos\vartheta_i+\langle G\rangle)}\right|^2\cdot \\ &\quad\cdot\left((k_0d_G)^2\cdot\langle(G-\langle G\rangle)^2\rangle_A\cdot e^{-(k_0\gamma d_G)^2/2}+k_0^4\gamma^4d_\zeta^2\cdot\langle\zeta^2\rangle_A\cdot e^{-(k_0\gamma d_\zeta)^2/2}\right)\end{aligned} \tag{12}$$

D.9 Scattering at the border of an absorbent half-plane

A hard half-plane and a locally reacting absorbent half-plane with the normalised surface admittance G have the y axis as common border line . A plane wave p_i is incident from the side of the hard half-plane under the polar angle ϑ_i and azimuthal angle φ_i (measured in the x,y plane relative to the x axis).

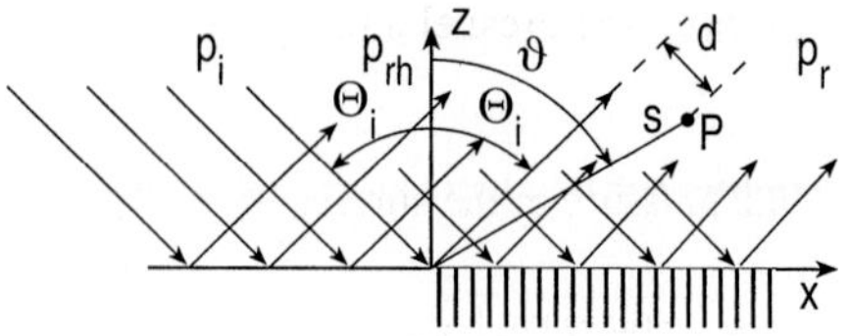

The problem becomes a two-dimensional one (in the x,z plane) by the substitutions

$$k_0^2 \to k^2 = k_0^2\left(1-\sin^2\vartheta_i \cdot \sin^2\varphi_i\right) \quad ; \quad k_x = k\cdot\sin\Theta_i \quad ; \quad k_z = k\cdot\cos\Theta_i$$

$$\sin\Theta_i = \frac{\sin\vartheta_i \cdot \cos\varphi_i}{\sqrt{1-\sin^2\vartheta_i\cdot\sin^2\varphi_i}} \quad ; \quad \cos\Theta_i = \frac{\cos\vartheta_i}{\sqrt{1-\sin^2\vartheta_i\cdot\sin^2\varphi_i}} \tag{1}$$

and a common factor $e^{-jk_y y}$ to each field quantity.

The sound field is composed as $p(x,z)= p_i(x,z)+ p_{rh}(x,z)+ p_s(x,z)$ with

- p_i= incident plane wave ;
- p_{rh}= reflected wave with “hard reflection”,
- p_s= scattered wave.

Combining p_i+p_{rh} , the sound field is :

$$p(x,z) = 2P_i\, e^{-jkx\cdot\sin\Theta_i}\cdot\cos(kz\cdot\cos\Theta_i) - jkG\int_0^\infty p(x_0,0)\cdot G(x,y\,|\,x_0,y_0)\,dx_0 \tag{2}$$

with Green’s function :

$$G(x,y\,|\,x_0,y_0) = -\frac{j}{4}\left[H_0^{(2)}(kR) + H_0^{(2)}(kR')\right]$$
$$R^2 = (x-x_0)^2 + (z-z_0)^2 \quad ; \quad R'^2 = (x-x_0)^2 + (z+z_0)^2 \tag{3}$$

containing Hankel functions of the second kind $H_0^{(2)}(z)$.

The far field of the sound pressure components $p_{rh}(s)+ p_s(s)$ is : (4)

$$p_{rh+s}(s) \xrightarrow[ks\to\infty]{} P_i\begin{cases} 1-\dfrac{G}{\cos\Theta_i+G}\left[(1-C(u)-S(u))+j(C(u)-S(u))\right] & ; \; d<0 \\ \dfrac{\cos\Theta_i-G}{\cos\Theta_i+G}+\dfrac{G}{\cos\Theta_i+G}\left[(1-C(u)-S(u))+j(C(u)-S(u))\right] & ; \; d>0 \end{cases}$$

with u from :

$$u^2 = \frac{1}{2} kd \frac{d}{s} = \frac{1}{2} kd \cdot \sin(\Theta - \Theta_i) = \frac{1}{2} ks \cdot \sin^2(\Theta - \Theta_i) \tag{5}$$

and C(u),S(u) the Fresnel's integrals : $C(u) = \sqrt{\frac{2}{\pi}} \int_0^u \cos(t^2)\,dt \quad ; \quad S(u) = \sqrt{\frac{2}{\pi}} \int_0^u \sin(t^2)\,dt$ (6)

The sound pressure in the surface at z=0 is :

$$p(x,0) = P_i\, e^{-jkx\cdot\sin\Theta_i} \left[2 - \frac{G}{G+\cos\Theta_i} \big(2 - U(-\Theta_i, kx) + jV(-\Theta_i, kx)\big) \right] \tag{7}$$

with the functions $U(\Theta,u)$, $V(\Theta,u)$ defined as the real and imaginary parts of :

$$U(\Theta,u) - jV(\Theta,u) = 1 - \cos\Theta \int_0^u \big(J_0(w) - jY_0(w)\big) \cdot [\cos(w\cdot\sin\Theta) - j\sin(w\cdot\sin\Theta)]\, dw \tag{8}$$

($J_0(z)$ and $Y_0(z)$ are Bessel and Neumann function, respectively). The evaluation of these integrals is described in [D.2] .

D.10 Absorbent strip in a hard baffle wall, with far field distribution

A locally reacting strip with normalised admittance G and width a , axial direction along the y axis, is placed in the plane x,y plane at (–a/2,+a/2). A plane sound wave p_i is incident from the direction –x under the polar angle Θ_i .

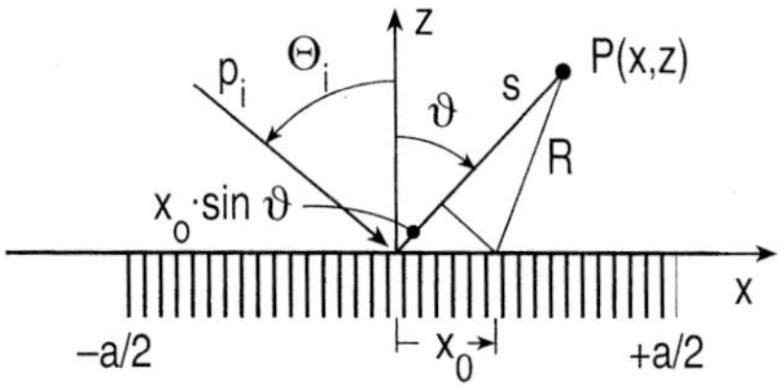

See the section D.9 for a possible component k_y of the wave vector in the y direction.

The sound field is composed as $p(x,z) = p_i(x,z) + p_{rh}(x,z) + p_s(x,z)$ with

- p_i= incident plane wave ;
- p_{rh}= reflected wave with "hard reflection",
- p_s= scattered wave.

Combining $p_i + p_{rh}$, the sound field is :

$$p(x,z) = 2P_i\, e^{-jkx\cdot\sin\Theta_i} \cdot \cos(kz\cdot\cos\Theta_i) - \frac{kG}{2} \int_0^\infty p(x_0,0) \cdot H_0^{(2)}(kR)\, dx_0 \tag{1}$$

with $R^2 = (x - x_0)^2 + z^2$.

In the far field :

$$p(x,z) = 2P_i\, e^{-jkx\cdot\sin\Theta_i} \cdot \cos(kz\cdot\cos\Theta_i) + \sqrt{\frac{j}{2\pi k}}\,\frac{e^{-jks}}{\sqrt{s}} \cdot V_z(k\sin\vartheta) \tag{2}$$

where $V_z(k\sin\vartheta)$ is the Fourier transform of the particle velocity distribution (in the z direction) at the plane z=0 . From the equivalent form of the scattered field p_s in the far field :

$$p_s(s) = -P_i\sqrt{\frac{j}{2\pi k}} \cdot \Phi_s(\Theta_i\,|\,\vartheta) \cdot \frac{e^{-jks}}{\sqrt{s}} \tag{3}$$

follows :

$$\Phi_s(\Theta_i\,|\,\vartheta) = \int_{-ka/2}^{+ka/2} G\cdot\frac{p(x_0,0)}{P_i}\cdot e^{+jkx_0\cdot\sin\vartheta}\, d(kx_0) = \int_{-\infty}^{+\infty}\frac{v_z(x_0,0)}{v_i}\cdot e^{+jkx_0\cdot\sin\vartheta}\, d(kx_0) \tag{4}$$

and the absorption cross section Q_a of the strip :

$$Q_a = \frac{2}{k}\mathrm{Re}\{\Phi_s(\Theta_i\,|\,\Theta_i)\} - \frac{1}{2\pi k}\int_{-\pi/2}^{+\pi/2} |\,\Phi_s(\Theta_i\,|\,\vartheta)\,|^2\, d\vartheta \tag{5}$$

The needed sound pressure distribution $p(x_0,0)$ has different possible approximations.

For small ka and low values of G , leading to $\Phi_s(\Theta_i\,|\,\vartheta)$, with si(z)=sin(z)/z ,

$$\begin{aligned} p(x_0,0) &\approx 2p_i(x_0,0) = 2P_i\, e^{-jkx_0\cdot\sin\Theta_i} \\ \Phi_s(\Theta_i\,|\,\vartheta) &= 2ka\,G\cdot \mathrm{si}\big(ka(\sin\vartheta - \sin\Theta_i)/2\big) \end{aligned} \tag{6}$$

and to Q_a :

$$\begin{aligned} \frac{Q_a}{a} &= 4\,\mathrm{Re}\{G\} - \frac{2}{\pi}ka\cdot|\,G\,|^2\int_{-\pi/2}^{+\pi/2}\mathrm{si}^2\big(ka(\sin\vartheta - \sin\Theta_i)/2\big)\, d\vartheta \\ &\xrightarrow[ka\ll 1]{} 4\,\mathrm{Re}\{G\} - 2ka\cdot|\,G\,|^2 \to 4\,\mathrm{Re}\{G\} \end{aligned} \tag{7}$$

Approximations for large ka and the corresponding $\Phi_s(\Theta_i\,|\,\vartheta)$ are :

$$\begin{aligned} p(x_0,0) &\approx P_i\,(1+r)\cdot e^{-jkx_0\cdot\sin\Theta_i} \\ &= 2P_i\frac{\cos\Theta_i}{G+\cos\Theta_i}\cdot e^{-jkx_0\cdot\sin\Theta_i} \\ \Phi_s(\Theta_i\,|\,\vartheta) &= 2\frac{G\cos\Theta_i}{G+\cos\Theta_i}ka\cdot \mathrm{si}\big(ka(\sin\vartheta - \sin\Theta_i)/2\big) \end{aligned} \tag{8}$$

The resulting Q_a is :

$$\frac{Q_a}{a} = 4\,\mathrm{Re}\{\frac{G\cos\Theta_i}{G+\cos\Theta_i}\} - \frac{2ka}{\pi}\left|\frac{G\cos\Theta_i}{G+\cos\Theta_i}\right|^2\int_{-\pi/2}^{+\pi/2}\mathrm{si}^2\big(ka(\sin\vartheta - \sin\Theta_i)/2\big)\, d\vartheta \tag{9}$$

D.11 Absorbent strip in a hard baffle wall, as a variational problem

Geometry and field composition as in section D.10 .

The variational principle is based on Helmholtz's theorem of superposition, which requires that the power of the "cross intensity" $p(x_0,0)\cdot v_z^a(x_0,0)$ of the desired field $p(x,z)$ and the particle velocity $v_z^a(x,z)$ of the adjoint field is zero at the plane $z=0$. The adjoint field is the solution for exchanged emission and immission points. The cross power is minimised by variation of the amplitude P of the estimate $P\cdot\exp(-jkx\cdot\sin\Theta_i)$. The expression to be minimised is :

$$4P_i\,kaG\cdot P-kaG\cdot P^2-\tfrac{1}{2}k^2G^2P^2\int\limits_{-a/2}^{+a/2}e^{jkx\cdot\sin\Theta_i}\,dx\cdot\int\limits_{-a/2}^{a/2}H_0^{(2)}(k|x-x_0|)\cdot e^{-jkx_0\cdot\sin\Theta_i}\,dx_0$$

$$=\mathrm{Min}(P) \qquad (1)$$

The partial derivative $\partial/\partial P$ is zero for :

$$P=2P_i\Bigg/\left\{1+\frac{kG}{2a}\int\limits_{-a/2}^{+a/2}e^{jkx\cdot\sin\Theta_i}\left(\int\limits_{-a/2}^{a/2}H_0^{(2)}(k|x-x_0|)\cdot e^{-jkx_0\cdot\sin\Theta_i}\,dx_0\right)dx\right\} \qquad (2)$$

Definition of auxiliary functions :

$$Z(\Theta,u)=\cos\Theta\int\limits_0^u\left(J_0(|w|)-j\,Y_0(|w|)\right)\cdot e^{-jw\cdot\sin\Theta}\,dw$$

$$=\begin{cases}-1+U(\Theta,u)-j\,V(\Theta,u) & ;\quad u<0\\ 1-U(\Theta,u)+j\,V(\Theta,u) & ;\quad u<0\end{cases} \qquad (3)$$

$$W(\Theta,ka)=\frac{1}{ka}\int\limits_0^{ka}Z(\Theta,u)\,du$$

(the functions $U(\Theta,u)$, $V(\Theta,u)$ are defined in section D.9).

The amplitude factor P becomes :

$$P=\frac{2P_i\cdot\cos\Theta_i}{\cos\Theta_i+\frac{G}{2}\left(W(\Theta_i,ka)+W(-\Theta_i,ka)\right)} \qquad (4)$$

and the sound pressure at $z=0$:

$$p(x,0) = 2P_i\left[1-\frac{G}{2}\,\frac{Z(\Theta_i,ka/2-kx)+Z(-\Theta_i,ka/2+kx)}{\cos\Theta_i+\frac{G}{2}\big(W(\Theta_i,ka)+W(-\Theta_i,ka)\big)}\right]\cdot e^{-jkx\cdot\sin\Theta_i}$$
$$\longrightarrow\begin{cases}2P_i\cdot e^{-jkx\cdot\sin\Theta_i} & ;\ |x|>>a\\ 2P_i\cdot e^{-jkx\cdot\sin\Theta_i}\,\dfrac{\cos\Theta_i}{G+\cos\Theta_i} & ;\ |kx|<<|ka|>>1\end{cases} \tag{5}$$

The angular far field distribution of the scattered field is :

$$\Phi_s(\Theta_i\,|\,\vartheta) = \frac{2\,ka\,G\cos\Theta_i}{\cos\Theta_i+\frac{G}{2}\big(W(\Theta_i,ka)+W(-\Theta_i,ka)\big)}\cdot si\big(ka\,(\sin\Theta-\sin\Theta_i)/2\big) \tag{6}$$

with $si(z)=\sin(z)/z$.

The absorption cross section Q_a of the strip follows with this from the previous sections.

Numerical examples for sound pressure distributions in the plane $z=0$:
(equivalencies for the parameters in the plot labels: "Theta" ~ Θ_i ; "F" ~ G ; "k0a" ~ k_0a);
The curve dashes become shorter for later entries in the parameter lists {…} .
Sound incidence is normal to the strip axis ($k_y=0$).

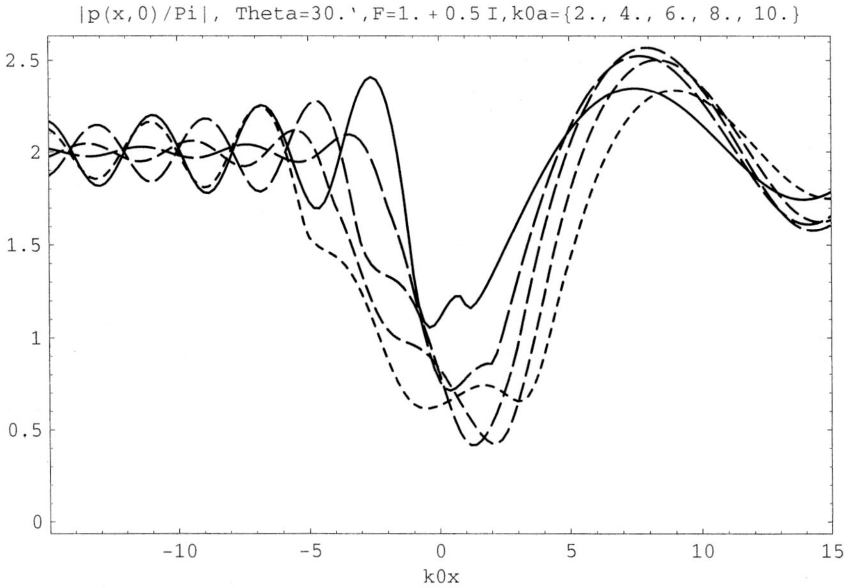

Sound pressure distribution for different k_0a .

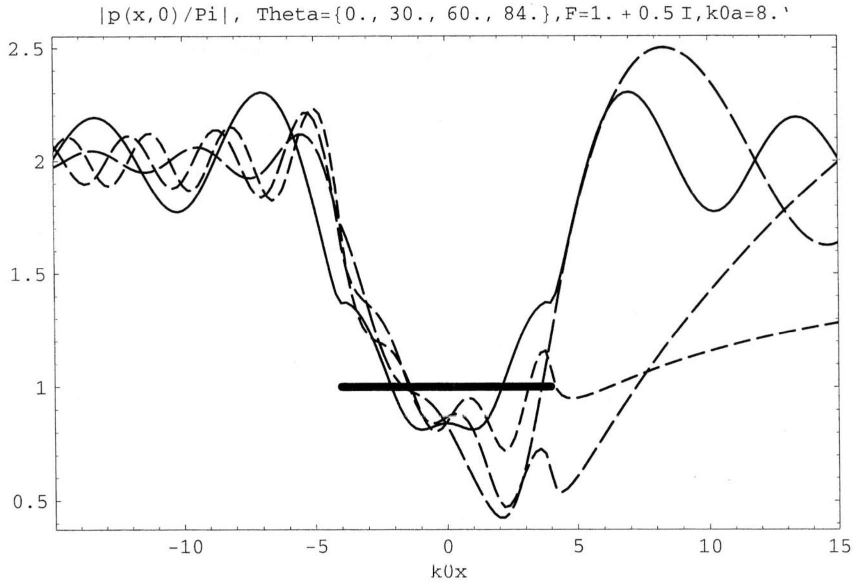

Sound pressure distributions for different angles of incidence Θ_i .

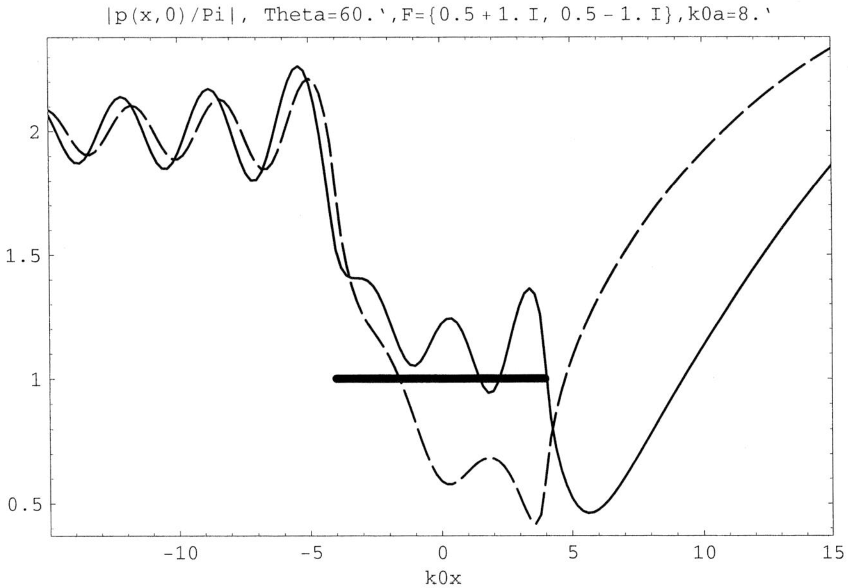

Sound pressure distributions for two normalised admittances F ~ G .

D.12 Absorbent strip in a hard baffle wall, with Mathieu functions

See Mechel, [D.3], for notations and relations of Mathieu functions.

The sound field around a locally reacting strip with normalised admittance G in a hard baffle wall can be formulated as a boundary value problem with exact solutions in elliptic-hyperbolic cylinder co-ordinates (ρ,ϑ). The co-ordinate curves are confocal ellipses and orthogonal confocal hyperbolic branches. The radial and azimuthal eigenfunctions in these co-ordinates are Mathieu functions.

Transformation between Cartesian and elliptic-hyperbolic co-ordinates:
The common foci are at $x=\pm c$.

$$x = x_1 = c\cdot\cosh\rho\cdot\cos\vartheta$$
$$y = x_2 = c\cdot\sinh\rho\cdot\sin\vartheta \qquad (1)$$
$$z = x_3 = z$$

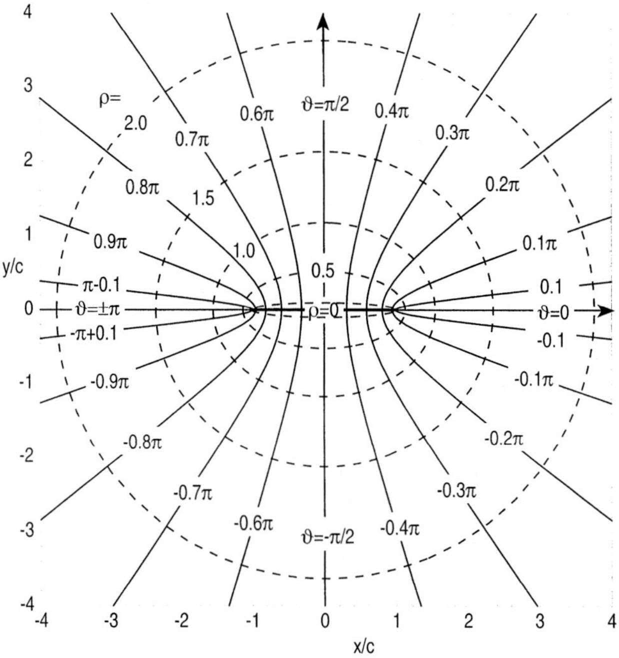

Co-ordinate lines in elliptic-hyperbolic cylinder co-ordinates.

The boundary surface of the absorbent strip is at $\rho=0$; the focus distance is $c=a/2$; the boundaries of the baffle wall are at $\vartheta=0$ and $\vartheta=\pi$. The Helmholtz differential equation (wave equation) $(\Delta + k_0^2)u = 0$ in elliptic-hyperbolic co-ordinates is :

$$\frac{\partial^2 u}{\partial\rho^2}+\frac{\partial^2 u}{\partial\vartheta^2}+\left(\cosh^2\rho-\cos^2\vartheta\right)\cdot\left(c^2\frac{\partial^2 u}{\partial z^2}+(k_0 c)^2 u\right)=0 \qquad (2)$$

For separated field functions $u(\rho,\vartheta)=T(\vartheta)\cdot R(\rho)$ this is equivalent to the pair of Mathieu differential equations :

$$\frac{d^2R(\rho)}{d\rho^2}-(\lambda-2q\cdot\cosh(2\rho))\cdot R(\rho)=0$$
$$\frac{d^2T(\vartheta)}{d\vartheta^2}+(\lambda-2q\cdot\cos(2\vartheta))\cdot T(\vartheta)=0 \tag{3}$$

The parameter q is determined by $q=(k_0c)^2/4$

The parameter λ stands for *characteristic values* of the Mathieu functions, for which the Mathieu differential equations have finite and periodic (in ϑ) solutions; they will be named $\lambda=ac_m$ for cos-like (symmetrical in ϑ) azimuthal Mathieu functions $T(\vartheta)=ce_m(\vartheta,q)$, and $\lambda=bc_m$ for sin-like (anti-symmetrical in ϑ) azimuthal Mathieu functions $T(\vartheta)=se_m(\vartheta,q)$. The azimuthal functions are associated, respectively, with radial Mathieu functions of the Bessel-type, $Jc_m(\rho)$, $Js_m(\rho)$, of the Neumann-type, $Yc_m(\rho)$, $Ys_m(\rho)$, and of the Hankel-type for outward propagating waves $Hc_m^{(2)}(\rho)=Jc_m(\rho)-j\cdot Yc_m(\rho)$ or $Hs_m^{(2)}(\rho)=Js_m(\rho)-j\cdot Ys_m(\rho)$.

A plane wave incident at an angle Θ against the major axis of the ellipses, i.e. against the plane of the strip and the baffle wall, is in Cartesian co-ordinates :

$$p_i(x,y)=e^{-jk_0(x\cos\Theta+y\sin\Theta)}=e^{-2jw\sqrt{q}}=p_i(\rho,\vartheta)$$
$$w=\cosh\rho\,\cos\vartheta\,\cos\Theta+\sinh\rho\,\sin\vartheta\,\sin\Theta \tag{4}$$

Its expansion in Mathieu functions is :

$$p_i(\rho,\vartheta)=2\sum_{m=0}^{\infty}(-j)^m ce_m(\Theta;q)\cdot Jc_m(\rho;q)\cdot ce_m(\vartheta;q)+$$
$$+2\sum_{m=1}^{\infty}(-j)^m se_m(\Theta;q)\cdot Js_m(\rho;q)\cdot se_m(\vartheta;q) \tag{5}$$

The sum of the incident wave p_i and of the reflected wave p_r with reflection at a hard plane containing the major axis of the ellipses (i.e. the baffle wall) is :

$$p_i(\rho,\vartheta)+p_r(\rho,\vartheta)=4\sum_{m=0}^{\infty}(-j)^m ce_m(\Theta;q)\cdot Jc_m(\rho;q)\cdot ce_m(\vartheta;q) \tag{6}$$

A scattered field $p_s(\rho,\vartheta)$ is added to these field components; it is formulated as a sum of terms as in p_i+p_r , but with yet undetermined term amplitudes a_m , and the Mathieu-Bessel functions $Jc_m(\rho;q)$ (which represent radial standing waves) replaced with Mathieu-Hankel functions $Hc_m^{(2)}(\rho;q)$; so it satisfies the boundary condition at the baffle wall ; thus :

$$p(\rho,\vartheta)=p_i(\rho,\vartheta)+p_r(\rho,\vartheta)+p_s(\rho,\vartheta)=$$
$$=4\sum_{m=0}^{\infty}(-j)^m ce_m(\Theta;q)\cdot ce_m(\vartheta;q)\cdot\left(Jc_m(\rho;q)+a_m\cdot Hc_m^{(2)}(\rho;q)\right) \tag{7}$$

The gradient in elliptic-hyperbolic co-ordinates is :

$$\mathrm{grad}\,u=\frac{\vec{e}_\rho}{c\sqrt{\sinh^2\rho+\sin^2\vartheta}}\frac{\partial u}{\partial\rho}+\frac{\vec{e}_\vartheta}{c\sqrt{\sinh^2\rho+\sin^2\vartheta}}\frac{\partial u}{\partial\vartheta}+\vec{e}_z\cdot\frac{\partial u}{\partial z} \tag{8}$$

and therefore the particle velocity v_ρ in the direction of the hyperbolic ρ-lines :

$$v_\rho=\frac{j}{k_0Z_0}\mathrm{grad}_\rho p=\frac{j}{k_0cZ_0}\frac{\partial p/\partial\rho}{\sqrt{\sinh^2\rho+\sin^2\vartheta}}$$

This is used for the boundary condition at the strip : $Z_0v_\rho(0,\vartheta)=-G\cdot p(0,\vartheta)$.

$$\xrightarrow[\rho=0]{}\frac{j}{k_0cZ_0}\frac{\partial p/\partial\rho}{|\sin\vartheta|} \qquad \xrightarrow[\vartheta=0,\ \vartheta=\pi]{}\frac{j}{k_0cZ_0}\frac{\partial p/\partial\rho}{\sinh\rho} \tag{9}$$

The boundary condition gives (a prime at the Mathieu functions indicates the derivative in ρ) :

$$\begin{aligned}&\sum_{m=0}^{\infty}(-j)^m ce_m(\Theta;q)\cdot ce_m(\vartheta;q)\cdot\left(Jc'_m(0;q)+a_m\cdot Hc'^{(2)}_m(0;q)\right)=\\&=jk_0cG\cdot|\sin\vartheta|\sum_{m=0}^{\infty}(-j)^m ce_m(\Theta;q)\cdot ce_m(\vartheta;q)\cdot\left(Jc_m(0;q)+a_m\cdot Hc^{(2)}_m(0;q)\right)\end{aligned} \tag{10}$$

The functions $ce_m(\vartheta;q)$ are orthogonal in ϑ over $(0,\pi)$ with the norms N_m :

$$\int_0^\pi ce_m(\vartheta;q)\cdot ce_n(\vartheta;q)\,d\vartheta=\delta_{m,n}\cdot N_m \tag{11}$$

Application of the orthogonality integral on both sides of the boundary condition gives, with the mode coupling coefficients :

$$T_{m,n}=\int_0^\pi ce_m(\vartheta;q)\cdot ce_n(\vartheta;q)\cdot|\sin\vartheta|\,d\vartheta \tag{12}$$

the linear, inhomogeneous system of equations for the amplitudes a_m :

$$\begin{aligned}&\sum_{m=0}^{\infty}a_m\cdot(-j)^m ce_m(\Theta;q)\cdot\left(jk_0cG\cdot T_{m,n}\cdot Hc^{(2)}_m(0;q)-\delta_{m,n}N_n\cdot Hc'^{(2)}_n(0;q)\right)=\\&=-\sum_{m=0}^{\infty}(-j)^m ce_m(\Theta;q)\cdot\left(jk_0cG\cdot T_{m,n}\cdot Jc_m(0;q)-\delta_{m,n}N_n\cdot Jc'_n(0;q)\right)\end{aligned} \tag{13}$$

The mode norms are $N_m=\pi/2$; the coupling coefficients $T_{m,n}$ can be expressed in terms of the Fourier coefficients of the Fourier series representation of $ce_m(\vartheta)$ (see [D.3, section 19.5]). After the solution of this (truncated) system of equations, the sound field is known.

Numerical examples :

(The correspondences for the parameters in the plot labels are: "theta" ~ Θ ; "k0a" ~ k_0a ; "Wn" ~ 1/G ; "I" ~ j ; "mhi"= upper limit of used orders m)

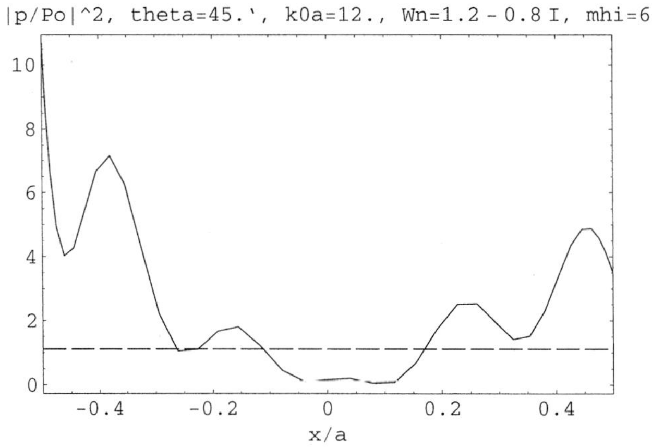

Distribution of sound pressure magnitude squared at the surface of an absorbent strip with mass-type reactance.

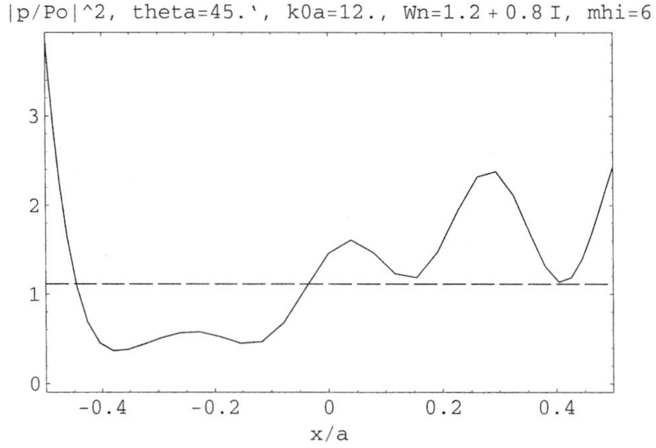

Distribution of sound pressure magnitude squared at the surface of an absorbent strip with spring-type reactance.

The horizontal dashed lines in the plots above are the squared sound pressure magnitudes at an absorber of infinite extend.

The *acoustic corner effect* :

Due to scattering at the borders of a finite-size absorber, its absorption in general is different from the absorption of an infinite, but otherwise equal, absorber. The *quantitative corner effect* is defined as the ratio of the effective power Π absorbed by the strip to the power Π_∞ absorbed by an area of the same size of an infinite, but otherwise equal absorber :

$$CE(\vartheta) = \frac{\Pi}{\Pi_\infty} = \frac{1}{4\,k_0 a \cdot \mathrm{Re}\{G\}} \cdot \left|1 + \frac{G}{\sin\vartheta}\right|^2 \cdot \sum_{m\geq 0} \left[4\, ce_m(\vartheta;q) \cdot \mathrm{Re}\left\{(-j)^m\, a_m^*\right\} + |a_m|^2\right] \qquad (14)$$

(In the diagram: "KE" ~ CE ; "theta" ~ ϑ ; "koa" ~ $k_0 a$; "Wn" ~ 1/G= impedance)

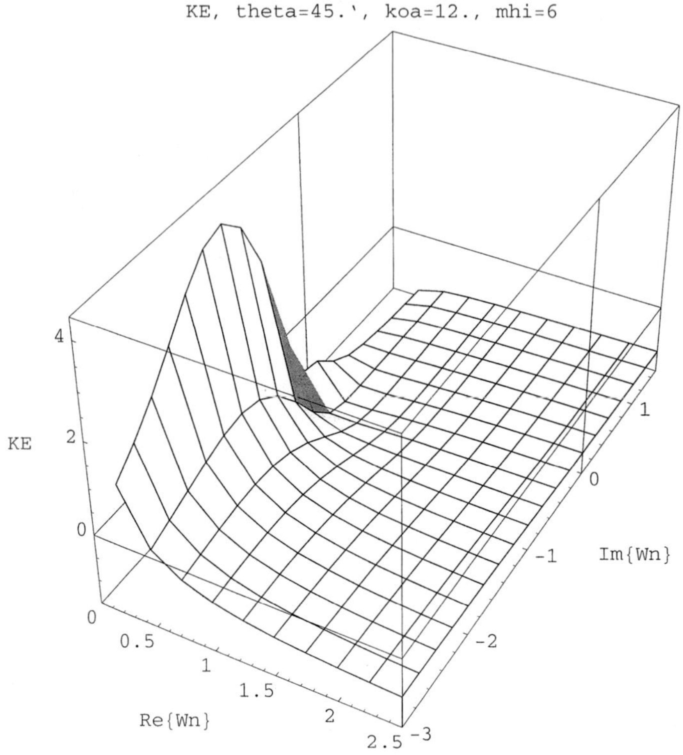

The corner effect may be positive or negative, depending on the sign of the reactance.

D.13 Absorption of finite-size absorbers, as a problem of radiation

Ref.: Mechel, [D.2]

The surface impedance Z_A of an infinite absorber is generally easily evaluated. The problem with finite-size absorbers in a baffle wall is the influence of border scattering. This influence can be taken into account by a simple equivalent network.

The absorbed effective power is :

$$\Pi_a = A \cdot \frac{|P_i|^2}{2Z_0} \frac{4\,\mathrm{Re}\{Z_A\}}{|Z_A + Z_s|^2} \tag{1}$$

The normalised absorption cross-section is :

$$\frac{Q_a}{A} = \frac{4\,\mathrm{Re}\{Z_A\}}{|Z_A + Z_s|^2} \tag{2}$$

Therein are :

A= area of the absorber ; P_i= amplitude of the incident plane wave ;

Z_A= surface impedance of the infinite absorber ;

Z_s= radiation impedance of a radiator with the size and shape of the absorber, when its surface oscillation pattern agrees with that of the exciting wave at the absorber surface.

This can be represented in an equivalent network :

- the pressure source has the amplitude $2P_i$;
- the radiation impedance Z_s is the internal source impedance;
- the impedance Z_A of the infinite absorber is the load impedance;
- the power in Z_A is the absorbed power Π_a .

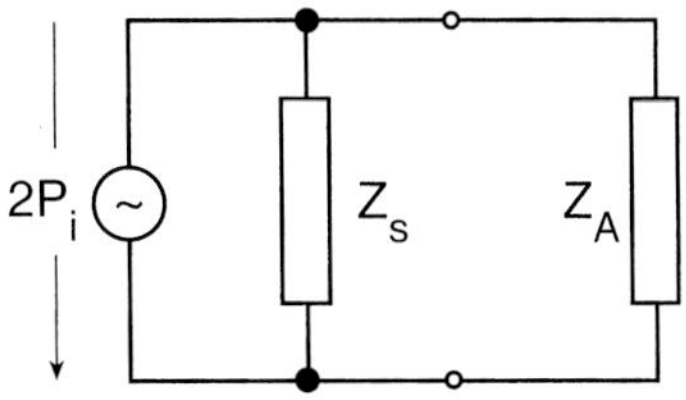

Thus the determination of the absorption by finite-size absorbers is reduced to the determination of their radiation impedance.

D.14 A monopole line source above an infinite, plane absorber; integration method

Ref.: Mechel, [D.4]

A monopole line source placed at Q is parallel to the absorber, with a normalised surface admittance G . S is the mirror-reflected point to Q . P is a field point.

The absorber may be locally or bulk reacting; it will be mentioned if results are valid for locally reacting absorbers only.

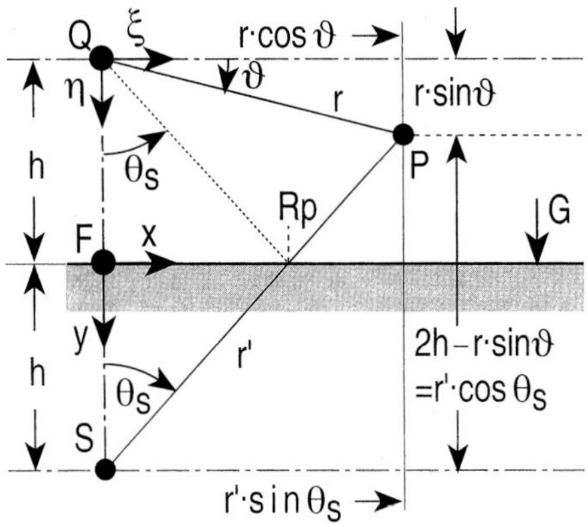

In the following, k is the wave vector component in the plane containing Q,S,P.

The field is composed as $p = p_Q + p_r$; p_Q= source free field ; p_r= reflected field .

Source free field (with unit amplitude) $p_Q(r) = H_0^{(2)}(kr)$ (1)

Field of plane wave incident under polar angle θ
(with reflection factor $R(\theta)$ for this angle of incidence, and $\Theta=\theta+\vartheta$) :

$$p_e + p_{er} = e^{-jkr\cdot\sin\Theta} + R(\Theta-\vartheta)\cdot e^{-2jkh\cdot\cos(\Theta-\vartheta)}\cdot e^{-jkr\cdot\sin(\Theta-2\vartheta)} \tag{2}$$

After application of the integral operation: $\frac{1}{\pi}\int\limits_{C(\Theta)}(p_e+p_{er})\,d\Theta$ (3)

the 1st term is the integral representation of the Hankel function with path $C(\Theta)$: $-j\infty \to 0 \to \pi \to \pi+j\infty$

$$H_0^{(2)}(kr) = \frac{1}{\pi}\int\limits_{C(\Theta)} e^{-jkr\cdot\sin\Theta}\,d\Theta \quad ; \quad \mathrm{Re}\{kr\}>0 \tag{4}$$

Thus the 2nd term yields :

$$p_r = \frac{1}{\pi}\int\limits_{C(\Theta)} R(\Theta-\vartheta)\cdot e^{-jkr\cdot(\sin(\Theta-2\vartheta)+2h/r\cdot\cos(\Theta-\vartheta))}\,d\Theta \tag{5}$$

and after a horizontal shift of the path, with $\varphi=\theta-\theta_S$ (see the sketch for θ_S and r') :

$$p_r = \frac{1}{\pi}\int\limits_{C(\varphi)} R(\varphi+\theta_s)\cdot e^{-jkr'\cdot\cos\varphi}\,d\varphi \quad ; \quad R(\varphi+\theta_s) = \frac{\cos(\varphi+\theta_s)-G}{\cos(\varphi+\theta_s)+G} \tag{6}$$

This is an exact representation for p_r ; the path $C(\varphi)$ is shown in the diagram, together with the shaded range, where poles of $R(\varphi+\theta_S)$ are possible (only for $\mathrm{Im}\{G\}>0$), and the "path of steepest descent" (pass way) Pw . If during the deformation $C(\varphi) \to$ Pw a pole is crossed, a "pole contribution" must be added to the integral of steepest descent; it has the form of a surface wave.

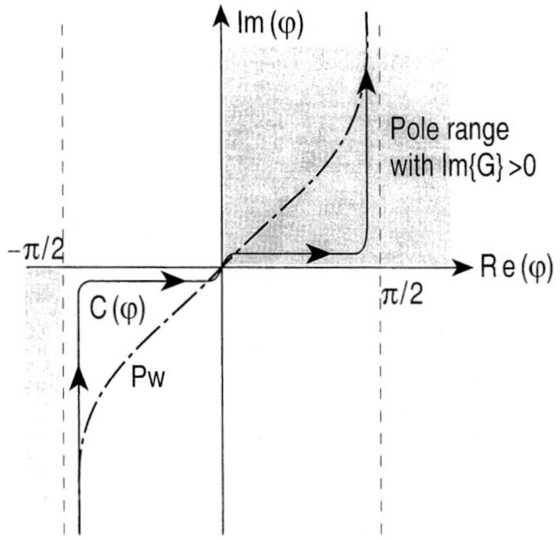

For direct numerical integration, use :

$$p_r = \frac{1}{\pi}\int\limits_{-\pi/2}^{+\pi/2}\frac{\cos(\varphi+\theta_s)-G}{\cos(\varphi+\theta_s)+G}\cdot e^{-jkr'\cdot\cos\varphi}\,d\varphi + \frac{2j}{\pi}\int\limits_0^{\infty}\frac{1+G^2-\cos^2\theta_s+\sinh^2\varphi''}{1-G^2-\cos^2\theta_s+\sinh^2\varphi''+2jG\cos\theta_s\cdot\sinh\varphi''}\cdot e^{-kr'\cdot\sinh\varphi''}\,d\varphi'' \tag{7}$$

The first integrand oscillates strongly for $kr'>>1$. Therefore use the method of integration along the steepest descent (also "saddle point integration" or "pass integration"). Some cases must be distinguished.

Saddle point at $\varphi_s=0$; on the pass way is $\varphi(s)=\pm\arccos(1-j\cdot s^2)$; $s \gtrless 0$, s being a running parameter on the pass way from $-\infty$ to $+\infty$; the saddle point is at $s=0$; slope of the pass way in the saddle point $d\varphi(0)/ds=1+j$.

No pole crossing, and no pole near the saddle point :

$$p_r = \sqrt{\frac{1}{\pi kr'}} e^{-jkr'}\left[\Phi(0)+\frac{1}{4kr'}\Phi''(0)+\ldots+\frac{1\cdot3\cdot5\cdot\ldots(2n-1)}{(2n)!\,(2kr')^n}\Phi^{(2n)}(0)\right] \tag{8}$$

(primes are derivatives with respect to s) with :

$$\Phi(s) = R(\varphi(s)+\theta_s)\cdot\frac{2j}{\sqrt{2j+s^2}} = \frac{\cos(\varphi(s)+\theta_s)-G}{\cos(\varphi(s)+\theta_s)+G}\cdot\frac{2j}{\sqrt{2j+s^2}} \tag{9}$$

With some derivatives performed (leaving $R^{(n)}(\theta_s)$ unevaluated) we have: (10)

$$p_r = \sqrt{\frac{2j}{\pi kr'}} e^{-jkr'}\cdot\left[\left(1+\frac{j}{8kr'}-\frac{9}{128(kr')^2}-\frac{75j}{1024(kr')^3}+\frac{3675}{32768(kr')^4}\right)\cdot R(\theta_s)+\right.$$
$$\left(\frac{j}{2}-\frac{5}{16\,kr'}-\frac{259j}{768(kr')^2}+\frac{3229}{6144(kr')^3}\right)\cdot\frac{R^{(2)}(\theta_s)}{kr'}-\left(\frac{1}{8}+\frac{35j}{192\,kr'}-\frac{329}{1024(kr')^2}\right)\cdot\frac{R^{(4)}(\theta_s)}{(kr')^2}-$$
$$\left.\left(\frac{j}{48}-\frac{7}{128\,kr'}\right)\cdot\frac{R^{(6)}(\theta_s)}{(kr')^3}+\frac{1}{384}\cdot\frac{R^{(8)}(\theta_s)}{(kr')^4}\right]$$

This form is valid for both locally and bulk reacting absorbers. The first terms in the parentheses collected give the *geometrical acoustic approximation* (or *mirror source approximation*):

$$p_r \xrightarrow[kr'\gg1]{} R(\theta_s)\cdot H_0^{(2)}(kr') \tag{11}$$

For locally reacting absorbers the derivatives $R^{(n)}(\theta_s)$ can be evaluated in advance (for bulk reacting absorbers they depend on the internal structure of the absorber) :

$$R(\theta_s) = \frac{\cos\theta_s - G}{\cos\theta_s + G}$$

$$R^{(2)}(\theta_s) = -2G\frac{2-\cos^2\theta_s + G\cdot\cos\theta_s}{(\cos\theta_s+G)^3} \tag{12}$$

$$R^{(4)}(\theta_s) =$$
$$2G\frac{(G-5\cos\theta_s)(\cos\theta_s+G)^2\cos\theta_s+4(\cos\theta_s+G)(2G-7\cos\theta_s)\sin^2\theta_s-24\sin^4\theta_s}{(\cos\theta_s+G)^5}$$

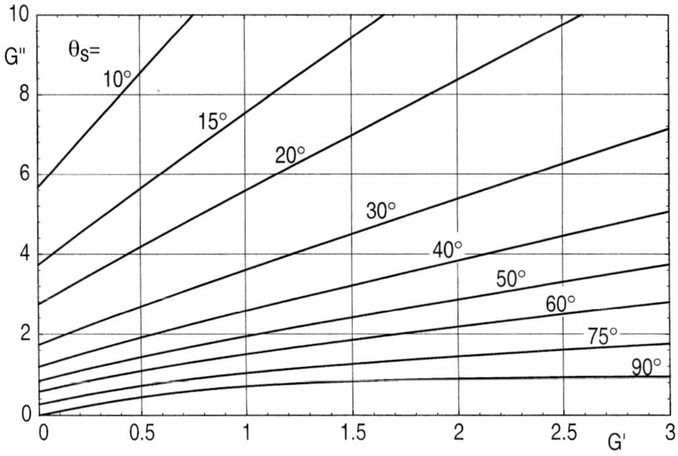

Limits for pole contributions with a locally reacting absorber.

A uniform pass integration :

It can be used also if the pole is near the saddle point; however, the simple pass integration above is preferable, if the pole is not near the saddle point.

Let the integral to be computed along the pass way Pw be of the form, with real $x>1$:

$$I = \int_{Pw} e^{x\cdot f(\varphi)} \cdot F(\varphi)\, d\varphi \tag{18}$$

It can be evaluated by:

$$\begin{aligned} I &= e^{x\cdot f(\varphi_s)} \cdot \left[\pm j\pi a \cdot W(\pm b\sqrt{x}) + \sqrt{\frac{\pi}{x}} \cdot T \right] \quad ; \quad Im\{b\} \gtrless 0 \\ &= e^{x\cdot f(\varphi_s)} \cdot \left[j\pi a \cdot W(b\sqrt{x}) + \sqrt{\frac{\pi}{x}} \cdot T - j\pi a \cdot e^{-x\cdot b^2} \right] \quad ; \quad \begin{cases} Im\{b\} = 0 \\ Re\{b\} \neq 0 \end{cases} \\ &= e^{x\cdot f(\varphi_s)} \cdot \sqrt{\frac{\pi}{x}} \cdot T \quad ; \quad b = 0 \end{aligned} \tag{19}$$

with the definitions :

$$a =: \lim_{\varphi\to\varphi_p} \left[(\varphi - \varphi_p) \cdot F(\varphi) \right] = N(\varphi_p) \Big/ \left. \frac{dD}{d\varphi} \right|_{\varphi=\varphi_p}$$

$$b =: \sqrt{f(\varphi_s) - f(\varphi_p)} \quad ; \quad b \xrightarrow[\varphi_p \to \varphi_s]{} \frac{\varphi_p - \varphi_s}{\gamma} \tag{20}$$

$$\gamma =: \sqrt{-2/f''(\varphi_s)} \quad ; \quad \arg(\gamma) = \left(\arg(d\varphi)\right)_{\varphi_s} \quad ; \quad \varphi \text{ along } Pw$$

$$T =: \gamma \cdot F(\varphi_s) + \frac{a}{b}$$

$$W(u) =: e^{-u^2} \cdot \mathrm{erfc}(-ju) \quad ; \quad \mathrm{erfc}(z) =: \frac{2}{\sqrt{\pi}} \int\limits_z^{\infty} e^{-y^2}\, dy \tag{20}$$

It is supposed that $F(\varphi)=N(\varphi)/D(\varphi)$ can be written as quotient of a numerator and denominator; thus a is the residue of $F(\varphi)$. The quantity b distinguishes cases of the relative position φ_p of the pole to the pass way Pw or to the saddle point φ_S. For $\mathrm{Im}\{b\}>0$ the pole is still outside Pw; for $\mathrm{Im}\{b\}<0$ it has been crossed by $C(\varphi) \to$ Pw, and $\mathrm{Im}\{b\}=0$ describes the situation that φ_p is on Pw. The addendum in the definition of b defines the sign of the root in b. The addendum in the definition of γ also serves to select the sign of the root; it demands that the argument of γ should agree with the argument of a step $d\varphi$ from the saddle point φ_S in the direction of the pass way. The function $W(u)$ is based on the complementary error function $\mathrm{erfc}(z)$.

The correspondences to the present integral along the pass way Pw as defined above are:

$$\begin{aligned} &x \Rightarrow kr' \\ &f(\varphi) \Rightarrow -j\cos\varphi \\ &F(\varphi) \Rightarrow R(\varphi+\theta_s) = \frac{\cos(\varphi+\theta_s)-G}{\cos(\varphi+\theta_s)+G} \end{aligned} \tag{21}$$

and the required quantities are:

$$\begin{aligned} &\varphi_S = 0 \quad ; \quad \varphi_p = \arccos(-G) - \theta_s \\ &\cos(\varphi_p+\theta_s) = -G \\ &\cos\varphi_p = -G\cdot\cos\theta_s + \sqrt{1-G^2}\cdot\sin\theta_s \quad ; \quad \mathrm{Im}\{\sqrt{1-G^2}\} \le 0 \end{aligned} \tag{22}$$

Other quantities in the above definitions are for a locally reacting absorber :

$$\begin{aligned} &a = -\frac{\cos(\varphi_p+\theta_s)-G}{\sin(\varphi_p+\theta_s)-G} = \frac{2G}{\sqrt{1-G^2}} \\ &b = \pm\sqrt{j}\sqrt{\cos\varphi_p - 1} \xrightarrow[\varphi_p\to\varphi_s=0]{} \mp\frac{1-j}{2}\varphi_S \\ &\quad = \pm(j)^{3/2}\sqrt{1+G\cdot\cos\theta_s - \sqrt{1-G^2}\cdot\sin\theta_s} \\ &\gamma = \sqrt{2j} \\ &T = \sqrt{2j}\cdot R(\theta_s) \mp \frac{2\sqrt{j}G}{\sqrt{1-G^2}}\frac{1}{\sqrt{1+G\cdot\cos\theta_s - \sqrt{1-G^2}\cdot\sin\theta_s}} \end{aligned} \tag{23}$$

The sign convention for γ is satisfied; the sign convention in b requires one to take the lower signs in b and T if the last root in b,T is evaluated with a positive real part. The desired field $p_r=I/\pi$ can be evaluated by insertion.

The diagrams below compare in 3D-plots (as "wire graphics") the magnitude of the sound pressure |p| over kx, ky from numerical integration of the exact integral (thick lines) with results from approximate methods (to which the pass integration belongs).

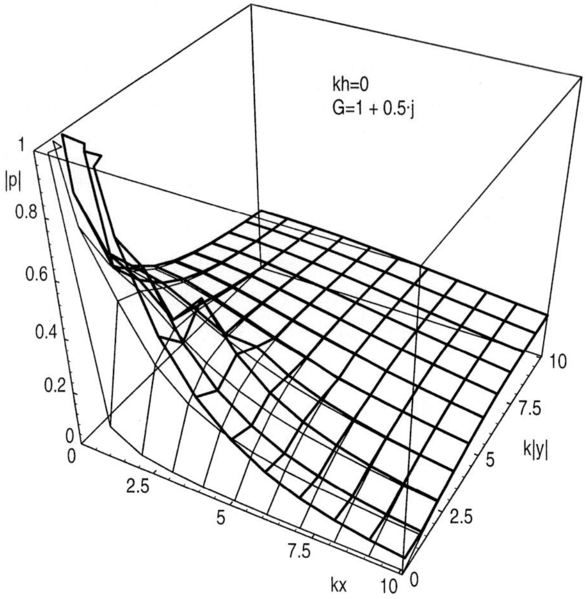

Comparison of numerical integration of the exact integral (thick lines) with the mirror source approximation (thin lines)

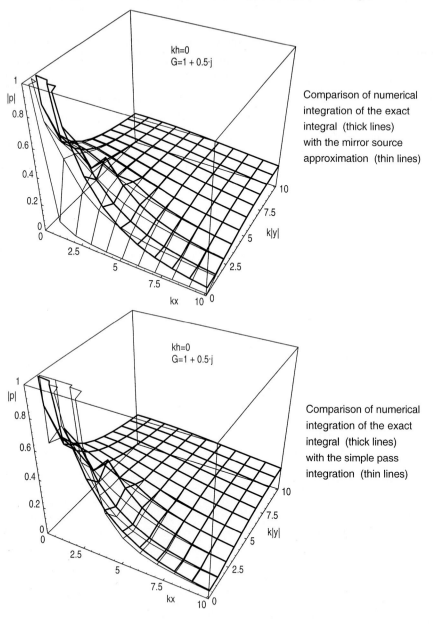

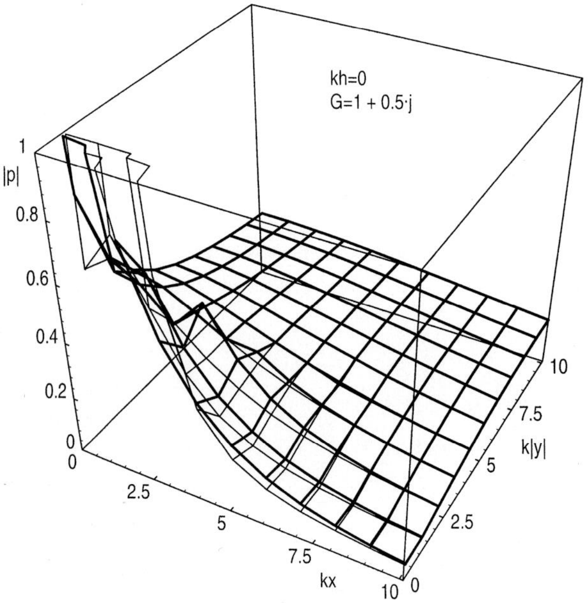

Comparison of numerical integration of the exact integral (thick lines) with the simple pass integration (thin lines)

D.15 A monopole line source above an infinite, plane absorber; with principle of superposition

Ref.: Mechel, [D.6];

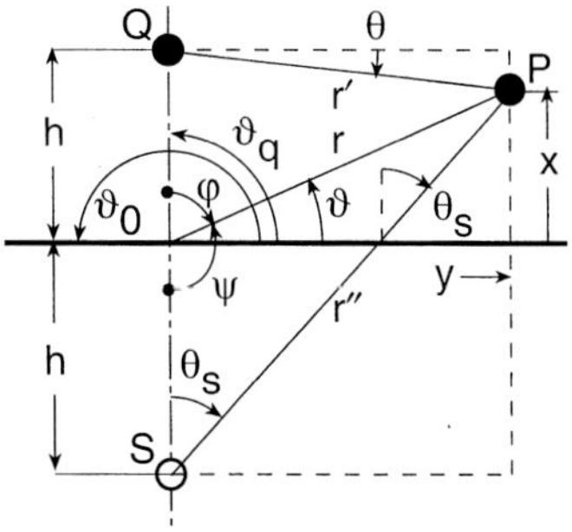

A monopole line source with volume flow q (per unit length) is placed at Q with a height h above a locally reacting plane with normalised surface admittance G. S is the mirror-reflected point to Q, P is a field point.

The sound field is formulated as

$p_a(\mathbf{r}) = A \cdot p_h(\mathbf{r}) + B \cdot p_s(\mathbf{r})$

with p_h the field above a hard plane,

and p_s the field above a soft plane, both satisfying individually the source condition.

The principle of superposition (3rd principle of superposition in section B.10) gives :

$$p_a(\mathbf{r}) = \frac{1}{1+G\cdot X(\mathbf{s})} \cdot \left[p_h(\mathbf{r}) + G\cdot X(\mathbf{s}) \cdot p_s(\mathbf{r})\right] \tag{1}$$

with $\mathbf{s}$ the projection of the field point P (along co-ordinate lines) on the plane, and the “cross-impedance”

$$X(\mathbf{s}) = \frac{p_h(\mathbf{s})}{Z_0 v_{sn}(\mathbf{s})} = -\frac{j k_0 \cdot p_h(\mathbf{s})}{\mathrm{grad}_n\, p_s(\mathbf{s})} \tag{2}$$

In the present task is $p_h = p_Q + p_{Sh}$; $p_w = p_Q + p_{Sw}$ with p_Q the free source field :

$$p_Q = P_0 \cdot H_0^{(2)}(k_0 r') = \frac{k_0 Z_0 \cdot q}{4} \cdot H_0^{(2)}(k_0 r') = \frac{p_Q(k_0 h)}{H_0^{(2)}(k_0 h)} \cdot H_0^{(2)}(k_0 r') \tag{3}$$

(the 2nd form replaces the amplitude P_0 by the source volume flow q ; the 3rd form describes the source strength by the free field sound pressure $p_Q(h)$ at the origin), and $p_h = p_Q + p_{Sh}$; $p_s = p_Q + p_{Ss}$, where p_{Sh} , p_{Ss} are the fields from the mirror sources in the case of a hard or soft plane, respectively, which for "ideal" reflection are exact forms of the scattered field:

$$p_{Sh} = P_0 \cdot H_0^{(2)}(k_0 r'') \quad ; \quad p_{Ss} = -P_0 \cdot H_0^{(2)}(k_0 r'') \tag{4}$$

One gets for the cross-impedance $X(y)$:

$$\frac{1}{X(y)} = \frac{-Z_0 v_{Qsx}}{p_{Qh}} = -j \cdot \frac{k_0 r_q}{2} \cdot \left(1 + \frac{H_2^{(2)}(k_0 r')}{H_0^{(2)}(k_0 r')}\right) = -j \cdot \frac{k_0 r_q}{2} \cdot \left(1 + \frac{H_2^{(2)}\left(\sqrt{(k_0 y)^2 + (k_0 r_q)^2}\right)}{H_0^{(2)}\left(\sqrt{(k_0 y)^2 + (k_0 r_q)^2}\right)}\right) \tag{5}$$

With $p_{Sw} = -p_{Sh}$ one can simplify to :

$$p_a(x,y) = p_Q(x,y) + \frac{1 - G \cdot X(y)}{1 + G \cdot X(y)} \cdot p_{Sh}(x,y) = p_Q(x,y) + \frac{1/X(y) - G}{1/X(y) + G} \cdot p_{Sh}(x,y) \tag{6}$$

Numerical comparison with saddle point integration (see section D.14) :

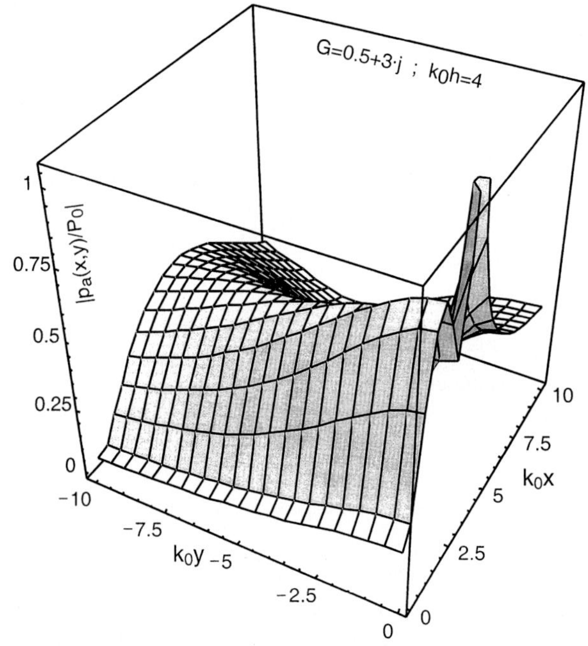

Sound pressure magnitude from a line source above an absorbing plane (on the x axis), evaluated with the principle of superposition.

The next diagram compares the above diagram (in a 3D-wire-plot) with results from the method of saddle point integration.

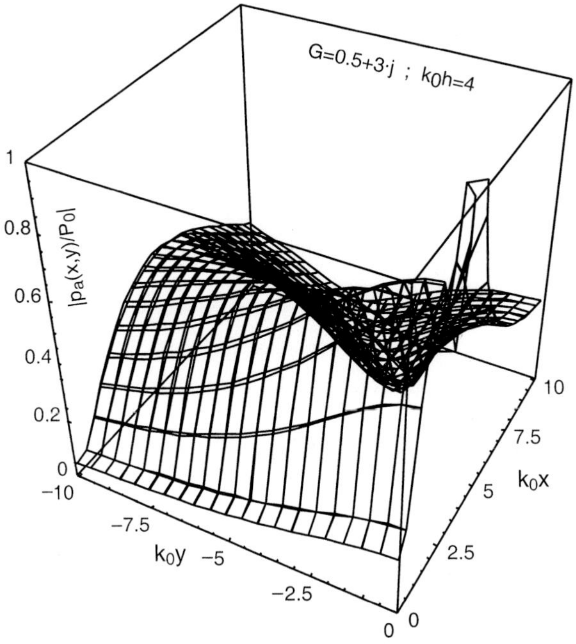

Comparison of the above figure with results from the saddle point integration.

D.16 A monopole point source above a bulk reacting plane, exact forms

Ref.: Mechel, [D.7]

A monopole point source is placed at a point Q at height h above an absorbent plane; a field point is at P . The plane may be bulk reacting. See section D.17 for a locally reacting plane. See [D.7] for references to the extensive literature about this problem. As an exception, the time factor in this section is $e^{-i\omega t}$, in order to facilitate the comparison with the literature, where this sign convention mostly is used.

The free field of the point source is
$$p_Q(r_1) = P_0 \frac{e^{i k_0 r_1}}{k_0 r_1} \tag{1}$$

the field p above the absorber is
$$\frac{p}{P_0} = \frac{e^{i k_0 r_1}}{k_0 r_1} + \frac{p_r}{P_0} \tag{2}$$

with r_1=dist (Q,P) and p_r the reflected field. The task is to find p_r .

An exact integral expression for p_r is :

$$\frac{p_r}{p_0} = i \int_0^{\pi/2-i\infty} J_0(k_0 r \cdot \sin\Theta) \cdot e^{i k_0 (z+h)\cdot\cos\Theta} \cdot R(\Theta) \cdot \sin\Theta \, d\Theta \tag{3}$$

where $R(\Theta)=(\cos\Theta-G)/(\cos\Theta+G)$ is the reflection factor of a plane wave incident under a polar angle Θ ; z is the co-ordinate normal to the plane directed into the half-space above the plane; r is the radius of P from the foot point of Q on the plane; $J_0(z)$ is the Bessel function of zero order. The path of integration in the complex Θ plane is: $0 \to \pi/2 \to \pi/2-i\cdot\infty$.

If the absorber is a half-space (indicated with index $\beta=2$, in contrast to index $\beta=1$ for half-space above the absorber) of a homogeneous, isotropic material the characteristic wave numbers and wave impedances in both half-spaces are k_β, Z_β , respectively, and the ratios $k=k_2/k_1$, $Z=Z_2/Z_1$. An exact formulation (SOMMERFELD) of the field in the upper half-space $\beta=1$ is :

$$\frac{p_1(r,z)}{p_0} = (1+kZ)\int_0^{\infty} \frac{y \cdot J_0(y k_1 r) \cdot e^{-k_1 z\sqrt{y^2-1}}}{\sqrt{y^2-k^2} + kZ\sqrt{y^2-1}} \, dy \tag{4}$$

This integral is used for numerical integration (as a reference for approximations), for which the interval of integration is subdivided in $(0,\infty) = (0,1)+(1,2)+(2,y_{hi})+(y_{hi},\infty)$, and the precision and convergence are checked separately in each sub interval.

In the case of two half-spaces, the integral above for p_r/p_0 can be transformed to (a form, which is suited for saddle point integration) :

$$\frac{p_r(r,\vartheta)}{p_0} = \frac{i}{2}\int_C H_0^{(1)}(k_1 r \cdot \sin\vartheta) \cdot e^{i k_1 H \cdot \cos\vartheta} \cdot R(\vartheta) \cdot \sin\vartheta \, d\vartheta \tag{5}$$

with the reflection factor :

$$R(\vartheta) = \frac{kZ\cos\vartheta - \sqrt{k^2 - \sin^2\vartheta}}{kZ\cos\vartheta + \sqrt{k^2 - \sin^2\vartheta}} \tag{6}$$

This form can be applied also for bulk reacting layers of finite thickness, if a corresponding reflection factor is used. The path of integration is $C = -\pi/2+i\cdot\infty \to -\pi/2 \to +\pi/2 \to +\pi/2-i\cdot\infty$. The cross-over from the positive bank of $\mathrm{Re}\{\vartheta\}$ to the negative bank is at $\mathrm{Re}\{\vartheta\}=0$.

Further exact forms (BUTOV) for the reflected field above a homogeneous half-space and the field in the lower half-space are :

$$\frac{p_r}{p_0} = \frac{i}{2\pi k_1} \int_{-\infty}^{\infty}\int_{-\infty}^{\infty} \frac{k_{2z} - kZ \cdot k_{1z}}{k_{2z} + kZ \cdot k_{1z}} \cdot e^{i(k_x x + k_y y)} \frac{e^{i k_{1z}|z+h|}}{k_{1z}} \, dk_x \, dk_y$$

$$\frac{p_2}{p_0} = \frac{i}{2\pi k_1} \int_{-\infty}^{\infty}\int_{-\infty}^{\infty} \frac{2kZ}{k_{2z} + kZ \cdot k_{1z}} \cdot e^{i(k_x x + k_y y)} e^{i k_{1z} h} e^{-i k_{2z} z} \, dk_x \, dk_y \qquad (7)$$

with wave number components $k_{\beta x}, k_{\beta y}, k_{\beta z}$ of k_β . The first line can be transformed to :

$$\frac{p_r}{p_0} = i \sum_{n=0}^{\infty} (-1)^n (4n+1) \cdot V_{2n} \cdot h_{2n}^{(1)}(k_1 r_2) \cdot P_{2n}(\cos\vartheta) \qquad (8)$$

with r_2=dist (mirror point of Q, P) ; $P_{2n}(z)$= Legendre polynomial ; $h_{2n}^{(1)}(z)$= spherical Hankel function of 1st kind ; and :

$$V_{2n} = \frac{1}{2}\int_{-1}^{1} V(x) \cdot P_{2n}(x)\, dx \quad ; \quad V(x) = \frac{kz \cdot x - \sqrt{k^2 - 1 + x^2}}{kz \cdot x + \sqrt{k^2 - 1 + x^2}} \qquad (9)$$

Although this form is elegant, it is not suited for numerical evaluations, because of problems of convergence caused by the spherical Hankel functions.

Another exact form for p_1 (BREKHOVSKIKH) above a homogeneous absorber half-space is :

$$\frac{p_1(r,z;h)}{p_0} = \frac{e^{i k_1 r_1}}{k_1 r_1} - \frac{e^{i k_1 r_2}}{k_1 r_2} + \frac{p_1(r,z+h;0)}{p_0}$$

$$\frac{p_1(r,z+h;0)}{p_0} = 2(1+kZ) \int_0^{\infty} J_0(y \cdot k_1 r) \frac{y \cdot e^{-k_1 H \sqrt{y^2-1}}}{\sqrt{y^2 - k^2} + kZ\sqrt{y^2-1}} \, dy \qquad (10)$$

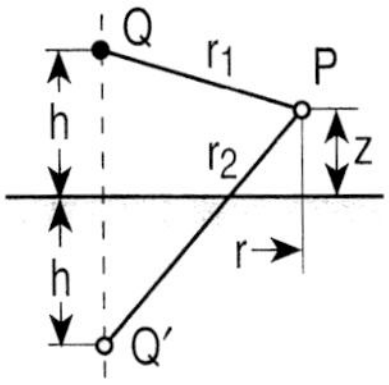

or with $\gamma = y \cdot k_1$:

$$\frac{p_1(r,z+h;0)}{p_0} = \frac{2(1+kZ)}{k_1} \int_0^{\infty} J_0(\gamma r) \frac{\gamma \cdot e^{-H\sqrt{\gamma^2 - k_1^2}}}{\sqrt{\gamma^2 - k_2^2} + kZ\sqrt{\gamma^2 - k_1^2}} \, d\gamma \qquad (11)$$

with r_1=dist(Q,p) ; r_2=dist (mirror point of Q, P) ; H=h+z= sum of heights of P and Q . The inclusion of the source height h in $p_1(r,z;h)$ indicates Brekhovskikh's rule: if one subtracts from the source free field the mirror source field, the remaining scattering term depends only on the sum of source and receiver heights. The 2nd form can be further modified to a form which is suited for saddle point integration :

$$\frac{p_1(r,H;0)}{p_0} = \frac{2(1+kZ)}{k_1} \int_{-\infty}^{\infty} \frac{\gamma \cdot e^{-H\sqrt{\gamma^2 - k_1^2} + i\gamma r}}{\sqrt{\gamma^2 - k_2^2} + kZ\sqrt{\gamma^2 - k_1^2}} \cdot H_0^{(1)}(\gamma r) \cdot e^{-\gamma r} \, d\gamma \qquad (12)$$

The path of integration is parallel to $Re\{\gamma\}$ with a small distance above this axis for $Re\{\gamma\}<0$ and a small distance below it for $Re\{\gamma\}>0$.

The exact form of VAN DER POL for two half-spaces is :

$$\frac{p_1(r,z)}{P_0}=\frac{e^{ik_1r_1}}{k_1r_1}+\frac{e^{ik_1r_2}}{k_1r_2}-\frac{1}{\pi k_1}\iiint_{V_2}\frac{\partial^2}{\partial z^2}\left(\frac{e^{ik_2r_1}}{r_1}\right)\frac{e^{ik_1r_2}}{r_2}\cdot r_0\,dr_0\,dz\,d\varphi \tag{13}$$

with integration over the half-space V_2 below the plane which contains the mirror-reflected point to Q , and r_0,φ determined from :

$$r_1^2=r_0^2+z^2\quad;\quad r_2^2=r^2-2rr_0\cos\varphi+r_0^2+(H+kZ\cdot z)^2 \tag{14}$$

D.17 A monopole point source above a locally reacting plane, exact forms

Ref.: Mechel, [D.7]

A monopole point source is placed at a point Q at height h above an absorbent plane; a field point is at P . The plane is locally reacting with a normalised admittance $G=1/Z$. See section D.16 for a bulk reacting plane; some of the forms for the field above the absorbent plane in that section can be used also for a locally reacting plane, if such forms apply the reflection factor R of a plane wave at the plane. See [D.7] for references to the extensive literature about this problem. As an exception, the time factor in this section is $e^{-i\omega t}$, in order to facilitate the comparison with the literature, where this sign convention mostly is used.

The free field of the point source is $$p_Q(r_1)=P_0\,\frac{e^{ik_0r_1}}{k_0r_1} \tag{1}$$

the field p above the absorber is $$\frac{p}{P_0}=\frac{e^{ik_0r_1}}{k_0r_1}+\frac{p_r}{P_0} \tag{2}$$

with r_1=dist (Q,P) and p_r the reflected field. The task is to find p_r .

The reflection factor of a plane wave incident under a polar angle Θ is (with $Z=1/G$)
$R(\Theta)=(\cos\Theta-G)/(\cos\Theta+G)=(Z\cdot\cos\Theta-1)/(Z\cdot\cos\Theta+1)$. (3)

An exact form of p_r is :

$$\frac{p_r}{p_0} = i \int_0^{\pi/2-i\infty} J_0(k_0 r \cdot \sin\vartheta) \cdot e^{i k_0 H \cdot \cos\vartheta} \cdot R(\vartheta) \cdot \sin\vartheta \, d\vartheta \tag{4}$$

H=z+h ; z is the co-ordinate normal to the plane directed into the half-space above the plane; r is the radius to P from the foot point of Q on the plane; $J_0(z)$ is the Bessel function of zero order. The path of integration in the complex ϑ plane is: $0 \to \pi/2 \to \pi/2 - i\cdot\infty$. Decomposition into real and imaginary parts with $\vartheta=\vartheta'+i\cdot\vartheta''$ of :

$$\begin{aligned} \sin\vartheta &= \sin\vartheta' \cdot \cosh\vartheta'' + i \cdot \cos\vartheta' \cdot \sinh\vartheta'' \\ \cos\vartheta &= \cos\vartheta' \cdot \cosh\vartheta'' - i \cdot \sin\vartheta' \cdot \sinh\vartheta'' \end{aligned} \tag{5}$$

gives :

$$\frac{p_r}{p_0} = i \int_0^1 J_0(k_0 r \cdot \sqrt{1-y^2}) \cdot e^{i k_0 H \cdot y} \frac{Z \cdot y - i}{Z \cdot y + i} dy + \int_0^\infty J_0(k_0 r \cdot \sqrt{1+y^2}) \cdot e^{-k_0 H \cdot y} \frac{Z \cdot y + i}{Z \cdot y - i} dy \tag{6}$$

Replacement of the Bessel function with the Hankel function yields :

$$\frac{p_r}{p_0} = \frac{i}{2} \int_C H_0^{(1)}(k_0 r \cdot \sin\vartheta) \cdot e^{i k_0 H \cdot \cos\vartheta} \cdot R(\vartheta) \cdot \sin\vartheta \, d\vartheta \tag{7}$$

with the path of integration $C = -\pi/2 + i\cdot\infty \to -\pi/2 \to +\pi/2 \to +\pi/2 - i\cdot\infty$.

A different exact form is :

$$\frac{p_r}{p_0} = \int_0^\infty J_0(k_0 r \cdot y) \cdot e^{-k_0 H \cdot \sqrt{y^2-1}} \frac{Z \cdot \sqrt{y^2-1} + i}{Z \cdot \sqrt{y^2-1} - i} \frac{y}{\sqrt{y^2-1}} dy \tag{8}$$

with a negative imaginary root in $0 \le y < 1$.

With a field composition as :

$$\frac{p(r,z;h)}{p_0} = \frac{e^{i k_0 r_1}}{k_0 r_1} - \frac{e^{i k_0 r_2}}{k_0 r_2} + \frac{p_1(r, z+h; 0)}{p_0} \tag{9}$$

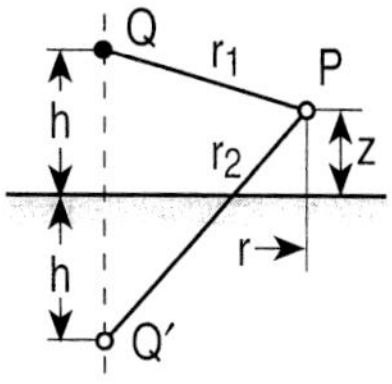

one gets :

$$\frac{p_1(r,H;0)}{p_0} = 2Z \int_0^\infty J_0(y \cdot k_0 r) \frac{y \cdot e^{-k_0 H \sqrt{y^2-1}}}{Z\sqrt{y^2-1} - i} dy \tag{10}$$

BUTOV's form for bulk reacting absorbers can be transformed so that it can be applied to locally reacting absorbers :

$$\frac{p_r}{p_0} = i \sum_{n=0}^\infty (-1)^n (4n+1) \cdot V_{2n} \cdot h_{2n}^{(1)}(k_1 r_2) \cdot P_{2n}(\cos\vartheta) \tag{11}$$

with $h_{2n}^{(1)}(z)$ spherical Hankel functions of the 1st kind ; $P_{2n}(z)$ Legendre polynomials ; r_2=dist(mirror point of Q, P), and

$$V_{2n=0} = \frac{1}{2Z}\int_{-Z}^{Z}\frac{y-1}{y+1}\,dy = 1-\frac{1}{Z}\ln\frac{1+Z}{1-Z} \tag{12}$$

$$V_{2n>0} = \frac{-2}{Z}Q_{2n}(1/Z) = \frac{-1}{Z}\left[P_{2n}(1/Z)\cdot\ln\frac{1+Z}{1-Z} - 4\sum_{m=0}^{[n-1/2]}\frac{2(n-m)-1}{(2m+1)(2n-m)}P_{2(n-m)-1}(1/Z)\right] \tag{13}$$

with $Q_{2n}(z)$ Legendre polynomials of the 2nd kind, and $[n-1/2]$ the highest integer $\le(n-1/2)$.

All integrals of the exact forms have oscillating integrands, and the interval of integration extends to infinity. If numerical integration is applied, the convergence must be improved. This is done according to the scheme :

$$I = \int_y^{\infty} f(x)\,dx = \int_y^{\infty} f_\infty(x)\,dx + \int_y^{\infty}\left(f(x)-f_\infty(x)\right)dx \tag{14}$$

where $f_\infty(x)$ is an asymptotic approximation to $f(x)$, and the analytical integral over $f_\infty(x)$ is known (it is dangerous to apply an approximation to that integral). In $\left(f(x)-f_\infty(x)\right)$ the oscillations at large x are reduced.

D.18 A monopole point source above a locally reacting plane, exact saddle point integration

Ref.: Mechel, [D.7]

The method of saddle point integration generally is considered as an approximate method to evaluate an integral which satisfies some criteria. In the present task, the saddle point integration can be applied so that it is an exact transformation of the integral, which makes it suited to numerical integration with high precision. If an integral as it appears in the present problem can be cast exactly into an integral over the path of steepest descent, it has the best possible form for a precise numerical evaluation.

Suppose the integral to be evaluated is of the form :

$$I = \int_C e^{a \cdot f(\vartheta)} \cdot F(\vartheta)\, d\vartheta \qquad (1)$$

This integral is suited for saddle point integration, if $a>>1$ is a large real number, and the path C goes to infinity on both sides.

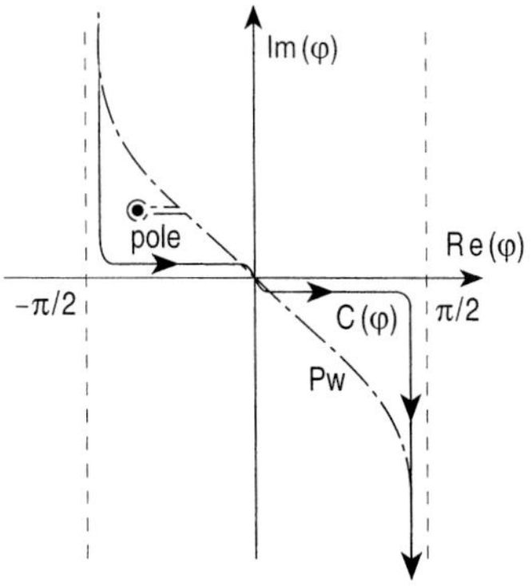

The saddle point integration in most of its applications is an approximation, because $f(\vartheta)$ is approximated as $f(\vartheta)\approx\alpha+\beta\cdot s^2$ with a real variable s on the pass way, and, more serious, $F(\vartheta)$ is expanded as a power series. If a pole is near the saddle point, the radius of convergence becomes small and the precision goes down.

The start integral I_0 in our problem comes from the 3rd integral of section D.17 , after multiplication and division of the integrand by $\exp(\pm ik_0r\cdot\sin\vartheta)$:

$$\frac{p_r}{P_0} = \frac{i}{2}\int_C e^{i\,k_0r_2\cdot\cos(\vartheta-\Theta_o)} \cdot \left[H_0^{(1)}(k_0r\cdot\sin\vartheta)\cdot e^{-i\,k_0r\cdot\sin\vartheta}\right]\cdot R(\vartheta)\cdot\sin\vartheta\; d\vartheta = \frac{i}{2}\cdot I_0 \qquad (2)$$

with the geometrical quantities as in the sketch and making use of

$$k_0\big((h+z)\cos\vartheta + r\sin\vartheta\big) = k_0r_2\cdot\cos(\vartheta-\Theta_0) \qquad (3)$$

The integration path $C(\varphi)$ and the path of steepest descent (pass way Pw) are shown in the sketch in the complex plane of $\varphi=\vartheta-\Theta_0$. If during the deformation $C(\varphi)\rightarrow Pw$ a pole of the reflection factor $R(\vartheta)$ is crossed, it is encircled as shown. This extra circle will give a “pole contribution”.

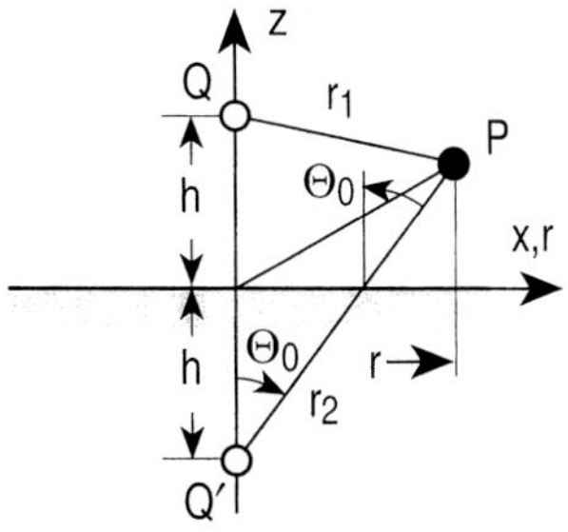

The oscillations of the term in brackets in the integral go to zero for large argument values, because the Hankel function oscillations are compensated by the exponential factor.

Comparing I_0 with the general integral I , correspondences are: $a\rightarrow k_0r_2$; $f(\vartheta)\rightarrow i\cdot\cos\varphi$. The saddle point ϑ_s with the maximum exponential factor (outside the brackets) follows from $df(\vartheta)/d\vartheta=0$, which in our case is $\varphi_s=0$, i.e. $\vartheta_s=\Theta_0$. The parameter form of the pass way equation is (with $\vartheta=\vartheta'+i\cdot\vartheta''$; values of ϑ on Pw are named ϑ_{Pw}) :

$$\cos(\vartheta'_{Pw}-\Theta_0)\cdot\cosh\vartheta''_{Pw}=1$$

$$\cos(\vartheta'_{Pw}-\Theta_0)=\frac{1}{\cosh\vartheta''_{Pw}} \tag{4}$$

$$\sin(\vartheta'_{Pw}-\Theta_0)=-\tanh\vartheta''_{Pw}$$

or equivalently :

$$\sin\vartheta'_{Pw}=\frac{\sin\Theta_0-\cos\Theta_0\cdot\sinh\vartheta''_{Pw}}{\cosh\vartheta''_{Pw}}$$

$$\cos\vartheta'_{Pw}=\frac{\cos\Theta_0+\sin\Theta_0\cdot\sinh\vartheta''_{Pw}}{\cosh\vartheta''_{Pw}} \tag{5}$$

and therefore the function $f(\vartheta)$ in the exponent:

$$f(\vartheta_{Pw})=-\tanh\vartheta''_{Pw}\cdot\sinh\vartheta''_{Pw}+i\xrightarrow[|\vartheta''_{Pw}|<<1]{}-(\vartheta''_{Pw})^2+i \tag{6}$$

All factors in the integrand of I_0 can be expressed as functions of ϑ''_{Pw} , especially :

$$R(\vartheta_{Pw})=R\langle\vartheta''_{Pw}\rangle=\frac{Z\cdot\cos\langle\vartheta''_{Pw}\rangle-1}{Z\cdot\cos\langle\vartheta''_{Pw}\rangle+1} \tag{7}$$

with the definitions :

$$\cos\langle\vartheta''_{Pw}\rangle=\cos\Theta_0+\sin\Theta_0\cdot\sinh\vartheta''_{Pw}-i\cdot\tanh\vartheta''_{Pw}\cdot\left(\sin\Theta_0-\cos\Theta_0\cdot\sinh\vartheta''_{Pw}\right)$$

$$\sin\langle\vartheta''_{Pw}\rangle=\sin\Theta_0-\cos\Theta_0\cdot\sinh\vartheta''_{Pw}+i\cdot\tanh\vartheta''_{Pw}\cdot\left(\cos\Theta_0+\sin\Theta_0\cdot\sinh\vartheta''_{Pw}\right) \tag{8}$$

With the transition $\vartheta\to\vartheta''_{Pw}$ the general integral I is transformed to :

$$I=\int_{+\infty}^{-\infty}e^{a\cdot g(\vartheta''_{Pw})}\cdot G(\vartheta''_{Pw})\,d\vartheta''_{Pw}$$

$$g(\vartheta''_{Pw})=i-\tanh\vartheta''_{Pw}\cdot\sinh\vartheta''_{Pw} \tag{9}$$

$$G(\vartheta''_{Pw})=F\big(\vartheta(\vartheta''_{Pw})\big)\frac{d\vartheta}{d\vartheta''_{Pw}}=F\big(\vartheta(\vartheta''_{Pw})\big)\frac{dg(\vartheta''_{Pw})/d\vartheta''_{Pw}}{df(\vartheta)/d\vartheta}$$

The last fraction becomes :

$$\frac{dg(\vartheta''_{Pw})/d\vartheta''_{Pw}}{df(\vartheta)/d\vartheta}=-\frac{\sinh\vartheta''_{Pw}\cdot(2-\tanh^2\vartheta''_{Pw})}{\tanh\vartheta''_{Pw}+i\cdot\sinh\vartheta''_{Pw}}=-\frac{2-\tanh^2\vartheta''_{Pw}}{i+1/\cosh\vartheta''_{Pw}} \tag{10}$$

The desired integral I_0 finally is (substitute for ease of writing $\vartheta''_{Pw}\to t$) :

$$I_0 = e^{i\,k_0 r_2}\int_0^{\infty} e^{-k_0 r_2\cdot\tanh t\cdot\sinh t}\,\frac{2-\tanh^2 t}{i+1/\cosh t}\cdot$$
$$\cdot\Big[R\langle t\rangle\cdot\sin\langle t\rangle\cdot H_0^{(1)}\big(k_0 r\,\sin\langle t\rangle\big)\cdot e^{-k_0 r\,\sin\langle t\rangle}+ \qquad (11)$$
$$+\,R\langle -t\rangle\cdot\sin\langle -t\rangle\cdot H_0^{(1)}\big(k_0 r\,\sin\langle -t\rangle\big)\cdot e^{-k_0 r\,\sin\langle -t\rangle}\Big]\,dt$$

In the special case $h=z=0$, i.e. $\Theta_0=\pi/2$, one gets :

$$\cos\langle\vartheta''_{Pw}\rangle=\sinh\vartheta''_{Pw}-i\cdot\tanh\vartheta''_{Pw}=-\cos\langle-\vartheta''_{Pw}\rangle$$
$$\sin\langle\vartheta''_{Pw}\rangle=1+i\cdot\tanh\vartheta''_{Pw}\cdot\sinh\vartheta''_{Pw}=\sin\langle-\vartheta''_{Pw}\rangle \qquad (12)$$
$$R\langle\vartheta''_{Pw}\rangle=\frac{Z\cdot\big(\sinh\vartheta''_{Pw}-i\cdot\tanh\vartheta''_{Pw}\big)-1}{Z\cdot\big(\sinh\vartheta''_{Pw}-i\cdot\tanh\vartheta''_{Pw}\big)+1}=1/R\langle-\vartheta''_{Pw}\rangle$$

and therewith :

$$I_0=\int_0^{\infty} e^{-k_0 r_2\cdot\tanh t\cdot\sinh t}\,\frac{2-\tanh^2 t}{i+1/\cosh t}\cdot(1+i\cdot\tanh t\cdot\sinh t)\cdot \qquad (13)$$
$$\cdot H_0^{(1)}\big(k_0 r\,(1+i\cdot\tanh t\cdot\sinh t)\big)\cdot\big(R\langle t\rangle+1/R\langle t\rangle\big)\,dt$$

The integrand in I_0 decreases quickly with increasing t . This is paid for with a complex argument of the Hankel function. The scattered field is $p_r/P_0=i/2\cdot I_0$.

If during the deformation $C(\varphi)\rightarrow Pw$ a pole of the reflection factor $R(\vartheta)$ is crossed, the pole contribution p_{rp} must be added to p_r :

$$\frac{p_{rp}}{P_0}=\frac{-2\pi}{Z}H_0^{(1)}(k_0 r\sqrt{1-1/Z^2})\cdot e^{-i\,k_0 H/Z}\quad ;\quad \mathrm{Re}\{\sqrt{1-1/Z^2}\}>0 \qquad (14)$$

D.19 A monopole point source above a locally reacting plane, approximations

Ref.: Mechel, [D.7]

See section D.17 for exact integral formulations of the solution, and see [D.7] for a discussion of the approximations and their precision.

A monopole point source is placed at a point Q at height h above an absorbent plane; a field point is at P with height z above the plane and horizontal distance r from Q. Q′ is the mirror-reflected point to Q. The plane is locally reacting with a normalised admittance G=1/Z . See [D.7] for a detailed discussion of the approximations.

As an exception, the time factor in this section is $e^{-i\omega t}$, in order to facilitate the comparison with the literature, where this sign convention mostly is used. Radii used are: r_1=dist(Q,P) ; r_2=dist(Q′,P) ; and the angle $\Theta_0 = \angle((Q',Q),(Q',P))$.

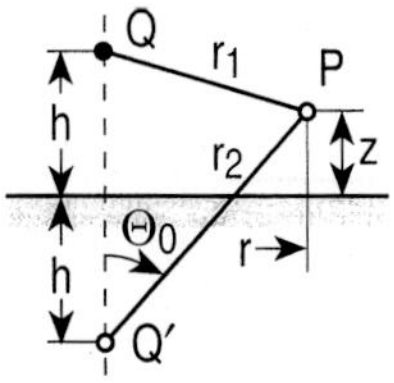

The start equation for a 1st approximation to the reflected field p_r is :

$$\frac{p_r}{P_0} = \frac{i}{2}\int_C e^{i k_0 r_2 \cdot \cos(\vartheta-\Theta_0)} \cdot \left[H_0^{(1)}(k_0 r \cdot \sin\vartheta) \cdot e^{-i k_0 r \cdot \sin\vartheta}\right] \cdot R(\vartheta) \cdot \sin\vartheta \, d\vartheta = \frac{i}{2} \cdot I_0 \qquad (1)$$

See section D.17 for definitions of r, r_2, $R\langle t \rangle$. An approximate saddle point integration is applied to I_0 (condition: a pole at $\cos\vartheta_p = -1/Z$ of the reflection factor $R(\vartheta)$ is not near to the saddle point $\vartheta_s = \Theta_0$ (see sketch in section D.18 for Θ_0)). The first order approximation is :

$$\frac{p_r(r,z)}{P_0} = \frac{e^{i k_0 r_2}}{k_0 r_2}\left[R(\Theta_0) - \frac{i}{2 k_0 r_2}\left(R'(\Theta_0)\cot\Theta_0 + R''(\Theta_0)\right)\right] + \left\langle \frac{p_{rp}}{P_0} \right\rangle \qquad (2)$$

The term $\langle p_{rp}/P_0 \rangle$ indicates a possible pole contribution p_{rp} (more, see below). In these equations is : r= horizontal distance between source Q and field point P ; z= height of field point P ; r_1= dist(Q,P) ; r_2= dist(Q′,P) ; $\Theta_0 = \angle((Q',Q),(Q',P))$; see section D.1 for $R(\Theta_0)$ (there $r(\Theta)$) and derivatives.

A higher approximation (condition: pole not near the saddle point) is :

$$\frac{p_r(r,z)}{P_0} = -\frac{e^{i k_0 r_2}}{k_0 r_2}\left[\left(\frac{1}{8 k_0 r_2} + i\right) \cdot F(\Theta_0) + \frac{1}{2 k_0 r_2} F''(\Theta_0)\right] + \left\langle \frac{p_{rp}}{P_0} \right\rangle \qquad (3)$$

with:

$$F(\Theta_0) = R(\Theta_0)\left[1 - \frac{\alpha}{\sin^2\Theta_0} - \frac{i\beta}{\sin\Theta_0}\right] \quad ; \quad \alpha = \frac{9}{128 (k_0 r)^2} \quad ; \quad \beta = \frac{1}{8 k_0 r}$$

$$F''(\Theta_0) = \frac{1}{4} R(\Theta_0)\left[\frac{1 - 3\sin^2\Theta_0}{\sin^2\Theta_0} - \alpha\frac{6 + 11\cos^2\Theta_0}{\sin^4\Theta_0} - i\beta\frac{2 + 3\cos^2\Theta_0}{\sin^3\Theta_0}\right] + \qquad (4)$$

$$+ R'(\Theta_0)\cos\Theta_0\left[\frac{1}{\sin\Theta_0} + \frac{3\alpha}{\sin^3\Theta_0} + \frac{i\beta}{\sin^2\Theta_0}\right] + R''(\Theta_0)\left[1 - \frac{\alpha}{\sin^2\Theta_0} - \frac{i\beta}{\sin\Theta_0}\right]$$

If the pole of the reflection factor is near the saddle point, the integral to be evaluated, and which can be transformed to the general form :

$$I = e^{a \cdot f(\vartheta_s)} \int_{-\infty}^{\infty} e^{-a \cdot f(\vartheta)} \cdot \Phi(s)\, ds \tag{5}$$

is modified further by separation of the simple pole in $\Phi(s)$, i.e. by setting :

$$\Phi(s) = \frac{\alpha}{s - \beta} + T(s) \tag{6}$$

with :

$$\alpha - \frac{2}{Z} H_0^{(1)}(k_0 r \sqrt{1 - 1/Z^2}) \cdot e^{-i k_0 r \sqrt{1-1/Z^2}} \quad ; \quad \mathrm{Re}\{\sqrt{1 - 1/Z^2}\} > 0$$

$$\beta = \sqrt{i\left(1 + \cos\Theta_0 / Z - \sin\Theta_0 \sqrt{1 - 1/Z^2}\right)} \quad ; \quad \mathrm{Im}\{\beta\} \begin{cases} > 0 \text{ ; pole above pass way} \\ = 0 \text{ ; pole on pass way} \\ < 0 \text{ ; pole below pass way} \end{cases} \tag{7}$$

The cases for $\mathrm{Im}\{\beta\}$ correspond to :

$$\mathrm{Im}\{\beta\} \gtreqless 0 \;\Leftrightarrow\; G'' \gtreqless \frac{-1}{\sin\Theta_0} \frac{(\cos\Theta_0 + G')(1 + \cos\Theta_0 \cdot G')}{\sqrt{1 + 2\cos\Theta_0\, G' + G'^2}} \tag{8}$$

One gets the approximation :

$$\frac{p_r}{P_0} = \frac{i \cdot e^{i k_0 r_2}}{2} \begin{cases} \pm i\pi\alpha \cdot W(\pm\beta\sqrt{k_0 r_2}) + \sqrt{\pi/(k_0 r_2)} \cdot T(0) \quad ; \quad \mathrm{Im}\{\beta\} \gtrless 0 \\ i\pi\alpha \cdot W(\beta\sqrt{k_0 r_2}) + \sqrt{\pi/(k_0 r_2)} \cdot T(0) - i\pi\alpha \cdot e^{-k_0 r_2 \cdot \beta^2} \quad ; \quad \mathrm{Im}\{\beta\} = 0 \end{cases} \tag{9}$$

with :

$$T(0) = \frac{\alpha}{\beta} + (1 - i) H_0^{(1)}(k_0 r \sin\Theta_0) \cdot e^{-i k_0 r \sin\Theta_0} \cdot \sin\Theta_0 \frac{\cos\Theta_0 - G}{\cos\Theta_0 + G}$$

$$W(u) = e^{-u^2} \cdot \mathrm{erfc}(-i u) \quad ; \quad \mathrm{erfc}(z) = \frac{2}{\sqrt{\pi}} \int_z^{\infty} e^{-x^2} dx \tag{10}$$

where erfc(z) is the complementary error function.

VAN MOORHEM's approximation :

$$\frac{p_r}{P_0} = \frac{e^{i k_0 r_2}}{k_0 r_2} \left[R(\Theta_0) + (1 - R(\Theta_0)) \cdot F(\Theta_0, k_0 r_2)\right] \tag{11}$$

with : (12)

The signs of other square roots are selected so that their real parts are positive.

This form is well suited for numerical integration; the results coincide with those of numerical integrations of other exact forms. It can be applied also for $h=z=0$, i.e. source and receiver on the plane.

An approximation which is derived from the last form of p_r/P_0 is for $k_0r_2 >>1$ and $|B|^2 >> |A|^2$ and $|B|^2 >> k_0r_2$:

$$\frac{p_r}{P_0}=\frac{e^{ik_0r_2}}{k_0r_2}\left[1-2k_0r_2G\frac{C}{B}\sum_{m=0}^{\infty}\frac{(2m)!}{(m!)^2(4B^2)^m}\cdot I_m\right]- \\ -\pi(1-C)G\cdot H_0^{(1)}(k_0r\sqrt{1-G^2})\cdot e^{-ik_0(h+z)\cdot G} \tag{25}$$

with iterative evaluation of the I_m :

$$I_0=\sqrt{\pi}\cdot e^{A^2}\cdot \text{erfc}(A) \quad ; \quad I_1=A+\left(\tfrac{1}{2}-A^2\right)\cdot I_0 \\ I_m=\left(m-\tfrac{1}{2}-A^2\right)\cdot I_{m-1}+(m-1)A^2\cdot I_{m-2} \tag{26}$$

This approximation computes very precisely in the mentioned range of conditions.

An approximation by NOBILE :

$$\frac{p_r}{P_0}=\frac{e^{ik_0r_2}}{k_0r_2}-\frac{4iG\cdot B}{G+\cos\Theta_0}\cdot e^{ik_0r_2}\sum_{n=0}^{\infty}(e_0\cdot E_n+K_n)\cdot T_n \tag{27}$$

with :

$$B=-i\sqrt{1+G\cos\Theta_0-\sin\Theta_0\sqrt{1-G^2}} \quad ; \quad \text{Re}\{\sqrt{\ldots}\}>0 \\ C=1+G\cos\Theta_0-\sin\Theta_0\sqrt{1-G^2} \quad ; \quad \text{Re}\{\sqrt{1-G^2}\}\geq 0 \\ T_n=\frac{1}{(2B)^n}\sum_{m=0}^{[n/2]}\binom{n-m}{m}\cdot a_{n-m}\cdot\left(\frac{-4B^2}{C}\right)^{n-m} \\ a_0=1 \quad ; \quad a_n=\frac{\frac{1}{2}-n}{n}\cdot a_{n-1} \\ e_0=\frac{1}{2}\sqrt{\frac{\pi}{ik_0r_2}}\cdot e^{-\lambda^2}\cdot\text{erfc}(-i\lambda) \\ \lambda=\sqrt{ik_0r_2}\sqrt{1+G\cos\Theta_0+\sin\Theta_0\sqrt{1-G^2}} \quad ; \quad \text{all Re}\{\sqrt{\ldots}\}>0 \tag{28}$$

and iterative evaluation of :

$$E_0 = 1 \quad ; \quad E_1 = -B \quad ; \quad E_n = -B \cdot E_{n-1} - i\frac{n-1}{2k_0 r_2} \cdot E_{n-2}$$
$$K_0 = 0 \quad ; \quad K_1 = -\frac{i}{2k_0 r_2} \quad ; \quad K_n = -B \cdot K_{n-1} - i\frac{n-1}{2k_0 r_2} \cdot K_{n-2} \tag{29}$$

An other approximation by Nobile, in which $\Re$ is a "reflection factor for a spherical wave" (see section D.20) :

$$\frac{p_r}{P_0} = \Re \frac{e^{i k_0 r_2}}{k_0 r_2} \quad ; \quad \Re = 1 + \frac{2G}{G + \cos\Theta_0} \sum_{n=0}^{\infty} (e_1 \cdot \overline{E}_n + \overline{K}_n) \cdot \overline{T}_n \tag{30}$$

with auxiliary quantities from above, except the newly defined quantities :

$$\overline{T}_n = \sum_{m=0}^{[n/2]} \binom{n-m}{m} \cdot a_{n-m} \cdot \left(\frac{-4B^2}{C}\right)^{n-m} \quad ; \quad e_1 = -2iBk_0 r_2 \cdot e_0 \tag{31}$$

$$\overline{E}_0 = 1 \quad ; \quad \overline{E}_1 = -\frac{1}{2} \quad ; \quad \overline{E}_n = -\frac{1}{2} \cdot \overline{E}_{n-1} - i\frac{n-1}{8k_0 r_2 B^2} \cdot \overline{E}_{n-2}$$
$$\overline{K}_0 = 0 \quad ; \quad \overline{K}_1 = -\frac{1}{2} \quad ; \quad \overline{K}_n = -\frac{1}{2} \cdot \overline{K}_{n-1} - i\frac{n-1}{8k_0 r_2 B^2} \cdot \overline{K}_{n-2} \tag{32}$$

In the finite sum the upper limits is $[x]$, the highest integer $\le x$.

Finally, the approximation obtained with the principle of superposition applied in section D.15 is :

$$\frac{p_r}{P_0} = \Re \frac{e^{i k_0 r_2}}{k_0 r_2} \quad ; \quad \Re = \frac{1/X(r) - G}{1/X(r) + G} \tag{33}$$

with the normalised cross-impedance $X(r)$ of the plane $z=0$ defined and given by :

$$\frac{1}{X(r)} = \frac{-Z_0 v_{sz}(r, z=0)}{p_h(r, z=0)} = \frac{i}{k_0} \frac{\partial\big(p_Q(r,0) - p_{Q'}(r,0)\big)/\partial z}{p_Q(r,0) + p_{Q'}(r,0)} \tag{34}$$

where $p_Q(r_1)$ is the source free field, and $p_{Q'}(r_2)$ the free field of a point source (of same strength) in the mirror-reflected point Q' to Q (p_h is the field, for which the plane $z=0$ is hard; p_s is the field for which that plane is soft). It is :

$$p_Q = \frac{e^{i k_0 r_1}}{k_0 r_1} \quad ; \quad p_{Q'} = \frac{e^{i k_0 r_2}}{k_0 r_2} \tag{35}$$
$$r_1 = \sqrt{(h-z)^2 + r^2} \quad ; \quad r_2 = \sqrt{(h+z)^2 + r^2}$$

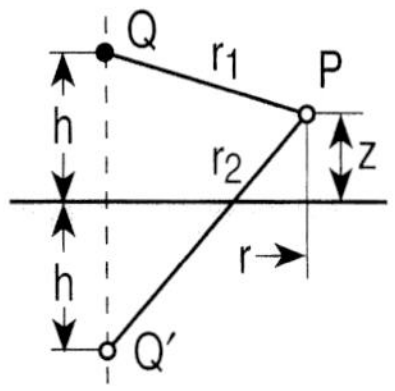

and therefore :

$$\frac{1}{X(r)} = \frac{k_0 h\left(i + k_0\sqrt{h^2 + r^2}\right)}{k_0^2\left(h^2 + r^2\right)}$$

$$\Re = \frac{i - G(k_0 h + r/h \cdot k_0 r) + k_0\sqrt{h^2 + r^2}}{i + G(k_0 h + r/h \cdot k_0 r) + k_0\sqrt{h^2 + r^2}} \xrightarrow[r=0]{} \frac{i + k_0 h(1-G)}{i + k_0 h(1+G)} \tag{36}$$

with the limit $\Re \to -1$ for $r \to \infty$. $\Re$ can be considered as a "reflection factor for spherical waves".

D.20 A monopole point source above a bulk reacting plane, approximations

Ref.: Mechel, [D.7]

See section D.16 for conventions used, and see [D.7] for a discussion of the approximations and their precision.

The object is a half-space with homogeneous, isotropic material having the characteristic wave number k_2 and wave impedance Z_2. The point source is in the upper half-space with k_1, Z_1 as characteristic wave number and wave impedance; it is at the source point Q with a height h above the plane. The field point P has a horizontal distance r of Q and a height z. The ratios $k=k_2/k_1$ and $Z=Z_2/Z_1$ are used; further $H=h+z$.

The approximation by DELANY / BAZLEY starts from VAN DER POL's exact form; it is :

$$\frac{p_r}{p_0} = \frac{e^{i k_0 r_2}}{k_0 r_2} + 2ik\int_0^\infty \frac{e^{i(k_1 r_3 + ky)}}{k_1 r_3}\,dy$$

$$k_1 r_3 = \sqrt{(k_1 r)^2 + (k_1 H + kZ \cdot y)^2} \quad ; \quad \mathrm{Im}\{\sqrt{\ldots}\} > 0 \tag{1}$$

The exponential function in the integrand decays exponentially, therefore this form is suited for numerical integration.

NORTON / RUDNICK propose a correction to this approximation :

$$kZ \to k^2 Z\sqrt{k^2 - \sin^2\Theta_0} \quad ; \quad \mathrm{Re}\{\sqrt{\ldots}\} > 0 \tag{2}$$

SOOMERFELD's approximation for the total field in the upper space :

$$\frac{p_1}{p_0}=\frac{e^{i k_1 r_1}}{k_1 r_1}+\frac{p_r}{p_0}$$
$$=2\pi C\cdot H_0^{(1)}(\gamma_p r)\cdot e^{-k_1 z\sqrt{(\gamma_p/k_1)^2-1}}+\frac{C_1(z)}{(k_1 r)^2}\cdot e^{i k_1 r}+\frac{C_2}{(k_1 r)^2}\cdot e^{i k_1 r\cdot k-k_1 z\sqrt{k^2-1}} \qquad (3)$$

with :

$$C=\frac{kZ}{1-kZ}\sqrt{\frac{k^2-1}{(kZ)^2-1}} \quad ; \quad C_1(z)=-2i(1+kZ)\left(\frac{kZ}{1-k^2}+\frac{kz}{\sqrt{1-k^2}}\right)$$
$$C_2=-2i(1+kZ)\frac{k}{(kZ)^2(k^2-1)} \quad ; \quad \gamma_p/k_1=k\sqrt{\frac{Z^2-1}{(kZ)^2-1}} \qquad (4)$$

The approximation by PAUL uses the notations $\delta=1/(kZ)$; $k_1H=k_1(h+z)$:

$$\frac{p_r}{p_0}=-\frac{e^{i k_0 r_2}}{k_0 r_2}+V(H,r) \quad ; \quad V(H,r)=V_1(H,r)+V_2(H,r) \qquad (5)$$

with :

$$V_1(H,r)=-2i\frac{1+\delta}{\delta^2(1-k^2)}\frac{e^{i k_1 r}}{(k_1 r)^2}\left[F_0(H)-i\frac{F_1(H)}{2k_1 r}-\frac{F_2(H)}{8(k_1 r)^2}\right] \qquad (6)$$

$$F_0(H)=1+k_1H\cdot\delta\cdot\sqrt{1-k^2}$$

$$F_1(H)=$$
$$\frac{1}{1-k^2}\left\{4-3(k_1H)^2+k^2\left(2+3(k_1H)^2\right)+k_1H\sqrt{1-k^2}\left[\delta\cdot\left(1-k_1H+k^2(2+k_1H)\right)-6/\delta\right]+\right.$$
$$\left.+k_1H\sqrt{1-k^2}\left[\delta\cdot\left(1-k_1H+k^2(2+k_1H)\right)-6/\delta\right]+6/\delta^2\right\}$$

$$F_2(H)=\frac{1}{(1-k^2)^2}\cdot$$
$$\cdot\left\{k_1H\delta\sqrt{1-k^2}\left[g+2(k_1H)^2+(k_1H)^4+k^2\left(36-14(k_1H)^2-(k_1H)^4\right)+12k^4(k_1H)^2\right]+\right.$$
$$+48-24(k_1H)^2+5(k_1H)^4+k^2\left(72-12(k_1H)^2-5(k_1H)^4\right)+36k^4(k_1H)^2- \qquad (7)$$
$$-\frac{k_1H}{\delta}\sqrt{1-k^2}\left[108-20(k_1H)^2+k^2\left(72+20(k_1H)^2\right)\right]-$$
$$\left.-\frac{1}{\delta}\left[168-60(k_1H)^2+k^2\left(72+60(k_1H)^2\right)\right]+120\frac{k_1H}{\delta^3}\sqrt{1-k^2}+\frac{120}{\delta^4}\right\}$$

$$V_2(H,r)=-2ik(1+\delta)\frac{e^{i(k_1 r\cdot k+k_1H\sqrt{1-k^2})}}{(k_1 r)^2}\left[G_0(H)-\frac{G_1(H)}{2k_1 r\cdot k}-\frac{G_2(H)}{8(k_1 r)^2\cdot k^2}\right]$$

$$G_0(H)=\frac{\delta}{1-k^2}$$
$$G_1(H)=-\frac{\delta}{(1-k^2)^2}\left[2+4k^2-6\delta^2k^2+3ik_1H\cdot k^2\sqrt{1-k^2}\right] \tag{8}$$
$$G_2(H)=-\frac{k^2\delta}{(1-k^2)^2}\Big[72+48k^2-ik_1H\left(36+39k^2\right)\sqrt{1-k^2}-15(k_1H)^2k^2(1-k^2)-$$
$$-\delta^2\left(72+168k^2-60ik_1H\cdot k^2\sqrt{1-k^2}\right)+120k^2\delta^4\Big]$$

An approximation given by ATTENBOROUGH / HAYEK /LAWTHER is :

$$\frac{p_r}{P_0}=\frac{e^{ik_1r_2}}{k_1r_2}\left[-1+2(1+kZ)\frac{\cos\Theta_0}{kZ\cdot\cos\Theta_0+\sqrt{k^2-\sin\Theta_0}}\left(1+\frac{iF}{k_1r_2}\right)\right] \tag{9}$$

with $\mathrm{Im}\{\sqrt{k^2-\sin\Theta_0}\}>0$ and : (10)

$$F=1-\frac{\sin^2\Theta_0}{k^2-\sin^2\Theta_0}\left(\frac{\cos\Theta_0+kZ\sqrt{k^2-\sin\Theta_0}}{kZ\cdot\cos\Theta_0+\sqrt{k^2-\sin\Theta_0}}\right)+\frac{kZ\cdot\sin^2\Theta_0}{\cos\Theta_0\cdot(kz\cdot\cos\Theta_0+\sqrt{k^2-\sin\Theta_0})}\cdot$$
$$\cdot\left\{1-\frac{\cos\Theta_0}{\sin^2\Theta_0}\left[\cos\Theta_0+\frac{1}{kZ\sqrt{k^2-\sin\Theta_0}}\left(\cos^2\Theta_0-\frac{3}{2}\sin^2\Theta_0+\frac{\cos^2\Theta_0\cdot\sin^2\Theta_0}{2(k^2-\sin\Theta_0)}\right)\right]\right\}$$

The denominators produce problems when they go to zero, i.e. for large flow resistivity values of a porous material in the lower half-space and at the same time $\Theta_0\to\pi/2$, i.e. source and receiver on the plane.

An approximation obtained by saddle point integration is :

$$\frac{p_r}{P_0}=\frac{e^{ik_1r_2}}{k_1r_2}\cdot\frac{-i}{\sqrt{\sin\Theta_0}}\left[\left(\frac{1}{8k_1r_2}+i\right)\cdot F(\Theta_0)+\frac{1}{2k_1r_2}F''(\Theta_0)\right] \tag{11}$$

with :

$$F(\Theta_0)=R(\Theta_0)\sqrt{\sin\Theta_0}\left[1-\frac{\alpha}{\sin^2\Theta_0}-\frac{i\beta}{\sin\Theta_0}\right]$$
$$F''(\Theta_0)=\frac{R(\Theta_0)}{4}\sqrt{\sin\Theta_0}\left[\frac{1-3\sin^2\Theta_0}{\sin^2\Theta_0}-\alpha\frac{6+11\cos^2\Theta_0}{\sin^4\Theta_0}-i\beta\frac{2+3\cos^2\Theta_0}{\sin^3\Theta_0}\right]+ \tag{12}$$
$$+R'(\Theta_0)\sqrt{\sin\Theta_0}\cos\Theta_0\left[\frac{1}{\sin\Theta_0}+\alpha\frac{3}{\sin^3\Theta_0}+i\beta\frac{1}{\sin^2\Theta_0}\right]+$$
$$+R''(\Theta_0)\sqrt{\sin\Theta_0}\left[1-\frac{\alpha}{\sin^2\Theta_0}-\frac{i\beta}{\sin\Theta_0}\right]$$

and the reflection factor and its derivatives :

$$R(\Theta_0)=\frac{kZ\cdot\cos\Theta_0-w}{kZ\cdot\cos\Theta_0+w}$$

$$R'(\Theta_0)=\frac{kZ}{w\cdot(kZ\cdot\cos\Theta_0+w)^2}\sin\Theta_0\cdot\left(1+\sin^2\Theta_0-2k^2\right)$$

$$R''(\Theta_0)=\frac{kZ}{w\cdot(kZ\cdot\cos\Theta_0+w)^2}\left[\cos\Theta_0\left(1+\sin^2\Theta_0-2k^2\right)+2\sin^2\Theta_0\cdot\cos\Theta_0+\right.$$
$$\left.+\left(1+\sin^2\Theta_0-2k^2\right)\cdot\left(\frac{\sin^2\Theta_0\cdot\cos\Theta_0}{w^2}+\frac{2\sin^2\Theta_0}{w}\frac{kZ\cdot w+\cos\Theta_0}{kZ\cdot\cos\Theta_0+w}\right)\right] \qquad (13)$$

$$\text{using : }\alpha=\frac{9}{128(k_1r)^2}\quad;\quad ß=\frac{1}{8k_1r}\quad;\quad w=\sqrt{k^2-\sin^2\Theta_0}\;;\operatorname{Im}\{w\}>0\,. \qquad (14)$$

For $k_1r_2 >> 1$ this approximation can be simplified to :

$$\frac{p_r}{P_0}=\frac{e^{ik_1r_2}}{k_1r_2}\left[R(\Theta_0)-\frac{i}{2k_1r_2}\left(R'(\Theta_0)\cdot\cot\Theta_0+R''(\Theta_0)\right)\right] \qquad (15)$$

References to part D :
Reflection of Sound

[D.1] MECHEL, F.P.
"Schallabsorber", Vol. I–III
S.Hirzel Verlag, Stuttgart, 1989, 1995, 1998

[D.2] MECHEL, F.P.
"Schallabsorber", Vol. I, Chapter 8:
"Plane absorbers with finite lateral dimensions"
S.Hirzel Verlag, Stuttgart, 1989

[D.3] MECHEL, F.P.
"Mathieu Functions; Formulas, Generation, Use"
S.Hirzel Verlag, Stuttgart, 1997

[D.4] MECHEL, F.P. Acta Acustica 2000
"A line source above a plane absorber"

[D.6] MECHEL, F.P. Acta Acustica, 2000
"Modified Mirror and Corner Sources with a Principle of Superposition"

[D.7] MECHEL, F.P.
"Schallabsorber", Vol. I, Chapter 13:
"Spherical waves over a flat absorber"
S.Hirzel Verlag, Stuttgart, 1989

E
Scattering of Sound

E.1 Plane wave scattering at cylinders

Ref.: Mechel, [E.1]

See section E.2 for a survey of formulas for cylinders and spheres.

The cylinder with diameter 2a is either bulk reacting, i.e. it consists of a homogeneous material with characteristic propagation constant Γ_a and wave impedance Z_a , or it consists of a similar material with same characteristic values, but locally reacting either in the axial direction or locally reacting in all directions. Local reaction is obtained either by a high flow resistivity Ξ of the porous material, or by thin partitions at mutual distances smaller than about $\lambda_0/4$. Sound incidence of the plane wave with unit amplitude is in the x,z plane.

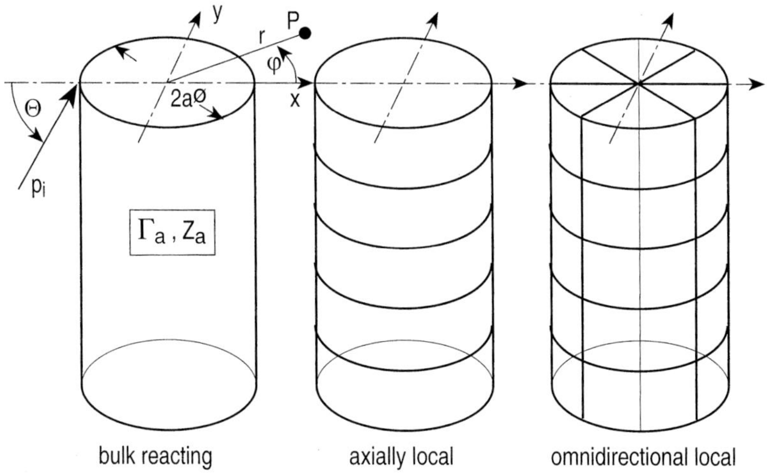

The field formulation will be given below for the bulk reacting cylinder; the fields for the other cylinders follow from that by simplifications.

Notations: • $\Gamma_{an}=\Gamma_a/k_0$; $Z_{an}=Z_a/Z_0$ normalised characteristic values;
- p_i= incident plane wave;
- p_s= scattered wave;
- $p=p_i+p_s$= total exterior field;
- p_a= interior field in the absorbing cylinder;

Expansion of the incident plane wave in Bessel functions :

$$p_i(r,\varphi,z)=e^{-jk_0z\cdot\sin\Theta}\sum_{m\ge0}\delta_m(-j)^m\cdot\cos(m\varphi)\cdot J_m(k_0r\cdot\cos\Theta) \quad ; \quad \delta_m=\begin{cases}1\,;\, m=0\\2\,;\, m>0\end{cases} \tag{1}$$

Formulation of the scattered field :

$$p_s(r,\varphi,z)=e^{-jk_0z\cdot\sin\Theta}\sum_{m\ge0}D_m\cdot\delta_m(-j)^m\cdot\cos(m\varphi)\cdot H_m^{(2)}(k_0r\cdot\cos\Theta) \tag{2}$$

Formulation of the interior field :

$$p_a(r,\varphi,z)=e^{-jk_0z\cdot\sin\Theta}\sum_{m\ge0}E_m\cdot\delta_m\cdot\cos(m\varphi)\cdot I_m(\Gamma_ar\cdot\cos\Theta_1) \quad ; \quad I_m(z)=(-j)^mJ_m(jz) \tag{3}$$

with Bessel functions $J_m(z)$, Hankel functions of the 2nd kind $H_m^{(2)}(z)$, modified Bessel functions $I_m(z)$.

From the boundary conditions of matching pressure and radial particle velocity :

$$jk_0\cdot\sin\Theta=\Gamma_a\cdot\sin\Theta_1 \quad ; \quad \Gamma_{an}\cdot\cos\Theta_1=\sqrt{1+\Gamma_{an}^2-\cos^2\Theta} \tag{4}$$

(Θ_1 is the refracted angle), and :

$$D_m=-\frac{J_m(\alpha)+jW_m\cdot J_m'(\alpha)}{H_m^{(2)}(\alpha)+jW_m\cdot H_m'^{(2)}(\alpha)}=-\frac{\left(\frac{\cos\Theta}{jW_m}+\frac{m}{k_0a}\right)\cdot J_m(\alpha)-\cos\Theta\cdot J_{m+1}(\alpha)}{\left(\frac{\cos\Theta}{jW_m}+\frac{m}{k_0a}\right)\cdot H_m^{(2)}(\alpha)-\cos\Theta\cdot H_{m+1}^{(2)}(\alpha)} \tag{5}$$

$$E_m=\frac{J_m(\alpha)+D_m\cdot H_m^{(2)}(\alpha)}{J_m(\beta)}$$

with the abbreviations: $\alpha=k_0a\cdot\cos\Theta \quad ; \quad \beta=jk_0a\cdot\Gamma_{an}\cdot\cos\Theta_1$ (6)

and the modal normalised surface impedances :

$$W_m=Z_{an}\frac{\cos\Theta}{\cos\Theta_1}\frac{I_m(\Gamma_aa\cdot\cos\Theta_1)}{I_m'(\Gamma_aa\cdot\cos\Theta_1)}=-jZ_{an}\frac{\cos\Theta}{\cos\Theta_1}\frac{J_m(y)}{J_m'(y)}$$

$$\frac{\cos\Theta}{jW_m}=-\frac{\Gamma_{an}\cdot\cos\Theta_1}{\Gamma_{an}Z_{an}}\left[\frac{J_{m+1}(y)}{J_m(y)}-\frac{m}{y}\right] \tag{7}$$

For the cylinder which is locally reacting in the axial direction :

- set $\Theta_1=0$;

for the cylinder which is locally reacting in all directions :

- retain in p_a only the term $m=0$;
- set $\Theta_1=0$;
- replace everywhere $W_m \to W_0$, in which case : $$\frac{\cos\Theta}{jW_m} \to \frac{\cos\Theta}{jW_0} = -\frac{1}{Z_{an}}\frac{J_1(jk_0a\cdot\Gamma_{an})}{J_0(jk_0a\cdot\Gamma_{an})} \quad (8)$$

The result then describes the scattering of a cylinder consisting of a (porous) material, and made locally reacting.

For a cylinder which is locally reacting in all directions, and described by a normalised surface admittance G :

- neglect p_a ;
- replace everywhere $W_m \to W_0= 1/G$, in which case : $$\frac{\cos\Theta}{jW_m} \to \frac{\cos\Theta}{jW_0} = -jG\cdot\cos\Theta \quad (9)$$

The incident plane wave is temporarily assumed to have an amplitude p_0 (which above was set $p_0=1$). The integrals $\oint$ below are taken at the cylinder surface (r=a). A star $*$ indicates the complex conjugate.

Scattering cross-section : (ratio of scattered power to incident intensity)

$$Q_s = \oint \mathrm{Re}\left\{\frac{p_s}{p_0}\cdot\frac{v_{rs}^*}{p_0/Z_0}\right\} dS \xrightarrow[p_0=1]{} \oint \mathrm{Re}\{p_s\cdot Z_0 v_{rs}^*\}\, dS \quad (10)$$

$$\frac{Q_s}{2a} = \frac{2}{k_0a}\sum_{m\geq 0}\delta_m\cdot|D_m|^2 \quad (11)$$

Absorption cross-section : (ratio of absorbed power to incident intensity)

$$Q_a = -\oint \mathrm{Re}\{p\cdot Z_0 v_r^*\}\, dS \quad (12)$$

$$\frac{Q_a}{2a} = \frac{-2}{k_0a}\sum_{m\geq 0}\delta_m\cdot\left(\mathrm{Re}\{D_m\}+|D_m|^2\right) \quad (13)$$

Extinction cross-section :

$$Q_e = Q_s + Q_a = -\oint \mathrm{Re}\{p_i^*\cdot v_r + p\cdot v_{ri}^*\}\, dS \quad (14)$$

$$\frac{Q_e}{2a} = \frac{-2}{k_0a}\sum_{m\geq 0}\delta_m\cdot\mathrm{Re}\{D_m\} \quad (15)$$

With the always possible separation of the scattered far field into an angular and a radial factor:

$$p_s(r,\vartheta,\varphi) \xrightarrow[k_0 r>>1]{} \Phi(\vartheta,\varphi)\cdot\frac{e^{-jk_0 r}}{r}+O(r^{-2}) \tag{16}$$

the extinction cross section is (extinction theorem) :

$$Q_e = -\frac{4\pi}{k_0}\,\mathrm{Im}\{\Phi(\vartheta_0,\varphi_0)\} \tag{17}$$

where a radius with the angles ϑ_0,φ_0 points in the forward direction of the incident wave.

Backscattering cross-section : (measures the strength of the backscattering to the source)

$$Q_r = 2\pi r\,\frac{|p_s(r,\vartheta_0+\pi,\varphi_0+\pi)|^2}{p_0^2} \quad ; \quad k_0 r >> 1 \tag{18}$$

Absorption cross-section for diffuse sound incidence

There exist several definitions in the literature (differing from each other in the reference intensity). Π_a= absorbed power ; I_i= intensity of an incident plane wave

1st definition :

$$Q_{a1} = \Pi_a/I_i$$
$$Q_{a1} = 4\pi\int_0^{\pi/2} Q_a(\Theta)\cdot\cos\Theta\; d\Theta \tag{19}$$

2nd definition :

$$Q_{a2} = \Pi_a/I_{i,dif} \quad ; \quad I_{i,dif} = \pi\cdot I_i$$
$$Q_{a2} = \frac{Q_{a1}}{\pi} = 4\int_0^{\pi/2} Q_a(\Theta)\cdot\cos\Theta\; d\Theta \tag{20}$$

3rd definition :

with $\Pi_{i,dif}$= incident power in a diffuse field on a cylinder of unit length and diameter

$$Q_{a3} = \Pi_a/\Pi_{i,dif} \quad ; \quad \Pi_{i,dif} = 4\pi\cdot I_i$$
$$Q_{a3} = \frac{Q_{a1}}{4\pi} = \frac{Q_{a2}}{4} = \int_0^{\pi/2} Q_a(\Theta)\cdot\cos\Theta\; d\Theta \tag{21}$$

E.2 Plane wave scattering at cylinders and spheres

Ref.: Mechel, [E.1]

See previous section E.1 for an oblique plane wave incident on a cylinder. This section briefly gives the fundamental relations for a plane wave incident on a sphere, and then collects equations for both spheres and cylinders (with normal incidence on the cylinder axis). In the case of a bulk reacting sphere, it consists of a homogeneous material with characteristic propagation constant Γ_a and wave impedance Z_a.

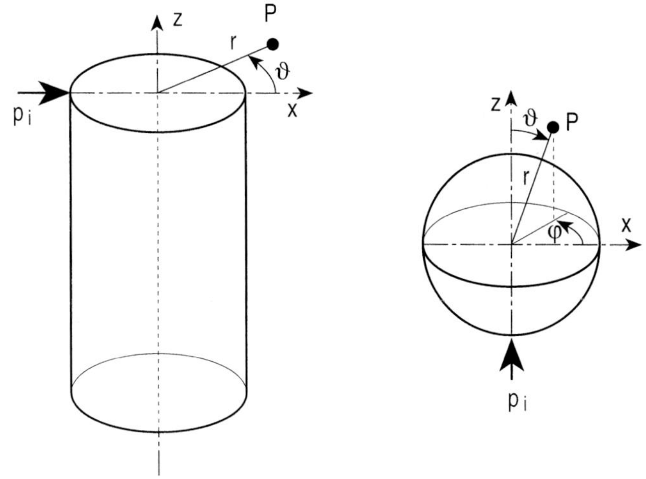

Diameter 2a ;
incident plane wave p_i ;
field point P .

Notations: • $\Gamma_{an}=\Gamma_a/k_0$; $Z_{an}=Z_a/Z_0$ normalised characteristic values;
- $p_i=$ incident plane wave (with amplitude $p_0=1$);
- $p_s=$ scattered wave;
- $p=p_i+p_s=$ total exterior field;
- $p_a=$ interior field in the absorbing cylinder or sphere;

Sound field formulations for the sphere :

Incident plane wave :

$$p_i(r,\vartheta)=e^{-jk_0r\cdot\cos\vartheta}=\sum_{m\ge 0}(2m+1)(-j)^m\cdot P_m(\cos\vartheta)\cdot j_m(k_0r) \qquad (1)$$

Scattered wave :

$$p_s(r,\vartheta)=\sum_{m\ge 0}D_m\cdot(2m+1)(-j)^m\cdot P_m(\cos\vartheta)\cdot h_m^{(2)}(k_0r) \qquad (2)$$

Total exterior field: $p=p_i+p_s$

with $P_m(z)=$ Legendre polynomial ; $j_m(z)=$ spherical Bessel function ; $h_m^{(2)}(z)=$ spherical Hankel function of 2nd kind.

The following table contains corresponding quantities for cylinders and spheres, of diameter 2a, which are *locally reacting* with a normalised surface admittance G . Hankel functions of the 2nd kind are written as $H_m(z)$. The argument k_0a of Bessel and Hankel functions is dropped. In some equations $W = 1/G = R + j\cdot X$ will be used. The amplitude factors C_m are $C_m = 2D_m - 1$.

Quantity	Symbol	Cylinder	Sphere
Factors	$D_m=$	$-\dfrac{(-jG+m/k_0a)J_m - J_{m+1}}{(-jG+m/k_0a)H_m - H_{m+1}}$	$-\dfrac{(-jG+m/k_0a)j_m - j_{m+1}}{(-jG+m/k_0a)h_m - h_{m+1}}$
	$C_m=$	$-\dfrac{(-jG+m/k_0a)H^*_m - H^*_{m+1}}{(-jG+m/k_0a)H_m - H_{m+1}}$	$-\dfrac{(-jG+m/k_0a)h^*_m - h^*_{m+1}}{(-jG+m/k_0a)h_m - h_{m+1}}$
	$T_m=$	$\cos(m\vartheta)$	$P_m(\cos\vartheta)$
	$\delta_m=$	$\begin{cases}1\,;\,m=0\\2\,;\,m>0\end{cases}$	$2m+1$
Cross-section	$S=$	$2a$	πa^2
Incident wave	$p_i(r,\vartheta)=$	$e^{-jk_0r\cdot\cos\vartheta}$	
		$\sum_{m\geq 0}\delta_m(-j)^m\cdot T_m\cdot J_m(k_0r)$	$\sum_{m\geq 0}\delta_m(-j)^m\cdot T_m\cdot j_m(k_0r)$
Scattered wave	$p_s(r,\vartheta)=$	$\sum_{m\geq 0}D_m\delta_m(-j)^m T_m J_m(k_0r)$	$\sum_{m\geq 0}D_m\delta_m(-j)^m T_m j_m(k_0r)$
Scattered far field	$p_s(r,\vartheta)\rightarrow$	$\rightarrow\sqrt{\dfrac{2j}{\pi}}\dfrac{e^{-jk_0r}}{\sqrt{k_0r}}\sum_{m\geq 0}D_m\delta_m T_m$ $\rightarrow\sqrt{\dfrac{2j}{\pi}}\Phi(\vartheta)\dfrac{e^{-jk_0r}}{\sqrt{k_0r}}$	$\rightarrow j\dfrac{e^{-jk_0r}}{k_0r}\sum_{m\geq 0}D_m\delta_m T_m$ $\rightarrow\Phi(\vartheta)\dfrac{e^{-jk_0r}}{k_0r}$
Total field	$p(r,\vartheta)=$	$\sum_{m\geq 0}\delta_m(-j)^m T_m\cdot$ $\cdot[J_m(k_0r)+D_m H_m(k_0r)]$	$\sum_{m\geq 0}\delta_m(-j)^m T_m\cdot$ $\cdot[j_m(k_0r)+D_m h_m(k_0r)]$
Scattering cross-section	$Q_s=$	$\oint \mathrm{Re}\{p_s\cdot v^*_{rs}\}dS$	
	$Q_s/S=$	$\dfrac{2}{k_0a}\sum_{m\geq 0}\delta_m\lvert D_m\rvert^2$	$\dfrac{4}{(k_0a)^2}\sum_{m\geq 0}\delta_m\lvert D_m\rvert^2$
	$=$	$\dfrac{2}{k_0a}\sum_{m\geq 0}\delta_m\lvert 1-C_m\rvert^2$	$\dfrac{4}{(k_0a)^2}\sum_{m\geq 0}\delta_m\lvert 1-C_m\rvert^2$

Table continued :

Quantity	Symbol	Cylinder	Sphere
Scattering cross-section; Approximations :			
$k_0a>>1$; $R=0$ or $=\infty$	$Q_s/S=$	2	2
$k_0a<<1$; $G=0$	$Q_s/S=$	$\frac{3\pi^2}{8}(k_0a)^3$	$\frac{7}{9}(k_0a)^4$
$k_0a<<1$; $\|G\|\to\infty$	$Q_s/S=$	$\frac{\pi^2}{2k_0a\ln^2(1/k_0a)}$	4
$k_0a<<1$; $G=$else	$Q_s/S=$	$5k_0a\,\|G\|^2$	$4(k_0a)^2\,\|G\|^2$
Absorption cross-section	$Q_a=$	$-\oint \mathrm{Re}\{p\cdot v_r^*\}\,dS$	
	$Q_a/S=$	$\frac{-2}{k_0a}\sum_{m\geq 0}\delta_m\left(\mathrm{Re}\{D_m\}+\|D_m\|^2\right)$	$\frac{-4}{(k_0a)^2}\sum_{m}\delta_m\left(\mathrm{Re}\{D_m\}+\|D_m\|^2\right)$
	$=$	$\frac{1}{2k_0a}\sum_{m\geq 0}\delta_m\left(1-\|C_m\|^2\right)$	$\frac{1}{(k_0a)^2}\sum_{m\geq 0}\delta_m\left(1-\|C_m\|^2\right)$
Absorption cross-section; Approximations :			
$k_0a<<1$; $\|X\|<<R$	$Q_a/S=$	$\frac{\pi}{R}$	$\frac{4}{R}$
$k_0a<<1$; $R<\|X\|/2$	$Q_a/S=$	$\frac{\pi R}{X^2}$	$\frac{4R}{X^2}$
Extinction cross-section with scattered far field	$Q_e=$	$-\oint \mathrm{Re}\{p_i^*\cdot v_r^*+p\cdot v_{ri}^*\}\,dS$	
	$Q_e/S=$	$\frac{-2}{k_0a}\sum_{m\geq 0}\delta_m T_m(0)\cdot\mathrm{Re}\{D_m\}$	$\frac{-4}{(k_0a)^2}\sum_{m\geq 0}\delta_m T_m(0)\cdot\mathrm{Re}\{D_m\}$
	$p_s\to$	$\sqrt{2j/\pi}\cdot\Phi(\vartheta)\cdot\frac{e^{-jk_0r}}{\sqrt{k_0r}}$	$j\cdot\Phi(\vartheta)\cdot\frac{e^{-jk_0r}}{k_0r}$
	$Q_e/S=$	$\frac{-2}{k_0a}\mathrm{Re}\{\Phi(0)\}$	$\frac{-4}{(k_0a)^2}\mathrm{Re}\{\Phi(0)\}$

Quantity	Symbol	Cylinder	Sphere
Backscatter cross-section	$Q_r=$	$2\pi r\cdot\lvert p_s(r,\pi)\rvert^2$	$4\pi r^2\cdot\lvert p_s(r,\pi)\rvert^2$
Backscatter cross-section; Approximations :			
$k_0a>>1$; $\lvert G\rvert\to 0$ or ∞	$Q_r/S=$	$\pi/2$	1
$k_0a<<1$; $\lvert G\rvert\to\infty$	$Q_r/S=$	$\dfrac{2\pi^2}{k_0a\left(\pi^2+4\ln^2(1.123/k_0a)\right)}$	4
$k_0a<1^{st}$ reson.; G else	$Q_r/S=$	$\dfrac{\pi^2}{2}k_0a\lvert G\rvert^2$	$4k_0a\lvert G\rvert^2$
Reactance at m-th resonance $k_0a<1$ ß=1.123	$X_0=$	$-k_0a\cdot\ln\dfrac{ß}{k_0a}\cdot\dfrac{1-\frac{1}{2}(k_0a)^2\ln\frac{ß}{k_0a}}{1-\frac{1}{2}(k_0a)^2\ln\frac{ß}{k_0a}}$	$\dfrac{-k_0a}{1+(k_0a)^2}$
	$X_1=$	$-k_0a\cdot\dfrac{1+2.14(k_0a)^2\ln\frac{ß}{k_0a}}{1-(k_0a)^2\ln\frac{ß}{k_0a}}$	$-k_0a\dfrac{1+2(k_0a)^2}{1+(k_0a)^2}$
	$X_m=$	$-\dfrac{k_0a}{m}\cdot\dfrac{1+\frac{(k_0a)^2}{4(m-1)}}{1+\frac{(m-2)(k_0a)^2}{4m(m-1)}}$	$-\dfrac{k_0a}{m}\cdot\dfrac{1}{1-\frac{(k_0a)^2}{2m^2}}$
Q_s at lowest resonance; $R=0$; $X=X_0$	$Q_s/S=$	$\dfrac{2}{k_0a}$	$\dfrac{4}{(k_0a)^2}$
Frequency of lowest back-scatter minimum	$k_0a=$	$\dfrac{2\lvert X\rvert}{2+3\lvert X\rvert^2}$ $\lvert X\rvert_{max}=\sqrt{2/3}$ $(k_0a)_{max}=1/\sqrt{6}$	$\dfrac{6\lvert X\rvert}{3+5\lvert X\rvert^2}$ $\lvert X\rvert_{max}=(k_0a)_{max}=\sqrt{3/10}$

The next table contains corresponding values for cylinders and spheres, with diameter $2a$, consisting of a *homogeneous (bulk reacting) material* with characteristic propagation constant Γ_a and wave impedance Z_a. Hankel functions of the 2nd kind are written as $H_m(z)$ and spherical Hankel functions of 2nd kind as $h_m(z)$; the arguments k_0a of Bessel and Hankel functions are dropped. The abbreviation $z=j\cdot\Gamma_a a$ will be used, and $\Gamma_{an}=\Gamma_a/k_0$, $Z_{an}=Z_a/Z_0$. A prime at functions indicates the derivative; a prime or double prime at Γ_{an}, Z_{an} indicates the real or imaginary part, respectively. An asterisk indicates the complex conjugate.

Quantity	Symbol	Cylinder	Sphere
Factors	$D_m=$	$-\dfrac{(-jG_m+m/k_0a)J_m-J_{m+1}}{(-jG_m+m/k_0a)H_m-H_{m+1}}$	$-\dfrac{(-jG_m+m/k_0a)j_m-j_{m+1}}{(-jG_m+m/k_0a)h_m-h_{m+1}}$
	$C_m=$	$-\dfrac{(-jG_m+m/k_0a)H^*_m-H^*_{m+1}}{(-jG_m+m/k_0a)H_m-H_{m+1}}$	$-\dfrac{(-jG_m+m/k_0a)h^*_m-h^*_{m+1}}{(-jG_m+m/k_0a)h_m-h_{m+1}}$
	$T_m=$	$\cos(m\vartheta)$	$P_m(\cos\vartheta)$
	$\delta_m=$	$\begin{cases}1\,;\,m=0\\2\,;\,m>0\end{cases}$	$2m+1$
Cross-section	$S=$	$2a$	πa^2
Modal admittance	$G_m=$ $(z=j\cdot\Gamma_a a)$	$j\dfrac{Z_0}{Z_a}\dfrac{J'_m(z)}{J_m(z)}$	$j\dfrac{Z_0}{Z_a}\dfrac{j'_m(z)}{j_m(z)}$
Incident wave	$p_i(r,\vartheta)=$	$e^{-jk_0r\cdot\cos\vartheta}$	$e^{-jk_0r\cdot\cos\vartheta}$
		$\sum\limits_{m\geq0}\delta_m(-j)^m\cdot T_m\cdot J_m(k_0r)$	$\sum\limits_{m\geq0}\delta_m(-j)^m\cdot T_m\cdot j_m(k_0r)$
Scattered wave	$p_s(r,\vartheta)=$	$\sum\limits_{m\geq0}D_m\cdot\delta_m(-j)^m\cdot T_m\cdot H_m(k_0r)$	$\sum\limits_{m\geq0}D_m\cdot\delta_m(-j)^m\cdot T_m\cdot h_m(k_0r)$
Scattered far field	$p_s(r,\vartheta)\to$	$-\sqrt{\dfrac{2j}{\pi k_0r}}e^{-jk_0r}\sum\limits_{m\geq0}C_m\cdot\delta_mT_m$	$\dfrac{j}{k_0r}e^{-jk_0r}\sum\limits_{m\geq0}C_m\cdot\delta_mT_m =: \dfrac{e^{-jk_0r}}{k_0r}\cdot\Phi(\vartheta)$

Table continued :

Quantity	Symbol	Cylinder	Sphere
Total ext. field	$p(r,\vartheta)=$	$\sum_{m\geq 0}\delta_m(-j)^m T_m\cdot \cdot\left(J_m(k_0r)+D_m H_m(k_0r)\right)$	$\sum_{m\geq 0}\delta_m(-j)^m T_m\cdot \cdot\left(j_m(k_0r)+D_m h_m(k_0r)\right)$
Scattering cross-section	$\frac{Q_s}{S}=$	$\frac{1}{2k_0a}\sum_m\delta_m\cdot\lvert 1-C_m\rvert^2$	$\frac{1}{(k_0a)^2}\sum_m\delta_m\cdot\lvert 1-C_m\rvert^2$
Absorption cross-section	$\frac{Q_a}{S}=$	$\frac{1}{2k_0a}\sum_m\delta_m\cdot\left(1-\lvert C_m\rvert^2\right)$	$\frac{1}{(k_0a)^2}\sum_m\delta_m\cdot\left(1-\lvert C_m\rvert^2\right)$
Approximations : $k_0a \ll 1$			
Modal admittance	$G_0=$	$\frac{k_0a}{2}\frac{\Gamma_{an}}{Z_{an}}$	$\frac{k_0a}{3}\frac{\Gamma_{an}}{Z_{an}}$
Scattering cross-section	$\frac{Q_s}{S}=$	$\frac{1}{2k_0a}\sum_m\delta_m\cdot\lvert 1-C_0\rvert^2 = \frac{\pi^2(k_0a)^3}{16}\cdot \cdot\left[1+\left(\left(\frac{\Gamma_{an}}{Z_{an}}\right)'-\left(\frac{\Gamma_{an}}{Z_{an}}\right)''\right)^2\right]$	$\frac{1}{(k_0a)^2}\sum_m\delta_m\cdot\lvert 1-C_0\rvert^2 = \frac{4(k_0a)^4}{9}\cdot\left[\left\lvert 1+j\frac{\Gamma_{an}}{Z_{an}}\right\rvert^2+ +3\left\lvert\frac{1+j\Gamma_{an}Z_{an}}{1-2j\Gamma_{an}Z_{an}}\right\rvert^2\right]$
Absorption cross-section, cylinder	$\frac{Q_a}{S}=$	$\frac{(1-\lvert C_0\rvert^2)}{2k_0a}=\frac{\pi k_0a}{2}\cdot\left(\frac{\Gamma_{an}}{Z_{an}}\right)'\cdot \cdot\frac{1+k_0a\cdot\ln(0.890\,k_0a)}{\left[1+k_0a\cdot\ln(0.890\,k_0a)\left(1+\frac{(k_0a)^2}{2}\left(\frac{\Gamma_{an}}{Z_{an}}\right)''\right)\right]^2}$	
Absorption cross-section, sphere	$\frac{Q_a}{S}=$	$\frac{(1-\lvert C_0\rvert^2)}{(k_0a)^2}=-\frac{4k_0a}{3}\cdot \mathrm{Im}\left\{1+j\frac{\Gamma_{an}}{Z_{an}}+3\frac{1+j\Gamma_{an}Z_{an}}{1-2j\Gamma_{an}Z_{an}}\right\}$	

Table continued :

Quantity	Symbol	Cylinder	Sphere
Approximations : $k_0 a \ll 1$			
Total ext. field at surface ; sphere	$p(a,\vartheta)=$	$1 - j k_0 a \frac{3\Gamma_{an} Z_{an}}{j+2\Gamma_{an} Z_{an}} \cdot \cos(\vartheta)$	
Radial particle velocity at surface ; sphere	$v_r(a,\vartheta)=$	$\frac{-k_0 a}{3} \frac{\Gamma_{an}}{Z_{an}} + \frac{3j}{j+2\Gamma_{an} Z_{an}} \cdot \cos(\vartheta)$	
Scattered field directivity in the far field, cylinder	$\Phi(\vartheta)=$	$\sqrt{\frac{-j\pi}{8k_0}}(k_0 a)^3 \left[-\left(1+j\frac{\Gamma_{an}}{Z_{an}}\right) + 2\frac{\Gamma_{an} Z_{an} - j}{\Gamma_{an} Z_{an} - j} \cdot \cos(\vartheta) \right]$	
Scattered field directivity in the far field, cylinder	$\Phi(\vartheta)=$	$\frac{k_0^2 a^3}{3} \left[-\left(1+j\frac{\Gamma_{an}}{Z_{an}}\right) + 2\frac{\Gamma_{an} Z_{an} - j}{\Gamma_{an} Z_{an} - j} \cdot \cos(\vartheta) \right]$	

Some numerical examples will illustrate the quantities and relations given above. The diagrams are taken immediately from the evaluation program *Mathematica* ; therefore correspondence of symbols in the plot labels to symbols used here is explained by "there" → here. If a plot contains several curves for different parameter values, these are given as a list {…}, and the dashes of the curves become shorter with increasing parameter value number in the list.

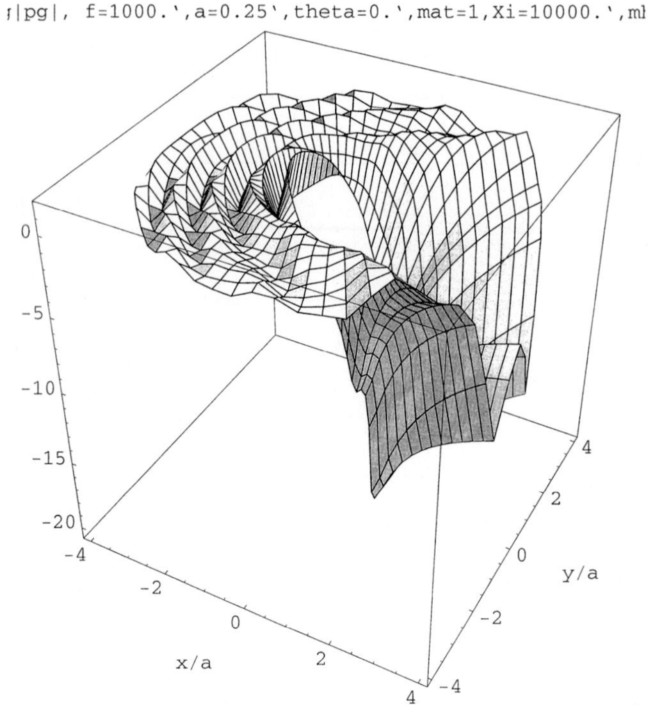

Total sound pressure level around a cylinder of homogeneous, bulk reacting, porous material .
Sound incidence from the left-hand side.
Correspondence:
"f" → frequency [Hz],
"a" → radius a in [m],
"theta" → Θ ,
"Xi" → "Ξ"= flow resistivity of material [Pa·s/m^2]
"mhi=32" → upper summation limit.

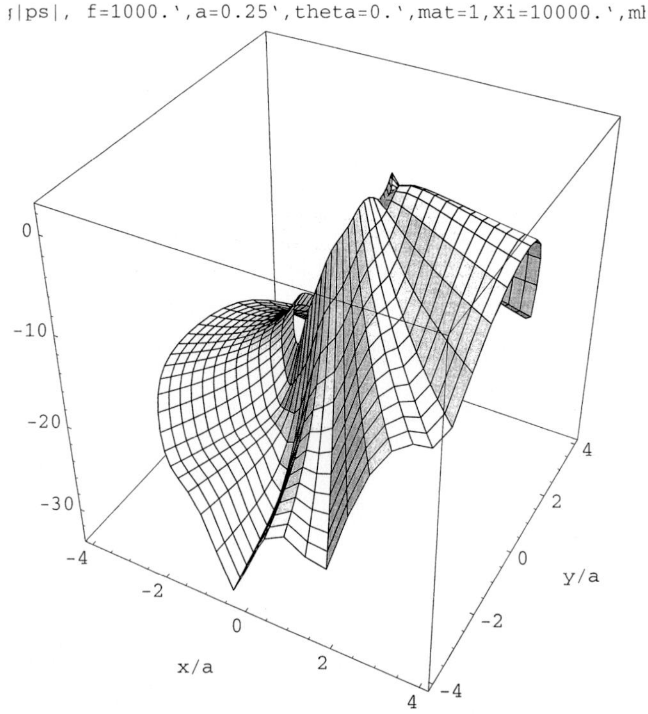

Scattered sound pressure level around a cylinder of homogeneous, bulk reacting, porous material .
Sound incidence from the left-hand side.

Correspondence:
"f" → frequency [Hz],
"a" → radius a in [m],
"theta" → Θ ,
"Xi" → "Ξ"= flow resistivity of material [Pa·s/m^2]
"mhi=32" → upper summation limit.

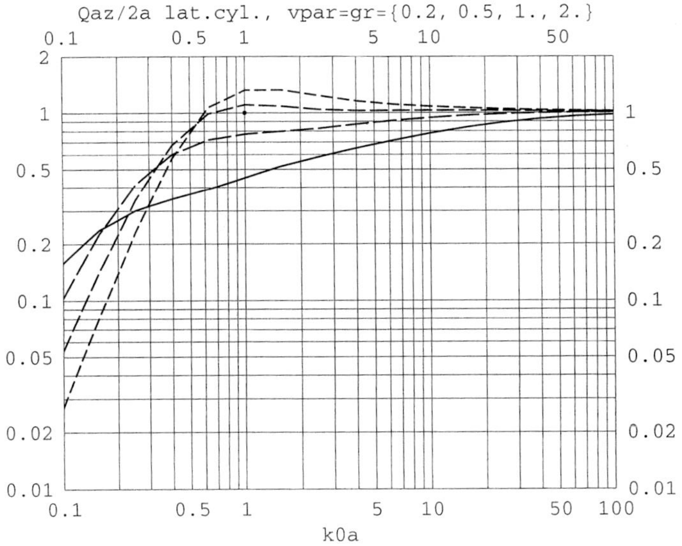

Normalised absorption cross section $Q_a/2a$ of a cylinder of homogeneous, bulk reacting, porous glass fibre material, with $\Theta=0$.

Correspondence:
"k0a" → k_0a,
"gr" → $R=\Xi\cdot a/Z_0$,
parameter values :
R={0.2 , 0.5 , 1. , 2.}

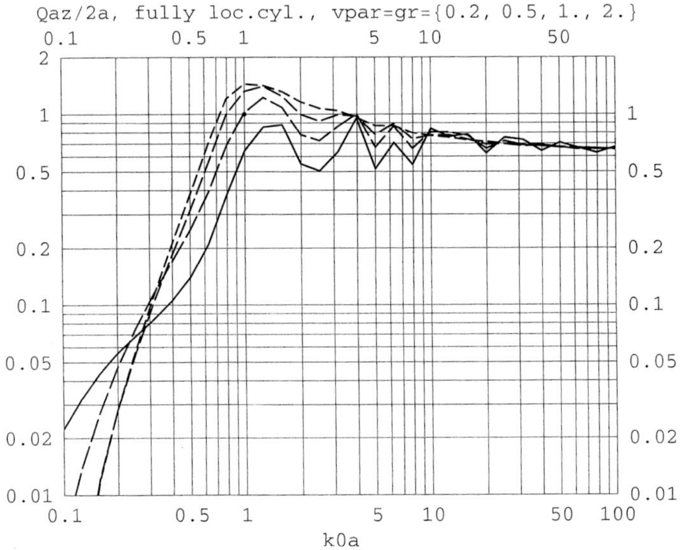

As above diagram, but sound incidence with $\Theta=45°$, and the cylinder made locally reacting .

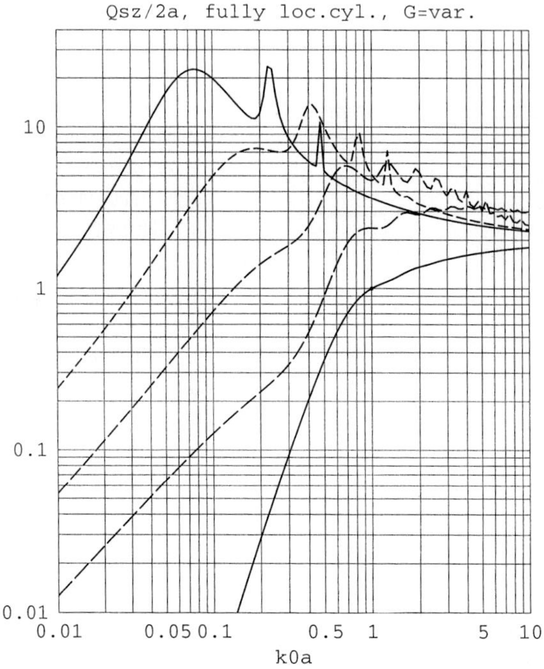

Normalised scattering cross section of a locally reacting cylinder with given values of the surface adimittance (normalised, from low to high) G={0 , 0.5j , 1j , 2j , 4j}.
This graph illustrates the scattering resonances (the exterior vibrating mass resonates with the resilience of the surface).

E.3 Multiple scattering at cylinders and spheres

Ref.: Mechel , [E.2]

Consider an "artificial medium" consisting of an arrangement (preferably random) of hard scatterers (cylindrical or spherical) with a root mean square average radius a and mutual distances such that the "massivity" μ of the arrangement (fraction of the space occupied by the scatterers) holds. A sound wave propagates through that medium with an effective (complex) wave number k_{eff} and wave impedance Z_{eff} given by :

$$\frac{k_{eff}^2}{k_0^2} = \frac{\rho_{eff}}{\rho_0} \cdot \frac{C_{eff}}{C_0} \quad ; \quad \frac{Z_{eff}^2}{Z_0^2} = \frac{\rho_{eff}}{\rho_0} / \frac{C_{eff}}{C_0} \tag{1}$$

with the effective density ρ_{eff} and compressibility C_{eff} :

$$\frac{\rho_{eff}}{\rho_0} = 1 + j\frac{8\mu}{\pi(k_0a)^2} \sum_{n=1,3,5\ldots} D_n \quad ; \quad \frac{C_{eff}}{C_0} = 1 + j\frac{8\mu}{\pi(k_0a)^2} \cdot [0.5\, D_0 + \sum_{n=2,4,6\ldots} D_n] \tag{2}$$

where the coefficients D_n are taken from the table in section E.2 , and $C_0 = 1/(\rho_0 c_0^2)$. (3)

E.4 Cylindrical wave scattering at cylinders

A line source at Q is parallel to the axis of a locally reacting cylinder with radius a and (normalised) surface admittance G. The field point is at P. The source distance r_q defines two radial zones (a),(b).

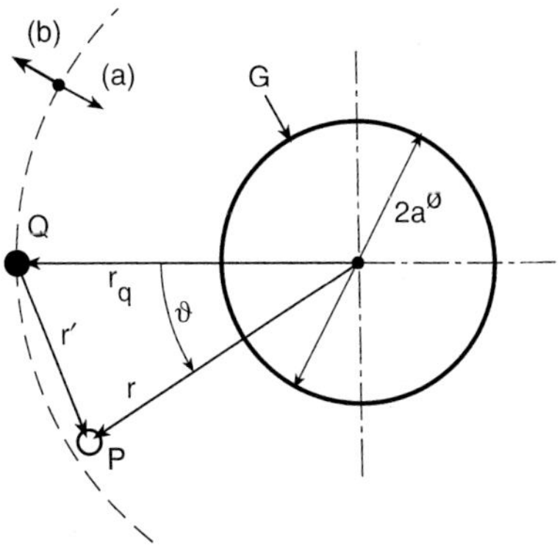

The sound field is composed as the sum of the source free field p_Q and the scattered field p_s :

$$p(r,\vartheta) = p_Q(r') + p_s(r,\vartheta) \tag{1}$$

The source free field p_Q is transformed with the addition theorem for Hankel functions to the co-ordinates (r,ϑ) :

$$p_Q(r') = P_0 \cdot H_0^{(2)}(k_0 r') = P_0 \cdot \begin{cases} \sum\limits_{m\geq 0} \delta_m \cdot J_m(k_0 r) \cdot H_m^{(2)}(k_0 r_q) \cdot \cos(m\vartheta) & ; \text{ in (a)} \\ \sum\limits_{m\geq 0} \delta_m \cdot J_m(k_0 r_q) \cdot H_m^{(2)}(k_0 r) \cdot \cos(m\vartheta) & ; \text{ in (b)} \end{cases} \tag{2}$$

$$\text{with :} \qquad \delta_m = \begin{cases} 1 \; ; & m=0 \\ 2 \; ; & m>0 \end{cases} \tag{3}$$

Formulation of the scattered field :

$$p_s(r,\vartheta) = P_0 \cdot \sum_{m\geq 0} a_m \cdot \delta_m \cdot J_m(k_0 r_q) \cdot H_m^{(2)}(k_0 r) \cdot \cos(m\vartheta) \tag{4}$$

The boundary condition $-Z_0\left(v_{Qr} + v_{sr}\right) \overset{!}{=} G \cdot \left(p_Q + p_s\right)$ gives the amplitudes :

$$\begin{aligned} a_m &= -\frac{j \cdot J'_m(k_0 a) + G \cdot J_m(k_0 a)}{j \cdot H'^{(2)}_m(k_0 a) + G \cdot H_m^{(2)}(k_0 a)} \cdot \frac{H_m^{(2)}(k_0 r_q)}{J_m(k_0 r_q)} \\ &\xrightarrow[G\to 0]{} -\frac{J'_m(k_0 a) \cdot H_m^{(2)}(k_0 r_q)}{J_m(k_0 r_q) \cdot H'^{(2)}_m(k_0 a)} = a_{hm} \\ &\xrightarrow[|G|\to\infty]{} -\frac{J_m(k_0 a) \cdot H_m^{(2)}(k_0 r_q)}{J_m(k_0 r_q) \cdot H_m^{(2)}(k_0 a)} = a_{sm} \end{aligned} \tag{5}$$

with the special cases $G \to 0$ (hard) and $|G| \to \infty$ (soft).

In a different notation, the scattered field is:

$$p_s(r,\vartheta) = -P_0 \cdot \sum_{m \ge 0} \delta_m \cdot c_m \cdot H_m^{(2)}(k_0 r_q) \cdot H_m^{(2)}(k_0 r) \cdot \cos(m\vartheta)$$

$$c_m = \frac{\left(G + \frac{m}{k_0 a}\right) \cdot J_m(k_0 a) - j \cdot J_{m+1}(k_0 a)}{\left(G + \frac{m}{k_0 a}\right) \cdot H_m^{(2)}(k_0 a) - j \cdot H_{m+1}^{(2)}(k_0 a)} \qquad (6)$$

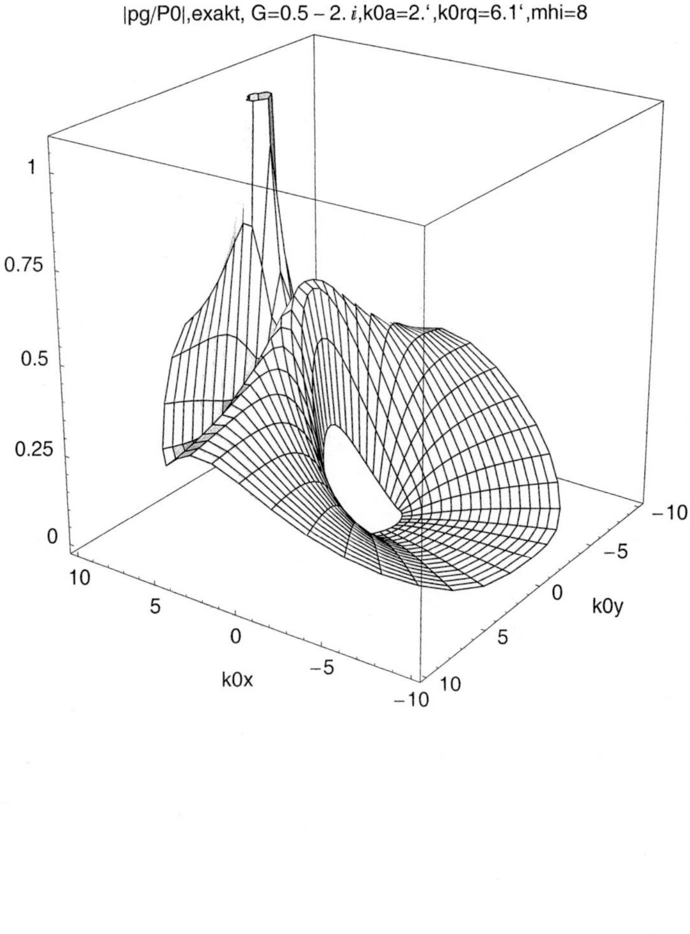

Sound pressure magnitude from a line source around a locally absorbing cylinder.
$G = 0.5-2 \cdot j$;
$k_0 a=2$; $k_0 r_q=6.1$;
upper summation limit $m_{hi}= 8$.

E.5 Cylindrical or plane wave scattering at a corner surrounded by a cylinder

Ref.: Mechel, [E.3]

The apex line of a corner with hard flanks at $\vartheta=0$ and $\vartheta=\vartheta_0\le 2\pi$ is surrounded by a locally reacting cylinder of radius a and (normalised) surface admittance G . The line source at Q has the co-ordinates (r_q, ϑ_q) .

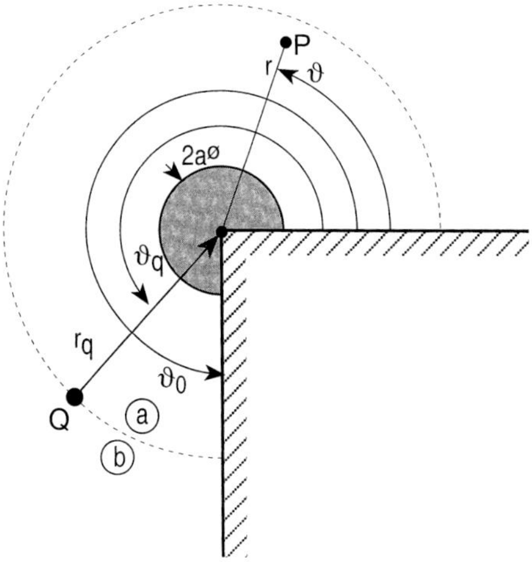

The sound field is formulated as a mode sum :

$$p(r,\vartheta,z) = Z(k_z z)\sum_\eta R_\eta(kr)\cdot T(\eta\vartheta) \qquad (1)$$

The factor $Z(k_z z)$ may be any of the functions $e^{\pm jk_z z}$, $\cos(k_z z)$, $\sin(k_z z)$ or a linear combination thereof. If $k_z\neq 0$, set $k^2 = k_0^2 - k_z^2$.

Below it will be supposed (for simplicity) that $k_z=0$, $Z(k_z z)=1$.

The azimuthal functions are $T(\eta\vartheta) = \cos(\eta\vartheta)$ and the azimuthal wave numbers satisfy the characteristic equation :

$$(\eta_n\vartheta_0)\cdot\tan(\eta_n\vartheta_0) = 0 \qquad (2)$$

with the solutions $\eta_n\vartheta_0=n\cdot\pi$; n=0,1,2,... .Field formulations in the two radial zones (a),(b) :

$$\begin{aligned} p_a(r,\vartheta) &= \sum_{n\ge 0} A_n\cdot H^{(2)}_{\eta_n}(kr_q)\cdot\left[H^{(1)}_{\eta_n}(kr) + r_n\cdot H^{(2)}_{\eta_n}(kr)\right]\cdot\cos(\eta_n\vartheta) \quad ; \quad a\le r\le r_q \\ p_b(r,\vartheta) &= \sum_{n\ge 0} A_n\cdot\left[H^{(1)}_{\eta_n}(kr_q) + r_n\cdot H^{(2)}_{\eta_n}(kr_q)\right]\cdot H^{(2)}_{\eta_n}(kr)\cdot\cos(\eta_n\vartheta) \quad ; \quad r_q\le r<\infty \end{aligned} \qquad (3)$$

with the modal reflection factors at the cylinder :

sound field at the later corner can approximately be assumed to be generated by a line source situated at the earlier corner. Thus the sound shielding by buildings can be evaluated by iteration over the surrounded corners.

The required intermediate values $L(kr,\vartheta=0)= 20\cdot\lg|p(kr,0)/p_Q(0)|$ can be evaluated by regression over kr for corner parameters ϑ_0 (r= distance between earlier and later corner) :

$$L(kr,0) = a_0 + a_1 \cdot x + +a_2 \cdot x^2 + \ldots \quad ; \quad x = \lg(kr)$$

$$a_i = \begin{cases} a_i(\vartheta_0, kr_q, \vartheta_q) \\ a_i(\vartheta_0, \vartheta_q) \end{cases} \quad ; \quad i = \begin{cases} 0,1,2 & ; \text{ line source} \\ 0,1,2,3 & ; \text{ plane wave} \end{cases} \tag{12}$$

In the case of a line source at (r_q,ϑ_q), the coefficients a_i are expanded as :

for $\vartheta_0 \neq 2\pi$:

$$\begin{aligned} a_i(kr_q, \vartheta_q) = {} & b_{0,0} + b_{1,0} \cdot z + b_{2,0} \cdot z^2 + b_{3,0} \cdot z^3 + b_{0,1} \cdot y + b_{0,2} \cdot y^2 + \\ & + b_{1,1} \cdot z \cdot y + b_{1,2} \cdot z \cdot y^2 + b_{2,1} \cdot z^2 \cdot y + b_{2,2} \cdot z^2 \cdot y^2 \\ z = \lg(kr_q) \quad ; \quad & y = (\vartheta_0 - \vartheta_q)_{rad} \quad ; \quad i = 0,1,2 \quad ; \text{ line source} \\ b_{m,n} = b_{m,n}(\vartheta_0, i) \quad ; \quad & \vartheta_0 \neq 2\pi \end{aligned} \tag{13}$$

for $\vartheta_0 = 2\pi$ (observe the change in sign of y):

$$\begin{aligned} a_i(kr_q, \vartheta_q) = {} & b_{0,0} + b_{1,0} \cdot z + b_{2,0} \cdot z^2 + b_{3,0} \cdot z^3 + \\ & + b_{0,1} \cdot y + b_{0,2} \cdot y^2 + b_{0,3} \cdot y^3 + \\ & + b_{1,1} \cdot z \cdot y + b_{1,2} \cdot z \cdot y^2 + b_{1,3} \cdot z \cdot y^3 + \\ & + b_{2,1} \cdot z^2 \cdot y + b_{2,2} \cdot z^2 \cdot y^2 + b_{2,3} \cdot z^2 \cdot y^3 \\ z = \lg(kr_q) \quad ; \quad & y = (\vartheta_q - \vartheta_0)_{rad} \quad ; \quad i = 0,1,2 \quad ; \text{ line source} \\ b_{m,n} = b_{m,n}(\vartheta_0, i) \quad ; \quad & \vartheta_0 = 2\pi \end{aligned} \tag{14}$$

In the case of a plane wave from ϑ_q, the coefficients a_i are expanded similarly. The coefficients are :

<u>*line source; ϑ_0=270°:*</u>

$$\begin{aligned} a_0 = {} & -3.5089\,14089 + 2.5221\,96950 \cdot z - 1.8833\,48105 \cdot z^2 + 0.49679\,54203 \cdot z^3 + \\ & + 0.025449\,89020 \cdot y + 0.83455\,44874 \cdot y^2 + \\ & + 0.36795\,33261 \cdot z \cdot y + 0.37770\,82514 \cdot z \cdot y^2 - \\ & - 0.090243\,91927 \cdot z^2 \cdot y - 0.19472\,45060 \cdot z^2 \cdot y^2 \end{aligned} \tag{15}$$

$$a_1 = -8.0485\,12868 - 0.0095415\,28038 \cdot z + 0.57694\,54995 \cdot z^2 - 0.30516\,73769 \cdot z^3 +$$
$$+ 0.38215\,48481 \cdot y + 0.26510\,97458 \cdot y^2 -$$
$$- 0.61597\,40550 \cdot z \cdot y + 3.2921\,01514 \cdot z \cdot y^2 +$$
$$+ 0.29582\,36606 \cdot z^2 \cdot y - 0.87248\,95618 \cdot z^2 \cdot y^2 \tag{15}$$

$$a_2 = -0.63106\,12470 - 0.14239\,72091 \cdot z + 0.016782\,28228 \cdot z^2 + 0.047564\,68413 \cdot z^3 -$$
$$- 0.14074\,94337 \cdot y - 0.084649\,13202 \cdot y^2 +$$
$$+ 0.50436\,46313 \cdot z \cdot y - 1.1078\,19931 \cdot z \cdot y^2 -$$
$$- 0.71819\,71540 \cdot z^2 \cdot y + 0.82565\,08092 \cdot z^2 \cdot y^2$$

line source; ϑ_0=225°:

$$a_0 = -0.095406\,93872 + 3.18259\,83742 \cdot z - 2.2114\,37627 \cdot z^2 + 0.57977\,54874 \cdot z^3 -$$
$$- 1.5214\,70673 \cdot y + 2.9650\,63818 \cdot y^2 +$$
$$+ 3.5125\,79747 \cdot z \cdot y - 4.4380\,49225 \cdot z \cdot y^2 -$$
$$- 1.7393\,92722 \cdot z^2 \cdot y + 2.1463\,07680 \cdot z^2 \cdot y^2 \tag{16}$$

$$a_1 = -7.3238\,33505 + 3.0205\,07185 \cdot z - 0.085537\,99890 \cdot z^2 - 0.48498\,23570 \cdot z^3 +$$
$$+ 3.2843\,16693 \cdot y - 4.2599\,41774 \cdot y^2 -$$
$$- 7.6016\,95001 \cdot z \cdot y + 16.521\,364584 \cdot z \cdot y^2 +$$
$$+ 4.2761\,59321 \cdot z^2 \cdot y - 7.4133\,41455 \cdot z^2 \cdot y^2$$

$$a_2 = -0.87657\,45279 - 1.2033\,62984 \cdot z + 0.35134\,58955 \cdot z^2 + 0.034014\,30131 \cdot z^3 -$$
$$- 1.5290\,12091 \cdot y + 1.9733\,88860 \cdot y^2 +$$
$$+ 4.0117\,26713 \cdot z \cdot y - 7.3478\,39244 \cdot z \cdot y^2 -$$
$$- 2.3524\,98508 \cdot z^2 \cdot y + 4.8096\,51759 \cdot z^2 \cdot y^2$$

line source; ϑ_0=360°: $(y=(\vartheta_q-\vartheta_0)_{rad})$

$$a_0 = -9.0746\,49329 + 1.3520\,53882 \cdot z - 1.1527\,52755 \cdot z^2 + 0.33068\,58082 \cdot z^3 -$$
$$- 0.14552\,94858 \cdot y + 0.59601\,96681 \cdot y^2 - 0.0080120\,63348 \cdot y^3 +$$
$$+ 0.62917\,07040 \cdot z \cdot y + 0.85327\,98218 \cdot z \cdot y^2 + 0.13257\,83868 \cdot z \cdot y^3 -$$
$$- 0.24818\,13528 \cdot z^2 \cdot y - 0.30784\,62069 \cdot z^2 \cdot y^2 - 0.046470\,02117 \cdot z^2 \cdot y^3 \tag{17}$$

$$a_1 = -8.7968\,78642 - 0.46051\,18745 \cdot z + 0.26898\,65167 \cdot z^2 - 0.035863\,57874 \cdot z^3 + \\ + 0.60928\,64775 \cdot y + 0.84163\,78006 \cdot y^2 + 0.13561\,84578 \cdot y^3 - \\ - 1.5301\,26829 \cdot z \cdot y - 1.5894\,22436 \cdot z \cdot y^2 - 0.66061\,64056 \cdot z \cdot y^3 + \\ + 0.71581\,76859 \cdot z^2 \cdot y + 0.62332\,22389 \cdot z^2 \cdot y^2 + 0.21759\,05849 \cdot z^2 \cdot y^3 \tag{17}$$

$$a_2 = -0.45505\,05284 + 0.13009\,65967 \cdot z - 0.069958\,29336 \cdot z^2 - 0.031078\,82535 \cdot z^3 - \\ - 0.24382\,01115 \cdot y - 0.30793\,36093 \cdot y^2 - 0.050234\,97076 \cdot y^3 + \\ + 0.78491\,93467 \cdot z \cdot y + 0.71812\,40230 \cdot z \cdot y^2 + 0.24418\,35343 \cdot z \cdot y^3 - \\ - 0.77826\,97727 \cdot z^2 \cdot y - 0.76045\,96554 \cdot z^2 \cdot y^2 - 0.21399\,67639 \cdot z^2 \cdot y^3$$

The corresponding expansions for plane wave incidence are :

plane wave; ϑ_0=270°:

$$a_0 = -2.3897\,74901 - 0.13952\,62072 \cdot y + 2.4146\,52403 \cdot y^2 - 2.0826\,38349 \cdot y^3 + \\ + 1.7277\,12526 \cdot y^4 - 0.50244\,23058 \cdot y^5$$

$$a_1 = -6.3871\,81695 + 1.4298\,99038 \cdot y - 6.2941\,35235 \cdot y^2 + 22.114\,36965\,37 \cdot y^3 - \\ - 19.769\,03930\,85 \cdot y^4 + 5.5652\,21750 \cdot y^5 \tag{18}$$

$$a_2 = -2.5521\,24597 - 2.3420\,05059 \cdot y + 14.786\,77851\,84 \cdot y^2 - 38.027\,36056\,53 \cdot y^3 + \\ + 36.173\,18669\,37 \cdot y^4 - 10.728\,44150\,31 \cdot y^5$$

$$a_3 = 0.58831\,91145 + 0.79537\,88225 \cdot y - 5.5243\,62056 \cdot y^2 + 13.784\,20931\,81 \cdot y^3 - \\ - 13.683\,72113\,07 \cdot y^4 + 4.3492\,90260 \cdot y^5$$

plane wave; ϑ_0=225°:

$$a_0 = 1.5680\,78212 + 0.046053\,30206 \cdot y + 0.99017\,80451 \cdot y^2 + 1.8882\,27817 \cdot y^3 - \\ - 1.2048\,73300 \cdot y^4 - 0.55725\,13587 \cdot y^5$$

$$a_1 = -3.0816\,59731 - 0.54132\,85417 \cdot y + 8.6690\,503927 \cdot y^2 - 24.294\,47775\,16 \cdot y^3 + \\ + 19.787\,94853\,53 \cdot y^4 + 0.77202\,84647 \cdot y^5 \tag{19}$$

$$a_2 = -3.7292\,26627 + 1.0808\,88001 \cdot y - 4.0768\,62638 \cdot y^2 + 52.423\,47912\,86 \cdot y^3 - \\ - 59.934\,10118\,99 \cdot y^4 + 11.328\,55522\,96 \cdot y^5$$

$$a_3 = 0.67106\,94247 - 0.54571\,49106 \cdot y + 3.3222\,27192 \cdot y^2 - 28.532\,94487\,42 \cdot y^3 + \\ + 41.134\,31302\,00 \cdot y^4 - 13.871\,41596\,90 \cdot y^5$$

<u>*plane wave;* ϑ_0=360°:</u> $(y=(\vartheta_0-\vartheta_q)_{rad})$

$$a_0=-8.6190\,29147-0.049142\,42523\cdot y+0.98893\,551678\cdot y^2-0.096779\,73083\cdot y^3+$$
$$+0.043145\,42296\cdot y^4-0.0070927\,45198\cdot y^5$$

$$a_1=-8.4954\,14675+0.63498\,41469\cdot y-0.94143\,82840\cdot y^2+0.90744\,14634\cdot y^3-$$
$$-0.031200\,88534\cdot y^4-0.031113\,06942\cdot y^5 \qquad (20)$$

$$a_2=-1.0515\,33957-2.3970\,34644\cdot y+6.1112\,32667\cdot y^2-5.9035\,08575\cdot y^3+$$
$$+2.0087\,47402\cdot y^4-0.20880\,99622\cdot y^5$$

$$a_3=0.21431\,16375+1.3968\,01316\cdot y-3.9023\,86481\cdot y^2+3.9492\,26446\cdot y^3-$$
$$-1.5499\,39902\cdot y^4+0.20338\,83155\cdot y^5$$

The following diagrams show the sound pressure level at the shadowed flank $(\vartheta=0)$, evaluated with mode sums (points) and from the above regressions (lines).

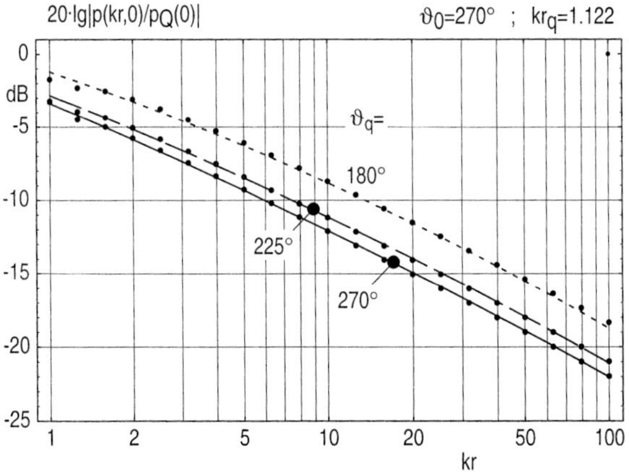

A line source at a small distance kr_q=1.122 .

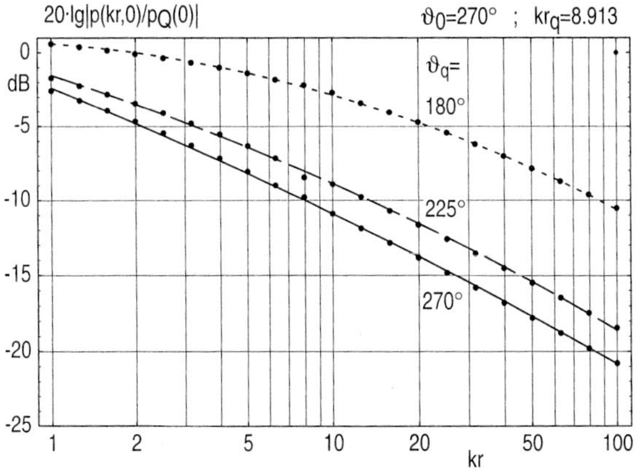

A line source at a medium distance kr_q=8.913 .

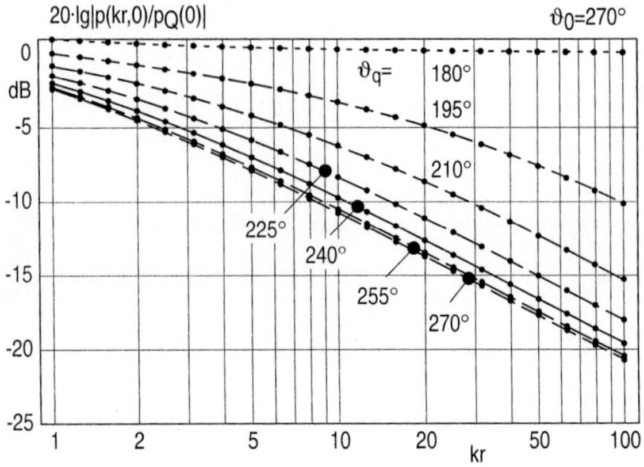

Plane wave incidence from different directions ϑ_q .

E.6 Plane wave scattering at a hard screen

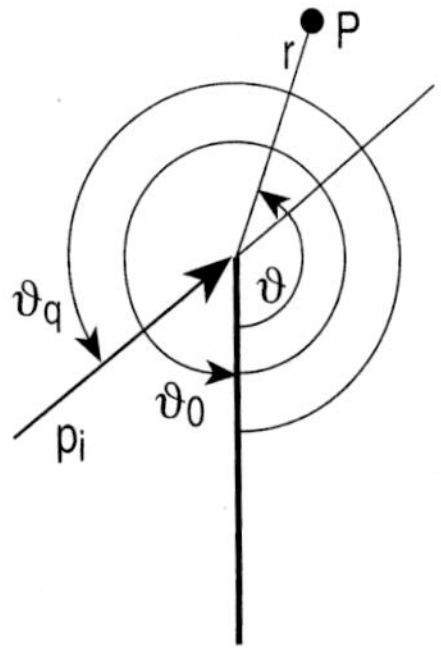

The hard screen is a special case with $\vartheta_0=2\pi$ of the section E.5 ; see also the next section E.7 .

A plane wave is incident under the direction ϑ_q . Its sound pressure at the screen corner is $p_Q(0)$. The radial wave number component is k (see section E.5)

The sound pressure around the screen is :

$$p(r,\vartheta) =$$

$$= p_Q(0)\frac{1+j}{2}\left\{e^{jkr\cos(\vartheta-\vartheta_q)}\left[\frac{1-j}{2}+C\left(\sqrt{2kr}\cos\frac{\vartheta-\vartheta_q}{2}\right)-jS\left(\sqrt{2kr}\cos\frac{\vartheta-\vartheta_q}{2}\right)\right]+\right. \tag{1}$$

$$\left. + e^{jkr\cos(\vartheta+\vartheta_q)}\left[\frac{1-j}{2}+C\left(\sqrt{2kr}\cos\frac{\vartheta+\vartheta_q}{2}\right)-jS\left(\sqrt{2kr}\cos\frac{\vartheta+\vartheta_q}{2}\right)\right]\right\}$$

with the Fresnel integrals defined by:

$$S(x) = \sqrt{\frac{2}{\pi}}\int_0^x \sin(t^2)\,dt \quad ; \quad C(x) = \sqrt{\frac{2}{\pi}}\int_0^x \cos(t^2)\,dt \tag{2}$$

E.7 Cylindrical or plane wave scattering at a screen with an elliptical cylinder atop

Ref.: See [E.5] for notation, formulas, and evaluation of Mathieu functions.

A hard, thin screen of height h has a locally absorbing, elliptical cylinder at its top; the surface admittance of the cylinder is G ; its long and short axes are 2a, 2b . The eccentricity of the ellipse is c .

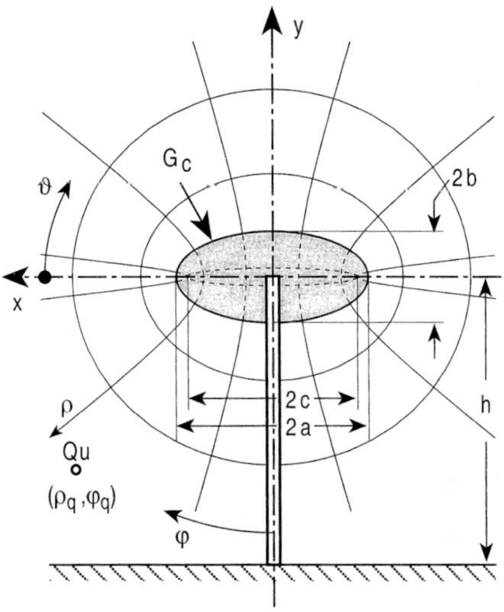

A line source parallel to the axis of the ellipse is at Qu with the elliptical co-ordinates (ρ_q,φ_q). For $\rho_q \to \infty$, a plane wave incidence prevails.

First, the height h is taken as $h \to \infty$; then the arrangement is mirror-reflected at y=–h , and the scattering of the field from the mirror-reflected source is evaluated, then both fields are superposed.

Transformation between Cartesian (x,y) and elliptic-hyperbolic (ρ,ϑ) co-ordinates :

$$x = c\cdot\cosh\rho\cdot\cos\vartheta \quad ; \quad y = c\cdot\sinh\rho\cdot\sin\vartheta \quad ; \quad z = z$$
$$x \pm jy = c\cdot\cosh(\rho \pm j\vartheta) \tag{1}$$

Also used are the co-ordinates (ρ,φ) with :

$$\varphi = \vartheta + \pi/2 \quad ; \quad \cos\vartheta = \sin\varphi \quad ; \quad \sin\vartheta = -\cos\varphi \tag{2}$$

Geometrical parameters (with ρ_c on the elliptical cylinder):

$$\frac{a}{c} = \cosh\rho_c \quad ; \quad \frac{b}{c} = \sinh\rho_c \quad ; \quad \frac{b}{a} = \tanh\rho_c \quad ; \quad c = \frac{a}{\cosh\rho_c} \tag{3}$$

With a separation $p(\rho,\vartheta,z)= R(\rho)\cdot T(\vartheta)\cdot Z(z)$ and an axial factor Z(z) proportional to either one or a linear combination of the functions $Z(z) = e^{\pm jk_z z}$; $\cos(k_z)$; $\sin(k_z z)$ given by a wave with a wave number k_z leading to the wave number k in the plane normal to the axis with $k^2 = k_0^2 - k_z^2$, the axial factor $Z(z)$ can be dropped; only $p(\rho,\vartheta)$ will be given.

Sound field from a line source:

The line source is placed at (ρ_q,ϑ_q) or (ρ_q,φ_q). Its polar distance to the origin is r_q with:

$$r_q^2 = x_q^2 + y_q^2 = c^2 \cdot \left[\cosh^2 \rho_q \cdot \sin^2 \varphi_q + \sinh^2 \rho_q \cdot \cos^2 \varphi_q\right] \tag{4}$$

When it has the volume flow q (per unit length), then it will produce the sound pressure in free space at the position of the origin:

$$p_Q(0) = \frac{1}{4} Z_0\, k_0 r_q \cdot q \cdot H_0^{(2)}(k r_q) \tag{5}$$

General field formulations in the two zones with $\rho_c \le \rho < \rho_q$ and $\rho_q < \rho < \infty$, respectively, separated from each other by the elliptic radius ρ_q of the line source position, are (integer summation index m):

$$\begin{aligned} p_1(\rho,\varphi) &= \sum_{m\ge 0} a_m \cdot Hc^{(2)}_{m/2}(\rho_q) \cdot \left[Hc^{(1)}_{m/2}(\rho) + r_m \cdot Hc^{(2)}_{m/2}(\rho)\right] \cdot ce_{m/2}(\varphi) \\ p_2(\rho,\varphi) &= \sum_{m\ge 0} a_m \cdot Hc^{(2)}_{m/2}(\rho) \cdot \left[Hc^{(1)}_{m/2}(\rho_q) + r_m \cdot Hc^{(2)}_{m/2}(\rho_q)\right] \cdot ce_{m/2}(\varphi) \end{aligned} \tag{6}$$

The term amplitudes a_m follow from the source condition; they are :

$$a_m = \frac{Z_0 q \cdot k_0 c}{2} \sum_{n\ge 0} ce_{n/2}(\varphi_q) \cdot Ic_{m,n}(\rho_q) \tag{7}$$

with the integrals (for more about them see below):

$$Ic_{m,n}(\rho) := \frac{1}{2\pi} \int_0^{2\pi} \sqrt{\sinh^2 \rho + \cos^2 \varphi} \cdot ce_{m/2}(\varphi) \cdot ce_{n/2}(\varphi)\, d\varphi \tag{8}$$

The modal reflection factors r_m (at the cylinder surface) are obtained from the boundary condition at the cylinder with the form $b_m = r_m \cdot a_m$ from the system of equations (with explicitly known a_m):

$$\begin{aligned} &\sum_{m\ge 0} b_m \cdot Hc^{(2)}_{m/2}(\rho_q) \cdot \left[G_c \cdot Hc^{(2)}_{m/2}(\rho_c) \cdot Ic_{m,n}(\rho_c) + \delta_{m,n} \cdot \frac{j}{2 k_0 c} Hc'^{(2)}_{m/2}(\rho_c)\right] = \\ &- \sum_{m\ge 0} a_m \cdot Hc^{(2)}_{m/2}(\rho_q) \cdot \left[G_c \cdot Hc^{(1)}_{m/2}(\rho_c) \cdot Ic_{m,n}(\rho_c) + \delta_{m,n} \cdot \frac{j}{2 k_0 c} Hc'^{(1)}_{m/2}(\rho_c)\right] \end{aligned} \tag{9}$$

($\delta_{m,n}$ the Kronecker symbol).

1st *special case* : In the first special case, the cylinder is supposed to be rigid, $G_c=0$. The system of equations for the r_m simplifies to:

$$b_m = r_m \cdot a_m = -a_m \cdot \frac{Hc'^{(1)}_{m/2}(\rho_c)}{Hc'^{(2)}_{m/2}(\rho_c)} \tag{10}$$

2nd *special case* : In the second special case with $\rho_c=0$, in which the cylinder degenerates to an absorbing strip of width $2c$, the equations formally remain unchanged, but the integrals $Ic_{m,n}(\rho_c)$ simplify drastically as will be seen below.

3rd *special case* : The third special case is a combination of the first and second: $G_c=0$ and $\rho_c=0$; the cylinder changes to a rigid strip. Then:

$$b_m = r_m \cdot a_m = -a_m \cdot \frac{Hc'^{(1)}_{m/2}(0)}{Hc'^{(2)}_{m/2}(0)} \tag{11}$$

There still remain the integrals $Ic_{m,n}(\rho)$ to be evaluated. They can be expressed in terms of the Fourier coefficients A_ν of the even azimuthal Mathieu functions $ce_\mu(\varphi)$, B_ν of the odd azimuthal Mathieu functions $se_\mu(\varphi)$, which are needed anyhow for the evaluation of such Mathieu functions.

When μ and ν are integers and both even, then:

$$Ic_{m,n}(\rho) = \frac{(-1)^{(\mu+\nu)/2}}{4\pi} \sum_{s,\sigma\ge 0} (-1)^{s+\sigma} A_{2s}(\mu) \cdot A_{2\sigma}(\nu) \cdot \left[I_{|s-\sigma|}(\rho) + I_{s+\sigma}(\rho) \right] \tag{12}$$

with :

$$\begin{aligned} I_i(\rho) &= \int_0^{2\pi} \cos(2i\varphi) \cdot \sqrt{\sinh^2\rho + \cos^2\varphi}\; d\varphi \\ &= 4 \int_0^{\pi/2} \cos(2i\varphi) \cdot \sqrt{\sinh^2\rho + \cos^2\varphi}\; d\varphi \quad ; \quad 2i = \text{even} \\ &= 0 \quad ; \quad 2i = \text{odd} \end{aligned} \tag{13}$$

and the values :

$$I_i(\rho) = 2\pi(-1)^{i+1} \cdot \cosh\rho \sum_{k\ge 0} \frac{(2(i+k)-3)!! \cdot (2(i+k)-1)!!}{k! \cdot (2i+k)!} \frac{1}{(2\cosh\rho)^{2(i+k)}} \tag{14}$$
$$i \ge 1 \;\; ; \;\; \rho > 0 \;\; ; \;\; (0)!! = (-1)!! = 1 \;\; ; \;\; (2n+1)!! = 1\cdot 3\cdot 5\cdot \ldots \cdot (2n+1)$$

with the special value: $I_0(\rho) = 4\cosh\rho \cdot E(1/\cosh\rho)$, ($E(z)$ the exponential integral), and for $\rho=0$:

$$I_i(0)=\begin{cases}\pi \;;\; 2i+1=0\\ 2\left[\dfrac{\sin(2i-1)\pi/2}{2i-1}+\dfrac{\sin(2i+1)\pi/2}{2i+1}\right] \;;\; 2i=\text{even}\\ 0\;;\; 2i=\text{odd}\neq 1\end{cases} \tag{15}$$

When μ and ν are integers and both odd, then:

$$Ic_{m,n}(\rho)=\frac{(-1)^{(\mu+\nu-2)/2}}{4\pi}\sum_{s,\sigma\geq 0}(-1)^{s+\sigma}B_{2s+1}(\mu)\cdot B_{2\sigma+1}(\nu)\cdot\left[I_{|s-\sigma|}(\rho)+I_{s+\sigma+1}(\rho)\right] \tag{16}$$

When μ and ν are integers, one even the other odd, then: $Ic_{m,n}(\rho)=0$.
When μ is half-valued, ν is an integer, or inversely, then: $Ic_{m,n}(\rho)=0$.
When μ , ν are both half-valued, with $\mu=\mu'+1/2$, $\nu=\nu'+1/2$, and both μ', ν' even or odd:

$$Ic_{m,n}(\rho)=\frac{1}{4\pi}\sum_{s,\sigma=-\infty}^{+\infty}(-1)^{s+\sigma}C_{2s}(\mu)\cdot C_{2\sigma}(\nu)\cdot I_{|(\mu'-\nu')/2+s-\sigma|}(\rho) \tag{17}$$

When μ , ν are both half-valued, with μ' even, ν' odd, or inversely:

$$Ic_{m,n}(\rho)=\frac{1}{4\pi}\sum_{s,\sigma=-\infty}^{+\infty}(-1)^{s+\sigma}C_{2s}(\mu)\cdot C_{2\sigma}(\nu)\cdot I_{|(\mu'+\nu'+1)/2+s+\sigma|}(\rho) \tag{18}$$

One gets in the limit $\rho\to\infty$ with: $\sqrt{\sinh^2\rho+\cos^2\varphi}=\sqrt{\cosh^2\rho-\sin^2\varphi}\to\cosh\rho$: (19)

$$I_i(\rho)\to 4\cosh\rho\int_0^{\pi/2}\cos(2i\varphi)\,d\varphi=\begin{cases}2\pi\cosh\rho \;;\; i=0\\ 0 \;;\; i\neq 0\,,\ \text{integer}\\ 2\pi\cosh\rho\cdot\sin(i\pi)/(i\pi)\;;\;\text{else}\end{cases} \tag{20}$$

This gives in the above cases of non-zero $Ic_{m,n}(\rho)$:

$$Ic_{m,n}(\rho)\to\frac{1}{2}(-1)^{(\mu+\nu)/2}\cosh\rho\cdot\left[\sum_{s,\sigma\geq 0;\ s=\sigma}A_{2s}(\mu)\cdot A_{2\sigma}(\nu)+A_0(\mu)\cdot A_0(\nu)\right]$$

$$Ic_{m,n}(\rho)\to\frac{1}{2}(-1)^{(\mu+\nu)/2-1}\cosh\rho\cdot\sum_{s,\sigma\geq 0;s=\sigma}B_{2s+1}(\mu)\cdot B_{2\sigma+1}(\nu)$$

$$Ic_{m,n}(\rho)=\frac{1}{2}\cosh\rho\cdot\sum_{\substack{s,\sigma=-\infty\\(\mu'-\nu')/2+s-\sigma=0}}^{+\infty}(-1)^{s+\sigma}C_{2s}(\mu)\cdot C_{2\sigma}(\nu) \tag{21}$$

$$Ic_{m,n}(\rho)=\frac{1}{2}\cosh\rho\cdot\sum_{\substack{s,\sigma=-\infty\\(\mu'+\nu'+1)/2+s+\sigma=0}}^{+\infty}(-1)^{s+\sigma}C_{2s}(\mu)\cdot C_{2\sigma}(\nu)$$

Sound field from a plane wave:

This case is treated as the limit $\rho_q \to \infty$. The polar radius r_q of the source position approaches $r_q \to c \cdot \cosh\rho_q$, whence $2\sqrt{q}\cosh\rho_q \to k\,r_q$. One further replaces

$$Z_0 q = \frac{4\,p_Q(0)}{k_0 r_q \cdot H_0^{(2)}(kr_q)} \tag{22}$$

Finally, one replaces above

$$a_m \cdot Hc_{m/2}^{(2)}(\rho_q) \xrightarrow[\rho_q \to \infty]{} 2\,p_Q(0) \cdot e^{j m\pi/4} \cdot \sum_{n \geq 0} ce_{n/2}(\varphi_q) \cdot Ic'_{m,n} \tag{23}$$

and in $b_m \cdot Hc_{m/2}^{(2)}(\rho_q) = r_m \cdot a_m \cdot Hc_{m/2}^{(2)}(\rho_q)$. (24)

Rigid screen with a mushroom-like hat :

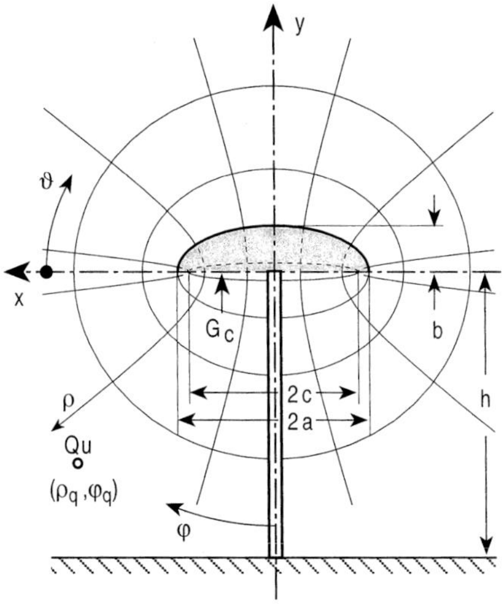

The cylindrical body atop the screen has an semi-elliptical shape; its surface is curved and rigid on the upper side, and is flat and absorbing on its lower side with the admittance G_c.

The boundary condition at the cylinder is :

$$G_c = \begin{cases} 0 & ; \ \pi/2 \leq \varphi \leq 3\pi/2 \\ G_c & ; \ \text{else} \end{cases}$$
$$\rho_c = \begin{cases} \rho_c & ; \ \pi/2 \leq \varphi \leq 3\pi/2 \\ 0 & ; \ \text{else} \end{cases} \tag{25}$$

i.e. :

$$\frac{j}{k_0 c\sqrt{\sinh^2\rho + \cos^2\varphi}} \frac{\partial p_1}{\partial \rho} = \begin{cases} 0 & ; \ \rho = \rho_c \ ; \ \pi/2 \leq \varphi \leq 3\pi/2 \\ -G_c \cdot p_1 & ; \ \rho = 0 \ \ ; \ \text{else} \end{cases} \tag{26}$$

The term amplitudes a_m remain as above; the system of equations for the $b_m = r_m \cdot a_m$ changes to :

$$\sum_{m\geq 0} b_m \cdot Hc^{(2)}_{m/2}(\rho_q) \cdot$$

$$\cdot \left[\left(Hc'^{(2)}_{m/2}(\rho_c) - Hc'^{(2)}_{m/2}(0) \right) \frac{1}{2\pi} \int_{\pi/2}^{3\pi/2} ce_{m/2}(\varphi) \cdot ce_{n/2}(\varphi)\, d\varphi - \right.$$

$$\left. - jk_0c \cdot G_c \cdot Hc^{(2)}_{m/2}(0) \frac{1}{2\pi} \int_{else} |\cos\varphi| \cdot ce_{m/2}(\varphi) \cdot ce_{n/2}(\varphi)\, d\varphi + \delta_{m,n} N_m Hc'^{(2)}_{m/2}(0) \right] =$$

$$= -\sum_{m\geq 0} a_m \cdot Hc^{(2)}_{m/2}(\rho_q) \cdot \tag{27}$$

$$\cdot \left[\left(Hc'^{(1)}_{m/2}(\rho_c) - Hc'^{(1)}_{m/2}(0) \right) \frac{1}{2\pi} \int_{\pi/2}^{3\pi/2} ce_{m/2}(\varphi) \cdot ce_{n/2}(\varphi)\, d\varphi - \right.$$

$$\left. - jk_0c \cdot G_c \cdot Hc^{(1)}_{m/2}(0) \frac{1}{2\pi} \int_{else} |\cos\varphi| \cdot ce_{m/2}(\varphi) \cdot ce_{n/2}(\varphi)\, d\varphi + \delta_{m,n} N_m Hc'^{(1)}_{m/2}(0) \right]$$

The N_m are the azimuthal mode norms :

$$N_m = \frac{1}{2\pi} \int_0^{2\pi} ce^2_{m/2}(\varphi)\, d\varphi \tag{28}$$

E.8 Uniform scattering at screens and dams

Ref.: Mechel, [E.4]; see [E.5] for notation, formulas, and evaluation of Mathieu functions

This section describes the plane wave scattering

- at a “high” absorbent dam, with the limit case of
  - a thin absorbent screen;
- at a “flat” absorbent dam, with the limit case of
  - a flat absorbent strip in a rigid baffle wall.

A semicircular absorbent dam could also be treated as a limiting case of this uniform theory (it will, however, be discussed separately in Section 10). All objects are situated on a hard ground. All absorbent objects are locally reacting with a normalised surface admittance G .

The distinction of the “high” and “flat” dam is necessary because of the orientation of the axes and co-ordinates of the elliptical cylinder, with which the objects are modelled.

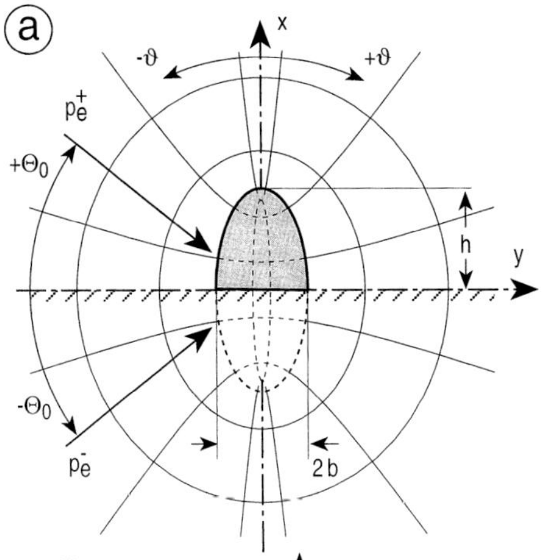

A high dam with its elliptical-hyperbolic co-ordinate system. The plane x=0 is the hard ground. p_e^+ is the incident plane wave; p_e^- is the mirror-reflected wave. The height of the dam is h ; its width on ground level is 2b .

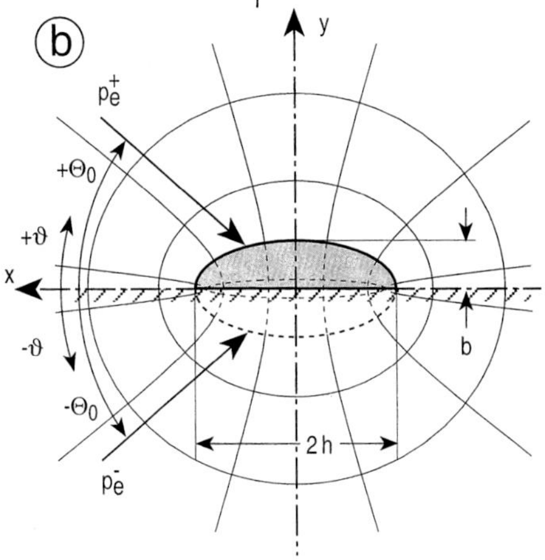

A flat dam with its elliptical-hyperbolic co-ordinate system. The plane y=0 is the hard ground. p_e^+ is the incident plane wave; p_e^- is the mirror-reflected wave. The height of the dam is b ; its width on ground level is 2h .

Limit cases are:

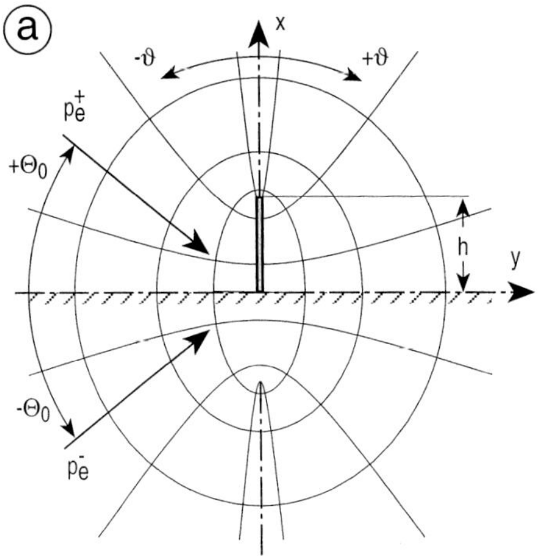

A thin, absorbent screen with its elliptical-hyperbolic co-ordinate system. The plane x=0 is the hard ground. p_e^+ is the incident plane wave; p_e^- is the mirror-reflected wave. The height of the screen is h.

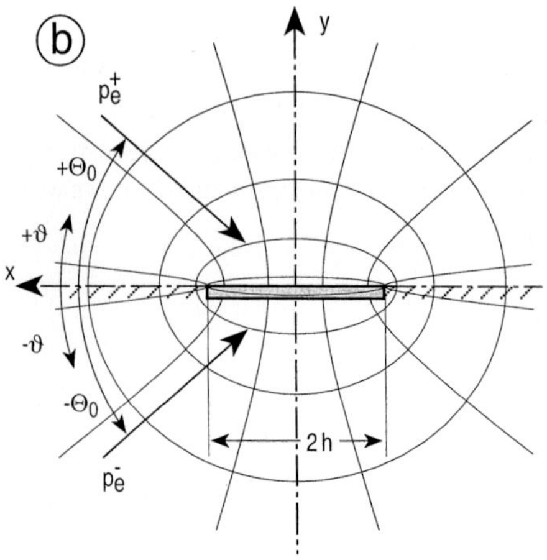

An absorbent strip in a hard baffle with its elliptical-hyperbolic co-ordinate system. The plane y=0 is the hard baffle wall. p_e^+ is the incident plane wave; p_e^- is the mirror-reflected wave. The width of the strip is 2h .

The co-ordinate transformation between Cartesian (x,y) and elliptic-hyperbolic co-ordinates (ρ,ϑ) is, with the eccentricity of the ellipses :

$$\left.\begin{aligned} x &= c\cdot\cosh\rho\cdot\cos\vartheta \\ y &= c\cdot\sinh\rho\cdot\sin\vartheta \end{aligned}\right\} \quad ; \quad 0\le\rho<\infty \quad ; \quad -\pi\le\vartheta\le+\pi \tag{1}$$

and in the backward direction :

$$\rho + j\vartheta = \operatorname{area}\cosh(\xi + j\eta) = \ln\left(\xi + j\eta \pm \sqrt{(\xi + j\eta)^2 - 1}\right) \tag{2}$$

with ξ=x/c , η=y/c . Geometrical parameters are with the elliptical radius ρ_c on the object:

$$\left.\begin{aligned} h/c &= \cosh\rho_c \\ b/c &= \sinh\rho_c \end{aligned}\right\} \quad ; \quad b/h = \tanh\rho_c \quad ; \quad c = \frac{h}{\cosh\rho_c} \tag{3}$$

The geometrical shadow limit for plane wave incidence with an angle Θ_0 is given by :

$$h/c - \operatorname{tg}\Theta_0\cdot\sinh\rho\cdot\sin\vartheta = \cosh\rho\cdot\cos\vartheta \tag{4}$$

with the special cases :

$$\begin{aligned} &\underline{\rho = 0:} \quad \vartheta = \arccos(h/c) \xrightarrow[h/c\to1]{} 0 \\ &\underline{\Theta_0 = 0:} \quad \vartheta = \arccos\frac{h/c}{\cosh\rho} \\ &\underline{\rho >> 1:} \quad \vartheta \approx -\arctan\frac{h/c}{\tan\Theta_0\tanh\rho} \end{aligned} \tag{5}$$

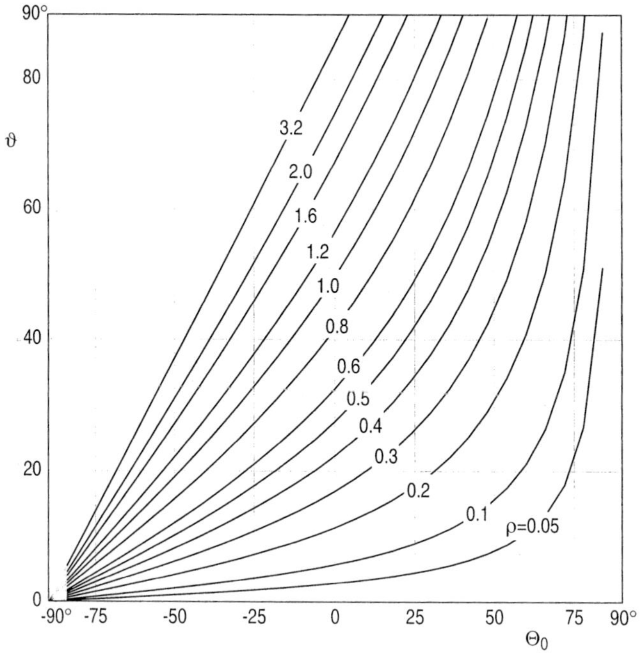

Geometrical shadow limits for h/c=1 .

Let $p_e^{\pm}$ be the incident wave and the mirror-reflected wave at the ground. $p_e = p_e^+ + p_e^-$ then is the "exciting" wave with hard ground, and $p_e = p_e^+ + \underline{r} \cdot p_e^-$ the exciting wave with absorbing ground having a reflection factor $\underline{r}$. The total field is composed as the sum $p= p_e+p_{rs}$ of the exciting wave p_e and a "reflected plus scattered" wave p_{rs} .

The following description uses the 2nd principle of superposition from section B.10 ; i.e. the task is subdivided in two sub-tasks (ß)=(h),(w) , in which the plane of symmetry through the scattering object (where it is sound transmissive) is considered first as hard (h), second as soft (w). The reflected plus scattered wave is marked and decomposed in both sub-tasks as $p_{rs}^{(ß)} = p_r^{(ß)} + p_s^{(ß)}$, where $p_r^{(ß)}$ is the reflected wave at the plane of symmetry, with hard reflection for (ß)=(h) and soft reflection for (ß)=(w), respectively, i.e. $p_r^{(w)}(y) = -p_r^{(h)}(y)$. The component $p_s^{(ß)}$ is the "truly" scattered wave. At the high dam, the co-ordinate normal to the plane of symmetry is $y \to \vartheta$. According to the principle of superposition the sound field on the front side (side of sound incidence) is :

$$\begin{aligned} p_{front}(\vartheta < 0) &= p_e(\vartheta) + \frac{1}{2}\left[p_s^{(h)}(\vartheta) + p_s^{(w)}(\vartheta)\right] \\ &= \frac{1}{2}\left[\left(p_e(\vartheta) + p_r^{(h)}(\vartheta)\right) + p_s^{(h)}(\vartheta) + \left(p_e(\vartheta) + p_r^{(w)}(\vartheta)\right) + p_s^{(w)}(\vartheta)\right] \end{aligned} \tag{6}$$

and the transmitted sound field on the back side is :

$$p_{back}(\vartheta>0) = p_e(\vartheta) + \frac{1}{2}\left[p_s^{(h)}(-\vartheta) - p_s^{(w)}(-\vartheta)\right]$$
$$= \frac{1}{2}\left[\left(p_e(-\vartheta) + p_r^{(h)}(-\vartheta)\right) + p_s^{(h)}(-\vartheta) - \left(p_e(-\vartheta) + p_r^{(w)}(-\vartheta)\right) - p_s^{(w)}(-\vartheta)\right] \tag{7}$$

The basis for the field analysis is the decomposition of a plane wave in Mathieu functions :

$$u(x,y) = e^{-jk_0(x\cos\alpha + y\sin\alpha)}$$
$$= 2\sum_{m=0}^{\infty}(-j)^m ce_m(\alpha)\cdot ce_m(\vartheta)\cdot Jc_m(\rho) + 2\sum_{m=1}^{\infty}(-j)^m se_m(\alpha)\cdot se_m(\vartheta)\cdot Js_m(\rho) \tag{8}$$

(α is the angle between the wave number vector and the positive x axis),
and the decomposition of the Hankel function of the 2nd kind in Mathieu functions :

$$H_0^{(2)}(k_0R) = 2\left[\sum_{m\ge 0} ce_m(\vartheta_0)\cdot ce_m(\vartheta)\cdot\begin{cases} Jc_m(\rho_0)\,Hc_m^{(2)}(\rho) & ; \ \rho>\rho_0 \\ Jc_m(\rho)\,Hc_m^{(2)}(\rho_0) & ; \ \rho<\rho_0\end{cases}\right.$$
$$\left. + \sum_{m\ge 1} se_m(\vartheta_0)\cdot se_m(\vartheta)\cdot\begin{cases} Js_m(\rho_0)\,Hs_m^{(2)}(\rho) & ; \ \rho>\rho_0 \\ Js_m(\rho)\,Hs_m^{(2)}(\rho_0) & ; \ \rho<\rho_0\end{cases}\right] \tag{9}$$

with the source of the Hankel function in the elliptical co-ordinates (ρ_0,ϑ_0). The parameter of the Mathieu differential equation is $q=(k_0c)^2/4$; $ce_m(\vartheta)$, $se_m(\vartheta)$ are even and odd azimuthal Mathieu functions; $Jc_m(\rho)$, $Yc_m(\rho)$, $Hc_m^{(2)}(\rho) = Jc_m(\rho) - j\cdot Yc_m(\rho)$ and $Js_m(\rho)$, $Ys_m(\rho)$, $Hs_m^{(2)}(\rho) = Js_m(\rho) - j\cdot Ys_m(\rho)$ are the associated radial Mathieu-Bessel, Mathieu-Neumann, and Mathieu-Hankel functions.

High dam :

Let the incident wave be a plane wave with $\alpha^{\pm}=\pi/2\pm\Theta_0$. The exciting wave on hard ground is:

$$p_e = p_e^+ + p_e^- = P_e e^{-jk_y y}\left(e^{+jk_x x} + e^{-jk_x x}\right) = 2P_e\cos k_x x\cdot e^{-jk_y y} \tag{10}$$

and with absorbing ground :

$$p_e = p_e^+ + \underline{r}\cdot p_e^- = P_e e^{-jk_y y}\left(e^{+jk_x x} + \underline{r}\cdot e^{-jk_x x}\right) \tag{11}$$

$$\text{with : } k_x = k_0\sin\Theta_0 \quad ; \quad k_y = k_0\cos\Theta_0 \ . \tag{12}$$

With hard ground, the exciting and reflected waves are in both sub-tasks :

$$p_e + p_r^{(h)} = 8P_e\sum_{r\ge 0}(-j)^{2r} ce_{2r}(\alpha^+)\cdot ce_{2r}(\vartheta)\cdot Jc_{2r}(\rho) \tag{13}$$

$$p_e + p_r^{(w)} = 8P_e\sum_{r\ge 0}(-j)^{2r+1} se_{2r+1}(\alpha^+)\cdot se_{2r+1}(\vartheta)\cdot Js_{2r+1}(\rho) \tag{14}$$

The scattered waves are formulated as :

$$p_s^{(h)} = P_e \sum_{r\ge 0} C_r \cdot ce_{2r}(\vartheta) \cdot Hc_{2r}^{(2)}(\rho) \tag{15}$$

$$p_s^{(w)} = P_e \sum_{r\ge 0} S_r \cdot se_{2r+1}(\vartheta) \cdot Hs_{2r+1}^{(2)}(\rho) \tag{16}$$

with still undetermined term amplitudes C_r, S_r. They follow from the boundary condition at the elliptic cylinder :

$$-Z_0\left(v_{e\rho} + v_{r\rho}^{(\beta)} + v_{s\rho}^{(\beta)}\right)_{\rho=\rho_c} \overset{!}{=} G\cdot\left(p_e + p_r^{(\beta)} + p_s^{(\beta)}\right)_{\rho=\rho_c} \tag{17}$$

using the integrals :

$$\begin{aligned} Ic_{m,\mu}(u) &:= \int_0^\pi ce_m(t)\cdot ce_\mu(t)\cdot\sqrt{u^2+\sin^2 t}\; dt \quad ; \quad u^2 = \sinh^2\rho \\ Is_{m,\mu}(u) &:= \int_0^\pi se_m(t)\cdot se_\mu(t)\cdot\sqrt{u^2+\sin^2 t}\; dt \end{aligned} \tag{18}$$

and the orthogonality relation :

$$\int_0^\pi ce_m(t)\cdot ce_\mu(t)\; dt = \int_0^\pi se_m(t)\cdot se_\mu(t)\; dt = \delta_{m,\mu}\cdot\pi/2 \tag{19}$$

One gets the following systems of equations for the term amplitudes:

for (ß)=(h) by: $\displaystyle\int_0^\pi \ldots\cdot ce_{2s}(t)\cdot\sqrt{u^2+\sin^2 t}\; dt \quad ; \quad s\ge 0$

$$\begin{aligned} &\sum_{r\ge 0} C_r\cdot\left[\delta_{r,s}\frac{j\pi}{2k_0c} Hc_{2r}'^{(2)}(\rho_c) + G\cdot Hc_{2r}^{(2)}(\rho_c)\cdot Ic_{2r,2s}\right] = \\ &= -8\sum_{r\ge 0}(-j)^{2r} ce_{2r}(\alpha^+)\cdot\left[\delta_{r,s}\frac{j\pi}{2k_0c} Jc_{2r}'(\rho_c) + G\cdot Jc_{2r}(\rho_c)\cdot Ic_{2r,2s}\right] \quad ; \quad s\ge 0 \end{aligned} \tag{20}$$

for (ß)=(w) by: $\displaystyle\int_0^\pi \ldots\cdot se_{2s+1}(t)\cdot\sqrt{u^2+\sin^2 t}\; dt \quad ; \quad s\ge 0$

$$\begin{aligned} &\sum_{r\ge 0} S_r\cdot\left[\delta_{r,s}\frac{j\pi}{2k_0c} Hs_{2r+1}'^{(2)}(\rho_c) + G\cdot Hs_{2r+1}^{(2)}(\rho_c)\cdot Is_{2r+1,2s+1}\right] = \\ &= -8\sum_{r\ge 0}(-j)^{2r+1} se_{2r+1}(\alpha^+)\cdot\left[\delta_{r,s}\frac{j\pi}{2k_0c} Js_{2r+1}'(\rho_c) + G\cdot Js_{2r+1}(\rho_c)\cdot Is_{2r+1,2s+1}\right] \quad ; \quad s\ge 0 \end{aligned} \tag{21}$$

With the Fourier coefficients $A_\nu(\mu)$ of the $ce_\mu(z)$ and $B_\nu(\mu)$ of the $se_\mu(z)$, the required integrals are :

$$Ic_{m,\mu} = \sum_{n,\nu\ge 0} A_n(m)\cdot A_\nu(\mu)\cdot\int_0^\pi \cos(n\vartheta)\cos(\nu\vartheta)\sqrt{u^2+\sin^2\vartheta}\;d\vartheta$$

$$Is_{m,\mu} = \sum_{n,\nu\ge 1} B_n(m)\cdot B_\nu(\mu)\cdot\int_0^\pi \sin(n\vartheta)\sin(\nu\vartheta)\sqrt{u^2+\sin^2\vartheta}\;d\vartheta \tag{22}$$

and :

$$Ic_{m,\mu} = \frac{1}{2}\sum_{n,\nu\ge 0} A_n(m)\cdot A_\nu(\mu)\cdot\left[I_{(n-\nu)/2}+I_{(n+\nu)/2}\right]$$

$$Is_{m,\mu} = \frac{1}{2}\sum_{n,\nu\ge 1} B_n(m)\cdot B_\nu(\mu)\cdot\left[I_{(n-\nu)/2}-I_{(n+\nu)/2}\right] \tag{23}$$

where :

$$I_i = I_i(u) := \int_0^\pi \cos(2i\vartheta)\sqrt{u^2+\sin^2\vartheta}\;d\vartheta \tag{24}$$

Special case of a hard high dam, i.e. G=0 :

has the explicit solutions :

$$C_r = -8(-1)^{2r}\, ce_{2r}(\alpha^+)\frac{Jc'_{2r}(\rho_w)}{Hc'^{(2)}_{2r}(\rho_w)} \quad ; \quad S_r = -8(-1)^{2r+1}\, se_{2r+1}(\alpha^+)\frac{Js'_{2r+1}(\rho_w)}{Hs'^{(2)}_{2r+1}(\rho_w)} \tag{25}$$

Special case of a thin screen, i.e. $\rho_c \to 0$:

has the special values :

$$Jc'_m(0) = 0 \quad ; \quad Js_m(0) = 0 \quad ; \quad Hc'^{(2)}_m(0) = -j\,Yc'_m(0) \quad ; \quad Hs^{(2)}_m(0) = -j\,Ys_m(0) \tag{26}$$

and if further G=0 , i.e. the screen is hard : C_r=0 .

With absorbent ground, having the reflection factor $\underline{r}$:

Exciting and reflected wave :

$$p_e + p_r^{(h)} = 4P_e \sum_{m\ge 0} (-j)^m ce_m(\alpha^+)\left(1+\underline{r}\cdot(-1)^m\right)\cdot ce_m(\vartheta)\cdot Jc_m(\rho)$$
$$p_e + p_r^{(w)} = 4P_e \sum_{m\ge 1} (-j)^m se_m(\alpha^+)\left(1-\underline{r}\cdot(-1)^m\right)\cdot se_m(\vartheta)\cdot Js_m(\rho) \qquad (27)$$

Formulation of the scattered waves for both sub-tasks:

$$p_s^{(h)} = P_e \sum_{m\ge 0} C_m \cdot ce_m(\vartheta)\cdot Hc_m^{(2)}(\rho)$$
$$p_s^{(w)} = P_e \sum_{m\ge 1} S_m \cdot se_m(\vartheta)\cdot Hs_m^{(2)}(\rho) \qquad (28)$$

The systems of equations for the term amplitudes C_m, S_m become :

$$\sum_{m\ge 0} C_m \cdot \left[\delta_{m,\mu}\frac{j\pi}{2k_0c} Hc_m'^{(2)}(\rho_c) + Z_0G\cdot Hc_m^{(2)}(\rho_c)\cdot Ic_{m,\mu}\right] = \qquad ;\ \mu\ge 0$$
$$= -4\sum_{m\ge 0} (-j)^m ce_m(\alpha^+)\left(1+\underline{r}\cdot(-1)^m\right)\cdot\left[\delta_{m,\mu}\frac{j\pi}{2k_0c} Jc_m'(\rho_c) + Z_0G\cdot Jc_m(\rho_c)\cdot Ic_{m,\mu}\right] \qquad (29)$$

$$\sum_{m\ge 1} S_m \cdot \left[\delta_{m,\mu}\frac{j\pi}{2k_0c} Hs_m'^{(2)}(\rho_c) + Z_0G\cdot Hs_m^{(2)}(\rho_c)\cdot Is_{m,\mu}\right] = \qquad ;\ \mu\ge 1$$
$$= -4\sum_{m\ge 1} (-j)^m se_m(\alpha^+)\left(1-\underline{r}\cdot(-1)^m\right)\cdot\left[\delta_{m,\mu}\frac{j\pi}{2k_0c} Js_m'(\rho_c) + Z_0G\cdot Js_m(\rho_c)\cdot Is_{m,\mu}\right] \qquad (30)$$

The orders m,μ in the integrals $Ic_{m,\mu}$, $Is_{m,\mu}$ have the same parity.

High dam and cylindrical incident wave:

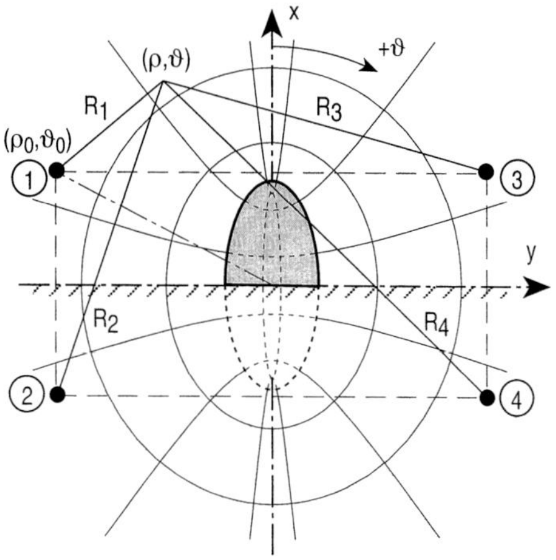

The original source (1) is at (ρ_0,ϑ_0) .
Some mirror sources (2)…(4) are used.

The original free field is :

$$p_e^+ = p_1 = P_e\cdot H_0^{(2)}(k_0R_1) \qquad (31)$$

The exciting wave with hard ground is: (32)

$$p_e = p_e^+ + p_e^- = P_e\cdot\left[H_0^{(2)}(k_0R_1) + H_0^{(2)}(k_0R_2)\right]$$

and with absorbent ground :

$$p_e = p_e^+ + \underline{r} \cdot p_e^- = P_e \cdot \left[H_0^{(2)}(k_0 R_1) + \underline{r} \cdot H_0^{(2)}(k_0 R_2) \right] \tag{33}$$

In the range $\rho<\rho_0$, and especially $\rho=\rho_c$ is with a hard ground :

$$p_e + p_r^{(h)} = 8 P_e \sum_{r \geq 0} ce_{2r}(\vartheta_0) \cdot ce_{2r}(\vartheta) \cdot Jc_{2r}(\rho) \cdot Hc_{2r}^{(2)}(\rho_0) \tag{34}$$

$$p_e + p_r^{(w)} = 8 P_e \sum_{r \geq 0} se_{2r+1}(\vartheta_0) \cdot se_{2r+1}(\vartheta) \cdot Js_{2r+1}(\rho) \cdot Hs_{2r+1}^{(2)}(\rho_0) \tag{35}$$

The formulations for the scattered waves of the sub-tasks (ß)=(h),(w) remain as above. The systems of equations for the term amplitudes are obtained from those above by the substitutions :

$$(ß) = \begin{cases} (h): & (-j)^m \cdot ce_m(\alpha^+) \rightarrow ce_m(\vartheta_0) \cdot Hc_m^{(2)}(\rho_0) \\ (w): & (-j)^m \cdot se_m(\alpha^+) \rightarrow se_m(\vartheta_0) \cdot Hs_m^{(2)}(\rho_0) \end{cases} \tag{36}$$

An absorbent ground is introduced as above.

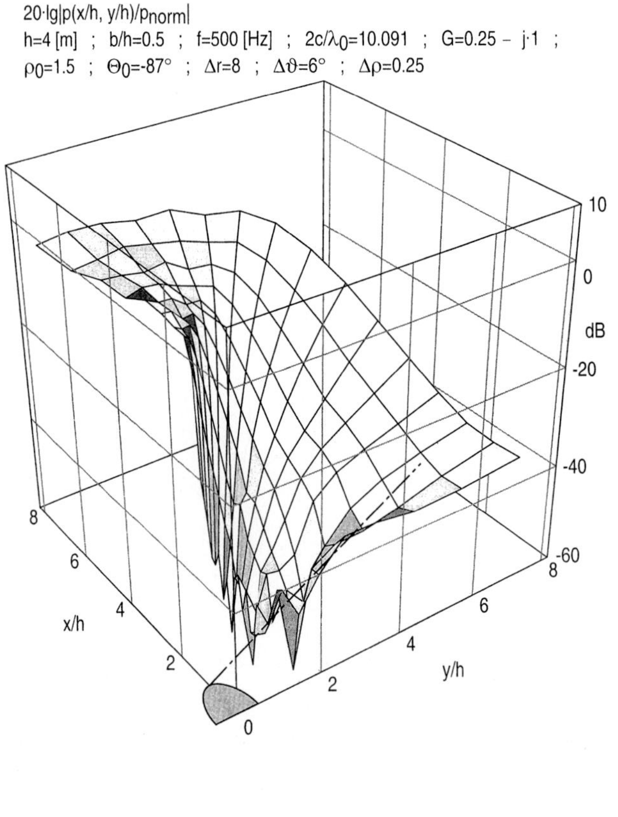

Sound pressure level on the shadow side behind a high dam; the line source is near to the ground at $\rho_0=1.5$; $\Theta_0=-87°$.

E.9 Scattering at a flat dam

The scattering at a flat dam is contained in a separate section, because the formulas are different from those for a high dam. See the previous section E.8 for the distinction between flat and high dams.

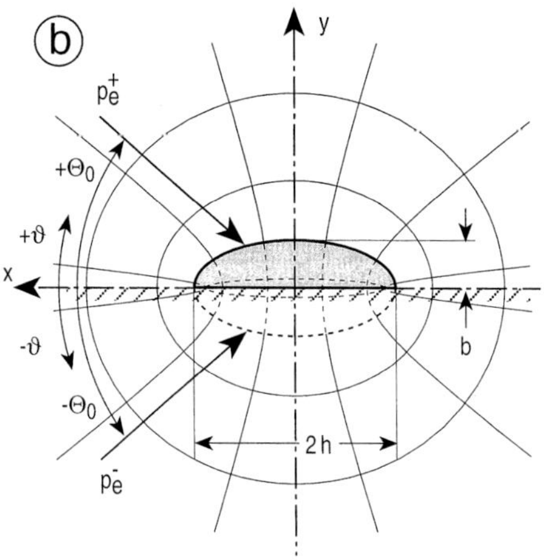

A flat dam with its elliptical-hyperbolic co-ordinate system. The plane y=0 is the hard ground. p_e^+ is the incident plane wave; p_e^- is the mirror-reflected wave. The height of the dam is b ; its width on ground level is 2h . The angle Θ_0 of sound incidence is measured towards ground. The dam is a semi-ellipse with eccentricity c . See the previous section E.8 for relations with other geometrical parameters.

The sound field is again evaluated with the principle of superposition (see previous section E.8). The exciting wave p_e , with the incident plane wave p_e^+ and the plane wave reflected at ground p_e^- (reflection factor $\underline{r}$ of the ground), is :

$$p_e = p_e^+ + \underline{r} \cdot p_e^- = P_e \cdot e^{jk_x x} \cdot \left(e^{jk_y y} + \underline{r}\, e^{-jk_y y}\right)$$
$$k_x = k_0 \cos\Theta_0 \quad ; \quad k_y = k_0 \sin\Theta_0 \tag{1}$$

The sound fields in front of (side of incidence) and behind the dam are :

$$p_{front}(x>0,y) = p_e(x,y) + \frac{1}{2}\left[p_s^{(h)}(x,y) + p_s^{(w)}(x,y)\right]$$
$$p_{back}(x<0,y) = p_e(x,y) + \frac{1}{2}\left[p_s^{(h)}(-x,y) - p_s^{(w)}(-x,y)\right] \tag{2}$$

with $p_s^{(ß)}(x,y)$ the scattered fields for the sub-task (ß)=(h) with hard plane x=0 and the sub-task (ß)=(w) with soft plane x=0 , respectively. The field in $x \geq 0$ in the two sub-tasks is made up as $p_e + p_r^{(ß)} + p_s^{(ß)}$, where $p_r^{(h)}$ is the exciting wave after hard reflection at the plane x=0 , and $p_r^{(w)}$ is the exciting wave after soft reflection at the plane x=0 . The sums $p_e + p_r^{(ß)}$ are :

$$p_e + p_r^{(h)} = 4P_e(1+\underline{r})\sum_{r\geq 0}(-j)^{2r} ce_{2r}(\alpha^+)ce_{2r}(\vartheta)Jc_{2r}(\rho) + 4P_e(1-\underline{r})\sum_{r\geq 1}(-j)^{2r} se_{2r}(\alpha^+)se_{2r}(\vartheta)Js_{2r}(\rho) \tag{3}$$

$$p_e + p_r^{(w)} = 4P_e(1+\underline{r})\sum_{r\geq 0}(-j)^{2r+1} ce_{2r+1}(\alpha^+)ce_{2r+1}(\vartheta)Jc_{2r+1}(\rho) + 4P_e(1-\underline{r})\sum_{r\geq 0}(-j)^{2r+1} se_{2r+1}(\alpha^+)se_{2r+1}(\vartheta)Js_{2r+1}(\rho) \tag{4}$$

The formulations for the scattered fields are, with still unknown term amplitudes $C_r^{(ß)}, S_r^{(ß)}$:

$$p_s^{(h)} = P_e\left[\sum_{r\geq 0} C_r^{(h)} ce_{2r}(\vartheta)Hc_{2r}^{(2)}(\rho) + \sum_{r\geq 1} S_r^{(h)} se_{2r}(\vartheta)Hs_{2r}^{(2)}(\rho)\right] \tag{5}$$

$$p_s^{(w)} = P_e\left[\sum_{r\geq 0} C_r^{(w)} ce_{2r+1}(\vartheta)Hc_{2r+1}^{(2)}(\rho) + \sum_{r\geq 0} S_r^{(w)} se_{2r+1}(\vartheta)Hs_{2r+1}^{(2)}(\rho)\right] \tag{6}$$

The boundary condition $-Z_0\left(v_{e\rho} + v_{r\rho}^{(ß)} + v_{s\rho}^{(ß)}\right)_{\rho=\rho_c} \overset{!}{=} G\cdot\left(p_e + p_r^{(ß)} + p_s^{(ß)}\right)_{\rho=\rho_c}$ at the dam surface gives for (ß)=(h) the two systems of equations ($\delta_{r,s}$= Kronecker symbol) :

$$\sum_{r\geq 0} C_r^{(h)}\left[\delta_{r,s}\frac{j\pi}{2k_0c}Hc_{2r}'^{(2)}(\rho_c) + G\cdot Hc_{2r}^{(2)}(\rho_c)\cdot Ic_{2r,2s}\right] = -4(1+\underline{r})\sum_{r\geq 0}(-j)^{2r} ce_{2r}(\alpha^+)\left[\delta_{r,s}\frac{j\pi}{2k_0c}Jc_{2r}'(\rho_c) + G\cdot Jc_{2r}(\rho_c)\cdot Ic_{2r,2s}\right] \quad ; \; s\geq 0 \tag{7}$$

$$\sum_{r\geq 1} S_r^{(h)}\left[\delta_{r,s}\frac{j\pi}{2k_0c}Hs_{2r}'^{(2)}(\rho_c) + G\cdot Hs_{2r}^{(2)}(\rho_c)\cdot Is_{2r,2s}\right] = -4(1-\underline{r})\sum_{r\geq 1}(-j)^{2r} se_{2r}(\alpha^+)\left[\delta_{r,s}\frac{j\pi}{2k_0c}Js_{2r}'(\rho_c) + G\cdot Js_{2r}(\rho_c)\cdot Ic_{2r,2s}\right] \quad ; \; s\geq 1 \tag{8}$$

and for (ß)=(w) two more systems of equations :

$$\sum_{r\geq 0} C_r^{(w)}\left[\delta_{r,s}\frac{j\pi}{2k_0c}Hc_{2r+1}'^{(2)}(\rho_c) + G\cdot Hc_{2r+}^{(2)}(\rho_c)\cdot Ic_{2r+1,2s+1}\right] = \quad ; \; s\geq 0$$
$$= -4(1+\underline{r})\sum_{r\geq 0}(-j)^{2r+1} ce_{2r+1}(\alpha^+)\left[\delta_{r,s}\frac{j\pi}{2k_0c}Jc_{2r+1}'(\rho_c) + G\cdot Jc_{2r+1}(\rho_c)\cdot Ic_{2r+1,2s+1}\right] \tag{9}$$

$$\sum_{r\geq 0} S_r^{(w)}\left[\delta_{r,s}\frac{j\pi}{2k_0c}Hs'^{(2)}_{2r+1}(\rho_c)+G\cdot Hs^{(2)}_{2r+1}(\rho_c)\cdot Is_{2r+1,2s+1}\right]= \quad ; \; s\geq 0$$
$$=-4(1-\underline{r})\sum_{r\geq 0}(-j)^{2r+1}se_{2r+1}(\alpha^+)\left[\delta_{r,s}\frac{j\pi}{2k_0c}Js'_{2r+1}(\rho_c)+G\cdot Js_{2r+1}(\rho_c)\cdot Ic_{2r+1,2s+1}\right] \tag{10}$$

The integrals $Ic_{m,n}$, $Is_{m,n}$ are described in section E.8 . The sound field is known after solution for the $C_r^{(\beta)}$, $S_r^{(\beta)}$.

E.10 Scattering at a semicircular absorbing dam on absorbing ground

Ref.: Mechel, [E.6]

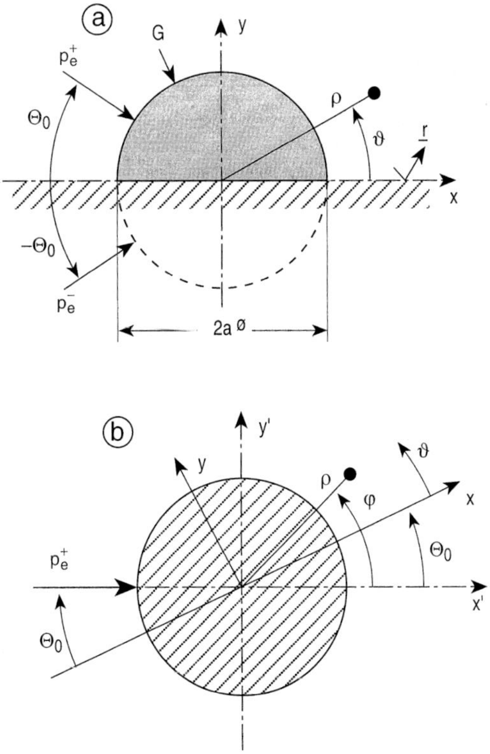

A semicircular, locally reacting dam, with radius a and normalised surface admittance G sits on an absorbent ground plane with reflection factor $\underline{r}$.

A plane or cylindrical wave p_e^+ is incident at an angle Θ_0 with the ground plane. The ground plane produces the mirror-reflected wave p_e^- . A field point is at the cylindrical co-ordinates (ρ,ϑ).

The second diagram (b) shows the co-ordinates as they are generally used in scattering problems at cylinders, such as in sections E.1 , E.2 .

The exciting wave is

$$p_e = p_e^+ + \underline{r}\cdot p_e^- \tag{1}$$

Incident plane wave :

The exciting wave expanded in Bessel functions :

$$p_e = P_e \cdot e^{-jk_x x}\left(e^{+jk_y y} + \underline{r} \cdot e^{-jk_y y}\right)$$
$$= P_e \sum_{m\geq 0} \delta_m (-j)^m \left[\cos\left(m(\vartheta+\Theta_0)\right) + \underline{r}\cdot\cos\left(m(\vartheta-\Theta_0)\right)\right]\cdot J_m(k_0\rho) \qquad (2)$$

with :

$$\delta_m = \begin{cases} 1 \; ; \; m=0 \\ 2 \; ; \; m\neq 0 \end{cases} \quad ; \quad k_x = k_0 \cos\Theta_0 \quad ; \quad k_y = k_0 \sin\Theta_0 \qquad (3)$$

Formulation of the scattered field with yet undetermined term amplitudes D_m :

$$p_s = P_e \sum_{m\geq 0} \delta_m (-j)^m \cdot D_m \cdot \left[\cos\left(m(\vartheta+\Theta_0)\right) + \underline{r}\cdot\cos\left(m(\vartheta-\Theta_0)\right)\right]\cdot H_m^{(2)}(k_0\rho) \qquad (4)$$

The boundary condition at the cylinder gives for the term amplitudes :

$$D_m = -\frac{J'_m(k_0h) - jZ_0G\cdot J_m(k_0h)}{H'^{(2)}_m(k_0h) - jZ_0G\cdot H^{(2)}_m(k_0h)} = -\frac{\left(m/k_0h - jZ_0G\right)\cdot J_m(k_0h) - J_{m+1}(k_0h)}{\left(m/k_0h - jZ_0G\right)\cdot H^{(2)}_m(k_0h) - H^{(2)}_{m+1}(k_0h)} \qquad (5)$$

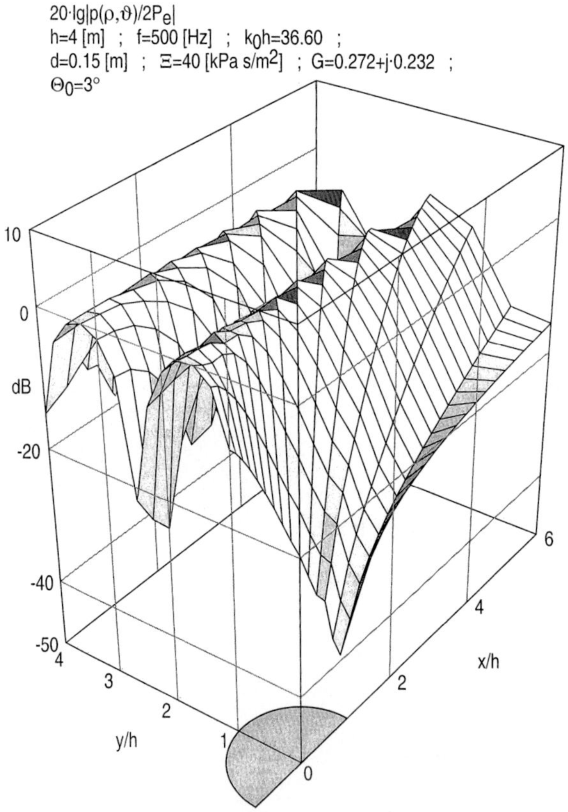

A semicircular absorbent dam on a hard ground plane. The absorption of the dam corresponds to that of a d=0.15 [m] thick glass fibre layer with a flow resistivity of Ξ=40 [kPa·s/m^2]. A plane wave is incident under an angle of elevation of Θ_0=3° . The diagram shows the sound pressure level in the shadow area.

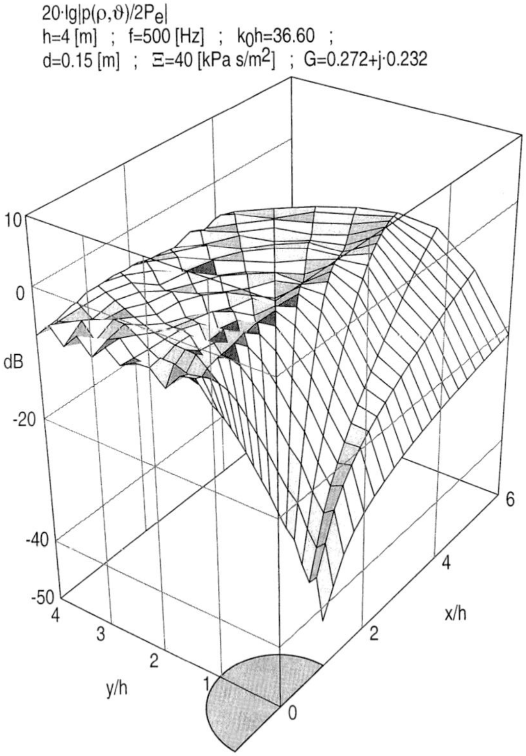

As above, but with a fully absorbent ground plane ($\underline{r}$=0) .

Incident cylindrical wave :

Let the line source Q of the cylindrical wave be at a distance ρ_0 from the dam axis under an elevation angle Θ_0 with the ground plane. The ground plane with the reflection factor $\underline{r}$ produces the mirror-reflected wave p_e^- to the original incident wave p_e^+ .

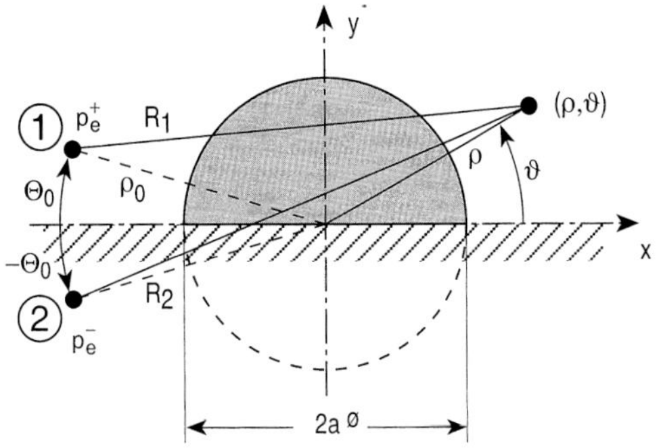

The exciting wave is in the radial range $\rho<\rho_0$:

$$p_e = P_e\left[H_0^{(2)}(k_0R_1)+\underline{r}\cdot H_0^{(2)}(k_0R_2)\right]$$
$$= P_e \sum_{m\geq 0}\delta_m(-1)^m \cdot H_m^{(2)}(k_0\rho_0)\cdot J_m(k_0\rho)\cdot\left[\cos\left(m(\vartheta+\Theta_0)\right)+\underline{r}\cdot\cos\left(m(\vartheta-\Theta_0)\right)\right] \quad (6)$$

The total field is $p = p_e + p_s$ with the scattered field :

$$p_s = P_e \sum_{m\geq 0} \delta_m (-1)^m \cdot D_m \cdot H_m^{(2)}(k_0\rho_0) \cdot H_m^{(2)}(k_0\rho) \cdot \left[\cos\left(m(\vartheta+\Theta_0)\right) + \underline{r} \cdot \cos\left(m(\vartheta-\Theta_0)\right)\right] \quad (7)$$

The term amplitudes D_m follow from the boundary condition at the dam surface $\rho=a$ as :

$$D_m = -\frac{J'_m(k_0a) - jG \cdot J_m(k_0a)}{H'^{(2)}_m(k_0a) - jG \cdot H_m^{(2)}(k_0a)} = -\frac{\left(m/k_0a - jG\right) \cdot J_m(k_0a) - J_{m+1}(k_0a)}{\left(m/k_0a - jG\right) \cdot H_m^{(2)}(k_0a) - H_{m+1}^{(2)}(k_0a)} \quad (8)$$

The component form $p_e = P_e\left[H_0^{(2)}(k_0R_1) + \underline{r} \cdot H_0^{(2)}(k_0R_2)\right]$ is valid for all $\rho \geq a$, like p_s.

The radii R_1, R_2 are given by :

$$\begin{aligned} k_0R_1 &= k_0\sqrt{(\rho\cos\vartheta + \rho_0\cos\Theta_0)^2 + (\rho\sin\vartheta - \rho_0\sin\Theta_0)^2} \\ k_0R_2 &= k_0\sqrt{(\rho\cos\vartheta + \rho_0\cos\Theta_0)^2 + (\rho\sin\vartheta + \rho_0\sin\Theta_0)^2} \end{aligned} \quad (9)$$

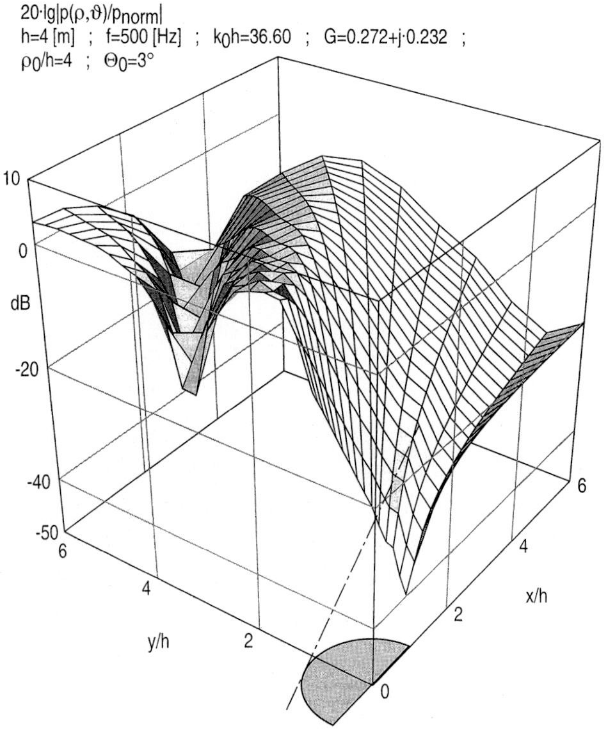

Sound pressure level in the shadow zone of a semi-circular absorbing dam on a hard ground plane for a cylindrical incident wave from the source position $\rho_0/h=4$; $\Theta_0=3°$.

E.11 Scattering in random media, general

Ref.: Mechel, [E.2]

This section brings general distinctions and concepts for the scattering of sound in random media. The composite medium consists of a fluid with randomly distributed scatterers.

The fluid
- may have no losses,
- may have viscous and thermal losses;

the scatterers
- may be different with respect to their shape, e.g. below:
  - spheres,
  - cylinders,
- may be different with respect to their consistence:
  - rigid or soft,
  - fluid, with or without losses,
  - elastic,
- may be different with respect to their dynamical behaviour:
  - not moving (though oscillating at their surface),
  - moving as a total under the influence of acoustical forces;

the composite medium
- may be disperse, i.e. multiple scattering negligible,
- may be dense, i.e. multiple scattering not negligible;

the scattering
- may retain the wave type (monotype scattering),
- may change the wave type into the triple of density, thermal, viscous waves (triple type scattering).

The Table 1 on the next page gives a survey of some of the different scattering processes. The upper rows belong to the monotype scattering: both the exciting wave Φ_e and the scattered wave Φ_s are of the same type. The lower rows describe the triple type scattering: the exciting wave Φ_e generates the triple of density (ρ), thermal (α), and viscous (ν) waves as scattered waves. The propagation of sound through the composite medium is composed of elementary scattering processes at single scatterers, which is indicated in the 1st column. In disperse media (2nd column) the multiple scattering (scattering of scattered fields) is neglected, i.e. the scattered waves propagate freely through the medium. In dense media (3rd column) multiple scattering must be taken into account. The exciting wave then is an "effective" wave Φ_E.

The scatterers will have a uniform random distribution in the composite medium. If there are inhomogeneities, e.g. holes or clusters, they are supposed to be randomly distributed as well and to form a sub-system of scatterers.

		1	2	3
	Type of Scattering	**Single Scatterer**	**Disperse Medium**	**Dense Medium**
	Monotype Scattering	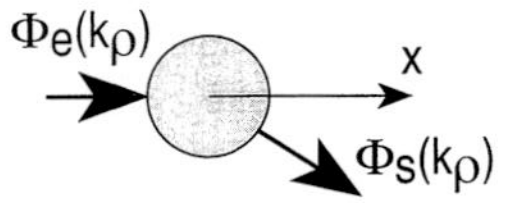	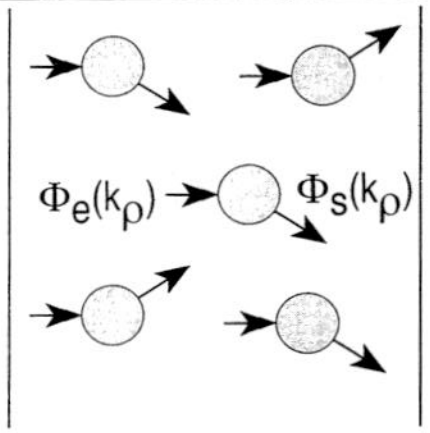	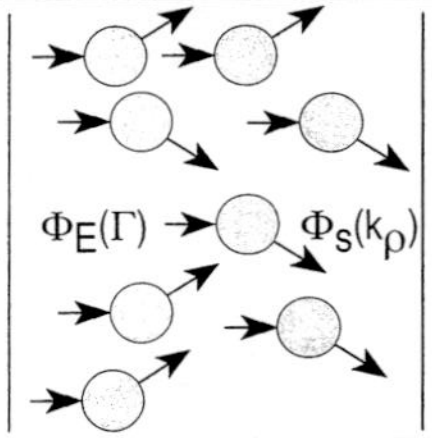
1	**Exciting Wave**	$\Phi_e(k_\rho x) = e^{-jk_\rho x}$	$\Phi_e(k_\rho) = e^{-jk_\rho x} + \sum_{j\neq i} \Phi_s(k_\rho r_j)$	$\Phi_E = e^{-\Gamma x} + \sum_{j\neq i} \Phi_s(k_\rho r_j)$
	Scattered Wave	$\Phi_s(k_\rho r)$	$\Phi_s(k_\rho r)$	$\Phi_s(k_\rho r)$
	Triple type Scattering	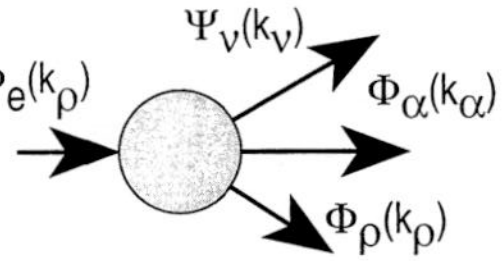	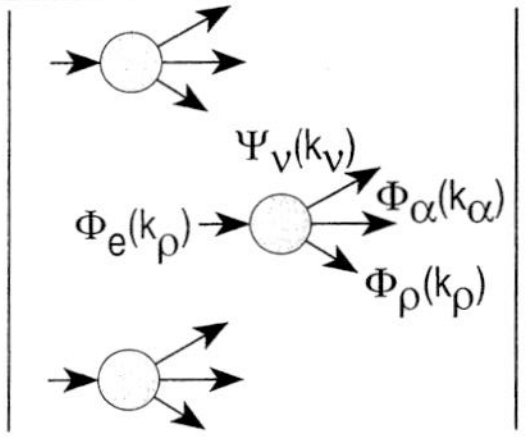	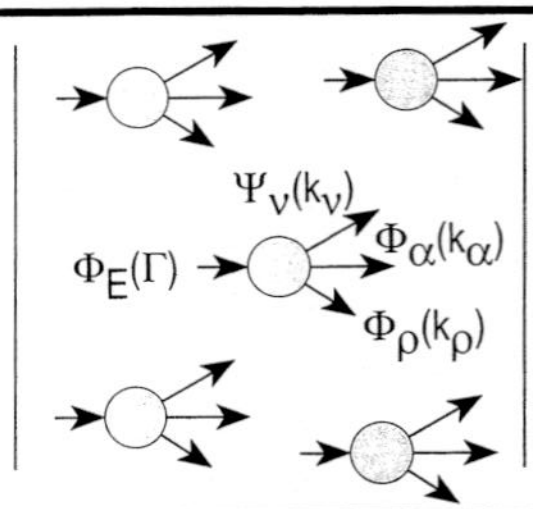
2	**Exciting Wave**	$\Phi_e(k_\rho x) = e^{-jk_\rho x}$	$\Phi_e(k_\rho x) = e^{-jk_\rho x} + \sum_{\beta, j\neq i} \Phi_s(k_\beta r_j)$	$\Phi_E = e^{-\Gamma x} + \sum_{\beta, j\neq i} \Phi_s(k_\beta r_j)$
	Scattered Wave	$\Phi_s(k_\beta r) = \{\Phi_\rho(k_\rho r), \Phi_\alpha(k_\alpha r), \Psi_\nu(k_\nu r)\}$	$\Phi_s(k_\beta r) = \{\Phi_\rho(k_\rho r), \Phi_\alpha(k_\alpha r), \Psi_\nu(k_\nu r)\}$	$\Phi_s(k_\beta r) = \{\Phi_\rho(k_\rho r), \Phi_\alpha(k_\alpha r), \Psi_\nu(k_\nu r)\}$

Table 1: Survey of scattering in random media.

REICHE's experiment:

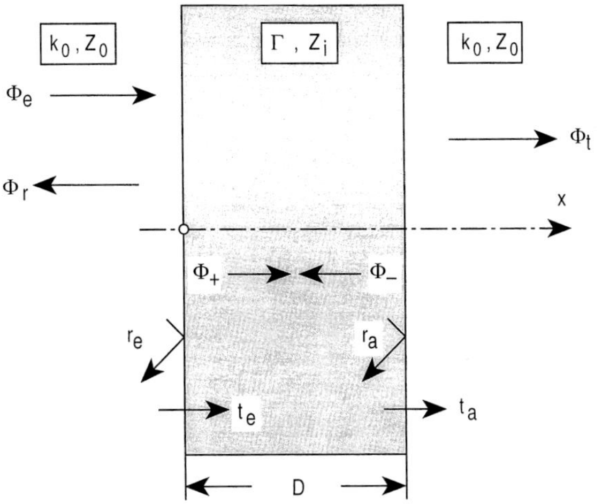

A simple experiment is the background of many theories for the determination of the characteristic propagation constant Γ and wave impedance Z_i inside a composite medium with random scatterers :

A plane wave Φ_e is incident on a layer of thickness D of the investigated material. A receiver on the front side "collects" the backscattered wave components from inside the layer as reflected wave Φ_r ;
a receiver on the back side "collects" the forward scattered wave components from inside the layer as transmitted wave Φ_t . The characteristic values of the material are determined from the reflection and transmission factors $\underline{r}_e, t_e$ on the front side, and $\underline{r}_a, t_a$ on the back side (reflection factors are underlined for distinction with the later used symbol r for radius and/or general co-ordinate). The forward and backward waves inside the layer are Φ_+, Φ_- , respectively. The fluid inside the layer (between the scatterers) equals the fluid outside the layer.

Monotype scattering :

$$\Phi_e(x) = e^{-jk_0 x} \quad ; \quad \Phi_r(x) = \underline{r}_e \cdot e^{+jk_0 x} \quad ; \quad \Phi_t(x) = t \cdot e^{-jk_0(x-D)}$$

$$\begin{aligned} &\underline{r}_e = \Phi_r(0)/\Phi_e(0) \quad ; \quad \underline{r}_a = \Phi_-(D)/\Phi_+(D) \\ &t_e = \Phi_+(0)/\Phi_e(0) \quad ; \quad t_a = \Phi_t(D)/\Phi_+(D) \\ &t = \Phi_t(D)/\Phi_e(0) \end{aligned} \qquad (1)$$

For transmission with $y = \Gamma D$:

$$\begin{aligned} &t_a = 1 + \underline{r}_a = \frac{2}{1 + Z_i/Z_0} \quad ; \quad t_e = \frac{1 - \underline{r}_a}{1 - \underline{r}_a^2 \cdot e^{-2y}} \\ &t = \frac{1 - \underline{r}_a^2}{1 - \underline{r}_a^2 \cdot e^{-2y}} \cdot e^{-y} \quad ; \quad t = t_e\, t_a\, e^{-y} \end{aligned} \qquad (2)$$

For reflection with $y = \Gamma D$:

$$\underline{r}_a = \frac{1 - Z_i/Z_0}{1 + Z_i/Z_0} \quad ; \quad \underline{r}_e = -\underline{r}_a \cdot \frac{1 - e^{-2y}}{1 - \underline{r}_a^2 \cdot e^{-2y}} \qquad (3)$$

Field inside the layer :

$$\Phi_I(x) = \Phi_+(x) + \Phi_-(x) = t_e \cdot \left(e^{-\Gamma x} + \underline{r}_a \cdot e^{+\Gamma(x-2D)}\right) \tag{4}$$

Field at a field point $\mathbf{r}$ (which may be well outside the layer) :

$$\Phi(\mathbf{r}) = \Phi_e(\mathbf{r}) + U(\mathbf{r}) \quad ; \quad U(\mathbf{r}) = \sum_s u_s(\mathbf{r} - \mathbf{r}_s) \tag{5}$$

where u_s is the scattered field of a single scatterer with running index s having its position at $\mathbf{r}_s$. This often used elementary decomposition implicitly supposes, that the exciting wave can reach the scatterers (even deeply inside the layer) without attenuation; this form therefore is restricted to disperse media. The single scatterer functions u_s are sums over Hankel functions of the 2nd kind $H_n^{(2)}(k_0 r)$ for cylindrical scatterers, or spherical Hankel functions of the 2nd kind $h_n^{(2)}(k_0 r) = \sqrt{\pi/(2k_0 r)}\, H_{n+1/2}^{(2)}(k_0 r)$ for spherical scatterers.

It is a further assumption in REICHE's experiment to choose receiving points for the transmitted and reflected fields at large distances, so that $k_0|x| >> 1$. Then the scattered far field can be written as a product of a factor $\Re(k_0 r)$ with only a radial variation, and an angular factor $g(\mathbf{o},\mathbf{i})$ which contains the angle between the direction $\mathbf{o}$ to the field point with the direction $\mathbf{i}$ of incidence on the scatterer. Replacing the summation over the index s of the scatterers by an integration, and taking into account the symmetry of the problem around the axis of propagation, one can write for the scattered fields outside the layer (at large distances of it) :

$$\begin{aligned} U_>(x) &= \Phi_e(k_0 x) \cdot C \int_0^D e^{jk_0\xi} \cdot G(\xi,\mathbf{i}) \cdot d\xi \quad ; \quad x > D \\ U_<(x) &= \Phi_e(-k_0 x) \cdot C \int_0^D e^{-jk_0\xi} \cdot G(\xi,-\mathbf{i}) \cdot d\xi \quad ; \quad x < 0 \end{aligned} \tag{6}$$

and inside the layer:

$$\Phi(x) = \Phi_e(k_0 x) \cdot \left[1 + C \int_0^x e^{jk_0\xi} \cdot G(\xi,\mathbf{i}) \cdot d\xi\right] + \Phi_e(-k_0 x) \cdot C \int_x^D e^{-jk_0\xi} \cdot G(\xi,-\mathbf{i}) \cdot d\xi \tag{7}$$

where C is a constant factor (depending on the far field decomposition) which will be given below, and

$$\begin{aligned} G(\xi,\mathbf{o}) &= g(\mathbf{o},\mathbf{i}) \cdot \Phi_+(0,\xi) + g(\mathbf{o},-\mathbf{i}) \cdot \Phi_-(\xi,D) \\ G(\xi,\pm\mathbf{i}) &= g(\pm\mathbf{i},\mathbf{i}) \cdot \Phi_+(0,\xi) + g(\pm\mathbf{i},-\mathbf{i}) \cdot \Phi_-(\xi,D) \end{aligned} \tag{8}$$

($\pm\mathbf{i}$ represent forward or backward direction, in or against the x direction) with :

$$\Phi(x) = \Phi_+(0,x) + \Phi_-(x,D)$$

$$\Phi_+(0,x) = e^{-jk_0 x} \cdot \left[1 + C \int_0^x e^{jk_0 \xi} \cdot G(\xi, \mathbf{i}) \cdot d\xi \right] \tag{9}$$

$$\Phi_-(x,D) = e^{+jk_0 x} \cdot C \int_x^D e^{-jk_0 \xi} \cdot G(\xi, -\mathbf{i}) \cdot d\xi$$

Introduce four symbols corresponding to the four possible combinations of forward and backward directions (and include, for ease, the constant C):

$$\begin{aligned} S_+ &= C\, g(\mathbf{i},\mathbf{i}) \quad ; \quad S_- = C\, g(-\mathbf{i},-\mathbf{i}) \\ R_+ &= C\, g(\mathbf{i},-\mathbf{i}) \quad ; \quad R_- = C\, g(-\mathbf{i},\mathbf{i}) \end{aligned} \tag{10}$$

assume further, that the scatterers have a forward-backward symmetry (as spheres and cylinders), then

$$\begin{aligned} g(-\mathbf{i},-\mathbf{i}) &= g(\mathbf{i},\mathbf{i}) \quad ; \quad g(-\mathbf{i},\mathbf{i}) = g(\mathbf{i},-\mathbf{i}) \\ S_+ &= S_- = S \quad ; \quad R_+ = R_- = R \end{aligned} \tag{11}$$

and two coupled integral equations are obtained:

$$\begin{aligned} \Phi_+(0,x) &= e^{-jk_0 x} \cdot \left[1 + \int_0^x e^{+jk_0 \xi} [S_+ \Phi_+(0,\xi) + R_+ \Phi_-(\xi,D)] \cdot d\xi \right] \\ \Phi_-(x,D) &= e^{+jk_0 x} \cdot \int_x^D e^{-jk_0 \xi} [R_- \Phi_+(0,\xi) + S_- \Phi_-(\xi,D)] \cdot d\xi \end{aligned} \tag{12}$$

The essential trick of REICHE's experiment is to use these intgral equations for obtaining the boundary conditions at the layer surfaces as well as a "wave equation". Differentiation of $\Phi_\pm$ with respect to x gives, for constant $S_\pm, R_\pm$:

$$\begin{aligned} \Phi'_\pm &= \mp jk_0\, \Phi_\pm \pm (S_\pm\, \Phi_\pm + R_\pm\, \Phi_\mp) = \pm T_\pm\, \Phi_\pm \pm R_\pm\, \Phi_\mp \\ \Phi''_\pm &+ (T_- - T_+)\, \Phi'_\pm + (R_+ R_- - T_+ T_-)\, \Phi_\pm = 0 \end{aligned} \tag{13}$$

with $T_\pm = -jk_0 + S_\pm$. The differential equation in the 2nd line is interpreted as a wave equation; then the parentheses of the last term contain the square of the effective wave number k_E with $\Gamma = jk_E$:

$$\begin{aligned} k_E^2 &= R_+ R_- - T_+ T_- \xrightarrow[\substack{R_+=R_-=R \\ S_+=S_-=S \\ T_+=T_-=T}]{} R^2 - T^2 = k_0^2 + R^2 - S^2 + 2jk_0 S \\ k_E^2 / k_0^2 &= 1 + \frac{R^2 - S^2}{k_0^2} + \frac{2j}{k_0} S \end{aligned} \tag{14}$$

(the last line contains the square of the index of refraction).

If the assumption of symmetrical scatterers cannot be made, formulate a general solution for $\Phi_\pm$:

$$\Phi_\pm = A_\pm\, e^{-jk_{E+}x} + B_\pm\, e^{+jk_{E-}x} \tag{15}$$

insert this in the derivatives above to obtain two homogeneous linear systems for the amplitudes $A_\pm, B_\pm$:

$$\begin{pmatrix} (T_+ + jk_{E+}) & R_+ \\ R_- & (T_- - jk_{E+}) \end{pmatrix} \cdot \begin{pmatrix} A_+ \\ A_- \end{pmatrix} = \begin{pmatrix} 0 \\ 0 \end{pmatrix} = \mathbf{M}(k_{E+}) \bullet \mathbf{A}$$

$$\begin{pmatrix} (T_+ - jk_{E-}) & R_+ \\ R_- & (T_- + jk_{E-}) \end{pmatrix} \cdot \begin{pmatrix} B_+ \\ B_- \end{pmatrix} = \begin{pmatrix} 0 \\ 0 \end{pmatrix} = \mathbf{M}(k_{E-}) \bullet \mathbf{B} \tag{16}$$

The condition of zero determinants gives the characteristic equations with solutions :

$$\begin{aligned} k_{E\pm} &= k_{E0} \pm \Delta \\ k_{E0} &= j\sqrt{\tfrac{1}{4}(T_+ + T_-)^2 - R_+R_-} \quad ; \quad \Delta = \tfrac{j}{2}(T_+ - T_-) \end{aligned} \tag{17}$$

The wave impedance is determined from the boundary conditions :

$$\begin{aligned} \Phi_+(0,0) &= A_+ + B_+ = 1 \\ \Phi_-(D,D) &= A_- \cdot e^{-jk_{E+}D} + B_- \cdot e^{+jk_{E-}D} = 0 \end{aligned} \tag{18}$$

which, together with the above systems of equations, give an inhomogeneous system of equations, the solutions of which are, with $k_{E+}+k_{E-}= 2k_{E0}$:

$$\begin{aligned} A_+ &= [1 - QQ'\, e^{-2jk_{E0}D}]^{-1} =: F \quad ; \quad A_- = -Q\, A_+ \\ B_- &= Q\, e^{-2jk_{E0}D} \cdot F \quad ; \quad B_+ = -Q'\, B_- \end{aligned} \tag{19}$$

with the abbreviations (besides of F) :

$$Q = R_-/(T_- - jk_{E+}) \quad ; \quad Q' = R_+/(T_+ - jk_{E-}) \tag{20}$$

The sound field inside the layer so becomes :

$$\Phi_I(x) = e^{-j\Delta\cdot x} \cdot \frac{1-Q}{1-QQ'\, e^{-2jk_{E0}D}} \cdot \left(e^{-jk_{E0}x} - Q\frac{1-Q'}{1-Q}\, e^{+jk_{E0}(x-2D)}\right) \tag{21}$$

In the special case of symmetrical scatterers :

$$S_+ = S_- = S \;;\; R_+ = R_- = R \;;\; T_+ = T_- = T$$
$$\Delta = 0 \;;\; k_{E+} = k_{E-} = k_{E0} = k_E \;;\; Q' = Q \;;\; F = \frac{1}{1 - Q^2\, e^{-2jk_E D}} \tag{22}$$

and :

$$\Phi_I(x) = \frac{1-Q}{1-Q^2\, e^{-2jk_E D}} \cdot (e^{-jk_E x} - Q\, e^{+jk_E (x-2D)}) \tag{23}$$

This can be made completely analogous to the field in a homogeneous medium layer by the equivalencies $Q \leftrightarrow \underline{r}_a$; $jk_E \leftrightarrow \Gamma$, from which follows the wave impedance of the layer material :

$$\frac{Z_i}{Z_0} = \frac{1-Q}{1+Q} \tag{24}$$

The reflected field in front of the layer and the transmitted field behind the layer are :

$$\begin{aligned} \Phi_r(x) &= U_<(x) = -Q\,(1 - e^{-2jk_E D})\, F \cdot e^{jk_0 x} \quad ; \quad x \le 0 \\ \Phi_t(x) &= U_>(x) = (1 - QQ')\, e^{-2jk_E D}\, F \cdot e^{-jk_0 x} \quad ; \quad x \ge D \end{aligned} \tag{25}$$

So the characteristic values of the composite material and the sound fields are known with this method (and with its restrictive assumptions), if the scattered far field functions $g_> = g(\mathbf{i},\mathbf{i})$ and $g_< = g(-\mathbf{i},\mathbf{i})$ are known from single scatterer evaluations, which are the scattered far fields in forward and backward directions. With them the characteristic values can be given other forms, using :

$$R = C\, g_< \;;\; T = C\, g_> - jk_0 \;;\; Q = \frac{C\, g_<}{C\, g_> - j(k_0 + k_E)} \tag{26}$$

and :

$$\frac{k_E}{k_0} = \sqrt{\frac{\rho_{eff}}{\rho_0} \cdot \frac{C_{eff}}{C_0}} \;;\; \frac{Z_i}{Z_0} = \sqrt{\frac{\rho_{eff}}{\rho_0} \Big/ \frac{C_{eff}}{C_0}} \tag{27}$$

$$\frac{\rho_{eff}}{\rho_0} = 1 + j\frac{C}{k_0}(g_> - g_<) \;;\; \frac{C_{eff}}{C_0} = 1 + j\frac{C}{k_0}(g_> + g_<) \tag{28}$$

For low scatterer number density $N \to 0$ will be $|\, jC \cdot g / k_0 \,| << 1$ and therefore :

$$\frac{k_E^2}{k_0^2} \approx 1 + 2j\frac{C}{k_0} \cdot g_> \;;\; \frac{Z_i^2}{Z_0^2} \approx 1 + 2j\frac{C}{k_0} \cdot g_< \tag{29}$$

In this approximation the wave number only depends on the forward scattering, and the wave impedance only on the backward scattering. With the extinction cross-section Q_e of a single scatterer :

$$Q_e = -2C \cdot \mathrm{Re}\{g_>\} \quad ; \quad C = \begin{cases} 2/k_0 & \text{Cylinder} \\ 2\pi/k_0^2 & \text{Sphere} \end{cases} \tag{30}$$

one finds the plausible result for the attenuation (in the present approximation) :

$$-2\,\mathrm{Im}\{k_E\} \approx N \cdot Q_e \tag{31}$$

i.e. the attenuation is proportional to the number density of the scatterers and to their extinction cross section.

E.12 Function tables for monotype scattering

This section gives tables of functions for monotype scattering to be applied in the general scheme of the previous section E.11 , i.e. incident and scattered waves are of the same type, to say: density waves.

It is not necessary to use potential functions Φ for the field with monotype scattering; therefore the field function here will be the sound pressure p . Time factor is, as usual, $e^{j\omega t}$. The incident plane wave has a unit amplitude.

The 1st table compiles radial functions $R_n(r)$ and azimuthal functions $T_n(\vartheta)$ for cylindrical and spherical scatterers. $Z_n(z)$ stands for either a Bessel, Neumann, or Hankel function; $K_n(z)$ stands for either a spherical Bessel, Neumann, of Hankel function. Because Hankel functions only are of the 2nd kind, the upper index $^{(2)}$ will be dropped (for ease of writing). The 2nd table gives formulations of the incident plane wave and of the scattered wave for both geometries of the scatterer. The 3rd table collects modal amplitudes and mode admittances (normalised with Z_0) for both locally reacting and bulk reacting scatterers. Bulk reacting scatterers are supposed to be of an isotropic, homogeneous material having the characteristic propagation constant Γ_σ and wave impedance Z_σ ; thus this type of scatterers can represent also fluids with losses. Hard and soft cylinders and spheres are treated as special cases of locally reacting scatterers. a is the radius of the scatterer. N is the number density of scatterers (number of scatterers per unit volume); μ is the massivity, i.e. the ratio of space occupied by the scatterers. The argument in radial functions is dropped if it is k_0a .

Some of the contents of these tables may be found also in the sections E.1, E.2 . But the tables here are completed with terms required in the previous section E.11 .

	Quantities	Cylinder	Sphere
1	**Radial Functions** $R_n(z)$	$Z_n(z) = \{J_n(z), Y_n(z), H_n(z)\}$ $H_n(z) = J_n(z) - j\,Y_n(z)$	$K_n(z) = \{j_n(z), y_n(z), h_n(z)\}$ $K_n(z) = \sqrt{\frac{\pi}{2z}} \cdot Z_{n+1/2}(z)$ $h_n(z) = j_n(z) - j\,y_n(z)$
2	$R'_n(z)$	$Z'_n(z) = \frac{n}{z} Z_n(z) - Z_{n+1}(z)$ $= Z_{n-1}(z) - \frac{n}{z} Z_n(z)$	$K'_n(z) = \frac{n}{z} K_n(z) - K_{n+1}(z)$ $= K_{n-1}(z) - \frac{n}{z} K_n(z)$
3	$J_n(z) \xrightarrow[\lvert z\rvert \ll 1]{}$ $j_n(z) \xrightarrow[\lvert z\rvert \ll 1]{}$	$\frac{(z/2)^n}{n!}$	$\frac{z^n}{1 \cdot 2 \cdot 3 \cdot \ldots \cdot (2n+1)}$
4	$J_n(z) \xrightarrow[\lvert z\rvert \gg 1]{}$ $j_n(z) \xrightarrow[\lvert z\rvert \gg 1]{}$	$\sqrt{\frac{2}{\pi z}} \cdot \cos(z - n\frac{\pi}{2} - \frac{\pi}{4})$	$\frac{1}{z} \cdot \cos(z - n\frac{\pi}{2})$
5	$Y_n(z) \xrightarrow[\lvert z\rvert \ll 1]{}$ $y_n(z) \xrightarrow[\lvert z\rvert \ll 1]{}$	$Y_0(z) \to \frac{2}{\pi} \ln z$ $Y_n(z) \to -\frac{(n-1)!}{\pi\,(z/2)^n}$; $n > 0$	$-\frac{1 \cdot 3 \cdot 5 \cdot \ldots \cdot (2n-1)}{z^{n+1}}$
6	$Y_n(z) \xrightarrow[\lvert z\rvert \gg 1]{}$ $y_n(z) \xrightarrow[\lvert z\rvert \gg 1]{}$	$\sqrt{\frac{2}{\pi z}} \cdot \sin(z - n\frac{\pi}{2} - \frac{\pi}{4})$	$\frac{1}{z} \cdot \sin(z - n\frac{\pi}{2})$
7	$H_n(z) \xrightarrow[\lvert z\rvert \ll 1]{}$ $h_n(z) \xrightarrow[\lvert z\rvert \ll 1]{}$	$H_0(z) \to -\frac{2j}{\pi} \ln z$ $H_n(z) \to \frac{j(n-1)!}{\pi\,(z/2)^n}$; $n > 0$	$j\,\frac{1 \cdot 3 \cdot 5 \cdot \ldots \cdot (2n-1)}{z^{n+1}}$
8	$H_n(z) \xrightarrow[\lvert z\rvert \gg 1]{}$ $h_n(z) \xrightarrow[\lvert z\rvert \gg 1]{}$	$\sqrt{\frac{2}{\pi z}} \cdot e^{-j(z - n\pi/2 - \pi/4)}$ $= \frac{j^n}{\sqrt{\pi/2}} \cdot \frac{e^{-jz}}{\sqrt{-jz}}$	$\frac{j}{z} \cdot e^{-j(z - n\pi/2)}$ $= j^n \cdot \frac{e^{-jz}}{(-jz)}$
9	$J_{n+1}(z) Y_n(z) -$ $Y_{n+1}(z) J_n(z)$	$\frac{2}{\pi z}$	$\frac{1}{z^2}$
10	**Polar Functions** $T_n(\vartheta)$	$\cos(n\,\vartheta)$	$P_n(\cos\vartheta)$
11	$T_0(\vartheta)$	1	1
12	$T_n(0)$	1	1
13	$T_n(\pi)$	$(-1)^n$	$(-1)^n$

Table 1 : Radial and polar functions for cylindrical and spherical co-ordinates

	Quantity:	Cylinder	Sphere
1	**Incident Wave**:	$p_e(r,\vartheta) = e^{-jk_0 r\cdot\cos\vartheta}$	
	$p_e(r,\vartheta) =$	$\sum_{n=0}^{\infty} \delta_n \cdot (-j)^n \cdot T_n(\vartheta) \cdot R_{en}(r)$	
2	$\delta_n =$	$\begin{cases} 1 \; ; \; n=0 \\ 2 \; ; \; n>0 \end{cases}$	$2n+1$
3	$T_n(\vartheta) =$	$\cos(n\vartheta)$	$P_n(\cos\vartheta)$
4	$R_{en}(r) =$	$J_n(k_0 r)$	$j_n(k_0 r)$
5	**Scattered Wave**: $p_s(r,\vartheta) =$	$\sum_{n=0}^{\infty} \delta_n \cdot (-j)^n \cdot D_n \cdot T_n(\vartheta) \cdot R_{sn}(r)$	
6	$R_{sn}(r) =$	$H_n(k_0 r)$	$h_n(k_0 r)$
7	$p_s(r,\vartheta) \xrightarrow[k_0 r \gg 1]{}$	$\Re(r) \cdot \Theta(\vartheta)$	
8	$\Re(r) =$	$\sqrt{\frac{2j}{\pi}} \frac{e^{-jk_0 r}}{\sqrt{k_0 r}}$	$j \frac{e^{-jk_0 r}}{k_0 r}$
9	$\Theta(\vartheta) = g(\mathbf{o},\mathbf{i}) =$	$\sum_{n=0}^{\infty} \delta_n D_n T_n(\vartheta)$	$\sum_{n=0}^{\infty} \delta_n D_n T_n(\vartheta)$
10	$\Theta(0) = g(\mathbf{i},\mathbf{i}) =$	$\sum_{n=0}^{\infty} \delta_n D_n T_n(0) = \sum_{n=0}^{\infty} \delta_n D_n$	$\sum_{n=0}^{\infty} \delta_n D_n T_n(0) = \sum_{n=0}^{\infty} \delta_n D_n$
11	$\Theta(\pi) = g(-\mathbf{i},\mathbf{i}) =$	$\sum_{n=0}^{\infty} (-1)^n \delta_n D_n$	$\sum_{n=0}^{\infty} (-1)^n \delta_n D_n$
12	$C =$	$\frac{2N}{k_0}$	$\frac{2\pi N}{k_0^2}$
13	Massivity $\mu =$	$N \cdot \pi a^2$	$N \cdot \frac{4}{3}\pi a^3$
14	$\frac{C}{k_0} =$	$\frac{2\mu}{\pi (k_0 a)^2}$	$\frac{3\mu}{2(k_0 a)^3}$

Table 2 : Incident and scattered waves.

	Quantity:	Cylinder	Sphere
1	**locally reacting**: G =	$-Z_0 \cdot v_r(a)/p(a) = G = \text{fix}$	
2	$D_n =$	$-\dfrac{(\frac{n}{k_0 a} - jG)\, J_n - J_{n+1}}{(\frac{n}{k_0 a} - jG)\, H_n - H_{n+1}}$	$-\dfrac{(\frac{n}{k_0 a} - jG)\, j_n - j_{n+1}}{(\frac{n}{k_0 a} - jG)\, h_n - h_{n+1}}$
3	**bulk reacting**: $G_n =$	$\dfrac{j}{Z_\sigma/Z_0}\,\dfrac{J'_n(j\Gamma_\sigma a)}{J_n(j\Gamma_\sigma a)}$	$\dfrac{j}{Z_\sigma/Z_0}\,\dfrac{j'_n(j\Gamma_\sigma a)}{j_n(j\Gamma_\sigma a)}$
4	$D_n =$	$-\dfrac{(\frac{n}{k_0 a} - jG_n)\, J_n - J_{n+1}}{(\frac{n}{k_0 a} - jG_n)\, H_n - H_{n+1}}$	$-\dfrac{(\frac{n}{k_0 a} - jG_n)\, j_n - j_{n+1}}{(\frac{n}{k_0 a} - jG_n)\, h_n - h_{n+1}}$

Table 3 : Modal admittances G (normalised) and amplitudes D .

The hard, not movable scatterer is a special case G=0 of a locally reacting scatterer :

$$D_n = \begin{cases} -\dfrac{J'_n(k_0 a)}{H'_n(k_0 a)} & \text{Cylinder} \\ -\dfrac{j'_n(k_0 a)}{h'_n(k_0 a)} & \text{Sphere} \end{cases} \tag{1}$$

The soft, not movable scatterer is a special case |G|→∞ of a locally reacting scatterer :

$$D_n = \begin{cases} -\dfrac{J(k_0 a)}{H(k_0 a)} & \text{Cylinder} \\ -\dfrac{j(k_0 a)}{h(k_0 a)} & \text{Sphere} \end{cases} \tag{2}$$

The bulk reacting scatterers of a homogeneous material are movable under the force of the sound field by formulation; the movement mainly is described by the mode n=1 . A hard, freely movable scatterer can be obtained from a bulk reacting scatterer by setting the compressibility of its material to $C_\sigma=0$ but with a finite density ρ_σ . This produces $\Gamma_\sigma \to 0$ and $Z_\sigma \to \infty$ so that a finite $Z_\sigma \cdot \Gamma_\sigma \to j\omega\rho_\sigma$ is obtained. The modal admittances for cylinders can be written (similarly for spheres) :

$$G_n = \frac{n}{jk_0 a}\,\frac{\rho_0}{\rho_\sigma} - \frac{j}{Z_\sigma/Z_0}\,\frac{J_{n+1}(j\Gamma_\sigma a)}{J_n(j\Gamma_\sigma a)} \tag{3}$$

The ratio of the Bessel functions is $j\Gamma_\sigma a/(2n+2)$ for cylinders and $j\Gamma_\sigma a/(2n+3)$ for spheres. So for $\Gamma_\sigma \to 0$ the 2nd term can be neglected, and in the 1st term the replacement $n \to n+1/2$ takes place for spheres.

E.13 Sound attenuation in a forest

The previous section E.12 is used for the evaluation of the attenuation of sound in a forest.

First, the forest is modelled as a random arrangement of trunks with average radius a. The escape of sound in the upward direction is neglected. The propagation constant $\Gamma = jk_E$ of the sound wave in the model forest is evaluated from :

$$\frac{\Gamma}{k_0} = j\sqrt{\frac{\rho_{eff}}{\rho_0} \cdot \frac{C_{eff}}{C_0}} \tag{1}$$

with :

$$\begin{aligned} \frac{\rho_{eff}}{\rho_0} &= 1 + j\frac{8\mu}{\pi(k_0a)^2} \sum_{n=1,3,5\ldots} D_n \\ \frac{C_{eff}}{C_0} &= 1 + j\frac{8\mu}{\pi(k_0a)^2} \cdot [0.5\, D_0 + \sum_{n=2,4,6\ldots} D_n] \end{aligned} \tag{2}$$

where μ is the massivity of the trunk arrangement, and D_n are the scattered mode amplitudes for hard cylinders from the previous section. The attenuation as level decrease per free field wave length λ_0 is :

$$(-\Delta L)_{\lambda_0} = 8.686 \cdot \mathrm{Re}\{\Gamma/k_0\} = \frac{8.686}{2\pi} \mathrm{Re}\{\Gamma\}\, \lambda_0 \quad [\mathrm{dB}] \tag{3}$$

$\mathrm{Re}\{\Gamma/k_0\}$ is plotted below over k_0a for $\mu = (2a/2R)^2$, where $2R$ is the average mutual distance of the trunks; up to $n_{max}=8$ scattered modes were used.

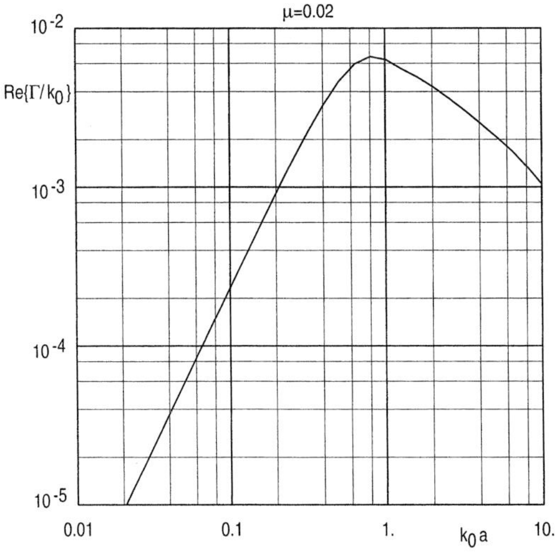

Attenuation constant in a model forest of hard cylindrical trunks with average radius a forming a massivity μ .

More instructive is the level decrease ΔL of sound penetrating a strip of forest of width D , with the front side and back side reflections at the forest borders included :

$$(-\Delta L)_D = -20 \cdot \log|t| = -20 \cdot (\log|t_e| + \log|t_a|) + 8.68\,\mathrm{Re}\{\Gamma\} D \quad [\mathrm{dB}] \tag{4}$$

This transmission loss $-\Delta L$ is plotted below for D=100 [m] and some trunk diameters.

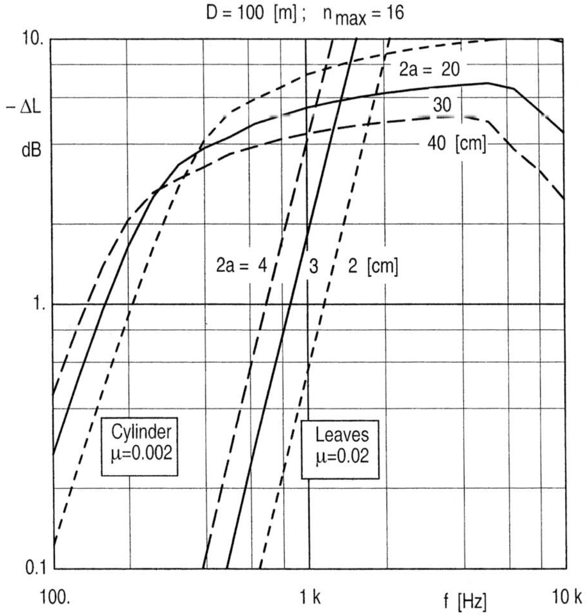

Transmission loss through a model forest, D=100 [m] wide.
The left-hand curves only consider cylindrical, hard trunks, the right-hand curves only consider the leaves, which are modelled as scattering spheres.

Leaves with an assumedly circular shape and average leave radius a_0 will have an effective leaf radius $a = a_0/3^{1/3} = 0.693 \cdot a_0$ normal to the direction of sound, because of their random orientation. The leaves are modelled with hard spheres of this effective radius. In C_{eff}/C_0 of the previous section E.12 the mode n=0 with mode amplitude D_0 dominates for small k_0a. The static compressibility of the composite model medium is reduced by $1 \to 1-\mu$; therefore one corrects $D_0 \to D_0/(1-\mu)$. In ρ_{eff}/ρ_0 the mode n=1 with the mode amplitude D_1 dominates. ρ_{eff} is corrected with the ratio of the oscillating mass of a free disk to that of a sphere. Thus :

$$D_0 \to D_0 \cdot \frac{1}{1-\mu} \quad ; \quad D_1 \to D_1 \cdot \frac{1+\frac{2}{\pi}\mu}{1+\frac{3}{2}\mu} \tag{5}$$

Higher order mode amplitudes D_n remain unchanged.

The curves in the above diagram show that the transmission loss for a parameter set a, μ, D can be evaluated from the transmission loss for a parameter set a_0,μ_0,D_0 by the transformation:

$$\Delta L(f, a, \mu, D) = \frac{a_0 \mu D}{a \mu_0 D_0} \cdot \Delta L(\frac{a}{a_0} f, a_0, \mu_0, D_0) \tag{6}$$

E.14 Mixed monotype scattering in random media

The fundamentals of this section are displayed in section E.11 . The difference of this section with respect to sections E.11 - E.13 , all dealing with monotype scattering, lies in the fact, that there not only is the exciting wave of the same type as the scattered wave (density wave), but also their free field wave numbers are supposed to agree with each other. For not too low scatterer densities N , however, the wave which excites a reference scatterer deeply in the layer of the composite material will have characteristic values different from those of the wave in free space. This section still makes the assumption that nearby neighbouring scatterers (to the reference scatterer) placed in forward direction are not shadowed by the reference scatterer with respect to nearby neighbouring scatterers in the backward direction (in front of the scatterer). This condition implies that :

$$\frac{N\,Q_s}{k_0} = \begin{cases} \dfrac{2\mu}{\pi\, k_0 a} \dfrac{Q_s}{S} \\[2ex] \dfrac{3\mu}{4\, k_0 a} \dfrac{Q_s}{S} \end{cases} \approx \begin{cases} \dfrac{3\pi}{4} \mu (k_0 a)^2 \\[2ex] \dfrac{1}{3} \mu (k_0 a)^3 \end{cases} \overset{!}{<<} 1 \quad ; \quad \begin{cases} \text{Cylinder} \\[2ex] \text{Sphere} \end{cases} \tag{1}$$

At the theoretically possible upper limit $\mu=1$ (which, however, would be in conflict with conditions for the application of monotype scattering), the limits above give

$k_0a \ll 0.65$ for the cylinder,

$k_0a \ll 1.44$ for the sphere.

N=	number density of scatterers;
μ=	massivity of composite material;
D=	material layer thickness
Q_s=	scattering cross section;
a=	radius of scatterer;
S=	cross section of scatterer;

The effective propagation constant and wave number will be symbolised with $\Gamma=jk_E$, the effective wave impedance with Z_E.

As in section E.11 , the scattered far field angular distribution $g(\mathbf{o},\mathbf{i})$ will be used (**o**= outward direction of the scattered field, **i**= inward direction of the exciting wave); but the different wave numbers k_0,k_E in both directions will also be indicated; and because only the forward and backward directions (parallel and antiparallel to the incident wave) will be relevant, one changes : $g(\mathbf{o},\mathbf{i}) \rightarrow g(\pm k_0, \pm k_E)$.

The sound field in the material layer is : $\Phi(x) = A\, e^{-jk_E x} + B\, e^{+jk_E x}$ (2)

with the relations between the amplitudes (see section E.11) :

$$A\, e^{-jk_E x} = e^{-jk_0 x} \cdot \left[1 + \int_0^x e^{jk_0\xi} \cdot (A \cdot S_+\, e^{-jk_E\xi} + B \cdot R_+\, e^{+jk_E\xi}) \cdot d\xi \right]$$
$$B\, e^{+jk_E x} = e^{+jk_0 x} \cdot \int_x^D e^{-jk_0\xi} \cdot (A \cdot R_-\, e^{-jk_E\xi} + B \cdot S_-\, e^{+jk_E\xi}) \cdot d\xi \qquad (3)$$

where the following abbreviations are used :

$$S_+ = C\, g(k_0, k_E) \quad ; \quad S_- = C\, g(-k_0, -k_E)$$
$$R_+ = C\, g(k_0, -k_E) \quad ; \quad R_- = C\, g(-k_0, k_E) \qquad (4)$$

The constant factor C can be taken from Table 2 in section E.12 . Integration yields the system of equations :

$$\frac{S_+}{k_E - k_0} - \frac{R_-}{k_E + k_0} = -j \quad ; \quad \frac{S_-}{k_E - k_0} - \frac{R_+}{k_E + k_0} = -j$$
$$\frac{A \cdot S_+}{k_E - k_0} - \frac{B \cdot R_-}{k_E + k_0} = -j \quad ; \quad \frac{B \cdot S_- \cdot e^{+jk_E D}}{k_E - k_0} - \frac{A \cdot R_- \cdot e^{-jk_E D}}{k_E + k_0} = 0 \qquad (5)$$

and the solutions of the last two equations :

$$A = (1-Q) \cdot F \quad ; \quad B = (1-Q')\, Q\, e^{-2jk_E D} \cdot F \qquad (6)$$

where (as in section E.11) the auxiliary quantities are used :

$$F = [1 - QQ' e^{-2jk_E D}]^{-1} \quad ; \quad Q = \frac{R_-}{S_+} \frac{k_E - k_0}{k_E + k_0} \quad ; \quad Q' = \frac{R_+}{S_-} \frac{k_E - k_0}{k_E + k_0} \tag{7}$$

For scatterers with front-to-back symmetry (in the statistical average) simplifications are :

$$\begin{aligned} &g(k_0, k_E) = g(-k_0, -k_E) \quad ; \quad g(-k_0, k_E) = g(k_0, -k_E) \\ &S_+ = S_- = S \quad ; \quad R_+ = R_- = R \quad ; \quad Q = Q' \\ &F = \frac{1}{1 - Q^2 e^{-2jk_E D}} \end{aligned} \tag{8}$$

For such scatterers the field inside the layer is :

$$\Phi_I(x) = \frac{1-Q}{1 - Q^2 e^{-2jk_E D}} \cdot \left(e^{-jk_E x} - Q\, e^{+jk_E (x-2D)} \right) \quad ; \quad 0 \le x \le D \tag{9}$$

the reflected field in front of the layer :

$$\Phi_r(x) = -Q\,(1 - e^{-2jk_E D})\, F \cdot e^{+jk_0 x} \quad ; \quad x \le 0 \tag{10}$$

the transmitted field behind the layer :

$$\Phi_t(x) = (1 - Q^2)\, e^{-2jk_E D}\, F \cdot e^{-jk_0 x} \quad ; \quad x \ge D \tag{11}$$

The analogy with a homogeneous layer gives the correspondences : $Q \leftrightarrow \underline{r}_a$; $jk_E \leftrightarrow \Gamma$ where $\underline{r}_a$ is the reflection factor of the internal plane wave at the back side of the material layer (see section E.11).

Despite the close analogy to the results of section E.11 (with pure monotype scattering) there are differences in the $g(\pm k_0, \pm k_E)$ (as compared to $g(\pm\mathbf{o}, \pm\mathbf{i})$, see below for values) and in the definition of Q. For ease of writing (and in close analogy to section E.11) the values of the scattered far field angular distribution are defined (for symmetrical scatterers) as :

$$\begin{aligned} g_>(k_0, k_E) &= g(+k_0, +k_E) = g(-k_0, -k_E) \\ g_<(k_0, k_E) &= g(-k_0, +k_E) = g(+k_0, -k_E) \end{aligned} \tag{12}$$

with which the wave impedance of the effective wave is :

$$\frac{Z_E}{Z_0} = \frac{1-Q}{1+Q} = \frac{\frac{k_E}{k_0}(g_> - g_<) + (g_> + g_<)}{\frac{k_E}{k_0}(g_> + g_<) + (g_> - g_<)} \tag{13}$$

and a square equation holds for the wave number :

$$\frac{k_E^2}{k_0^2} - \frac{k_E}{k_0} \cdot j\frac{C}{k_0}(g_> - g_<) - \left(1 + j\frac{C}{k_0}(g_> + g_<) \right) = 0 \tag{14}$$

Since both $g_>$, $g_<$ contain k_E, Z_E the equation for k_E must be solved numerically, in general.

The still sought $g_>, g_<$ follow from the solution of the scattering task at a single scatterer. The scatterer shall consist of a homogeneous material with a characteristic propagation constant $\Gamma_\sigma = jk_\sigma$ and a characteristic wave impedance Z_σ. The scattered field is formulated as in section E.11, i.e. with the scattered mode amplitudes D_n, except for the substitution $k_0 \to k_E$. The interior field is formulated as :

$$p_\sigma(r,\vartheta) = \sum_{n=0}^{\infty} \delta_n \, (-j)^n \, E_n \, T_n(\vartheta) \, R_{\sigma n}(r) \tag{15}$$

with δ_n and the azimuthal functions $T_n(\vartheta)$ taken from Table 1 of section E.11 , and the radial functions :

$$R_{\sigma n}(r) = \begin{cases} J_n(k_\sigma r) & ; \text{ Cylinder} \\ j_n(k_\sigma r) & ; \text{ Sphere} \end{cases} \tag{16}$$

The boundary conditions at the surface give for the scattered mode amplitudes D_n and the interior mode amplitudes E_n (below for a cylinder, similarly for a sphere) :

$$\begin{aligned} D_n &= \frac{1}{X}\Big[\frac{1}{Z_\sigma/Z_0} J_n(k_E a)\, J'_n(k_\sigma a) - \frac{1}{Z_E/Z_0} J_n(k_\sigma a)\, J'_n(k_E a)\Big] \\ E_n &= \frac{1}{X}\Big[J_n(k_E a)\, H'^{(2)}_n(k_0 a) - \frac{1}{Z_E/Z_0} H^{(2)}_n(k_0 a)\, J'_n(k_E a)\Big] \end{aligned} \tag{17}$$

with the abbreviation :

$$X = J_n(k_\sigma a)\, H'^{(2)}_n(k_0 a) - \frac{1}{Z_\sigma/Z_0} H^{(2)}_n(k_0 a)\, J'_n(k_\sigma a) \tag{18}$$

(a prime indicates the derivative with respect to the argument). With the modal (normalised) admittances G_n the scattered mode amplitudes can be written as :

$$\begin{aligned} G_n &= \frac{j}{Z_\sigma/Z_0} \frac{J'_n(j\Gamma_\sigma a)}{J_n(j\Gamma_\sigma a)} \\ D_n &= \frac{-1}{Z_E/Z_0} \frac{\left(\frac{n}{k_E a} - j\frac{Z_E}{Z_0} G_n\right)\cdot J_n(k_E a) - J_{n+1}(k_E a)}{\left(\frac{n}{k_0 a} - jG_n\right)\cdot H^{(2)}_n(k_0 a) - H^{(2)}_{n+1}(k_0 a)} \end{aligned} \tag{19}$$

A locally reacting scatterer with a (normalised) surface admittance G is obtained by the substitution $G_n \to G$.

The usual writing of the characteristic values k_E, Z_E of the composite medium

$$\frac{k_E}{k_0} = \sqrt{\frac{\rho_{eff}}{\rho_0} \cdot \frac{C_{eff}}{C_0}} \quad ; \quad \frac{Z_E}{Z_0} = \sqrt{\frac{\rho_{eff}}{\rho_0} / \frac{C_{eff}}{C_0}} \tag{20}$$

with the effective density ρ_{eff} and compressibility C_{eff} is possible with :

$$\frac{\rho_{eff}}{\rho_0} = \frac{1}{1 - \frac{k_0}{k_E} \cdot j \frac{C}{k_0} (g_> - g_<)} \quad ; \quad \frac{C_{eff}}{C_0} = 1 + j \frac{C}{k_0} (g_> + g_<) \tag{21}$$

In the special case of a composite medium consisting of hard, parallel cylinders with radius a, and the plane wave incident normally on the cylinders (a forest of trunks, see section E.13), the expressions needed in the equations for the characteristic values are :

$$\begin{aligned} g_>(k_0,k_E) + g_<(k_0,k_E) &= D_0(k_0,k_E) + 2 \sum_{n=2,4,\ldots}^{n_{max}} \delta_n \, D_n(k_0,k_E) \\ g_>(k_0,k_E) - g_<(k_0,k_E) &= 2 \sum_{n=1,3,5,\ldots}^{n_{max}} \delta_n \, D_n(k_0,k_E) \end{aligned} \tag{22}$$

The diagram shows the attenuation coefficient $\mathrm{Re}\{\Gamma_E/k_0\}$ in such a medium where the scatterers have a massivity μ=0.02 (compare with section E.13).

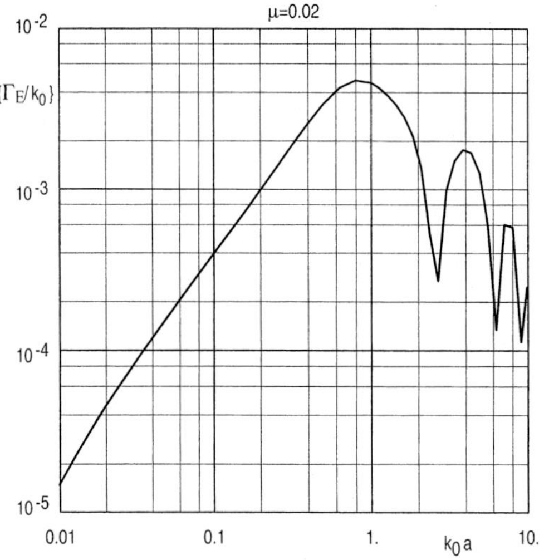

Attenuation coefficient in a composite medium of parallel, hard cylinders of radius a, forming a massivity μ=0.02, evaluated with the method of mixed monotype scattering.

E.15 Multiple triple-type scattering in random media

See section E.11 for general distinctions and notations. The sections about sound in capillaries in the chapter "Duct Acoustics" contain fundamentals about sound fields with thermal and viscous losses.

A sound wave with a scalar potential function Φ_e is incident on a scatterer in the composite medium. If the mutual distances of the scatterers are larger than the thickness of the shear boundary layer at a scatterer, it will be a density wave; otherwise it will be an "effective" wave Φ_E (i.e. influenced by the three wave types). The scatterer produces a scattered density wave Φ_ρ, a temperature wave Φ_α, and a viscous wave $\vec{\Psi}$. $\Phi_ß$; ß= e,ρ,α are scalar potentials with particle velocities $\vec{v}_ß = -\text{grad}\,\Phi_ß$, and $\vec{\Psi}$ is a vector potential, so that the total particle velocity is

$$\vec{v} = -\sum_ß \text{grad}\,\Phi_ß + \text{rot}\,\vec{\Psi}\,. \qquad (1)$$

The component fields obey wave equations :

$$(\Delta + k_ß^2)\Phi_ß = 0 \quad ; \quad (\Delta + k_\nu^2)\vec{\Psi} = 0 \qquad (2)$$

with characteristic wave numbers (given as squares) :

$$k_\rho^2 \approx k_0^2 = \left(\frac{\omega}{c_0}\right)^2 \quad ; \quad k_\nu^2 = -j\frac{\omega}{\nu} \quad ;$$
$$k_\alpha^2 \approx k_{\alpha 0}^2 = -j\frac{\kappa\omega}{\alpha} = \kappa\,\text{Pr}\cdot k_\nu^2 \qquad (3)$$

c_0= adiabatic sound velocity;
ρ_0= air density;
C_0=air compressibility;
$\nu = \eta/\rho_0$= kinematic viscosity;
η= dynamic viscosity
$\alpha = \Lambda/(\rho_0 c_p)$= temperature conductivity;
Λ= heat conductivity;
c_p= specific heat at constant pressure;
$\text{Pr} = \nu/\alpha$= Prandtl number;
ω= angular frequency;
P_0= atmospheric pressure;

The sound pressure p in the scattered field is :

$$p = P_0\Pi_\rho \cdot \Phi_\rho + P_0\Pi_\alpha \cdot \Phi_\alpha \qquad (4)$$

The coefficients $\Pi_ß$ are given in the mentioned sections about capillaries. The ratio Π_E/Π_ρ for an effective wave Φ_E can be expressed by the effective density ρ_{eff} of the composite medium:

$$P_0\Pi_E = jk_E\,Z_E = j\omega\,\rho_{eff} \quad ; \quad \frac{P_0\Pi_E}{P_0\Pi_\rho} = \frac{k_E}{k_0}\frac{Z_i}{Z_0} = \frac{\rho_{eff}}{\rho_0} \qquad (5)$$

If the composite medium is statistically homogeneous, the scattered vector potentials $\vec{\Psi}$ compensate each other in the forward and backward directions of scattering; thus the scattered far field angular distributions $g(\mathbf{o},\mathbf{i})$ do not contain the viscous wave in those directions. A similar compensation for the thermal wave Φ_α does not exist; it can be neglected in the propagating wave Φ_e only, if the immediate neighbours of a reference scatterer are outside the boundary layer.

The scatterer is assumed below to be either a cylinder or a sphere of a fluid with thermal and viscous losses (hard, soft, or locally reacting scatterers can be treated as special cases of this general assumption). Field quantities and material parameters inside the scatterer are marked with a prime. Only Hankel functions of the 2nd kind will appear; the upper index $^{(2)}$ therefore will be dropped (for ease of writing). The exciting wave Φ_e is supposed to have unit amplitude: $\Phi_e = e^{-jk_e x} = e^{-jk_e r\cos\vartheta}$. (6)

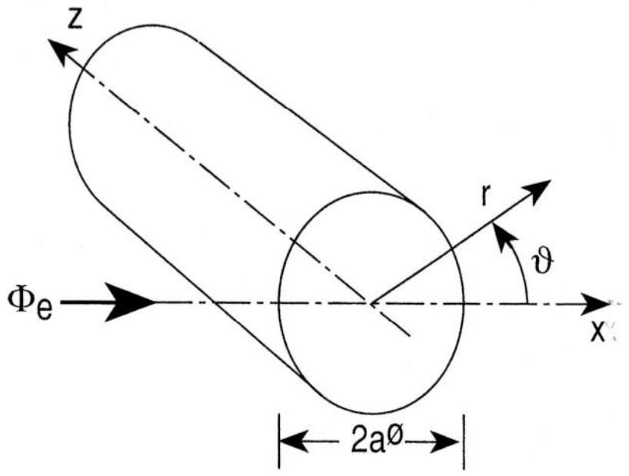

The particle velocities are :

$$\vec{v} = -\text{grad}\sum_{ß=e,\rho,\alpha}\Phi_ß + \text{rot}\,\Psi_v \quad ; \quad \text{outside}$$

$$\vec{v}' = -\text{grad}\sum_{ß'=\rho',\alpha'}\Phi_{ß'} + \text{rot}\,\Psi_{v'} \quad ; \quad \text{inside} \qquad (7)$$

The vector potential of the viscous wave has the components : (8)

$$\Psi_ß = \begin{cases}\{0,0,\Psi_{ßz}\} & ; \text{ Cylinder} \\ \{0,0,\Psi_{ß\varphi}\} & ; \text{ Sphere}\end{cases} \quad ; \quad ß = v, v'$$

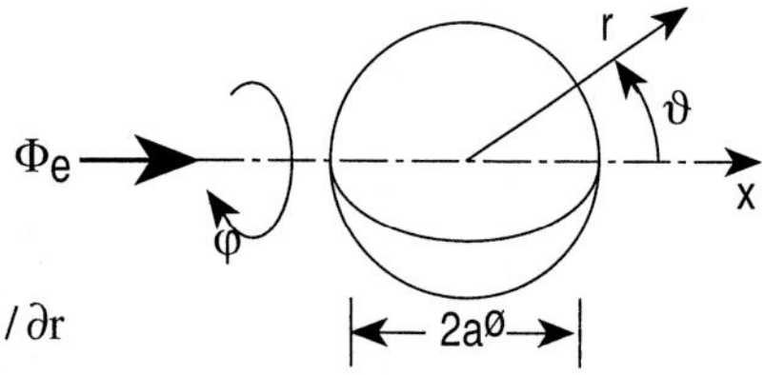

The boundary conditions are : (9)

(a): $v_r = v'_r$; (b): $v_\vartheta = v'_\vartheta$

(c): $T = T'$; (d): $\Lambda\cdot\partial T/\partial r = \Lambda'\cdot\partial T'/\partial r$

(e): $p_{rr} = p'_{rr}$; (f): $p_{r\vartheta} = p'_{r\vartheta}$

(a),(b) fit the radial and tangential particle velocities; (c),(d) fit the (alternating) temperature and heat flow; (e),(f) fit the radial and tangential tensions, which are (i,j standing for co-ordinates; η= dynamic viscosity):

$$p_{ii} = p + 2\eta\cdot\frac{\partial v_i}{\partial x_i} \quad ; \quad p_{ij} = \eta\cdot\left(\frac{\partial v_j}{\partial x_i} + \frac{\partial v_i}{\partial x_j}\right) \quad ; \quad i \neq j \qquad (10)$$

The pressure field is (p_0= atmospheric pressure) :

$$p = p_0\cdot\sum_ß \Pi_ß\,\Phi_ß \quad ; \quad ß = \begin{cases} e,\rho,\alpha & \text{outside} \\ \rho',\alpha' & \text{inside}\end{cases} \qquad (11)$$

The following tables give:

- strain velocities in cylindrical and spherical co-ordinates;
- vector components of grad and rot in both systems;
- field formulations of the incident wave, the scattered wave, and the wave inside the scatterer;
- terms appearing in the boundary conditions.

	Cylinder	Sphere
1	$\dot{s}_{rr} = \frac{\partial v_r}{\partial r}$	$\dot{s}_{rr} = \frac{\partial v_r}{\partial r}$
2	$\dot{s}_{\vartheta\vartheta} = \frac{1}{r}(v_r + \frac{\partial v_\vartheta}{\partial \vartheta})$	$\dot{s}_{\vartheta\vartheta} = \frac{1}{r}(v_r + \frac{\partial v_\vartheta}{\partial \vartheta})$
3	$\dot{s}_{zz} = \frac{\partial v_z}{\partial z}$	$\dot{s}_{\varphi\varphi} =$ $\frac{1}{\sin\vartheta}[v_r \sin\vartheta + v_\vartheta \cos\vartheta + \frac{\partial v_\varphi}{\partial \varphi}]$
4	$\dot{s}_{r\vartheta} = \dot{s}_{\vartheta r} =$ $\frac{1}{2}[\frac{\partial v_\vartheta}{\partial r} - \frac{v_\vartheta}{r} + \frac{1}{r}\frac{\partial v_r}{\partial \vartheta}]$	$\dot{s}_{r\vartheta} = \dot{s}_{\vartheta r} =$ $\frac{1}{2r}[r\frac{\partial v_\vartheta}{\partial r} - v_\vartheta + \frac{\partial v_r}{\partial \vartheta}]$
5	$\dot{s}_{z\vartheta} = \dot{s}_{\vartheta z} =$ $\frac{1}{2}[\frac{1}{r}\frac{\partial v_z}{\partial \vartheta} + \frac{\partial v_\vartheta}{\partial z}]$	$\dot{s}_{r\varphi} = \dot{s}_{\varphi r} =$ $\frac{1}{2r\sin\vartheta}[\frac{\partial v_r}{\partial \varphi} + r\frac{\partial v_\varphi}{\partial r}\sin\vartheta - v_\varphi \sin\vartheta]$
6	$\dot{s}_{zr} = \dot{s}_{rz} =$ $\frac{1}{2}[\frac{\partial v_r}{\partial z} + \frac{\partial v_z}{\partial r}]$	$\dot{s}_{\vartheta\varphi} = \dot{s}_{\varphi\vartheta} =$ $\frac{1}{2r\sin\vartheta}[\frac{\partial v_\varphi}{\partial \vartheta}\sin\vartheta - v_\varphi \cos\vartheta + \frac{\partial v_\vartheta}{\partial \varphi}]$

Table 1 : Components of the strain velocity in cylindrical and spherical co-ordinates

Table 2 : Components of grad and rot in cylindrical and spherical co-ordinates.

	Component	Cylinder	Sphere
1		grad U	
2	r	$\frac{\partial U}{\partial r}$	$\frac{\partial U}{\partial r}$
3	ϑ	$\frac{1}{r}\frac{\partial U}{\partial \vartheta}$	$\frac{1}{r}\frac{\partial U}{\partial \vartheta}$
4	z, φ	$\frac{\partial U}{\partial z}$	$\frac{1}{r\sin\varphi}\frac{\partial U}{\partial \varphi}$
5		rot **V**	
6	r	$\frac{1}{r}\frac{\partial V_z}{\partial \vartheta} - \frac{\partial V_\vartheta}{\partial z}$	$\frac{1}{r\sin\vartheta}\left(\frac{\partial}{\partial \vartheta}(\sin\vartheta\, V_\varphi) - \frac{\partial V_\vartheta}{\partial \varphi}\right)$
7	ϑ	$\frac{\partial V_r}{\partial z} - \frac{\partial V_z}{\partial r}$	$\frac{1}{r\sin\vartheta}\left(\frac{\partial V_r}{\partial \varphi} - \frac{1}{r}\frac{\partial}{\partial r}(r\, V_\varphi)\right)$
8	z, φ	$\frac{1}{r}(\frac{\partial (rV_\vartheta)}{\partial r} - \frac{\partial V_r}{\partial \vartheta})$	$\frac{1}{r}\frac{\partial}{\partial r}(r\, V_\vartheta) - \frac{1}{r}\frac{\partial V_r}{\partial \vartheta}$

	Quantity	Cylinder	Sphere
1	**Incident Wave:** $\Phi_e(r,\vartheta) =$	$\Phi_e(r,\vartheta) = e^{-jk_e r\cdot\cos\vartheta}$ $= \sum_{n=0}^{\infty} \delta_n \cdot (-j)^n \cdot T_n(\vartheta) \cdot R_{en}(r)$	
2	$\delta_n =$	$\begin{cases} 1 \ ; \ n=0 \\ 2 \ ; \ n>0 \end{cases}$	$2n+1$
3	$T_n(\vartheta) =$	$\cos(n\vartheta)$	$P_n(\cos\vartheta)$
4	$R_{en}(r) =$	$J_n(k_e r)$	$j_n(k_e r)$
5	**Scattered Wave** $\Phi_ß(r,\vartheta) =$	$\sum_{n=0}^{\infty} \delta_n \cdot (-j)^n \cdot A_{ßn} \cdot T_n(\vartheta) \cdot R_{an}(r)$	
6	$T_n(\vartheta) =$	$\cos(n\vartheta)$	$P_n(\cos\vartheta)$
7	$\Psi_{\nu z(\varphi)}(r,\vartheta) =$	$\sum_{n=0}^{\infty} \delta_n \cdot (-j)^n \cdot A_{\nu n} \cdot T_n^1(\vartheta) \cdot R_{an}(r) \ ; \ A_{\nu 0} = 0$	
8	$T_n^1(\vartheta) =$	$\sin(n\varphi)$	$P_n^1(\cos\vartheta)$
9	$R_{an}(r) =$	$H_n(k_ß r) \ ; \ ß = \rho, \alpha, \nu$	$h_n(k_ß r) \ ; \ ß = \rho, \alpha, \nu$
10	**Interior Wave** $\Phi_{ß'}(r,\vartheta) =$	$\sum_{n=0}^{\infty} \delta_n \cdot (-j)^n \cdot A_{ß'n} \cdot T_n(\vartheta) \cdot R_{in}(r)$	
11	$T_n(\vartheta) =$	$\cos(n\vartheta)$	$P_n(\cos\vartheta)$
12	$\Psi_{\nu' z(\varphi)}(r,\vartheta) =$	$\sum_{n=0}^{\infty} \delta_n \cdot (-j)^n \cdot A_{\nu' n} \cdot T_n^1(\vartheta) \cdot R_{in}(r) \ ; \ A_{\nu' 0} = 0$	
13	$T_n^1(\vartheta) =$	$\sin(n\varphi)$	$P_n^1(\cos\vartheta)$
14	$R_{in}(r) =$	$J_n(k_{ß'} r) \ ; \ ß' = \rho', \alpha', \nu'$	$j_n(k_{ß'} r) \ ; \ ß' = \rho', \alpha', \nu'$

Table 3 : Field formulations of the incident wave, the scattered wave and the interior wave.

	Quantity	Cylinder	Sphere
1	$v_r =$	$-\frac{\partial}{\partial r}\sum_\beta \Phi_\beta + \frac{1}{r}\frac{\partial \Psi_z}{\partial \vartheta}$	$-\frac{\partial}{\partial r}\sum_\beta \Phi_\beta + \frac{1}{r\sin\vartheta}\frac{\partial}{\partial \vartheta}(\sin\vartheta\, \Psi_\varphi)$
2	$v_\vartheta =$	$-\frac{1}{r}\frac{\partial}{\partial \vartheta}\sum_\beta \Phi_\beta - \frac{\partial \Psi_z}{\partial r}$	$-\frac{1}{r}\frac{\partial}{\partial \vartheta}\sum_\beta \Phi_\beta - \frac{\Psi_\varphi}{r} - \frac{\partial \Psi_\varphi}{\partial r}$
3	$\frac{T}{T_0} =$	$\sum_\beta \Theta_\beta \Phi_\beta$	$\sum_\beta \Theta_\beta \Phi_\beta$
4	$\frac{\partial}{\partial r}\frac{T}{T_0} =$	$\sum_\beta \Theta_\beta \frac{\partial \Phi_\beta}{\partial r}$	$\sum_\beta \Theta_\beta \frac{\partial \Phi_\beta}{\partial r}$
5	$P_{rr} =$	$p_0 \sum_\beta \Pi_\beta \Phi_\beta + 2\eta[-\sum_\beta \frac{\partial^2 \Phi_\beta}{\partial r^2} + \frac{\partial}{\partial \vartheta}(-\frac{\Psi_z}{r^2} + \frac{1}{r}\frac{\partial \Psi_z}{\partial r})]$	$p_0 \sum_\beta \Pi_\beta \Phi_\beta + 2\eta\{-\sum_\beta \frac{\partial^2 \Phi_\beta}{\partial r^2} + \frac{1}{\sin\vartheta}\frac{\partial}{\partial \vartheta}[\sin\vartheta(-\frac{\Psi_\varphi}{r^2} + \frac{1}{r}\frac{\partial \Psi_\varphi}{\partial r})]\}$
6	$P_{r\vartheta} =$	$\eta[-2\sum_\beta \frac{\partial}{\partial \vartheta}(\frac{1}{r}\frac{\partial \Phi_\beta}{\partial r} - \frac{\Phi_\beta}{r^2}) - (\frac{\partial^2 \Psi_z}{\partial r^2} - \frac{1}{r}\frac{\partial \Psi_z}{\partial r}) + \frac{1}{r^2}\frac{\partial^2 \Psi_z}{\partial \vartheta^2}]$	$\eta\{-2\sum_\beta \frac{\partial}{\partial \vartheta}(\frac{1}{r}\frac{\partial \Phi_\beta}{\partial r} - \frac{\Phi_\beta}{r^2}) - (\frac{\partial^2 \Psi_\varphi}{\partial r^2} - \frac{2}{r^2}\Psi_\varphi) + \frac{1}{r^2}\frac{\partial}{\partial \vartheta}[\frac{1}{\sin\vartheta}\frac{\partial}{\partial \vartheta}(\sin\vartheta\,\Psi_\varphi)]\}$

Table 4 : Terms in the boundary conditions

In Table 3 are

- $P_n(z)$ Legendre polynomials,
- $P_n^1(z)$ associate Legendre functions

with the useful relations :

$$P_n^1(z) = -\sqrt{1-z^2}\,\frac{dP_n(z)}{dz} = -\sqrt{1-z^2}\,P_n{}'(z)$$
$$P_n^1(\cos\vartheta) = -\sin\vartheta\, P_n{}'(\cos\vartheta) = \frac{d}{d\vartheta}P_n(\cos\vartheta) \tag{12}$$

$$\frac{1}{\sin\vartheta}\frac{d}{d\vartheta}(\sin\vartheta\cdot P_n^1(\cos\vartheta))=\sin^2\vartheta\cdot P_n''(\cos\vartheta)-2\cos\vartheta\cdot P_n'(\cos\vartheta)=-n(n+1)\cdot P_n(\cos\vartheta) \tag{13}$$

and the recursive evaluation :

$$P_{n+1}(z)=\frac{1}{n+1}[(2n+1)\,z\,P_n(z)-n\,P_{n-1}(z)] \quad ; \quad P_0(z)=1 \quad ; \quad P_1(z)=z \tag{14}$$

from which can be evaluated :

$$P_n^1(\cos\vartheta)=\frac{n}{\sin\vartheta}[\cos\vartheta\,P_n(\cos\vartheta)-P_{n-1}(\cos\vartheta)] \tag{15}$$

The following Table 5 contains the equations of the boundary conditions with the above field formulations. Use was made of the derivatives of the radial functions :

$$z\cdot R_n'(z)=n\cdot R_n(z)-z\cdot R_{n+1}(z) \tag{16}$$

and for the second derivatives of the cylindrical radial functions $Z_n(z)$ and spherical radial functions $K_n(z)$:

$$\begin{aligned} z^2\cdot Z_n''(z)&=(n^2-n-z^2)\cdot Z_n(z)+z\cdot Z_{n+1}(z)\\ z^2\cdot K_n''(z)&=(n^2-n-z^2)\cdot K_n(z)+2z\cdot K_{n+1}(z) \end{aligned} \tag{17}$$

The following terms, appearing in the table, can be simplified as :

$$\begin{aligned} &\frac{\rho_0 a^2\Pi_\rho}{\eta}=-(k_v a)^2\frac{k_\rho^2}{k_0^2}\frac{1-(k_\rho/k_{\alpha0})^2}{1-\kappa(k_\rho/k_{\alpha0})^2}\approx-(k_v a)^2\\ &\frac{\rho_0 a^2\Pi_\alpha}{\eta}\approx-(k_v a)^2(1-\frac{4}{3}\kappa\,\mathrm{Pr}) \quad ; \quad \frac{\rho_0 a^2\Pi_E}{\eta}=-(k_v a)^2\frac{\rho_{eff}}{\rho_0} \end{aligned} \tag{18}$$

Further, the coefficients $\Theta_E,\Theta_\rho,\Theta_\alpha$ can be written as :

$$\begin{aligned} &\Theta_E=\frac{\kappa}{\alpha}\frac{k_E^2}{k_{\alpha0}^2}\frac{\kappa-1}{1-\kappa\cdot k_E^2/k_{\alpha0}^2}\approx\frac{\kappa(\kappa-1)}{\alpha}\frac{k_E^2}{k_{\alpha0}^2}\\ &\Theta_\beta=\frac{\kappa}{\alpha}\frac{k_\beta^2}{k_{\alpha0}^2}\frac{\kappa-1}{1-\kappa\cdot k_\beta^2/k_{\alpha0}^2}\approx\begin{cases}\dfrac{\kappa(\kappa-1)}{\alpha}\dfrac{k_\rho^2}{k_{\alpha0}^2} & ;\ \beta=\rho\\[2ex] \dfrac{-\kappa}{\alpha} & ;\ \beta=\alpha\end{cases} \end{aligned} \tag{19}$$

$$\frac{\Theta_\beta}{\Theta_\alpha}\approx-(\kappa-1)\frac{k_\beta^2}{k_{\alpha0}^2} \quad ; \quad \beta=\rho,E \quad ; \quad \Lambda\,\Theta_\beta\approx\begin{cases}\rho_0 c_p(\kappa-1)\dfrac{k_\beta^2}{k_{\alpha0}^2} & ;\ \beta=\rho,E\\[2ex] -\rho_0 c_p & ;\ \beta=\alpha\end{cases} \tag{20}$$

Bound. condit.	**Equations for Cylinder**
$v_r = v'_r$	$n\,J_n(k_e a) - k_e a\, J_{n+1}(k_e a) +$ $+ \sum_{\beta=\rho,\alpha} A_{\beta n}\,[n\,H_n(k_\beta a) - k_\beta a\, H_{n+1}(k_\beta a)] - A_{\nu n}\, n\, H_n(k_\nu a) =$ $= \sum_{\beta'=\rho',\alpha'} A_{\beta' n}\,[n\,J_n(k_{\beta'} a) - k_{\beta'} a\, J_{n+1}(k_{\beta'} a)] - A_{\nu' n}\, n\, J_n(k_{\nu'} a)$
$v_\vartheta = v'_\vartheta$	$n\,J_n(k_e a) + \sum_{\beta=\rho,\alpha} A_{\beta n}\, n\, H_n(k_\beta a) - A_{\nu n}\,[n\,H_n(k_\nu a) - k_\nu a\, H_{n+1}(k_\nu a)] =$ $= \sum_{\beta'=\rho',\alpha'} A_{\beta' n}\, n\, J_n(k_{\beta'} a) - A_{\nu' n}\,[n\,J_n(k_{\nu'} a) - k_{\nu'} a\, J_{n+1}(k_{\nu'} a)]$
$\frac{T}{T_0} = \frac{T'}{T_0}$	$\Theta_e\, J_n(k_e a) + \sum_{\beta=\rho,\alpha} A_{\beta n}\,\Theta_\beta\, H_n(k_\beta a) = \sum_{\beta'=\rho',\alpha'} A_{\beta' n}\,\Theta_{\beta'}\, J_n(k_{\beta'} a)$
$\Lambda \frac{\partial}{\partial r}\frac{T}{T_0} =$ $\Lambda' \frac{\partial}{\partial r}\frac{T'}{T_0}$	$\Lambda \left[\Theta_e\,[n\,J_n(k_e a) - k_e a\, J_{n+1}(k_e a)] + \sum_{\beta=\rho,\alpha} A_{\beta n}\,\Theta_\beta\,[n\,H_n(k_\beta a) - k_\beta a\, H_{n+1}(k_\beta a)] \right]$ $= \Lambda' \left[\sum_{\beta'=\rho',\alpha'} A_{\beta' n}\,\Theta_{\beta'}\,[n\,J_n(k_{\beta'} a) - k_{\beta'} a\, J_{n+1}(k_{\beta'} a)] \right]$
$P_{rr} = P'_{rr}$	$\eta \Big[[\frac{p_0 a^2}{\eta}\Pi_e + 2(n - n^2 + (k_e a)^2)]\, J_n(k_e a) - 2\,k_e a\, J_{n+1}(k_e a) +$ $\sum_{\beta=\rho,\alpha} A_{\beta n} \left([\frac{p_0 a^2}{\eta}\Pi_\beta + 2(n - n^2 + (k_\beta a)^2)]\, H_n(k_\beta a) - 2\,k_\beta a\, H_{n+1}(k_\beta a) \right) +$ $+ A_{\nu n}\, 2n\,[(n-1)\, H_n(k_\nu a) - k_\nu a\, H_{n+1}(k_\nu a)] \Big] =$ $= \eta' \Big[A_{\nu' n}\, 2n\,[(n-1)\, J_n(k_{\nu'} a) - k_{\nu'} a\, J_{n+1}(k_{\nu'} a)] +$ $\sum_{\beta'=\rho',\alpha'} A_{\beta' n} \left([\frac{p_0 a^2}{\eta'}\Pi_{\beta'} + 2(n - n^2 + (k_{\beta'} a)^2)]\, J_n(k_{\beta'} a) - 2\,k_{\beta'} a\, J_{n+1}(k_{\beta'} a) \right) \Big]$
$P_{r\vartheta} = P'_{r\vartheta}$	$\eta \Big[2n\,[(n-1)\, J_n(k_e a) - k_e a\, J_{n+1}(k_e a)] +$ $+ 2n \sum_{\beta=\rho,\alpha} A_{\beta n}\,[(n-1)\, H_n(k_\beta a) - k_\beta a\, H_{n+1}(k_\beta a)] -$ $- A_{\nu n}\,[(2n(n-1) - (k_\nu a)^2)\, H_n(k_\nu a) + 2\,k_\nu a\, H_{n+1}(k_\nu a)] \Big] =$ $= \eta' \Big[2n \sum_{\beta'=\rho',\alpha'} A_{\beta' n}\,[(n-1)\, J_n(k_{\beta'} a) - k_{\beta'} a\, J_{n+1}(k_{\beta'} a)] -$ $- A_{\nu' n}\,[(2n(n-1) - (k_{\nu'} a)^2)\, J_n(k_{\nu'} a) + 2\,k_{\nu'} a\, J_{n+1}(k_{\nu'} a)] \Big]$

Bound. condit.	Equations for Sphere
$v_r = v'_r$	$n\, j_n(k_e a) - k_e a\, j_{n+1}(k_e a) + \sum_{\beta=\rho,\alpha} A_{\beta n}\,[n\, h_n(k_\beta a) - k_\beta a\, h_{n+1}(k_\beta a)] + A_{\nu n}\, n(n+1)\, h_n(k_\nu a) = \sum_{\beta'=\rho',\alpha'} A_{\beta' n}\,[n\, j_n(k_{\beta'} a) - k_{\beta'} a\, j_{n+1}(k_{\beta'} a)] + A_{\nu' n}\, n(n+1)\, j_n(k_{\nu'} a)$
$v_\vartheta = v'_\vartheta$	$j_n(k_e a) + \sum_{\beta=\rho,\alpha} A_{\beta n}\, h_n(k_\beta a) + A_{\nu n}\,[(n+1)\, h_n(k_\nu a) - k_\nu a\, h_{n+1}(k_\nu a)] = \sum_{\beta'=\rho',\alpha'} A_{\beta' n}\, j_n(k_{\beta'} a) + A_{\nu' n}\,[(n+1)\, j_n(k_{\nu'} a) - k_{\nu'} a\, j_{n+1}(k_{\nu'} a)]$
$\frac{T}{T_0} = \frac{T'}{T_0}$	$\Theta_e\, j_n(k_e a) + \sum_{\beta=\rho,\alpha} A_{\beta n}\,\Theta_\beta\, h_n(k_\beta a) = \sum_{\beta'=\rho',\alpha'} A_{\beta' n}\,\Theta_{\beta'}\, j_n(k_{\beta'} a)$
$\Lambda \frac{\partial}{\partial r}\frac{T}{T_0} = \Lambda' \frac{\partial}{\partial r}\frac{T'}{T_0}$	$\Lambda \left[\Theta_e\,[n\, j_n(k_e a) - k_e a\, j_{n+1}(k_e a)] + \sum_{\beta=\rho,\alpha} A_{\beta n}\,\Theta_\beta\,[n\, h_n(k_\beta a) - k_\beta a\, h_{n+1}(k_\beta a)]\right] = \Lambda'\left[\sum_{\beta'=\rho',\alpha'} A_{\beta' n}\,\Theta_{\beta'}\,[n\, j_n(k_{\beta'} a) - k_{\beta'} a\, j_{n+1}(k_{\beta'} a)]\right]$
$P_{rr} = P'_{rr}$	$\eta \left[[\frac{p_0 a^2}{\eta}\Pi_e + 2(n - n^2 + (k_e a)^2)]\, j_n(k_e a) - 4 k_e a\, j_{n+1}(k_e a) + \sum_{\beta=\rho,\alpha} A_{\beta n}\left([\frac{p_0 a^2}{\eta}\Pi_\beta + 2(n - n^2 + (k_\beta a)^2)]\, h_n(k_\beta a) - 4 k_\beta a\, h_{n+1}(k_\beta a)\right) - A_{\nu n}\, 2n(n+1)\,[(n-1)\, h_n(k_\nu a) - k_\nu a\, h_{n+1}(k_\nu a)] \right] = \eta' \left[-A_{\nu' n}\, 2n(n+1)\,[(n-1)\, j_n(k_{\nu'} a) - k_{\nu'} a\, j_{n+1}(k_{\nu'} a)] + \sum_{\beta'=\rho',\alpha'} A_{\beta' n}\left([\frac{p_0 a^2}{\eta'}\Pi_{\beta'} + 2(n - n^2 + (k_{\beta'} a)^2)]\, j_n(k_{\beta'} a) - 4 k_{\beta'} a\, j_{n+1}(k_{\beta'} a)\right)\right.$
$P_{r\vartheta} = P'_{r\vartheta}$	$\eta \left[2\,[(n-1)\, j_n(k_e a) - k_e a\, j_{n+1}(k_e a)] + 2 \sum_{\beta=\rho,\alpha} A_{\beta n}\,[(n-1)\, h_n(k_\beta a) - k_\beta a\, h_{n+1}(k_\beta a)] + A_{\nu n}\,[(2(n^2-1) - (k_\nu a)^2)\, h_n(k_\nu a) + 2 k_\nu a\, h_{n+1}(k_\nu a)] \right] = \eta' \left[2 \sum_{\beta'=\rho',\alpha'} A_{\beta' n}\,[(n-1)\, j_n(k_{\beta'} a) - k_{\beta'} a\, j_{n+1}(k_{\beta'} a)] + A_{\nu' n}\,[(2(n^2-1) - (k_{\nu'} a)^2)\, j_n(k_{\nu'} a) + 2 k_{\nu'} a\, j_{n+1}(k_{\nu'} a)] \right]$

The six boundary equations in the above tables (for each shape of the scatterer) originally contained common factors on both sides; they have been divided out. If these boundary equations are to be used for other types of scatterers than fluid cylinders or spheres, the factors must be included again before the modification of the equations. These factors are contained in the following Table 6 , the 1st column of which indicates the number of the corresponding row in the above tables and the index letter used above for the boundary conditions.

	Quantity	Factors	
		Cylinder	Sphere
1 (a)	v_r	$-(-j)^n\delta_n \dfrac{\cos(n\vartheta)}{a}$	$-(-j)^n\delta_n \dfrac{P_n(\cos\vartheta)}{a}$
2 (b)	v_ϑ	$(-j)^n\delta_n \dfrac{\sin(n\vartheta)}{a}$	$-(-j)^n\delta_n \dfrac{d\,P_n(\cos\vartheta)/d\vartheta}{a}$
3 (c)	$\dfrac{T_1}{T_0}$	$(-j)^n\delta_n \cos(n\vartheta)$	$(-j)^n\delta_n P_n(\cos\vartheta)$
4 (d)	$\Lambda\dfrac{\partial}{\partial r}\dfrac{T_1}{T_0}$	$(-j)^n\delta_n \dfrac{\cos(n\vartheta)}{a}$	$(-j)^n\delta_n \dfrac{P_n(\cos\vartheta)}{a}$
5 (e)	p_{rr}	$(-j)^n\delta_n \dfrac{\cos(n\vartheta)}{a^2}$	$(-j)^n\delta_n \dfrac{P_n(\cos\vartheta)}{a^2}$
6 (f)	$p_{r\vartheta}$	$(-j)^n\delta_n \dfrac{\sin(n\vartheta)}{a^2}$	$-(-j)^n\delta_n \dfrac{d\,P_n(\cos\vartheta)/d\vartheta}{a^2}$

Table 6 : Common factors on both sides of the boundary equations.

Special types of scatterers :

Elastic scatterers :

Replace the dynamic viscosity η' with the shear modulus G' or the Lamé constant μ' and the sound velocity c'_0 with the compression modulus K' or the 2nd Lamé constant λ' according to :

$$\begin{aligned} &\eta' \to \mu'/j\omega = G'/j\omega \\ &c'_0 \to \sqrt{K'/\rho'_0} \quad ; \quad K' = \lambda' + \tfrac{2}{3}\mu' \end{aligned} \qquad (21)$$

Rigid scatterer at rest :

Set $v'_r = v'_\vartheta = 0 \;;\; k_{\rho'} = k_{\alpha'} \to 0 \;;\; R_{n>0}(k_{\beta'}a) = 0 \;;\; R_0(k_{\beta'}a) = 1$ (22)

Delete the boundary conditions (e),(f) ; in (a)-(d) delete on the right-hand sides all interior wave terms except the temperature wave term. In the special case of an isothermal surface delete (d) and set in (c) the right-hand side to zero. In this special case only a set of amplitudes $A_{\beta n}$ with $\beta=\rho,\alpha,\nu$ must be determined.

Porous scatterers :

Mostly $|k_\rho a|^2, |k_E a|^2 << 1$ and $|k_\rho / k_\beta|^2 << 1 ; \beta = \alpha, \nu$. Then the radial functions $R_n(k_\rho a)$ can be approximated with the 1st term of their power series, and terms with $\Theta_\rho/\Theta_\alpha$ and Π_α/Π_ρ can be neglected.

Isothermal, hard scatterer freely oscillating :

This case will here be fully formulated. The oscillation is in the x direction ($\delta_1=2$ is retained from the general formulations) :

$$\Phi_x = (-j)^1 \delta_1 \cdot A_x \frac{x}{a} = (-j)^1 \delta_1 \cdot A_x \frac{r}{a} \cos\vartheta$$
$$v_x = j\delta_1 \frac{A_x}{a} \;;\; v_r = j\delta_1 A_x \frac{\cos\vartheta}{a} \;;\; v_\vartheta = -j\delta_1 A_x \frac{\sin\vartheta}{a} \tag{23}$$

On the right-hand sides of the boundary conditions (a),(b) all terms with $n\neq1$ vanish; and for $n=1$ there will appear A_x . The right-hand side of the boundary condition (c) disappears, and also the complete boundary condition (d). The two last boundary conditions (e), (f) are replaced by an equation for the balance of force K_x (which is the integral of the stresses in the x direction over the scatterer surface) :

$$K_x = \oint (p_{rr} \cos\vartheta - p_{r\vartheta} \sin\vartheta) \cdot dA \overset{!}{=} j\omega M \cdot v_x = j\omega M \cdot j\delta_1 \frac{A_x}{a} \tag{24}$$

where $M=\rho'_0 V_0$ is the scatterer's mass (per unit length of the cylinder). The integrals over the terms with $p_{rr}, p_{r\vartheta}$ lead to, respectively :

$$I_{rr} = \int_0^\pi P_n(\cos\vartheta) \cos\vartheta \sin\vartheta \cdot d\vartheta = \int_{-1}^{+1} y \cdot P_n(y)\, dy = \frac{2}{3} \delta_{1,n}$$
$$I_{r\vartheta} = \int_0^\pi \frac{\partial}{\partial\vartheta}(P_n(\cos\vartheta)) \sin^2\vartheta \cdot d\vartheta = -2\, I_{rr} = -\frac{4}{3} \delta_{1,n} \tag{25}$$

with the Kronecker symbol $\delta_{m,n}$.

The boundary equations for an isothermal, movable, hard scatterer are given in the next table.

Bound. condit.	Equations
Cylinder	
$v_r = v_x \cos\vartheta$	$\sum\limits_{\beta=\rho,\alpha} A_{\beta n}\,[n\,H_n(k_\beta a) - k_\beta a\,H_{n+1}(k_\beta a)] - A_{\nu n}\,n\,H_n(k_\nu a) - A_x\,\delta_{1,n} =$ $= -[n\,J_n(k_e a) - k_e a\,J_{n+1}(k_e a)]$
$v_\vartheta = -v_x \sin\vartheta$	$\sum\limits_{\beta=\rho,\alpha} A_{\beta n}\,n\,H_n(k_\beta a) - A_{\nu n}\,[n\,H_n(k_\nu a) - k_\nu a\,H_{n+1}(k_\nu a)] - A_x\,\delta_{1,n} =$ $= -n\,J_n(k_e a)$
$\frac{T}{T_0} = 0$	$\sum\limits_{\beta=\rho,\alpha} A_{\beta n}\,\Theta_\beta\,H_n(k_\beta a) = -\Theta_e\,J_n(k_e a)$
$K_x = j\omega M v_x$	$\sum\limits_{\beta=\rho,\alpha} A_{\beta 1}\,[\frac{p_0 a^2}{\eta}\Pi_\beta + 2\,(k_\beta a)^2]\,H_1(k_\beta a) -$ $- A_{\nu 1}\,(k_\nu a)^2\,H_1(k_\nu a) - A_x\,(k_\nu a)^2\,\frac{\rho'_0}{\rho_0} =$ $= -[\frac{p_0 a^2}{\eta}\Pi_e + 2\,(k_e a)^2]\,J_1(k_e a)$
Sphere	
$v_r = v_x \cos\vartheta$	$\sum\limits_{\beta=\rho,\alpha} A_{\beta n}\,[n\,h_n(k_\beta a) - k_\beta a\,h_{n+1}(k_\beta a)] + A_{\nu n}\,n(n+1)\,h_n(k_\nu a) - A_x\,\delta_{1,n} =$ $= -[n\,j_n(k_e a) - k_e a\,j_{n+1}(k_e a)]$
$v_\vartheta = -v_x \sin\vartheta$	$\sum\limits_{\beta=\rho,\alpha} A_{\beta n}\,h_n(k_\beta a) + A_{\nu n}\,[(n+1)\,h_n(k_\nu a) - k_\nu a\,h_{n+1}(k_\nu a)] - A_x\,\delta_{1,n} =$ $= -j_n(k_e a)$
$\frac{T}{T_0} = 0$	$\sum\limits_{\beta=\rho,\alpha} A_{\beta n}\,\Theta_\beta\,h_n(k_\beta a) = -\Theta_e\,j_n(k_e a)$
$K_x = j\omega M v_x$	$\sum\limits_{\beta=\rho,\alpha} A_{\beta 1}\,[\frac{p_0 a^2}{\eta}\Pi_\beta + 2\,(k_\beta a)^2)]\,h_1(k_\beta a) +$ $+ A_{\nu 1}\,2(k_\nu a)^2\,h_1(k_\nu a) - A_x\,(k_\nu a)^2\,\frac{\rho'_0}{\rho_0} =$ $= -[\frac{p_0 a^2}{\eta}\Pi_e + 2\,(k_e a)^2)]\,j_1(k_e a)$

Table 7 : Boundary conditions for an isothermal, hard, freely movable scatterer.

E.16 Plane wave scattering at elastic cylindrical shell

Ref.: Paniklenko / Rybak, [E.7]

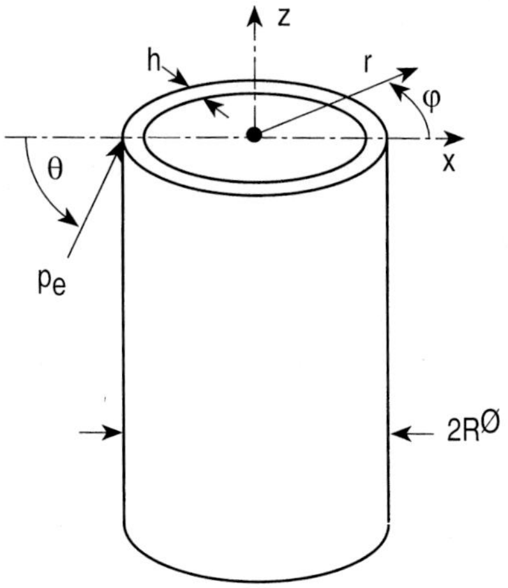

A plane wave p_e is incident (under an angle θ with the radius) on a cylindrical shell with outer radius R and thickness h .

The exterior sound field is composed as

$p = p_e + p_r + p_s$

with p_r= scattered field from a hard cylinder,

p_s= additional scattering due to elasticity.

Parameters of the surrounding medium:

ρ_0, c_0, k_0, Z_0 = density, sound speed, free field wave number, free field wave impedance.

Parameters of the shell:

ρ, σ, E, η= density, Poisson ratio, Young's modulus, loss factor;

c_D, k_D= speed and wave number of the dilatational wave in a plate of thickness h;

$\underline{E} = E\cdot(1+j\cdot\eta)$= complex Young's modulus.

Abbreviations:

$$\Omega_0 = k_0 R \; ; \; \Omega = k_D R \; ; \; \xi = k_0 r\cdot\cos\theta \; ; \; \delta_0 = 1 \; ; \; \delta_{n>0} = 2 \; ; \tag{1}$$

Field component formulations :

$$p_e(r,\varphi) = e^{-j(k_0 r\cdot\cos\theta\cdot\cos\varphi + k_0 z\cdot\sin\theta)}$$

$$p_r(r,\varphi) = \sum_{n\geq 0} \frac{J'_n(\xi)}{H_n^{(2)}(\xi)} \tag{2}$$

$$p_s(r,\varphi) = \frac{-2Z_0}{\pi\Omega_0\cos\theta} e^{-jk_0 z\cdot\sin\theta} \sum_{n\geq 0} \frac{\delta_n(-j)^n H_n^{(2)}(\xi)\cdot\cos(n\varphi)}{\left(H_n'^{(2)}(\Omega_0\cos\theta)\right)^2 \left(Z_{mn}+Z_{sn}\right)}$$

with radiation impedance Z_{sn} of the n-th mode of the shell:

$$Z_{sn} = \frac{-jZ_0}{\cos\theta} \frac{H_n^{(2)}(\Omega_0\cos\theta)}{H_n'^{(2)}(\Omega_0\cos\theta)} \tag{3}$$

and the mechanical impedance Z_{mn} of the n-th shell mode :

$$Z_{mn} = \frac{-j\rho h}{\Omega^2}\frac{D}{D_1} \quad ; \quad D = \mathrm{Det}\{A_{ik}\} \quad ; \quad D_1 = A_{11}\cdot A_{22} - A_{12}\cdot A_{21} \quad ; \quad i,k = 1,\ldots,3 \tag{4}$$

with matrix coefficients :

$$\begin{aligned} &A_{11} = \Omega^2 - (\Omega_0 \sin\theta)^2 - (1-\sigma)n^2/2 \quad ; \quad A_{12} = -A_{21} = -j(1+\sigma)\Omega_0 \sin(n\theta/2) \\ &A_{13} = -A_{31} = -j\sigma\Omega_0 \sin\theta \\ &A_{22} = \Omega^2 - (1-\sigma)\frac{(\Omega_0 \sin\theta)^2}{2} - n^2 \quad ; \quad A_{23} = A_{32} = -n \\ &A_{33} = \Omega^2 - h^2\left((\Omega_0 \sin\theta)^2 + n^2\right)/12 - 1 \end{aligned} \tag{5}$$

Asymptotic form of p_s (for $k_0 r >> 1$) :

$$\begin{aligned} &p_s \approx \sqrt{\frac{-2j}{\pi\xi}}\, e^{-j(k_0\xi + k_0 z\cdot\sin\theta)} \cdot \Phi_s(\Omega_0,\varphi,\theta) \\ &\Phi_s(\Omega_0,\varphi,\theta) = \sum_{n\ge 0}\Phi_n = \frac{-2Z_0}{\pi\Omega_0\cos\theta}\sum_{n\ge 0}\frac{\delta_n\cdot\cos(n\varphi)}{\left(H_n^{(2)}(\Omega_0\cos\theta)\right)^2\left(Z_{mn}+Z_{sn}\right)} \end{aligned} \tag{6}$$

Shell resonances without shell losses :

Resonance condition: $\mathrm{Im}\{Z_{mn}+Z_{sn}\} = 0$. (7)

For $n \ge 2$ the resonances with fluid load are about at the resonances of the shell without fluid load.

Approximation for low frequencies $\Omega_0 << 2n+1$:

$$Z_{sn} \approx \frac{Z_0}{\cos\theta}\left[\frac{4\pi}{(n!)^2}\left(\frac{\Omega_0\cos\theta}{2}\right)^{2n+1} + j\frac{\Omega_0\cos\theta}{n}\right] \tag{8}$$

Far field angular distribution of radiating mode in resonance :

$$\Phi_n^{res}(\varphi) \approx -\delta_n \cos(n\varphi) \tag{9}$$

Scattered far field of the n-th mode in resonance :

$$p_s \approx -\sqrt{\frac{-2j}{\pi\xi}}\, e^{-j(k_0\xi + k_0 z\cdot\sin\theta)} \cdot \delta_n \cos(n\varphi) \tag{10}$$

Scattering cross-section in resonance :

$$Q_s = \frac{2\delta_n^2}{\pi k_0 \cos\theta}\int_0^{2\pi}\cos^2(n\varphi)\, d\varphi = \frac{8}{k_0\cos\theta} \tag{11}$$

Quality factor q_n of the resonance of the n-th mode (without shell losses):

$$q_n = \frac{\mathrm{Im}\{Z_{sn}\} + \rho h}{\mathrm{Re}\{Z_{sn}\}} = \frac{(n!)^2}{2\pi n}\left(\frac{2}{\Omega_0}\right)^{2n}\left(1 + \frac{\rho h}{\rho_0 R} n\right) \tag{12}$$

Shell resonances with shell losses :

With $E \to E\cdot(1+j\eta)$; $c_D^2 \to c_D^2(1+j\eta)$ (13)

Mechanical shell mode impedance :

$$Z_{mn} = j\omega\rho h\left[1 - (n^2-1)^2 \frac{h^2}{12R^2}\left(\frac{c_D}{c_0}\right)^2 \frac{1}{\Omega_0^2}\right] \tag{14}$$

Resonances at :

$$\Omega_{0,res}^2(n) = \frac{n B_n}{1 + n\dfrac{\rho h}{\rho_0 R}} \quad ; \quad B_n := \frac{\rho h}{\rho_0 R}\left(\frac{c_D}{c_0}\right)^2 \frac{h^2}{12R^2}(n^2-1)^2 \tag{15}$$

For $n \geq 2$, with losses $Z_{mn,\eta}$, without losses $Z_{mn,0}$:

$$Z_{mn,\eta} \approx Z_{mn,0} + \frac{\eta Z_0 B_n}{\Omega_0} \tag{16}$$

Ratio of radiation loss to internal loss :

$$\frac{\mathrm{Re}\{Z_{sn}\}}{\mathrm{Re}\{Z_{mn}\}} = \frac{8\pi}{\eta(n!)^2}\frac{1}{B_n}\left(\frac{\Omega_0}{2}\right)^{2n+1} \tag{17}$$

Far field angular distribution in resonance with losses and $\theta=0$ for $\Omega_{0,res} \ll 1$:

$$\Phi_n^{res} \approx \frac{-2\delta_n \cos\varphi}{\pi(n!/2)^2(2/\Omega_0)^{2(n+1)}\,\eta B_n} \tag{18}$$

Relation to far field angular distribution Φ_{nh} of a hard cylinder :

$$\frac{\Phi_n^{res}}{\Phi_{nh}} = \frac{2}{\eta\left(1 + n\dfrac{\rho h}{\rho_0 R}\right)} \tag{19}$$

E.17 Plane wave backscattering by a liquid sphere

Ref.: Johnson, R.K. [E.8]

Consider a fluid sphere with radius a and ρ_1, c_1, k_1 for density, sound speed, free field wave number, respectively, of the sphere fluid in an outer medium with ρ_0, c_0, k_0, respectively. Ratios of densities: $g=\rho_1/\rho_0$; of sound velocities $\gamma=c_1/c_0$.

The backscattering cross-section σ for an incident plane wave is :

$$\frac{\sigma}{\pi a^2}=\frac{2}{k_0 a}\left|\sum_{m\geq 0}\frac{(-1)^m(2m+1)}{1+j\cdot C_m}\right|$$

$$C_m=\frac{\dfrac{\alpha'_m}{\alpha_m}\dfrac{y_m(k_0 a)}{j_m(k_1 a)}-\dfrac{\beta_m}{\alpha_m}g\gamma}{\dfrac{\alpha'_m}{\alpha_m}\dfrac{j_m(k_0 a)}{j_m(k_1 a)}-g\gamma} \tag{1}$$

$$\alpha_m=m\cdot j_m(k_0 a)-(m+1)\cdot j_{m+1}(k_0 a)\quad ;\quad \alpha'_m=m\cdot j_m(k_1 a)-(m+1)\cdot j_{m+1}(k_1 a)$$

$$\beta_m=m\cdot y_m(k_0 a)-(m+1)\cdot y_{m+1}(k_0 a)\quad ;\quad \beta'_m=m\cdot y_m(k_1 a)-(m+1)\cdot y_{m+1}(k_1 a)$$

with $j_m(z)$ spherical Bessel functions, $y_m(z)$ spherical Neumann functions.

Approximation for $k_0 a \ll 1$:

$$\frac{\sigma}{\pi a^2}\approx 4(k_0 a)^4\left[\frac{1-g\gamma^2}{3g\gamma^2}+\frac{1-g}{1+2g}\right]^2 \tag{2}$$

Special case: air bubble in water :

$$\frac{\sigma}{\pi a^2}\approx 4\Big/\left\{\left[(f_0/f)^2-1\right]^2+\delta^2\right\}\quad ;\quad f_0=\frac{1}{2\pi a}\sqrt{3\kappa P/\rho_0} \tag{3}$$

with f_0 the 1st bubble resonance frequency; κ= adiabatic exponent of air; P= static pressure; $\delta\approx 1/5$ an attenuation exponent of the bubble oscillation.

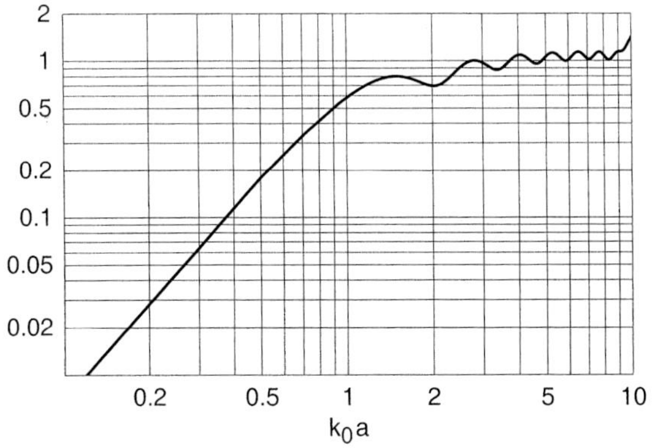

Normalised backscatter cross-section $\sigma/(\pi a^2)$ for an air bubble in water (exact form).

E.18 Spherical wave scattering at a perfectly absorbing wedge

Ref.: Rawlins, [E.9]

Attention:
The time factor here is $e^{-i\omega t}$.
A point source at $Q=\{r_q,\Phi_q,z_q\}$ sends a spherical wave onto a wedge with half wedge angle Ω.
The object treated is an idealised model for an absorbing wedge; the scattered field is half the sum of the fields for a hard and a soft wedge. Thus the wedge here is perfectly absorbing for all directions of incident sound.

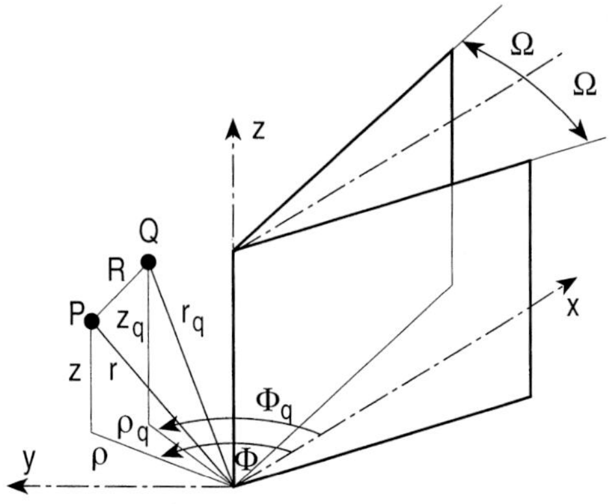

Field composition: $p(\rho,\Phi,z)=p_i(\rho,\Phi,z)+p_s(\rho,\Phi,z)$ (1)

with incident wave $p_i=\dfrac{e^{ik_0R}}{k_0R}$

General solution :

$$p=\frac{1}{2\pi i\nu}\int_{C_1+C_2}\frac{e^{ik_0R(\alpha)}}{k_0R(\alpha)}\cdot\cot\frac{\pi-\alpha-\Phi-\Phi_q}{2\nu}\,d\alpha$$

$$R=\sqrt{\rho^2+\rho_q^2-2\rho\rho_q\cos(\Phi-\Phi_q)+(z-z_q)^2}$$

$$R(\alpha)=\sqrt{\rho^2+\rho_q^2+2\rho\rho_q\cos\alpha+(z-z_q)^2} \qquad (2)$$

$$\nu=2(\pi-\Omega)/\pi$$

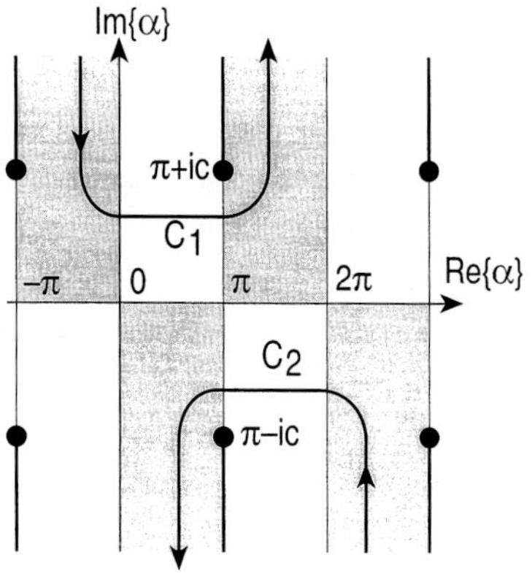

The path of integration circumvents the branch points at $\pi\pm ic$ with

$$c=2\cosh^{-1}\frac{R}{2\sqrt{\rho\rho_q}} \qquad (3)$$

Near field :

$$p = \frac{1}{\nu} \sum_{n \ge 0} \delta_n \cdot \cos \frac{n(\Phi - \Phi_q)}{\nu} \cdot S_{n/\nu} \quad ; \quad \delta_0 = 1 \quad ; \quad \delta_{n>0} = 2$$

$$S_\tau = \frac{i}{2k_0} \int_{-\infty}^{+\infty} e^{i t(z-z_q)} \cdot \begin{Bmatrix} J_\tau\left(\sqrt{k_0^2 - t^2}\right) \\ H_\tau^{(1)}\left(\sqrt{k_0^2 - t^2}\right) \end{Bmatrix} dt \quad ; \quad \begin{cases} \rho < \sqrt{k_0^2 - t^2} \\ \rho > \sqrt{k_0^2 - t^2} \end{cases} \tag{4}$$

with $J_\tau(z)$= Bessel function; $H_\tau^{(1)}(z)$= Hankel function of 1st kind.

Approximation for $k_0\rho << 1$:

$$p \approx \frac{i}{\nu} h_0^{(1)}\left(k_0\sqrt{\rho^2 + \rho_q^2 + (z - z_q)^2} \right) +$$

$$+ \frac{2i}{\Gamma(1/\nu)} \left(k_0\rho\rho_q / 2\right)^{1/\nu} \cdot \frac{h_{1/\nu}^{(1)}\left(k_0\sqrt{\rho^2 + \rho_q^2 + (z - z_q)^2} \right)}{\left(\rho^2 + \rho_q^2 + (z - z_q)^2\right)^{1/(2\nu)}} \cdot \cos \frac{\Phi - \Phi_q}{\nu} + \tag{5}$$

$$+ O\left((k_0\rho)^{\min(2/\nu, 2)}\right)$$

with $h_n^{(1)}(z)$= spherical Hankel function of 1st kind; $\Gamma(z)$= Gamma function.

Far field , $k_0\rho\rho_q/R_1 >> 1$:

$$p \approx \sum_{n,m} \frac{e^{i k_0 R(\alpha_{nm})}}{k_0 R(\alpha_{nm})} + \left\{ V(-\pi - \Phi + \Phi_q) - V(\pi - \Phi + \Phi_q) \right\} \tag{6}$$

with summation over all n,m with $|\Phi - \Phi_q + 2nm\pi\nu| < \pi$, and :

$$\alpha_{nm} = \pi - \Phi + \Phi_q - 2nm\pi\nu \quad ; \quad R_1 = \sqrt{(\rho + \rho_q)^2 + (z - z_q)^2}$$

$$V(ß) = \frac{1}{2\pi\nu} \int_0^\infty \frac{e^{i k_0 R(it)}}{k_0 R(it)} \, \frac{\sin(ß / \nu)}{\cosh(t / \nu) - \cos(ß / \nu)} dt \tag{7}$$

Approximation for the scattered far field :

$$p_s \approx \frac{e^{i(k_0 R_1 + \pi/4)}}{\sqrt{2\pi k_0 R_1}} \, \frac{1}{k_0\sqrt{\rho\rho_q}} \, \frac{1}{\nu} \, \frac{\sin(\pi / \nu)}{\cos(\pi / \nu) - \cos((\Phi - \Phi_q) / \nu)} \tag{8}$$

E.19 Impulsive spherical wave scattering at a hard wedge

Ref.: Biot / Tolstoy, [E.10]; Ouis, [E.11]

See also Sections E.5 , E.6 . This section will give exact solutions in the time domain for an impulsive point source, and approximations for a point source with harmonic signal.

A hard wedge has its apex line on the z axis of a cylindrical co-ordinate system (r,ϑ,z) and its flanks at $\vartheta=0$, $\vartheta=\Theta_0$. The wedge may be convex ($\Theta_0>\pi$) or concave ($\Theta_0<\pi$).

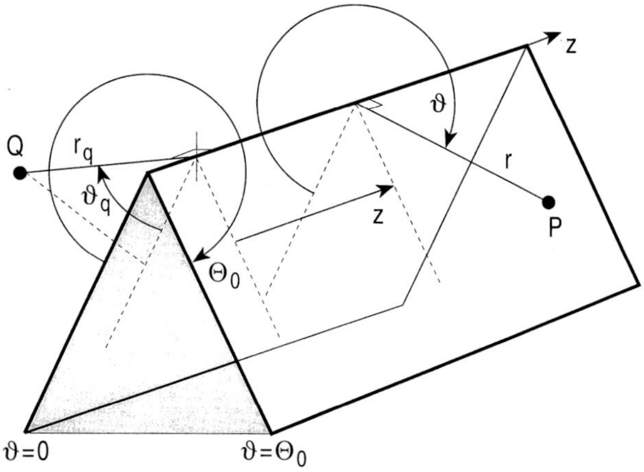

A point source Q with volume flow q is at $(r_q,\vartheta_q,0)$; the observer point P is in (r,ϑ,z) .

The point source sends a delta pulse

$$p_q(r') = \frac{\rho_0 q}{4\pi r'}\cdot\delta(t-r'/c_0) \tag{1}$$

with t= time; r′= distance from Q ; δ(z)= Dirac delta function.

Composition of the field :

$$p = p_q(R_q) + \sum_s p_s(R_s) + p_d \tag{2}$$

with p_q= direct source contribution; p_s= mirror source contribution; p_d= diffracted wave. Some or all of the contributions may vanish, depending on the time interval and the geometrical situation.

In $t<t_0$; $t_0=R_0/c_0$; $R_0=\sqrt{(r-r_q)^2+z^2}$ no signal is received, p=0 .

In $t_0<t<\tau_0$; $\tau_0=R_a/c_0$; $R_a=\sqrt{(r+r_q)^2+z^2}$ the shortest distance between Q and P passing the apex line, only $p_q(R_q)$ and (possibly) mirror source contributions $p_s(R_s)$ are received. $p_q(R_q)$ is obtained by the substitution

$r' \to R_q = \sqrt{r^2 + r_q^2 + z^2 - 2 r r_q \cos(\vartheta - \vartheta_q)}$

in $p_q(r')$, and $p_S(R_S)$ is obtained by a similar substitution in $p_q(r')$ with $r_q \to r_S$; $\vartheta_q \to \vartheta_S$ where r_S, ϑ_S are the co-ordinates of the mirror source.

Mirror sources (of different orders) represent specular reflections at the wedge flanks.

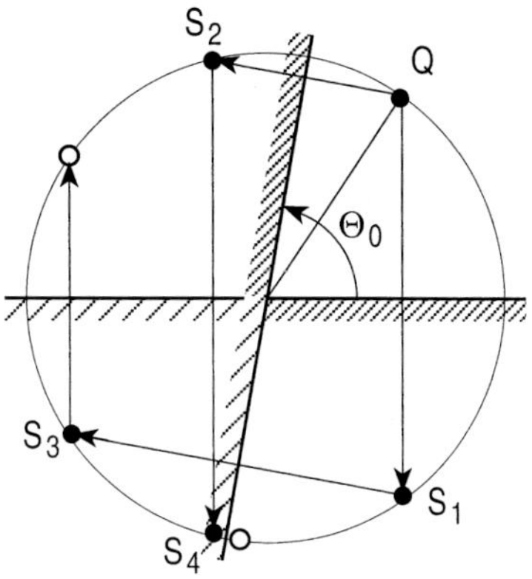

The original source Q and a mirror source S produce a new mirror source at one of the flanks only if they are on the "field side" of that flank (which for that decision is extended to infinity). The black dots represent possible image sources, the open circles are excluded. In general there are conditions in which no image source contribution exists.

For a contribution $p_S(R_S)$ the condition

$$\arccos \frac{r_s^2 + r_q^2 + z^2 - (c_0 t)^2}{2 r_s r_q} \le \pi \quad \text{must hold.} \qquad (3)$$

The diffracted wave p_d *in time* :

The diffracted wave is received for $t > \tau_0$. It vanishes if the wedge angle is an integer fraction of π : $\Theta_0 = \pi/m$. Its time function is :

$$p_d(t) = \frac{-q Z_0}{4\pi\Theta_0} \cdot \{ß\} \cdot \frac{e^{-\pi y/\Theta_0}}{r r_q \cdot \sinh(y)} \qquad (4)$$

$$y = \operatorname{arccosh} \frac{(c_0 t)^2 - (r^2 + r_q^2 + z^2)}{2 r r_q}$$

$$\{ß\} = \frac{\sin\left(\pi(\pi \pm \vartheta \pm \vartheta_q)/\Theta_0\right)}{1 - 2e^{-\pi y/\Theta_0} \sin\left(\pi(\pi \pm \vartheta \pm \vartheta_q)/\Theta_0\right) + e^{-2\pi y/\Theta_0}} \qquad (5)$$

where $\{ß\}$ is the sum of terms with the four possible combinations of signs.

E.20 Spherical wave scattering at a hard screen

Ref.: Biot / Tolstoy, [E.10]; Ouis, [E.11]

The hard screen is the special case $\Theta_0 = 2\pi$ of the previous Section E.19 . The diffracted wave $p_d(t)$ in this case can be given an alternative form, valid for $z=0$ (see sketch in E.19):

$$p_d(t) = \frac{-q\rho_0}{4\pi^2 c_0}\sqrt{\frac{t_+^2 - t_-^2}{t^2 - t_+^2}} \cdot \left\{\left\{ \frac{\cos\left((\vartheta \pm \vartheta_q)/2\right)}{t^2 - t_+^2 + (t_+^2 - t_-^2)\cos^2\left((\vartheta \pm \vartheta_q)/2\right)} \right\}\right\} \quad ; \quad t_\pm = (r \pm r_q)/c_0 \qquad (1)$$

where $\{\{\ldots\}\}$ is the abbreviation for the sum of two terms corresponding to different signs in the argument of the trigonometric function. This form is suited for (approximate) Fourier transformation :

$$p_d(\omega) = \int_{\tau_0}^{\infty} p_d(t) \cdot e^{j\omega t}\, dt \quad ; \quad \tau_0 = \frac{r + r_q}{c_0} \qquad (2)$$

Sound field for a harmonic point source:

The sound field for a harmonic point source with angular frequency $\omega = 2\pi f$ is obtained by a Fourier transformation. The contributions p_q , p_s are the values of the spherical wave

$$p(R) = \frac{jk_0^2 q Z_0}{4\pi} \frac{e^{-jk_0 R}}{k_0 R} \qquad (3)$$

with q= volume flow amplitude, and $R \to R_q$; $R \to R_s$, respectively (see E.19).

Approximations of different orders $p_{di}(f)$ will be given below for the diffracted field $p_d(f)$ in the frequency range.

1st order : development of $p_d(\tau)$ for $\tau = t - \tau_0 << \tau_0$

$$p_{d1}(\omega) = \frac{-q\rho_0}{4\pi^2 c_0} \frac{1}{\sqrt{2t_+(t_+^2 - t_-^2)}} \cdot \left\{\left\{ \frac{1}{\cos\left((\vartheta \pm \vartheta_q)/2\right)} \right\}\right\} \frac{1+j}{2\sqrt{f}} \cdot e^{j\omega\tau_0} \qquad (4)$$

2nd order : in the range of the 1st order, but improved

$$p_{d2}(\omega) = \frac{-q\rho_0}{4\pi^2 c_0}\sqrt{\frac{t_+^2 - t_-^2}{2t_+}} \frac{\pi e^{j\omega\tau_0}}{2t_+} \cdot \left\{\left\{ \frac{\cos\left((\vartheta \pm \vartheta_q)/2\right) \cdot e^{-j\omega a_\pm}}{\sqrt{a_\pm}} \cdot \operatorname{erfc}\left(\sqrt{-j\omega a_\pm}\right) \right\}\right\} \qquad (5)$$

$$a_\pm = (t_+^2 - t_-^2)\cos^2\left((\vartheta \pm \vartheta_q)/2\right)\big/(2t_+) \ge 0$$

with the complementary error function erfc(z) .

3rd order : $\tau^2 + 2t_+\tau \ll (t_+^2 - t_-^2)\cos^2\left((\vartheta \pm \vartheta_q)/2\right)$ (6)

$$p_{d3}(\omega) = \frac{-q\rho_0}{4\pi^2 c_0}\frac{e^{j\omega(\tau_0 - t_+)}}{\sqrt{t_+^2 - t_-^2}} \cdot \left\{\left\{\frac{1}{\cos\left((\vartheta \pm \vartheta_q)/2\right)}\right\}\right\} \cdot K_0(-j\omega t_+) \tag{7}$$

with $K_0(z)$ the modified Bessel function of 2nd kind and order zero.

4th order : $\sqrt{\tau(\tau + 2t_+)} \approx \sqrt{2t_+\tau}$ (8)

$$p_{d4}(\omega) = \frac{-q\rho_0}{4\pi^2 c_0}\sqrt{\frac{t_+^2 - t_-^2}{2t_+}}\frac{\pi e^{j\omega\tau_0}}{2} \cdot \left\{\left\{\frac{\cos\left((\vartheta \pm \vartheta_q)/2\right)}{\sqrt{\Delta_\pm}}\left[\left[\frac{e^{-j\omega\tau_{1,2}}}{\sqrt{\tau_{1,2}}}\operatorname{erfc}\left(\sqrt{-j\omega\tau_{1,2}}\right)\right]\right]\right\}\right\} \tag{9}$$

$$\Delta_\pm = t_+^2 - \left(t_+^2 - t_-^2\right)\cos^2\left((\vartheta \pm \vartheta_q)/2\right) \ge 0 \quad ; \quad \tau_{1,2} = t_+ \mp \sqrt{\Delta_\pm} \ge 0 \tag{10}$$

with [[...]] standing for the difference of the term with index 1 minus term with index 2 .

5th order : After expansion of

$$\frac{1}{\sqrt{\tau + 2t_+}} = \frac{1}{\sqrt{2t_+}}\sum_{n \ge 0}\binom{-1/2}{n}\left(\frac{\tau}{2t_+}\right)^n \quad ; \quad \tau \le 2t_+ \tag{11}$$

and using three series terms :

$$\begin{aligned} p_{d5}(\omega) = {} & \frac{-q\rho_0}{4\pi^2 c_0}\sqrt{\frac{t_+^2 - t_-^2}{2t_+}}\frac{\pi e^{j\omega\tau_0}}{2} \cdot \left\{\left\{\frac{\cos\left((\vartheta \pm \vartheta_q)/2\right)}{\sqrt{\Delta_\pm}}\right.\right. \cdot \\ & \cdot \left[\left[\left(\frac{1}{\sqrt{\tau_{1,2}}} + \frac{\sqrt{\tau_{1,2}}}{4t_+}\right) \cdot e^{-j\omega\tau_{1,2}}\operatorname{erfc}\left(-j\omega\tau_{1,2}\right) - \right.\right. \\ & \left.\left.\left.\left. - \frac{1}{4t_+\sqrt{-j\pi\omega}} + \frac{3}{64\pi}\frac{\tau_{1,2}^{3/2}}{t_+^2}\left(K_0(-j\omega\tau_{1,2}/2) - K_1(-j\omega\tau_{1,2}/2)\left(1 + \frac{1}{j\omega\tau_{1,2}}\right)\right)\right]\right]\right\}\right\} \end{aligned} \tag{12}$$

With more terms :

$$p_{d5}(\omega)=\frac{-q\rho_0}{4\pi^2 c_0}\sqrt{\frac{t_+^2-t_-^2}{2t_+}}\,\frac{\pi e^{j\omega\tau_0}}{2}\cdot\left\{\left\{\frac{\cos\left((\vartheta\pm\vartheta_q)/2\right)}{\sqrt{\Delta_\pm}}\cdot\right.\right. \tag{13}$$

$$\cdot\left[\left[\frac{e^{-j\omega\tau_{1,2}}}{\sqrt{\tau_{1,2}}}\cdot\operatorname{erfc}\left(-j\omega\tau_{1,2}\right)+\sum_{n\ge1}\binom{-1/2}{n}\frac{1}{(2t_+)^n}\cdot\left(\quad(-1)^n\tau_{1,2}^{n-1/2}e^{-j\omega\tau_{1,2}}\cdot\operatorname{erfc}\left(\sqrt{-j\omega\tau_{1,2}}\right)+\right.\right.\right.$$

$$\left.\left.\left.\left.\left.+2^{1-n}\sqrt{\frac{2}{\pi}}(-2j\omega)^{1/2-n}\sum_{m=0}^{n-1}(2n-2m-3)!!\cdot(-2j\omega\tau_{1,2})^m\right)\right]\right]\right\}\right\}$$

6th *order* : with development of the fraction in $\{\{\ldots\}\}$ of $p_d(t)$

$$p_{d6}(\omega)=\frac{-q\rho_0}{4\pi^2 c_0}\frac{e^{j\omega\tau_0}}{\sqrt{t_+^2-t_-^2}}\sum_{n\ge0}(-2t_+)^n\Gamma(n+\tfrac{1}{2})\cdot U(n+\tfrac{1}{2},n+1;-2j\omega t_+)\cdot$$

$$\cdot\left\{\left\{\frac{1}{\cos\left((\vartheta\pm\vartheta_q)/2\right)}\sum_{m=0}^{n}\frac{1}{\tau_1^m\cdot\tau_2^{n-m}}\right\}\right\} \tag{14}$$

where $U(a,b;z)=1/z^a\cdot{}_2F_0(a,1+a-b,-1/z)$ is the Tricomi function.

References to part E :

Scattering of Sound

[E.1] MECHEL, F.P.
"Schallabsorber", Vol. I, Chapter 6:
"Cylindrical sound absorbers"
S.Hirzel Verlag, Stuttgart, 1989

[E.2] MECHEL, F.P.
"Schallabsorber", Vol. II, Chapter 14:
"Characteristic valucs of composite media"
S.Hirzel Verlag, Stuttgart, 1995

[E.3] MECHEL, F.P. J. Sound Vibr. 219 (1999) 559-579
"Improvement of Corner Shielding by an Absorbing Cylinder"

[E.4] MECHEL, F.P. Acta Acustica 83 (1997) 260-283
"A Uniform Theory of Sound Screens and Dams"

[E.5] MECHEL, F.P.
"Mathieu Functions; Formulas, Generation, Use"
S. Hirzel Verlag, Stuttgart, 1997

[E.6] MECHEL, F.P.
"Schallabsorber", Vol. III, Chapter 22:
"Semicircular absorbing dam on absorbing ground"
S.Hirzel Verlag, Stuttgart, 1998

[E.7] PANIKLENKO, A.P.; RYBAK, S.A. Sov.Phys.Acoust. 30 (1984) 148-151

[E.8] JOHNSON, R.K. J.Acoust.Soc.Amer. 61 (1977) 375-377

[E.9] RAWLINS J.Sound Vibr. 41 (1975) 391-393

[E.10] BIOT, M.A.; TOLSTOY, I. J.Acoust.Soc.Amer. 29 (1957) 381-391
"Formulation of Wave Propagation in Infinite Media by Normal Coordinates with an Application to Diffraction"

[E.11] OUIS, D. Report TVBA-3094, 1997, Lund Inst.of Technology
"Theory and Experiment of the Diffraction by a Hard Half Plane"

F
Radiation of Sound

Radiation of sound takes place, not only if a surface is driven by an internal force, but also if the surface is set in vibration by an incident sound wave. Then radiation is the back reaction of the surface to the incident sound in the process of reflection and/or scattering. Part of the power which the vibrating surface produces with the exciting sound pressure is radiated as effective power to infinity; this gives rise to the *radiation loss* of the surface. Part of the reaction is contained in non-radiating near fields; they will influence the tuning of resonating surfaces by the inertia of their *oscillating mass*. This oscillating mass can be represented as the mass contained in a prism with the cross-section of the vibrating surface (e.g. an orifice) and the length of an *end correction*. The advantage of the concept of the oscillating mass and of the end correction is the possibility to include them as members in equivalent networks (they are determined just so that this is possible).

The distinction between "mechanical impedance", "impedance", and "flow impedance" from the section A.3 Conventions is reminded.

F.1 Definition of radiation impedance and end corrections

Ref.: Mechel, [F.1]

Let $v_n(s)$ be the oscillating velocity in a surface A with the co-ordinate s in A , and directed normal to the surface towards the side, on which a sound pressure $p(s)$ exists. The time average sound power produced is :

$$\Pi = \Pi' + j \cdot \Pi'' = \frac{1}{2} \iint_A p(s) \cdot v_n^*(s)\, dA = \iint_A I_n(s)\, dA \tag{1}$$

with the normal time average sound intensity $I_n(s)$. The *radiation impedance* $Z_r = Z'_r + j \cdot Z''_r$ is defined by :

$$\Pi := \frac{1}{2} Z_r \cdot \iint_A |v_n(s)|^2\, dA \tag{2}$$

The *mechanical radiation impedance* Z_{mr} (which is suitable for a small surface A and/or conphase excitation) is defined by :

$$\Pi := \frac{1}{2} Z_{mr} \cdot \langle |v_n(s)|^2 \rangle_A \tag{3}$$

where $\langle \ldots \rangle_A$ stands for the average over A . It is evident that : $Z_{mr}=A \cdot Z_r$.

A normal component $Z_{Fn}(s)$ of a *field impedance* can be defined by: $p(s)= Z_{Fn}(s) \cdot v_n(s)$ on A. Then :

$$\Pi = \frac{1}{2} \iint_A Z_{Fn}(s) \cdot |v_n(s)|^2 \, dA \tag{4}$$

Special case : $Z_{Fn}(s)$=const(s) : $\qquad Z_r=Z_{Fn}$ (5)

Special case : $|v_n(s)|$=const(s) : $\qquad Z_r = \frac{1}{A} \iint_A Z_{Fn}(s) \, dA = \frac{1}{A} \iint_A \frac{dA}{G_n}$ (6)

with the field admittance component $G_n=1/Z_{Fn}$.

Special case : $v_n(s)$=const(s) in magnitude and phase : $Z_r = \frac{\langle p(s) \rangle_A}{v_n}$ (7)

Special case : p(s)=const(s) $\qquad \Pi = \frac{1}{2} p \iint_A v_n^*(s) \, dA = \frac{1}{2} p \cdot q^*$ (8)

with q= volume flow of the surface A .

Related quantities :

The *radiation efficiency* σ is defined as the ratio of the real (effective) power radiated by A to the effective power, which a section of size A of an infinite surface with constant surface velocity v_n would radiate :

$$\sigma = \Pi' \Big/ \frac{1}{2} \iint_A |v_n(s)|^2 \, dA = \frac{Z'_r}{Z_0} \tag{9}$$

The *oscillating mass* M_r is given by : $Z''_{mr} = j\omega \cdot M_r$ or a mass surface density m_r given by
$Z''_r = j\omega \cdot m_r$ with $M_r = A \cdot m_r$.

The *end correction* $\Delta\ell$ is the height of a prism of cross-section A containing the oscillating mass M_r :

$$\Delta\ell = \frac{M_r}{\rho_0 A} = \frac{m_r}{\rho_0} = \frac{Z''_{mr}}{\omega \rho_0 A} = \frac{Z''_r}{\omega \rho_0} = \frac{Z''_r}{k_0 Z_0} \quad ; \quad \frac{\Delta\ell}{a} = \frac{Z''_r}{k_0 a \cdot Z_0} \tag{10}$$

The non-dimensional form $\Delta\ell/a$ contains any length a, mostly the radius of the surface A.

Also used is the *radiation factor* S, which is the ratio of the power Π of A to the power Π_0 of a small spherical radiator with the same square average of the volume flow density $\langle|q|^2\rangle_A$ as the considered surface A:

$$\Pi = S\cdot\Pi_0 \quad ; \quad \Pi_0 = \frac{Z_0}{2}\frac{k_0^2}{4\pi}\langle|q|^2\rangle_A \quad ; \quad Z_r = Z_0\frac{k_0^2 A}{4\pi}\cdot S \qquad (11)$$

F.2 Some methods to evaluate the radiation impedance

The simplest radiators are piston radiators and "breathing" radiators with constant normal particle velocity over the radiator surface A: $v_n(s)$=const. According to section F.1 only the average sound pressure $\langle p(s)\rangle_A$ at the surface must be evaluated.

Also simple are radiators with a surface A which is on a co-ordinate surface of a co-ordinate system in which the wave equation is separable (e.g. spheres, cylinders, ellipsoids etc.) and if the vibration pattern agrees with an eigen function (mode) in that system, because then the modal field impedance of the vibration is constant over A, so it agrees with the radiation impedance (see section F.1).

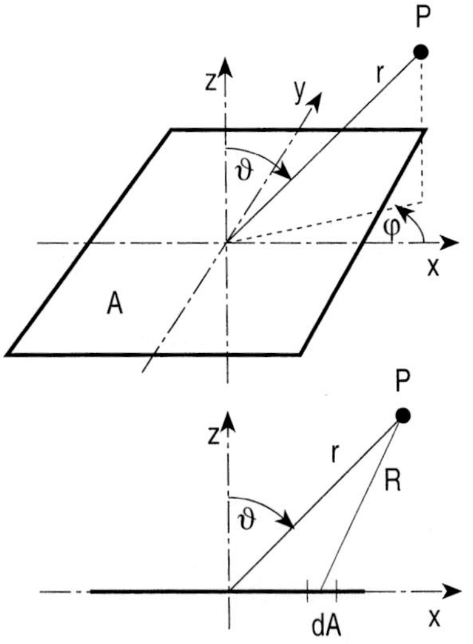

An important family of radiators are plane surfaces A in a surrounding plane baffle wall. Let the normal particle velocity at points (x_0,y_0) of A be $v(x_0,y_0)$. The sound pressure at a field point P(x,y,z) is then:

$$p(x,y,z) = = \frac{jk_0Z_0}{2\pi}\iint_A v(x_0,y_0)\frac{e^{-jk_0R}}{R}\,dx_0\,dy_0 \qquad (1)$$

$$= \frac{jk_0Z_0}{2\pi}\iint_A v(x_0,y_0)\cdot G(x,y,z\,|\,x_0,y_0,0)\,dx_0\,dy_0$$

with Green's function $G(x,y,z\,|\,x_0,y_0,0)$.

One gets with the Fourier transform of $v(x_0,y_0)$ (in the hard baffle wall z=0):

$$V(k_1,k_2)=\int\int_{-\infty}^{+\infty} v(x_0,y_0)\cdot e^{-j(k_1x_0+k_2y_0)}\,dx_0\,dy_0 \tag{2}$$

for the complex power :

$$\Pi=\Pi'+j\cdot\Pi''=\frac{k_0Z_0}{8\pi^2}\int\int_{-\infty}^{+\infty}\frac{|V(k_1,k_2)|^2}{\sqrt{k_0^2-k_1^2-k_2^2}}\,dk_1\,dk_2 \tag{3}$$

and therefore for the radiation impedance :

$$Z_r=k_0Z_0\int\int_{-\infty}^{+\infty}\frac{|V(k_1,k_2)|^2}{\sqrt{k_0^2-k_1^2-k_2^2}}\,dk_1\,dk_2\Bigg/\int\int_{-\infty}^{+\infty}|V(k_1,k_2)|^2\,dk_1\,dk_2 \tag{4}$$

The sound pressure in the far field is given by :

$$p(x,y,z)=\frac{k_0Z_0}{4\pi^2}\int\int_{-\infty}^{+\infty}\frac{V(k_1,k_2)}{\sqrt{k_0^2-k_1^2-k_2^2}}\cdot e^{-j(k_1x+k_2y+z\sqrt{k_0^2-k_1^2-k_2^2})}\,dk_1\,dk_2 \tag{5}$$

Special case :

The surface A is a strip with the strip axis on the y axis and $v(x_0,y_0)=\mathrm{const}(y_0)$:

$$p(x,z)=\frac{k_0Z_0}{2\pi}\int_{-\infty}^{+\infty}\frac{V(k_1)}{\sqrt{k_0^2-k_1^2}}\cdot e^{-j(k_1x+z\sqrt{k_0^2-k_1^2})}\,dk_1 \tag{6}$$

$$\Pi'=\frac{k_0Z_0}{4\pi}\int_{-k_0}^{+k_0}\frac{|V(k_1)|^2}{\sqrt{k_0^2-k_1^2}}\,dk_1 \tag{7}$$

$$Z_r=\frac{k_0Z_0}{2\pi A\langle|v_n|^2\rangle_A}\int_{-\infty}^{+\infty}\frac{|V(k_1)|^2}{\sqrt{k_0^2-k_1^2}}\,dk_1 \tag{8}$$

(Π' and Z_r per unit strip length; A= strip width).

Special case :

The plane surface A and the velocity $v(r)$ have a radial symmetry.

The role of the Fourier transform of $v(r)$ is taken over by a Hankel transform:

$$V(k_r)=2\pi\int_0^{\infty}v(r_0)\cdot J_0(k_rr_0)\cdot r_0\,dr_0 \tag{9}$$

One gets for the sound pressure far field:

$$p(r,\vartheta)=\frac{jk_0Z_0}{2\pi}\frac{e^{-jk_0r}}{r}\cdot V(k_0\sin\vartheta) \tag{10}$$

and for the effective sound power Π' and the radiation impedance Z_r :

$$\Pi'=\frac{k_0^2Z_0}{4\pi}\int_0^{\pi/2}|V(k_0\sin\vartheta)|^2\cdot\sin\vartheta\,d\vartheta=\frac{k_0Z_0}{4\pi}\int_0^{k_0}\frac{|V(k_r)|^2}{\sqrt{k_0^2-k_r^2}}\cdot k_r\,dk_r \tag{11}$$

$$Z_r=\frac{k_0Z_0}{2\pi A\langle|v_n|^2\rangle_A}\int_{-\infty}^{+\infty}\frac{|V(k_r)|^2}{\sqrt{k_0^2-k_r^2}}\cdot k_r\,dk_r \tag{12}$$

BOUWKAMP, [F.3], evaluates the radiation impedance of a plane piston radiator with particle velocity distribution v(x,y)=const as :

$$Z_r=\frac{Z_0k_0^2A}{4\pi^2}\int_0^{2\pi}d\varphi\int_0^{\pi/2+j\infty}|D(\vartheta,\varphi)|^2\cdot\sin\vartheta\,d\vartheta \tag{13}$$

where $D(\vartheta,\varphi)$ is the far field directivity function of the radiated sound (directivity pattern with unit value in the maximum). The integration over $\vartheta=0\to\vartheta=\pi/2$ returns the real part of Z_r ; the integration $\vartheta=\pi/2+j\cdot 0\to\vartheta=\pi/2+j\cdot\infty$ returns the imaginary part of Z_r .

F.3 Spherical radiators

Ref.: Mechel, [F.1]

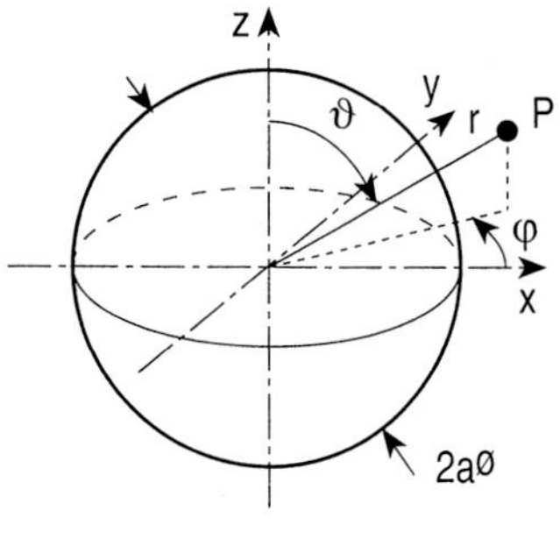

Let $v(\vartheta,\varphi)$ be the pattern of the normal (outward) particle velocity on the sphere with radius a .

The pattern is synthesised with spherical modes :

$$v(\vartheta,\varphi)=\sum_{n=0}^{\infty}\sum_{m=0}^{n}V_{m,n}\cdot P_n^m(\cos\vartheta)\cdot\cos(m\varphi) \tag{1}$$

with associate Legendre functions :

$$P_n^m(x)=(1-x^2)^{m/2}\frac{d^mP_n(x)}{dx^m}\;;\;m\ge 1\quad;\quad P_n(x)=P_n^0(x)=\frac{1}{2^n\,n!}\frac{d^n}{dx^n}(x^2-1)^n \tag{2}$$

defined via the Legendre polynomials; some special values :

$$P_0(x) = 1 \quad ; \quad P_1(x) = x$$
$$P_2(x) = (3x^2 - 1)/2 \quad ; \quad P_3(x) = (5x^3 - 3x)/2 \tag{3}$$

The modal velocity amplitudes are :

$$V_{m,n} = \frac{1}{N_{m,n}} \int_0^{2\pi} d\varphi \int_0^{\pi} v(\vartheta,\varphi) \cdot P_n^m(\cos\vartheta) \cdot \cos(m\varphi) \cdot \sin\vartheta \, d\vartheta \tag{4}$$

with the mode norms :

$$N_{m,n} = \int_0^{2\pi} \cos^2(m\varphi)\, d\varphi \int_{-1}^{1} \left(P_n^m(x)\right)^2 dx = \frac{2\pi}{\delta_m} \frac{2}{2n+1} \frac{(n+m)!}{(n-m)!} \quad ; \quad \delta_m = \begin{cases} 1 ; m=0 \\ 2 ; m>0 \end{cases} \tag{5}$$

The sound pressure at the surface of the sphere is :

$$p(a,\vartheta,\varphi) = \sum_{n=0}^{\infty} \sum_{m=0}^{n} Z_n \cdot V_{m,n} \cdot P_n^m(\cos\vartheta) \cdot \cos(m\varphi) \tag{6}$$

where Z_n are the modal impedances at the sphere surface (directed inward) :

$$Z_n = -j\rho_0 c_0 \frac{h_n^{(2)}(k_0 a)}{h_n'^{(2)}(k_0 a)} \tag{7}$$

with the spherical Hankel functions of the 2nd kind $h_n^{(2)}(z)$.

If the sphere oscillates in a single mode, the modal impedance is the radiation impedance (see section F.1).

Special case :

The vibration pattern $v(\vartheta,\varphi)$=const(φ) , i.e. the oscillation is symmetrical around the z axis.

$$v(\vartheta) = \sum_{n=0}^{\infty} V_n \cdot P_n(\cos\vartheta)$$
$$V_n = \left(n + \frac{1}{2}\right) \int_0^{\pi} v(\vartheta) \cdot P_n(\cos\vartheta) \cdot \sin\vartheta \, d\vartheta \tag{8}$$

$$p(r,\vartheta) = -j\rho_0 c_0 \sum_{n=0}^{\infty} V_n \cdot P_n(\cos\vartheta) \frac{h_n^{(2)}(k_0 r)}{h_n'^{(2)}(k_0 a)} \xrightarrow[r \to a]{} \sum_{n=0}^{\infty} Z_n V_n \cdot P_n(\cos\vartheta) \tag{9}$$

Special case :

Breathing sphere: $V_{n>0}=0$; $V_0=v(\vartheta,\varphi)=$ const(ϑ,φ) . (10)

Radiation impedance (= zero mode impedance):

$$Z_{r0} = \rho_0 c_0 \frac{j k_0 a}{1 + j k_0 a} = \rho_0 c_0 \frac{(k_0 a)^2 + j k_0 a}{1 + (k_0 a)^2} \tag{11}$$

Oscillating mass :

$$M_{r0} = A \cdot \frac{Z''_{r0}}{\omega} = \frac{\rho_0 \cdot 4\pi a^3}{1 + (k_0 a)^2} \xrightarrow[k_0 a \ll 1]{} \rho_0 \cdot 4\pi a^3 = \rho_0 \cdot 3\,\mathrm{Vol} \tag{12}$$

Special case :

Oscillating rigid sphere: $V_{n \neq 1} = 0$; $v(\vartheta) = V_1 \cdot \cos\vartheta$. (13)

Radiation impedance (= 1st order mode impedance) :

$$Z_{r1} = \frac{j\rho_0 c_0}{\dfrac{2}{k_0 a} - \dfrac{h_0^{(2)}(k_0 a)}{h_1^{(2)}(k_0 a)}} = \rho_0 c_0 \frac{(k_0 a)^4 + j k_0 a\left(2 + (k_0 a)^2\right)}{4 + (k_0 a)^4}$$

$$\xrightarrow[k_0 a \ll 1]{} \rho_0 c_0 \frac{(k_0 a)^4}{4} + j\omega\rho_0 a/2 \tag{14}$$

$$M_{r1} \xrightarrow[k_0 a \ll 1]{} \rho_0 \cdot \tfrac{3}{2}\mathrm{Vol}$$

In general, for the n-th mode oscillation (n>0) :

$$Z'_{rn} = \rho_0 c_0 \frac{(k_0 a)^{2n+2}}{(n+1)^2 [1 \cdot 3 \cdot \ldots \cdot (2n-1)]^2} \quad ; \quad (k_0 a)^2 \ll |2n-1|$$

$$= \rho_0 c_0 \quad ; \quad k_0 a \gg n^2 + 1 \tag{15}$$

$$Z''_{rn} = \rho_0 c_0 \frac{k_0 a}{n+1} \quad ; \quad (k_0 a)^2 \ll |2n-1|$$

$$= \rho_0 c_0 / k_0 a \quad ; \quad k_0 a \gg n^2 + 1 \tag{16}$$

$$M_{rn} \xrightarrow[k_0 a \ll 1]{} \rho_0 \cdot \tfrac{3}{n+1}\mathrm{Vol}$$

Correspondence in the graphs below: “Zsn” → Z_{rn}/Z_0 ; “k0a” → $k_0 a$.

The maximum of $\mathrm{Im}\{Z_{rn}/Z_0\}$ is at about $k_0 a = n$.

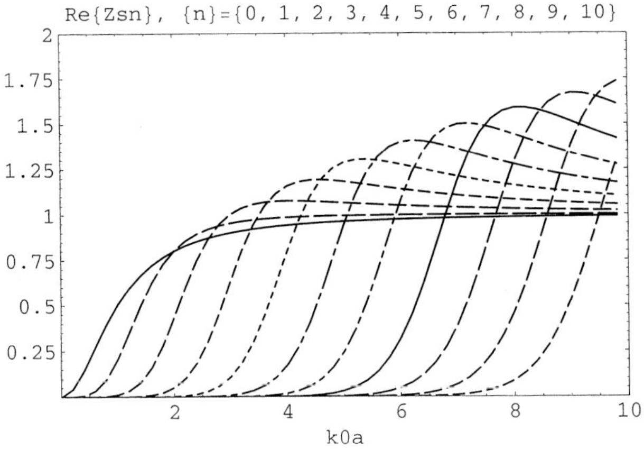
Re{Zsn}, {n}={0, 1, 2, 3, 4, 5, 6, 7, 8, 9, 10}
2
1.75
1.5
1.25
1
0.75
0.5
0.25
2
4
6
8
10
k0a

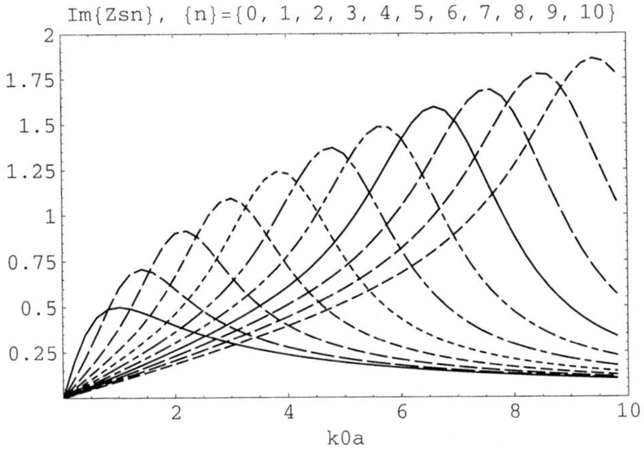
Im{Zsn}, {n}={0, 1, 2, 3, 4, 5, 6, 7, 8, 9, 10}
2
1.75
1.5
1.25
1
0.75
0.5
0.25
2
4
6
8
10
k0a

F.4 Cylindrical radiators

Ref.: Mechel, [F.1]

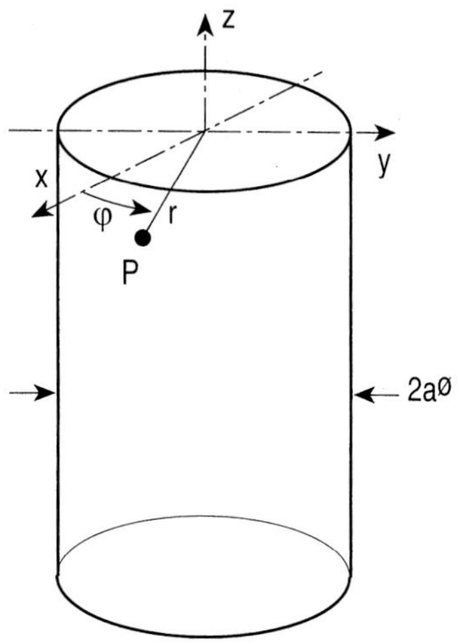

Let $v(\vartheta,\varphi)$ be the pattern of the normal (outward) particle velocity on the cylinder with radius a.

The pattern is synthesised with cylindrical modes :

$$v(\varphi,z) = \sum_{m,n\geq 0} V_{m,n} \cdot \cos(n\varphi) \cdot \cos(k_m z) \tag{1}$$

The sound pressure field is then : (2)

$$p(r,\varphi,z) = \sum_{m,n\geq 0} V_{m,n} \cdot Z_{m,n}(r) \cdot \cos(n\varphi) \cdot \cos(k_m z)$$

with the modal field impedances (in radial direction) :

$$Z_{m,n}(r) = \frac{-jk_0 Z_0}{k_{rm}} \frac{H_n^{(2)}(k_{rm} r)}{H_n'^{(2)}(k_{rm} r)} \quad ; \quad k_{rm}^2 = k_0^2 - k_m^2 \tag{3}$$

($H_n^{(2)}(z)$ Hankel functions of the 2nd kind).

The modal velocity amplitudes $V_{m,n}$ are obtained from the integral transformation of the given pattern $v(\vartheta,\varphi)$:

$$V_{m,n} = \frac{\delta_m \delta_n}{4\pi} \lim_{L\to\infty} \int_{-L}^{L} dz \int_0^{2\pi} v(\varphi,z) \cdot \cos(n\varphi) \cdot \cos(k_m z)\, d\varphi \quad ; \quad \delta_m = \begin{cases} 1 ; m = 0 \\ 2 ; m > 0 \end{cases} \tag{4}$$

Special case :

The cylinder surface oscillates with only one azimuthal mode m and one axial wave number k_m. Then (according to section F.1) the modal wave impedance $Z_{m,n}(a)$ is the radiation impedance Z_r :

$$Z_r = Z_{m,n}(a) = \frac{-jk_0 Z_0}{k_{rm}} \frac{H_n^{(2)}(k_{rm} a)}{H_n'^{(2)}(k_{rm} a)} \tag{5}$$

For an axially conphase oscillation (k_m=0) : $Z_{rn} = Z_{0,n}(a) = -j\rho_0 c_0 \frac{H_n^{(2)}(k_0 a)}{H_n'^{(2)}(k_0 a)}$ (6)

For thin cylinders (k_0a<<1) and n>0 : $\frac{Z'_{rn}}{\rho_0 c_0} \approx \pi k_0 a \frac{(k_0 a)^{2n}}{(n!)^2 \cdot 2^{2n-1}} \quad ; \quad \frac{Z''_{rn}}{\rho_0 c_0} \approx \frac{k_0 a}{n}$ (7)

For n=0 :
$$\frac{Z_{r0}}{\rho_0 c_0} \approx \frac{\pi k_0 a}{2} - j k_0 a \cdot \ln(k_0 a) \tag{8}$$

Special case :
A slow mode in the axial direction : $k_m^2 > k_0^2$

The modal radiation impedances then are :

$$Z_{m,n}(a) = j\rho_0 c_0 \frac{k_0}{\sqrt{k_m^2 - k_0^2}} \Bigg/ \left[\frac{K_{n+1}\left(a\sqrt{k_m^2 - k_0^2}\right)}{K_n\left(a\sqrt{k_m^2 - k_0^2}\right)} - \frac{n}{a\sqrt{k_m^2 - k_0^2}} \right] \tag{9}$$

with $K_n(z)$ modified Bessel functions of the 2nd kind.

Correspondence in the graphs below: "Zsn" → Z_{rn}/Z_0 ; "k0a" → k_0a . For $k_m=0$.

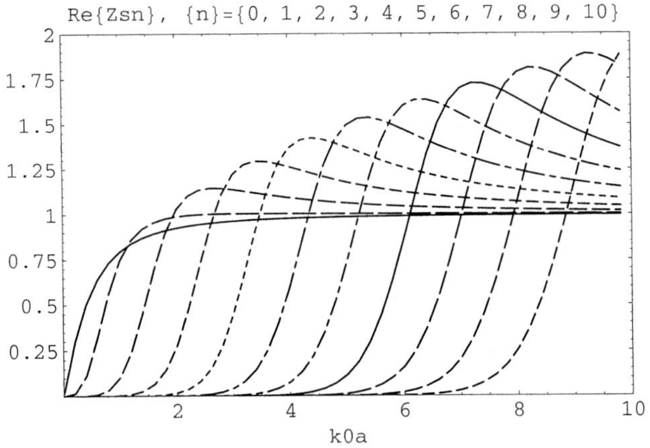

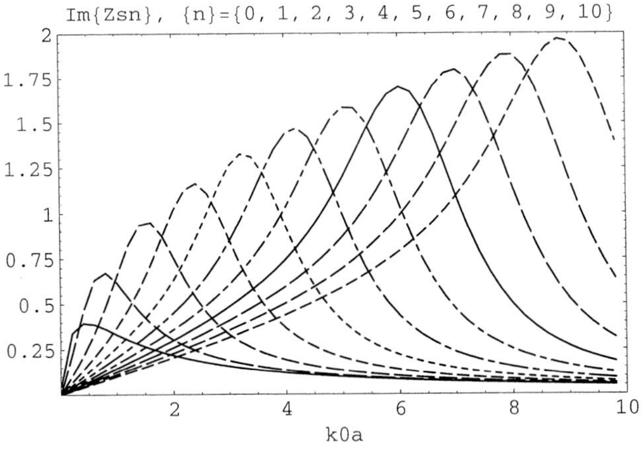

F.5 Piston radiator on a sphere

Ref.: Mechel, [F.1]

This case corresponds to the classical Helmholtz resonator.

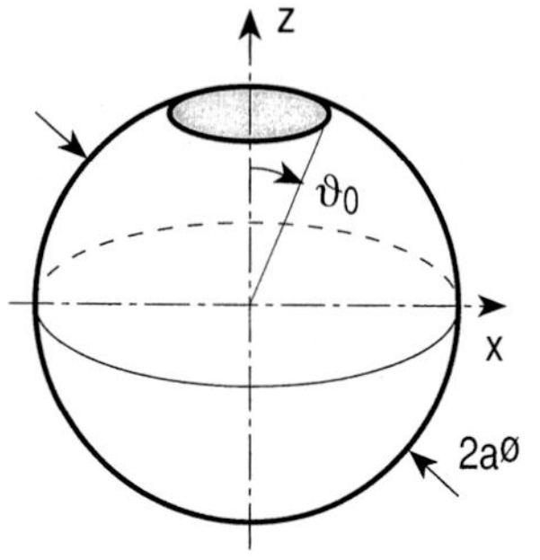

A hollow hard sphere with radius a has a circular hole which subtendeds an angle ϑ_0 with the z axis.

Let the particle velocity be constant in the hole :

$$v(\vartheta)=\begin{cases} v_0 & ; \; 0\le\vartheta<\vartheta_0 \\ 0 & ; \; \vartheta_0<\vartheta\le\pi \end{cases}$$

Modal velocity amplitudes at r=a :

$$V_n=(n+1/2)\cdot v_0 \int_{\cos\vartheta_0}^{1} P_n(x)\,dx=\frac{v_0}{2}\left[P_{n-1}(\cos\vartheta_0)-P_{n+1}(\cos\vartheta_0)\right] \tag{1}$$

with $P_n(z)$ Legendre polynomials, and $P_{-1}(z)=1$.
Radial particle velocity and sound pressure at r=a :

$$v(a,\vartheta)=\sum_{n=0}^{\infty} V_n\cdot P_n(\cos\vartheta) \quad ; \quad p(a,\vartheta)=\sum_{n=0}^{\infty} Z_n(a)\cdot V_n\cdot P_n(\cos\vartheta) \tag{2}$$

using the modal (radial) impedances :

$$Z_n(a)=-j\rho_0 c_0\frac{h_n^{(2)}(k_0a)}{h_n'^{(2)}(k_0a)} \tag{3}$$

with the spherical Hankel functions of the 2nd kind $h_n^{(2)}(z)$.

Because $v(\vartheta)$=const over the hole, its radiation impedance is given by the average sound pressure and the particle velocity (see section F.1) with the radiator surface :

$$A=2\pi a^2\int_0^{\vartheta_0}\sin\vartheta\, d\vartheta=2\pi a^2(1-\cos\vartheta_0) \tag{4}$$

$$\langle p(a,\vartheta)\rangle_A=v_0\frac{\pi a^2}{A}\sum_{n=0}^{\infty}\frac{Z_n(a)}{2n+1}\left[P_{n-1}(\cos\vartheta_0)-P_{n+1}(\cos\vartheta_0)\right]^2 \tag{5}$$

This gives the radiation impedance :

$$Z_r = \frac{1}{2(1-\cos\vartheta_0)} \sum_{n=0}^{\infty} \frac{Z_n(a)}{2n+1} \left[P_{n-1}(\cos\vartheta_0) - P_{n+1}(\cos\vartheta_0)\right]^2 \tag{6}$$

In the limit of low frequencies :

$$\frac{Z_r}{\rho_0 c_0} \approx \frac{1+\cos\vartheta_0}{2} \frac{Z_{r0}}{\rho_0 c_0} = \frac{1+\cos\vartheta_0}{2} \frac{jk_0 a}{1+jk_0 a} \tag{7}$$

Correspondence in the diagrams below: “Zs” $\rightarrow Z_r/\rho_0 c_0$; “theta0” $\rightarrow \vartheta_0$; “k0a” $\rightarrow k_0 a$.
The dashes become shorter for higher list entries of ϑ_0 .

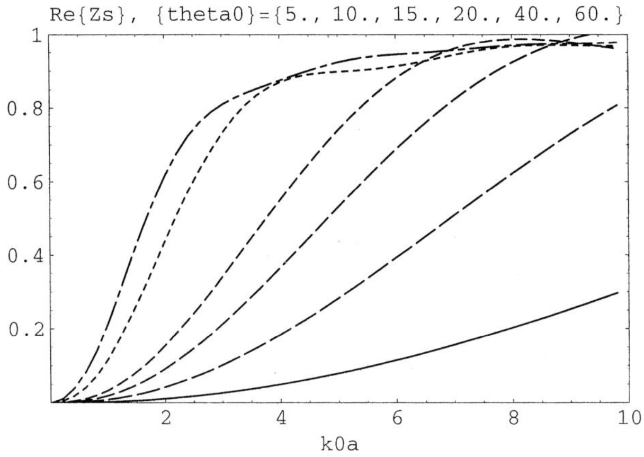

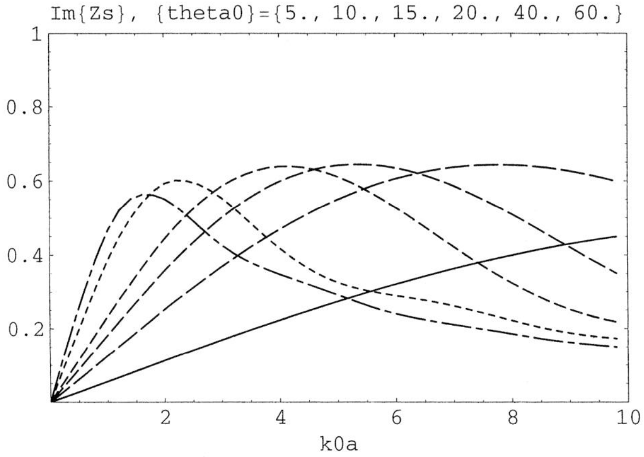

F.6 Strip-shaped radiator on cylinder

Ref.: Mechel, [F.1]

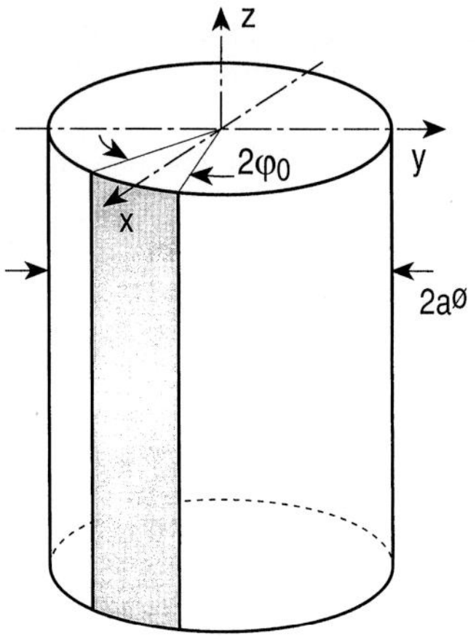

A hard cylinder with radius a has a vibrating strip on its surface, which subtends an angle φ_0 with the x axis.

The radial particle velocity be constant in azimuthal direction and may have a propagating or standing wave pattern in the axial direction :

$$v(a,\varphi,z)=\begin{cases} v_0\cdot g(k_m z) & ; \ -\varphi_0\le\varphi\le\varphi_0 \\ 0 & ; \ \varphi_0<\varphi<2\pi-\varphi_0 \end{cases} \quad (1)$$

The modal particle velocity amplitudes are :

$$V_{m,n}=\frac{\delta_m}{\pi}v_0\varphi_0\frac{\sin(n\varphi_0)}{n\varphi_0} \quad ; \quad \delta_m=\begin{cases}1\,;\,m=0\\2\,;\,m>0\end{cases} \quad ; \quad k_r^2=k_0^2-k_m^2 \qquad (2)$$

The radiation impedance is evaluated as :

$$Z_r=\iint_A p\cdot v^* dA\Big/\iint_A |v|^2\,dA=\frac{\varphi_0}{\pi}\sum_{n=0}^{\infty}\delta_n Z_{m,n}(a)\left(\frac{\sin(n\varphi_0)}{n\varphi_0}\right)^2$$
$$=\rho_0 c_0\frac{-j\varphi_0}{\pi}\frac{k_0 a}{k_r a}\sum_{n=0}^{\infty}\delta_n\frac{H_n^{(2)}(k_r a)}{H_n'^{(2)}(k_r a)}\left(\frac{\sin(n\varphi_0)}{n\varphi_0}\right)^2 \qquad (3)$$

At high frequencies : $Z_r\to\rho_0 c_0\cdot k_0/k_r$. (4)

Correspondence in the diagrams below: "Zs" $\to Z_r/\rho_0 c_0$; "phi0" $\to\varphi_0$; "k0a" $\to k_0 a$. The dashes become shorter for higher list entries of φ_0 . The axial wave number there is k_m=0.

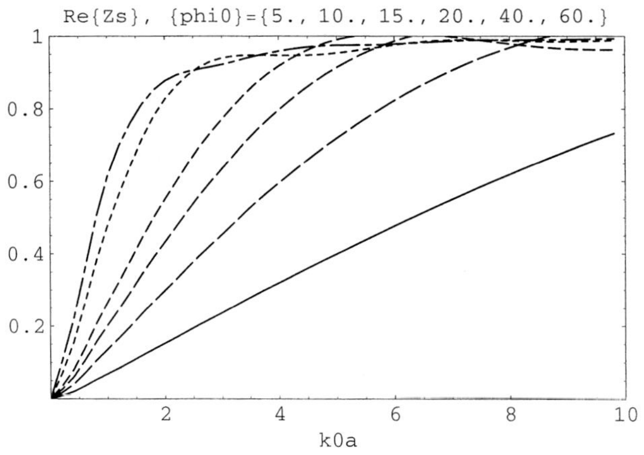

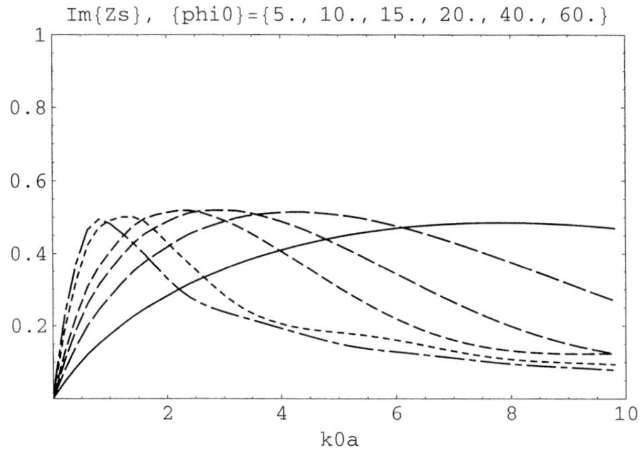

F.7 Plane piston radiators

Ref.: Mechel, [F.1]

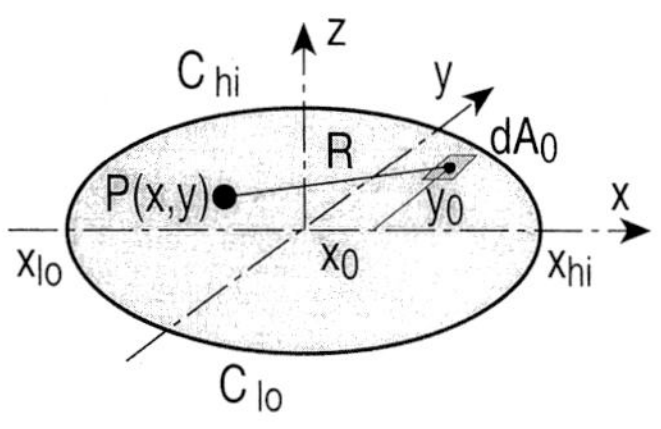

A plane surface A , surrounded by a plane, hard baffle wall, oscillates with a constant velocity v .

A general scheme of evaluation for the radiation impedance Z_r can be designed for surfaces A with convex border lines.

The evaluation applies the field impedance $Z_F(x,y)$ on the radiating surface.

$$\frac{Z_F}{Z_0} = \frac{j}{2\pi} \iint_{k_0^2 A} \frac{e^{-jk_0R}}{R} \, d(k_0^2 A) = \frac{1}{2\pi} \int_{k_0 x_{lo}}^{k_0 x_{hi}} d(k_0 x_0) \int_{k_0 C_{lo}(x_0)}^{k_0 C_{hi}(x_0)} \left[\frac{\sin k_0 R}{k_0 R} + j \frac{\cos k_0 R}{k_0 R} \right] d(k_0 y_0) \quad (1)$$

$$\frac{Z_r}{Z_0} = \frac{1}{2\pi k_0^2 A} \int_{k_0 x_{lo}}^{k_0 x_{hi}} d(k_0 x) \int_{k_0 C_{lo}(x)}^{k_0 C_{hi}(x)} \frac{Z_F(x,y)}{Z_0} \, d(k_0 y) \quad (2)$$

Circular piston radiator with radius a :

$$\frac{Z_r}{Z_0} = 1 - \frac{J_1(2k_0 a)}{k_0 a} + j \frac{S_1(2k_0 a)}{k_0 a} \quad (3)$$

with $J_1(z)$ Bessel function, $S_1(z)$ Struve function.

Approximation for low $k_0 a$ (with $x = 2k_0 a$; for about $x < 4$; range depends on number of terms) :

$$\frac{Z_r'}{Z_0} = \frac{x^2}{2 \cdot 4} - \frac{x^4}{2 \cdot 4^2 \cdot 6} + \frac{x^6}{2 \cdot 4^2 \cdot 6^2 \cdot 8} - + \ldots \quad ; \quad \frac{Z_r''}{Z_0} = \frac{4}{\pi} \left[\frac{x}{3} - \frac{x^3}{3^2 \cdot 5} + \frac{x^5}{3^2 \cdot 5^2 \cdot 7} - + \ldots \right] \quad (4)$$

Approximation for high $k_0 a$ (with $x = 2k_0 a$, for about $x > 4$) :

$$\frac{Z_r'}{Z_0} = 1 - \frac{2}{x} \sqrt{\frac{2}{\pi x}} \cdot \sin(x - \pi/4) \quad ; \quad \frac{Z_r''}{Z_0} = \frac{4}{\pi x} \left[1 - \sqrt{\frac{2}{x}} \cdot \sin(x + \pi/4) \right] \quad (5)$$

Correspondence in the diagram below: "Zs" $\rightarrow Z_r/\rho_0 c_0$; "k0a" $\rightarrow k_0 a$.
Full line: $\mathrm{Re}\{Z_r/\rho_0 c_0\}$, dashed line: $\mathrm{Im}\{Z_r/\rho_0 c_0\}$.

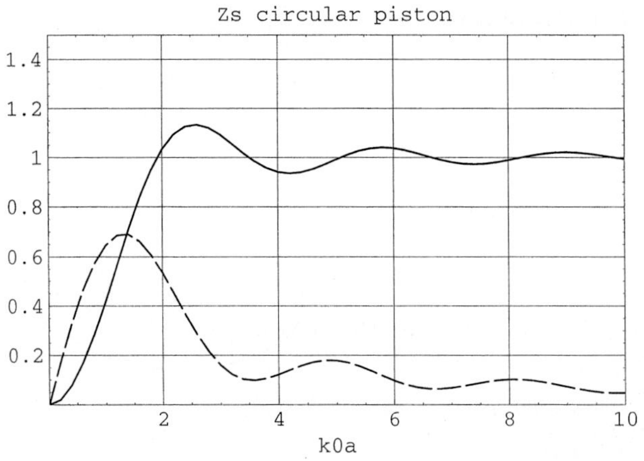

Oscillating free circular disk with radius a ; oscillation normal to disk :

The sound field is described in oblate spheroidal co-ordinates (ρ,ϑ,φ) (which are generated by rotation of the elliptic-hyperbolic cylinder co-ordinates (ρ,ϑ) around the short axis of the ellipses), with relation to the Cartesian co-ordinates :

$$z = a\cdot\sinh\rho\cdot\cos\vartheta \quad ; \quad \begin{matrix} x \\ y \end{matrix} = a\cdot\cosh\rho\cdot\sin\vartheta\cdot\begin{matrix}\cos \\ \sin\end{matrix}\varphi \tag{6}$$

The co-ordinate value $\rho=0$ describes a circular disk with radius a normal to the z axis.

$$\frac{Z_r}{Z_0} = \frac{-8jk_0a}{9}\sum_{n=1,3,\ldots}^{\infty}\left[\frac{he_{0n}(-jk_0a, j\sinh\rho)}{d\,he_{0n}(-jk_0a, j\sinh\rho)/d\rho}\right]_{\rho=0}\frac{d_1(-jk_0a\,|\,0,n)}{\Delta_{0n}}S_{0n}(-jk_0a,\cos\vartheta) \tag{7}$$

with : $S_{0n}(\rho,\vartheta)$= azimuthal spheroidal function; $he_{0n}(\rho,z)$= even radial spheroidal function of the 3rd kind; the term $d_1(-jk_0a|0,n)$ comes from the expansion of $S_{0n}(\rho,\vartheta)$ in associate Legendre functions :

$$S_{0n}(\rho,\vartheta) = \sum_{m=1,3,\ldots}^{\infty} d_m(\rho\,|\,0,1)\cdot T_m^0(\vartheta) \tag{8}$$

and Δ_{0n} from :

$$\Delta_{0n} = \int_{-1}^{+1} S_{0n}^2(\rho,\vartheta)\,d\vartheta \tag{9}$$

Approximation for low k_0a :

$$\frac{Z_r}{Z_0} \approx \frac{16}{27\pi^2}(k_0a)^4 + j\frac{8}{3\pi}k_0a \tag{10}$$

Elliptic piston in a baffle wall :

The ellipse has a long axis 2a and a short axis 2b ; the ratio of the axes is ß=b/a .

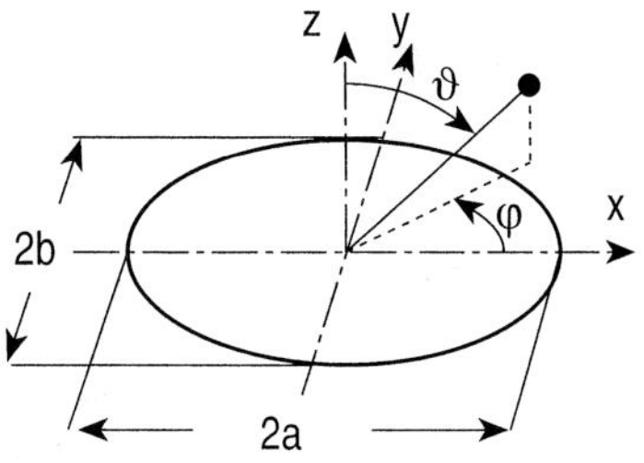

Some evaluations in the literature start from the BOUWKAMP integral (see section F.2) with the far field directivity function of the radiated sound :

$$D(\vartheta,\varphi) = 2\frac{J_1(k_0\sin\vartheta\sqrt{a^2\cos^2\varphi + b^2\sin^2\varphi})}{k_0\sin\vartheta\sqrt{a^2\cos^2\varphi + b^2\sin^2\varphi}} \tag{11}$$

One solution for the real part of the radiation impedance $Z_r = Z_r' + j\cdot Z_r''$ is :

$$\frac{Z_r'}{Z_0} = k_0a\cdot k_0b\sum_{m=0}^{\infty}\frac{(k_0a)^{2m}}{(m+1)!\,(m+2)!}\cdot {}_2F_1\left(-m\,;\tfrac{1}{2}\,;1\,;\mu^2\right) \quad ; \quad \mu^2 = 1-ß^2 \tag{12}$$

with ${}_2F_1(\alpha;\beta;\gamma;z)$ the hypergeometric function. The numerical errors become large for $k_0a \gg 1$.

A solution suited for numerical integration is :

$$\frac{Z_r'}{Z_0} = 1 - \frac{2}{\pi} k_0^2 ab \int_0^{\pi/2} \frac{J_1(2B)}{B^3}\, d\varphi \quad ; \quad B = k_0 a \sqrt{\cos^2\varphi + ß^2 \sin^2\varphi}$$

$$\frac{Z_r''}{Z_0} = \frac{2}{\pi} k_0^2 ab \int_0^{\pi/2} \frac{S_1(2B)}{B^3}\, d\varphi \tag{13}$$

with $J_1(z)$= Bessel function; $S_1(z)$= Struve function. The numerical integration can be avoided by an expansion of the integrands. This leads to the iterative evaluation :

$$\frac{Z_r'}{Z_0} = ß\left[(k_0a)^2/2 + \sum_{n=2}^{n_{hi}} c_n' \cdot I_n'\right]$$

$$c_1' = (k_0a)^2/2 \quad ; \quad c_n' = \frac{-(k_0a)^2}{n\cdot(n+1)} \cdot c_{n-1}' \quad ; \quad I_0' = 1/ß \quad ; \quad I_1' = 1 \quad ; \quad I_n' = 2I_n'/\pi \tag{14}$$

$$\frac{Z_r''}{Z_0} = \frac{4k_0b}{\pi^2}\left[\frac{4}{3}I_0'' - \frac{16}{45}(k_0a)^2 \cdot I_1'' + \sum_{n=2}^{n_{hi}} c_n'' \cdot I_n''\right]$$

$$c_1'' = -16(k_0a)^2/45 \quad ; \quad c_n'' = -4(k_0a)^2/((2n+1)(2n+3)) \tag{15}$$

$$I_0'' = K(\mu^2) \quad ; \quad I_1'' = E(\mu^2) \quad ; \quad I_n'' = \frac{2n-2}{2n-1}(1+ß^2)\cdot I_{n-1}'' - \frac{2n-3}{2n-1}ß^2 \cdot I_{n-2}''$$

with $K(z)$, $E(z)$ the complete elliptic integrals of the 1st and 2nd kind. The upper summation limit should be $n_{hi} \ge 2(k_0a+1)$.

A further solution for the real component of Z_r is :

$$\frac{Z_r'}{Z_0} = 1 - \frac{J_1(2k_0a)}{k_0a} - (1-ß)\cdot J_2(2k_0a) - \frac{ß}{k_0a}\sum_{n=2}^{n_{hi}} \hat{c}_n \cdot \hat{I}_n \cdot J_{1+n}(2k_0a)$$

$$\hat{c}_1 = (1-ß^2)\cdot k_0a \quad ; \quad \hat{c}_n = \frac{(1-ß^2)k_0a}{n} \cdot \hat{c}_{n-1} \tag{16}$$

$$\hat{I}_0 = 1/ß \quad ; \quad \hat{I}_1 = 1/(ß(1+ß)) \quad ; \quad \hat{I}_n = \left(\frac{1}{1-ß^2} + \frac{2n-3}{2n-2}\right)\cdot \hat{I}_{n-1} - \frac{1}{1-ß^2}\frac{2n-3}{2n-2}\cdot \hat{I}_{n-2}$$

with $J_n(z)$= Bessel functions.

Correspondence in the diagrams below: "Zs" $\to Z_r/Z_0$; "k0a" $\to k_0a$; dashes become shorter for higher positions in the parameter list $\{ß\}=\{0.25 , 0.5 , 1\}$.

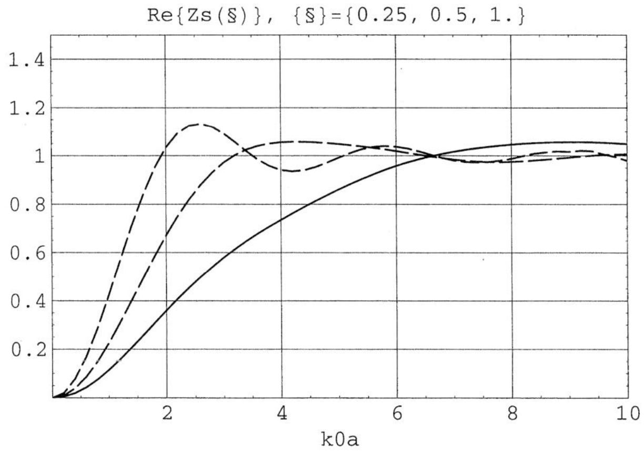

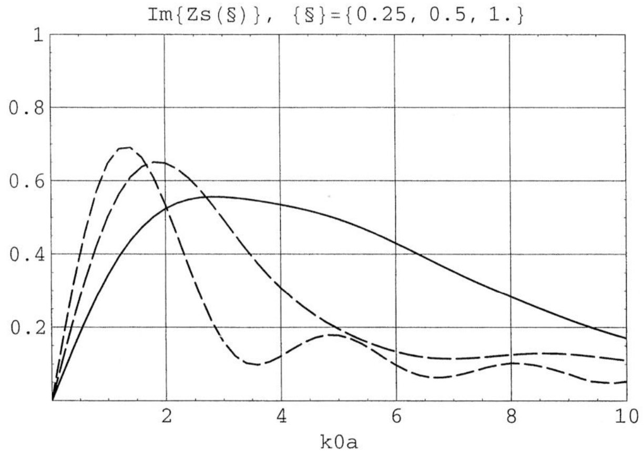

Rectangular piston in a baffle wall :

The rectangle has a long side a and a short side b ; the side length ratio is ß=b/a .

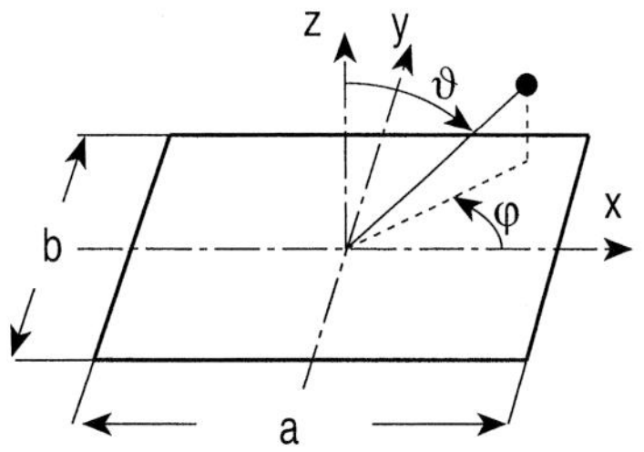

Some evaluations in the literature start from the BOUWKAMP integral (see section F.2) with the far field directivity function of the radiated sound (with si(z)=(sinz)/z):

$$D(\vartheta,\varphi) = \mathrm{si}\left(k_0 a/2 \cdot \sin\vartheta \cdot \cos\varphi\right) \cdot \mathrm{si}\left(k_0 b/2 \cdot \sin\vartheta \cdot \sin\varphi\right) \tag{17}$$

A first form of the radiation impedance $Z_r = Z_r' + j \cdot Z_r''$ is : (18)

$$\frac{Z_r'}{Z_0} =$$

$$1 + \frac{ß}{\pi}\left[\mathrm{Ci}(k_0 a) - \frac{\sin k_0 a}{k_0 a} + \frac{\cos k_0 a - 1}{(k_0 a)^2}\right] + \frac{1}{\pi ß}\left[\mathrm{Ci}(k_0 b) - \frac{\sin k_0 b}{k_0 b} + \frac{\cos k_0 b - 1}{(k_0 b)^2}\right] - \frac{2ß}{\pi} I_1(k_0 a, ß)$$

$$\frac{Z_r''}{Z_0} =$$

$$-\frac{ß}{\pi}\left[\mathrm{Si}(k_0 a) + \frac{\sin k_0 a}{(k_0 a)^2} + \frac{\cos k_0 a - 2}{k_0 a}\right] - \frac{1}{\pi ß}\left[\mathrm{Si}(k_0 b) + \frac{\sin k_0 b}{(k_0 b)^2} + \frac{\cos k_0 b - 2}{k_0 b}\right] - \frac{2ß}{\pi} I_2(k_0 a, ß)$$

with Ci(z), Si(z) the integral cosine and sine functions, and the integrals :

$$I_1(k_0 a, ß) = \int_0^1 \left[\mathrm{Ci}(k_0 a\sqrt{x^2 + 1/ß^2}) + \frac{1}{ß^2}\mathrm{Ci}(k_0 b\sqrt{x^2 + ß^2})\right] \cdot (1-x)\,dx$$

$$I_2(k_0 a, ß) = \int_0^1 \left[\mathrm{Si}(k_0 a\sqrt{x^2 + 1/ß^2}) + \frac{1}{ß^2}\mathrm{Si}(k_0 b\sqrt{x^2 + ß^2})\right] \cdot (1-x)\,dx \tag{19}$$

A second form of the radiation impedance $Z_r = Z_r' + j \cdot Z_r''$ is :

$$\frac{Z_r'}{Z_0} = 1 - \frac{2}{\pi ß (k_0 a)^2} \cdot$$

$$\left[1 + \cos(k_0 a\sqrt{1+ß^2}) + k_0 a\sqrt{1+ß^2} \cdot \sin(k_0 a\sqrt{1+ß^2}) - \cos(k_0 a) - \cos(k_0 b)\right] +$$

$$+ \frac{2}{\pi\sqrt{ß}} \cdot I_a(k_0 a, ß) \tag{20}$$

$$\frac{Z_r''}{Z_0} = \frac{2}{\pi ß (k_0 a)^2} \cdot$$

$$\left[\sin(k_0 a\sqrt{1+ß^2}) - k_0 a\sqrt{1+ß^2} \cdot \cos(k_0 a\sqrt{1+ß^2}) + k_0 a(1 + 1/ß) - \sin(k_0 a) - \sin(k_0 b)\right] -$$

$$- \frac{2}{\pi\sqrt{ß}} \cdot I_b(k_0 a, ß)$$

with the integrals :

$$I_a(k_0a,\beta)=\int_{\sqrt{\beta}}^{\sqrt{\beta+1/\beta}}\sqrt{1-\beta/x^2}\cdot\cos(x\,k_0a\sqrt{\beta})\,dx+\beta\int_{1/\sqrt{\beta}}^{\sqrt{\beta+1/\beta}}\sqrt{1-1/(\beta x)^2}\cdot\cos(x\,k_0a\sqrt{\beta})\,dx$$

$$I_b(k_0a,\beta)=\int_{\sqrt{\beta}}^{\sqrt{\beta+1/\beta}}\sqrt{1-\beta/x^2}\cdot\sin(x\,k_0a\sqrt{\beta})\,dx+\beta\int_{1/\sqrt{\beta}}^{\sqrt{\beta+1/\beta}}\sqrt{1-1/(\beta x)^2}\cdot\sin(x\,k_0a\sqrt{\beta})\,dx$$

(21)

A modification thereof leads to a fast numerical evaluation:

$$\frac{Z_r'}{Z_0}=1-\frac{2}{\pi k_0^2ab}\cdot$$
$$\cdot\left[1+\cos(k_0\sqrt{a^2+b^2})+k_0\sqrt{a^2+b^2}\cdot\sin(k_0\sqrt{a^2+b^2})-\cos(k_0a)-\cos(k_0b)\right]+\frac{2}{\pi}\cdot\hat{I}_a$$

$$\frac{Z_r''}{Z_0}=\frac{2}{\pi k_0^2ab}\cdot$$
$$\cdot\left[k_0(a+b)+\sin(k_0\sqrt{a^2+b^2})-k_0\sqrt{a^2+b^2}\cdot\cos(k_0\sqrt{a^2+b^2})-\sin(k_0a)-\sin(k_0b)\right]-\frac{2}{\pi}\cdot\hat{I}_b$$

(22)

with the integrals :

$$\hat{I}_a=\int_1^{\sqrt{1+(b/a)^2}}\sqrt{1-1/x^2}\cdot\cos(x\,k_0a)\,dx+\int_1^{\sqrt{1+(a/b)^2}}\sqrt{1-1/x^2}\cdot\cos(x\,k_0b)\,dx$$

$$\hat{I}_b=\int_1^{\sqrt{1+(b/a)^2}}\sqrt{1-1/x^2}\cdot\sin(x\,k_0a)\,dx+\int_1^{\sqrt{1+(a/b)^2}}\sqrt{1-1/x^2}\cdot\sin(x\,k_0b)\,dx$$

(23)

The component integrals are of the forms :

$$\tilde{I}_a(A,B)=\int_1^B\sqrt{1-1/x^2}\cdot\cos(Ax)\,dx\quad;\quad\tilde{I}_b(A,B)=\int_1^B\sqrt{1-1/x^2}\cdot\sin(Ax)\,dx \qquad (24)$$

They can be evaluated iteratively :

$$\tilde{I}_a(A,B)=I_{-1}+\sum_{n=1}^{\infty}(-1)^n\frac{A^{2n}}{(2n)!}\cdot I_{2n-1}\quad;\quad\tilde{I}_b(A,B)=A\cdot I_0+\sum_{n=1}^{\infty}(-1)^n\frac{A^{2n+1}}{(2n+1)!}\cdot I_{2n} \qquad (25)$$

with start values and recursion for the I_m :

$$I_m=\frac{B^{m-1}}{m+2}(B^2-1)^{3/2}+\frac{m-1}{m+2}\cdot I_{m-2}$$

$$I_{-1}=\sqrt{B^2-1}-\arccos(1/B)\quad;\quad I_0=\frac{B}{2}\sqrt{B^2-1}-\frac{1}{2}\ln(B+\sqrt{B^2-1})$$

(26)

An approximation for large k_0a (>5) and not too small b/a is :

$$\frac{Z_r'}{Z_0}=1-\sqrt{\frac{2}{\pi}}\left[\frac{\cos(k_0a-\pi/4)}{(k_0a)^{3/2}}+\frac{\cos(k_0b-\pi/4)}{(k_0b)^{3/2}}\right]-\frac{2}{\pi k_0^2ab}\left[1-\cos k_0a-\cos k_0b\right]-$$
$$-\frac{9}{8}\sqrt{\frac{2}{\pi}}\left[\frac{\sin(k_0a-\pi/4)}{(k_0a)^{5/2}}+\frac{\sin(k_0b-\pi/4)}{(k_0b)^{5/2}}\right]+\frac{2(a^2+b^2)^{3/2}}{\pi(k_0ab)^3}\sin(k_0\sqrt{a^2+b^2}) \qquad (27)$$

$$\frac{Z_r''}{Z_0}=\frac{2(a+b)}{\pi k_0ab}+\sqrt{\frac{2}{\pi}}\left[\frac{\sin(k_0a-\pi/4)}{(k_0a)^{3/2}}+\frac{\sin(k_0b-\pi/4)}{(k_0b)^{3/2}}\right]-\frac{2}{\pi k_0^2ab}\left[\sin k_0a+\sin k_0b\right]-$$
$$-\frac{9}{8}\sqrt{\frac{2}{\pi}}\left[\frac{\cos(k_0a-\pi/4)}{(k_0a)^{5/2}}+\frac{\cos(k_0b-\pi/4)}{(k_0b)^{5/2}}\right]+\frac{2(a^2+b^2)^{3/2}}{\pi(k_0ab)^3}\cos(k_0\sqrt{a^2+b^2}) \qquad (28)$$

Correspondence in the diagrams below: "Zs" → Z_r/Z_0 ; "k0a" → k_0a ; dashes become shorter for higher positions in the parameter list {ß}={0.25 , 0.5 , 1} .

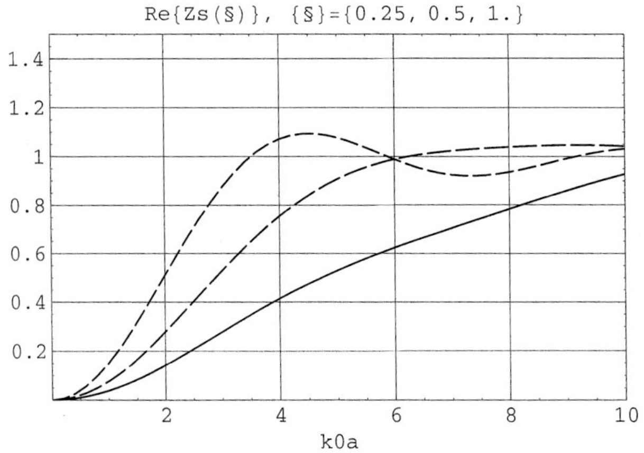

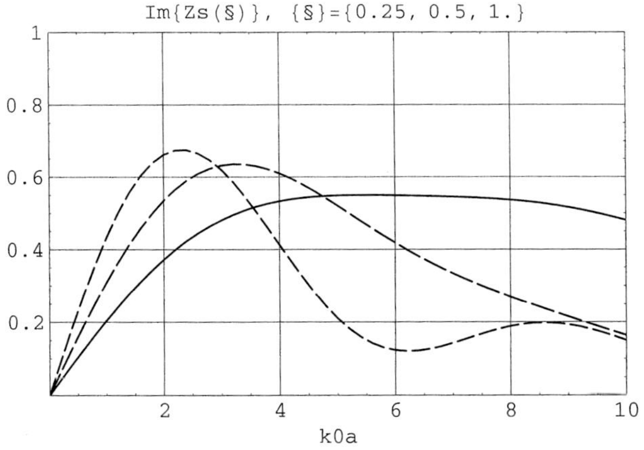
Im{Zs(§)}, {§}={0.25, 0.5, 1.}
1
0.8
0.6
0.4
0.2
2
4
6
8
10
k0a

F.8 Uniform end correction of plane piston radiators

Ref.: Mechel, [F.1]

The normalised end correction of a radiator is defined from its radiation reactance Z_r'' by :

$$\frac{\Delta\ell}{a} = \frac{Z_r''}{k_0 a \cdot Z_0} \tag{1}$$

where a is any side length. Thus $\Delta\ell/a$ equals the tangent of the curve of Z_r''/Z_0 over k_0a at the origin $k_0a=0$.

If one takes $a= A^{3/4} \cdot U^{1/2}$ (2)

with A= area , U= periphery of the piston surface, then the curves of Z_r''/Z_0 over k_0a coincide at the origin $k_0a=0$ for different shapes of the surface, supposed its border line is convex. So one can deduce end corrections for piston shapes with unknown solutions for Z_r from end corrections of shapes with known solutions.

F.9 Narrow strip-shaped, field excited radiator

Ref.: Mechel, [F.4]

A plane radiator is called "field excited", if its vibration pattern agrees with the pattern of an obliquely incident plane wave at the surface.

The object here is an infinitely long strip of width a in a hard baffle wall, which is excited by a plane wave with polar angle Θ of incidence and azimuthal angle Φ with the strip axis.
Either $\Phi=0$ (then a unlimited), or $\Phi\neq 0$, then $a \ll \lambda_0$; the oscillation velocity of the strip surface can be assumed to be constant across the strip :

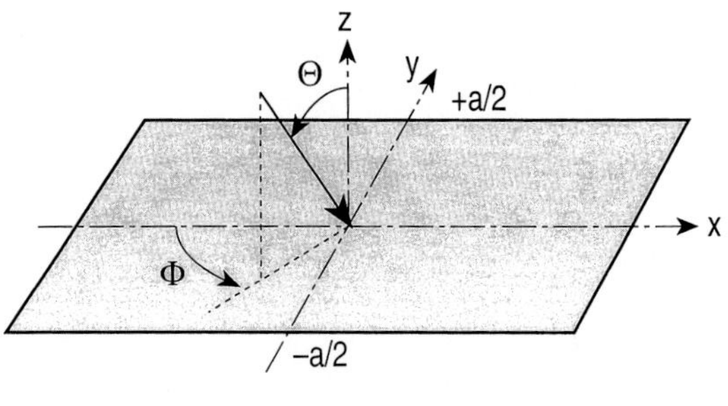

$$v(x,y) = V_0 \cdot e^{-jk_x x} \quad ; \quad k_x = k_0 \cdot \sin\Theta \cdot \cos\Phi \tag{1}$$

According to section F.1 , because of $|v|$=const : $Z_r = \frac{1}{A}\iint_A Z_F \, dA = \frac{1}{V_0 a}\int_{-a/2}^{+a/2} p(y,0)\, dy$ (2)

with the field impedance $Z_F = p(x,y,0)/v(x,y)$ and $p(x,y,z) = p(y,z) \cdot e^{-jk_x x}$. The lateral sound pressure distribution is :

$$p(y,z) = \frac{k_0 Z_0 V_0 a}{2\pi}\int_{-\infty}^{+\infty} \frac{\sin(k_y a/2)}{k_y a/2} \frac{e^{-j(k_y y + z\sqrt{k_0^2 - k_x^2 - k_y^2})}}{\sqrt{k_0^2 - k_x^2 - k_y^2}} dk_y \tag{3}$$

and therewith the radiation impedance :

$$\frac{Z_r}{Z_0} = \frac{k_0 a}{2\pi}\int_{-\infty}^{+\infty}\left(\frac{\sin(k_y a/2)}{k_y a/2}\right)^2 \frac{dk_y}{\sqrt{k_0^2 - k_x^2 - k_y^2}} \tag{4}$$

In a different form :

$$\frac{Z_r}{Z_0} = \frac{k_0}{k^2 a}\int_0^{ka}(ka - |u|) \cdot H_0^{(2)}(|u|)\, du \quad ; \quad k^2 = k_0^2 - k_x^2 = k_0^2(1 - \sin^2\Theta \cdot \cos^2\Phi) \tag{5}$$

with the Hankel function of the 2nd kind $H_0^{(2)}(z)$. After analytical solution of the integral :

$$\frac{Z_r}{Z_0} = k_0 a\left\{H_0^{(2)}(ka) - \frac{H_1^{(2)}(ka)}{ka} + \frac{2j}{\pi(ka)^2} + \frac{\pi}{2}\left[H_1^{(2)}(ka) \cdot S_0(ka) - H_0^{(2)}(ka) \cdot S_1(ka)\right]\right\} \tag{6}$$

or as real and imaginary parts :

$$\begin{aligned} \frac{Z_r'}{Z_0} &= k_0 a\left\{J_0(ka) - \frac{J_1(ka)}{ka} + \frac{\pi}{2}\left[J_1(ka) \cdot S_0(ka) - J_0(ka) \cdot S_1(ka)\right]\right\} \\ \frac{Z_r''}{Z_0} &= -k_0 a\left\{Y_0(ka) - \frac{Y_1(ka)}{ka} - \frac{2}{\pi(ka)^2} + \frac{\pi}{2}\left[Y_1(ka) \cdot S_0(ka) - Y_0(ka) \cdot S_1(ka)\right]\right\} \end{aligned} \tag{7}$$

with $J_n(z)$= Bessel function; $Y_n(z)$= Neumann function; $S_n(z)$= Struve function.

Approximation for small ka (with c=0.57721 , Euler's constant):

$$\frac{Z_r'}{Z_0} = \tag{8}$$

$$k_0 a\left\{\left(1 - \frac{(ka)^2}{6} + \frac{(ka)^4}{64}\right)\left(1 - \frac{(ka)^2}{3} + \frac{(ka)^4}{45}\right) - \left(\frac{1}{2} - \frac{(ka)^2}{16} + \frac{(ka)^4}{192}\right)\left(1 - (ka)^2 + \frac{(ka)^4}{9}\right)\right\}$$

$$\frac{Z_r''}{Z_0} = \frac{2k_0 a}{\pi}\left\{1 - \frac{(ka)^2}{9} + \frac{(ka)^4}{225} - \right.$$
$$-\left(1 - \frac{(ka)^2}{3} + \frac{(ka)^4}{45}\right)\left[\left(\ln\frac{ka}{2} + c\right)\left(1 - \frac{(ka)^2}{4} + \frac{(ka)^4}{64}\right) + (ka)^2\left(\frac{1}{4} - \frac{3(ka)^2}{128}\right)\right] + \tag{9}$$
$$\left. + \left(1 - (ka)^2 + \frac{(ka)^4}{9}\right)\left[\left(\ln\frac{ka}{2} + c\right)\left(\frac{1}{2} - \frac{(ka)^2}{16} + \frac{(ka)^4}{192}\right) - \frac{1}{4} + \frac{5(ka)^2}{64} - \frac{10(ka)^4}{2304}\right]\right\}$$

F.10 Wide strip-shaped, field excited radiator

Ref.: Mechel, [F.4]

A plane radiator is called "field excited", if its vibration pattern agrees with the pattern of an obliquely incident plane wave at the surface.

The object here is an infinitely long strip of width a in a hard baffle wall, which is excited by a plane wave with polar angle ϑ of incidence and azimuthal angle φ with the normal to the strip axis.

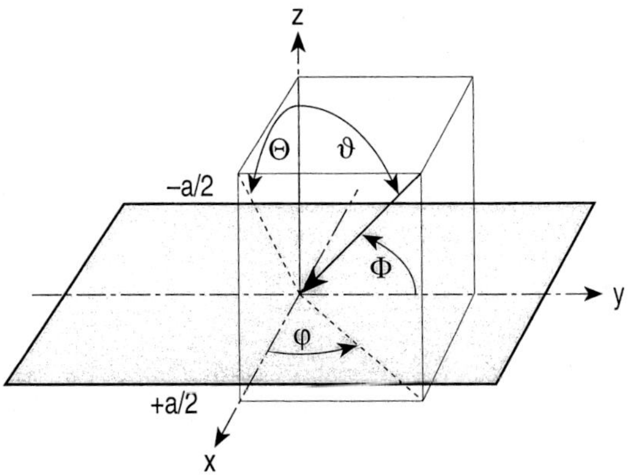

Notice the different co-ordinates and angles as compared to section F.9 .

$$\cos\Phi = \sin\varphi \cdot \sin\vartheta$$
$$\cos\Theta = \cos\vartheta / \sqrt{1 - \sin^2\varphi \cdot \sin^2\vartheta}$$
$$(1)$$
$$\cos\vartheta = \sin\Phi \cdot \cos\Theta$$
$$\sin\varphi = \cos\Phi / \sqrt{1 - \sin^2\Phi \cdot \cos^2\Theta}$$

Radiation impedance :

$$\frac{Z_r}{Z_0} = \frac{C}{\sin\Phi} \quad ; \quad b = k_0 a \cdot \sin\Phi$$
$$C = A + jB = \int_0^b \left(1 - \frac{x}{b}\right) \cdot \cos(x \cdot \sin\Theta) \cdot H_0^{(2)}(x)\, dx \tag{2}$$

with $H_0^{(2)}(x)$ = Hankel function of the 2nd kind. After power series expansion of the factor to the Hankel function in the integrand :

$$A = \sum_{n=0}^{\infty} (-1)^n \frac{\sin^{2n}\Theta \cdot b^{2n+1}}{(2n)!} \cdot$$
$$\cdot \left[\frac{1}{2n+1}\, {}_1F_2\!\left(1/2+n\,;1,3/2+n\,;-b^2/4\right) - \frac{1}{2n+2}\, {}_1F_2\!\left(1+n\,;1,2+n\,;-b^2/4\right) \right] \tag{3}$$

$$B = \frac{-1}{\pi} \sum_{n=0}^{\infty} (-1)^n \frac{\sin^{2n}\Theta \cdot b^{2n+1}}{(2n)!} \cdot$$
$$\cdot \left[\ln\!\left(\frac{4}{b^2}\right) \cdot \left(\frac{1}{2n+2}\, {}_1F_2\!\left(1+n\,;1,2+n\,;-b^2/4\right) - \frac{1}{2n+1}\, {}_1F_2\!\left(1/2+n\,;1,3/2+n\,;-b^2/4\right) \right) + \right.$$
$$+ \frac{2}{(2n+2)^2} \cdot {}_2F_3\!\left(1+n,1+n\,;1,2+n,2+n\,;-b^2/4\right) -$$
$$\left. - \frac{2}{(2n+1)^2} \cdot {}_2F_3\!\left(1/2+n,1/2+n\,;1,3/2+n,3/2+n\,;-b^2/4\right) \right]$$

with hypergeometric functions ${}_1F_2(a_1\,; b_1,b_2\,; z)$ and ${}_2F_3(a_1,a_2\,; b_1,b_2,b_3\,; z)$.

Correspondence in the diagram below: "Zs" → Z_r/Z_0 ; "theta" → ϑ ; "phi" → φ ; "k0a" → k_0a ; full line: real part ; dashed line: imaginary part.

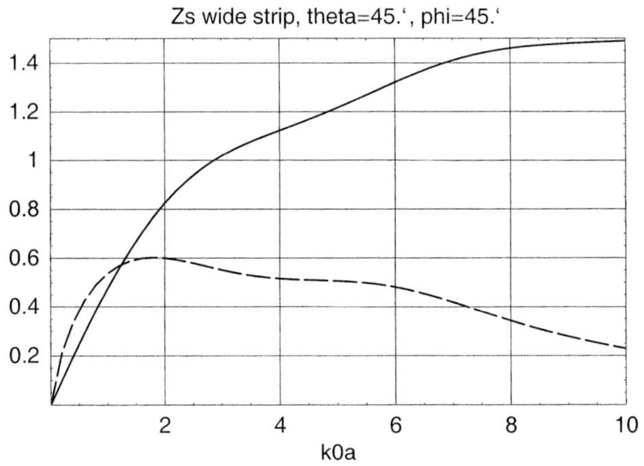

F.11 Wide rectangular, field excited radiator

Ref.: Mechel, [F.4]

A plane radiator is called "field excited", if its vibration pattern agrees with the pattern of an obliquely incident plane wave at the surface.

The object here is a rectangle A with side lengths a,b in a hard baffle wall, which is excited by a plane wave with polar angle ϑ_i of incidence and azimuthal angle φ_i with the axis parallel to side a .

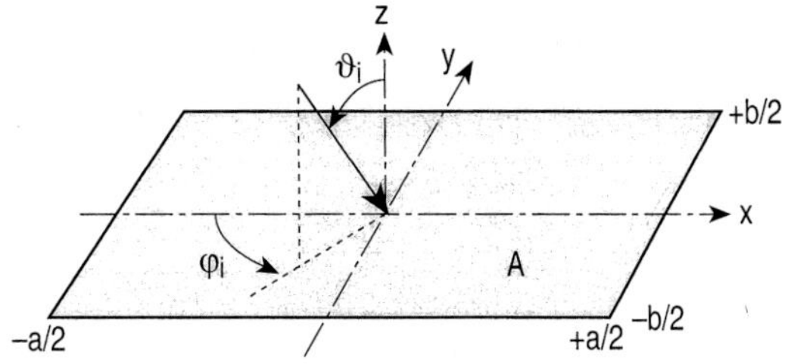

Velocity pattern on A :

$$v(x,y)=V_0\cdot e^{-j(k_x x+k_y y)}$$
$$k_x=k_0\cdot\sin\vartheta_i\cdot\cos\varphi_i=k_0\cdot\mu_x \qquad (1)$$
$$k_y=k_0\cdot\sin\vartheta_i\cdot\sin\varphi_i=k_0\cdot\mu_y$$

The sound pressure field is :

$$p(x,y,z)=\frac{jk_0Z_0}{2\pi}\iint_A v(x_0,y_0)\frac{e^{-jk_0R}}{R}\,dA_0 \quad ; \quad R=\sqrt{(x-x_0)^2+(y-y_0)^2} \qquad (2)$$

The definition of the radiation impedance Z_r with the radiated power gives a first form :

$$Z_r=\frac{jk_0Z_0}{2\pi A}\iint_A dA\iint_A \frac{e^{-jk_0R}}{R}e^{-j(k_x(x_0-x)+k_y(y_0-y))}\,dA_0 \quad ; \quad R^2=(x_0-x)^2+(y_0-y)^2 \qquad (3)$$

The fact that $|v(x,y)|$=const on A and that, therefore, the radiation impedance follows from the average field impedance with the Fourier transform of the velocity distribution leads to a form with fewer integrations :

$$\begin{aligned} V(k_1,k_2)&=\iint_A v(x_0,y_0)e^{-j(k_1x_0+k_2y_0)}\,dx_0dy_0 \\ &=V_0ab\frac{\sin((k_1+k_x)a/2)}{(k_1+k_x)a/2}\,\frac{\sin((k_2+k_y)b/2)}{(k_2+k_y)b/2} \end{aligned} \qquad (4)$$

using $\alpha_1=k_1/k_0$; $\alpha_2=k_2/k_0$:

$$\frac{Z_r}{Z_0}=\frac{k_0a\,k_0b}{4\pi^2}\int_{-\infty}^{+\infty}\!\!\int\left(\frac{\sin((\mu_x-\alpha_1)k_0a/2)}{(\mu_x-\alpha_1)k_0a/2}\right)^2\left(\frac{\sin((\mu_y-\alpha_2)k_0b/2)}{(\mu_y-\alpha_2)k_0b/2}\right)^2\frac{d\alpha_1\,d\alpha_2}{\sqrt{1-\alpha_1^2-\alpha_2^2}} \qquad (5)$$

The 3rd form starts from the BOUWKAMP integral (see section F.2)

$$\frac{Z_r}{Z_0}=\frac{k_0a\cdot k_0b}{4\pi^2}\int_0^{2\pi}d\varphi\int_0^{\pi/2+j\infty}|D(\vartheta,\varphi)|^2\cdot\sin\vartheta\,d\vartheta \qquad (6)$$

with the far field directivity function :

$$D(\vartheta,\varphi)=\frac{\sin\left(\frac{k_0a}{2}(\sin\vartheta_i\cos\varphi_i-\sin\vartheta\cos\varphi)\right)}{\frac{k_0a}{2}(\sin\vartheta_i\cos\varphi_i-\sin\vartheta\cos\varphi)}\cdot\frac{\sin\left(\frac{k_0b}{2}(\sin\vartheta_i\sin\varphi_i-\sin\vartheta\sin\varphi)\right)}{\frac{k_0b}{2}(\sin\vartheta_i\sin\varphi_i-\sin\vartheta\sin\varphi)} \qquad (7)$$

The 2nd form can be transformed to :

$$\frac{Z_r}{Z_0}=\frac{2j}{\pi k_0a\cdot k_0b}\int_{x=0}^{k_0a}dx\int_0^{k_0b}(k_0a-x)(k_0b-y)\cos(\mu_x x)\cos(\mu_y y)\frac{e^{-j\sqrt{x^2+y^2}}}{\sqrt{x^2+y^2}}\,dy \qquad (8)$$

This becomes for normal sound incidence with $\vartheta_i=0$; $\mu_x=\mu_y=0$:

$$\frac{Z_r}{Z_0}=\frac{2j}{\pi k_0a\cdot k_0b}\int_{x=0}^{k_0a}dx\int_0^{k_0b}(k_0a-x)(k_0b-y)\frac{e^{-j\sqrt{x^2+y^2}}}{\sqrt{x^2+y^2}}\,dy \qquad (9)$$

The double integral can be transformed by substitution of variables to :

$$\frac{Z_r}{Z_0}=\frac{2j}{\pi k_0a\cdot k_0b}\left[\int_0^{\operatorname{arctg}(b/a)}I(k_0a/\cos\varphi)\,d\varphi+\int_{\operatorname{arctg}(b/a)}^{\pi/2}I(k_0b/\sin\varphi)\,d\varphi\right] \qquad (10)$$

with the intermediate integrals :

$$I(R)=\int_0^R\left(U+V\cdot r+W\cdot r^2\right)\cos(\alpha r)\cdot\cos(\beta r)\cdot e^{-jr}\,dr$$

$$U=k_0a\cdot k_0b\quad;\quad V=-(k_0a\cdot\sin\varphi+k_0b\cdot\cos\varphi)\quad;\quad W=\sin\varphi\cdot\cos\varphi \qquad (11)$$

$$\alpha=\mu_x\cdot\cos\varphi\quad;\quad \beta=\mu_y\cdot\sin\varphi$$

See the reference for an analytical procedure to solve the integrals contained in I(R) .

Correspondence in the following diagrams: “Zs” → Z_r/Z_0 ; “thetai” → ϑ_i ; “k0a” → k_0a ; parameters: b/a=3 ; $\varphi_i=0$; the dashes become shorter with increasing position of the parameter value of ϑ_i in the list $\{\vartheta_i\}$.

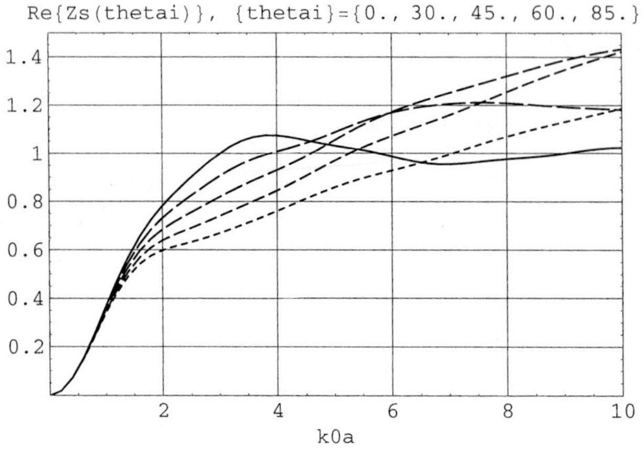
Re{Zs(thetai)}, {thetai}={0., 30., 45., 60., 85.}
1.4
1.2
1
0.8
0.6
0.4
0.2
2
4
6
8
10
k0a

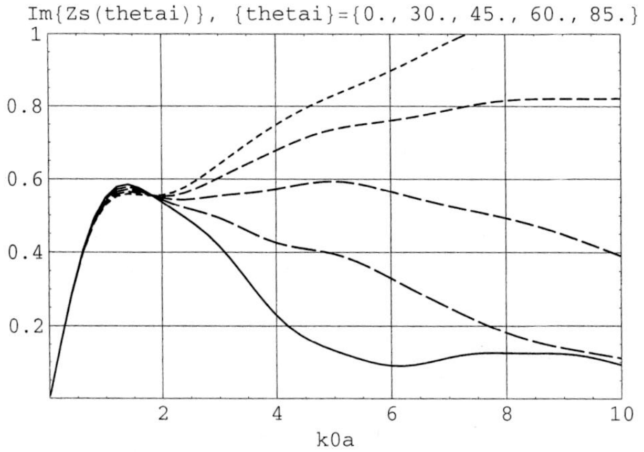
Im{Zs(thetai)}, {thetai}={0., 30., 45., 60., 85.}
1
0.8
0.6
0.4
0.2
2
4
6
8
10
k0a

F.12 End corrections :

Ref.: Mechel, [F.2]

See section F.1 for the definition of end corrections. End corrections represent the inertial near fields at expansions (orifices) of the cross-section available for the sound wave. End corrections are mostly of interest for small k_0a , where a is a characteristic lateral dimension of the orifice. End corrections are influenced by the shape of the orifice and of the space which is available for the sound wave behind the orifice. Therefore in general the orifices on both sides of a "neck" must be distinguished (exterior and interior end correction). The relations of the end correction $\Delta\ell$ of an orifice with area A to the radiation impedance $Z_r = Z'_r + j\cdot Z''_r$ and the oscillating mass M_r are :

$$\Delta\ell = \frac{M_r}{\rho_0 A} = \frac{m_r}{\rho_0} = \frac{Z''_{mr}}{\omega\rho_0 A} = \frac{Z''_r}{\omega\rho_0} = \frac{Z''_r}{k_0 Z_0} \quad ; \quad \frac{\Delta\ell}{a} = \frac{Z''_r}{k_0 a\cdot Z_0} \qquad (1)$$

Object		M_r	Remarks
Monopole sphere	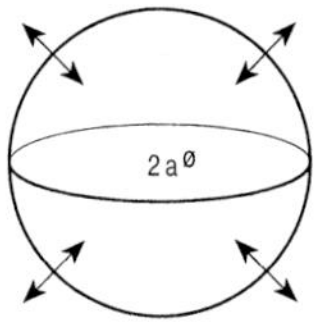	$M_r = 3\rho_0 V \frac{1}{1+(k_0a)^2} \xrightarrow[k_0a\ll 1]{} 3\rho_0 V$	$A=4\pi a^2$ $V=4\pi a^3/3$
Oscillating sphere	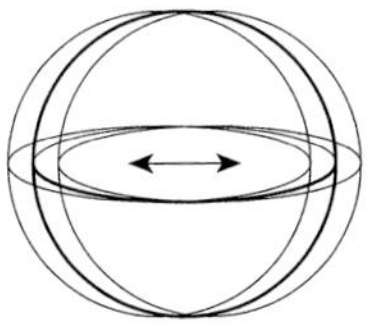	$M_r = 3\rho_0 V \frac{2+(k_0a)^2}{4+(k_0a)^4} \xrightarrow[k_0a\ll 1]{} \frac{3}{2}\rho_0 V$	
Sphere in n-th mode	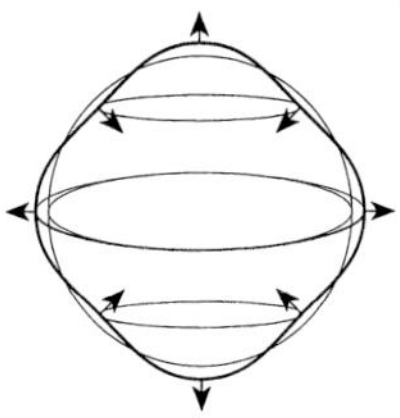	$M_r \xrightarrow[k_0a\ll\lvert 2n-1\rvert]{} 3\rho_0 V \frac{1}{n+1}$ $\xrightarrow[k_0a\gg n^2+1]{} 3\rho_0 V \frac{1}{(k_0a)^2}$	

Object		M_r	Remarks
Monopole cylinder	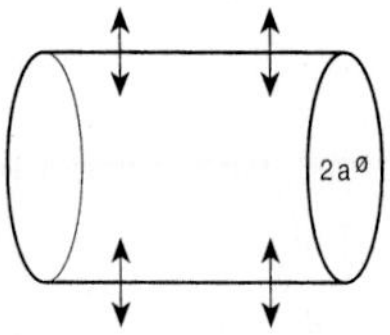	$M_r \xrightarrow[k_0a\ll 1]{} -2\rho_0 V \ln(k_0 a)$	$A=2\pi a$ $V=\pi a^2$
Oscillating cylinder	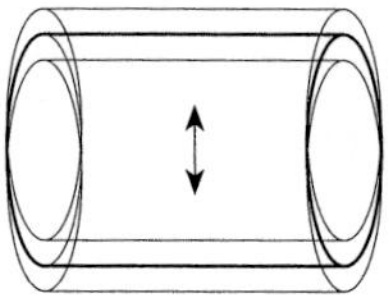	$M_r \xrightarrow[k_0a\ll 1]{} 2\rho_0 V$	
Cylinder in n-th mode	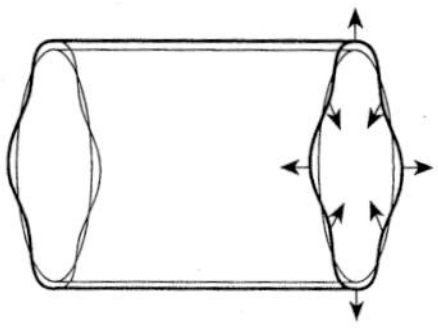	$M_r \xrightarrow[k_0a\ll 1]{} 2\rho_0 V \frac{1}{n}$	

Object		$\frac{\Delta\ell}{a}$	Remarks
Circle in baffle wall	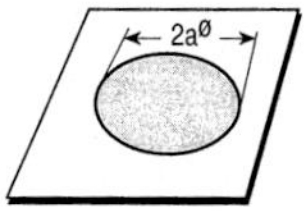	$0.785 = \pi/4 < \Delta\ell/a \le 8/3\pi = 0.85$ $\frac{\Delta\ell}{a} = \frac{8}{3\pi}[1 - \frac{2}{15}(k_0a)^2 + \frac{8}{525}(k_0a)^4]$	a= radius
Tube orifice in free space	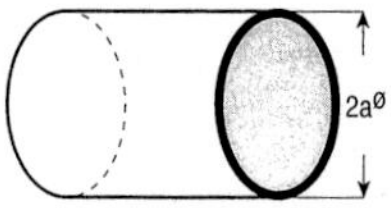	$(0.65 \text{ to } 0.69)\cdot a^2/\lambda_0$	a= radius
Half monopole sphere in baffle wall	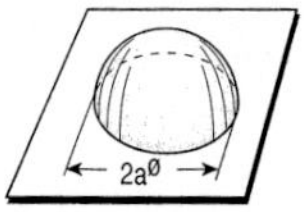	$\Delta\ell/a = 2/[1+(k_0a)^2] \to 2$	a= radius
Orifice on sphere	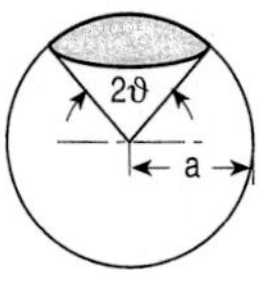	$\frac{\Delta\ell}{a} = [1+\cos\vartheta]/[2\,(1+(k_0a)^2)]$	a= radius $a\cdot\sin\vartheta$= orifice radius

Object		$\frac{\Delta\ell}{a}$	**Remarks**
Elliptical orifice in baffle wall	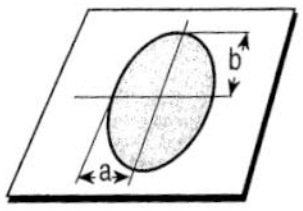	$\frac{\Delta\ell}{a}=\frac{16}{3\pi^2}K(1-ß^2)$; $\frac{\Delta\ell}{b}=ß\cdot\frac{\Delta\ell}{a}$ $K(1-ß^2)\approx\frac{4+ß^2}{8}\ln\frac{16}{ß^2}-\frac{ß^2}{4}$ $0<ß\le 0.641$ $K(1-ß^2)\approx\frac{\pi}{2}\frac{11+5ß^2}{7+9ß^2}$ $0.641\le ß\le 1$	a= small b= large half axis ß= a/b <1
Orifice in tube wall	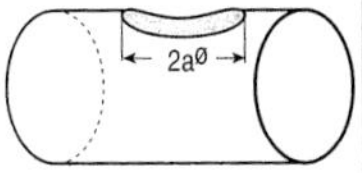	$\Delta\ell\approx U/8+(\lambda_0/2\pi)\cdot\chi_0(2k_0\sqrt{S/\pi})$ $\chi_0(x)=\frac{4}{\pi}\int_0^{\pi/2}\sin(x\cos\alpha)\sin^2\alpha\,d\alpha$ $\approx x^2/8$; x«1	U=2πa= periphery S= πa^2=area of orifice
Circular fence in tube	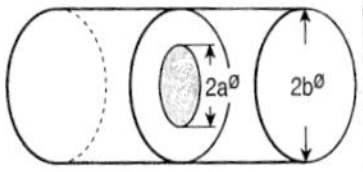	$\Delta l/a\approx -0.0445728-0.728326\,x-$ $-0.177078\,x^2+0.0339531\,y+$ $+0.00810471\,y^2-0.00100762\,xy$ $\sigma=(a/b)^2$; $x=\lg\sigma$; $y=\lg(b/\lambda_0)$	a= fence radius b= tube radius
Free circular disk	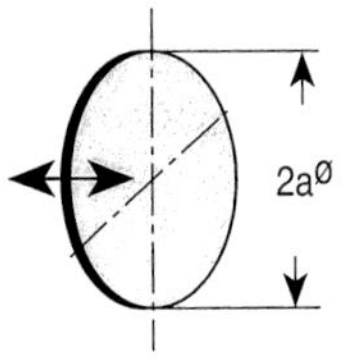	$\Delta l/a\approx 8/3\pi$	a= radius
Rectangular orifice in baffle	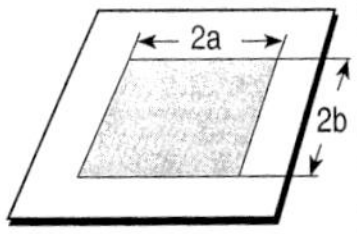	$\frac{\Delta l}{a}=\frac{2}{3\pi}[ß+\frac{1-(1+ß^2)^{3/2}}{ß^2}]+$ $+\frac{2}{\pi}[\frac{1}{ß}\ln(ß+\sqrt{1+ß^2})+$ $+\ln(\frac{1}{ß}(1+\sqrt{1+ß^2}))]$	2a, 2b = sides ß = a/b ≤ 1

Object		$\frac{\Delta\ell}{a}$	**Remarks**
Slit on a cylinder	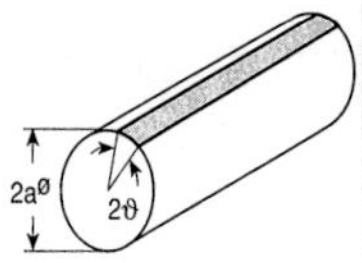	$\frac{\Delta l}{2a\vartheta}=\frac{1}{\pi}\sum_{n=1}\frac{1}{n}\left(\frac{\sin n\vartheta}{n\vartheta}\right)^2-\frac{\ln(k_0a)}{2\pi}$	a = cylinder radius ϑ = angle of slit
Rectangular orifice in baffle wall	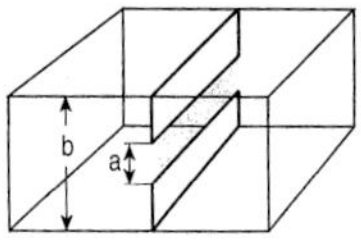	$\Delta\ell/a=\frac{1}{\pi}\ln[\frac{1}{2}\mathrm{tg}(\frac{\pi\sigma}{4})+\frac{1}{2}\cot(\frac{\pi\sigma}{4})]$ $\approx\frac{1}{\pi}\ln[\sin(\frac{\pi\sigma}{2})]\,;\ \sigma<1$ $\approx\frac{\pi}{8}(1-\sigma)^2\ ;\ b-a\ll b$ $\approx-0.395450\,x+0.346161\,x^2+$ $0.141928\,x^3+0.0200128\,x^4$	a= slit width b= duct width; σ=a/b ; x=lgσ

Expansion of a flat duct	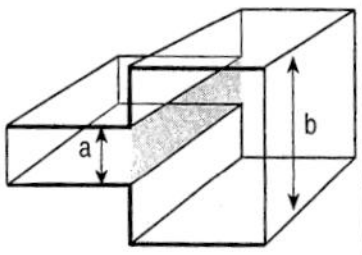	$\Delta\ell/a=\frac{1}{\pi}[\frac{(1-\beta)^2}{2\beta}\ln\frac{1+\beta}{1-\beta}+\ln\frac{(1+\beta)^2}{4\beta}]$
		$\approx\frac{1}{\pi}[1-\ln(4\beta)]\quad;\ a\ll b$
		$\approx\frac{1}{4\pi}(1-\beta)^2[1-\ln((1-\beta)/2)]\quad;\ a\approx b$
			a= narrow duct height;
			b= wide duct height;
			ß=a/b

Grid of slits	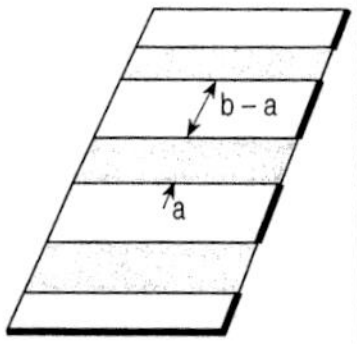	$\Delta\ell/a=\sigma\sum_{n=1}^{\infty}\frac{\sin^2(n\pi\sigma)}{(n\pi\sigma)^3}$ $\approx-\sqrt{2}/\pi\cdot\ln[\sin(\pi\sigma/2)]\quad;\quad 0{,}1<\sigma<0{,}7$	a= slit width L= slit distance σ=a/L
Hole grid; square arrangement	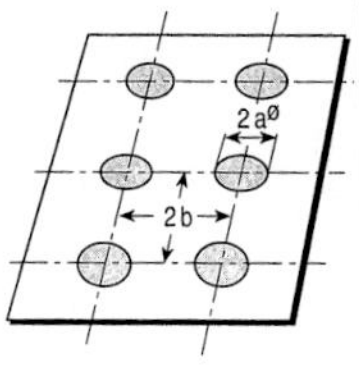	$\Delta\ell/a=0.79(1-1.47\sqrt{\sigma}+0.47\,\sigma^{3/2})$	a= hole radius; 2b= hole distance; $\sigma=\pi a^2/(2b)^2$

Object		$\frac{\Delta\ell}{a}$	Remarks
Hole grid; hexagonal arrangement	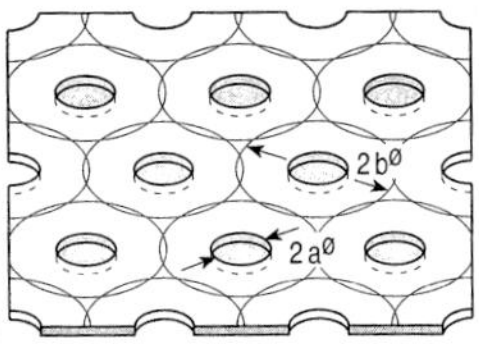	$\Delta\ell/a \approx -0.0454728 - 0.728326\,x - 0.177078\,x^2 + 0.0339531\,y + 0.00810471\,y^2 - 0.00100762\,xy$; $\sigma = (a/b)^2$; $x = \lg\sigma$; $y = \lg(b/\lambda_0)$	a= hole radius b= hole distance

End correction of a slit in a grid of parallel slits :

Alternative representations; a= width of slit; L=a+b= slit centre distance

$$\frac{\Delta\ell}{a} = \frac{a}{L}\cdot\sum_{n=1}^{\infty}\frac{\sin^2(n\pi a/L)}{(n\pi a/L)^3} \tag{2}$$

$$\frac{\Delta\ell}{a} \approx \frac{1}{\pi}\ln\left[\frac{1}{2}\tan\left(\frac{\pi}{4}\frac{a}{L}\right) + \frac{1}{2}\cot\left(\frac{\pi}{4}\frac{a}{L}\right)\right] \tag{3}$$

from regression:

$$\frac{\Delta\ell}{a} = -0.395450\cdot x + 0.346161\cdot x^2 + 0.141928\cdot x^3 + 0.0200128\cdot x^4 \tag{4}$$

Radiation reactance (used below for reference):

$$\frac{Z''_{r0}}{Z_0} = \frac{2a}{L}\sum_{n>0}\frac{1}{\sqrt{\left(n\frac{\lambda_0}{L}\right)^2 - 1}}\left(\frac{\sin(n\pi\, a/L)}{n\pi\, a/L}\right)^2 \tag{5}$$

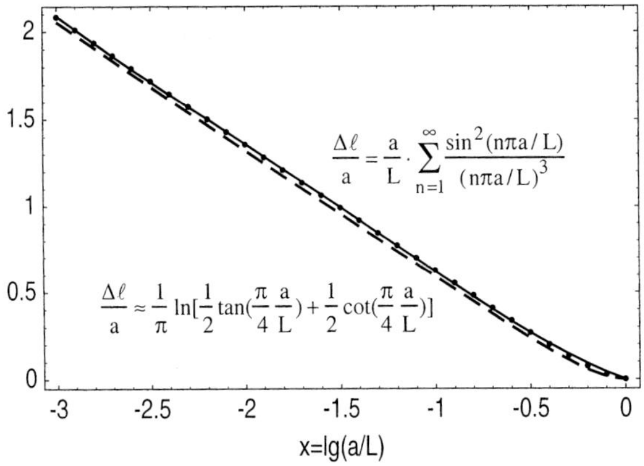

Influence of higher modes in the neck of a slit grid plate :

Width and distance of slits as above: the slits are in a plate of thickness d ;
radiation reactance of a back orifice :

$$Z''_{rb} = Z''_{r0} \cdot (1 + \frac{\Delta Z''_{rb}}{Z''_{r0}}) = Z''_{rb} \cdot (1 - 10^{F(x,y)})$$

$$F(x,y) = \lg(-\frac{\Delta Z''_{sh}}{Z''_{sh0}}) = f(x) \cdot (1 + g(y)) \qquad (6)$$

$$x = \lg(a/L) \quad ; \quad y = \lg(d/a)$$

with (in $-3 \le x < 0$ and $-1 \le y \le 1$):

$$f(x) = -1.73968 + 1.48435\,(x+1.5) - 1.84230\,(x+1.5)^2 + \\ + 0.292538\,(x+1.5)^3 + 0.428402\,(x+1.5)^4$$

$$g(y) = H(-y) \cdot [0.00259355\,y - 0.0758181\,y^2 + \\ + 0.330845\,y^3 + 0.226933\,y^4] \qquad (7)$$

$$H(-y) = \begin{cases} 1; & y \le 0 \\ 0; & y > 0 \end{cases}$$

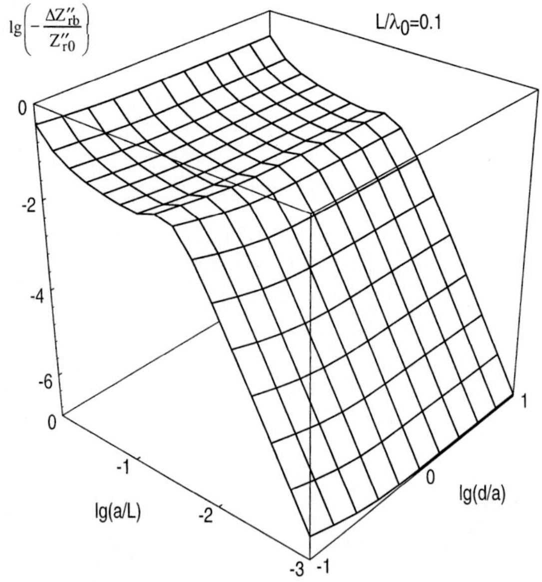

Relative change of the radiation reactance of a slit in a slit grid due to higher modes in the neck of the slit plate.

Interior end correction of the slit orifice in a slit resonator array :

No losses and only a plane wave in the slit (i.e. narrow slit).

The resonators repeat in the y direction with a period length L=a+b .

Lateral wave numbers in the volume :

$$\gamma_0 = jk_0 \; ; \quad \gamma_n = k_0\sqrt{(n\frac{\lambda_0}{L})^2 - 1} \qquad (8)$$

$$\mathrm{Re}\{\gamma_n\} \ge 0 \quad \text{or} \quad \mathrm{Im}\{\gamma_n\} \ge 0$$

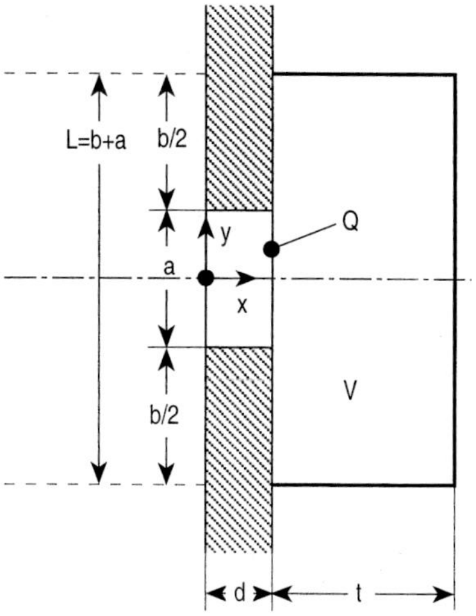

Impedance of the back orifice :

$$\frac{Z_{sh}}{Z_0} = -j\frac{a/L}{\tan(k_0 t)} + j2\frac{a}{L}\sum_{i>0}\frac{k_0}{\gamma_i}\frac{s_i^2}{\tanh(\gamma_i t)} \qquad (9)$$

$$s_0 = 1 \quad ; \quad s_i = \frac{\sin(i\pi\, a/L)}{i\pi\, a/L} \qquad (10)$$

The 1st term (outside the sum) is the spring reactance of the volume; thus the sum term is the mass reactance at the interior orifice. The back side end correction therefore is :

$$\frac{\Delta\ell_b}{a} = \frac{2}{L}\sum_{i>0}\frac{1}{\gamma_i}\frac{s_i^2}{\tanh(\gamma_i t)} \qquad (11)$$

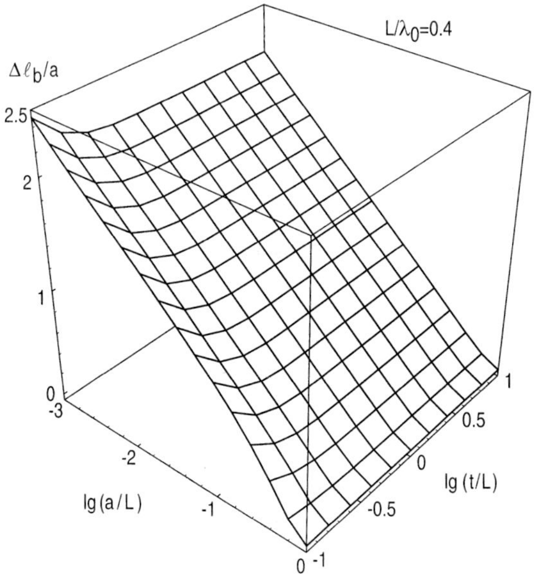

Influence of the shape parameter t/L on the interior end correction of the slit in a slit resonator.

Interior end correction of the slit orifice in a slit resonator array with higher modes in the neck :

Geometrical parameters as above. $\Delta\ell_{b0}/a$ interior end correction from above with only plane wave in the neck.

$$\frac{\Delta\ell_b}{a} \approx \frac{\Delta\ell_{b0}}{a}(x)\cdot[1+f(y)]\cdot[1+g(z)]$$

$$x = \lg\frac{a}{L} \quad ; \quad y = \lg\frac{d}{a} \quad ; \quad z = \lg\frac{t}{L} \qquad (12)$$

$$f(y) = 0.00144829\cdot y + 0.00255510\cdot y^2 + 0.03430510\cdot y^3 + 0.01568299\cdot y^4$$

$$g(z) = -0.000932290\cdot z - 0.00767204\cdot z^2 - 0.01925972\cdot z^3 - 0.01804839\cdot z^4$$

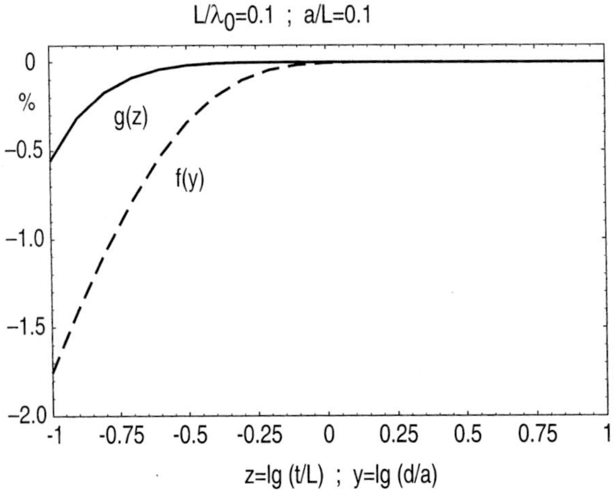

Influence (in per cent) of the shape factors d/a and t/L on the interior end correction, with higher neck modes taken into account.

Interior orifice impedance of a slit in a slit array, with viscous and caloric losses in the neck taken into account :

Let Z_{b0} be the back orifice impedance without losses. The back orifice impedance $Z_b = Z_b' + j\cdot Z_b''$ can be approximated with :

$$\frac{Z_b'}{Z_0} = \frac{Z_{b0}'}{Z_0}\left(1+\frac{10^{F'(x)}}{\sqrt{a_{[m]}}\cdot\sqrt[3]{a/L}}\right) \quad ; \quad \frac{Z_b''}{Z_0} = \frac{Z_{b0}''}{Z_0}\left(1+\frac{10^{F''(x)}}{\sqrt{a_{[m]}}\cdot\sqrt[3]{a/L}}\right) \quad ; \quad x = \lg\frac{f_{[Hz]}a_{[m]}}{(a/L)^{3/2}} ;$$

$$F'(x) = -4.64106 + 0.435993\,x + 0.0142851\,x^2 + 0.000461347\,x^3 \qquad (13)$$

$$F''(x) = -2.26665 - 0.492331\,x - 0.000719182\,x^2 - 0.0010208\,x^3$$

Interior orifice impedance of a slit in a slit array in contact with a porous absorber layer :

Let the characteristic propagation constant and wave impedance of the porous material be Γ_a , Z_a . Air gap thickness t=0 .
Ξ= flow resistivity of the porous material.

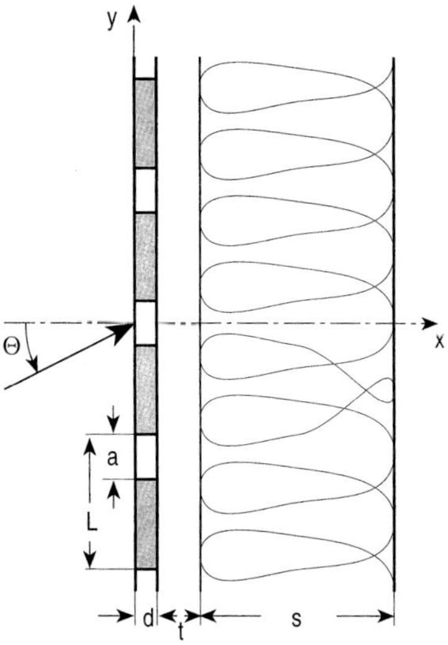

$$\varepsilon_n = k_0 \sqrt{(\sin\Theta + n\lambda_0 / L)^2 + (\Gamma_a / k_0)^2} \qquad (14)$$

Impedance Z_b of the back slit orifice : (15)

$$\frac{Z_b}{Z_0} = 2\frac{a}{L}\frac{\Gamma_a Z_a}{k_0 Z_0}\sum_{n>0}\frac{k_0}{\varepsilon_n}\left(\frac{\sin(n\pi a/L)}{n\pi a/L}\right)^2 \coth(\varepsilon_n s)$$

Back orifice end correction : (16)

$$\frac{\Delta\ell_b}{a} = -2j\,\frac{s}{L}\frac{\Gamma_a Z_a}{k_0 Z_0}\sum_{n>0}\left(\frac{\sin(n\pi a/L)}{n\pi a/L}\right)^2 \frac{\coth(\varepsilon_n s)}{\varepsilon_n s}$$

Interior end correction of a slit in a slit array in contact with a porous absorber layer :

$$\frac{\Delta\ell_b}{a} = j\,\frac{\Gamma_a Z_a}{k_0 Z_0}\cdot$$
$$\begin{aligned}&(0.0389998 + 0.454066\cdot x - 0.345328\cdot x^2 - 0.125386\cdot x^3 -\\ &0.0143782\cdot y + 0.00418541\cdot y^2 + 0.0170766\cdot y^3 -\\ &0.0142094\cdot z - 0.0715597\cdot z^2 + 0.0915584\cdot z^3 -\\ &0.0115326\cdot x\cdot y - 0.0195509\cdot x\cdot z - 0.0595634\cdot y\cdot z)\end{aligned} \qquad (17)$$
$$x = \lg(a/L)\ ;\ \ y = \lg(\Xi s / Z_0)\ ;\ \ z = \lg(s/L)$$

Interior orifice impedance of a slit in a slit array with an air gap between the slit plate and a porous absorber layer :

Geometrical and material parameters as well as ε_n as above.

$$\gamma_0 = jk_0\cos\Theta\ ;\ \ \gamma_n = k_0\sqrt{(\sin\theta + n\lambda_0/L)^2 - 1} \qquad (18)$$

Impedance Z_b of the back side orifice :

$$\frac{Z_b}{Z_0} = \frac{a}{L}\left[\frac{1+r_0\,e^{-2jk_0 t}}{1-r_0\,e^{-2jk_0 t}} + 2j\sum_{n>0}\frac{k_0}{\gamma_n}\left(\frac{\sin(n\pi a/L)}{n\pi a/L}\right)^2\frac{1+r_n\,e^{-2\gamma_n t}}{1-r_n\,e^{-2\gamma_n t}}\right] \qquad (19)$$

$$r_n = \frac{1 - j\dfrac{k_0 Z_0}{\Gamma_a Z_a}\dfrac{\varepsilon_n}{\gamma_n}\tanh(\varepsilon_n s)}{1 + j\dfrac{k_0 Z_0}{\Gamma_a Z_a}\dfrac{\varepsilon_n}{\gamma_n}\tanh(\varepsilon_n s)} \tag{20}$$

The 1st term in the brackets is the front side impedance of the porous layer transformed to the plane of the back side orifices of the slit plate. Therefore the 2nd term (sum term) is the mass impedance Z_{bm} of the oscillating mass of the back side orifice. The r_n are the modal reflection factors at the front side of the porous layer. The end correction of the back slit orifice is :

$$\frac{\Delta\ell_b}{a} = \frac{-j}{k_0 a}\frac{Z_{bm}}{Z_0} = 2\sum_{n>0}\frac{1}{\gamma_n L}\left(\frac{\sin(n\pi a/L)}{n\pi a/L}\right)^2 \frac{1 + r_n\, e^{-2\gamma_n t}}{1 - r_n\, e^{-2\gamma_n t}} \tag{21}$$

Correspondence and parameters in the following diagrams: "Δlb/a" → $\Delta\ell_b/a$; "F" → L/λ_0 ; parameters: a/L=0.25 ; d/a=1 ; s/L=1 ; R=Ξ·s/Z_0=1 ; porous layer of glass fibres.

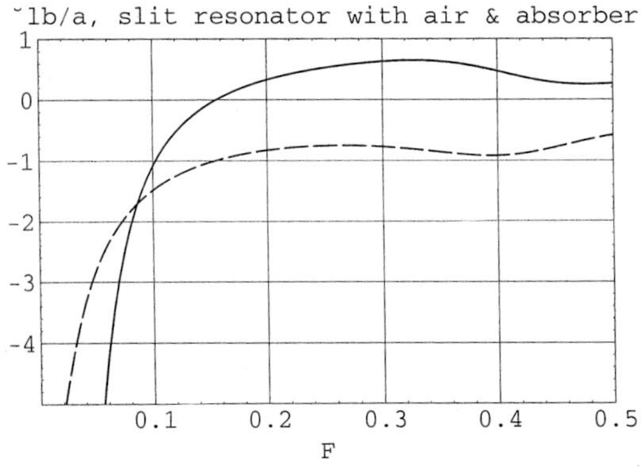

Real (full line) and imaginary (dashed line) part of the interior end correction $\Delta\ell_b/a$ for t/a=0.01 (other parameters given above) . The real part represents a mass reactance if it is positive; at negative values it represents the influence of the porous material on the spring reactance of the volume. The negative imaginary part represents a flow resistance.

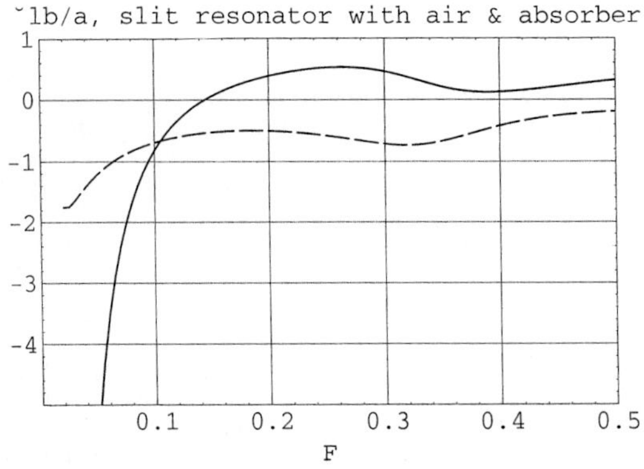

As above, but with a larger distance t/a= 1 between plane of orifices and absorber layer.

F.13 Piston radiating into a hard tube

Ref.: Mawardi, [F.5]

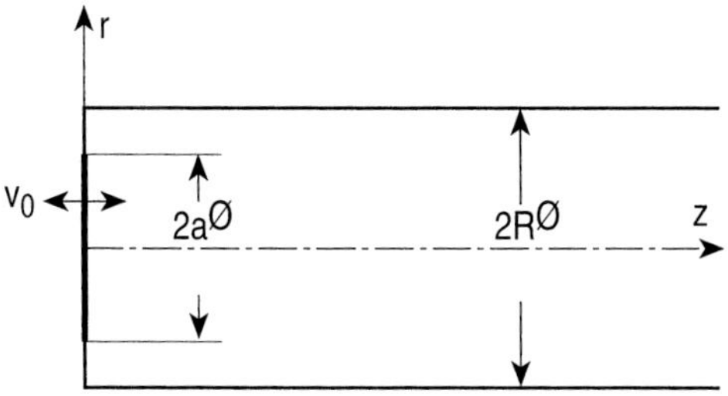

A circular piston with diameter 2a oscillates with a velocity amplitude v_0 in a hard end surface of a hard, circular tube with diameter 2R .
$S=\pi a^2$ = piston area.

Sound pressure on the piston surface z=0:

$$p(r,z=0)=v_0\left[Z_0+j\frac{\omega\rho_0}{S}\sum_{n\ge1}\frac{J_0(k_{0n}r)\cdot J_1(k_{0n}a)\cdot a}{2k_{0n}\sqrt{k_{0n}^2-k_0^2}\cdot J_0^2(k_{0n}a)}\right] \tag{1}$$

with $k_0=\omega/c_0$; k_{mn}= n-th root of $J'_m(k_{mn}R)=0$.

The 2nd term vanishes in the special case a=R with $J_1(k_{0n}a)=0$.

The radiation impedance Z_s is :

$$Z_s=Z_0+j\omega\rho_0\sum_{n\ge1}\frac{J_1^2(k_{0n}a)}{k_{0n}^2\sqrt{k_{0n}^2-k_0^2}\cdot J_0^2(k_{0n}a)} \tag{2}$$

F.14 Oscillating mass of a fence in a hard tube

Ref.: Iwanov-Schitz / Rschevkin, [F.6]

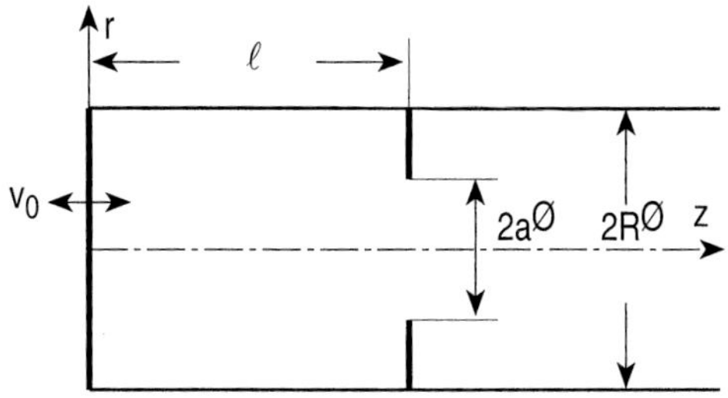

A hard tube with diameter 2R is driven by a plane wave with velocity amplitude v_0 from a piston in a distance ℓ to a thin fence with aperture diameter 2a ; $S=\pi a^2$.

M_I , M_{II} are the oscillating masses of

the fence orifice towards the piston and towards the tube, respectively.

$$M_I = 4S\rho_0 R \sum_{m\geq 1} \frac{J_1^2(x_m a)\cdot \coth(x_m \ell)}{(x_m R)^3 \cdot J_0^2(x_m R)} \quad ; \quad M_{II} = 4S\rho_0 R \sum_{m\geq 1} \frac{J_1^2(x_m a)}{(x_m R)^3 \cdot J_0^2(x_m R)} \tag{1}$$

with x_m the roots of $J_0'(x_m R) = 0$. In the limit $M_I \xrightarrow[\ell\to\infty]{} M_{II}$.

F.15 A ring-shaped piston in a baffle wall

Ref.: Antonov / Putyrev, [F.7]

A ring with interior radius r_0 and exterior radius r_1 oscillates in a baffle wall. The ring surface area is $S_R = \pi(r_1^2 - r_0^2)$; the interior circle areas are $S_0 = \pi r_0^2$; $S_1 = \pi r_1^2$; the radius ratio $\alpha = r_0/r_1$ with $0 \leq \alpha < 1$;the area ratio $\beta = S_R/S_0 = (r_1/r_0)^2 - 1$.

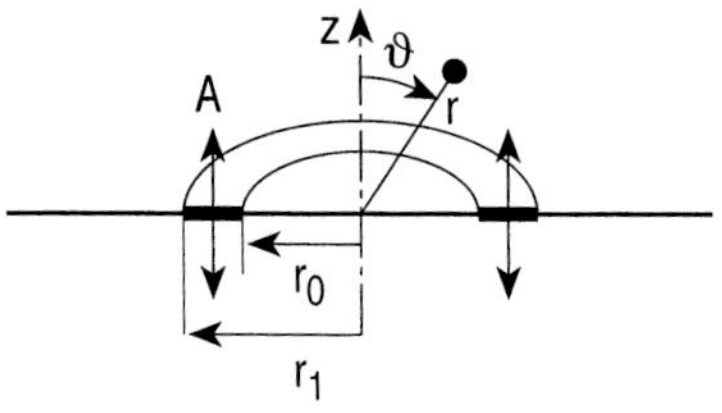

The mechanical radiation impedance $Z_s = Z'_s + j\cdot Z''_s$ (force/velocity) is evaluated by :

$$Z_s = \frac{jk_0 Z_0}{2\pi} \iint_{S_R} dS_1 \iint_{S_R} \frac{e^{-jk_0 r}}{r} dS$$

$$= 2\pi Z_0 \left[\int_{r_0}^{r_1} \left(1 - J_0(2k_0 r) - \tfrac{4}{\pi} I_s\right)\cdot r\,dr + j\int_{r_0}^{r_1} \left(S_0(2k_0 r) - \tfrac{4}{\pi} I_c\right)\cdot r\,dr \right] \tag{1}$$

with $J_0(z)$ the Bessel function , $S_0(z)$ the Struve function of zero order, and the integrals :

$$\left.\begin{matrix} I_s \\ I_c \end{matrix}\right\} = \int_0^{\arcsin(r_0/r)} \left.\begin{matrix} \sin(k_0 r\cdot\cos\vartheta) \\ \cos(k_0 r\cdot\cos\vartheta) \end{matrix}\right\} \cdot \sin\left(k_0\sqrt{r_0^2 - r^2\sin^2\vartheta} \right) d\vartheta \tag{2}$$

Approximation for low frequencies $k_0 r_0 \ll 1$ and $k_0 r_1 \ll 1$:

$$\frac{Z_s}{Z_0} \approx S_0 \left[\frac{\beta^2}{2}(k_0 r_0)^2 + j\frac{8}{3\pi}\frac{k_0 r_0}{\alpha^2}\left((1+\alpha^2)(1-E(\alpha)) + (1-\alpha^2)K(\alpha)\right) \right] \tag{3}$$

with $E(\alpha)$, $K(\alpha)$ the complete elliptic integrals of 1st and 2nd kinds, respectively.

For low frequencies $k_0 r_0 \ll 1$ and $k_0 r_1 \ll 1$ and a slender ring $0 < \beta < 0.6$:

$$\frac{Z''_s}{Z_0} \approx \frac{S_0 ß^2}{2\pi} k_0 r_0 \left[(1 - 0.25ß) \ln\frac{16}{ß} + \frac{3}{2} \right] \tag{4}$$

Special case of a small circular piston radiator, i.e. $r_0 \to 0$ and $k_0 r_1 << 1$:

$$\frac{Z_s}{Z_0} \approx S_1 \left[\frac{(k_0 r_1)^2}{2} + j\frac{8}{3\pi} k_0 r_1 \right] \tag{5}$$

Far field of a ring-shaped piston radiator with elongation amplitude A :

$$p(r,\vartheta) \approx -\frac{1}{2} A Z_0 \frac{r_1}{r} k_0 r_1 \left[2\frac{J_1(k_0 r_1 \sin\vartheta)}{k_0 r_1 \sin\vartheta} - 2\alpha^2 \frac{J_1(k_0 r_0 \sin\vartheta)}{k_0 r \sin\vartheta} \right] \cdot e^{-jk_0 r} \tag{6}$$

F.16 Measures of radiation directivity

Let $p(r,\vartheta,\varphi)$ be the sound pressure generated by a radiator in the far field, $k_0 r >> 1$.

Directivity factor :

$$D_0(\vartheta,\varphi) = \frac{p(r,\vartheta,\varphi)}{p(r,\vartheta_0,\varphi_0)} \quad \left(\text{or} \quad = \frac{|p(r,\vartheta,\varphi)|}{|p(r,\vartheta_0,\varphi_0)|} \right) \tag{1}$$

where $p(r,\vartheta_0,\varphi_0)$ is the sound pressure in a reference direction (mostly the direction of some axis of symmetry of the radiator).

Directivity coefficient :

$$D_0^2(\vartheta,\varphi) == \frac{|p(r,\vartheta,\varphi)|^2}{|p(r,\vartheta_0,\varphi_0)|^2} \tag{2}$$

Directivity value :

$$D_m(\vartheta,\varphi) = \frac{|p(r,\vartheta,\varphi)|^2}{\langle |p(r,\vartheta,\varphi)|^2 \rangle_{\vartheta,\varphi}} \tag{3}$$

Directivity index :

$$D_{L0}(\vartheta,\varphi) = 10 \cdot \lg \frac{|p(r,\vartheta,\varphi)|^2}{|p(r,\vartheta_0,\varphi_0)|^2} \tag{4}$$

Directivity :

$$D_{Lm}(\vartheta,\varphi) = 10 \cdot \lg \frac{|p(r,\vartheta,\varphi)|^2}{\langle |p(r,\vartheta,\varphi)|^2 \rangle_{\vartheta,\varphi}} \tag{5}$$

Sharpness of directivity pattern ß is given as the angle between the normal to the radiator and the direction for which the intensity decreases to 1/2 of the maximum value.

F.17 Directivity of radiator arrays

Ref.: Skudrzyk, [F.8]

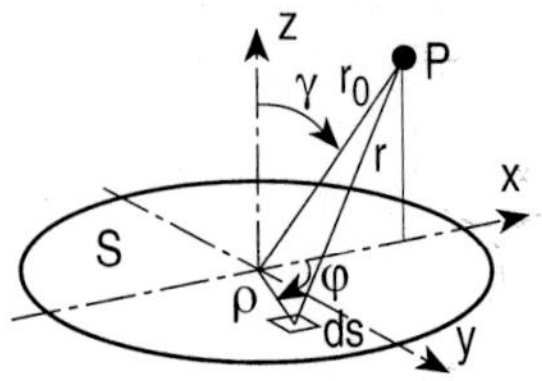

The far field p of a plane radiator with area S and normal velocity distribution V(x,y) in an infinite baffle wall can be evaluated with the Huygens-Rayleigh integral :

$$p = \frac{jk_0Z_0}{2\pi}\int_S \frac{V(x,y)\cdot e^{-jk_0r}}{r}ds = \frac{jk_0Z_0}{2\pi}\frac{e^{-jk_0r_0}}{r_0}\int_S V(x,y)\cdot e^{+jk_0(x\cos(r_0,x)+y\cos(r_0,y))}ds \quad (1)$$

where r_0 is the radius from a reference point on the radiator to the field point P . If the velocity V has the same phase on the radiator, the sound pressure attains its maximum P_0 in the direction normal to the radiator :

$$P_0 = \frac{jk_0Z_0}{2\pi}\frac{e^{-jk_0r_0}}{r_0}\cdot Q \quad ; \quad Q = \int_S V\,ds \quad (2)$$

We describe the sound pressure in other directions with the directivity factor D : $p = D\cdot P_0$ with:

$$\begin{aligned} D &= \frac{1}{Q}\int_S e^{+jk_0(x\cos(r_0,x)+y\cos(r_0,y))}dQ \quad ; \quad dQ = V\cdot ds \\ &= \frac{1}{Q}\int_S e^{+jk_0\rho\cos\varphi\sin\gamma}dQ \xrightarrow[\text{symm.}]{} \frac{1}{Q}\int_S \cos(k_0\rho\cos\varphi\sin\gamma)dQ \end{aligned} \quad (3)$$

(see the graph for γ,φ). The last relation holds for a radiator with a central axis of symmetry.

In the case of an array with small elementary radiators having conphase volume flows Q_n the integral is replaced by a sum :

$$D = \frac{1}{Q}\sum_n Q_n\cdot e^{+jk_0(x_n\cos(r_n,x_n)+y_n\cos(r_n,y_n))} \quad ; \quad Q = \sum_n Q_n \quad (4)$$

Two point sources with equal volume flow Q_i *at* $x=0$ *and* $x=d$:

$$D = e^{jk_0d/2\cdot\cos(r,x)}\cdot\cos\left(k_0d/2\cdot\cos(r,x)\right) = e^{jk_0d/2\cdot\sin\gamma}\cdot\cos\left(k_0d/2\cdot\sin\gamma\right)$$

Maxima of $|D|$ are at angles γ with $d \sin\gamma = 2\nu \cdot \lambda_0 / 2 \quad ; \quad \nu = 1,2,3,\ldots$

minima occur at odd multiples of $\lambda_0/2$.

Point sources equally spaced along a line :

The n point sources spaced at intervals d again are conphase and of equal strength.

$$D = \frac{1}{n} \sum_{\nu=0}^{n-1} e^{j\nu k_0 d \sin\gamma} = e^{j(n-1)\Delta} \frac{\sin(n\Delta)}{n \cdot \sin\Delta} \quad ; \quad \Delta = \tfrac{1}{2} k_0 d \sin\gamma \tag{5}$$

Zeroes of the directivity are at angles γ with $\sin\gamma = \nu\lambda_0/nd$; the principal maximum (with unit value) is at $\gamma = 0$; the angles for the following maxima are at :

$$\sin\gamma = \frac{(2\nu+1)\pi}{2nd} \quad ; \quad \nu = 1,2,3,\ldots \tag{6}$$

with values at the maxima :

$$D_\nu = \frac{1}{n \sin\Delta} = \frac{1}{n \sin((2\nu+1)\pi/(2n))} \tag{7}$$

Densely packed linear array :

With $\ell = n d$ the length of the array :

$$D = e^{j\frac{1}{2}k_0 \ell \sin\gamma} \frac{\sin(\frac{1}{2} k_0 \ell \sin\gamma)}{\frac{1}{2} k_0 \ell \sin\gamma} \tag{8}$$

Densely packed circular array :

The circle has the radius a ; the elementary volume flow $dQ = Q_0 ds = Q_0 \cdot a d\varphi$ is constant along the circle.

$$D = \frac{1}{2\pi} \int_0^{2\pi} e^{j k_0 a \sin\gamma \cos\varphi} \, d\varphi = J_0(\Delta) \quad ; \quad \Delta = k_0 a \sin\gamma \tag{9}$$

Sources at constant intervals along a circle :

Let n point sources with equal volume flow Q be distributed with equal intervals on a circle with radius a . r_0= radius from circle centre to field point P ; γ= angle between circle axis and r_0 ; φ= angle between the x axis in the plane of the circle and the projection of r_0 on the circle plane.

$$D = J_0(k_0 a \sin\gamma) + 2 j^n J_n(k_0 a \sin\gamma) \cdot \cos(n\varphi) + 2 j^{2n} J_{2n}(k_0 a \sin\gamma) \cdot \cos(2n\varphi) + \ldots \tag{10}$$

Circular piston in a baffle wall :

The piston radius is a . Elementary volume flow $dQ= Q_0 \cdot r\, dr\, d\varphi$; $x= r \cdot \cos\varphi$; $y= r \cdot \sin\varphi$.

$$D = \frac{1}{\pi a^2} \int_0^a \int_0^{2\pi} \cos(k_0 r \cos\varphi \sin\gamma)\, r\, dr\, d\varphi = \frac{2}{a^2} \int_0^a r\, J_0(k_0 r \sin\gamma)\, dr = 2 \frac{J_1(k_0 a \sin\gamma)}{k_0 a \sin\gamma} \tag{11}$$

Rectangular piston in a baffle wall :

The side lengths are 2a, 2b ; the elementary volume flow $dQ= Q_0 \cdot dx\, dy$.

$$D = D_1 \cdot D_2$$

$$D_1 = \frac{1}{2a} \int_{-a}^{+a} e^{j k_0 x \cos(r,x)}\, dx = \frac{\sin\left(k_0 a \cos(r_0,x)\right)}{k_0 a \cos(r_0,x)} \tag{12}$$

$$D_2 = \frac{1}{2b} \int_{-b}^{+b} e^{j k_0 y \cos(r,y)}\, dy = \frac{\sin\left(k_0 b \cos(r_0,y)\right)}{k_0 b \cos(r_0,y)}$$

Rectangular plate, clamped at opposite edges, in its fundamental mode :

Let the plate be in a one-dimensional vibration with (approximate) velocity distribution :

$$V(y) = V_0 \cdot \left(1 - y^2 / b^2\right) \tag{13}$$

where 2a is the length of the supported edges and 2b that of the other two edges. Average velocities:

$$\langle V \rangle = \tfrac{2}{3} V_0 \quad ; \quad \langle V^2 \rangle = \tfrac{8}{15} V_0^2 \tag{14}$$

$$D = \frac{3}{\Delta^2} \left(\frac{\sin\Delta}{\Delta} - \cos\Delta \right) \quad ; \quad \Delta = k_0 b \sin\gamma \tag{15}$$

Rectangular plate, free at opposite edges, in its fundamental resonance :

Let the plate be in a one-dimensional vibration with (approximate) velocity distribution :

$$V(y) = V_0 \cdot \left(1 - 2y^2 / b^2\right) \tag{16}$$

The nodal lines (V= 0) are at $y = \pm b / \sqrt{2}$. The average velocities are $\langle V \rangle = \tfrac{1}{3} V_0$; $\langle V^2 \rangle = \tfrac{7}{15} V_0^2$. (17)

$$D = \frac{12}{\Delta^2} \left(\frac{\sin\Delta}{\Delta} - \cos\Delta \right) - \frac{3\sin\Delta}{\Delta} \quad ; \quad \Delta = k_0 b \sin\gamma \tag{18}$$

Circular membrane and plate :

Let the radius be a . The velocity distribution of the fundamental mode can be represented by a power series :

$$V(\rho) = V_0 + V_1\left(1-\rho^2/a^2\right) + V_2\left(1-\rho^2/a^2\right)^2 + \ldots \tag{19}$$

$$D = \left[V_0 + \frac{1}{2}V_1 + \frac{1}{3}V_2 + \ldots + \frac{1}{n+1}V_n\right] \cdot \left[2V_0\frac{J_1(\Delta)}{\Delta} + 2\cdot 1!\cdot V_1\frac{J_2(\Delta)}{\Delta^2} + \ldots + 2^{n+1}\cdot n!\cdot V_n\frac{J_{n+1}(\Delta)}{\Delta^{n+1}}\right]^{-1} \quad ; \quad \Delta = k_0 a \sin\gamma \tag{20}$$

For a velocity distribution $V(\rho)= V_0\cdot J_0(k_B\rho)$ (21)

with the bending wave number k_B on the radiator :

$$D = \frac{1}{a^2}\frac{k_B a}{J_1(k_B a)}\frac{a}{k_B^2 - k_0^2 \sin\gamma}\left[k_B J_0(k_0 a \sin\gamma)\, J_1(k_B a) - k_0 \sin\gamma\, J_1(k_0 a \sin\gamma)\, J_0(k_B a)\right] \tag{22}$$

If the membrane or plate is supported at its edge, i.e. $J_0(k_B a)= 0$:

$$D = \frac{k_B^2}{k_B^2 - k_0^2 \sin\gamma} J_0(k_0 a \sin\gamma) \tag{23}$$

Circular radiator with radial and azimuthal nodal lines :

Develop the velocity distribution into a Fourier series : $V(\rho,\varphi_0) = \sum_{m\geq 0} V_m(\rho)\cdot\cos(m\varphi_0)$ (24)

with radial nodal lines for integer $m>0$, and circular nodal lines at $V_m(\rho)= 0$.

Write the far field pressure as : $p(r,\gamma,\varphi) = \frac{e^{-jk_0 r}}{r}\sum_{m\geq 0} K_m(\gamma,\varphi)$ (25)

The directivity factor of a sum term is then :

$$D_m(\gamma,\varphi) = \frac{p_m(r,\gamma,\varphi)}{p_0(r,0,\varphi)} = \frac{K_m(\gamma,\varphi)}{K_0(0,\varphi)} = \frac{2\pi\cdot K_m(\gamma,\varphi)}{Q} = \frac{2\pi\cdot K_m(\gamma,\varphi)}{\langle V\rangle S} \quad ; \quad S = \pi a^2 \tag{26}$$

$$K_m(\gamma,\varphi) = \cos(m\varphi)\cdot e^{jm\pi/2}\int_0^a J_m(k_0\rho\sin\gamma)\cdot V_m(\rho)\cdot\rho\, d\rho \tag{27}$$

Introducing the integral transform (which is tabulated for many $V_{cm}(\rho)$) :

$$f_m(\lambda) = \frac{1}{a^2}\int_0^a J_m(\lambda\rho)\cdot V_m(\rho)\cdot\rho\, d\rho \tag{28}$$

one gets :

$$K_m(\gamma,\varphi) = a^2\cdot j^m\cdot f_m(k_0\sin\gamma)\cdot\cos(m\varphi) \tag{29}$$

The directivity factor $D(\gamma,\vartheta)$ is the sum of the $D_m(\gamma,\vartheta)$.

Particle velocity: $$v = v_r = \frac{p(r)}{Z_0}\left(1 - \frac{j}{k_0 r}\right) \quad (2)$$

Energy density: $$w = \frac{\rho_0}{(4\pi r^2)^2}\,|q|^2\left((k_0 r)^2 + \tfrac{1}{2}\right) \quad (3)$$

Effective intensity: $$I = I_r = \frac{|p(r)|^2}{2Z_0} \quad (4)$$

Radiated (effective) power: $$\Pi = 4\pi r^2 \cdot I_r = \frac{\rho_0 \omega^2}{8\pi c_0}|q|^2 \quad (5)$$

Radiant energy in a shell of unit thickness: $$E' = \frac{\rho_0 k_0^2}{4\pi}|q|^2 \quad (6)$$

Reactive energy outside the radius r : $$E'' = \frac{\rho_0}{8\pi r}|q|^2 \quad (7)$$

If the source has a finite radius $a \ll \lambda_0$:

Surface impedance (outward): $$Z_s = \frac{p(a)}{v_r(a)} = \frac{Z_0}{1 - j/k_0 a} = Z_0 \frac{k_0 a(k_0 a + j)}{1 + (k_0 a)^2} \quad (8)$$

Let a monopole source with volume flow amplitude q be at $\mathbf{r_0} = (x_0, y_0, z_0)$:

Sound pressure in $\mathbf{r} = (x,y,z)$: $$p(\mathbf{r}) = j k_0 Z_0 \cdot q \cdot g(\mathbf{r}\,|\,\mathbf{r_0}) \quad (9)$$

with: $$g(\mathbf{r}\,|\,\mathbf{r_0}) = \frac{e^{-jk_0 R}}{4\pi R} \quad ; \quad R^2 = |\mathbf{r} - \mathbf{r_0}|^2 = (x - x_0)^2 + (y - y_0)^2 + (z - z_0)^2 \quad (10)$$

Dipole :

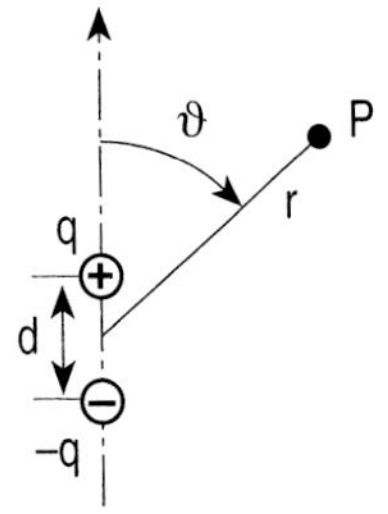

Two monopoles with opposite sign of the volume flow q at a mutual distance $d \ll \lambda_0$.

Dipole strength: $D = q \cdot d$ (11)

Sound pressure:

$$p(\mathbf{r}) = j k_0 Z_0 \cdot q \cdot \left(g(\mathbf{r}\,|\,\tfrac{1}{2}\mathbf{d}) - g(\mathbf{r}\,|\,-\tfrac{1}{2}\mathbf{d})\right) = -k_0^2 Z_0 D \frac{e^{-jk_0 r}}{4\pi r}\left(1 - \frac{j}{k_0 r}\right) \cdot \cos\vartheta \quad (12)$$

Velocity components:

$$v_r = -k_0^2 D\frac{e^{-jk_0r}}{4\pi r}\left(1-\frac{j}{k_0r}-\frac{2}{(k_0r)^2}\right)\cdot\cos\vartheta \quad ; \quad v_\vartheta = -jk_0D\frac{e^{-jk_0r}}{4\pi r^2}\left(1-\frac{j}{k_0r}\right)\cdot\sin\vartheta \tag{13}$$

Effective energy density :
$$w = \rho_0\left(\frac{k_0^2 D}{4\pi r}\right)^2\left[\cos^2\vartheta+\frac{1}{2(k_0r)^2}+\frac{1+3\cos^2\vartheta}{2(k_0r)^4}\right] \tag{14}$$

Effective intensity :
$$I_r = \frac{Z_0}{2}\left(\frac{k_0^2D}{4\pi r}\right)^2\cos^2\vartheta \tag{15}$$

Effective power :
$$\Pi = \frac{Z_0}{2}\frac{4\pi^2}{3\lambda_0^4}|D|^2 = \frac{\rho_0\omega^4}{24\pi c_0^3}|D|^2 \tag{16}$$

Radiant energy in a shell of unit thickness:
$$E' = \frac{\rho_0 k_0^2}{12\pi}|D|^2 \tag{17}$$

Reactive energy outside the radius r :
$$E'' = \frac{\rho_0}{12\pi r^3}|D|^2 \tag{18}$$

A dipole corresponds to a small hard sphere with radius $a \ll \lambda_0$ oscillating back and forth in the direction of the dipole axis with a maximum surface velocity U_d in that direction.

Maximum velocity:
$$U_d = \frac{D}{2\pi a^3}e^{-jk_0a}\left(1+jk_0a-\tfrac{1}{2}(k_0a)^2\right) \tag{19}$$

Driving force:
$$F_d = \frac{jk_0Z_0\cdot D}{3}e^{-jk_0a}\left(1-jk_0a\right) \tag{20}$$

Mechanical driving impedance:
$$Z_d = \frac{F_d}{U_d} = \frac{2\pi a^3k_0Z_0}{3}\frac{(k_0a+j)}{\left(1+jk_0a-\frac{1}{2}(k_0a)^2\right)} \tag{21}$$

A dipole centred at the point $\mathbf{r_0}=(x_0,y_0,z_0)$ with dipole strength vector $\mathbf{D}=(D_x,D_y,D_z)$; with $\mathbf{R}=\mathbf{r}-\mathbf{r_0}$ and $\mathbf{R}$ having the spherical angles ϑ_R, φ_R has the sound pressure field :

$$p(\mathbf{r}) = jk_0Z_0\cdot\mathbf{D}\cdot\mathbf{g}(\mathbf{r}|\mathbf{r_0}) \quad ; \quad \mathbf{g}(\mathbf{r}|\mathbf{r_0}) = (g_x,g_y,g_z) \tag{22}$$

$$g_x = \sin\vartheta_R\cos\varphi_R\cdot|g_\omega| \quad ; \quad g_y = \sin\vartheta_R\sin\varphi_R\cdot|g_\omega| \quad ; \quad g_z = \cos\vartheta_R\cdot|g_\omega|$$
$$|g_\omega| = \frac{jk_0}{4\pi}\frac{e^{-jk_0R}}{R}\left(1-\frac{j}{k_0R}\right) \tag{23}$$

Lateral quadrupole :

For $d \ll \lambda_0$, with $D_{xy} = q\cdot d^2$ (24)

Sound pressure field :

$$p = -jk_0^3Z_0\cdot D_{xy}\frac{xye^{-jk_0r}}{4\pi r^3}\left(1-\frac{3j}{k_0r}-\frac{3}{(k_0r)^2}\right) \tag{25}$$

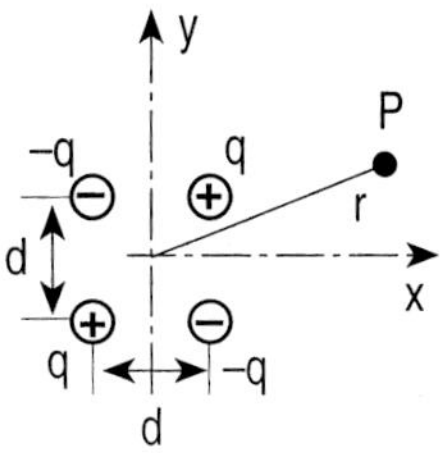

Linear quadrupole :

The two central monopoles collapse to a volume flow $-2q$. For $d \ll \lambda_0$, with $D_{xx} = q \cdot d^2$ (26)

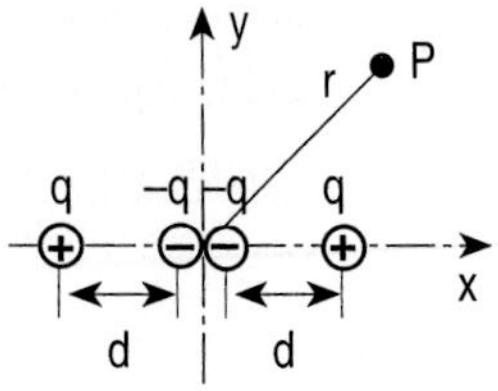

sound pressure field :

$$p = -jk_0^3 Z_0 \cdot D_{xx} \frac{e^{-jk_0 r}}{4\pi r} \left[\left(\frac{x}{r} \right)^2 - \frac{3x^2 - r^2}{r^2} \left(\frac{j}{k_0 r} + \frac{1}{(k_0 r)^2} \right) \right] \quad (27)$$

F.20 Plane radiator in a baffle wall

Ref.: Heckl, [F.5]

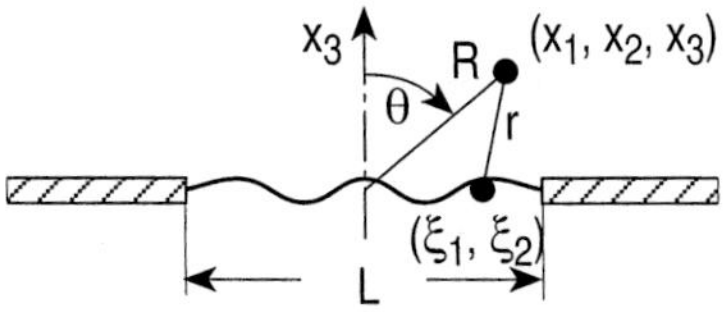

A plane radiator with either dimensions L×B in Cartesian co-ordinates (x_1,x_2,x_3) or radius a in polar co-ordinates (R,θ,φ) is contained in a hard baffle wall.
A point on the radiator is at (ξ_1,ξ_2). The radiator area is, respectively, $S = L \cdot B = \pi \cdot a^2$.

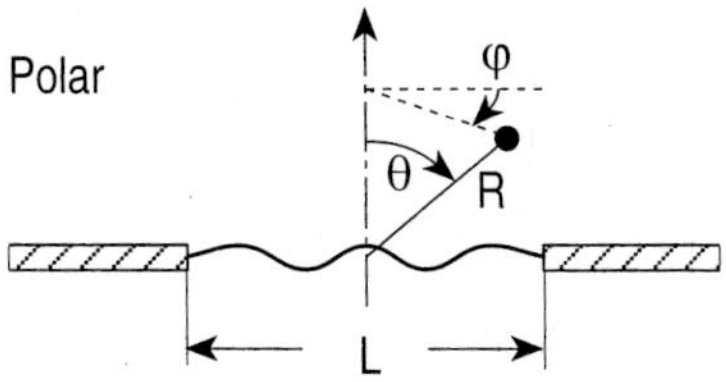

Geometrical relations :

$$r^2 = (x_1 - \xi_1)^2 + (x_2 - \xi_2)^2 + x_3^2$$
$$R^2 = x_1^2 + x_2^2 + x_3^2 \quad (1)$$

$$x_1 = R \cdot \sin\theta \cdot \cos\varphi \quad ; \quad x_2 = R \cdot \sin\theta \cdot \sin\varphi \quad ; \quad x_3 = R \cdot \cos\theta \quad (2)$$

Quantities :

$v(\xi_1,\xi_2)$	given velocity distribution of the radiator;
$\hat{v}(k_1,k_2)$	Fourier transform of $v(\xi_1,\xi_2)$;
$p(x_1,x_2,x_3)$	sound pressure in a field point;
Π	effective sound power radiated to one side;
$k_b = 2\pi/\lambda_b$	wave number of radiator bending wave;
k_1, k_2	bending wave number components in directions x_1, x_2 ;
r_q, χ	polar co-ordinates r,φ of a point on the source in polar co-ordinates;
p_L, Π_L	sound pressure and effective power radiated by a line source;
v_0	velocity amplitude of the radiator;
Z_s	radiation impedance;

σ	radiation efficiency;
m_W	oscillating medium mass;
η	bending wave loss factor;
g(x)	envelope of the radiator velocity distribution;

Sound pressure in a far field point, i.e. $k_0 L^2/R \ll 1$ or $R\cdot\lambda_0 > L^2$:

Cartesian:
$$p(x_1,x_2,x_3) = \frac{j k_0 Z_0}{2\pi} \iint_S v(\xi_1,\xi_2) \frac{e^{-jk_0 r}}{r} d\xi_1 \, d\xi_2 \tag{3}$$

polar:
$$p(R,\theta,\varphi) = \frac{j k_0 Z_0}{2\pi} \frac{e^{-jk_0 R}}{R} \iint_S v(\xi_1,\xi_2)\cdot e^{j k_0 \sin\theta(\xi_1\cos\varphi+\xi_2\sin\varphi)} d\xi_1 \, d\xi_2 \tag{4}$$

Using the wave number spectrum of the radiator pattern :

$$p(x_1,x_2,x_3) = \frac{k_0 Z_0}{4\pi^2} \int\!\!\int_{-\infty}^{+\infty} \frac{\hat{v}(k_1,k_2)}{\sqrt{k_0^2-k_1^2-k_2^2}} \cdot e^{j(k_1 x_1+k_2 x_2)} \cdot e^{-j x_3 \sqrt{k_0^2-k_1^2-k_2^2}} \, dk_1 \, dk_2 \tag{5}$$

Long source (in x_2 direction) :

$$p_L(x_1,x_3) = \frac{k_0 Z_0}{2\pi} \int_{-\infty}^{+\infty} \frac{\hat{v}(k_1)}{\sqrt{k_0^2-k_1^2}} \cdot e^{j k_1 x_1} \cdot e^{-j x_3 \sqrt{k_0^2-k_1^2}} \, dk_1 \tag{6}$$

Radiator with radial symmetry (index r):

$$p_r(R,\theta) = \frac{j k_0 Z_0}{2\pi R} \cdot \hat{v}_r(k_0 \sin\theta)\cdot e^{-jk_0 R} \tag{7}$$

Wave number spectrum of radiator velocity pattern:

rectangular (Fourier transform) :

$$\hat{v}(k_1,k_2) = \int\!\!\int_{-\infty}^{+\infty} v(\xi_1,\xi_2)\cdot e^{-j(k_1\xi_1+k_2\xi_2)} d\xi_1 \, d\xi_2 = \iint_S v(\xi_1,\xi_2)\cdot e^{-j(k_1\xi_1+k_2\xi_2)} d\xi_1 \, d\xi_2$$
$$v(\xi_1,\xi_2) = \frac{1}{4\pi^2} \int\!\!\int_{-\infty}^{+\infty} \hat{v}(k_1,k_2)\cdot e^{+j(k_1\xi_1+k_2\xi_2)} dk_1 \, dk_2 \tag{8}$$

with radial symmetry (Hankel transform) :

$$\hat{v}(k_1,k_2) \to \hat{v}_r(k_r) = 2\pi \int_0^{\infty} v(r_q)\cdot J_0(k_r r_q)\cdot r_q \, dr_q \quad ; \quad k_r = \sqrt{k_1^2+k_2^2} \quad ; \quad \begin{matrix} \xi_1 = r_q \cdot \cos\chi \\ \xi_2 = r_q \cdot \sin\chi \end{matrix} \tag{9}$$

Effective sound power Π radiated to one side :

$$\Pi = \frac{1}{2Z_0} \int_0^{\pi/2} \int_0^{2\pi} |p(R,\theta,\varphi)|^2 \cdot R^2 \sin\theta \, d\varphi \, d\theta \tag{10}$$

$$\begin{aligned} \Pi &= \frac{k_0^2 Z_0}{8\pi^2} \int_0^{\pi/2} \int_0^{2\pi} |\hat{v}(-k_0 \sin\theta\cos\varphi, -k_0 \sin\theta \sin\varphi)|^2 \sin\theta \, d\varphi \, d\theta \\ &= \frac{k_0 Z_0}{8\pi^2} \mathrm{Re}\left\{ \int_{-\infty}^{+\infty} \int |\hat{v}(k_1,k_2)|^2 \frac{dk_1 \, dk_2}{\sqrt{k_0^2 - k_1^2 - k_2^2}} \right\} \end{aligned} \tag{11}$$

Line source :

$$\Pi_L = \frac{k_0 Z_0}{4\pi} \int_{-k_0}^{+k_0} |\hat{v}(k_1)|^2 \frac{dk_1}{\sqrt{k_0^2 - k_1^2}} = \frac{k_0 Z_0}{4\pi} \int_{-\pi/2}^{+\pi/2} |\hat{v}(k_0 \cos\psi)|^2 \, d\psi \tag{12}$$

Source with radial symmetry :

$$\Pi_r = \frac{k_0^2 Z_0}{4\pi} \int_0^{\pi/2} |\hat{v}_r(k_0 \sin\theta)|^2 \cdot \sin\theta \, d\theta \tag{13}$$

Radiation impedance Z_s (Π is complex power; $\langle\ldots\rangle$ indicates average) :

Definition : $$\Pi = \frac{S}{2} \cdot Z_s \cdot \langle v^2 \rangle_{\xi_1,\xi_2} \tag{14}$$

$$Z_s = \frac{k_0 Z_0}{4\pi \langle v^2 \rangle_{\xi_1,\xi_2}} \int_{-\infty}^{+\infty} \int |\hat{v}(k_1,k_2)|^2 \frac{dk_1 \, dk_2}{\sqrt{k_0^2 - k_1^2 - k_2^2}} \tag{15}$$

Radiation efficiency σ :

Definition : $$\sigma = \frac{\mathrm{Re}\{Z_s\}}{Z_0} \tag{16}$$

$$\sigma = k_0 \cdot \iint_{k_1^2 + k_2^2 < k_0^2} |\hat{v}(k_1,k_2)|^2 \frac{dk_1 \, dk_2}{\sqrt{k_0^2 - k_1^2 - k_2^2}} \Bigg/ \int_{-\infty}^{+\infty} \int |\hat{v}(k_1,k_2)|^2 \, dk_1 \, dk_2 \tag{17}$$

Oscillating mass m_w :

Definition : $$m_w = \frac{S \cdot \mathrm{Im}\{Z_s\}}{\omega} \tag{18}$$

$$m_w = \rho_0 S \iint\limits_{k_1^2+k_2^2>k_0^2} |\hat{v}(k_1,k_2)|^2 \frac{dk_1\,dk_2}{\sqrt{-k_0^2+k_1^2+k_2^2}} \Bigg/ \iint\limits_{-\infty}^{+\infty} |\hat{v}(k_1,k_2)|^2\,dk_1\,dk_2 \tag{19}$$

Useful substitutions for evaluation :

Set $k_1 \to k_0 \cosh(z\cos\varphi)$; $k_2 \to k_0 \cosh(z\sin\varphi)$

makes :
$$\frac{dk_1\,dk_2}{\sqrt{k_1^2+k_2^2-k_0^2}} = k_0 \cdot \cosh z \cdot dz\,d\varphi \tag{20}$$

For line sources, set $k_1 \to k_0 \cosh(z)$; $k_2 \to 0$; $S \to B$

makes :
$$\frac{dk_1}{\sqrt{k_1^2-k_0^2}} = dz \tag{21}$$

Fourier transforms of some 1-dimensional velocity patterns:

Pattern:	Range:	Fourier transform:
$v(\xi_1) = v_0$	$\lvert\xi_1\rvert < L/2$	$\frac{\hat{v}(k_1)}{v_0} = L\frac{\sin(k_1L/2)}{k_1L/2}$
$v(\xi_1) = v_0\left(1-2\lvert\xi_1/L\rvert\right)$	$\lvert\xi_1\rvert < L/2$	$\frac{\hat{v}(k_1)}{v_0} = \frac{L}{2}\left(\frac{\sin(k_1L/4)}{k_1L/4}\right)^2$
$v(\xi_1) = v_0\left(1-(2\xi_1/L)^2\right)^2$	$\lvert\xi_1\rvert < L/2$	$\frac{\hat{v}(k_1)}{v_0} = L\left[\left(24\left(\frac{2}{k_1L}\right)^5 - 8\left(\frac{2}{k_1L}\right)^3\right)\cdot\sin\frac{k_1L}{2} - 24\left(\frac{2}{k_1L}\right)^4\cdot\cos\frac{k_1L}{2}\right]$
$v(\xi_1) = v_0\left(2\xi_1/L\right)$	$\lvert\xi_1\rvert < L/2$	$\frac{\hat{v}(k_1)}{v_0} = -jL\left[\frac{\sin(k_1L/2)}{(k_1L/2)^2} - \frac{\cos(k_1L/2)}{k_1L/2}\right]$
$v(\xi_1) = v_0\left[3(2\xi_1/L)^2-1\right]/2$	$\lvert\xi_1\rvert < L/2$	$\frac{\hat{v}(k_1)}{v_0} = L\left[3\frac{\cos(k_1L/2)}{(k_1L/2)^2} + \left(\frac{2}{k_1L} - 3\left(\frac{2}{k_1L}\right)^3\right)\sin\frac{k_1L}{2}\right]$
$v(\xi_1) = v_0\,e^{-\alpha\lvert\xi_1\rvert}$	$\lvert\xi_1\rvert < \infty$	$\frac{\hat{v}(k_1)}{v_0} = \frac{2\alpha}{\alpha^2+\xi_1^2}$

Hankel transforms of some velocity patterns with radial symmetry :

Pattern:	Range:	Hankel transform:
$v(r_q)=v_0$	$r_q<a$	$\dfrac{\hat{v}_r(k_r)}{v_0\,a^2}=2\pi\dfrac{J_1(k_r a)}{k_r a}$
$v(r_q)=v_0\,J_0(k_b r_q)$	$r_q<a$	$\dfrac{\hat{v}_r(k_r)}{v_0\,a^2}=2\pi\dfrac{k_r a\,J_0(k_b a)J_1(k_r a)-k_b a\,J_0(k_r a)J_1(k_b a)}{(k_r a)^2-(k_b a)^2}$
$v(r_q)=v_0\,e^{-\alpha\,r_q}$	$r_q<\infty$	$\dfrac{\hat{v}_r(k_r)}{v_0\,a^2}=2\pi\dfrac{\alpha}{(\alpha^2+k_r^2)^{3/2}}$
$v(r_q)=v_0\,e^{-p^2\,r_q^2}$	$r_q<\infty$	$\dfrac{\hat{v}_r(k_r)}{v_0\,a^2}=\dfrac{2\pi}{2p^2}e^{-k_r^2/4p^2}$
$v(r_q)=v_0/r_q$	$r_q<\infty$	$\dfrac{\hat{v}_r(k_r)}{v_0\,a^2}=\dfrac{2\pi}{k_r}$
$v(r_q)=v_0/r_q^{\kappa}$	$r_q<\infty$	$\dfrac{\hat{v}_r(k_r)}{v_0\,a^2}=\pi\,2^{\kappa+2}\dfrac{\Gamma(1+\kappa/2)}{\Gamma(-\kappa/2)}k_r^{-(\kappa+2)}$

Radiator with *nearly periodic velocity pattern* $v_M(x_1)$:

$$v_M(x_1)=v_0\cdot g(x_1)\cdot\cos(k_b x_1)$$

$$\hat{v}_M(k_1)=\frac{v_0}{2}\int\hat{g}(\mu)[\delta(k_1-k_b-\mu)+\delta(k_1+k_b-\mu)]d\mu \qquad (22)$$

$$=\frac{v_0}{2}[\hat{g}(k_1-k_b)+\hat{g}(k_1+k_b)]$$

with $\delta(x)$ the Dirac delta function, and $\hat{g}(k)$ the Fourier transform of the envelope $g(x_1)$. For a pattern with $\sin(k_b x_1)$ multiply $\hat{v}_M(k_1)$ with j and subtract the 2nd function in [...].

F.21 Ratio of radiation and excitation efficiencies of plates

Ref.: Heckl, [F.6]

Consider two "experiments" :

- plate excited at a point with a force F radiates a sound power Π ;
- plate excited by a diffuse sound field with pressure p vibrates with a velocity v .

Define the radiation efficiency a by $\Pi_{eff} = a \cdot F_{eff}^2$; (1)

define the excitation efficiency b by $v_{eff}^2 = b \cdot p_{eff}^2$ (2)

Then:

$$\frac{a}{b} = \frac{Z_0 k_0^2}{4\pi} \tag{3}$$

Consider two "experiments" :

- plate excited by a line source with a force F_L radiates a sound power Π_L ;
- plate excited by a diffuse sound field with pressure p vibrates with a velocity v .

Define the radiation efficiency α by $\Pi_{L,eff} = \alpha \cdot F_{L,eff}^2$; (4)

define the excitation efficiency ß by $v_{eff}^2 = ß \cdot p_{eff}^2$ (5)

Then:

$$\frac{\alpha}{ß} = \frac{Z_0 k_0}{4} \tag{6}$$

F.22 Radiation of plates with special excitations

Ref.: Ver, [F.7]

Let S be the area of a plate in a baffle wall ; $\langle v^2 \rangle_S$ the average of the squared vibration velocity ; Π the sound power radiated into one half-space ; σ with $\Pi = Z_0 S \cdot \sigma \cdot \langle v^2 \rangle_S / 2$ the radiation efficiency ; f_{cr} the critical frequency ; σ_P Poisson ratio ; c_D dilatation wave velocity ; h plate thickness ; $m'' = \rho_p h$ surface mass density.

Infinite plate with a free bending wave k_B at $f > f_{cr}$:

$$\sigma = 1/\sqrt{1-(k_B/k_0)^2} = 1/\sqrt{1-(f_{cr}/f)} \tag{1}$$

Infinite plate (without losses) excited by a point force with amplitude F for $f \ll f_{cr}$:
(with $m''=\rho_p h$ surface mass density)

$$\Pi \approx \frac{\rho_0}{2\pi c_0}\left(\frac{F}{2m''}\right)^2\left[1-\frac{\rho_0}{k_0 m''}\tan^{-1}\frac{k_0 m''}{\rho_0}\right] \approx \begin{cases} \dfrac{\rho_0}{2\pi c_0}\left(\dfrac{F}{2m''}\right)^2 & ; \quad \dfrac{k_0 m''}{\rho_0} \gg 1 \\ \dfrac{(k_0 F)^2}{24\pi Z_0} & ; \quad \dfrac{k_0 m''}{\rho_0} \ll 1 \end{cases} \tag{2}$$

Infinite plate (without losses) with a point velocity source with amplitude v for $k_0 m''/\rho_0 \gg 1$:

$$\Pi \approx \frac{2}{\pi^3}\frac{\rho_0 c_0^3}{f_{cr}^2}\, v^2 \tag{3}$$

Radius a of the equivalent ideal piston radiator (with same $\langle v^2\rangle_S$ and unit efficiency) :

$$a = \sqrt{8/\pi^3}\cdot\lambda_{cr} = 0.286\cdot\lambda_{cr} \quad ; \quad \lambda_{cr} = c_0/f_{cr} \tag{4}$$

Far field for a point force acting on an infinite plate, with force amplitude F :
(R= radius from excitation point to field point; ϑ its polar angle)

For a thin plate without losses ($f< 0.7\cdot f_{cr}$) :

$$p(R,\vartheta) = \frac{jk_0}{2\pi}\cdot F\cdot\frac{e^{-jk_0R}}{R}\cdot\frac{\cos\vartheta}{1+j\dfrac{k_0 m''}{\rho_0}\cos\vartheta\cdot\left(1-(f/f_{cr})^2\sin^4\vartheta\right)} \tag{5}$$

For a thick plate without losses ($f> 0.7\cdot f_{cr}$) :

$$p(R,\vartheta) = \frac{jk_0}{2\pi}\cdot F\cdot\frac{e^{-jk_0R}}{R}\cdot\frac{[1+\varphi(\vartheta)]\cos\vartheta}{[1+\varphi(\vartheta)]+j\dfrac{k_0 m''}{\rho_0}\left\{1+\left[1-\dfrac{1-\sigma_P}{24}\dfrac{(\pi c_D\sin\vartheta)^2}{c_0^2}\right]\cdot\varphi(\vartheta)\right\}} \tag{6}$$

$$\varphi(\vartheta) = \frac{2(k_0 h)^2}{\pi^2(1-\sigma_P)}\left(\sin^2\vartheta-(c_0/c_D)^2\right)$$

For plates with a loss factor η substitute $c_D \to c_D\cdot(1+j\cdot\eta/2)$; $f_{cr} \to f_{cr}\cdot(1+j\cdot\eta/2)$. (7)

References to part F :

Radiation of Sound

[F.1] MECHEL, F.P.
"Schallabsorber", Vol. I, Chapter 9:
"Radiation impedances"
S.Hirzel Verlag, Stuttgart, 1989

[F.2] MECHEL, F.P.
"Schallabsorber", Vol. II, Chapter 22:
"Collection of end corrections"
S.Hirzel Verlag, Stuttgart, 1995

[F.3] BOUWKAMP, C.J. Philipps Research Report 1 (1945/46) 251-277
"Diffraction Theory"

[F.4] MECHEL, F.P.
"Schallabsorber", Vol. I, Chapter 10:
"Radiation impedance of field pattern excited radiators"
S.Hirzel Verlag, Stuttgart, 1989

[F.5] MAWARDI, O.K. J.Acoust.Soc.Amer. 23 (1951) 571-576

[F.6] IWANOV-SCHITZ, K.M.; RSCHEVKIN, S.N. Acustica 13 (1963) 403-406

[F.7] ANTONOV, S.N.; PUTYREV, V.A. Sovj.Phys.Acoust. 30 (1984) 429-432

[F.8] SKUDRZYK, E.
"The Foundations of Acoustics", Chapt. 26
Springer Verlag, 1971, N.Y.

[F.9] SKUDRZYK, E.
"The Foundations of Acoustics", Chapt. 21
Springer Verlag, 1971, N.Y.

[F.10] MORSE, P.M.; INGARD, K.U.
"Theoretical Acoustics", Chapt. 7
McGraw-Hill, 1968, N.Y.

[F.5] HECKL, M. Acustica 37 (1977) 155-166

[F.6] HECKL, M. Frequenz 18 (1964) 299-304

[F.7] VER, I.
in "Noise and Vibration Control", chapter 11
McGraw-Hill, N.Y., 1971

G
Porous Absorbers

If no extra reference is given in the sections of this chapter, see [G.1].

The aims of the sections in this chapter are twofold: first to derive the characteristic propagation constant Γ_a and the characteristic wave impedance Z_a of a plane wave in the porous material as functions of structure data, second to derive the flow resistivity Ξ of the material as function of structure data, because the flow resistivity is the most useful material parameter for the evaluation of Γ_a, Z_a. Different models of a porous material will be presented; special attention will be given to fibrous materials. See also the sections about scattering in random media in the chapter "Scattering of Sound" for propagation constant and wave impedance in fibrous and granular media.

G.1 Structure parameters of porous materials

The definition of structure parameters depends on the model theory in which they are applied (see the relevant sections for specific definitions). Especially, see the chapter about BIOT's theory for special parameters of that theory. This section describes the structure parameters which come from the theory of the "quasi-homogeneous material" (see next section), because they are most often used to describe qualitatively porous absorber materials; the theory of the quasi-homogeneous material is the simplest theory.

Volume porosity σ_V , *massivity* μ :

The volume porosity σ_V is the ratio of air volume contained in the porous material to the total volume; it is given by $\sigma_V = 1-\rho_a/\rho_m$ with ρ_a= bulk density of the porous absorber material, ρ_m= density of the (dense) matrix material. For glass or mineral fibre materials a value $\rho_m \approx 2250$ [kg/m^3] may be used.

For some considerations it is advantageous to apply the massivity $\mu=1-\sigma$.

The following table gives ranges of σ_V for some materials.

Material	σ_V from	σ_V to
Mineral fibre materials	0.92	0.99
Foams	0.95	0.995
Felts	0.83	0.95
Wood-fibre board	0.65	0.80
Wood-wool board	0.50	0.65
Porous render	0.60	0.65
Pumice concrete	0.25	0.50
Pumice fill	0.65	0.85
Gravel and stone chip fill	0.25	0.45
Ceramic filtres	0.33	0.42
Brick	0.25	0.30
Sinter metal	0.10	0.25
Fire-clay	0.15	0.35
Sand stone	0.02	0.06
Marble		ca. 0.005

Structure factor χ :

The structure factor is the most ambiguous quantity in porous material theories. It is defined in the theory of the quasi-homogeneous material as $\chi = \sigma_V/\sigma_S$ with σ_S= surface porosity of a cut through the material. Its value depends on the type of the pore shapes. Because the pore volume V_p is given by the integral over the pore surface S_p over a distance x normal to the considered pore surface :

$$V_p = \int_x S_p(x)\,dx \quad ; \quad \chi = \frac{\langle \sigma_S \rangle_x}{\sigma_S} \qquad (1)$$

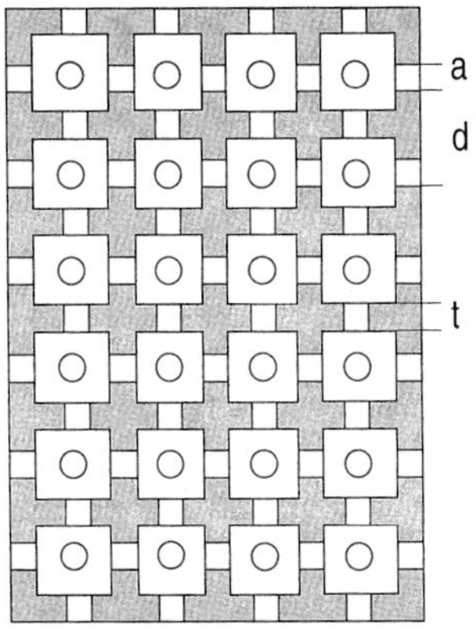

A model of an open-cellular foam consisting of cubic cells with connecting pores has the values :

$$\sigma_V \approx 1 - 3\frac{t}{d} \quad ; \quad \chi = \frac{(1+\pi a/2)(d-3t)}{\pi a^2} \qquad (2)$$

Two end corrections $\Delta\ell = \pi a/2$ have been added to the neck length t .

The following table gives the volume and surface porosities σ_V, σ_S and the structure factor χ of some regular model structures; a is the width of the pores, d is their distance.

Structure	σ_V	σ_S	χ
Flat pores, longitudinal or inclined	a/d	a/d	1
Square pores, longitudinal or inclined	$(a/d)^2$	$(a/d)^2$	1
Round pores, longitudinal or inclined	$\pi(a/d)^2/4$	$\pi(a/d)^2/4$	1
Square fibres, longitudinal or inclined	$2a/d-(a/d)^2$	$2a/d-(a/d)^2$	1
Round fibres, longitudinal or inclined	$1-\pi(a/d)^2/4$	$1-\pi(a/d)^2/4$	1
Square fibres, transversal	$2a/d-(a/d)^2$	a/d	$2-a/d$
Round fibres, transversal	$1-\pi(a/d)^2/4$	$1-a/d$	$1+a/d+(1-\pi/4)(a/d)^2$
Array of cubes	$3a/d-3(a/d)^2+(a/d)^3$	$2a/d-(a/d)^2$	$3/2-3a/(4d)+(a/d)^2/8$

Flow resistivity Ξ ; *absorber variable* E :

The flow resistivity Ξ of a porous material is its flow resistance per unit thickness for stationary flow with low velocity V (about V=0.05 [cm/s]). For a material test sample of thickness dx :

$$\Xi = -\frac{1}{dx}\frac{dP}{V} \tag{3}$$

with dP= static pressure difference across the sample in the flow direction.

Theories mostly determine the interior velocity V_i in a pore for a given pressure difference. This "internal" flow resistivity Ξ_i is related to the flow resistivity Ξ of the sample by $\Xi = \Xi_i/\sigma$.

A suitable non-dimensional quantity R is the ratio of the flow resistance $\Xi \cdot d$ of a layer of thickness d with the free field wave impedance Z_0 : $R = \Xi \cdot d/Z_0$. According to :

$$R = \frac{\Xi d}{\omega\rho_0 \cdot \lambda_0 / 2\pi} \tag{4}$$

this is the ratio of the flow resistance to the mass reactance of a layer of air with thickness $\lambda_0/2\pi$. An important non-dimensional quantity for the evaluation of the characteristic data Γ_a, Z_a is the "absorber variable" (f= frequency) :

$$E = \frac{\rho_0 f}{\Xi} \tag{5}$$

For fibrous materials consisting of parallel fibres, one distinguishes between the flow resistivity $\Xi_{||}$ if the flow is parallel to the fibres, and $\Xi_{\perp}$ if the flow is transversal to the fibres. Empirical data obtained by SULLIVAN for parallel fibre materials with mono-valued fibre radii a lead to the relations :

$$\Xi_{||} = 3.94 \cdot \frac{\eta}{a^2}\frac{\mu^{1.413}}{1-\mu}[1+27\mu^3] \tag{6}$$

a= fibre radius;
η= dynamic viscosity;
μ=1–σ= massivity

and

$$\Xi_{\perp} = \begin{cases} 10.56\dfrac{\eta}{a^2}\dfrac{\mu^{1.531}}{(1-\mu)^3} & ; \quad a \approx 6-10\,[\mu m] \\ 6.8\dfrac{\eta}{a^2}\dfrac{\mu^{1.296}}{(1-\mu)^3} & ; \quad a \approx 20-30\,[\mu m] \end{cases} \tag{7}$$

Semi-empirical data (analytical relation fitted to experimental values) for fibre materials with mono-valued fibre radii a and random fibre orientation give :

$$\Xi = 4 \cdot \frac{\eta}{a^2}[0.55\frac{\mu^{4/3}}{(1-\mu)} + \sqrt{2}\frac{\mu^2}{(1-\mu)^3}] \tag{8}$$

Empirical data for fibrous materials with random fibre radius distribution and random fibre orientation can be approximated by :

$$\Xi = \frac{\eta}{\langle a^2 \rangle} \cdot \begin{cases} 3.2\,\mu^{1.42} & ;\text{ glass fibre material} \\ 4.4\,\mu^{1.59} & ;\text{ mineral fibre material} \end{cases} \tag{9}$$

G.2 Theory of the quasi-homogeneous material

An equivalent network is designed for a homogeneous material, taking into account a finite volume porosity σ_V, a structure factor χ of randomly oriented pores, and a relaxation time constant τ for heat exchange between air and matrix material. A possible vibration of the matrix, induced by the friction between air and matrix, can also be included.

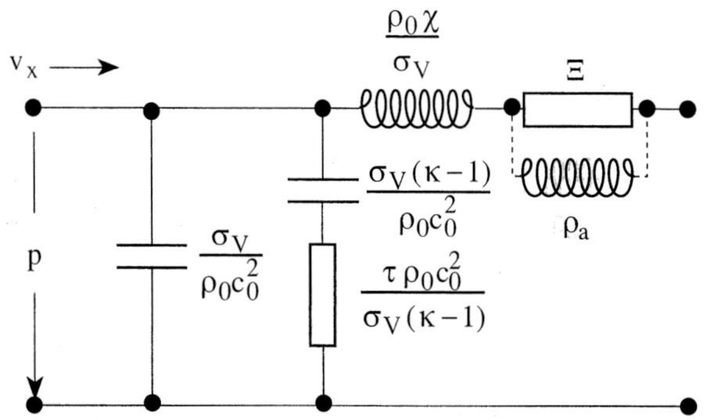

Equivalent network for unit length of a porous material.

The 1st transversal branch represents the compressibility of the air in the material; the 2nd transversal branch represents the relaxation due to heat exchange with the matrix; the 1st longitudinal element represents the inertia of the air in the pores (modified by the structure factor χ); the 2nd longitudinal element represents the friction. If matrix vibration is included, the bulk density ρ_a of the material is parallel to Ξ. This can be taken into account by an effective resistivity :

$$\Xi \to \Xi_{eff} = \frac{j\omega\rho_a \cdot \Xi}{j\omega\rho_a + \Xi} . \tag{10}$$

Characteristic values :

$$\frac{\Gamma_a}{k_0} = j\sqrt{\chi\,(1+\frac{\kappa-1}{1+j\omega\tau})\cdot(1-j\frac{\sigma_V\Xi}{\omega\rho_0\chi})} \tag{11}$$

$$\frac{Z_a}{Z_0} = \frac{1}{\sigma_V}\sqrt{\chi\,(1-j\frac{\sigma_V\Xi}{\omega\rho_0\chi})/(1+\frac{\kappa-1}{1+j\omega\tau})} \tag{12}$$

k_0= free field wave number;
Z_0= free field wave impedance;
ρ_0= density of air;
ω= angular frequency;
κ= adiabatic exponent of air;
Ξ= flow resistivity;
χ= structure factor;
σ_V= volume porosity;
τ= heat relaxation time constant;

Introducing $E=\rho_0 f/\Xi$ and $E_0=\rho_0 f_0/\Xi$ with $f_0=$ relaxation frequency, the characteristic values are :

$$\frac{\Gamma_a}{k_0}=j\sqrt{\frac{\kappa+jE/E_0}{1+jE/E_0}\cdot(\chi-j\frac{\sigma_v}{2\pi E})}\quad;\quad\frac{Z_a}{Z_0}=\frac{1}{\sigma_v}\sqrt{\frac{1+jE/E_0}{\kappa+jE/E_0}\cdot(\chi-j\frac{\sigma_v}{2\pi E})}\tag{13}$$

$E_0\to 0$ belongs to an isothermal sound wave in the material; $E_0\to\infty$ belongs to an adiabatic sound wave; $E_0=0.1$ is a typical value for mineral fibre materials.

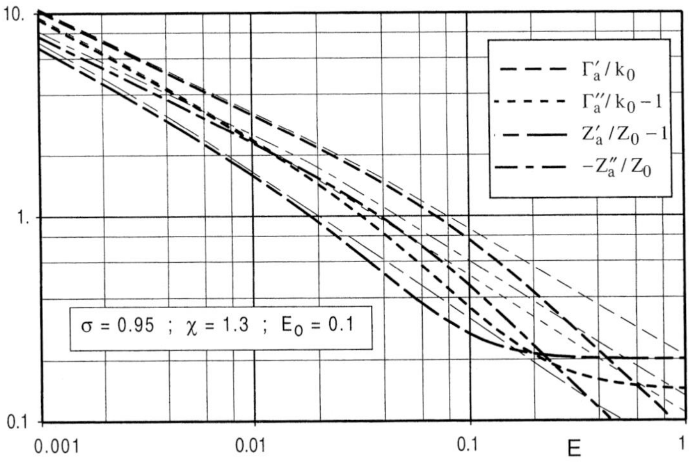

Thick lines: theory of quasi-homogeneous material; thin lines: measured values.

G.3 RAYLEIGH model with round capillaries

The Rayleigh model has for a long time been the only existing model for porous materials. This model consists of parallel circular capillaries with radius a and mutual distance d in a block of the matrix material. The sound propagation in the capillary is determined with viscous and thermal losses at the capillary wall taken into account. See the chapter "Duct Acoustics" for sound propagation in capillaries. The arrangement and mutual distance of the capillaries is supposed to be such, that a prescribed porosity σ is obtained (e.g. $\sigma=\pi a^2/d^2$ for a square arrangement), even if the value of the porosity cannot be realised physically (in a square arrangement the realisable porosity is $\sigma\le\pi/4=0.785$), because only a single capillary is considered. The theory gives the propagation constant Γ_a of the density wave (which is considered to be the propagation constant of sound in the porous material), and the "interior" axial wave impedance Z_i of the density wave in a capillary. Its relation to the desired wave impedance Z_a of the porous material is $Z_a=Z_i/\sigma$.

The normalised characteristic values of a cell are :

$$\frac{\Gamma_a}{k_0}=j\sqrt{\frac{\rho_{eff}}{\rho_0}\cdot\frac{C_{eff}}{C_0}} \quad ; \quad \frac{Z_i}{Z_0}=\sqrt{\frac{\rho_{eff}}{\rho_0}\Big/\frac{C_{eff}}{C_0}} \tag{1}$$

with the ratios of the effective air density and air compressibility (index 0 indicates free field values) :

$$\frac{\rho_{eff}}{\rho_0}=\frac{1}{1-J_{1,0}(k_\nu a)} \quad ; \quad \frac{C_{eff}}{C_0}=1+(\kappa-1)\cdot J_{1,0}(k_{\alpha0}a) \tag{2}$$

and the function :

$$J_{1,0}(z)=2\frac{J_1(z)}{z\cdot J_0(z)}=2\cdot\frac{1}{2-}\,\frac{z^2}{4-}\,\frac{z^2}{6-}\,\frac{z^2}{8-\ldots}\cdots\frac{z^2}{2\cdot n-\ldots} \quad ; n=1,2,3\ldots \tag{3}$$

with Bessel functions $J_n(z)$ and the continued-fraction expansion of their ratio in the last term (the expansion must continue until $|z|^2/(2n)<<1$). The asymptotic approximation

$$J_{1,0}(z)\approx\frac{2}{z}\,\frac{\tan(z-\frac{\pi}{4})+\frac{3}{8z}}{1+\frac{1}{8z}\tan(z-\frac{\pi}{4})} \tag{4}$$

may be applied for large arguments $|z|$.
The squares of the wave numbers used are

$$k_\nu^2=-j\omega/\nu \quad ; \quad k_{\alpha0}^2=k_\nu^2\cdot\sqrt{\kappa\,Pr}\,. \tag{5}$$

κ= adiabatic exponent;
ν= kinematic viscosity;
η= dynamic viscosity;
Pr= Prandtl number;
μ= cell massivity;
C_0= $1/(\rho_0 c_0^2)$;
Ξ= flow resistivity

The argument $k_\nu a$ should be replaced by an argument that can also be applied for material with random pore radius distribution. This is the “absorber variable” $E=\rho_0 f/\Xi$. For its application, the flow resistivity Ξ_i in a pore must be determined.

The flow velocity profile in a circular capillary is, with the static pressure gradient dP/dx :

$$V(r)=-\frac{1}{4\eta}\frac{dP}{dx}(a^2-r^2) \tag{6}$$

The flow resistivity Ξ_i is defined with the volume flow Q through the capillary cross-section area S :

$$\Xi_i=\frac{-dP/dx\cdot S}{Q}=\frac{8\eta}{a^2} \quad ; \quad Q=-\frac{\pi a^4}{8\eta}\frac{dP}{dx} \tag{7}$$

Thus the desired relations between the arguments are :

$$(k_\nu a)^2=-16\pi j\,E \quad ; \quad (k_{\alpha0}a)^2=-16\pi j\,\kappa\,Pr\,E \tag{8}$$

If in a material test the (exterior) flow resistivity Ξ is determined, its relation to the interior flow resistivity is $\Xi = \Xi_i / \sigma$ with the porosity σ of the material.

The continued-fraction approximation (up to the 5th fraction) of the characteristic values is with the variable E :

$$\left(\frac{\Gamma_i}{k_0}\right)^2 =$$

$$-\kappa \frac{75 + 30\pi j\,[8 + Pr(5+3\kappa)]\,E - \pi^2[90 + Pr(480+288\kappa) + Pr^2\,\kappa(80+10\kappa)]\,E^2 - \ldots}{2\pi E\,\{75j - 40\pi(1+6\kappa\,Pr)\,E - 2\pi^2 j\,\kappa\,Pr(64+45\kappa\,Pr)\,E^2 + \ldots} \tag{9}$$

$$\frac{-4\pi^3 j\,Pr[45+27\kappa+\kappa\,Pr(84+8\kappa)]\,E^3 + 12\pi^4\kappa\,Pr^2(8+\kappa)\,E^4}{+48\pi^3(\kappa\,Pr)^2\,E^3\}}$$

$$\left(\frac{Z_i}{Z_0}\right)^2 =$$

$$\frac{1}{\kappa}\,\frac{225 + 720\pi j\,(1+\kappa\,Pr)\,E - 18\pi^2[15 + \kappa\,Pr(128+15\kappa\,Pr)]\,E^2 - \ldots}{2\pi E\,\{225j - 30\pi[4+Pr(15+9\kappa)]\,E - 6\pi^2 j\,Pr[40+24\kappa+\kappa\,Pr(40+5\kappa)]\,E^2 + \ldots} \tag{10}$$

$$\frac{-864\pi^3 j\kappa\,Pr(1+\kappa\,Pr)\,E^3 + 324\pi^4(\kappa\,Pr)^2\,E^4}{+16\pi^3\kappa\,Pr^2(8+\kappa)\,E^3\}}$$

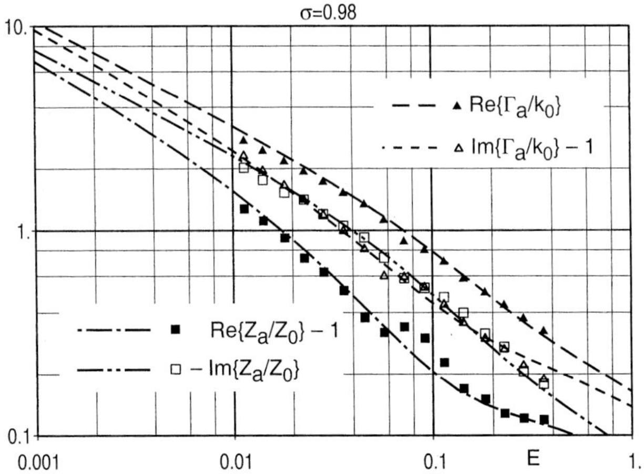

Comparison of the characteristic values from the Rayleigh model (curves) with measurements at a technical mineral fibre absorber (points).

G.4 Model with flat capillaries

The good agreement between characteristic values from measurements with fibrous absorber materials and the theory of the Rayleigh model is rather surprising, because the Rayleigh model has a somewhat "inverted" geometry as compared to a fibre material. Therefore the simpler model with flat capillaries (instead of round capillaries) may also have a chance.

The model consists of parallel flat capillaries, 2h wide and at a mutual distance d, in a block of the matrix material. The sound propagation in the capillary is determined with viscous and thermal losses at the capillary wall taken into account. See the chapter "Duct Acoustics" for sound propagation in capillaries. The arrangement and mutual distance of the capillaries are supposed to be such, that a prescribed porosity σ is obtained ($\sigma = 2h/d$). The theory gives the propagation constant Γ_a of the density wave (which is considered to be the propagation constant of sound in the porous material), and the "interior" axial wave impedance Z_i of the density wave in a capillary. Its relation to the desired wave impedance Z_a of the porous material is $Z_a = Z_i/\sigma$.

The normalised characteristic values of a capillary are :

$$\frac{\Gamma_a}{k_0} = j\sqrt{\frac{\rho_{eff}}{\rho_0}\cdot\frac{C_{eff}}{C_0}} \quad ; \quad \frac{Z_i}{Z_0} = \sqrt{\frac{\rho_{eff}}{\rho_0} \Big/ \frac{C_{eff}}{C_0}} \tag{1}$$

with the ratios of the effective air density and air compressibility (index 0 indicates free field values) :

$$\frac{\rho_{eff}}{\rho_0} = \frac{1}{1-\dfrac{\tan(k_v h)}{k_v h}} \quad ; \quad \frac{C_{eff}}{C_0} = 1+(\kappa-1)\frac{\tan(k_{\alpha 0} h)}{k_{\alpha 0} h} \tag{2}$$

The squares of the wave numbers used are

$$k_v^2 = -j\omega/\nu \quad ; \quad k_{\alpha 0}^2 = k_v^2 \cdot \kappa \Pr . \tag{3}$$

κ= adiabatic exponent;
ν= kinematic viscosity;
η= dynamic viscosity;
Pr= Prandtl number;
$C_0 = 1/(\rho_0 c_0^2)$;
Ξ= flow resistivity

The argument $k_v h$ should be replaced by an argument that can also be applied for a material with a random pore width distribution. This is the "absorber variable" $E = \rho_0 f / \Xi$. For its application, the flow resistivity Ξ_i in a pore must be determined.

The flow velocity profile in a flat capillary is, with the static pressure gradient dP/dx and y the transversal co-ordinate :

$$V(y) = -\frac{1}{2\eta}\frac{dP}{dx}(h^2 - y^2) \tag{4}$$

The flow resistivity Ξ_i is defined with the volume flow Q through the capillary cross-section:

$$\Xi_i = \frac{-dP/dx \cdot 2h}{Q} = \frac{3\eta}{h^2} \quad ; \quad Q = 2\int_0^h V(y) \cdot dy = -\frac{2}{3\eta}\frac{dP}{dx}h^3 \tag{5}$$

Thus the desired relations between the arguments are :

$$|k_v h|^2 = 6\pi E \quad ; \quad (k_v h)^2 = -6\pi j\, E \quad ; \quad (k_{\alpha 0} h)^2 = -6\pi j\, \kappa \Pr E \tag{6}$$

If in a material test the (exterior) flow resistivity Ξ is determined, its relation to the interior flow resistivity is $\Xi = \Xi_i / \sigma$ with the porosity σ of the material.

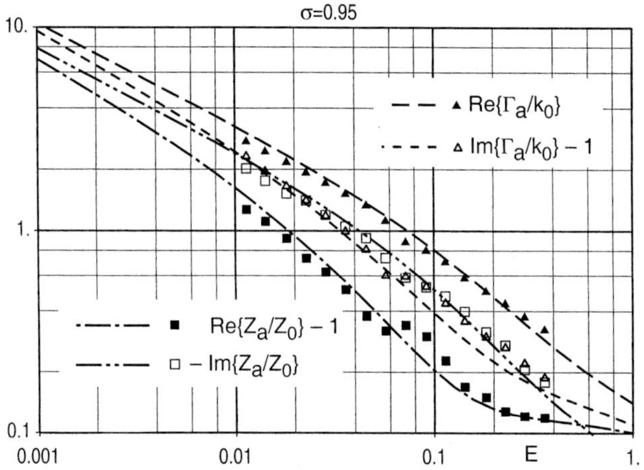

Comparison of the characteristic values from the model with flat capillaries (curves) with measurements at a technical mineral fibre absorber (points).

G.5 Longitudinal flow resistivity in parallel fibres

The geometry of the fibre bundle models agrees better with the geometry of real fibrous materials. First a regular arrangement of the fibres is used in the theory, then the diameters and distances of the fibres are randomised. The direction of flow and sound relative to the fibres must be treated separately, depending on being parallel to the fibres or transversal.

Regular arrangement :

Fibres of equal radius a are arranged in a regular array with cells of radius R around each fibre. The flow velocity V_i is parallel to the fibres.

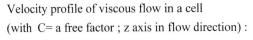

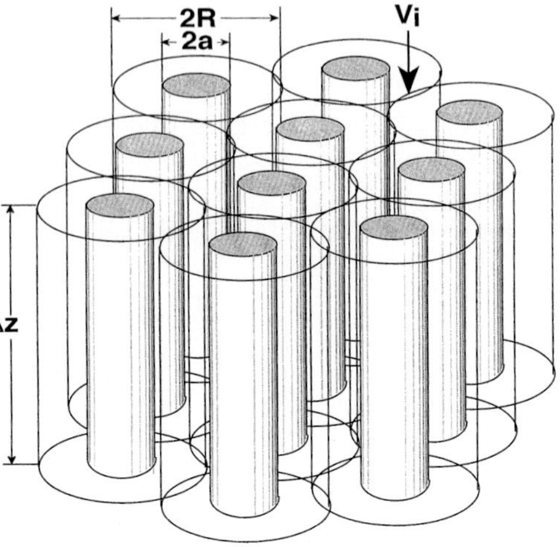

The cell radius R is adjusted to give the porosity σ :

$$\sigma = 1-(a/R)^2 = 1 - N\cdot\pi a^2 \qquad (1)$$

with N= number of fibres in unit area.

Velocity profile of viscous flow in a cell (with C= a free factor ; z axis in flow direction) :

$$v_z(r) = C\left[\frac{R^2}{2}\ln\frac{r}{a} - \frac{1}{4}(r^2-a^2)\right] \qquad (2)$$

Flow resistivity, with definition $\Xi_{\|} = (-\partial p/\partial z)/\langle v_z\rangle$ (3)

$$\Xi_{\|} = (-\partial p/\partial z)/\langle v_z\rangle = \frac{4\eta}{a^2}\,\frac{\mu}{2\mu - \ln\mu - \frac{1}{2}\mu^2 - 1{,}5} \qquad (4)$$

with η= dynamic viscosity of air; μ=1–σ= massivity.

Random arrangement :

A material sample contains I cells, i=1,2,…,I , with random variation of the fibre radius a_i and/or the cell cross-section area $S_i = \pi R_i^2$ which gives a massivity $\mu_i = (a_i/R_i)^2$ of each cell. The resulting flow resistivity is :

$$\Xi_{\|} = \frac{4\eta}{\sum\limits_{i=1}^{I} R_i^2 \cdot \frac{S_i}{\sum_i S_i}\cdot[2\mu_i - 0{,}5\mu_i^2 - \ln\mu_i - 1{,}5]} \qquad (5)$$

One can collect the fibre radii a_i in groups around radius group values a_m with a relative frequency q_m , and correspondingly the cell radii R_i in groups around group radii R_n with relative frequency p_n . The flow resistivity is :

$$\Xi_{\|} = \frac{4\eta < a_m^2 >}{\mu \sum_{m,n} q_m p_n \cdot R_n^4 \left[2(\frac{a_m}{R_n})^2 - \frac{1}{2}(\frac{a_m}{R_n})^4 - 2\ln(\frac{a_m}{R_n}) - 1,5 \right]} \tag{6}$$

Technical fibrous absorbers mostly have Poisson distributions of the a_m and R_n .

G.6 Longitudinal sound in parallel fibres

Consider a fibre model as in section G.4 with sound propagation parallel to the fibres. The sound field in each cell is evaluated with viscous and thermal losses (at the isothermal fibre surface) taken into account. The propagation constant of the density wave is Γ_a ; the axial wave impedance of the density wave is Z_i (interior wave impedance); its relation to the characteristic impedance Z_a (exterior wave impedance) of a material sample is $Z_a = Z_i/\sigma$, with σ= porosity of the material.

The normalised characteristic values of a cell are :

$$\frac{\Gamma_a}{k_0} = j\sqrt{\frac{\rho_{eff}}{\rho_0} \cdot \frac{C_{eff}}{C_0}} \quad ; \quad \frac{Z_i}{Z_0} = \sqrt{\frac{\rho_{eff}}{\rho_0} \Big/ \frac{C_{eff}}{C_0}} \tag{1}$$

with the ratios of the effective air density and air compressibility (index 0 indicates free field values) :

$$\frac{\rho_{eff}}{\rho_0} = \frac{1}{1 - H_{1,0}(k_\nu a;\mu)} \quad ; \quad \frac{C_{eff}}{C_0} = 1 + (\kappa - 1) \cdot H_{1,0}(k_{\alpha,0} a;\mu) \tag{2}$$

κ= adiabatic exponent; ν= kinematic viscosity Pr= Prandtl number; μ= cell massivity; $C_0 = 1/(\rho_0 c_0^2)$

and the function :

$$H_{1,0}(x,\mu) = \frac{-2\mu}{x(1-\mu)} \frac{J_1(x/\sqrt{\mu}) \cdot Y_1(x) - J_1(x) \cdot Y_1(x/\sqrt{\mu})}{J_1(x/\sqrt{\mu}) \cdot Y_0(x) - J_0(x) \cdot Y_1(x/\sqrt{\mu})} \tag{3}$$

with Bessel functions $J_n(z)$ and Neumann functions $Y_n(z)$. The squares of the wave numbers used are $k_\nu^2 = -j\omega/\nu$; $k_{\alpha 0}^2 = k_\nu^2 \cdot \sqrt{\kappa \, Pr}$. (4)

The replacement of the argument $k_\nu a$ by the absorber variable $E = \rho_0 f / \Xi$ is performed with the result of the previous section G.5 :

$$E = \frac{1}{8\pi} |k_\nu a|^2 \frac{2\mu - \ln\mu - \frac{1}{2}\mu^2 - 1.5}{\mu} \approx \frac{1}{8\pi} |k_\nu a|^2 \frac{2\mu - \ln\mu - 1.5}{\mu} \tag{5}$$

$$|k_v a|^2 = 8\pi E \frac{\mu}{2\mu - \ln\mu - \frac{1}{2}\mu^2 - 1.5} \approx 8\pi E \frac{\mu}{2\mu - \ln\mu - 1.5} \qquad (6)$$

The present model also permits us to determine the relaxation (angular) frequency ω_0 of the heat transfer between the sound wave and the matrix (ω_0 was needed in section G.2 , but it was not possible there to determine its magnitude). The relaxation is marked by a deep minimum of $\mathrm{Im}\{C_{eff}/C_0\}$ in the plane of $|k_v a|^2$ and massivity μ . Its position gives the relations:

$$|k_v a|^2 = \frac{\omega_0}{\nu} a^2 = 10^{(1.69843+2.58454\,\log\mu+0.541408(\log\mu)^2+0.0726279(\log\mu)^3)}$$

$$\omega_0 = \frac{\nu}{a^2}|k_v a|^2 = \frac{\nu}{a^2}\cdot 10^{(1.69843+2.58454\,\log\mu+0.541408(\log\mu)^2+0.0726279(\log\mu)^3)} \qquad (7)$$

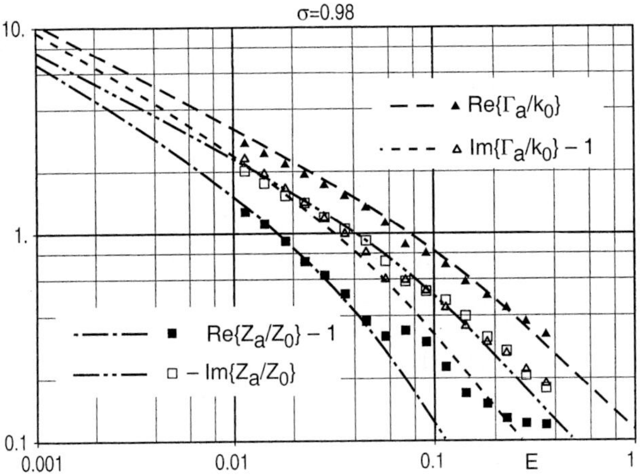

Comparison of the characteristic values from the model of a regular arrangement of longitudinal fibres (curves) with measured data from a technical mineral fibre absorber.

The model of parallel fibres with longitudinal sound propagation can be easily randomised with respect to the fibre radii a_i and cell radii R_i or cell areas $A_i = \pi R_i^2$. Their random values are counted in groups of interval widths $\Delta a, \Delta A$ around average group values a_m, A_n . If their relative frequencies q_m, p_n have the forms of Poisson distributions with distribution parameters λ, Λ , the relations hold :

$$q_m = e^{-\lambda}\frac{\lambda^m}{m!} \quad ; \quad p_n = e^{-\Lambda}\frac{\Lambda^n}{n!} \quad ; \quad m, n = 0, 1, 2, \ldots$$

$$\langle a_m^2 \rangle = (\lambda^2 + 2\lambda + 1/4)\,\Delta a^2 \quad ; \quad \langle A_n \rangle = (\Lambda + 1/2)\,\Delta A \qquad (8)$$

with the total and the group massivities :

$$\mu = \pi \frac{<a_m^2>}{<A_n>} \quad ; \quad \mu_{mn} = \mu \frac{\Lambda + 1/2}{\lambda^2 + 2\lambda + 1/4} \frac{(m+1/2)^2}{n+1/2} \tag{9}$$

The effective group densities and compressibilities then are :

$$\frac{\rho_{mn}}{\rho_0} = \frac{1}{1 - H_{1,0}(k_v a_m, \mu_{mn})} \quad ; \quad \frac{C_{mn}}{C_0} = 1 + (\kappa - 1) H_{1,0}(k_{\alpha 0} a_m, \mu_{mn}) \tag{10}$$

The characteristic values are evaluated from the statistically relevant density ρ_s and compressibility C_s by :

$$\frac{\Gamma_a}{k_0} = j\sqrt{\frac{\rho_s}{\rho_0} \cdot \frac{C_s}{C_0}} \quad ; \quad \frac{Z_i}{Z_0} = \sqrt{\frac{\rho_s}{\rho_0} / \frac{C_s}{C_0}} \tag{11}$$

with :

$$1/\frac{\rho_s}{\rho_0} = \frac{\sum\limits_{m,n}^{<} q_m\, p_n \,/ \frac{\rho_{mn}}{\rho_0}}{\sum\limits_{m,n}^{<} q_m\, p_n} \quad ; \quad \frac{C_s}{C_0} = \frac{\sum\limits_{m,n}^{<} q_m\, p_n \frac{C_{mn}}{C_0}}{\sum\limits_{m,n}^{<} q_m\, p_n} \tag{12}$$

where the upper symbol $<$ at the summation sign indicates that the summation is performed only if $\mu_{mn}<1$; otherwise it is skipped. Typical values of the distribution parameters are $\lambda=2.5$ (deviations thereof have little influence on the result) and $\Lambda=4.5$.

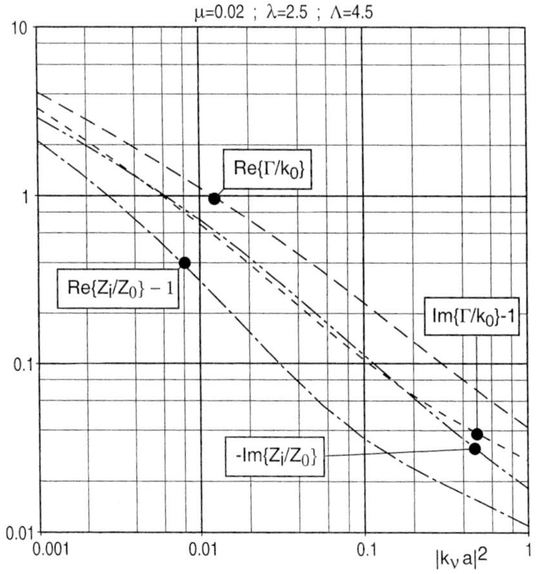

Characteristic values from the model with parallel fibres and longitudinal sound propagation, if the fibre radii and the cell radii are Poisson distributed.

G.7 Transversal flow resistivity in parallel fibres

The model is as depicted in section G.5 , but the flow velocity V_i is normal to the fibres. Analytical studies (see [G.1]) have shown, that a cell model, in which the cell "walls" are immaterial symmetry surfaces for the disturbance flow field (u,v), can also be used for this transversal flow as an approximation (in fact there is a windward-to-leeward unbalance, but it is small for massivity values as found in technical fibre absorbers).

The cell and the incoming flow U are as in the sketch. The flow field is $\vec{V} = U \cdot (1+u, v)$. The disturbance flow $\vec{v}$ and its radial derivative are supposed to be zero on the cell surface. Besides the Cartesian co-ordinates (x,y) cylindrical co-ordinates (r,φ) are also used (φ=0 on the x axis).

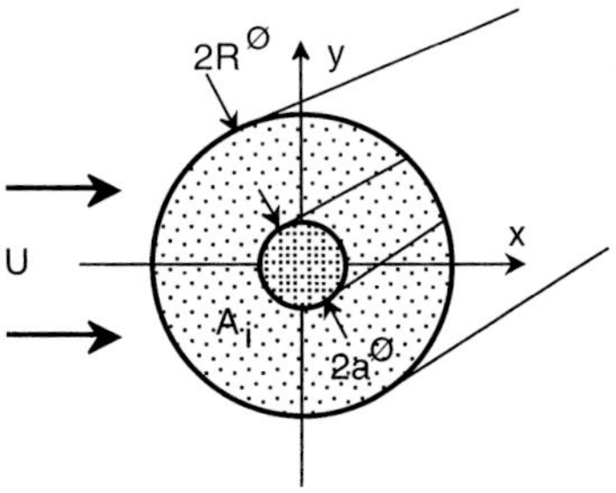

General solutions with the present conditions of symmetry of the Navier-Stokes equation are:

$$u(kr,\varphi) = \vec{v}_x / U =$$

$$= -\frac{C_0 \cos\varphi}{kr} - \sum_{m\geq 1} m\,[C_m\,(kr)^{m-1} \cdot \cos(m-1)\varphi - D_m\,(kr)^{-m-1} \cdot \cos(m+1)\varphi] \; -$$

$$-\frac{1}{2} e^{kr\cos\varphi} \cdot \sum_{n\geq 0} A_n \cdot [(\cos n\varphi + \frac{n}{kr}\cos(n+1)\varphi)\, I_n(kr) - \cos\varphi \cos n\varphi\, I_{n-1}(kr)] + \qquad (1)$$

$$+ B_n \cdot [(\cos n\varphi + \frac{n}{kr}\cos(n+1)\varphi)\, K_n(kr) - \cos\varphi \cos n\varphi\, K_{n-1}(kr)]$$

$$v(kr,\varphi) = \vec{v}_y / U =$$

$$= -\frac{C_0 \sin\varphi}{kr} - \sum_{m\geq 1} m\,[C_m\,(kr)^{m-1} \cdot \sin(m-1)\varphi + D_m\,(kr)^{-m-1} \cdot \sin(m+1)\varphi] \; -$$

$$-\frac{1}{2} e^{kr\cos\varphi} \cdot \sum_{n\geq 0} A_n \cdot [\frac{n}{kr}\sin(n+1)\varphi\, I_n(kr) \;\; - \sin\varphi \cos n\varphi\, I_{n-1}(kr)] + \qquad (2)$$

$$+ B_n \cdot [\frac{n}{kr}\sin(n+1)\varphi)\, K_n(kr) + \sin\varphi \cos n\varphi\, K_{n-1}(kr)]$$

with the flow parameter k=U/(2ν) (ν= kinematic viscosity) , the modified Bessel functions $I_m(z)$, $K_m(z)$ of the 1st and 2nd kinds, and yet undetermined coefficients A_n, B_n, C_n, D_n . These are determined from the boundary conditions :

$$u(ka,\varphi)\big|_{kr=ka}=-1 \;;\quad v(ka,\varphi)\big|_{kr=ka}=0 \;;$$
$$u_r(kr,\varphi)\big|_{kr=kR}=0 \;;\quad \left.\frac{\partial u_\varphi(kr,\varphi)}{\partial kr}\right|_{kr=kR}=0 \tag{3}$$

which, with a summation up to m,n=3 , give a linear system of equations of the form :

$$\begin{bmatrix} c_{1,1} & c_{1,2}\cdots c_{1,14} & c_{1,15} \\ c_{2,1} & c_{2,2}\cdots c_{2,14} & c_{2,15} \\ \vdots & \vdots \quad\; \vdots & \vdots \\ c_{15,1} & c_{15,2}\cdots c_{15,14} & c_{15,15} \end{bmatrix} \cdot \begin{bmatrix} A_0 \\ A_1 \\ \vdots \\ D_3 \end{bmatrix} = \begin{bmatrix} -1 \\ 0 \\ \vdots \\ 0 \end{bmatrix} \tag{4}$$

with the list of unknown coefficients :

$$\{A_0, A_1, A_2, A_3, B_0, B_1, B_2, B_3, C_0, C_1, C_2, C_3, D_1, D_2, D_3,\} \tag{5}$$

and the matrix coefficient rows: (6)

$c_1 = \{-(1/2 + ka^2/8)\cdot I_0(ka) + (ka/4 + ka^3/32)\cdot I_1(ka)\;,\; (1/4 + 3\,ka^2/32)\cdot I_0(ka) - (5\,ka/16 + ka^3/32)\cdot I_1(ka)\;,\; (ka/8 + ka^3/48)\cdot I_1(ka) - ka^2/12\cdot I_2(ka)\;,\; ka^2/32\cdot I_2(ka) - ka^3/96\cdot I_3(ka)\;,\; -(1/2 + ka^2/8)\cdot K_0(ka) - (ka/4 + ka^3/32)\cdot K_1(ka)\;,\; -(1/4 + 3\,ka^2/32)\cdot K_0(ka) - (5\,ka/16 + ka^3/32)\cdot K_1(ka), -(ka/8 + ka^3/48)\cdot K_1(ka) - ka^2/12\cdot K_2(ka)\;,\; -ka^2/32\cdot K_2(ka) - ka^3/96\cdot K_3(ka)\;,\; 0\;,\; -1\;,\; 0\;,\; 0\;,\; 0\;,\; 0\;,\; 0\}\;;$

$c_2 = \{-(ka/2 + ka^3/16)\cdot I_0(ka) + (1/2 + 3\,ka^2/16)\cdot I_1(ka)\;,\; (3\,ka/8 + 5\,ka^3/96)\cdot I_0(ka) - (3/4 + 11\,ka^2/48)\cdot I_1(ka)\;,\; (1/4 + ka^2/8)\cdot I_1(ka) - (3\,ka/8 + ka^3/24)\cdot I_2(ka)\;,\; (ka/8 + 5\,ka^3/192)\cdot I_2(ka) - 3\,ka^2/32\cdot I_3(ka)\;,\; -(ka/2 + ka^3/16)\cdot K_0(ka) - (1/2 + 3\,ka^2/16)\cdot K_1(ka)\;,\; -(3\,ka/8 + 5\,ka^3/96)\cdot K_0(ka) + (-3/4 - 11\,ka^2/48)\cdot K_1(ka)\;,\; (-1/4 - ka^2/8)\cdot K_1(ka) - (3\,ka/8 + ka^3/24)\cdot K_2(ka)\;,\; -(ka/8 + 5\,ka^3/192)\cdot K_2(ka) - 3\,ka^2/32\cdot K_3(ka)\;,\; -1/ka\;,\; 0\;,\; -2\,ka\;,\; 0\;,\; 0\;,\; 0\;,\; 0\}\;;$

$c_3 = \{-(ka^2/8\cdot I_0(ka)) + (ka/4 + ka^3/24)\cdot I_1(ka)\;,\; (1/4 + ka^2/8)\cdot I_0(ka) - (1/(2\,ka) + 3\,ka/8 - ka^3/24)\cdot I_1(ka)\;,\; (ka/4 + 7\,ka^3/192)\cdot I_1(ka) - (1 + 3\,ka^2/16)\cdot I_2(ka)\;,\; (1/4 + 3\,ka^2/32)\cdot I_2(ka) - (7\,ka/16 + ka^3/32)\cdot I_3(ka)\;, - ka^2/8\cdot K_0(ka) + (-ka/4 - ka^3/24)\cdot K_1(ka)\;,\; -(-1/4\,ka^2/8)\cdot K_0(ka) - (1/(2\,ka) + 3\,ka/8 + ka^3/24)\cdot K_1(ka)\;,\; -(ka/4 + 7\,ka^3/192)\cdot K_1(ka) - (1 + 3\,ka^2/16)\cdot K_2(ka)\;,\; -(1/4 + 3\,ka^2/32)\cdot K_2(ka) - (7\,ka/16 + ka^3/32)\cdot K_3(ka)\;,\; 0\;,\; 0\;,\; 0\;,\; -3\,ka^2\;,\; ka^{-2}\;,\; 0\;,\; 0\};$

$c_4 = \{-ka^3/48\cdot I_0(ka) + ka^2/16\cdot I_1(ka)\;,\; (ka/8 + 5\,ka^3/192)\cdot I_0(ka) - (1/4 + 3\,ka^2/32)\cdot I_1(ka)\;,\; (1/4 + 3\,ka^2/32)\cdot I_1(ka) - (ka^{-1} + ka/2 + ka^3/32)\cdot I_2(ka)\;,\; (ka/4 + ka^3/32)\cdot I_2(ka) - (5/4 + 7\,ka^2/32)\cdot I_3(ka)\;,\; -ka^3/48\cdot K_0(ka) - ka^2/16\cdot K_1(ka)\;,\; -(ka/8 + 5\,ka^3/192)\cdot K_0(ka) - -(1/4 + 3\,ka^2/32)\cdot K_1(ka)\;, -(1/4 + 3\,ka^2/32)\cdot K_1(ka) - (ka^{-1} + ka/2 + ka^3/32)\cdot K_2(ka)\;,\; -(ka/4 + ka^3/32)\cdot K_2(ka) - (5/4 + 7\,ka^2/32)\cdot K_3(ka)\;,\; 0\;,\; 0\;,\; 0\;,\; 0\;,\; 0\;,\; 2/ka^3\;,\; 0\};$

$c_5 = \{ka^3/96\cdot I_1(ka)\ ,\ ka^2\cdot/32\cdot I_0(ka) - (ka/16 + ka^3/96)\cdot I_1(ka)\ ,\ (ka/8 + ka^3/48)\cdot I_1(ka)$
$- (1/2 + ka^2/8)\cdot I_2(ka)\ ,\ (1/4 + 3\ ka^2/32)\cdot I_2(ka) - (3/(2\ ka) + 5\ ka/8 + ka^3/32)\cdot I_3(ka)\ ,$
$- ka^3/96\cdot K_1(ka)\ ,\ - ka^2/32\cdot K_0(ka) - (ka/16 + ka^3/96)\cdot K_1(ka)\ ,\ - (ka/8 + ka^3/48)\cdot K_1(ka)$
$- (1/2 + ka^2/8)\cdot K_2(ka)\ ,\ - (1/4 + 3\ ka^2/32)\cdot K_2(ka) - (3/(2\ ka) + 5\ ka/8 + ka^3/32)\cdot K_3(ka)\ ,$
$0\ ,\ 0\ ,\ 0\ ,\ 0\ ,\ 0\ ,\ 0\ ,\ 3/ka^4\};$

$c_6 = \{(1/2 + ka^2/16)\cdot I_1(ka)\ ,\ (ka/8 + ka^3/96)\cdot I_0(ka) - (1/4 + ka^2/48)\cdot I_1(ka)\ ,$
$- I_1(ka)/4\ - ka/8\cdot I_2(ka)\ ,\ - (ka/8 + ka^3/192)\cdot I_2(ka) - ka^2/32\cdot I_3(ka)\ ,$
$- (1/2 + ka^2/16)\cdot K_1(ka)\ ,\ - (ka/8 + ka^3/96)\cdot K_0(ka) - (1/4 + ka^2/48)\cdot K_1(ka)\ ,$
$K_1(ka)/4 - ka/8\cdot K_2(ka)\ ,\ (ka/8 + ka^3/192)\cdot K_2(ka) - ka^2/32\cdot K_3(ka)\ ,\ -1/ka\ ,\ 0\ ,\ 2\ ka\ ,$
$0\ ,\ 0\ ,\ 0\ ,\ 0\};$

$c_7 = \{(ka/4 + ka^3/48)\cdot I_1(ka)\ ,\ (1/4 + ka^2/16)\cdot I_0(ka) - (1/(2\ ka) + ka/8)\cdot I_1(ka)\ ,$
$ka^3\cdot/192\cdot I_1(ka) - (1/2 + ka^2/16)\cdot I_2(ka)\ ,\ - (1/4 + ka^2/32)\cdot I_2(ka) - 3\ ka/16\cdot I_3(ka)\ ,$
$- (ka/4 + ka^3/48)\cdot K_1(ka)\ ,\ - (1/4 + ka^2/16)\cdot K_0(ka) - (1/(2\ ka) + ka/8)\cdot K_1(ka)\ ,$
$- ka^3/192\cdot K_1(ka) - (1/2 + ka^2/16)\cdot K_2(ka)\ ,\ (1/4 + ka^2/32)\cdot K_2(ka) - 3\ ka/16\cdot K_3(ka)\ ,\ 0\ ,$
$0\ ,\ 0\ ,\ 3\ ka^2\ ,\ 1/ka^2\ ,\ 0\ ,\ 0\};$

$c_8 = \{ka^2/16\cdot I_1(ka)\ ,\ (ka/8 + ka^3/64)\cdot I_0(ka) - (1/4 + ka^2/32)\cdot I_1(ka)\ ,\ (1/4 + ka^2/32)\cdot I_1(ka)$
$- (1/ka + ka/4)\cdot I_2(ka)\ ,\ - (3/4 + 3\ ka^2/32)\cdot I_3(ka)\ ,\ - ka^2/16\cdot K_1(ka)\ ,\ - (ka/8$
$+ ka^3/64)\cdot K_0(ka) - (1/4 + ka^2/32)\cdot K_1(ka)\ ,\ - (1/4 + ka^2/32)\cdot K_1(ka) - (1/ka + ka/4)\cdot K_2(ka)\ ,$
$- (3/4 + 3\ ka^2/32)\cdot K_3(ka)\ ,\ 0\ ,\ 0\ ,\ 0\ ,\ 0\ ,\ 0\ ,\ 2/ka^3\ ,\ 0\};$

$c_9 = \{(1/2 + kR^2/8)\cdot I_1(kR) - (kR/4 + kR^3/32)\cdot I_0(kR)\ ,\ (kR/4 + kR^3/32)\cdot I_0(kR)$
$- (1/2 + kR^2/8)\cdot I_1(kR)\ ,\ kR^2/16\cdot I_1(kR) - (kR/4 + kR^3/48)\cdot I_2(kR)\ ,\ kR^3/96\cdot I_2(kR)$
$- kR^2/16\cdot I_3(kR)\ ,\ - (1/2 + kR^2/8)\cdot K_1(kR) - (kR/4 + kR^3/32)\cdot K_0(kR)\ ,\ - (kR/4 +$
$kR^3/32)\cdot K_0(kR) - (1/2 + kR^2/8)\cdot K_1(kR)\ ,\ - kR^2/16\cdot K_1(kR) - (kR/4 + kR^3/48)\cdot K_2(kR)\ ,$
$- kR^3/96\cdot K_2(kR) - kR^2/16\cdot K_3(kR)\ ,\ -1/kR\ ,\ 0\ ,\ 0\ ,\ 0\ ,\ 0\ ,\ 0\ ,\ 0\};$

$c_{10} = \{(kR/2 + kR^3/16)\cdot I_1(kR) - (1/2 + 3\ kR^2/16)\cdot I_0(kR)\ ,\ (1/2 + 3\ kR^2/16)\cdot I_0(kR)$
$- (1/(2\ kR) + 9\ kR/16 + 5\ kR^3/96)\cdot I_1(kR)\ ,\ (kR/4 + kR^3/24)\cdot I_1(kR) - (3/4 +$
$5\ kR^2/24)\cdot I_2(kR)\ ,\ kR^2/16\cdot I_2(kR) - (5\ kR/16 + 5\ kR^3/192)\cdot I_3(kR)\ ,$
$- (kR/2 + kR^3/16)\cdot K_1(kR) - (1/2 + 3\ kR^2/16)\cdot K_0(kR)\ ,\ - (1/2 + 3\ kR^2/16)\cdot K_0(kR)$
$- (1/(2\ kR) + 9\ kR/16 + 5\ kR^3/96)\cdot K_1(kR)\ ,\ - (kR/4 + kR^3/24)\cdot K_1(kR) - (3/4$
$+ 5\ kR^2/24)\cdot K_2(kR)\ ,\ - kR^2/16\cdot K_2(kR) - (5\ kR/16 + 5\ kR^3/192)\cdot K_3(kR)\ ,$
$0\ ,\ -1\ ,\ 0\ ,\ 0\ ,\ 1/kR^2\ ,\ 0\ ,\ 0\};$

$c_{11} = \{kR^2/8\cdot I_1(kR) - (kR/4 + kR^3/24)\cdot I_0(kR)\ ,\ (kR/4 + kR^3/24)\cdot I_0(kR)$
$- (1/2 + kR^2/6)\cdot I_1(kR)\ ,\ (1/2 + kR^2/8)\cdot I_1(kR) - (1/kR + kR/2 + 7\ kR^3/192)\cdot I_2(kR)\ ,$
$(kR/4 + kR^3/32)\cdot I_2(kR) - (1 + 3\ kR^2/16)\cdot I_3(kR)\ ,\ - kR^2/8\cdot K_1(kR) - (kR/4 +$
$kR^3/24)\cdot K_0(kR)\ ,\ - (kR/4 + kR^3/24)\cdot K_0(kR) - (1/2 + kR^2/6)\cdot K_1(kR)\ ,$

$-(1/2 + kR^2/8)\cdot K_1(kR) - (1/kR + kR/2 + 7\,kR^3/192)\cdot K_2(kR)$, $-(kR/4 + kR^3/32)\cdot K_2(kR) - (1 + 3\,kR^2/16)\cdot K_3(kR)$, 0 , 0 , $-2\,kR$, 0 , 0 , $2/kR^3$, 0};

c_{12} = {$kR^3/48\cdot I_1(kR) - kR^2/16\cdot I_0(kR)$, $kR^2/16\cdot I_0(kR) - (3\,kR/16 + 5\,kR^3/192)\cdot I_1(kR)$, $(kR/4 + kR^3/32)\cdot I_1(kR) - (3/4 + 5\,kR^2/32)\cdot I_2(kR)$, $(1/2 + kR^2/8)\cdot I_2(kR) - (3/(2\,kR) + 5\,kR/8 + kR^3/32)\cdot I_3(kR)$, $-kR^3/48\cdot K_1(kR) - kR^2/16\cdot K_0(kR)$, $-kR^2/16\cdot K_0(kR) - (3\,kR/16 + 5\,kR^3/192)\cdot K_1(kR)$, $-(kR/4 + kR^3/32)\cdot K_1(kR) - (3/4 + 5\,kR^2/32)\cdot K_2(kR)$, $-(1/2 + kR^2/8)\cdot K_2(kR) - (3/(2\,kR) + 5\,kR/8 + kR^3/32)\cdot K_3(kR)$, 0 , 0 , 0 , $-3\,kR^2$, 0 , 0 , $3/kR^4$};

c_{13} = {$kR/8\cdot I_0(kR) + (1/2 + kR^2/16)\cdot I_1(kR)$, $(-1/(2\,kR) + kR/16 + kR^3/96)\cdot I_0(kR) + (1/kR^2 + kR^2/48)\cdot I_1(kR)$, $-(3/4 + kR^2/24)\cdot I_1(kR) + 3/(2\,kR)\cdot I_2(kR)$, $-(5\,kR/16 + kR^3/192)\cdot I_2(kR) + 5/8\cdot I_3(kR)$, $kR/8\cdot K_0(kR) - (1/2 + kR^2/16)\cdot K_1(kR)$, $(1/(2\,kR) - kR/16 - kR^3/96)\cdot K_0(kR) + (1/kR^2 + kR^2/48)\cdot K_1(kR)$, $(3/4 + kR^2/24)\cdot K_1(kR) + 3/(2\,kR)\cdot K_2(kR)$, $(5\,kR/16 + kR^3/192)\cdot K_2(kR) + 5/8\cdot K_3(kR)$, 0 , 0 , 0 , 0 , $-2/kR^3$, 0 , 0};

c_{14} = {$(1/4 + kR^2/16)\cdot I_0(kR) + (kR/4 + kR^3/48)\cdot I_1(kR)$, $(kR^2\cdot I_0(kR))/24 + (kR\cdot I_1(kR))/24$, $(-1/kR - kR/4 + kR^3/192)\cdot I_1(kR) + (1/4 + 3/kR^2 + kR^2/192)\cdot I_2(kR)$, $(-1 - kR^2/8)\cdot I_2(kR) + (3/kR + kR/8)\cdot I_3(kR)$, $(1/4 + kR^2/16)\cdot K_0(kR) + (-kR/4 - kR^3/48)\cdot K_1(kR)$, $-(kR^2\cdot K_0(kR))/24 + (kR\cdot K_1(kR))/24$, $(1/kR + kR/4 - kR^3/192)\cdot K_1(kR) + (1/4 + 3/kR^2 + kR^2/192)\cdot K_2(kR)$, $(1 + kR^2/8)\cdot K_2(kR) + (3/kR + kR/8)\cdot K_3(kR)$, 0 , 0 , 2 , 0 , 0 , $-6/kR^4$, 0};

c_{15} = {$(kR\cdot I_0(kR))/8 + (kR^2\cdot I_1(kR))/16$, $(kR/16 + kR^3/64)\cdot I_0(kR) + (kR^2\cdot I_1(kR))/32$, $(-1/4 - kR^2/32)\cdot I_1(kR) + I_2(kR)/(2\,kR)$, $(-3/(2\,kR) - (3\,kR)/8)\cdot I_2(kR) + (3/4 + 6/kR^2)\cdot I_3(kR)$, $(kR\cdot K_0(kR))/8 - (kR^2\cdot K_1(kR))/16$, $(-kR/16 - kR^3/64)\cdot K_0(kR) + (kR^2\cdot K_1(kR))/32$, $(1/4 + kR^2/32)\cdot K_1(kR) + K_2(kR)/(2\,kR)$, $(3/(2\,kR) + (3\,kR)/8)\cdot K_2(kR) + (3/4 + 6/kR^2)\cdot K_3(kR)$, 0 , 0 , 0 , $6\,kR$, 0 , 0 , $-12/kR^5$};

The diagram below shows the field of magnitude $|\vec{V}/U|$ of the total velocity $\vec{V}$ to the incoming velocity U for ka= 0.001 ; R/a= 10 .

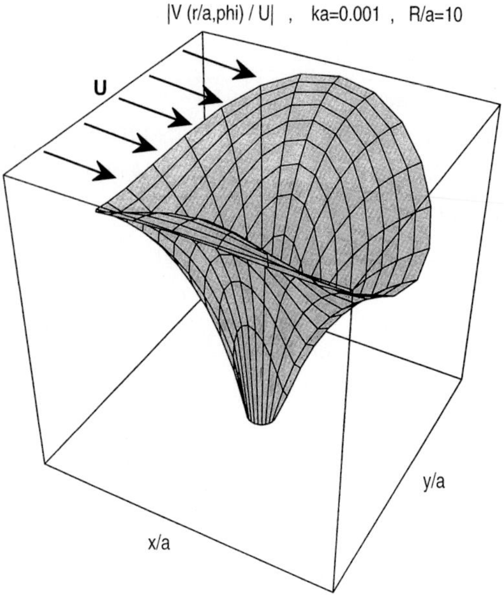

Field of the total velocity magnitude around a fibre in a regular fibre array with transversal incoming flow U .

On each fibre a pressure force F_{xp} and a viscous force F_{xv} act in the x direction :

$$F_{xp} = -\int_0^{2\pi} a\, p\, d\varphi \quad ; \quad F_{xv} = -\int_0^{2\pi} a\eta \cdot \left.\frac{\partial \vec{V}_\varphi}{\partial r}\right|_{r=a} \cdot \cos\vartheta \cdot d\varphi \tag{7}$$

Their values are :

$$\frac{F_{xp}}{\frac{1}{2}\rho U^2 \cdot 2a} = \pi \cdot \left(\frac{C_0}{ka} + 2\, ka\, C_2\right) \tag{8}$$

$$\frac{F_{xv}}{\frac{1}{2}\rho U^2 \cdot 2a} =$$

$$\begin{aligned} &-\pi \cdot \{ [ka\, (\frac{1}{16} + \frac{ka^2}{192} + \frac{ka^4}{6144})\, I_0(ka) + (\frac{1}{4} + \frac{ka^2}{32} + \frac{ka^4}{768} + \frac{ka^6}{36864})\, I_1(ka)] \cdot A_0 + \\ &+ [(-\frac{1}{4ka} + \frac{ka}{32} + \frac{ka^3}{256} + \frac{5ka^5}{36864})\, I_0(ka) + (\frac{1}{2ka^2} + \frac{ka^2}{128} + \frac{5ka^4}{9216})\, I_1(ka)] \cdot A_1 + \\ &+ [-(\frac{3}{8} + \frac{ka^2}{48} + \frac{ka^4}{3072} - \frac{ka^6}{92160})\, I_1(ka) + (\frac{3}{4ka} - \frac{ka^3}{1536} + \frac{ka^5}{23040})\, I_2(ka)] \cdot A_2 + \\ &+ [-(\frac{5ka}{32} + \frac{13ka^3}{1536} + \frac{11ka^5}{61440})\, I_2(ka) + (\frac{5}{16} - \frac{11ka^4}{30720})\, I_3(ka)] \cdot A_3 + \end{aligned} \tag{9}$$

$$+[ka\,(\frac{1}{16}+\frac{ka^2}{192}+\frac{ka^4}{6144})\,K_0(ka)-(\frac{1}{4}+\frac{ka^2}{32}+\frac{ka^4}{768}+\frac{ka^6}{36864})\,K_1(ka)]\cdot B_0+$$
$$+[(\frac{1}{4ka}-\frac{ka}{32}-\frac{ka^3}{256}-\frac{5ka^5}{36864})\,K_0(ka)+(\frac{1}{2ka^2}+\frac{ka^2}{128}+\frac{5ka^4}{9216})\,K_1(ka)]\cdot B_1+$$
$$+[(\frac{3}{8}+\frac{ka^2}{48}+\frac{ka^4}{3072}-\frac{ka^6}{92160})\,K_1(ka)+(\frac{3}{4ka}-\frac{ka^3}{1536}+\frac{ka^5}{23040})\,K_2(ka)]\cdot B_2+ \tag{9}$$
$$+[(\frac{5ka}{32}+\frac{13ka^3}{1536}+\frac{11ka^5}{61440})\,K_2(ka)+(\frac{5}{16}-\frac{11ka^4}{30720})\,K_3(ka)]\cdot B_3-\frac{1}{ka^3}\cdot D_1\}$$

For small $ka \ll 1$ a power series development of the modified Bessel functions in F_{xv} gives the approximation :

$$\frac{F_{xv}}{\frac{1}{2}\rho U^2\cdot 2a}= \tag{10}$$

$$-\pi\cdot\{[0.1875\,ka+0.0520833\,ka^2]\cdot A_0+[2.63111\cdot 10^{-8}\,ka+0.0130208\,ka^3]\cdot A_1+$$
$$+[-0.093750\,ka-0.0260416\,ka^3]\cdot A_2+[1.31556\cdot 10^{-7}\,ka-0.013021\,ka^3]\cdot A_3+$$
$$+[-0.25/ka-0.076978\,ka+0.0116033\,ka^3-(0.1875\,ka+0.0520833\,ka^3)\ln ka]\cdot B_0+$$
$$+[0.5/ka^3-0.125/ka+0.0312499\,ka+0.00198205\,ka^3+$$
$$+(2.63111\cdot 10^{-8}\,ka+0.0130208\,ka^3)\ln ka]\cdot B_1+$$
$$+[1.5/ka^3+4.500\cdot 10^{-8}/ka+0.0502076\,ka-0.00766781\,ka^3+$$
$$+(0.09375008\,ka+0.0260416\,ka^3)\ln ka]\cdot B_2+$$
$$+[2.5/ka^3+7.500\cdot 10^{-8}/ka-0.02500025\,ka-0.00235060\,ka^3+$$
$$+(1.315556\cdot 10^{-7}\,ka-0.0130210\,ka^3)\ln ka\,]\cdot B_3-\frac{1}{ka^3}\cdot D_1\}$$

The test sample of a model material with regular fibre arrangement is supposed to have the dimensions D_x in the flow direction and D_y normal to the flow and to the fibres. The flow resistivity $\Xi_\perp$ is given by the sum of the viscous forces on the fibres in the sample :

$$\Xi_\perp=\frac{2\eta\cdot\sum_{i=1}^{I} ka_i\,\frac{F_{xv,i}(ka_i,kR_i)}{a_i\,\rho_0 U^2}}{D_x D_y(1-\mu)} \tag{11}$$

with η= dynamic viscosity; μ= $1-\sigma$= massivity of the material; $D_x D_y=I\cdot\pi\langle R_i^2\rangle$. The terms under the sum in the numerator are nearly independent of ka_i ; so they are a function $c(R_i/a_i)$ of the remaining parameter R_i/a_i . This function can be evaluated by (a regression through the values from the cell model) :

$$c(\frac{R}{a}) = e^{f(x)} \quad ; \quad x = \ln\frac{R}{a}$$

$$f(x) = 0.865823 + \frac{0.0250214}{x^3} - \frac{0.322560}{x^2} + \frac{1.78839}{x} - \\ - 0.530524\,x + 0.0604543\,x^2 - 0.00312698\,x^3 \tag{12}$$

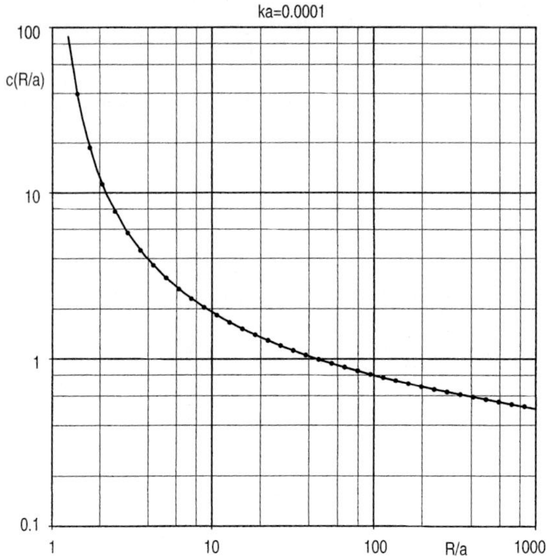

Function c(R/a) of the sum terms in the numerator of $\Xi_\perp$. Points: from the model theory; curve: regression. The curves coincide if ka is varied by powers of ten.

If the fibre and cell radii are randomised with relative frequencies q_m of the fibre radius a_i in the counting group with mean value a_m , and correspondingly p_n the relative frequency of R_i in the counting group around R_n , the flow resistivity $\Xi_\perp$ is :

$$\Xi_\perp \frac{<a^2>}{\eta} = \frac{2\mu}{\pi(1-\mu)} \cdot \sum_{m,n} q_m\, p_n \cdot c(\frac{R_n}{a_m}) \tag{13}$$

If the relative frequencies have Poisson distributions :

$$q_m = e^{-\lambda} \cdot \frac{\lambda^m}{m!} \quad ; \quad p_n = e^{-\Lambda} \cdot \frac{\Lambda^n}{n!} \quad ; \quad m,n = 0,1,2,\ldots \tag{14}$$

the required radius ratios are :

$$\frac{R_n}{a_m} = \frac{n+1/2}{m+1/2}\,\frac{1}{\sqrt{\mu}}\sqrt{\frac{\lambda^2+2\lambda+1/4}{\Lambda^2+2\Lambda+1/4}} \tag{15}$$

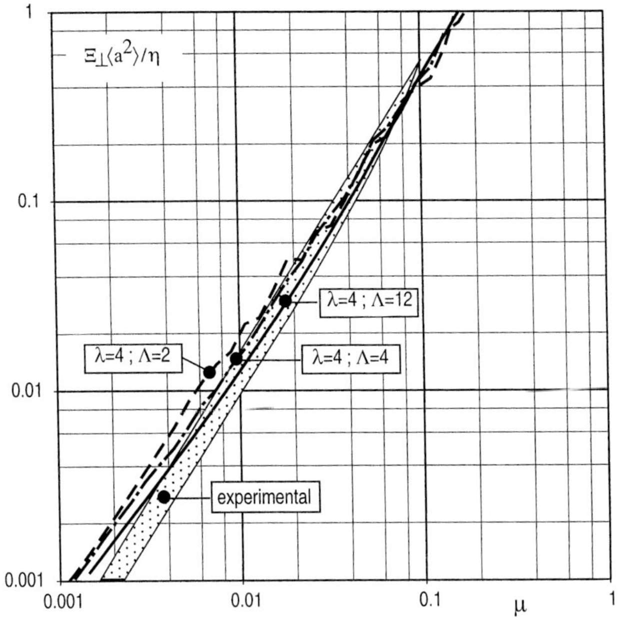

Flow resistivity $\Xi_{\perp}$ of a bundle of parallel fibres for transversal flow. Computed curves with different parameters of the Poisson distributions, and shaded range of experimental data.

G.8 Transversal sound in parallel fibres

The model of the porous material is illustrated in section G.5 , but the sound propagation is transversal to the fibres, which are assumed to be at rest. An elementary cell and the co-ordinates used are shown here.

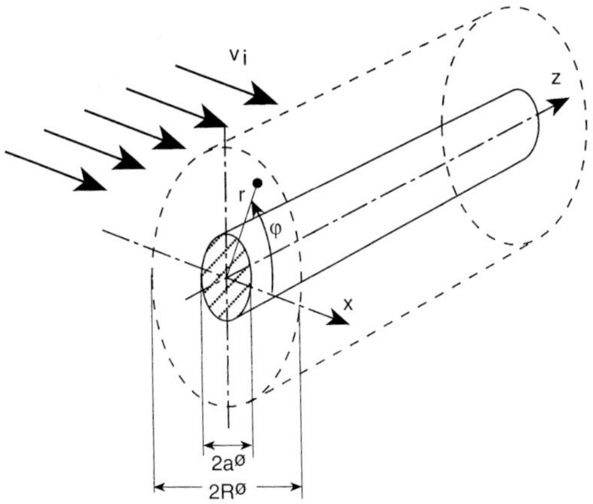

The direction of the incident wave is v_i . The sound field in a cell is evaluated with density, viscous and thermal waves taken into account.

The cell radius R is determined so that, with the given fibre radius a , the desired massivity μ is obtained.

In a first model (*closed cell model*) the cell surfaces are adiabatic surfaces of symmetry for the field scattered at a fibre (the scattered fields do not penetrate the cell surface, but the surface is transparent for the incident wave); in a second model the scattered field freely propagates through the fibre bundle (*open cell model*); this model is finally used to evaluate multiple scattering between fibres (*multiple scattering model*). The advantage of the multiple scattering model is that it permits random fibre distances.

The characteristic propagation constant Γ_i and wave impedance Z_i of the density wave in the fibre bundle are determined from the effective density ρ_{eff} and effective compressibility C_{eff} by :

$$\frac{\Gamma_i}{k_0} = j\sqrt{\frac{\rho_{eff}}{\rho_0} \cdot \frac{C_{eff}}{C_0}} \quad ; \quad \frac{Z_i}{Z_0} = \sqrt{\frac{\rho_{eff}}{\rho_0} / \frac{C_{eff}}{C_0}} \tag{1}$$

which in turn are evaluated by the integrals :

$$\frac{\rho_{eff}}{\rho_0} = \frac{\oint [\Phi_i + \Phi_\rho + \frac{\Pi_\alpha}{\Pi_\rho}\Phi_\alpha] \cdot \cos\varphi \cdot ds}{\oint [\Phi_i + \Phi_\rho + \Phi_\alpha] \cdot \cos\varphi \cdot ds - \int_0^{2\pi} d\varphi \int_a^R (\frac{\partial \Psi_z}{\partial \varphi}\cos\varphi + r\frac{\partial \Psi_z}{\partial r}\sin\varphi) \cdot dr} \tag{2}$$

$$\frac{C_{eff}}{C_0} = \frac{R\int_0^{2\pi} v_{ir}(R,\varphi) \cdot d\varphi}{k_\rho^2 \iint_{A_0} [(\Phi_e + \Phi_\rho) + \frac{\Pi_\alpha}{\Pi_\rho}\Phi_\alpha] \cdot dA} \tag{3}$$

Therein Φ_i is the potential function of the incident density wave; Φ_ρ, Φ_α are the potentials of the scattered density and thermal waves, Ψ_z is the z component of the vector potential of the scattered viscous wave; k_ρ, k_α, k_v are the free field wave numbers of the density, thermal and viscous waves, respectively; for the factors Π_ρ, Π_α see the Sections J2 and J2. The integrals $\oint \ldots ds$ are taken over the surface of the cell; the integral over A_0 is over the cell area. The field components of the scattered field are formulated as Fourier series over φ ; the index $\beta=\rho, \alpha$ stands for the density and thermal wave types.

Closed cell model :

Fourier series formulations of the component fields:

$$\Phi_\beta(r,\varphi) = \sum_{n=0} [A_{\beta n} H_n^{(2)}(k_\beta r) + B_{\beta n} J_n(k_\beta r)] \cdot \cos(n\varphi) = \sum_{n=0} \Phi_{\beta rn}(r) \cdot \cos(n\varphi) \quad ; \quad \beta = \rho, \alpha \tag{4}$$

$$\vec{\Psi} = \{0, 0, \Psi_z\},$$
$$\Psi_z(r,\varphi) = \sum_{n=0} [A_{vn} H_n^{(2)}(k_v r) + B_{vn} J_n(k_v r)] \cdot \sin(n\varphi) = \sum_{n=0} \Psi_{vrn}(r) \cdot \sin(n\varphi) \tag{5}$$

The boundary conditions at the cell and fibre surfaces are (for each Fourier series term) :

$$k_\rho R\,\Phi'_{\rho rn}(R)+k_\alpha R\,\Phi'_{\alpha rn}(R)-n\,\Psi_{zrn}(R)=0$$
$$-n[\Phi_{\rho rn}(R)+\Phi_{\alpha rn}(R)]+n[k_\rho R\,\Phi'_{\rho rn}(R)+k_\alpha R\,\Phi'_{\alpha rn}(R)]-(k_v R)^2\,\Psi''_{zrn}(R)=0 \qquad (6)$$
$$k_\rho R\,\Phi'_{\rho rn}(R)+\frac{\Theta_\alpha}{\Theta_\rho}k_\alpha R\,\Phi'_{\alpha rn}(R)=0$$

$$n[\Phi_{\rho rn}(a)+\Phi_{\alpha rn}(a)]-k_v a\,\Psi'_{zrn}(a)=-n\,\Phi_{ern}(a)$$
$$k_\rho R\,\Phi'_{\rho rn}(a)+k_\alpha R\,\Phi'_{\alpha rn}(a)-n\,\Psi_{zrn}(a)=-k_\rho a\,\Phi'_{ern}(a) \qquad (7)$$
$$\Phi_{\rho rn}(a)+\frac{\Theta_\alpha}{\Theta_\rho}\Phi_{\alpha rn}(a)=-\Phi_{ern}(a)$$

(for the factors Θ_ρ, Θ_α see Sections J1, J2; the prime indicates the derivative with respect to the argument of the radial functions in the field terms). The inhomogeneous linear systems of equations can be solved for the amplitudes of the Fourier terms (see reference [G.1, chapter 12] for more details). The integrals for ρ_{eff} and C_{eff} need only the terms $n=0$ for the effective compressibility, and $n=1$ for the effective density (other integrals vanish); so the final equations are (A_i is the arbitrary amplitude of the incident density wave):

$$\frac{\rho_{eff}}{\rho_0}=\frac{(-2jA_i+B_{\rho 1})[J_1(k_\rho R)-\frac{a}{R}J_1(k_\rho a)]+A_{\rho 1}[H_1^{(2)}(k_\rho R)-\frac{a}{R}H_1^{(2)}(k_\rho a)]+\ldots}{(-2jA_i+B_{\rho 1})[J_1(k_\rho R)-\frac{a}{R}J_1(k_\rho a)]+A_{\rho 1}[H_1^{(2)}(k_\rho R)-\frac{a}{R}H_1^{(2)}(k_\rho a)]+\ldots}$$
$$\frac{\ldots+\frac{\Pi_\alpha}{\Pi_\rho}\left[A_{\alpha 1}[H_1^{(2)}(k_\alpha R)-\frac{a}{R}H_1^{(2)}(k_\alpha a)]+B_{\alpha 1}[J_1(k_\alpha R)-\frac{a}{R}J_1(k_\alpha a)]\right]}{\ldots+A_{\alpha 1}[H_1^{(2)}(k_\alpha R)-\frac{a}{R}H_1^{(2)}(k_\alpha a)]+B_{\alpha 1}[J_1(k_\alpha R)-\frac{a}{R}J_1(k_\alpha a)]-\ldots} \qquad (8)$$
$$\frac{\ldots}{-A_{v1}[H_1^{(2)}(k_v R)-\frac{a}{R}H_1^{(2)}(k_v a)]-B_{v1}[J_1(k_v R)-\frac{a}{R}J_1(k_v a)]}$$

$$\frac{C_{eff}}{C_0}=\frac{k_\rho R\,J_1(k_\rho R)\,A_i}{(A_i+B_{\rho 0})[yJ_1(y)]_{k_\rho a}^{k_\rho R}+A_{\rho 0}[yH_1^{(2)}(y)]_{k_\rho a}^{k_\rho R}+\ldots}$$
$$\frac{\ldots}{\ldots+\frac{\Pi_\alpha}{\Pi_\rho}\frac{k_\rho^2}{k_\alpha^2}\left[A_{\alpha 0}[yH_1^{(2)}(y)]_{k_\alpha a}^{k_\alpha R}+B_{\alpha 0}[yJ_1(y)]_{k_\alpha a}^{k_\alpha R}\right]} \qquad (9)$$

The free field wave numbers k_ρ, k_α, k_v can be reduced to $k_0=\omega/c_0$ (c_0= adiabatic sound velocity) and $k_v^2=-j\,\omega/\nu$ (ν= kinematic viscosity; see Sections J1, J2), and k_v in turn can be related to the flow resistivity Ξ of a transversal fibre bundle with the absorber variable $E=\rho_0 f/\Xi$ by the relation :

$$|k_v a|^2 = \frac{4\mu}{1-\mu} \cdot c(\frac{R}{a}) \cdot E \tag{10}$$

with the massivity μ and the function $c(R/a)$ from section G.7 .

The real part of the characteristic impedance from the transversal fibre closed cell model typically shows a resonance because of the assumedly identical interactions of the fibres with the cell surfaces in all cells (the closed cell model cannot handle the time lag of scattering during the propagation through the bundle).

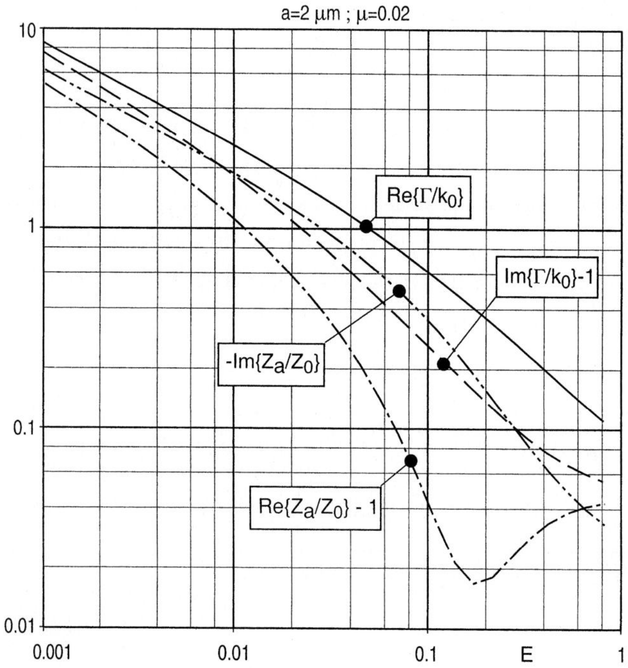

Characteristic values of a transversal fibre bundle evaluated with the model of regular arrangement.

Open cell model :

The potentials of the incident density wave Φ_i and of the scattered waves Φ_β ; $\beta = \alpha, \nu$ (scattered density and temperature wave) as well as the scattered viscosity wave Ψ are formulated as :

$$\Phi_i(r,\varphi) = A_i \cdot e^{-jk_\rho x}$$

$$= A_i \sum_{n=0} \delta_n (-j)^n \cdot J_n(k_\rho r) \cdot \cos(n\varphi) = \sum_{n=0} \Phi_{irn}(r) \cdot \cos(n\varphi) \quad ; \quad \delta_n = \begin{cases} 1; & n=0 \\ 2; & n>0 \end{cases} \tag{11}$$

$$\Phi_\beta(r,\varphi) = \sum_{n=0} A_{\beta n} H_n^{(2)}(k_\beta r)\cdot\cos(n\varphi) = \sum_{n=0} \Phi_{\beta rn}(r)\cdot\cos(n\varphi) \quad ; \quad \beta=\rho,\alpha \tag{12}$$

$$\vec{\Psi} = \{0, 0, \Psi_z\}$$

$$\Psi_z(r,\varphi) = \sum_{n=0} A_{\nu n} H_n^{(2)}(k_\nu r)\cdot\sin(n\varphi) = \sum_{n=0} \Psi_{\nu rn}(r)\cdot\sin(n\varphi) \tag{13}$$

The scattered field amplitudes follow from the system of linear equations (n=0,1) which represent the boundary conditions :

$$\sum_{\beta=\rho,\alpha} A_{\beta n} n\, H_n^{(2)}(k_\beta a) - A_{\nu n}\,[k_\nu a\, H_{n-1}^{(2)}(k_\nu a) - n\, H_n^{(2)}(k_\nu a)] = -n\,\delta_n\,(-j)^n J_n(k_\rho a)\,A_i$$

$$\begin{aligned}\sum_{\beta=\rho,\alpha} A_{\beta n}[k_\beta a\, H_{n-1}^{(2)}(k_\beta a) - n\, H_n^{(2)}(k_\beta a)] - A_{\nu n}\, n\, H_n^{(2)}(k_\nu a) = \\ = -\delta_n\,(-j)^n[k_\rho a\, J_{n-1}(k_\rho a) - n\, J_n(k_\rho a)]\,A_i\end{aligned} \tag{14}$$

$$A_{\rho n}\, H_n^{(2)}(k_\rho a) + \frac{\Theta_\alpha}{\Theta_\rho} A_{\alpha n}\, H_n^{(2)}(k_\alpha a) = -\delta_n\,(-j)^n J_n(k_\rho a)\,A_i$$

and with them the effective density ρ_{eff} and effective compressibility C_{eff} are evaluated from :

$$\frac{\rho_{eff}}{\rho_0} = \frac{-2jA_i[J_1(k_\rho R) - \frac{a}{R}J_1(k_\rho a)] + A_{\rho 1}[H_1^{(2)}(k_\rho R) - \frac{a}{R}H_1^{(2)}(k_\rho a)] + \ldots}{-2jA_i[J_1(k_\rho R) - \frac{a}{R}J_1(k_\rho a)] + A_{\rho 1}[H_1^{(2)}(k_\rho R) - \frac{a}{R}H_1^{(2)}(k_\rho a)] + \ldots}$$

$$\frac{\ldots + \frac{\Pi_\alpha}{\Pi_\rho} A_{\alpha 1}[H_1^{(2)}(k_\alpha R) - \frac{a}{R}H_1^{(2)}(k_\alpha a)]}{\ldots + A_{\alpha 1}[H_1^{(2)}(k_\alpha R) - \frac{a}{R}H_1^{(2)}(k_\alpha a)] - A_{\nu 1}[H_1^{(2)}(k_\nu R) - \frac{a}{R}H_1^{(2)}(k_\nu a)]} \tag{15}$$

$$\frac{C_{eff}}{C_0} = \frac{A_i\, k_\rho R\, J_1(k_\rho R) + A_{\rho 0}\, k_\rho R\, H_1^{(2)}(k_\rho R) + A_{\alpha 0}\, k_\alpha R\, H_1^{(2)}(k_\alpha R)}{A_i\,[yJ_1(y)]_{k_\rho a}^{k_\rho R} + A_{\rho 0}\,[yH_1^{(2)}(y)]_{k_\rho a}^{k_\rho R} + \frac{\Pi_\alpha}{\Pi_\rho}\frac{k_\rho^2}{k_\alpha^2} A_{\alpha 0}\,[yH_1^{(2)}(y)]_{k_\alpha a}^{k_\alpha R}} \tag{16}$$

with $[f(y)]_a^b = f(b) - f(a)$.

The characteristic values Γ and Z_i of the propagation constant and wave impedance of the sound wave propagating through the open cell model fibre bundle approach realistic values only at relatively high values of $|k_\nu a|^2 = \omega a^2/\nu$; at low values the presence of the fibres is under-weighted. This is the consequence of the assumedly missing scattering interaction between neighbouring fibres at low frequencies, where the viscous boundary layer at the fibres fills the whole interspace between the fibres in reality. Nevertheless, the open cell model is needed for the evaluation of the scattered field in the multiple scattering model.

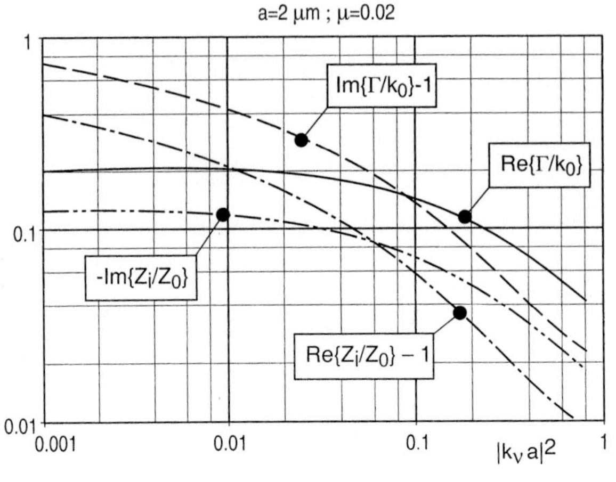

Characteristic values in a transversal fibre bundle evaluated with the open cell model, which contains only single scattering at a fibre.

Multiple scattering model :

This model considers the scattering at a reference fibre (at position P_i) at which not only the incident wave p_i is scattered, but also the scattered fields coming from neighbouring fibres (at positions P_j).

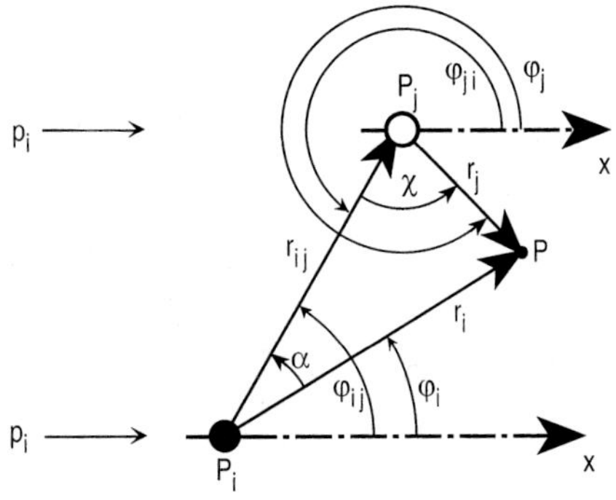

A minimum distance d between the reference fibre and the neighbouring fibres is assumed. The fibres outside the range d may be arranged randomly with a porosity σ or massivity μ or fibre number density N of the transversal fibre bundle:

$$\sigma = 1 - \mu = 1 - N \cdot \pi a^2 \tag{17}$$

The "radiating" scattered fields, i.e. the scattered fields which assumedly propagate freely through the bundle are indicated with a prime (with r,φ the co-ordinates centred in the scattering fibre) :

$$\begin{aligned} \Phi'_{\beta}(r,\varphi) &= \sum_{n=0}^{\infty} A_{\beta n} \cdot H_n^{(2)}(k_\beta r) \cdot \cos n\varphi \quad ; \quad \beta = \rho, \alpha \\ \Psi'_z(r,\varphi) &= \sum_{n=0}^{\infty} A_{\nu n} \cdot H_n^{(2)}(k_\nu r) \cdot \sin n\varphi \quad ; \quad A_{\nu 0} = 0 \end{aligned} \tag{18}$$

The total scattered field at a point of immission P is indicated with a double prime; it is obtained by integration over the contributions of all neighbouring fibres to the reference fibre. If the point of immission P is at the position P_i of the reference fibre, the total scattered field is :

$$\begin{aligned}\Phi''_ß(r_i,\varphi_i) &= \int_0^{2\pi} d\varphi_{ij} \int_a^{\infty} N(r_{ij})\cdot\Phi'_ß(r_j,\varphi_j)\, r_{ij}\cdot dr_{ij} \\ &= N\sum_{n=0} A_{ßn}\cdot \int_0^{2\pi}\int_d^{\infty} H_n^{(2)}(k_ß r_j)\, r_{ij}\cdot \cos n\varphi_j \cdot dr_{ij}\cdot d\varphi_{ij}\end{aligned} \tag{19}$$

$$\begin{aligned}\Psi''_z(r_i,\varphi_i) &= \int_0^{2\pi} d\varphi_{ij} \int_a^{\infty} N(r_{ij})\cdot\Psi'_z(r_j,\varphi_j)\, r_{ij}\cdot dr_{ij} \\ &= N\sum_{n=0} A_{vn}\cdot \int_0^{2\pi}\int_d^{\infty} H_n^{(2)}(k_v r_j)\, r_{ij}\cdot \sin n\varphi_j \cdot dr_{ij}\cdot d\varphi_{ij}\end{aligned} \tag{20}$$

(the first expressions also permit a variable fibre density function $N(r_{ij})$). After application of the addition theorem for Hankel functions, to transform all co-ordinates to those of the reference fibre, one gets for the total scattered field around the reference fibre :

$$\begin{aligned}\Phi_ß(r_i,\varphi_i) &= \sum_{n=0} A_{ßn}\cdot [H_n^{(2)}(k_ß r_i)\cdot\cos(n\varphi_i) + \\ &+(-1)^n N \sum_{m=-\infty}^{\infty} \int_0^{2\pi} d\varphi_{ij} \int_d^{\infty} r_{ij}\cdot H_{n+m}^{(2)}(k_ß r_{ij})\cdot J_m(k_ß r_i)\cdot\cos((n+m)\varphi_{ij}-m\varphi_i)\cdot dr_{ij}\,] \\ \Psi_z(r_i,\varphi_i) &= \sum_{n=0} A_{vn}\cdot [H_n^{(2)}(k_v r_i)\cdot\sin(n\varphi_i) + \\ &+(-1)^n N \sum_{m=-\infty}^{\infty} \int_0^{2\pi} d\varphi_{ij} \int_d^{\infty} r_{ij}\cdot H_{n+m}^{(2)}(k_v r_{ij})\cdot J_m(k_v r_i)\cdot\sin((n+m)\varphi_{ij}-m\varphi_i)\cdot dr_{ij}\,]\end{aligned} \tag{21}$$

with the geometrical relations :

$$\begin{aligned}\varphi_{ji} &= \varphi_{ij}+\pi \\ r_j &= \sqrt{r_i^2+r_{ij}^2-2\,r_i\,r_{ij}\cos(\varphi_{ij}-\varphi_i)} \\ \varphi_j &= \varphi_{ji}+\operatorname{arctg}\frac{r_i\sin(\varphi_{ij}-\varphi_i)}{r_{ij}-r_i\cos(\varphi_{ij}-\varphi_i)}\end{aligned} \tag{22}$$

In the above expressions for $\Phi_ß, \Psi_z$ all integrals over φ_{ij} disappear except those with m+n=0; then the integral returns the value 2π . So the total scattered field at the reference fibre becomes :

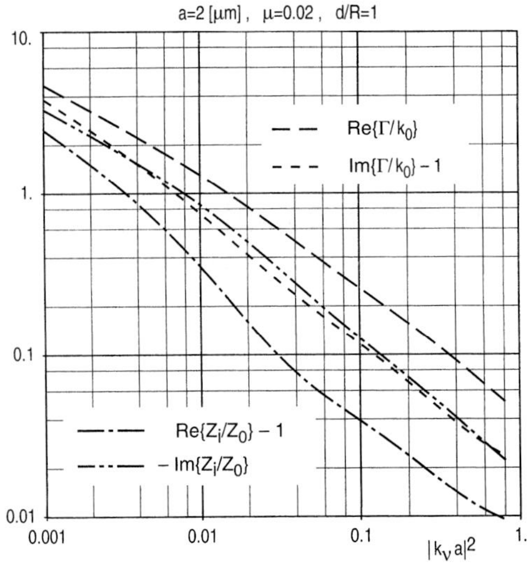

Components of the characteristic values of a transversal fibre bundle, evaluated with multiple scattering.

It still remains to determine the connection of the independent variable $|k_\nu a|^2$ to the flow resistivity variable $E = \rho_0 f / \Xi$. This is given by (see section G.7) :

$$|k_\nu a|^2 = \frac{4\mu}{1-\mu} \cdot c(x) \cdot E$$

$$c(\frac{R}{a}) = e^{f(x)} \quad ; \quad x = \ln\frac{R}{a} = -0.5\,\ln(\mu) \tag{33}$$

$$f(x) = 0.865823 + \frac{0.0250214}{x^3} - \frac{0.322560}{x^2} + \frac{1.78839}{x} - 0.530524 \cdot x +$$
$$+ 0.0604543 \cdot x^2 - 0.00312698 \cdot x^3$$

The multiple scattering model of a transversal fibre bundle only contains the function $K_{1,0}(k_\beta a,\ \mu)$ in the characteristic values. This function can be approximately represented by:

$$\mathrm{Re}\{K_{1,0}(x,y)\} = 0{,}51 - 0{,}49 \cdot \tanh(2{,}9 \cdot d_r(x,y))]$$

$$\mathrm{Im}\{K_{1,0}(x,y)\} = \frac{h(x,y)}{\cosh^{3/2}(2{,}5 \cdot d_i(x,y))}$$

$$d_r(x,y) = -0.5903666 - 0.8220386 \cdot x + 0.5694317 \cdot y \tag{34}$$

$$d_i(x,y) = -0.494366 - 0.802776 \cdot x + 0.59628 \cdot y$$

$$h(x,y) = -0.4959784 - 0.1499322 / x + 0.003230626 \cdot x$$

With these relations and $(k_\nu a)^2 = -j\,|k_\nu a|^2 \quad ; \quad (k_\alpha a)^2 \approx \kappa\,\mathrm{Pr} \cdot (k_\nu a)^2$ (35)

(Pr= Prandtl number) the characteristic values of a transversal fibre bundle (with multiple scattering) can easily be evaluated. See the comparison below between exact evaluation and approximation.

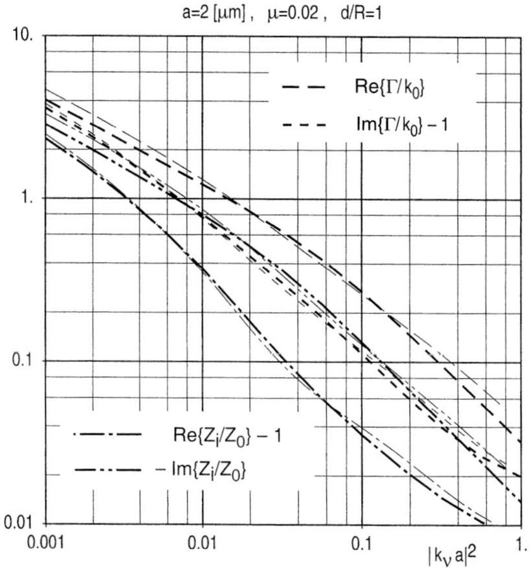

Comparison of the components of the characteristic values in a transversal fibre bundle from exact evaluation (thick lines) and approximation (thin lines).

G.9 Effective wave multiple scattering in transversal fibre bundle

As in section G.8 (with which this section has a number of similarities) the porous material consists of parallel fibres (at rest) with fibre radius a and fibre number density N . A sound wave propagates inside the fibre bundle normal to the fibres. Whereas it was assumed in section G.8 that the propagating wave is a density wave, here it is supposed to be a less specified "effective" wave with potential Φ_E with propagation constant $\Gamma = jk_e$. The propagating wave (exciting wave) may be some hybrid wave of a density wave, a thermal wave and a viscous wave. The propagation constant and wave impedance Z_e of this wave will be determined. See section G.8 for the assumed co-ordinates. As before, the index $\beta = \rho, \alpha$ indicates a density or thermal wave with scalar potential functions Φ_β , and Ψ_z is the z component of the vector potential of a viscous wave. It is further assumed, as before, that the nearest neighbours of a reference fibre at the position (r_i, φ_i) have a distance d from it.

Exciting wave formulation :

$$\begin{aligned} \Phi_E(r,\varphi) &= A_E \cdot e^{-\Gamma x} = A_E \cdot e^{-jk_e r \cos\varphi} = A_E \cdot \sum_{n=0} (-j)^n \, \delta_n \, J_n(k_e r) \cos(n\varphi) \\ &= A_E \cdot \sum_{n=0} \Phi_{Ern}(r) \cdot \cos(n\varphi) \quad ; \quad \delta_n = \begin{cases} 1 \, ; n = 0 \\ 2 \, ; n > 0 \end{cases} \end{aligned} \tag{1}$$

Scattered field formulation (marked with a prime) :

$$\begin{aligned} \Phi'_\beta(r,\varphi) &= \sum_{n=0} A_{\beta n} \, H_n^{(2)}(k_\beta r) \cos(n\varphi) = \sum_{n=0} \Phi'_{\beta n}(r) \cdot \cos(n\varphi) \quad ; \quad \beta = \rho, \alpha \\ \Psi'_z(r,\varphi) &= \sum_{n=0} A_{\nu n} \, H_n^{(2)}(k_\nu r) \sin(n\varphi) = \sum_{n=0} \Psi'_{\nu n}(r) \cdot \sin(n\varphi) \quad ; \quad A_{\nu 0} = 0 \end{aligned} \tag{2}$$

Total scattered field at a reference fibre at (r_i, φ_i) (marked with a double prime):

$$\begin{aligned} \Phi''_\beta(r_i,\varphi_i) &= N \sum_{n=0} A_{\beta n} \int_0^{2\pi} d\varphi_{ij} \int_d^\infty e^{-jk_e r_{ij} \cos\varphi_{ij}} \cdot H_n^{(2)}(k_\beta r_j) \cos(n\varphi_j) \cdot r_{ij} \cdot dr_{ij} \\ \Psi''_z(r_i,\varphi_i) &= N \sum_{n=0} A_{\nu n} \int_0^{2\pi} d\varphi_{ij} \int_d^\infty e^{-jk_e r_{ij} \cos\varphi_{ij}} \cdot H_n^{(2)}(k_\nu r_j) \sin(n\varphi_j) \cdot r_{ij} \cdot dr_{ij} \end{aligned} \tag{3}$$

and after application of the addition theorem for Hankel functions and integration :

$$\begin{aligned}\Phi_\beta(r_i,\varphi_i) &= \\ &= \sum_{n=0} \cos(n\varphi_i)\cdot\{A_{\beta n}\, H_n^{(2)}(k_\beta r_i) - \frac{2\pi N}{k_\beta^2-k_e^2}(-j)^n\, J_n(k_\beta r_i)\cdot \sum_{m=0}(j)^m\, B_{\beta mn}\cdot A_{\beta m}\} \\ &= \sum_{n=0}\Phi_{\beta rn}(r_i)\cdot\cos(n\varphi_i)\end{aligned} \tag{4}$$

$$\begin{aligned}\Psi_z(r_i,\varphi_i) &= \\ &= \sum_{n=0} \sin(n\varphi_i)\cdot\{A_{\nu n}\, H_n^{(2)}(k_\nu r_i) + \frac{2\pi N}{k_\nu^2-k_e^2}(-j)^n\, J_n(k_\nu r_i)\cdot \sum_{m=0}(j)^m\, B_{\nu mn}\cdot A_{\nu m}\} \\ &= \sum_{n=0}\Psi_{zrn}(r_i)\cdot\sin(n\varphi_i)\end{aligned} \tag{5}$$

with the abbreviations for both $\beta=\rho,\alpha$:

$$\begin{aligned}B_{\beta mn} &= k_e d\cdot[H^{(2)}_{m+n}(k_\beta d)\, J_{m+n-1}(k_e d) + H^{(2)}_{m-n}(k_\beta d)\, J_{m-n-1}(k_e d)] - \\ &\quad - k_\beta d\cdot[H^{(2)}_{m+n-1}(k_\beta d)\, J_{m+n}(k_e d) + H^{(2)}_{m-n-1}(k_\beta d)\, J_{m-n}(k_e d)] \\ B_{\nu mn} &= k_e d\cdot[H^{(2)}_{m+n}(k_\nu d)\, J_{m+n-1}(k_e d) - H^{(2)}_{m-n}(k_\nu d)\, J_{m-n-1}(k_e d)] - \\ &\quad - k_\nu d\cdot[H^{(2)}_{m+n-1}(k_\nu d)\, J_{m+n}(k_e d) - H^{(2)}_{m-n-1}(k_\nu d)\, J_{m-n}(k_e d)]\end{aligned} \tag{6}$$

These scattered waves plus the exciting wave have to satisfy the boundary conditions at the reference fibre (which is assumed to be isothermal) :

$$v_\varphi(a,\varphi_i)=0: \quad \begin{aligned}& n\cdot[A_{\rho n}\, H_n^{(2)}(k_\rho a) - \frac{2\pi N}{k_\rho^2-k_e^2}(-j)^n\, J_n(k_\rho a)\cdot\sum_{m=0} j^m\, B_{\rho mn}\cdot A_{\rho m}] + \\ & +n\cdot[A_{\alpha n}\, H_n^{(2)}(k_\alpha a) - \frac{2\pi N}{k_\alpha^2-k_e^2}(-j)^n\, J_n(k_\alpha a)\cdot\sum_{m=0} j^m\, B_{\alpha mn}\cdot A_{\alpha m}] - \\ & -k_\nu a\cdot[A_{\nu n}\, H_n'^{(2)}(k_\nu a) + \frac{2\pi N}{k_\nu^2-k_e^2}(-j)^n\, J_n'(k_\nu a)\cdot\sum_{m=0} j^m\, B_{\nu mn}\cdot A_{\nu m}] = \\ & = -n\,\delta_n\,(-j)^n\, J_n(k_e a)\cdot A_E\end{aligned} \tag{7}$$

$$v_r(a,\varphi_i)=0: \quad \begin{aligned}& k_\rho a\cdot[A_{\rho n}\, H_n'^{(2)}(k_\rho a) - \frac{2\pi N}{k_\rho^2-k_e^2}(-j)^n\, J_n'(k_\rho a)\cdot\sum_{m=0} j^m\, B_{\rho mn}\cdot A_{\rho m}] + \\ & +k_\alpha a\cdot[A_{\alpha n}\, H_n'^{(2)}(k_\alpha a) - \frac{2\pi N}{k_\alpha^2-k_e^2}(-j)^n\, J_n'(k_\alpha a)\cdot\sum_{m=0} j^m\, B_{\alpha mn}\cdot A_{\alpha m}] - \\ & -n\cdot[A_{\nu n}\, H_n^{(2)}(k_\nu a) + \frac{2\pi N}{k_\nu^2-k_e^2}(-j)^n\, J_n(k_\nu a)\cdot\sum_{m=0} j^m\, B_{\nu mn}\cdot A_{\nu m}] = \\ & = -\delta_n\,(-j)^n\, k_e a\, J_n'(k_e a)\cdot A_E\end{aligned} \tag{8}$$

$$[A_{\rho n}\, H_n^{(2)}(k_\rho a) - \frac{2\pi\, N}{k_\rho^2 - k_e^2}(-j)^n\, J_n(k_e a)\cdot \sum_{m=0}(j)^m B_{\rho mn}\cdot A_{\rho m}] +$$

$$\frac{T(a,\varphi_i)}{T_0} = 0 :\; +\frac{\Theta_\alpha}{\Theta_\rho}\cdot[A_{\alpha n}\, H_n^{(2)}(k_\alpha a) - \frac{2\pi\, N}{k_\alpha^2 - k_e^2}(-j)^n\, J_n(k_\alpha a)\cdot \sum_{m=0}(j)^m B_{\alpha mn}\cdot A_{\alpha m}] = \quad (9)$$

$$= -\delta_n\,(-j)^n\, J_n(k_e a)\cdot A_E$$

With a solution for the amplitudes $A_{\beta n}, A_{\nu n}$ evaluate :

$$\frac{\rho_{eff}}{\rho_0} = \frac{[r\cdot(\Phi_{Er1}(r) + \Phi_{\rho r1}(r) + \frac{\Pi_\alpha}{\Pi_\rho}\,\Phi_{\alpha r1}(r))]_a^R}{[r\cdot(\Phi_{Er1}(r) + \Phi_{\rho r1}(r) + \Phi_{\alpha r1}(r) - \Psi_{zr1}(r))]_a^R}$$

$$\frac{C_{eff}}{C_0} = \frac{[\hat{\Phi}_{Er0}(r) + \hat{\Phi}_{\rho r0}(r) + \hat{\Phi}_{\alpha r0}(r)]_a^R}{[\frac{\rho_{eff}}{\rho_0}\frac{k_\rho^2}{k_e^2}\,\hat{\Phi}_{Er0}(r) + \hat{\Phi}_{\rho r0}(r) + \frac{\Pi_\alpha\, k_\rho^2}{\Pi_\rho\, k_\alpha^2}\,\hat{\Phi}_{\alpha r0}(r)]_a^R} \quad (10)$$

with $[f(y)]_a^b = f(b) - f(a)$ and $\hat{\Phi}_\beta(r) = \int^r k_\beta^2 r\cdot\Phi_\beta(r)\, dr$.

The evaluation must proceed iteratively, because the equations from the boundary conditions contain (besides the unknown amplitudes) the unknown wave number k_e . We begin the iteration with an approximation $\Gamma = jk_e$ from section G.8 ; solve for a 1st approximation of $A_{\beta n}$, $A_{\nu n}$; insert into the expressions for ρ_{eff}, C_{eff} ; evaluate the next approximation for the propagating wave from :

$$\frac{\Gamma}{k_0} = j\sqrt{\frac{\rho_{eff}}{\rho_0}\cdot\frac{C_{eff}}{C_0}} \quad ; \quad \frac{Z_e}{Z_0} = \sqrt{\frac{\rho_{eff}}{\rho_0} / \frac{C_{eff}}{C_0}} \quad (11)$$

and resume the cycle of iteration. Apply in the evaluations, with the massivity μ :

$$\frac{a}{R} = \sqrt{\mu} \quad ; \quad \frac{2\pi N}{k_\beta^2 - k_e^2} = \frac{2\mu}{(k_\beta a)^2 - (k_e a)^2} \quad (12)$$

The following diagram compares the components of the characteristic values from the present iterative evaluation (with an “effective” propagating wave) with values from section G.8 (with the density wave as the propagating wave). Up to $m_{max}=4$ orders of scattering were used.

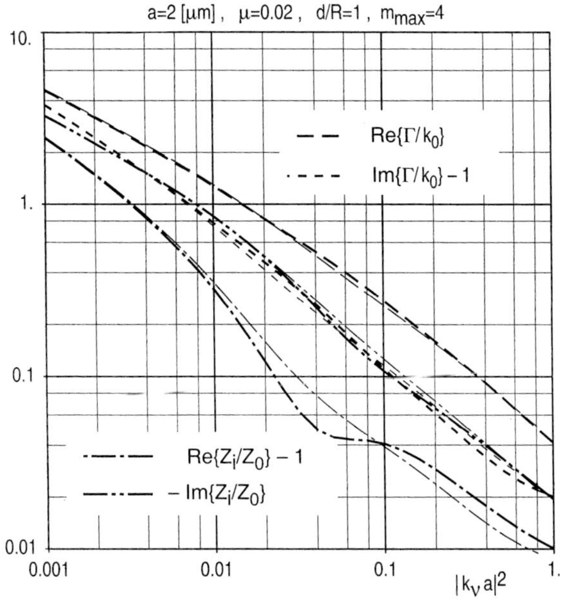

Comparison of the components of characteristic values in a transversal fibre bundle, evaluated iteratively for an effective propagating wave (thick lines, from this section), and with a propagating density wave (thin lines, from section G.8).

The present method, although much more complicated numerically than the method of section G.8 , is well suited to evaluating particle velocity profiles around a fibre in a transversal fibre bundle.

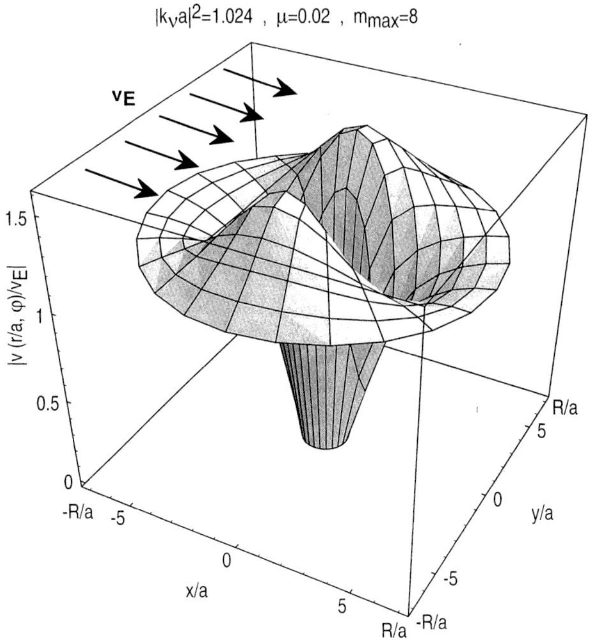

Profile of the magnitude of the total particle velocity around a fibre in a fibre bundle with transversal sound propagation.

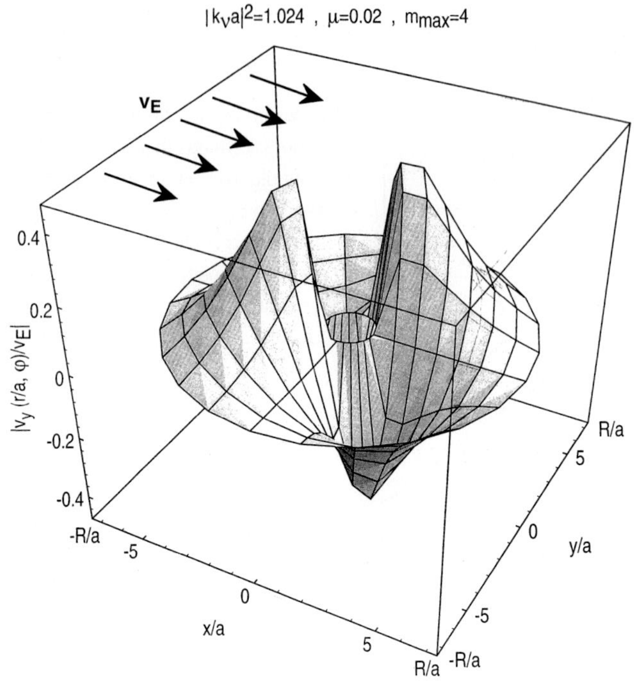

Profile of the real component of the transversal particle velocity around a fibre in a fibre bundle with transversal sound propagation.

G.10 BIOT's theory of porous absorbers

Ref.: BIOT's papers in [G.2], a survey in [G1]

Whereas in other sections of this chapter the matrix of the porous material is assumed to be rigid, it may be elastic in BIOT's theory. The consequence is the onset of many wave types by the coupling between the matrix and the enclosed fluid in the pores. BIOT's theory is not a "terminated" theory; he took some fundamental relations between flow and sound from the theory of circular capillaries, which was available at his time, instead of results from possibly better suited models, which are now available. The price for the wider range of application of BIOT's theory is a number of specially defined material parameters, for which BIOT has given prescriptions how to measure them.

Fundamental assumptions :

- The matrix is homogeneous and isotropic at scales which are larger than the scale of the pores.
- The pores are interconnected.
- The size of the pores is small compared to volume elements dV considered and small compared to the wavelength.

Fundamental equations :

Let $\vec{u}_s$ be the movement (elongation) of the solid, averaged over a volume element dV ; let $\vec{u}_f$ be the average movement of the fluid in dV .

Equations of motion :

$$\rho_{ss}\frac{\partial^2\vec{u}_s}{\partial t^2}+\rho_{sf}\frac{\partial^2\vec{u}_f}{\partial t^2}=P\cdot\text{grad div}\,\vec{u}_s+Q\cdot\text{grad div}\,\vec{u}_f-N\cdot\text{rot rot}\,\vec{u}_s+bF(\omega)\cdot\left(\frac{\partial\vec{u}_f}{\partial t}-\frac{\partial\vec{u}_s}{\partial t}\right)\quad(1)$$

$$\rho_{ff}\frac{\partial^2\vec{u}_f}{\partial t^2}+\rho_{sf}\frac{\partial^2\vec{u}_s}{\partial t^2}=R\cdot\text{grad div}\,\vec{u}_f+Q\cdot\text{grad div}\,\vec{u}_s-bF(\omega)\cdot\left(\frac{\partial\vec{u}_f}{\partial t}-\frac{\partial\vec{u}_s}{\partial t}\right)\quad(2)$$

Strain-stress equations :

$$\tau^s_{ij}=[(P-2N)\cdot\text{div}\,\vec{u}_s+Q\cdot\text{div}\,\vec{u}_f\,]\delta_{ij}+N\cdot\left(\frac{\partial u_{si}}{\partial x_j}+\frac{\partial u_{sj}}{\partial x_i}\right)$$

$$\tau^f_{ij}=-\sigma\,p\,\delta_{ij}=[R\cdot\text{div}\,\vec{u}_f+Q\cdot\text{div}\,\vec{u}_s\,]\delta_{ij}\quad(3)$$

Here (e.g.) u_{si} is the component of $\vec{u}_s$ in the direction of the co-ordinate x_i ; τ^s is the tension on the matrix absorber in a unit area; τ^f is the tension on the fluid in a unit area; i,j= 1,2,3 stand for co-ordinates; e.g. τ^s_{33} is the x_3 component of the tension on the matrix acting on a surface normal to x_3 , and τ^s_{13} means a shear tension in the x_1 direction on a surface normal to x_3 . δ_{ij} is the Kronecker symbol. ρ_{mn} are effective mass densities if m=n, and they are coupling coefficients between solid and fluid if m≠n . P,Q,R,N are elastic constants, introduced by BIOT. A corresponds to the 1st Lamé constant of the material matrix, N to its 2nd Lamé constant. In most cases P=A+2N holds. Q is evidently a coupling coefficient between matrix and fluid; it also determines the coefficient R by the relation $R=-Q\cdot e/\varepsilon$, where $e=\operatorname{div}\vec{u}_s$ and $\varepsilon=\operatorname{div}\vec{u}_f$ are the strains of the solid and the fluid, respectively. There are three coupling coefficients between solid and fluid in the equations : ρ_{sf}, $bF(\omega)$, Q .

With σ= volume porosity of the porous material, the effective densities are :

$$\rho_{ss}=(1-\sigma)\rho_s-\rho_{sf} \quad ; \quad \rho_{ff}=\sigma\rho_0-\rho_{sf} \tag{4}$$

where ρ_s is the density of the (compact) solid, and $\rho_0=\rho_f$ is the density of the fluid. The coupling density ρ_{sf} describes the extra inertia of a relative motion between solid and fluid :

$$\rho_{sf}=-(\chi-1)\sigma\rho_0 \tag{5}$$

The term χ represents the *tortuosity* of the pores in the material (it corresponds to the structure factor in older theories; see below for its determination); it is a pure form factor.

The term $bF(\omega)$ is a coupling factor of the viscous forces; it is associated with the relative velocity $\partial\vec{u}_f/\partial t-\partial\vec{u}_s/\partial t$ of fluid and solid. It is mainly determined by the flow resistivity Ξ of the material; it is frequency dependent, whereby the transition from an approximately parabolic flow profile at low frequencies to an approximately rectangular profile at high frequencies can be described. BIOT has taken the quantity $bF(\omega)$ from the observation that the effective densities ρ_{eff} in flat and in circular capillaries have about the same frequency dependence, up to a "stretching" of the frequency axis with a factor c (see the Sections J1, J2; see there also for the definition of $J_{1,0}(z)$). So he took from circular capillaries (a= pore radius in the capillary model; $k_\nu=-j\sqrt{(\omega/\nu)}$; ν= kinematic viscosity) :

$$bF(\omega)=-\frac{\sigma^2\Xi}{4}\,\frac{ck_\nu a\cdot J_{1,0}(ck_\nu a)}{1-\dfrac{2}{ck_\nu a}J_{1,0}(ck_\nu a)} \tag{6}$$

The equivalent radius a of the capillary for a given porous material can be taken from :

$$\Xi=\frac{8\eta}{a^2}\frac{\chi}{\sigma} \qquad (\eta=\text{ dynamic viscosity}) \tag{7}$$

with the measured flow resistivity Ξ, porosity σ, tortuosity χ. The matching factor c in the argument of $J_{1,0}(z)$ changes between c=1 for cylindrical pores, and $c=(4/3)^{1/2}$ for flat pores; values for triangular and square pores have been evaluated.

A different method (from JOHNSON et al.) to determine $bF(\omega)$ is :

$$bF(\omega) = \sigma^2 \Xi \left(1 + \frac{4j\chi^2 \eta \rho_0 \omega}{\Xi^2 \Lambda^2 \sigma^2}\right)^{1/2} \quad ; \quad \Lambda = c\left(\frac{8\eta\chi}{\Xi\sigma}\right)^{1/2} \tag{8}$$

The coupling factor Q is called the *potential coupling factor*. It takes into account that the pressure in the pores may change, even if there $\mathrm{div}\,\vec{u}_f=0$, by a dilatation of the matrix.

If only shear stresses act on the absorber, no dilatation takes place; the strain-stress equations reduce to :

$$\tau^s_{ij} = N\left[\frac{\partial u_{si}}{\partial x_j} + \frac{\partial u_{sj}}{\partial x_i}\right] \quad ; \quad \tau^f_{ij} = 0 \tag{9}$$

thus N is the shear modulus of the matrix.

Suppose a sample of the porous material to be coated with a thin, limp foil, allowing a connection between the inside and outside space, and exerting a static pressure P_s on the coated sample. Whereas the matrix may be deformed, the pressure in the pores remains constant.

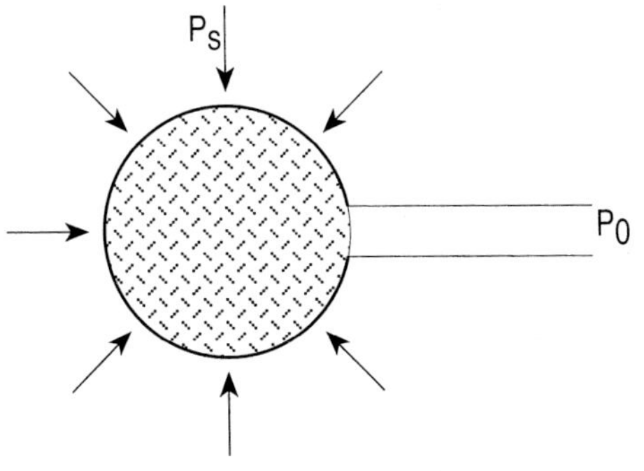

The strain-stress equations simplify to :

$$-P_s = (P - \frac{4}{3}N)\cdot \mathrm{div}\,\vec{u}_s + Q\cdot \mathrm{div}\,\vec{u}_f \quad ; \quad 0 = R\cdot \mathrm{div}\,\vec{u}_f + Q\cdot \mathrm{div}\,\vec{u}_s \tag{10}$$

After elimination of $\vec{u}_f$:

$$K_b = P - \frac{4}{3}N - \frac{Q^2}{R} \tag{11}$$

where $K_b = -P_s / \mathrm{div}\,\vec{u}_s$ is the bulk compression modulus of the matrix. (12)

Suppose further that a material sample is subjected (in a tank) to a hydrostatic pressure p_f . The forces acting on the matrix and the pores are $(1-\sigma)\cdot p_f$ and $\sigma\cdot p_f$, respectively. The strain-stress equations become :

$$-(1-\sigma)p_f = (P - \frac{4}{3}N)\mathrm{div}\,\vec{u}_s + Q\,\mathrm{div}\,\vec{u}_f \quad ; \quad -\sigma p_f = R\,\mathrm{div}\,\vec{u}_f + Q\,\mathrm{div}\,\vec{u}_s \tag{13}$$

With K_s the compression modulus of the compact matrix material, and K_f the compression modulus of the fluid in the pores, one gets:

$$P=\frac{(1-\sigma)(1-\sigma-K_b/K_s)K_s+(K_s/K_f)K_b}{1-\sigma-K_b/K_s+\sigma K_s/K_f}+\frac{4}{3}N$$
$$Q=\frac{(1-\sigma-K_b/K_s)\,\sigma K_s}{1-\sigma-K_b/K_s+\sigma K_s/K_f} \qquad (14)$$
$$R=\frac{\sigma^2 K_s}{1-\sigma-K_b/K_s+\sigma K_s/K_f}$$

In many materials $K_s \gg K_b$ and $K_s \gg K_f$; then the equations simplify to :

$$P=K_b+\frac{4}{3}N+\frac{(1-\sigma)^2}{\sigma}K_f \quad ; \quad Q=(1-\sigma)K_f \quad ; \quad R=\sigma K_f \qquad (15)$$

The compression modulus K_f of the fluid in the pores changes from isothermal compression at low frequencies (if the heat capacity and heat conduction of the matrix material are much larger than those of the fluid) to adiabatic compression at high frequencies. This transition is taken from the model of circular capillary pores (with the same fitting parameter c from above for other pore shapes) :

$$K_f=\rho_0 c_0^2/[1+(\kappa-1)J_{1,0}(c\,k_{\alpha 0}a)] \qquad (16)$$

(a= equivalent pore radius; κ= adiabatic exponent; $k_{\alpha 0}=\kappa\mathrm{Pr}\cdot k_v$; Pr= Prandtl number; k_v= wave number of viscous wave; $J_{1,0}(z)$ see the chapter "Duct Acoustics").

Wave equations :

The solution of the fundamental equations in principle consists of a triple of two longitudinal compressional waves and a transversal shear wave. The strain fields $\vec{u}_s, \vec{u}_f$ in the solid and the fluid are described by scalar and vector potentials :

$$\vec{u}_s=\mathrm{grad}\,\Phi+\mathrm{rot}\,\vec{H} \quad ; \quad \vec{u}_f=\mathrm{grad}\,\Psi+\mathrm{rot}\,\vec{G} \qquad (17)$$

Insertion in the fundamental equations gives :

$$\rho_{ss}\frac{\partial^2\Phi}{\partial t^2}+\rho_{sf}\frac{\partial^2\Psi}{\partial t^2}=P\cdot\mathrm{div\,grad}\,\Phi+Q\cdot\mathrm{div\,grad}\,\Psi+bF(\omega)\cdot\left(\frac{\partial\psi}{\partial t}-\frac{\partial\Phi}{\partial t}\right)$$
$$\rho_{ff}\frac{\partial^2\Psi}{\partial t^2}+\rho_{sf}\frac{\partial^2\Phi}{\partial t^2}=R\cdot\mathrm{div\,grad}\,\Psi+Q\cdot\mathrm{div\,grad}\,\Phi-bF(\omega)\cdot\left(\frac{\partial\Psi}{\partial t}-\frac{\partial\Phi}{\partial t}\right) \qquad (18)$$

and :

$$\rho_{ss}\frac{\partial^2\vec{H}}{\partial t^2}+\rho_{sf}\frac{\partial^2\vec{G}}{\partial t^2}=N\cdot\operatorname{grad}\operatorname{div}\vec{H}+bF(\omega)\cdot\left(\frac{\partial\vec{G}}{\partial t}-\frac{\partial\vec{H}}{\partial t}\right)$$
$$\rho_{ff}\frac{\partial^2\vec{G}}{\partial t^2}+\rho_{sf}\frac{\partial^2\vec{H}}{\partial t^2}=-bF(\omega)\cdot\left(\frac{\partial\vec{G}}{\partial t}-\frac{\partial\vec{H}}{\partial t}\right) \tag{19}$$

With a time factor $e^{j\omega t}$ and the abbreviations :

$$\tilde{\rho}_{ss}=\rho_{ss}-j\frac{bF(\omega)}{\omega}\quad;\quad\tilde{\rho}_{ff}=\rho_{ff}-j\frac{bF(\omega)}{\omega}\quad;\quad\tilde{\rho}_{sf}=\rho_{sf}+j\frac{bF(\omega)}{\omega}\quad;\quad\tilde{\chi}=\chi-j\frac{bF(\omega)}{\omega\sigma\rho_0} \tag{20}$$

one gets :

$$-\omega^2(\tilde{\rho}_{ss}\Phi+\tilde{\rho}_{sf}\Psi)=P\cdot\Delta\Phi+Q\cdot\Delta\Psi$$
$$-\omega^2(\tilde{\rho}_{ff}\Psi+\tilde{\rho}_{sf}\Phi)=R\cdot\Delta\Psi+Q\cdot\Delta\Phi \tag{21}$$

Elimination of $\Delta\Psi$ yields :

$$\Psi=\frac{(PR-Q^2)\cdot\Delta\Phi+\omega^2(\tilde{\rho}_{ss}R-\tilde{\rho}_{sf}Q)\cdot\Phi}{\omega^2(\tilde{\rho}_{ff}Q-\tilde{\rho}_{sf}R)}$$
$$(PR-Q^2)\cdot\Delta^2\Phi+\omega^2(\tilde{\rho}_{ss}R+\tilde{\rho}_{ff}P-2\tilde{\rho}_{sf}Q)\cdot\Delta\Phi+\omega^4(\tilde{\rho}_{ss}\tilde{\rho}_{ff}-\tilde{\rho}_{sf}^2)\cdot\Phi=0 \tag{22}$$

The last equation is formally interpreted as a product of two wave equations $(\Delta-k_1^2)\cdot(\Delta-k_2^2)\Phi=0$ with a solution $\Phi=\Phi_1+\Phi_2$ whereof the sum terms obey : (23)

$$(\Delta-k_1^2)\Phi_1=0\quad;\quad(\Delta-k_2^2)\Phi_2=0$$
$$k_{1,2}^2=\frac{\omega^2}{2(PR-Q^2)}[(\tilde{\rho}_{ss}R+\tilde{\rho}_{ff}P-2\tilde{\rho}_{sf}Q)\pm\sqrt{D}\,] \tag{24}$$
$$D=(\tilde{\rho}_{ss}R+\tilde{\rho}_{ff}P-2\tilde{\rho}_{sf}Q)^2-4(PR-Q^2)(\tilde{\rho}_{ss}\tilde{\rho}_{ff}-\tilde{\rho}_{sf}^2)$$

There are two compressional waves $\Phi_{1,2}$ in the solid with characteristic wave numbers $k_{1,2}$

The scalar potential Ψ for the sound wave in the fluid can be written as :

$$\Psi=\mu_1\Phi_1+\mu_2\Phi_2$$
$$\mu_{1,2}=\frac{\tilde{\rho}_{ss}R-\tilde{\rho}_{sf}Q-(PR-Q^2)k_{1,2}^2/\omega^2}{\tilde{\rho}_{ff}Q-\tilde{\rho}_{sf}R} \tag{25}$$

The vector potentials $\vec{G},\vec{H}$ can be derived similarly. One finds :

$$\vec{G}=-\frac{\tilde{\rho}_{sf}}{\tilde{\rho}_{ff}}\cdot\vec{H}=\mu_3\cdot\vec{H}\quad;\quad\Delta\vec{H}-\frac{\omega^2}{N}\left(\frac{\tilde{\rho}_{ss}\tilde{\rho}_{ff}-\tilde{\rho}_{sf}^2}{\tilde{\rho}_{ff}}\right)\cdot\vec{H}=\left(\Delta-k_3^2\right)\vec{H}=0$$
$$k_3^2=\frac{\omega^2}{N}\frac{\tilde{\rho}_{ss}\tilde{\rho}_{ff}-\tilde{\rho}_{sf}^2}{\tilde{\rho}_{ff}}\quad;\quad\mu_3=-\frac{\tilde{\rho}_{sf}}{\tilde{\rho}_{ff}} \tag{26}$$

Special case: weak coupling, densities of solid and fluid very different :

In many porous absorber materials the porosity is $\sigma \approx 1$; the tortuosity is $1 \le \chi \le 2$; the density of the solid material is $\tilde{\rho}_{ss} >> \tilde{\rho}_{ff}$ and also $\tilde{\rho}_{ss} >> \tilde{\rho}_{sf}$. Then BIOT's parameters simplify to :

$$P \approx K_b + \frac{4}{3}N \quad ; \quad Q \approx 0 \quad ; \quad R \approx K_f \tag{27}$$

the wave numbers of the compressional waves approximately become :

$$k_1^2 \approx \omega^2 [\frac{\tilde{\rho}_{ss}}{P} - \frac{\tilde{\rho}_{ss}\tilde{\rho}_{ff} - \tilde{\rho}_{sf}^2}{R\tilde{\rho}_{ss}}] \quad ; \quad k_2^2 \approx \omega^2 [\frac{\tilde{\rho}_{ff}}{R} - \frac{\tilde{\rho}_{sf}^2}{R\tilde{\rho}_{ss}}] \tag{28}$$

and the amplitude ratio of the shear wave to the compressional waves :

$$\mu_1 \approx -\frac{P}{R}\left(\frac{\tilde{\rho}_{ff}}{\tilde{\rho}_{sf}} - \frac{\tilde{\rho}_{sf}}{\tilde{\rho}_{ss}}\right) \quad ; \quad \mu_2 \approx -\frac{\tilde{\rho}_{ss}}{\tilde{\rho}_{sf}} + \frac{P}{R}\left(\frac{\tilde{\rho}_{ff}}{\tilde{\rho}_{sf}} - \frac{\tilde{\rho}_{sf}}{\tilde{\rho}_{ss}}\right) \tag{29}$$

The condition for this weak coupling is satisfied at frequencies above $f = \sigma^2 \Xi / 2\pi(1-\sigma)\rho_s$.

The effective densities under the conditions mentioned are :

$$\tilde{\rho}_{ss} \approx (1-\sigma)\rho_s - j\frac{bF(\omega)}{\omega} \quad ; \quad \tilde{\rho}_{ff} \approx \sigma\rho_0 - j\frac{bF(\omega)}{\omega} \quad ; \quad \tilde{\rho}_{sf} \approx j\frac{bF(\omega)}{\omega} \tag{30}$$

One gets for not too high flow resistivities Ξ :

$$k_1^2 \approx \omega^2 [\frac{\tilde{\rho}_{ss}}{P} + \frac{\tilde{\rho}_{sf}^2}{\tilde{\rho}_{ss}R - \tilde{\rho}_{ff}P}] \quad ; \quad k_2^2 \approx \omega^2 [\frac{\tilde{\rho}_{ff}}{R} - \frac{\tilde{\rho}_{sf}^2}{\tilde{\rho}_{ss}R - \tilde{\rho}_{ff}P}] \tag{31}$$

The amplitude ratios are then :

$$\begin{aligned} \mu_1 &\approx \frac{\tilde{\rho}_{ss} - P k_1^2/\omega^2}{-\tilde{\rho}_{sf}} \approx \frac{P\tilde{\rho}_{sf}}{\tilde{\rho}_{ss}R - \tilde{\rho}_{ff}P} \\ \mu_2 &\approx \frac{\tilde{\rho}_{ss} - P k_2^2/\omega^2}{-\tilde{\rho}_{sf}} \approx \frac{(\tilde{\rho}_{ss}R - \tilde{\rho}_{ff}P)^2 + PR\tilde{\rho}_{sf}}{(\tilde{\rho}_{ss}R - \tilde{\rho}_{ff}P)R\tilde{\rho}_{sf}} \end{aligned} \tag{32}$$

The wave number of the shear wave approximates the value for that wave in the evacuated matrix :

$$k_3^2 \approx \frac{\omega^2}{N}(1-\sigma)\rho_s \quad ; \quad \mu_3 \approx 0 \tag{33}$$

Input impedance of a porous layer with rigid backing (as an example) :

A plane wave is incident under a polar angle Θ on a plane layer of a porous material with thickness d . The graph shows schematically the three excited waves in the layer.

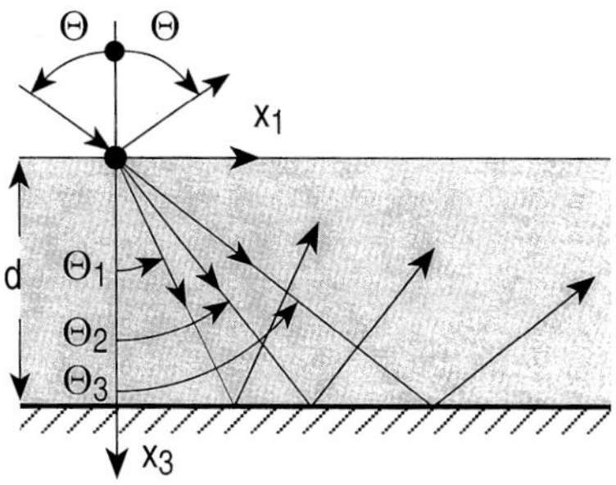

The three waves are formulated as three displacement potential functions; φ_3 is the non-vanishing component of the vector potential $\vec{H}$ of the shear wave :

in the forward direction :

$$\varphi_n = A_n \cdot e^{-jk_n(x_1 w_{n1} + x_3 w_{n3})} \quad ; \quad n = 1, 2, 3 \tag{34}$$

waves reflected at the hard wall :

$$\psi_n = B_n \cdot e^{-jk_n(x_1 w_{n1} - x_3 w_{n3})} \quad ; \quad n = 1, 2, 3 \tag{35}$$

From SNELL's law , with k_n= wave numbers of the BIOT waves :

$$k_0 \sin\Theta = w_{n1} k_n \quad ; \quad w_{n3} = \sqrt{1 - w_{n1}} \tag{36}$$

Let the layer be in tight contact with the hard wall at x_3=d ; thus

$$u_{s1}(x_1, d) = u_{s3}(x_1, d) = u_{f3}(x_1, d) = 0 \tag{37}$$

Special case of an open front side : i.e. no cover sheet on the absorber layer

Let p be the sound pressure, and v_3 be the particle velocity in the x_3 direction in front of the absorber; therefore (with the common factor in x_1 dropped and an unit amplitude of the incident wave supposed) :

$$v_3 = \frac{\cos\Theta}{Z_0}(1 - R) \quad ; \quad p = 1 + R \tag{38}$$

with the reflection factor R of the layer. Boundary conditions at the front side are :

$$v_3 = j\omega[(1-\sigma)u_{s3} + \sigma u_{f3}] \quad ; \quad -(1-\sigma)p = \tau_{33}^s \quad ; \quad -\sigma p = \tau_{33}^f \tag{39}$$

The field formulations, when inserted in the boundary conditions at the back and front sides, give the system of six equations :

$$\begin{aligned}
u_{s1} &= -jk_1 w_{11}(A_1 e^{-jk_1 w_{13} d} + B_1 e^{+jk_1 w_{13} d}) - \\
&\quad jk_2 w_{21}(A_2 e^{-jk_2 w_{23} d} + B_2 e^{+jk_2 w_{23} d}) + \\
&\quad jk_3 w_{33}(A_3 e^{-jk_3 w_{33} d} - B_3 e^{+jk_3 w_{33} d}) = 0 \\
u_{s3} &= -jk_1 w_{13}(A_1 e^{-jk_1 w_{13} d} - B_1 e^{+jk_1 w_{13} d}) - \\
&\quad jk_2 w_{23}(A_2 e^{-jk_2 w_{23} d} - B_2 e^{+jk_2 w_{23} d}) - \\
&\quad jk_3 w_{31}(A_3 e^{-jk_3 w_{33} d} + B_3 e^{+jk_3 w_{33} d}) = 0 \\
u_{f3} &= \mu_1[-jk_1 w_{13}(A_1 e^{-jk_1 w_{13} d} - B_1 e^{+jk_1 w_{13} d})] + \\
&\quad \mu_2[-jk_2 w_{23}(A_2 e^{-jk_2 w_{23} d} - B_2 e^{+jk_2 w_{23} d})] + \\
&\quad \mu_3[-jk_3 w_{31}(A_3 e^{-jk_3 w_{33} d} + B_3 e^{+jk_3 w_{33} d})] = 0 \\
\tau_{33}^s &= (P-2N)\,\mathrm{div}\,\vec{u}_s + Q\,\mathrm{div}\,\vec{u}_f + 2N\frac{\partial u_{s3}}{\partial x_3} \\
&= (P-2N+\mu_1 Q)(-k_1^2(A_1+B_1)) + \\
&\quad (P-2N+\mu_2 Q)(-k_2^2(A_2+B_2)) + \\
&\quad 2N[-k_1^2 w_{13}^2(A_1+B_1) - k_2^2 w_{23}^2(A_2+B_2) - k_3^2 w_{31} w_{33}(A_3-B_3)] \\
&= -(1-\sigma)p \\
\tau_{33}^f &= Q\,\mathrm{div}\,\vec{u}_s + R\,\mathrm{div}\,\vec{u}_f \\
&= (Q+\mu_1 R)(-k_1^2 w_{13}^2(A_1+B_1)) + \\
&\quad (Q+\mu_2 R)(-k_2^2 w_{23}^2(A_2+B_2)) \\
&= -\sigma p \\
\tau_{13}^s &= N\left(\frac{\partial u_{s1}}{\partial x_3} + \frac{\partial u_{s3}}{\partial x_1}\right) \\
&= N[-k_1^2 w_{11} w_{13}(A_1-B_1)) - k_2^2 w_{21} w_{23}(A_2-B_2)) - k_3^2(A_3+B_3)] = 0
\end{aligned} \tag{40}$$

Together with the relation for the reflection factor they are a system of seven equations for A_1, B_1, A_2, B_2, A_3, B_3 and R. After a numerical solution, the input impedance Z of the absorber layer can be evaluated from :

$$\frac{Z}{Z_0} = \frac{1}{\cos\Theta}\frac{1+R}{1-R} \tag{41}$$

Special case of an adhesive foil or membrane on the front side :

Let ρ be the surface mass density of the foil, S its bending stiffness, and T its tension in the case of a membrane. Let P_1 be immediately in front of the absorber, P_2 a point in the surface of the absorber, and P_3 a point immediately behind the front side of the absorber. The equations of motion of the foil are :

$$-\omega^2\rho\, u_3(P_2) = \tau_{33}^{s}(P_3) + \tau_{33}^{f}(P_3) - \tau_{33}^{f}(P_1) + T\frac{\partial^2 u_3(P_2)}{\partial x_3^2}$$
$$-\omega^2\rho\, u_1(P_2) = \tau_{13}^{s}(P_3) + S\frac{\partial^2 u_1(P_2)}{\partial x_3^2} \qquad (42)$$

With $\tau_{33}^{f}(P_1) = -p$ these equations can be transformed to :

$$-\omega^2\rho_1\, u_3(P_2) = \tau_{33}^{s}(P_3) + \tau_{33}^{f}(P_3) - \tau_{33}^{f}(P_1)$$
$$-\omega^2\rho_2\, u_1(P_2) = \tau_{13}^{s}(P_3) \qquad (43)$$

with :

$$\rho_1 = \rho - T k_0^2 \sin^2\Theta/\omega^2 \quad ; \quad \rho_2 = \rho - S k_0^2 \sin^2\Theta/\omega^2 \qquad (44)$$

The boundary conditions at the front side now become :

$$u_{s3}(P_3) = u_{f3}(P_3) = u_3(P_2) = u_3(P_1) \quad ; \quad u_{s1}(P_3) = u_1(P_2) \qquad (45)$$

Together with the boundary conditions at the back side one has seven equations ($(u_i)_n$ is the component in the direction x_n of u_i) :

$$(j\omega\rho_1 - Z)\cdot j\omega u_3(P_2) - \tau_{33}^{s}(P_3) - \tau_{33}^{f}(P_3) = 0$$
$$-\omega^2\rho_2\cdot (u_1)_1(P_3) - \tau_{13}^{s}(P_3) = 0$$
$$(u_1)_3(P_2) = (u_2)_3(P_2) \qquad (46)$$
$$(u_1)_1 = 0 \quad ; \quad (u_2)_3 = 0 \quad ; \quad (u_1)_3 = 0 \quad \text{at} \quad x_3 = d$$
$$Z = \frac{-\tau_{33}^{f}(P_1)}{j\omega\, u_3(P_1)}$$

The numerical solution of the system of equations, after insertion of the field formulations, gives the amplitudes of the wave components and the input impedance Z .

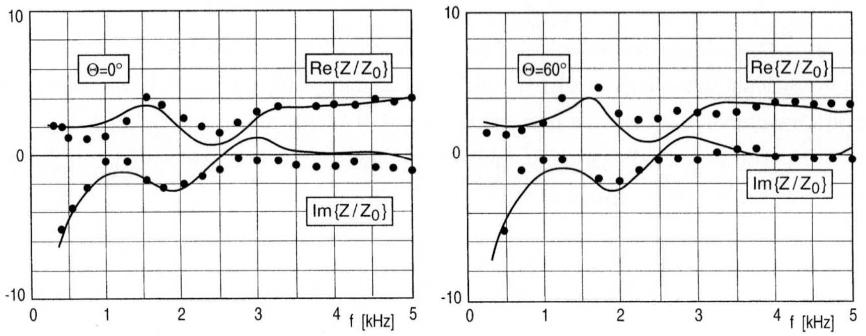

Measured (points) and evaluated (curves) components of the input impedance Z of an open layer of PU foam, d=2 [cm] thick.

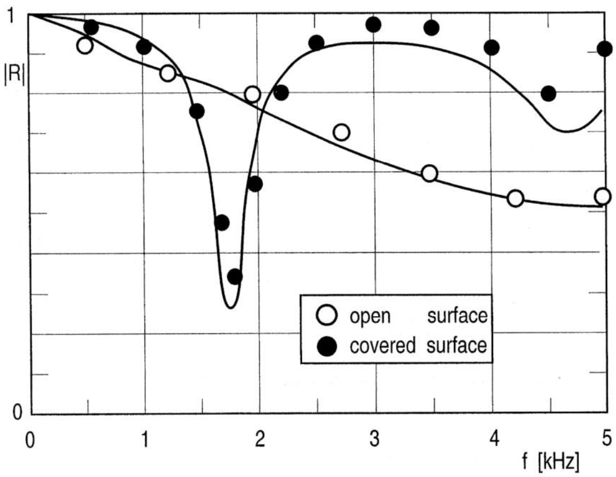

Measured (points) and evaluated (curves) magnitude of the reflection factor R of a d=2 [cm] thick PU foam, once with open surface, once with a cover foil.

G.11 Empirical relations for characteristic values of fibre absorbers

Ref. Mechel, [G.1] ; Mechel, Grundmann, [G.3]

For a great number of glass fibre and mineral fibre absorber material from different producers and with a wide range of bulk densities the characteristic values, i.e. the propagation constant Γ_a and wave impedance Z_a , as well as the flow resistivity Ξ were carefully measured in [G.3]. The materials could be subdivided (from an acoustical point of view) into three product groups:

- glass fibre products,
- mineral fibre products (rockwool),
- basalt wool products.

This section shows experimental values for the components of the normalised characteristic values $\Gamma_{an}=\Gamma_a/k_0=\Gamma'_{an}+j\cdot\Gamma''_{an}$ and $Z_{an}= Z_{an}/Z_0= Z'_{an}+j\cdot Z''_{an}$ over the "absorber variable" $E=\rho_0 f/\Xi$ (ρ_0= density of air ; f= frequency ; Ξ= flow resistivity) . It is advantageous (according to a proposal by Delany-Bazley) to plot Γ'_{an} , $\Gamma''_{an}-1$, $Z'_{an}-1$, $-Z''_{an}$ as functions of E , because the experimental data then group around simple curves, and to derive empirical relations for these quantities by regressions through the data.

The following table gives for the three product groups average values of

- the fibre diameter d, • the distribution parameter Λ of a Poisson distribution of the diameters, to which the empirical distribution can be best matched,
- the shot content (per weight, for shot with diameters >100 [µm]),
- the content of organic binder (per weight).

Product group	**Average fibre diameter** d µm	**Distribution parameter** Λ (Δd=1 µm)	**Shot content** % (> 100 µm Ø)	**Binder content** %
Glass fibre	5,2	5,3	< 1	4,8
Basalt wool	3,9	4,0	32,4	1,0
Mineral fibre	4,0	2,8	27,1	2,1

The following diagrams show (for three product groups) experimental points, thick full curves from a floating average, thin dashed lines from the Delany-Bazley formula (see below).

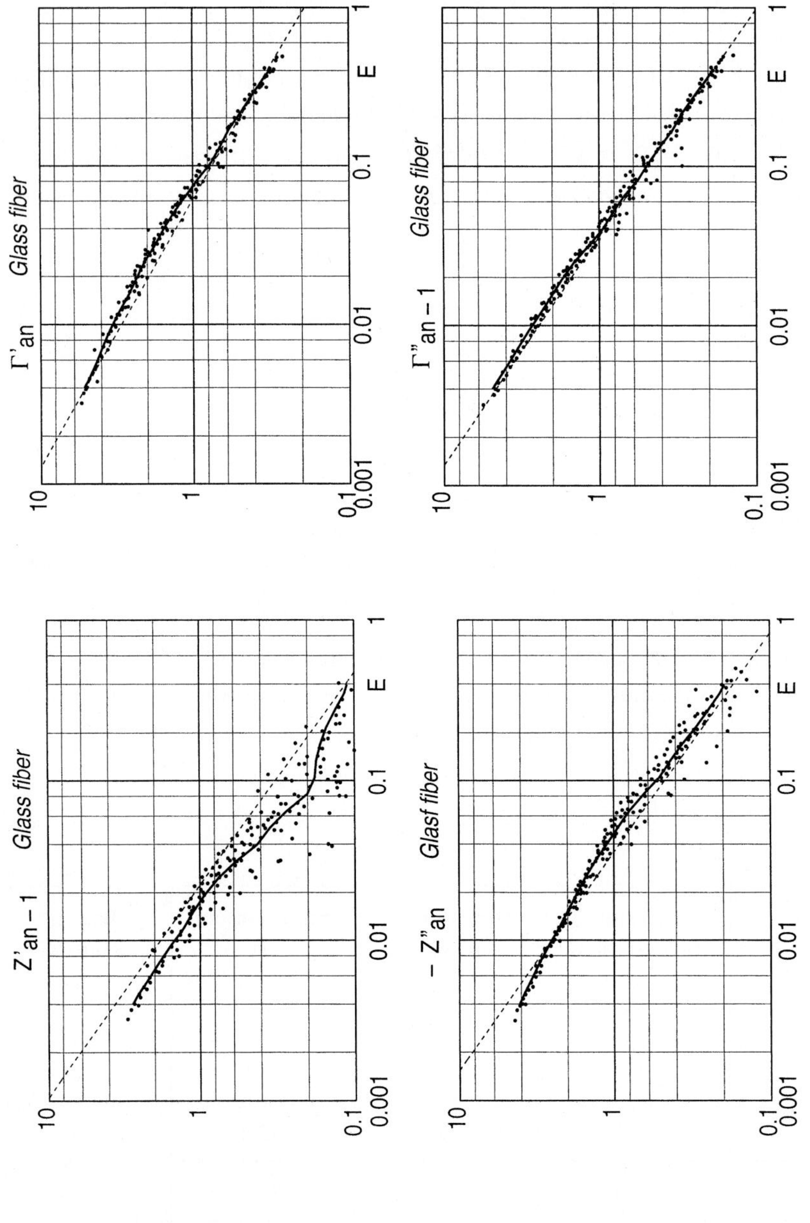

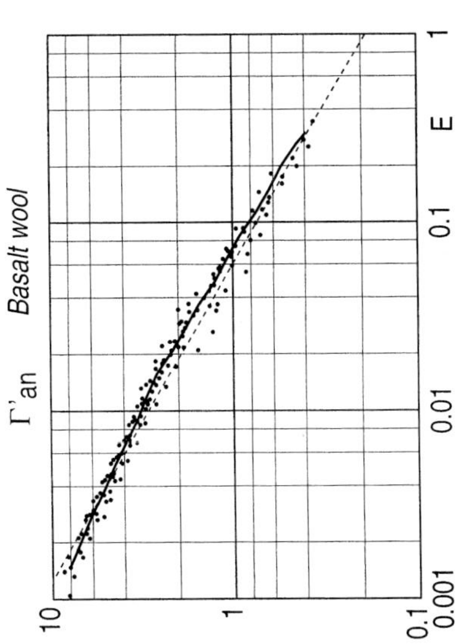
Γ'an
Basalt wool
E
0.001
0.01
0.1
1
10
1
0.1

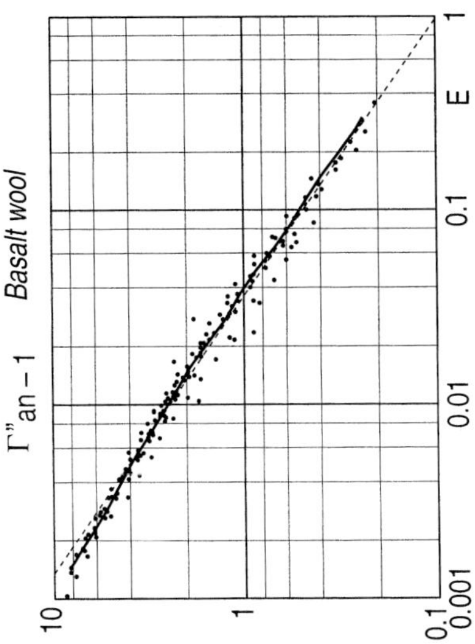
Γ"an − 1
Basalt wool
E
0.001
0.01
0.1
1
10
1
0.1

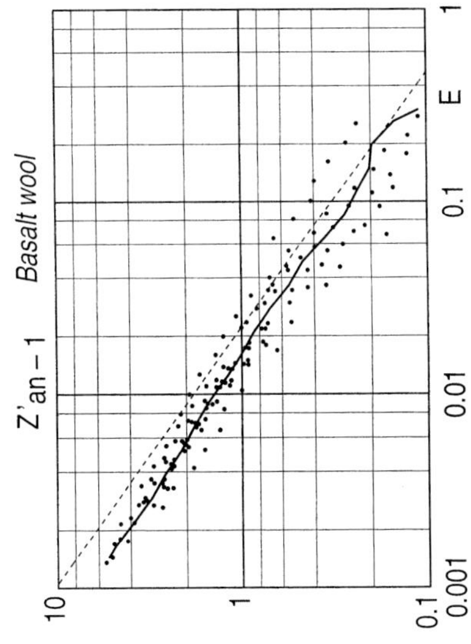
Z'an − 1
Basalt wool
E
0.001
0.01
0.1
1
10
1
0.1

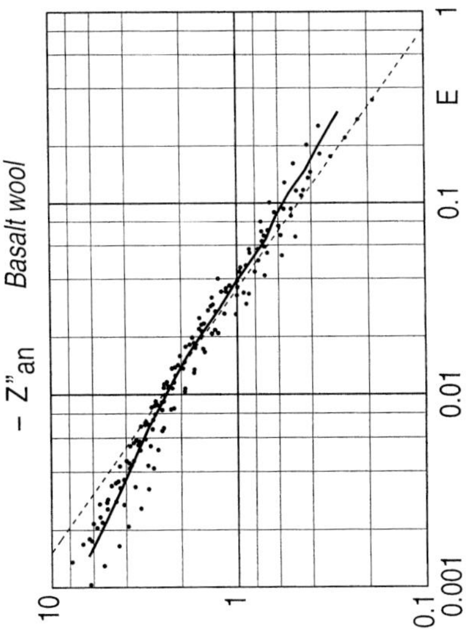
− Z"an
Basalt wool
E
0.001
0.01
0.1
1
10
1
0.1

Coefficients	Γ_{an} Mineral fibre / Glass fibre	Z_{an} Mineral fibre / Glass fibre
$ß_{-1}$	$-0.003\,557\,57 - j\,0.000\,016\,489\,7$ $-0.004\,518\,36 + j\,0.000\,541\,333$	$0.002\,678\,6 + j\,0.003\,857\,61$ $-0.001\,713\,87 + j\,0.001\,194\,89$
$ß_{-1/2}$	$0.421\,329 + j\,0.342\,011$ $0.421\,987 + j\,0.376\,270$	$0.135\,298 - j\,0.394\,160$ $0.283\,876 - j\,0.292\,168$
$ß_0$	$-0.507\,733 + j\,0.086\,655$ $-0.383\,809 - j\,0.353\,780$	$0.946\,702 + j\,1.476\,53$ $-0.463\,860 + j\,0.188\,081$
$ß_{1/2}$	$-0.142\,339 + j\,1.259\,86$ $-0.610\,867 + j\,2.599\,22$	$-1.452\,02 - j\,4.562\,33$ $3.127\,36 + j\,0.941\,600$
$ß_1$	$1.290\,48 - j\,0.082\,0811$ $1.133\,41 - j\,1.,748\,19$	$4.031\,71 + j\,7.560\,31$ $-2.109\,2\,0 - j\,1.323\,98$
$ß_{3/2}$	$-0.771\,857 - j\,0.668\,050$ 0	$-2.869\,93 - j\,4.904\,37$ 0

Table 3: Coefficients for the relations ().

The regression is good in the "range of definition" $0.003 \le E \le 0.4$, but, as with all regressions, the errors become large if this range is exceeded.

G.12 Characteristic values from theoretical models fitted to experimental data

Simple numerical regressions through experimental data may return nonsense results, if the range of definition of the regression is aborted. Some extension beyond that range is possible, if theoretical models for porous materials are matched to experimental data by suitably fitting the parameters of the model theory, because physical aspects can be taken into account when the parameters are fitted. The model theory must be simple enough to do that.

The independent variable is $E = \rho_0 f / \Xi$

(ρ_0= density of air ; f= frequency ; Ξ= flow resistivity)

Theory of the quasi-homogeneous material fitted to experimental data :
(see section G.2)

The characteristic values are evaluated from :

$$\Gamma_{an} = \frac{\Gamma}{k_0} = j\sqrt{\frac{\rho_{eff}}{\rho_0}\cdot\frac{C_{eff}}{C_0}} \quad ; \quad Z_{an} = \frac{Z_a}{Z_0} = \frac{1}{\sigma}\sqrt{\frac{\rho_{eff}}{\rho_0} / \frac{C_{eff}}{C_0}}$$

$$\frac{\rho_{eff}}{\rho_0} = \chi - j\frac{\sigma\cdot g(E)}{2\pi E} \quad ; \quad \frac{C_{eff}}{C_0} = \frac{\kappa + \alpha_1 \cdot jE/E_0}{1 + jE/E_0} \tag{1}$$

$$g(E) = \gamma_0 + \gamma_1\, E + \gamma_2\, E^2$$

with fitted parameters for three product groups; $\sigma = 1-\mu$ = porosity:

Parameters	**Glass fibre**	**Basalt wool**	**Rockwool**
χ	1.3	1.3	1.3
μ	0.02	0.02	0.02
κ	1.40 + 0.15 j	1.40 + 0.10 j	1.40 + 0.15 j
α_1	1	1.1	1.1
E_0	0.125	0.10 (0.07 for Z_{an})	0.125
γ_0	1.098 72	0.976 206	1.121 40
γ_1	0.333 239	2.184 74	1.499 53
γ_2	1.626 42	−1.262 75	0.468 552

The next diagram shows, as an example, the normalised propagation constant in glass fibre materials; the points are floating averages over experimental data, the thick, full lines come from the fitted theory, the thin dashed curves belong to the original theory.

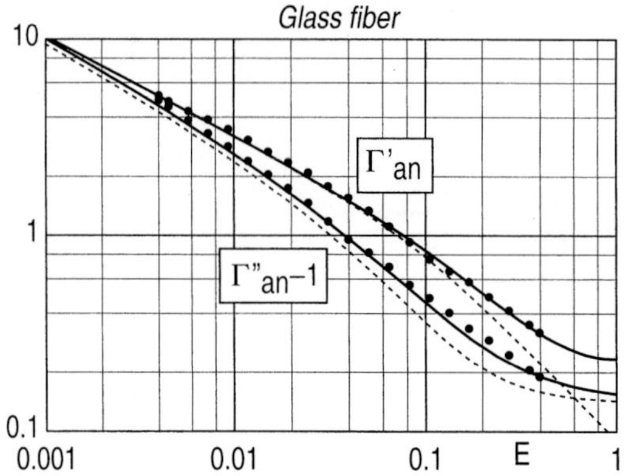

Components of the normalised propagation constant in glass fibre materials;
points: floating average of experimental data,
thick lines: fitted theory,
thin lines: original theory.

Fitted model of flat capillaries :
(see section G.4)

The characteristic values are evaluated from :

$$\frac{\Gamma}{k_0} = \Gamma_{an} = j\sqrt{\frac{\rho_{eff}}{\rho_0}\cdot\frac{C_{eff}}{C_0}} \quad ; \quad \frac{Z_a}{Z_0} = Z_{an} = \frac{1}{\sigma}\sqrt{\frac{\rho_{eff}}{\rho_0} / \frac{C_{eff}}{C_0}}$$

$$\frac{\rho_{eff}}{\rho_0} = \frac{1}{1 - \dfrac{\tan\sqrt{-6\pi j\, a_1\, E}}{\sqrt{-6\pi j\, a_1\, E}}} \quad ; \quad \frac{C_{eff}}{C_0} = 1 + (\kappa - 1)\,\frac{\tan\sqrt{-6\pi j\, \kappa\, Pr\, a_2\, E}}{\sqrt{-6\pi j\, \kappa\, Pr\, a_2\, E}} \qquad (2)$$

with Pr= Prandtl number ; $\sigma = 1-\mu$= porosity and fitted parameters for three product groups:

Material	**Char. Value**	μ	κ	a_1	a_2
Glass fibre	Γ_{an}	–	1.60 + 0.1 j	1	1
	Z_{an}	0.05	1.40 + 0.1 j	1	1.5
Basalt wool	Γ_{an}	–	1.40 + 0.15 j	1	1
	Z_{an}	0.05	1.40 + 0.1 j	0.9	1.7
Rockwool	Γ_{an}	–	1.70 + 0.1 j	1	0.6
	Z_{an}	0,05	1.40 + 0.1 j	0.9	1.7

The diagrams show the characteristic values for rockwool fibre materials (points: from floating averages over experimental data; curves: from fitted capillary model).

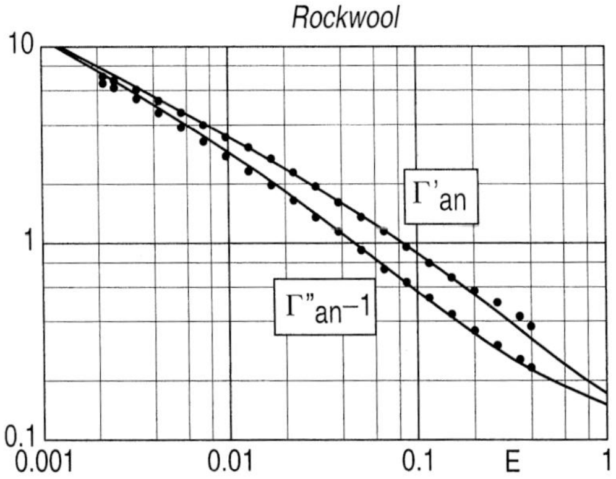

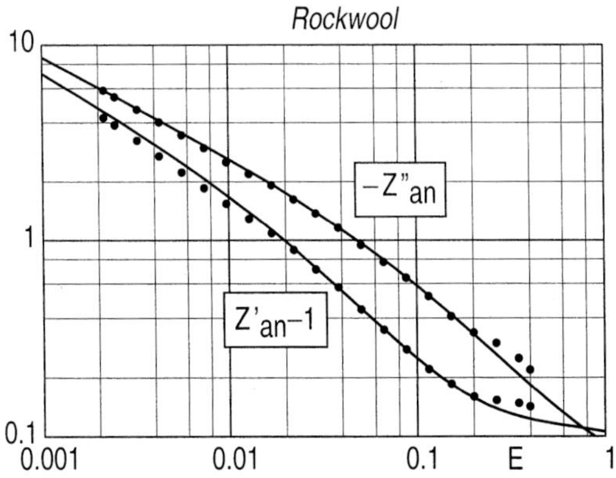

References to part G :

Porous Absorbers

[G.1] MECHEL, F.P.
"Schallabsorber", Vol. II,
S.Hirzel Verlag, Stuttgart, 1995

[G.2] TOLSTOY, I., Editor
"Acoustics, Elasticity, and Thermodynamics of Porous Media;
Twenty-one Papers by M.A. Biot"
Acoustical Society of America, Amer.Inst.of Physics, N.Y., 1992

[G.3] MECHEL, F.P., GRUNDMANN, R.
"Akustische Kennwerte von Faserabsorbern", Vol. I, Bericht BS 85/83, 1983;
"Materialdaten", Vol. II, Bericht BS 75/82, 1982
Berichte des Fraunhofer-Instituts für Bauphysik, Stuttgart

H
Compound Absorbers

Sound absorbers, except simple porous layers, are compound absorbers, i.e. they consist of elements in special arrangements, such as air volumes, foils (either limp or elastic, either tight or porous), membranes, plates (either stiff or elastic, either tight or porous), mostly with perforations ("necks") in the shape of e.g. slits or circular holes, porous absorber layers, etc. The aim mostly is to evaluate the input admittance G of such absorbers, or impedance $Z=1/G$, because it is this quantity with which absorbers enter into acoustical computations.

Many compound absorbers, in turn, are arrays of elementary absorbers, such as arrays of Helmholtz resonators, and they have an inhomogeneous surface, e.g. the neck areas of Helmholtz resonators and the hard plate between the necks. One must distinguish what the input admittance stands for, either for the neck area or for the whole array. If the lateral dimensions of an array element are small compared to the wavelength λ_0 (typically $< \lambda_0/4$), the performance of an absorber in most applications can be equivalently described by an average admittance (or "homogenised" admittance), which is the average of the local admittances in an array. Otherwise the array must be treated as a periodic structure with the admittance profile along the surface explicitly taken into account.
A further distinction concerns the radiation impedance $Z_r=Z'_r+j\cdot Z''_r$ of the absorber, or more distinctly the radiation resistance Z'_r, whether it is included in the absorber impedance Z, or not. This distinction comes from the general equivalent network of a source and an absorber.

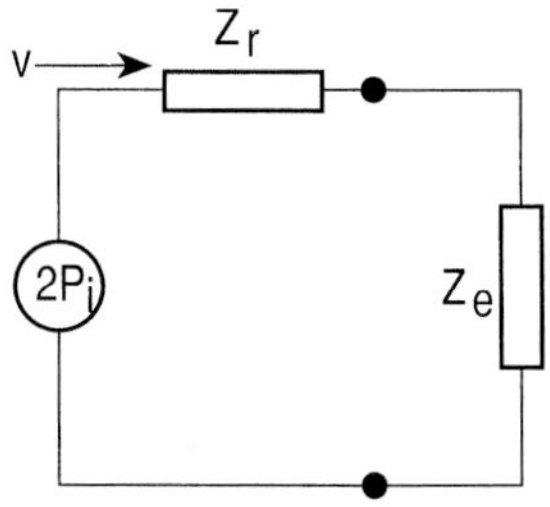

The network consists of a pressure source, with P_i the sound pressure of the incident wave and the internal source impedance Z_r, and the absorber with input impedance Z_e. For a plane wave with polar angle Θ of incidence the radiation resistance is

$$Z'_r = Z_0 / \cos\Theta$$

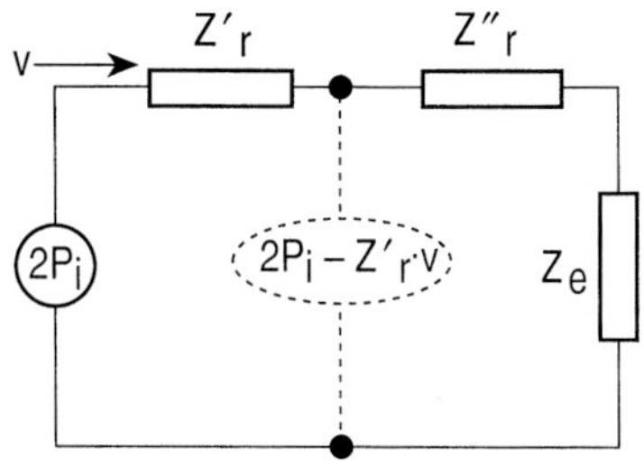

It does not contain any information about the absorber. The radiation reactance, however, contains the oscillating mass of necks and therefore influences the tuning of resonators.

So it makes sense to attribute Z_r' to the source and Z_r'' to the absorber.

It should be noted, that this attribution is a matter of convention, and therefore one must examine absorber formulas for the convention used.

It will be indicated at the end of some sections in the present chapter about absorber elements how the absorber element is introduced into the equivalent chain network of a multi-layer absorber. Most technical sound absorbers can be described by such an equivalent network (see section C.5). Some sections below will mainly give chains of equations which lead finally to the desired input admittance G or input impedance Z by iterated insertion.

H.1 Absorber of flat capillaries

Ref.: Mechel, [H.1]

See section J.1 for sound in flat capillaries.

A plane sound wave is incident on a layer of thickness d (with hard backing) consisting of thin plates with mutual distance 2h . The arrangement has a surface porosity σ .

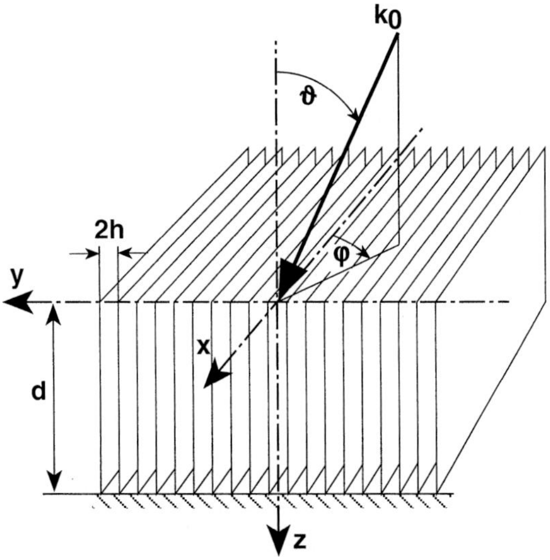

The plates are first assumed to be normal to the back wall.

Viscous and thermal losses are considered in the capillaries between the plates.

With reflection factor r at the front side, and transmission factor t from outside to inside of the capillaries :

Sound wave in front of the absorber :

$$p = p_e + p_r = A \cdot e^{-j(k_x x + k_y y)} (e^{-jk_z z} + r\, e^{+jk_z z})$$
$$k_x = k_0 \sin\vartheta \cos\varphi \quad ; \quad k_y = k_0 \sin\vartheta \sin\varphi \quad ; \quad k_z = k_0 \cos\vartheta \qquad (1)$$

Sound wave inside the capillaries :

$$p_a = A \cdot t \cdot e^{-\Gamma_{ax} x} \cdot \cosh(\Gamma_{az}(z-d)) \cdot e^{-jk_y y}$$

$$(\Delta - \Gamma_a^2) p_a = 0 \quad ; \quad \Gamma_{ax}^2 + \Gamma_{az}^2 = \Gamma_a^2$$

$$\Gamma_{ax} = jk_x = jk_0 \sin\vartheta \cos\varphi \quad ; \quad \frac{\Gamma_{az}}{k_0} = \sqrt{\left(\frac{\Gamma_a}{k_0}\right)^2 - \sin^2\vartheta \cos^2\varphi} \tag{2}$$

$$v_{az} = \frac{-1}{\Gamma_a Z_a} \frac{\partial p_a}{\partial z} \quad ; \quad v_{ax} \neq 0 \quad ; \quad v_{ay} = 0$$

with Γ_a= propagation constant in a flat capillary; $Z_a = Z_i/\sigma$; Z_i= wave impedance in a flat capillary (see section J.1).

Input impedance :

$$\frac{Z}{Z_0} = \frac{Z_a}{Z_0} \frac{\Gamma_a / k_0}{\Gamma_{az} / k_0} \coth(k_0 d \cdot \Gamma_{az} / k_0)$$

$$\xrightarrow[k_0 d \ll 1]{} \frac{1}{k_0 d} \frac{Z_a/Z_0 \cdot \Gamma_a/k_0}{(\Gamma_a/k_0)^2 - \sin^2\vartheta \cos^2\varphi} = \begin{cases} \dfrac{1}{k_0 d} \dfrac{Z_a/Z_0}{\Gamma_a/k_0} & ; \quad \vartheta = 0 \text{ or } \varphi = \pm\pi/2 \\ \dfrac{1}{k_0 d} \dfrac{Z_a/Z_0 \cdot \Gamma_a/k_0}{(\Gamma_a/k_0)^2 - \sin^2\vartheta} & ; \quad \varphi = 0 \end{cases} \tag{3}$$

$$\xrightarrow[k_0 d \gg 1]{} \frac{Z_a/Z_0 \cdot \Gamma_a/k_0}{\sqrt{(\Gamma_a/k_0)^2 - \sin^2\vartheta \cos^2\varphi}} = \begin{cases} Z_a/Z_0 & ; \quad \vartheta = 0 \text{ or } \varphi = \pm\pi/2 \\ \dfrac{Z_a/Z_0 \cdot \Gamma_a/k_0}{\sqrt{(\Gamma_a/k_0)^2 - \sin^2\vartheta}} & ; \quad \varphi = 0 \end{cases}$$

Reflection factor r and absorption coefficient α as usual :

$$r = \frac{\cos\vartheta \dfrac{Z}{Z_0} - 1}{\cos\vartheta \dfrac{Z}{Z_0} + 1} \quad ; \quad \alpha(\vartheta,\varphi) = 1 - |r|^2$$

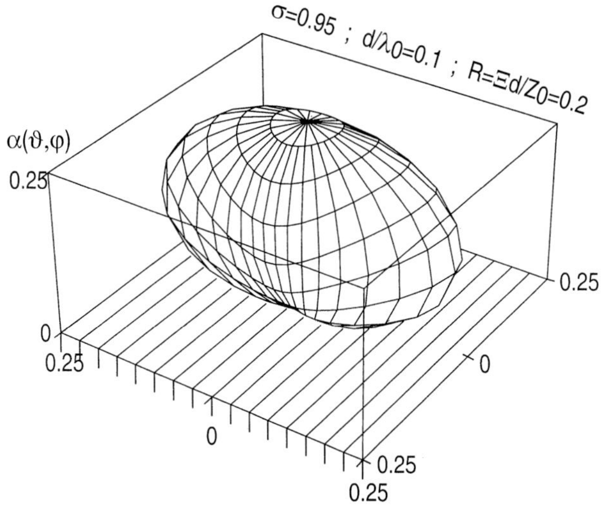

Absorption coefficient $\alpha(\vartheta,\varphi)$ as a function of the direction of sound incidence.
The lamellae distance 2h is given by the normalised flow resistance R of the arrangement.

Next, the lamellae are assumed to be inclined with ϑ_0 .

The effective depth changes to :

$$d_{eff} = d / \cos\vartheta_0$$

the outside wave impedance changes to

$$Z_0 = Z_0 / \cos\vartheta_0$$

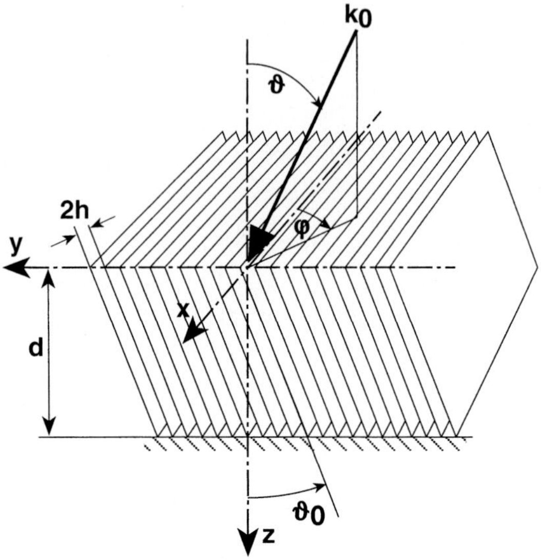

Thus : (4)

$$\frac{Z(\vartheta_0 = 0, d)}{Z_0} \to \frac{1}{\cos\vartheta_0} \frac{Z(\vartheta_0 = 0, d_{eff})}{Z_0}$$

and : (5)

$$r = \frac{Z/Z_0 - 1}{Z/Z_0 + 1} \to \frac{\dfrac{Z(0,d_{eff})}{Z_0} \dfrac{\cos\vartheta}{\cos\vartheta_0} - 1}{\dfrac{Z(0,d_{eff})}{Z_0} \dfrac{\cos\vartheta}{\cos\vartheta_0} + 1}$$

The inclination of the lamellae (indicated in the next diagram) has no immediate influence on the sound absorption $\alpha(\vartheta,\varphi)$ as a function of angles of incidence :

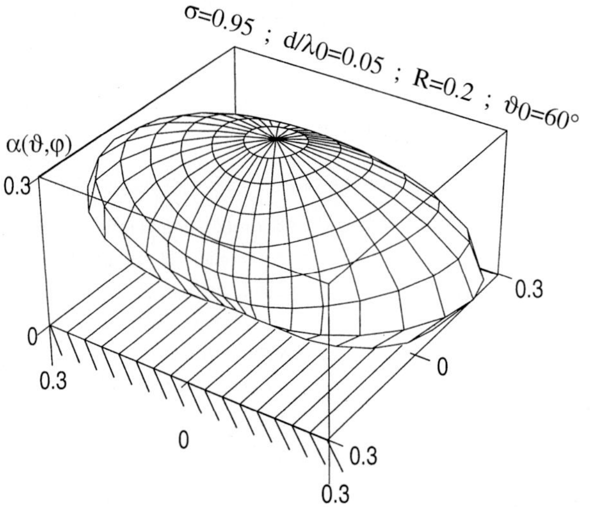

Absorption coefficient $\alpha(\vartheta,\varphi)$ as a function of the direction of sound incidence, with inclined lamellae.
The lamellae distance 2h is given by the normalised flow resistance R of the arrangement.

H.2 Plate with narrow slits

Ref. Mechel, [H.2]

"Narrow" slits mean that only plane waves are considered in the neck channels (in contrast to "wide" slits in the next section, where higher modes are assumed in the necks), but they are still wide enough, so that viscous and thermal losses in the necks may be neglected.

Consider an array of parallel slits of width a and mutual distance L in a (rigid) plate of thickness d .

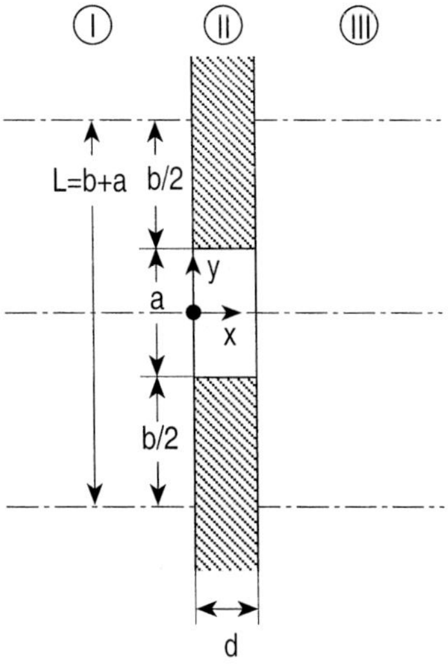

Excitation is by a plane wave with normal incidence and amplitude A_e .

There are three sound zones I, II, III.

Field formulation in zone II :

$$p_{II}(x,y) = Be^{-jk_0x} + Ce^{jk_0x}$$
$$v_{IIx}(x,y) = \frac{1}{Z_0}[Be^{-jk_0x} - Ce^{jk_0x}] \qquad (1)$$
$$v_{IIy}(x,y) = 0$$

Field in zone I in front of the plate :

$$p_I(x,y) = A_e e^{-jk_0x} + A_0 e^{jk_0x} + 2\sum_{n>0} A_n\, e^{\gamma_n x}\cos(\eta_n y)$$
$$v_{Ix}(x,y) = \frac{1}{Z_0}[A_e e^{-jk_0x} - A_0 e^{jk_0x} + 2j\sum_{n>0} A_n \frac{\gamma_n}{k_0} e^{\gamma_n x}\cos(\eta_n y)] \qquad (2)$$
$$v_{Iy}(x,y) = -\frac{2j}{Z_0}\sum_{n>0} A_n \frac{\eta_n}{k_0} e^{\gamma_n x}\sin(\eta_n y)$$

Field in zone III behind the plate :

$$p_{III}(x,y) = D_0 e^{-jk_0x} + 2\sum_{n>0} D_n\, e^{-\gamma_n x}\cos(\eta_n y)$$
$$v_{IIIx}(x,y) = \frac{1}{Z_0}[D_0 e^{-jk_0x} - 2j\sum_{n>0} D_n \frac{\gamma_n}{k_0} e^{-\gamma_n x}\cos(\eta_n y)] \qquad (3)$$
$$v_{IIIy}(x,y) = -\frac{2j}{Z_0}\sum_{n>0} D_n \frac{\eta_n}{k_0} e^{-\gamma_n x}\sin(\eta_n y)$$

Wave numbers and propagation constants :

$$\eta_0 = 0 \quad ; \quad \eta_n = \frac{2\pi n}{L} = k_0 \cdot n\frac{\lambda_0}{L} \quad ; \quad n = 1,2,\ldots$$

$$\gamma_0 = jk_0 \; ; \quad \gamma_n = \sqrt{\eta_n^2 - k_0^2} = k_0\sqrt{(n\frac{\lambda_0}{L})^2 - 1} \quad ; \quad n = 1,2,\ldots \qquad (4)$$

From particle velocity boundary conditions follows :

$$A_0 = A_e - \frac{a}{L}\cdot(B - C)$$

$$A_n = -j\frac{a}{L}\frac{k_0}{\gamma_n}\,si_n \cdot (B - C) \qquad (5)$$

$$D_n = j\frac{a}{L}\frac{k_0}{\gamma_n}\,si_n\, e^{\gamma_n d}\cdot(B e^{-jk_0 d} - C e^{+jk_0 d}) \quad ; \quad n = 0, 1, 2,\ldots$$

with $\quad si_n = \frac{\sin(n\pi a/L)}{n\pi a/L}; \quad si_0 = 1$

From matching average sound pressures in the slit orifices follows :

$$\frac{B}{A_e} = \frac{(1+S)\,e^{+jk_0 d}}{2S\cos(k_0 d) + j(1+S^2)\sin(k_0 d)}$$

$$\frac{C}{A_e} = \frac{-(1-S)\,e^{-jk_0 d}}{2S\cos(k_0 d) + j(1+S^2)\sin(k_0 d)} \qquad (6)$$

$$\frac{B-C}{A_e} = 2\frac{\cos(k_0 d) + jS\sin(k_0 d)}{2S\cos(k_0 d) + j(1+S^2)\sin(k_0 d)}$$

$$\frac{B\,e^{-jk_0 d} - C\,e^{+jk_0 d}}{A_e} = \frac{2}{2S\cos(k_0 d) + j(1+S^2)\sin(k_0 d)} \qquad (7)$$

with the abbreviation : $S = \frac{a}{L}[1 + 2j\sum_{i>0}\frac{k_0}{\gamma_i}s_i^2]$ (8)

Front side orifice impedance Z_{sf} :

$$\frac{Z_{sf}}{Z_0} = \frac{\langle p_{II}(0,y)\rangle}{Z_0\langle v_{IIx}(0,y)\rangle} = \frac{B+C}{B-C} = \frac{S + j\tan(k_0 d)}{1 + jS\cdot\tan(k_0 d)} \qquad (9)$$

The last expression has the typical form of a (normalised) impedance (here S) which is transformed by a transmission line of length d .

Back side orifice impedance Z_{sb} :

$$\frac{Z_{sb}}{Z_0} = \frac{\langle p_{II}(d,y)\rangle}{Z_0\langle v_{IIx}(d,y)\rangle} = \frac{Be^{-jk_0 d} + Ce^{+jk_0 d}}{Be^{-jk_0 d} - Ce^{+jk_0 d}} = S \qquad (10)$$

$$\frac{Z_{sb}}{Z_0} = \frac{a}{L}[1 + 2j\sum_{n>0} \frac{1}{\sqrt{(n\frac{\lambda_0}{L})^2 - 1}} (\frac{\sin(n\pi\, a/L)}{n\pi\, a/L})^2] \tag{11}$$

The 1st term in the brackets represents the radiation resistance.

If one subtracts in the front side orifice the sound pressure of the equivalent source $2P_i = 2A_e$:

$$\frac{\langle p_{II}(0,y) - 2A_e \rangle}{Z_0 \langle v_{IIx}(0,y) \rangle} = \frac{B + C - 2A_e}{B - C} = -S \tag{12}$$

i.e. the orifice impedances on both sides are symmetrical.

End correction of the orifice :

$$\frac{\Delta\ell}{a} = \frac{Z''_{sb}}{k_0 a\, Z_0} = \frac{1}{k_0 a} S'' \approx \frac{1}{\pi}\sum_{n=1}^{\infty} \frac{S_n^2}{n} = \frac{a}{L} \cdot \sum_{n=1}^{\infty} \frac{\sin^2(n\pi a/L)}{(n\pi a/L)^3} \tag{13}$$

If the summation is approximated by an integration :

$$\frac{\Delta\ell}{a} \approx \frac{1}{\pi} \ln[\frac{1}{2}\tan(\frac{\pi}{4}\frac{a}{L}) + \frac{1}{2}\cot(\frac{\pi}{4}\frac{a}{L})] \tag{14}$$

and from a numerical regression :

$$x = \lg(a/L)$$

$$\frac{\Delta\ell}{a} = -0.395450 \cdot x + 0.346161 \cdot x^2 + 0.141928 \cdot x^3 + 0.0200128 \cdot x^4 \tag{15}$$

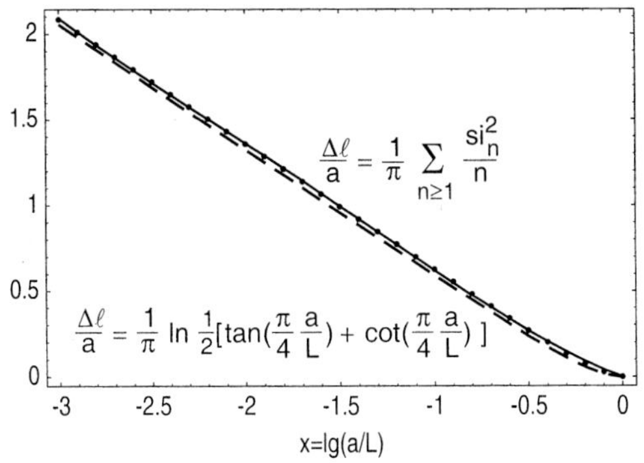

End correction of a slit in an array;
points: summation;
full line: regression;
dashed line: integration.

H.3 Plate with wide slits

Ref. Mechel, [H.2]

See section H.2 for the arrangement, co-ordinates and field zones.
In contrast to section H.2 , higher modes are assumed in the neck channels. The field formulations in the three zones are :

Zone I :

$$p_I(x,y) = A_e e^{-jk_0x} + \sum_{n=0} \delta_n A_n \, e^{\gamma_n x} \cos(\eta_n y)$$

$$Z_0 v_{Ix}(x,y) = A_e e^{-jk_0x} + j\sum_{n=0} \delta_n A_n \frac{\gamma_n}{k_0} e^{\gamma_n x} \cos(\eta_n y) \qquad (1)$$

$$Z_0 v_{Iy}(x,y) = -2j\sum_{n>0} A_n \frac{\eta_n}{k_0} e^{\gamma_n x} \sin(\eta_n y)$$

Zone II :

$$p_{II}(x,y) = \sum_{n=0} [B_n e^{-j\kappa_n x} + C_n e^{j\kappa_n x}] \cos(\varepsilon_n y)$$

$$Z_0 v_{IIx}(x,y) = \sum_{n=0} \frac{\kappa_n}{k_0} [B_n e^{-j\kappa_n x} - C_n e^{j\kappa_n x}] \cos(\varepsilon_n y) \qquad (2)$$

$$Z_0 v_{IIy}(x,y) = -j\sum_{n=0} \frac{\varepsilon_n}{k_0} [B_n e^{-j\kappa_n x} + C_n e^{j\kappa_n x}] \sin(\varepsilon_n y)$$

Zone III :

$$p_{III}(x,y) = \sum_{n=0} \delta_n D_n \, e^{-\gamma_n x} \cos(\eta_n y)$$

$$Z_0 v_{IIIx}(x,y) = -j\sum_{n=0} \delta_n D_n \frac{\gamma_n}{k_0} e^{-\gamma_n x} \cos(\eta_n y) \qquad (3)$$

$$Z_0 v_{IIIy}(x,y) = -2j\sum_{n>0} D_n \frac{\eta_n}{k_0} e^{-\gamma_n x} \sin(\eta_n y)$$

with $\delta_n = \begin{cases} 1\,;\, n=0 \\ 2\,;\, n>0 \end{cases} \quad ; \quad \delta_{mn} = \begin{cases} 1\,;\, m=n \\ 0\,;\, m\neq n \end{cases}$ (4)

and lateral wave numbers :

$$\eta_0 = 0 \;\; ; \;\; \eta_n = \frac{2\pi n}{L} = k_0 \cdot n \frac{\lambda_0}{L} \;\; ; \;\; \varepsilon_0 = 0 \;\; ; \;\; \varepsilon_n = \frac{2\pi n}{a} = k_0 \cdot n \frac{\lambda_0}{a} \qquad (5)$$

as well as axial propagation constants :

$$\gamma_0=jk_0\;;\quad \gamma_n=\sqrt{\eta_n^2-k_0^2}=k_0\sqrt{(n\frac{\lambda_0}{L})^2-1}\quad;\quad \mathrm{Re}\{\gamma_n\}\ge 0\quad\text{or}\quad \mathrm{Im}\{\gamma_n\}\ge 0$$
$$\kappa_0=k_0\;;\quad \kappa_n=\sqrt{k_0^2-\varepsilon_n^2}=k_0\sqrt{1-(n\frac{\lambda_0}{a})^2}\quad;\quad \mathrm{Im}\{\kappa_n\}\le 0\quad\text{or}\quad \mathrm{Re}\{\kappa_n\}\ge 0 \tag{6}$$

Mode coupling coefficients in the orifice planes are :

$$\begin{aligned} s_{m,n} &= \frac{1}{a}\int_{-a/2}^{+a/2}\cos(\eta_m y)\cos(\varepsilon_n y)\,dy \quad;\quad m=0,1,2,\ldots \\ &= \frac{1}{2}\left[\frac{\sin((\eta_m-\varepsilon_n)\,a/2)}{(\eta_m-\varepsilon_n)\,a/2}+\frac{\sin((\eta_m+\varepsilon_n)\,a/2)}{(\eta_m+\varepsilon_n)\,a/2}\right] \\ &= (-1)^n\frac{m\pi\,a/L}{(m\pi\,a/L)^2-(n\pi)^2}\sin(m\pi\,a/L) \\ &= \frac{(-1)^n}{\pi}\,\frac{m\frac{a}{L}\cdot\sin(m\pi\frac{a}{L})}{(m\frac{a}{L})^2-n^2}\quad;\quad m\frac{a}{L}\ne n\ne 0 \end{aligned} \tag{7}$$

and the special cases :

$$s_{m,n}=\tfrac{1}{2}\quad\text{for}\quad m\frac{a}{L}=n\ne 0\quad;\quad s_{0,0}=1\quad;\quad s_{m,0}=\frac{\sin(m\pi\,a/L)}{m\pi\,a/L}\quad;\quad s_{0,n>0}=0 \tag{8}$$

The boundary conditions for the particle velocities at the zone limits give (m=0,1,2,...) :

$$A_m=-j\frac{k_0}{\gamma_m}[-\delta_{0,m}A_e+\frac{a}{L}\sum_{n=0}\frac{\kappa_n}{k_0}s_{m,n}\cdot(B_n-C_n)]$$
$$D_m\,e^{-\gamma_m d}=j\frac{a}{L}\frac{k_0}{\gamma_m}\sum_{n=0}\frac{\kappa_n}{k_0}s_{m,n}\cdot(B_n e^{-j\kappa_n d}-C_n e^{+j\kappa_n d})] \tag{9}$$

the boundary conditions for the sound pressure yield (m=0,1,2,...) :

$$\frac{1}{\delta_m}(B_m+C_m)=\delta_{0,m}A_e+\sum_{n=0}\delta_n\,s_{n,m}\cdot A_n$$
$$\frac{1}{\delta_m}(B_m e^{-j\kappa_m d}+C_m e^{+j\kappa_m d})=\sum_{n=0}\delta_n\,e^{-\gamma_n d}\,s_{n,m}\cdot D_n \tag{10}$$

Instead of solving these systems for A_m,B_m,C_m,D_m the auxiliary quantities X_n,Y_n are introduced :

$$X_{n\pm}:=B_n\pm C_n\quad;\quad Y_{n\pm}:=B_n e^{-j\kappa_n d}\pm C_n e^{+j\kappa_n d} \tag{11}$$

with intrinsic relations :

$$\begin{aligned} X_{n+} &= X_{n-}\cdot\frac{1+e^{-2j\kappa_n d}}{1-e^{-2j\kappa_n d}} - 2Y_{n-}\cdot\frac{e^{-j\kappa_n d}}{1-e^{-2j\kappa_n d}} \\ Y_{n+} &= 2X_{n-}\cdot\frac{e^{-j\kappa_n d}}{1-e^{-2j\kappa_n d}} - Y_{n-}\cdot\frac{1+e^{-2j\kappa_n d}}{1-e^{-2j\kappa_n d}} \end{aligned} \tag{12}$$

The B_n, C_n follow from :

$$\begin{aligned} B_n &= \tfrac{1}{2}(X_{n+}+X_{n-}) = \tfrac{1}{2}(Y_{n+}+Y_{n-})e^{+j\kappa_n d} \\ C_n &= \tfrac{1}{2}(X_{n+}-X_{n-}) = \tfrac{1}{2}(Y_{n+}-Y_{n-})e^{-j\kappa_n d} \end{aligned} \tag{13}$$

A coupled systems of equations for X_{n-}, Y_{n-} is obtained with the form :

$$\begin{aligned} &\sum_{n=0} a_{mn}\cdot X_{n-} + c_m\cdot Y_{m-} = b_m\cdot A_e \quad ; \quad m = 0, 1, 2, \ldots \\ &c_m\cdot X_{m-} + \sum_{n=0} a_{mn}\cdot Y_{n-} = 0 \quad ; \quad m = 0, 1, 2, \ldots \end{aligned} \tag{14}$$

and the coefficients :

$$\begin{aligned} a_{m,n} &= j\frac{a}{L}\frac{\kappa_n}{k_0}S_{m,n} + \frac{\delta_{m,n}}{\delta_m}\frac{1+e^{-2j\kappa_m d}}{1-e^{-2j\kappa_m d}} \\ c_m &= -\frac{2}{\delta_m}\frac{e^{-j\kappa_m d}}{1-e^{-2j\kappa_m d}} \quad ; \quad b_m = \delta_{0,m} + s_{0,m} = 2\delta_{0,m} \end{aligned} \tag{15}$$

with the abbreviations :

$$\begin{aligned} S_{m,n} &= \sum_{i=0}\delta_i\frac{k_0}{\gamma_i}s_{i,m}\cdot s_{i,n} \quad ; \quad m,n = 0, 1, 2, \ldots \\ &= -j\,\delta_{0m}\delta_{0n} + 2\cdot\sum_{i>0}\frac{k_0}{\gamma_i}s_{i,m}\cdot s_{i,n} \end{aligned} \tag{16}$$

The normalised backside orifice impedance is :

$$\frac{Z_{sb}}{Z_0} = \frac{\langle p_{II}(d,y)\rangle}{Z_0\langle v_{IIx}(d,y)\rangle} = \frac{Y_{0+}}{Y_{0-}} = j\frac{a}{L}\Big[\sum_{i=0}\delta_i\frac{k_0}{\gamma_i}s_{i0}^2 + \sum_{n>0}\frac{\kappa_n}{k_0}\frac{Y_{n-}}{Y_{0-}}\cdot\sum_{i=0}\delta_i\frac{k_0}{\gamma_i}s_{i0}s_{in}\Big] \tag{17}$$

The 1st term in the brackets is just the orifice impedance of a neck with only plane waves in it; thus the 2nd term is a correction term for the influence of higher modes in the neck.

The normalised front side orifice impedance is :

$$\begin{aligned} \frac{Z_{sf}}{Z_0} &= \frac{\langle p_{II}(0,y)\rangle}{Z_0\langle v_{IIx}(0,y)\rangle} = \frac{X_{0+}}{X_{0-}} \\ &= \frac{2A_e}{B_0-C_0} - j\frac{a}{L}\Big[\sum_{i=0}\delta_i\frac{k_0}{\gamma_i}s_{i0}^2 + \sum_{n>0}\frac{\kappa_n}{k_0}\frac{X_{n-}}{X_{0-}}\cdot\sum_{i=0}\delta_i\frac{k_0}{\gamma_i}s_{i0}s_{in}\Big] \end{aligned} \tag{18}$$

After subtraction of the sound pressure of the equivalent source (the 1st term in the last expression) the orifice impedances on both sides remain symmetrical.

The slit impedances above were defined with the average sound pressure and axial particle velocity. The slit radiation impedances, which are defined with the radiated power, see section F.1) are :

$$\frac{Z_{rb}}{Z_0}=\frac{\int_a p_{II}(d,y)\cdot Z_0 v_{IIx}^*(d,y)\,dy}{Z_0^2\int_a v_{IIx}(d,y)\cdot v_{IIx}^*(d,y)\,dy}=\frac{\sum_{n=0}\frac{1}{\delta_n}\frac{\kappa_n^*}{k_0}Y_{n+}Y_{n-}^*}{\sum_{n=0}\frac{1}{\delta_n}\left|\frac{\kappa_n}{k_0}\right|^2\left|Y_{n-}\right|^2}$$

$$\frac{Z_{rf}}{Z_0}=\frac{\int_a p_{II}(0,y)\cdot Z_0 v_{IIx}^*(0,y)\,dy}{Z_0^2\int_a v_{IIx}(0,y)\cdot v_{IIx}^*(0,y)\,dy}=\frac{\sum_{n=0}\frac{1}{\delta_n}\frac{\kappa_n^*}{k_0}X_{n+}X_{n-}^*}{\sum_{n=0}\frac{1}{\delta_n}\left|\frac{\kappa_n}{k_0}\right|^2\left|X_{n-}\right|^2} \qquad (19)$$

Relative change of the orifice reactance Z''_{sb} (and therefore also for the end correction) due to higher modes as compared with the orifice reactance Z''_{sb0} with only plane waves in the neck (see section H.2) :

$$Z''_{sb}=Z''_{sb0}\cdot\left(1+\frac{\Delta Z''_{sb}}{Z''_{sb0}}\right)=Z''_{sb0}\cdot\left(1-10^{F(x,y)}\right)$$

$$F(x,y)=\lg\left(-\frac{\Delta Z''_{sb}}{Z''_{sb0}}\right)=f(x)\cdot(1+g(y)) \qquad (20)$$

$$x=\lg(a/L)\quad ;\quad y=\lg(d/a)$$

with functions f(x) and g(y) in the ranges $-3\le x<0$ and $-1\le y\le 1$:

$$f(x)=-1.73968+1.48435\,(x+1.5)-1{,}84230\,(x+1.5)^2+ \\ 0.292538\,(x+1.5)^3+0.428402\,(x+1.5)^4$$

$$g(y)=H(-y)\cdot[0.00259355\,y-0.0758181\,y^2+ \\ 0.330845\,y^3+0.226933\,y^4]\quad ;\quad H(-y)=\begin{cases}1; & y\le 0\\ 0; & y>0\end{cases} \qquad (21)$$

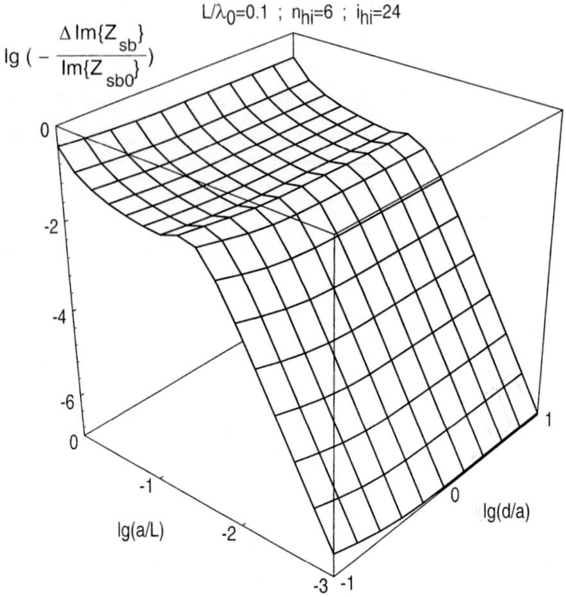

Relative change of the imaginary part of the neck orifice impedance Z_{sb} due to higher modes in the neck, as compared to the impedance Z_{sb0} with only plane waves in the neck.

The influence of higher modes is small if $a/L< 0.25$, and only if $d/a<<1$ does the plate thickness become sensible.

H.4 Dissipationless slit resonator

Ref. Mechel, [H.2]

Parallel slits in a neck plate and air volumes V behind them form an array of slit resonators.

Excitation is by a plane wave with normal incidence and amplitude A_e .

First, higher modes will be assumed in the necks, then the special case of only plane waves in the neck will be treated.

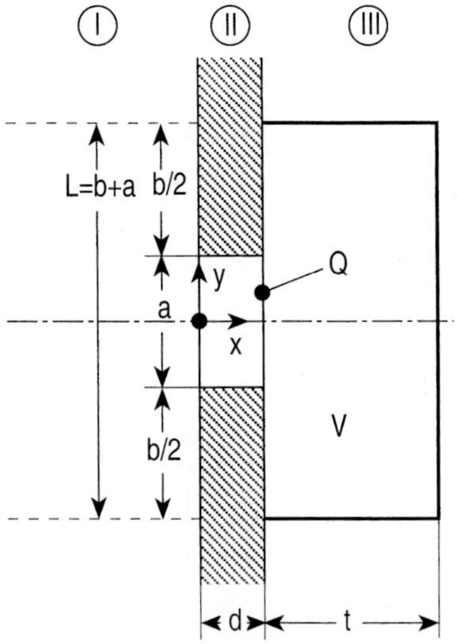

The field formulations *with higher modes* remain as in section H.3 , except in the volumes in zone III :

$$p_{III}(x,y) = \sum_{n=0} \delta_n D_n \cosh(\gamma_n(x-d-t))\cos(\eta_n y)$$

$$Z_0 v_{IIIx}(x,y) = j\sum_{n=0} \delta_n D_n \frac{\gamma_n}{k_0}\sinh(\gamma_n(x-d-t))\cos(\eta_n y) \qquad (1)$$

$$Z_0 v_{IIIy}(x,y) = -2j\sum_{n>0} D_n \frac{\eta_n}{k_0}\cosh(\gamma_n(x-d-t))\sin(\eta_n y)$$

(with wave numbers and propagation constants from section H.3).

The boundary conditions give for the auxiliary quantities $X_{n\pm}, Y_{n\pm}$:

$$X_{n\pm} := B_n \pm C_n \quad ; \quad Y_{n\pm} := B_n e^{-j\kappa_n d} \pm C_n e^{+j\kappa_n d}$$

$$B_n = \tfrac{1}{2}(X_{n+} + X_{n-}) = \tfrac{1}{2}(Y_{n+} + Y_{n-})e^{+j\kappa_n d}$$

$$C_n = \tfrac{1}{2}(X_{n+} - X_{n-}) = \tfrac{1}{2}(Y_{n+} - Y_{n-})e^{-j\kappa_n d} \qquad (2)$$

$$X_{n+} = X_{n-}\cdot\frac{1+e^{-2j\kappa_n d}}{1-e^{-2j\kappa_n d}} - 2Y_{n-}\cdot\frac{e^{-j\kappa_n d}}{1-e^{-2j\kappa_n d}}$$

$$Y_{n+} = 2X_{n-}\cdot\frac{e^{-j\kappa_n d}}{1-e^{-2j\kappa_n d}} - Y_{n-}\cdot\frac{1+e^{-2j\kappa_n d}}{1-e^{-2j\kappa_n d}}$$

a coupled system of linear equations :

$$\sum_{n=0} a_{mn}\cdot X_{n-} + c_m\cdot Y_{m-} = b_m\cdot A_e \quad ; \quad m = 0,1,\ldots$$

$$c_m\cdot X_{m-} + \sum_{n=0} d_{mn}\cdot Y_{n-} = 0 \quad ; \quad m = 0,1,\ldots \qquad (3)$$

with coefficients :

$$a_{m,n} = j\frac{a}{L}\frac{\kappa_n}{k_0}S_{m,n} + \frac{\delta_{m,n}}{\delta_m}\frac{1+e^{-2j\kappa_m d}}{1-e^{-2j\kappa_m d}}$$

$$d_{m,n} = j\frac{a}{L}\frac{\kappa_n}{k_0}T_{m,n} + \frac{\delta_{m,n}}{\delta_m}\frac{1+e^{-2j\kappa_m d}}{1-e^{-2j\kappa_m d}} \qquad (4)$$

$$c_m = -\frac{2}{\delta_m}\frac{e^{-j\kappa_m d}}{1-e^{-2j\kappa_m d}} \quad ; \quad b_m = \delta_{0,m} + s_{0,m} = 2\delta_{0,m}$$

wherein the $S_{m,n}$ are defined as in section H.3 , as well as the $s_{m,n}$, and :

$$T_{m,n} = \sum_{i=0} \delta_i \frac{k_0}{\gamma_i}\frac{si_{i,m}\cdot si_{i,n}}{\tanh(\gamma_n t)} \quad ; \quad m,n = 0,1,2,\ldots \qquad (5)$$

The amplitudes A_m, D_m are given by :

$$A_m = -j\frac{k_0}{\gamma_m}[-\delta_{0,m}A_e + \frac{a}{L}\sum_{n=0}\frac{\kappa_n}{k_0}s_{m,n}\cdot X_{n-}]$$

$$D_m = j\frac{a}{L}\frac{k_0}{\gamma_m\cdot\sinh(\gamma_m t)}\sum_{n=0}\frac{\kappa_n}{k_0}s_{m,n}\cdot Y_{n-} \tag{6}$$

The back side orifice impedance Z_{sb} is :

$$\frac{Z_{sb}}{Z_0} = j\frac{a}{L}[T_{0,0} + \sum_{n>0}\frac{\kappa_n}{k_0}\frac{Y_{n-}}{Y_{0-}}T_{0,n}]$$

$$= -j\frac{a/L}{\tan(k_0 t)} + j\frac{a}{L}[2\sum_{i>0}\frac{k_0}{\gamma_i}\frac{s_{i,0}^2}{\tanh(\gamma_i t)} + \sum_{n>0}\frac{\kappa_n}{k_0}\frac{Y_{n-}}{Y_{0-}}\sum_{i=0}\delta_i\frac{k_0}{\gamma_i}\frac{s_{i,0}s_{i,n}}{\tanh(\gamma_i t)}] \tag{7}$$

The 1st term in the 2nd line is the spring reactance of the resonator volume when it is driven by a piston of width a. Therefore the 2nd term in the 2nd line is the mass reactance of the back side orifice.

With *only plane waves* in the neck (i.e. narrow necks and/or low frequencies) substitute

$$B_0 = B \;;\; C_0 = C \;;\; B_{n>0} = C_{n>0} = 0 \tag{8}$$

to get for the back side orifice impedance Z_{sb} :

$$\frac{Z_{sb}}{Z_0} = j\frac{a}{L}T_{0,0} = -j\frac{a/L}{\tan(k_0 t)} + j2\frac{a}{L}\sum_{i>0}\frac{k_0}{\gamma_i}\frac{s_{i,0}^2}{\tanh(\gamma_i t)} \tag{9}$$

The front side orifice impedance Z_{sf} has the form of the impedance of a free plate (see section H.2 ; see there for S):

$$\frac{Z_{sf}}{Z_0} = \frac{\langle p_{II}(0,y)\rangle}{Z_0\langle v_{IIx}(0,y)\rangle} = \frac{B+C}{B-C} = \frac{2A_e}{B-C} - S = \frac{Z_{sb}/Z_0 + j\tan(k_0 d)}{1 + j(Z_{sb}/Z_0)\tan(k_0 d)} \tag{10}$$

The back side end correction $\Delta\ell_b/a$ can be defined from the back side neck reactance Z''_{sb} by:

$$\frac{\Delta\ell_b}{a} = \frac{1}{k_0 a}(\frac{Z''_{sb}}{Z_0} + j\frac{a/L}{\tan(k_0 t)}) \tag{11}$$

The influence of the shape parameter t/L is of interest.

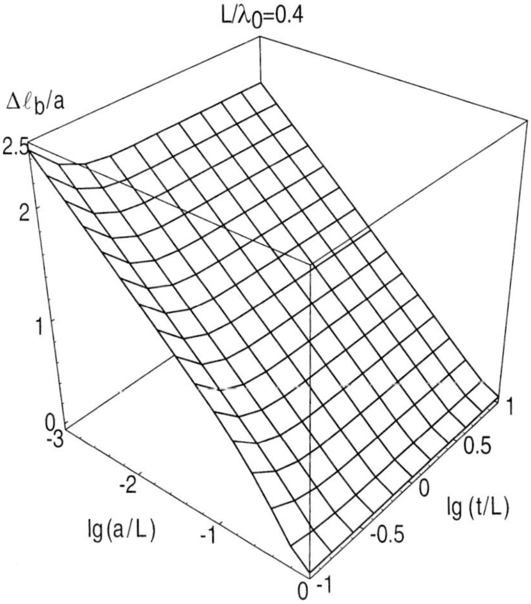

Influence of the shape parameter t/L on the end correction of the back side neck orifice (towards the resonator volume).

The back side end correction differs sensibly from the end correction of a free plate (i.e. from the front side end correction) only for rather small values of t/L ; then it is larger than the front side end correction.

For slit resonators *with higher modes* in the neck the back orifice end correction is (in the range $-1 \le y, z \le 0$):

$$\frac{\Delta\ell_b}{a} \approx \frac{\Delta\ell}{a}(x)\cdot[1+f(y;z_0,x_0)]\cdot[1+g(z;y_0,x_0)]$$

$$x = \lg\frac{a}{L} \quad ; \quad y = \lg\frac{d}{a} \quad ; \quad z = \lg\frac{t}{L}$$

$$f(y;z_0,x_0) \approx g(z;y_0,x_0) \approx 0 \qquad \text{for} \quad y \ge y_0 \quad ; \quad z \ge z_0 \tag{12}$$

$$x_0 = -1 \quad ; \quad y_0 = z_0 = 0{,}3 \approx \lg(2)$$

$$f(y;z_0,x_0) = 0.00144829\cdot y + 0.00255510\cdot y^2 + 0.03430510\cdot y^3 + 0.01568299\cdot y^4$$

$$g(z;y_0,x_0) = -0.000932290\cdot z - 0.00767204\cdot z^2 - 0.01925972\cdot z^3 - 0.01804839\cdot z^4$$

where $\Delta\ell/a$ is the end correction for a free slit in an array.

H.5 Resonance frequencies and radiation loss of slit resonators

Ref. Mechel, [H.2]

The slit resonator is a special form of a Helmholtz resonator. Let V be the volume of the resonator, Q the cross-section area of the neck; then the resilience F, the oscillating mass M and the angular resonance frequency ω_0 of the resonator usually are given as (ρ_0= density of air; c_0= sound velocity) are :

$$F=\frac{V}{Q^2\rho_0 c_0^2} \quad ; \quad M=\rho_0 Q\,(d+\Delta\ell+\Delta\ell_b)$$
$$\omega_0=\frac{1}{\sqrt{F\cdot M}}=c_0\sqrt{\frac{Q}{V\cdot(d+\Delta\ell+\Delta\ell_b)}}\approx c_0\sqrt{\frac{Q}{V\cdot(d+2\Delta\ell)}} \tag{1}$$

This formula is known to return seriously false results for some parameter combinations.

If the resonance condition is defined by zero reactance of the front side orifice impedance Z_{sf}, then it is given by (for slit resonators with only plane waves in the neck) :

$$\mathrm{Im}\{\frac{Z_{sv}}{Z_0}+S\}=\mathrm{Im}\{\frac{S+T+j(1+ST)\tan k_0 d}{1+jT\tan k_0 d}\}\overset{!}{=}0$$
$$S=\frac{a}{L}[1+2j\sum_{n>0}\frac{k_0}{\gamma_n}s_n^2] \quad ; \quad s_n=\frac{\sin(n\pi a/L)}{n\pi a/L}; \quad s_0=1 \tag{2}$$
$$T=j\frac{a}{L}\sum_{n=0}\frac{k_0}{\gamma_n}\frac{s_n^2}{\tanh(\gamma_n t)}=j\frac{a}{L}[-\frac{a/L}{\tan(k_0 t)}+2\sum_{n>0}\frac{k_0}{\gamma_n}\frac{s_n^2}{\tanh(\gamma_n t)}]$$

Because of the periodicity of $\tan(k_0 d)$ one must further demand that the zero value is crossed with positive slopes (in order to avoid anti-resonances), i.e. transition from spring to mass type reactance. The resonance condition is then :

$$k_0 a\,(\frac{\Delta\ell}{a}+\frac{\Delta\ell_b}{a})-\frac{a/L}{\tan(k_0 t)}+[1-(k_0 a)^2\cdot\frac{\Delta\ell}{a}\frac{\Delta\ell_b}{a}+\frac{a}{L}\frac{\Delta\ell}{a}\frac{k_0 a}{\tan(k_0 t)}]\cdot\tan(k_0 d)=0 \tag{3}$$

For low frequencies with $\tan(k_0 d)\approx k_0 d$, $\tan(k_0 t)\approx k_0 t$:

$$(2\pi\frac{a}{\lambda_0})^2[\frac{t}{a}(\frac{d}{a}+2\frac{\Delta\ell}{a})+\frac{a}{L}\frac{d}{a}\frac{\Delta\ell}{a}]-(2\pi\frac{a}{\lambda_0})^4\frac{d}{a}(\frac{\Delta\ell}{a})^2-\frac{a}{L}=0 \tag{4}$$

and neglecting further the term with $(a/\lambda_0)^4$, the lowest resonance is approximately:

$$\frac{L}{\lambda_0} \approx \frac{1}{2\pi\sqrt{\frac{t}{L}(\frac{d}{a}+2\frac{\Delta\ell}{a})+(\frac{a}{L})^2\frac{d}{a}\frac{\Delta\ell}{a}}} \tag{5}$$

A better approximation is obtained with a continued fraction expansion of $\tan z$:

$$\frac{L}{\lambda_0} \approx \frac{1}{2\pi\sqrt{\frac{t}{L}(\frac{d}{a}+2\frac{\Delta\ell}{a}+\frac{1}{3}\frac{t}{L})}} \tag{6}$$

or, with volume V, volume cross-section area Q_V, neck cross-section area Q :

$$\omega_0 \approx c_0\sqrt{\frac{Q}{V\cdot(d+\Delta\ell+\Delta\ell_b+\frac{1}{3}QV/Q_V^2)}} \tag{7}$$

The form of this resonance formula may be compared with the form of the traditional formula.

A resonance formula for the lowest resonance with a one step higher precision is :

$$\frac{L}{\lambda_0} \approx \frac{L/t}{2\pi}\sqrt{\frac{v-\sqrt{v^2-4uw}}{2u}}$$

$$u=\frac{(a/L)^2}{t/L}\frac{\Delta\ell}{a}(1+\frac{\Delta\ell/a}{t/L}) \quad ; \quad v=1+\frac{1}{3}\frac{t/L}{d/a}+\frac{\Delta\ell}{a}(\frac{2}{d/a}+3\frac{(a/L)^2}{t/L}) \quad ; \quad w=\frac{t/L}{d/a} \tag{8}$$

A slit resonator in an array has a *radiation loss*, corresponding to its back radiation (reflection). Its loss factor η is given by :

$$\eta=\frac{R}{\omega_0 M}=R\cdot\omega_0 F=\frac{R}{\sqrt{\frac{M}{F}}} \tag{9}$$

or with approximations for the circuit elements at resonance :

$$R=\rho_0 c_0\frac{a^2}{L} \quad ; \quad F=\frac{Lt}{a^2\rho_0 c_0^2} \quad ; \quad M=\rho_0 a^2(\frac{d}{a}+2\frac{\Delta l}{a}+\frac{1}{3}\frac{t}{L})$$

$$\eta=\frac{R}{\sqrt{M/F}}=\sqrt{\frac{t/L}{d/a+2\Delta l/a+\frac{1}{3}t/L}} \tag{10}$$

H.6 Slit array with viscous and thermal losses :

Ref. Mechel, [H.3]

The object is an array of slits in a free plate with thickness d , the width of the slits is a , and their mutual distance L .

The sketch shows the combination with a resonator volume V for the next section; it further shows zones of the sound field, field quantities for which boundary conditions exist, and mode amplitudes as well as mode wave numbers of the field formulations.

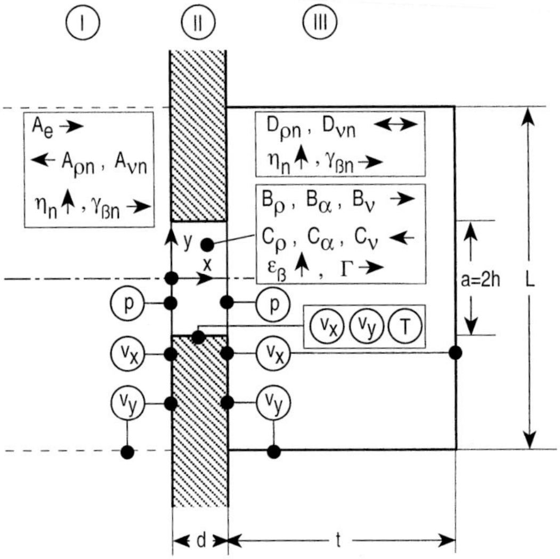

A plane sound wave with amplitude A_e is incident normally on the plate.

A simplification is applied: the thermal wave component is neglected in the reflected field in the zone I and in the transmitted field in the zone III (however, viscous waves are considered in these zones). The full triple of density wave ($\beta=\rho$), viscous wave ($\beta=\nu$), and thermal wave ($\beta=\alpha$) is applied in the necks. These waves satisfy the wave equations (see the Sections J1, J2) :

$$(\Delta + k_\beta^2)\Phi_\beta = 0 \quad , \quad \beta = \rho, \alpha \quad ; \quad (\Delta + k_\nu^2)\vec{\Psi} = 0$$
$$k_\nu^2 = -j\frac{\omega}{\nu} \quad ; \quad k_\alpha^2 \approx k_{\alpha 0}^2 = -j\kappa\frac{\omega}{\alpha} = \kappa\,\mathrm{Pr}\cdot k_\nu^2 \quad ; \quad k_\rho^2 \approx k_0^2 = \frac{\omega}{c_0} \tag{1}$$

The particle velocity $\vec{v}$, the sound pressure p and the oscillating (absolute) temperature are :

$$\vec{v} = -\mathrm{grad}(\Phi_\rho + \Phi_\alpha) + \mathrm{rot}\vec{\Psi}$$
$$\frac{p}{P_0} = \Pi_\rho \cdot \Phi_\rho + \Pi_\alpha \cdot \Phi_\alpha \tag{2}$$
$$\frac{T}{T_0} = \Theta_\rho \cdot \Phi_\rho + \Theta_\alpha \cdot \Phi_\alpha$$

ρ_0=	density;
c_0=	adiabatic sound velocity;
κ=	adiabatic exponent;
ν=	kinematic viscosity;
η=	dynamic viscosity;
α=	temperature conductivity;
Pr=	Prandtl number;
P_0,T_0=	atmospheric pressure and temperature;

with scalar potential functions $\Phi_ß$; ß= α,ν; and a vector potential $\vec{\Psi}$. See the sections about capillaries in the Chapter J for the coefficients $\Pi_ß,\Theta_ß$.

Particle velocity components for boundary conditions are :

$$v_x = -\frac{\partial(\Phi_\rho + \Phi_\alpha)}{\partial x} + \frac{\partial \Psi_z}{\partial y} \quad ; \quad v_y = -\frac{\partial(\Phi_\rho + \Phi_\alpha)}{\partial y} - \frac{\partial \Psi_z}{\partial x} \tag{3}$$

Numerical coefficients used :

$$\delta_n = \begin{cases} 1 \,; n = 0 \\ 2 \,; n > 0 \end{cases} \quad ; \quad \delta_{m,n} = \begin{cases} 1 \,; m = n \\ 0 \,; m \neq n \end{cases} \tag{4}$$

Incident plane wave (formulated as a potential function) : $\Phi_e = A_e \cdot e^{-jk_\rho x}$ (5)

Field potential formulations

in *Zone I*:

$$\Phi_I(x,y) = A_e \cdot e^{-jk_\rho x} + \sum_{n\geq 0} \delta_n\, A_{\rho n} \cdot e^{+\gamma_{\rho n} x} \cdot \cos(\eta_n y)$$
$$\Psi_I(x,y) = \sum_{n\geq 0} \delta_n\, A_{\nu n} \cdot e^{+\gamma_{\nu n} x} \cdot \sin(\eta_n y) \quad ; \quad A_{\nu 0} = 0 \tag{6}$$

in *Zone II*:

$$\Phi_{II}(x,y) = \sum_{ß=\rho,\alpha} [B_ß \cdot e^{-\Gamma x} + C_ß \cdot e^{+\Gamma x}] \cdot \cos(\varepsilon_ß y)$$
$$\Psi_{II}(x,y) = [B_\nu \cdot e^{-\Gamma x} + C_\nu \cdot e^{+\Gamma x}] \cdot \sin(\varepsilon_\nu y) \tag{7}$$

in *Zone III*:

$$\Phi_{III}(x,y) = \sum_{n\geq 0} \delta_n\, D_{\rho n} \cdot e^{-\gamma_{\rho n} x} \cdot \cos(\eta_n y)$$
$$\Psi_{III}(x,y) = \sum_{n\geq 0} \delta_n\, D_{\nu n} \cdot e^{-\gamma_{\nu n} x} \cdot \sin(\eta_n y) \quad ; \quad D_{\nu 0} = 0 \tag{8}$$

with wave numbers and propagation constants :

$$\eta_n = n\frac{2\pi}{L} \quad ; \quad n = 0, 1, 2\ldots \quad ; \quad \gamma_{ßn}^2 = \eta_n^2 - k_ß^2 \quad ; \quad ß = \rho,\alpha \quad ; \quad \gamma_{ß0} = jk_ß$$
$$\varepsilon_ß^2 = \Gamma^2 + k_ß^2 \quad ; \quad ß = \rho,\alpha,\nu \tag{9}$$

and Γ the known solution (see sections about capillaries) of the characteristic equation :

$$(\Gamma h)^2 (\frac{\Theta_\rho}{\Theta_\alpha} - 1)\frac{\tan(\varepsilon_\nu h)}{\varepsilon_\nu h} + \varepsilon_\rho h \cdot \tan(\varepsilon_\rho h) - \frac{\Theta_\rho}{\Theta_\alpha} \varepsilon_\alpha h \cdot \tan(\varepsilon_\alpha h) = 0 \tag{10}$$

Relations between amplitudes :

$$B_\alpha = -\frac{\Theta_\rho}{\Theta_\alpha}\frac{\cos(\varepsilon_\rho h)}{\cos(\varepsilon_\alpha h)} B_\rho \quad ; \quad B_\nu = -\frac{\Gamma}{\varepsilon_\nu}\frac{\cos(\varepsilon_\rho h)}{\cos(\varepsilon_\nu h)}(1 - \frac{\Theta_\rho}{\Theta_\alpha}) B_\rho \tag{11}$$

$$C_\alpha = -\frac{\Theta_\rho}{\Theta_\alpha}\frac{\cos(\varepsilon_\rho h)}{\cos(\varepsilon_\alpha h)}C_\rho \quad ; \quad C_v = +\frac{\Gamma}{\varepsilon_v}\frac{\cos(\varepsilon_\rho h)}{\cos(\varepsilon_v h)}(1-\frac{\Theta_\rho}{\Theta_\alpha})C_\rho \tag{12}$$

So only B_ρ, C_ρ must be determined in the set of $B_ß, C_ß$.

Mode coupling coefficients :

$$s_n = \frac{1}{a}\int_{-h}^{+h}\cos(\eta_n y)\,dy = \frac{\sin(\eta_n h)}{\eta_n h} = \frac{\sin(n\pi a/L)}{n\pi a/L} \quad ; \quad s_0 = 1$$

$$S_{ßn} = \frac{1}{a}\int_{-a/2}^{+a/2}\cos(\varepsilon_ß y)\cdot\cos(\eta_n y)\,dy = \frac{1}{2}[\frac{\sin(\varepsilon_ß-\eta_n)h}{(\varepsilon_ß-\eta_n)h}+\frac{\sin(\varepsilon_ß+\eta_n)h}{(\varepsilon_ß+\eta_n)h}]$$

$$= \frac{\eta_n h\cdot\sin(\eta_n h)\cdot\cos(\varepsilon_ß h)-\varepsilon_ß h\cdot\cos(\eta_n h)\cdot\sin(\varepsilon_ß h)}{(\eta_n^2-\varepsilon_ß^2)h^2} \tag{13}$$

$$R_{ßn} = \frac{1}{a}\int_{-a/2}^{+a/2}\sin(\varepsilon_ß y)\cdot\sin(\eta_n y)\,dy = \frac{1}{2}[\frac{\sin(\varepsilon_ß-\eta_n)h}{(\varepsilon_ß-\eta_n)h}-\frac{\sin(\varepsilon_ß+\eta_n)h}{(\varepsilon_ß+\eta_n)h}]$$

$$= \frac{\varepsilon_ß h\cdot\sin(\eta_n h)\cdot\cos(\varepsilon_ß h)-\eta_n h\cdot\cos(\eta_n h)\cdot\sin(\varepsilon_ß h)}{(\eta_n^2-\varepsilon_ß^2)h^2}$$

with special cases :

$$S_{ßn} = \frac{\sin(\varepsilon_ß h)}{\varepsilon_ß h}; n=0 \quad ; \quad S_{ßn} = \frac{1}{2}(1+\frac{\sin(\eta_n a)}{\eta_n a}); \eta_n = \varepsilon_ß \quad ; \quad S_{ßn} = 1; n = \varepsilon_ß = 0$$

$$R_{ßn} = 0 \;;\begin{cases} n = 0 \\ \varepsilon_ß = 0 \end{cases} \quad ; \quad R_{ßn} = \frac{1}{2}(1-\frac{\sin(\eta_n a)}{\eta_n a}); \eta_n = \varepsilon_ß \tag{14}$$

The auxiliary quantities : $X_\pm = B_\rho \pm C_\rho \quad ; \quad Y_\pm = B_\rho e^{-\Gamma d} \pm C_\rho e^{+\Gamma d}$ (15)

with intrinsic relations :

$$X_+ = X_-\cdot\frac{1+e^{-2\Gamma d}}{1-e^{-2\Gamma d}} - 2Y_-\cdot\frac{e^{-\Gamma d}}{1-e^{-2\Gamma d}}$$

$$Y_+ = 2X_-\cdot\frac{e^{-\Gamma d}}{1-e^{-2\Gamma d}} - Y_-\cdot\frac{1+e^{-2\Gamma d}}{1-e^{-2\Gamma d}} \tag{16}$$

from which follow :

$$B_\rho = \tfrac{1}{2}(X_+ + X_-) = \tfrac{1}{2}(Y_+ + Y_-)\,e^{+\Gamma d}$$

$$C_\rho = \tfrac{1}{2}(X_+ - X_-) = \tfrac{1}{2}(Y_+ - Y_-)\,e^{-\Gamma d} \tag{17}$$

are solutions of the coupled system of equations : (18)

$$X_-\cdot[\frac{1+e^{-2\Gamma d}}{1-e^{-2\Gamma d}}(1-U\sum_{n\ge0}\delta_n s_n V_n) - U\sum_{n\ge0}\delta_n s_n W_n\,] - Y_-\cdot 2\frac{e^{-\Gamma d}}{1-e^{-2\Gamma d}}(1-U\sum_{n\ge0}\delta_n s_n V_n) =$$

$$= 2U\cdot A_e$$

$$X_- \cdot 2\frac{e^{-\Gamma d}}{1-e^{-2\Gamma d}}(1-U\sum_{n\geq 0}\delta_n s_n V_n) - Y_- \cdot [\frac{1+e^{-2\Gamma d}}{1-e^{-2\Gamma d}}(1-U\sum_{n\geq 0}\delta_n s_n V_n) - U\sum_{n\geq 0}\delta_n s_n W_n\,] = 0$$

The other mode amplitudes follow from :

$$\begin{aligned} A_{\rho n} &= \delta_{0,n}\cdot A_e + V_n\cdot X_+ + W_n \cdot Y_- \\ D_{\rho n} e^{-\gamma_{\rho n} d} &= V_n \cdot Y_+ - W_n \cdot Y_- \end{aligned} \tag{19}$$

with the coefficients :

$$U = [\,S_{\rho 0} - S_{\alpha 0}\frac{\Pi_\alpha}{\Pi_\rho}\frac{\Theta_\rho}{\Theta_\alpha}\frac{\cos(\varepsilon_\rho h)}{\cos(\varepsilon_\alpha h)}\,]^{-1}$$

$$V_n = \frac{a}{L}\frac{\eta_n}{\eta_n^2-\gamma_{\rho n}\gamma_{\nu n}}[\varepsilon_\rho R_{\rho n} - \varepsilon_\alpha R_{\alpha n}\frac{\Theta_\rho}{\Theta_\alpha}\frac{\cos(\varepsilon_\rho h)}{\cos(\varepsilon_\alpha h)} - \frac{\Gamma^2}{\varepsilon_\nu}R_{\nu n}(1-\frac{\Theta_\rho}{\Theta_\alpha})\frac{\cos(\varepsilon_\rho h)}{\cos(\varepsilon_\nu h)}] \tag{20}$$

$$W_n = \frac{a}{L}\frac{\gamma_{\nu n}\Gamma}{\eta_n^2-\gamma_{\rho n}\gamma_{\nu n}}[S_{\rho n} - S_{\alpha n}\frac{\Theta_\rho}{\Theta_\alpha}\frac{\cos(\varepsilon_\rho h)}{\cos(\varepsilon_\alpha h)} - S_{\nu n}(1-\frac{\Theta_\rho}{\Theta_\alpha})\frac{\cos(\varepsilon_\rho h)}{\cos(\varepsilon_\nu h)}]$$

Introducing the abbreviations :

$$\begin{aligned} e_n &= S_{\rho n} - S_{\alpha n}\frac{\Theta_\rho}{\Theta_\alpha}\frac{\cos(\varepsilon_\rho h)}{\cos(\varepsilon_\alpha h)} - S_{\nu n}(1-\frac{\Theta_\rho}{\Theta_\alpha})\frac{\cos(\varepsilon_\rho h)}{\cos(\varepsilon_\nu h)} \\ d_n &= \varepsilon_\rho R_{\rho n} - \varepsilon_\alpha R_{\alpha n}\frac{\Theta_\rho}{\Theta_\alpha}\frac{\cos(\varepsilon_\rho h)}{\cos(\varepsilon_\alpha h)} - \frac{\Gamma^2}{\varepsilon_\nu}R_{\nu n}(1-\frac{\Theta_\rho}{\Theta_\alpha})\frac{\cos(\varepsilon_\rho h)}{\cos(\varepsilon_\nu h)}\;;\quad d_0 = 0 \end{aligned} \tag{21}$$

the coefficients can be written as :

$$V_n = \frac{a}{L}\frac{\eta_n}{\eta_n^2-\gamma_{\rho n}\gamma_{\nu n}}d_n \quad;\quad W_n = \frac{a}{L}\frac{\gamma_{\nu n}\Gamma}{\eta_n^2-\gamma_{\rho n}\gamma_{\nu n}}e_n \tag{22}$$

The back side slit impedance Z_{sb} is then :

$$\frac{Z_{sb}}{Z_0} = -j\frac{k_\rho^2}{k_0\Gamma}\frac{1-k_\rho^2/k_{\alpha 0}^2}{1-\kappa k_\rho^2/k_{\alpha 0}^2}\frac{1}{U e_0}\frac{U\sum\limits_{n\geq 0}\delta_n s_n W_n}{1-U\sum\limits_{n\geq 0}\delta_n s_n V_n} \tag{23}$$

and the front side slit impedance Z_{sf} :

$$\frac{Z_{sf}}{Z_0} = +j\frac{k_\rho^2}{k_0\Gamma}\frac{1-k_\rho^2/k_{\alpha 0}^2}{1-\kappa k_\rho^2/k_{\alpha 0}^2}\frac{1}{U e_0}\frac{\tanh(\Gamma d)\cdot[1-U\sum\limits_{n\geq 0}\delta_n s_n V_n] - U\sum\limits_{n\geq 0}\delta_n s_n W_n}{[1-U\sum\limits_{n\geq 0}\delta_n s_n V_n] - \tanh(\Gamma d)\cdot U\sum\limits_{n\geq 0}\delta_n s_n W_n} \tag{24}$$

Using the approximations, which are possible for $|\varepsilon_\rho h| << 1\,;|\Gamma h|^2 << |k_ß h|^2\;;ß = \alpha,\nu$:

$$S_{\rho 0} = \frac{\sin(\varepsilon_\rho h)}{\varepsilon_\rho h} \approx 1 \quad ;$$
$$S_{\beta 0}\frac{\cos(\varepsilon_\rho h)}{\cos(\varepsilon_\beta h)} \approx S_{\rho 0}\frac{\tan(\varepsilon_\beta h)}{(\varepsilon_\beta h)} \xrightarrow[|\varepsilon_\beta h|\gg 1]{} S_{\rho 0}\frac{1-j}{\sqrt{2}\,|k_\beta h|} \approx \frac{1-j}{\sqrt{2}\,|k_\beta h|} \tag{25}$$

$$U \approx \frac{1}{S_{\rho 0}} \quad ; \quad W_0 \approx -\frac{a}{L}\frac{\Gamma}{\gamma_{\rho 0}}e_0 = -\frac{a}{L}\frac{\Gamma}{jk_0}e_0 \quad ; \quad e_0 \approx S_{\rho 0}[1-\frac{1-j}{\sqrt{2}\,|k_\beta h|}] \approx 1-\frac{1-j}{\sqrt{2}\,|k_\beta h|} \tag{26}$$

one gets :

$$\frac{Z_{sb}}{Z_0} \approx \frac{a}{L} - \frac{jk_0}{\Gamma}\frac{\sqrt{2}\,|k_\nu h|}{\sqrt{2}\,|k_\nu h|-(1-j)}\frac{2\sum\limits_{n>0} s_n W_n}{1-2\sum\limits_{n>0} s_n V_n} \tag{27}$$

The 1st term is the normalised radiation resistance; in the 2nd term the first two fractions have about unit value for not too narrow slits. The end correction can be evaluated from :

$$\frac{\Delta\ell}{a} = \frac{\mathrm{Im}\{Z_{sb}\}}{Z_0\, k_0 a} \tag{28}$$

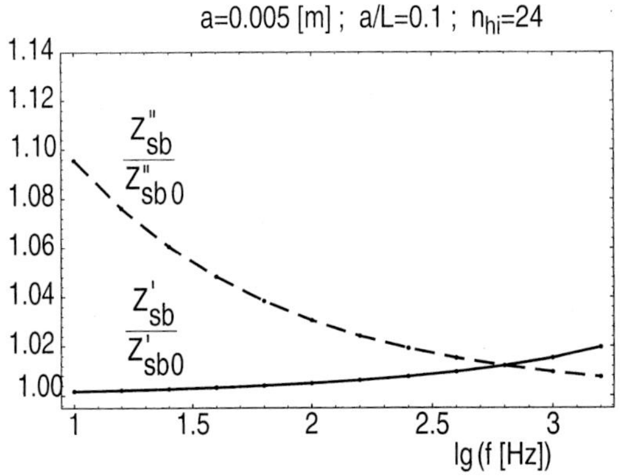

Ratios of the components of the slit impedance Z_{sb} with losses to these of the slit impedance Z_{sb0} without losses.

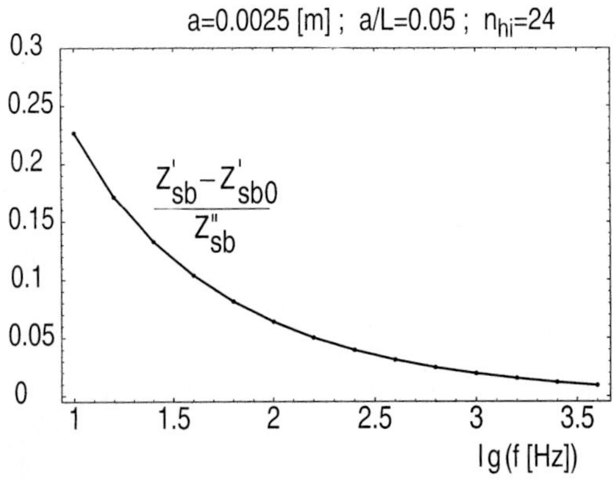

Loss factor of the oscillating mass of a slit, when viscous and thermal losses in the neck are taken into account.

The components of the slit impedance $Z_{sb}=Z'_{sb}+j\cdot Z''_{sb}$ can approximately be evaluated from those of the slit impedance $Z_{sb0}=Z'_{sb0}+j\cdot Z''_{sb0}$ without losses by :

$$\frac{Z'_{sb}}{Z_0}=\frac{Z'_{sb0}}{Z_0}\cdot\left(1+\frac{10^{F'(x)}}{\sqrt{a_{[m]}}\cdot\sqrt[3]{a/L}}\right) \quad ; \quad x=\lg\frac{f_{[Hz]}a_{[m]}}{(a/L)^{3/2}} \; ;$$

$$\frac{Z''_{sb}}{Z_0}=\frac{Z''_{sb0}}{Z_0}\cdot\left(1+\frac{10^{F''(x)}}{\sqrt{a_{[m]}}\cdot\sqrt[3]{a/L}}\right) \quad ; \qquad (29)$$

$$F'(x)=-4.64106+0.435993\,x+0.0142851\,x^2+0.000461347\,x^3$$

$$F''(x)=-2.26665-0.492331\,x-0.000719182\,x^2-0.0010208\,x^3$$

H.7 Slit resonator with viscous and thermal losses

Ref. Mechel, [H.3]

See the schematic drawing in the Section H.6 .

The field formulations remain as in the previous section H.6, except for the field in

Zone III :

$$\Phi_{III}(x,y)=\sum_{n=0}\delta_n\,D_{\rho n}\cdot\cosh(\gamma_{\rho n}(x-d-t))\cdot\cos(\eta_n y)$$

$$\Psi_{III}(x,y)=\sum_{n=0}\delta_n\,D_{\nu n}\cdot\sinh(\gamma_{\nu n}(x-d-t))\cdot\sin(\eta_n y) \quad ; \quad D_{\nu 0}=0 \qquad (1)$$

The system of equations for the auxiliary quantities X_-, Y_- (see H.6) is : (2)

$$X_-\cdot\left[\frac{1+e^{-2\Gamma d}}{1-e^{-2\Gamma d}}\left(1-U\sum_{n=0}\delta_n s_n V_n\right)-U\sum_{n=0}\delta_n s_n W_n\right]-Y_-\cdot 2\frac{e^{-\Gamma d}}{1-e^{-2\Gamma d}}\left(1-U\sum_{n=0}\delta_n s_n V_n\right)=$$
$$=2U\cdot A_e$$

$$X_-\cdot 2\frac{e^{-\Gamma d}}{1-e^{-2\Gamma d}}\left(1-U\sum_{n=0}\delta_n s_n V'_n\right)-Y_-\cdot\left[\frac{1+e^{-2\Gamma d}}{1-e^{-2\Gamma d}}\left(1-U\sum_{n=0}\delta_n s_n V'_n\right)-U\sum_{n=0}\delta_n s_n W'_n\right]=0$$

with the new coefficients :

$$V'_n=\frac{a}{L}\frac{\eta_n}{\eta_n^2-\gamma_{\rho n}\gamma_{\nu n}\cdot\tanh(\gamma_{\rho n}t)/\tanh(\gamma_{\nu n}t)}\,d_n \quad ; \quad V'_0=0$$

$$W'_n=\frac{a}{L}\frac{\Gamma\,\gamma_{\nu n}/\tanh(\gamma_{\nu n}t)}{\eta_n^2-\gamma_{\rho n}\gamma_{\nu n}\cdot\tanh(\gamma_{\rho n}t)/\tanh(\gamma_{\nu n}t)}\,e_n \qquad (3)$$

and all other terms as in H.6 .

The back and front orifice impedances Z_{sb}, Z_{sf} become :

$$\frac{Z_{sb}}{Z_0}=-j\frac{k_\rho^2}{k_0\Gamma}\frac{1-k_\rho^2/k_{\alpha 0}^2}{1-\kappa k_\rho^2/k_{\alpha 0}^2}\frac{1}{U e_0}\frac{U\sum\limits_{n\ge 0}\delta_n s_n W'_n}{1-U\sum\limits_{n\ge 0}\delta_n s_n V'_n}$$

$$\frac{Z_{sf}}{Z_0}=+j\frac{k_\rho^2}{k_0\Gamma}\frac{1-k_\rho^2/k_{\alpha 0}^2}{1-\kappa k_\rho^2/k_{\alpha 0}^2}\frac{1}{U e_0}\frac{\tanh(\Gamma d)\cdot[1-U\sum\limits_{n\ge 0}\delta_n s_n V'_n]-U\sum\limits_{n\ge 0}\delta_n s_n W'_n}{[1-U\sum\limits_{n\ge 0}\delta_n s_n V'_n]-\tanh(\Gamma d)\cdot U\sum\limits_{n\ge 0}\delta_n s_n W'_n} \qquad (4)$$

Compared with the results of section H.6 only the substitutions $V_n \to V'_n$, $W_n \to W'_n$ take place, which correspond to the substitutions $\gamma_{\rho n} \to \gamma_{\rho n}\cdot\tanh(\gamma_{\rho n}t)$ and $\gamma_{\rho n} \to \gamma_{\rho n}\cdot\tanh(\gamma_{\rho n}t)$.

Let Z''_{sM} be the mass reactance part of the back orifice impedance Z_{sb} with losses, and let Z''_{sb0} be the mass reactance part of the *free slit plate without losses* (see section H.2) , then the relative change of the reactance can be evaluated with :

$$x=\lg(f[Hz]) \;;\; y=\lg(a[m]) \;;\; z=\lg(a/L) \;;\; u=\lg(t/L) \qquad (5)$$

from :

$$\begin{aligned}\lg\left(\frac{Z''_{sM}}{Z''_{sb0}}-1\right)= &\\ &-2.240408\\ &-0.1580984\cdot x \quad +0.00688292\cdot x^2 \quad +0.0225970\cdot x^3\\ &-0.7868117\cdot y \quad +0.3117230\cdot y^2 \quad +0.0739239\cdot y^3\\ &+0.7621584\cdot z \quad +0.4961154\cdot z^2 \quad +0.1579759\cdot z^3\\ &-1.113747\cdot u \quad +1.609799\cdot u^2 \quad -2.026946\cdot u^3\\ &+0.2694603\cdot x\cdot y \quad +0.1078516\cdot x\cdot y^2 \quad +0.0741470\cdot x^2\cdot y\\ &+0.1401039\cdot x\cdot z \quad +0.00720527\cdot x\cdot z^2 \quad -0.0424421\cdot x^2\cdot z\\ &+0.0937094\cdot u\cdot x \quad -0.0519085\cdot u\cdot x^2 \quad +0.7279337\cdot u^2\cdot x\\ &-0.1959382\cdot y\cdot z \quad -0.00180315\cdot y^2\cdot z \quad -0.0587445\cdot y\cdot z^2\\ &-1.014977\cdot u\cdot y \quad -0.1716795\cdot u\cdot y^2 \quad +1.373450\cdot u^2\cdot y\\ &+0.1977607\cdot u\cdot z \quad -0.1665151\cdot u^2\cdot z \quad +0.0690112\cdot u\cdot z^2\end{aligned} \qquad (6)$$

The loss factor η can be evaluated with :

$$x=\lg(a[m]) \;;\; y=\lg(a/L) \;;\; z=\lg(t/L) \;;\; u=\lg(d/a) \qquad (7)$$

in the range :

$0.0025 \le a \le 0.02$ [m] ; $0.025 \le a/L \le 0.4$; $0.25 \le t/L \le 2.0$; $0.25 \le d/a \le 4.0$

from :

$$
\begin{aligned}
\lg(\eta) = \; & -3.42990 \\
& - 0.567811\cdot x - 0.405786\cdot y + 0.395143\cdot z \\
& + 0.0811464\cdot z^2 - 0.0337095\cdot u + 0.0871987\cdot u^2 \\
& + 0.0168052\cdot u\cdot x + 0.0184409\cdot u^2\cdot x - 0.225751\cdot u\cdot y \\
& + 0.0404207\cdot u^2\cdot y - 0.143725\cdot u\cdot z - 0.0130437\cdot u^2\cdot z \\
& - 0.132369\cdot u\cdot z^2 - 0.114934\cdot x\cdot y - 0.0195440\cdot x\cdot z \\
& - 0.000512528\cdot x\cdot z^2 + 0.0682123\cdot y\cdot z + 0.0335370\cdot y\cdot z^2 \\
& + 0.0534482\cdot u\cdot x\cdot y + 0.0314014\cdot u^2\cdot x\cdot y - 0.00696663\cdot u\cdot x\cdot z \\
& + 0.0156990\cdot u^2\cdot x\cdot z - 0.0805588\cdot u\cdot y\cdot z - 0.00456461\cdot u^2\cdot y\cdot z \\
& - 0.0409866\cdot x\cdot y\cdot z + 0.0150972\cdot u\cdot x\cdot y\cdot z + 0.0183208\cdot u^2\cdot x\cdot y\cdot z \\
& + 0.597378\cdot u^2\cdot z^2 - 0.0122410\cdot u\cdot x\cdot z^2 + 0.00565404\cdot u^2\cdot x\cdot z^2 \\
& - 0.0586390\cdot u\cdot y\cdot z^2 + 0.454270\cdot u^2\cdot y\cdot z^2 - 0.00757341\cdot x\cdot y\cdot z^2 \\
& - 0.000668200\cdot u\cdot x\cdot y\cdot z^2 + 0.00446479\cdot u^2\cdot x\cdot y\cdot z^2
\end{aligned} \tag{8}
$$

A regression for the lowest resonance frequency f_0[Hz] is with the same variables and range :

$$
\begin{aligned}
\lg(f_0\,[\mathrm{Hz}]) = \; & 1.624\,303 - 0.321\,020\cdot u - 0.128\,558\cdot u^2 \\
& - 1.046\,357\cdot x + 0.011\,080\,6\cdot u\cdot x + 0.010\,078\,7\cdot u^2\cdot x \\
& + 1.041\,716\cdot y - 0.092\,742\,1\cdot u\cdot y + 0.003\,998\,00\cdot u^2\cdot y \\
& - 0.084\,163\,8\cdot x\cdot y + 0.025\,871\,8\cdot u\cdot x\cdot y + 0.017\,817\,2\cdot u^2\cdot x\cdot y \\
& - 0.623\,277\cdot z + 0.136\,128\cdot u\cdot z - 0.022\,086\,6\cdot u^2\cdot z \\
& - 0.010\,705\,2\cdot x\cdot z - 0.001\,128\,57\cdot u\cdot x\cdot z - 0.000\,679\,359\cdot u^2\cdot x\cdot z \\
& - 0.057\,692\cdot y\cdot z + 0.066\,738\,9\cdot u\cdot y\cdot z - 0.026\,252\,4\cdot u^2\cdot y\cdot z \\
& - 0.020\,474\,4\cdot x\cdot y\cdot z - 0.001\,982\,88\cdot u\cdot x\cdot y\cdot z + 0.001\,241\,23\cdot u^2\cdot x\cdot y\cdot z \\
& - 0.080\,625\,9\cdot z^2 + 0.083\,414\cdot u\cdot z^2 + 0.003\,182\,64\cdot u^2\cdot z^2 \\
& + 0.002\,644\,97\cdot x\cdot z^2 - 0.001\,501\,23\cdot u\cdot x\cdot z^2 - 0.002\,655\,46\cdot u^2\cdot x\cdot z^2 \\
& - 0.021\,889\,3\cdot y\cdot z^2 + 0.036\,204\,5\cdot u\cdot y\cdot z^2 - 0.007\,951\,19\cdot u^2\cdot y\cdot z^2 \\
& + 0.005\,702\,69\cdot x\cdot y\cdot z^2 - 0.005\,804\,03\cdot u\cdot x\cdot y\cdot z^2 - 0.003\,461\,27\cdot u^2\cdot x\cdot y\cdot z^2
\end{aligned} \tag{9}
$$

H.8 Free plate with an array of circular holes, with losses

Ref. Mechel, [H.4]

The object is a (rigid) plate of thickness d, containing circular holes of diameter 2a in a hexagonal arrangement.

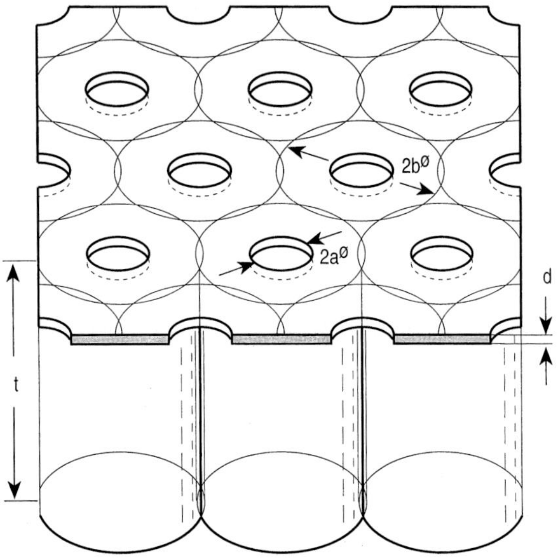

A plane wave with normal incidence has the amplitude A_e .

The sketch shows the arrangement with a symmetry cell behind each hole In this section the length of the cell is t= ∞ . The radius b of the cell is fixed so that the cells cover all the backside area of the plate (so square arrays can also be treated with this model).

Zones I and III are assumed to be dissipationless; the necks (zone II) have viscous and thermal losses, or the neck walls may be absorbent.

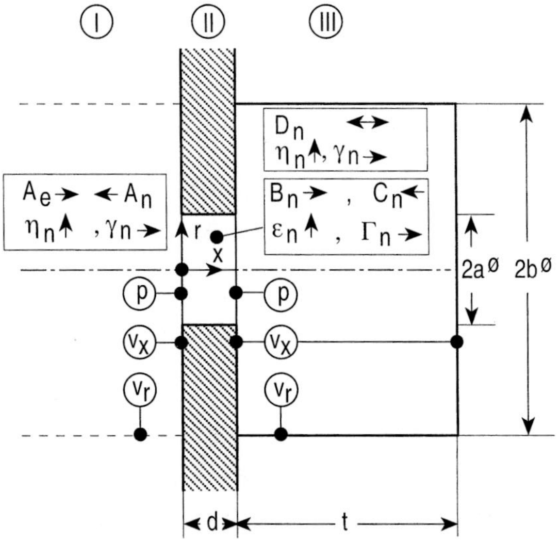

The sketch (with a resonator volume for the next section) shows field quantities which have to satisfy boundary conditions, mode amplitudes in the zones, and wave numbers.

The fundamentals of this section correspond closely to those of section H.7 for slit arrays with losses, except that the losses in zones I and III are neglected here (their effect is not important).

Sound field formulations

in *Zone I*: (1)

$$p_I(x,r) = A_e e^{-jk_0x} + \sum_{n\ge 0} A_n\, e^{\gamma_n x} J_0(\eta_n r)$$

$$Z_0 v_{Ix}(x,r) = A_e e^{-jk_0x} + j\sum_{n\ge 0} A_n\, \frac{\gamma_n}{k_0}\, e^{\gamma_n x} J_0(\eta_n r)$$

in *Zone II*: (2)

$$p_{II}(x,r) = \sum_{n\ge 0} [B_n e^{-\Gamma_n x} + C_n e^{\Gamma_n x}]\, J_0(\varepsilon_n r)$$

$$Z_0 v_{IIx}(x,r) = -j\sum_{n\ge 0} \frac{\Gamma_n}{k_0} [B_n e^{-\Gamma_n x} - C_n e^{\Gamma_n x}]\, J_0(\varepsilon_n r)$$

in *Zone III*: (3)

$$p_{III}(x,r) = \sum_{n\ge 0} D_n\, e^{-\gamma_n x} J_0(\eta_n r)$$

$$Z_0 v_{IIIx}(x,r) = -j\sum_{n\ge 0} D_n\, \frac{\gamma_n}{k_0}\, e^{-\gamma_n x} J_0(\eta_n y)$$

The sound field in the neck (zone II) is formulated as a mode sum. Some cases can be treated as follows:

1st Use the formulation as it is, if the neck walls are absorbent with an admittance G and higher neck modes are to be considered. Solve the characteristic equation

$$\varepsilon_n a\, \frac{J_1(\varepsilon_n a)}{J_0(\varepsilon_n a)} = j\, k_0 a\, Z_0 G \tag{4}$$

for a sufficiently large set of wave numbers ε_n and determine the axial propagation constants Γ_n from $\Gamma_n^2 = \varepsilon_n^2 - k_0^2$.

2nd The neck wall is absorbent, but the neck is narrow, so that only the fundamental neck mode must be retained: determine ε_0, Γ_0 as above and set $B_{n>0}=0$ and $C_{n>0}=0$.

3rd The neck wall is hard, and higher neck modes will be considered : Proceed as in 1st , but with the characteristic equation for $G=0$.

4th The neck wall is hard, and only a plane wave is assumed in the neck : Proceed as in 2nd , but with ε_0, Γ_0 for $G=0$.

5th The neck wall is hard, and only the fundamental capillary mode will be considered (the neck is very narrow, and viscous and thermal losses in it must be considered) : Take for Γ_0 the propagation constant in circular capillaries (see Sections J1, J2 about capillaries in the chapter “Duct Acoustics”) and evaluate ε_0 from $\varepsilon_0^2 = \Gamma_0^2 + k_0^2$. (5)

6th A somewhat exotic model assumes a very narrow neck, but with higher capillary modes. Either solve the characteristic equation of circular capillaries for a set of higher mode propagation constants (which is not easy), or solve

$$\varepsilon_n a \frac{J_1(\varepsilon_n a)}{J_0(\varepsilon_n a)} = j\, k_0 a\, Z_0 G \tag{6}$$

with an equivalent Z_0G :

$$Z_0 G = \frac{j}{k_0 a} \frac{(k_\rho a)^2 (1 - \frac{\Theta_\rho}{\Theta_\alpha}) \frac{J_1(k_\nu a)}{k_\nu a\, J_0(k_\nu a)} - \frac{\Theta_\rho}{\Theta_\alpha} k_\alpha a \frac{J_1(k_\alpha a)}{J_0(k_\alpha a)}}{1 - 2(1 - \frac{\Theta_\rho}{\Theta_\alpha}) \frac{J_1(k_\nu a)}{k_\nu a\, J_0(k_\nu a)}} \tag{7}$$

for a set of ε_n and then Γ_n from $\Gamma_n^2 = \varepsilon_n^2 - k_0^2$. (8)

Relations between wave numbers and propagation constants are :

$$\gamma_n^2 = \eta_n^2 - k_0^2 \quad ; \quad \Gamma_n^2 = \varepsilon_n^2 - k_0^2 \tag{9}$$

$z_n = \eta_n b$ are solutions (n=0,1,2,...) of $J_1(z)=0$ with $z_0=0$.

Mode coupling coefficients :

$$T_{n,m} = \frac{1}{a^2} \int_0^a J_0(\varepsilon_n r)\, J_0(\eta_m r) \cdot r\, dr = \frac{\varepsilon_n a\, J_1(\varepsilon_n a)\, J_0(z_m \frac{a}{b}) - z_m \frac{a}{b} J_0(\varepsilon_n a)\, J_1(z_m \frac{a}{b})}{(\varepsilon_n a)^2 - z_m^2 (\frac{a}{b})^2}$$

$$= \frac{J_0(\varepsilon_n a)\, J_0(\eta_m a)}{(\varepsilon_n a)^2 - (\eta_m)^2} [j k_0 a Z_0 G - \eta_m a \frac{J_1(\eta_m a)}{J_0(\eta_m a)}] \tag{10}$$

$$T_{m,0} = \frac{1}{a^2} \int_0^a J_0(\varepsilon_m r) \cdot r\, dr = \frac{J_1(\varepsilon_m a)}{\varepsilon_m a} = \frac{J_0(\varepsilon_m a)}{(\varepsilon_m a)^2} \cdot j k_0 a Z_0 G$$

$$T_{m,i} T_{n,i} = J_0^2(\eta_i a) \frac{J_0(\varepsilon_m a)}{(\varepsilon_m a)^2 - (\eta_i a)^2} \frac{J_0(\varepsilon_n a)}{(\varepsilon_n a)^2 - (\eta_i a)^2} (j k_0 a Z_0 G - \eta_i a \frac{J_1(\eta_i a)}{J_0(\eta_i a)})^2$$

and :

$$R_m = \frac{1}{a^2} \int_0^a J_0(\varepsilon_n r)\, J_0(\varepsilon_m r) \cdot r\, dr = \tfrac{1}{2} J_0^2(\varepsilon_m a) [1 - \frac{(k_0 a)^2}{(\varepsilon_m a)^2} (Z_0 G)^2] \tag{11}$$

The boundary conditions lead to a coupled system of equations for the auxiliary quantities :

$$X_{n\pm} := B_n \pm C_n \quad ; \quad Y_{n\pm} := B_n e^{-\Gamma_n d} \pm C_n e^{\Gamma_n d} \tag{12}$$

with intrinsic relations :

$$X_{n+}=X_{n-}\cdot\frac{1+e^{-2\Gamma_n d}}{1-e^{-2\Gamma_n d}}-2Y_{n-}\cdot\frac{e^{-\Gamma_n d}}{1-e^{-2\Gamma_n d}}$$
$$Y_{n+}=2X_{n-}\cdot\frac{e^{-\Gamma_n d}}{1-e^{-2\Gamma_n d}}-Y_{n-}\cdot\frac{1+e^{-2\Gamma_n d}}{1-e^{-2\Gamma_n d}} \tag{13}$$

The system of equations has the form (m=0,1,…):

$$\sum_{n\ge 0} a_{mn}\cdot X_{n-}+c_m\cdot Y_{m-}=b_m\cdot A_e$$
$$c_m\cdot X_{m-}+\sum_{n\ge 0} a_{mn}\cdot Y_{n-}=0 \tag{14}$$

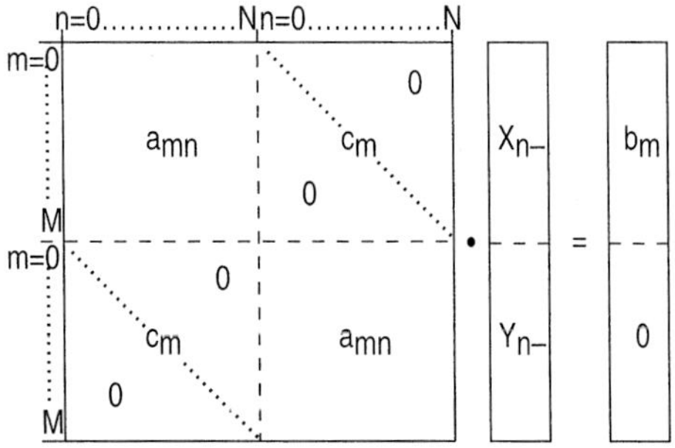

with coefficients :

$$a_{m,n}=2\left(\frac{a}{b}\right)^2\frac{\Gamma_n}{k_0}\sum_{i\ge 0}\frac{k_0}{\gamma_i J_0^2(z_i)}T_{m,i}T_{n,i}+\delta_{m,n}\frac{1+e^{-2\Gamma_m d}}{1-e^{-2\Gamma_m d}}$$
$$c_m=-2R_m\frac{e^{-\Gamma_m d}}{1-e^{-2\Gamma_m d}}\quad;\quad b_m=T_{m,0}\,(1+j\frac{k_0}{\gamma_0}) \tag{15}$$

The mode amplitudes follow from solutions with : (16)

$$A_n=\frac{k_0}{\gamma_n J_0^2(z_n)}[\delta_{0,n}\cdot jA_e-2(\frac{a}{b})^2\sum_{i\ge 0}\frac{\Gamma_i}{k_0}T_{i,n}\cdot X_{i-}]$$
$$B_n=\tfrac{1}{2}(X_{n+}+X_{n-})=\tfrac{1}{2}(Y_{n+}+Y_{n-})e^{+\Gamma_n d}\quad;\quad C_n=\tfrac{1}{2}(X_{n+}-X_{n-})=\tfrac{1}{2}(Y_{n+}-Y_{n-})e^{-\Gamma_n d}$$
$$D_n=2(\frac{a}{b})^2\frac{k_0\,e^{\gamma_n d}}{\gamma_n J_0^2(z_n)}\sum_{i=0}\frac{\Gamma_i}{k_0}T_{i,n}\cdot Y_{i-}$$

The front side and back side orifice impedances Z_{sf}, Z_{sb} are obtained from :

$$\frac{Z_{sf}}{Z_0}=\frac{<p_{II}(0,r)>_a}{Z_0<v_{IIx}(0,r)>_a}=j\frac{\sum_{n\ge 0}T_{n,0}\cdot X_{n+}}{\sum_{n\ge 0}\frac{\Gamma_n}{k_0}T_{n,0}\cdot X_{n-}}$$
$$\frac{Z_{sb}}{Z_0}=\frac{<p_{II}(d,r)>_a}{Z_0<v_{IIx}(d,r)>_a}=j\frac{\sum_{n\ge 0}T_{n,0}\cdot Y_{n+}}{\sum_{n\ge 0}\frac{\Gamma_n}{k_0}T_{n,0}\cdot Y_{n-}} \tag{17}$$

If only the fundamental mode n=0 is retained in the neck, the system of equations becomes :

$$a_{0,0}\cdot X_{0-}+c_0\cdot Y_{0-}=b_0\cdot A_e\quad;\quad c_0\cdot X_{0-}+a_{0,0}\cdot Y_{0-}=0 \tag{18}$$

or:
$$X_{0-} = \frac{b_0\, a_{0,0}}{a_{0,0}^2 - c_0^2} \quad ; \quad Y_{0-} = \frac{-b_0\, c_0}{a_{0,0}^2 - c_0^2} \tag{19}$$

with :
$$a_{0,0} = 2(\frac{a}{b})^2 \frac{\Gamma_0}{k_0} \sum_{i\geq 0} \frac{k_0}{\gamma_i J_0^2(z_i)} T_{0,i} T_{0,i} + \frac{1+e^{-2\Gamma_0 d}}{1-e^{-2\Gamma_0 d}}$$
$$b_0 = T_{0,0}(1 + j\frac{k_0}{\gamma_0}) \quad ; \quad c_0 = -2R_0 \frac{e^{-\Gamma_0 d}}{1-e^{-2\Gamma_0 d}} \tag{20}$$

The end correction $\Delta\ell/a$ of an orifice follows from $\dfrac{\Delta\ell}{a} = \dfrac{\mathrm{Im}\{Z_{sb}/Z_0\}}{k_0 a}$. (21)

Analytical results for necks with viscous and thermal losses in a free plate can be represented by the regression :

$$\frac{\Delta\ell}{a} = -0.0454728 - 0.728326\, x - 0.177078\, x^2 + 0.0339531\, y + 0.00810471\, y^2 - 0.00100762\, xy \tag{22}$$
$$x = \lg(\sigma) = \lg(a^2/b^2) \quad ; \quad y = \lg(b/\lambda_0)$$

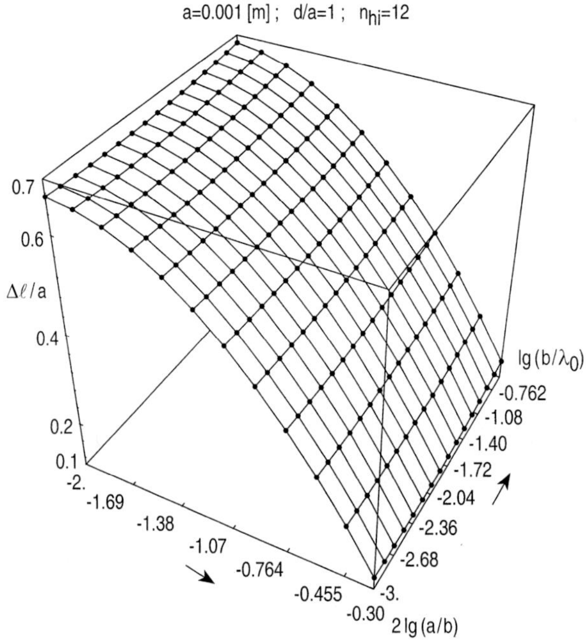

End correction $\Delta\ell/a$ of an orifice of an array of circular necks in a free plate, with losses in the neck;
points: analytical solution;
curves: regression.

H.9 Array of Helmholtz resonators with circular necks

Ref. Mechel, [H.4]

The arrangement is as shown in section H.8 , but now with a finite length t of the cells behind the necks. The fields in zones I and III are formulated as in section H.8 . The field formulation in the zone II is now :

$$p_{III}(x,r)=\sum_{n\geq 0} D_n \cosh(\gamma_n(x-d-t))\, J_0(\eta_n r)$$
$$Z_0 v_{IIIx}(x,r)=j\sum_{n\geq 0} D_n \frac{\gamma_n}{k_0}\sinh(\gamma_n(x-d-t))\, J_0(\eta_n y) \qquad (1)$$

with wave numbers as in section H.8 . The auxiliary quantities X_{n-}, Y_{n-} of that section now are solutions of the coupled system of equations (m=0,1,2,...) :

$$\sum_{n\geq 0} a_{mn}\cdot X_{n-} + c_m\cdot Y_{m-} = b_m\cdot A_e$$
$$c_m\cdot X_{m-} + \sum_{n\geq 0} d_{mn}\cdot Y_{n-} = 0 \qquad (2)$$

with the coefficients $a_{m,n}$, b_m , c_m from section H.8 , and :

$$d_{m,n}=2(\frac{a}{b})^2\frac{\Gamma_n}{k_0}\sum_{i\geq 0}\frac{k_0\coth(\gamma_i t)}{\gamma_i J_0^2(z_i)}T_{m,i}T_{n,i}+\delta_{m,n}\frac{1+e^{-2\Gamma_m d}}{1-e^{-2\Gamma_m d}} \qquad (3)$$

The new mode amplitudes D_n are evaluated with solutions by :

$$D_n=2(\frac{a}{b})^2\frac{k_0}{\gamma_n\sinh(\gamma_n d)\cdot J_0^2(z_n)}\sum_{i=0}\frac{\Gamma_i}{k_0}T_{i,n}\cdot Y_{i-} \qquad (4)$$

The front side and back side orifice impedances Z_{sf} , Z_{sb} are :

$$\frac{Z_{sf}}{Z_0}=\frac{<p_{II}(0,r)>_a}{Z_0<v_{IIx}(0,r)>_a}=j\frac{\sum\limits_{n\geq 0}T_{n,0}\cdot X_{n+}}{\sum\limits_{n\geq 0}\frac{\Gamma_n}{k_0}T_{n,0}\cdot X_{n-}}$$
$$\frac{Z_{sb}}{Z_0}=\frac{<p_{II}(d,r)>_a}{Z_0<v_{IIx}(d,r)>_a}=j\frac{\sum\limits_{n\geq 0}T_{n,0}\cdot Y_{n+}}{\sum\limits_{n\geq 0}\frac{\Gamma_n}{k_0}T_{n,0}\cdot Y_{n-}} \qquad (5)$$

If losses can be neglected in the neck, i.e. $\varepsilon_n a = z_n$ with z_n the solutions of $J_1(z_n)=0$; $n=0,1,2,\ldots$; $z_0=0$, then :

$$\eta_n b = z_n \quad ; \quad \eta_0 = 0 \quad ; \quad \gamma_n^2 = \eta_n^2 - k_0^2 \quad ; \quad \gamma_0 = jk_0$$
$$\varepsilon_n a = z_n \quad ; \quad \varepsilon_0 = 0 \quad ; \quad \Gamma_n^2 = \varepsilon_n^2 - k_0^2 \quad ; \quad \Gamma_0 = jk_0 \tag{6}$$

and :

$$R_m = \frac{1}{a^2}\int_0^a J_0^2(\varepsilon_m r)\cdot r\,dr = \tfrac{1}{2} J_0^2(\varepsilon_m a)$$

$$T_{n,m} = \frac{1}{a^2}\int_0^a J_0(\varepsilon_n r)\, J_0(\eta_m r)\cdot r\,dr = \frac{\varepsilon_n a\, J_1(\varepsilon_n a)\, J_0(z_m \frac{a}{b}) - z_m \frac{a}{b} J_0(\varepsilon_n a)\, J_1(z_m \frac{a}{b})}{(\varepsilon_n a)^2 - z_m^2 (\frac{a}{b})^2}$$

$$= -\frac{\eta_m a\, J_0(\varepsilon_n a)\, J_1(\eta_m a)}{(\varepsilon_n a)^2 - (\eta_m)^2} \tag{7}$$

$$T_{n,0} = \frac{1}{a^2}\int_0^a J_0(\varepsilon_n r)\cdot r\,dr = \frac{J_1(\varepsilon_n a)}{\varepsilon_n a} = 0 \quad ; \quad n>0$$

$$T_{0,m} = \frac{1}{a^2}\int_0^a J_0(\eta_m r)\cdot r\,dr = \frac{J_1(\eta_m a)}{\eta_m a} \quad ; \quad m>0$$

$$T_{0,0} = \frac{1}{2}$$

further: $b_m = \delta_{0,m}$. The orifice impedances then become :

$$\frac{Z_{sb}}{Z_0} = \frac{\langle p_{II}(d,r) \rangle_a}{Z_0 \langle v_{IIx}(d,r) \rangle_a} = \frac{Y_{0+}}{Y_{0-}} \quad ; \quad \frac{Z_{sf}}{Z_0} = \frac{\langle p_{II}(0,r) \rangle_a}{Z_0 \langle v_{IIx}(0,r) \rangle_a} = \frac{X_{0+}}{X_{0-}} \tag{8}$$

The equivalent network of a Helmholtz resonator can be conceived as in the diagram.

Z_R is the radiation resistance of the front orifice;

Z_1, Z_2 represent the neck;

Z_F is the spring reactance of the resonator volume.

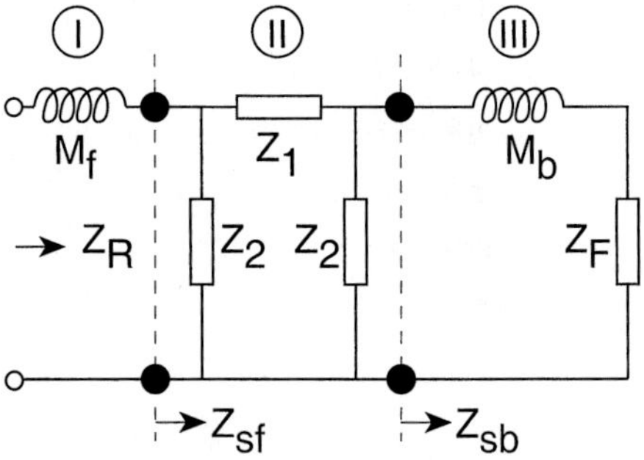

Then M_f, M_b represent the oscillating masses of the front and back orifices.

$$Z_1 = jZ_0 \sin(k_0 d) \quad ; \quad Z_2 = -jZ_0 \frac{\sin(k_0 d)}{1-\cos(k_0 d)} \quad ; \quad Z_F = -j\sigma \cot(k_0 t) \tag{9}$$

with $\sigma = (a/b)^2$ the surface porosity of the neck plate.

The end corrections are given by : $\dfrac{\Delta\ell}{a} = \dfrac{\mathrm{Im}\{Z_M/Z_0\}}{k_0 a} = \dfrac{\omega M/Z_0}{k_0 a}$. (10)

The front side end correction $\Delta\ell_f/a$ is that of the front side orifice of a free plate (see section H.8). With the back side orifice impedance Z_{sb} written as :

$$\frac{Z_{sb}}{Z_0} = \frac{Z_{Mb}}{Z_0} - j\sigma\cot(k_0 t) \tag{11}$$

the interior end correction can be represented for $t/b \geq 0.5$ by : (12)

$$x = \lg\sigma = 2\lg(a/b) \quad ; \quad y = \lg(b/\lambda_0)$$

$$\frac{\Delta\ell_b}{a} =$$

$$-0.0481939 - 0.731823x - 0.179629x^2 + 0.0342687y + 0.00818059y^2 - 0.00101281xy$$

The resonance (angular) frequency ω_0 follows from the resonance condition :

$$k_0 a\left(\frac{\Delta\ell_f}{a} + \frac{\Delta\ell_b}{a}\right) - \sigma\cot(k_0 t) + [1 - (k_0 a)^2 \frac{\Delta\ell_f}{a}\frac{\Delta\ell_b}{a} + \sigma k_0 a \frac{\Delta\ell_f}{a}\cot(k_0 t)] \cdot \tan(k_0 d) = 0 \tag{13}$$

with an approximation for $k_0 d \ll 1$ and $k_0 t \ll 1$
(with S_a= neck cross-section area; V= resonator volume) :

$$(k_0 a)^2 \approx \frac{\sigma \frac{d}{t}}{\frac{d}{a}\left[\frac{d}{a} + \sigma\frac{d}{t}\frac{\Delta\ell_f}{a} + \frac{\Delta\ell_f}{a} + \frac{\Delta\ell_b}{a}\right]} \quad ; \quad \omega_0 \approx c_0\sqrt{\frac{S_a}{V\,(d + \Delta\ell_f + \Delta\ell_b) + S_a d\,\Delta\ell_f}} \tag{14}$$

H.10 Slit resonator array with porous layer in the volume, fields

Ref. Mechel, [H.5]

The arrangement consists of a stiff plate at a distance t+s from a hard wall, which contains an array of parallel slits of width a at a mutual distance L . Behind the plate is a porous layer of thickness d (backed by the wall) at a distance t from the plate.

A plane wave with amplitude A_e is obliquely incident under a polar angle Θ .

Special cases, such as $\Theta=0$; t=0 ; $s=\infty$, will be considered. A great interest lies in the influence of the porous layer on the back side end correction.

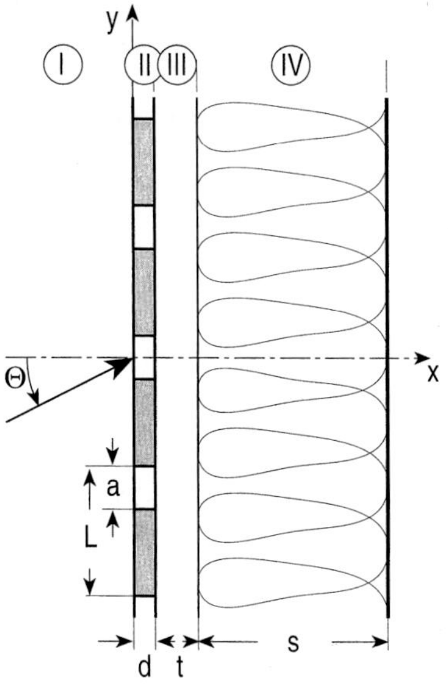

The arrangement is treated as a periodic structure with period length L . The field in the necks is composed as mode sums in a hard duct (viscous and thermal losses can generally be neglected compared with the losses introduced by the porous layer). As a special case, only plane waves in the necks will be also assumed.

Below is $\delta_m = \begin{cases} 1\,; m=0 \\ 2\,; m>0 \end{cases}$ (1)

Oblique incidence ; t >0 ; higher neck modes :

Field formulation in the *Zone I* :

$$p_I(x,y) = A_e\, e^{-j(k_x x + k_y y)} + \sum_{n=-\infty}^{+\infty} A_n \cdot e^{\gamma_n x} \cdot e^{-j\beta_n y}$$

$$Z_0\, v_{Ix}(x,y) = A_e \frac{k_x}{k_0} e^{-j(k_x x + k_y y)} + j \sum_{n=-\infty}^{+\infty} A_n \frac{\gamma_n}{k_0} \cdot e^{\gamma_n x} \cdot e^{-j\beta_n y} \qquad (2)$$

$$k_x = k_0 \cos\Theta \quad ; \quad k_y = k_0 \sin\Theta$$

with : $$\beta_0 = k_y = k_0 \sin\Theta \quad ; \quad \beta_n = \beta_0 + n\frac{2\pi}{L} = k_0(\sin\Theta + n\frac{\lambda_0}{L}) \tag{3}$$

$$\gamma_n^2 = \beta_n^2 - k_0^2 \quad ; \quad \gamma_0 = jk_x = jk_0 \cos\Theta \quad ; \quad \gamma_n = k_0\sqrt{(\sin\theta + n\lambda_0 / L)^2 - 1}$$

The necks in *Zone II* are numbered with $\nu = 0, \pm 1, \pm 2, \ldots$, beginning with the neck which contains the x axis. The local co-ordinate in the ν-th neck is $y_\nu = y - \nu \cdot L$ with $|y_\nu| \le a/2$.

$$p_{II}(x, y_\nu) = e^{-j\beta_0 \nu L} \sum_{m=0}^{\infty} (B_m e^{-j\kappa_m x} + C_m e^{+j\kappa_m x}) \cos(m\pi(\frac{y_\nu}{a} - \frac{1}{2}))$$

$$Z_0 v_{IIx}(x, y_\nu) = e^{-j\beta_0 \nu L} \sum_{m=0}^{\infty} \frac{\kappa_m}{k_0} (B_m e^{-j\kappa_m x} - C_m e^{+j\kappa_m x}) \cos(m\pi(\frac{y_\nu}{a} - \frac{1}{2})) \tag{4}$$

with : $$\kappa_m = \begin{cases} \sqrt{k_0^2 - (m\pi/a)^2} & ; \; m \le m_g \\ -j\sqrt{(m\pi/a)^2 - k_0^2} & ; \; m > m_g \end{cases} \quad ; \quad m_g = \mathrm{INT}(k_0 a/\pi) = \mathrm{INT}(2a/\lambda_0) \tag{5}$$

The index limit m_g defines the transition from propagating modes to cut-off modes.

In the air gap of *Zone III* (with wave numbers as in Zone I) :

$$p_{III}(x, y) = \sum_{n \ge 0} (D_n e^{-\gamma_n y} + E_n e^{+\gamma_n x}) \cdot e^{-j\beta_n y}$$

$$Z_0 v_{IIIx}(x, y) = -j \sum_{n \ge 0} \frac{\gamma_n}{k_0} (D_n e^{-\gamma_n y} - E_n e^{+\gamma_n x}) \cdot e^{-j\beta_n y} \tag{6}$$

In the absorber layer, *Zone IV* :

$$p_{IV}(x, y) = \sum_{n \ge 0} F_n \cosh(\varepsilon_n (x - d - t - s)) \cdot e^{-j\beta_n y}$$

$$Z_0 v_{IVx}(x, y) = -\frac{k_0 Z_0}{\Gamma_a Z_a} \sum_{n \ge 0} \frac{\varepsilon_n}{k_0} F_n \sinh(\varepsilon_n (x - d - t - s)) \cdot e^{-j\beta_n y} \tag{7}$$

with the characteristic propagation constant Γ_a and wave impedance Z_a of the porous material

and : $$\varepsilon_n = \sqrt{\beta_n^2 + \Gamma_a^2} = k_0 \sqrt{(\sin\Theta + n\lambda_0 / L)^2 + (\Gamma_a / k_0)^2} \; . \tag{8}$$

Auxiliary amplitudes are introduced :

$$X_{m\pm} =: B_m \pm C_m \quad ; \quad Y_{m\pm} =: B_m e^{-j\kappa_m d} \pm C_m e^{+j\kappa_m d} \tag{9}$$

with intrinsic relations :

$$
\begin{aligned}
X_{m+} &= X_{m-}\frac{1+e^{-2j\kappa_m d}}{1-e^{-2j\kappa_m d}} - 2Y_{m-}\frac{e^{-j\kappa_m d}}{1-e^{-2j\kappa_m d}} \\
Y_{m+} &= 2X_{m-}\frac{e^{-j\kappa_m d}}{1-e^{-2j\kappa_m d}} - Y_{m-}\frac{1+e^{-2j\kappa_m d}}{1-e^{-2j\kappa_m d}}
\end{aligned} \tag{10}
$$

and giving the other amplitudes by :

$$
\begin{aligned}
B_m &= \tfrac{1}{2}(X_{m+}+X_{m-}) = \tfrac{1}{2}(Y_{m+}+Y_{m-})e^{+j\kappa_m d} \\
C_m &= \tfrac{1}{2}(X_{m+}-X_{m-}) = \tfrac{1}{2}(Y_{m+}-Y_{m-})e^{-j\kappa_m d} \\
A_n &= j\frac{k_0}{\gamma_n}\Big[\delta_{0,n}\cos\Theta - \frac{a}{2L}\sum_{m=0}^{\infty} X_{m-}\frac{\kappa_m}{k_0}s_{m,n}\Big] \\
D_n &= j\frac{a}{2L}\frac{k_0}{\gamma_n}\frac{e^{+\gamma_n d}}{1-r_n\,e^{-2\gamma_n t}}\sum_{m=0}^{\infty} Y_{m-}\frac{\kappa_m}{k_0}s_{m,n} \\
E_n &= D_n\cdot r_n\,e^{-2\gamma_n(d+t)} \\
F_n &= \frac{1}{\cosh(\varepsilon_n s)}\left(D_n\,e^{-\gamma_n(d+t)} + E_n\,e^{+\gamma_n(d+t)}\right)
\end{aligned} \tag{11}
$$

with the modal reflection factors at the zone boundary III–IV :

$$
r_n = \frac{1-j\dfrac{k_0 Z_0}{\Gamma_a Z_a}\dfrac{\varepsilon_n}{\gamma_n}\tanh(\varepsilon_n s)}{1+j\dfrac{k_0 Z_0}{\Gamma_a Z_a}\dfrac{\varepsilon_n}{\gamma_n}\tanh(\varepsilon_n s)} \tag{12}
$$

The X_{n-}, Y_{n-} are solutions of the coupled system of linear equations :

$$
\sum_{n=0}^{\infty} a_{m,n}X_{n-} + c_m Y_{m-} = b_m \quad ; \quad \sum_{n=0}^{\infty} d_{m,n}Y_{n-} + c_m X_{m-} = 0 \tag{13}
$$

with the coefficients :

$$
\begin{aligned}
a_{m,n} &= j\frac{a}{2L}\frac{\kappa_n}{k_0}\sum_{i=-\infty}^{\infty}\frac{k_0}{\gamma_i}s_{m,i}s_{n,i} + 2(-1)^m\frac{\delta_{m,n}}{\delta_m}\frac{1+e^{-2j\kappa_m d}}{1-e^{-2j\kappa_m d}} \\
d_{m,n} &= j\frac{a}{2L}\frac{\kappa_n}{k_0}\sum_{i=-\infty}^{\infty}\frac{k_0}{\gamma_i}\frac{1+r_i\,e^{-2\gamma_i t}}{1-r_i\,e^{-2\gamma_i t}}s_{m,i}s_{n,i} + 2(-1)^m\frac{\delta_{m,n}}{\delta_m}\frac{1+e^{-2j\kappa_m d}}{1-e^{-2j\kappa_m d}} \\
c_m &= -\frac{4(-1)^m}{\delta_m}\frac{e^{-j\kappa_m d}}{1-e^{-2j\kappa_m d}} \quad ; \quad b_m = 2\,s_{m,0}
\end{aligned} \tag{14}
$$

where

$$s_{m,n} = e^{jm\pi/2}\left[\frac{\sin((m\pi-\beta_n a)/2)}{(m\pi-\beta_n a)/2} + (-1)^m \frac{\sin((m\pi+\beta_n a)/2)}{(m\pi+\beta_n a)/2}\right]$$

$$= 2\frac{\beta_n a/2}{(\beta_n a/2)^2-(m\pi/2)^2}\cdot\begin{cases}\sin(\beta_n a/2) & ;\ m=\text{even}\\ -j\cos(\beta_n a/2) & ;\ m=\text{odd}\end{cases} = \begin{cases}2 & ;\ \beta_n a=m\pi\,,\quad m=0\\ e^{jm\pi/2} & ;\ \beta_n a=m\pi\,,\quad m\neq 0\end{cases}$$

$$= \begin{cases}2 & ;\ -\beta_n a=m\pi\,,\quad m=0\\ (-1)^m e^{jm\pi/2} & ;\ -\beta_n a=m\pi\,,\quad m\neq 0\end{cases}$$

and $s_{-m,n}= s_{m,n}$. (15)

Oblique incidence ; t >0 ; higher neck modes ; infinite layer thickness $s\to\infty$:

Change the field formulation in *Zone IV* to :

$$p_{IV}(x,y) = \sum_{n\geq 0} F_n\, e^{-\varepsilon_n x}\, e^{-j\beta_n y}$$

$$Z_0\, v_{IVx}(x,y) = \frac{k_0 Z_0}{\Gamma_a Z_a}\sum_{n\geq 0}\frac{\varepsilon_n}{k_0} F_n\, e^{-\varepsilon_n x}\, e^{-j\beta_n y} \qquad (16)$$

substitute in r_n : $\tanh(\varepsilon_n s)\to 1$

and evaluate the amplitudes from : $F_n = e^{\varepsilon_n (d+t)}\left(D_n\, e^{-\gamma_n (d+t)} + E_n\, e^{+\gamma_n (d+t)}\right)$ (17)

the other expressions remain .

Normal incidence ; t >0 ; higher neck modes :

It is not advisable to treat this case as a special case $\Theta=0$ of the above results, because the anti-symmetrical modes for oblique incidence (odd m) will vanish, and the matrix will get a banded structure.

Field in *Zone I* :

$$p_I(x,y) = A_e\, e^{-jk_0 x} + \sum_{n\geq 0}\delta_n\, A_n\, e^{\gamma_n x}\cos(\eta_n y)$$

$$Z_0\, v_{Ix}(x,y) = A_e\, e^{-jk_0 x} + j\sum_{n\geq 0}\delta_n \frac{\gamma_n}{k_0} A_n\, e^{\gamma_n x}\cos(\eta_n y) \qquad (18)$$

with: $\delta_n = \begin{cases}1 & ;\ n=0\\ 2 & ;\ n>0\end{cases}$; $\eta_n = n\dfrac{2\pi}{L}$; $\gamma_n = \sqrt{\eta_n^2-k_0^2}$; $\gamma_0 = jk_0$ (19)

Field in *Zone II* :

$$p_{II}(x,y) = \sum_{m\geq 0}(B_m\, e^{-j\kappa_m x} + C_m\, e^{+j\kappa_m x})\cos(2m\pi y/a)$$

$$Z_0\, v_{IIx}(x,y) = \sum_{m\geq 0}\frac{\kappa_m}{k_0}(B_m\, e^{-j\kappa_m x} - C_m\, e^{+j\kappa_m x})\cos(2m\pi y/a) \tag{20}$$

with:
$$\kappa_m = \begin{cases}\sqrt{k_0^2 - (2m\pi/a)^2} & ;\ m \leq m_g \\ -j\sqrt{(2m\pi/a)^2 - k_0^2} & ;\ m > m_g\end{cases} \quad ;\ \kappa_0 = k_0 \tag{21}$$

and the limit index for cut-off : $m_g = \mathrm{INT}(a/\lambda_0)$.

Field in *Zone III* :

$$p_{III}(x,y) = \sum_{n\geq 0}\delta_n\,(D_n\, e^{-\gamma_n x} + E_n\, e^{+\gamma_n x})\cos(\eta_n y)$$

$$Z_0\, v_{IIIx}(x,y) = -j\sum_{n\geq 0}\delta_n\frac{\gamma_n}{k_0}(D_n\, e^{-\gamma_n x} - E_n\, e^{+\gamma_n x})\cos(\eta_n y) \tag{22}$$

Field in *Zone IV* :

$$p_{IV}(x,y) = \sum_{n\geq 0}\delta_n\, F_n\cosh(\varepsilon_n(x-d-t-s))\cos(\eta_n y)$$

$$Z_0\, v_{IVx}(x,y) = -\frac{k_0 Z_0}{\Gamma_a Z_a}\sum_{n\geq 0}\delta_n\frac{\varepsilon_n}{k_0}F_n\sinh(\varepsilon_n(x-d-t-s))\cos(\eta_n y) \tag{23}$$

with:
$$\varepsilon_n = \sqrt{\eta_n^2 + \Gamma_a^2}\quad ;\ \varepsilon_0 = \Gamma_a \tag{24}$$

Mode coupling coefficients :

$$\begin{aligned}
S_{m,n} &:= \frac{1}{a}\int_{-a/2}^{+a/2}\cos(2m\pi y/a)\cos(\eta_n y)\,dy \\
&= \frac{1}{2}\left[\frac{\sin(m\pi - n\pi a/L)}{m\pi - n\pi a/L} + \frac{\sin(m\pi + n\pi a/L)}{m\pi + n\pi a/L}\right] \ ;\ m,n>0 \\
&= \frac{-(-1)^m}{\pi}\frac{na/L}{m^2 - (na/L)^2}\sin(n\pi a/L)\ ;\ m,n>0\ ,\ m\neq na/L \\
&= \frac{\sin(n\pi a/L)}{n\pi a/L}\ ;\ m=0\ ,\ n\geq 0 \\
&= 0\ ;\ m>0\ ,\ n=0 \\
&= 1\ ;\ m=n=0 \\
&= \tfrac{1}{2}\ ;\ m = na/L \neq 0
\end{aligned} \tag{25}$$

The auxiliary amplitudes X_{n-} , Y_{n-} from above are again solutions of two coupled systems of equations as above, but with coefficients :

$$a_{m,n} = j\frac{a}{L}\frac{\kappa_n}{k_0}\sum_{i\geq 0}\delta_i\frac{k_0}{\gamma_i}S_{m,i}\,S_{n,i} + \frac{\delta_{m,n}}{\delta_m}\frac{1+e^{-2j\kappa_m d}}{1-e^{-2j\kappa_m d}}$$

$$d_{m,n} = j\frac{a}{L}\frac{\kappa_n}{k_0}\sum_{i\geq 0}\delta_i\frac{k_0}{\gamma_i}S_{m,i}\,S_{n,i}\frac{1+r_i\,e^{-2\gamma_i t}}{1-r_i\,e^{-2\gamma_i t}} + \frac{\delta_{m,n}}{\delta_m}\frac{1+e^{-2j\kappa_m d}}{1-e^{-2j\kappa_m d}} \tag{26}$$

$$c_m = -\frac{2}{\delta_m}\frac{e^{-j\kappa_m d}}{1-e^{-2j\kappa_m d}} \quad ; \quad b_m = 2\delta_{0,m}$$

The mode amplitudes follow from solutions of this system as :

$$A_m = j\frac{k_0}{\gamma_m}[\delta_{0,m} - \frac{a}{L}\sum_{n=0}\frac{\kappa_n}{k_0}S_{n,m}\,X_{n-}]$$

$$B_m = \tfrac{1}{2}(X_{m+} + X_{m-}) = \tfrac{1}{2}(Y_{m+} + Y_{m-})e^{+j\kappa_m d}$$

$$C_m = \tfrac{1}{2}(X_{m+} - X_{m-}) = \tfrac{1}{2}(Y_{m+} - Y_{m-})e^{-j\kappa_m d}$$

$$D_m = j\frac{a}{L}\frac{e^{+\gamma_m d}}{1-r_m\,e^{-2\gamma_m t}}\frac{k_0}{\gamma_m}\sum_{n=0}\frac{\kappa_n}{k_0}S_{n,m}\,Y_{n-} \tag{27}$$

$$E_m = r_m\,e^{-2\gamma_m(d+t)}\,D_m$$

$$F_m = \frac{1}{\cosh(\varepsilon_m s)}(D_m\,e^{-\gamma_m(d+t)} + E_m\,e^{+\gamma_m(d+t)})$$

Normal incidence ; t >0 ;only plane waves in the neck :

The system of equations to be solved simplifies to :

$$a_{0,0}\cdot X_{0-} + c_0\cdot Y_{0-} = b_0 \quad ; \quad c_0\cdot X_{0-} + d_{0,0}\cdot Y_{0-} = 0 \tag{28}$$

or :

$$X_{0-} = \frac{b_0 d_{0,0}}{a_{0,0}d_{0,0} - c_0^2} \quad ; \quad Y_{0-} = \frac{-b_0 c_0}{a_{0,0}d_{0,0} - c_0^2} \quad ; \quad \frac{X_{0-}}{Y_{0-}} = -\frac{d_{0,0}}{c_0} \tag{29}$$

with coefficients :

$$a_{0,0} = j\frac{a}{L}\sum_{i\geq 0}\delta_i\frac{k_0}{\gamma_i}\left(\frac{\sin(i\pi a/L)}{i\pi a/L}\right)^2 - j\cot(k_0 d)$$

$$d_{0,0} = j\frac{a}{L}\sum_{i\geq 0}\delta_i\frac{k_0}{\gamma_i}\left(\frac{\sin(i\pi a/L)}{i\pi a/L}\right)^2\frac{1+r_i\,e^{-2\gamma_i t}}{1-r_i\,e^{-2\gamma_i t}} - j\cot(k_0 d) \tag{30}$$

$$c_0 = \frac{j}{\sin(k_0 d)} \quad ; \quad b_0 = 2$$

The modal reflection factors are :

$$r_n = \frac{1 - j\dfrac{k_0 Z_0}{\Gamma_a Z_a}\dfrac{\varepsilon_n}{\gamma_n}\,\mathrm{Tanh}(\varepsilon_n s)}{1 + j\dfrac{k_0 Z_0}{\Gamma_a Z_a}\dfrac{\varepsilon_n}{\gamma_n}\,\mathrm{Tanh}(\varepsilon_n s)} \tag{31}$$

and the mode amplitudes :

$$\begin{aligned}
A_n &= j\frac{k_0}{\gamma_n}[\delta_{0,n} - \frac{a}{L}\frac{\sin(n\pi a/L)}{n\pi a/L}X_{0-}] \\
B_0 &= \tfrac{1}{2}(X_{0+} + X_{0-}) = \tfrac{1}{2}(Y_{0+} + Y_{0-})e^{+jk_0 d} \\
C_0 &= \tfrac{1}{2}(X_{0+} - X_{0-}) = \tfrac{1}{2}(Y_{0+} - Y_{0-})e^{-jk_0 d} \\
D_n &= j\frac{a}{L}\frac{e^{+\gamma_n d}}{1 - r_n e^{-2\gamma_n t}}\frac{k_0}{\gamma_n}\frac{\sin(n\pi a/L)}{n\pi a/L}Y_{0-} \\
E_n &= r_n e^{-2\gamma_n (d+t)} D_n \\
F_n &= \frac{1}{\cosh(\varepsilon_n s)}(D_n e^{-\gamma_n (d+t)} + E_n e^{+\gamma_n (d+t)})
\end{aligned} \tag{32}$$

Normal incidence ; t =0 ; higher neck modes :

If the neck plate is in contact with the porous layer, Zone III is obsolete; the amplitudes D_n, E_n are not needed. The axial function in the Zone IV changes to $\cosh(\varepsilon_n(x-d-s))$; $\sinh(\varepsilon_n(x-d-s))$. The system of equations for X_{n-}, Y_{n-} has the coefficients :

$$\begin{aligned}
a_{m,n} &= j\frac{a}{L}\frac{\kappa_n}{k_0}\sum_{i\ge 0}\delta_i\frac{k_0}{\gamma_i}S_{m,i}S_{n,i} + \frac{\delta_{m,n}}{\delta_m}\frac{1+e^{-2j\kappa_m d}}{1-e^{-2j\kappa_m d}} \\
d_{m,n} &= j\frac{a}{L}\frac{\kappa_n}{k_0}\frac{\Gamma_a Z_a}{k_0 Z_0}\sum_{i\ge 0}\delta_i\frac{k_0}{\varepsilon_i}S_{m,i}S_{n,i}\coth(\varepsilon_i s) + \frac{\delta_{m,n}}{\delta_m}\frac{1+e^{-2j\kappa_m d}}{1-e^{-2j\kappa_m d}} \\
c_m &= -\frac{2}{\delta_m}\frac{e^{-j\kappa_m d}}{1-e^{-2j\kappa_m d}} \quad ; \quad b_m = 2\delta_{0,m}
\end{aligned} \tag{33}$$

The amplitudes F_m follow from :

$$F_n = \frac{a}{L}\frac{\Gamma_a Z_a}{k_0 Z_0}\frac{k_0}{\varepsilon_n \cdot \sinh(\varepsilon_n s)}\sum_{m=0}\frac{\kappa_m}{k_0}S_{m,n}\cdot Y_{m-} \tag{34}$$

with the other amplitudes as above.

Normal incidence ; t = 0 ; only plane waves in the neck :

The coefficients of the two equations for X_{0-} and Y_{0-} become :

$$a_{0,0} = j\frac{a}{L}\sum_{i\ge 0}\delta_i \frac{k_0}{\gamma_i}\left(\frac{\sin(i\pi a/L)}{i\pi a/L}\right)^2 + \frac{1+e^{-2jk_0d}}{1-e^{-2jk_0d}}$$

$$d_{0,0} = \frac{a}{L}\frac{\Gamma_a Z_a}{k_0 Z_0}\sum_{i\ge 0}\delta_i \frac{k_0}{\varepsilon_i}\left(\frac{\sin(i\pi a/L)}{i\pi a/L}\right)^2 \coth(\varepsilon_i s) + \frac{1+e^{-2jk_0d}}{1-e^{-2jk_0d}} \qquad (35)$$

$$c_0 = -2\frac{e^{-jk_0d}}{1-e^{-2jk_0d}} \quad ; \quad b_0 = 2$$

H.11 Slit resonator array with porous layer in the volume, impedances

Ref. Mechel, [H.5]

The object is the same as in the previous section H.10 .

The intention in this section is to evaluate the average impedance Z (without radiation resistance Z_R) of an array of slit Helmholtz resonators with a porous absorber layer in the resonator volume by a chain of equations, which represent a simple equivalent network. Some elements of the network are described with the help of the field evaluations in the previous section H.10.

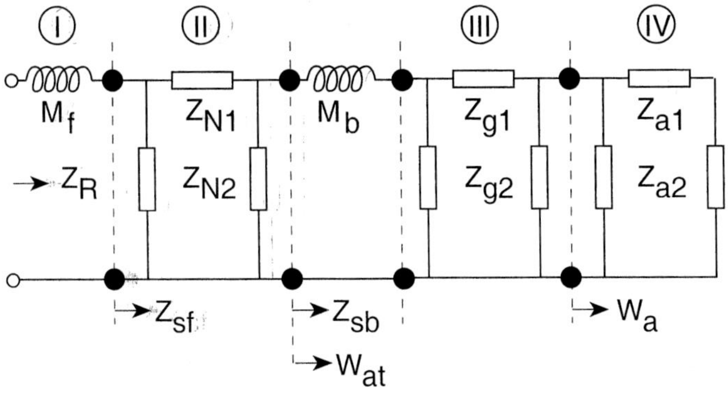

The symmetrical Π-fourpoles with Z_{N1},Z_{N2} ; Z_{g1},Z_{g2} ; Z_{a1},Z_{a2} represent, respectively, the neck, the air gap between the neck plate and absorber layer, and the absorber layer. W_a is the input impedance of the absorber layer; W_{at} is that impedance transformed to the back side surface of the neck plate; Z_{sf}, Z_{sb} are the orifice impedances of the front side orifice and back side orifice. M_f, M_b are the oscillating masses of the two orifices.

The equations are :

$$\frac{Z}{Z_0}=\frac{L}{a}[j k_0 a\frac{\Delta\ell}{a}+\frac{Z_{sf}}{Z_0}]$$
$$\frac{Z_{sf}}{Z_0}=\frac{j\tan(k_0 d)+Z_{sb}/Z_0}{1+j\,Z_{sb}/Z_0\cdot\tan(k_0 d)}$$
$$\frac{Z_{sb}}{Z_0}=\frac{Z_{Msb}}{Z_0}+\frac{a}{L}\frac{W_{at}}{Z_0}\quad;\quad\frac{Z_{Msb}}{Z_0}=j k_0 a\frac{\Delta\ell_b}{a} \tag{1}$$
$$\frac{W_{at}}{Z_0}=\frac{j\tan(k_0 t)+W_a/Z_0}{1+j\,W_a/Z_0\cdot\tan(k_0 t)}\quad;\quad\frac{W_a}{Z_0}=\frac{Z_a/Z_0}{\tanh(\Gamma_a s)}$$

$\Delta\ell/a$ is the front side end correction; the 3rd line is a defining equation for the back side end correction $\Delta\ell_b/a$ using the back side orifice impedance Z_{sb} evaluated from the sound field as given in section H.10 for different conditions. Some of them will be considered below. There $\Gamma_{an}=\Gamma_a/k_0$; $Z_{an}=Z_a/Z_0$ are normalised characteristic values of the porous material.

Normal sound incidence, absorber layer in contact with neck plate, t=0, plane wave in the neck

The back side orifice impedance Z_{sb} is (with quantities from section H.10) :

$$\begin{aligned}\frac{Z_{sb}}{Z_0}&=\frac{\langle p_{II}(d,y)\rangle_a}{Z_0\langle v_{IIx}(d,y)\rangle_a}=\frac{Y_{0+}}{Y_{0-}}\\&=-2\frac{e^{-jk_0 d}}{1-e^{-2jk_0 d}}\frac{d_{0,0}}{c_0}-\frac{1+e^{-2jk_0 d}}{1-e^{-2jk_0 d}}=d_{0,0}-\frac{1+e^{-2jk_0 d}}{1-e^{-2jk_0 d}}\\&=\frac{a}{L}\frac{\Gamma_a Z_a}{k_0 Z_0}\sum_{n\geq 0}\delta_n\frac{k_0}{\varepsilon_n}(\frac{\sin(n\pi a/L)}{n\pi a/L})^2\coth(\varepsilon_n s)\\&=\frac{a}{L}[\frac{Z_a/Z_0}{\tanh(\Gamma_a s)}+2\frac{\Gamma_a Z_a}{k_0 Z_0}\sum_{n>0}\frac{k_0}{\varepsilon_n}(\frac{\sin(n\pi a/L)}{n\pi a/L})^2\coth(\varepsilon_n s)\,]\end{aligned} \tag{2}$$

For the absorber layer in contact with the neck plate, as supposed here, $W_a=W_{at}$; therefore evidently :

$$\frac{Z_{Mb}}{Z_0}=2\frac{a}{L}\frac{\Gamma_a Z_a}{k_0 Z_0}\sum_{n>0}\frac{k_0}{\varepsilon_n}(\frac{\sin(n\pi a/L)}{n\pi a/L})^2\coth(\varepsilon_n s) \tag{3}$$

and :

$$\frac{\Delta\ell_b}{a} = \frac{-j}{k_0 a}\frac{Z_{Mb}}{Z_0}$$

$$= -2j\,\frac{s}{L}\frac{\Gamma_a Z_a}{k_0 Z_0}\sum_{n>0}\left(\frac{\sin(n\pi a/L)}{n\pi a/L}\right)^2 \frac{\coth(\varepsilon_n s)}{\varepsilon_n s} = 2\frac{\sigma}{\sigma_a}\frac{\rho_{eff}}{\rho_0}\frac{s}{a}\sum_{n>0}\left(\frac{\sin(n\pi a/L)}{n\pi a/L}\right)^2 \frac{\coth(\varepsilon_n s)}{\varepsilon_n s}$$

with : $$\frac{\Gamma_a Z_a}{k_0 Z_0} = \frac{j}{\sigma_a}\frac{\rho_{eff}}{\rho_0} \qquad (4)$$

where σ_a= porosity of the porous material ; ρ_{eff}= its (acoustical) effective density .

A set of variables for $\Delta\ell_b/a$ is :

$$F = \frac{L}{\lambda_0} \;;\; R = \frac{\Xi s}{Z_0} \;;\; E = \frac{\rho_0 f}{\Xi} = \frac{s}{L}\frac{\Gamma}{R} \;;\; \frac{a}{L} \;;\; \frac{d}{a} \;;\; \frac{t}{L} \;;\; \frac{s}{L} \qquad (5)$$

A regression through analytically evaluated end correction values for the absorber layer in contact with the neck plate is :

$$\begin{aligned}\frac{\Delta\ell_b}{a} = j\frac{\Gamma_a Z_a}{k_0 Z_0}\cdot(&0.0389998 + 0.454066\cdot x - 0.345328\cdot x^2 - 0.125386\cdot x^3 - \\ &0.0143782\cdot y + 0.00418541\cdot y^2 + 0.0170766\cdot y^3 - \\ &0.0142094\cdot z - 0.0715597\cdot z^2 + 0.0915584\cdot z^3 - \\ &0.0115326\cdot x\cdot y - 0.0195509\cdot x\cdot z - 0.0595634\cdot y\cdot z)\end{aligned} \qquad (6)$$

$x = \lg(a/L)$; $y = \lg(R) = \lg(\Xi s/Z_0)$; $z = \lg(s/L)$

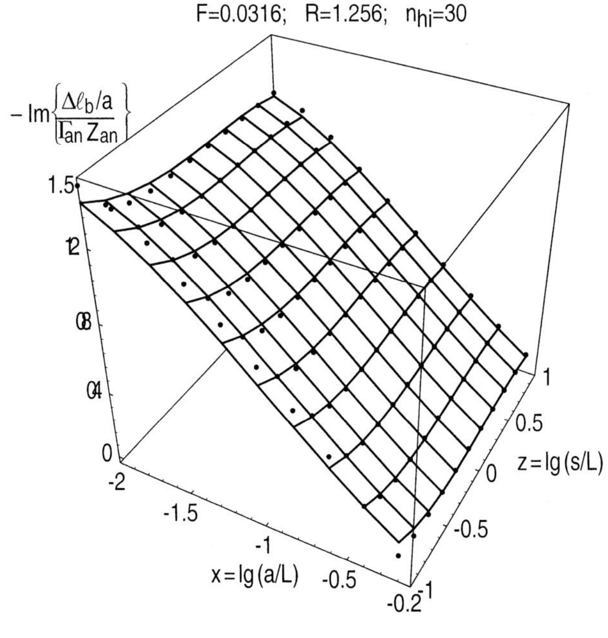

Back side orifice end correction $\Delta\ell_b/a$ if the absorber layer is in contact with the neck plate ; points: analytic evaluation; curves: regression.

Normal sound incidence, t>0, plane wave in the neck :

The back side orifice impedance Z_{sb} is (with quantities from the previous section H.10) :

$$\begin{aligned}\frac{Z_{sb}}{Z_0} &= \frac{\langle p_{II}(d,y)\rangle_a}{Z_0\langle v_{IIx}(d,y)\rangle_a} = \frac{Y_{0+}}{Y_{0-}} \\ &= -2\frac{e^{-jk_0d}}{1-e^{-2jk_0d}}\frac{d_{0,0}}{c_0} - \frac{1+e^{-2jk_0d}}{1-e^{-2jk_0d}} = d_{0,0} - \frac{1+e^{-2jk_0d}}{1-e^{-2jk_0d}} \\ &= j\frac{a}{L}\sum_{n\geq 0}\delta_n\frac{k_0}{\gamma_n}\left(\frac{\sin(n\pi a/L)}{n\pi a/L}\right)^2\frac{1+r_n\,e^{-2\gamma_n t}}{1-r_n\,e^{-2\gamma_n t}} \\ &= \frac{a}{L}\left[\frac{1+r_0\,e^{-2jk_0t}}{1-r_0\,e^{-2jk_0t}} + 2j\sum_{n>0}\frac{k_0}{\gamma_n}\left(\frac{\sin(n\pi a/L)}{n\pi a/L}\right)^2\frac{1+r_n\,e^{-2\gamma_n t}}{1-r_n\,e^{-2\gamma_n t}}\right]\end{aligned} \tag{7}$$

and :

$$\frac{Z_{Mb}}{Z_0} = 2j\frac{a}{L}\sum_{n>0}\frac{k_0}{\gamma_n}\left(\frac{\sin(n\pi a/L)}{n\pi a/L}\right)^2\frac{1+r_n\,e^{-2\gamma_n t}}{1-r_n\,e^{-2\gamma_n t}} \tag{8}$$

The variation of Z_{Mb} in the parameter space makes it not suited for the definition of an end correction with a representation by regression in this case.

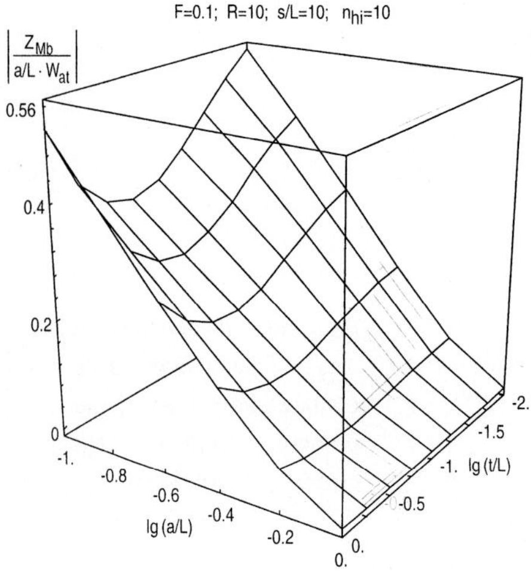

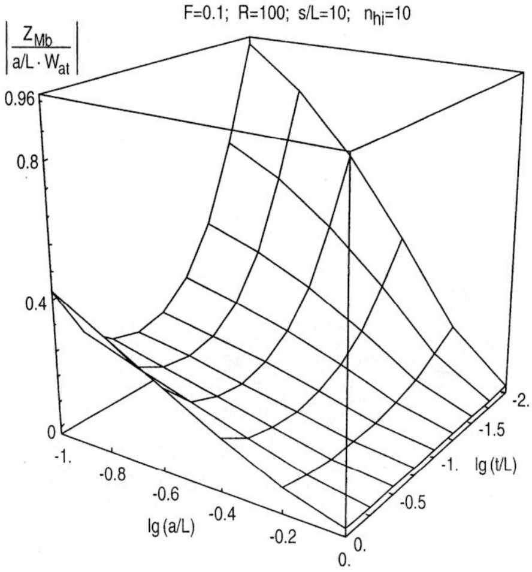

Normal sound incidence, infinite absorber layer in contact with neck plate, t=0; s=∞, plane wave in the neck :

The impedance Z_{Mb} of the back orifice oscillating mass is (with quantities from the previous section H.10) :

$$\frac{Z_{Mb}}{Z_0} = 2\frac{a}{L}\Gamma_{an}Z_{an}\sum_{n>0}\left(\frac{\sin(n\pi a/L)}{n\pi a/L}\right)^2\frac{1}{\sqrt{(n\lambda_0/L)^2+\Gamma_{an}^2}} \tag{9}$$

from the comparison with the corresponding impedance Z_{Mb0} of a free neck plate, and assuming $n\cdot\lambda_0/L >> |\Gamma_{an}|^2$; $n\cdot\lambda_0/L >> 1$ for all $n\geq 1$, one gets :

$$\frac{Z_{Mb}}{Z_0} \approx -j\Gamma_{an}Z_{an}\frac{Z_{Mb0}}{Z_0} = k_0 a\Gamma_{an}Z_{an}\frac{\Delta\ell}{a} \tag{10}$$

With a higher order approximation for $\lambda_0/L >> 1$:

$$\begin{aligned}\frac{Z_{Mb}}{Z_0} &\approx -j\Gamma_{an}Z_{an}\frac{Z_{Mb0}}{Z_0} + 2\frac{a}{L}\left(\frac{\sin(\pi a/L)}{\pi a/L}\right)^2\Gamma_{an}Z_{an}\left(\frac{1}{\sqrt{(\lambda_0/L)^2+\Gamma_{an}^2}}-\frac{1}{\sqrt{(\lambda_0/L)^2-1}}\right) \\ &\approx k_0 a\Gamma_{an}Z_{an}\left[\frac{\Delta\ell}{a}+\frac{1}{\pi}\left(\frac{\sin(\pi a/L)}{\pi a/L}\right)^2\left(\frac{1}{\sqrt{1-(L/\lambda_0\cdot\Gamma_{an})^2}}-1\right)\right]\end{aligned} \tag{11}$$

Therein $\Delta\ell/a$ is the front side orifice end correction.

Normal sound incidence, infinite absorber layer, s=∞, air gap between neck plate and porous layer, t> 0 , plane wave in the neck :

The back orifice oscillating mass impedance is :

$$\frac{Z_{Mb}}{Z_0} = 2j\frac{a}{L}\sum_{n>0}\frac{k_0}{\gamma_n}\left(\frac{\sin(n\pi a/L)}{n\pi a/L}\right)^2\frac{1+r_n\,e^{-2\gamma_n t}}{1-r_n\,e^{-2\gamma_n t}} \tag{12}$$

and the modal reflection factors in this case : $r_n = \left(1-\frac{j}{\Gamma_{an}Z_{an}}\frac{\varepsilon_n}{\gamma_n}\right)/\left(1+\frac{j}{\Gamma_{an}Z_{an}}\frac{\varepsilon_n}{\gamma_n}\right)$ (13)

with the wave number ratio :

$$\frac{\varepsilon_n}{\gamma_n} = \sqrt{\frac{(n\lambda_0/L)^2+\Gamma_{an}^2}{(n\lambda_0/L)^2-1}} \tag{14}$$

consequently :

$$\frac{1+r_n\,e^{-2\gamma_n t}}{1-r_n\,e^{-2\gamma_n t}} = \frac{1+\frac{j}{\Gamma_{an}Z_{an}}\frac{\varepsilon_n}{\gamma_n}\tanh(\gamma_n t)}{\tanh(\gamma_n t)+\frac{j}{\Gamma_{an}Z_{an}}\frac{\varepsilon_n}{\gamma_n}} \tag{15}$$

Under the condition and ensuing approximations :

$$(n\lambda_0/L)^2 \gg \begin{cases} 1 \\ |\Gamma_{an}|^2 \end{cases} \quad ; \quad \frac{\varepsilon_n}{\gamma_n} \approx 1 \quad ; \quad \gamma_n t \approx k_0 t \cdot n\lambda_0/L = 2\pi n\, t/L \tag{16}$$

one gets :

$$\frac{1+r_n\, e^{-2\gamma_n t}}{1-r_n\, e^{-2\gamma_n t}} \approx \frac{1+\dfrac{j}{\Gamma_{an} Z_{an}} \tanh(2\pi n\, t/L)}{\tanh(2\pi n\, t/L)+\dfrac{j}{\Gamma_{an} Z_{an}}} \xrightarrow[\pi t/L>1]{} \frac{1+\dfrac{j}{\Gamma_{an} Z_{an}}}{1+\dfrac{j}{\Gamma_{an} Z_{an}}} = 1 \tag{17}$$

Thus for $t/L > 1/\pi$: $\dfrac{Z_{Mb}}{Z_0} \approx \dfrac{Z_{Mb0}}{Z_0}$.

An approximation of higher order is :

$$\frac{Z_{msh}}{Z_0} \xrightarrow[\substack{s\to\infty \\ \lambda_0/L \gg 1}]{}$$

$$jk_0 a \left[\frac{\Delta\ell}{a} + \frac{1}{\pi}\left(\frac{\sin(\pi a/L)}{\pi a/L}\right)^2 \frac{\left(1-\dfrac{j}{\Gamma_{an} Z_{an}} \sqrt{\dfrac{(\lambda_0/L)^2+\Gamma_{an}^2}{(\lambda_0/L)^2-1}}\right)(1-\tanh(2\pi t/L))}{\tanh(2\pi t/L)+\dfrac{j}{\Gamma_{an} Z_{an}} \sqrt{\dfrac{(\lambda_0/L)^2+\Gamma_{an}^2}{(\lambda_0/L)^2-1}}} \right] \tag{18}$$

$$\approx jk_0 a \left[\frac{\Delta\ell}{a} + \frac{1}{\pi}\left(\frac{\sin(\pi a/L)}{\pi a/L}\right)^2 \frac{\left(1-\dfrac{j}{\Gamma_{an} Z_{an}} \sqrt{1+(\Gamma_{an} L/\lambda_0)^2}\right)(1-\tanh(2\pi t/L))}{\tanh(2\pi t/L)+\dfrac{j}{\Gamma_{an} Z_{an}} \sqrt{1+(\Gamma_{an} L/\lambda_0)^2}} \right]$$

H.12 Slit resonator array with porous layer on back orifice

Ref. Mechel, [H.6]

The object here is similar to that in sections H.10, H.11 , but a (possibly thin) absorber layer covers the back side orifices and allows an air space deeper in the resonator volume. This is a common method to apply an additional loss to Helmholtz resonators.

A (stiff) plate of thickness d contains an array of parallel slits, width a , mutual distance L , each of which is backed with a resonator volume V of depth t .
A porous material layer of thickness s<t is in contact with the back side of the plate.
The characteristic values of the layer material are Γ_a, Z_a , and in normalised form $\Gamma_{an}= \Gamma_a/k_0$; $Z_{an}= Z_a/Z_0$. If the layer becomes a cloth, wire mesh, felt, etc , the limit transition $s \to 0$ is made with its flow resistance $R= \Xi s/Z_0$ kept constant.

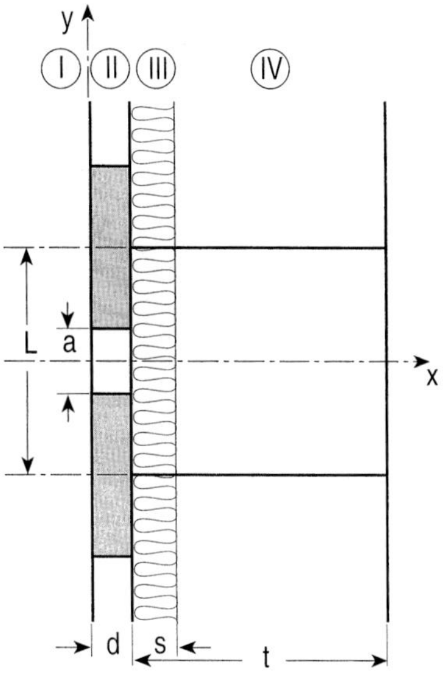

A normally incident plane wave with amplitude A_e is assumed.

The field formulations in the zones I and II are taken from section H.10 .

Field formulation in *Zone III* :

$$\begin{aligned} p_{III}(x,y) &= \sum_{n\geq 0} \delta_n \left(D_n\, e^{-\varepsilon_n x} + E_n\, e^{+\varepsilon_n x}\right) \cos(\eta_n y) \\ Z_0\, v_{IIIx}(x,y) &= \frac{k_0 Z_0}{\Gamma_a Z_a} \sum_{n\geq 0} \delta_n \frac{\varepsilon_n}{k_0} \left(D_n\, e^{-\varepsilon_n x} - E_n\, e^{+\varepsilon_n x}\right) \cos(\eta_n y) \end{aligned} \tag{1}$$

Field formulation in *Zone IV* :

$$\begin{aligned} p_{IV}(x,y) &= \sum_{n\geq 0} \delta_n\, F_n\, \cosh(\gamma_n (x-d-t)) \cdot \cos(\eta_n y) \\ Z_0 v_{IVx}(x,y) &= j \sum_{n\geq 0} \delta_n \frac{\gamma_n}{k_0} F_n\, \sinh(\gamma_n (x-d-t)) \cdot \cos(\eta_n y) \end{aligned} \tag{2}$$

with wave numbers and propagation constants :

$$\eta_n = 2n\pi/L \;\; ; \;\; \gamma_n = \sqrt{\eta_n^2 - k_0^2} \;\; ; \;\; \gamma_0 = jk_0$$

$$\kappa_m = \begin{cases} \sqrt{k_0^2 - (2m\pi a/L)^2} & ; \; m \le m_g \\ -j\sqrt{(2m\pi a/L)^2 - k_0^2} & ; \; m > m_g \end{cases} \;\; ; \;\; m_g = \mathrm{INT}(a/\lambda_0) \;\; ; \;\; \kappa_0 = k_0 \tag{3}$$

$$\varepsilon_n = \sqrt{\eta_n^2 + \Gamma_a^2} \;\; ; \;\; \varepsilon_0 = \Gamma_a$$

The limit order m_g separates cut-on and cut-off (i.e. radiating and non-radiating) spatial harmonics.

Modal reflection factors at the back surface of the porous layer are :

$$r_n = \frac{\dfrac{j}{\Gamma_{an} Z_{an}} \dfrac{\varepsilon_n}{\gamma_n} \coth(\gamma_n (t-s)) - 1}{\dfrac{j}{\Gamma_{an} Z_{an}} \dfrac{\varepsilon_n}{\gamma_n} \coth(\gamma_n (t-s)) + 1} \tag{4}$$

The auxiliary amplitudes $X_{n\pm}, Y_{n\pm}$ are defined as in section H.10 . The X_{n-}, Y_{n-} are solutions of the coupled system of equations :

$$\sum_{n=0}^{\infty} a_{m,n} X_{n-} + c_m Y_{m-} = b_m \;\; ; \;\; \sum_{n=0}^{\infty} d_{m,n} Y_{n-} + c_m X_{m-} = 0 \tag{5}$$

with the coefficients :

$$a_{m,n} = j \frac{a}{L} \frac{\kappa_n}{k_0} \sum_{i \ge 0} \delta_i \frac{k_0}{\gamma_i} S_{m,i} S_{n,i} + \frac{\delta_{m,n}}{\delta_m} \frac{1 + e^{-2j\kappa_m d}}{1 - e^{-2j\kappa_m d}}$$

$$d_{m,n} = \frac{a}{L} \frac{\Gamma_a Z_a}{k_0 Z_0} \frac{\kappa_n}{k_0} \sum_{i \ge 0} \delta_i \frac{k_0}{\varepsilon_i} S_{m,i} S_{n,i} \frac{1 + r_i e^{-2\varepsilon_i s}}{1 - r_i e^{-2\varepsilon_i s}} + \frac{\delta_{m,n}}{\delta_m} \frac{1 + e^{-2j\kappa_m d}}{1 - e^{-2j\kappa_m d}} \tag{6}$$

$$c_m = \frac{-2}{\delta_m} \frac{e^{-j\kappa_m d}}{1 - e^{-2j\kappa_m d}} \;\; ; \;\; b_m = 2\delta_{0,m}$$

$$\text{wherein :} \quad \delta_m = \begin{cases} 1 ; m = 0 \\ 2 ; m > 0 \end{cases} \;\; ; \;\; \delta_{m,n} = \begin{cases} 1 ; m = n \\ 0 ; m \ne n \end{cases} \tag{7}$$

and $S_{m,n}$ is found in section H.10 .

The field term amplitudes follow from the solutions X_{n-}, Y_{n-} as :

$$A_n = -j \frac{k_0}{\gamma_n} \Big[\frac{a}{L} \sum_{m=0} \frac{\kappa_m}{k_0} S_{m,n} \cdot X_{m-} - \delta_{0,n} A_e \Big] \tag{8}$$

$$B_m = \tfrac{1}{2}(X_{m+} + X_{m-}) = \tfrac{1}{2}(Y_{m+} + Y_{m-}) e^{+j\kappa_m d} \tag{9}$$

$$C_m = \tfrac{1}{2}(X_{m+} - X_{m-}) = \tfrac{1}{2}(Y_{m+} - Y_{m-})e^{-j\kappa_m d}$$

$$D_n = \frac{a}{L}\frac{\Gamma_a Z_a}{k_0 Z_0}\frac{k_0}{\varepsilon_n}\frac{e^{\varepsilon_n d}}{1 - e^{-\varepsilon_n s}[1 + r_n e^{-2\varepsilon_n s} - e^{-\varepsilon_n s}} \sum_{m\geq 0} \frac{\kappa_m}{k_0} S_{m,n} \cdot Y_{m-} \tag{10}$$

$$E_n = D_n e^{-2\varepsilon_n (d+s)}[1 - e^{\varepsilon_n s} + r_n e^{-\varepsilon_n s}]$$

$$F_n = \frac{1}{\cosh(\gamma_n (t-s))}[D_n e^{-\varepsilon_n (d+s)} + E_n e^{+\varepsilon_n (d+s)}]$$

The back orifice impedance Z_{sb} becomes :

$$\frac{Z_{sb}}{Z_0} = \frac{a}{L} Z_{an} [\frac{1 + r_0 e^{-2\Gamma_a s}}{1 - r_0 e^{-2\Gamma_a s}} + 2\Gamma_{an} \sum_{i>0} (\frac{\sin(i\pi a/L)}{i\pi a/L})^2 \frac{k_0}{\varepsilon_i} \frac{1 + r_i e^{-2\varepsilon_i s}}{1 - r_i e^{-2\varepsilon_i s}}] \tag{11}$$

and the impedance Z_{Mb} of the oscillating mass at the back orifice :

$$\frac{Z_{Mb}}{Z_0} = 2\frac{a}{L}\Gamma_{an} Z_{an} \sum_{i>0} (\frac{\sin(i\pi a/L)}{i\pi a/L})^2 \frac{k_0}{\varepsilon_i} \frac{1 + r_i e^{-2\varepsilon_i s}}{1 - r_i e^{-2\varepsilon_i s}} \tag{12}$$

In the limit $s \to 0$ for thin orifice covers :

$$\frac{Z_{sb}}{Z_0} = \frac{a}{L}[R - j\cot(k_0 t) + 2j\sum_{i>0} (\frac{\sin(i\pi a/L)}{i\pi a/L})^2 \frac{k_0}{\gamma_i} \coth(\gamma_i t)] \tag{13}$$

This is the value for an empty resonator volume except for the term R for the flow resistance of the thin cover. If R has no reactive component, the tuning of the resonator is not changed by the additional resistance, nor the end correction of the interior orifice.

If the porous foil can freely oscillate, the substitution $R \to R_{eff}$ should be made, with :

$$R_{eff} = \frac{1}{Z_0}\frac{j\omega m_f \cdot \Xi s}{j\omega m_f + \Xi s} = \frac{jk_0 s \cdot \rho_f/\rho_0 \cdot R}{jk_0 s \cdot \rho_f/\rho_0 + R} \tag{14}$$

where m_f is the surface mass density of the foil, and ρ_f the density of the foil material.

H.13 Slit resonator array with porous layer on front orifice

Ref. Mechel, [H.6]

Sections H.10 – H.12 deal with additional damping of Helmholtz resonators with porous layers. This section considers an arrangement of the porous layer which gives the possibility to combine the broad-band absorption of a porous layer with the peak absorption of a resonator.

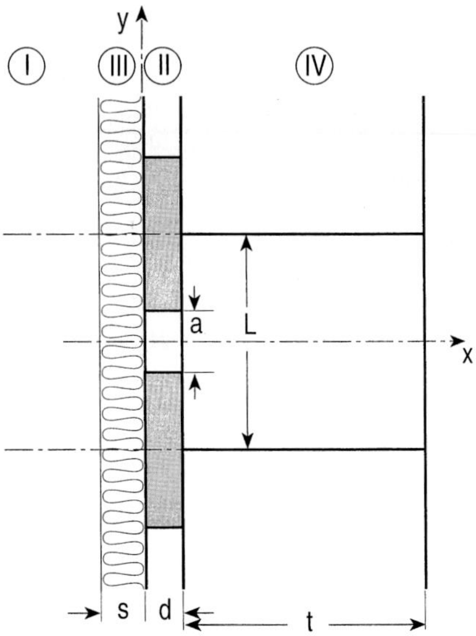

In contrast to the previous sections, the porous layer is now placed on the front side of the resonator array.

A plane wave with amplitude A_e is normally incident.

The characteristic values of the layer material are Γ_a, Z_a , and in normalised form $\Gamma_{an}=\Gamma_a/k_0$; $Z_{an}=Z_a/Z_0$. If the layer becomes a cloth, wire mesh, felt, etc. , the limit transition $s\to 0$ is made with its flow resistance $R=\Xi s/Z_0$ kept constant.

The field formulations in the zones are :

Zone I :

$$\begin{aligned} p_I(x,y) &= A_e\,e^{-jk_0x} + \sum_{n\ge 0}\delta_n A_n\,e^{\gamma_n x}\cos(\eta_n y) \\ Z_0 v_{Ix}(x,y) &= A_e\,e^{-jk_0x} + j\sum_{n\ge 0}\delta_n\frac{\gamma_n}{k_0}A_n\,e^{\gamma_n x}\cos(\eta_n y) \end{aligned} \tag{1}$$

Zone III :

$$\begin{aligned} p_{III}(x,y) &= \sum_{n\ge 0}\delta_n(D_n\,e^{-\varepsilon_n x}+E_n\,e^{+\varepsilon_n x})\cos(\eta_n y) \\ Z_0 v_{IIIx}(x,y) &= \frac{k_0 Z_0}{\Gamma_a Z_a}\sum_{n\ge 0}\delta_n\frac{\varepsilon_n}{k_0}(D_n\,e^{-\varepsilon_n x}-E_n\,e^{+\varepsilon_n x})\cos(\eta_n y) \end{aligned} \tag{2}$$

Zone II :

$$\begin{aligned} p_{II}(x,y) &= \sum_{m\ge 0}(B_m\,e^{-j\kappa_m x}+C_m\,e^{+j\kappa_m x})\cos(2m\pi y/a) \\ Z_0 v_{IIx}(x,y) &= \sum_{m\ge 0}\frac{\kappa_m}{k_0}(B_m\,e^{-j\kappa_m x}-C_m\,e^{+j\kappa_m x})\cos(2m\pi y/a) \end{aligned} \tag{3}$$

Zone IV :

$$\begin{aligned} p_{IV}(x,y) &= \sum_{n\ge 0}\delta_n F_n\cosh(\gamma_n(x-d-t))\cos(\eta_n y) \\ Z_0 v_{IVx}(x,y) &= j\sum_{n\ge 0}\delta_n\frac{\gamma_n}{k_0}F_n\sinh(\gamma_n(x-d-t))\cos(\eta_n y) \end{aligned} \tag{4}$$

The wave numbers and propagation constants are :

$$\eta_n = 2n\pi/L \;;\; \gamma_n = \sqrt{\eta_n^2 - k_0^2} \;;\; \gamma_0 = j k_0$$

$$\kappa_m = \begin{cases} \sqrt{k_0^2 - (2m\pi a/L)^2} & ; \; m \le m_g \\ -j\sqrt{(2m\pi a/L)^2 - k_0^2} & ; \; m > m_g \end{cases} \;;\; m_g = \mathrm{INT}(a/\lambda_0) \;;\; \kappa_0 = k_0 \tag{5}$$

$$\varepsilon_n = \sqrt{\eta_n^2 + \Gamma_a^2} \;;\; \varepsilon_0 = \Gamma_a$$

The boundary conditions give the set of equations for the mode amplitudes :

$$\begin{aligned}
&\delta_{0,n} A_e\, e^{+jk_0 s} + j\frac{\gamma_n}{k_0} A_n\, e^{-\gamma_n s} = \frac{k_0 Z_0}{\Gamma_a Z_a}\frac{\varepsilon_n}{k_0}(D_n\, e^{+\varepsilon_n s} - E_n\, e^{-\varepsilon_n s}) \\
&\frac{k_0 Z_0}{\Gamma_a Z_a}\frac{\varepsilon_n}{k_0}(D_n - E_n) = \frac{a}{L}\sum_{m\ge 0}\frac{\kappa_m}{k_0} S_{m,n}(B_m - C_m) \\
&\frac{\gamma_n}{k_0} F_n \sinh(\gamma_n t) = j\frac{a}{L}\sum_{m\ge 0}\frac{\kappa_m}{k_0} S_{m,n}(B_m\, e^{-j\kappa_m d} - C_m\, e^{+j\kappa_m d}) \\
&\delta_{0,n} A_e\, e^{+jk_0 s} + A_n\, e^{-\gamma_n s} = (D_n\, e^{+\varepsilon_n s} + E_n\, e^{-\varepsilon_n s}) \\
&\frac{1}{\delta_m}(B_m + C_m) = \sum_{n\ge 0}\delta_n S_{m,n}(D_n + E_n) \\
&\frac{1}{\delta_m}(B_m\, e^{-j\kappa_m d} + C_m\, e^{+j\kappa_m d}) = \sum_{n\ge 0}\delta_n S_{m,n} F_n \cosh(\gamma_n t)
\end{aligned} \tag{6}$$

with the mode coupling coefficients :

$$\begin{aligned}
S_{m,n} &:= \frac{1}{a}\int_{-a/2}^{+a/2}\cos(2m\pi y/a)\cos(\eta_n y)\, dy \\
&= \frac{1}{2}\left[\frac{\sin(m\pi - n\pi a/L)}{m\pi - n\pi a/L} + \frac{\sin(m\pi + n\pi a/L)}{m\pi + n\pi a/L}\right] \;;\; m,n>0 \\
&= \frac{-(-1)^m}{\pi}\frac{na/L}{m^2 - (na/L)^2}\sin(n\pi a/L) \;;\; m,n>0\,,\; m \ne na/L \\
&= \frac{\sin(n\pi a/L)}{n\pi a/L} \;;\; m=0\,,\; n\ge 0 \\
&= 0 \;;\; m>0\,,\; n=0 \\
&= 1 \;;\; m=n=0 \\
&= \tfrac{1}{2} \;;\; m = na/L \ne 0
\end{aligned} \tag{7}$$

From now on only plane waves are assumed to exist in the necks, i.e. $B_{m>0} = C_{m>0} = 0$.

The equations of the boundary conditions simplify to :

$\Delta\ell_f$= front side end correction;
$\Delta\ell_b$= end correction of neck orifice towards volume;
Z_1= entrance impedance of the 1st neck;
G_2= admittance of the 2nd neck and air gap in parallel;
G_{f2}= entrance admittance of 2nd neck;
Z_{sb}= output impedance of back orifice of 2nd neck;
G_y= entrance admittance of air gap.

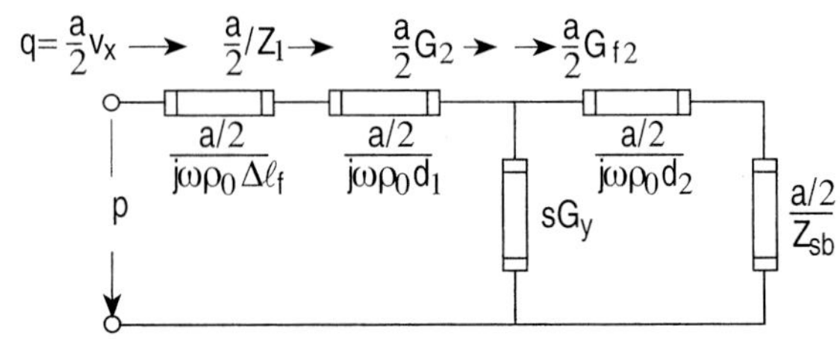

The chain of equations is (G is evaluated without radiation resistance) :

$$Z_0G = \frac{a/L}{jk_0a\dfrac{\Delta\ell_f}{a} + \dfrac{Z_1}{Z_0}} \quad ; \quad \frac{Z_1}{Z_0} = \frac{1 + j\tan(k_0d_1)\cdot Z_0G_2}{j\tan(k_0d_1) + Z_0G_2}$$

$$Z_0G_2 = Z_0G_{f2} + 2\frac{s}{a}Z_0G_y \quad ; \quad Z_0G_{f2} = \frac{1 + j\tan(k_0d_2)\cdot Z_{sb}/Z_0}{j\tan(k_0d_2) + Z_{sb}/Z_0} \qquad (1)$$

$$\frac{Z_{sb}}{Z_0} = \frac{-ja/L}{\tan(k_0t)} + jk_0a\frac{\Delta\ell_b}{a} + R_{res} \quad ; \quad G_y = \frac{1}{Z_y}\tanh(\Gamma_y\ell)$$

$\Delta\ell_f$, $\Delta\ell_b$ may be taken from a previous section about end corrections in a resonator array; R_{res} stands for a possibly added (normalised) resistance representing additional losses in the back side orifice and/or in the resonator volume. Γ_y, Z_y are the characteristic propagation constant and wave impedance, respectively, in a flat capillary of width s (see sections J1, J2 about capillaries in the chapter "Duct Acoustics").

If a poro-elastic foil (see later sections about foils) with effective surface mass density m_{eff} tightly covers the entrance orifice of the 2nd neck, and if the air gap length ℓ is different on both sides of the neck, $\ell \to \ell_1, \ell_2$, then evaluate G_{f2} and G_2 from :

$$\frac{1}{Z_0G_{f2}} = j\frac{\omega m_{eff}}{Z_0} + \frac{j\tan(k_0d_2) + Z_{sb}/Z_0}{1 + j\tan(k_0d_2)\cdot Z_{sb}/Z_0}$$

$$Z_0G_2 = Z_0G_{f2} + \frac{s}{a}(Z_0G_{y1} + Z_0G_{y2}) \qquad (2)$$

H.15 Array of slit resonators with subdivided neck plate and floating foil in the gap

Ref.: Mechel, [H.7]

The object is similar to that in the previous section H.14 , but the analysis is rather different. A freely floating poro-elastic foil (see later sections about foils) with effective surface mass density m_{eff} is placed in the air gap between the parts of the neck plate. From a technical point of view this is a possibility to protect mechanicaly sensitive foils; from an analytical point of view, sound transmission through the foil over all of its length must be considered. The analysis assumes (possibly different) air gap thicknesses s_1,s_2 in front of and behind the foil.

The necks have the shapes shown only for analytical reasons: they indicate that no shift of the co-ordinates in front of and behind the neck plate is expressly considered.

Schematic arrangement :

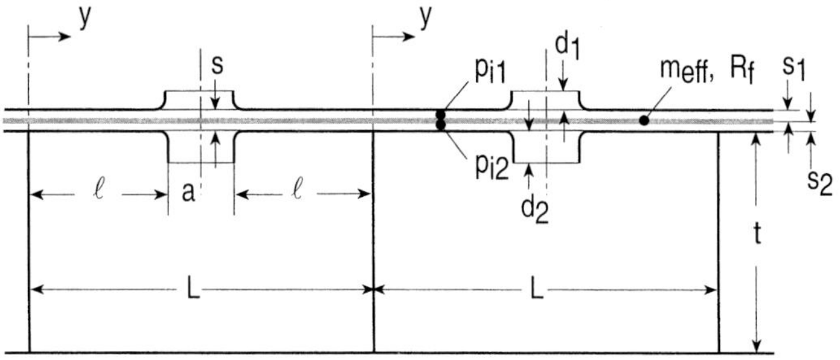

Schematic network :

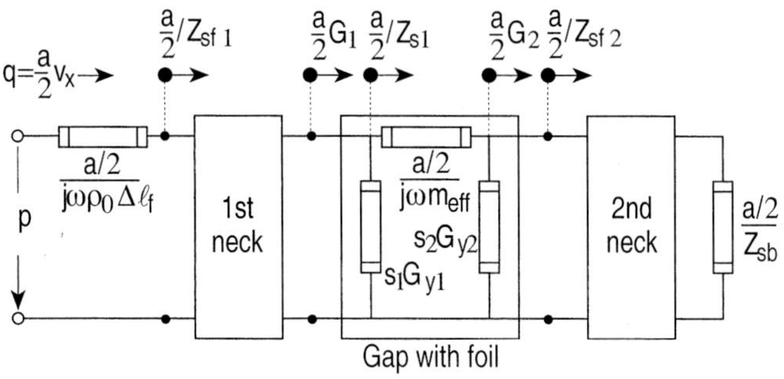

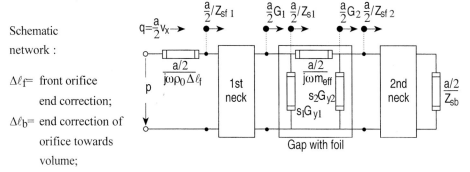

$\Delta\ell_f=$ front orifice end correction;

$\Delta\ell_b=$ end correction of orifice towards volume;

$Z_{sf1}=$ entrance impedance of 1st neck; $Z_{sf2}=$ entrance impedance of 2nd neck;

$Z_{sb}=$ exit impedance of 2nd neck;

G_1= admittance at exit of 1st neck; G_2= load admittance of foil;

G_{y1}, G_{y2}= input admittances of air gaps.

Detail in the air gaps :

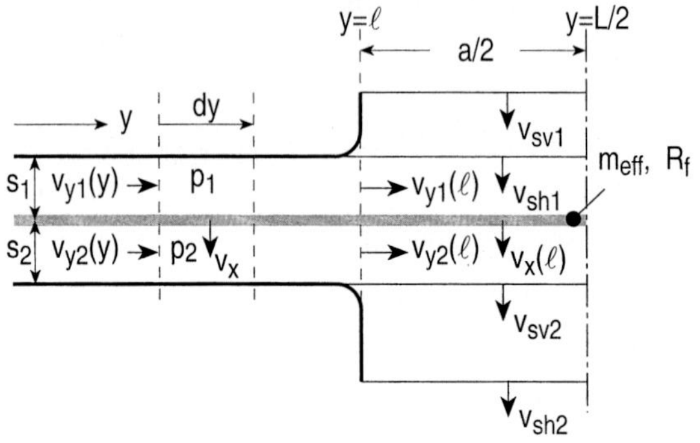

The chain of equations for the average surface admittance $G = 1/Z$ of the arrangement is, with the assumption (for the moment) that sound transmission through the foil in the gaps can be neglected :

$$\frac{1}{Z_0 G} = (jk_0 a \cdot \Delta\ell_f / a + Z_{sf1} / Z_0) \cdot L / a$$

$$\frac{Z_{sf1}}{Z_0} = \frac{1 + j\tan(k_0 d_1) \cdot Z_0 G_1}{j\tan(k_0 d_1) + Z_0 G_1}$$

$$Z_0 G_1 = \frac{Z_0}{Z_{s1}} + 2\frac{s_1}{a} \cdot Z_0 G_{y1}$$

$$\frac{Z_{s1}}{Z_0} = \frac{j\omega\, m_{eff}}{Z_0} + \frac{1}{Z_0 G_2} \qquad (1)$$

$$Z_0 G_2 = \frac{Z_0}{Z_{sf2}} + 2\frac{s_2}{a} \cdot Z_0 G_{y2}$$

$$\frac{Z_{sf2}}{Z_0} = \frac{j\tan(k_0 d_2) + Z_{sh} / Z_0}{1 + j\tan(k_0 d_2) \cdot Z_{sh} / Z_0}$$

$$\frac{Z_{sb}}{Z_0} = \frac{-ja / L}{\tan(k_0 t)} + jk_0 a \cdot \Delta\ell_b / a + R_{res}$$

Γ_{y1}, Γ_{y2}=	capillary propagation constants in the gaps;
Z_{y1}, Z_{y2}=	capillary wave impedances in the gaps;
$\Delta\ell_f/a$=	end correction of front side orifice;
$\Delta\ell_b/a$=	end correction of orifice towards resonator;
m_{eff}=	effective surface mass density of foil:
R_{res}=	normalised resistance for possible additional loss in the volume;

$$G_{y1} = \frac{1}{Z_{y1}} \tanh(\Gamma_{y1}\ell) \quad ; \quad G_{y2} = \frac{1}{Z_{y2}} \tanh(\Gamma_{y2}\ell) \qquad (2)$$

The two gaps on each side in fact are coupled wave guides; the differential equations for which are :

$$p_1 - p_2 = j\omega m \cdot v_x$$

$$-\frac{\partial v_{y1}}{\partial y} - \frac{v_x}{s_1} = j\omega C_{eff,1} \cdot p_1 \quad ; \quad -\frac{\partial v_{y2}}{\partial y} + \frac{v_x}{s_2} = j\omega C_{eff,2} \cdot p_2 \tag{3}$$

$$-\frac{\partial p_1}{\partial y} = j\omega\rho_{eff} \cdot v_{y1} \quad ; \quad -\frac{\partial p_2}{\partial y} = j\omega\rho_{eff} \cdot v_{y2}$$

where the effective air densities $\rho_{eff,i}$ and air compressibilities $C_{eff,i}$ follow from the capillary propagation constants $\Gamma_{y,i}$ and wave impedances $Z_{y,i}$ (i=1,2) by :

$$\frac{\Gamma_{y,i}}{k_0} = j\sqrt{\frac{\rho_{eff,i}}{\rho_0} \cdot \frac{C_{eff,i}}{C_0}} \quad ; \quad \frac{Z_{y,i}}{Z_0} = \sqrt{\frac{\rho_{eff,i}}{\rho_0} \Big/ \frac{C_{eff,i}}{C_0}} \tag{4}$$

One gets two coupled, inhomogeneous wave equations for the fields in the air gaps :

$$\begin{aligned} \frac{\partial^2 p_1}{\partial y^2} + [\omega^2 \rho_{eff,1} C_{eff,1} - \frac{\rho_{eff,1}}{m s_1}] \cdot p_1 &= -\frac{\rho_{eff,1}}{m s_1} \cdot p_2 \\ \frac{\partial^2 p_2}{\partial y^2} + [\omega^2 \rho_{eff,2} C_{eff,2} - \frac{\rho_{eff,2}}{m s_2}] \cdot p_2 &= -\frac{\rho_{eff,2}}{m s_2} \cdot p_1 \end{aligned} \tag{5}$$

with solutions satisfying the symmetry conditions :

$$p_i(y) = A_i \cdot \cosh(\Gamma_a y) + B_i \cdot \cosh(\Gamma_b y) \quad ; \quad i = 1, 2 \tag{6}$$

The characteristic equation of the system has the solutions :

$$\Gamma_a^2 = \Gamma_b^2 = \frac{\Gamma_{y1}^2 \dfrac{\rho_{eff,2}}{m s_2} - \Gamma_{y2}^2 \dfrac{\rho_{eff,1}}{m s_1}}{\dfrac{\rho_{eff,2}}{m s_2} - \dfrac{\rho_{eff,1}}{m s_1}} := \Gamma^2 \tag{7}$$

Therefore the gap fields are :

$$p_i(y) = A_i \cdot \cosh(\Gamma y) \quad ; \quad v_{yi}(y) = -A_i \frac{\Gamma}{j\omega\rho_{eff,i}} \cdot \sinh(\Gamma y) \quad ; \quad i = 1, 2 \tag{8}$$

and the gap input admittances :

$$G_{yi} = -\frac{v_{yi}(\ell)}{p_i(\ell)} = \frac{\Gamma}{j\omega\rho_{eff,i}} \tanh(\Gamma\ell) \tag{9}$$

The load admittance G_1 of the back orifice of the 1st neck (which is needed in the 2nd line of the chain of equations) follows as : (10)

$$Z_0 G_1 = j\frac{Z_0}{\omega m_{eff}} \left[\frac{1}{1 - (\dfrac{\omega}{\omega_2})^2 + j\dfrac{\omega m_{eff}}{Z_0}(Z_0 G_{sv2} + 2\dfrac{s_2}{a} G_{y2})} - (1 - (\frac{\omega}{\omega_1})^2 + j\frac{\omega m_{eff}}{Z_0} \cdot 2\frac{s_1}{a} G_{y1}) \right]$$

If different gap lengths ℓ_1,ℓ_2 are used on opposite sides of the neck, substitute :

$$2\frac{s_i}{a} Z_0 G_{yi}(\ell) \rightarrow \frac{s_i}{a}\left(Z_0 G_{yi}(\ell_1) + Z_0 G_{yi}(\ell_2)\right) \quad ; \quad i=1,2 \tag{11}$$

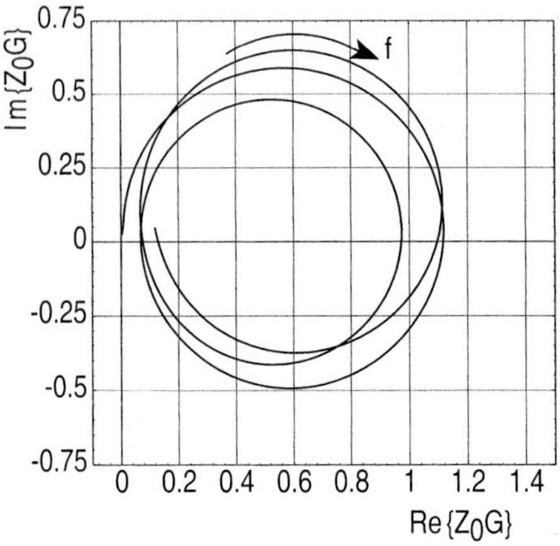

Normalised surface admittance Z_0G of a slit resonator array with subdivided neck plate and a floating poro-elastic foil in the gap, for increasing frequency f .
For parameters see below.

R_f=0.35 ; $f_{cr}\cdot d$=12 [Hz·m] ; ρ_f=2750 [kg/m^2] ; d_f=0.0001 [m] ;
L=0.05 [m] ; a=0.02 [m] ; s=0.002 [m] ; d=0.02 [m] ; t=0.1 [m] ;
d_1/d=0.5 ; s_1/s=0.45 ; Ξ=125 [Pa·s/m2] ;

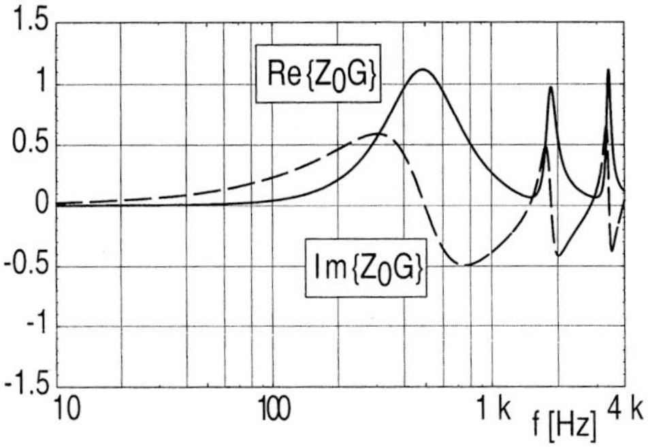

Frequency response curves of the real and imaginary parts of Z_0G for the above arrangement.

The parameters of the above example are $s=s_1+s_2$; $d=d_1+d_2$; R_f= normalised flow resistance of the poro-elastic foil ; d_f= foil thickness ; ρ_f= foil material density (the foil is of aluminium); $f_{cr}\cdot d$= product of foil thickness and critical frequency.

H.16 Array of slit resonators covered with a foil

Ref.: Mechel, [H.7]

An array of slit resonators is covered with a poro-elastic foil of effective surface mass density m_{eff} (for ease of writing also $m_{eff} \to m$). The width of the air gaps between the foil and the neck plate may be different on both sides of a neck, and also the length of the gaps.

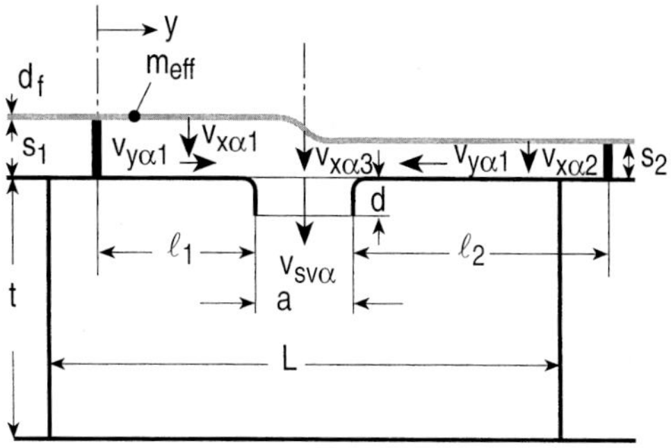

$s = (s_1+s_2)/2$; $L = \ell_1+\ell_2+a$

The resonator neck may be excited in three ways :

- via the left-hand side foil (α=1);
- via the right-hand side foil (α=2);
- via the foil in the orifice range (α=3).

A separate index ß=1,2,3 indicates the foil ranges (left, right, centre). The three ways of excitation can be considered as three sources. According to Helmholtz's source superposition theorem the general state is a superposition of three states, in which only one source is active and the two others are short-circuited. The following graph shows a schematical p/q-network of the arrangement with the three sources.

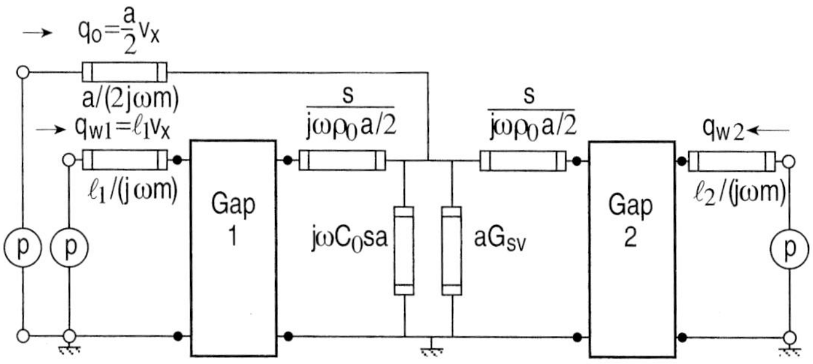

Let $\rho_{eff,ß}$, $C_{eff,ß}$ be the effective air density and air compressibility in a flat capillary with hard and rigid walls having lateral dimension $s_ß$; they can be evaluated from the capillary propagation constant $\Gamma_{yß}$ and wave impedance $Z_{yß}$ (see sections J1, J2 about sound in capillaries in the chapter "Duct Acoustics") by :

$$\Gamma_{y\beta}^2 = -\omega^2 \rho_{eff,\beta} C_{eff,\beta} \quad ; \quad j\omega \rho_{eff,\beta} = \Gamma_{y\beta} Z_{y\beta} \tag{1}$$

The supports of the foil in the above sketch need not be solid supports; their acoustical role can be played by positions of sound field symmetry.

The differential equations in the gaps are :

$$\begin{aligned} -\frac{\partial v_{y\alpha\beta}}{\partial y} + \frac{v_{x\alpha\beta}}{s_\beta} &= j\omega C_{eff,\beta} \cdot p_{i\alpha\beta} \\ -\frac{\partial p_{i\alpha\beta}}{\partial y} &= j\omega \rho_{eff,\beta} \cdot v_{y\alpha\beta} \quad ; \quad \alpha = 1,2,3 \; ; \; \beta = 1,2 \; ; \; i = 1,2 \end{aligned} \tag{2}$$

They lead to the wave equations :

$$\begin{aligned} &\frac{\partial^2 p_{i\alpha\beta}}{\partial y^2} + \omega^2 \rho_{eff,\beta} C_{eff,\beta} \cdot p_{i\alpha\beta} + j\omega \rho_{eff,\beta} \frac{v_{x\alpha\beta}}{s_\beta} = \\ &\quad \frac{\partial^2 p_{i\alpha\beta}}{\partial y^2} - \Gamma_{y\beta}^2 \cdot p_{i\alpha\beta} + j\omega \rho_{eff,\beta} \frac{v_{x\alpha\beta}}{s_\beta} = 0 \end{aligned} \tag{3}$$

Or, with the Kronecker symbol $\delta_{\alpha,\beta}$:

$$\begin{aligned} &\delta_{\alpha,\beta} \cdot p - p_{i\alpha\beta} = j\omega m_{eff} \cdot v_{x\alpha\beta} \quad ; \quad v_{x\alpha\beta} = \frac{1}{j\omega m_{eff}} (\delta_{\alpha,\beta} \cdot p - p_{i\alpha\beta}) \\ &\frac{\partial^2 p_{i\alpha\beta}}{\partial y^2} - (\Gamma_{y\beta}^2 + \frac{\rho_{eff,\beta}}{m_{eff}\, s_\beta}) \cdot p_{i\alpha\beta} = -\delta_{\alpha,\beta} \frac{\rho_{eff,\beta}}{m_{eff}\, s_\beta} \cdot p \quad ; \quad \alpha = 1,2,3 \; ; \; \beta = 1,2 \; ; \; i = 1,2 \end{aligned} \tag{4}$$

Suitable solutions are :

$$\begin{aligned} p_{i\alpha\beta}(y) &= A_{\alpha\beta} \cosh(\Gamma_\beta y) + \delta_{\alpha,\beta} \cdot B_\beta \cdot p \quad ; \quad \alpha = 1,2,3 \; ; \; \beta = 1,2 \\ v_{y\alpha\beta}(y) &= -\frac{\Gamma_\beta}{j\omega \rho_{eff,\beta}} A_{\alpha\beta} \sinh(\Gamma_\beta y) = -\frac{\Gamma_\beta}{\Gamma_{y\beta} Z_{y\beta}} A_{\alpha\beta} \sinh(\Gamma_\beta y) \end{aligned} \tag{5}$$

with :

$$\begin{aligned} \Gamma_\beta^2 &= \Gamma_{y\beta}^2 + \frac{\rho_{eff,\beta}}{s_\beta m_{eff}} = \Gamma_{y\beta}^2 + \frac{\Gamma_{y\beta} Z_{y\beta}}{j\omega s_\beta m_{eff}} \\ B_\beta &= \frac{\rho_{eff,\beta} / s_\beta m_{eff}}{\Gamma_{y\beta}^2 + \rho_{eff,\beta} / s_\beta m_{eff}} = \frac{1}{1 + j \dfrac{\omega m_{eff}}{Z_0} \dfrac{\Gamma_{y\beta}\, s_\beta}{Z_{y\beta} / Z_0}} \end{aligned} \tag{6}$$

The boundary conditions of continuity in the orifice range lead to :

$$\frac{A_{11}}{p}=-\frac{B_1}{2}\frac{2s_1U_2+WV_2}{U_1V_2+U_2V_1} \quad ; \quad \frac{A_{12}}{p}=+\frac{B_1}{2}\frac{2s_1U_1-WV_1}{U_1V_2+U_2V_1}$$
$$\frac{A_{21}}{p}=+\frac{B_2}{2}\frac{2s_2U_2-WV_2}{U_1V_2+U_2V_1} \quad ; \quad \frac{A_{22}}{p}=-\frac{B_2}{2}\frac{2s_2U_1+WV_1}{U_1V_2+U_2V_1} \qquad (7)$$
$$\frac{A_{31}}{p}=\frac{\cosh(\Gamma_2\ell_2)}{U_1\cosh(\Gamma_2\ell_2)+U_2\cosh(\Gamma_1\ell_1)} \quad ; \quad \frac{A_{32}}{p}=\frac{\cosh(\Gamma_1\ell_1)}{U_1\cosh(\Gamma_2\ell_2)+U_2\cosh(\Gamma_1\ell_1)}$$

with :

$$s=(s_1+s_2)/2 \quad ; \quad \omega_0^2=\frac{1}{m_{eff}\,sC_0}=\frac{\rho_0c_0^2}{m_{eff}\,s}$$
$$W=1-\left(\frac{\omega}{\omega_0}\right)^2+j\omega m_{eff}\,G_{sv}$$
$$U_ß=j\frac{s_ß}{a}\omega m_{eff}\frac{\Gamma_ß}{\Gamma_{yß}Z_{yß}}\sinh(\Gamma_ß\ell_ß)+\frac{1}{2}W\cosh(\Gamma_ß\ell_ß) \quad ; \quad ß=1,2 \qquad (8)$$
$$V_ß=j\frac{sa}{2}\omega\rho_0\frac{\Gamma_ß}{\Gamma_{yß}Z_{yß}}\sinh(\Gamma_ß\ell_ß)+s_ß\cosh(\Gamma_ß\ell_ß) \quad ; \quad ß=1,2$$

The average surface admittance of the arrangement is defined as :

$$G=\frac{1}{L}\frac{1}{p}\left[\int_0^{\ell_1}(v_{x11}+v_{x21}+v_{x31})dy+a\cdot(v_{x13}+v_{x23}+v_{x33})+\int_0^{\ell_2}(v_{x12}+v_{x22}+v_{x32})dy\right] \qquad (9)$$

One gets with the above solutions :

$$Z_0G=j\frac{Z_0}{\omega m_{eff}}\left[-1+\frac{L_1}{L}B_1+\frac{L_2}{L}B_2+\left(\frac{A_{11}}{p}+\frac{A_{21}}{p}+\frac{A_{31}}{p}\right)\cdot\left[\frac{\sinh(\Gamma_1\ell_1)}{\Gamma_1\ell_1}+\frac{a}{2L}\cosh(\Gamma_1\ell_1)\right]+\right.$$
$$\left.+\left(\frac{A_{12}}{p}+\frac{A_{22}}{p}+\frac{A_{32}}{p}\right)\cdot\left[\frac{\sinh(\Gamma_2\ell_2)}{\Gamma_2\ell_2}+\frac{a}{2L}\cosh(\Gamma_2\ell_2)\right]\right]$$

In the *special case of symmetrical gaps*, i.e. $s_1=s_2=s$; $\ell_1=\ell_2=\ell$:

$$Z_0G=j\frac{Z_0}{\omega m}\left[-1+B+\left(\frac{A_{11}}{p}+\frac{A_{21}}{p}+\frac{A_{31}}{p}\right)\cdot\left[2\frac{\sinh(\Gamma\ell)}{\Gamma\ell}+\frac{a}{L}\cosh(\Gamma\ell)\right]\right] \qquad (11)$$
$$\frac{A_{11}}{p}=-\frac{B}{4}\frac{2sU+WV}{UV} \quad ; \quad \frac{A_{21}}{p}=+\frac{B}{4}\frac{2sU-WV}{UV} \quad ; \quad \frac{A_{31}}{p}=\frac{1}{2U}$$

with :

$$\Gamma^2=\Gamma_y^2+\frac{\rho_{eff}}{s\,m_{eff}}=\Gamma_y^2+\frac{\Gamma_yZ_y}{j\omega s\,m_{eff}} \quad ; \quad B=\frac{\rho_{eff}/sm_{eff}}{\Gamma_y^2+\rho_{eff}/sm_{eff}}=\frac{1}{1+j\dfrac{\omega m_{eff}}{Z_0}\dfrac{\Gamma_y s}{Z_y/Z_0}} \qquad (12)$$

$$W = 1 - \left(\frac{\omega}{\omega_0}\right)^2 + j\omega m_{eff}\, G_{sv}$$

$$U = j\frac{s}{a}\omega m_{eff}\frac{\Gamma}{\Gamma_y Z_y}\sinh(\Gamma\ell) + \frac{1}{2}W\cosh(\Gamma\ell) \qquad (13)$$

$$V = j\frac{sa}{2}\omega\rho_0\frac{\Gamma}{\Gamma_y Z_y}\sinh(\Gamma\ell) + s\cdot\cosh(\Gamma\ell)$$

and : $$\frac{A_{11}}{p} + \frac{A_{21}}{p} + \frac{A_{31}}{p} = \frac{1 - BW}{2U}$$

so that finally in this special case :

$$\begin{aligned} Z_0 G &= j\frac{Z_0}{\omega m_{eff}}\left[-1 + B + (1 - BW)\frac{\frac{a}{L} + 2\frac{\tanh(\Gamma\ell)}{\Gamma\ell}}{W + 2\frac{\ell}{a}(1 + j\omega s m_{eff}\frac{\Gamma_y}{Z_y})\frac{\tanh(\Gamma\ell)}{\Gamma\ell}}\right] \\ &= j\frac{Z_0}{\omega m_{eff}}\left[-1 + B + (1 - BW)\frac{\frac{a}{L} + 2\frac{\tanh(\Gamma\ell)}{\Gamma\ell}}{W + 2\frac{\ell}{a}(1 - \omega^2/\omega_{eff}^2)\frac{\tanh(\Gamma\ell)}{\Gamma\ell}}\right] \end{aligned} \qquad (14)$$

with $\omega_{eff}^2 = 1/(m_{eff} s \cdot C_{eff})$ the square of the foil resonance (angular) frequency of the cover foil with the effective air compressibility in the gap.

See [H.7] for a discussion of the result and of a possible parameter non-linearity in the gaps.

H.17 Poro-elastic foils

Ref.: Mechel, [H.7] ; Mechel, [H.8]

Poro-elastic foils may be • tight or porous, • limp or elastic. Their common features are
• lateral homogeneity in scales which are comparable with the free bending wavelength,
• incompressibility (at least approximate). So the description below of poro-elastic foils uniformly covers materials like limp metal or plastic foils, thin porous layers, clothes, gauzes, felts, wire mesh, perforated metal sheets, elastic plates (with or without perforations). Poro-elastic foils need not be plane, they may also have the form of curved shells.

It is assumed that the foil is placed in a co-ordinate surface of a co-ordinate system in which the wave equation is separable. We apply orthogonal co-ordinates $\{x_1, x_2, x_3\}$ for a general survey, and assume that the foil occupies a co-ordinate surface $\{x_1, x_2\}$ and that the co-ordinate x_3 is normal to the foil, which is at the position $x_3=\xi$. It seems that CREMER has introduced the term *Trennimpedanz* (partition impedance) for the quantity Z_T defined by:

$$Z_T = \frac{\Delta p}{v} = \frac{p_{front}(x_1,x_2,\xi) - p_{back}(x_1,x_2,\xi)}{v(x_1,x_2)} \tag{1}$$

where p_{front}, p_{back} are the sound pressures in front and behind the foil, respectively, and v is the velocity of the foil which is counted positive in the direction front → back.

The boundary conditions to be applied at a foil are :

$$v_{front} = v_{back} = v \quad ; \quad p_{front} - p_{back} = Z_T \cdot v \tag{2}$$

Begin with a *tight and limp foil* : $Z_T = j\omega\, m_f = j\omega\rho_f d_f$

where m_f = surface mass density of the foil ; d_f = foil thickness ; ρ_f = foil material density.

1st generalisation : *Porous limp foil* :

The flow resistance $\Sigma = \Xi \cdot d_f$ (Ξ = flow resistivity of the foil material) acts in parallel with the mass reactance of the foil :

$$Z_t = \frac{j\omega\, m_f \cdot \Sigma}{j\omega\, m_f + \Sigma} = j\omega\, m_f \frac{1}{1 + \dfrac{j\omega\, m_f}{\Sigma}} = j\omega\, m_p \tag{3}$$

So this 1st generalisation is performed by the substitution : $m_f \to m_{eff,p} = \dfrac{m_f}{1 + \dfrac{j\omega\, m_f}{\Sigma}}$. (4)

2nd generalisation : *Tight elastic foil* :

The oscillation of the foil (which is indeed a thin plate) obeys the bending wave equation :

$$\left[\Delta_{x_1,x_2}\Delta_{x_1,x_2} - k_B^4\right] v = \frac{j\omega}{B} \cdot \Delta p \tag{5}$$

where $\Delta_{x,y}$ is the Laplace operator in the indicated co-ordinates, k_B is the wave number of the free bending wave on the plate, B is the bending stiffness, and $\Delta p = p_{front} - p_{back}$ is the driving sound pressure difference. With

$$k_B^4 = \omega^2 \frac{m_f}{B} \quad ; \quad \frac{k_0}{k_B} = \sqrt{\frac{f}{f_{cr}}} \tag{6}$$

where $\omega = 2\pi f$ and f_{cr} = (critical) coincidence frequency, one immediately gets :

$$Z_T = \frac{\Delta p}{v_p} = j\omega\, m_f \cdot \left[1 - \left(\frac{f}{f_{cr}} \right)^2 \frac{1}{k_0^4} \cdot \frac{\Delta_{x_1,x_2} \Delta_{x_1,x_2} v}{v} \right] \tag{7}$$

$$\frac{Z_T}{Z_0} = j k_0 \frac{m_f}{\rho_0} \cdot \left[1 - \left(\frac{f}{f_{cr}} \right)^2 \frac{1}{k_0^4} \cdot \frac{\Delta_{x_1,x_2} \Delta_{x_1,x_2} v}{v} \right] = j k_0 \frac{m_f}{\rho_0} \cdot \left[1 - \left(\frac{f}{f_{cr}} \right)^2 \sin^4 \chi \right] \tag{8}$$

with an effective angle χ of sound incidence on the plate (see below).

3rd generalisation : *Elastic foil with bending wave losses* :

Energy dissipation in the foil can be taken into account by a loss factor η introducing a complex modulus $B \to B \cdot (1+j\eta)$. This leads to:

$$\begin{aligned} \frac{Z_T}{Z_0} &= Z_m F \cdot \left[\left(1 - F^2 \sin^4 \chi \right) - j\eta F^2 \sin^4 \chi \right] \\ &= Z_m F \cdot \left[1 - \left(\frac{f}{f_c} \right)^2 - j\eta \left(\frac{f}{f_c} \right)^2 \right] \quad ; \quad Z_m = j \frac{\omega_{cr} m_f}{Z_0} \quad ; \quad F = \frac{f}{f_{cr}} \end{aligned} \tag{9}$$

where Z_m is the normalised inertial impedance of the plate at the critical frequency f_{cr}, and f_c is the coincidence frequency at the incidence angle χ, with $f_{cr} = f_c \cdot \sin^2 \chi$.

So elasticity and bending losses of the foil can be taken into account by using an effective surface mass density :

$$m_f \to m_{eff,e} = m_f \cdot \left[1 - \left(\frac{f}{f_c} \right)^2 - j\eta \left(\frac{f}{f_c} \right)^2 \right] \tag{10}$$

4th generalisation : *combine porosity and elasticity effects* :

Substitute : $$m_f \to m_{eff,p,e} = \frac{m_{eff,e}}{1 + \dfrac{j\omega\, m_{eff,e}}{\Sigma}} \tag{11}$$

Cylindrical shell :

The co-ordinate system is $\{x_1, x_2, x_3\} \to \{\vartheta, z, r\}$. The value $\xi = a$ is the radius of the shell. The sound fields separate into factors (v_p = velocity of the shell) :

$$p(r,\vartheta,z) = R(r) \cdot T(\vartheta) \cdot U(z) \quad ; \quad v_p(a,\vartheta,z) = A \cdot T(\vartheta) \cdot U(z) \tag{12}$$

The form of U(z) may be any of, or a linear combination of :

$$U(z)=\begin{cases} e^{\pm j k_z z} \\ \cos(k_z z) \\ \sin(k_z z) \end{cases} \tag{13}$$

The shape of R(r) may be one of the cylinder functions

$$Z_n(k_r r)=\left\{J_n(k_r r),\, Y_n(k_r r),\, H_n^{(1)}(k_r r),\, H_n^{(2)}(k_r r)\right\} \tag{14}$$

Then $T(\vartheta)$ for fields which are periodic in ϑ is :

$$T(\vartheta)=\begin{cases} \cos(n\vartheta) \\ \sin(n\vartheta) \end{cases} \tag{15}$$

The Laplace operators in cylindrical co-ordinates are:

$$\Delta=\frac{\partial^2}{\partial r^2}+\frac{1}{r}\frac{\partial}{\partial r}+\frac{1}{r^2}\frac{\partial^2}{\partial\vartheta^2}+\frac{\partial^2}{\partial z^2} \quad ; \quad \Delta_{\vartheta,z}=\frac{1}{a^2}\frac{\partial^2}{\partial\vartheta^2}+\frac{\partial^2}{\partial z^2} \tag{16}$$

The bending wave equation is satisfied with the above field factors, when the secular equation holds :

$$k_0^2=k_z^2+k_r^2 \quad ; \quad 1=(k_z/k_0)^2+(k_r/k_0)^2=\sin^2\Theta+\cos^2\Theta \tag{17}$$

The angle Θ is between the wave vector and the radius. The two-dimensional Laplace operator gives together with the Bessel differential equation for the $Z_n^{(i)}(k_r r)$:

$$\Delta_{\vartheta,z}\, p(a,\vartheta,z)=-\left(\frac{n^2}{a^2}+k_z^2\right)\cdot p(a,\vartheta,z) \tag{18}$$

Therefore:

$$\frac{\Delta_{\vartheta,z}\Delta_{\vartheta,z}v_p}{v_p}=\left(\frac{n^2}{a^2}+k_z^2\right)^2=k_0^4\left(\frac{n^2}{(k_0a)^2}+\sin^2\Theta\right)^2 \tag{19}$$

Comparing this with the form for Z_T/Z_0 leads to the effective angle χ of incidence:

$$\sin\chi=\left(\frac{n^2}{(k_0a)^2}+\sin^2\Theta\right)^{1/2} \tag{20}$$

Spherical shell :

Spherical co-ordinate system $\{r,\vartheta,\varphi\}$ and the correspondence $\{x_1,x_2,x_3\}\rightarrow\{\vartheta,\varphi,r\}$. The factors of the field are :

$$p(r,\vartheta,\varphi)=R(r)\cdot T(\vartheta)\cdot P(\varphi) \quad ; \quad v_p(a,\vartheta,z)=A\cdot T(\vartheta)\cdot P(\varphi) \tag{21}$$

with R(r)= spherical Bessel functions, $T(\vartheta)$= (associated) Legendre functions of the first and second kind :

$$R(r) = z_m(k_0 r) = \left\{ j_m(k_0 r),\, y_m(k_0 r),\, h_m^{(1)}(k_0 r),\, h_m^{(2)}(k_0 r) \right\} \tag{22}$$

$$T(\vartheta) = \begin{cases} P_m^n(\cos\vartheta) \\ Q_m^n(\cos\vartheta) \end{cases} \qquad P(\varphi) = \begin{cases} \cos(n\varphi) \\ \sin(n\varphi) \end{cases} \tag{23}$$

The Laplace operators are:

$$\begin{aligned} \Delta &= \frac{\partial^2}{\partial r^2} + \frac{2}{r}\frac{\partial}{\partial r} + \frac{1}{r^2}\frac{\partial^2}{\partial \vartheta^2} + \frac{1}{r^2 \mathrm{tg}\,\vartheta}\frac{\partial}{\partial \vartheta} + \frac{1}{r^2 \sin^2\vartheta}\frac{\partial^2}{\partial \varphi^2} \\ \Delta_{\vartheta,\varphi} &= \frac{1}{a^2}\frac{\partial^2}{\partial \vartheta^2} + \frac{1}{a^2 \mathrm{tg}\,\vartheta}\frac{\partial}{\partial \vartheta} + \frac{1}{a^2 \sin^2\vartheta}\frac{\partial^2}{\partial \varphi^2} \end{aligned} \tag{24}$$

and therefore:

$$\frac{\Delta_{\vartheta,z}\Delta_{\vartheta,z} v_p}{v_p} = k_0^4 \left(\frac{m(m+1)}{(k_0 a)^2} \right)^2 \tag{25}$$

with the effective angle χ of incidence given by:

$$\sin\chi = \left(\frac{m(m+1)}{(k_0 a)^2} \right)^{1/2} \tag{26}$$

Because of $P_m^n(\cos\vartheta) \equiv 0$; $n > m$, the angle of incidence is $\chi=0$ for the "breathing sphere" m=n=0, which is plausible.

See [H.8] for elliptic-cylindrical and hyperbolic-cylindrical shells.

Partition impedance of membranes :

Membranes get their bending stiffness from the tension in their plane (only plane membranes in the (x,y) plane; the method could also be applied to blown-up balloons). The inhomogeneous wave equation of a membrane is:

$$\left(\Delta + k_m^2\right) v_m = \frac{j\omega}{T}\left(p_{front} - p_{back}\right) \quad ; \quad k_m = \omega\sqrt{M/T} \tag{27}$$

with the surface mass density M and the tension T of the membrane. The partition impedance follows immediately:

$$Z_T = \frac{p_{front} - p_{back}}{v_m} = \frac{T}{j\omega}\left(\frac{\Delta v_m}{v_m} + k_m^2 \right) = \frac{jT}{\omega}\left(k_{mx}^2 + k_{my}^2 - k_m^2 \right) \tag{28}$$

for a pattern of the membrane velocity:

$$v_m(x,y) = A \cdot e^{\pm j k_{mx} x} \cdot e^{\pm j k_{my} y} \tag{29}$$

H.18 Foil resonator

Ref.: Mechel, [H.7]

A foil resonator consists of a foil having a (effective) surface mass density m_f at a distance t from a hard wall; the interspace may be filled with air or (partially) with a porous material.

The surface impedance Z for normal sound incidence is :

$$\frac{Z}{Z_0} = jk_0t\frac{m_f}{\rho_0 t} + \begin{cases} -j\cot(k_0t) & ; \text{ air} \\ \dfrac{Z_{an}}{\tanh(\Gamma_{an}\, k_0t)} & ; \text{ porous material} \end{cases} \tag{1}$$

with $\Gamma_{an}=\Gamma_a/k_0$, $Z_{an}=Z_a/Z_0$ normalised characteristic values of the porous material.

The (angular) resonance frequency with air in the volume is, under the condition $k_0t \ll 1$, :

$$\omega_0 = c_0\sqrt{\rho_0/(m_f\, t)} \quad ; \quad f_0 \approx 600/\sqrt{m_f\, t} \quad (f_0 \text{ in Hz; m in kg/m}^2\text{; t in cm}) \tag{2}$$

If $k_0t \ll 1$ does not hold, the resonance equation is (for the n-th resonance) :

$$\frac{\omega_n t}{c_0}\cdot\tan\frac{\omega_n t}{c_0} = \frac{\rho_0 t}{m_f} \tag{3}$$

An approximation for the lowest resonance solution is :

$$\left(\frac{\omega_0 t}{c_0}\right)^2 = \frac{105+45(\rho_0 t/m_f) - \sqrt{11025+5250(\rho_0 t/m_f)+1605(\rho_0 t/m_f)^2}}{20+2(\rho_0 t/m_f)} \tag{4}$$

For oblique sound incidence with the polar angle of incidence Θ, substitute $Z_0 \to Z_0/\cos\Theta$; $k_0 \to k_0\cdot\cos\Theta$.

As long as $k_0t\cdot\cos\Theta \ll 1$, the resonance changes to $\omega_0 \to \omega_0/\cos\Theta$.

For oblique incidence on a foil resonator with porous material in the volume :

$$\frac{Z}{Z_0} = jk_0t\frac{m_f}{\rho_0 t} + \frac{\Gamma_{an}Z_{an}}{\sqrt{\Gamma_{an}^2+\sin^2\Theta}\cdot\tanh(k_0t\sqrt{\Gamma_{an}^2+\sin^2\Theta})} \tag{5}$$

The lowest three resonances $\omega_0, \omega_1, \omega_2$ (with air in the volume) can be read from the following nomograph. Enter the nomograph on its vertical axis with the value m of the foil surface mass density; proceed horizontally to the line for the distance t ; proceed vertically to one of the (thick) lines for $\omega_0, \omega_1, \omega_2$ and from there horizontally to the line for the value of m ;

then vertically to the horizontal axis, where you can read the resonance frequency f_0 (or f_1 , f_2). (The example in the nomograph is for m= 0.5 [kg/m^2] ; t= 2 [cm] ; giving f_0= 700 [Hz])

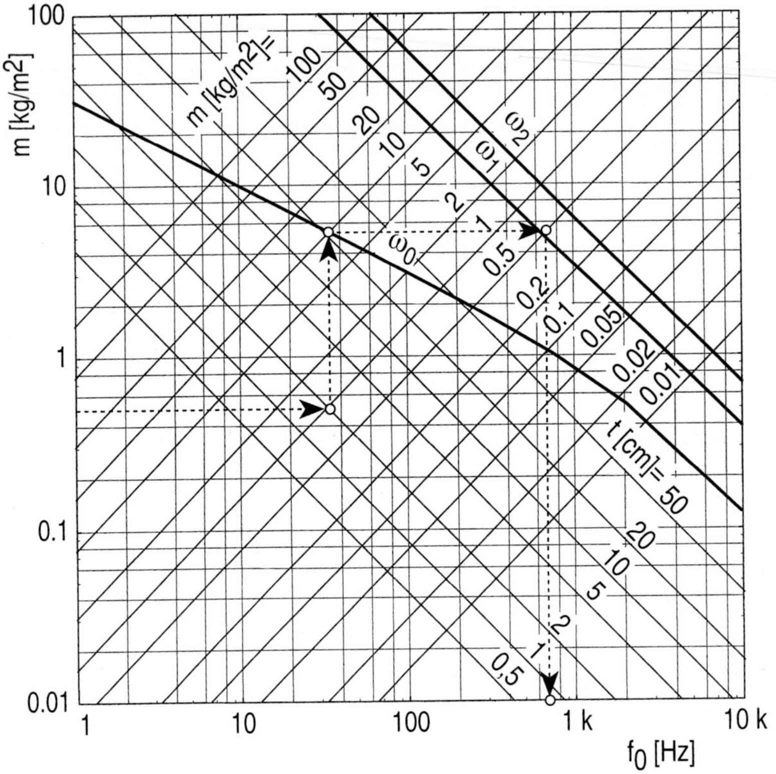

H.19 Ring resonator

Ring resonators are used, for example, in mufflers and in low-frequency silencer sections, e.g. in gas turbine run-up and test cells.

The aim is, to evaluate the orifice input admittance G . The input orifice may be covered with a poro-elastic foil (see section H.17) with a partition impedance Z_S (normalised). The normalised admittance G is then:

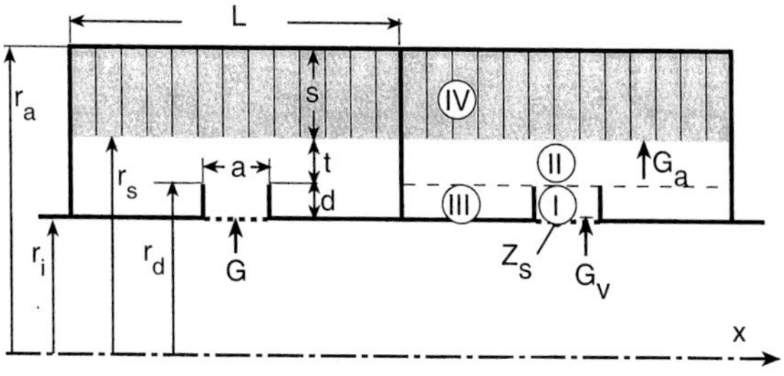

$$G = \frac{G_V}{1+Z_S G_V} \tag{1}$$

where G_V is the (normalised) input admittance of the ring-shaped neck. A porous absorber layer in zone IV ($r_S \le r \le r_a$) is made locally reacting by partitions.

Corresponding to the low-frequency application, only the fundamental cylindrical mode is assumed to propagate in the field zones.

Field formulations in the zones (with Bessel, Neumann, and Hankel functions) :

Zone I:
$$\begin{aligned} p_I &= a_0 \cdot H_0^{(2)}(k_0 r) + b_0 \cdot H_0^{(1)}(k_0 r) = (a_0+b_0)\cdot J_0(k_0 r) + j\cdot(a_0-b_0)\cdot Y_0(k_0 r) \\ Z_0 v_{rI} &= -j\cdot(a_0+b_0)\cdot J_1(k_0 r) + (a_0-b_0)\cdot Y_1(k_0 r) \end{aligned} \tag{2}$$

Zone II:
$$\begin{aligned} p_{II} &= c_0 \cdot J_0(k_0 r) + d_0 \cdot Y_0(k_0 r) \\ Z_0 v_{rII} &= -j\cdot[c_0 \cdot J_1(k_0 r) + d_0 \cdot Y_1(k_0 r)] \end{aligned} \tag{3}$$

Zone III:
$$\begin{aligned} p_{III} &= e_0 \cdot J_0(k_0 r) + f_0 \cdot Y_0(k_0 r) \\ Z_0 v_{rIII} &= -j\cdot[e_0 \cdot J_1(k_0 r) + f_0 \cdot Y_1(k_0 r)] \end{aligned} \tag{4}$$

Zone IV:
$$\begin{aligned} p_{IV} &= g_0 \cdot J_0(-j\Gamma_{an} r) + h_0 \cdot Y_0(-j\Gamma_{an} r) \\ Z_0 v_{rIV} &= \frac{j}{Z_{an}}[g_0 \cdot J_1(-j\Gamma_{an} r) + h_0 \cdot Y_1(-j\Gamma_{an} r)] \end{aligned} \tag{5}$$

with Γ_{an}, Z_{an} the normalised characteristic values of the porous material.

The radial input admittance (normalised) G_a of the absorber layer (zone IV) is :

$$G_a = \frac{j}{\Gamma_{an}} \frac{J_1(j\Gamma_{an}k_0r_s)\cdot Y_1(j\Gamma_{an}k_0r_a) - J_1(j\Gamma_{an}k_0r_a)\cdot Y_1(j\Gamma_{an}k_0r_s)}{J_0(j\Gamma_{an}k_0r_s)\cdot Y_1(j\Gamma_{an}k_0r_a) - J_1(j\Gamma_{an}k_0r_a)\cdot Y_0(j\Gamma_{an}k_0r_s)} \tag{6}$$

Boundary conditions :

the interior neck orifice is additionally loaded with the mass impedance $Z_m = j\cdot k_0 a\cdot\Delta\ell_i/a$ of the interior orifice end correction.

$r = r_s$:

$$-j[c_0\cdot J_1(k_0r_s) + d_0\cdot Y_1(k_0r_s)] \stackrel{!}{=} G_a\cdot[c_0\cdot J_0(k_0r_s) + d_0\cdot Y_0(k_0r_s)] \tag{7}$$

$r = r_d$; I–II :

velocity :

$$-j[c_0\cdot J_1(k_0r_d) + d_0\cdot Y_1(k_0r_d)] \stackrel{!}{=} \frac{a}{L}[-j(a_0+b_0)\cdot J_1(k_0r_d) + (a_0-b_0)\cdot Y_1(k_0r_d)] - j\left(1-\frac{a}{L}\right)[e_0\cdot J_1(k_0r_d) + f_0\cdot Y_1(k_0r_d)] \tag{8}$$

pressure :

$$(a_0+b_0)\cdot J_0(k_0r_d) + j(a_0-b_0)\cdot Y_0(k_0r_d) - -Z_m\cdot[-j(a_0+b_0)\cdot J_1(k_0r_d) + (a_0-b_0)\cdot Y_1(k_0r_d)] \stackrel{!}{=} c_0\cdot J_0(k_0r_d) + d_0\cdot Y_0(k_0r_d) \tag{9}$$

$r = r_d$; II–III :

pressure :

$$e_0\cdot J_0(k_0r_d) + f_0\cdot Y_0(k_0r_d) \stackrel{!}{=} c_0\cdot J_0(k_0r_d) + d_0\cdot Y_0(k_0r_d) \tag{10}$$

$r = r_i$:

$$e_0\cdot J_1(k_0r_i) + f_0\cdot Y_1(k_0r_i) \stackrel{!}{=} 0 \tag{11}$$

Setting the arbitrary amplitude $a_0=1$, the neck input admittance is :

$$G_v = -j\frac{(1+b_0)\cdot J_1(k_0r_i) + j(1-b_0)\cdot Y_1(k_0r_i)}{(1+b_0)\cdot J_0(k_0r_i) + j(1-b_0)\cdot Y_0(k_0r_i)} \tag{12}$$

So one must solve the boundary condition equations for $b_0=: -N/D$ with : (13)

$$
\begin{aligned}
N = {} & \\
\Big\{ & -J_i(k_0 r_i)\big(G_a J_0(k_0 r_s) + j J_1(k_0 r_s)\big)\big(J_1(k_0 r_i)\, Y_0(k_0 r_d) - J_0(k_0 r_d)\, Y_1(k_0 r_i)\big) \cdot \\
& \cdot \big(J_1(k_0 r_s)\, Y_0(k_0 r_d) - J_0(k_0 r_d)\, Y_1(k_0 r_s)\big)\Big(-j\, H_0^{(1)}(k_0 r_d) + Z_m H_1^{(1)}(k_0 r_d)\Big) + \\
& + \Big[G_a \big(J_0(k_0 r_d)\, Y_0(k_0 r_s) - J_0(k_0 r_s)\, Y_0(k_0 r_d)\big) + j\big(J_0(k_0 r_d)\, Y_1(k_0 r_s) - J_1(k_0 r_s)\, Y_0(k_0 r_d)\big)\Big] \cdot \\
& \cdot \Big[-j a/L \cdot J_0(k_0 r_d)\, J_1(k_0 r_i)\, H_1^{(1)}(k_0 r_d)\big(J_1(k_0 r_i)\, Y_0(k_0 r_d) - J_0(k_0 r_d)\, Y_1(k_0 r_i)\big) - \\
& \quad - J_1(k_0 r_i)\big(J_1(k_0 r_i)\big(J_1(k_0 r_s)\, Y_0(k_0 r_d) + (a/L - 1)\, J_0(k_0 r_d)\, Y_1(k_0 r_d)\big) - \\
& \quad - J_0(k_0 r_d)\, Y_1(k_0 r_i)\big(J_1(k_0 r_s) + (a/L - 1)\, J_1(k_0 r_d)\big)\big) \cdot \Big(-j\, H_0^{(1)}(k_0 r_d) + Z_m H_1^{(1)}(k_0 r_d)\Big)\Big]\Big\}
\end{aligned}
$$

$$
\begin{aligned}
D = {} & \\
\Big\{ & J_1(k_0 r_i)\big(G_a J_0(k_0 r_s) + j J_1(k_0 r_s)\big)\big(J_1(k_0 r_i)\, Y_0(k_0 r_d) - J_0(k_0 r_d)\, Y_1(k_0 r_i)\big) \cdot \\
& \cdot \big(J_1(k_0 r_s)\, Y_0(k_0 r_d) - J_0(k_0 r_d)\, Y_1(k_0 r_s)\big)\Big(j\, H_0^{(2)}(k_0 r_d) - Z_m H_1^{(2)}(k_0 r_d)\Big) + \\
& + \Big[G_a \big(J_0(k_0 r_d)\, Y_0(k_0 r_s) - J_0(k_0 r_s)\, Y_0(k_0 r_d)\big) - j\big(J_1(k_0 r_s)\, Y_0(k_0 r_d) - J_0(k_0 r_d)\, Y_1(k_0 r_s)\big)\Big] \cdot \\
& \cdot \Big[-j a/L \cdot J_0(k_0 r_d)\, J_1(k_0 r_i)\, H_1^{(2)}(k_0 r_d)\big(J_1(k_0 r_i)\, Y_0(k_0 r_d) - J_0(k_0 r_d)\, Y_1(k_0 r_i)\big) + \\
& \quad + J_1(k_0 r_i)\big(J_1(k_0 r_i)\big(J_1(k_0 r_s)\, Y_0(k_0 r_d) + (a/L - 1)\, J_0(k_0 r_d)\, Y_1(k_0 r_d)\big) - \\
& \quad - J_0(k_0 r_d)\, Y_1(k_0 r_i)\big(J_1(k_0 r_s) + (a/L - 1)\, J_1(k_0 r_d)\big)\big) \cdot \Big(j\, H_0^{(2)}(k_0 r_d) - Z_m H_1^{(2)}(k_0 r_d)\Big)\Big]\Big\}
\end{aligned}
\tag{14}
$$

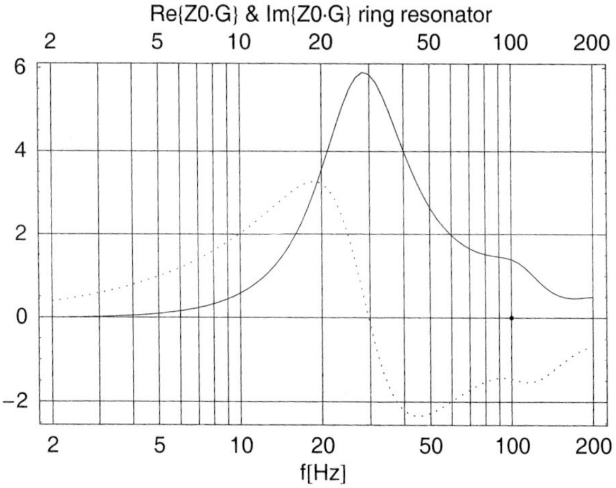

Example of Re{Z_0G} , Im{Z_0G} of a low-tuned ring resonator for a turbine test cell; with input parameters for computation :

```
(*Duct*)
ri[m]=3., ra[m]=4.9
(*Cell*)
L[m]=1., a[m]=0.25, d[m]=0.2, t[m]=0.2, s[m]=1.5
(*Absorber*)
s[m]=1.5, Ξ[Pas/m^2]=500.
```

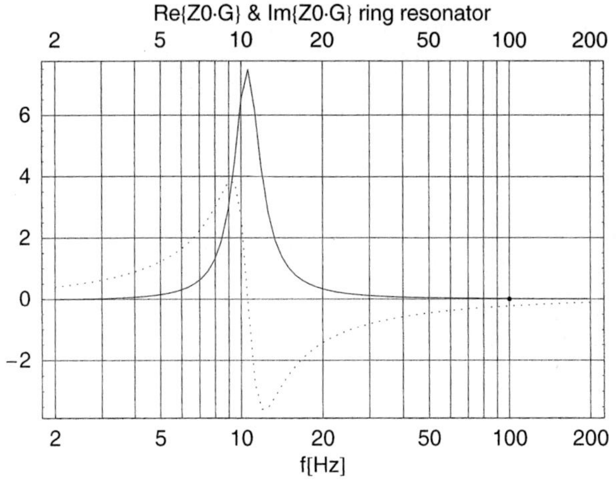

H.20 Wide-angle absorber, scattered far field

Ref.: Mechel, [H.9] ; Mechel, [H.10] ; Schroeder & Gerlach, [H.11]

The objects of this section originally were conceived as wide-angle diffusers, [H.11], but the unavoidable losses make them effective absorbers; and "diffuser" and "absorber" are contradictions per se . This section is more concerned with the "diffuser", in that it describes mainly the scattered far field (see the next section H.21 for other field ranges).

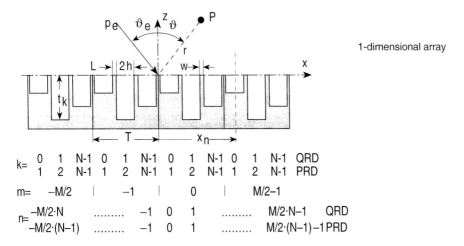

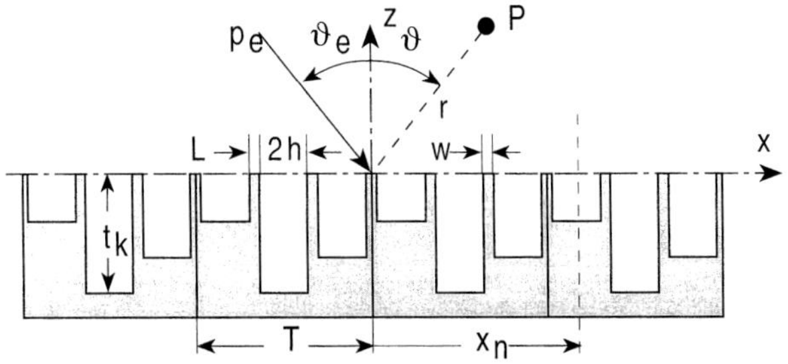

1-dimensional array

k=	0	1	N-1	0	1	N-1	0	1	N-1	0	1	N-1	QRD
	1	2	N-1	1	2	N-1	1	2	N-1	1	2	N-1	PRD

m= –M/2 | –1 | 0 | M/2–1

n= –M/2·N .... –1 0 1 .... M/2·N–1 QRD

n= –M/2·(N–1) .... –1 0 1 .... M/2·(N–1)–1 PRD

The objects are 1-dimensional or 2-dimensional arrays of $\lambda/4$ resonators. The depth t_k of the resonators varies in one of two possible pseudo-random manners (see below). Mostly, the arrangement is composed of groups of resonators, and the pseudo-random variation of t_k is within the group; then the object has a periodic structure. Three indices will be used: k for the number of a resonator in the group; m for the group ; n for the resonator within the arrangement.

The "classical" arrangements are, with N a prime number:

QRD: *quadratic residue diffuser*

1-dimensional :
$$t_k = \frac{\pi c_0 \,\mathrm{mod}(k^2, N)}{N\omega_r} \quad ; \quad k = 0,\ 1, \ldots,\ N-1 \tag{1}$$

2-dimensional :
$$t_{k,\ell} = \frac{\pi c_0 \,\mathrm{mod}(k^2+\ell^2, N)}{N\omega_r} \quad ; \quad k,\ell = 0,\ 1, \ldots,\ N-1 \tag{2}$$

with: $\omega_r = 2\pi f_r$ = "working (angular) frequency" ; mod(a,b)= "modulo function"= remainder of a/b .

PRD: *primitive root diffuser*

1-dimensional :
$$t_k = \frac{\pi c_0 \, \mathrm{mod}(\rho^k, N)}{N\omega_r} \quad ; \quad k = 1,\ldots,\ N-1 \tag{3}$$

with ρ= "primitive root" of N ($\mathrm{mod}(\rho^k,N)$ produces the numbers 1,2,…,N–1 in irregular sequence, if k=1,2,…,N–1).

The required Helmholtz numbers are for 1-dimensional diffusers :

$$k_0 t_k = \frac{f}{f_r} \cdot \begin{cases} \frac{\pi}{N}\mathrm{mod}(k^2,N) \ ; \ k=0,\ 1,\ldots,\ N-1 \ ; \ \mathrm{QRD} \\ \frac{\pi}{N}\mathrm{mod}(\rho^k,N) \ ; \ k=1,\ldots,\ N-1 \ ; \ \mathrm{PRD} \end{cases} \tag{4}$$

Cell centre co-ordinates are :

QRD:
$$x_n = x_{k,m} = (n+\tfrac{1}{2})L = mT + (k+\tfrac{1}{2})L \ ; \ \begin{cases} n = -MN/2,\ldots,+MN/2-1 \\ m = -M/2,\ldots,+M/2-1 \\ k = 0,\ldots,N-1 \end{cases} \tag{5}$$

PRD:
$$x_n = x_{k,m} = (n+\tfrac{1}{2})L = mT + (k-\tfrac{1}{2})L \ ; \ \begin{cases} n = -M(N-1)/2,\ldots,+M(N-1)/2-1 \\ m = -M/2,\ldots,+M/2-1 \\ k = 1,\ldots,N-1 \end{cases}$$

The input admittance (normalised) $G(x_k)$ of a chamber is :

$$G(x_k) = \frac{\tanh(\Gamma_{an} k_0 t_k)}{Z_{an}} \tag{6}$$

where Γ_{an}, Z_{an} are the normalised propagation constant and wave impedance in the cell; the fundamental mode values in capillaries are used, if viscous and thermal losses are to be taken into account (see sections J1, J2 about capillaries in the chapter "Duct Acoustics").

Scattered far field for 2-dimensional diffuser :

The scattered far field p_s for a 2-dimensional diffuser always can be factorised as:

$$p_s(r,\vartheta,\varphi) = P_i \cdot \Phi_s(\vartheta_e,\varphi_e \mid \vartheta,\varphi) \cdot \frac{e^{-jk_0 r}}{k_0 r} \tag{7}$$

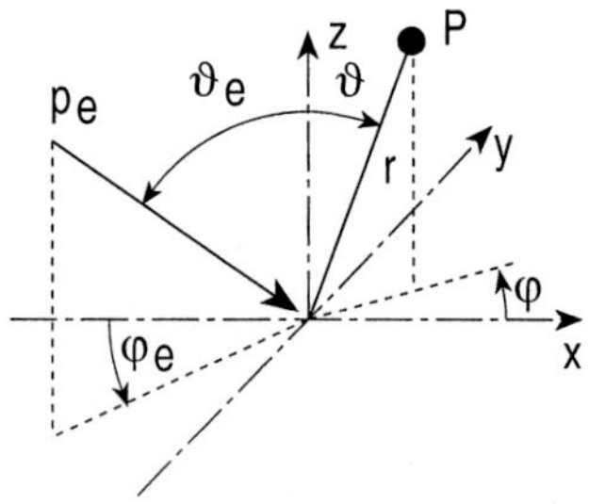

The angular distribution Φ_s is evaluated from :

$$\Phi_s(\vartheta_e,\varphi_e\,|\,\vartheta,\varphi)=\frac{-jk_0^2\cos\vartheta_e}{2\pi(\cos\vartheta_e+\langle G\rangle)}\iint_A G(x,y)\cdot e^{-j(\mu_x x+\mu_y y)}\cdot dx\,dy \tag{8}$$

with :
$$\begin{aligned}\mu_x&=k_0(\sin\vartheta_e\cos\varphi_e-\sin\vartheta\cos\varphi)\\ \mu_y&=k_0(\sin\vartheta_e\sin\varphi_e-\sin\vartheta\sin\varphi)\end{aligned} \tag{9}$$

If the surface $A=(MT)^2$ of the arrangement would have the homogeneous, averaged admittance $\langle G\rangle$:

$$\Phi_s=\frac{-jk_0^2\cos\vartheta_e\cdot<G>}{2\pi(\cos\vartheta_e+<G>)}(MT)^2\,\frac{\sin(\mu_x MT/2)}{\mu_x MT/2}\,\frac{\sin(\mu_y MT/2)}{\mu_y MT/2} \tag{10}$$

For the 2-dimensional QRD with cell centre co-ordinates :

$$\begin{aligned}x_m&=x_{k,m_g}=(m+\tfrac{1}{2})L=m_gT+(k+\tfrac{1}{2})L\ ;\quad\begin{cases}m=-MN/2,\dots,+MN/2\\ m_g=-M/2,\dots,+M/2-1\\ k=0,\dots,N-1\end{cases}\\ y_n&=x_{\ell,n_g}=(n+\tfrac{1}{2})L=n_gT+(\ell+\tfrac{1}{2})L\ ;\quad\begin{cases}n=-MN/2,\dots,+MN/2\\ n_g=-M/2,\dots,+M/2-1\\ \ell=0,\dots,N-1\end{cases}\end{aligned} \tag{11}$$

(m_g, n_g= group indices ; k,ℓ= cell indices), the scattered far field distribution is :

$$\begin{aligned}\Phi_s&=\frac{-j(k_0h)^2\cos\vartheta_e}{\pi(\cos\vartheta_e+<G>)}\cdot\frac{\sin\mu_x h}{\mu_x h}\cdot\frac{\sin\mu_y h}{\mu_y h}\cdot e^{-j(\mu_x+\mu_y)L/2}\cdot\\ &\cdot\sum_{m_g,n_g}e^{-j(\mu_x m_g+\mu_y n_g)T}\cdot\sum_{k,\ell}G(k,\ell)e^{-j(\mu_x k+\mu_y\ell)L}\end{aligned} \tag{12}$$

The difference of the QRD , as compared with a surface A of equal size and same average admittance $\langle G\rangle$, is mainly produced by the factor of the last sum.

In the following examples of $|\Phi_s|$ the cell orifices may contain a normalised flow resistance R (e.g. a wire mesh; see parameter list). The QRD with N=11 is exceptional, in that it shows a strong backscattering.

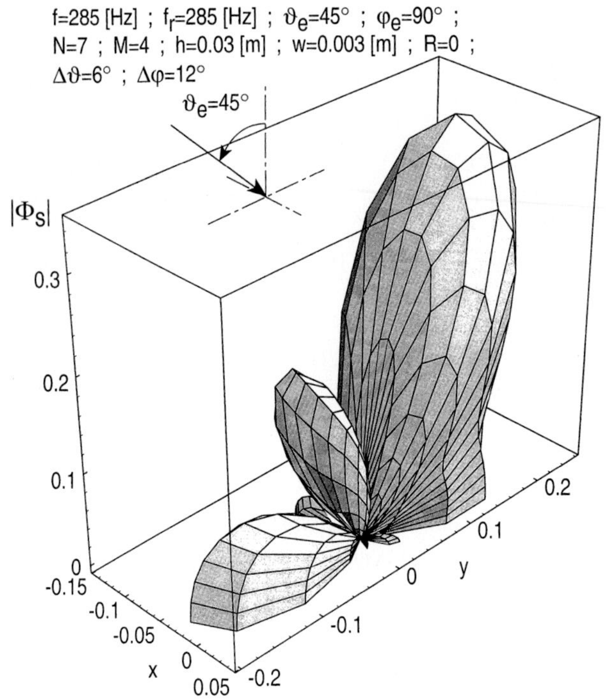

Angular distribution of the scattered far field with M= 4 .

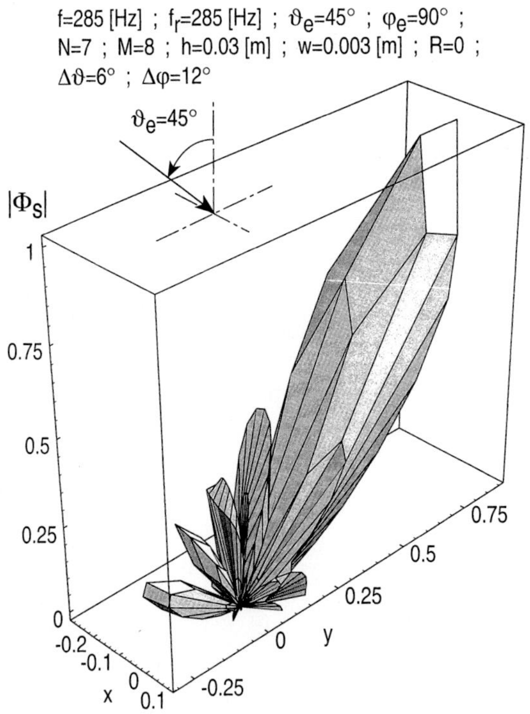

As above, but the number M of groups is increased to M=8 in both directions.

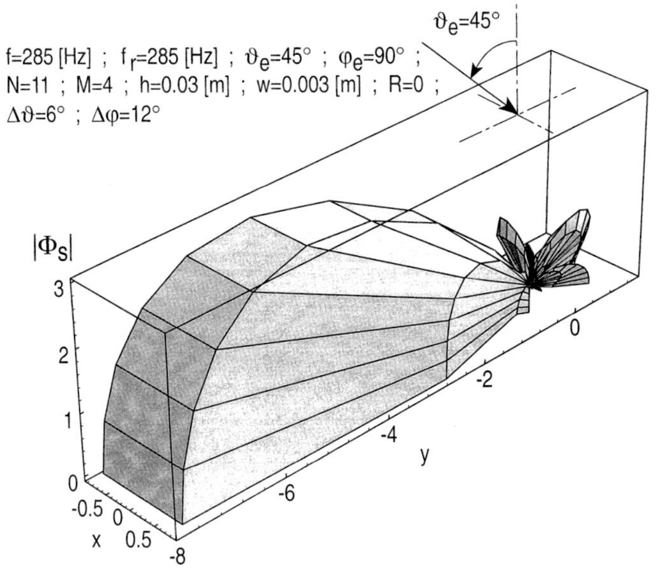

Scattered far field directivity of a QRD with N=11 .

Scattered far field for 1-dimensional diffuser : ($\varphi_e = 0$)

Scattered far field :

$$p_s(r,\vartheta,\varphi) = -P_i\sqrt{\frac{j}{2\pi}}\cdot\Phi_s(\vartheta_e\,|\,\vartheta)\cdot\frac{e^{-jk_0r}}{\sqrt{k_0r}} \tag{13}$$

Directivity function (A= width of diffuser):

$$\Phi_s(\vartheta_e\,|\,\vartheta) = \int_{-k_0A/2}^{+k_0A/2} G(x)\,\frac{p(x,0)}{P_i}\,e^{jk_0x\sin\vartheta}\,d(k_0x) \tag{14}$$

with sound pressure at the surface and surface admittance $G(x)$:

$$\frac{p(x,0)}{P_i} = 2\,e^{-jk_0x\sin\vartheta_e} - \\ -\frac{\cos\vartheta_e}{\cos\vartheta_e + \langle G\rangle}\,e^{-jk_0x\sin\vartheta_e}\int_{-k_0A/2}^{+k_0A/2} G(x')\,H_0^{(2)}(k_0\,|\,x-x'|)\,e^{-jk_0(x'-x)\sin\vartheta_e}\,d(k_0x') \tag{15}$$

So Φ_s has the form :

$$\Phi_s(\vartheta_e\,|\,\vartheta) = 2\int\limits_{-k_0A/2}^{+k_0A/2} G(x)\; e^{jk_0x(\sin\vartheta-\sin\vartheta_e)}\; d(k_0x) - \frac{\cos\vartheta_e}{\cos\vartheta_e + <G>}\cdot$$

$$\cdot\iint\limits_{-k_0A/2}^{+k_0A/2} G(x)G(y)\; e^{jk_0(x\sin\vartheta-y\sin\vartheta_e)}\cdot H_0^{(2)}(k_0\,|\,x-y\,|)\; d(k_0x)\; d(k_0y) \tag{16}$$

$$:= I_1 - \frac{\cos\vartheta_e}{\cos\vartheta_e + \langle G\rangle}\cdot I_2$$

The 1st integral I_1 is for a QRD (for a PRD : summation $k=1,\ldots,N-1$, and change sign in the exponent of the last factor in the 3rd line below):

$$I_1 = 4k_0h\frac{\sin\left(k_0h(\sin\vartheta-\sin\vartheta_e)\right)}{k_0h(\sin\vartheta-\sin\vartheta_e)}\sum_n G(x_n)e^{jk_0x_n(\sin\vartheta-\sin\vartheta_e)}$$

$$= 4k_0h\frac{\sin\left(k_0h(\sin\vartheta-\sin\vartheta_e)\right)}{k_0h(\sin\vartheta-\sin\vartheta_e)}e^{j\,k_0L/2\cdot(\sin\vartheta-\sin\vartheta_e)}\,. \tag{17}$$

$$\cdot\sum_{m=-M/2}^{M/2-1} e^{jmk_0T(\sin\vartheta-\sin\vartheta_e)}\sum_{k=0}^{N-1} G(x_k)e^{jk\,k_0L(\sin\vartheta-\sin\vartheta_e)}$$

The 2nd integral I_2 is :

$$I_2 = \sum_{n,n'} G(x_n)G(x_{n'})\cdot e^{jk_0(x_n\sin\vartheta-x_{n'}\sin\vartheta_e)}\cdot I_{n,n'}$$

$$I_{n,n'} := \iint\limits_{-k_0h}^{k_0h} e^{j(x\sin\vartheta-y\sin\vartheta_e)}\cdot H_0^{(2)}(|\,n-n'\,|k_0L + x - y)\; dx\; dy \tag{18}$$

See [H.9] for the integration of $I_{n,n'}$.

H.21 Wide-angle absorber, near field and absorption

Ref.: Mechel, [H.9] ; Mechel, [H.10]

The object is the same as in the previous section H.20 . This section is mainly concerned with the field analysis near the absorber (diffuser) and its absorption. The parts of this section are :

- 1-dimensional absorber :
    - exterior field without losses; in the cells fundamental capillary mode;
    - exterior field without losses; in the cells higher capillary modes;
    - exterior field with losses;
- 2-dimensional absorber .

A plane wave p_e is incident with a polar angle ϑ_e .

1-dimensional absorber :

The absorber is composed of cell groups with width T= N·L (for QRD) or T= (N–1)·L (for PRD). The cell raster is L= 2h+w ; 2h= cell width ; w= thickness of walls between cells. The absorber is treated as a periodic structure with period length T .

Fundamental capillary mode in the cells :

Field in front of the absorber :

$$p(x,z) = p_e(x,z) + p_s(x,z) \quad ; \quad \begin{aligned} p_e(x,z) &= P_e \cdot e^{j(-xk_x + zk_z)} \\ p_s(x,z) &= \sum_{n=-\infty}^{+\infty} A_n \cdot e^{-\gamma_n z} \cdot e^{-j\beta_n x} \end{aligned} \tag{1}$$

with :

$$\begin{aligned} &k_x = k_0 \sin\vartheta_e \quad ; \quad k_z = k_0 \cos\vartheta_e \\ &\beta_n = \beta_0 + n\frac{2\pi}{T} \quad ; \quad \frac{\gamma_n}{k_0} = \sqrt{(\sin\vartheta_e + n\cdot\lambda_0/T)^2 - 1} \quad ; \quad \gamma_0 = jk_z = jk_0\cos\vartheta_e \end{aligned} \tag{2}$$

Index range of radiating space harmonics (the other harmonics are surface waves) :

$$-\frac{T}{\lambda_0}(1+\sin\vartheta_e) \le n_s \le \frac{T}{\lambda_0}(1-\sin\vartheta_e) \tag{3}$$

At the surface :

$$p(x,0) = [P_e + A_0 + \sum_{n\ne 0} A_n \cdot e^{-jn\frac{2\pi}{T}x}]\, e^{-jk_x x} \tag{4}$$

$$-Z_0 v_z(x,0) = [(P_e - A_0)\cos\vartheta_e + j\sum_{n\neq 0} A_n \frac{\gamma_n}{k_0}\cdot e^{-jn\frac{2\pi}{T}x}]\, e^{-jk_x x} \tag{5}$$

Boundary condition :

$$(P_e - A_0)\cos\vartheta_e + j\sum_{n\neq 0} A_n \frac{\gamma_n}{k_0}\cdot e^{-jn\frac{2\pi}{T}x} = G(x)\cdot[P_e + A_0 + \sum_{n\neq 0} A_n \cdot e^{-jn\frac{2\pi}{T}x}] \tag{6}$$

or, with the Fourier analysis of $G(x)$:

$$G(x) = \sum_{n=-\infty}^{+\infty} g_n \cdot e^{-jn\frac{2\pi}{T}x} \quad ; \quad g_n = \frac{1}{T}\int_0^T G(x)\cdot e^{+jn\frac{2\pi}{T}x}\, dx \tag{7}$$

the system of equations for the A_n ($\delta_{m,n}$= Kronecker symbol) :

$$\sum_{n=-n_{hi}}^{+n_{hi}} A_n \cdot [g_{-m-n} - j\delta_{m,-n}\frac{\gamma_n}{k_0}] = P_e\,(\delta_{m,0}\cos\vartheta_e - g_{-m}) \quad ; \quad m = -n_{hi},\ldots,+n_{hi} \tag{8}$$

With the admittance profile :

$$G(x_k) = \frac{\tanh(\Gamma_{an}\, k_0 t_k)}{Z_{an}} \tag{9}$$

in the k-th cell with :

$$k_0 t_k = \frac{f}{f_r}\frac{\pi}{N}\cdot\begin{cases}\mathrm{mod}(k^2,N) & ; \quad \text{QRD} \\ \mathrm{mod}(\rho^k,N) & ; \quad \text{PRD}\end{cases} \tag{10}$$

where Γ_{an}, Z_{an} are the normalised capillary propagation constant and wave impedance, the Fourier components are :

$$\begin{aligned} g_n &= \frac{2h}{T}\sum_{k=0}^{N-1} G(x_k)\, e^{-jn\pi(2k+1)/N}\,\frac{\sin(2n\pi h/T)}{2n\pi h/T} \quad ; \quad \text{QRD} \\ g_n &= \frac{2h}{T}\sum_{k=1}^{N-1} G(x_k)\, e^{-jn\pi(2k-1)/(N-1)}\,\frac{\sin(2n\pi h/T)}{2n\pi h/T} \quad ; \quad \text{PRD} \end{aligned} \tag{11}$$

The absorption coefficient $\alpha(\vartheta_e)$ is :

$$\alpha(\vartheta_e) = 1 - \left|\frac{A_0}{P_e}\right|^2 - \frac{1}{\cos\vartheta_e}\sum_{n_s\neq 0}\left|\frac{A_{n_s}}{P_e}\right|^2 \sqrt{1-(\sin\vartheta_e + n_s\lambda_0/T)^2} \tag{12}$$

where the summation index n_s spans the range of radiating space harmonics, but not n_s=0 . The 2nd term $|A_0/P_e|^2$ represents the geometrical reflection $|r_g|^2$; therefore the 3rd term represents the non-geometrical reflection $|r_s|^2$ by scattering.

The following diagram shows $\alpha(\vartheta_e)$ over f ; 1st: without losses in the exterior space taken into account (thick, full line), 2nd: the exterior losses considered by the substitution $k_0 \to k_\rho$, where k_ρ is the free field wave number of viscous and heat conducting air (thin, dashed line). This substitution surely substantially under-estimates the real losses; however, the strong modification of $\alpha(\vartheta_e)$ by the substitution indicates the sensitivity for losses.

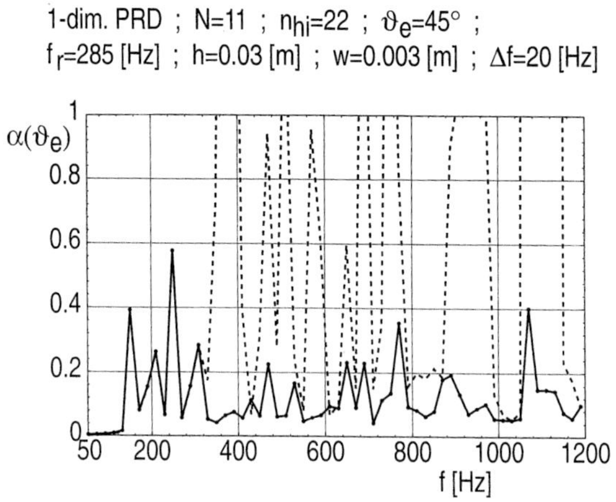

Absorption coefficient $\alpha(\vartheta_e)$ of a 1-dimensional PRD ;
full: no losses in exterior space;
dashed: with free field wave number k_ρ of lossy air used in the exterior space.

This result suggests that with a small (normalised) additional flow resistance in the orifices (e.g. by a wire mesh) a good absorber could be realised.

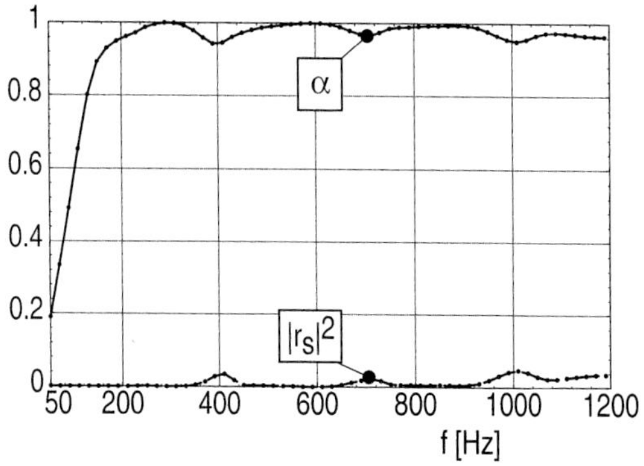

A 1-dimensional PRD as above, but with an additional flow resistance R=0.4 in the cell orifices.

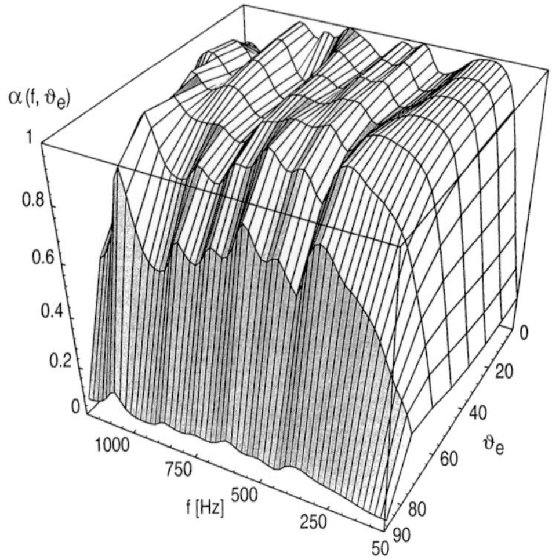

Absorption coefficient $\alpha(\vartheta_e)$ of the 1-dimensional PRD from above, plotted over f and ϑ_e.

Losses and higher modes in the cells of a 1-dimensional absorber :

Assume density waves of lossy air in the exterior space, i.e. with a free field wave number $k_\rho \approx k_0$, and assume capillary wave modes in the cells. The field formulation in the k-th cell is :

$$p_k(x,z) = e^{-j\beta_0 x_k} \sum_{n=0}^{\infty} B_{k,n} \cdot \cosh(\Gamma_n(z-t_k)) \cdot q_n(x-x_k) \tag{13}$$

with $\beta_0 = k_\rho \cdot \sin\vartheta_e$ and the mode profiles

$$q_n(x-x_k) = \begin{cases} \cos(\varepsilon_n(x-x_k)) \ ; \ n = n_e \\ \sin(\varepsilon_n(x-x_k)) \ ; \ n = n_o \end{cases} \tag{14}$$

for the symmetrical even modes (n_e= 0,2,4…) and the anti-symmetrical odd modes (n_o= 1,3,5…) . The characteristic equation for the $\varepsilon_n = \varepsilon_{\rho n}$ of *symmetrical modes* is :

$$\begin{aligned} &[(\varepsilon_\rho h)^2 - (k_\rho h)^2](\frac{\Theta_\rho}{\Theta_\alpha} - 1)\frac{\tan\sqrt{(\varepsilon_\rho h)^2 - (k_\rho h)^2 + (k_v h)^2}}{\sqrt{\ldots}} + \varepsilon_\rho h \cdot \tan\varepsilon_\rho h - \\ &- \frac{\Theta_\rho}{\Theta_\alpha}\sqrt{(\varepsilon_\rho h)^2 - (k_\rho h)^2 + (k_\alpha h)^2} \cdot \tan\sqrt{\ldots} = 0 \end{aligned} \tag{15}$$

where $\sqrt{\ldots}$ stands for the nearest root, and :

$$\frac{\Theta_\rho}{\Theta_\alpha} = \frac{(k_\rho h)^2}{(k_\alpha h)^2} \frac{1 - \kappa (k_\alpha h)^2 / (k_{\alpha 0} h)^2}{1 - \kappa (k_\rho h)^2 / (k_{\alpha 0} h)^2} \tag{16}$$

with the free field wave numbers :

$$(k_0h)^2=(\frac{\omega h}{c_0})^2 \;;\; (k_\nu h)^2=-j\frac{\omega}{\nu}h^2 \;;\; (k_{\alpha 0}h)^2=\kappa\,Pr\cdot(k_\nu h)^2$$

$$\left.\begin{matrix}(k_\rho h)^2\\ (k_\alpha h)^2\end{matrix}\right\}=\frac{[\frac{1}{(k_0h)^2}+\frac{4}{3(k_\nu h)^2}+\frac{\kappa}{(k_{\alpha 0}h)^2}]\mp\sqrt{[\ldots]^2-2\cdot\{\ldots\}}}{\{\frac{2\kappa}{(k_{\alpha 0}h)^2}(\frac{1}{\kappa(k_0h)^2}+\frac{4}{3(k_\nu h)^2})\}} \qquad (17)$$

where [...] and {...} under the root repeat the corresponding expressions from outside the root. The characteristic equation for the $\varepsilon_n=\varepsilon_{\rho n}$ of *anti-symmetrical modes* is obtained by the substitutions $\cos\rightarrow\sin$; $\sin\rightarrow-\cos$; $\tan\rightarrow-\cot$. Start values for the numerical solution of the characteristic equation are for even modes $\varepsilon_0h=\varepsilon_\rho h$; $\varepsilon_{n>0}h=n\pi/2$, and for odd modes $\varepsilon_{n>0}h=n\pi/2$.

The field in the exterior space is formulated as :

$$p(x,z)=p_e(x,z)+p_s(x,z) \;;\; \begin{matrix}p_e(x,z)=P_e\cdot e^{j(-xk_x+zk_z)}\\ p_s(x,z)=\sum_{n=-\infty}^{+\infty}A_n\cdot e^{-\gamma_n z}\cdot e^{-j\beta_n x}\end{matrix} \qquad (18)$$

where p_s is periodic in x with a period length T , and with :

$$\begin{gathered}k_x=k_0\sin\vartheta_e \;;\; k_z=k_0\cos\vartheta_e\\ \beta_n=\beta_0+n\frac{2\pi}{T} \;;\; \frac{\gamma_n}{k_0}=\sqrt{(\sin\vartheta_e+n\cdot\lambda_0/T)^2-1} \;;\; \gamma_0=jk_z=jk_0\cos\vartheta_e\end{gathered} \qquad (19)$$

The boundary conditions at the surface give a linear system of equations for the amplitudes A_n of the reflected space harmonics :

$$\begin{aligned}\sum_{n=-n_{hi}}^{+n_{hi}}A_n\Bigg[&\frac{T}{h}\delta_{m,n}[j\delta_{m,0}\cos\vartheta_e+(1-\delta_{m,0})\frac{\gamma_m}{k_0}]-\\ &-\overset{\circ}{\sum_k^{N-1}}e^{j(m-n)2\pi x_k/T}\sum_{i=0}^{i_{hi}}(-1)^i\frac{\Gamma_i}{k_\rho}\tanh(\Gamma_i t_k)\frac{S_{m,i}\cdot S_{n,i}}{Q_i}\Bigg]=\\ =\Bigg[j\frac{T}{h}&\delta_{m,0}\cos\vartheta_e+\overset{\circ}{\sum_k^{N-1}}e^{jm2\pi x_k/T}\sum_{i=0}^{i_{hi}}(-1)^i\frac{\Gamma_i}{k_\rho}\tanh(\Gamma_i t_k)\frac{S_{m,i}\cdot S_{0,i}}{Q_i}\Bigg]\cdot P_e\end{aligned} \qquad (20)$$

where $\delta_{m,n}$ is the Kronecker symbol, and the circle o at $\circ\sum$ indicates that cells with depth $t_k=0$ (which exist for a QRD) are excluded from the summation. The amplitudes $B_{k,n}$ follow, with a set of solutions A_n , from :

$$B_{k,m} = \frac{(P_e + A_0)R_{0,m} + \sum_{n\neq 0} A_n\, e^{-jn2\pi x_k/T} R_{n,m}}{Q_m \cosh(\Gamma_m t_k)} \tag{21}$$

These equations use mode norms and coupling coefficients :

$$Q_n := \frac{1}{h}\int_{-h}^{+h} q_n^2(y)\,dy = \begin{cases} 1+\dfrac{\sin 2\varepsilon_n h}{2\varepsilon_n h} & ;\; n = n_e = 0,2,\ldots \\ 1-\dfrac{\sin 2\varepsilon_n h}{2\varepsilon_n h} & ;\; n = n_o = 1,3,\ldots \end{cases} \tag{22}$$

$$S_{m,n} := \frac{1}{h}\int_{-h}^{+h} e^{j\beta_m y}\cdot\begin{Bmatrix}\cos(\varepsilon_n y)\\ \sin(\varepsilon_n y)\end{Bmatrix}\cdot dy \;;\; \begin{Bmatrix} n = n_e \\ n = n_o \end{Bmatrix} \;;\; m = 0,\pm 1,\pm 2,\ldots$$

$$= \frac{1}{h(\beta_m^2-\varepsilon_n^2)}[(\beta_m+\varepsilon_n)\sin((\beta_m-\varepsilon_n)h)+(\beta_m-\varepsilon_n)\sin((\beta_m+\varepsilon_n)h)] \;;\; n = n_e \tag{23}$$

$$= \frac{1}{h(\beta_m^2-\varepsilon_n^2)}[(\beta_m+\varepsilon_n)\sin((\beta_m-\varepsilon_n)h)-(\beta_m-\varepsilon_n)\sin((\beta_m+\varepsilon_n)h)] \;;\; n = n_o$$

$$R_{m,n} := \frac{1}{h}\int_{-h}^{+h} e^{-j\beta_m y}\cdot\begin{Bmatrix}\cos(\varepsilon_n y)\\ \sin(\varepsilon_n y)\end{Bmatrix}\cdot dy \;;\; \begin{Bmatrix} n = n_e \\ n = n_o \end{Bmatrix} \;;\; m = 0,\pm 1,\pm 2,\ldots = \begin{cases} S_{m,n_e} & ;\; n = n_e \\ -S_{m,n_o} & ;\; n = n_o \end{cases} \tag{24}$$

Limit values for $\beta_m \to \pm\varepsilon_n$ are : $\quad S_{m,n_e} \to Q_{n_e} \;;\; S_{m,n_o} \to S_{m,n_o} \to \pm j\, Q_{n_o}$. (24)

See [H.9] for field evaluation with 2-dimensional absorbers.

H.22 Tight panel absorber, rigorous solution

Ref.: Mechel, [H.12], [H.13]

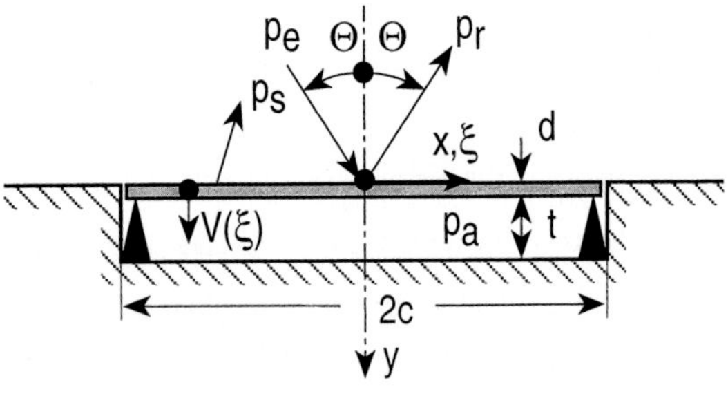

A tight, long, elastic panel is simply supported at its borders at $x = \pm c$. Its thickness is d , the plate material density ρ_p , the elastic parameter for bending $f_{cr}d$, with the critical frequency f_{cr} , the bending loss factor is η . The panel covers a back volume of depth t . The characteristic propagation constant and wave impedance in the back volume are Γ_a , Z_a (thus the back volume may be filled with air, i.e. $\Gamma_a \to j\cdot k_0$; $Z_a \to Z_0$, if t is not too small, or Γ_a , Z_a from a flat capillary for small t , or Γ_a , Z_a from porous materials if the back volume is filled with such material). The front side of the arrangement is flush with a hard baffle wall. A plane wave p_e is incident (normal to the z axis) with a polar angle Θ .

Field formulation in front of the absorber:

$$p(x,y) = p_e(x,y) + p_r(x,y) + p_s(x,y) \tag{1}$$

with p_r the reflected wave after reflection at a hard plane y=0 ; p_s the scattered wave.

$$\begin{gathered} p_e(x,y) = P_e \cdot e^{-jk_x x} \cdot e^{-jk_y y} \quad ; \quad p_r(x,y) = P_e \cdot e^{-jk_x x} \cdot e^{+jk_y y} \\ k_x = k_0 \sin\Theta \quad ; \quad k_y = k_0 \cos\Theta \end{gathered} \tag{2}$$

Field p_a in the back volume, with the wave and impulse equations

$$\left(\Delta - \Gamma_a^2\right) p_a = 0 \quad ; \quad v_a = \frac{-1}{\Gamma_a Z_a} \operatorname{grad} p_a \tag{3}$$

as sum of volume modes :

$$p_a(x,y) = \sum_{k\ge 0} a_k \cdot p_{ak}(x) \cdot \cos\left(\kappa_k (y-t)\right) \tag{4}$$

$$\kappa_k c = j\sqrt{(\Gamma_a c)^2 + \gamma_k^2}$$

$$p_{ak}(\xi) = \begin{cases} \cos\left(k\pi\xi/2\right) = \cos(\gamma_k \xi) & ; \; k = 0,2,4,\ldots \\ \sin\left(k\pi\xi/2\right) = \sin(\gamma_k \xi) & ; \; k = 1,3,5,\ldots \end{cases} \quad ; \quad \gamma_k = k\pi/2 \tag{5}$$

$$v_{ay}(\xi, y=0) = \frac{-1}{\Gamma_a Z_a} \sum_{k\ge 0} a_k \kappa_k \cdot p_{ak}(\xi) \cdot \sin\left(\kappa_k t\right) \tag{6}$$

Plate vibration velocity V(x), or $V(\xi)$ with $\xi = x/c$:

$$\begin{gathered} V(\xi) = \sum_{n\ge 1} V_n \cdot v_n(\xi) \\ v_n(x) = \begin{cases} \cos\left(n\pi\xi/2\right) = \cos(\gamma_n \xi) & ; \; n = 1,3,5,\ldots \\ \sin\left(n\pi\xi/2\right) = \sin(\gamma_n \xi) & ; \; n = 2,4,6,\ldots \end{cases} \quad ; \quad \gamma_n = n\pi/2 \end{gathered} \tag{7}$$

The feature of $p_s(x,0) = p_s(\xi)$ to have a finite normal particle velocity $v_{sy}(x,0)$ in $-c \le x \le +c$, and zero normal velocity outside suggests the use of elliptic-hyperbolic cylinder co-ordinates (ρ,ϑ) for the formulation of that component field. These co-ordinates follow from the Cartesian co-ordinates (x,y) by the transformation :

$$x = c \cdot \cosh\rho \cdot \cos\vartheta \qquad y = c \cdot \sinh\rho \cdot \sin\vartheta \tag{8}$$

$x = \pm c$ are the positions of the common foci of the ellipses and hyperbolic branches.

At $\rho = 0$:

$\xi = x/c = \cos\vartheta$.

$$v_\rho \xrightarrow[\rho\to 0]{} -v_y$$

$$v_\vartheta \xrightarrow[\rho\to 0]{} v_x$$

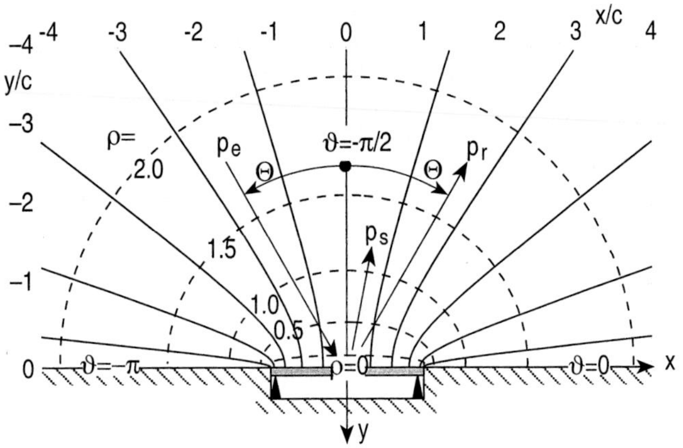

$$\text{grad}\, p \xrightarrow[\rho\to 0]{} \frac{1}{c \sin\vartheta}\left[\frac{\partial p}{\partial \rho}\vec{n}_\rho + \frac{\partial p}{\partial \vartheta}\vec{n}_\vartheta\right] \tag{9}$$

The wave equation in these co-ordinates writes as:

$$\frac{\partial^2 p}{\partial \rho^2} + \frac{\partial^2 p}{\partial \vartheta^2} + (k_0 c)^2\left(\cosh^2\rho - \cos^2\vartheta\right) \cdot p(\rho,\vartheta) = 0 \tag{10}$$

It separates for $p(\rho,\vartheta) = U(\vartheta)\cdot W(\rho)$ into the two Mathieu differential equations:

$$\frac{d^2U(z)}{dz^2} + \left(b - 4q\cos^2 z\right)\cdot U(z) = 0$$
$$\frac{d^2W(z)}{dz^2} - \left(b - 4q\cosh^2 z\right)\cdot W(z) = 0 \tag{11}$$

with $q = (k_0 c)^2/4$ and b a separation constant. Solutions are the Mathieu functions (see [H.13] for these functions). The sum $p_e + p_r$ can be expanded in Mathieu functions $ce_m(\vartheta)$, $Jc_m(\rho)$:

$$p_e(\rho,\vartheta) + p_r(\rho,\vartheta) = 4P_e \sum_{m=0}^{\infty} (-j)^m ce_m(\alpha)\cdot Jc_m(\rho)\cdot ce_m(\vartheta) \tag{12}$$

A formulation of p_s which has the mentioned features for each term is :

$$p_s(\rho,\vartheta) = 4\sum_{m=0}^{\infty} D_m(-j)^m ce_m(\alpha)\cdot Hc_m^{(2)}(\rho)\cdot ce_m(\vartheta) \qquad (13)$$

$$\alpha = \pi/2-\Theta \quad ; \quad q=(k_0c)^2/4$$

The $ce_m(\vartheta)$ are „azimuthal Mathieu functions" which are even in ϑ at $\vartheta=0$, and the $Hc_m^{(2)}(\rho) = Jc$ the „Mathieu-Bessel" function $Jc_m(\rho)$ and the „Mathieu-Neumann" function $Yc_m(\rho)$ like the cylindrical Hankel function of the 2nd kind. The Mathieu functions depend on the parameter q. It is noted for later use that $ce_m(\vartheta)$, $Jc_m(\rho)$, $Yc_m(\rho)$ are real functions, and $Jc'_m(0)=0$, $Hc'^{(2)}_m(0) = -j\cdot Yc'_m(0)$ (where the primes indicate derivatives with respect to ρ). It will be important for later evaluations that the $ce_m(\vartheta)$ are generated as a Fourier series :

$$ce_m(\vartheta) = \sum_{s=0}^{+\infty} A_{2s+p}\cdot\cos((2s+p)\vartheta) \quad ; \quad m=2r+p \quad ; \quad \begin{cases} r=0,1,2,\ldots \\ p=0,1 \end{cases} \qquad (14)$$

so the real Fourier coefficients A_{2s+p} are delivered by the computing program which generates the Mathieu function.

The plate vibration modes $v_n(\xi)$, the back volume modes $p_{ak}(\xi)$, and the Mathieu functions $ce_m(\vartheta)$ are orthogonal functions in the range of the plate with norms :

$$N_{pn} = \int_{-1}^{1} v_n^2(\xi)\,d\xi = 1 \quad ; \quad N_{ak} = \int_{-1}^{1} p_{ak}^2(\xi)\,d\xi = \begin{cases} 2\,; k=0 \\ 1\,; k>0 \end{cases}$$
$$N_{sm} = \int_{-\pi}^{0} ce_m^2(\vartheta)\,d\vartheta = \int_{0}^{\pi} ce_m^2(\vartheta)\,d\vartheta = \frac{\pi}{2} \qquad (15)$$

The remaining boundary conditions to be satisfied are :

$$v_{sy}(\xi) = V(\xi)$$
$$v_{ay}(\xi) = V(\xi) \qquad (16)$$
$$p_e(\xi)+p_r(\xi)+p_s(\xi)-p_a(\xi) = \sum_{n\geq 1} V_n Z_{Tn}\cdot v_n(\xi)$$

where, in the last condition, Z_{Tn} are modal partition impedances of the panel :

$$\frac{Z_{Tn}}{Z_0} = 2\pi Z_m F\left[\eta F^2\left(\frac{\gamma_n}{k_0c}\right)^4 + j\left(1-F^2\left(\frac{\gamma_n}{k_0c}\right)^4\right)\right] \quad ; \quad F=\frac{f}{f_{cr}} \quad ; \quad Z_m = \frac{f_{cr}d}{Z_0}\rho_p \qquad (17)$$

The last condition assumes that the left-hand side is expanded in plate modes $v_n(\xi)$ and that the condition holds term-wise.

Multiplication of the 1st condition by $\sin\vartheta \cdot ce_m(\vartheta)$ and integration with respect to ϑ over $(-\pi,0)$ gives :

$$D_m = \frac{-k_0 c}{2\pi(-j)^m} \frac{1}{ce_m(\alpha)\cdot Yc'_m(0)} \sum_{n\geq 1} Z_0 V_n \cdot Q_{m,n} \tag{18}$$

Multiplication of the 2nd condition by $p_{ak}(\xi)$ and integration over $-1 \leq \xi \leq +1$ gives :

$$a_k = \frac{-\Gamma_a Z_a}{\kappa_k \cdot N_{ak} \cdot \sin(\kappa_k t)} \sum_{n\geq 1} V_n \cdot S_{k,n} \quad ; \quad k \geq 0 \tag{19}$$

Multiplication of the last condition with $v_n(\xi)$ and integration over $-1 \leq \xi \leq +1$ gives :

$$\begin{aligned} &N_{pn} \frac{Z_{Tn}}{Z_0} \cdot Z_0 V_n = \\ &4 \sum_{m\geq 0} (-j)^m ce_m(\alpha)\cdot Q_{m,n} \cdot \left[P_e Jc_m(0) + D_m Hc_m^{(2)}(0)\right] - \sum_{k\geq 0} a_k \cdot S_{k,n} \cdot \cos(\kappa_k t) \end{aligned} \tag{20}$$

and after insertion of D_m, a_k the linear system of equations for $Z_0 V_n$ (ν= 1,2,3,...):

$$\begin{aligned} &\sum_{n\geq 1} Z_0 V_n \cdot \\ &\cdot \left\{ \delta_{n,\nu} N_{p\nu} - \frac{k_0 c Z_0}{Z_{T\nu}} \left[\frac{2j}{\pi} \sum_{m\geq 0} \frac{Hc_m^{(2)}(0)}{Hc_m'^{(2)}(0)} \cdot Q_{m,\nu} Q_{m,n} + \frac{\Gamma_a Z_a}{k_0 Z_0} \sum_{k\geq 0} \frac{S_{k,\nu} \cdot S_{k,n} / N_{ak}}{\kappa_k c \cdot \tan(\kappa_k t)} \right] \right\} = \\ &= 4 P_e \frac{Z_0}{Z_{T\nu}} \sum_{m\geq 0} (-j)^m ce_m(\alpha) \cdot Q_{m,\nu} \cdot Jc_m(0) \end{aligned} \tag{21}$$

with the Kronecker symbol $\delta_{n,\nu}$. After its solution, the amplitudes D_m , a_k follow from above.

These equations use the mode coupling coefficients :

$$S_{k,n} := \int_{-1}^{+1} p_{ak}(\xi) \cdot v_n(\xi)\, d\xi$$

$$p_{ak}(\xi) = \begin{cases} \cos(k\pi\xi/2) = \cos(\gamma_k \xi) & ; \quad k = 0,2,4,\ldots \\ \sin(k\pi\xi/2) = \sin(\gamma_k \xi) & ; \quad k = 1,3,5,\ldots \end{cases} \quad ; \quad \gamma_k = k\pi/2 \tag{22}$$

$$v_n(\xi) = \begin{cases} \cos(n\pi\xi/2) = \cos(\gamma_n \xi) & ; \quad n = 1,3,5,\ldots \\ \sin(n\pi\xi/2) = \sin(\gamma_n \xi) & ; \quad n = 2,4,6,\ldots \end{cases} \quad ; \quad \gamma_n = n\pi/2$$

with the values :

$$S_{k,n} = \begin{cases} 0 \quad ; \quad k_e \;\&\; n_e \\ 0 \quad ; \quad k_o \;\&\; n_o \\ \dfrac{2}{\pi}\left(\dfrac{(-1)^{(k_o-n_e-1)/2}}{k_o - n_e} + \dfrac{(-1)^{(k_o+n_e-1)/2}}{k_o + n_e} \right) \quad ; \quad k_o \;\&\; n_e \\ \dfrac{2}{\pi}\left(\dfrac{(-1)^{(k_e-n_o-1)/2}}{k_e - n_o} - \dfrac{(-1)^{(k_e+n_o-1)/2}}{k_e + n_o} \right) \quad ; \quad k_e \;\&\; n_o \end{cases} \tag{23}$$

and the coupling coefficients :

$$Q_{m,n} := \int_{-1}^{+1} ce_m(\arccos\xi) \cdot v_n(\xi)\, d\xi = \int_0^{\pi} \sin\vartheta \cdot ce_m(\vartheta) \cdot v_n(\cos\vartheta)\, d\vartheta$$

$$v_n(\xi) = \begin{cases} \cos(n\pi\xi/2) = \cos(\gamma_n \xi) \quad ; \quad n = 1,3,5,\ldots \\ \sin(n\pi\xi/2) = \sin(\gamma_n \xi) \quad ; \quad n = 2,4,6,\ldots \end{cases} \quad ; \quad \gamma_n = n\pi/2 \tag{24}$$

$$ce_m(\vartheta) = \sum_{s=0}^{+\infty} A_{2s+p} \cdot \cos((2s+p)\vartheta) \quad ; \quad m = 2r + p \quad ; \quad \begin{cases} r = 0,1,2,\ldots \\ p = 0,\, 1 \end{cases}$$

with zero values if both m and n are even or odd, and for different parities of m, n :

$$Q_{2r,n} = \sqrt{\frac{2\pi}{\gamma_n}} \sum_{s\geq 0} A_{2s} \left[J_{1/2}(\gamma_n) + (1-\delta_{0,s}) \sum_{i=1}^{s} (-1)^i \frac{i!}{(2i)!} \left(\frac{2}{\gamma_n} \right)^i \cdot J_{i+1/2}(\gamma_n) \prod_{k=0}^{i-1} \left(4s^2 - 4k^2 \right) \right] \tag{25}$$

$$Q_{2r+1,n} = \sqrt{\frac{2\pi}{\gamma_n}} \sum_{s\geq 0} A_{2s+1} \left[J_{3/2}(\gamma_n) + (1-\delta_{0,s}) \sum_{i=1}^{s} (-1)^i \frac{i!}{(2i)!} \prod_{k=1}^{i} \left((2s+1)^2 - (2k-1)^2 \right) \left(\frac{2}{\gamma_n} \right)^i \cdot J_{i+3/2}(\gamma_n) \right]$$

with Bessel functions of half-integer orders $J_{i+1/2}(z)$.

The sound absorption coefficient $\alpha(\Theta) = \Pi_a/\Pi_e$ is evaluated with the effective incident power Π_e (per unit panel length)

$$\Pi_e = \frac{c \cdot \cos\Theta}{Z_0} |P_e|^2 \tag{26}$$

and the absorbed effective sound power :

$$\Pi_a = \frac{c}{2}\mathrm{Re}\left\{\int_{-1}^{+1} (p_e + p_r + p_s)\cdot v_{sy}^* \, d\xi\right\}$$
$$= \frac{-4c\pi}{k_0 c Z_0}\sum_{m\ge 0} ce_m^2(\alpha)\cdot Yc'_m(0)\cdot \mathrm{Re}\left\{\left(P_e\cdot Jc_m(0) + D_m\cdot Hc_m^{(2)}(0)\right)\cdot D_m^*\right\} \tag{27}$$

Special case : The back volume is locally reacting

i.e. its input impedance (at $y=0$) is : $Z_b = Z_a\cdot\coth(\Gamma_a t)$ (28)

The boundary conditions then become :

$$p_e(\xi) + p_r(\xi) + p_s(\xi) = \sum_{n\ge 1} V_n (Z_{Tn} + Z_b)\cdot v_n(\xi)$$
$$v_{sy}(\xi) = \sum_{n\ge 1} V_n \cdot v_n(\xi) \tag{29}$$

The system of equations for the plate mode amplitudes $Z_0 V_n$ will be ($\nu = 1,2,3,\dots$):

$$\sum_{n\ge 1} Z_0 V_n \cdot\left[\delta_{n,\nu} - \frac{2jk_0 c}{\pi}\,\frac{Z_0}{Z_{T\nu}+Z_b}\sum_{m\ge 0} Q_{m,\nu}\cdot Q_{m,n}\cdot\frac{Hc_m^{(2)}(0)}{Hc_m'^{(2)}(0)}\right] =$$
$$= 4P_e\frac{Z_0}{Z_{T\nu}+Z_b}\sum_{m\ge 0}(-j)^m ce_m(\alpha)\cdot Jc_m(0)\cdot Q_{m,\nu} \tag{30}$$

The amplitudes D_m of the scattered field are evaluated as above, and the absorption coefficient also. An alternative form for the sound absorption coefficient in this special case is :

$$\Pi_a = \frac{c}{2}\int_{-1}^{+1}\mathrm{Re}\left\{\sum_{n\ge 1} V_n (Z_{Tn}+Z_b)\cdot v_n(\xi)\cdot\sum_{n\ge 1} V_n^*\cdot v_n(\xi)\right\} d\xi$$
$$= \frac{c}{2Z_0}\sum_{n\ge 1} N_{pn}\,\mathrm{Re}\left\{\frac{Z_{Tn}+Z_b}{Z_0}\right\}\cdot |Z_0 V_n|^2 \tag{31}$$
$$\alpha(\Theta) = \frac{1}{2\cos\Theta}\sum_{n\ge 1}\mathrm{Re}\left\{\frac{Z_{Tn}+Z_b}{Z_0}\right\}\cdot |Z_0 V_n / P_e|^2$$

The numerical examples for a plywood panel absorber use as constant parameters $d = 6$ mm ; $\rho_p = 700$ kg/m^3 ; $f_{cr}d = 20$ Hz·m ; $\eta = 0.02$; the back volume with depth $t = 10$ cm is filled with glass fibre material having a flow resistivity $\Xi = 2500$ Pa·s/m^2 . The angle of sound incidence is $\Theta = 45°$. The upper limits used for the modes are $n_{hi} = 10$; $k_{hi} = m_{hi} = 8$. The plots of $\alpha(\Theta)$ over the frequency f also contain (as dashed curves, for orientation) the absorption coefficient for an infinite panel.

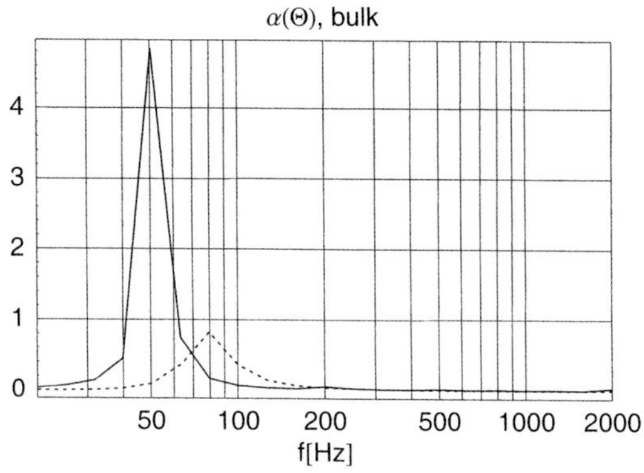

Sound absorption coefficient α(Θ) for Θ= 45° of a plywood panel absorber with c= 0.2 m; modal analysis: full line; infinite panel: dashed line.

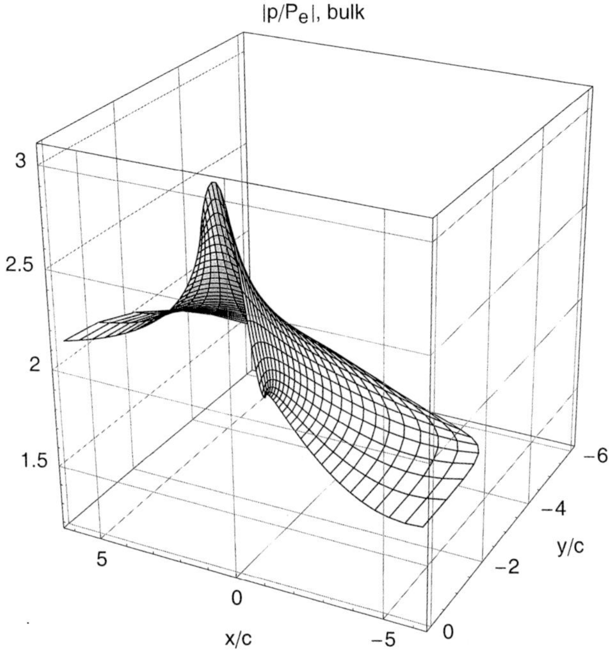

Magnitude of the sound pressure field for the above absorber, at f= 50 Hz ; sound incidence is from the side of negative x/c values.

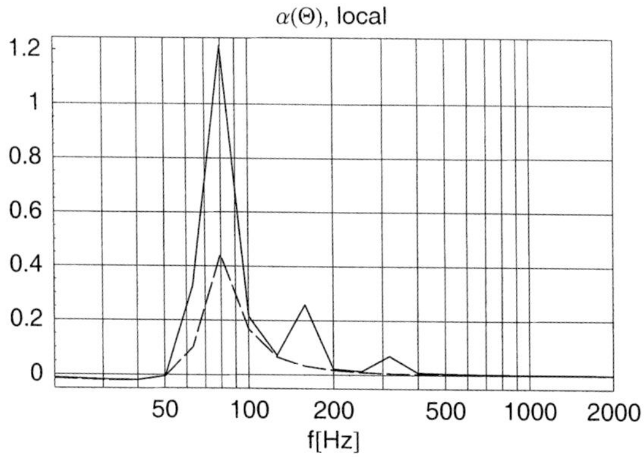

Sound absorption coefficient α(Θ) for Θ= 45° of a plywood panel absorber with c= 0.5 m and locally reacting back volume; modal analysis: full line; infinite panel: dashed line.

H.23 Tight panel absorber, approximations

Ref.: Mechel, [H.12]

The object and the symbols used are as in the previous section H.22. This section describes approximations which avoid the evaluation of Mathieu functions.

The principal step in such approximations is the subdivision of the boundary value problem into two sub-tasks. The 1st sub-tasks finds the plate mode amplitudes with the assumption that $p_s(\xi)$ can be neglected compared to $p_e(\xi)+p_r(\xi)$. This sum is assumed to be the driving force on the front side for the plate motion. The assumption is plausible if the surface impedance of the plate is not too small (i.e. outside resonances). The 2nd step then evaluates the absorbed power with the plate mode amplitudes V_n found in the 1st step.

If the back volume is assumed to be bulk reacting (i.e. possible sound propagation parallel to the plate) the boundary conditions of the 1st sub-task are :

$$p_e(\xi)+p_r(\xi)-p_a(\xi)=2P_e\cdot e^{-jk_x c\xi}-p_a(\xi)=\sum_{n\geq 1} V_n Z_{Tn}\cdot v_n(\xi)$$

$$v_{ay}(\xi)=\sum_{n\geq 1} V_n\cdot v_n(\xi) \tag{1}$$

where in the 1st equation the left-hand side is assumed to be expanded in plate modes $v_n(\xi)$ so that modal plate partition impedances Z_{Tn} (see section H.22) can be applied. It should be noted that this equation describes the excitation of the plate by a distributed force without radiation load on the side of excitation. Multiplication of the 1st equation by $v_\nu(\xi)$ (ν= 1,2,3,...) and integration over $-1\leq\xi\leq+1$ yields the equations :

$$V_\nu Z_{T\nu}N_{p\nu}=2P_e\cdot R_\nu-\sum_{k\geq 0} a_k\cdot S_{k,\nu}\cdot\cos(\kappa_k t) \tag{2}$$

with the mode coupling coefficients $S_{k,n}$ from the previous section, and the new coefficients:

$$R_n:=\int_{-1}^{+1} e^{-jk_x c\xi}\cdot v_n(\xi)\,d\xi=\begin{cases}\dfrac{4n\pi(-1)^n\cos(k_x c)}{(n\pi)^2-4(k_x c)^2} & ;\ n=\text{odd}\\[2ex] \dfrac{-4jn\pi(-1)^{n/2}\sin(k_x c)}{(n\pi)^2-4(k_x c)^2} & ;\ n=\text{even}\end{cases} \tag{3}$$

Insertion of the back volume mode amplitudes a_k which follow from the 2nd boundary conditions as in the previous Section 22 :

$$a_k = \frac{-\Gamma_a Z_a}{\kappa_k \cdot N_{ak} \cdot \sin(\kappa_k t)} \sum_{n\geq 1} V_n \cdot S_{k,n} \quad ; \quad k \geq 0 \tag{4}$$

gives the system of equations to be solved for V_n if the back volume is bulk reacting :

$$\sum_{n\geq 1} V_n \cdot \left[\delta_{n,\nu} \cdot N_{p\nu} - \frac{\Gamma_a Z_a}{Z_{T\nu}} \sum_{k\geq 0} \frac{S_{k,n} \cdot S_{k,\nu} \cdot \cot(\kappa_k t)}{\kappa_k \cdot N_{ak}} \right] = 2P_e \cdot \frac{R_\nu}{Z_{T\nu}} \tag{5}$$

($\delta_{n,\nu}$ is the Kronecker symbol).

If the back volume is locally reacting, the only remaining boundary condition becomes :

$$p_e(\xi) + p_r(\xi) = \sum_{n\geq 1} V_n (Z_{Tn} + Z_b) \cdot v_n(\xi) \tag{6}$$

Multiplication and integration as before gives the explicit expressions for V_n :

$$V_n = \frac{2P_e \cdot R_n}{(Z_{Tn} + Z_b) N_{pn}} \tag{7}$$

The 2nd sub-task determines the absorbed sound power, assuming that the plate velocity $V(\xi)$, expanded in $V_n \cdot v_n(\xi)$, is a given oscillation (i.e. again without consideration of a possible back reaction of radiation on the oscillation).

In a first variant of this step one applies the product $(p_e(\xi)+p_r(\xi)) \cdot V^*(\xi)$ for the evaluation of the power which $(p_e(\xi)+p_r(\xi))$ feeds into the plate. So one makes the same error twice, because $(p_e(\xi)+p_r(\xi))$ is not the true exciting pressure.

$$\Pi_{a1} = \frac{c}{2} \int_{-1}^{+1} \mathrm{Re}\left\{ (p_e(\xi) + p_r(\xi)) \cdot \sum_{n\geq 1} V_n^* \cdot v_n(\xi) \right\} d\xi = P_e \cdot c \sum_{n\geq 1} \mathrm{Re}\left\{ V_n^* \cdot R_n \right\} \tag{8}$$

In a second variant one takes into account the sound pressure that the plate with the given velocity profile $V(\xi)$ in a baffle wall radiates. We write p_s for the radiated sound ($V(\xi)$ is counted positive in the direction oriented into the plate; thus the plate in fact is a sink for the energy of p_s). The absorbed power is given by the integral over $(p_e(\xi) + p_r(\xi)) \cdot V^*(\xi) + p_s(\xi) \cdot V^*(\xi)$. The 1st term gives the power contribution Π_{a1} ; the 2nd term is the absorbed effective power Π_{as} due to p_s . This can be obtained by (see [H.14]) :

$$\Pi_{as} = \frac{c}{2} \int_{-1}^{+1} \mathrm{Re}\left\{ p_s(\xi) \cdot V^*(\xi) \right\} d\xi = \frac{k_0 Z_0}{4\pi} \int_{-k_0}^{k_0} \frac{|\hat{v}(k_1)|^2}{\sqrt{k_0^2 - k_1^2}} dk_1 = \frac{k_0 Z_0}{4\pi} \int_{-\pi/2}^{\pi/2} |\hat{v}(k_0 \cos\psi)|^2 \, d\psi \tag{9}$$

where $\hat{v}(k_1)$ is the wave number spectrum of $V(x)$, which follows by a Fourier transform ($L = 2c$ the plate width):

$$\hat{v}(k_1) = \int_{-\infty}^{+\infty} V(x)\cdot e^{-jk_1x}\,dx = \int_L V(x)\cdot e^{-jk_1x}\,dx = c\int_{-1}^{+1} V(\xi)\cdot e^{-jk_1c\cdot\xi}\,d\xi \tag{10}$$

and the back transformation :

$$V(x) = \frac{1}{2\pi}\int_{-\infty}^{+\infty} \hat{v}(k_1)\cdot e^{+jk_1x}\,dk_1 \tag{11}$$

In the present application $V(\xi)$ is the sum of terms $V_n\cdot v_n(\xi)$. The wave number spectrum $\hat{v}_n(k_1)$ of $v_n(\xi)$ is :

$$\hat{v}_n(k_1) = c\int_{-1}^{+1} e^{-jk_1c\xi}\cdot v_n(\xi)\,d\xi = \begin{cases} c\dfrac{4n\pi(-1)^n\cos(k_1c)}{(n\pi)^2-4(k_1c)^2} & ;\quad n=\text{odd} \\[2ex] c\dfrac{-4jn\pi(-1)^{n/2}\sin(k_1c)}{(n\pi)^2-4(k_1c)^2} & ;\quad n=\text{even} \end{cases} \tag{12}$$

and the contribution to the absorbed power becomes:

$$\Pi_{as} = \frac{k_0Z_0}{4\pi}\int_{-\pi/2}^{\pi/2}\left|\sum_{n\geq 1} V_n\cdot\hat{v}_n(k_0\cos\psi)\right|^2 d\psi \tag{13}$$

The integral must be evaluated numerically.

A 3rd variant of the 2nd sub-task completes the absorbed intensity to $(p_e(\xi)+p_r(\xi)+p_s(\xi))\cdot V^*(\xi)$, but

$$p_s(x,y) = \sum_{n\geq 1} d_n\cdot v_n(x)\cdot f_n(y) \tag{14}$$

A plausible form for $f_n(y)$ representing outgoing waves is $f_n(y)= \exp(j\varepsilon_n y)$; the terms satisfy the wave equation and Sommerfeld's condition if $(\varepsilon_n c)^2 = (k_0c)^2 - \gamma_n^2\,;\,\mathrm{Im}\{\varepsilon_n\}\leq 0$. From $v_{sy}(\xi) = V(\xi)$ one gets :

$$d_n = -\frac{k_0}{\varepsilon_n}Z_0V_n \tag{15}$$

The absorbed power is :

$$\begin{aligned} \Pi_a &= \frac{c}{2}\int_{-1}^{+1}\mathrm{Re}\left\{(p_e(\xi)+p_r(\xi)+p_s(\xi))\cdot\sum_{n\geq 1}V_n^*\cdot v_n(\xi)\right\}d\xi = \Pi_{a1}+\Pi_{as} \\ \Pi_{as} &= \frac{c}{2}\int_{-1}^{+1}\mathrm{Re}\left\{\sum_{n\geq 1}d_n\cdot v_n(\xi)\cdot\sum_{n\geq 1}V_n^*\cdot v_n(\xi)\right\}d\xi = \frac{-c}{2Z_0}\sum_{n\geq 1}\mathrm{Re}\left\{\frac{k_0}{\varepsilon_n}\right\}\cdot N_{pn}\cdot|Z_0V_n|^2 \end{aligned} \tag{16}$$

The approximation results are acceptable, except in or near plate resonances.

A very simple approximation, serving more for orientation than as an approximation, assumes the plate to be infinitely wide $(L \to \infty)$. Then the absorption coefficient $\alpha(\Theta)$ follows from the reflection factor R as $\alpha(\Theta)= 1-|R|^2$ with :

$$R = \frac{(Z_T + Z_b)\cdot \cos\Theta - Z_0}{(Z_T + Z_b)\cdot \cos\Theta + Z_0} \tag{17}$$

where Z_T is the plate partition impedance of an infinite panel :

$$\frac{Z_T}{Z_0} = 2\pi Z_m F\left[\eta F^2 \sin^4\Theta + j\left(1 - F^2 \sin^4\Theta\right)\right] \quad ; \quad F = \frac{f}{f_{cr}} \quad ; \quad Z_m = \frac{f_{cr}d}{Z_0}\rho_p \tag{18}$$

and Z_b is the input impedance of the back volume, which is given, for a locally reacting volume, by :

$$Z_b = Z_a \cdot \coth(\Gamma_a t) \tag{19}$$

and for a bulk reacting volume by :

$$\frac{Z_b}{Z_0} = \frac{\Gamma_{an} Z_{an}}{\Gamma_{an}\cos\theta_1}\cdot \coth(k_0 t\, \Gamma_{an}\cos\theta_1) \quad ; \quad \Gamma_{an} = \Gamma_a / k_0 \, ; Z_{an} = Z_a / Z_0$$
$$\Gamma_{an}\cos\theta_1 = \sqrt{\Gamma_{an}^2 + \sin^2\Theta} \tag{20}$$

H.24 Porous panel absorber, rigorous solution

Ref.: Mechel, [H.12]

The object is like the absorber in section H.22 , except now the panel is assumed to be perforated with a porosity σ .
Field formulations and symbols will be taken from H.22.

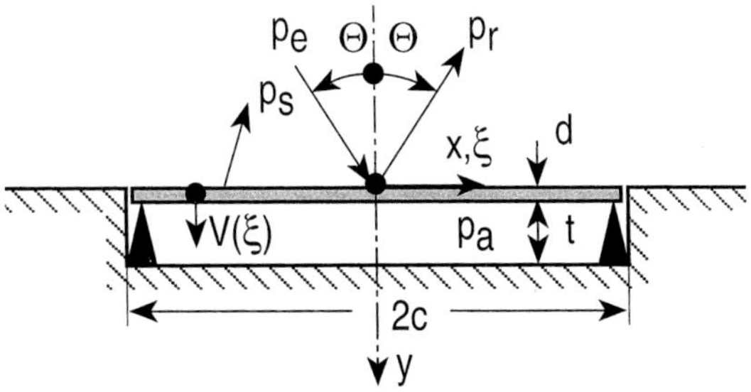

A tight, long, elastic panel is simply supported at its borders at $x= \pm c$.
Its thickness is d , the plate material density ρ_p , the elastic parameter for bending $f_{cr}d$, with the critical frequency f_{cr} , the bending loss factor is η . The panel covers a back volume of depth t . The characteristic propagation constant and wave impedance in the back volume are Γ_a , Z_a (thus the back volume may be filled with air, i.e. $\Gamma_a \to j\cdot k_0$; $Z_a \to Z_0$, if t is

not too small, or Γ_a, Z_a from a flat capillary for small t, or Γ_a, Z_a from porous materials if the back volume is filled with such material). The front side of the arrangement is flush with a hard baffle wall. A plane wave p_e is incident (normal to the z axis) at a polar angle Θ.

To make the perforation tractable in the analysis, it is assumed that a "micro-structured" perforation is applied. This means that the diameter of the perforations and their distances are small compared with both the sound wave length and the panel width. We further assume a homogeneous distribution of the perforation over the panel (except possibly for narrow border areas).

One first has to fix how the acoustic qualities of the perforation and of the perforated panel have to be defined. The perforation changes the mechanical parameters, effective plate material density ρ_p and bending modulus B of the plate:

$$\rho_p \to \rho_p(1-\sigma) \quad ; \quad B \to B\cdot(1-\sqrt{\sigma}) \quad ; \quad f_{cr}d \to f_{cr}d\sqrt{\frac{1-\sigma}{1-\sqrt{\sigma}}} \tag{1}$$

The indicated change in B is for approximately square holes and perforation raster; more sophisticated relations can be derived for other geometries. The symbol Z_T will be used for the partition impedance of an equivalent tight plate, evaluated with these parameters and defined by $\Delta p = Z_T \cdot v_p$ where v_p is the velocity of this tight panel. The pores are characterised by an impedance $Z_r = Z_r' + j\cdot Z_r''$ determined by $\Delta p = Z_r \cdot v_r$, where Δp is the pressure difference driving the average velocity v_r through the perforated plate at rest. Preferably one determines Z_r' experimentally (because the technical roughness of hole walls and the effect of rounding of the hole corners are difficult to describe analytically; an exception could be straight, very fine holes for which the real part Z'_r also can be determined precisely from the theory of capillaries), and the imaginary part Z_r'' by evaluation from :

$$\frac{Z_r''}{Z_0} = \frac{k_0 a}{\sigma}\left(\frac{d}{a} + \frac{\Delta\ell_e}{a} + \frac{\Delta\ell_i}{a}\right) \tag{2}$$

where a is any representative hole dimension (mostly its radius), and $\Delta\ell_e$, $\Delta\ell_i$ are the exterior and interior end corrections, respectively, where the important distinction should made for the interior end correction whether the interior orifice ends in air or on a porous material.

Z_r can be realised either by the friction force and end corrections of the pores themselves and/ or by the impedance of thin porous sheets (e.g. a fine wire mesh) covering the perforation orifice. In the latter case it must be distinguished whether the sheet is force-locking with the plate or not.

The assumption of the microstructure of the perforation implies that the sound pressure distributions along the panel surfaces do not have significant ripples due to the perforation pattern. The formulations of the component fields therefore remain as in section H.22 . For the determination of the unknown amplitudes a_k, D_m, V_n one needs three boundary conditions as in section H.22, but now modified for the parallel volume flow through the panel and the pores.

The sketch indicates, in a representative area element S, the distribution of the velocities on the plate and the holes.

The average velocity is :

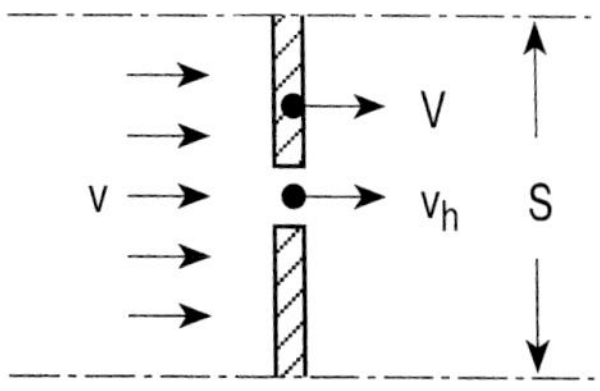

$$v = (1-\sigma)\cdot V + \sigma\cdot v_h$$
$$= (1-\sigma)\cdot V + v_r \qquad (3)$$
$$= (1-\sigma)\cdot V + \Delta p / Z_r$$

The 2nd sketch shows the equivalent network for the perforated panel.

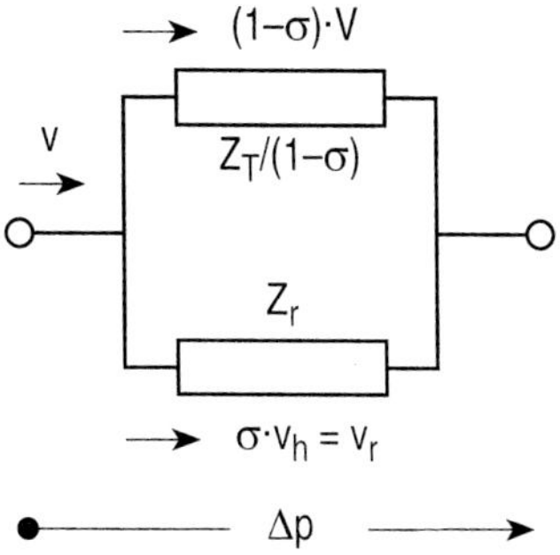

The effective impedance is :

$$Z_{eff} = \frac{Z_r \cdot Z_T / (1-\sigma)}{Z_r \cdot Z_T + (1-\sigma)} \qquad (4)$$

and the 1st boundary condition becomes :

$$\Delta p = Z_{eff} \cdot v$$
$$= \frac{Z_r \cdot Z_T / (1-\sigma)}{Z_r + Z_T / (1-\sigma)} \left[(1-\sigma)\cdot V + \Delta p / Z_r\right] \qquad (5)$$

or with a transformation, if $\sigma \neq 1$:

$$\Delta p \frac{Z_r}{Z_r + Z_T / (1-\sigma)} = \frac{Z_r \cdot Z_T}{Z_r + Z_T / (1-\sigma)} \cdot V \qquad (6)$$

This corresponds to the boundary condition with tight plates in section H.22 if we use the effective plate partition impedance :

$$Z_{Teff} = \frac{Z_r \cdot Z_T}{Z_r + Z_T / (1-\sigma)} \qquad (7)$$

In these relations is $\Delta p = p_e + p_r + p_s - p_a$. (8)

The other boundary conditions of matching velocities become :

$$v = (1-\sigma)\cdot V + \Delta p / Z_r \overset{!}{=} \begin{cases} v_{sy} \\ v_{ay} \end{cases} \qquad (9)$$

[H.10] MECHEL, F.P. Acustica, 81 (1995
"The Wide-angle Diffuser – a Wide-angle Absorber ?"

[H.11] SCHROEDER, M.R., GERLACH, R.E. 9th ICA, Madrid, 19
"Diffuse sound reflection surfaces"

[H.12] MECHEL, F.P. submitted to J.Soun
"Panel Absorber"

[H.13] MECHEL, F.P.
„Mathieu Functions; Formulas, Generation, Use"
S.Hirzel Verlag, Stuttgart, 1997

[H.14] HECKL, M. Acustica 37 (1977)
„Abstrahlung von ebenen Schallquellen"

I
Sound Transmission

This chapter deals with the sound transmission through objects like porous absorber material layers ("noise barriers"), slits and holes in walls, wide passages which nevertheless are sound insulating ("noise sluices"), plates and multiple leaf walls, suspended ceilings, office fences, etc.

I.1 "Noise barriers"

Ref.: Mechel, [I.1]

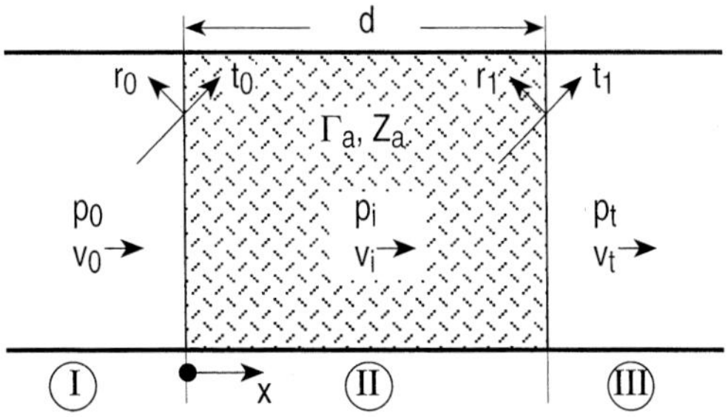

Flanking ducts, e.g. cable ducts or plenum ducts of suspended ceilings, may be the critical path for sound transmission between neighbouring rooms. It may be difficult to install a well fitting partition wall in the duct, but it is easy to fill the duct to some length d with a "plug" of porous absorber material. Let Γ_a, Z_a be the characteristic propagation constant and wave impedance of the material.

Sound fields for *normal incidence* :

in Zone I :
$$\begin{aligned} p_0 &= e^{-jk_0x} + r_0\, e^{+jk_0x} \\ Z_0 v_0 &= e^{-jk_0x} - r_0\, e^{+jk_0x} \end{aligned} \tag{1}$$

in Zone II :
$$\begin{aligned} p_i &= A\cdot\cosh\Gamma_a x + B\cdot\sinh\Gamma_a x \\ Z_0 v_i &= \frac{-1}{Z_a/Z_0}[A\cdot\sinh\Gamma_a x + B\cdot\cosh\Gamma_a x] \end{aligned} \tag{2}$$

in Zone III :
$$\begin{aligned} p_t &= t\,e^{-jk_0x} \\ Z_0 v_t &= t\,e^{-jk_0x} \end{aligned} \tag{3}$$

The boundary conditions give : $A = 1 + r_0 \;;\; B = -\frac{Z_a}{Z_0}(1 - r_0)$ (4)

and for the sound transmission factor : $t\,e^{-jk_0 d} = \frac{2}{2\cosh\Gamma_a d + (\frac{Z_a}{Z_0} + \frac{Z_0}{Z_a})\sinh\Gamma_a d}$ (5)

The transmission coefficient $\tau = |t\,e^{-jk_0 d}|^2$ and from that the transmission loss $R = -10\cdot\lg\tau$ is :

$$R = -20\cdot\lg\left|\frac{2}{2\cosh\Gamma_a d + (Z_a/Z_0 + Z_0/Z_a)\sinh\Gamma_a d}\right| \;[\text{dB}] \quad (6)$$

It is advisable to take for the evaluation of Γ_a, Z_a an effective absorber variable E_{eff}, which takes the vibration of the material matrix (with bulk density RG) into account :

$$E_{eff} = \rho_0 f/\Xi_{eff} = E - \frac{j}{2\pi}\frac{\rho_0}{RG} \quad (7)$$

ρ_0=	density of air;
RG=	bulk density of porous material;
Ξ=	flow resistivity of material;
E=	$\rho_0 f/\Xi$;
f=	frequency;

The diagram compares sound transmission loss values R for layers of basalt wool with different thickness d, from measurements (points) and from the present evaluation.

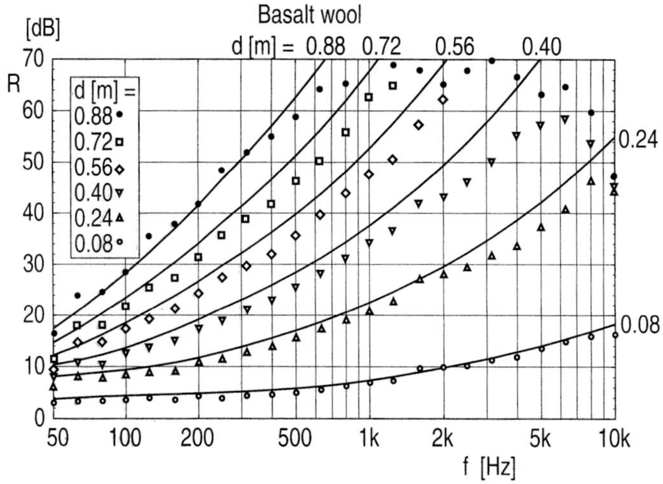

Oblique sound incidence : (at polar angle Θ)

Internal angle Θ_i :

(with $\Gamma_{an}= \Gamma_a/k_0$; $Z_{an}= Z_a/Z_0$)

$$\sin\Theta_i = \frac{j}{\Gamma_{an}}\sin\Theta$$
$$\Gamma_{an}\cos\Theta_i = \sqrt{\Gamma_{an}^2 + \sin^2\Theta} \qquad (8)$$

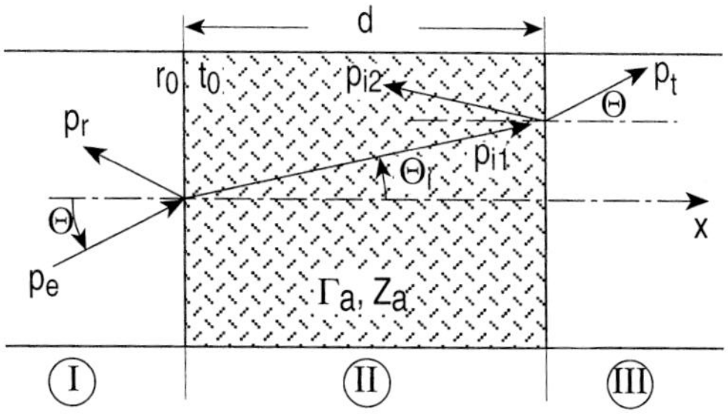

Sound fields :

$$p_e = P_e \cdot e^{-jk_0(x\cos\Theta + y\sin\Theta)}$$
$$p_r = r_0 P_e \cdot e^{-jk_0(-x\cos\Theta + y\sin\Theta)}$$
$$p_i = P_{i1} \cdot e^{-\Gamma_a(x\cos\Theta_i + y\sin\Theta_i)} + P_{i2} \cdot e^{-\Gamma_a(-x\cos\Theta_i + y\sin\Theta_i)} \qquad (9)$$
$$p_t = P_t \cdot e^{-jk_0(x\cos\Theta + y\sin\Theta)}$$

Reflection and transmission factors :

$$r_1 = \frac{1-z}{1+z} \quad ; \qquad r_0 = -r_1\frac{1-e^{-2y}}{1-r_1^2 e^{-2y}} = -\frac{(1-z^2)(1-e^{-2y})}{(1+z)^2-(1-z)^2 e^{-2y}}$$
$$t_1 = 1 + r_1 = \frac{2}{1+z} \quad ; \quad t_0 = \frac{1-r_1}{1-r_1^2 e^{-2y}} = \frac{2z^2}{(1+z)^2-(1-z)^2 e^{-2y}} \qquad (10)$$
$$t = t_0 t_1 e^{-y} = e^{-y}\frac{4z}{(1+z)^2-(1-z)^2 e^{-2y}}$$

with abbreviations : $y := \Gamma_a d\cos\Theta_i = k_0 d\cdot\Gamma_{an}\cdot\cos\Theta_i \quad ; \quad z := Z_{an}\dfrac{\cos\Theta}{\cos\Theta_i}$ (11)

For a given transmission coefficient $\tau(\Theta)= |t(\Theta)|^2$, the sound transmission coefficient with *diffuse* sound incidence is

in 3 dimensions : $$\tau_{3D-diff} = 2\int_0^{\Theta_{max}} \tau(\Theta)\cos\Theta\sin\Theta\, d\Theta \qquad (12)$$

in 2 dimensions : $$\tau_{2D-diff} = \int_0^{\Theta_{max}} \tau(\Theta)\cos\Theta\, d\Theta \qquad (13)$$

Transmission loss values for different kinds of excitation (with $\Gamma_a = \Gamma_a' + j\cdot\Gamma_a''$):

Plane wave excitation with normal incidence:

$$R_{\perp} \approx 8.68 \cdot \Gamma_a' d + 20 \log\left|\tfrac{1}{2}[1 + \tfrac{1}{2}(\frac{Z_a}{Z_0} + \frac{Z_0}{Z_a})]\right| \tag{14}$$

Conphase excitation with given v_0 :

$$R_{v_0} \approx 8.68 \cdot \Gamma_a' d + 20 \log\left|\tfrac{1}{2}[1 + \frac{Z_0}{Z_a}]\right| \tag{15}$$

Conphase excitation with given p_0 :

$$R_{p_0} \approx 8.68 \cdot \Gamma_a' d + 20 \log\left|\tfrac{1}{2}[1 + \frac{Z_a}{Z_0}]\right| \tag{16}$$

I.2 Sound transmission through a slit in a wall

Ref.: Mechel, [I.2]

Let a slit of width $2a$ be in a hard wall of thickness d . A plane wave p_e is incident at the angles indicated in the graph.

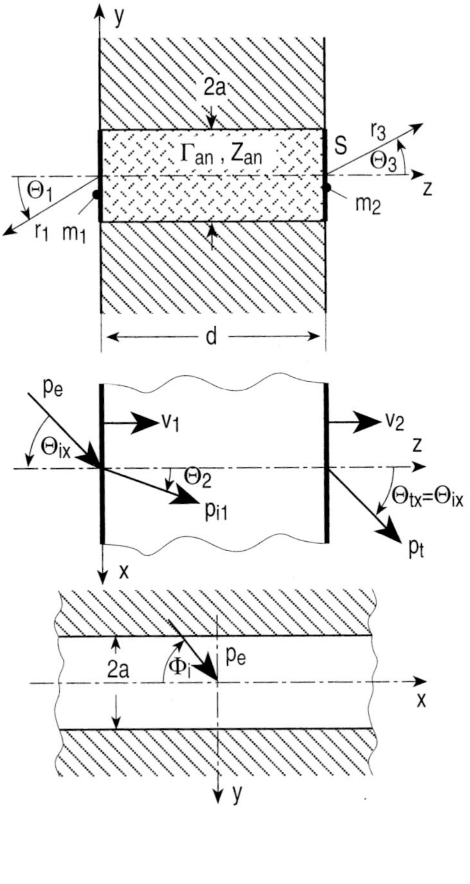

The slit is possibly filled with a porous material having the normalised characteristic values Γ_{an}, Z_{an} . For air in the slit : $\Gamma_{an} \to j$; $Z_{an} \to 1$.

The slit orifices may be covered with (poro-elastic) foils with surface mass densities m_1, m_2 . These can represent plastic sealing masses.

The sound field on the front side has the components :

$$p_1 = p_e + p_r + p_s \tag{1}$$

p_e= incident plane wave,

p_r= reflected wave from a hard wall,

p_s= scattered wave.

with formulations :

$$p_e(x,y,z) = P_e \cdot e^{-jk_0(x\cdot\sin\Theta_i\ \cos\Phi_i + y\cdot\sin\Theta_i\ \sin\Phi_i + z\cdot\cos\Theta_i)}$$
$$p_r(x,y,z) = P_e \cdot e^{-jk_0(x\cdot\sin\Theta_i\ \cos\Phi_i + y\cdot\sin\Theta_i\ \sin\Phi_i - z\cdot\cos\Theta_i)} \tag{2}$$

and :

$$p_s(x,y,z) = \frac{-j\omega\rho_0\ V_1\cdot 2a}{2\pi}\cdot e^{-jk_x x}\cdot\int_{-\infty}^{+\infty}\frac{\sin\alpha a}{\alpha a}\frac{e^{-j\alpha y + z\sqrt{\alpha^2-\gamma^2}}}{\sqrt{\alpha^2-\gamma^2}}\,d\alpha \tag{3}$$

where V_1 is the average particle velocity in the entrance orifice, and

$$k_x = k_0\cdot\sin\Theta_i\cdot\cos\Phi_i$$
$$\gamma^2 = k_0^2 - k_x^2 \tag{4}$$

The waves in the slit are :

$$p_{i1}(x,z) = P_{i1}\cdot e^{-\Gamma_a\ (x\cdot\sin\Theta_2 + z\cdot\cos\Theta_2)}$$
$$p_{i2}(x,z) = P_{i2}\cdot e^{-\Gamma_a\ (x\cdot\sin\Theta_2 - z\cdot\cos\Theta_2)} \tag{5}$$

with the internal angle Θ_2 from :

$$\frac{\sin\Theta_2}{\sin\Theta_i} = \frac{jk_0\ \cos\Phi_i}{\Gamma_a} = \frac{j\cos\Phi_i}{\Gamma_{an}}\quad ;\quad \Gamma_{an}\cos\Theta_2 = \sqrt{\Gamma_{an}^2 + \sin^2\Theta_i\ \cos^2\Phi_i} \tag{6}$$

The transmitted wave is :

$$p_t(x,y,z') = \frac{j\omega\rho_0\ V_2\cdot 2a}{2\pi}\cdot e^{-jk_x x}\int_{-\infty}^{+\infty}\frac{\sin\alpha a}{\alpha a}\frac{e^{-j\alpha y - z'\sqrt{\alpha^2-\gamma^2}}}{\sqrt{\alpha^2-\gamma^2}}\,d\alpha \tag{7}$$

with V_2= average particle velocity amplitude in the exit orifice, and shifted z co-ordinate z' ($z' = z-d$). Both p_s and p_t satisfy the boundary condition at the hard wall surfaces.

Let $Z_1 = j\omega m_1 + Z_{r1}$; $Z_2 = j\omega m_2 + Z_{r2}$ be the sums of the sealing impedance and radiation impedance Z_r of the orifices (see chapter "Radiation Impedance and End Corrections"). Setting the arbitrary amplitude P_e=1 , the boundary conditions give the system of equations :

$$\begin{pmatrix} 1 & 1 & Z_1 & 0 \\ 1 & -1 & -Z_a/\cos\Theta_2 & 0 \\ e^{-\Gamma_a d\cos\Theta_2} & -e^{+\Gamma_a d\cos\Theta_2} & 0 & -Z_a/\cos\Theta_2 \\ e^{-\Gamma_a d\cos\Theta_2} & e^{+\Gamma_a d\cos\Theta_2} & 0 & -Z_2 \end{pmatrix}\cdot\begin{pmatrix} P_{i1} \\ P_{i2} \\ V_1 \\ V_2 \end{pmatrix} = \begin{pmatrix} 2\,\mathrm{si}(k_0 a\sin\Theta_i\ \sin\Phi_i) \\ 0 \\ 0 \\ 0 \end{pmatrix} \tag{8}$$

with si(z)= (sin z)/z . The average particle velocity amplitude in the exit orifice V_2 is :

$$V_2 = \frac{2\dfrac{Z_a}{\cos\Theta_2}\cdot si(k_0 a\ \sin\Theta_i\ \sin\Phi_i)}{\dfrac{Z_a}{\cos\Theta_2}(Z_1+Z_2)\cosh(\Gamma_a d\cos\Theta_2)+[(\dfrac{Z_a}{\cos\Theta_2})^2+Z_1 Z_2]\sinh(\Gamma_a d\cos\Theta_2)} \tag{9}$$

The sound transmission coefficient $\tau(\Theta_i,\Phi_i)$ is then :

$$\tau(\Theta_i,\Phi_i) = \frac{Z_0}{\cos\Theta_i}\cdot \mathrm{Re}\{Z_{r2}\}\cdot\left|\frac{V_2}{P_e}\right|^2 \tag{10}$$

The sound transmission coefficient τ_{dif} for diffuse sound incidence follows by integration :

$$\tau_{dif} = \frac{1}{\pi}\int_0^{2\pi} d\Phi \int_0^{\pi/2} \tau(\Theta,\Phi)\cos\Theta\sin\Theta\ d\Theta = \frac{4}{\pi}\int_0^{\pi/2} d\Phi \int_0^{\pi/2} \tau(\Theta,\Phi)\cos\Theta\sin\Theta\ d\Theta \tag{11}$$

The normalised radiation impedance $Z_{rn}=Z_r/Z_0$ is (with $u=2\gamma a$) :

$$Z_{rn} = 2k_0 a\left\{H_0^{(2)}(u)+\frac{\pi}{2}[H_1^{(2)}(u)S_0(u)-H_0^{(2)}(u)S_1(u)]-\frac{1}{u}H_1^{(2)}(u)+\frac{2j}{\pi u^2}\right\} \tag{12}$$

where $H_n^{(2)}(u)$ are Hankel functions of the 2nd kind, and $S_n(u)$ are Struve functions. The real and imaginary parts of Z_{rn} are :

$$\begin{aligned} Z'_{rn} &= 2k_0 a\left\{J_0(u)-\frac{J_1(u)}{u}+\frac{\pi}{2}[J_1(u)S_0(u)-J_0(u)S_1(u)]\right\} \\ Z''_{rn} &= -2k_0 a\left\{Y_0(u)-\frac{Y_1(u)}{u}-\frac{2}{\pi u^2}+\frac{\pi}{2}[Y_1(u)S_0(u)-Y_0(u)S_1(u)]\right\} \end{aligned} \tag{13}$$

and an approximation for small u (with C=0.577216 Euler's constant) :

$$\begin{aligned} Z_{rn} = k_0 a\Bigg\{&\left[1-u^2/24+u^4/960-u^6/64512+u^8/6635520\right]+ \\ &+\frac{j}{\pi}\Big[3-19u^2/144+7u^4/1800-353u^6/5419008+413u^8/597196800- \\ &-(\ln\frac{u}{2}+C)\left(2-u^2/12+u^4/480-u^6/32256+u^8/3317760\right)\Big]\Bigg\} \end{aligned} \tag{14}$$

Approximations (for $\Phi_i=0$):

Use the set of non-dimensional parameters (with $m_1=m_2=m$) :

$$F = f d/c_0 = d/\lambda_0\ ;\quad A = 2a/d\ ;\quad X = \Xi d/Z_0\ ;\quad M = m/\rho_0 d \tag{15}$$

For the analytical discussion below, $\Gamma_{an} \to j$ and $Z_{an} \to 1$ are used for an empty slit; in numerical evaluations it is better to use the propagation constant and wave impedance of a flat capillary for Γ_a, Z_a (see chapter "Duct Acoustics").

Neglecting terms $(FA\cdot\cos\Theta)^n$ with $n>1$:

$$\tau(\Theta) = \frac{\pi}{\cos\Theta}\left|\{\pi + j[4 - 2C - 2\ln(\pi FA\cos\Theta) + 2\pi\frac{M}{A\cos\Theta}]\}\cdot \right.$$
$$\left. \cdot\{1 + \sin^2\Theta - 2\pi^2F^2 + j\pi\kappa FX\} + \frac{\pi}{A\cos\Theta}\{\frac{1}{2\pi\sigma}\frac{X}{F} + \frac{j}{\kappa\sigma} - j4\pi^2\kappa\sigma F^2M^2\}\right|^{-2} \tag{16}$$

with κ = adiabatic exponent of air ; σ = porosity of absorber material. For an empty slit, set $X=0$ and $\kappa\sigma \to 1$. For a very narrow and empty slit :

$$\tau(\Theta) = \frac{A}{\pi F}\left|1 + 2M - 4\pi^2F^2M(1+M)\cos^2\Theta\right|^{-2} \tag{17}$$

For a very narrow, empty slit without sealing : $$\tau(\Theta) = \frac{A}{\pi F} \tag{18}$$

For an empty, narrow slit with sealing, in a thin wall (or at low frequencies, $F<<1$) : $$\tau(\Theta) = \frac{A}{\pi F(1+2M)^2} \tag{19}$$

For a narrow slit with porous material, in a thin wall (or at low frequencies, $F<<1$) : $$\tau(\Theta) = 4\pi\sigma^2\cdot FA/X^2 \tag{20}$$

The frequency response curves of $R(\Theta) = -\lg\tau(\Theta)$ evidently can have very different shapes.

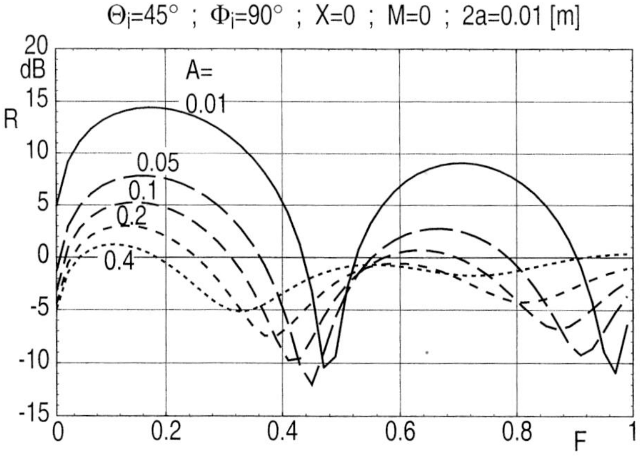

Empty slit, oblique incidence

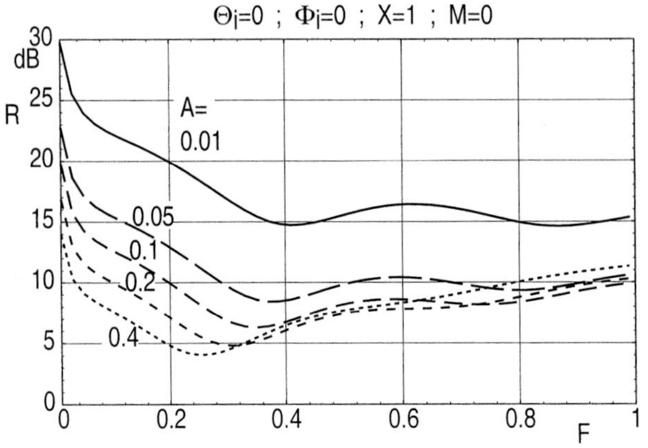

Filled slit, normal incidence.

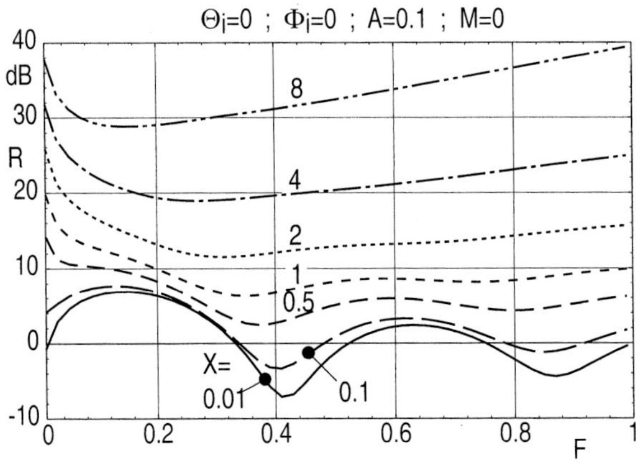

Slits with different absorber material fill; normal incidence.

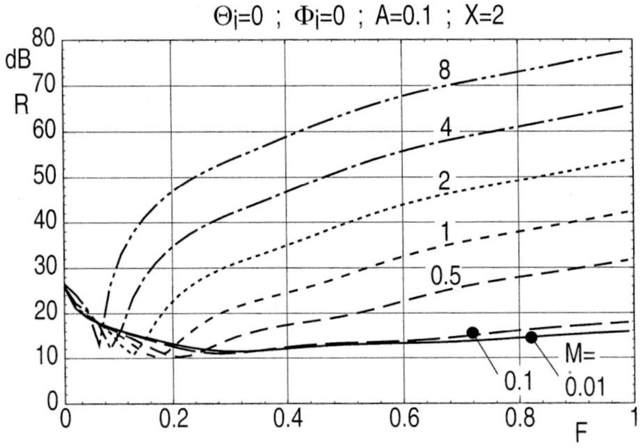

Slits with absorber fill and different sealing.

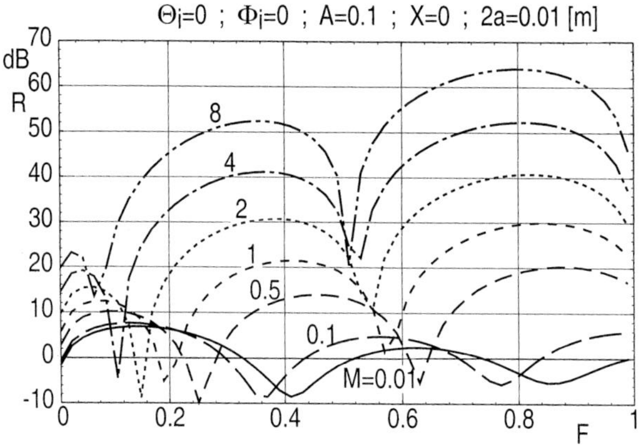

Empty slit with different sealing.

I.3 Sound transmission through a hole in a wall

Ref.: Mechel, [I.2]

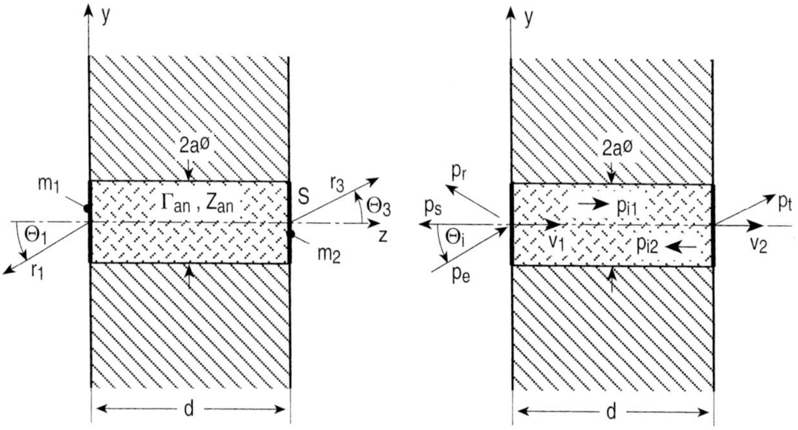

A circular hole with diameter $2a$ is in a wall of thickness d. The hole is possibly filled with a porous material having a normalised propagation constant Γ_{an} and a normalised wave impedance Z_{an}. Possibly poro-elastic foils with effective surface mass densities m_1, m_2 seal the hole orifices. A plane wave p_e is incident at a polar angle Θ_i. The graphs show the co-ordinates and wave components used. The sound field p_1 on the front side is composed as $p_1 = p_e + p_r + p_s$, where p_r is the plane wave after reflection at a hard wall, and p_s is the scattered field.

If the hole is "empty", i.e. filled with air, substitute for analytical discussions $\Gamma_{an} \to j$ and $Z_{an} \to 1$, and for numerical evaluations use the propagation constant and wave impedance for sound propagation in a circular capillary (see chapter "Duct Acoustics").

Field formulations :

$$p_e(x,y) = P_e \cdot e^{-jk_0(z\cos\Theta_i + y\sin\Theta_i)}$$

$$p_r(x,y) = P_e \cdot e^{-jk_0(-z\cos\Theta_i + y\sin\Theta_i)} \qquad (1)$$

$$p_s(r_1,\Theta_1) = j(k_0 a)^2 Z_0 \cdot V_1 \cdot \frac{e^{-jk_0 r_1}}{k_0 r_1} \frac{J_1(k_0 a \sin\Theta_1)}{k_0 a \sin\Theta_1}$$

where V_1 is the average particle velocity in the front side orifice. Further, in the hole:

$$p_{i1}(z) = P_{i1} \cdot e^{-\Gamma_a z} \quad ; \quad p_{i2}(z) = P_{i2} \cdot e^{+\Gamma_a z} \qquad (2)$$

and for the transmitted wave :

$$p_t(r_3,\Theta_3) = j(k_0 a)^2 Z_0 \cdot V_2 \cdot \frac{e^{-jk_0 r_3}}{k_0 r_3} \frac{J_1(k_0 a \sin\Theta_3)}{k_0 a \sin\Theta_3} \qquad (3)$$

where V_2 is the average particle velocity in the back side orifice. p_s and p_t satisfy the boundary condition at the wall.

The boundary conditions for matching the field components give the system of equations :

$$\begin{pmatrix} 1 & 1 & j\omega m_1 + Z_{r1} & 0 \\ 1 & -1 & -Z_a & 0 \\ e^{-\Gamma_a d} & -e^{+\Gamma_a d} & 0 & -Z_a \\ e^{-\Gamma_a d} & e^{+\Gamma_a d} & 0 & -j\omega m_2 - Z_{r2} \end{pmatrix} \cdot \begin{pmatrix} P_{i1} \\ P_{i2} \\ V_1 \\ V_2 \end{pmatrix} = \begin{pmatrix} 2 \\ 0 \\ 0 \\ 0 \end{pmatrix} \qquad (4)$$

where Z_{r1}, Z_{r2} are the radiation impedances of circular piston radiators in a baffle wall (see chapter F). The solutions are, with $Z_1 = j\omega m_1 + Z_{r1}$, $Z_2 = j\omega m_2 + Z_{r2}$:

$$\begin{aligned} P_{i1}/P_e &= 2Z_a(Z_a + Z_2)e^{+\Gamma_a d}/D \\ P_{i2}/P_e &= 2Z_a(Z_a - Z_2)e^{-\Gamma_a d}/D \\ V_1/P_e &= 4(Z_a \cosh\Gamma_a d + Z_2 \sinh\Gamma_a d)/D \\ V_2/P_e &= 4Z_a/D \end{aligned} \qquad (5)$$

with the determinant of the matrix :

$$D = 2\,[Z_a(Z_1 + Z_2)\cosh\Gamma_a d + (Z_a^2 + Z_1 Z_2)\sinh\Gamma_a d] \qquad (6)$$

The transmission loss R of the hole follows from the transmission coefficient τ , which is the ratio of the transmitted effective power Π_t to the incident effective power Π_e :

$$R = -10\lg\tau \;\; [\text{dB}] \quad ; \quad \tau(\Theta_i) = \frac{\Pi_t}{\Pi_e(\Theta_i)}$$

$$\Pi_e(\Theta_i) = \tfrac{1}{2} S \cdot \cos\Theta_i \cdot \frac{|P_e|^2}{Z_0} \quad ; \quad \Pi_t = \tfrac{1}{2} S \cdot \mathrm{Re}\{Z_{r2}\} \cdot |V_2|^2 \tag{7}$$

(S= orifice area). Thus :

$$\tau(\Theta_i) = \frac{Z_0}{\cos\Theta_i} \cdot \mathrm{Re}\{Z_{r2}\} \cdot \left|\frac{V_2}{P_e}\right|^2$$

$$= \frac{Z_0}{\cos\Theta_i} \cdot \mathrm{Re}\{Z_{r2}\} \cdot \left|\frac{2Z_a}{Z_a(Z_1+Z_2)\cosh\Gamma_a d + (Z_a^2 + Z_1 Z_2)\sinh\Gamma_a d}\right|^2 \tag{8}$$

The sound transmission coefficient τ_{dif} for diffuse sound incidence from the whole half-space is simply : $\tau_{dif} = \pi \cdot \tau(0)$

because of : $\tau(\Theta_i) = \tau(0)/\cos\Theta_i$. (9)

Approximations and special cases :

(normal incidence $\Theta_i=0$, and $m_1 = m_2$; index n at impedances means normalisation)

Use the set of non-dimensional parameters :

$$F = f\,d/c_0 = d/\lambda_0 \quad ; \quad A = 2a/d \quad ; \quad X = \Xi d/Z_0 \quad ; \quad M = m/\rho_0 d \tag{10}$$

where Ξ= flow resistivity of the porous fill material (needed for the evaluation of Γ_{an}, Z_{an}).

For $2k_0a \ll 1$ and $|\Gamma_a d| \ll 1$:

$$\tau(0) \approx \frac{\mathrm{Re}\{Z_{rn}\}}{\left|Z_{1n} + (Z_{an}^2 + Z_{1n}^2)\,k_0 d \dfrac{\Gamma_{an}}{2Z_{an}}\right|^2} \tag{11}$$

and with the leading term of the radiation impedance :

$$\tau(0) \approx \frac{1}{2}\left(\frac{\pi A F}{\frac{\pi^2}{2} A^2 F^2 + X/(2\sigma)}\right)^2 \approx 2(\pi\sigma\, AF/X)^2 \tag{12}$$

where σ= porosity of the absorber fill. If additionally the hole is empty (X=0) :

$$\tau(0) \approx \frac{1}{8}\left(\frac{A}{\frac{4}{3\pi}A + M}\right)^2 \tag{13}$$

Empty and unsealed hole (M=0 ; X=0) at low frequencies :

I.4 Hole transmission with equivalent network

Ref.: Mechel, [I.2]

The sound transmission through holes and slits in a wall is derived in the previous sections I.2, I.3 as a boundary value problem. Equal results are obtained by application of the method of equivalent networks.

The equivalent network of a sealed and filled hole (or slit) in a wall is in the pU-analogy and in the pI-analogy :

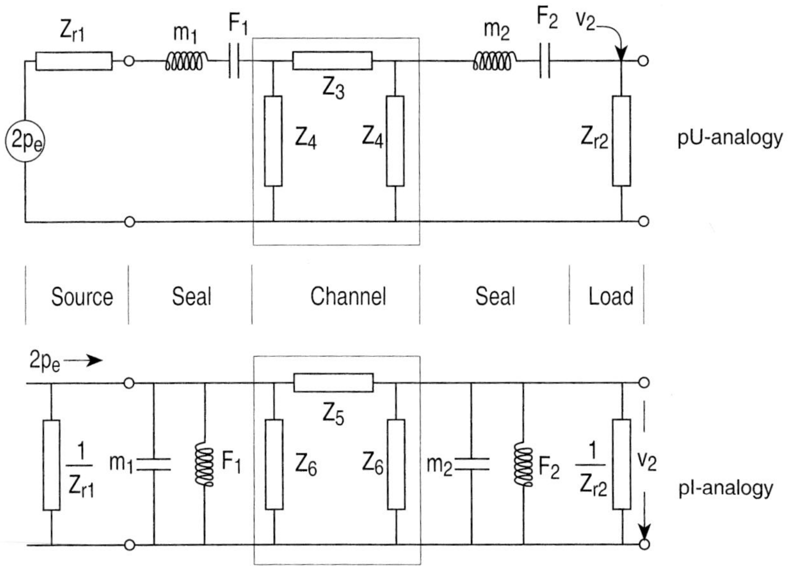

The elements are :

m_1, F_1 ; $m2, F2$ — surface mass densities and resilience values of the orifice seals;

Z_{r1}, Z_{r2} — radiation impedances of the orifices;

$$Z_3 = Z_a \cdot \sinh \Gamma_a d \quad ; \quad \frac{1}{Z_4} = \frac{1}{Z_a} \frac{\cosh \Gamma_a d - 1}{\sinh \Gamma_a d}$$
$$Z_5 = \frac{1}{Z_a} \cdot \sinh \Gamma_a d \quad ; \quad \frac{1}{Z_6} = Z_a \frac{\cosh \Gamma_a d - 1}{\sinh \Gamma_a d} \tag{1}$$

Combine :

$$Z_1 = Z_{r1} + j\omega m_1 + \frac{1}{j\omega F_1} \quad ; \quad Z_2 = Z_{r2} + j\omega m_2 + \frac{1}{j\omega F_2} \tag{2}$$

and :

$$Z_I = Z_1 + \frac{1}{Z_6} = Z_{r1} + j\omega m_1 + \frac{1}{j\omega F_1} + Z_a \frac{\cosh \Gamma_a d - 1}{\sinh \Gamma_a d}$$
$$Z_{II} = Z_2 + \frac{1}{Z_6} = Z_{r2} + j\omega m_2 + \frac{1}{j\omega F_2} + Z_a \frac{\cosh \Gamma_a d - 1}{\sinh \Gamma_a d} \tag{3}$$

Then :

$$V_2 = \frac{2P_e}{Z_I + Z_{II} + Z_5 \cdot Z_I \cdot Z_{II}} \tag{4}$$

and the transmission coefficient of the *hole* :

$$\tau(\Theta_i) = \frac{Z_0}{\cos\Theta_i} \cdot \mathrm{Re}\{Z_{r2}\} \cdot \left| \frac{2}{Z_I + Z_{II} + Z_5 \cdot Z_I \cdot Z_{II}} \right|^2 \tag{5}$$

Insertion of the abbreviations leads to the result of the boundary value problem.

For a *slit* apply the substitutions :

$$\Gamma_a \to \Gamma_a \cos\Theta_2 \quad ; \quad Z_a \to Z_a / \cos\Theta_2$$
$$Z_{r,circle} \to Z_{r,strip} \tag{6}$$
$$2P_e \to 2P_e \; si(k_0 a \sin\Theta_i \; \sin\Phi_i)$$

with si(z)= sin(z)/z . This returns the result of section I.2 .

I.5 Sound transmission through lined slits in a wall

Ref.: Mechel, [I.2]

Sometimes joints of wall elements which form slits in the wall cannot be sealed or filled, e.g. the lower gap at a door.

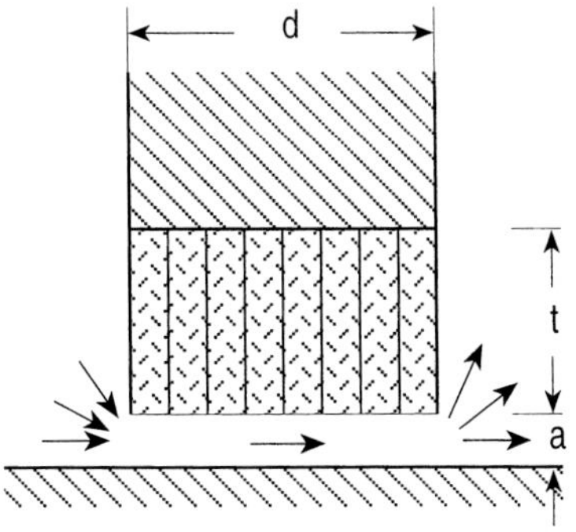

The quantities and relations are those of section I.2 , except that Γ_a, Z_a are the propagation constant and wave impedance of the least attenuated mode in a flat (silencer) duct of height a , which is lined with an absorber having a surface admittance G_y .

The lining is preferably made locally reacting (if necessary with thin partitions). Because $k_0a<<1$ holds in general, low-frequency approximations can be used for the determination of Γ_a.

Channel wave components :
(with substitutions $\Gamma_a \to \Gamma_s$; $Z_a \to Z_s$ in order to avoid confusion with material data)

$$p_{i1}(x,y,z) = P_{i1} \cdot e^{-\Gamma_s\,(x\cdot\sin\Theta_2 + z\cdot\cos\Theta_2)} \cdot \cos\varepsilon_y y$$
$$p_{i2}(x,y,z) = P_{i2} \cdot e^{-\Gamma_s\,(x\cdot\sin\Theta_2 - z\cdot\cos\Theta_2)} \cdot \cos\varepsilon_y y \qquad \varepsilon_y^2 = \Gamma_s^2 + k_0^2 \tag{1}$$

$\varepsilon_y a$ is a solution of the characteristic equation for a *locally reacting lining* :

$$\varepsilon_y a \cdot \tan\varepsilon_y a = j \cdot k_0 a \cdot Z_0 G_y := j \cdot U \tag{2}$$

The low-frequency approximation is applicable :

$$(\varepsilon_y a)^2 = \frac{105 + 45jU \pm \sqrt{11025 + 5250jU - 1605U^2}}{20 + 2jU} \tag{3}$$

The sign of the root is chosen so that the real part of is a minimum. The wave impedance Z_s of the least attenuated mode is

$$\Gamma_s a = \sqrt{(\varepsilon_y a)^2 - (k_0 a)^2} \tag{4}$$

$$\frac{Z_s}{Z_0} = \frac{j}{\Gamma_s / k_0} \tag{5}$$

If the lining is a *bulk reacting* homogeneous, porous absorber layer of thickness t with characteristic values Γ_a, Z_a of the material, the equation to be solved is :

$$\varepsilon_y a \cdot \tan\varepsilon_y a = -j\left(\frac{\Gamma_a}{k_0}\frac{Z_a}{Z_0}\right)^{-1} \sqrt{(\varepsilon_y a)^2 - (\eta a)^2} \cdot \tan\left(\frac{t}{a}\sqrt{(\varepsilon_y a)^2 - (\eta a)^2}\right)$$
$$(\eta a)^2 := (\Gamma_a a)^2 + (k_0 a)^2 \tag{6}$$

A continued fraction approximation for low frequencies leads to a polynomial equation :

$$a_0 + a_1 \cdot (\varepsilon_y a)^2 + a_2 \cdot (\varepsilon_y a)^4 + a_3 \cdot (\varepsilon_y a)^6 + a_4 \cdot (\varepsilon_y a)^8 = 0 \tag{7}$$

with coefficients :

$$a_0 = -11025(\eta a)^2 - 1050(t/a)^2(\eta a)^4$$
$$a_1 = 11025 + (\eta a)^2(4725 + 2100(t/a)^2) + 450(t/a)^2(\eta a)^4 - \\ - A \cdot [11025/(t/a)^2 + 4725(\eta a)^2 + 105(t/a)^2(\eta a)^4] \tag{8}$$

$$a_2 = -\{4725 + 1050(t/a)^2 + (\eta a)^2(105 + 900(t/a)^2) + 10(t/a)^2(\eta a)^4 + \\ + A \cdot [1050/(t/a)^2 + 4725 + (\eta a)^2(450 + 210(t/a)^2) + 10(t/a)^2(\eta a)^4]\}$$

$$a_3 = 105 + 450(t/a)^2 + 20(t/a)^2(\eta a)^2 - A\cdot[450 + 105(t/a)^2 + 20(t/a)^2(\eta a)^2]$$
$$a_4 = 10(t/a)^2(A-1) \tag{9}$$

and the abbreviation :
$$A := j\frac{t}{a}\frac{\Gamma_a}{k_0}\frac{Z_a}{Z_0} \tag{10}$$

The least attenuated mode in the slit channel has a cosine profile; the matching to the exterior sound fields has to be performed with the average values of sound pressure and axial particle velocity.

The boundary conditions in the orifices lead to the system of equations :

$$\begin{aligned}
si(\varepsilon_y a)\cdot(P_{i1} + P_{i2}) + Z_{r1}V_1 + 0 &= 2P_e\, si(k_y a)\\
j\beta\, si(\varepsilon_y a)\cdot(P_{i1} - P_{i2}) + Z_0 V_1 + 0 &= 0\\
si(\varepsilon_y a)\cdot(P_{i1}e^{-\gamma} + P_{i2}e^{+\gamma}) + 0 - Z_{r2}V_2 &= 0\\
j\beta\, si(\varepsilon_y a)\cdot(P_{i1}e^{-\gamma} - P_{i2}e^{+\gamma}) + 0 + Z_0 V_2 &= 0
\end{aligned} \tag{11}$$

with $si(z) = \sin(z)/z$ and the abbreviations :

$$\begin{aligned}
\beta &:= \Gamma_s/k_0\cdot\cos\Theta_2 = \sqrt{(\varepsilon_y/k_0)^2 - 1 + \sin^2\Theta_i\,\cos^2\Phi_i}\\
\gamma &:= \Gamma_s d\cdot\cos\Theta_2 = k_0 d\sqrt{(\varepsilon_y/k_0)^2 - 1 + \sin^2\Theta_i\,\cos^2\Phi_i}
\end{aligned} \tag{12}$$

The matrix determinant is :

$$D = 2\,si^2(\varepsilon_y a)\cdot[j\beta Z_0(Z_{r1}+Z_{r2})\cdot\cosh\gamma + (\beta^2 Z_{r1}Z_{r2} - Z_0^2)\cdot\sinh\gamma] \tag{13}$$

and the wanted average axial particle velocity V_2 in the exit orifice :

$$\frac{V_2}{P_e} = \frac{2j\beta Z_0\, si(k_y a)}{j\beta Z_0(Z_{r1}+Z_{r2})\cdot\cosh\gamma + (\beta^2 Z_{r1}Z_{r2} - Z_0^2)\cdot\sinh\gamma} \tag{14}$$

The coefficient of transmission is, with the normalised radiation impedances Z_{r1n}, Z_{r2n} of the orifices (see section I.2) :

$$\tau(\Theta_i,\Phi_i) = \frac{\mathrm{Re}\{Z_{r2n}\}}{\cos\Theta_i}\left|\frac{2j\beta\cdot si(k_0 a\sin\Theta_i\,\sin\Phi_i)}{j\beta(Z_{r1n}+Z_{r2n})\cdot\cosh\gamma + (\beta^2 Z_{r1n}Z_{r2n} - 1)\cdot\sinh\gamma}\right|^2 \tag{15}$$

A set of non-dimensional parameters for a slit with a lining consisting of a layer of porous material (flow resistivity Ξ, made locally reacting by partitions), which is covered with a foil of surface mass density m, is :

$$F = fd/c_0 = d/\lambda_0 \;;\quad A = 2a/d \;;\quad X = \Xi t/Z_0 \;;\quad M_s = m/\rho_0 d \;;\quad T = t/d \tag{16}$$

The frequency response curves of the sound transmission loss $R = -10\cdot\lg(\tau)$ have a great variety of forms, depending on the parameter values. A few examples will be given below.

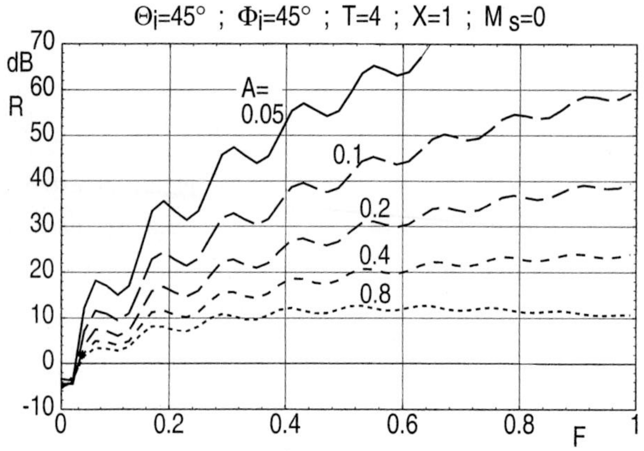
Θi=45° ; Φi=45° ; T=4 ; X=1 ; M s=0
dB
R
A=
0.05
0.1
0.2
0.4
0.8
F

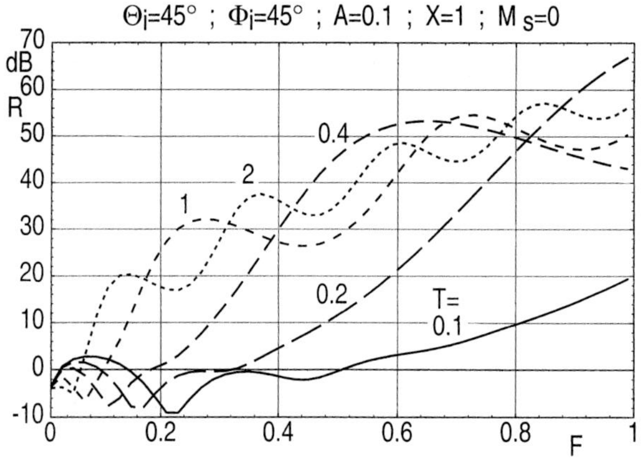
Θi=45° ; Φi=45° ; A=0.1 ; X=1 ; M s=0
dB
R
0.4
2
1
0.2
T=
0.1
F

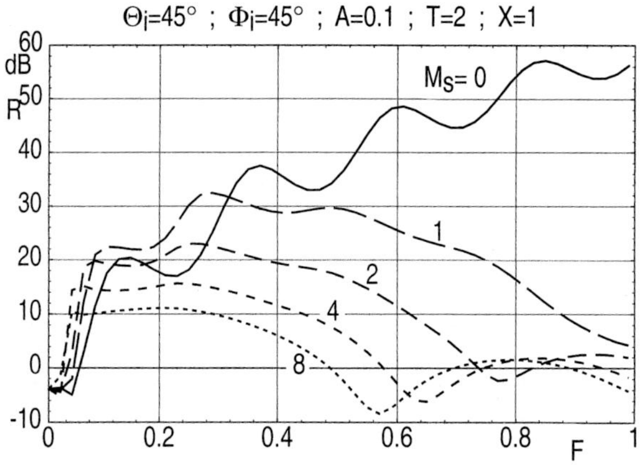
Θi=45° ; Φi=45° ; A=0.1 ; T=2 ; X=1
dB
R
Ms= 0
1
2
4
8
F

I.6 Chambered joint

Ref.: Mechel, [I.2]

Some joints between construction elements have a cross-section which, in principle, is a sequence of chambers (e.g. joints of facade elements, joints of the window frame, etc.).

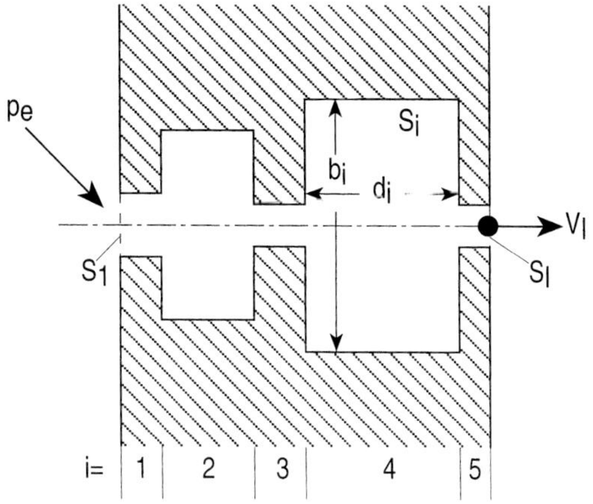

Let a plane sound wave p_e be incident with the wave vector in the plane normal to the wall and the length of the joint, at a polar angle Θ_i.

Let S_i ; i= 1,2,…,I ; be the cross-section areas of the duct elements, in which the joint can be subdivided.

The sound transmission coefficient $\tau(\Theta_i)$ will be :

$$\tau(\Theta_i) = \frac{S_I}{S_1}\frac{\mathrm{Re}\{Z_{rI}/Z_0\}}{\cos\Theta_i}\cdot\left|\frac{Z_0 V_I}{P_e}\right|^2 = 4\frac{S_I}{S_1}\frac{\mathrm{Re}\{Z_{rI}/Z_0\}}{\cos\Theta_i}\cdot\left|(\frac{Z_0}{S_I}\, S_I V_I)/(2P_e)\right|^2 \qquad (1)$$

where Z_{rI} is the radiation impedance of the exit orifice SI, and Z_{r1} the radiation impedance of the entrance orifice i=1 .

The composed duct can be described with an equivalent chain network, in which the duct sections are represented by Π-fourpoles which are separated in the longitudinal branch by mass reactances $Z_{mi,k}$ of an internal orifice i if it enters into a wider duct section k .

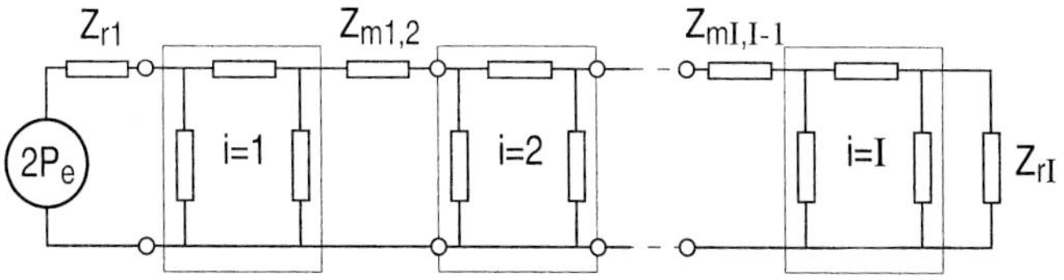

with limit values for $\varepsilon_m a \to \pm k_0 a \cdot \sin\Theta$: $S_{m_{sy}} \to N_{m_{sy}}$; $S_{m_{as}} \to \pm j N_{m_{as}}$

and if $\varepsilon_m a \to 0$: $N_{m_{sy}} \to 1$; $N_{m_{as}} \to 0$ (7)

Field matching at the sluice entrance gives for the mode amplitudes :

$$\frac{A_{m_{sy}}}{P_e} = \frac{S_{m_{sy+}}}{N_{m_{sy}}}\left(1 + r_0\, e^{-j k_0 a \sin\Theta}\right) \quad ; \quad \frac{A_{m_{as}}}{P_e} = \frac{S_{m_{as+}}}{N_{m_{as}}}\left(1 - r_0\, e^{-j k_0 a \sin\Theta}\right) \tag{8}$$

Effective sound power of the incident wave p_e : $\Pi_e(\Theta) = \frac{a}{Z_0}\left|P_e\right|^2 \cos\Theta$ (9)

Effective modal sound power incident in the duct on the exit orifice :

$$\Pi_m(d) = \frac{1}{2}\,\mathrm{Re}\left\{\int_{-a}^{a} p_m(y,d)\cdot v_{zm}^{*}(y,d)\,dy\right\} \tag{10}$$

Assuming a good radiation efficiency of the modes at the duct exit (because the duct is wide), the transmission coefficient of the noise sluice is : (11)

$$\tau(\Theta) = \frac{\sum_m \Pi_m(d)}{\Pi_e(\Theta)} = \frac{1}{2\cos\Theta}\sum_m \frac{\Gamma_m''}{k_0} e^{-2\Gamma_m' d}\left|\frac{S_m}{N_m}(1 \pm r_0\, e^{-j k_0 a \sin\Theta})\right|^2 \left[\frac{\sinh 2\varepsilon_m'' a}{2\varepsilon_m'' a} \pm \frac{\sin 2\varepsilon_m' a}{2\varepsilon_m' a}\right]$$

(with $\Gamma_m = \Gamma_m' + j\cdot\Gamma_m''$; $\varepsilon_m = \varepsilon_m' + j\cdot\varepsilon_m''$). The summation is over the symmetrical (with the upper sign in $\pm$) and anti-symmetrical modes (with the lower sign in $\pm$).

In the following examples of the transmission loss $R = -\lg(\tau(\Theta))$ the lining of the duct and of the wall in front of the duct is assumed to be a layer of glass fibre material, if necessary made locally reacting by internal partitions; the flow resistivity of the material is Ξ .

The mode orders m in the summation are taken in a range $\mathrm{Max}(1, m_0 - \Delta m) \le m \le m_0 + \Delta m$ with an interval width $2\Delta m$ and the central order m_0 selected so that $\left|\mathrm{Re}\{\varepsilon_m a - k_0 a \cdot \sin\Theta\}\right| = \min(m)$.

Θ=45° ; a=1 [m] ; d=3 [m] ; t=0.25 [m] ; t_0=0.25 [m] ; Ξ=10 [kPas/m^2]

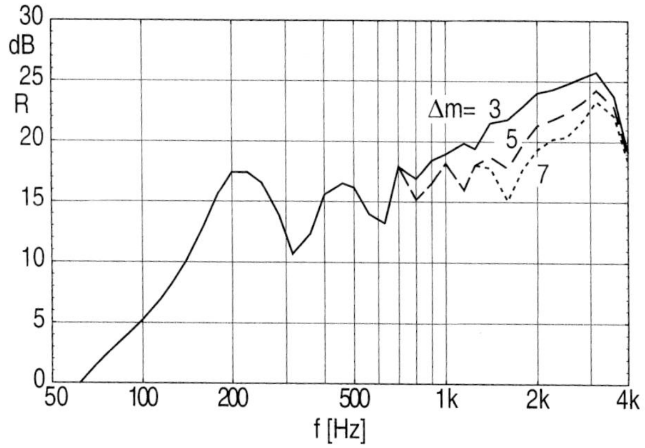

Influence of the order interval width Δm on $R(\Theta)$.

Θ=45° ; a=1 [m] ; d=3 [m] ; t=0.25 [m] ; t_0=0.25 [m] ; Δm=5

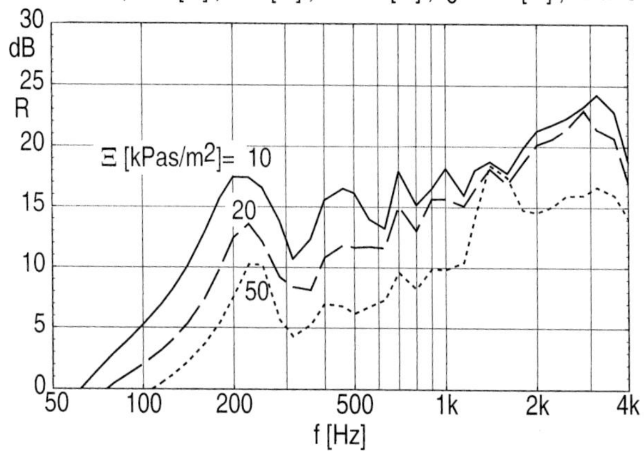

Influence of lining resistivity Ξ on $R(\Theta)$.

a=1 [m] ; d=3 [m] ; t=0.25 [m] ; t_0=0.25 [m] ; Ξ=10 [kPas/m^2] ; Δm=5

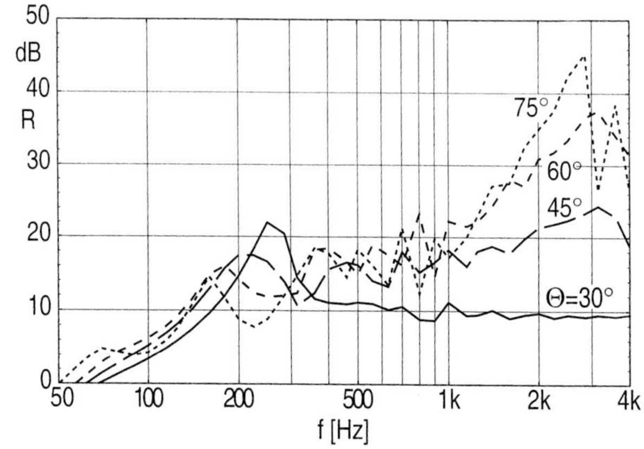

Influence of angle of incidence on $R(\Theta)$.

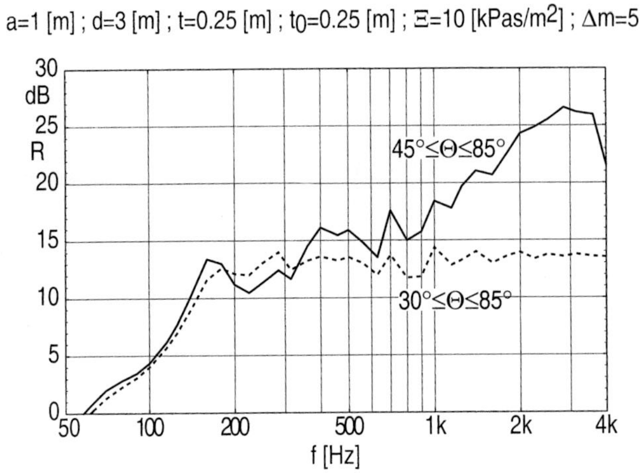

Sound transmission loss R of a noise sluice for quasi-diffuse sound incidence with two ranges of incidence angle.

Measures to avoid sound incidence at small Θ :

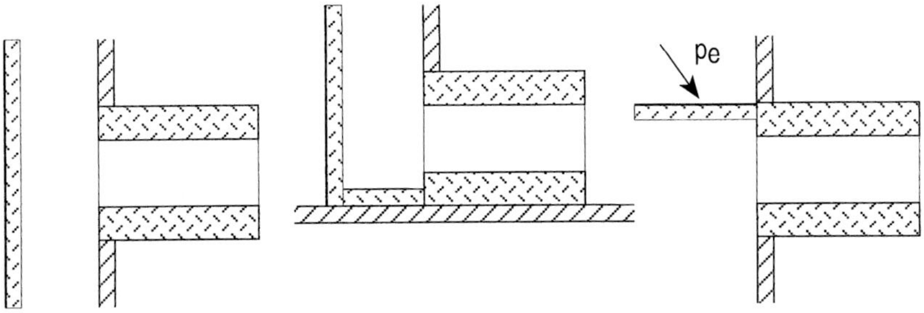

I.8 Sound transmission through plates, some fundamentals

Ref.: Mechel, [I.4]

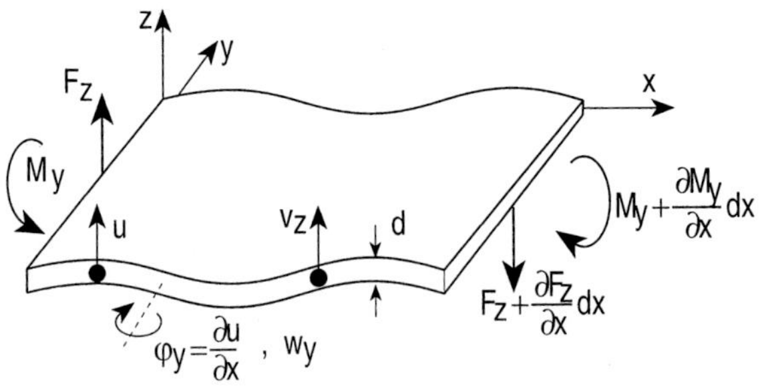

Kinematic and dynamic quantities at a plate section, in Cartesian co-ordinates.

u : elongation;

v_z : velocity;

φ : rotation angle with y as axis;

w_z : rotational velocity;

F_z : transversal force;

M_y : torsional moment with y as axis.

Bending wave equation : $$[\Delta\Delta - k_B^4]\,v_z = \frac{j\omega}{B}\,\delta p \tag{1}$$

with : Δ= Laplace operator ; ω= angular frequency ; B= bending stiffness ; δp= $p_{front}-p_{back}$= driving sound pressure difference ; k_B= free bending wave number .

$$k_B = \frac{\omega}{c_B} = \frac{2\pi}{\lambda_B} = \sqrt{\omega}\,\sqrt[4]{m/B} = \sqrt{\frac{\omega}{c_L d}}\,\sqrt[4]{12(1-\sigma^2)} = \frac{1}{d}\sqrt{k_L d}\,\sqrt[4]{12(1-\sigma^2)}$$
$$= k_0\sqrt{f_{cr}/f} \xrightarrow[\sigma=0.35]{} 4.515\sqrt{\frac{f}{c_L d}} \tag{2}$$

with: c_B= bending wave velocity ; λ_B= bending wavelength ; m= surface mass density ; d= plate thickness ; k_L= longitudinal wave number ; c_L= longitudinal wave velocity ; σ= Poisson's ratio ; f= frequency ; f_{cr}= critical coincidence frequency .

$$f_{cr} = \frac{c_0^2}{2\pi}\sqrt{\frac{m}{B}} = \frac{c_0^2}{2\pi d}\sqrt{\frac{12\rho(1-\sigma^2)}{E}} = \frac{c_0^2}{2\pi d\,c_L}\sqrt{12(1-\sigma^2)} \xrightarrow[\sigma=0.35]{} \frac{60761}{d\,c_L} \tag{3}$$

Coincidence frequency for angle Θ of incidence : $$f_c = \frac{f_{cr}}{\sin^2\Theta}\ . \tag{4}$$

Wave type	**Speed**	**Remark**
Shear wave, Torsional wave	$c_S = \sqrt{S/\rho}$	ρ= plate material density
Compressional wave in a bar	$c_{L,St} = \sqrt{E/\rho}$	
Compressional wave, Longitudinal wave	$c_D = \sqrt{D/\rho}$ $= c_0 = \gamma P_0$ in a gas	γ =adiabatic exponent P_0=static pressure
Bar bending wave	$c_{B,St} = \sqrt[4]{\frac{\omega^2 B_{St}}{m'}}$	m'=mass per length
Plate bending wave	$c_{B,Pl} = c_B = \sqrt{\omega}\,\sqrt[4]{\frac{B}{m}}$	m=mass per area (surface mass density)
Rayleigh wave	$c_{Rayl} \approx 0.92\cdot c_s$	

Table 1: Characteristic wave speeds.

If a plate is infinite in its lateral extension (in practice: if its smallest lateral dimension is large compared to the bending wavelength λ_B, so that plate resonances can be neglected in their influence on the transmission loss) the most important quantity of a plate is its partition impedance Z_T (see below and section H.17).

If finite plate dimensions must be considered, the boundary conditions at the plate boundaries must be taken into account. "Classical" boundary conditions are given in the Table 2 . They can be expressed also with *force impedances* Z_F and *momentum impedances* Z_M at the boundary $\{x_R, y_R\}$. These impedances are defined by :

$$\begin{aligned} F(x_R, y_R) &= Z_K(x_R, y_R) \cdot v_z(x_R, y_R) \\ M_i(x_R, y_R) &= Z_M(x_R, y_R) \cdot w_i(x_R, y_R) \end{aligned} \tag{5}$$

Additional *corner impedances* Z_E apply at plate corners $\{x_E, y_E\}$:

$$F_E(x_E, y_E) = Z_E(x_E, y_E) \cdot v_z(x_E, y_E) \tag{6}$$

Fixation	Condition	Symbol
Simply supported	$v(\xi,1) = \dfrac{\partial^2 v(\xi,1)}{\partial \eta^2} = 0$ $1/Z_K = Z_M = 0$	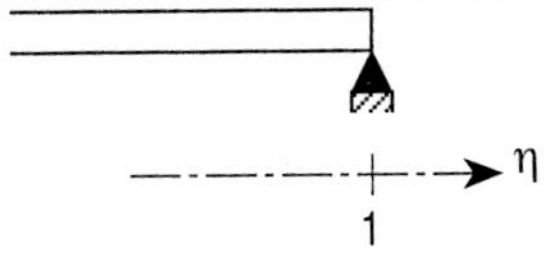
Clamped	$v(\xi,1) = \dfrac{\partial v(\xi,1)}{\partial \eta} = 0$ $1/Z_K = 1/Z_M = 0$	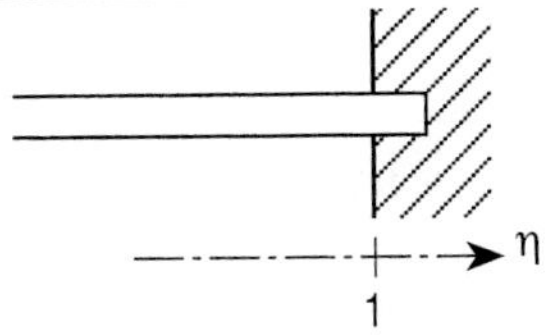
Free	$\dfrac{\partial^2 v}{\partial \eta^2} + \sigma ß^2 \dfrac{\partial^2 v}{\partial \xi^2} = 0$ $\dfrac{\partial^3 v}{\partial \eta^3} + (2-\sigma) ß^2 \dfrac{\partial^3 v}{\partial \eta \partial \xi^2} = 0$ $Z_K = Z_M = 0$	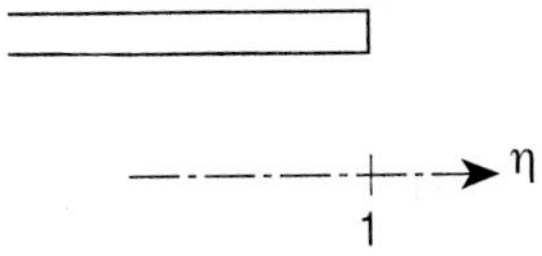 ß= b/a
Hinged	$Z_K = 1/Z_M = 0$	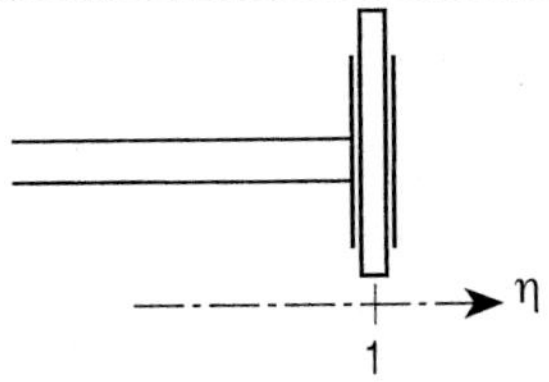

Table 2: Classical boundary conditions for plates.

Partition impedance Z_T :

Let a thin, i.e. incompressible, plate in an orthogonal co-ordinate system $\{x_1,x_2,x_3\}$ be on the co-ordinate surface $x_3=\xi$, and let $p_f(x_1,x_2,\xi_3)$ be the sound pressure at the plate on its front side (side of excitation), and $p_b(x_1,x_2,\xi_3)$ the sound pressure at the plate on its back side; further let $\Delta p = p_f(x_1,x_2,\xi_3) - p_b(x_1,x_2,\xi_3)$ be the driving pressure difference, which generates a plate velocity $v(x_1,x_2,\xi_3)$ (counted positive in the direction front → back).

The boundary conditions at the plate are :

$$v_f = v_b = v \quad ; \quad p_f - p_b = Z_T \cdot v \tag{7}$$

where the last condition demands proportionality between Δp and v with a constant factor Z_T if the plate is homogeneous in x_1,x_2 (i.e., for example, the plate is large or a closed shell with constant curvature). It is assumed that p_f,p_b satisfy the wave equation in air (and possible other boundary conditions at surfaces other than the plate). The bending wave equation for v

$$\left[\Delta_{x_1,x_2}\Delta_{x_1,x_2} - k_B^4\right]v = \frac{j\omega}{B}\cdot\Delta p \tag{8}$$

and the last boundary condition can be combined to :

$$\frac{\Delta_{x_1,x_2}\Delta_{x_1,x_2}v}{v} - k_B^4 = \frac{j\omega}{B}\cdot\frac{\Delta p}{v} = \frac{j\omega}{B}\cdot Z_T \tag{9}$$

For stationary sound fields and homogeneous plates the function $v(x_1,x_2)$ is, up to a constant factor, the same as $p_f(x_1,x_2,\xi_3)$, $p_b(x_1,x_2,\xi_3)$. So the 1st term in the last equation is known for given sound field formulations.

For a plane plate in the plane $z=0$ and a plane wave p_e incident in the x,z plane at a polar angle Θ , i.e. with a trace wave number $k_x = k_0 \cdot \sin\Theta$, the partition impedance for a plate with surface mass density m is:

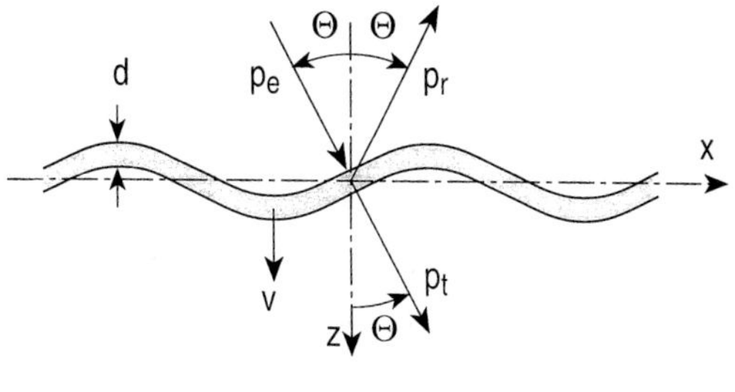

$$Z_T = \frac{\Delta p}{v} = \frac{B}{j\omega}\left(k_x^4 - k_B^4\right) = j\omega m\frac{k_B^4 - k_x^4}{k_B^4} = j\omega m\left(1 - (\frac{f}{f_c})^2\right) = j\omega m\left(1 - (\frac{f}{f_{cr}})^2 \sin^4\Theta\right) \tag{10}$$

If the plate has bending losses, i.e. $B \to B(1+j\eta)$ with the loss factor η , then :

$$\frac{Z_T}{Z_0} = \frac{\omega m}{Z_0}\left[\eta(\frac{f}{f_{cr}})^2 \sin^4\Theta + j\left(1-(\frac{f}{f_{cr}})^2 \sin^4\Theta\right)\right]$$

$$= 2\pi Z_m F\left[\eta F^2 \sin^4\Theta + j\left(1-F^2 \sin^4\Theta\right)\right] \qquad F := \frac{f}{f_{cr}} \; ; \; Z_m := \frac{f_{cr} m}{Z_0} = \frac{f_{cr} d}{Z_0}\rho \quad (11)$$

$$= 2\pi Z_m F\left[\eta(\frac{f}{f_c})^2 + j\left(1-(\frac{f}{f_c})^2\right)\right]$$

where ρ is the plate material mass density.

As long as the compressibility of the plate can be neglected, the TIMOSHENKO-MINDLIN theory of thick plates gives in an analogous way for the partition impedance :

$$Z_T = j\omega m \frac{k_B^4 - (k_x^2 - k_L^2)(k_x^2 - k_R^2)}{k_B^4 - k_R^2(k_x^2 - k_L^2)} \; ; \; k_R = \frac{\sqrt{12}}{\pi} k_S \qquad (12)$$

with k_x= trace wave number of the exciting field ; k_B= free bending wave number ; k_L= longitudinal wave number ; k_S= shear wave number.

Material	**Density** ρ [kg/m^3]	**E modulus** E [MN/m^2]	$f_{cr}d$ [Hz·m]	Z_m [-]	**Loss fact.** η [-]
Construction Materials:					
Concrete	2100-2300	25-40·10^3	15-18.5	77-104	0.05
Lean concrete	2000	15 000	23	112	
Light concrete	800-1400	1.5-3·10^3	37-48	72-164	0.015
Porous concrete	600-700	1.4-2·10^3	37-45	54-77	0.01
Cement floor	2200	30 000	17	91	
Xylolith floor	1600	6 000	32.5	127	0.03
Asphalt floor	2200	6-15·10^3	24-42	129-225	0.03-0.3
Plaster floor	1200	20 000	15.5-16	45-47	0.006
Gypsum panel	1000-1200	3.5-7·10^3	24-35	58-102	0.004
Plaster board	1000	3 200	31-35	85	0.03
Fibre cement board	2000-2100	20-30·10^3	16.5-20	80-102	0.01
Brick wall	1700-1800	9-25·10^3	16-27	66-118	0.04
Glass	2500	60-80·10^3	11-13	67-79	0.001
Chip board	600-1000	2-5·10^3	23-36	34-88	0.03
Plywood	600-800	5-12·10^3	14-34.5	20-65	0.02
Oak wood	700	200-1000	18-32	31-55	0.01

Pine wood	480	100-500	20-32	23-37	0.01
Hard board	1000	$3\text{-}4.5\cdot 10^3$	29.5-36.5	72-89	0.015
Plastics:					
Acryl glass	1200	5600	29	85	0.06
Polypropylene	1100	3000	38	102	0.1
Polyester	1200	4500	32.5	95	0.14
PVC, hard	1300	2700	43.5	138	0.04
PVC, 30% softener	1250		48	1220	
Polyethylene, hard	950	1700	47	109	0.04
Polyethylene, soft	920	400	95.5	214	0.1

Table 3: Density and elastic constants of materials.

Material	**Density** ρ [kg/m^3]	**E modulus** E [MN/m^2]	$f_{cr}d$ [Hz·m]	Z_m [-]	**Loss fact.** η [-]
Polystyrene	1070	3000	37.5	98	0.01
Polystyrene+30% glass fibre	1450	8000	27	95	
Polyester+glass fibre	2200	11500	27.5	147	0.02
Metals:					
Aluminium	2700	74 000	12	79	$7\cdot 10^{-5}$
Lead	11400	18000	48.5	1348	0.02
Copper	8900	125000	17	369	
Brass	8500	96000	18.6	386	0.001
Steel, Cast steel	7800	200000	12.3	234	$1\cdot 10^{-4}$
Malleable iron	7500	170000	13.2	241	
Cast iron with spher.graphite	7250	120000	15.4	272	0.01
Cast iron with lamell.graphite	7250	120000	15.4	272	0.02
Zinc	7130	13000	46.5	809	
Tin	7280	4400	81	1438	

Table 3: Density and elastic constants of materials (continued).

Table 3 above gives the required elastic data of materials. Of special interest are the density ρ and the product $f_{cr}\cdot d$ which is a material constant.

Quantity	**Symbol , Relations**	**Remark**
Lame constants	λ, μ $-p_{ii} = \lambda \cdot \mathrm{div}\,\vec{s} + 2\mu \cdot \varepsilon_{ii}$ $-p_{ik} = 2\mu \cdot \varepsilon_{ik}$	
Poisson number	$\sigma = -\varepsilon_{yy}/\varepsilon_{xx} = \dfrac{\lambda}{2(\lambda+\mu)}$ $= \dfrac{E}{2S} - 1$ $\dfrac{\mu}{\lambda} = \dfrac{1-2\sigma}{2\sigma}; \quad 0 \le \sigma < 0.5$	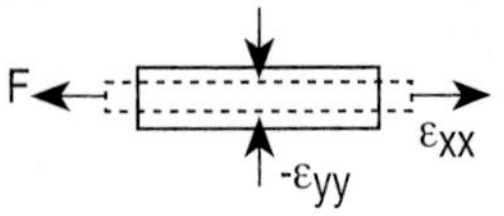 incompressible: σ=0.5
Shear modulus	$S = \mu \quad [\mathrm{Pa}]$ $= \dfrac{1}{2}\dfrac{1}{1+\sigma}E; \quad (\sigma < 0.4)$	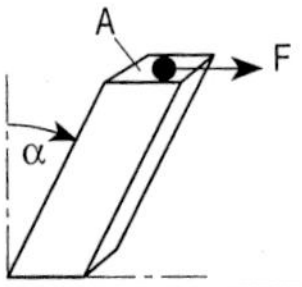 $p = F/A$ $S = p/\alpha$
E modulus, Young's modulus	$E = \mu\dfrac{3\lambda+2\mu}{\lambda+\mu} \quad [\mathrm{Pa}]$ $= 2\mu/1+\sigma) = 2S(1+\sigma)$	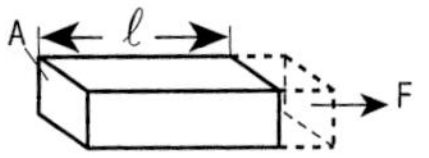 $-p=F/A=E\cdot\Delta\ell/\ell=E\cdot s_{xx}$
Compression modulus	$K = \lambda + \frac{2}{3}\mu \quad [\mathrm{Pa}]$ $= (\dfrac{2}{3} + \dfrac{2\sigma}{1-2\sigma}) \cdot S$ $= \dfrac{E}{3(1-2\sigma)} = \dfrac{1}{\kappa}$	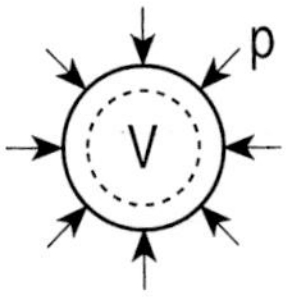 $-p=K\cdot dV/V$ κ=compressibility
Dilation modulus	$D = \lambda + 2\mu \quad [\mathrm{Pa}]$ $= 2\dfrac{1-\sigma}{1-2\sigma} \cdot S = \dfrac{1-\sigma}{(1+\sigma)(1-2\sigma)} \cdot E$ $= \dfrac{1}{3}\dfrac{1-\sigma}{1+\sigma} \cdot K$	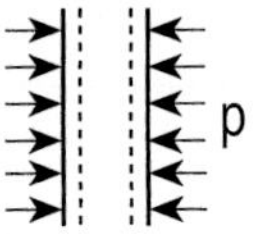 $-p=D\cdot s_{xx}$ for gas: $D = \rho_0 c_0^2 = \gamma P$
Bar bending modulus	$B_{st} = \dfrac{bd^3}{12} \cdot E \quad [\mathrm{Pa} \cdot \mathrm{m}^4]$	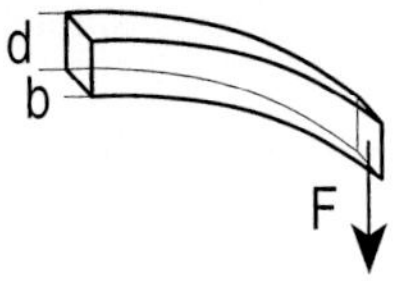

Quantity	Symbol , Relations	Remark
Plate bending modulus	$B = B_{pl} = I \cdot E \quad [Pa \cdot m^3]$ $= \frac{d^3}{3}\frac{\mu(\lambda+\mu)}{\lambda+2\mu} = \frac{d^3}{12(1-\sigma^2)} \cdot E$ $= \frac{d^3}{6(1-\sigma)} \cdot S$	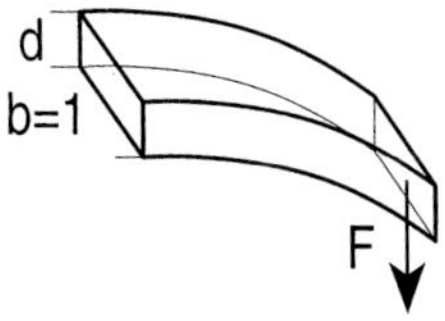 Moment of inertia $I = \frac{d^3}{12(1-\sigma^2)}$
Plate dilation modulus	$D_{pl} = \frac{E}{1-\sigma^2} \quad [Pa]$	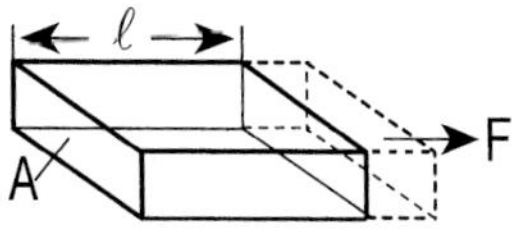

Table 4: Elastic constants

I.9 Sound transmission through a simple plate

Ref.: Mechel, [I.4]

A plate is "simple" if it is incompressible, homogeneous (except possibly in a micro-scale), isotropic, and unbounded (or at least with large dimensions, so that boundary effects can be neglected).

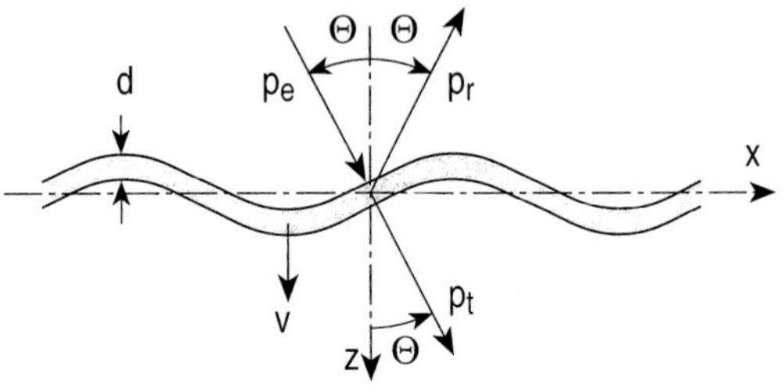

Transmission as a boundary value problem :

Field formulations :

$$p_e(x,z) = P_e \cdot e^{-jk_x x} \cdot e^{-jk_z z}$$
$$p_r(x,z) = r \cdot P_e \cdot e^{-jk_x x} \cdot e^{+jk_z z} \qquad k_x^2 + k_z^2 = k_0^2 \quad ; \quad k_x = k_0 \cdot \sin\Theta \quad ; \quad k_z = k_0 \cdot \cos\Theta \tag{1}$$
$$p_t(x,z) = P_t \cdot e^{-jk_x x} \cdot e^{-jk_z z}$$

Plate velocity :
$$v(x) = V \cdot e^{-jk_x x} \tag{2}$$

Plate driving pressure :
$$\delta p(x) = p_e(x,0) + p_r(x,0) - p_t(x,0) \tag{3}$$

From the bending wave equation :
$$\begin{gathered}[\frac{\partial^4}{\partial x^4} - k_B^4]\, v_z(x) = \frac{j\omega}{B}\delta p(x) \\ (k_x^4 - k_B^4)\cdot V \cdot e^{-jk_x x} = \frac{j\omega}{B}\left(P_e(1+r) - P_t\right)\cdot e^{-jk_x x}\end{gathered} \tag{4}$$

the partition impedance (boundary condition for pressure) follows :

$$\begin{aligned} Z_T &= \frac{\delta p}{v_z} = \frac{\left(P_e(1+r) - P_t\right)}{V} = \frac{B}{j\omega}\left(k_x^4 - k_B^4\right) = j\omega m \frac{k_B^4 - k_x^4}{k_B^4} \\ &= j\omega m\left(1 - (\frac{f}{f_c})^2\right) = j\omega m\left(1 - (\frac{f}{f_{cr}})^2 \sin^4\Theta\right) \end{aligned} \tag{5}$$

With boundary conditions for

particle velocity :

$$v_z \stackrel{!}{=} v_{tz}(x,0) = \frac{k_z}{k_0 Z_0} P_t \cdot e^{-jk_x x}$$
$$v_z \stackrel{!}{=} v_{ez}(x,0) + v_{rz}(x,0) = P_e \frac{k_z}{k_0 Z_0}(1-r)\cdot e^{-jk_x x} \tag{6}$$

the system of equations follows :

$$\begin{pmatrix} Z_T & -1 & 1 \\ 1 & 0 & -\cos\Theta/Z_0 \\ 1 & \cos\Theta/Z_0 & 0 \end{pmatrix} \cdot \begin{pmatrix} V \\ rP_e \\ P_t \end{pmatrix} = \begin{pmatrix} 1 \\ 0 \\ \cos\Theta/Z_0 \end{pmatrix} \tag{7}$$

with solution :

$$\frac{P_t}{P_e} = \left(1 + \frac{1}{2}\frac{Z_T \cos\Theta}{Z_0}\right)^{-1} \tag{8}$$

Sound transmission coefficient :

$$\tau(\Theta) = \left|\frac{P_t}{P_e}\right|^2 = \left|1 + \frac{1}{2}\frac{Z_T}{Z_0}\cos\Theta\right|^{-2} \tag{9}$$

After insertion :

$$\frac{1}{\tau(\Theta)} = 1 + \left(\frac{\cos\Theta}{2Z_0}\omega m[1 - (\frac{k_0}{k_B}\sin\Theta)^2]\right)^2 = 1 + \left(\frac{\cos\Theta}{2Z_0}\omega m[1 - (\frac{f}{f_c})^2]\right)^2$$
$$= 1 + \left(\pi Z_m F \cdot \cos\Theta \cdot (1 - F^2 \sin^4\Theta)\right)^2 = 1 + \left(\pi Z_m \sqrt{F(F-y)} \cdot (1-y^2)\right)^2 \tag{10}$$

with non-dimensional parameters :

$$y := \frac{f}{f_c} \; ; \; F := \frac{f}{f_{cr}} \; ; \; Z_m := \frac{f_{cr} m}{Z_0} = \frac{f_{cr} d}{Z_0}\rho \tag{11}$$

Transmission coefficients for diffuse sound incidence in three and two dimensions :

$$\tau_{3-dif} = \frac{\int_0^{\Theta_{hi}} \tau(\Theta)\cos\Theta\sin\Theta\, d\Theta}{\int_0^{\Theta_{hi}} \cos\Theta\sin\Theta\, d\Theta} = \frac{2}{\sin^2\Theta_{hi}}\int_0^{\Theta_{hi}} \tau(\Theta)\cos\Theta\sin\Theta\, d\Theta$$
$$\tau_{2-dif} = \frac{\int_0^{\Theta_{hi}} \tau(\Theta)\cos\Theta\, d\Theta}{\int_0^{\Theta_{hi}} \cos\Theta\, d\Theta} = \frac{1}{\sin\Theta_{hi}}\int_0^{\Theta_{hi}} \tau(\Theta)\cos\Theta\, d\Theta \tag{12}$$

(Θ_{hi}= upper limit, $\le\pi/2$, of sound incidence angles).

Approximations and special cases :

Berger's law for $f \ll f_{cr}$:
$$\frac{1}{\tau(\Theta)} \approx \left(\omega m \frac{\cos\Theta}{2Z_0}\right)^2 \tag{13}$$

$\dfrac{f \cdot \sin\Theta}{f_c} \gg 1$:
$$\frac{1}{\tau(\Theta)} \approx 1 + \left(\frac{\omega m}{2Z_0}\cos\Theta\,\sin^2\Theta (\frac{f}{f_c})^2\right)^2 \tag{14}$$

Normal or grazing incidence :
$$\tau(0) = \frac{1}{1+(\pi Z_m F)^2} \quad ; \quad \tau(\pi/2) = 1 \tag{15}$$

Diffuse incidence :
$$\tau_{3-dif} = \frac{2}{F\sin^2\Theta_{hi}} \int_0^{F\sin^2\Theta_{hi}} \frac{dy}{1+\pi^2 Z_m^2\, F(F-y)(1-y^2)^2} \tag{16}$$

When $R_{dif} = -10 \cdot \lg(\tau_{dif})$ is plotted over F, it depends on the single parameter F_m (which is a material constant) (besides the parameter Θ_{hi} of the test conditions).

Transmission with equivalent circuit :

The equivalent circuit method is adequate for the sound transmission through a simple plate.

The immediate result is, as above :

$$\tau(\Theta) = \left|\frac{P_t}{P_e}\right|^2 = \left|1 + \frac{1}{2}\frac{Z_T}{Z_0}\cos\Theta\right|^{-2} \tag{17}$$

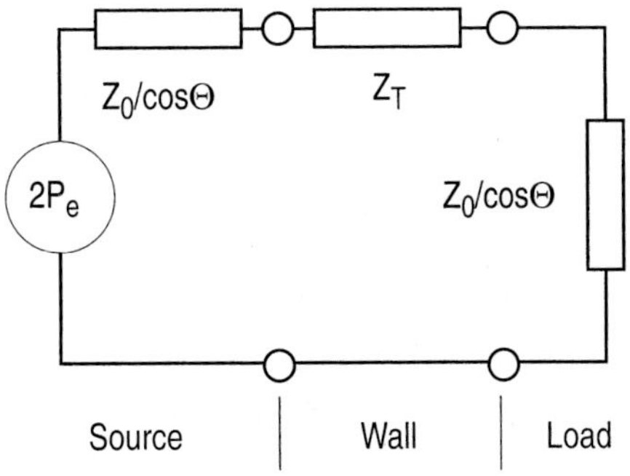

Plate with bending losses :

With bending loss factor η , use above :

$$\frac{Z_T}{Z_0} = \frac{\omega m}{Z_0}\left[\eta(\frac{f}{f_{cr}})^2\sin^4\Theta + j\left(1-(\frac{f}{f_{cr}})^2\sin^4\Theta\right)\right]$$
$$= 2\pi Z_m F\left[\eta F^2\sin^4\Theta + j\left(1-F^2\sin^4\Theta\right)\right] \tag{18}$$
$$= 2\pi Z_m F\left[\eta(\frac{f}{f_c})^2 + j\left(1-(\frac{f}{f_c})^2\right)\right]$$

Transmission coefficient for diffuse sound incidence :

$$\tau_{3-dif} = \frac{2}{F\sin^2\Theta_{hi}} \int_0^{F\sin^2\Theta_{hi}} \frac{dy}{[1+\pi\eta Z_m\sqrt{F(F-y)}\,y^2]^2 + \pi^2 Z_m^2 F(F-y)(1-y^2)^2} \tag{19}$$

Thick plates :

Use above the partition impedance from the TIMOSHENKO-MINDLIN theory :

$$Z_T = j\omega m \frac{k_B^4 - (k_x^2 - k_L^2)(k_x^2 - k_R^2)}{k_B^4 - k_R^2(k_x^2 - k_L^2)} \quad ; \quad k_R = \frac{\sqrt{12}}{\pi} k_S \tag{20}$$

with k_x= trace wave number of the exciting field ; k_B= free bending wave number ; k_L= longitudinal wave number ; k_S= shear wave number.

Alternatively, use in $Z_T = j\omega m[1-(\frac{k_s}{k_B})^4]$ the "corrected" bending wave number k_B :

$$k_B^2 = \frac{4.43}{24}\frac{\omega^2 m d^2}{B} + \sqrt{\left(\frac{4.43}{24}\frac{\omega^2 m d^2}{B}\right)^2 - \frac{m\omega^2}{B}\left(\frac{0.26\,\omega^2 md}{E} - 1\right)} \tag{21}$$

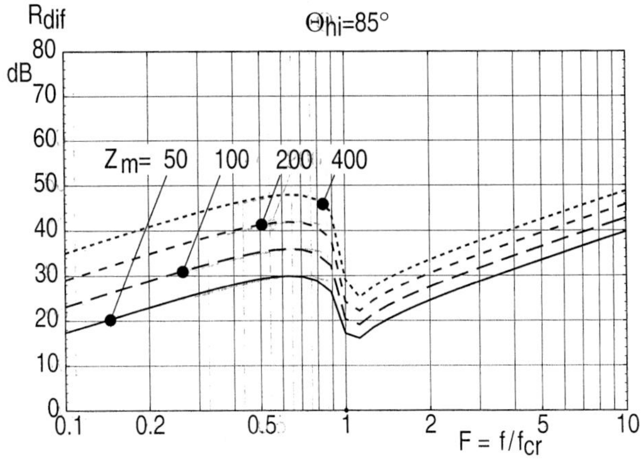

Transmission loss for diffuse sound incidence of simple plates with different values of Z_m.

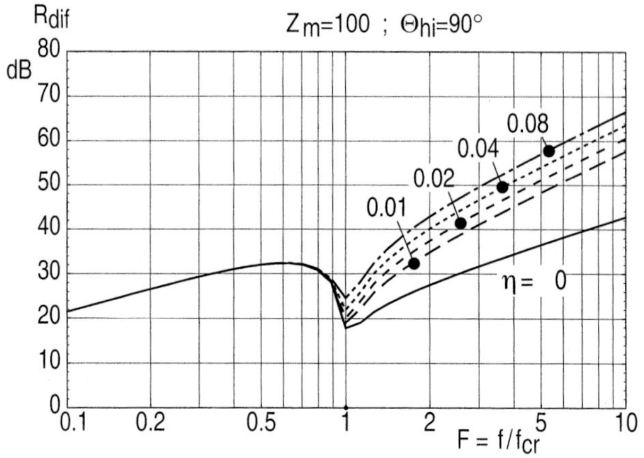

Transmission loss for diffuse sound incidence of simple plates with different values of bending loss factor η .

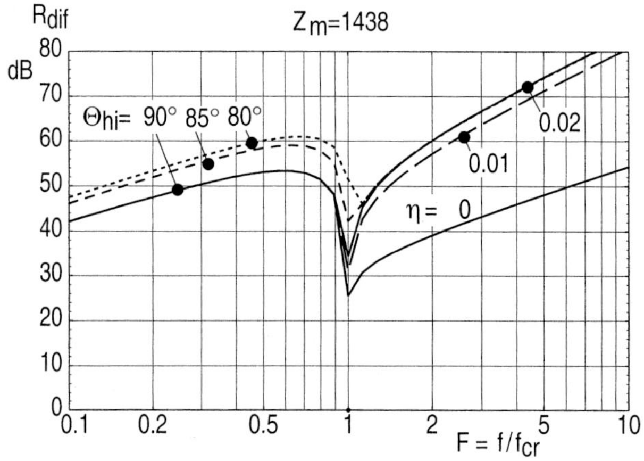

Highest possible transmission loss values of a simple plate for diffuse sound incidence; the plate consists of tin .

I.10 Infinite double-shell wall with absorber fill

Ref.: Mechel, [I.5]

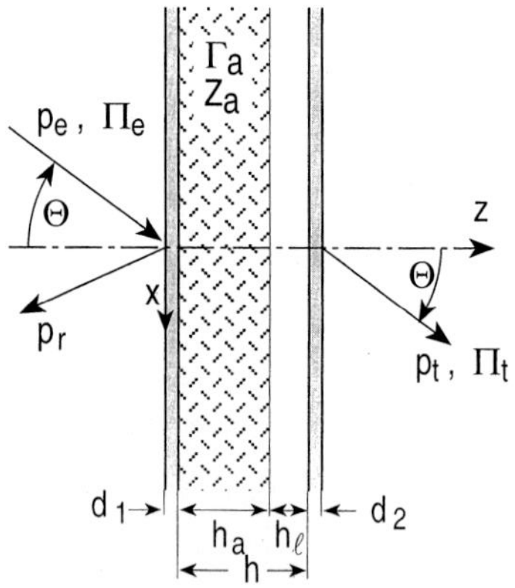

Two simple plates with thicknesses d_1, d_2 form an interspace which is filled with a fraction $\beta = h_a/h$ with a bulk reacting porous material having the characteristic values Γ_a, Z_a .

A plane wave p_e is incident at a polar angle of incidence Θ .

The sound transmission is evaluated with the method of equivalent networks.

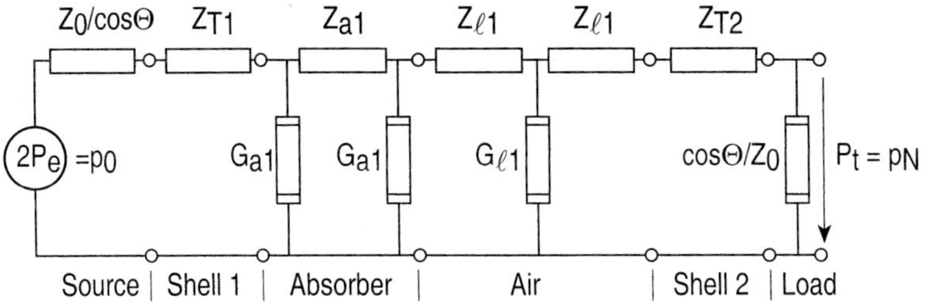

Z_{Ti} ; i=1,2; are partition impedances of the plates; network elements of the absorber layer :

$$\frac{Z_{a1}}{Z_0}=\frac{Z_a}{Z_0}\,\frac{\sinh(\Gamma_a h_a\cos\Theta_a)}{\cos\Theta_a}\quad ;\quad Z_0G_{a1}=\frac{\cos\Theta_a}{Z_a/Z_0}\,\frac{\cosh(\Gamma_a h_a\cos\Theta_a)-1}{\sinh(\Gamma_a h_a\cos\Theta_a)} \tag{1}$$

and of the air layer :

$$\frac{Z_{\ell 1}}{Z_0}=j\,\frac{1-\cos(k_0h_\ell\cos\Theta)}{\cos\Theta\,\sin(k_0h_\ell\cos\Theta)}\quad ;\quad Z_0G_{\ell 1}=j\cos\Theta\,\sin(k_0h_\ell\cos\Theta) \tag{2}$$

with internal angle in the absorber layer from : $\cos\Theta_a=\sqrt{1+\dfrac{\sin^2\Theta}{(\Gamma_a/k_0)^2}}$ (3)

The sound transmission coefficient is : $\tau(\Theta)=\left|\dfrac{P_t}{P_e}\right|^2=4\left|\dfrac{p_N}{p_0}\right|^2$ (4)

The numerical evaluation may apply the iterative method of section C.5 , or analytically with:

$$\frac{p_N}{p_0}=\frac{P_t}{2P_e}=\frac{1}{z_e+(b+g_{a1}z_e)z_a} \tag{5}$$

and the auxiliary quantities :

$$\begin{aligned}
&z_a=z_{T1}+1/\cos\Theta\;;\; z_b=z_{\ell 1}+z_{T2}\;;\; z_c=1+z_b\cos\Theta\\
&z_d=z_c+a\cdot z_{\ell 1}\;;\; z_e=z_d+b\cdot z_{a1}\\
&a=\cos\Theta+g_{\ell 1}z_c\;;\; b=a+g_{a1}z_d
\end{aligned} \tag{6}$$

in which z,g are normalised (to Z_0) impedances and admittances.

If the *interspace is full* (with absorber material) :

$$\begin{aligned}
&\frac{p_N}{p_0}=\frac{P_t}{2P_e}=\frac{1}{z_c+(a+g_{a1}z_c)z_a}\\
&z_a=z_{T1}+1/\cos\Theta\;;\; z_b=1+z_{T2}\cos\Theta\;;\; z_c=z_b+a\cdot z_{a1}\\
&a=\cos\Theta+g_{a1}\cdot z_b
\end{aligned} \tag{7}$$

If the *interspace is empty* (i.e. no absorber) :

$$\frac{p_N}{p_0} = \frac{P_t}{2P_e} = \frac{1}{z_a + (\cos\Theta + g_{\ell 1} z_a) z_b} \tag{8}$$

$$z_a = 1 + (z_{\ell 1} + z_{T2})\cos\Theta \quad ; \quad z_b = z_{\ell 1} + z_{T1} + 1/\cos\Theta$$

or after insertion :

$$\frac{p_N}{p_0} = \frac{P_t}{2P_e} = \frac{1}{2 + (2z_{\ell 1} + z_{T1} + z_{T2})\cos\Theta + g_{\ell 1}(\frac{1}{\cos\Theta} + z_{\ell 1} + z_{T1})(\frac{1}{\cos\Theta} + z_{\ell 1} + z_{T2})} \tag{9}$$

The *double-shell resonance* for an empty interspace is at :

$$f_0(\Theta) = \frac{c_0}{2\pi\cos\Theta}\sqrt{\frac{\rho_0}{h}\left(\frac{1}{m_1} + \frac{1}{m_2}\right)} \tag{10}$$

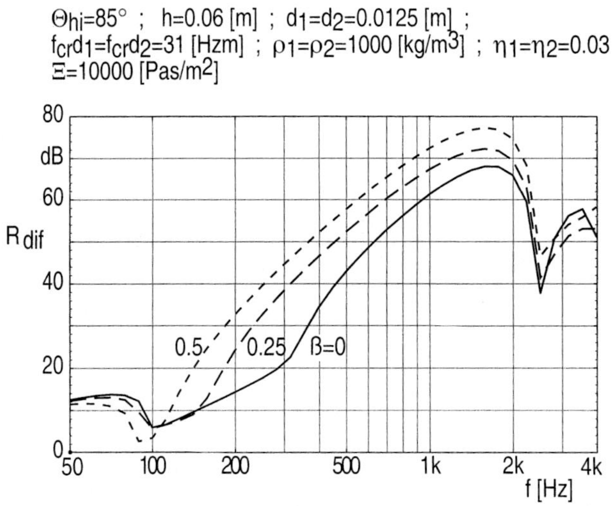

Sound transmission loss for diffuse sound incidence of a double-shell wall of plaster board shells, with different fill factors $ß = h_a/h$ of the absorber layer.

I.11 Double-shell wall with thin air gap

Ref.: Mechel, [I.5]

The present object is formally treated as a double-shell wall, completely filled with absorber material, from the previous section I.10 , but using for Γ_a, Z_a the characteristic values in a flat capillary. Finally the limit transition $\Gamma_a h_a \to 0$ is applied.

$$\frac{p_N}{p_0} = \frac{P_t}{2P_e} = \frac{1}{z_c + (a + g_{a1} z_c) z_a}$$

$$z_a = z_{T1} + 1/\cos\Theta \;\; ; \;\; z_b = 1 + z_{T2}\cos\Theta \;\; ; \;\; z_c = z_b + a \cdot z_{a1} \tag{1}$$

$$a = \cos\Theta + g_{a1} \cdot z_b$$

In the limit :
$$z_{a1} \to \frac{\Gamma_a Z_a}{k_0 Z_0} k_0 h_a \frac{\cos\Theta_a}{\cos\Theta} \;\; ; \;\; g_{a1} \to \frac{1}{z_{a1}} - \frac{1}{z_{a1}} = 0 \tag{2}$$

and therewith the transmission factor :

$$\frac{p_N}{p_0} \approx \frac{1}{2 + (z_{T1} + z_{T2})\cos\Theta + \dfrac{\Gamma_a Z_a}{k_0 Z_0} k_0 h_a \cos\Theta_a} \tag{3}$$

In the 3rd term of the denominator for very small h_a (with η= dynamic viscosity of air) :

$$\frac{\Gamma_a Z_a}{k_0 Z_0} \approx \frac{12\eta}{\omega \rho_0 h_a^2} \tag{4}$$

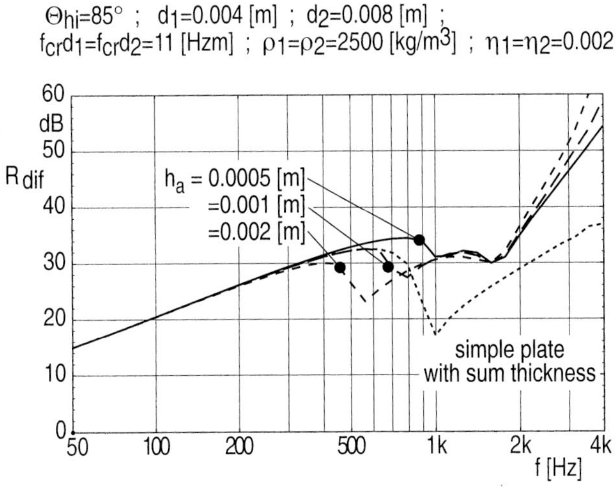

Double glass pane with thin air gap of different thickness.

I.12 Plate with absorber layer

Ref.: Mechel, [I.5]

It is assumed that the structure-borne sound transmission between the plate and the porous absorber layer is negligible (no or only loose contact between them). The evaluation uses the equivalent network method (it produces identical results with the solution of a boundary value problem). Γ_a, Z_a are the characteristic values of the porous material; $\Gamma_{an} = \Gamma_a/k_0$, $Z_{an} = Z_a/Z_0$; z,g are normalised (with Z_0) impedances or admittances.

Positions of the absorber layer in front of or behind the plate give the same transmission coefficients.

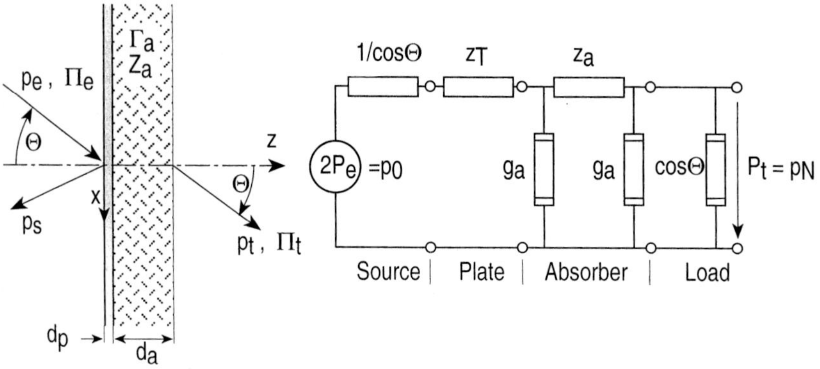

The network elements are :

$$z_a = \frac{Z_{an}}{\cos\Theta_a}\sinh(\Gamma_{an}\,k_0 d_a \cos\Theta_a)$$

$$g_a = \frac{\cos\Theta_a}{Z_{an}}\,\frac{\cosh(\Gamma_{an}\,k_0 d_a \cos\Theta_a) - 1}{\sinh(\Gamma_{an}\,k_0 d_a \cos\Theta_a)} \tag{1}$$

$$g_a z_a = \cosh(\Gamma_{an}\,k_0 d_a \cos\Theta_a) - 1 \quad ; \quad \cos\Theta_a = \sqrt{1 + \frac{\sin^2\Theta}{\Gamma_{an}^2}}$$

$$z_T = 2\pi Z_m F[\eta F^2 \cdot \sin^4\Theta + j(1 - F^2 \cdot \sin^4\Theta)] \quad ; \quad Z_m = \frac{f_{cr} d_p}{Z_0}\rho \quad ; \quad F = \frac{f}{f_{cr}}$$

(f_{cr}= critical frequency of the plate ; ρ= its density ; η= its bending loss factor).

Transmission loss : $R(\Theta) = 10\lg(1/\tau(\Theta)) = 10 \cdot \lg(0.25\,|\,p_0/p_N\,|^2)$ (2)

with :

$$\frac{p_N}{p_0} = \frac{P_t}{2P_e} = \frac{1}{z_2 + (a + g_a \cdot z_2)/\cos\Theta} \tag{3}$$

$$z_1 = 1 + z_T\cos\Theta \quad ; \quad a = \cos\Theta + g_a \cdot z_1 \quad ; \quad z_2 = z_1 + a \cdot z_a$$

and after insertion :

$$\frac{p_0}{p_N} = \frac{2P_e}{P_t} \tag{4}$$

$$= 2 + g_a \cdot (2z_a + 2z_T + g_a z_a z_T) + (z_a + z_T + g_a z_a z_T) \cdot \cos\Theta + ga \cdot (2 + g_a z_a)/\cos\Theta$$

I.13 Sandwich panels

Ref.: Mechel, [I.6]

Sandwich panels are combinations of sheets with high E- and shear modulus G (index 2) with boards having lower E- and shear moduli (index 1). The layers are combined with an adhesive layer, either very thin (or at least with no shear), or of thickness δ , the adhesive having a shear modulus G (without index).

One must distinguish between boards which are tight, and boards which are porous.

Tight boards :

It is sufficient to know the effective bending stiffness B of the sandwich. The required partition impedance Z_T is then obtained from section I.8 .

No.	Sandwich	Connection
1	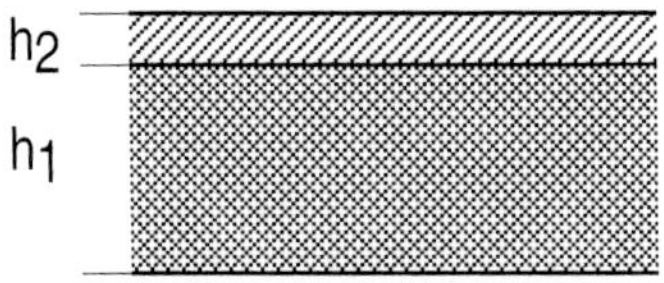	fix connection
2	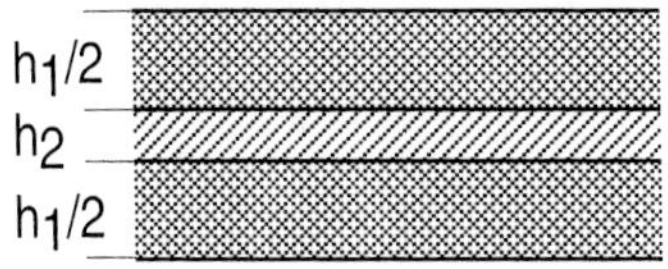	fix connection
3	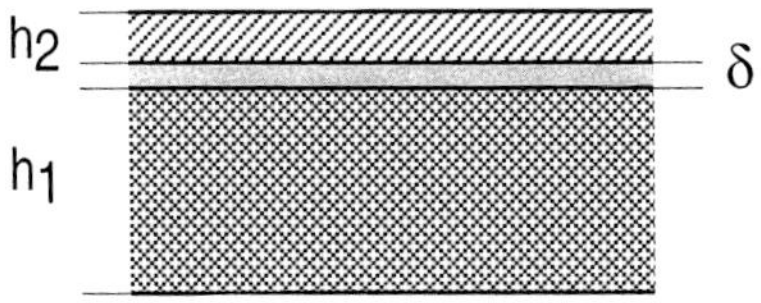	connection with shear $\delta \ge \frac{3.5\cdot 10^{-3}}{h_1}\frac{G}{E_1} - 1.3\cdot 10^{-12}$ [m] $E_2 h_2 \ge 2\cdot 10^7$ [Pa m]
4	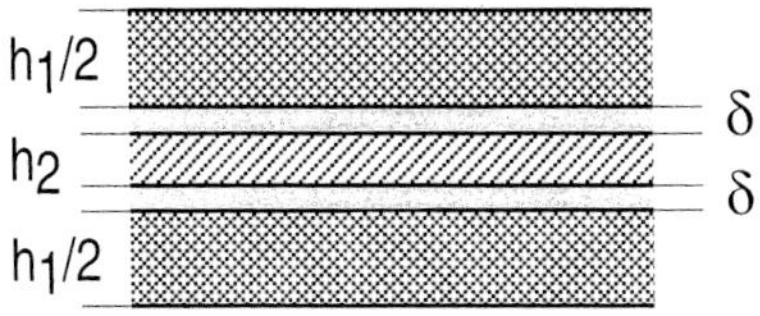	connection with shear $\delta \ge \frac{0.25\cdot 10^{-3}}{h_1}\frac{G}{E_1}$ [m] h_i in [m]; G, E_i in [Pa]

Table 1: Sandwich panels

No.	Effective bending modulus	Remark
1	$B=B_2\dfrac{1+2\dfrac{E_1}{E_2}\left[2(\dfrac{h_1}{h_2})+3(\dfrac{h_1}{h_2})^2+2(\dfrac{h_1}{h_2})^3\right]+(\dfrac{h_1}{h_2})^2(\dfrac{E_1}{E_2})^4}{1+\dfrac{h_1}{h_2}\dfrac{E_1}{E_2}}$	
2	$B\approx B_1\left(1+\dfrac{h_2}{h_1}\right)^3+B_2$	
3	$B\approx B_1+B_2+3G\delta\dfrac{h_1^2}{4}+\dfrac{E_1h_1E_2h_2(h_1/2+h_2/2)^2\cdot g}{E_1h_1+g\cdot(E_1h_1+E_2h_2)}-$ $-3G\delta\dfrac{h_1+h_2}{4}\dfrac{E_1h_1^2+2gE_2h_1h_2}{E_1h_1+g\cdot(E_1h_1+E_2h_2)}$	$\delta\ll h_1$ $g=\dfrac{G}{\delta E_1h_1\omega}\sqrt{B/m}$
4	$B=B_2+2E_1\dfrac{(h_1/2)^3}{12\,(1-\sigma_1^2)}$	

Table 2: Effective bending muduli B for sandwiches from Table 1 ..

Sandwich with porous board on front side :

A porous layer of thickness h propagates both sound waves in the pores (index 2) and dilatational waves in the matrix (index 1). Their coupling with each other leads to two characteristic propagation constants $\Gamma_\pm$.

The boundary conditions between board and sheet neglect shear stresses at the boundary II-III .

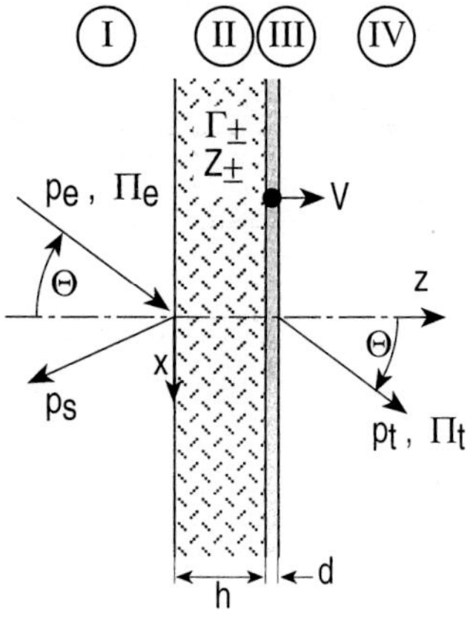

Field formulations (without common factor e^{-jk_xx}):

$$p_e(z)=P_e\cdot e^{-jk_zz}$$
$$p_s(z)=P_s\cdot e^{+jk_zz}\quad;\quad P_s=r\cdot P_e$$
$$v_{ez}(0)+v_{sz}(0)=P_e\frac{k_z}{k_0Z_0}(1-r)$$

$$k_x^2+k_z^2=k_0^2 \qquad (1)$$
$$k_x=k_0\sin\Theta\quad;\quad k_z=k_0\cos\Theta$$

$$p_t(z) = P_t \cdot e^{-jk_z(z-h-d)} \quad ; \quad v_{tz}(z) = \frac{k_z}{k_0 Z_0} P_t \cdot e^{-jk_z(z-h-d)} \tag{2}$$

Sound waves in the matrix (index 1) and in the pores (index 2) of the board :

$$\begin{aligned}
p_1(z) &= \frac{\Gamma_{z+}}{\Gamma_+} P_{12+}[A \cdot e^{-\Gamma_{z+}z} + B \cdot e^{+\Gamma_{z+}z}] + \frac{\Gamma_{z-}}{\Gamma_-} P_{12-}[C \cdot e^{-\Gamma_{z-}z} + D \cdot e^{+\Gamma_{z-}z}] \\
Z_0 v_{1z}(z) &= \frac{\Gamma_{z+}}{\Gamma_+} \frac{V_{12+}}{Z_{22+}}[A \cdot e^{-\Gamma_{z+}z} - B \cdot e^{+\Gamma_{z+}z}] + \frac{\Gamma_{z-}}{\Gamma_-} \frac{V_{12-}}{Z_{22-}}[C \cdot e^{-\Gamma_{z-}z} - D \cdot e^{+\Gamma_{z-}z}] \\
p_2(z) &= A \cdot e^{-\Gamma_{z+}z} + B \cdot e^{+\Gamma_{z+}z} + C \cdot e^{-\Gamma_{z-}z} + D \cdot e^{+\Gamma_{z-}z} \\
Z_0 v_{2z}(z) &= \frac{\Gamma_{z+}/\Gamma_+}{Z_{22+}}[A \cdot e^{-\Gamma_{z+}z} - B \cdot e^{+\Gamma_{z+}z}] + \frac{\Gamma_{z-}/\Gamma_-}{Z_{22-}}[C \cdot e^{-\Gamma_{z-}z} - D \cdot e^{+\Gamma_{z-}z}]
\end{aligned} \tag{3}$$

with : $\quad \Gamma_\pm^2 = \Gamma_{x\pm}^2 + \Gamma_{z\pm}^2 \;;\; \Gamma_{x\pm} = jk_x \;;\; \Gamma_{z\pm}^2 = \Gamma_\pm^2 + k_x^2 = \Gamma_\pm^2 + k_0^2 \sin^2\Theta$

The free field propagation constants $\Gamma_\pm$ are solutions of the characteristic equation :

$$\begin{aligned}
&(\frac{\Gamma}{k_0})^4 \cdot \frac{K_1}{K_0}\frac{K_2}{K_0} + (\frac{\Gamma}{k_0})^2 \cdot [\chi\frac{K_1}{K_0} + \frac{K_2}{K_0}(\chi - 1 + \frac{\rho_1}{\rho_0}) - \frac{j\sigma}{2\pi E}(\frac{K_1}{K_0} + \frac{K_2}{K_0})] + \\
&\quad + \sigma(\chi - 1) + \chi\frac{\rho_1}{\rho_0} - \frac{j\sigma}{2\pi E}(\frac{\rho_1}{\rho_0} + \sigma) = 0
\end{aligned} \tag{4}$$

($\pm$ indicates the sign of the root in the solution Γ^2), or in an approximation for $\sigma \approx 1$; $\chi \approx 1$; $\rho_1/\rho_0 \ll 1$:

$$(\frac{\Gamma}{k_0})^4 \cdot \frac{K_1}{K_0}\frac{K_2}{K_0} + (\frac{\Gamma}{k_0})^2 \cdot [\chi\frac{K_1}{K_0} + \frac{K_2}{K_0}\frac{\rho_1}{\rho_0} - \frac{j\sigma}{2\pi E}(\frac{K_1}{K_0} + \frac{K_2}{K_0}) + \frac{\rho_1}{\rho_0}(\chi - \frac{j\sigma}{2\pi E}) = 0 \tag{5}$$

The coefficients $Z_{22\pm}$, $P_{12\pm}$, $V_{12\pm}$ are :

$$\begin{aligned}
Z_{22\pm} &= \frac{p_2}{Z_0 v_2} = -j\frac{\Gamma_\pm}{k_0}\frac{K_2}{K_0} \frac{2\pi E \cdot (\chi - \sigma) - j\sigma}{2\pi E \cdot [\chi - 1 + (\frac{\Gamma_\pm}{k_0})^2 \frac{K_2}{K_0}\frac{\sigma - 1}{\sigma}] - j\sigma} \\
P_{12\pm} &= \frac{p_1}{p_2} = \frac{2\pi E \cdot (\chi + (\frac{\Gamma_\pm}{k_0})^2 \frac{K_2}{K_0}) - j\sigma}{2\pi E \cdot (\chi - \sigma) - j\sigma} \\
V_{12\pm} &= \frac{v_1}{v_2} = \frac{2\pi E \cdot [\chi + (\frac{\Gamma_\pm}{k_0})^2 \frac{K_2}{K_0}] - j\sigma}{2\pi E \cdot [\chi - 1 + (\frac{\Gamma_\pm}{k_0})^2 \frac{K_2}{K_0}\frac{\sigma - 1}{\sigma}] - j\sigma}
\end{aligned} \tag{6}$$

with the compression modulus of the air in the pores :

$$\frac{K_2}{K_0} = \frac{1 + jE/E_0}{\kappa + jE/E_0} \quad ; \quad E_0 = \frac{\rho_0 f_0}{\Xi} \quad ; \quad 2\pi f_0 \cdot \tau_0 = 1 \tag{7}$$

See the inset fràme for the meaning of other symbols.

The velocity V of the sheet is defined with its partition impedance Z_T and the driving pressure difference Δp :

$$V \cdot Z_T = p_1(h) + p_2(h) - p_t(h+d) \tag{8}$$

The seven unknown amplitudes P_s (or r), P_t , A, B, C, D, V are determined from the boundary conditions (one boundary condition is contained in the definition of V) :

at I-II :

$$\begin{aligned} & p_1(0) = (1-\sigma) P_e (1+r) \\ & p_2(0) = \sigma P_e (1+r) \\ & Z_0[(1-\sigma) v_{1z}(0) + \sigma v_{2z}(0)] = Z_0\left(v_e(0) + v_s(0)\right) = \frac{k_z}{k_0} P_e (1-r) \end{aligned} \tag{9}$$

at II-III : $v_{1z}(h) = V \quad ; \quad v_{2z}(h) = V$ (10)

at III-IV : $v_{tz}(d+h) = V$ (11)

The system of equations :

$$(\text{Matrix}) \bullet \begin{pmatrix} P_s / P_e \\ P_t / P_e \\ A / P_e \\ B / P_e \\ C / P_e \\ D / P_e \\ Z_0 V / P_e \end{pmatrix} = \begin{pmatrix} 1-\sigma \\ \sigma \\ k_z / k_0 \\ 0 \\ 0 \\ 0 \\ 0 \end{pmatrix}$$

- ρ_0= density of air ;
- c_0= adiabatic sound velocity ;
- κ= adiabatic exponent of air ;
- k_0= ω/c_0 ;
- K_0= $\rho_0 c_0^2$ = adiabatic compression modulus of air ;
- ρ_1= density of matrix material ;
- K_1= dilation modulus of matrix material ;
- K_2= compression modulus of air in the pores;
- σ= porosity of porous material ;
- χ= tortuosity of porous material ;
- Ξ= flow resistivity of porous material ;
- E= $\rho_0 f/\Xi$= absorber variable of porous material ;
- E_0= value of E at relaxation frequency f_0 in the pores ;

has a matrix with columns :

$$\{P_s/P_e\} = \{-(1-\sigma),\ -\sigma,\ k_z/k_0,\ 0,\ 0,\ 0,\ 0\}$$

$$\{P_t/P_e\} = \{0,\ 0,\ 0,\ 0,\ 0,\ k_z/k_0,\ 1\}$$

$$\begin{aligned} \{A/P_e\} = \{ & \Gamma_{z+} P_{12+}/\Gamma_+,\ 1,\ \Gamma_{z+}(\sigma + V_{12+}(1-\sigma))/(\Gamma_+ Z_{22+}), \\ & \Gamma_{z+} V_{12+}/(e^{\Gamma_{z+}h}\,\Gamma_+ Z_{22+}),\ \Gamma_{z+}/(e^{\Gamma_{z+}h}\,\Gamma_+ Z_{22+}),\ 0, \\ & -(1+\Gamma_{z+} P_{12+}/\Gamma_+)/e^{\Gamma_{z+}h} \} \end{aligned} \tag{13}$$

$$\begin{aligned} \{B/P_e\} = \{ & \Gamma_{z+} P_{12+}/\Gamma_+,\ 1,\ -\Gamma_{z+}(\sigma + V_{12+}(1-\sigma))/(\Gamma_+ Z_{22+}), \\ & -\Gamma_{z+} V_{12+} e^{\Gamma_{z+}h}/(\Gamma_+ Z_{22+}),\ -\Gamma_{z+} e^{\Gamma_{z+}h}/(\Gamma_+ Z_{22+}),\ 0, \\ & -(1+\Gamma_{z+} P_{12+}/\Gamma_+) \cdot e^{\Gamma_{z+}h} \} \end{aligned} \tag{14}$$

$$\{C/P_e\} = \{\Gamma_{z-}P_{12-}/\Gamma_-,\ 1,\ \Gamma_{z-}(\sigma + V_{12-}(1-\sigma))/(\Gamma_- Z_{22-}),$$
$$\Gamma_{z-}V_{12-}/(e^{\Gamma_{z-}h}\,\Gamma_- Z_{22-}),\ \Gamma_{z-}/(e^{\Gamma_{z-}h}\,\Gamma_- Z_{22-}),\ 0,$$
$$-(1+\Gamma_{z-}P_{12-}/\Gamma_-)/e^{\Gamma_{z-}h}\}$$

$$\{D/P_e\} = \{\Gamma_{z-}P_{12-}/\Gamma_-,\ 1,\ -\Gamma_{z-}(\sigma + V_{12-}(1-\sigma))/(\Gamma_- Z_{22-}),$$
$$-\Gamma_{z-}V_{12-}e^{\Gamma_{z-}h}/(\Gamma_- Z_{22-}),\ -\Gamma_{z-}e^{\Gamma_{z-}h}/(\Gamma_- Z_{22-}),\ 0,$$
$$-(1+\Gamma_{z-}P_{12-}/\Gamma_-)\cdot e^{\Gamma_{z-}h}\} \tag{15}$$

$$\{Z_0 V/P_e\} = \{0,\ 0,\ 0,\ -1,\ -1,\ -1,\ Z_T/Z_0\}$$

The finally desired transmission coefficient $\tau(\Theta)$ follows from $$\tau(\Theta) = |P_t/P_e|^2 \tag{16}$$

The following example is for a sandwich with a front-side glass fibre board, having different bulk densities $RG= (1-\sigma)\cdot\rho_1$ (with $\rho_1= 2500$ [kg/m^3]) and flow resistivities Ξ , and a back-side plaster board, d= 12.5 [mm] thick .

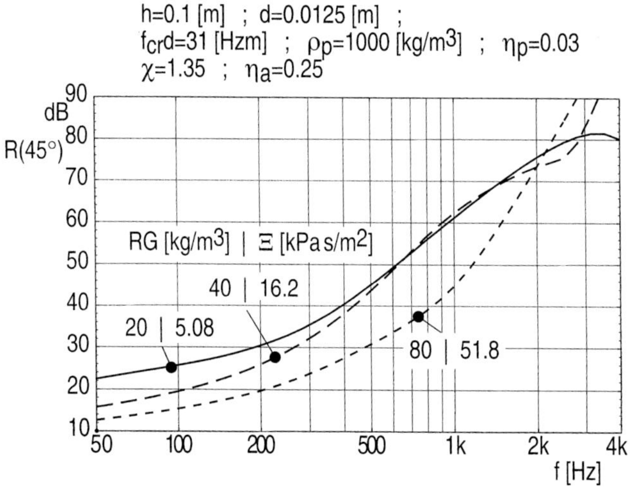

Sound transmission loss for oblique sound incidence of a sandwich with a front-side glass fibre board of different bulk densities RG and flow porosities Ξ , and a back-side plaster board, d= 12.5 [mm] thick.

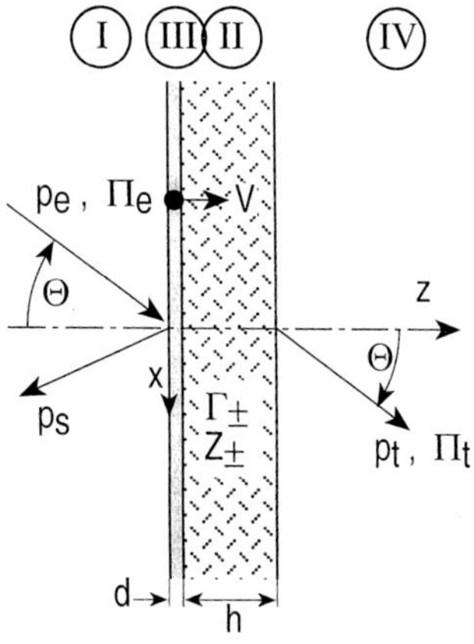

Sandwich with porous board on back side :

The field definitions remain as in the previous arrangement.

The boundary conditions now are :

plate :

$$V\cdot Z_T = p_e(0) + p_s(0) - (p_1(d) + p_2(d)) \tag{17}$$

$$\text{at I-III:}\quad v_{ez}(0)+v_{sz}(0)=\frac{k_z}{k_0 Z_0}(P_e-P_s)\overset{!}{=}V \tag{18}$$

$$\text{at III-II:}\quad v_{1z}(d)=V \;;\; v_{2z}(d)=V \tag{19}$$

at II-IV :

$$\begin{aligned} &p_1(d+h)=(1-\sigma)P_t \\ &p_2(d+h)=\sigma P_t \\ &Z_0[(1-\sigma)v_{1z}(d+h)+\sigma v_{2z}(d+h)]=Z_0 v_{tz}(d+h)=\frac{k_z}{k_0}P_t \end{aligned} \tag{20}$$

The system of equations to be solved is :

$$(\text{Matrix})\bullet\begin{pmatrix} P_s/P_e \\ P_t/P_e \\ A/P_e \\ B/P_e \\ C/P_e \\ D/P_e \\ Z_0V/P_e \end{pmatrix}=\begin{pmatrix} -\cos\Theta \\ 0 \\ 0 \\ 0 \\ 0 \\ 0 \\ -1 \end{pmatrix} \tag{21}$$

with the matrix columns :

$$\{P_s/P_e\}=\{-\cos\Theta,\ 0,\ 0,\ 0,\ 0,\ 0,\ 1\}$$

$$\{P_t/P_e\}=\{0,\ 0,\ 0,\ -(1-\sigma),\ -\sigma,\ -\cos\Theta,\ 0\}$$

$$\begin{aligned} \{A/P_e\}=\Big\{&0,\ \Gamma_{z+}V_{12+}/(\Gamma_+ Z_{22+}e^{\Gamma_{z+}d}),\ \Gamma_{z+}/(\Gamma_+ Z_{22+}e^{\Gamma_{z+}d}), \\ &\Gamma_{z+}P_{12+}/(\Gamma_+ e^{\Gamma_{z+}(d+h)}),\ 1/e^{\Gamma_{z+}(d+h)}, \\ &\Gamma_{z+}/(\Gamma_+ Z_{22+}e^{\Gamma_{z+}(d+h)})\cdot[\sigma+(1-\sigma)V_{12+}],\ -(1+\Gamma_{z+}P_{12+}/\Gamma_+)/e^{\Gamma_{z+}d}\Big\} \end{aligned} \tag{22}$$

$$\begin{aligned} \{B/P_e\}=\Big\{&0,\ -\Gamma_{z+}V_{12+}e^{\Gamma_{z+}d}/(\Gamma_+ Z_{22+}),\ -\Gamma_{z+}e^{\Gamma_{z+}d}/(\Gamma_+ Z_{22+}), \\ &\Gamma_{z+}P_{12+}e^{\Gamma_{z+}(d+h)}/\Gamma_+,\ e^{\Gamma_{z+}(d+h)}, \\ &-\Gamma_{z+}e^{\Gamma_{z+}(d+h)}/(\Gamma_+ Z_{22+})\cdot[\sigma+V_{12+}(1-\sigma)],\ -e^{\Gamma_{z+}d}\cdot(1+\Gamma_{z+}P_{12+}/\Gamma_+)\Big\} \end{aligned} \tag{23}$$

$$\begin{aligned} \{C/P_e\}=\Big\{&0,\ \Gamma_{z-}V_{12-}/(\Gamma_- Z_{22-}e^{\Gamma_{z-}d}),\ \Gamma_{z-}/(\Gamma_- Z_{22-}e^{\Gamma_{z-}d}), \\ &\Gamma_{z-}P_{12-}/(\Gamma_- e^{\Gamma_{z-}(d+h)}),\ 1/e^{\Gamma_{z-}(d+h)}, \\ &\Gamma_{z-}/(\Gamma_- Z_{22-}e^{\Gamma_{z-}(d+h)})\cdot[\sigma+(1-\sigma)V_{12-}],\ -(1+\Gamma_{z-}P_{12-}/\Gamma_-)/e^{\Gamma_{z-}d}\Big\} \end{aligned} \tag{24}$$

$$\{D/P_e\} = \{0,\ -\Gamma_{z-}V_{12-}e^{\Gamma_{z-}d}/(\Gamma_-Z_{22-}),\ \ -\Gamma_{z-}e^{\Gamma_{z-}d}/(\Gamma_-Z_{22-}),$$
$$\Gamma_{z-}P_{12-}e^{\Gamma_{z-}(d+h)}/\Gamma_-,\ \ e^{\Gamma_{z-}(d+h)}, \tag{25}$$
$$-\Gamma_{z-}e^{\Gamma_{z-}(d+h)}/(\Gamma_-Z_{22-})\cdot[\sigma+(1-\sigma)V_{12-}],\ \ -e^{\Gamma_{z-}d}\cdot(1+\Gamma_{z-}P_{12-}/\Gamma_-)\}$$

$$\{Z_0V/P_e\} = \{-1,\ -1,\ -1,\ 0,\ 0,\ 0,\ -Z_T\} \tag{26}$$

The finally desired transmission coefficient $\tau(\Theta)$ follows from $\qquad \tau(\Theta) = |P_t/P_e|^2 \qquad (27)$

The following example is for the same object as above. Both the example and the equations show that the position of the absorber board modifies the sound transmission loss.

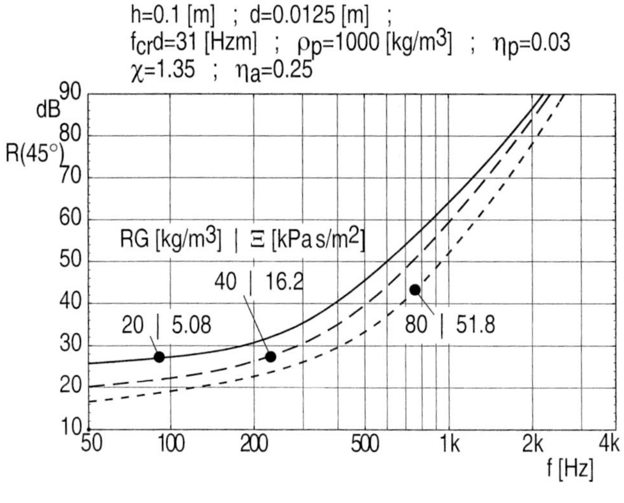

Same sandwich as above, but with a reversed arrangement of the glass fibre board and the plaster board.

Sandwich with a porous layer as core :

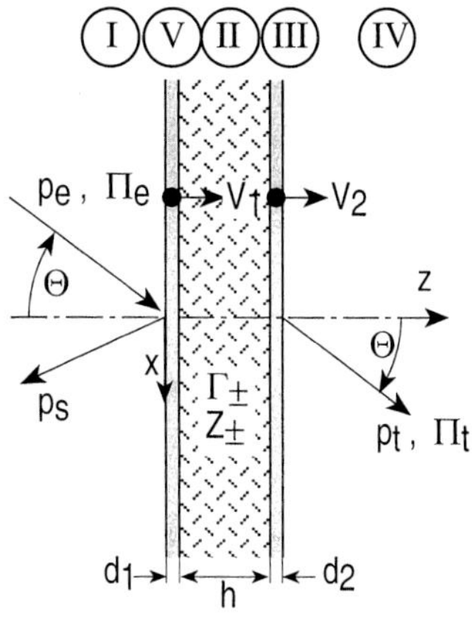

The transmitted wave p_t is modified :

$$p_t(z) = P_t\cdot e^{-jk_z(z-h-d_1-d_2)}$$
$$v_{tz}(z) = \frac{k_z}{k_0Z_0}P_t\cdot e^{-jk_z(z-h-d_1-d_2)} \tag{28}$$

The equations for the cover plates are :

$$V_1\cdot Z_{T1} = p_e(0)+p_s(0)-(p_1(d_1)+p_2(d_1))$$
$$V_2\cdot Z_{T2} = p_1(d_1+h)+p_2(d_1+h)-p_t(d_1+d_2+h) \tag{29}$$

The boundary conditions are now :

at I-V : $$v_{ez}(0)+v_{sz}(0)=\frac{k_z}{k_0 Z_0}(P_e-P_s)\overset{!}{=}V_1 \tag{30}$$

at V-II : $$v_{1z}(d_1)=V_1 \quad ; \quad v_{2z}(d_1)=V_1 \tag{31}$$

at II-III : $$v_{1z}(d_1+h)=V_2 \quad ; \quad v_{2z}(d_1+h)=V_2 \tag{32}$$

at III-V : $$v_{tz}(d_1+d_2+h)=\frac{k_z}{k_0 Z_0}P_t \overset{!}{=} V_2 \tag{33}$$

The system of equations to be solved is :

$$(\text{Matrix})\bullet\begin{pmatrix} P_s/P_e \\ P_t/P_e \\ A/P_e \\ B/P_e \\ C/P_e \\ D/P_e \\ Z_0V_1/P_e \\ Z_0V_2/P_e \end{pmatrix}=\begin{pmatrix} -\cos\Theta \\ 0 \\ 0 \\ 0 \\ 0 \\ 0 \\ -1 \\ 0 \end{pmatrix} \tag{34}$$

with the matrix columns :

$$\begin{aligned}&\{P_s/P_e\}=\{-\cos\Theta,\ 0,\ 0,\ 0,\ 0,\ 0,\ 1,\ 0\}\\&\{P_t/P_e\}=\{0,\ 0,\ 0,\ 0,\ 0,\ \cos\Theta,\ 0,\ -1\}\end{aligned} \tag{35}$$

$$\begin{aligned}\{A/P_e\}=\Big\{&0,\ \Gamma_{z+}V_{12+}/(\Gamma_+Z_{22+}e^{\Gamma_{z+}d_1}),\ \Gamma_{z+}/(\Gamma_+Z_{22+}e^{\Gamma_{z+}d_1}),\\&\Gamma_{z+}V_{12+}/(\Gamma_+Z_{22+}e^{\Gamma_{z+}(d_1+h)}),\ \Gamma_{z+}/(\Gamma_+Z_{22+}e^{\Gamma_{z+}(d_1+h)}),\ 0,\\&-(1+\Gamma_{z+}P_{12+}/\Gamma_+)/e^{\Gamma_{z+}d_1},\ (1+\Gamma_{z+}P_{12+}/\Gamma_+)/e^{\Gamma_{z+}(d_1+h)}\Big\}\\\{B/P_e\}=\Big\{&0,\ -\Gamma_{z+}V_{12+}e^{\Gamma_{z+}d_1}/(\Gamma_+Z_{22+}),\ -\Gamma_{z+}e^{\Gamma_{z+}d_1}/(\Gamma_+Z_{22+}),\\&-\Gamma_{z+}V_{12+}e^{\Gamma_{z+}(d_1+h)}/(\Gamma_+Z_{22+}),\ -\Gamma_{z+}e^{\Gamma_{z+}(d_1+h)}/(\Gamma_+Z_{22+}),\ 0,\\&-e^{\Gamma_{z+}d_1}\cdot(1+\Gamma_{z+}P_{12+}/\Gamma_+),\ e^{\Gamma_{z+}(d_1+h)}\cdot(1+\Gamma_{z+}P_{12+}/\Gamma_+)\Big\}\end{aligned} \tag{36}$$

$$\begin{aligned}\{C/P_e\}=\Big\{&0,\ \Gamma_{z-}V_{12-}/(\Gamma_-Z_{22-}e^{\Gamma_{z-}d_1}),\ \Gamma_{z-}/(\Gamma_-Z_{22-}e^{\Gamma_{z-}d_1}),\\&\Gamma_{z-}V_{12-}/(\Gamma_-Z_{22-}e^{\Gamma_{z-}(d_1+h)}),\ \Gamma_{z-}/(\Gamma_-Z_{22-}e^{\Gamma_{z-}(d_1+h)}),\ 0,\\&-(1+\Gamma_{z-}P_{12-}/\Gamma_-)/e^{\Gamma_{z-}d_1},\ (1+\Gamma_{z-}P_{12-}/\Gamma_-)/e^{\Gamma_{z-}(d_1+h)}\Big\}\end{aligned} \tag{37}$$

$$\{D/P_e\} = \{0,\ -\Gamma_{z-}V_{12-}e^{\Gamma_{z-}d_1}/(\Gamma_-Z_{22-}),\ -\Gamma_{z-}e^{\Gamma_{z-}d_1}/(\Gamma_-Z_{22-}),$$
$$-\Gamma_{z-}V_{12-}e^{\Gamma_{z-}(d_1+h)}/(\Gamma_-Z_{22-}),\ -\Gamma_{z-}e^{\Gamma_{z-}(d_1+h)}/(\Gamma_-Z_{22-}),\ 0, \tag{38}$$
$$-e^{\Gamma_{z-}d_1}\cdot(1+\Gamma_{z-}P_{12-}/\Gamma_-),\ e^{\Gamma_{z-}(d_1+h)}\cdot(1+\Gamma_{z-}P_{12-}/\Gamma_-)\}$$

$$\{Z_0V_1/P_e\} = \{-1,\ -1,\ -1,\ 0,\ 0,\ 0,\ -Z_{T1},\ 0\}$$
$$\{Z_0V_2/P_e\} = \{0,\ 0,\ 0,\ -1,\ -1,\ -1,\ 0,\ -Z_{T2}\} \tag{39}$$

The finally desired transmission coefficient $\tau(\Theta)$ follows from $\tau(\Theta) = |P_t/P_e|^2$ (40)

The example is for a sandwich with a glass fibre core for different bulk densities RG and flow resistivities Ξ of the glass fibre material.

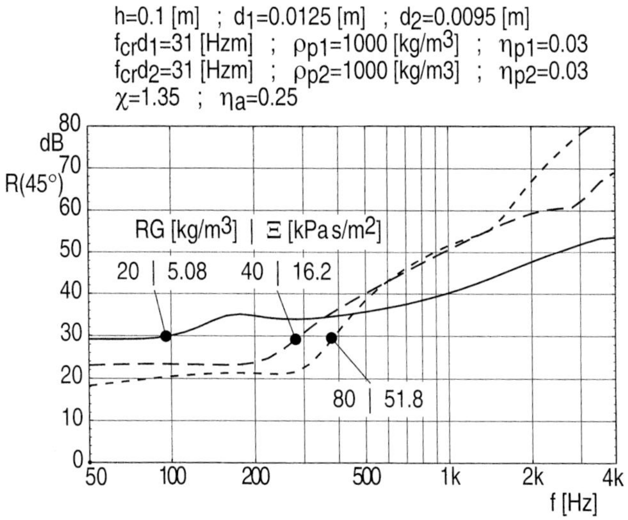

Sandwich with a glass fibre core and plaster board cover plates, for different bulk densities RG and flow resistivities Ξ of the core material.

I.14 Finite size plate

Ref.: Mechel, [I.7]

Let the plate be two-dimensional (for ease of writing, a three-dimensional plate is treated similarly), infinite in the y direction, and with supported borders at $x=\pm h$. The non-dimensional co-ordinate $\xi = x/h$ will be used. In this section solutions $v_n(\xi) = v_n(\gamma_n \xi)$ of the homogeneous bending wave equation

$$\left(\frac{\partial^4}{\partial \xi^4} - \gamma_n^4\right) v_n(\xi) = 0 \tag{1}$$

are given, which satisfy the boundary conditions of different kinds of boundary support. These solutions are *plate modes* which will be used to synthesise plate velocity patterns :

$$V(\xi) = \sum_n V_n \cdot v_n(\xi) \tag{2}$$

The bending wave equation :

$$\left(\frac{\partial^4}{\partial \xi^4} - (k_B h)^4\right) V(\xi) = h^4 \frac{j\omega}{B} \delta p(\xi) \tag{3}$$

then gives :

$$\sum_n V_n\left(\gamma_n^4 - (k_B h)^4\right) v_n(\xi) = h^4 \frac{j\omega}{B} \delta p(\xi) \tag{4}$$

which can be written as :

$$\sum_n V_n Z_{Tn} \cdot v_n(\xi) = \delta p(\xi) \tag{5}$$

therewith defining the *modal partition impedances* Z_{Tn} :

$$Z_{Tn} = \frac{B}{j\omega}\left((\gamma_n / h)^4 - k_B^4\right) = j\omega m\left[1 - \left(\frac{\gamma_n}{k_B h}\right)^4\right] = j\omega m\left[1 - \left(\frac{k_{bn}}{k_B}\right)^4\right] \tag{6}$$

In the last expression is $k_{bn} = \gamma/h$ is the modal bending wave number. For a plate with bending losses (η= bending loss factor) correspondingly :

$$Z_{Tn} = \omega m\left[\eta\left(\frac{\gamma_n}{k_B h}\right)^4 + j\left(1-\left(\frac{\gamma_n}{k_B h}\right)^4\right)\right]$$

$$\frac{Z_{Tn}}{Z_0} = 2\pi Z_m F\left[\eta F^2\left(\frac{\gamma_n}{k_0 h}\right)^4 + j\left(1-F^2\left(\frac{\gamma_n}{k_0 h}\right)^4\right)\right] \qquad F=\frac{f}{f_{cr}}\ ;\ Z_m=\frac{f_{cr}m}{Z_0}=\frac{f_{cr}d}{Z_0}\rho \qquad (7)$$

Depending on the pattern of excitation symmetrical modes $v_n^{(s)}(\xi)$ and anti-symmetrical modes $v_n^{(a)}(\xi)$ will be excited, even for symmetrical support at $\xi=\pm 1$ (symmetry defined with respect to $\xi=0$). With different supports at both sides, the plate velocity pattern can be written as a mode sum over symmetrical and anti-symmetrical modes.

For the "classical" supports (see section I.8) the plate modes are orthogonal :

$$\int_{-1}^{+1} v_m^{(\beta)}(\xi)\cdot v_n^{(\beta)}(\xi)\,d\xi = \delta_{mn}\cdot N_{Pn}^{(\beta)} \quad ; \quad \delta_{mn}=\begin{cases}1\ ; & m=n\\ 0\ ; & m\neq n\end{cases} \qquad (8)$$

δ_{mn}= Kronecker symbol; $N_{Pn}^{(\beta)}$= norm of the plate mode. Modes with different types ß of symmetry are evidently always orthogonal to each other.

Simply supported plate :

(the most probable classical support for many technical fixations of construction panels)

Boundary conditions :

$$v_n(\xi)=\frac{\partial^2 v_n(\xi)}{\partial\xi^2}=0 \quad ; \quad \xi=\pm 1 \qquad (9)$$

Solutions :

$$v_n^{(\beta)}(\xi)=\begin{cases}\cos(\gamma_n^{(s)}\xi) & ;\ \gamma_n^{(s)}=n_o\dfrac{\pi}{2}\ ;\ n_o=1,3,5,\ldots\\ \sin(\gamma_n^{(a)}\xi) & ;\ \gamma_n^{(a)}=n_e\dfrac{\pi}{2}\ ;\ n_e=2,4,6,\ldots\end{cases} \qquad (10)$$

Polynomial solutions are excluded. Mode norms :

$$N_{Pn}^{(\beta)}=\int_{-1}^{+1}[v_n^{(\beta)}(\xi)]^2\,d\xi=1 \qquad (11)$$

Clamped plate : ß= s,a or ß= 0,1 respectively

Boundary conditions :

$$v_n(\pm 1)=\frac{\partial v_n(\pm 1)}{\partial\xi}=0 \qquad (12)$$

General solutions :

$$v_n^{(ß)}(\xi) = \begin{cases} \cos(\gamma_n^{(s)}\xi) + C_n^{(s)} \cosh(\gamma_n^{(s)}\xi) \\ \sin(\gamma_n^{(a)}\xi) + C_n^{(a)} \sinh(\gamma_n^{(a)}\xi) \end{cases} \tag{13}$$

where $C_n^{(ß)}$ are solutions of :

$$C_n^{(ß)} = (-1)^ß \frac{\sin\gamma_n^{(ß)}}{\sinh\gamma_n^{(ß)}} = -\frac{\cos\gamma_n^{(ß)}}{\cosh\gamma_n^{(ß)}} \tag{14}$$

The 2nd equation gives the characteristic equation for the $\gamma_n^{(ß)}$:

$$\tan\gamma_n^{(ß)} = \mp\tanh\gamma_n^{(ß)} \quad ; \quad ß = \begin{cases} s \\ a \end{cases} \tag{15}$$

with approximate solutions : $\gamma_n^{(ß)} \approx \pi(n \mp 1/4) \; ; \; n = 1,2,3,\ldots$ (16)

n	$\gamma_{n,ex}^{(s)}$	$\gamma_{n,apr}^{(s)}$	$\gamma_{n,ex}^{(a)}$	$\gamma_{n,apr}^{(a)}$
1	2.36502	2.35619	3.92660	3.92699
2	5.49780	5.49779	7.06858	7.06858
3	8.63938	8.63938	10.21018	10.21018
4	11.78097	11.78097	13.35177	13.35177
≥5	14.92257	$\gamma_n^{(s)} \approx \pi(n-1/4)$	16.49336	$\gamma_n^{(a)} \approx \pi(n+1/4)$

Table 1: Characteristic values γ_n (exact and approximations) for a clamped plate.

Mode norms : (17)

$$N_{Pn}^{(ß)} = \int_{-1}^{+1} [v_n^{(ß)}(\gamma_n^{(ß)}\xi)]^2 \, d\xi =$$

$$= 1 \pm \frac{\sin(2\gamma_n^{(ß)})}{2\gamma_n^{(ß)}} \pm [C_n^{(ß)}]^2 \left(1 \pm \frac{\sinh(2\gamma_n^{(ß)})}{2\gamma_n^{(ß)}}\right) + 2\frac{C_n^{(ß)}}{\gamma_n^{(ß)}}\left(\sin\gamma_n^{(ß)} \cdot \cosh\gamma_n^{(ß)} \pm \cos\gamma_n^{(ß)} \cdot \sinh\gamma_n^{(ß)}\right)$$

$$= 1 \pm \frac{\sin(2\gamma_n^{(ß)})}{2\gamma_n^{(ß)}} \pm [C_n^{(ß)}]^2 \left(1 \pm \frac{\sinh(2\gamma_n^{(ß)})}{2\gamma_n^{(ß)}}\right)$$

n	$N^{(s)}_{Pn,ex}$	$N^{(s)}_{Pn,apr}$	$N^{(a)}_{Pn,ex}$	$N^{(a)}_{Pn,apr}$
1	1.017651	1.029835	0.9992230	0.9995197
2	1.000034	1.000043	0.9999986	0.9999989
3	1.000000	1.000000	1.000000	1.000000
4	1.000000	1.000000	1.000000	1.000000

Table 2: Mode norms (exact and approximate) for a clamped plate.

Free plate : ß= s,a or ß= 0,1 respectively

Boundary conditions :

$$\frac{\partial^2 v_n(\xi)}{\partial \xi^2} = \frac{\partial^3 v_n(\xi)}{\partial \xi^3} = 0 \quad ; \quad \xi = \pm 1 \tag{18}$$

General solutions :

$$v_n^{(ß)}(\xi) = \begin{cases} \cos(\gamma_n^{(s)}\xi) + C_n^{(s)} \cosh(\gamma_n^{(s)}\xi) \\ \sin(\gamma_n^{(a)}\xi) + C_n^{(a)} \sinh(\gamma_n^{(a)}\xi) \end{cases} \tag{19}$$

where $C_n^{(ß)}$ are solutions of :

$$C_n^{(ß)} = -(-1)^ß \frac{\sin\gamma_n^{(ß)}}{\sinh\gamma_n^{(ß)}} = \frac{\cos\gamma_n^{(ß)}}{\cosh\gamma_n^{(ß)}} \tag{20}$$

Characteristic equation for the $\gamma_n^{(ß)}$:

$$\tan\gamma_n^{(ß)} = \mp\tanh\gamma_n^{(ß)} \quad ; \quad ß = \begin{cases} s \\ a \end{cases} \tag{21}$$

with the same solutions and approximations as for the clamped plate. Additionally a piston-like oscillation (with index n=0) of the form $v_0^{(s)}(\xi) = 1 + C_0^{(s)} = \text{const}$ is possible with the characteristic value $\gamma_0^{(s)} = 0$. With the choice $C_0^{(s)} = 1$, this mode has a norm with unit value. Further, an anti-symmetrical mode is possible with :

$$v_0^{(a)}(\xi) = C_0^{(a)} \cdot \xi \quad ; \quad \gamma_0^{(a)} = 0 \quad ; \quad C_0^{(a)} = 1 \quad ; \quad n = 0 \tag{22}$$

(the choice $C_0^{(s)} = 1$ is arbitrary). These additional modes are called *polynomial modes*.

The mode norms are the same as for the clamped plate, except for the polynomial modes :

$$N_{P0}^{(s)} = 2(1+C_0^{(s)})^2 = 8 \quad ; \quad N_{P0}^{(a)} = \frac{2}{3} C_0^{(a)2} = \frac{2}{3} \tag{23}$$

The polynomial modes are orthogonal to the other modes.

I.15 Single plate across a flat duct

Ref.: Mechel, [I.7]

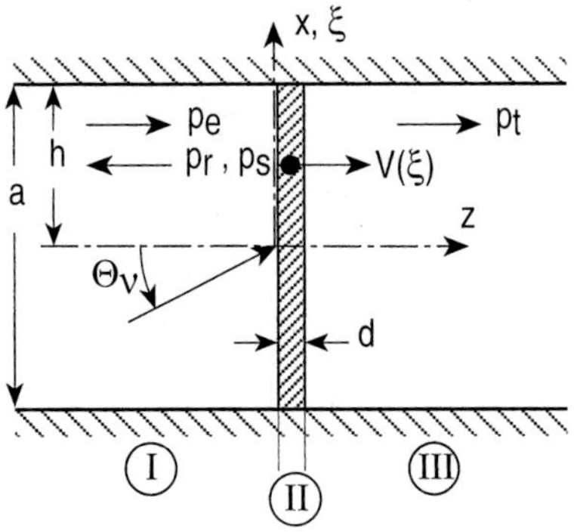

A flat (2-dimensional) duct of width $a = 2h$ with hard walls is subdivided by a plate of thickness d, the plate will have different "classical" fixations at the duct walls in $x=\pm h$, i.e. in $\xi = x/h = 1$.

The excitation will first be by a single duct mode of order μ and (arbitrary) amplitude P_e. Later "mode mixtures" will be considered.

The field in front of the plate (i.e. in zone I) is formulated as a sum

$$p_I(\xi,z) = p_e(\xi,z) + p_r(\xi,z) + p_s(\xi,z) \tag{1}$$

with $p_e(\xi,z)$= incident duct mode; $p_r(\xi,z)$= hardly reflected duct mode; $p_s(\xi,z)$= scattered field. The scattered field p_s and the transmitted wave p_t are formulated as sums of duct modes. The velocity pattern of the plate is formulated as a sum of plate modes (see previous section I.14 , especially for γ_n and $C_n^{(ß)}$).

$$\begin{aligned} p_e(\xi,z) &= P_e \cdot q_\mu^{(ß)}(\xi) \cdot e^{-jk_{\mu z}z} \\ Z_0 v_{ez}(\xi,z) &= \cos\Theta_\mu P_e \cdot q_\mu^{(ß)}(\xi) \cdot e^{-jk_{\mu z}z} \end{aligned} \tag{2}$$

$$\begin{aligned} p_r(\xi,z) &= P_e \cdot q_\mu^{(ß)}(\xi) \cdot e^{+jk_{\mu z}z} \\ Z_0 v_{rz}(\xi,z) &= -\cos\Theta_\mu P_e \cdot q_\mu^{(ß)}(\xi) \cdot e^{+jk_{\mu z}z} \end{aligned} \tag{3}$$

$$\begin{aligned} p_s(\xi,z) &= \sum_\nu P_{s\nu} \cdot q_\nu^{(ß)}(\xi) \cdot e^{+jk_{\nu z}z} \\ Z_0 v_{sz}(\xi,z) &= -\sum_\nu \cos\Theta_\nu P_{s\nu} \cdot q_\nu^{(ß)}(\xi) \cdot e^{+jk_{\nu z}z} \end{aligned} \tag{4}$$

$$p_t(\xi,z)=\sum_\nu P_{t\nu}\cdot q_\nu^{(\beta)}(\xi)\cdot e^{-jk_{\nu z}(z-d)}$$
$$Z_0 v_{tz}(\xi,z)=\sum_\nu \cos\Theta_\nu P_{t\nu}\cdot q_\nu^{(\beta)}(\xi)\cdot e^{-jk_{\nu z}(z-d)} \tag{5}$$
$$V(\xi)=\sum_n V_n\cdot v_n^{(\beta)}(\xi)$$

with the duct mode lateral profiles for symmetrical (ß=s) and anti-symmetrical (ß=a) modes:

$$q_\nu^{(\beta)}(\xi)=\begin{cases}\cos(\kappa_\nu^{(s)}\xi) & \kappa_\nu^{(s)}=\nu_e\frac{\pi}{2}\ ;\ \nu_e=0,2,4,\ldots\\ \sin(\kappa_\nu^{(a)}\xi) & \kappa_\nu^{(a)}=\nu_o\frac{\pi}{2}\ ;\ \nu_o=1,3,5,\ldots\end{cases} \tag{6}$$

having the duct mode norms :

$$N_\nu^{(\beta)}=\int_{-1}^{+1}[q_\nu^{(\beta)}(\xi)]^2\,d\xi=1\pm\frac{\sin(2\kappa_\nu^{(\beta)})}{2\kappa_\nu^{(\beta)}}=\frac{2}{\delta_\nu}\begin{cases}\text{symm.}\\ \text{anti-symm.}\end{cases}\ ;\quad \delta_i=\begin{cases}1\ ;\ i=0\\ 2\ ;\ i>0\end{cases} \tag{7}$$

Setting $k_{\nu x}^{(\beta)}=\kappa_\nu^{(\beta)}/h$, one can introduce duct mode angles $\Theta_\nu^{(\beta)}$, using the wave equation, by :

$$k_0^2=[k_{\nu x}^{(\beta)}]^2+[k_{\nu z}^{(\beta)}]^2=k_0^2[\sin^2\Theta_\nu^{(\beta)}+\cos^2\Theta_\nu^{(\beta)}]$$
$$k_{\nu x}^{(\beta)}=k_0\sin\Theta_\nu^{(\beta)}\ ;\ k_{\nu z}^{(\beta)}=k_0\cos\Theta_\nu^{(\beta)}=k_0\sqrt{1-\sin^2\Theta_\nu^{(\beta)}}\ ;\ \begin{cases}\mathrm{Re}\{\sqrt{\ldots}\}\ge 0\\ \mathrm{Im}\{\sqrt{\ldots}\}\ge 0\end{cases} \tag{8}$$

They represent the angle with the z axis of plane waves, from which the duct modes can be composed.

Coupling coefficients between duct modes and plate modes : $S_{\nu n}^{(\beta)}=\int_{-1}^{+1}q_\nu^{(\beta)}(\xi)\cdot v_n^{(\beta)}(\xi)\,d\xi$ (9)

The remaining boundary conditions of zone matching are :

$$P_{s\nu}=-P_{t\nu}$$
$$\sum_\nu P_{t\nu}\cos\Theta_\nu\cdot q_\nu(\xi)=\sum_n Z_0 V_n\cdot v_n(\xi) \tag{10}$$
$$2\sum_\nu[\delta_{\mu\nu}P_e-P_{t\nu}]\cdot q_\nu(\xi)=\sum_n Z_{Tn}V_n\cdot v_n(\xi)$$

They lead to the system of equations for Z_0V_n :

$$\sum_{n(\beta_\mu)} Z_0V_n\cdot\left[\delta_{mn}N_{Pm}\frac{Z_{Tm}}{Z_0}+\sum_{\nu(\beta_\nu)}\delta_\nu\frac{S_{\nu m}S_{\nu n}}{\cos\Theta_\nu}\right]=2S_{\mu m}\cdot P_e \tag{11}$$

$n(\text{ß}_\mu)$, $\nu(\text{ß}_\mu)$ means: the summation over n and m are in the range of indices belonging to the symmetry type of the incident duct mode, and ν is in the range of plate modes for the type of plate fixation used. The transmitted duct mode amplitudes $P_{t\nu}$ are evaluated with:

$$P_{t\nu} = \frac{\delta_\nu}{2\cos\Theta_\nu} \sum_{n(\text{ß}_\mu)} S_{\nu n} \cdot Z_0 V_n \tag{12}$$

and the scattered duct mode amplitudes $P_{s\nu}$ from the 1st boundary condition. The sound transmission coefficient τ_μ for the incident μ-th duct mode is :

$$\tau_\mu = \frac{\delta_\mu}{4\cos\Theta_\mu} \sum_{\nu(\text{ß}_\mu)}^{\nu_{gr}} \frac{\delta_\nu}{\cos\Theta_\nu} \cdot \left| \sum_{n(\text{ß}_\mu)}^{n_{ob}} S_{\nu n} \frac{Z_0 V_n}{P_e} \right|^2 \tag{13}$$

The range of duct mode indices used is the range of cut-on modes, with the limit :

$$n_{con} = 1 - \text{ß}/2 + \frac{k_0 a}{2\pi} \tag{14}$$

(ß=0 for symmetrical duct modes; ß=1 for anti-symmetrical duct modes)

Simply supported plate :

$$\begin{aligned} S_{\nu n}^{(s)} &= \int_{-1}^{+1} \cos(\nu_e \frac{\pi}{2}\xi) \cdot \cos(n_o \frac{\pi}{2}\xi)\, d\xi \\ &= \frac{\sin(\nu_e - n_o)\pi/2}{(\nu_e - n_o)\pi/2} + \frac{\sin(\nu_e + n_o)\pi/2}{(\nu_e + n_o)\pi/2} = \frac{4}{\pi}(-1)^{(\nu_e + n_o - 1)/2} \frac{n_o}{n_o^2 - \nu_e^2} \\ S_{\nu n}^{(a)} &= \int_{-1}^{+1} \sin(\nu_o \frac{\pi}{2}\xi) \cdot \sin(n_e \frac{\pi}{2}\xi)\, d\xi \\ &= \frac{\sin(\nu_o - n_e)\pi/2}{(\nu_o - n_e)\pi/2} - \frac{\sin(\nu_o + n_e)\pi/2}{(\nu_o + n_e)\pi/2} = \frac{4}{\pi}(-1)^{(n_e + \nu_o - 1)/2} \frac{n_e}{\nu_o^2 - n_e^2} \end{aligned} \tag{15}$$

Clamped plate :

$$S_{\nu n}^{(\text{ß})} = \int_{-1}^{+1} \left\{ \begin{matrix} \cos(\nu_e \frac{\pi}{2}\xi) \\ \sin(\nu_o \frac{\pi}{2}\xi) \end{matrix} \right\} \cdot v_n^{(\text{ß})}(\gamma_n^{(\text{ß})}\xi)\, d\xi \quad ; \quad \left\{ \begin{matrix} (\text{ß}) = (s) \; ; \; \nu = \nu_e = 0,2,4,\ldots \\ (\text{ß}) = (a) \; ; \; \nu = \nu_o = 1,3,5,\ldots \end{matrix} \right. \tag{16}$$

$$S_{\nu n}^{(\beta)} = \begin{cases} 8(-1)^{\nu_e/2}\gamma_n^{(s)}\left(\dfrac{\sin\gamma_n^{(s)}}{16(\gamma_n^{(s)})^4-(\nu_e\pi)^4}+C_n^{(s)}\dfrac{\sinh\gamma_n^{(s)}}{16(\gamma_n^{(s)})^4+(\nu_e\pi)^4}\right) \\ -8(-1)^{(\nu_o-1)/2}\gamma_n^{(a)}\left(\dfrac{\cos\gamma_n^{(a)}}{16(\gamma_n^{(a)})^4-(\nu_o\pi)^4}-C_n^{(a)}\dfrac{\cosh\gamma_n^{(a)}}{16(\gamma_n^{(a)})^4+(\nu_o\pi)^4}\right) \end{cases}$$

$$= \begin{cases} 64(-1)^{\nu_e/2}\dfrac{(\gamma_n^{(s)})^3\sin\gamma_n^{(s)}}{16(\gamma_n^{(s)})^4-(\nu_e\pi)^4} \\ -64(-1)^{(\nu_o-1)/2}\dfrac{(\gamma_n^{(a)})^3\cos\gamma_n^{(a)}}{16(\gamma_n^{(a)})^4-(\nu_o\pi)^4} \end{cases} \qquad (17)$$

Free plate :

$$S_{\nu n}^{(\beta)} = \begin{cases} 16(-1)^{\nu_e/2}(\nu_e\pi)^2\dfrac{\gamma_n^{(s)}\sin\gamma_n^{(s)}}{16(\gamma_n^{(s)})^4-(\nu_e\pi)^4} \\ -16(-1)^{(\nu_o-1)/2}(\nu_o\pi)^2\dfrac{\gamma_n^{(a)}\cos\gamma_n^{(a)}}{16(\gamma_n^{(a)})^4-(\nu_o\pi)^4} \end{cases} \qquad (18)$$

Coupling coefficients of polynomial plate modes :

$$S_{\nu_e 0}^{(s)} = 2(1+C_0^{(s)})\frac{\sin(\nu_e\pi/2)}{\nu_e\pi/2} = \begin{cases} 2(1+C_0^{(s)})=4 & ; \ \nu_e=0 \\ 0 & ; \ \nu_e>0 \end{cases}$$

$$S_{\nu_o 0}^{(a)} = C_0^{(a)}\frac{8}{\nu_o\pi}(-1)^{(\nu_o-1)/2} = \frac{8}{\nu_o\pi}(-1)^{(\nu_o-1)/2} \qquad (19)$$

Relation between coupling coefficients for free and clamped plates :

$$S_{\nu n,\,free}^{(\beta)} = \frac{\pi^2}{4}\left(\frac{\nu^{(\beta)}}{\gamma_n^{(\beta)}}\right)^2 \cdot S_{\nu n,\,clamped}^{(\beta)} \approx \left(\frac{\nu^{(\beta)}/2}{n\mp 1/4}\right)^2 \cdot S_{\nu n,\,clamped}^{(\beta)} \qquad (20)$$

Mixture of incident duct modes :
(written for a 3-dimensional, rectangular duct with mode angles $\Theta_{m,n}$ and mode amplitudes $A_{m,n}$ of incident modes, with an arbitrary reference pressure p_0)

Sound transmission coefficient τ for an arbitrary mixture of incident duct modes, each of which has a modal transmission coefficient $\tau(\Theta_{m,n})$:

$$\tau = \frac{\sum_{m,n} \frac{\tau(\Theta_{mn})}{\delta_m \delta_n} \cdot \left| \frac{A_{mn}}{p_0} \right|^2 \cdot \cos\Theta_{mn}}{\sum_{m,n} \frac{1}{\delta_m \delta_n} \cdot \left| \frac{A_{mn}}{p_0} \right|^2 \cdot \cos\Theta_{mn}} \quad ; \quad \delta_m = \begin{cases} 1 \,;\, m = 0 \\ 2 \,;\, m > 0 \end{cases} \tag{21}$$

In the special case that all incident duct modes have the same energy density (a model that corresponds best to the diffuse sound incidence of room acoustics) :

$$\left| \frac{A_{mn}}{p_0} \right|^2 = \frac{\delta_m \delta_n}{\sum_{m,n} \mathrm{Re}\{\cos\Theta_{mn}\}} \tag{22}$$

Therefore :

$$\tau = \frac{\sum_{m,n} \tau(\Theta_{mn}) \cos\Theta_{mn}}{\sum_{m,n} \cos\Theta_{mn}} \tag{23}$$

(this is, up to the discretisation of the angle of incidence, the relation for diffuse sound incidence).

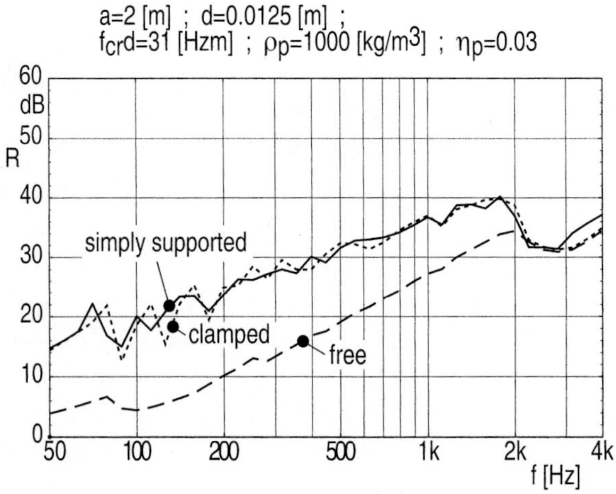

Sound transmission loss R of a plaster board across a duct, a= 2 [m] wide, for all propagating duct modes incident with same energy density, and three different boundary fixations of the plate.

I.16 Single plate in a wall niche

Ref.: Mechel, [I.8]

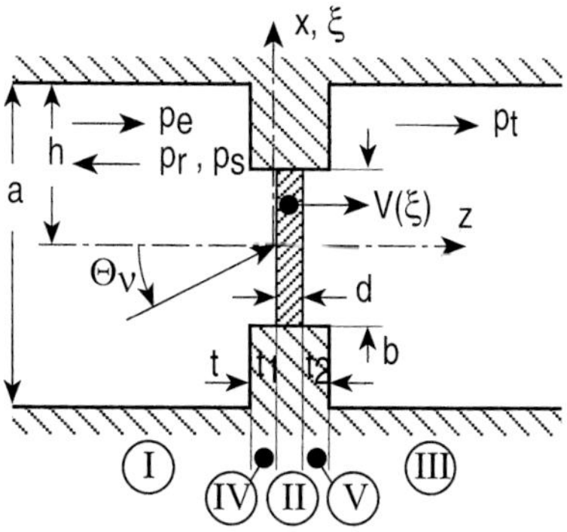

The influence of mounting a wall in a niche of the partition wall (baffle wall) between an emission and a receiving room has played for a while as a "niche effect" some rolc in the discussion of sound transmission tests.

The object and the sound field formulations here are similar to those in the previous section I.15 , except for the new fields in the niches with depths t_1, t_2 , which are formulated as sums of "niche modes". The plate and the niche have the width $b = 2h'$.

The niche is centred in the baffle wall. The baffle wall is hard and rigid. Only a freely supported plate will be considered below (most parts of the formulas below can be used for other types of fixation also; then the pertinent values for γ_n and $S^{(\beta)}_{i,n}$ must be used). The incident wave p_e is the μ-th (propagating) duct mode (symmetrical or anti-symmetrical). Because of the central position of the niche, all other modes have the same symmetry as the incident mode.

The field on the front side is composed as : $p_I = p_e + p_r + p_s$ (1)

with: $p_e =$ incident duct mode;

$p_r =$ incident duct mode after hard reflection;

$p_s =$ scattered field.

p_r, p_s and the transmitted wave p_t are formulated as sums of duct modes. The plate vibration is formulated as a sum of plate modes. So far the field formulations are similar to those in section I.15 ; here additional mode sums of "niche modes" will be used for the fields p_1, p_2 in the front and back niche, respectively.

Non-dimensional lateral co-ordinates : $\xi = x/h \,;\, \xi' = x/b = x/(2h')$ (2)

Field formulations :

Zone I: (emission side)

$$\begin{aligned} p_e(\xi,z) &= P_e \cdot q_\mu(\xi)\cdot e^{-jk_{\mu z}z} \\ Z_0 v_{ez}(\xi,z) &= \cos\Theta_\mu\, P_e \cdot q_\mu(\xi)\cdot e^{-jk_{\mu z}z} \end{aligned} \tag{3}$$

$$\begin{aligned} p_r(\xi,z) &= P_e \cdot q_\mu(\xi)\cdot e^{+jk_{\mu z}z} \\ Z_0 v_{rz}(\xi,z) &= -\cos\Theta_\mu\, P_e \cdot q_\mu(\xi)\cdot e^{+jk_{\mu z}z} \end{aligned} \tag{4}$$

$$\begin{aligned} p_s(\xi,z) &= \sum_\nu P_{s\nu} \cdot q_\nu(\xi)\cdot e^{+jk_{\nu z}z} \\ Z_0 v_{sz}(\xi,z) &= -\sum_\nu \cos\Theta_\nu\, P_{s\nu} \cdot q_\nu(\xi)\cdot e^{+jk_{\nu z}z} \end{aligned} \tag{5}$$

Zone III: (transmission side) :

$$\begin{aligned} p_t(\xi,z) &= \sum_\nu P_{t\nu} \cdot q_\nu(\xi)\cdot e^{-jk_{\nu z}(z-t)} \\ Z_0 v_{tz}(\xi,z) &= \sum_\nu \cos\Theta_\nu\, P_{t\nu} \cdot q_\nu(\xi)\cdot e^{-jk_{\nu z}(z-t)} \end{aligned} \tag{6}$$

Zone IV: (front side niche) :

$$\begin{aligned} p_1(\xi',z) &= \sum_i \left(A_i\, e^{-jg_{iz}(z-t_1)} + B_i\, e^{+jg_{iz}(z-t_1)}\right)\cdot \varphi_i^{(1)}(\xi') \\ Z_0 v_{1z}(\xi',z) &= \sum_i \cos\Phi_i \left(A_i\, e^{-jg_{iz}(z-t_1)} - B_i\, e^{+jg_{iz}(z-t_1)}\right)\cdot \varphi_i^{(1)}(\xi') \end{aligned} \tag{7}$$

Zone V:
(back side niche, if its width h'' is different from that of the front side niche; $\xi''= x/h''$) :

$$\begin{aligned} p_2(\xi'',z) &= \sum_i \left(C_i\, e^{-jg_{iz}^{(2)}(z-t_1-d)} + D_i\, e^{+jg_{iz}^{(2)}(z-t_1-d)}\right)\cdot \varphi_i^{(2)}(\xi'') \\ Z_0 v_{2z}(\xi'',z) &= \sum_i \cos\Phi_i^{(2)} \left(C_i\, e^{-jg_{iz}^{(2)}(z-t_1-d)} - D_i\, e^{+jg_{iz}^{(2)}(z-t_1-d)}\right)\cdot \varphi_i^{(2)}(\xi'') \end{aligned} \tag{8}$$

For $h'=h''$ is $\xi''=\xi'$ and the upper niche indices (1),(2) are not needed.

Zone II (plate) :

$$Z_0 V(\xi') = \sum_n Z_0 V_n \cdot v_n(\xi') \tag{9}$$

Mode profiles :

Duct modes :

$$q_\nu^{(\beta)}(\xi) = \begin{cases} \cos(\kappa_\nu^{(s)}\xi)\,;\,\text{symmetrical} & \kappa_\nu^{(s)} = \nu_e \dfrac{\pi}{2} \quad ; \quad \nu_e = 0, 2, 4, \ldots \\ \sin(\kappa_\nu^{(a)}\xi)\,;\,\text{anti-symmetrical} & \kappa_\nu^{(a)} = \nu_o \dfrac{\pi}{2} \quad ; \quad \nu_o = 1, 3, 5, \ldots \end{cases} \tag{10}$$

having the duct mode norms :

$$N_{K\nu}^{(\beta)} = \int_{-1}^{+1} [q_\nu^{(\beta)}(\xi)]^2\, d\xi = 1 \pm \frac{\sin(2\kappa_\nu^{(\beta)})}{2\kappa_\nu^{(\beta)}} = \frac{2}{\delta_\nu} \begin{cases} \text{symm.} \\ \text{anti-symm.} \end{cases} \quad ; \quad \delta_n = \begin{cases} 1 \;;\; n=0 \\ 2 \;;\; n>0 \end{cases} \tag{11}$$

Niche modes :

$$\varphi_i(\xi') = \begin{cases} \cos(i_e \dfrac{\pi}{2}\xi') \;;\; \beta = s = 0 \;;\; i_e = 0,2,4\ldots \\ \sin(i_o \dfrac{\pi}{2}\xi') \;;\; \beta = a = 1 \;;\; i_o = 1,3,5\ldots \end{cases} \tag{12}$$

with axial wave numbers and mode angles :

$$g_{iz} = k_0 \cos\Phi_i = k_0\sqrt{1-\sin^2\Phi_i} \quad ; \quad \begin{cases} \mathrm{Re}\sqrt{\ldots} \ge 0 \\ \mathrm{Im}\sqrt{\ldots} \le 0 \end{cases} \quad ; \quad \sin\Phi_i = i^{(\beta)}\frac{\lambda_0}{2b} \tag{13}$$

having niche mode norms :

$$N_{Ni}^{(\beta)} = \frac{2}{\delta_i} \tag{14}$$

Plate modes :

$$v_n^{(\beta)}(\xi) = \begin{cases} \cos(\gamma_n^{(s)}\xi) \;;\; \gamma_n^{(s)} = n_o \dfrac{\pi}{2} \;;\; n_o = 1,3,5,\ldots \\ \sin(\gamma_n^{(a)}\xi) \;;\; \gamma_n^{(a)} = n_e \dfrac{\pi}{2} \;;\; n_e = 2,4,6,\ldots \end{cases} \tag{15}$$

having plate mode norms :

$$N_{Pn}^{(\beta)} = \int_{-1}^{+1} [v_n^{(\beta)}(\xi)]^2\, d\xi = 1 \tag{16}$$

Auxiliary amplitudes :

$$X_{i\pm} = A_i \pm B_i \qquad X_{i+} = \frac{j}{\tan(g_{iz}t_1)}\cdot X_{i-} - \frac{j}{\sin(g_{iz}t_1)}\cdot Y_{i-}$$
$$Y_{i\pm} = A_i\, e^{+jg_{iz}t_1} \pm B_i\, e^{-jg_{iz}t_1} \qquad Y_{i+} = \frac{j}{\sin(g_{iz}t_1)}\cdot X_{i-} - \frac{j}{\tan(g_{iz}t_1)}\cdot Y_{i-} \tag{17}$$

$$U_{i\pm} = C_i \pm D_i \qquad U_{i+} = \frac{-j}{\tan(g_{iz}t_2)}\cdot U_{i-} + \frac{j}{\sin(g_{iz}t_2)}\cdot W_{i-}$$
$$W_{i\pm} = C_i\, e^{-jg_{iz}t_2} \pm D_i\, e^{+jg_{iz}t_2} \qquad W_{i+} = \frac{-j}{\sin(g_{iz}t_2)}\cdot U_{i-} + \frac{j}{\tan(g_{iz}t_2)}\cdot W_{i-} \tag{18}$$

The boundary conditions of field matching at the zone limits lead to two coupled systems of equations for Y_{1-}, W_{1-} :

$$\begin{aligned} &\{M_{11}\}\circ\{Y_{\iota-}\}+\{M_{12}\}\circ\{W_{\iota-}\}=2P_e\cdot\{Q_{\mu i}\}_\mu \\ &\{M_{21}\}\circ\{Y_{\iota-}\}+\{M_{22}\}\circ\{W_{\iota-}\}=0 \end{aligned} \tag{19}$$

with matrices :

$$\begin{aligned} &\{M_{11}\}=-j\,\{I_{ii}\}*\{N_{Ni}\cos(g_{iz}t_1)\}+\frac{b}{a}\{\sin(g_{iz}t_1)\}*\{T_{\nu i}\}^t\circ\{T_{\nu\iota}\}*\left\{\frac{\cos\Phi_\iota}{N_{K\nu}\cos\Theta_\nu}\right\} \\ &\{M_{12}\}=+j\,\{I_{ii}\}*\{N_{Ni}\cos(g_{iz}t_2)\}-\frac{b}{a}\{\sin(g_{iz}t_2)\}*\{T_{\nu i}\}^t\circ\{T_{\nu\iota}\}*\left\{\frac{\cos\Phi_\iota}{N_{K\nu}\cos\Theta_\nu}\right\} \\ &\{M_{21}\}=j\{I_{ii}\}*\left\{\frac{N_{Ni}}{\sin(g_{iz}t_1)}\right\} \\ &\{M_{22}\}=j\{I_{ii}\}*\left\{\frac{N_{Ni}}{\sin(g_{iz}t_2)}\right\}-\{G_{ii'}\}\circ\{H_{i'\iota}\} \end{aligned} \tag{20}$$

and right-side vector :

$$\{Q_{\mu i}\}_\mu=\{\sin(g_{iz}t_1)T_{\mu i}\}_\mu \tag{21}$$

where $\circ$ indicates a matrix multiplication; $\{a_m\}*\{c_m\}=\{a_m c_m\}$ indicates a term-wise multiplication of two vectors, and correspondingly $\{c_m x_{mn}\}=\{c_m\}*\{x_{mn}\}$ indicates the multiplication of the m-th row of the matrix $\{x_{mn}\}$ with the element c_m of the vector $\{c_m\}$, and $\{d_n x_{mn}\}=\{d_n\}*\{x_{mn}\}^t\,\{\{d_n\}*\{x_{mn}\}^t\}^t$ indicates the multiplication of the n-th column of $\{x_{mn}\}$ with d_n ; $\{x_{mn}\}^t$ is the transposed matrix; $\{I_{ii}\}$ is the unit matrix.

The coupling coefficients $T_{\nu i}=T_{\nu i}^{(ß)}$ between duct modes and niche modes (where the 2nd form indicates the symmetry type ß= s,a) are :

$$\begin{aligned} T_{\nu i}&=\int_{-1}^{+1} q_\nu\!\left(\frac{b}{a}\xi'\right)\cdot\varphi_i(\xi')\,d\xi' \\ T_{\nu i}^{(ß)}&=\int_{-1}^{+1}\begin{matrix}\cos(\nu_e\pi/2\cdot b/a\cdot\xi')\cdot\cos(i_e\pi/2\cdot\xi')\\ \sin(\nu_o\pi/2\cdot b/a\cdot\xi')\cdot\sin(i_o\pi/2\cdot\xi')\end{matrix}\cdot d\xi'\;;\quad ß=\begin{cases}s\\a\end{cases} \\ &=\frac{\sin((i-\nu b/a)\pi/2)}{(i-\nu b/a)\pi/2}\pm\frac{\sin((i+\nu b/a)\pi/2)}{(i+\nu b/a)\pi/2}\;;\quad ß=\begin{cases}s\;;\;\nu_e,i_e\\a\;;\;\nu_o,i_o\end{cases} \end{aligned} \tag{22}$$

Abbreviation used :

$$\{G_{i\iota}\} := \left\{ j\{I_{ii}\} * \left\{ N_{Ni} \frac{\sin(g_{iz}(t_1+t_2))}{\sin(g_{iz}t_1)\cdot\sin(g_{iz}t_2)} \right\} - \{S_{in}\} \circ \{S_{\iota n}\}^t * \left\{ \frac{Z_{Tn}\cdot \mathrm{os}\,\Phi_\iota}{N_{Pn}} \right\} \right\} \tag{23}$$

with which :

$$\{G_{i\iota}\} \circ \{U_{\iota -}\} = j \left\{ \frac{N_{Ni} Y_{i-}}{\sin(g_{iz}t_1)} + \frac{N_{Ni} W_{i-}}{\sin(g_{iz}t_2)} \right\} \tag{24}$$

The symbol Z_{Tn} stands for the n-th modal partition impedance of the plate (see section I.15).

With the solutions Y_{1-}, W_{1-} the other mode amplitudes in the duct are evaluated from :

$$P_{s\nu} = \frac{-b/a}{N_{K\nu}\cos\Theta_\nu} \sum_i \cos\Phi_i T_{\nu i} \cdot Y_{i-}$$
$$P_{t\nu} = \frac{b/a}{N_{K\nu}\cos\Theta_\nu} \sum_i \cos\Phi_i T_{\nu i} \cdot W_{i-} \tag{25}$$

If one is only interested in the $P_{t\nu}$, a simplified system of equations is :

$$\left\{ -\{M_{11}\} \circ \{M_{21}\}^{-1} \circ \{M_{22}\} + \{M_{12}\} \right\} \circ \{W_{\iota -}\} = 2P_e \cdot \{Q_{\mu i}\}_\mu \tag{26}$$

The transmission coefficient τ_μ for a single incident (propagating) duct mode is :

$$\tau_\mu = \frac{\sum_\nu \Pi'_{t\nu}}{b/a\cdot \Pi'_{e\mu}} = \frac{1}{b/a\cdot N_{K\mu}\cos\Theta_\mu} \sum_\nu N_{K\nu}\cos\Theta_\nu \left| \frac{P_{t\nu}}{P_e} \right|^2$$
$$= \frac{b/a}{N_{K\mu}\cos\Theta_\mu} \sum_\nu \frac{1}{N_{K\nu}\cos\Theta_\nu} \left| \{T_{\nu i}\} \circ \left\{ \cos\Phi_i \frac{W_{i-}}{P_e} \right\} \right|^2 \tag{27}$$
$$= \frac{b/a}{N_{K\mu}\cos\Theta_\mu} \sum_\nu \frac{1}{N_{K\nu}\cos\Theta_\nu} \left| \sum_i T_{\nu i} \cos\Phi_i \frac{W_{i-}}{P_e} \right|^2$$

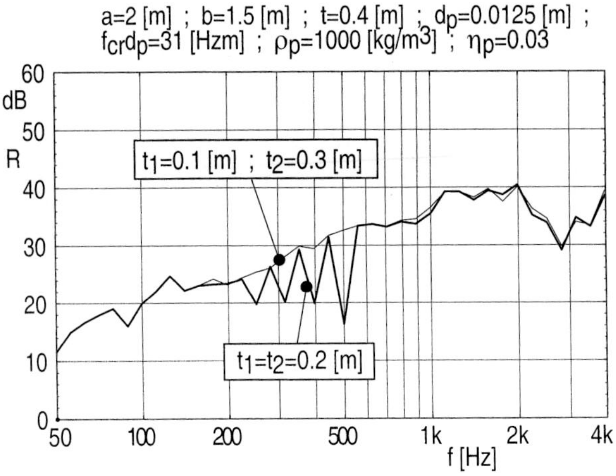

Transmission loss R through a plaster board plate in a niche in the partition wall between test rooms. All propagating modes of the emission side duct are incident with equal energy density. The example shows the singularity of a central position of the test object in a niche.

I.17 Strip-shaped wall in infinite baffle wall

Ref.: Mechel, [I.9]

A wall of thickness d and width $a=2c$ is placed in a hard baffle wall, also of thickness d.

Elliptic-hyperbolic cylindrical systems of co-ordinates ρ,ϑ are used with focus positions at $x=c$.

A plane wave p_e is assumed to be incident at a polar angle Θ ; it becomes p_r after hard reflection at the front side surface. An additional scattered wave p_s is needed to formulate the front side sound field :

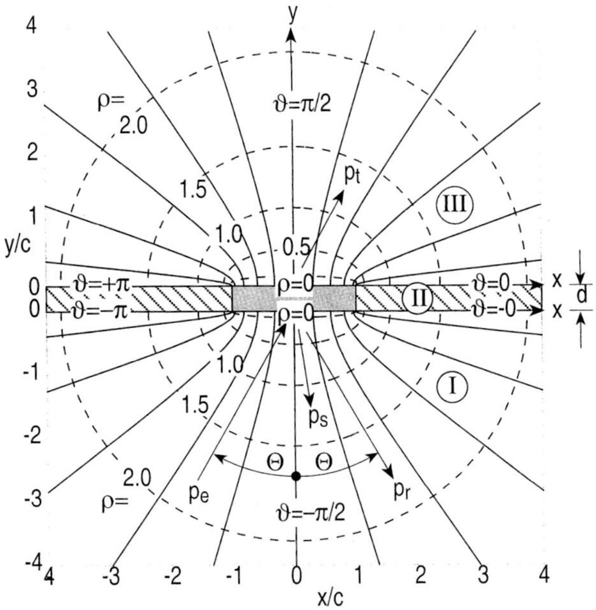

$$p_I = p_e + p_r + p_s \tag{1}$$

The transmitted field is p_t . Both p_s and p_t are formulated as sums of Mathieu functions. (see [D.3] for notations, formulas and generation of Mathieu functions). The velocity pattern of the plate is formulated as a sum of plate modes in the normalised co-ordinate $\xi = x/c$.

The sound transmitting wall here is assumed to be a simply supported single plate. Other types of walls are treated correspondingly.

Transformation between Cartesian and elliptic-hyperbolic co-ordinates :

$$\left.\begin{array}{l} x = c\cdot\cosh\rho\cdot\cos\vartheta \\ z = c\cdot\sinh\rho\cdot\sin\vartheta \end{array}\right\} \quad ; \quad 0 \le \rho < \infty \quad ; \quad -\pi \le \vartheta \le +\pi \tag{2}$$

Field formulations :

$$p_e(\rho,\vartheta) + p_r(\rho,\vartheta) = 4P_e \sum_{m=0}^{\infty} (-j)^m\, ce_m(\alpha)\cdot Jc_m(\rho)\cdot ce_m(\vartheta) \tag{3}$$

$$p_t(\rho,\vartheta) = -p_s(\rho,\vartheta) = 2\sum_{m=0}^{\infty} D_m (-j)^m\, ce_m(\alpha)\cdot Hc_m^{(2)}(\rho)\cdot ce_m(\vartheta) \tag{4}$$

$$\left.\begin{array}{l} Z_0 v_{t\rho}(0,\vartheta>0) \\ Z_0 v_{s\rho}(0,\vartheta<0) \end{array}\right\} = \frac{\pm j}{\beta\,\sin\vartheta} \sum_{m=0} D_m (-j)^m\, ce_m(\alpha)\; Hc_m'^{(2)}(0)\cdot ce_m(\vartheta) \tag{5}$$

where : P_e= amplitude of incident plane wave ; $\alpha = \pi/2-\Theta$; $\beta = k_0 c/2$; $ce_m(\vartheta)$= even azimuthal Mathieu functions ; $Hc_m^{(2)}(\rho)$= radial Hankel-Mathieu functions of the 2nd kind (associated with the $ce_m(\vartheta)$) ; D_m= mode amplitudes.

Plate velocity :

$$V(\xi) = \sum_n V_n\cdot v_n^{(s)}(\xi) \tag{6}$$

with (symmetrical , σ=s, and anti-symmetrical , σ=a) plate modes :

$$v_n^{(\sigma)}(\xi) = \begin{cases} \cos(\gamma_n^{(s)}\xi) \;;\; \gamma_n^{(s)} = n_o\pi/2 \;;\; n_o = 1,3,5,\ldots \;;\; (\sigma)=(s)=\text{symmetrical} \\ \sin(\gamma_n^{(a)}\xi) \;;\; \gamma_n^{(a)} = n_e\pi/2 \;;\; n_e = 2,4,6,\ldots \;;\; (\sigma)=(a)=\text{anti-symmetrical} \end{cases} \tag{7}$$

having mode norms : $$N_{Pn} = \int_{-1}^{+1} v_n^{(\sigma)2}(\xi)\, d\xi = 1 \tag{8}$$

and modal plate partition impedances :

$$\frac{Z_{Tn}}{Z_0} = 2\pi Z_m F\left[\eta F^2\left(\frac{\gamma_n}{k_0 c}\right)^4 + j\left(1-F^2\left(\frac{\gamma_n}{k_0 c}\right)^4\right)\right] \quad ; \quad F = \frac{f}{f_{cr}} \quad ; \quad Z_m = \frac{f_{cr} m}{Z_0} = \frac{f_{cr} d}{Z_0}\rho_P \tag{9}$$

where: f= frequency ; f_{cr}= critical frequency ; m=d·ρp= plate surface mass density ; η= plate loss factor.

The boundary conditions give a linear system of equations for the plate mode amplitudes V_n :

$$\sum_n Z_0 V_n \left[\delta_{n,\nu} \cdot N_{P\nu} \frac{Z_{T\nu}}{Z_0} + \frac{8\beta}{j\pi} \sum_{m=0} \frac{Hc_m^{(2)}(0)}{Hc_m'^{(2)}(0)} Q_{m,\nu} Q_{m,n} \right] = \tag{10}$$
$$= 4 P_e \sum_{m=0} (-j)^m ce_m(\alpha)\, Jc_m(0)\, Q_{m,\nu} \quad ; \quad \nu = 1,2,3,\ldots$$

and with its solutions the mode amplitudes D_m of the transmitted wave :

$$D_m = \frac{2\beta (j)^{m-1}}{\pi\, ce_m(\alpha) Hc_m'^{(2)}(0)} \sum_n Q_{m,n} \cdot Z_0 V_n \tag{11}$$

Therein: $\delta_{n,\nu}$= Kronecker symbol, and mode coupling coefficients :

$$Q_{m,n} := \int_{-1}^{+1} ce_m(\arccos\xi; \beta^2) \cdot v_n^{(\sigma)}(\gamma_n \xi)\, d\xi = \int_0^{\pi} \sin\vartheta \cdot ce_m(\vartheta; \beta^2) \cdot v_n^{(\sigma)}(\gamma_n \cos\vartheta)\, d\vartheta \tag{12}$$

which are evaluated for symmetrical plate modes, for which m=2r , by : (13)

$$Q_{2r,n} = \sqrt{\frac{2\pi}{\gamma_n}} \sum_{s=0} A_{2s} \left[J_{1/2}(\gamma_n) + (1-\delta_{0,s}) \sum_{i=1}^{s} (-1)^i \frac{i!}{(2i)!} \left(\frac{2}{\gamma_n}\right)^i \cdot J_{i+1/2}(\gamma_n) \prod_{k=0}^{i-1} \left(4s^2 - 4k^2\right) \right]$$

and for anti-symmetrical plate modes, for which m=2r+1 : (14)

$$Q_{2r+1,n} =$$
$$\sqrt{\frac{2\pi}{\gamma_n}} \sum_{s=0} A_{2s+1} \left[J_{3/2}(\gamma_n) + (1-\delta_{0,s}) \sum_{i=1}^{s} (-1)^i \frac{i!}{(2i)!} \prod_{k=1}^{i} \left((2s+1)^2 - (2k-1)^2\right) \left(\frac{2}{\gamma_n}\right)^i \cdot J_{i+3/2}(\gamma_n) \right]$$

Therein $J_n(z)$ are Bessel functions, and A_n are the Fourier series components needed for the evaluation of the azimuthal Mathieu functions.

The sound transmission coefficient $\tau(\Theta)$ for oblique incidence finally is :

$$\tau(\Theta) = \frac{1}{\beta\cos\Theta} \sum_{m=0} \left| \frac{D_m}{P_e} \right|^2 ce_m^2(\pi/2 - \Theta) \tag{15}$$

and for diffuse sound incidence (two-dimensional) :

$$\tau_{2-dif} = \frac{2}{\beta} \int_{-\pi/2}^{+\pi/2} \tau(\Theta) \cos\Theta\, d\Theta \tag{16}$$

I.18 Finite size plate with a front side absorber layer

Ref.: Mechel, [I.10]

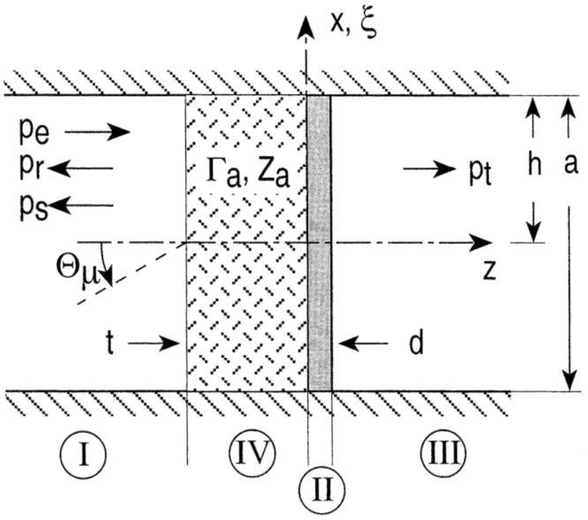

A simply supported plate of thickness d with a porous layer of thickness t and characteristic values Γ_a, Z_a of the material on its front side is placed across a flat duct with lateral dimension $a=2h$. There is no (or only loose) mechanical contact between the plate and the layer.

The μ-th (propagating) duct mode is assumed to be the incident wave; the index ß= s,a indicates whether the incident duct mode is symmetrical or anti-symmetrical. All other fields are of the same symmetry type. The field p_I on the front side is composed as $p_I = p_e + p_r + p_s$ of the incident wave p_e, of its hard reflection p_r (at x=0), and of a scattered wave. The scattered wave p_s and the transmitted wave p_t, as well as the field p_a in the absorber layer are formulated as duct mode sums. The plate vibration $V(\xi)$ is a plate mode sum with $\xi = x/h$.

Field formulations :

$$p_e(\xi,z) = P_e \cdot q_\mu^{(ß)}(\xi) \cdot e^{-jk_{\mu z}z}$$
$$Z_0 v_{ez}(\xi,z) = \cos\Theta_\mu P_e \cdot q_\mu^{(ß)}(\xi) \cdot e^{-jk_{\mu z}z} \quad ; \tag{1}$$

$$p_r(\xi,z) = P_e \cdot q_\mu^{(ß)}(\xi) \cdot e^{+jk_{\mu z}z}$$
$$Z_0 v_{rz}(\xi,z) = -\cos\Theta_\mu P_e \cdot q_\mu^{(ß)}(\xi) \cdot e^{+jk_{\mu z}z} \quad ; \tag{2}$$

$$p_s(\xi,z) = \sum_\nu P_{s\nu} \cdot q_\nu^{(ß)}(\xi) \cdot e^{+jk_{\nu z}z}$$
$$Z_0 v_{sz}(\xi,z) = -\sum_\nu \cos\Theta_\nu P_{s\nu} \cdot q_\nu^{(ß)}(\xi) \cdot e^{+jk_{\nu z}z} \quad ; \tag{3}$$

$$p_t(\xi,z) = \sum_\nu P_{t\nu} \cdot q_\nu^{(ß)}(\xi) \cdot e^{-jk_{\nu z}(z-d)}$$
$$Z_0 v_{tz}(\xi,z) = \sum_\nu \cos\Theta_\nu P_{t\nu} \cdot q_\nu^{(ß)}(\xi) \cdot e^{-jk_{\nu z}(z-d)} \quad ; \tag{4}$$

$$V(\xi) = \sum_n V_n \cdot v_n^{(\beta)}(\xi) \; ; \tag{5}$$

$$\begin{aligned} p_a(\xi,z) &= \sum_\nu P_{a\nu} \cdot q_\nu(\xi) \cdot \left[e^{-\Gamma_\nu z} + r_\nu e^{+\Gamma_\nu z} \right] \\ Z_0 v_{az}(\xi,z) &= \sum_\nu \frac{\Gamma_\nu}{\Gamma_a Z_{an}} P_{a\nu} \cdot q_\nu(\xi) \cdot \left[e^{-\Gamma_\nu z} - r_\nu e^{+\Gamma_\nu z} \right] \end{aligned} \tag{6}$$

with :

$$q_\nu^{(\beta)}(\xi) = \begin{cases} \cos(\kappa_\nu^{(s)}\xi) \\ \sin(\kappa_\nu^{(a)}\xi) \end{cases} \; ; \quad \kappa_\nu^{(\beta)} = k_{\nu x}^{(\beta)} h = \left. \begin{matrix} \nu_e \\ \nu_o \end{matrix} \right\} \frac{\pi}{2} \quad ; \quad \begin{cases} \nu_e = 0, 2, 4\ldots \\ \nu_o = 1, 3, 5,\ldots \end{cases}$$

$$\begin{aligned} \frac{k_{\nu z}}{k_0} &= \cos\Theta_\nu = \sqrt{1 - \left(\frac{\nu\pi}{2k_0 h} \right)^2} \quad ; \quad \begin{cases} \mathrm{Re}\{\sqrt{}\} \geq 0 \\ \mathrm{Im}\{\sqrt{}\} \leq 0 \end{cases} \\ \frac{\Gamma_\nu}{k_0} &= \sqrt{\left(\frac{\Gamma_a^2}{k_0} \right) + \left(\frac{\nu\pi}{2k_0 h} \right)^2} \quad ; \quad \mathrm{Re}\{\sqrt{}\} \geq 0 \quad ; \quad N_{K\nu}^{(\beta)} = 2/\delta_\nu \end{aligned} \tag{7}$$

and :

$$v_n^{(\beta)}(\xi) = \begin{cases} \cos(\gamma_n^{(s)}\xi) \quad ; \quad \gamma_n^{(s)} = n_o \dfrac{\pi}{2} \quad ; \quad n_o = 1,3,5,\ldots \\ \sin(\gamma_n^{(a)}\xi) \quad ; \quad \gamma_n^{(a)} = n_e \dfrac{\pi}{2} \quad ; \quad n_e = 2,4,6,\ldots \end{cases} \tag{8}$$

The boundary conditions give a linear system of equations for the plate mode amplitudes V_n :

$$\begin{aligned} &\sum_n Z_0 V_n \cdot \left[\delta_{m,n} N_{Pm} \frac{Z_{Tm}}{Z_0} + \sum_\nu \frac{\delta_\nu}{2} \frac{S_{\nu m} S_{\nu n}}{\cos\Theta_\nu} \frac{(1+C_\nu)^2 - (1-C_\nu)^2 e^{-2\Gamma_\nu t}}{(1+C_\nu) - (1-C_\nu) e^{-2\Gamma_\nu t}} \right] = \\ &= \frac{4 C_\mu S_{\mu m} e^{-\Gamma_\mu t}}{(1+C_\mu) - (1-C_\mu) e^{-2\Gamma_\mu t}} \cdot P_e \quad ; \quad m \end{aligned} \tag{9}$$

with the plate mode norms $N_{Pm}=1$; the abbreviations :

$$C_\nu := \Gamma_{an} Z_{an} \cos\Theta_\nu \frac{k_0}{\Gamma_\nu} \quad ; \quad \Gamma_{an} = \Gamma_a / k_0 \quad ; \quad Z_{an} = Z_a / Z_0 \tag{10}$$

and the mode coupling coefficients :

$$\begin{aligned} S_{\nu n}^{(s)} &= \int_{-1}^{+1} \cos(\nu_e \frac{\pi}{2}\xi) \cdot \cos(n_o \frac{\pi}{2}\xi) \, d\xi \\ &= \frac{\sin(\nu_e - n_o)\pi/2}{(\nu_e - n_o)\pi/2} + \frac{\sin(\nu_e + n_o)\pi/2}{(\nu_e + n_o)\pi/2} = \frac{4}{\pi}(-1)^{(\nu_e + n_o - 1)/2} \frac{n_o}{n_o^2 - \nu_e^2} \end{aligned} \tag{11}$$

$$S_{\nu n}^{(a)} = \int_{-1}^{+1} \sin(\nu_o \frac{\pi}{2}\xi)\cdot \sin(n_e \frac{\pi}{2}\xi)\, d\xi$$

$$= \frac{\sin(\nu_o - n_e)\pi/2}{(\nu_o - n_e)\pi/2} - \frac{\sin(\nu_o + n_e)\pi/2}{(\nu_o + n_e)\pi/2} = \frac{4}{\pi}(-1)^{(n_e+\nu_o-1)/2}\frac{n_e}{\nu_o^2 - n_e^2} \tag{12}$$

One computes, with the solutions Z_0V_n, the transmitted duct mode amplitudes P_{t_ν} by :

$$P_{t\nu} = \frac{\delta_\nu}{2\cos\Theta_\nu}\sum_n S_{\nu n}\cdot Z_0 V_n \tag{13}$$

or directly the transmission coefficient :

$$\tau_\mu = \frac{\delta_\mu}{4\cos\Theta_\mu}\sum_\nu^{\nu_{lim}} \frac{\delta_\nu}{\cos\Theta_\nu}\cdot\left|\sum_n^{n_{hi}} S_{\nu n}\frac{Z_0 V_n}{P_e}\right|^2 \tag{14}$$

where the upper summation limit ν_{lim} is the index limit for propagating duct modes, and n_{hi} is the upper index limit for the plate modes used, which is set by the convergence of the system of equations for Z_0V_n .

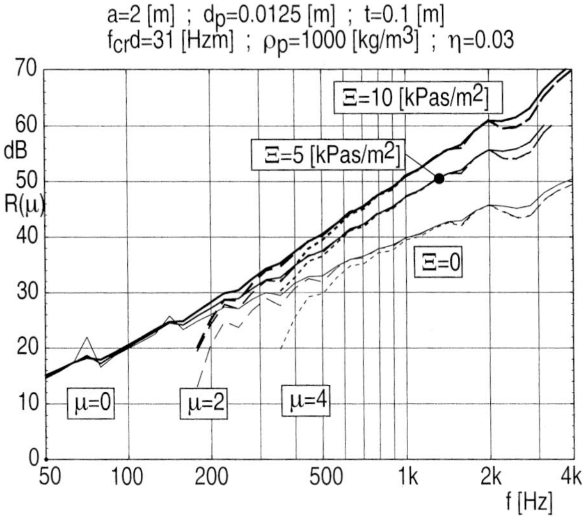

Sound transmission loss $R(\mu)$ for the μ-th duct mode through a plaster board with a front side glass fibre layer (if $\Xi>0$) having flow resistivity values Ξ .

I.19 Finite size plate with a back side absorber layer

Ref.: Mechel, [I.10]

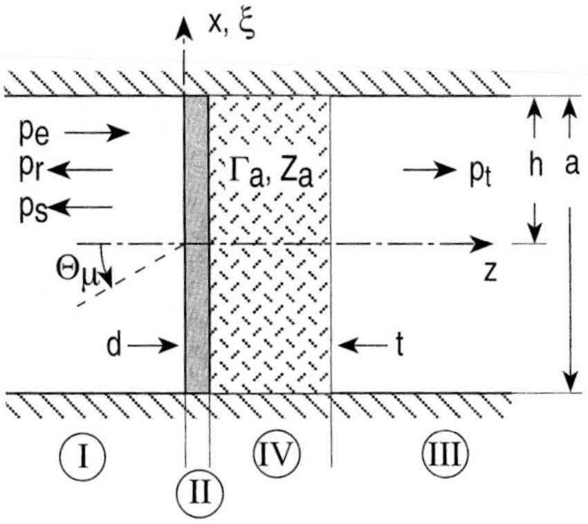

See the previous section I.18 for the duct, the plate, its mounting and the field composition.

The incident wave is again the μ-th propagating duct mode.

The system of equations for the plate mode amplitudes V_n now reads :

$$\sum_n Z_0 V_n \cdot \left[\delta_{m,n} N_{Pn} \frac{Z_{Tm}}{Z_0} + \sum_\nu \frac{\delta_\nu}{2\cos\Theta_\nu} \left(1 + C_\nu \frac{1+r_\nu}{1-r_\nu}\right) \cdot S_{\nu n} S_{\nu m} \right] = 2 S_{\mu m} \cdot P_e \qquad (1)$$

with the modal reflection factors at the back side of the absorber layer :

$$r_\nu = e^{-2\Gamma_\nu t} \frac{1 - C_\nu}{1 + C_\nu} \qquad (2)$$

After solution for $Z_0 V_n$, the transmitted duct mode amplitudes are evaluated from :

$$P_{t\nu} = \delta_\nu \frac{C_\nu}{1 + C_\nu} \frac{e^{-\Gamma_\nu t}}{(1 - r_\nu)\cos\Theta_\nu} \sum_n S_{\nu n} \cdot Z_0 V_n \qquad (3)$$

or the sound transmission coefficient directly from :

$$\tau_\mu = \frac{\delta_\mu}{\cos\Theta_\mu} \sum_\nu^{\nu_{lim}} \frac{\delta_\nu}{\cos\Theta_\nu} e^{-2\Gamma'_\nu t} \left| \frac{C_\nu}{(1 - r_\nu)(1 + C_\nu)} \right|^2 \cdot \left| \sum_n^{n_{hi}} S_{\nu n} \cdot \frac{Z_0 V_n}{P_e} \right|^2 \qquad (4)$$

with $\Gamma' = \mathrm{Re}\{\Gamma\}$; see the previous section for the mode order limits ν_{lim} and n_{hi} .

The numerical results are the same as with a front side absorber layer (see previous section I.18), except for some details around the coincidence frequency of the plate.

I.20 Finite size double wall with an absorber core

Ref.: Mechel, [I.11]

A double wall with a porous absorber layer as a core is mounted in a flat, hard duct. The plate borders are simply supported at the duct walls (as an example; other supports only need other plate mode wave numbers γ_n, mode norms Np_n, and modal partition impedance Z_{Tn}). There is no (or only a loose) mechanical contact between the plates and the absorber layer. The characteristic values of the porous material are Γ_a, Z_a, or in normalised forms : $\Gamma_{an} = \Gamma_a/k_0$, $Z_{an} = Z_a/Z_0$.

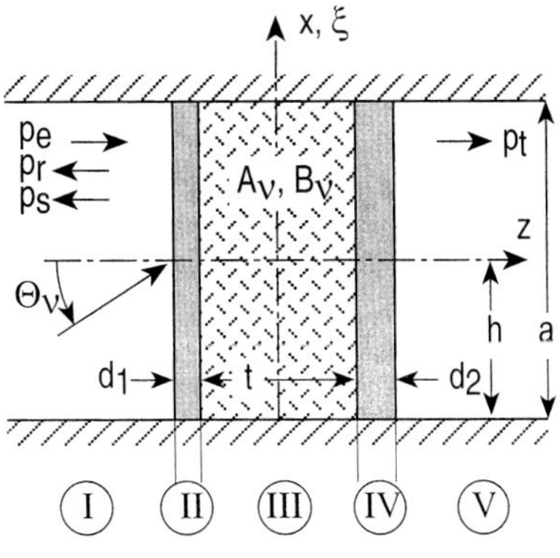

The incident wave p_e is the μ-th (symmetrical or anti-symmetrical) propagating duct mode. The sound field on the front side is composed as : $p_I = p_e + p_r + p_s$ (1)
where p_r is the incident mode after hard reflection at $z = -(t/2 + d_1)$; p_s is the scattered field. The scattered and the transmitted wave p_t are formulated as duct mode sums, as well as the sound field p_a in the absorber layer. The plate velocity patterns $V^{(i)}(\xi)$; i=1,2; are plate mode sums with $\xi = x/h$.

Field formulations :

$$\begin{aligned} p_e(\xi,z) &= P_e \cdot q_\mu(\xi) \cdot e^{-jk_{\mu z}(z+t/2+d_1)} \\ Z_0 v_{ez}(\xi,z) &= \cos\Theta_\mu \cdot p_e(\xi,z) \end{aligned} \tag{2}$$

$$\begin{aligned} p_r(\xi,z) &= P_e \cdot q_\mu(\xi) \cdot e^{+jk_{\mu z}(z+t/2+d_1)} \\ Z_0 v_{rz}(\xi,z) &= -\cos\Theta_\mu \cdot p_r(\xi,z) \end{aligned} \tag{3}$$

$$\begin{aligned} p_s(\xi,z) &= \sum_\nu P_{s\nu} \cdot q_\nu(\xi) \cdot e^{+jk_{\nu z}(z+t/2+d_1)} \\ Z_0 v_{sz}(\xi,z) &= -\sum_\nu \cos\Theta_\nu \, P_{s\nu} \cdot q_\nu(\xi) \cdot e^{+jk_{\nu z}(z+t/2+d_1)} \end{aligned} \tag{4}$$

$$\begin{aligned} p_t(\xi,z) &= \sum_\nu P_{t\nu} \cdot q_\nu(\xi) \cdot e^{-jk_{\nu z}(z-t/2-d_2)} \\ Z_0 v_{tz}(\xi,z) &= \sum_\nu \cos\Theta_\nu \, P_{t\nu} \cdot q_\nu(\xi) \cdot e^{-jk_{\nu z}(z-t/2-d_2)} \end{aligned} \tag{5}$$

I.21 Plenum modes

Ref.: Mechel, [I.12]

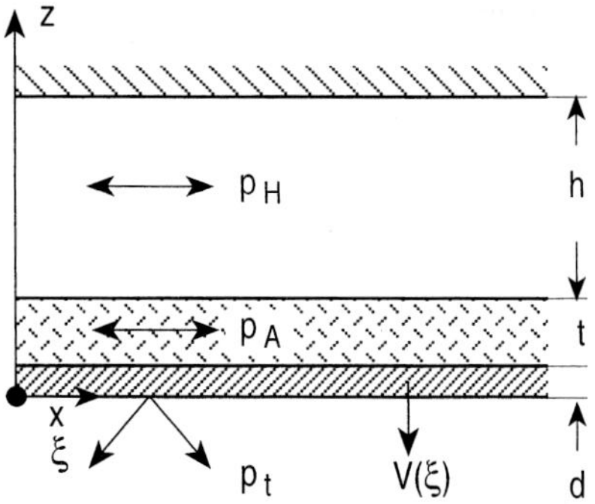

This section serves as preparation for the next section I.22 . It deals with characteristic solutions in the plenum of suspended ceilings.

The object is a flat (2-dimensional) duct with one hard wall (at z= d+t+h), an air space of height h ($d+t \le z \le d+t+h$), a porous absorber layer of thickness t ($d \le z \le d+t$), and an elastic plate of thickness d ($0 \le z \le d$). The free space $z \le 0$ belongs to the object.

The porous absorber material is described by its characteristic propagation constant Γ_a and wave impedance Z_a (or in normalised form: $\Gamma_{an}=\Gamma_a/k_0$, $Z_{an}=Z_a/Z_0$). The elastic plate is described by its partition impedance Z_T for a polar angle of incidence Φ .

Elementary solutions are sought, which obey the wave equations in air and in the absorber material :

$$\left(\Delta+k_0^2\right)p_H=0 \quad ; \quad \left(\Delta+k_0^2\right)p_t=0 \quad ; \quad \left(\Delta-\Gamma_a^2\right)p_A=0 \tag{1}$$

and the bending wave equation of the plate (which is guaranteed by using the partition impedance), as well as the boundary conditions. The solutions are called "plenum modes", with mode index n . A normalised longitudinal co-ordinate may be used with some reference length a : $\xi = x/a$.

Formulations of the component fields :

$$\begin{aligned} p_{Hn}(\xi,z) &= P_{Hn}\cdot e^{\pm\Gamma_n a\,\xi}\cdot\cos\left(\varepsilon_n\,(z-h-t-d)\right) \\ Z_0 v_{Hnz}(\xi,z) &= -j\frac{\varepsilon_n}{k_0}P_{Hn}\cdot e^{\pm\Gamma_n a\,\xi}\cdot\sin\left(\varepsilon_n\,(z-h-t-d)\right) \end{aligned} \tag{2}$$

$$\begin{aligned} p_{tn}(\xi,z) &= P_{tn}\cdot e^{\pm\Gamma_n a\,\xi}\cdot e^{-\Gamma_n a\,\xi}\cdot e^{j\kappa_n z} \\ Z_0 v_{tnz}(\xi,z) &= -\frac{\kappa_n}{k_0}p_{tn}(\xi,z) \end{aligned} \tag{3}$$

$$p_{An}(\xi,z) = e^{\pm\Gamma_n a\,\xi}\left[C_n\cdot e^{-\gamma_n (z-d)}+D_n\cdot e^{+\gamma_n (z-d)}\right] \tag{4}$$

From the wave equation in the plenum space it follows that :

$$\Gamma_n^2 - \varepsilon_n^2 + k_0^2 = 0$$

$$\left(\frac{\Gamma_n}{jk_0}\right)^2 + \left(\frac{\varepsilon_n}{k_0}\right)^2 = 1 = \sin^2\Phi_n + \cos^2\Phi_n \quad ; \quad \begin{cases} \sin\Phi_n = \Gamma_n / jk_0 \\ \cos\Phi_n = \varepsilon_n / k_0 \end{cases} \tag{5}$$

$$\Gamma_n h = \sqrt{(\varepsilon_n h)^2 - (k_0 h)^2} \quad ; \quad \mathrm{Re}\{\Gamma_n h\} \geq 0$$

which defines a modal angle of incidence Φ_n . The corresponding equations in the free space $z \leq 0$ lead to $\varepsilon_n = \pm\kappa_n$. The wave equation in the absorber material is satisfied with :

$$\frac{\gamma_n}{k_0} = \sqrt{\Gamma_{an}^2 + 1 - (\varepsilon_n / k_0)^2} \quad ; \quad \mathrm{Re}\{\sqrt{\ldots}\} \geq 0 \tag{6}$$

An abbreviation used later is :

$$G_n := \Gamma_{an} Z_{an} \frac{\varepsilon_n}{\gamma_n} \tan(\varepsilon_n h) \tag{7}$$

The modal partition impedance of the plate is (see section I.9) :

$$\frac{Z_{Tn}}{Z_0} = 2\pi Z_m F\left[\eta F^2 \sin^4\Phi_n + j\left(1 - F^2 \sin^4\Phi_n\right)\right] \quad ; \quad F := \frac{f}{f_{cr}} \quad ; \quad Z_m := \frac{f_{cr} m}{Z_0} = \frac{f_{cr} d}{Z_0}\rho$$

$$\sin^4\Phi_n = \left(1 - (\varepsilon_n / k_0)^2\right)^2 \tag{8}$$

The remaining boundary conditions give the homogeneous linear system of equations for the mode amplitudes C_n, D_n :

$$C_n e^{-\gamma_n t} \cdot \left[j\frac{\varepsilon_n}{k_0} \tan(\varepsilon_n h) - \frac{\gamma_n / k_0}{\Gamma_{an} Z_{an}} \right] + D_n e^{+\gamma_n t} \cdot \left[j\frac{\varepsilon_n}{k_0} \tan(\varepsilon_n h) + \frac{\gamma_n / k_0}{\Gamma_{an} Z_{an}} \right] = 0$$

$$C_n \cdot \left[1 + \frac{\gamma_n / k_0}{\Gamma_{an} Z_{an}} \left(\frac{k_0}{\kappa_n} + \frac{Z_{Tn}}{Z_0} \right) \right] + D_n \cdot \left[1 - \frac{\gamma_n / k_0}{\Gamma_{an} Z_{an}} \left(\frac{k_0}{\kappa_n} + \frac{Z_{Tn}}{Z_0} \right) \right] = 0 \tag{9}$$

A non-trivial solution exists, if the determinant of the coefficient matrix vanishes; this gives the characteristic equation for the wave numbers of the plenum modes : (10)

$$\cosh(\gamma_n t) \cdot$$

$$\left\{ j\frac{\varepsilon_n}{k_0} \tan(\varepsilon_n h) \cdot \left[\tanh(\gamma_n t) + \frac{\gamma_n / k_0}{\Gamma_{an} Z_{an}} \left(\frac{k_0}{\kappa_n} + \frac{Z_{Tn}}{Z_0} \right) \right] + \frac{\gamma_n / k_0}{\Gamma_{an} Z_{an}} \cdot \left[1 + \frac{\gamma_n / k_0}{\Gamma_{an} Z_{an}} \left(\frac{k_0}{\kappa_n} + \frac{Z_{Tn}}{Z_0} \right) \right] \right\} = 0$$

The leading factor can be assumed to be $\cosh(\gamma_n t) \neq 0$; thus the expression in the curled brackets must be zero. Taking $z = \kappa_n h$ as the quantity for which solutions shall be found, the equation reads :

$$\begin{aligned}&\mathrm{j}z\cdot\tan z\cdot\left[z\cdot\tanh\left(k_0 t\frac{\gamma_n}{k_0}\right)+\frac{1}{\Gamma_{an}Z_{an}}\frac{\gamma_n}{k_0}\left(1+z\frac{Z_{Tn}}{Z_0}\right)\right]+\\&+\frac{1}{\Gamma_{an}Z_{an}}\frac{\gamma_n}{k_0}\left[z+\frac{1}{\Gamma_{an}Z_{an}}\frac{\gamma_n}{k_0}\left(1+z\frac{Z_{Tn}}{Z_0}\right)\right]=0\end{aligned} \tag{11}$$

(γ_n and Z_{Tn} are functions of z). A method of solution for a set z_n of modes is described in [I.12] .

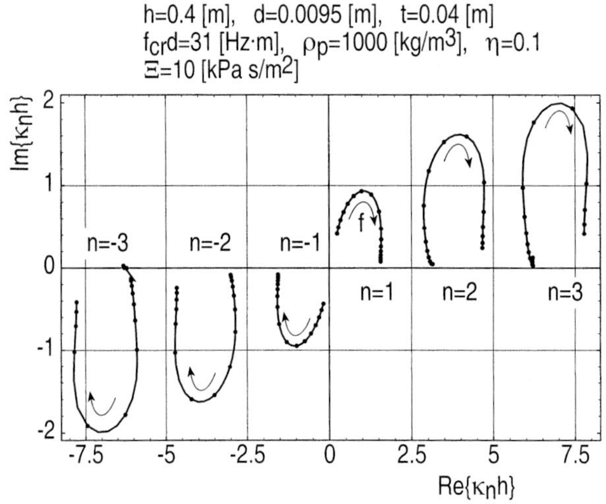

Example of plenum mode solutions $\kappa_n h$ for a plaster board as the elastic plate, and with a t= 4 [cm] thick glass fibre mat as absorber layer.

I.22 Sound transmission through suspended ceilings

Ref.: Mechel, [I.12]

A typical set-up of a suspended ceiling is taken from the previous section I.21 , from where also notations for the component fields are used. The next graph shows the arrangement of an emission room of width a_s (index s on the emission side from the German *Sendeseite*) and a receiving room of width a_e (index e on the receiver side from the German *Empfangsseite*) . The suspended ceiling spans over both rooms; the partition wall between the two rooms is rigid. The target quantity is the flanking transmission loss $R_f = -10\cdot\lg(\tau)$ of sound transmission through the suspended ceiling. First, the transmission coefficient τ_μ for the μ-th propagating mode of the emission room as incident wave p_e will be given; then the transmission loss for all propagating emission room modes (with equal energy densities) follows.

The non-dimensional axial co-ordinates $\xi_s = x/a_s$; $\xi_e = -x/a_e$ will be used. The back walls of the plenum are assumed to have reflection factors r_s, r_e .

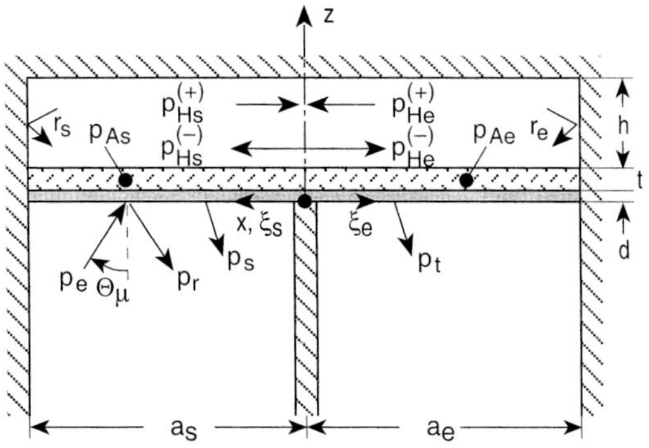

The sound field in the emission room is composed as : $p_e+p_r+p_s$ where p_e is the μ-th propagating room mode of the emission room, p_r is this mode after hard reflection at the lower surface of the suspended ceiling, p_s is a (back-) scattered field. Both p_s and the transmitted sound p_t are composed as mode sums of the room modes in the relevant rooms.

The sound fields in the plenum spaces and in the absorber layer are composed as sums of plenum modes. A lower index ß= s,e indicates to which side the sound wave belongs; an upper index (±) indicates the direction of a mode.

The incident and reflected emission room modes are (μ= 0,1,2,...) :

$$p_e(\xi_s,z) = P_e \cdot \cos(\mu\pi\xi_s) \cdot e^{-j k_{\mu z} z} \quad ; \quad Z_0 v_{ez}(\xi_s,z) = \cos\Theta_\mu \cdot p_e(\xi_s,z)$$
$$p_r(\xi_s,z) = P_e \cdot \cos(\mu\pi\xi_s) \cdot e^{+j k_{\mu z} z} \quad ; \quad Z_0 v_{rz}(\xi_s,z) = -\cos\Theta_\mu \cdot p_r(\xi_s,z) \tag{1}$$

with modal angles of incidence :

$$\sin\Theta_\mu = \frac{k_{\mu x}}{k_0} = \frac{\mu\pi}{k_0 a_s} \quad ; \quad \cos\Theta_\mu = \frac{k_{\mu z}}{k_0} = \sqrt{1-\left(\frac{\mu\pi}{k_0 a_s}\right)^2} \quad ; \quad \begin{cases} \mathrm{Re}\{\sqrt{\ldots}\} \ge 0 \\ \mathrm{Im}\{\sqrt{\ldots}\} \le 0 \end{cases} \tag{2}$$

and similarly modal angle Θ_ν in the receiving room.

The transmitted sound field is (ν= 0,1,2,...) : (3)

$$p_t(\xi_e,z) = \sum_\nu P_{t\nu} \cdot \cos(\nu\pi\xi_e) \cdot e^{j k_{\nu z} z} \quad ; \quad Z_0 v_{tz}(\xi_e,z) = -\sum_\nu P_{t\nu} \cos\Theta_\nu \cdot \cos(\nu\pi\xi_e) \cdot e^{j k_{\nu z} z}$$

The desired sound transmission coefficient is :

$$\tau_\mu = \frac{\delta_\mu}{\cos\Theta_\mu} \sum_\nu^{\nu_{lim}} \frac{\cos\Theta_\nu}{\delta_\nu} \left|\frac{P_{t\nu}}{P_e}\right|^2 \tag{4}$$

with ν_{lim} the mode order limit for propagating modes, given by the condition $\nu \le \dfrac{k_0 a_e}{\pi} = \dfrac{2a_e}{\lambda_0}$ (5)

The modal components in the plenum space are formulated as (ß= s,e) :

$$\begin{aligned} p_{Hn}(\xi_ß,z) &= \left[P_{Hßn}^{(+)}\cdot e^{+\Gamma_n a\,\xi_ß} + P_{Hßn}^{(-)}\cdot e^{-\Gamma_n a\,\xi_ß}\right]\cdot\cos\left(\varepsilon_n\,(z-h-t-d)\right) \\ Z_0 v_{Hnz}(\xi_ß,z) &= -j\frac{\varepsilon_n}{k_0}\left[P_{Hßn}^{(+)}\cdot e^{+\Gamma_n a\,\xi_ß} + P_{Hßn}^{(-)}\cdot e^{-\Gamma_n a\,\xi_ß}\right]\cdot\sin\left(\varepsilon_n\,(z-h-t-d)\right) \end{aligned} \tag{6}$$

$$\begin{aligned} p_{An}(\xi_ß,z) = {} & e^{+\Gamma_n a\,\xi_ß}\left[C_{ßn}{}^{(+)}\cdot e^{-\gamma_n\,(z-d)} + D_{ßn}{}^{(+)}\cdot e^{+\gamma_n\,(z-d)}\right]+ \\ & + e^{-\Gamma_n a\,\xi_ß}\left[C_{ßn}{}^{(-)}\cdot e^{-\gamma_n\,(z-d)} + D_{ßn}{}^{(-)}\cdot e^{+\gamma_n\,(z-d)}\right] \end{aligned} \tag{7}$$

and in the absorber layer :

$$\begin{aligned} p_{An}(\xi_ß,z) = {} & e^{+\Gamma_n a\,\xi_ß}\left[C_n{}^{(+)}\cdot e^{-\gamma_n\,(z-d)} + D_n{}^{(+)}\cdot e^{+\gamma_n\,(z-d)}\right]+ \\ & + e^{-\Gamma_n a\,\xi_ß}\left[C_n{}^{(-)}\cdot e^{-\gamma_n\,(z-d)} + D_n{}^{(-)}\cdot e^{+\gamma_n\,(z-d)}\right] \end{aligned} \tag{8}$$

The reflection at the back walls of the plenum with given reflection factors $r_ß$ can serve to eliminate some sets of amplitudes :

$$P_{Hßn}^{(+)} = r_ß\cdot e^{-2\Gamma_n a_ß}\cdot P_{Hßn}^{(-)} \tag{9}$$

and from the matching of fields at the surface between the plenum space and absorber layer :

$$C_{ßn}^{(\pm)} = \frac{1}{2}P_{Hßn}^{(\pm)}\cdot e^{+\gamma_n t}\cdot\cos(\varepsilon_n h)\cdot\left(1+jG_n\right)\;;\quad D_{ßn}^{(\pm)} = \frac{1}{2}P_{Hßn}^{(\pm)}\cdot e^{-\gamma_n t}\cdot\cos(\varepsilon_n h)\cdot\left(1-jG_n\right) \tag{10}$$

The field matching in the plane x=0 leads to two coupled systems of linear equations for the $P_{Hßn}^{(-)}$:

$$\begin{aligned} &\sum_n\left[P_{Hsn}^{(-)}\left(1+r_s\,e^{-2\Gamma_n a_s}\right) - P_{Hen}^{(-)}\left(1+r_e\,e^{-2\Gamma_n a_e}\right)\right]\cdot M_{mn} = \\ &\quad = jhS_{\mu m}\cdot P_{Hs\mu}^{(h)} - \frac{t}{\Gamma_{an}Z_{an}}\left(R_{\mu m}^{(-)}\cdot A_{s\mu} + R_{\mu m}^{(+)}\cdot B_{s\mu}\right) \end{aligned} \tag{11}$$

$$\sum_n\frac{\Gamma_n}{k_0}\left[P_{Hsn}^{(-)}\left(1-r_s\,e^{-2\Gamma_n a_s}\right) + P_{Hen}^{(-)}\left(1-r_e\,e^{-2\Gamma_n a_e}\right)\right]\cdot M_{mn} = 0 \tag{12}$$

With the solutions one evaluates $P_{Hßn}^{(+)}$ and with these $C_{ßn}^{(\pm)}$, $D_{ßn}^{(\pm)}$. The amplitudes $P_{t\nu}$, which are needed for the transmission coefficient, are evaluated from :

$$P_{t\nu} = -\frac{\delta_\nu}{\cos\Theta_\nu}\sum_n\frac{\gamma_n/k_0}{\Gamma_{an}Z_{an}}\left[\left(C_{en}^{(+)} - D_{en}^{(+)}\right)T_{\nu n}^{(+)} + \left(C_{en}^{(-)} - D_{en}^{(-)}\right)T_{\nu n}^{(-)}\right] \tag{13}$$

In the above equations we have :

$$A_\mu = \frac{1}{2} e^{+\chi_\mu t}\left[\cos(k_{\mu z}h) + j\, C_\mu \sin(k_{\mu z}h)\right]\cdot P_H$$
$$B_\mu = \frac{1}{2} e^{-\chi_\mu t}\left[\cos(k_{\mu z}h) - j\, C_\mu \sin(k_{\mu z}h)\right]\cdot P_H \tag{14}$$

and :

$$P_H = \frac{4 P_e \cdot e^{-\chi_\mu t}}{\cos(k_{\mu z}h)\left[1 + e^{-2\chi_\mu t} + 1/C_\mu \cdot \left(1 + Z_{T\mu}/Z_0 \cdot \cos\Theta_\mu\right)\left(1 - e^{-2\chi_\mu t}\right)\right] + \ldots}$$
$$\frac{\ldots}{\ldots + j\sin(k_{\mu z}h)\left[C_\mu\left(1 - e^{-2\chi_\mu t}\right) + \left(1 + Z_{T\mu}/Z_0 \cdot \cos\Theta_\mu\right)\left(1 + e^{-2\chi_\mu t}\right)\right]} \tag{15}$$

with :

$$k_{\mu z}h = k_0 h \cdot \cos\Theta_\mu \quad ; \quad \frac{\chi_\mu}{k_0} = \sqrt{\Gamma_{an}^2 + \left(\frac{\mu\pi}{k_0 a}\right)^2} \quad ; \quad \mathrm{Re}\{\sqrt{\ldots}\} \ge 0$$
$$C_\mu := \Gamma_{an} Z_{an} \cos\Theta_\mu \frac{k_0}{\chi_\mu} \tag{16}$$

Further, the weight factors $M_{m,n}$ of the inter-orthogonality of plenum mode factors are used; they are defined by :

$$\frac{1}{jk_0 Z_0}\int_H p_{Hm}(z)\cdot p_{Hn}(z)\, dz + \frac{1}{\Gamma_a Z_a}\int_A p_{Am}(z)\cdot p_{An}(z)\, dz := \frac{M_{m,n}}{k_0 Z_0} \tag{17}$$

(if the plate of the suspended ceiling is rigid, then $M_{m,n} = \delta_{m,n}$ = Kronecker symbol); with evaluation by :

$$M_{m,n} = \frac{h}{2j}\left[\frac{\sin(\varepsilon_m - \varepsilon_n)h}{(\varepsilon_m - \varepsilon_n)h} + \frac{\sin(\varepsilon_m + \varepsilon_n)h}{(\varepsilon_m + \varepsilon_n)h}\right] + \frac{\cos(\varepsilon_m h)\cos(\varepsilon_n h)}{4\Gamma_{an} Z_{an}} t +$$
$$+\left[(1 + jG_m)(1 + jG_n)\frac{1 - e^{-(\gamma_m + \gamma_n)t}}{(\gamma_m + \gamma_n)t} - (1 - jG_m)(1 - jG_n)\frac{1 - e^{+(\gamma_m + \gamma_n)t}}{(\gamma_m + \gamma_n)t} + \right.$$
$$\left. + (1 + jG_m)(1 - jG_n)\frac{1 - e^{-(\gamma_m - \gamma_n)t}}{(\gamma_m - \gamma_n)t} - (1 - jG_m)(1 + jG_n)\frac{1 - e^{+(\gamma_m - \gamma_n)t}}{(\gamma_m - \gamma_n)t}\right] \tag{18}$$

Other factors are mode coupling factors, between directly transmitted field and plenum mode field in the absorber layer :

$$R_{\mu n}^{(\pm)} := \frac{1}{t}\int_{d}^{d+t} e^{\pm\chi_n(z-d)} \cdot p_{An}^{(\pm)}(z)\,dz$$
$$= \frac{\cos(\varepsilon_n h)}{2t(\gamma_n^2-\chi_n^2)}\Big[e^{\gamma_n t}\cdot(1+jG_n)(\gamma_n \pm \chi_n) - e^{-\gamma_n t}\cdot(1-jG_n)(\gamma_n \mp \chi_n) + $$
$$+ 2e^{\pm\chi_n t}(j\gamma_n G_n \pm \chi_n)\Big] \tag{19}$$

between directly transmitted and plenum mode field in the plenum space :

$$S_{\mu n} := \frac{1}{h}\int_{t+d}^{h+t+d} \cos\big(k_{\mu z}(z-h-t-d)\big)\cdot\cos\big(\varepsilon_n(z-h-t-d)\big)\,dz$$
$$= \frac{1}{2}\left[\frac{\sin(k_{\mu z}-\varepsilon_n)h}{(k_{\mu z}-\varepsilon_n)h} + \frac{\sin(k_{\mu z}+\varepsilon_n)h}{(k_{\mu z}+\varepsilon_n)h}\right] \tag{20}$$

and between the room modes and the absorber field along the common surface :

$$T_{\nu n}^{(\pm)} := \int_0^1 q_\nu(\xi)\cdot e^{\pm\Gamma_n a\xi}\,d\xi = \frac{\pm\Gamma_n a}{(\Gamma_n a)^2+(\nu\pi)^2}\Big[(-1)^\nu\cdot e^{\pm\Gamma_n a}-1\Big] \tag{21}$$

The simulation of diffuse sound incidence assumes that all propagating room modes have the same sound energy density and are incident modes. The sound transmission coefficient for diffuse sound incidence is evaluated by :

$$\tau_{dif} = \sum_{\mu=0}^{\mu_{lim}} \tau_\mu \cos\Theta_\mu \Big/ \sum_{\mu=0}^{\mu_{lim}} \cos\Theta_\mu \tag{22}$$

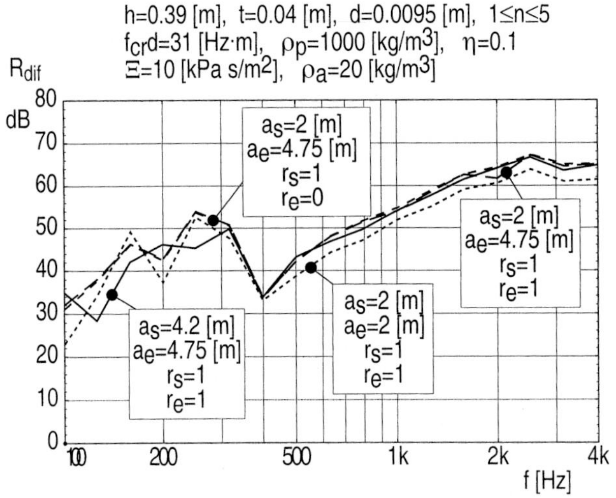

Sound transmission loss for diffuse sound incidence through a suspended ceiling with a plaster board plate, a t=4 [cm] glass fibre layer on it, a plenum height of h= 39 [cm], for different situations of room sizes a_s, a_e and plenum back wall reflections factors r_s,r_e .

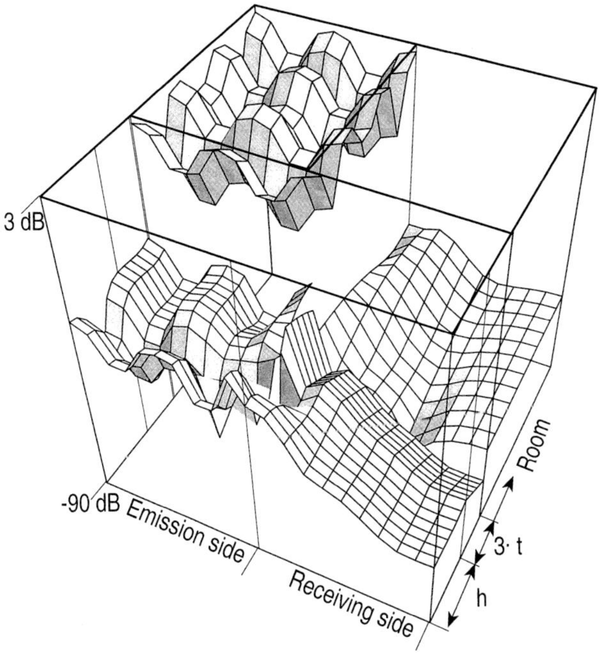

Sound pressure level profile below and in the plenum of a suspended ceiling for the 1st higher room mode as incident mode.
Left-behind: emission room;
right-behind: receiving room;
left-front: plenum above emission,
right-front: plenum above receiving room.
The space occupied by the absorber layer is drawn with a 3-fold magnification.
The suspended ceiling consists of a d=9.5 [mm] plaster board, covered with a t=8 [cm] glass fibre mat; the plenum is h=35 [cm] high; the room sizes are $a_s = a_e = 4$ [m].
The plenum back walls are hard.

I.23 Office fences

Ref.: Mechel, [I.13]

In a two-dimensional room of height H with a hard floor and an absorbent ceiling is placed a (thin) absorbent and sound transmitting fence with its upper corner at y=h and (possibly) a gap of height b towards the floor. The rooms are anechoic in the ±x direction.

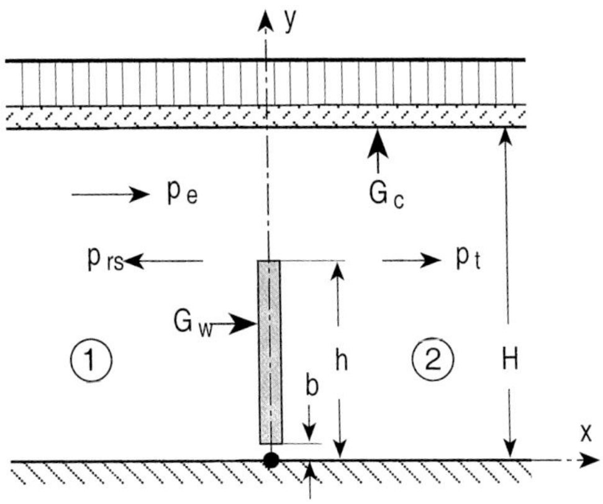

The ceiling is assumed (as an approximation) to be locally reacting with an admittance G_c. The surface admittance G_w of the fence may be different on both sides, G_{w1}, G_{w2}.

First, a non-transmissive fence without a lower gap $(b=0)$ will be considered; the modifications of the results for other conditions will be explained below.

Let the incident wave be the μ-th mode of the room :

$$p_e(x,y) = P_e \cdot \cos(\varepsilon_\mu y) \cdot e^{-\Gamma_\mu x} \tag{1}$$

where $\varepsilon_\mu H$ is a solution of the characteristic equation :

$$\varepsilon_\mu H \cdot \tan(\varepsilon_\mu H) = j\, k_0 H \cdot Z_0 G_c \tag{2}$$

and $$\Gamma_\mu^2 = \varepsilon_\mu^2 - k_0^2 \tag{3}$$

The field on the emission side (1) is composed as $p_1 = p_e + p_{rs}$ with the backscattered field formulated as a sum of room modes :

$$p_{rs}(x,y) = \sum_n B_n \cdot \cos(\varepsilon_n y) \cdot e^{+\Gamma_n x} \tag{4}$$

Similarly the transmitted field p_t is a sum of room modes :

$$p_t(x,y) = \sum_n D_n \cdot \cos(\varepsilon_n y) \cdot e^{-\Gamma_n x} \tag{5}$$

The matching of the fields to the fence admittance and to each other leads to two coupled systems of linear equations for B_n, D_n :

$$\begin{aligned} &\sum_n B_n \cdot j\frac{\Gamma_n}{k_0}\left(\delta_{m,n} N_m - S_{m,n}\right) - \sum_n D_n \cdot \left(Z_0 G_{w2} - \delta_{m,n}\, j\frac{\Gamma_m}{k_0} N_m \right) = \\ &= jP_e \frac{\Gamma_\mu}{k_0}\left(\delta_{\mu,m} N_m - S_{\mu,m}\right) \end{aligned} \tag{6}$$

$$\begin{aligned} &\sum_n B_n \cdot \left(j\frac{\Gamma_n}{k_0} S_{m,n} - \delta_{m,n} N_m Z_0 G_{w1} \right) - Z_0 G_{w1} \sum_n D_n \cdot \left(S_{m,n} - \delta_{m,n} N_m\right) = \\ &= P_e \left(j\frac{\Gamma_\mu}{k_0} S_{\mu,m} + \delta_{\mu,m} N_\mu Z_0 G_{w1} \right) \end{aligned} \tag{7}$$

Therein are: $\delta_{m,n}$ the Kronecker symbol; N_m the mode norms :

$$\begin{aligned} \frac{1}{H}\int_0^H \cos(\varepsilon_m y) \cdot \cos(\varepsilon_n y)\, dy &= \delta_{m,n} \cdot \frac{1}{2}\left(1 + \frac{\sin(2\varepsilon_m H)}{2\varepsilon_m H} \right) = \delta_{m,n} \cdot N_m \\ &\xrightarrow[\varepsilon_m = \varepsilon_n = 0]{} 1 \end{aligned} \tag{8}$$

and $S_{m,n}$ the mode coupling coefficients :

$$\frac{1}{H}\int_0^h \cos(\varepsilon_m y)\cdot\cos(\varepsilon_n y)\,dy = \frac{h}{2H}\left(\frac{\sin(\varepsilon_m-\varepsilon_n)h}{(\varepsilon_m-\varepsilon_n)h}+\frac{\sin(\varepsilon_m+\varepsilon_n)h}{(\varepsilon_m+\varepsilon_n)h}\right) =: S_{m,n}$$

$$\xrightarrow[\varepsilon_m=\varepsilon_n]{} \frac{h}{2H}\left(1+\frac{\sin(2\varepsilon_m h)}{2\varepsilon_m h}\right) \tag{9}$$

If the fence is sound transmissive, its surface admittances G_{w1}, G_{w2} are determined with a free space termination of the fence. If there is a bottom gap $(b\neq 0)$, the mode coupling coefficients $S_{m,n}$ are evaluated for the interval (b,h) of integration, i.e. everywhere $S_{m,n}(0,h)$ is replaced by $S_{m,n}(0,h) \to S_{m,n}(b,h)$.

I.24 Office fences, with 2nd principle of superposition

Ref.: Mechel, [I.13]

The object is the same as in the previous section I.23 , but it will be treated here with the 2nd principle of superposition (PSP) from section B.10 . The advantage of the PSP is, that it halves the size of the system of equations to be solved, and it makes the field formulations more plausible. The PSP can be applied, if the object has a plane of symmetry S (which is x=0 in our case). It splits the task in two sub-tasks: 1st the sound transmissive parts of S are assumed to be hard (upper index ß=(h)), 2nd the sound transmissive parts of S are assumed to be soft (upper index ß=(w)). The surface of the fence has the admittances $G_w^{(\beta)}$ in both sub-tasks (i.e. with hard or soft termination at x=0).

The incident wave p_e in both sub-tasks is assumed to be the μ-th mode of the room, which is lined on one side with a locally reacting ceiling :

$$p_e(x,y) = P_e\cdot\cos(\varepsilon_\mu y)\cdot e^{-\Gamma_\mu x} \tag{1}$$

It is associated on the front side (1) (see graph in section I.23) with the mode field $p_r^{(\beta)}(x,y)$ after hard or soft reflection, respectively, at x=0 :

$$p_r^{(\beta)}(x,y) = \pm P_e\cdot\cos(\varepsilon_\mu y)\cdot e^{+\Gamma_\mu x} \quad ; \quad (\beta)=(h),(w) \tag{2}$$

and a scattered wave

$$p_s^{(\beta)}(x,y) = \sum_n C_n^{(\beta)}\cdot\cos(\varepsilon_n y)\cdot e^{+\Gamma_n x} \tag{3}$$

which is a sum of duct modes. The sound fields in front of and behind the screen are then :

$$\begin{aligned} p_1(x<0,y) &= p_e(x,y) + \frac{1}{2}\left[p_s^{(h)}(x,y) + p_s^{(w)}(x,y)\right] \\ p_2(x>0,y) &= p_e(x,y) + \frac{1}{2}\left[p_s^{(h)}(-x,y) - p_s^{(w)}(-x,y)\right] \end{aligned} \tag{4}$$

Application of the boundary conditions in the surface plane of the screen gives two systems of linear equations for the $C_n^{(\beta)}$:

$$\begin{aligned} &\sum_n C_n^{(h)} \cdot \left[Z_0 G_w^{(h)} \cdot S_{m,n} - \delta_{m,n} \cdot j\frac{\Gamma_m}{k_0} N_m \right] = -2 Z_0 G_w^{(h)} S_{\mu,m} \cdot P_e \\ &\sum_n C_n^{(w)} \cdot \left[j\frac{\Gamma_n}{k_0} \cdot S_{m,n} - \delta_{m,n} \cdot Z_0 G_w^{(s)} N_m \right] = 2j\frac{\Gamma_\mu}{k_0} S_{\mu,m} \cdot P_e \end{aligned} \tag{5}$$

Therein the N_m are norms of the duct modes :

$$\frac{1}{H}\int_0^H \cos(\varepsilon_m y)\cdot\cos(\varepsilon_n y)\,dy = \delta_{m,n}\cdot\frac{1}{2}\left(1+\frac{\sin(2\varepsilon_m H)}{2\varepsilon_m H}\right) = \delta_{m,n}\cdot N_m \xrightarrow[\varepsilon_m=\varepsilon_n=0]{} 1 \tag{6}$$

and the $S_{m,n}= S_{m,n}(b,h)$ are mode coupling coefficients :

$$\frac{1}{H}\int_b^h \cos(\varepsilon_m y)\cdot\cos(\varepsilon_n y)\,dy =: S_{m,n}(b,h) \tag{7}$$

The method can be applied for any sound source on the emission side, if its source profile (either a given pressure or a given velocity) can be synthesised with room modes. Then the above evaluation will be performed mode-wise and the results superimposed.

The numerical examples below for the sound pressure level change $20\cdot\lg(|p/p_e|)$ due to the fence assume the following parameter values : frequency f= 500 [Hz] ; H= 3.5 [m] ; h= 2 [m] ; b= 0.1 [m] (if there is a gap); *ceiling* (from low to high): d= 2 [cm] boards of compressed mineral fibre; bulk density RG= 400 [kg/m^3] ; flow resistance $5\cdot Z_0$; elastic constant $f_{cr}\cdot d$= 70 ; plus a 5 [cm] thick mineral fibre felt with flow resistivity Ξ= 10 [kPa·s/m^2] and bulk density RG= 15 [kg/m^3]; plus a locally reacting air layer 40 [m] thick below the hard construction ceiling; *fence* : a mineral fibre board, 10 [cm] thick, flow resistivity Ξ= 10 [kPa·s/m^2], covered (on both sides) with a perforated metal sheet, 1.5 [mm] thick, porosity 36 %, round perforations 4 [mm] wide (there is no (or only loose) mechanical contact between the metal sheet and the mineral fibre board). If the fence is assumed to be non-transmissive, a heavy metal sheet may be assumed in its centre. The incident wave is the fundamental room mode μ=1 .

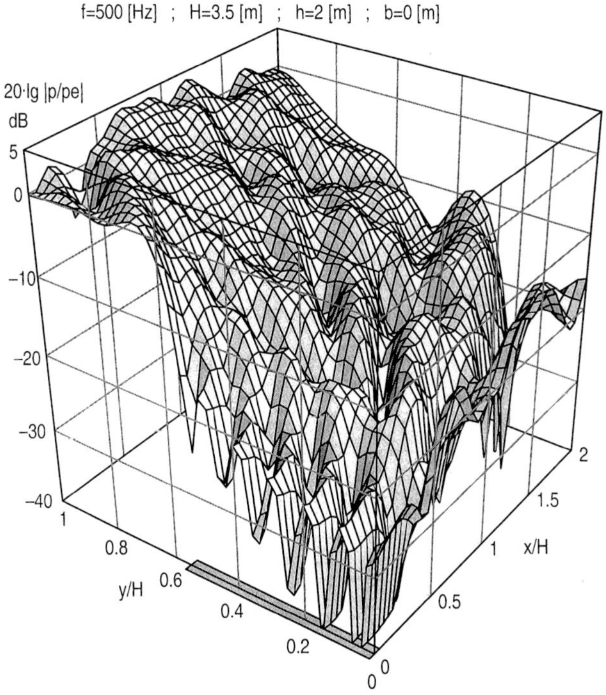

Sound pressure level change behind the fence;
the fence is non-transmissive and has no gap at its foot.

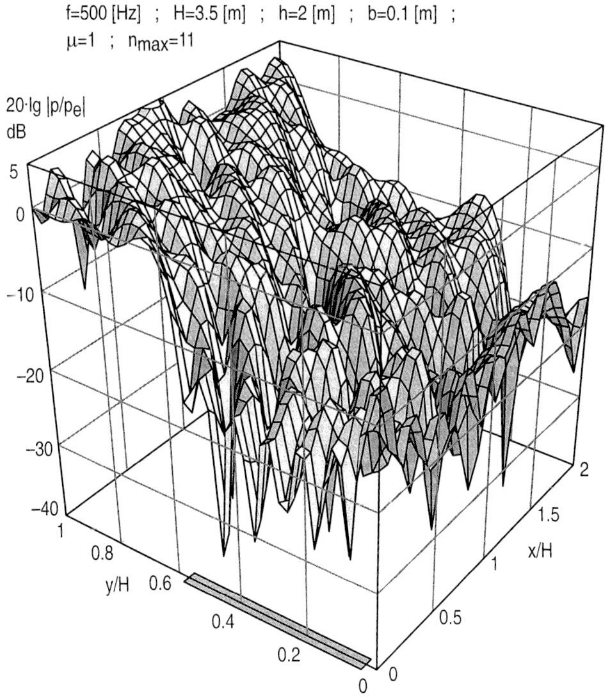

As above, but fence has a gap between foot and floor, b= 10 [cm] wide.

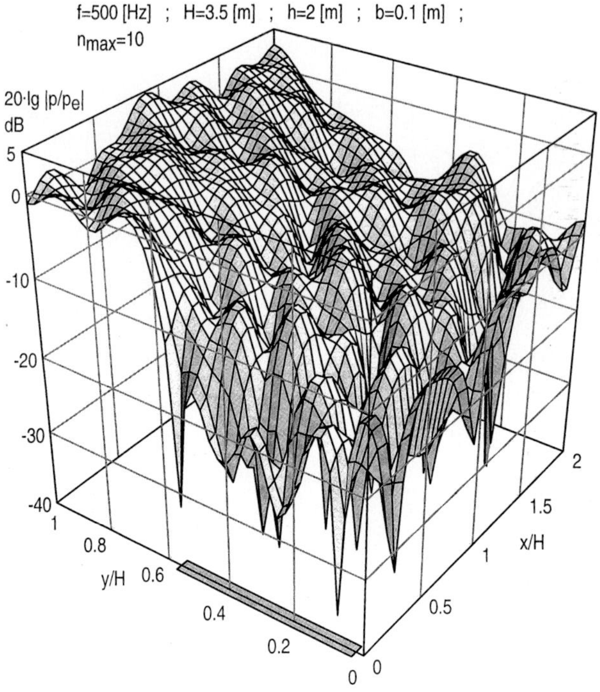

As above (i.e. transmissive fence with gap), but the incident wave is a plane wave.

I.25 Infinite plate between two different fluids

Ref.: Alekseev / Dianov [I.14]

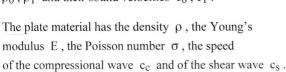

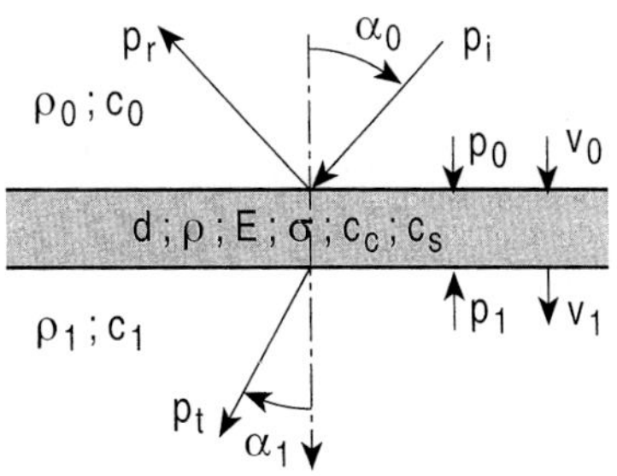

An infinite plate of thickness d is placed between two fluids with index 0 on the side of incidence of a plane wave p_i, and index 1 on the side of the transmitted wave p_t.

The fluids are characterised by their densities ρ_0, ρ_1 and their sound velocities c_0, c_1.

The plate material has the density ρ, the Young's modulus E, the Poisson number σ, the speed of the compressional wave c_c and of the shear wave c_s.

Sound pressures at the plate surfaces : $p_0 = p_i + p_r \; ; \; p_1 = p_t$ (1)

v_0, v_1 are the corresponding plate surface velocities.

Definitions :

Normal components of the fluid wave impedances: $Z_{0n} = \dfrac{\rho_0 c_0}{\cos\alpha_0} \; ; \; Z_{1n} = \dfrac{\rho_1 c_1}{\cos\alpha_1}$ (2)

Reflection factor: $R = \dfrac{p_r}{p_i}$ (3)

Transmission factor: $T = \dfrac{p_t}{p_i}$ (4)

Pressure ratio: $K = \dfrac{p_t}{p_i + p_r} = \dfrac{p_1}{p_0} = \dfrac{T}{1+R}$ (5)

Plate input impedance: $Z = \dfrac{p_0}{v_0} = Z_{0n}\dfrac{1+R}{1-R}$ (6)

Symmetrical impedance
(impedance for plate compression): $Z_s = \dfrac{p_0 + p_1}{v_0 - v_1}$ (7)

Anti-symmetrical impedance
(impedance for plate bending): $Z_a = \dfrac{p_0 - p_1}{v_0 + v_1}$ (8)

From the relation of the effective powers for a plate without losses $\mathrm{Re}\{p_0 v_0^*\} = \mathrm{Re}\{p_1 v_1^*\}$:

$$|K|^2 = Z_{1n} \cdot \mathrm{Re}\{1/Z\} \tag{9}$$

In the special case $Z_{0n} \ll |Z|$ & $|Z_s| \gg |Z_a|$ & $|Z_s| \gg Z_{1n}$ is $K \cdot Z \approx Z_{1n}$.

If the medium on the side of incidence is nearly soft, i.e. Z_{0n} is negligible :

$$K(\alpha_1) = \frac{Z_{1n}(Z_s - Z_a)}{2Z_s Z_a + Z_{1n}(Z_s + Z_a)} \approx D(\alpha_1)/2$$
$$Z(\alpha_1) = \frac{2Z_s Z_a + Z_{1n}(Z_s + Z_a)}{2Z_{1n} + Z_s + Z_a} \approx Z_{0n} \frac{1}{1 - R(\alpha_1)} \tag{10}$$

The symmetrical and anti-symmetrical impedances Z_s , Z_a follow from :

$$Z_s = -j\left[W_c \cdot \cot\frac{a_1}{2} + W_s \cdot \cot\frac{b_1}{2}\right] \quad ; \quad Z_a = +j\left[W_c \cdot \tan\frac{a_1}{2} + W_s \cdot \tan\frac{b_1}{2}\right] \tag{11}$$

with :

$$W_c = \frac{\rho c_c}{\cos\alpha} \cdot \cos^2(2ß) \quad ; \quad W_s = \frac{\rho c_s}{\cos ß} \cdot \sin^2(2ß)$$

$$\sin\alpha = \frac{c_c}{c_1} \cdot \sin\alpha_1 \quad ; \quad \sin ß = \frac{c_s}{c_1} \cdot \sin\alpha_1$$
$$a_1 = \frac{\omega d}{c_c} \cdot \cos\alpha \quad ; \quad b_1 = \frac{\omega d}{c_s} \cdot \cos ß \tag{12}$$
$$c_c = \sqrt{\frac{E(1-\sigma)}{\rho(1+\sigma)(1-2\sigma)}} \quad ; \quad c_s = \sqrt{\frac{E}{2\rho(1+\sigma)}}$$

α is the diffracted angle of the compressional wave in the half-infinite plate material; ß is the diffracted angle of the shear wave in the half-infinite plate material.

The sound transmission can be represented by an equivalent circuit. The transformer has the transformer ratio

$$N = \frac{Z_s + Z_a}{Z_s - Z_a} \tag{13}$$

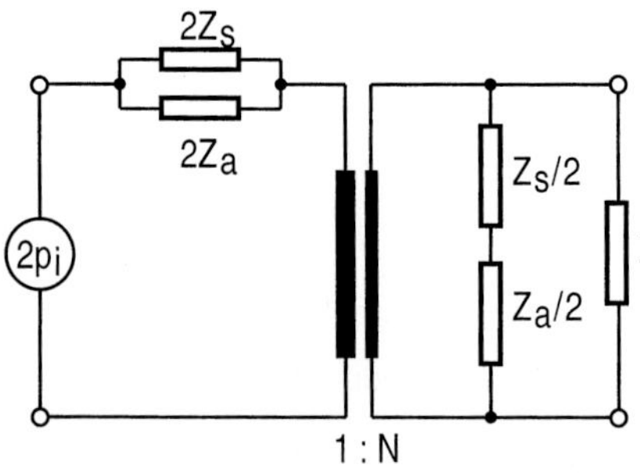

Special cases :

a) *Normal radiation* $\alpha_1 = 0$:

$$K(0)=\frac{1}{\cos\frac{\omega d}{c_c}+j\frac{\rho c_c}{\rho_1 c_1}\sin\frac{\omega d}{c_c}} \quad ; \quad N=\cos\frac{\omega d}{c_c} \tag{14}$$

b) *Coincidence of the incident wave with the anti-symmetrical wave in the free plate* $Z_a=0$:

$$K(\alpha_1)=1 \quad ; \quad Z(\alpha_1)=\frac{Z_1\cdot Z_s}{2Z_1+Z_s} \quad ; \quad N=1 \tag{15}$$

c) *Coincidence of the incident wave with the symmetrical wave in the free plate* i.e. $Z_s=0$:

$$K(\alpha_1)=-1 \quad ; \quad Z(\alpha_1)=\frac{Z_1\cdot Z_a}{2Z_1+Z_a} \quad ; \quad N=-1 \tag{16}$$

d) *Inpermeable plate* i.e. $Z_s=Z_a$:

$$K(\alpha_1)=0 \quad ; \quad Z(\alpha_1)=Z_s=Z_a \quad ; \quad N\to\infty \tag{17}$$

e) *Incidence at 1st critical angle* i.e. $Z_s\to\infty$:

$$K(\alpha_1)=\frac{Z_{1n}}{Z_{1n}+2Z_a} \quad ; \quad Z(\alpha_1)=Z_{1n}+2Z_a \quad ; \quad N=1 \tag{18}$$

f) *Only near field on the receiver side,* i.e. $\sin\alpha_1\to\infty$; $\cos\alpha_1\to j\infty$:

$$K(\alpha_1)=0 \quad ; \quad Z(\alpha_1)\to -2j\rho c_c\left(\frac{c_s}{c_c}\right)^2\frac{c_c^2-c_s^2}{c_c\cdot c_1}\sin\alpha_1\to -j\infty \tag{19}$$

If, additionally, the plate is thin :

$$Z(\alpha_1)\to -16j\frac{c_s}{\omega d}\left[1-\left(c_s/c_c\right)^2\right] \tag{20}$$

I.26 Sandwich plate with an elastic core

Ref.: Beshenkov, [I.15]

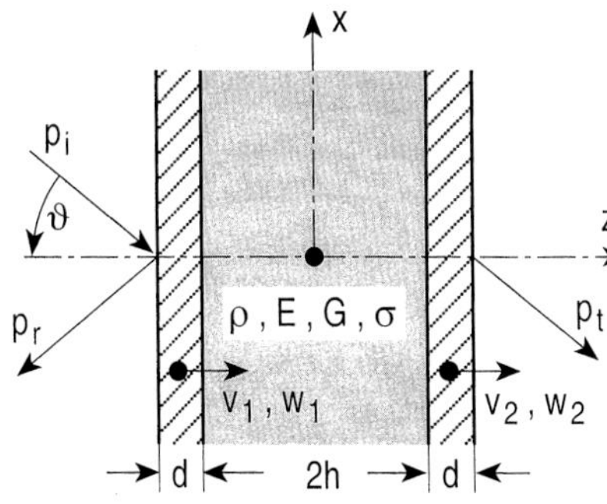

An elastic layer of thickness 2h and ρ= density , E= Young's modulus, G= shear modulus, σ= Poisson ratio of its material is covered on both sides with identical thin sheets of thickness d and ρ_1= density, E_1= Young's modulus, σ_1= Poisson ratio of the sheet material.

A plane wave p_i is incident at a polar angle ϑ. The moduli may be complex with loss factors η_E, η_G for the core, and η_{E1} for the cover sheets.

v_1, w_1 and v_2, w_2 are the velocities and elongations of the sheets; p_r is the reflected wave; p_t is the transmitted wave. Trace wave number and normal field impedance :

$$k_\vartheta = k_0 \cdot \sin\vartheta \quad ; \quad Z_\vartheta = Z_0 / \cos\vartheta \tag{1}$$

Decomposition in compressional (symmetrical, index s) and translational (anti-symmetrical, index a) parts:

$$\begin{aligned} v_s &= \tfrac{1}{2}(v_1 - v_2) \quad ; \quad v_a = \tfrac{1}{2}(v_1 + v_2) \\ w_s &= \tfrac{1}{2}(v_1 - v_2) \quad ; \quad w_a = \tfrac{1}{2}(v_1 + v_2) \\ p_s &= \tfrac{1}{2}\left[(p_i + p_r)_{z=-(h+d)} + (p_t)_{z=+(h+d)}\right] \quad ; \quad p_a = \tfrac{1}{2}\left[(p_i + p_r)_{z=-(h+d)} - (p_t)_{z=+(h+d)}\right] \end{aligned} \tag{2}$$

Impedances :

$$Z_s = \frac{p_s}{v_s} = Z'_s + jZ''_s \quad ; \quad Z_a = \frac{p_a}{v_a} = Z'_a + jZ''_a \tag{3}$$

Transmission coefficient τ_ϑ :

$$\tau_\vartheta = Z_\vartheta^2 \frac{(Z'_s - Z'_a)^2 + (Z''_s - Z''_a)^2}{\left[(Z_\vartheta + Z'_s)^2 + Z''^2_s\right] \cdot \left[(Z_\vartheta + Z'_a)^2 + Z''^2_a\right]} \tag{4}$$

The impedances Z_s, Z_a are evaluated from :

$$\begin{aligned} Z_s &= \frac{j}{2}\left[\frac{k_\vartheta^2}{\omega}\frac{A_1}{A_2} + \frac{a_7}{\omega} - \omega \cdot a_6\right] \quad ; \quad Z_a = \frac{j}{2}\left[\frac{k_\vartheta^2}{\omega}\frac{B_1}{B_2} - \omega \cdot c_3\right] \\ A_1 &= k_\vartheta^4\left(c_1 \cdot a_4 + a_1^2\right) - k_\vartheta^2\left(a_4 \cdot a_5 + 2a_1 \cdot a_3\right) + k_\vartheta^2\omega^2\left(c_2 \cdot a_4 + c_1 \cdot c_3 + 2a_1 \cdot a_2\right) + \\ &\quad + \omega^4\left(c_2 \cdot c_3 + a_2^2\right) - \omega^2\left(c_3 \cdot a_5 + 2a_2 \cdot a_3\right) + a_3^2 \\ A_2 &= k_\vartheta^2 \cdot a_4 + \omega^2 c_3 \\ B_1 &= k_\vartheta^4\left(c_1 \cdot b_4 - b_1^2\right) - k_\vartheta^2\left(b_3 \cdot (b_4 + c_1 + 2b_1)\right) + k_\vartheta^2\omega^2\left(c_2 \cdot b_4 + c_1 \cdot b_5 - 2b_1 \cdot b_2\right) + \\ &\quad + \omega^4\left(c_2 \cdot b_5 - b_2^2\right) - \omega^2\left(b_3 \cdot (b_5 + c_2 + 2b_2)\right) \\ B_2 &= k_\vartheta^2 \cdot b_4 + b_3 + \omega^2 b_5 \end{aligned} \tag{5}$$

The coefficients a_i, b_i are evaluated with the dilatation moduli D , D_1 of the core and cover sheets, respectively :

$$D = E\frac{1-\sigma}{(1+\sigma)(1-2\sigma)} \quad ; \quad D_1 = E_1\frac{1}{1-\sigma_1^2} \tag{6}$$

from the relations :

$$\begin{aligned} &a_1 = D_1 \cdot d^2 \quad ; \quad a_2 = -\rho_1 \cdot d^2 \quad ; \quad a_3 = -2D\frac{\sigma}{1-\sigma} \\ &a_4 = 2D_1 \cdot d + 2D \cdot h \quad ; \quad a_5 = \tfrac{2}{3}G \cdot h \quad ; \quad a_6 = -2\left(\rho_1 d + \tfrac{1}{3}\rho h\right) \quad ; \quad a_7 = -\frac{2D}{h} \end{aligned} \tag{7}$$

$$b_1 = -a_1 \cdot h \quad ; \quad b_2 = -a_2 \cdot h \quad ; \quad b_3 = -2Gh$$

$$b_4 = -2h^2\left(E_1 d + \tfrac{1}{3}Eh\right) \quad ; \quad b_5 = -a_6 \cdot h^2$$

$$c_1 = -\tfrac{2}{3}E_1 d^3 \quad ; \quad c_2 = \tfrac{2}{3}\rho_1 d^3 \quad ; \quad c_3 = -2(\rho h + \rho_1 d)$$

If the core is a viscous fluid with density ρ, sound speed c, kinematic viscosity ν, the equivalent elastic constants are :

$$D \to \rho c^2 \quad ; \quad G \to j\omega\nu \quad ; \quad \sigma \to \frac{1}{2}\left(1 - j\frac{\omega\nu}{\rho c^2}\right) \tag{8}$$

and the coefficients become (if the mass ρh of the core can be neglected compared to the surface mass density $m_1 = \rho_1 d$ of the cover sheet) :

$$\begin{aligned} &a_1 = D_1 \cdot d^2 \quad ; \quad a_2 = -\rho_1 \cdot d^2 \quad ; \quad a_3 = -2\rho c^2\left(1 - j\frac{\omega\nu}{\rho c^2}\right) \Big/ \left(1 + j\frac{\omega\nu}{\rho c^2}\right) \\ &a_4 = 2D_1 \cdot d + 2\rho c^2 \cdot h \quad ; \quad a_5 = \tfrac{2}{3}j\omega\nu \cdot h \quad ; \quad a_6 = -2m_1 \quad ; \quad a_7 = -2\frac{\rho c^2}{h} \\ &b_1 = -a_1 \cdot h \quad ; \quad b_2 = -a_2 \cdot h \quad ; \quad b_3 = -2j\omega\nu h \\ &b_4 = -2D_1 \cdot h^2 - \tfrac{2}{3}\rho c^2 \cdot h^2 \quad ; \quad b_5 = -a_6 \cdot h^2 \\ &c_1 = -\tfrac{2}{3}E_1 d^3 \quad ; \quad c_2 = \tfrac{2}{3}m_1 d^2 \quad ; \quad c_3 = -2m_1 \end{aligned} \tag{9}$$

I.27 Wall of multiple sheets with air interspaces

Ref.: Sharp / Beauchamps, [I.16]

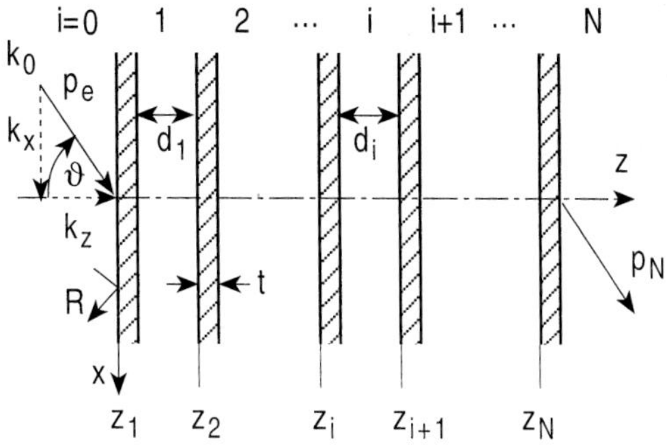

N elastic plates $i=1,\ldots,N$ with thicknesses t_i and possibly different materials are at mutual distances d_i.

A plane wave p_e is incident at a polar angle ϑ .

Incident wave p_e :

$$p_e(x,z) = p_x(x) \cdot e^{-jk_z z}$$
$$p_x(x) = P \cdot e^{-jk_x x} \quad ; \quad k_x = k_0 \cdot \sin\vartheta \quad ; \quad k_z = k_0 \cdot \cos\vartheta \tag{1}$$

Sound pressure fields :

$$p_0 = p_x(x) \cdot \left(e^{-jk_z z} + R \cdot e^{+jk_z z} \right) \quad ; \quad z < 0$$

$$\vdots$$

$$p_i = p_x(x) \cdot \left(A_i \cdot e^{-jk_z (z-z_i)} + B_i \cdot e^{+jk_z (z-z_i)} \right) \quad ; \quad z_i < z < z_{i+1} \tag{2}$$

$$\vdots$$

$$p_N = p_x(x) \cdot T \cdot e^{-jk_z (z-z_N)} \quad ; \quad z > z_N$$

with R= front side reflection factor ; T= (total) transmission factor. R_i , T_i are reflection and transmission factors at the i-th (single) plate.

Boundary conditions:

i= 1 :

$$A_1 = T_1 + B_1 \cdot R_1 \quad ; \quad R = R_1 + B_1 \cdot T_1 \tag{3}$$

i= 2,…,N–1 :

$$A_i = T_i \cdot A_{i-1} \cdot e^{-jk_z d_{i-1}} + R_i \cdot B_i \quad ; \quad B_{i-1} \cdot e^{+jk_z d_{i-1}} = B_i \cdot T_i + R_i \cdot A_{i-1} \cdot e^{-jk_z d_{i-1}} \tag{4}$$

i= N :

$$T = T_N \cdot A_{N-1} \cdot e^{-jk_z d_{N-1}} \quad ; \quad B_{N-1} \cdot e^{+jk_z d_{N-1}} = R_N \cdot A_{N-1} \cdot e^{-jk_z d_{N-1}} \tag{5}$$

Whence follows a matrix equation:

$$\begin{Bmatrix} T \\ 0 \end{Bmatrix} = C \bullet \begin{Bmatrix} 1 \\ R \end{Bmatrix} \quad ; \quad C = \prod_{i=1}^{N} \frac{1}{T_i} \cdot C^{(i)} \quad ; \quad C^{(i)} = \begin{Bmatrix} c_{11}^{(i)} & c_{12}^{(i)} \\ c_{21}^{(i)} & c_{22}^{(i)} \end{Bmatrix} \tag{6}$$

with coefficients:

i= 1 :

$$c_{11}^{(1)} = T_1^2 - R_1^2 \quad ; \quad c_{12}^{(1)} = R_1 \quad ; \quad c_{21}^{(1)} = -R_1 \quad ; \quad c_{22}^{(1)} = 1 \tag{7}$$

i= 2,…,N–1 :

$$c_{11}^{(i)} = (T_i^2 - R_i^2) e^{-jk_z d_{i-1}} \; ; \; c_{12}^{(i)} = R_i e^{+jk_z d_{i-1}} \; ; \; c_{21}^{(i)} = -R_i e^{-jk_z d_{i-1}} \; ; \; c_{22}^{(i)} = e^{+jk_z d_{i-1}} \tag{8}$$

i= N :

$$c_{11}^{(N)} = T_N^2 e^{+jk_z d_{N-1}} \quad ; \quad c_{12}^{(N)} = 0 \quad ; \quad c_{21}^{(N)} = -R_N e^{-jk_z d_{N-1}} \quad ; \quad c_{22}^{(N)} = e^{+jk_z d_{N-1}} \tag{9}$$

The required single plate reflection and transmission factors R, T (index i is dropped) are:

$$R = -F \cdot \left[1 + Z_B \cdot Z_E \frac{\cos^2\vartheta}{4Z_0^2} \right] \quad ; \quad T = -F \cdot (Z_B + Z_E) \frac{\cos\vartheta}{2Z_0} \tag{10}$$

with the auxiliary quantity :

$$F = \frac{1}{\left(1 + Z_B \frac{\cos\vartheta}{2Z_0}\right)\left(1 - Z_E \frac{\cos\vartheta}{2Z_0}\right)} \tag{11}$$

and the bending wave impedance Z_B , longitudinal wave impedance Z_E of the plate, which are

for thin plates :

$$Z_B = j\left(\omega m'' - \frac{B \cdot k_x^4}{\omega}\right) \quad ; \quad Z_E = j\frac{4m'' \cdot c_s^2}{\omega t^2 (1-\sigma)} \frac{2k_x^2 - (1-\sigma)\cdot \omega^2 / c_s^2}{k_x^2 - \omega^2 / c_d^2} \tag{12}$$

for thick plates :

$$Z_B = -j\frac{8m'' c_s^4}{t\omega^3}\left[ßk_x^2 \tanh(ßt/2) - \left(k_x^2 - \omega^2/(2c_s^2)\right)\tanh(\alpha t/2)/\alpha \right]$$

$$Z_E = -j\frac{8m'' c_s^4}{t\omega^3}\left[-ßk_x^2 \coth(ßt/2) + \left(k_x^2 - \omega^2/(2c_s^2)\right)\coth(\alpha t/2)/\alpha \right] \tag{13}$$

$$\alpha^2 = k_x^2 - \omega^2 / c_d^2 \quad ; \quad ß^2 = k_x^2 - \omega^2 / c_s^2$$

where $m'' = \rho_p t$ the surface mass density; ρ_p= plate material density; E= Young's modulus; B= bending modulus; σ= Poisson ratio; velocities c_s , c_d of shear and dilatational waves :

$$B = \frac{Et^3}{12(1-\sigma^2)} \quad ; \quad c_s = \sqrt{\frac{E}{2\rho_p(1+\sigma)}} \quad ; \quad c_d = \sqrt{\frac{E(1-\sigma)}{\rho_p(1+\sigma)(1-2\sigma)}} \tag{14}$$

Transmission loss with oblique incidence : $R_\vartheta = -10\cdot\log|T|^2$; transmission loss for diffuse incidence :

$$R = -10 \cdot \log\tau \quad ; \quad \tau = \frac{\int_0^{\vartheta_{lim}} |T|^2 \sin(2\vartheta)\, d\vartheta}{\int_0^{\vartheta_{lim}} \sin(2\vartheta)\, d\vartheta} \tag{15}$$

Special case : *Double sheet*

$$T = \frac{T_1 T_2\, e^{-jk_z d}}{1 - R_1 R_2\, e^{-2jk_z d}} \tag{16}$$

Special case : *Triple sheet*

$$T = \frac{-T_1 T_3\, e^{+jk_z(d_1+d_2)}}{R_1 R_3 \left\{ T_2 - \frac{1}{T_2}\left[\frac{e^{+2jk_z d_2}}{R_3} - R_2\right]\left[\frac{e^{+2jk_z d_1}}{R_1} - R_2\right]\right\}} \tag{17}$$

References to part I :

Sound Transmission

[I.1] MECHEL, F.P.
"Schallabsorber", Vol. III, Chapter 7:
"The absorber barrier"
S.Hirzel Verlag, Stuttgart, 1998

[I.2] MECHEL, F.P.
"Schallabsorber", Vol. III, Chapter 8:
"Sound transmission through gaps and holes (openings) in a wall"
S.Hirzel Verlag, Stuttgart, 1998

[I.3] MECHEL, F.P.
"Schallabsorber", Vol. III, Chapter 9:
"The noise sluice"
S.Hirzel Verlag, Stuttgart, 1998

[I.4] MECHEL, F.P.
"Schallabsorber", Vol. III, Chapter 10:
"Sound transmission through infinite, simple plates"
S.Hirzel Verlag, Stuttgart, 1998

[I.5] MECHEL, F.P.
"Schallabsorber", Vol. III, Chapter 11:
"Sound transmission through infinite, multiple plates"
S.Hirzel Verlag, Stuttgart, 1998

[I.6] MECHEL, F.P.
"Schallabsorber", Vol. III, Chapter 12:
"Infinite sandwich plate"
S.Hirzel Verlag, Stuttgart, 1998

[I.7] MECHEL, F.P.
"Schallabsorber", Vol. III, Chapter 14:
"Sound transmission through a finite wall in a duct"
S.Hirzel Verlag, Stuttgart, 1998

[I.8] MECHEL, F.P.
"Schallabsorber", Vol. III, Chapter 15:
"Simple plate in a niche of the baffle wall"
S.Hirzel Verlag, Stuttgart, 1998

[I.9] MECHEL, F.P.
"Schallabsorber", Vol. III, Chapter 16:
"A strip of a plate in an infinita baffle wall"
S.Hirzel Verlag, Stuttgart, 1998

[I.10] MECHEL, F.P.
"Schallabsorber", Vol. III, Chapter 17:
"Sound transmission through a finite plate combimed with an absorber layer"
S.Hirzel Verlag, Stuttgart, 1998

[I.11] MECHEL, F.P.
"Schallabsorber", Vol. III, Chapter 18:
"Multi-layer finite walls"
S.Hirzel Verlag, Stuttgart, 1998

[I.12] MECHEL, F.P.
"Schallabsorber", Vol. III, Chapter 19:
"Flanking sound transmission through suspended ceilings"
S.Hirzel Verlag, Stuttgart, 1998

[I.13] MECHEL, F.P.
"Schallabsorber", Vol. III, Chapter 24:
"Fences in a room"
S.Hirzel Verlag, Stuttgart, 1998

[I.14] ALEKSEEV ; DIANOV Sovj.Phys.Acoust. 22 (1976) 181-183

[I.15] BESHENKOV, S.N. Sovj.Phys.Acoust. 29 (1974) 115-117

[I.16] SHARP ; BEAUCHAMPS J.Sound Vibr. 9 (1969) 383

J
Duct Acoustics

This chapter deals with sound propagation in ducts. It begins with hard and smooth ducts, in which viscous and thermal losses at the walls are taken into account; this is important in narrow ducts (capillaries). The remainder deals with lined ducts of different cross sections and different linings. Sometimes the duct is assumed to be infinitely long, sometimes it has a finite length, but is still long enough to neglect reflections from the duct exit at the duct entrance. This assumption makes the contents of this chapter different from the contents of the chapter "K. Acoustic Mufflers", where the reflections at both ends of duct sections play a dominant role. A section at the end of this chapter will discuss the influence of flow on sound attenuation in lined ducts in an approximation which is precise enough for most technical applications. A more sophisticated discussion of the influence of flow will be given in the chapter "N. Flow Acoustics".

J.1 Flat capillary with isothermal boundaries

Ref.: Mechel, [J.1]

For the fundamental relations and notations used see section B.1 .

The duct has a width 2h ; x is the axial co-ordinate, y the transversal co-ordinate normal to the walls; the duct is infinite in the z direction (its dimension in this direction is much larger than that in the y direction). The co-ordinate origin is placed in the middle of the height.

The specific heat and the heat conduction of the wall material is assumed to be much higher than those of air; therefore the isothermal boundary condition holds; the temperature fluctuation of the walls is zero.

The scalar potentials for the density wave (index ρ) and the temperature wave (index α), as well as for the vector wave potential of the viscosity wave (index ν), are formulated with a common axial propagation constant Γ as :

$$\begin{aligned}\Phi_{\rho,\alpha}(x,y) &= A_{\rho,\alpha}\, e^{-\Gamma x}\cos(\varepsilon_{\rho,\alpha} y)\\ \Psi_z(x,y) &= A_\nu\, e^{-\Gamma x}\sin(\varepsilon_\nu y)\end{aligned} \tag{1}$$

The wave equations for the three types of waves then give the secular equations :

$$\varepsilon^2_{\rho,\alpha,\nu} = \Gamma^2 + k^2_{\rho,\alpha,\nu} \tag{2}$$

Wave number definitions used are :

$$k_0^2 = \left(\frac{\omega}{c_0}\right)^2 \quad ; \quad k_\nu^2 = -j\frac{\omega}{\nu} \quad ; \quad k_{\alpha 0}^2 = -j\frac{\kappa\omega}{\alpha} = \kappa\, \mathrm{Pr}\cdot k_\nu^2 \tag{3}$$

The boundary conditions at the walls lead to the system of equations (in matrix form) for the amplitudes :

$$\begin{pmatrix} \Gamma h\cos(\varepsilon_\rho h) & \Gamma h\cos(\varepsilon_\alpha h) & \varepsilon_\nu h\cos(\varepsilon_\nu h)\\ \varepsilon_\rho h\sin(\varepsilon_\rho h) & \varepsilon_\alpha h\sin(\varepsilon_\alpha h) & \Gamma h\sin(\varepsilon_\nu h)\\ \Theta_\rho\cos(\varepsilon_\rho h) & \Theta_\alpha\cos(\varepsilon_\alpha h) & 0 \end{pmatrix}\cdot\begin{pmatrix}A_\rho\\ A_\alpha\\ A_\nu\end{pmatrix} = \begin{pmatrix}0\\0\\0\end{pmatrix} \tag{4}$$

(see B.1 for $\Theta_\rho,\Theta_\alpha$). For the existence of a non-trivial solution the determinant must be zero :

$$(\Gamma h)^2\left(\frac{\Theta_\rho}{\Theta_\alpha}-1\right)\frac{\tan(\varepsilon_\nu h)}{\varepsilon_\nu h} + \varepsilon_\rho h\cdot\tan(\varepsilon_\rho h) - \frac{\Theta_\rho}{\Theta_\alpha}\varepsilon_\alpha h\cdot\tan(\varepsilon_\alpha h) = 0 \tag{5}$$

This is the exact characteristic equation for Γ . A good explicit approximation to the solution is under the condition $|\varepsilon_\rho h| \ll 1$:

$$\left(\frac{\Gamma}{k_0}\right)^2 \approx -\frac{1+(\kappa-1)\dfrac{\tan(k_{\alpha 0}h)}{k_{\alpha 0}h}}{1-\dfrac{\tan(k_\nu h)}{k_\nu h}} \tag{6}$$

The characteristic axial wave impedance Z is (with the same degree of approximation) :

$$\frac{Z}{Z_0} = \frac{\langle p(x,y)\rangle_y}{Z_0\langle v_x(x,y)\rangle_y} \approx j\frac{k_0}{\Gamma}\,\frac{\dfrac{\tan(\varepsilon_\rho h)}{\varepsilon_\rho h}}{\dfrac{\tan(\varepsilon_\rho h)}{\varepsilon_\rho h} - \dfrac{\tan(k_\nu h)}{k_\nu h}} \tag{7}$$

or with some transformations:

$$\frac{\Gamma}{k_0} = j\sqrt{\left[1+(\kappa-1)\frac{\tan(k_{\alpha 0}h)}{k_{\alpha 0}h}\right]\Big/\left[1-\frac{\tan(k_\nu h)}{k_\nu h}\right]} \tag{8}$$

$$\frac{Z}{Z_0}=\frac{1}{\sqrt{[1+(\kappa-1)\frac{\tan(k_{\alpha 0}h)}{k_{\alpha 0}h}]\cdot[1-\frac{\tan(k_\nu h)}{k_\nu h}]}} \tag{9}$$

Amplitude ratios :

$$\frac{A_\alpha}{A_\rho}=-\frac{\Theta_\rho}{\Theta_\alpha}\frac{\cos(\varepsilon_\rho h)}{\cos(\varepsilon_\alpha h)} \quad ; \quad \frac{A_\nu}{A_\rho}=(\frac{\Theta_\rho}{\Theta_\alpha}-1)\frac{\Gamma h}{\varepsilon_\nu h}\frac{\cos(\varepsilon_\rho h)}{\cos(\varepsilon_\nu h)} \tag{10}$$

Axial particle velocity profile (relative to the axial velocity of the density wave in the centre) :

$$\begin{aligned}\frac{v_x(x,y)}{v_{x\rho}(x,0)}&=\cos(\varepsilon_\rho h)\cdot\left[\frac{\cos(\varepsilon_\rho y)}{\cos(\varepsilon_\rho h)}-\frac{\Theta_\rho}{\Theta_\alpha}\frac{\cos(\varepsilon_\alpha y)}{\cos(\varepsilon_\alpha h)}+(\frac{\Theta_\rho}{\Theta_\alpha}-1)\frac{\cos(\varepsilon_\nu y)}{\cos(\varepsilon_\nu h)}\right]\\ &\approx 1-\frac{\Theta_\rho}{\Theta_\alpha}\frac{\cos(k_{\alpha 0}y)}{\cos(k_{\alpha 0}h)}+(\frac{\Theta_\rho}{\Theta_\alpha}-1)\frac{\cos(k_\nu y)}{\cos(k_\nu h)} \quad \approx 1-\frac{\cos(k_\nu y)}{\cos(k_\nu h)}\end{aligned} \tag{11}$$

Transversal particle velocity profile (relative to the axial velocity of the density wave in the centre) :

$$\begin{aligned}\frac{v_y(x,y)}{v_{x\rho}(x,0)}&=\frac{\varepsilon_\rho h}{\Gamma h}\sin(\varepsilon_\rho y)-\frac{\varepsilon_\alpha h}{\Gamma h}\frac{\Theta_\rho}{\Theta_\alpha}\frac{\cos(\varepsilon_\rho h)}{\cos(\varepsilon_\alpha h)}\sin(\varepsilon_\alpha y)+\frac{\Gamma h}{\varepsilon_\nu h}(\frac{\Theta_\rho}{\Theta_\alpha}-1)\frac{\cos(\varepsilon_\rho h)}{\cos(\varepsilon_\nu h)}\sin(\varepsilon_\nu y)\\ &\approx\frac{k_0 h}{\Gamma/k_0}[\frac{y}{h}(1+(\frac{\Gamma}{k_0})^2)+\frac{\kappa-1}{k_{\alpha 0}h}\frac{\sin(k_{\alpha 0}y)}{\cos(k_{\alpha 0}h)}-(\frac{\Gamma}{k_0})^2\frac{1}{k_\nu h}\frac{\sin(k_\nu y)}{\cos(k_\nu h)}]\end{aligned} \tag{12}$$

Example of particle velocity profiles (with $2h=4\cdot 10^{-4}$ [m]):

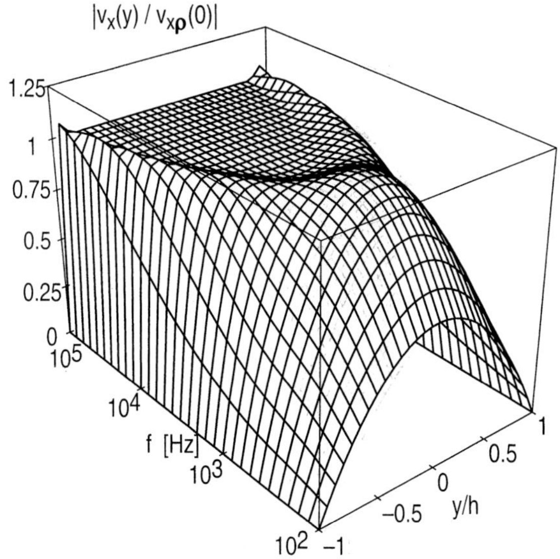

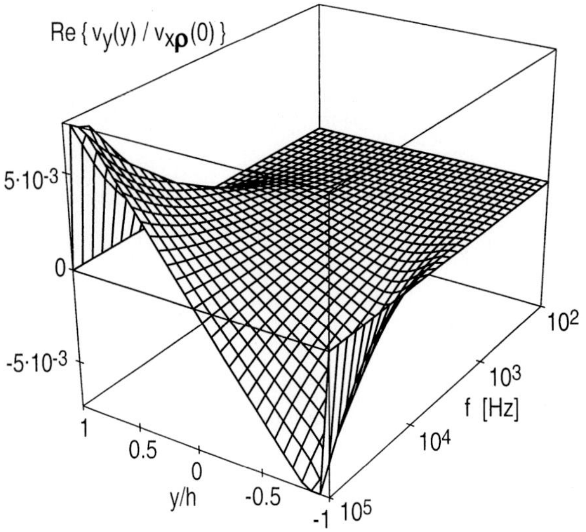

J.2 Flat capillary with adiabatic boundaries

Ref.: Mechel, [J.1]

For the fundamental relations and notations used see section B.1 .

No heat exchange takes place between the medium in the capillary and the walls.

Field formulation and secular equations as in section D.1 . The boundary conditions are :

$$\begin{pmatrix} \Gamma h\cos(\varepsilon_\rho h) & \Gamma h\cos(\varepsilon_\alpha h) & \varepsilon_\nu h\cos(\varepsilon_\nu h) \\ \varepsilon_\rho h\sin(\varepsilon_\rho h) & \varepsilon_\alpha h\sin(\varepsilon_\alpha h) & \Gamma h\sin(\varepsilon_\nu h) \\ \Theta_\rho\varepsilon_\rho h\sin(\varepsilon_\rho h) & \Theta_\alpha\varepsilon_\alpha h\sin(\varepsilon_\alpha h) & 0 \end{pmatrix} \cdot \begin{pmatrix} A_\rho \\ A_\alpha \\ A_\nu \end{pmatrix} = \begin{pmatrix} 0 \\ 0 \\ 0 \end{pmatrix} \tag{1}$$

The characteristic equation from det=0 becomes :

$$\begin{aligned} &(\Gamma h)^2 \frac{\tan(\varepsilon_\nu h)}{\varepsilon_\nu h}\left[\frac{\Theta_\rho}{\Theta_\alpha}\varepsilon_\rho h\tan(\varepsilon_\rho h) - \varepsilon_\alpha h\tan(\varepsilon_\alpha h)\right] - \\ &-\left(\frac{\Theta_\rho}{\Theta_\alpha} - 1\right)\cdot\varepsilon_\rho h\tan(\varepsilon_\rho h)\cdot\varepsilon_\alpha h\tan(\varepsilon_\alpha h) = 0 \end{aligned} \tag{2}$$

Approximate solutions for the propagation constant :

$$\left(\frac{\Gamma}{k_0}\right)^2 \approx \frac{-1}{1+\dfrac{1}{\dfrac{\Theta_\rho}{\Theta_\alpha}-1}\dfrac{\tan k_v h}{k_v h}} \approx \frac{-1}{1-\dfrac{1}{1+(\kappa-1)\dfrac{(k_0 h)^2}{(k_{\alpha 0} h)^2}}\dfrac{\tan k_v h}{k_v h}} \approx \frac{-1}{1-\dfrac{\tan k_v h}{k_v h}} \tag{3}$$

Approximate solution for the wave impedance :

$$\frac{Z_i}{\rho_0 c_0} \approx \frac{j}{\Gamma/k_0}\frac{1}{1-\dfrac{\tan(k_v h)}{k_v h}} \approx -\frac{j}{\Gamma/k_0}(\Gamma/k_0)^2 \approx -j\frac{\Gamma}{k_0} \tag{4}$$

With adiabatic boundary conditions, the normalised wave impedance is approximately the rotated normalised propagation constant.

Amplitude ratios of the component waves :

$$\frac{A_\alpha}{A_\rho} = -\frac{\Theta_\rho}{\Theta_\alpha}\frac{\varepsilon_\rho h}{\varepsilon_\alpha h}\frac{\sin(\varepsilon_\rho h)}{\sin(\varepsilon_\alpha h)}$$
$$\frac{A_v}{A_\rho} = \frac{\Gamma h}{\varepsilon_v h}\frac{\cos(\varepsilon_\rho h)}{\cos(\varepsilon_v h)}\cdot\left[\frac{\Theta_\rho}{\Theta_\alpha}\frac{\varepsilon_\rho h}{\varepsilon_\alpha h}\frac{\tan(\varepsilon_\rho h)}{\tan(\varepsilon_\alpha h)}-1\right] \tag{5}$$

J.3 Circular capillary with isothermal boundary

Ref.: Mechel, [J.1]

For the fundamental relations and used notations see section B.1 .

The capillary has the radius a . The temperature at the wall is constant.

Formulation of the scalar potentials for the density wave (index ρ) and the temperature wave (index α), as well as for the vector wave potential of the viscosity wave (index ν) with a common axial propagation constant Γ :

$$\Phi_{\rho,\alpha}(r,z) = A_{\rho,\alpha}\cdot e^{-\Gamma x}\cdot J_0(\varepsilon_{\rho,\alpha} r)$$
$$\Psi_\varphi(r,z) = A_v\cdot e^{-\Gamma x}\cdot J_1(\varepsilon_v r) \tag{1}$$

with Bessel functions $J_n(z)$. Secular equations $\varepsilon^2_{\rho,\alpha,\nu} = \Gamma^2 + k^2_{\rho,\alpha,\nu}$. (2)

The boundary conditions (in matrix form) :

$$\begin{pmatrix} \Gamma\cdot J_0(\varepsilon_\rho a) & \Gamma\cdot J_0(\varepsilon_\alpha a) & \varepsilon_\nu\cdot J_0(\varepsilon_\nu a) \\ \varepsilon_\rho\cdot J_1(\varepsilon_\rho a) & \varepsilon_\alpha\cdot J_1(\varepsilon_\alpha a) & \Gamma\cdot J_1(\varepsilon_\nu a) \\ \Theta_\rho\cdot J_0(\varepsilon_\rho a) & \Theta_\alpha\cdot J_0(\varepsilon_\alpha a) & 0 \end{pmatrix}\cdot\begin{pmatrix} A_\rho \\ A_\alpha \\ A_\nu \end{pmatrix}=\begin{pmatrix} 0\\0\\0 \end{pmatrix} \tag{3}$$

Characteristic equation for Γ :

$$(\Gamma a)^2(\frac{\Theta_\rho}{\Theta_\alpha}-1)\frac{J_1(\varepsilon_\nu a)}{\varepsilon_\nu a\cdot J_0(\varepsilon_\nu a)}+\varepsilon_\rho a\frac{J_1(\varepsilon_\rho a)}{J_0(\varepsilon_\rho a)}-\frac{\Theta_\rho}{\Theta_\alpha}\varepsilon_\alpha a\frac{J_1(\varepsilon_\alpha a)}{J_0(\varepsilon_\alpha a)}=0 \tag{4}$$

Approximate solution :

$$(\frac{\Gamma}{k_0})^2\approx\frac{\dfrac{\Theta_\rho}{\Theta_\alpha(k_0 a)^2}\cdot k_{\alpha 0}a\dfrac{J_1(k_{\alpha 0}a)}{J_0(k_{\alpha 0}a)}-\dfrac{1}{2}}{(\dfrac{\Theta_\rho}{\Theta_\alpha}-1)\dfrac{J_1(k_\nu a)}{k_\nu a\cdot J_0(k_{\alpha 0}a)}+\dfrac{1}{2}}\approx$$
$$\approx-\frac{1+(\kappa-1)\cdot 2\dfrac{J_1(k_{\alpha 0}a)}{k_{\alpha 0}a\cdot J_0(k_{\alpha 0}a)}}{1-2\dfrac{J_1(k_\nu a)}{k_\nu a\cdot J_0(k_\nu a)}}=-\frac{1+(\kappa-1)\cdot J_{1,0}(k_{\alpha 0}a)}{1-J_{1,0}(k_\nu a)} \tag{5}$$

Wave impedance Z with: $$J_{1,0}(x)=\frac{2}{x}\frac{J_1(x)}{J_0(x)} \tag{6}$$

$$\frac{Z}{Z_0}=j\frac{(k_\rho a)^2}{k_0 a\cdot\Gamma a}\frac{1-\dfrac{(k_\rho a)^2}{(k_{\alpha 0}a)^2}}{1-\kappa\dfrac{(k_\rho a)^2}{(k_{\alpha 0}a)^2}}\frac{J_{1,0}(\varepsilon_\rho a)-\dfrac{\Pi_\alpha}{\Pi_\rho}\dfrac{\Theta_\rho}{\Theta_\alpha}J_{1,0}(\varepsilon_\alpha a)}{J_{1,0}(\varepsilon_\rho a)-\dfrac{\Theta_\rho}{\Theta_\alpha}J_{1,0}(\varepsilon_\alpha a)+(\dfrac{\Theta_\rho}{\Theta_\alpha}-1)J_{1,0}(\varepsilon_\nu a)}$$
$$\approx\frac{j}{\Gamma/k_0}\frac{J_{1,0}(\varepsilon_\rho a)-(1-\dfrac{4\kappa\Pr}{3})(\kappa-1)\dfrac{(k_0 a)^2}{(k_{\alpha 0}a)^2}J_{1,0}(\varepsilon_\alpha a)}{J_{1,0}(\varepsilon_\rho a)-(\kappa-1)\dfrac{(k_0 a)^2}{(k_{\alpha 0}a)^2}J_{1,0}(\varepsilon_\alpha a)-(1+(\kappa-1)\dfrac{(k_0 a)^2}{(k_{\alpha 0}a)^2})J_{1,0}(\varepsilon_\nu a)}$$
$$\approx\frac{j}{\Gamma/k_0}\frac{J_{1,0}(\varepsilon_\rho a)}{J_{1,0}(\varepsilon_\rho a)-J_{1,0}(k_\nu a)}\approx\frac{j}{\Gamma/k_0}\frac{1}{1-J_{1,0}(k_\nu a)} \tag{7}$$

Combined solutions for numerical applications:

$$\frac{\Gamma}{k_0}=j\sqrt{\frac{1+(\kappa-1)J_{1,0}(k_{\alpha 0}a)}{1-J_{1,0}(k_\nu a)}}$$
$$\frac{Z}{Z_0}=\frac{1}{\sqrt{[1+(\kappa-1)J_{1,0}(k_{\alpha 0}a)]\cdot[1-J_{1,0}(k_\nu a)]}} \tag{8}$$

Effective density ρ_{eff}: $$\frac{\rho_{eff}}{\rho_0}=-j\frac{\Gamma}{k_0}\cdot\frac{Z_i}{Z_0}=\frac{1}{1-J_{1,0}(k_\nu a)} \tag{9}$$

Effective compressibility C_{eff}

$$\frac{C_{eff}}{C_0} = -j\frac{\Gamma}{k_0} / \frac{Z_i}{Z_0} = 1 + (\kappa - 1)\, J_{1,0}(k_{\alpha 0} a) \qquad (10)$$

(ρ_0 , C_0 values for air without losses).

Amplitude ratios of component waves :

$$\frac{A_\alpha}{A_\rho} = -\frac{\Theta_\rho}{\Theta_\alpha}\frac{J_0(\varepsilon_\rho a)}{J_0(\varepsilon_\alpha a)} \quad ; \quad \frac{A_\nu}{A_\rho} = (\frac{\Theta_\rho}{\Theta_\alpha} - 1)\frac{\Gamma a}{\varepsilon_\nu a}\frac{J_0(\varepsilon_\rho a)}{J_0(\varepsilon_\nu a)} \qquad (11)$$

Approximations from the literature to the propagation constant in capillaries (for more details, see [D.1]) ; the thick curve represents the solution of the exact characteristic equation; the approximation given above nearly coincides with the exact curve; it agrees with the approximation by ZWIKKER / KOSTEN, although it is differently derived :

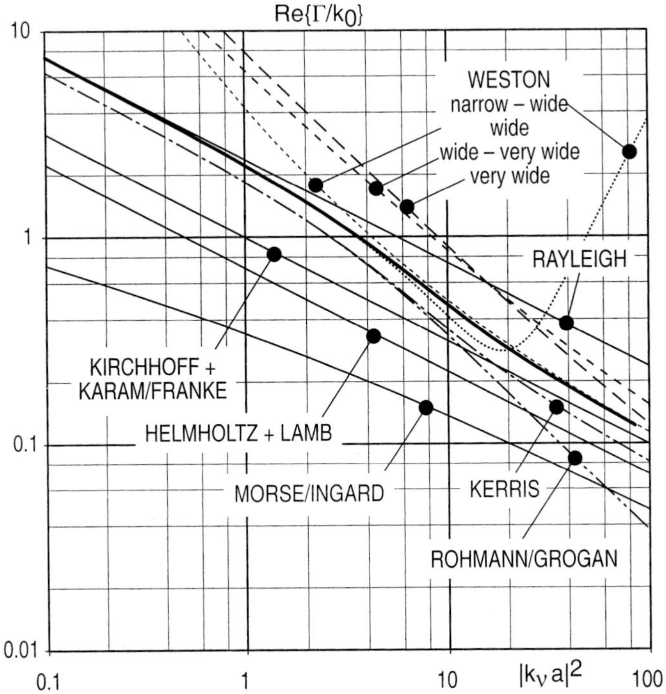

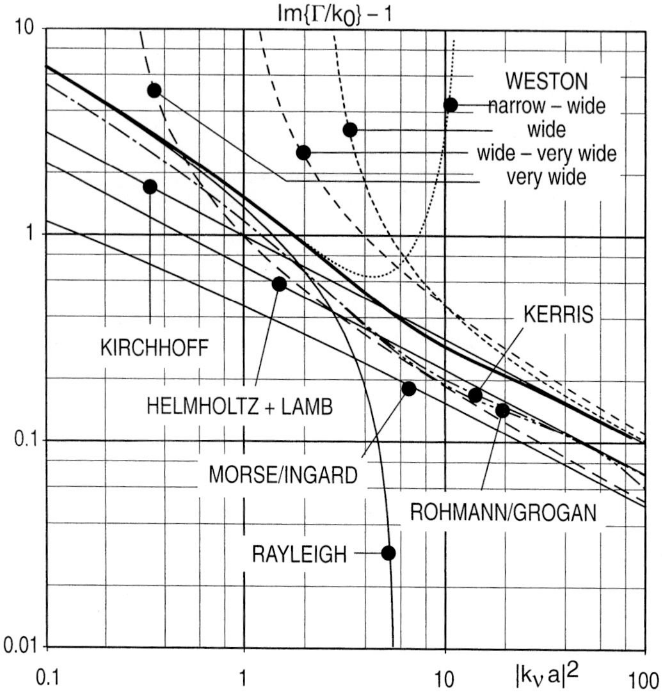

J.4 Lined ducts, general

In general, the interior space of lined ducts is prismatic or cylindrical, and the surfaces of the lining can be assumed to lie on co-ordinate surfaces of co-ordinate systems in which the wave equation is separable. In Cartesian co-ordinates, for example, a solution of the wave equation (without flow, see later sections for flow superposition) with the axis in the x direction is :

$$p(x,y,z) = \left(a\cdot\cos(\varepsilon_m y)+b\cdot\sin(\varepsilon_m y)\right)\cdot\left(\alpha\cdot\cos(\eta_n z)+\beta\cdot\sin(\eta_n z)\right)\cdot\left(c\cdot e^{-\Gamma_{m,n}x}+d\cdot e^{+\Gamma_{m,n}x}\right) \tag{1}$$

with the *secular equation* :

$$\Gamma_{m,n}^2 = \varepsilon_m^2+\eta_n^2-k_0^2 \tag{2}$$

If the x axis is in the duct centre, then the 1st terms in the parentheses (with a,α) describe symmetrical fields, the 2nd terms (with b,β) describe anti-symmetrical fields. The lateral wave numbers ε_m,η_n are solutions of a *characteristic equation* which follows from the boundary conditions at the lining surfaces. For locally reacting linings (see later sections for other types of linings) with surface admittances G_y, G_z on both sides of the y direction and z direction, respectively, with half-duct heights h_y, h_z in these directions, the characteristic equations for the symmetrical field components are :

$$\begin{aligned} \varepsilon_m h_y \cdot \tan(\varepsilon_m h_y) &= j\, k_0 h_y \cdot Z_0 G_y =: j U_y \\ \eta_n h_z \cdot \tan(\eta_n h_z) &= j\, k_0 h_z \cdot Z_0 G_z =: j U_z \end{aligned} \tag{3}$$

The right-hand sides, and therefore U_y, U_z are known values for given linings. The equations have an infinite number m,n= (0),1,2,... of solutions because of the periodicity of the tan function. They are *mode solutions*. It is important that the task of finding mode solutions is the same, whether the duct is 2-dimensional or 3-dimensional; in the latter case the same task has to be solved twice, and only in the evaluation of the axial propagation constant from

$$\Gamma_{m,n}^2 = \varepsilon_m^2 + \eta_n^2 - k_0^2 \tag{4}$$

the dimensionality enters. That is why in Cartesian co-ordinates mostly 2-dimensional (*flat*) ducts are considered.

Mostly the linings on opposite sides of the duct are the same; the duct is symmetrical. If further the sound excitation is symmetrical, it is sufficient to consider symmetrical modes only (it is important in this context that the *least attenuated mode* is a symmetrical mode).

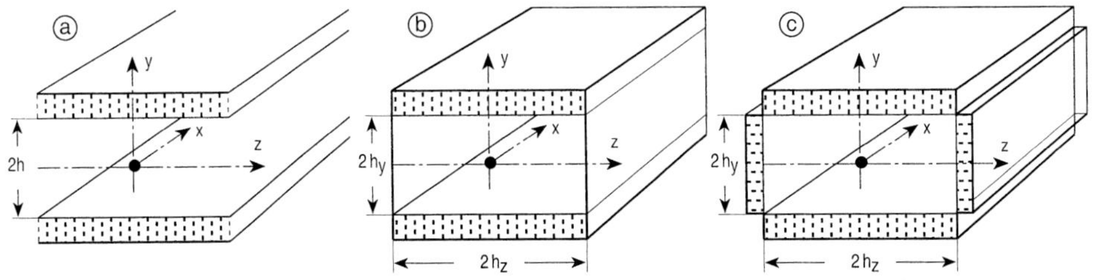

A "standard form" of a rectangular lined duct therefore is a flat duct with a hard wall at y=0 (plane of symmetry) and a lining surface at y=h . The secular equation then becomes :

$$\Gamma_m^2 = \varepsilon_m^2 - k_0^2 \tag{5}$$

It is automatically satisfied, if a *modal angle* is introduced by :

$$\begin{gathered} 1 = (\varepsilon_n / k_0)^2 - (\Gamma_n / k_0)^2 = (\varepsilon_n / k_0)^2 + (\Gamma_n / j k_0)^2 = \sin^2\Theta_n + \cos^2\Theta_n \\ \sin\Theta_n = \frac{\varepsilon_n}{k_0} \quad ; \quad \cos\Theta_n = \frac{\Gamma_n}{j k_0} \end{gathered} \tag{6}$$

With the present choice of association, the modal angle Θ_n is the angle which the wave vector of the plane waves include with the x axis, which by their superposition form the trigonometric lateral mode profile :

$$\left.\begin{matrix}\cos(\varepsilon_y y)\\ \sin(\varepsilon_y y)\end{matrix}\right\} = \frac{1}{2}\left[e^{+j\varepsilon_y y} \pm e^{-j\varepsilon_y y}\right] \tag{7}$$

The mode angles are defined, even when they become complex quantities.

The most important target quantity of a silencer of finite length L (L sufficiently large, so that end reflections can be neglected; see later sections for end effects) is its transmission loss $D_L = 8.68 \cdot \mathrm{Re}\{\Gamma L\} = L/h \cdot D_h$ with $D_h = 8.68 \cdot \mathrm{Re}\{\Gamma h\}$. D_h is the preferred quantity for the presentation of the silencer attenuation, because it follows immediately from the secular and characteristic equations, and it permits a better comparison between silencers with different linings.

Suppose we have a set of computed D_h curves, plotted in a double-logarithmic scale over f·h [Hz·m] and a required transmission loss D_L plotted in a double-logarithmic scale over f [Hz], and one of the diagrams be on a transparent foil (or the diagrams are plotted in a graphics computer program which permits drawing and moving of graphs in different levels). Then select a suitable silencer with the following procedure.

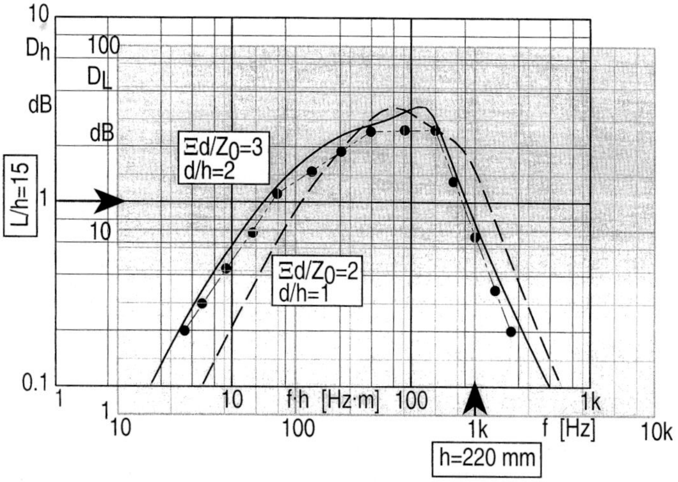

The required D_L over f is plotted as points in the shaded diagram. The computed D_h over f·h are plotted as lines in the other diagram. Move one of the diagrams so, that the D_L values are just below a computed D_h. Read opposite to f=1 kHz the value of h in mm, and opposite to D_h=1 the value of L/h . The other values needed are taken from the parameter list.

The main sub-task in the determination of D_h is the solution of the characteristic equation. Important tools are *Muller's procedure* for the solution of transcendent complex equations, and the *continued fraction* representation of the tan(z), cot(z) functions as well as of ratios of Bessel functions.

Muller's procedure for the solution of the equation f(z)=0 needs three start values z_{i-2}, z_{i-1}, z_i and the associated function values $f_{i-2} = f(z_{i-2})$, $f_{i-1} = f(z_{i-1})$, $f_i = f(z_i)$. A new approximation to the solution is :

$$z_{i+1} = z_i + \lambda_{i+1} \cdot (z_i - z_{i-1}) \tag{8}$$

where λ_{i+1} is a solution of the quadratic equation :

$$\lambda_{i+1}^2 \cdot \lambda_i h_i + \lambda_{i+1} \cdot g_i + f_i \delta_i = 0 \tag{9}$$

with the abbreviations :

$$\begin{aligned}
\lambda_i &= \frac{z_i - z_{i-1}}{z_{i-1} - z_{i-2}} \\
\delta_i &= 1 + \lambda_i = \frac{z_i - z_{i-2}}{z_{i-1} - z_{i-2}} \\
h_i &= f_{i-2}\lambda_i - f_{i-1}\delta_i + f_i \\
g_i &= f_{i-2}\lambda_i^2 - f_{i-1}\delta_i^2 + f_i(\lambda_i + \delta_i)
\end{aligned} \tag{10}$$

Therefore :

$$\lambda_{i+1} = \frac{1}{2\lambda_i h_i}\left[-g_i \pm \sqrt{g_i^2 - 4 f_i \lambda_i \delta_i h_i}\,\right] = \frac{-2 f_i \delta_i}{g_i \pm \sqrt{g_i^2 - 4 f_i \lambda_i \delta_i h_i}} \tag{11}$$

The sign of the root is selected so that the denominator in the 2nd form has the maximum magnitude. Special cases are :

$$\begin{aligned}
\lambda_i = 0: &\quad \lambda_{i+1} = -\frac{f_i}{f_i - f_{i-1}} \\
h_i = 0: &\quad \lambda_{i+1} = -\frac{f_i \delta_i}{g_i} \\
g_i = 0: &\quad \lambda_{i+1} = \pm j\sqrt{\frac{f_i \delta_i}{\lambda_i h_i}} \\
\text{radicand} = 0: &\quad \lambda_{i+1} = -\frac{g_i}{2\lambda_i h_i} = -\frac{2 f_i \delta_i}{g_i}
\end{aligned} \tag{12}$$

Approximations z_i must not coincide. The iteration is terminated if either or both

$$|f(z_{i+1})| \le \delta^2 \quad \text{and/or} \quad |1 - z_i/z_{i+1}| < \delta \tag{13}$$

with a small number δ ($\approx 10^{-8}$). One can, to some degree, influence the direction of the search for the solution by the arrangement of the starters z_{i-2}, z_{i-1}, z_i .

Continued fractions Cf may be written in one of the forms :

$$Cf = b_0 + \cfrac{a_1}{b_1 + \cfrac{a_2}{b_2 + \cfrac{a_3}{b_3 + \ldots}}} = b_0 + \frac{a_1}{b_1 +}\,\frac{a_2}{b_2 +}\,\frac{a_3}{b_3 +}\ldots \tag{14}$$

The evaluation "from behind" is fast if one knows where to truncate the expansion. An evaluation in the opposite direction uses the recursion :

$$
\begin{aligned}
&Cf_n = \frac{A_n}{B_n} = b_0 + \frac{a_1}{b_1 +}\frac{a_2}{b_2 +}\frac{a_3}{b_3 +}\dots\frac{a_n}{b_n} \\
&A_{-1} \equiv 1 \;;\quad A_0 = b_0 \;;\quad B_{-1} \equiv 0 \;;\quad B_0 \equiv 1 \\
&A_n = b_n A_{n-1} + a_n A_{n-2} \\
&B_n = b_n B_{n-1} + a_n B_{n-2}
\end{aligned}
\tag{15}
$$

Tests of convergence can be performed repeatedly after some number of steps.

J.5 Modes in rectangular ducts with locally reacting lining

Ref.: Mechel, [J.2]

Let the axial co-ordinate x be in the centre of the duct with heights $2h_y$, $2h_z$. Let the linings on opposite walls be equal and have the surface admittances G_y, G_z on the walls normal to the y and z axis, respectively. Modes (i.e. solutions of the wave equation and of the boundary conditions) have the form :

$$p(x,y,z) = q_y(y)\cdot q_z(z)\cdot e^{-\Gamma x} \tag{1}$$

with lateral profiles :

$$
\begin{aligned}
q_y(y) &= \begin{cases} \cos(\varepsilon_y y) \;;\; \text{symmetrical mode} \\ \sin(\varepsilon_y y) \;;\; \text{anti-symmetrical mode} \end{cases} \\
q_z(z) &= \begin{cases} \cos(\varepsilon_z z) \;;\; \text{symmetrical mode} \\ \sin(\varepsilon_z z) \;;\; \text{anti-symmetrical mode} \end{cases}
\end{aligned}
\tag{2}
$$

The wave equation is satisfied if the secular equation holds :

$$\Gamma^2 = \varepsilon_y^2 + \varepsilon_z^2 - k_0^2 \;;\quad Re\{\Gamma\} \ge 0 \;;\quad Im\{\Gamma\} \ge 0 \tag{3}$$

(the 1st sign convention has priority; the 2nd convention holds if $Re\{\dots\}=0$). The boundary conditions at the lining surfaces give the characteristic equations :

symmetrical modes :

$$
\begin{aligned}
\varepsilon_y h_y \cdot \tan(\varepsilon_y h_y) &= j k_0 h_y \cdot Z_0 G_y =: jU_y \\
\varepsilon_z h_z \cdot \tan(\varepsilon_z h_z) &= j k_0 h_z \cdot Z_0 G_z =: jU_z
\end{aligned}
\tag{4}
$$

anti-symmetrical modes :

$$
\begin{aligned}
\varepsilon_y h_y \cdot \cot(\varepsilon_y h_y) &= -j k_0 h_y \cdot Z_0 G_y =: -jU_y \\
\varepsilon_z h_z \cdot \cot(\varepsilon_z h_z) &= -j k_0 h_z \cdot Z_0 G_z =: -jU_z
\end{aligned}
\tag{5}
$$

U_y, U_z are known quantities for a given lining. If two opposite walls are hard, e.g. $G_z=0$:

$$\varepsilon_z h_z = \begin{cases} n\pi & ; \; n=0,1,2,\ldots \; ; \; \text{symmetrical modes} \\ (n+1/2)\pi & ; \; n=0,1,2,\ldots \; ; \; \text{anti-symmetrical modes} \end{cases} \tag{6}$$

Modes in locally lined ducts are orthogonal to each other over the duct height :

$$\frac{1}{2h_y}\int_{-h_y}^{h_y} q_{ym}(y)\cdot q_{yn}(y)\,dy = \delta_{m,n}\cdot N_{yn} = \frac{\delta_{m,n}}{2}\left[1 \pm \frac{\sin 2\varepsilon_m h_y}{2\varepsilon_m h_y}\right] \; ; \; \begin{cases}\text{symm.} \\ \text{anti-symm.}\end{cases} \tag{7}$$

N_{yn} are the mode norms. The characteristic equations for locally reacting linings have the form, with $\varepsilon h \to z$:

$$z\cdot \tan z = jU \quad ; \; \text{symm.} \; ; \quad z/\tan z = -jU \; ; \; \text{anti-symm.} \tag{8}$$

with U a known (in general complex) number with positive real part. They induce a transformation between z and U . If one plots for given real or imaginary parts of $z = z'+j\cdot z''$ with running second part the evaluated value of U in the complex plane, one gets a type of "Morse chart".

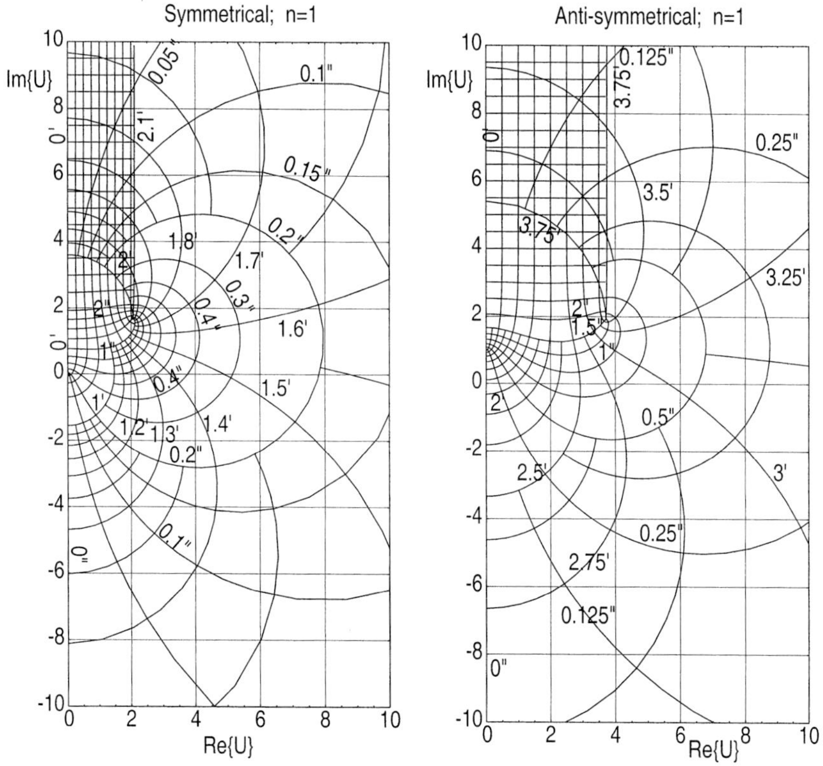

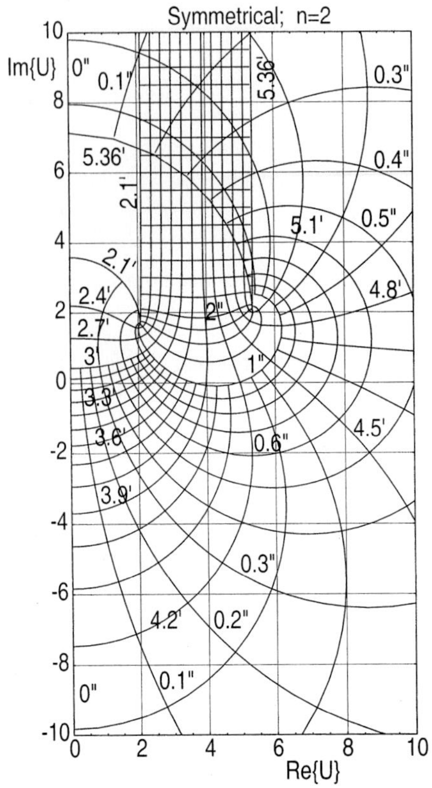

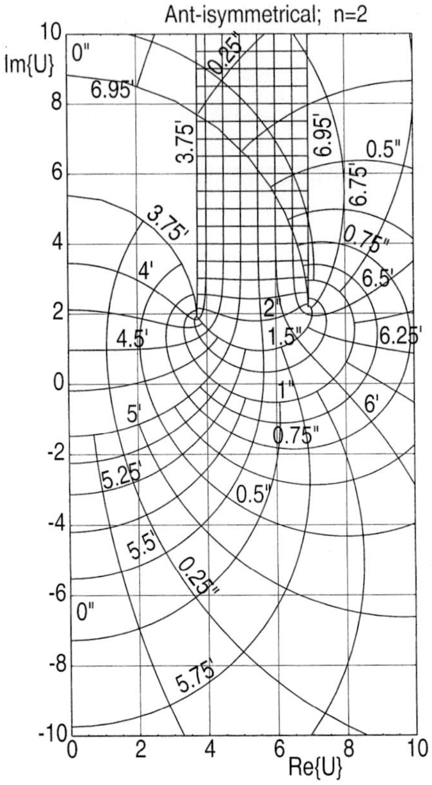

If one plots the above charts over the complex U plane and introduces a 3rd dimension $\mathrm{Re}\{z\}$, one gets more instructive three-dimensional charts, for example for symmetrical modes:

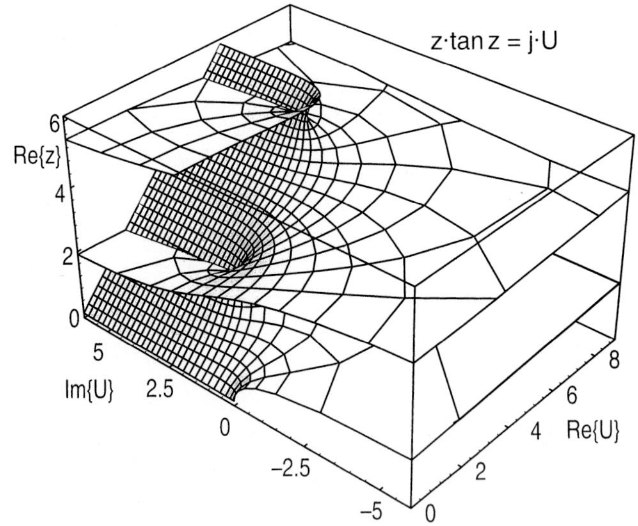

Evidently there are two types of modes: one set of modes with curved chart lines, and a single mode with nearly rectilinear chart lines.
The modes with curved lines have correspondences in a hard duct; the mode with the rectilinear lines is a surface wave; there is no corresponding solution in the hard duct.

J.6 Least attenuated mode in rectangular, locally lined ducts

Ref.: Mechel, [J.2]

Designing a silencer with the attenuation of the least attenuated mode is a "safe" design, because the least attenuated mode is one of the modes to excite easily (see later section about excitation efficiency of modes), and for a sufficiently long silencer other possibly excited modes will have decayed at the silencer exit, so that the least attenuated mode determines the exit sound pressure level.

The least attenuated mode is among the two lowest symmetrical modes. A number of methods have been described for its evaluation. The lateral wave number $z=\varepsilon h$ is a solution of the characteristic equation

$$z\cdot \tan z = jU \qquad (1)$$

with

$$U = k_0 h\cdot Z_0 G\ . \qquad (2)$$

The characteristic equation is even in z, so approximations for z^2 are normal.

An approximate solution z_{ap} can be tested as follows: let run the parts of $z_{ex}=z'_{ex}+j\cdot z''_{ex}$ and evaluate the associated U from the characteristic equation; then solve with this value of U for the approximation z_{ap}; plot the lines for z_{ex} in the complex plane of z_{ap}. If the approximation is good, it reproduces (approximately) the co-ordinate grid of that plane.

1st approximation :

Expand tan(z) as a power series around z=0 and retain the 1st term; this gives

$$z^2 \approx j\cdot U\ . \qquad (3)$$

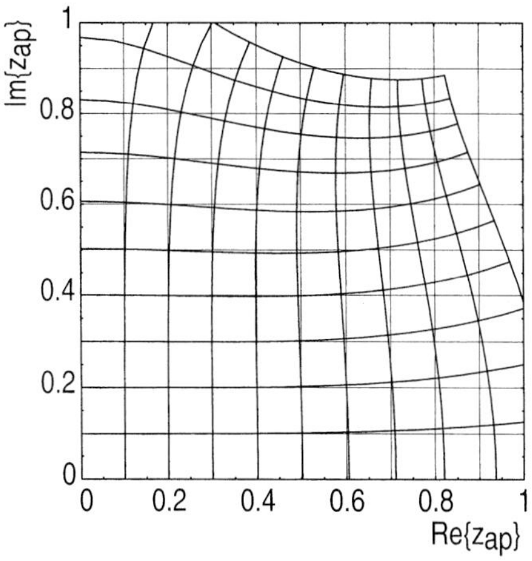

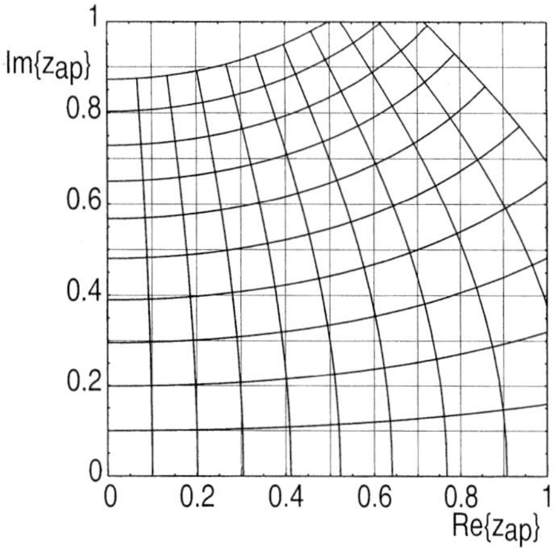

2nd approximation :

The power series expansion up to terms z^4 gives the approximation :

$$z^2 \approx 3/2\cdot\left(-1+\sqrt{1+4jU/3}\right) \qquad (4)$$

$$z^4 - (15 + 6jU)z^2 + 15jU = 0$$

$$z^2 = \frac{1}{2}\left(15 + 6jU \pm \sqrt{225 + 120jU - 36U^2}\right) \tag{19}$$

$$(10 + jU)z^4 - (105 + 45jU)z^2 + 105jU = 0 \tag{20}$$

$$\boxed{z^2 = \frac{105 + 45jU \pm \sqrt{11025 + 5250jU - 1505U^2}}{20 + 2jU}} \tag{21}$$

$$z^6 - (105 + 15jU)z^4 + (945 + 420jU)z^2 - 945jU = 0$$

$$(21 + jU)z^6 - (1260 + 210jU)z^4 + (10395 + 4725jU)z^2 - 10395jU = 0 \tag{22}$$

$$z^8 - (378 + 28jU)z^6 + (17325 + 3150jU)z^4 - (135135 + 62370jU)z^2 + \\ + 135135jU = 0 \tag{23}$$

$$(36 + jU)z^8 - (6930 + 630jU)z^6 + (270270 + 51975jU)z^4 - \\ - (2027025 + 945945jU)z^2 + 2027025jU = 0 \tag{24}$$

The precision test of the framed approximation is shown in the graph at the side. In the grey area the sign of the root was chosen so as to make the real part of the root positive, and in the other range negative. The limit line passes through the 1st branch point (where the lines are curved).

Evidently this approximation can be used also for parts of the 2nd "Morse chart", i.e. for the 2nd mode.

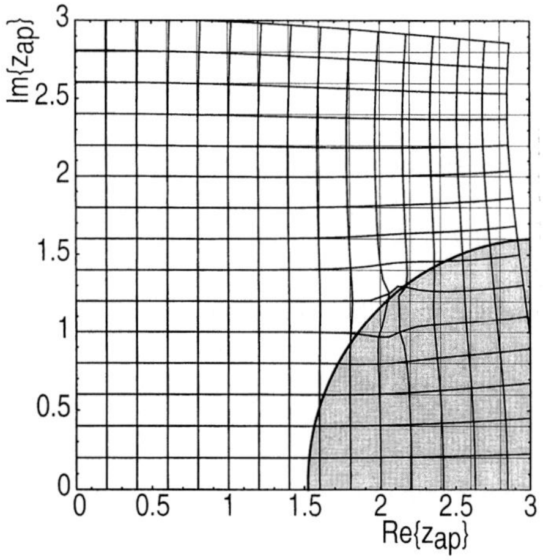

Higher degree polynomials of the continued fraction expansion give more than just one solution for z^2. They belong (with different precision) to the lower order modes. To find the solution for the least attenuated mode, exclude all approximations z which are not in the 1st quadrant, and take from the remaining approximations that which makes $\mathrm{Re}\{\Gamma h\}$ a minimum.

FROMMHOLD has modified the coefficients of the continued fraction approximations to move the range of application more towards the range of technical values of U ; he proposes :

If $0 \le \mathrm{Re}\{Z_0 G\} \le 3$; $-1.5 \le \mathrm{Im}\{Z_0 G\} \le 1.5$:

$$z^2 \approx \frac{(2.74-0.52\cdot j)\,jU}{2.88-0.55\cdot j+jU} \tag{25}$$

If $2 \le \mathrm{Re}\{Z_0G\} \le 5$; $3 \le \mathrm{Im}\{Z_0G\} \le 6$:

$$z^2 \approx \frac{(78.94-5.43\cdot j)+(34.47-2.2\cdot j)\,jU \pm \sqrt{\ldots}}{(16.1-1.11\cdot j)+2\,jU}$$
$$\sqrt{\ldots} = \sqrt{(6203-857\cdot j)+(2887.3-372\cdot j)\,jU-(867.4-130\cdot j)\,U^2} \tag{26}$$

The sign of the root is determined with the criterion $\mathrm{Re}\{\Gamma h\}=$ minimum .

J.7 Sets of mode solutions in rectangular, locally lined ducts

Ref.: Mechel, [J.2]

The charts of the transformation $z \to U$ which is induced by the characteristic equation show branch points. The evaluation of sets of mode solutions begins with the determination of these branch points z_b and the associated values U_b follow from the characteristic equations :

$$\begin{aligned} &z\cdot\tan z = jU \quad ; \quad \text{symmetrical modes} \\ &z/\tan z = -jU \quad ; \quad \text{anti-symmetrical modes} \end{aligned} \tag{1}$$

The branch points are solutions of :

$$\tan z + \frac{z}{\cos^2 z} = 0 \;\; ; \;\; \text{symmetrical} \;\; ; \;\; \cot z - \frac{z}{\sin^2 z} = 0 \;\; ; \;\; \text{anti-symmetrical} \tag{2}$$

Approximations of the functions $z''_b = f(z'_b)\,;\, U''_b = g(U'_b)\,;\, z'_b(m)\,;\, U'_b(m)$
(with $z_b = z'_b + j\cdot z''_b\,;\, U_b = U'_b + j\cdot U''_b$ and m= 0,1,2,... the mode order) are : (3)

$$z''_b = 0.702568\cdot(z'_b)^{1/3} + 0.216438\cdot(z'_b)^{1/2} - 0.036625\cdot z'_b + 0.000143119\cdot z'^2_b \;\; ; \;\; \text{symm.}$$
$$z''_b = 0.0232164 + 0.829796\cdot(z'_b)^{1/2} - 0.0827732\cdot z'_b + 0.000351925\cdot z'^2_b \;\; ; \;\; \text{anti-symm.}$$

$$\begin{aligned} U''_b &= 2.02599\cdot(U'_b)^{1/3} - 0.655631\cdot(U'_b)^{1/2} + 0.00631631\cdot U'_b + \\ &\quad + 0.000002508 27\cdot U'^2_b \;\; ; \;\; \text{symm.} \\ U''_b &= 1.0 + 0.553673\cdot(U'_b)^{1/2} - 0.0412894\cdot U'_b + 0.000120216\cdot U'^2_b \;\; ; \;\; \text{anti-symm.} \end{aligned} \tag{4}$$

$$z'_b(m) = -1.4403\,\sqrt{m} + 3.76029 \cdot m - 0.0284415 \cdot m^2 + 0.000620241 \cdot m^3 \quad ; \ \text{symm.}$$
$$z'_b(m) = 3.39478 \cdot m - 0.023865 \cdot m^2 + 0.000669072 \cdot m^3 \quad ; \ \text{anti-symm.} \tag{5}$$

$$U'_b(m) = -1.50237\sqrt{m} + 3.76029 \cdot m - 0.0284415 \cdot m^2 + 0.000620241 \cdot m^3 \quad ; \ \text{symm.}$$
$$U'_b(m) = 3.39042 \cdot m - 0.023312 \cdot m^2 + 0.000651624 \cdot m^3 \quad ; \ \text{anti-symm}$$

	symmetrical		anti-symmetrical	
m	z_b	U_b	z_b	U_b
0	0.+j 0.	0.+j 0.	0.+j 0.	0.+j 1.
1	2.1062 + j 1.12536	2.05998 + j 1.65061	3.74884 + j 1.38434	3.71944 + j 1.89528
2	5.35627 + j 1.55157	5.33471 + j 2.05785	6.94998 + j 1.6761	6.93297 + j 2.18022
3	8.53668 + j 1.77554	8.52264 + j 2.27847	10.1193 + j 1.85838	10.1073 + j 2.36058
4	11.6992 + j 1.9294	11.6888 + j 2.43112	13.2773 + j 1.99157	13.2681 + j 2.49295
5	14.8541 + j 2.04685	14.8458 + j 2.54799	16.4299 + j 2.09663	16.4224 + j 2.59758
6	18.0049 + j 2.14189	17.9981 + j 2.64271	19.5794 + j 2.1834	19.5731 + j 2.6841
7	21.1534 + j 2.22172	21.1476 + j 2.72234	22.7270 + j 2.25732	22.7216 + j 2.75786
8	24.3003 + j 2.29055	24.2952 + j 2.79103	25.8734 + j 2.32171	25.8686 + j 2.82214
9	27.4462 + j 2.35105	27.4417 + j 2.85144	29.0188 + j 2.37876	29.0146 + j 2.87911
10	30.5913 + j 2.40501	30.5872 + j 2.90533	32.1636 + j 2.42996	32.1598 + j 2.93025
11	33.7358 + j 2.45372	33.7321 + j 2.95399	35.3079 + j 2.4764	35.3044 + j 2.97665
12	36.8799 + j 2.4981	36.8765 + j 2.99833	38.4518 + j 2.5189	38.4486 + j 3.01911
13	40.0236 + j 2.53887	40.0205 + j 3.03906	41.5954 + j 2.55807	41.5924 + j 3.05825
14	43.1671 + j 2.57656	43.1642 + j 3.07673	44.7387 + j 2.59439	44.7359 + j 3.09455
15	46.3103 + j 2.61161	46.3076 + j 3.11176	47.8819 + j 2.62825	47.8793 + j 3.1284
16	49.4534 + j 2.64436	49.4509 + j 3.1445	51.0248 + j 2.65997	51.0224 + j 3.1601
17	52.5963 + j 2.6751	52.5939 + j 3.17522	54.1677 + j 2.68979	54.1654 + j 3.18991
18	55.7390 + j 2.70407	55.7368 + j 3.20417	57.3104 + j 2.71794	57.3082 + j 3.21804
19	58.8817 + j 2.73144	58.8796 + j 3.23154	60.4530 + j 2.74459	60.4509 + j 3.24468
20	62.0242 + j 2.7574	62.0222 + j 3.25748	63.5955 + j 2.76988	63.5935 + j 3.26997

Table 1: Branch points z_b and associated U_b for symmetrical and anti-symmetrical modes in flat ducts with locally reacting lining.

The characteristic equations transform the U-plane (with $\mathrm{Re}\{U\} \geq 0$) into a strip in the 1st quadrant of the z-plane. A one-to-one correspondence of a z-strip and the U-plane, with limit curves which can be evaluated (!), is shown in the graph below. The z-strip is limited by lower limit curves $z_{gl}(m)$ (g= *Grenze* in German), which are vertical lines from the branch point $z_b(m)$ to the axis $\mathrm{Re}\{z\}$, and by upper limit curves $z_{gu}(m)$, which are quarter ellipses between $z_b(m)$ and $z''_{g0}(m)$ on the axis $\mathrm{Im}\{z_b\}$. The associated limit curves in the U-plane are shown in the lower graph. In the shaded range of the U-plane the surface wave mode is evaluated.

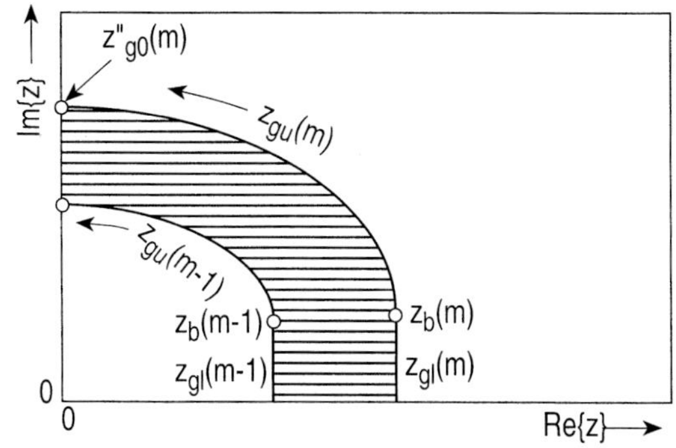

A strip in the z-plane into which the U-plane (with Re{U}≥0) is transformed. It is limited by vertical lower curves through the branch points, and by upper elliptic arcs.

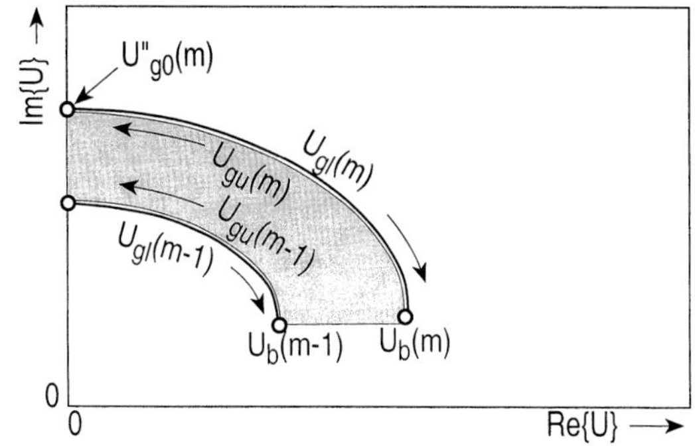

Transformation of the limits of the z-strip into the U-plane. If U is in the shaded range, the surface wave mode is evaluated.

The equation for the limit curve in the U-plane in the form $U'_g(m) = f\left(U''_g(m)\right)$ for both U_{gu} and U_{gl} is :

$$U'_g(m) = U'_b(m)\sqrt{1-\left(\frac{U''_g(m)-U''_b(m)}{\dfrac{z'_b(m)\cdot\sin(2z'_b(m))}{1+\cos(2z'_b(m))}+U''_b(m)}\right)^2} \tag{7}$$

The equation for the elliptic limit curve in the z-plane in the form $z'_{gu}(m) = f\left(z''_{gu}(m)\right)$ is :

$$z'_{gu}(m) = z'_b(m)\sqrt{1-\left(\frac{z''_{gu}(m)-z''_b(m)}{z''_{g0}(m)-z''_b(m)}\right)^2} \quad ; \quad z''_b(m) \le z''_{gu}(m) \le z''_{g0}(m) \tag{8}$$

and in the form $z''_{gu}(m) = f\left(z'_{gu}(m)\right)$:

$$z''_{gu}(m) = z''_b(m) + \left(z''_{g0}(m) - z''_b(m)\right)\sqrt{1 - \left(\frac{z'_{gu}(m)}{z'_b(m)}\right)^2} \quad ; \quad 0 \le z'_{gu}(m) < z'_b(m) \tag{9}$$

Required values of $z''_{g0}(m)\,;U''_{g0}(m)$ are contained in the next table :

m	symmetrical $z''_{g0}(m)$ & $U''_{g0}(m)$	anti-symmetrical $z''_{g0}(m)$ & $U''_{g0}(m)$
0	0.00000	0.00000
1	3.55637	5.39512
2	7.13605	8.83041
3	10.4985	12.1485
4	13.788	15.4182
5	17.0412	18.6597
6	20.276	21.8853
7	23.4928	25.0985
8	26.7013	28.2986
9	29.8984	31.4972
10	33.0907	34.6867
11	36.2800	37.8715
12	39.4623	41.0533
13	42.6456	44.2313
14	45.8196	47.4114
15	48.9979	50.5792
16	52.1657	53.7582
17	55.3359	56.9203
18	58.5124	60.0893
19	61.6746	63.2566
20	64.8482	66.4240

Table 2: Values of endpoints of elliptic arcs on imaginary axis.

It should be noted, that the branch points, range limit curves and endpoints of the elliptic arcs coincide with the origin for $m=0$.

The modes are counted as $m= 1,2,3,\ldots$ in the procedure for the evaluation of a mode solution $z=\varepsilon h$ for given values of U and m (usually counting is $m= 0,1,2,\ldots$). The procedure works with the following steps :

1st: special case $U=0$ *?*

Take
$$z=\varepsilon h=\begin{cases}(m-1)\pi & ;\ \text{symmetrical}\\ (m-1/2)\pi & ;\ \text{anti-symmetrical}\end{cases} \tag{10}$$

2nd: U in the surface wave range ?

That is, U is in the range limited by • the curve $U''_b=g(U'_b)$ which connects the branch point images, • the imaginary axis, • the curves $U'_g(n)=f\left(U''_g(n)\right)$ for $n=m-1$ and $n=m$.

In that case use the fast converging (because $z\approx U$) iteration $i=0,1,2,\ldots$

$$z_{i+1}=jU/\tan(z_i)\,;\text{symm.}\quad ;\quad z_{i+1}=-jU/\cot(z_i)\,;\text{anti-symm.} \tag{11}$$

with $z_0=U$.

3rd: Else :

Expand the characteristic equation in a continued fraction, using the periodicity of $\tan(z)$, $\cot(z)$:

$$\begin{aligned} z\cdot\tan(z-m\pi)&=\frac{z(z-m\pi)}{1-}\,\frac{(z-m\pi)^2}{3-}\,\frac{(z-m\pi)^2}{5-}\ldots=jU \quad ;\ \text{symm.}\\ z\cdot\cot(z-m\pi)&=\frac{z}{z-m\pi}\left(1-\frac{(z-m\pi)^2}{3-}\,\frac{(z-m\pi)^2}{5-}\ldots\right)=-jU \quad ;\ \text{anti-symm.}\end{aligned} \tag{12}$$

and derive a polynomial equation in z^2 by truncation. Take the solution for which $z''_b(m)\le z''<z''_b(m+1)$. Produce two start solutions for MULLER's procedure by truncating the continued fraction at different depth, and take as 3rd starter the mean value between them. Then solve the characteristic equation (in its original form) with these starters and MULLER's procedure (for many applications the solution of the polynomial equation, if the degree of the polynomial is not too low, is already sufficiently precise).

Because the curves $U''_{gu}(m)=f\left(U'_{gu}(m)\right);z''_{gu}(m)=f\left(z'_{gu}(m)\right)$ are not exact transforms of each other, it may happen that no solution of the polynomial equation has

$$z''_b(m)\le z''<z''_b(m+1)\ .$$

Then the desired solution is in or near the range of the surface wave mode. Take in this (seldom) instance the solution from the iteration above.

The (important) advantage of the procedure is that no mode in a mode set is missed or returned twice; the disadvantage is that the surface wave mode (if any exists) is subdivided and the "pieces" are attributed to different modes, so mode solution curves in the z plane do not look "nice"; but this feature does not disturb modal analysis computations. In the numerical example shown below, the surface wave solution (the arc which approximately agrees with the curve $U(k_0h)$) is subdivided; a mode solution jumps whenever this solution crosses a limit of a z-strip.

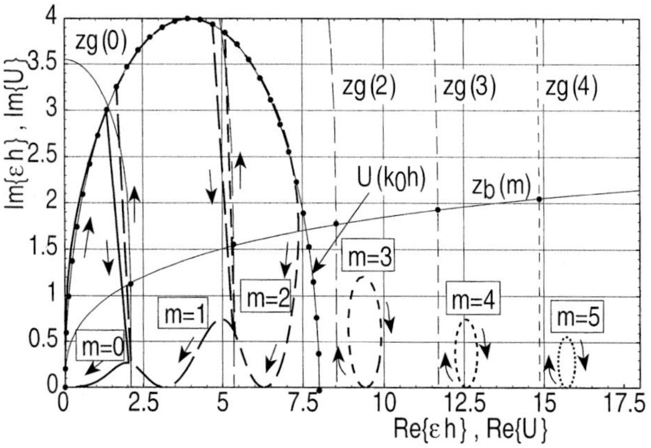

The graph shows (in the U plane) the curve $U(k_0h)$ (full line with dots) ; the curve $z_b(m)$ connecting the branch points (thin full line); the limit curves $z_g(m)$ of the z-strips (thin dashed curves); and the mode solutions for the modes m= 0,1,…,5 . The direction of increasing frequency is indicated by arrows.

A procedure for a set of mode solutions and a list of k_0h values, which avoids the subdivision of the surface wave mode, which returns continuous mode solutions (for a variation of k_0h) and which is relatively robust against "mode jumping", proceeds as follows. It is supposed that the list $\{k_0h\}$ begins with low values (if necessary, prepend such values which you may drop later), and the difference Δk_0h is not too large ($\Delta k_0h \leq 0.1$; in the maximum ≤ 0.2).

The procedure is first described for symmetrical modes.

Begin to work through the list (i=1,2,…) of k_0h with starting values $z_{s1} = m\pi$; $z_{s2} = m\pi + 0.01\cdot j$; $z_{s3} = m\pi + 0.01 + 0.01\cdot j$ for MULLER's procedure and find $z_{i=1}$. Write the characteristic equation as :

$$z^2 \approx m\pi\cdot z + j\cdot k_0h\cdot Z_0G\cdot\left[1-\frac{(z-m\pi)^2}{3-}\,\frac{(z-m\pi)^2}{5-\ldots}\cdots\frac{(z-m\pi)^2}{(n_{hi}-2)^2-\dfrac{(z-m\pi)^2}{n_{hi}^2}}\right] \qquad (13)$$

$$\xrightarrow[z\to m\pi]{} (m\pi)^2 + j\cdot k_0h\cdot Z_0G \xrightarrow[k_0h\to 0 \text{ and / or } G\to 0]{} (m\pi)^2$$

and use on the right-hand side $z \to z_{i-1}$ to find a starter approximation z_{s3} for z_i . The other starters for MULLER's procedure are $z_{s2}= z_{i-2}$; $z_{s1}= z_{i-3}$ (in the 2nd step i=2 take the previous solution as z_{s2} and the mean value of z_{s2}, z_{s3} as z_{s1}). The order of the starters is important, and do not use the NEWTON-RAPHSON method; otherwise mode jumping will happen.

The next graph shows again the mode solutions from above, but now with this method; in the second graph the range has been enlarged over which the surface wave mode spans; the mode index m=0 is attributed to the surface wave mode.

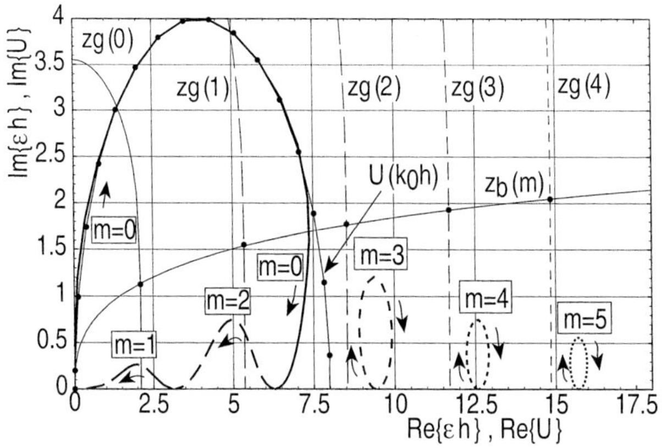

Mode solutions as above, but with a method returning steady curves for the modes.

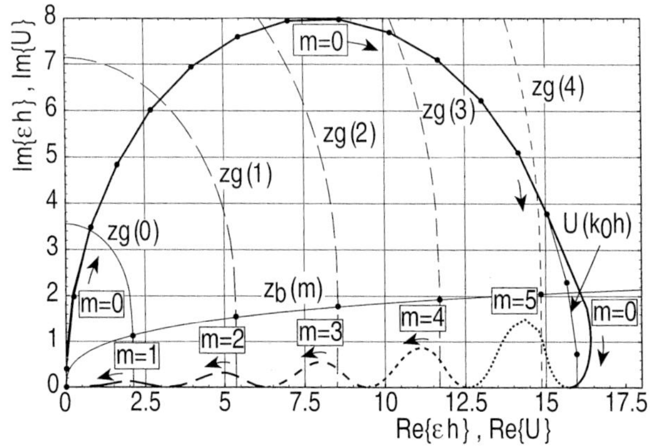

Mode solutions as above, but with an extended span of the surface wave mode (m=0) .

In the case of anti-symmetrical modes proceed as above, but replace $m \to (m+1/2)$ (because of $\cot(z)= -\tan(z-(m+1/2)\pi)$).

J.8 Flat duct with a bulk reacting lining

Ref.: Mechel, [J.3]

A flat duct with axial co-ordinate x and lateral co-ordinate y has a hard wall at y=0 , and is lined with a bulk reacting absorber at y= h . The lining is a layer of thickness d of a porous material with characteristic values Γ_a, Z_a (or in normalised form: $\Gamma_{an}=\Gamma_a/k_0$, $Z_{an}=Z_a/Z_0$), possibly covered with a poro-elastic foil having a partition impedance Z_s (s from *series impedance*).

Fundamental relations are contained in Table1.

Relation	Free duct	Absorber layer
Wave equation	$\left(\Delta+k_0^2\right)p(x,y)=0$	$\left(\Delta-\Gamma_a^2\right)p_a(x,y)=0$
Field formulation	$p(x,y,z)=P_0\,q(y)\cdot s(z)\cdot e^{-\Gamma x}$	$p_a(x,y,z)=P_a\,q_a(y)\cdot s(z)\cdot e^{-\Gamma x}$
Profile $\begin{cases}\text{symm.}\\ \text{antisymm.}\end{cases}$ in y	$q(y)=\begin{cases}\cos(\varepsilon_y y)\\ \sin(\varepsilon_y y)\end{cases}$	$q_a(y)=\cos\left(\varepsilon_{ay}(\pm y-h-d)\right)$
Profile $\begin{cases}\text{symm.}\\ \text{antisymm.}\end{cases}$ in z	$s(z)=\begin{cases}\cos(\varepsilon_z z)\\ \sin(\varepsilon_z z)\end{cases}$	
Secular equation	$\varepsilon_y^2=\Gamma^2+k_0^2-\varepsilon_z^2$	$\varepsilon_{ay}^2=\Gamma^2-\Gamma_a^2-\varepsilon_z^2$
Velocity in y direction	$v_y=\dfrac{j}{k_0 Z_0}\dfrac{\partial p}{\partial y}$	$v_{ay}=\dfrac{-1}{\Gamma_a Z_a}\dfrac{\partial p_a}{\partial y}$
Lateral admittance y=h symmetrical	$G_y=-j\dfrac{\varepsilon_y h}{k_0 h Z_0}\tan\varepsilon_y h$	$G_{ay}=-\dfrac{\varepsilon_{ay} d}{\Gamma_a d Z_a}\tan\varepsilon_{ay} d$
Lateral admittance y=h anti-symmetrical	$G_y=j\dfrac{\varepsilon h}{k_0 h Z_0}\cot\varepsilon h$	$G_{ay}=-\dfrac{\varepsilon_{ay} d}{\Gamma_a d Z_a}\tan\varepsilon_{ay} d$

Table 1 : Relations in the free rectangular duct and in the absorber layer.

The boundary condition is the agreement of the lateral admittances on both sides of y= h :

$$G_y \stackrel{!}{=} 1/\left(Z_s+1/G_{ay}\right) \tag{1}$$

This gives the characteristic equations

for *symmetrical modes* :
$$\varepsilon_y h \cdot \tan \varepsilon_y h = jk_0 h \Big/ \left(\frac{Z_s}{Z_0} - \frac{\Gamma_a}{\varepsilon_{ay}} \frac{Z_a}{Z_0} \cot \varepsilon_{ay} d \right) := jU \qquad (2)$$

for *anti-symmetrical modes* :
$$\varepsilon_y h \cdot \cot \varepsilon_y h = -jk_0 h \Big/ \left(\frac{Z_s}{Z_0} - \frac{\Gamma_a}{\varepsilon_{ay}} \frac{Z_a}{Z_0} \cot \varepsilon_{ay} d \right) = -jU \qquad (3)$$

From the secular equations :
$$(\varepsilon_{ay} h)^2 = (\varepsilon_y h)^2 - (\Gamma_a h)^2 - (k_0 h)^2 \qquad (4)$$

The function U now contains the solution $\varepsilon_y h$ (in contrast to locally reacting linings), and the form of the characteristic equation and the method of its solution thus depends on the structure of the lining. Without a cover of the absorber layer (i.e. $Z_s = 0$), the function U is :

$$U = -\frac{h}{d} \frac{k_0}{\Gamma_a} \frac{Z_0}{Z_a} \varepsilon_a d \cdot \tan \varepsilon_a d \qquad (5)$$

If the absorber layer is made locally reacting (e.g. either by a high low resistivity or by internal partition walls normal to the surface), the characteristic equation for symmetrical modes becomes :

$$\varepsilon h \cdot \tan \varepsilon h = jk_0 h \frac{\tanh(\Gamma_a d)}{Z_a / Z_0} \qquad (6)$$

If, on the other hand, the term Z_s/Z_0 is large compared with the 2nd term in the parentheses of the characteristic equation : $U \to \frac{k_0 h}{Z_s / Z_0}$, i.e. the lining behaves like a locally reacting lining.

The characteristic equation is even in εh (as for locally reacting absorbers), but now Re{U} < 0 is possible (in contrast to locally reacting absorbers); therefore solutions εh are no longer necessarily in the 1st quadrant. "Morse charts" for the solutions cannot be drawn.

Modes in a duct with a bulk reacting lining (terminated with a hard wall towards the outer space) are orthogonal to each other, if the lateral field profile within the outer walls is written as $q(y) = q^{(1)}(y) + q^{(2)}(y)$, where (i)=(1) stands for the free duct, and (i)=(2) for the absorber layer. The orthogonality relation is :

$$\frac{1}{jk_0 Z_0} \iint_{A_1} q_m^{(1)} \cdot q_n^{(1)} \, dA_1 + \frac{1}{\Gamma_a Z_a} \iint_{A_2} q_m^{(2)} \cdot q_n^{(2)} \, dA_2 = \delta_{m,n} \cdot N_m \qquad (7)$$

where $\delta_{m,n}$ is the Kronecker symbol, and N_m the mode norm :

$$N_m = \frac{1}{jk_0 Z_0} \iint_{A_1} \left(q_m^{(1)}\right)^2 dA_1 + \frac{1}{\Gamma_a Z_a} \iint_{A_2} \left(q_m^{(2)}\right)^2 dA_2 \qquad (8)$$

For multi-layer absorbers an additional term $i>2$ will appear on the left-hand sides for each layer.

J.9 Flat duct with an anisotropic, bulk reacting lining

Ref.: Mechel, [J.3]

The object is the same as in the previous section J.8 , but the porous material layer now is assumed to be anisotropic, i.e. with characteristic values Γ_{ax}, Z_{ax} and Γ_{ay}, Z_{ay} in the co-ordinate directions x,y . The relations in the free duct and in the absorber layer are presented in the following Table 1.

Relation	Free duct	Absorber layer
Wave equation	$\left(\Delta + k_0^2\right) p(x,y) = 0$	$\left(\partial^2/(\Gamma_{ax}\partial x)^2 + \partial^2/(\Gamma_{ay}\partial y)^2 - 1\right) p_a(x,y) = 0$
Field formulation	$p(x,y) = P_0\, q(y) \cdot e^{-\Gamma x}$	$p_a(x,y) = P_a\, q_a(y) \cdot e^{-\Gamma x}$
Profile $\begin{cases}\text{symm.}\\ \text{antisymm.}\end{cases}$	$q(y) = \begin{cases}\cos(\varepsilon_y y)\\ \sin(\varepsilon_y y)\end{cases}$	$q_a(y) = \cos\left(\varepsilon_{ay}(\pm y - h - d)\right)$
Secular equation	$\varepsilon_y^2 = \Gamma^2 + k_0^2$	$\varepsilon_{ay}^2/\Gamma_{ay}^2 = \Gamma^2/\Gamma_{ax}^2 - 1$
Velocity in y direct.	$v_y = \frac{j}{k_0 Z_0}\frac{\partial p}{\partial y}$	$v_{ay} = \frac{-1}{\Gamma_{ay} Z_{ay}}\frac{\partial p_a}{\partial y}$
Lateral admittance symmetrical	$G_y = -j\frac{\varepsilon_y h}{k_0 h Z_0}\tan \varepsilon_y h$	$G_{ay} = -\frac{\varepsilon_{ay} d}{\Gamma_{ay} d Z_{ay}}\tan \varepsilon_{ay} d$
Lateral admittance anti-symmetrical	$G_y = j\frac{\varepsilon h}{k_0 h Z_0}\cot \varepsilon h$	$G_{ay} = -\frac{\varepsilon_{ay} d}{\Gamma_{ay} d Z_{ay}}\tan \varepsilon_{ay} d$

Table 1 : Relations in the free duct and in the anisotropic absorber layer.

The characteristic equations for duct modes are now :

symmetrical mode :
$$\varepsilon_y h \cdot \tan \varepsilon_y h = j k_0 h \Big/ \left(\frac{Z_s}{Z_0} - \frac{\Gamma_{ay}}{\varepsilon_{ay}}\frac{Z_{ay}}{Z_0}\cot \varepsilon_{ay} d\right) := jU \qquad (1)$$

anti-symmetrical mode : $$\varepsilon_y h \cdot \cot \varepsilon_y h = -j k_0 h \Big/ \left(\frac{Z_s}{Z_0} - \frac{\Gamma_{ay}}{\varepsilon_{ay}} \frac{Z_{ay}}{Z_0} \cot \varepsilon_{ay} d \right) = -jU \tag{2}$$

secular equation : $$\varepsilon_{ay}^2 = \Gamma^2 \cdot \left(\Gamma_{ay}^2 / \Gamma_{ax}^2 \right) - \Gamma_{ay}^2 = \left(\varepsilon_y^2 - k_0^2 \right) \cdot \left(\Gamma_{ay}^2 / \Gamma_{ax}^2 \right) - \Gamma_{ay}^2 \tag{3}$$

axial propagation constant : $$\Gamma h = \sqrt{(\varepsilon_y h)^2 - (k_0 h)^2} \tag{4}$$

No principally new features are introduced by the anisotropy, only the amount of computation is somewhat increased.

J.10 Mode solutions in a flat duct with bulk reacting lining

Ref.: Mechel, [J.3]

Because no transformation between $z = \varepsilon h$ and a meaningful known variable, like U with locally reacting linings, can be defined, a method of solution as with such absorbers which uses the branch points cannot be designed.

Continued fraction expansion, symmetrical mode (flat duct, $\varepsilon_z = 0$) :
(applicable for the least attenuated mode only)

Transform the characteristic equation to :

$$\varepsilon_y h \cdot \tan \varepsilon_y h = j \varepsilon_{ay} d \cdot \tan \varepsilon_{ay} d \Big/ \left(\frac{Z_s}{k_0 h \cdot Z_0} \cdot \varepsilon_{ay} d \cdot \tan \varepsilon_{ay} d - \frac{d}{h} \frac{\Gamma_a}{k_0} \frac{Z_a}{Z_0} \right) \tag{1}$$

with : $$\varepsilon_{ay} d = \sqrt{(d/h)^2 \left((\varepsilon_y h)^2 - (k_0 h)^2 \right) - (\Gamma_a d)^2} \tag{2}$$

and apply the continued fraction expansion to $z \cdot \tan(z)$ (with $z = \varepsilon_y h$ and $a = (1 + \Gamma_{an}^2)(k_0 h)^2$; $Z_{sn} = Z_s / Z_0$)

Even with the rather low precision of expansion up to the 2nd fraction and the special case $Z_s = 0$ (i.e. no cover foil on the porous layer), the polynomial equation becomes somewhat lengthy :

$$-(z^2)^6(d/h)^4(k_0h)^2+(z^2)^5(d/h)^4(k_0h)^2(6+4a-\Gamma_{an}^2Z_{an}^2)- \qquad (3)$$
$$-(z^2)^4(d/h)^2(k_0h)^2(-6\Gamma_{an}^2Z_{an}^2+(d/h)^2(9+6a^2-2a(-12+\Gamma_{an}^2Z_{an}^2)))+$$
$$+(z^2)^3(k_0h)^2(-9\Gamma_{an}^2Z_{an}^2-6(d/h)^2a\cdot\Gamma_{an}^2Z_{an}^2+(d/h)^4a\cdot(36+4a^2-a\cdot(-36+\Gamma_{an}^2Z_{an}^2)))-$$
$$-(z^2)^2(d/h)^4a\cdot(k_0h)^2(54+24a+a^2)+(z^2)\cdot 6(d/h)^4a^3(k_0h)^2(6+a)-$$
$$-9(d/h)^4a^4(k_0h)^2=0$$

With a cover foil (i.e. $Z_s\neq 0$) with the same depth of expansion, the equation becomes for $z=\varepsilon h$: (4)

$$z^{14}\cdot 9(d/h)^4Z_{sn}^2+z^{13}\cdot 6j(d/h)^4k_0hZ_{sn}-z^{12}\cdot(d/h)^4((k_0h)^2+36aZ_{sn}^2)-$$
$$-z^{11}\cdot 6j(d/h)^4k_0h(3+4*a)Z_{sn}+$$
$$+z^{10}\cdot(d/h)^4(k_0h)^2(6-\Gamma_{an}^2Z_{an}^2+2a(2+27aZ_{sn}^2/(k_0h)^2))+$$
$$+z^9\cdot 36j(d/h)^4ak_0hZ_{sn}(2+a)-$$
$$-z^8\cdot(d/h)^2(k_0h)^2(-6\Gamma_{an}^2Z_{an}^2+(d/h)^2(9-2a(-12+\Gamma_{an}^2Z_{an}^2)+6a^2(1+6aZ_{sn}^2/(k_0h)^2))-$$
$$-z^7\cdot 12j(d/h)^4a^2k_0hZ_{sn}(9+2a)+$$
$$+z^6\cdot(k_0h)^2(-9\Gamma_{an}^2Z_{an}^2-6(d/h)^2a\Gamma_{an}^2Z_{an}^2+(d/h)^4a(36-a(-36+\Gamma_{an}^2Z_{an}^2)+$$
$$a^2(4+9aZ_{sn}^2/(k_0h)^2)))+$$
$$+z^5\cdot 6j(d/h)^4a^3k_0hZ_{sn}(12+a)-z^4\cdot(d/h)^4a^2(k_0h)^2(54+24a+a^2)-$$
$$-z^3\cdot 18j(d/h)^4a^4k_0hZ_{sn}+z^2\cdot 6(d/h)^4a^3(k_0h)^2(6+a)-9(d/h)^4a^4(k_0h)^2=0$$

The problem is, how to find the right solution among the roots. The root z in the 1st quadrant and with smallest magnitude is often the right choice.

Iteration through a list of $\{k_0h\}$:

At very low k_0h the admittance of all linings becomes small. Then the mode solution with the corresponding absorber, in which all layers are assumed to be locally reacting, are suitable starters for MULLER's procedure. For later entries of the k_0h list, take previous solutions as starters. It may be necessary to make the steps Δk_0h very small.

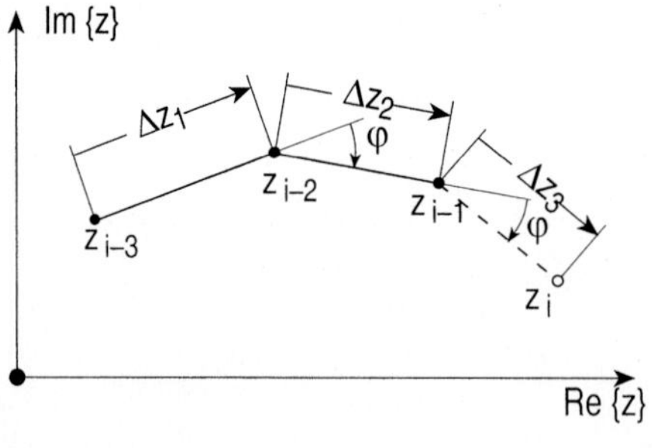

It is helpful (and permits larger Δk_0h values) to use as the 3rd starter for the solution z_i an extrapolated value. Let z_{i-1}, z_{i-2}, z_{i-3} be the previous solutions for the previous k_0h values.

The extrapolated z_i (which is the 3rd start value together with z_{i-1} and z_{i-2}) is evaluated from (with $\angle z$ the argument of the complex z) :

$$\begin{aligned} &z_i = |\Delta z_3| \cdot e^{j\angle\Delta z_3} \quad ; \quad |\Delta z_3| = c\cdot|\Delta z_2| = c\cdot|z_{i-2} - z_{i-1}| \quad ; \\ &\angle\Delta z_3 = \angle\Delta z_2 + \varphi = 2\cdot\angle\Delta z_2 - \angle\Delta z_1 \quad ; \\ &-\varphi = \angle\Delta z_1 - \angle\Delta z_2 \quad ; \\ &\Delta z_2 = z_{i-1} - z_{i-2} \quad ; \quad \Delta z_1 = z_{i-2} - z_{i-3} \end{aligned} \tag{5}$$

Start the numerical solution with mode values for the locally reacting absorber :

Sometimes it is proposed, to take the mode solution for the absorber with all layers made locally reacting (and some values nearby as the two other starters) as start values for MULLER's method, not only for the lowest entries of a k_0h list, as above, but for all k_0h values. This method fails, except in very harmless cases (because the numerical procedure may pass on its way from the starters to the true value through apparent resonances of the lining, which do not exist).

Iteration through the modal angle :

Define modal angles from the secular equation (below for three dimensions) in its form :

$$\begin{aligned} 1 &= (\varepsilon_y/k_0)^2 + \left((\varepsilon_z/k_0)^2 - (\Gamma/k_0)^2\right) \\ &= \cos^2\varphi + \sin^2\varphi\cdot\left(\sin^2\psi + \cos^2\psi\right) \end{aligned} \tag{6}$$

If one associates the terms as $\quad \cos\varphi = \varepsilon_y/k_0 \quad ; \quad \sin\varphi = \sqrt{1-(\varepsilon_y/k_0)^2} \qquad (7)$

then φ is the angle of incidence on the absorber. The surface admittance of a bulk reacting lining can be written as a function of $\cos\varphi$, $\sin\varphi$ (which may be complex). The admittance of the locally reacting absorber is obtained for normal incidence. Start the evaluation by finding the mode solution for the locally reacting absorber with the method described in section J.7 . Evaluate with it the mode angle φ as above. Insert this into the admittance formula of Z_0G_{ay} of the lining, and solve for the next approximation with the method for locally reacting absorbers (i.e. φ is kept constant during the performance of MULLER's procedure). Repeat until the solution becomes stationary.

The advantage of this *method of φ-iteration* is its robustness against mode jumping. It can also be applied for multi-layer absorbers, where other methods mostly run into problems. After about 8 iterations the result is mostly stationary in its first 4 to 5 decimals.

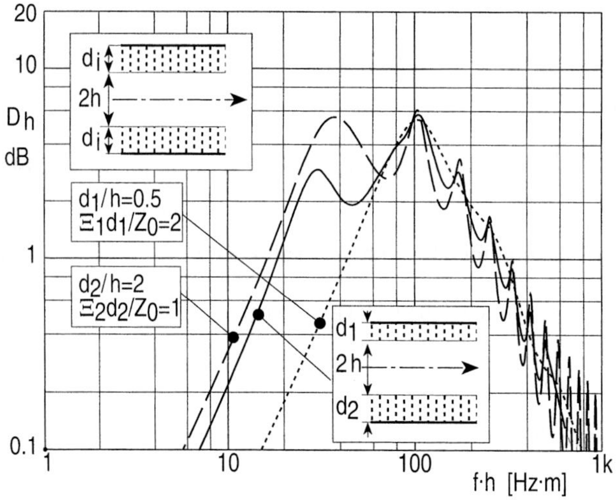

Attenuation D_h of the least attenuated mode in a duct with unsymmetrical, locally reacting lining (full line), compared with D_h in ducts with symmetrical linings (dashed lines).

J.12 Flat duct with an unsymmetrical, bulk reacting lining

Ref.: Mechel, [J.4]

The object is as in the previous section J.11 , but the lining consists of bulk reacting layers of a porous material (mineral fibre in the numerical examples) having the normalised characteristic values $\Gamma_{an,i}, Z_{an,i}$ on the duct sides i=1,2 , possibly covered with a foil having the partition impedance Z_{si} .

The secular and characteristic equations for the lateral wave number $z=\varepsilon h$ are :

$$\Gamma h=\sqrt{z^2-(k_0h)^2}$$

$$\left[z\cdot\tan z-jU_s\right]\cdot\left[z\cdot\cot z+jU_s\right]=U_a^2 \tag{1}$$

$$U_s=\frac{1}{2}(U_1+U_2)\quad ;\quad U_a=\frac{1}{2}(U_1-U_2)$$

with :

$$U_i=k_0h\frac{y_i\cdot\tan y_i}{Z_{si}/Z_0\cdot y_i\cdot\tan y_i-\frac{d_i}{h}k_0h\cdot\Gamma_{an,i}Z_{an,i}}\xrightarrow{Z_{si}\to 0}\frac{-y_i\cdot\tan y_i}{\frac{d_i}{h}\cdot\Gamma_{an,i}Z_{an,i}} \tag{2}$$

$$y_i^2=(d_i/h)^2\left(z^2-(k_0h)^2(1+\Gamma_{an,i}^2\right)\quad ;\quad i=1,2$$

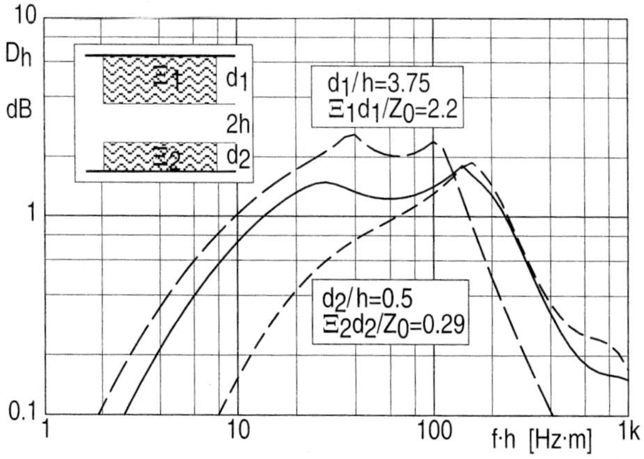

Attenuation D_h of the least attenuated mode in a flat duct with unsymmetrical, bulk reacting lining (full line), and in ducts with symmetrical lining (dashed lines).

J.13 Round duct with a locally reacting lining

Ref.: Mechel, [J.5]

The lining is defined by a surface admittance G . The form of a mode is :

$$p(r,\theta,x) = P \cdot q(r,\theta) \cdot e^{-\Gamma x} \tag{1}$$

The lateral profile $q(r,\theta)$ must satisfy the Bessel differential equation : (2)

$$\left[\frac{\partial^2}{\partial r^2} + \frac{1}{r}\frac{\partial}{\partial r} + \frac{1}{r^2}\frac{\partial^2}{\partial\theta^2} + \Gamma^2 + k_0^2\right] q(r,\theta) = 0$$

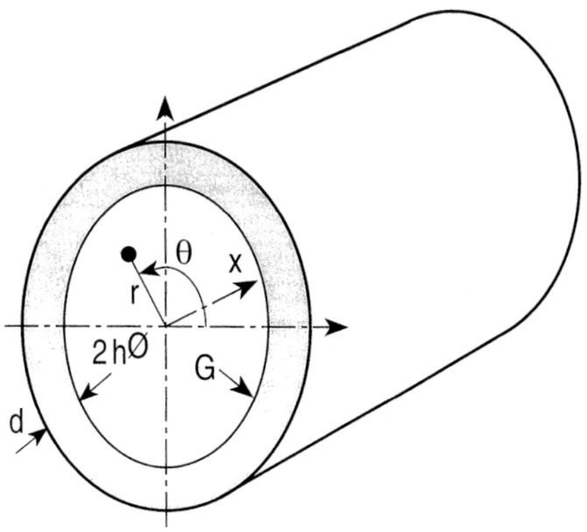

Solutions have the general form :

$$q(r,\theta) = \cos(m\theta)\left[J_m(\varepsilon_m r) + b \cdot Y_m(\varepsilon_m r)\right] \quad ; \quad m = 0,1,2,\ldots \tag{3}$$

with Bessel functions $J_m(z)$ and Neumann functions $Y_m(z)$. If the origin $r=0$ belongs to the field area, the Neumann function must be excluded, because it is singular there. So, mode profiles in round ducts are :

$$q(r,\theta) = \cos(m\theta) \cdot J_m(\varepsilon_m r) \quad ; \quad m = 0,1,2,\ldots \tag{4}$$

The Bessel differential equation requires (secular equation) : $\varepsilon_m^2 = \Gamma^2 + k_0^2$ (5)

with $\Gamma \to \Gamma_m$.

The radial particle velocity is :

$$v_r = \frac{j}{k_0 Z_0}\frac{\partial p}{\partial r} = \frac{j\varepsilon_m}{k_0 Z_0} P\cos(m\theta)\cdot J'_m(\varepsilon_m r)\cdot e^{-\Gamma x} \tag{6}$$

With it the boundary condition gives the characteristic equation for $\varepsilon_m h$:

$$(\varepsilon_m h)\frac{J'_m(\varepsilon_m h)}{J_m(\varepsilon_m h)} = -jk_0 h\cdot Z_0 G =: -j\cdot U \tag{7}$$

or, with $\quad J'_m(z) = J_{m-1}(z) - \frac{m}{z}J_m(z)$, (8)

$$(\varepsilon_m h)\frac{J_{m-1}(\varepsilon_m h)}{J_m(\varepsilon_m h)} = m - jU \tag{9}$$

The function $\quad F_m(z) := z\frac{J_{m-1}(z)}{J_m(z)}\quad$ can be expanded in a continued fraction : (10)

$$F_m(z) = 2m - \frac{z^2}{2(m+1)-}\,\frac{z^2}{2(m+2)-}\,\frac{z^2}{2(m+3)-}\cdots \tag{11}$$

Therewith the characteristic equation can be written as (with the abbreviations $z=\varepsilon_m h$) :

$$\frac{z^2}{2(m+1)-}\,\frac{z^2}{2(m+2)-}\,\frac{z^2}{2(m+3)-}\cdots = jU + m \tag{12}$$

It becomes for the fundamental azimuthal mode m=0 :

$$\frac{z^2}{2-}\,\frac{z^2}{4-}\,\frac{z^2}{6-}\,\frac{z^2}{8-}\cdots = jU \tag{13}$$

The solution with the indicated length of the fraction is : (14)

$$(\varepsilon_0 h)^2 \approx \frac{96+36jU \pm \sqrt{9216+2304jU-912U^2}}{12+jU}$$

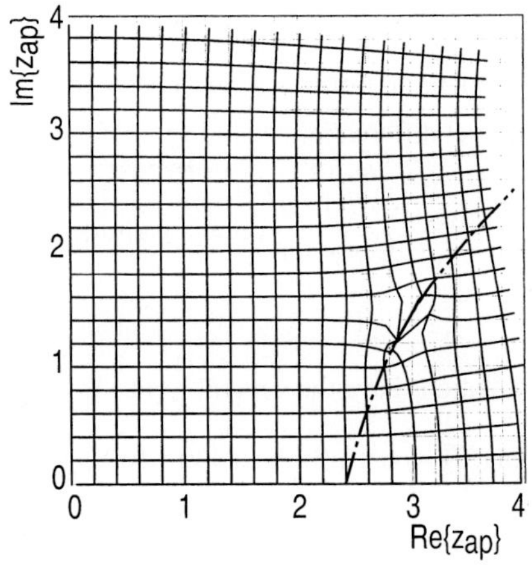

Its precision test is contained in the diagram opposite. The root is evaluated with a negative real part in the range above the dash-dotted limit curve, and with a positive real part below that curve.

The coefficients a_i for polynomial approximations

$$a_0 + a_1\cdot z^2 + a_2\cdot z^4 + \ldots + a_i\cdot z^{2i} = 0 \tag{15}$$

of the characteristic equation with increasing depth of the expansion are given in Table 1 .

Table 1: Coefficients of the polynomial approximations to the characteristic equation for azimuthal modes of orders m=0,1,2,3,4

m	a_0	a_1	a_2	a_3	a_4	a_5
0	8 jU	–(4 + jU)	–	–	–	–
	–384 jU	24 (8 + 3 jU)	–(12 + jU)	–	–	–
	46080 jU	1920 j (12j – 5U)	96 (20 + 3 jU)	–(24 + jU)	–	–
	–10321920 jU	322560 (16 + 7 jU)	17280 j (28 j – 5U)	800 (12 + jU)	–(40 + jU)	–
	3715891200 jU	92897280 j (20 j – 9U)	5160960 (36 + 7 jU)	94080 j (48 j – 5U)	600 (56 + 3 jU)	–(60 + jU)
1	24 (1 + jU)	–(7 + jU)	–	–	–	–
	1920 j (j – U)	48 (13 + 3 jU)	–(17 + jU)	–	–	–
	322560 (1 + jU)	5760 j (19 j – 5U)	480 (9 + jU)	–(31 + jU)	–	–
	92897280 j (j – U)	1290240 (25 + 7 jU)	40320 j (37 j – 5U)	1200 (15 + jU)	–(49 + jU)	–
	40874803200 (1 + jU)	464486400 j (31 j – 9U)	15482880 (47 + 7 jU)	188160 j (59 j – 5U)	840 (67 + 3 jU)	–(71 + jU)
2	48 (2 + jU)	–(10 + jU)	–	–	–	–
	5760 j (2 j – U)	240 (6 + jU)	–(22 + jU)	–	–	–
	1290240 (2 + jU)	13440 j (26 j – 5U)	240 (34 + 3 jU)	–(38 + jU)	–	–
	464486400 j (2 j – U)	3870720 (34 + 7 jU)	80640 j (46 j – 5U)	1680 (18 + jU)	–(58 + jU)	–
	245248819200 (2 + jU)	5109350400 j (14 j – 3U)	38707200 (58 + 7 jU)	1693440 j (14 j – U)	3360 (26 + jU)	–(82 + jU)
3	80 (3 + jU)	–(13 + jU)	–	–	–	–
	13440 j (3 j – U)	120 (23 + 3 jU)	–(27 + jU)	–	–	–
	3870720 (3 + jU)	26880 j (33 j – 5U)	336 (41 + 3 jU)	–(45 + jU)	–	–
	1703116800 j (3 j – U)	9676800 (43 + 7 jU)	725760 j (11 j – U)	2240 (21 + jU)	–(67 + jU)	–
	1062744883200 (3 + jU)	5109350400 j (53 j – 9U)	85155840 (69 + 7 jU)	564480 j (81 j – 5U)	1440 (89 + 3 jU)	–(93 + jU)
4	120 (4 + jU)	–(16 + jU)	–	–	–	–
	26880 j (4 j – U)	168 (28 + 3 jU)	–(32 + jU)	–	–	–
	9676800 (4 + jU)	241920 j (8 j – U)	1344 (16 + jU)	–(52 + jU)	–	–
	5109350400 j (4 j – U)	21288960 (52 + 7 jU)	241920 j (64 j – 5U)	2880 (24 + jU)	–(76 + jU)	–
	3719607091200 (4 + jU)	13284311040 j (64 j – 9U)	170311680 (80 + 7 jU)	887040 j (92 j – 5U)	1800 (100 + 3 jU)	–(104 + jU)

Mode charts can be plotted as lines of Re{z}=const and Im{z}=const in the complex U plane for azimuthal mode orders m= 0,1,2,… and radial mode orders n=0,1,2,… .

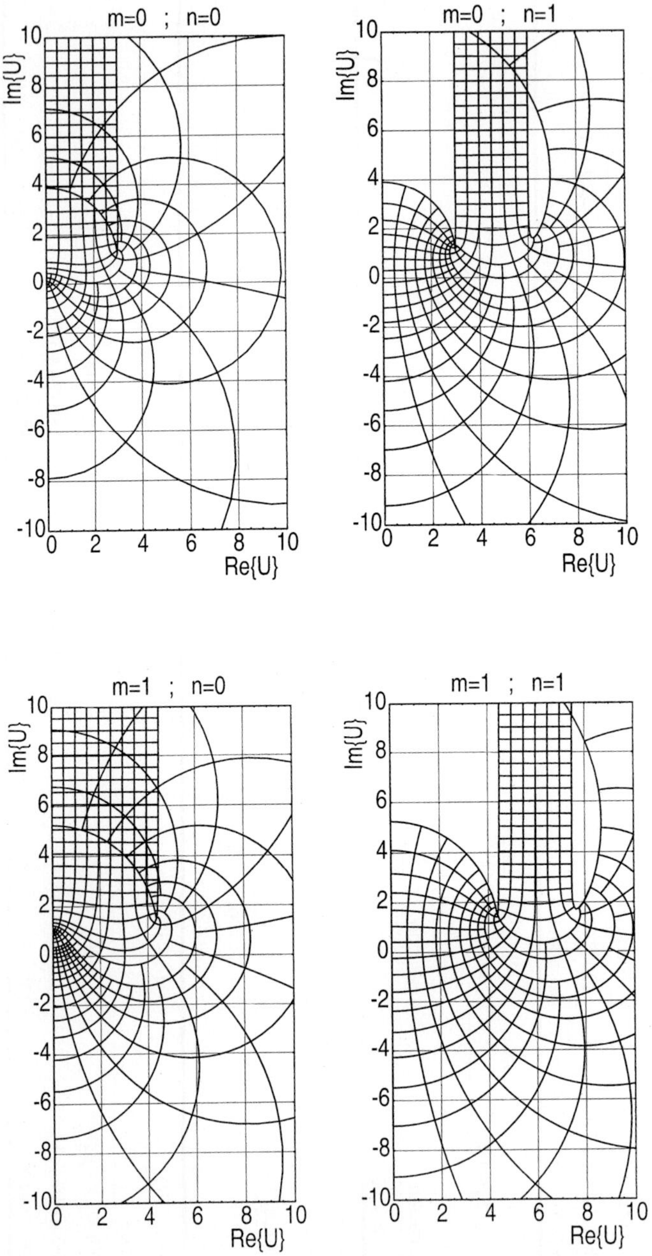

Three-dimensional mode charts are created, if one plots the mesh points at a height $Re\{z\}$ above the U plane for any azimuthal order m .

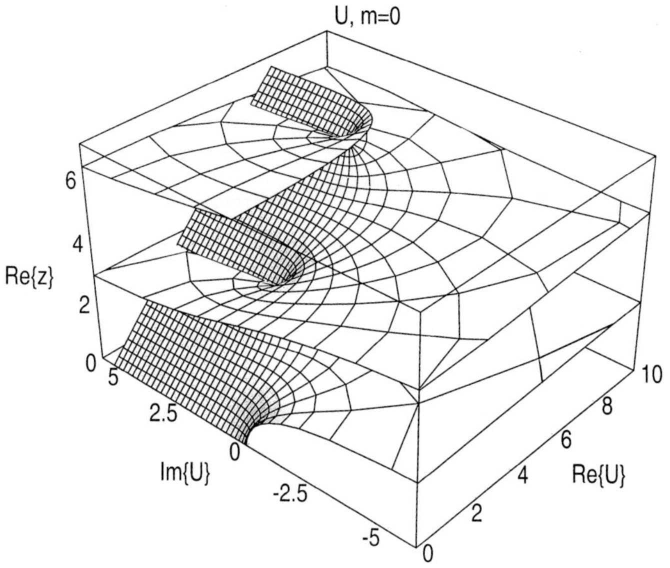

3D mode chart for the azimuthal mode m=0 and some radial modes.

The evaluation of sets of mode solutions $z=\varepsilon h$ is similar to the task in section J.7 . It needs the branch points z_b and the transforms U_b of them. The z_b are solutions of the equation :

$$\frac{J_{m-1}(z)}{J_m(z)}\left(2m - z\frac{J_{m-1}(z)}{J_m(z)}\right) - z \overset{!}{=} 0 \quad ; \quad z \neq 0 \tag{16}$$

The following tables contain branch points $z_b(m,n)$ and associated $U_b(m,n)$ for azimuthal orders m= 0,1,…,10 and radial orders n= 0,1,…,20 . The entries of the tables are the real and imaginary parts.

Further the coefficients are given (again with their real and imaginary parts) in tables for the representation of $z_b(m,n)$, $U_b(m,n)$ as functions of n for given m :

$$\begin{aligned} z_b(m,n) &= a_m + b_m \cdot n^{1/2} + c_m \cdot n + d_m \cdot n^2 + e_m \cdot n^3 \\ U_b(m,n) &= A_m + B_m \cdot n^{1/2} + C_m \cdot n + D_m \cdot n^2 + E_m \cdot n^3 \end{aligned} \tag{17}$$

and coefficients for the functions $z''_b(m,n) = f(z'_b(m,n))$; $U''_b(m,n) = f(U'_b(m,n))$ as :

$$\begin{aligned} z''_b &= \bar{a}_m + \bar{b}_m \cdot z'^{1/2}_b + \bar{c}_m \cdot z'_b + \bar{d}_m \cdot z'^{3/2}_b + \bar{e}_m \cdot z'^2_b + \bar{f}_m \cdot z'^3_b \\ U''_b &= \overline{A}_m + \overline{B}_m \cdot U'^{1/2}_b + \overline{C}_m \cdot U'_b + \overline{D}_m \cdot U'^{3/2}_b + \overline{E}_m \cdot U'^2_b + \overline{F}_m \cdot U'^3_b \end{aligned} \tag{18}$$

Table 2a : Branch points $z_b(m,n)$ for azimuthal orders m and radial orders n in round ducts with a locally reacting lining.

m	$z_b(m,0)$	$z_b(m,1)$	$z_b(m,2)$	$z_b(m,3)$	$z_b(m,4)$	$z_b(m,5)$	$z_b(m,6)$	$z_b(m,7)$	$z_b(m,8)$	$z_b(m,9)$	$z_b(m,10)$
0	0.	2.9803824	6.17515307	9.3419610	12.4985071	15.650104	18.7989117	21.945980	25.0918858	28.2369731	31.381461
	0.	1.2796025	1.61871738	1.8188728	1.96145954	2.0723098	2.16301098	2.2397725	2.30631281	2.36503612	2.4175870
1	0.	4.4662985	7.69410395	10.874574	14.0388913	17.195565	20.3479682	23.497724	26.6457175	29.7924754	32.938331
	0.	1.4674704	1.72697154	1.8949433	2.02006280	2.1199462	2.20312418	2.2744062	2.33677883	2.39222622	2.4421349
2	0.	5.8168507	9.11848718	12.334640	15.5203741	18.691419	21.8541613	25.011725	28.1658298	31.3175030	34.467400
	0.	1.6000118	1.81551915	1.9613351	2.07308877	2.1640392	2.24083200	2.3073267	2.36597916	2.41845412	2.4659342
3	0.	7.1009646	10.4837004	13.743724	16.9575046	20.148156	23.3254221	26.494193	29.6572168	32.8161611	35.972101
	0.	1.7069432	1.89206035	2.0209608	2.12187295	2.2052780	2.27651937	2.3387612	2.39405346	2.44380805	2.4890417
4	0.	8.3439388	11.8075223	15.114516	18.3597199	21.573052	24.7675211	27.949815	31.1237603	34.2917166	37.455223
	0.	1.7983610	1.96027390	2.0754868	2.16727350	2.2441437	2.31047269	2.3688878	2.42111618	2.46836325	2.5115077
5	0.	9.5583600	13.1003688	16.455227	19.7335515	22.971391	26.1848108	29.382231	32.5685480	35.7468196	38.919064
	0.	1.8791146	2.02226408	2.1259798	2.20988517	2.2809907	2.34291347	2.3978501	2.44726287	2.49218466	2.5333769
6	0.	10.751546	14.3689547	17.771528	21.0837394	24.347158	27.5806698	30.794337	33.9940832	37.1836547	40.365546
	0.	1.9519719	2.07937182	2.1731742	2.25014251	2.3160901	2.37401800	2.4257663	2.47257456	2.51532886	2.5546894
7	0.	11.928184	15.6179031	19.067527	22.4138509	25.703439	28.9577821	32.188476	35.4024288	38.6040478	41.796294
	0.	2.0186875	2.13251552	2.2176033	2.28837560	2.3496560	2.40392982	2.4527351	2.49712076	2.53784550	2.5754810
8	0.	13.091486	16.8505539	20.346310	23.7266496	27.042689	30.3183214	33.566582	36.7953083	40.0095430	43.212698
	0.	2.0804554	2.18235758	2.2596692	2.32484276	2.3818611	2.43276785	2.4788405	2.52096170	2.55977848	2.5957841
9	0.	14.243760	18.0694113	21.610265	25.0243271	28.366891	31.6640769	34.930267	38.1741788	41.4014603	44.615958
	0.	2.1381313	2.22939479	2.2996848	2.35975104	2.4128473	2.46063214	2.5041544	2.54415004	2.58116683	2.6156280
10	0.	15.386730	19.2764083	22.861283	26.308655	29.677677	32.9965422	36.280895	39.5402853	42.7809382	46.007117
	0.	2.1923528	2.27401119	2.3378991	2.39326958	2.4427332	2.48760786	2.5287392	2.56673216	2.60204547	2.6350391

Table 2b : Branch points $z_b(m,n)$ for azimuthal orders m and radial orders n in round ducts with a locally reacting lining (continued).

m	$z_b(m,11)$	$z_b(m,12)$	$z_b(m,13)$	$z_b(m,14)$	$z_b(m,15)$	$z_b(m,16)$	$z_b(m,17)$	$z_b(m,18)$	$z_b(m,19)$	$z_b(m,20)$
0	34.525496 2.4651401	37.6691784 2.50856370	40.812583 2.5485179	43.9557618 2.58551614	47.098757 2.6199657	50.2415986 2.65219513	53.384312 2.6824733	56.5269161 2.71102288	59.6694268 2.73803069	62.811857 2.7636547
1	36.083507 2.4875121	39.2281563 2.52911295	42.372390 2.5675181	45.5162895 2.60318383	48.659915 2.6364750	51.8033147 2.66768817	54.946524 2.6970676	58.0895726 2.72481678	61.2324837 2.75110719	64.375276 2.7760845
2	37.615957 2.5092906	40.7634791 2.54918414	43.910183 2.5861281	47.0562300 2.62052925	50.201740 2.6527155	53.3468057 2.68295513	56.491500 2.7114703	59.6358794 2.73844721	62.7799907 2.76404342	65.923871 2.7883934
3	39.125763 2.5305124	42.2776539 2.56880112	45.428140 2.6043630	48.5774940 2.63756183	51.725920 2.6686927	54.8735766 2.69799875	58.020588 2.7256823	61.1670524 2.75191372	64.3130496 2.77683800	67.458644 2.8005793
4	40.615330 2.5512115	43.7727783 2.58798682	46.928107 2.6222383	50.0817171 2.65429176	53.233915 2.6844132	56.3849379 2.71282291	59.534972 2.7397055	62.6841652 2.76521697	65.8326389 2.78949068	68.980491 2.8126413
5	42.086671 2.5714189	45.2506280 2.60676298	48.411660 2.6397692	51.5703113 2.67072952	54.726996 2.6998839	57.8820393 2.72743228	61.035696 2.7535428	64.1881718 2.77835854	67.3396329 2.80200202	70.490216 2.8245792
6	43.541487 2.5911633	46.7127217 2.62515010	49.880164 2.6569703	53.0445061 2.68688559	56.206278 2.7151124	59.3658956 2.74183199	62.523689 2.7671977	65.6799238 2.79134056	68.8348159 2.81437322	71.988543 2.8363935
7	44.981236 2.6104708	48.1603706 2.64316737	51.334805 2.6738558	54.5053795 2.70277034	57.672744 2.7301061	60.8374082 2.75602747	63.999779 2.7806740	67.1601857 2.80416564	70.3188954 2.82660595	73.476130 2.8480852
8	46.407174 2.6293655	49.5947157 2.66083276	52.776622 2.6904391	55.9538830 2.71839384	59.127268 2.7448727	62.2973818 2.77002428	65.464711 2.7939758	68.6296464 2.81683668	71.7925113 2.83870220	74.953571 2.8596555
9	47.820395 2.6478694	51.0167569 2.67816304	54.206533 2.7067329	57.3908614 2.73376582	60.570629 2.7594194	63.7465385 2.78382802	66.919153 2.8071071	70.0889297 2.82935680	73.2562454 2.85066423	76.421412 2.8711060
10	49.221856 2.6660028	52.4273767 2.69517382	55.625353 2.7227494	58.8170688 2.74889562	62.003528 2.7737536	65.1855288 2.79744428	68.363712 2.8200725	71.5386020 2.84172925	74.7106281 2.86249447	77.880149 2.8824385

Table 3a : Branch points $U_b(m,n)$ for modes with azimuthal order m and radial order n in a round duct with a locally reacting lining.

m	$U_b(m,0)$	$U_b(m,1)$	$U_b(m,2)$	$U_b(m,3)$	$U_b(m,4)$	$U_b(m,5)$	$U_b(m,6)$	$U_b(m,7)$	$U_b(m,8)$	$U_b(m,9)$	$U_b(m,10)$
0	0.	2.9803824	6.17515307	9.3419610	12.4985071	15.650104	18.7989117	21.945980	25.0918858	28.2369731	31.381461
	0.	1.2796025	1.61871738	1.8188728	1.96145954	2.0723098	2.16301098	2.2397725	2.30631281	2.36503612	2.4175870
1	0.	4.3645604	7.63203406	10.829870	14.0039585	17.166901	20.3236666	23.476634	26.6270897	29.7757956	32.923231
	1.	1.5016772	1.74101667	1.9027655	2.02510184	2.1234860	2.20575852	2.2764495	2.33841359	2.39356630	2.4432550
2	0.	5.4907943	8.90541529	12.175571	15.3932865	18.585549	21.7634172	24.932316	28.0952333	31.2539566	34.409622
	2.	1.6950243	1.85895745	1.9869593	2.09020428	2.1763663	2.25017531	2.3146755	2.37192429	2.42337139	2.4700748
3	0.	6.4813156	10.0605876	13.419744	16.6943117	19.926292	23.1335608	26.325136	29.5060904	32.6795116	35.847388
	3.	1.8701363	1.97163375	2.0697508	2.15532518	2.2298322	2.29539999	2.3537805	2.40631550	2.45402684	2.4977011
4	0.	7.3835846	11.1309878	14.586504	17.9251176	21.203156	24.4452811	27.664212	30.8672365	34.0588524	37.241995
	4.	2.0322668	2.07941812	2.1506167	2.21981999	2.2832936	2.34092956	2.3933440	2.44123699	2.48523973	2.5258873
5	0.	8.2221186	12.1365161	15.691562	19.0982555	22.426342	25.7070081	28.956619	32.1846863	35.3971529	38.597938
	5.	2.1845043	2.18286741	2.2294453	2.28339614	2.3364279	2.38645997	2.4330944	2.47645099	2.51680343	2.5544540
6	0.	9.0116093	13.0901833	16.745849	20.2228660	23.603546	26.9252866	30.207976	33.4633108	36.6986717	39.918969
	6.	2.3288533	2.28250429	2.3062807	2.34592951	2.3890567	2.43180353	2.4728523	2.51179287	2.54856962	2.5832688
7	0.	9.7617169	14.0010662	17.757338	21.3058567	24.740752	28.1053201	31.422838	34.7071219	37.9669654	41.208261
	7.	2.4667050	2.37877745	2.3812247	2.40738075	2.4410835	2.47684338	2.5124977	2.54714695	2.58043032	2.6122325
8	0.	10.479213	14.8758052	18.732066	22.3526020	25.842719	29.2513290	32.604957	35.9194713	39.2050406	42.468523
	8	2.5990743	2.47206344	2.4543972	2.46775429	2.4924595	2.52150723	2.5519495	2.58243119	2.61230611	2.6412699
9	0.	11.169070	15.7194383	19.674737	23.3673841	26.913312	30.3667933	33.757471	37.1031938	40.4154669	43.702092
	9.	2.7267290	2.56267754	2.5259193	2.52707712	2.5431644	2.56575149	2.5911533	2.61758702	2.64413811	2.6703241
10	0.	11.835078	16.5358998	20.589105	24.3536822	27.955719	31.45462389	34.883035	38.2607134	41.6004613	44.911000
	10.	2.8502677	2.65088497	2.5959057	2.58538746	2.5931956	2.6095514	2.6300728	2.65257265	2.67588250	2.6993511

Table 3b: Branch points $U_b(m,n)$ for modes with azimuthal order m and radial order n in a round duct with a locally reacting lining (continued).

m	$U_b(m,11)$	$U_b(m,12)$	$U_b(m,13)$	$U_b(m,14)$	$U_b(m,15)$	$U_b(m,16)$	$U_b(m,17)$	$U_b(m,18)$	$U_b(m,19)$	$U_b(m,20)$
0	34.525496 2.4651401	37.6691784 2.50856370	40.812583 2.5485179	43.9557618 2.58551614	47.098757 2.6199657	50.2415986 2.65219513	53.384312 2.6824733	56.5269161 2.71102288	59.6694268 2.73803069	62.811857 2.7636547
1	36.069713 2.4884634	39.2154611 2.52993170	42.360632 2.5682308	45.5053390 2.60381027	48.649669 2.6370303	51.7936874 2.66818403	54.937445 2.6975133	58.0809835 2.72521973	61.2243340 2.75147339	64.367523 2.7764189
2	37.562987 2.5128291	40.7145778 2.55224590	43.864770 2.5888055	47.0138402 2.62289203	50.161996 2.6548173	53.3093968 2.68483785	56.456167 2.7131673	59.6024038 2.73998526	62.7481871 2.76544436	65.893580 2.7896752
3	39.011063 2.5379526	42.1714748 2.57526884	45.329301 2.6100418	48.4850439 2.64259107	51.639082 2.6731805	54.7917072 2.70203008	57.943149 2.7293250	61.0935882 2.75522285	64.2431718 2.77985839	67.392019 2.8033480
4	40.418666 2.5636249	43.5902778 2.59882201	46.757859 2.6317861	49.9221743 2.66277443	53.083807 2.6920041	56.2432072 2.71965912	59.400731 2.7458970	62.5566605 2.77085312	65.7112253 2.79464478	68.864612 2.8173741
5	41.789737 2.5896899	44.9744709 2.62276930	48.153544 2.6539191	51.3280090 2.68333713	54.498673 2.7111952	57.6661623 2.73764260	60.830973 2.7628097	63.9935030 2.78681032	67.1540736 2.80974448	70.312949 2.8317003
6	43.127608 2.6160297	46.3270288 2.64700563	49.519028 2.6763473	52.7049613 2.70419550	55.885870 2.7306788	59.0625683 2.75591320	62.235703 2.7800025	65.4057936 2.80303968	68.5732632 2.82510783	71.738459 2.8462813
7	44.435123 2.6425538	47.6505153 2.67144898	50.856634 2.6989962	54.0551440 2.72528223	57.247328 2.7503941	60.4341950 2.77441551	63.616548 2.7974250	66.7950354 2.81949525	69.9701856 2.84069288	73.142435 2.8610789
8	45.714733 2.6691925	48.9471592 2.69603480	52.168396 2.7218066	55.3804194 2.74654278	58.584759 2.7702909	61.7826203 2.79310361	64.974968 2.8150351	68.1625812 2.83613827	71.3460989 2.85646395	74.526049 2.8760600
9	46.968572 2.6958912	50.2189128 2.72071188	53.456107 2.7447305	56.6824395 2.76793265	59.899688 2.7903279	63.1092582 2.81193925	66.312276 2.8327973	69.5096543 2.85293593	72.7021446 2.87239063	75.890368 2.8911966
10	48.198507 2.7226073	51.4674976 2.74543936	54.721357 2.7677292	57.9626776 2.78941536	61.193485 2.8104709	64.4153818 2.83089039	67.629659 2.8506816	70.8373648 2.86986025	74.0393625 2.88844680	77.236369 2.9064642

m	a_m	b_m	c_m	d_m	e_m
0	–0.368205765 0.1469519108	0.27627919216 1.39952627045	3.071531336840 –0.26965854932	0.0019178022618 0.0064183572442	–0.0000317773239 –0.0001024222785
1	0.9879138803 0.6540921622	0.45297970214 0.96167589506	3.024301642897 –0.15014414119	0.0032892893190 0.0028975202864	–0.0000552435395 –0.0000419843633
2	2.0521650936 0.9577515405	0.83345985094 0.73558052419	2.929014637243 –0.09439767730	0.0058440123552 0.0014452291919	–0.0000970698817 –0.0000186834770
3	3.0330022534 1.1785983848	1.22534943787 0.58911235561	2.839114176362 –0.06130501868	0.0080091011338 0.0006764141953	–0.0001305099148 –7.14670392·10^–6
4	3.9850498138 1.3540592956	1.59208568134 0.48335506807	2.762104033654 –0.03920700769	0.0096590276873 0.0002173543069	–0.0001543296062 –7.29673305·10^–7
5	4.9274383020 1.5007561805	1.92754209316 0.40182259263	2.697648032647 –0.02333700567	0.0108708424789 –0.000077651008	–0.0001704204923 3.0844587418·10–6
6	5.8677962944 1.6275460840	2.23322708646 0.33611110602	2.643937503965 –0.01134703166	0.0117393813727 –0.000277093023	–0.0001807192236 5.445314093·10^–6
7	6.8093247657 1.7397127229	2.51247480112 0.28141377516	2.599110725057 –0.00193820478	0.0123445094137 –0.000417131079	–0.0001867537156 6.942356734·10^–6
8	7.7533394623 1.8406628730	2.76875711191 0.23475325470	2.561568904705 0.005668071753	0.0127481558357 –0.000518443632	–0.0001896588790 7.902688174·10^–6
9	8.7003048861 1.9327222066	3.00519948413 0.19417363167	2.530009158087 0.011967067444	0.0129972690267 –0.000593571691	–0.0001902652854 8.518568157·10^–6
10	9.6502912998 2.0175497001	3.22448082853 0.15833007251	2.503386726405 0.017288702225	0.0131273172995 –0.000650474842	–0.0001891806589 8.90808346238·^–6

Table 4: Coefficients for $z_b(m,n)=f(n)$ for $n \geq 1$.

m	A_m	B_m	C_m	D_m	E_m
0	–0.368205759	0.27627918188	3.071531341247	0.0019178020235	–0.0000317773180
	0.1469519014	1.39952628507	–0.26965855525	0.0064183575291	–0.0001024222849
1	0.7282201609	0.66715771889	2.967519068383	0.0049072430267	–0.0000825306369
	0.7715863448	0.84366874624	–0.11529468602	0.0017887755870	–0.0000222567162
2	1.2844860115	1.42244565712	2.779892426397	0.0098790246486	–0.0001633416420
	1.2656576981	0.43704401766	–0.00823731455	–0.001223792151	0.00002813268974
3	1.6588783877	2.21884022363	2.597005607418	0.0142874040691	–0.0002314514703
	1.6855162704	0.11061198552	0.074449811796	–0.003448411178	0.00006448705901
4	1.9527849797	2.98917578557	2.432549760658	0.0179043227384	–0.0002845999367
	2.0573696067	–0.1665669779	0.142748671627	–0.005231565146	0.00009321965254
5	2.2049928625	3.71899899875	2.286842692153	0.0208394175184	–0.0003256545970
	2.3955102277	–0.4107674316	0.201761528312	–0.006743458033	0.00011740354356
6	2.4329866599	4.40793395345	2.157483380377	0.0232363066293	–0.0003576065424
	2.7085693213	–0.6314780024	0.254380707608	–0.008076867207	0.00013867336588
7	2.6455580966	5.05948744153	2.041808793319	0.0252172122947	–0.0003828178735
	3.0021502783	–0.8346524594	0.302371835308	–0.009286287303	0.00015796641044
8	2.8475035353	5.67793946463	1.937478158243	0.0268771251663	–0.0004030388285
	3.2800950094	–1.0242412011	0.346875537688	–0.010405619456	0.00017585217640
9	3.0416180708	6.26736142134	1.842563478661	0.0282880484561	–0.0004195460889
	3.5451532755	–1.2029842154	0.388662803041	–0.011456954116	0.00019269335736
10	3.2296254730	6.83134065850	1.755509313657	0.0295043406055	–0.0004332708985
	3.7993627839	–1.3728521687	0.428274604376	–0.012455254269	0.00020873028611

Table 5: Coefficients for $U_b(m,n)=f(n)$ for $n \geq 1$.

Table 6: Coefficients for $z''_b(m,n) = f(z'(m,n))$; $U''_b(m,n) = f(U'(m,n))$ of the curve connecting the branch points .

m	to:	$\bar{a}_m$ & $\overline{A}_m$	$\bar{b}_m$ & $\overline{B}_m$	$\bar{c}_m$ & $\overline{C}_m$	$\bar{d}_m$ & $\overline{D}_m$	$\bar{e}_m$ & $\overline{E}_m$	$\bar{f}_m$ & $\overline{F}_m$
0	z″	0.	1.03255472049	–0.209891280205	0.02650507820	–0.001547034821	2.717121859·10^–6
	U″	0.	1.03255471568	–0.209891275705	0.02650507684	–0.001547034677	2.717121329·10^–6
1	z″	0.	1.03768613394	–0.212546972815	0.02699886920	–0.001581192667	2.777630939·10^–6
	U″	1.	0.01695130841	0.1842150467024	–0.0437173781	0.0035595268032	–8.96778277·10^–6
2	z″	0.	1.04988152225	–0.218290124515	0.02796403550	–0.001641020525	2.857957129·10^–6
	U″	2.	–0.7586897835	0.4206222861074	–0.0775339878	0.0055645102340	–0.0000120934311
3	z″	0.	1.07021984409	–0.228003625757	0.02963847580	–0.001749250152	3.031500895·10^–6
	U″	3.	–1.4149143941	0.5869873503596	–0.0967221457	0.0064302118295	–0.0000125320673
4	z″	0.	1.09724952002	–0.240732875512	0.03180605385	–0.001887999641	3.252218427·10^–6
	U″	4.	–1.9974527414	0.7142045954165	–0.1084914003	0.0067714996070	–0.0000119578544
5	z″	0.	1.12979055127	–0.255733684141	0.03430556310	–0.002044526252	3.490226516·10^–6
	U″	5,	–2.5297526272	0.8170922590251	–0.1160930652	0.0068500426717	–0.0000110096253
6	z″	0.	1.16700545337	–0.272501726616	0.03703335580	–0.002211096670	3.729591742·10^–6
	U″	6.	–3.02571125991	0.9038948758619	–0.1212266312	0.0067942928935	–9.96235424·10^–6
7	z″	0.	1.20830030824	–0.290694642818	0.03992283817	–0.002383059269	3.962241644·10^–6
	U″	7.	–3.4943027473	0.9796199915083	–0.1248598502	0.0066724572248	–8.93938211·10^–6
8	z″	0.	1.25324561024	–0.310074608988	0.04293030500	–0.002557583186	4.184329860·10^–6
	U″	8.	–3.9416611884	1.0474900263916	–0.1275745816	0.0065223770294	–7.99458282·10^–6
9	z″	0.	1.30152387453	–0.330472119525	0.04602637876	–0.002732925622	4.394274371·10^–6
	U″	9.	–4.3721534847	1.1096650362265	–0.1297334263	0.0063654705151	–7.14831915·10^–6
10	z″	0.	1.35289583687	–0.351763534989	0.04919089721	–0.002908009942	4.591708949·10^–6
	U″	10.	–4.7889854598	1.1676373868072	–0.1315676758	0.0062138410826	–6.40447321·10^–6

Above the curve $U''_b(m,n) = f\big(U'_b(m,n)\big)$ connecting the branch points in the U-plane for the surface wave mode it is evident that : $z \approx U$. This permits an iterative approximation to a solution with the iteration scheme : (19)

$$z \approx z_1 = \frac{(m-jU)U}{F_m(U)}$$

$$\approx z_2 = (m-jU)\frac{J_m(z_1)}{J_{m-1}(z_1)} = \frac{(m-jU)z_1}{F_m(z_1)}$$

$$\approx z_3 = (m-jU)\frac{J_m(z_2)}{J_{m-1}(z_2)} = \frac{(m-jU)z_2}{F_m(z_2)}$$

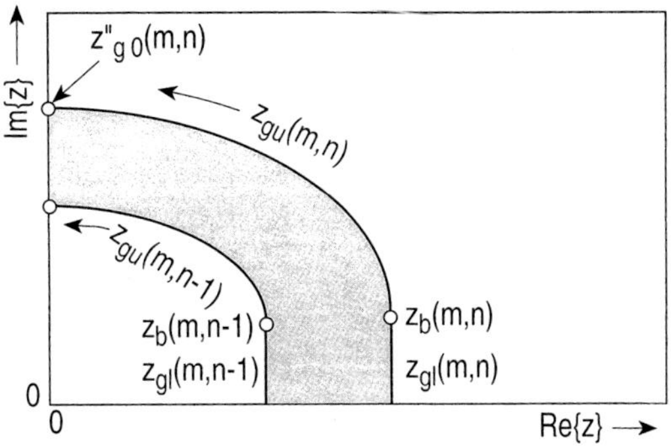

The radial modes n to a given azimuthal mode order m subdivide the z-plane into "mode strips". The limits pass through the branch points $z_b(m,n)$. The lower limit branches $z_{gl}(m,n)$ are vertical lines down to the real axis $\mathrm{Re}\{z\}$; the upper branches of the limits $z_{gu}(m,n)$ are quarter ellipses. The transforms of these limit branches are nearly coincident quarter elliptic arcs with the forms (prime and double prime indicate real and imaginary parts) :

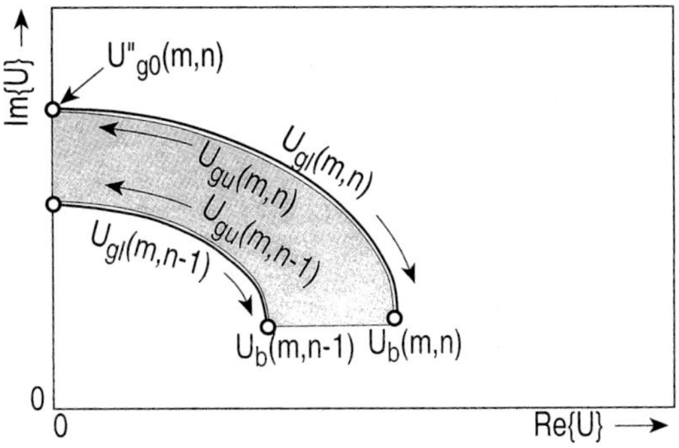

$$U''_g(m,n) = U''_b(m,n) + \big(U''_{g0}(m,n) - U''_b(m,n)\big)\sqrt{1 - U'^2_g(m,n)/U'^2_b(m,n)} \tag{20}$$

where $U''_{g0}(m,n)$ are the end points of the arcs on the imaginary axis $\mathrm{Im}\{U\}$; they have the same values as $z''_{g0}(m,n)$:

$$U''_{g0}(m,n) = z''_{g0}(m,n) = \mathrm{Im}\{j \cdot F_m(\mathrm{Re}\{z_b(m,n)\}) - m\} \tag{21}$$

In the following procedure for evaluation of a value $z(m,n)=\varepsilon_{m,n}h$ of a desired mode set for a given value U , the radial mode counting is n=0,1,2,... and the azimuthal mode counting is m=0,1,2,... . A solution of radial mode order n is sought.

1st: special case U= 0 *?*

Take the (n+1)-th (non-zero) root of $J'_m(z) = 0$. (22)

2nd: U in the surface wave range ?

That is, U is in the range limited by
• the curve $U''_b = g(U'_b)$ which connects the branch point images U_b ,

• the imaginary axis,
• the curves $U'_g(m,n) = f\big(U''_g(m,n)\big)$ for n and n+1 .

Then evaluate z with the above iteration.

3rd: Else :

The solution belongs to the lower part of the z-strip, where continued fraction expansions converge quickly. Take as starters z_{si} ; i=1,2,3; for MULLER's procedure solutions of

$$z_{s1}^2 = \frac{4(1+m)(2+m)(m+jU)}{4+3m+jU}$$

$$z_{s3}^2 = \frac{2}{12+5m+jU}\Big[\ (2+m)(3+m)(8+5m+3jU) \pm$$

$$\pm\Big(\ (2+m)(3+m)\cdot(384+608m+294m^2+57m^3+5m^4+ \tag{23}$$

$$+6j(2+m)(8+3m+m^2)U-(38+25m+5m^2)U^2)\ \Big)^{1/2}\Big]$$

$$z_{s2} = (z_{s1}+z_{s3})/2$$

Select the sign of the root in z_{s3} so that it lies in the lower part of the z-strip (i.e $z'_b(m,n) < z'_{s3} \le z'_b(m,n+1)$ and $0 \le z''_{s3} \le f\big(z'_b(m,n)\big)$, with which the curve connecting the branch points $z_b(m,n)$ is indicated.

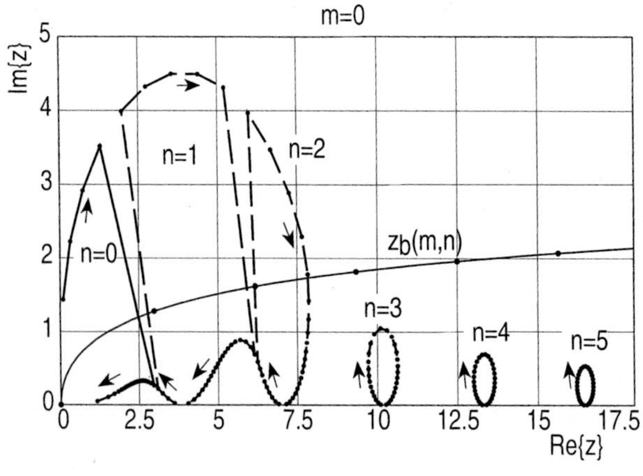

A set of mode solutions in a round duct with locally reacting lining, evaluated with the method described. A surface wave mode, if it exists, is distributed over some mode orders. The solution with the large arc is a surface wave mode.

The next graph shows a negative example of what may happen, if either imprecise start values and/or an unfavourable method of numerical solution are used; the returned solutions are "hopping" between modes (mode hopping). This would have bad (if not catastrophic) consequences in a modal field analysis tried with such results (some solutions are missing, others are returned several times).

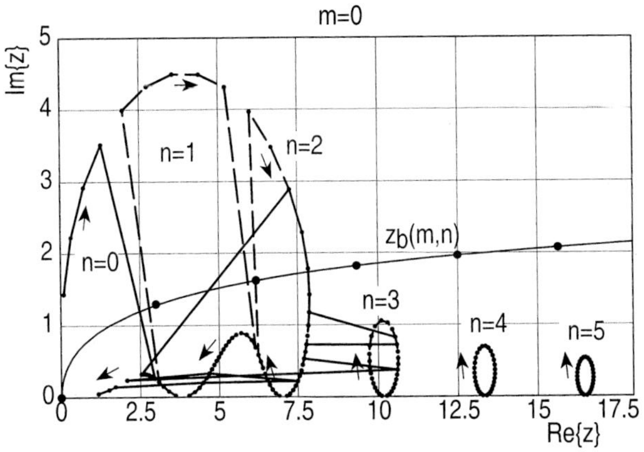

An example of “mode hopping”.

J.14 Admittance of annular absorbers approximated with flat absorbers

Ref.: Mechel, [J.5]

The evaluation of the surface admittance G of annular absorbers may be tedious. This is illustrated with a simple porous layer of thickness d having characteristic values Γ_a, Z_a . Let its interior surface be at the radius r= h .

If the layer is made locally reacting by cellular partitions (i.e. locally reacting in all directions), its surface admittance is :

$$Z_0G=\frac{j\,k_0}{Z_{an}}\,\frac{J_1(-j\Gamma_a h)\cdot Y_1(-j\Gamma_a(h+d))-Y_1(-j\Gamma_a h)\cdot J_1(-j\Gamma_a(h+d))}{J_0(-j\Gamma_a h)\cdot Y_1(-j\Gamma_a(h+d))-Y_0(-j\Gamma_a h)\cdot J_1(-j\Gamma_a(h+d))} \tag{1}$$

If the layer has ring-shaped partitions, its surface admittance for the m-th azimuthal mode is :

$$Z_0G=\frac{j\,k_0}{Z_{an}}\,\frac{J'_m(-j\Gamma_a h)\cdot Y'_m(-j\Gamma_a(h+d))-Y'_m(-j\Gamma_a h)\cdot J'_m(-j\Gamma_a(h+d))}{J_m(-j\Gamma_a h)\cdot Y'_m(-j\Gamma_a(h+d))-Y_m(-j\Gamma_a h)\cdot J'_m(-j\Gamma_a(h+d))} \tag{2}$$

The next graph shows sound attenuation curves D_h (for the least attenuated mode) in a round duct with a simple glass fibre layer with cellular partitions, first evaluated with the admittance of ring-shaped absorbers, then with the admittance of the same, but plane absorber.

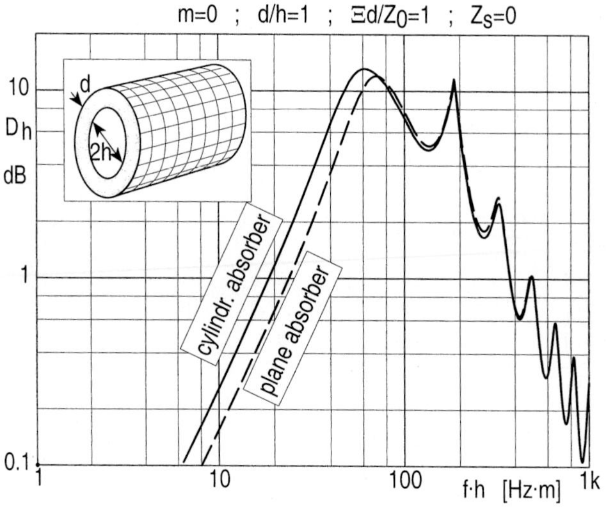

Attenuation D_h in a round duct with a locally reacting porous layer, either evaluated as a cylindrical layer or as a plane layer.

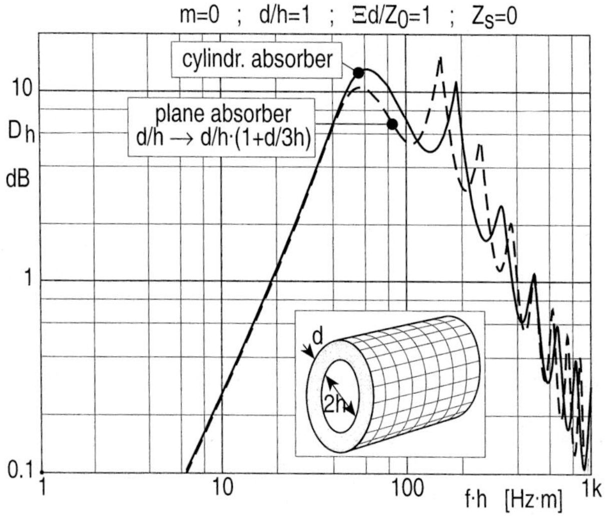

As above, but the thickness of the plane absorber increased in the evaluation.

The rule which can be taken from this example (checked with other examples, too) is :

- For frequencies up to the 1st maximum in the D_h-curve use the admittance of a plane absorber after having increased the thickness of all air or porous layers in the absorber by

$$d/h \to d/h\cdot(1+\tfrac{1}{3}d/h)\ . \qquad (3)$$

- For higher frequencies use the plane absorber with the original thicknesses.

J.15 Round duct with a bulk reacting lining

Ref.: Mechel, [J.5]

A round duct, 2h wide, is lined with a layer of thickness d of porous material having characteristic values Γ_a, Z_a, or in normalised form $\Gamma_{an} = \Gamma_a/k_0$, $Z_{an} = Z_a/Z_0$. The layer is possibly covered with a foil having a partition impedance Z_s.

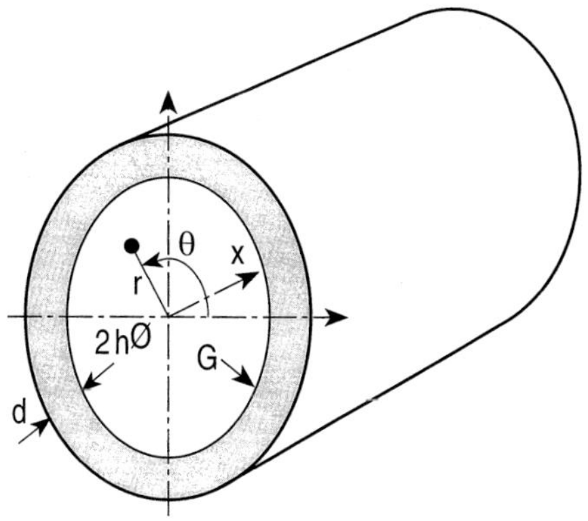

The analysis goes some way in parallel with the analysis in section J.13 , but the surface admittance G of the lining now becomes field dependent.

The field formulation in the free duct for the m-th azimuthal mode is :

$$p(r,\theta,x) = P\cdot\cos(m\theta)\cdot J_m(\varepsilon_m r)\cdot e^{-\Gamma_m x} \quad ; \quad m = 0,1,2,\ldots$$

$$v_r = \frac{j}{k_0 Z_0}\frac{\partial p}{\partial r} = \frac{j\varepsilon_m}{k_0 Z_0} P\cos(m\theta)\cdot J'_m(\varepsilon_m r)\cdot e^{-\Gamma_m x} \tag{1}$$

$$\varepsilon_m^2 = \Gamma_m^2 + k_0^2$$

The characteristic equation for $\varepsilon_m h$ is :

$$(\varepsilon_m h)\frac{J'_m(\varepsilon_m h)}{J_m(\varepsilon_m h)} = -j k_0 h\cdot Z_0 G \tag{2}$$

or :

$$(\varepsilon_m h)\frac{J_{m-1}(\varepsilon_m h)}{J_m(\varepsilon_m h)} = m - jU \quad ; \quad U =: k_0 h\cdot Z_0 G \tag{3}$$

So far the analysis is as in section J.13 . To obtain the surface admittance G , the field in the absorber layer is formulated as :

$$p_a(r,\theta,x) = \left[B J_m(\varepsilon_{am} r) + C Y_m(\varepsilon_{am} r)\right]\cdot\cos(m\theta)\cdot e^{-\Gamma_m x}$$

$$v_{ar}(r,\theta,x) = \frac{-\varepsilon_{am}}{\Gamma_a Z_a}\left[B J'_m(\varepsilon_{am} r) + C Y'_m(\varepsilon_{am} r)\right]\cdot\cos(m\theta)\cdot e^{-\Gamma_m x} \tag{4}$$

with : $$\varepsilon_{am}^2 = \Gamma^2 - \Gamma_a^2 = \varepsilon_m^2 - k_0^2\left(1+\Gamma_{an}^2\right) . \tag{5}$$

The boundary condition at the hard outer duct wall gives : $$C = -B\frac{J'_m(\varepsilon_{am}(h+d))}{Y'_m(\varepsilon_{am}(h+d))} \tag{6}$$

The surface admittance of the lining (without Z_s) becomes :

$$Z_0G = \frac{-\varepsilon_{am}}{\Gamma_a Z_{an}}\frac{J'_m(\varepsilon_{am}h)\cdot Y'_m(\varepsilon_{am}(h+d)) - Y'_m(\varepsilon_{am}h)\cdot J'_m(\varepsilon_{am}(h+d))}{J_m(\varepsilon_{am}h)\cdot Y'_m(\varepsilon_{am}(h+d)) - Y_m(\varepsilon_{am}h)\cdot J'_m(\varepsilon_{am}(h+d))} \tag{7}$$

and with Z_s : $$Z_0G \to 1/\left(Z_s/Z_0 + 1/Z_0G\right) \tag{8}$$

With the derivatives of the Bessel and Neumann functions substituted, the function U becomes for $Z_s= 0$ and $y= \varepsilon_{am}h$, for abbreviation :

$$U = \frac{1}{\Gamma_{an}Z_{an}}\cdot$$

$$\cdot\frac{\left[m J_m(y) - y J_{m-1}(y)\right]\cdot\left[m Y_m(y(1+d/h)) - y(1+d/h)Y_{m-1}(y(1+d/h))\right] - \ldots}{m\cdot\left[J_m(y)\cdot Y_m(y(1+d/h)) - Y_m(y)\cdot J_m(y(1+d/h))\right] + \ldots} \tag{9}$$

$$\frac{\ldots - \left[m Y_m(y) - y Y_{m-1}(y)\right]\cdot\left[m J_m(y(1+d/h)) - y(1+d/h)J_{m-1}(y(1+d/h))\right]}{\ldots + y(1+d/h)\cdot\left[Y_m(y)\cdot J_{m-1}(y(1+d/h)) - J_m(y)\cdot Y_{m-1}(y(1+d/h))\right]}$$

and for $Z_s= 0$ and $m= 0$:

$$U = k_0 h Z_0 G = \frac{y}{\Gamma_{an}Z_{an}}\frac{J_1(y)\cdot Y_1(y(1+d/h)) - Y_1(y)\cdot J_1(y(1+d/h))}{J_0(y)\cdot Y_1(y(1+d/h)) - Y_0(y)\cdot J_1(y(1+d/h))} \tag{10}$$

The function U of the characteristic equation contains in a complicated manner the desired solution $z= \varepsilon_m h$ of that equation.

It is supposed for the following methods of solution that modes will be determined for a list of k_0h values which begins at low values (if a mode for a single k_0h value is needed, possibly a list must be prepended).

1st method : iteration of layer resistance :

This method makes use of the fact that a mode-safe method for locally reacting linings exists, and that at low frequency and/or high flow resistance values $R= \Xi\cdot d/Z_0$ (Ξ= layer flow resistivity) the bulk reacting absorber becomes nearly locally reacting. So start the iteration i=1,2,... through the list k_0h for i=1,2,3 and begin for each i the evaluation of the mode for a locally reacting absorber with a high value R_k (about $R_k > 8$) ; evaluate three solutions for $R_{k=1}, R_{k=2}, R_{k=3}$ with the tendency of $R_k \to R$ for increasing k. Take these values as start values in MULLER'S procedure for the solution of the characteristic equation with the bulk reacting absorber, but with $R_{k=3}$. Then iterate with this task through R_k, taking previous solutions z_k as new starters. For values $i > 2$ take previous solutions z_{i-2}, z_{i-1} as two starters for the new z_i, and take the approximation from the iteration through R_k as the 3rd starter (it helps to avoid mode hopping, which happens when only previous z_i are used, even with small steps $\Delta k_0 h$).

2nd method : start with approximations for low frequencies :

The iteration through R_k in the above method may be time consuming.

If the absorber is *locally reacting* , U is independent of z. The characteristic equation reads :

$$\begin{aligned} &z \cdot J'_m(z) + jU \cdot J_m(z) = 0 \quad ; \quad m > 0 \\ &z \cdot J_1(z) - jU \cdot J_0(z) = 0 \quad ; \quad m = 0 \end{aligned} \tag{11}$$

The method makes use of the fact that for $k_0 h \to 0$ the function for every absorber decreases at least with the square of $k_0 h$. Thus starter solutions at low frequency (for the radial modes n) are the solutions $z_{m,n}$ of $J'_m(z) = 0$ for $m > 0$, and of $J_1(z) = 0$ for $m = 0$. If $n \geq 1$, this value and nearby values with small shifts $-\Delta z'$ and $+j \cdot \Delta z''$ are used as starters for MULLER's procedure. If $n=0$ (i.e. the fundamental radial mode) these approximations are not precise enough; then approximations from the continued fraction expansion should be applied. From the solution $i=4$ on (of the list of $k_0 h$) use previous solutions together with an extrapolated estimate of the new solution (see section J.10).

If the absorber is *bulk reacting*, the statement $U \xrightarrow[k_0 h \to 0]{} O\left((k_0 h)^i\right) \; ; \; i > 2$ still holds at low frequencies. For $n>0$ the same starters can be used as above. But for $n=0$ we must be more precise. Either use in this case the "iteration through R" from above, or derive continued fraction approximations for the characteristic equation.

For the *least attenuated mode* one determines solutions for $n=0$ and $n=1$ and takes the solution with minimum $\mathrm{Re}\{\Gamma_n h\}$.

J.16 Annular ducts

Ref.: Mechel, [J.6]

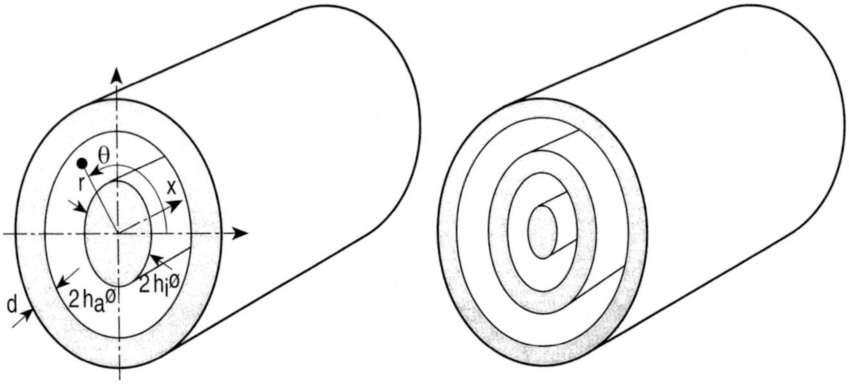

Such ducts are in principle different from round ducts, because now the Neumann functions appear.

We distinguish with indices i,a radii h_i, h_a and absorber functions $U_i = k_0 h_i \cdot Z_0 G_i$ and $U_a = k_0 h_a \cdot Z_0 G_a$ on the interior or outer side, respectively, of the free duct which spans over $h_i \le r \le h_a$.

Field formulation for the m-th azimuthal mode :

$$\begin{aligned} p(r,\theta,x) &= [A \cdot J_m(\varepsilon r) + B \cdot Y_m(\varepsilon r)] \cdot \cos(m\theta) \cdot e^{-\Gamma x} \\ v_r(r,\theta,x) &= \frac{j\varepsilon}{k_0 Z_0}[A \cdot J'_m(\varepsilon r) + B \cdot Y'_m(\varepsilon r)] \cdot \cos(m\theta) \cdot e^{-\Gamma x} \end{aligned} \tag{1}$$

(we neglect for the moment the indexing of ε, Γ with m). The boundary conditions give the system of equations :

$$\begin{aligned} &A \cdot [j\varepsilon \cdot J'_m(\varepsilon h_a) - k_0 Z_0 G_a \cdot J_m(\varepsilon h_a)] + B \cdot [j\varepsilon \cdot Y'_m(\varepsilon h_a) - k_0 Z_0 G_a \cdot Y_m(\varepsilon h_a)] = 0 \\ &A \cdot [j\varepsilon \cdot J'_m(\varepsilon h_i) + k_0 Z_0 G_i \cdot J_m(\varepsilon h_i)] + B \cdot [j\varepsilon \cdot Y'_m(\varepsilon h_i) + k_0 Z_0 G_i \cdot Y_m(\varepsilon h_i)] = 0 \end{aligned} \tag{2}$$

A solution must null the determinant :

$$\begin{aligned} &[j\varepsilon \cdot J'_m(\varepsilon h_a) - k_0 Z_0 G_a \cdot J_m(\varepsilon h_a)] \cdot [j\varepsilon \cdot Y'_m(\varepsilon h_i) + k_0 Z_0 G_i \cdot Y_m(\varepsilon h_i)] - \\ &[j\varepsilon \cdot J'_m(\varepsilon h_i) + k_0 Z_0 G_i \cdot J_m(\varepsilon h_i)] \cdot [j\varepsilon \cdot Y'_m(\varepsilon h_a) - k_0 Z_0 G_a \cdot Y_m(\varepsilon h_a)] = 0 \end{aligned} \tag{3}$$

With the recurrence relations for derivatives of Bessel and Neumann functions, and using the abbreviations $z=\varepsilon h_a$; $\alpha=h_i/h_a$ a different form of the equation is :

$$\begin{aligned}&[z\cdot J_{m-1}(z)+(jU_a-m)\cdot J_m(z)]\cdot[\alpha z\cdot Y_{m-1}(\alpha z)-(jU_i+m)\cdot Y_m(\alpha z)]-\\&-[\alpha z\cdot J_{m-1}(\alpha z)-(jU_i+m)\cdot J_m(\alpha z)]\cdot[z\cdot Y_{m-1}(z)+(jU_a-m)\cdot Y_m(z)]=0\end{aligned} \tag{4}$$

In the special case $m=0$ with $J_0'(z)=-J_1(z);\ Y_0'(z)=-Y_1(z)$ it becomes :

$$\begin{aligned}&[z\cdot J_1(z)-jU_a\cdot J_0(z)]\cdot[\alpha z\cdot Y_1(\alpha z)+jU_i\cdot Y_0(\alpha z)]-\\&-[\alpha z\cdot J_1(\alpha z)+jU_i\cdot J_0(\alpha z)]\cdot[z\cdot Y_1(z)-jU_a\cdot Y_0(z)]=0\end{aligned} \tag{5}$$

and after multiplication :

$$\begin{aligned}&\alpha z^2\cdot[J_1(z)\cdot Y_1(\alpha z)-J_1(\alpha z)\cdot Y_1(z)]-jU_a\,\alpha z\cdot[J_0(z)\cdot Y_1(\alpha z)-J_1(\alpha z)\cdot Y_0(z)]+\\&+jU_i\,z\cdot[J_1(z)\cdot Y_0(\alpha z)-J_0(\alpha z)\cdot Y_1(z)]+U_aU_i\cdot[J_0(z)\cdot Y_0(\alpha z)-J_0(\alpha z)\cdot Y_0(z)]=0\end{aligned} \tag{6}$$

An interior porous layer of thickness $d_i< h_i$ has the surface admittance (Γ_i,Z_i the characteristic values of the material) :

$$Z_0G_i=\frac{-j\,k_0}{Z_i}\,\frac{J_1(j\Gamma_ih_i)\cdot Y_1(j\Gamma_i(h_i-d_i))-Y_1(j\Gamma_ih_i)\cdot J_1(j\Gamma_i(h_i-d_i))}{J_0(j\Gamma_ih_i)\cdot Y_1(j\Gamma_i(h_i-d_i))-Y_0(j\Gamma_ih_i)\cdot J_1(j\Gamma_i(h_i-d_i))} \tag{7}$$

and if $d_i=h_i$, i.e. a central absorber (locally reacting) :

$$Z_0G_i=\frac{-j\,k_0}{Z_i}\,\frac{J_1(j\Gamma_ih_i)}{J_0(j\Gamma_ih_i)} \tag{8}$$

An outer porous layer has the admittance :

$$Z_0G_a=\frac{j\,k_0}{Z_a}\,\frac{J_1(j\Gamma_ah_a)\cdot Y_1(j\Gamma_a(h_a+d_a))-Y_1(j\Gamma_ah_a)\cdot J_1(j\Gamma_a(h_a+d_a))}{J_0(j\Gamma_ah_a)\cdot Y_1(j\Gamma_a(h_a+d_a))-Y_0(j\Gamma_ah_a)\cdot J_1(j\Gamma_a(h_a+d_a))} \tag{9}$$

For the numerical solution, an iterative scheme over a list of k_0h , which begins at low values, is recommended.

For $m=0$ *and locally reacting absorbers* :

For low k_0h the last three terms in the last form of the characteristic equation disappear with highest order in k_0h ; therefore an approximate solution is found from :

$$[J_1(z)\cdot Y_1(\alpha z)-J_1(\alpha z)\cdot Y_1(z)]=0 \tag{10}$$

Such solutions are tabulated in the literature. A regression for $z_{m=0,n=1}$ over $0\le\alpha\le 0.8$ is :

$$\begin{aligned}z_{0,1}\approx\ &3.8050757+2.65957569\cdot\alpha-21.764629204\alpha^2+\\&+135.27002785\cdot\alpha^3-249.58953616\cdot\alpha^4+172.92523266\cdot\alpha^5\end{aligned} \tag{11}$$

(use small shifts $-\Delta z'$ and $+j\cdot\Delta z''$ for the two other starters of MULLER's procedure). A starter at low k_0h for the case n=0 (i.e. lowest radial mode) is taken from the power series expansion of the characteristic equation :

$$\begin{aligned}
&\frac{2}{\pi}\left[j(U_a+U_i)+U_aU_i\ln\alpha\right]- \\
&-\frac{z^2}{2\pi}\left\{\left(1-\alpha^2\right)\left[2+j(U_a-U_i)+U_aU_i\right]+\ln\alpha\cdot\left[(U_a-2j)U_i+\alpha^2(U_i+2j)U_a\right]\right\}+ \\
&+\frac{z^4}{64\pi}\left\{\left(1-\alpha^2\right)\left[8+2jU_a-10jU_i+3U_aU_i+\alpha^2(8+10jU_a-2jU_i+3U_aU_i)\right]+\right. \\
&\qquad\left.+2\ln\alpha\cdot\left[(U_a-4j)U_i+4\alpha^2(U_a-2j)(U_i+2j)+\alpha^4U_a(U_i+4j)\right]\right\}=0
\end{aligned} \tag{12}$$

Take in the solution for z^2 the root with a negative real part. Further take the solution of this equation with 4th power term z^4 neglected as a 2nd starter, and the mean value of both as the 3rd starter.

If the absorber is bulk reacting and m=0 :

Find initial starters (at the lower end of the k_0h-list) with the assumption that the absorber is approximately locally reacting. Take these solutions as starters with the equation for the bulk reacting absorber, and for later k_0h-values take previous solutions and an extrapolated new solution (see section J.10) .

If the absorber is bulk reacting and m>0 :

For n>0 take the solution of the term without U_a, U_i in the expanded characteristic equation as starter. For n=0 solve the power series expansion of the characteristic equation, now with expansion of U_a, U_i , for a starter.

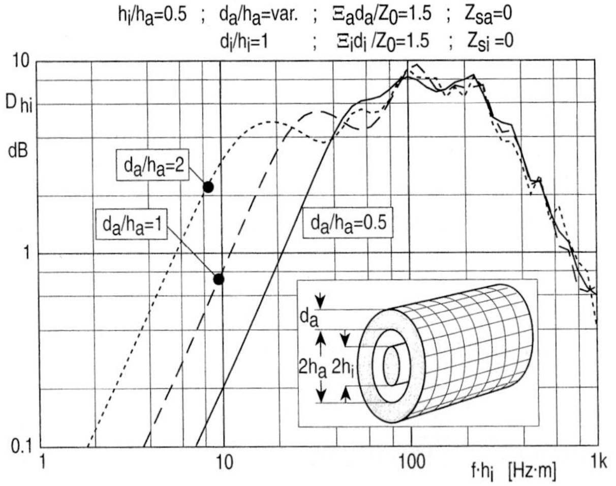

Attenuation curves D_{hi} for ring-shaped silencers with locally reacting glass fibre layers for different thickness ratios of the outer layer.

J.17 Duct with a cross-layered lining

Ref.: Mechel, [J.7]

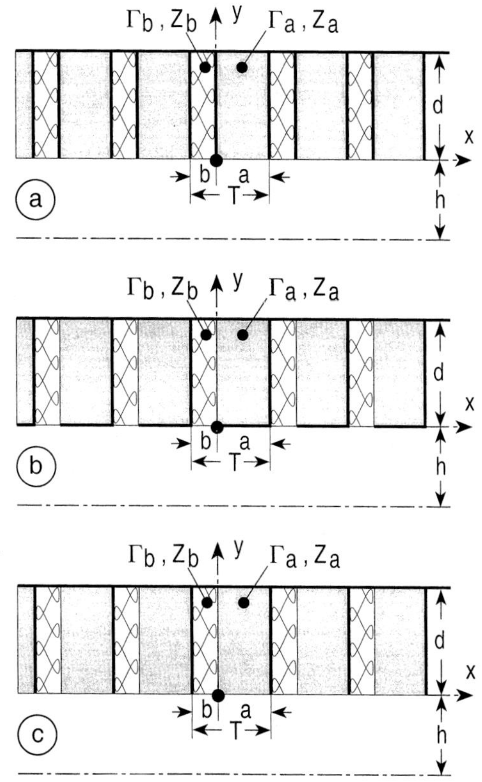

Often silencers need low values of the (normalised) flow resistance $R= \Xi \cdot d/Z_0$ of porous layers to obtain high attenuation values D_h, but the layer thickness d must not be small, otherwise the lower limit frequency of attenuation would be high. The necessarily low flow resistance values Ξ lead to low bulk densities or to coarse fibres; both measures reduce the mechanical stability of the absorber. A remedy can be to place layers side by side, one of the layers just being a "place holder" made out of (e.g.) scrambled wire mats.

If characteristic values Γ_α, Z_α ; α= a,b ; are normalised with k_0, Z_0, respectively, this is indicated with an additional index n .

In case (a) , in which both layers are separated from each other, if both layer thicknesses a,b are small compared to the wavelength, the effective lining admittance is the weighted average

$$Z_0 G_y = \frac{a/T}{Z_{an}} \tanh(\Gamma_a d) + \frac{b/T}{Z_{bn}} \tanh(\Gamma_b d) \qquad (1)$$

In case (b) a couple of layers forms a lined cross-directed duct, one of the layer heads is covered towards the main duct.

In case (c) layer couples form cross-directed, lined ducts, and both layer heads are open towards the main duct.

In cases (b),(c) first the propagation constant Γ_s in the side ducts is determined. For the arrangement (b) the average wall admittance seen from the main duct is approximately :

$$Z_0 G_y = \frac{b/T}{Z_{bn}} \frac{\Gamma_s}{\Gamma_b} \tanh(\Gamma_s d) \tag{2}$$

For the arrangement (c) the average admittance of the cross-layered lining of the main duct is:

$$Z_0 G_y = \left[\frac{a/T}{\Gamma_a Z_{an}} + \frac{b/T}{\Gamma_b Z_{bn}} \right] \cdot \Gamma_s \tanh(\Gamma_s d) \tag{3}$$

Average admittance values G_y are sufficient only for small T/λ_0. Otherwise the main duct should be treated as a duct with an axially periodic lining. Formulas will be given below, 1st with the assumption of only the fundamental mode in the side duct, 2nd with higher modes in the side duct. Sometimes the layer indices will be collected as $\alpha = a,b$.

Only the fundamental mode in the side duct :

Wave and impulse equation in the layers :

$$\left(\Delta - \Gamma_\alpha^2\right) p_\alpha(x,y) = 0 \quad ; \quad v_{\alpha,x}(x,y) = \frac{-1}{\Gamma_\alpha Z_\alpha} \frac{\partial p_\alpha}{\partial x} \tag{4}$$

Field formulations in the layers (α may also assume the values of a,b) :

$$p_\alpha(x,y) = P_\alpha \cdot \frac{\cosh(\Gamma_s(y-d))}{\cosh(\Gamma_s d)} \cdot \cos(\varepsilon_\alpha(x \mp \alpha)) \quad ; \quad \alpha = \begin{cases} a \ ; & x \in a \\ b \ ; & x \in b \end{cases}$$

$$v_{x\alpha}(x,y) = \frac{1}{Z_\alpha} P_\alpha \cdot \frac{\cosh(\Gamma_s(y-d))}{\cosh(\Gamma_s d)} \cdot \frac{\varepsilon_\alpha}{\Gamma_\alpha} \sin(\varepsilon_\alpha(x \mp \alpha)) \tag{5}$$

$$v_{y\alpha}(x,y) = \frac{-1}{Z_\alpha} P_\alpha \cdot \frac{\Gamma_s}{\Gamma_\alpha} \frac{\sinh(\Gamma_s(y-d))}{\cosh(\Gamma_s d)} \cdot \cos(\varepsilon_\alpha(x \mp \alpha))$$

and from the wave equation : $\varepsilon_\alpha^2 = \Gamma_s^2 - \Gamma_\alpha^2$ (6)

If the main duct carries higher modes in the z direction, this changes to : $\varepsilon_\alpha^2 = \Gamma_s^2 - \Gamma_\alpha^2 - \varepsilon_z^2$ (7) (will be neglected below).

The characteristic equation of the side duct is :

$$\frac{-\varepsilon_a}{\Gamma_a Z_a} \tan(\varepsilon_a a) = \frac{\varepsilon_b}{\Gamma_b Z_b} \tan(\varepsilon_b b) \tag{8}$$

or with the abbreviations :

$$z_b = \varepsilon_b b \quad ; \quad A = \frac{b}{a} \frac{\Gamma_b Z_b}{\Gamma_a Z_a} \quad ; \quad B = (\Gamma_b b)^2 - (\Gamma_a b)^2 \tag{9}$$

it is :

$$z_b \cdot \tan(z_b) + A \cdot \left(\frac{a}{b}\sqrt{z_b^2 + B}\right) \cdot \tan\left(\frac{a}{b}\sqrt{z_b^2 + B}\right) = 0 \tag{10}$$

An approximate equation $$C_0 + C_1 \cdot z_b^2 - C_2 \cdot z_b^4 + C_3 \cdot z_b^6 - C_4 \cdot z_b^8 = 0 \tag{11}$$
is obtained by continued fraction expansion with the coefficients : (12)

$$C_0 = 525(a/b)^2 A \cdot B\left(21 - 2(a/b)^2 B\right)$$

$$C_1 = 15\left\{5(a/b)^2 A\left[147 - 7\left(9 + 4(a/b)^2\right)B + 6(a/b)^2 B^2\right] + 7\left(105 - 45(a/b)^2 B + (a/b)^4 B^2\right)\right\}$$

$$C_2 = 5\left\{210 + 3(a/b)^2[315 + 7A(45 - B) - 30B)] - 2(a/b)^4\left[(21 - B)B - A(105 - 90B + B^2\right]\right\}$$

$$C_3 = 5(a/b)^2\left\{90 + A\left[21 + (a/b)^2(90 - 4B)\right] + (a/b)^2(21 - 4B)\right\}$$

$$C_4 = 10(1 + A)(a/b)^4$$

One gets with its solution :

$$\Gamma_s d = \frac{d}{b}\sqrt{z_b^2 + (k_0 b)^2 \Gamma_{bn}^2} \tag{13}$$

or in the case of a higher mode in the z direction :

$$\Gamma_s d = \frac{d}{b}\sqrt{z_b^2 + (k_0 b)^2 \Gamma_{bn}^2 + (\varepsilon_z b)^2} \tag{14}$$

Select from the polynomial solutions the solution with minimum $\mathrm{Re}\{\Gamma_s d\} > 0$.

The normalised surface admittances of the layer heads are :

$$Z_0 G_{\alpha y} = \frac{\Gamma_s d}{k_0 d\, \Gamma_{\alpha n} Z_{\alpha n}} \tanh(\Gamma_s d) \tag{15}$$

Field formulation in the main duct with spatial harmonics :

$$\begin{aligned} p(x,y) &= \sum_{n=-\infty}^{+\infty} P_n \cdot \frac{\cos\left(\varepsilon_n (y + h)\right)}{\cos \varepsilon_n h} \cdot e^{-j\beta_n x} \\ Z_0 v_y(x,y) &= -j\sum_n P_n \cdot \frac{\varepsilon_n}{k_0} \frac{\sin\left(\varepsilon_n (y + h)\right)}{\cos \varepsilon_n h} \cdot e^{-j\beta_n x} \end{aligned} \tag{16}$$

with axial wave numbers :

$$\beta_n = \beta_0 + n\frac{2\pi}{T} \quad ; \quad \mathrm{Im}\{\beta_0\} \le 0 \quad ; \quad \varepsilon_n^2 = k_0^2 - \beta_n^2 \tag{17}$$

The wall admittance at $y=0$ is periodic with period $T= a+b$ and has the axial profile :

$$Z_0G(x)=\begin{cases}Z_0G_{by} & ;\ -b\le x<0\\ Z_0G_{ay} & ;\ 0<x\le a\end{cases} \tag{18}$$

As a Fourier series :

$$Z_0G(x)=\sum_{\nu} g_\nu\cdot e^{+j2\nu\pi\cdot x/T}\quad ;\quad g_\nu=\frac{1}{T}\int_T Z_0G(x)\cdot e^{-j2\nu\pi\cdot x/T}\,dx \tag{19}$$

with coefficients :

$$\begin{aligned} g_\nu&=\frac{1}{a+b}\left[a\cdot Z_0G_{ay}\frac{\sin(\nu\pi a/T)}{\nu\pi a/T}\cdot e^{-j\nu\pi a/T}+b\cdot Z_0G_{by}\frac{\sin(\nu\pi b/T)}{\nu\pi b/T}\cdot e^{+j\nu\pi b/T}\right]\\ g_0&=\frac{1}{a+b}\left[a\cdot Z_0G_{ay}+b\cdot Z_0G_{by}\right]=Z_0G_y\end{aligned} \tag{20}$$

(notice in general $g_{-\nu}\neq g_{+\nu}$).

One splits the main duct field into a periodic factor and a propagation factor :

$$p(x,y)=P(x,y)\cdot e^{-j\beta_0 x}\quad ;\quad v_y(x,y)=V_y(x,y)\cdot e^{-j\beta_0 x} \tag{21}$$

The boundary conditions at $y=0$ give the linear, homogeneous system of equations :

$$\sum_n P_n\cdot\left[\delta_{m,n}\cdot j\frac{\varepsilon_n}{k_0}\tan(\varepsilon_n h)+g_{n-m}\right]=0\quad ;\quad m,\ n=0,\pm1,\pm2,\ldots \tag{22}$$

The determinant set to zero represents the characteristic equation for β_0 . A simplified boundary condition requires the matching of the periodic factor $V_y(x,0)$ with the particle velocity $V_{abs}(x,0)$ in the surface of the lining :

$$-jP_n\frac{\varepsilon_n}{k_0}\tan(\varepsilon_n h)=\frac{1}{T}\int_T Z_0V_{abs}(x,0)\cdot e^{+j2n\pi x/T}\,dx \tag{23}$$

This leads to :

$$\sum_n\left\{\delta_{m,n}-jk_0h\frac{g_{n-m}}{\varepsilon_n h\cdot\tan(\varepsilon_n h)}\right\}\cdot\frac{1}{T}\int_T Z_0V_{abs}(x,0)\cdot e^{+j2n\pi x/T}\,dx=0 \tag{24}$$

If only one mode exists in the side duct, and this is supposed to have a plane velocity profile, so that $V_{abs}=G\cdot P$ for an exciting pressure P , this equation becomes :

$$\begin{aligned}&\sum_n\left\{\delta_{m,n}-jk_0h\frac{g_{n-m}}{\varepsilon_n h\cdot\tan(\varepsilon_n h)}\right\}\cdot g_{-n}=0 && ;\quad m\neq0\\ &g_0-jk_0h\sum_n\frac{g_n\cdot g_{-n}}{\varepsilon_n h\cdot\tan(\varepsilon_n h)}=0 && ;\quad m=0\end{aligned} \tag{25}$$

The equation for $m=0$ is the characteristic equation for $\beta_0 h$ which is contained in the $\varepsilon_n h$ (the leading term plus the term with $n=0$ in the sum form the characteristic equation for a duct with a homogeneous lining). One can take the mode wave numbers $\varepsilon_n h$ in a duct with the average admittance from $Z_0 G_{ay}$, $Z_0 G_{by}$ as start approximations in the numerical solution with MULLER's procedure for $(\beta_0 h)^2 = (k_0 h)^2 - (\varepsilon_0 h)^2$.

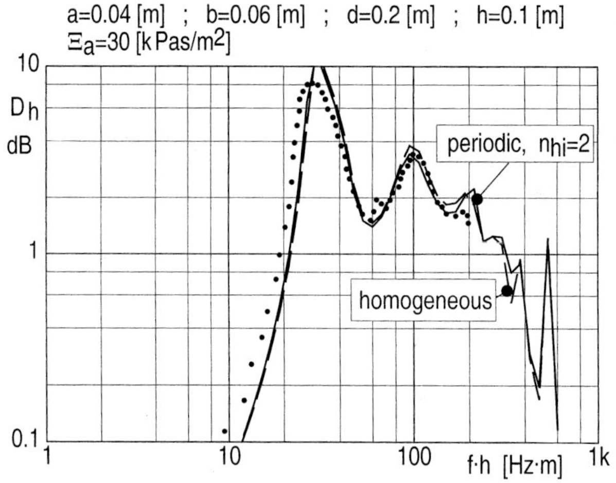

Mineral fibre boards, a=4 [cm] thick, with mutual distance b=6 [cm] form a cross-layered lining of d= 20 [cm] thickness (air in the layers α=a) of a h= 10 [cm] wide duct.
Points: measured;
full line: periodic duct with spatial harmonics;
dashed: homogeneous duct with average admittance of the layers treated as side ducts.

The assumption of a plane mode profile, made above, can be dropped. One obtains with the true mode profile :

$$Z_0 V_{abs}(x,0) = \begin{cases} Z_0 v_{ay}(x,0) = \dfrac{\Gamma_s \tanh(\Gamma_s d)}{\Gamma_a Z_{an}} P_a \cos(\varepsilon_a a) \dfrac{\cos\big(\varepsilon_a (x-a)\big)}{\cos(\varepsilon_a a)} & ; \ x \in a \\ Z_0 v_{by}(x,0) = \dfrac{\Gamma_s \tanh(\Gamma_s d)}{\Gamma_b Z_{bn}} P_b \cos(\varepsilon_b b) \dfrac{\cos\big(\varepsilon_b (x+b)\big)}{\cos(\varepsilon_b b)} & ; \ x \in b \end{cases} \tag{26}$$

and (notice the definition of the abbreviation γ_n) :

$$\frac{1}{T}\int_T Z_0 V_{abs}(x,0)\cdot e^{+j2n\pi x/T}\, dx \ := \ P\cdot\gamma_n =$$

$$= P\cdot\left\{ \frac{1}{\Gamma_a d Z_{an}} \frac{a/T}{(\varepsilon_a a)^2 - (2n\pi a/T)^2} \left[j2n\pi \frac{a}{T}\left(\frac{e^{j2n\pi a/T}}{\cos(\varepsilon_a a)} - 1 \right) + \varepsilon_a a \cdot \tan(\varepsilon_a a) \right] + \right. \tag{27}$$

$$\left. + \frac{1}{\Gamma_b d Z_{bn}} \frac{b/T}{(\varepsilon_b b)^2 - (2n\pi b/T)^2} \left[j2n\pi \frac{b}{T}\left(1 - \frac{e^{-j2n\pi b/T}}{\cos(\varepsilon_b b)} \right) + \varepsilon_b b \cdot \tan(\varepsilon_b b) \right] \right\}$$

Therewith the equation to be solved is :

$$\gamma_0 - jk_0 h \sum_n \frac{g_n \cdot \gamma_n}{\varepsilon_n h \cdot \tan(\varepsilon_n h)} = 0 \tag{28}$$

The larger amount of computations often does not pay for.

Higher modes in the side duct : (mode index σ in the side ducts)

Field formulation in the side ducts $\alpha = a,b$:

$$p_\alpha(x,y) = \sum_\sigma P_{\alpha\sigma} \cdot \frac{\cosh\left(\Gamma_{s\sigma}(y-d)\right)}{\cosh(\Gamma_{s\sigma} d)} \cdot \frac{\cos\left(\varepsilon_{\alpha\sigma}(x \mp \alpha)\right)}{\cos\left(\varepsilon_{\alpha\sigma}\alpha\right)} \quad ; \quad \alpha = \begin{cases} a \; ; \; x \in a \\ b \; ; \; x \in b \end{cases}$$

$$v_{x\alpha}(x,y) = \frac{1}{Z_\alpha} \sum_\sigma P_{\alpha\sigma} \cdot \frac{\cosh\left(\Gamma_{s\sigma}(y-d)\right)}{\cosh(\Gamma_{s\sigma} d)} \cdot \frac{\varepsilon_{\alpha\sigma}}{\Gamma_\alpha} \frac{\sin\left(\varepsilon_{\beta\sigma}(x \mp \alpha)\right)}{\cos\left(\varepsilon_{\alpha\sigma}\alpha\right)} \tag{29}$$

$$v_{y\alpha}(x,y) = \frac{-1}{Z_\alpha} \sum_\sigma P_{\alpha\sigma} \cdot \frac{\Gamma_{s\sigma}}{\Gamma_\alpha} \frac{\sinh\left(\Gamma_{s\sigma}(y-d)\right)}{\cosh(\Gamma_{s\sigma} d)} \cdot \frac{\cos\left(\varepsilon_{\alpha\sigma}(x \mp \alpha)\right)}{\cos\left(\varepsilon_{\alpha\sigma}\alpha\right)}$$

with : $\varepsilon_{\alpha\sigma}^2 = \Gamma_{s\sigma}^2 - \Gamma_\alpha^2$ and the characteristic equation :

$$\frac{-\varepsilon_{a\sigma}}{\Gamma_a Z_a} \tan(\varepsilon_{a\sigma} a) = \frac{\varepsilon_{b\sigma}}{\Gamma_b Z_b} \tan(\varepsilon_{b\sigma} b) \tag{30}$$

With the transversal mode profiles :
$$q_{\alpha\sigma}(x) = \frac{\cos\left(\varepsilon_{\alpha\sigma}(x \mp \alpha)\right)}{\cos\left(\varepsilon_{\alpha\sigma}\alpha\right)} \tag{31}$$

the orthogonality relation in the side ducts :

$$\frac{1}{\Gamma_{an} Z_{an}} \int_0^a q_{a\sigma}(x) \cdot q_{a\tau}(x)\, dx + \frac{1}{\Gamma_{bn} Z_{bn}} \int_{-b}^0 q_{b\sigma}(x) \cdot q_{b\tau}(x)\, dx = \delta_{\sigma,\tau} \cdot T N_\sigma \tag{32}$$

gives the mode norms N_σ :

$$T \cdot N_\sigma = \frac{a/2}{\Gamma_{an} Z_{an} \cos^2(\varepsilon_{a\sigma} a)} \left[1 + \frac{\sin(2\varepsilon_{a\sigma} a)}{2\varepsilon_{a\sigma} a}\right] + \frac{b/2}{\Gamma_{bn} Z_{bn} \cos^2(\varepsilon_{b\sigma} b)} \left[1 + \frac{\sin(2\varepsilon_{b\sigma} b)}{2\varepsilon_{b\sigma} b}\right] \tag{33}$$

The field formulation in the main duct remains as above.

Matching the sound fields in the plane $y=0$ leads to the homogeneous linear system of equations for the amplitudes P_n of the space harmonics :

$$\sum_n P_n \left[\delta_{m,n} \cdot \varepsilon_m h \cdot \tan(\varepsilon_m h) - j\frac{h}{d} \sum_\sigma \Gamma_{s\sigma} d \tanh(\Gamma_{s\sigma} d) \cdot \frac{R_{n,\sigma} R_{-m,\sigma}}{N_\sigma}\right] = 0 \tag{34}$$

with the coupling coefficients $R_{n,\sigma}$ between side duct modes and main duct spatial harmonics :

$$T\cdot R_{n,\sigma}:=\frac{1}{\Gamma_{an}Z_{an}}\int_0^a e^{-j\beta_n x}\cdot\frac{\cos(\varepsilon_{a\sigma}(x-a))}{\cos(\varepsilon_{a\sigma}a)}\,dx+\frac{1}{\Gamma_{bn}Z_{bn}}\int_{-b}^0 e^{-j\beta_n x}\cdot\frac{\cos(\varepsilon_{b\sigma}(x+b))}{\cos(\varepsilon_{b\sigma}b)}\,dx$$

$$\begin{aligned}&=\frac{j/2}{\Gamma_{an}Z_{an}\cos(\varepsilon_{a\sigma}a)}\cdot\\&\cdot\left[\frac{-1}{\beta_n}+\frac{2}{\beta_n}\frac{2\varepsilon_{a\sigma}^2-\beta_n^2}{4\varepsilon_{a\sigma}^2-\beta_n^2}e^{-j\beta_n a}+\frac{1}{4\varepsilon_{a\sigma}^2-\beta_n^2}\left(\beta_n\cos(2\varepsilon_{a\sigma}a)-2j\varepsilon_{a\sigma}\sin(2\varepsilon_{a\sigma}a)\right)\right]+\\&+\frac{j/2}{\Gamma_{bn}Z_{bn}\cos(\varepsilon_{b\sigma}b)}\cdot\\&\cdot\left[\frac{1}{\beta_n}-\frac{2}{\beta_n}\frac{2\varepsilon_{b\sigma}^2-\beta_n^2}{4\varepsilon_{b\sigma}^2-\beta_n^2}e^{+j\beta_n b}-\frac{1}{4\varepsilon_{b\sigma}^2-\beta_n^2}\left(\beta_n\cos(2\varepsilon_{b\sigma}b)+2j\varepsilon_{b\sigma}\sin(2\varepsilon_{b\sigma}b)\right)\right]\end{aligned}\qquad(35)$$

The determinant of the above system of equations set to zero is the characteristic equation of the system. It can be simplified, if only the periodic factor of the main duct field is matched to the field in the side ducts :

$$\sum_n P_n\left[\delta_{m,n}\cdot\varepsilon_m h\cdot\tan(\varepsilon_m h)-j\frac{h}{d}\sum_\sigma\Gamma_{s\sigma}d\,\tanh(\Gamma_{s\sigma}d)\cdot\frac{S_{n,\sigma}S_{-m,\sigma}}{N_\sigma}\right]=0\qquad(36)$$

where the $S_{n,\sigma}$ are obtained from the $R_{n,\sigma}$ by the substitution $\beta_n\to 2n\pi/T$, especially for n=0 :

$$\begin{aligned}&T\cdot S_{0,\sigma}=\frac{a/2}{\Gamma_{an}Z_{an}\cos(\varepsilon_{a\sigma}a)}\left[1+\frac{\sin(2\varepsilon_{a\sigma}a)}{2\varepsilon_{a\sigma}a}\right]+\frac{b/2}{\Gamma_{bn}Z_{bn}\cos(\varepsilon_{b\sigma}b)}\left[1+\frac{\sin(2\varepsilon_{b\sigma}b)}{2\varepsilon_{b\sigma}b}\right]\\&\frac{S_{0,\sigma}\cdot S_{0,\sigma}}{N_\sigma}\xrightarrow[|\varepsilon_{a\sigma}a|,|\varepsilon_{b\sigma}b|\ll 1]{}\frac{a/T}{\Gamma_{an}Z_{an}}+\frac{b/T}{\Gamma_{bn}Z_{bn}}\end{aligned}\qquad(37)$$

The determinant equation of the 2nd system can be approximated by :

$$\varepsilon_0 h\cdot\tan(\varepsilon_0 h)-j\frac{h}{d}(\Gamma_{s0}d)\cdot\tanh(\Gamma_{s0}d)\left(\frac{a/T}{\Gamma_{an}Z_{an}}+\frac{b/T}{\Gamma_{bn}Z_{bn}}\right)=0\qquad(38)$$

which is just the characteristic equation with the average head admittance.

A special form of ducts with cross-layered linings is the “pine-tree silencer”. These silencers are used in air flows with heavy dust load. Their axis is vertical. The intention is that dust deposits on the branches of the trees will slide down due to the vibrations of the structure.

The pine-tree baffles mostly belong to the type (b) of the initial graph of this section.

Because of the inclination of the branches, the layer thicknesses change to $a \to a\cdot\cos\varphi$, $b \to b\cdot\cos\varphi$ and the admittances $G_{\alpha y} \to G_{\alpha y}\cdot\cos\varphi$.

Sometimes the air-filled channels (layers with b in the sketch) are terminated near the "trunk" with an absorber layer having a reflection factor r for the incident fundamental side duct mode. The surface admittance at y=0 then becomes approximately :

$$Z_0 G_{by} = \frac{\Gamma_s}{\Gamma_b Z_{bn}} \frac{1 - r\cdot e^{-2\Gamma_s s}}{1 + r\cdot e^{-2\Gamma_s s}} \cos\varphi \qquad (39)$$

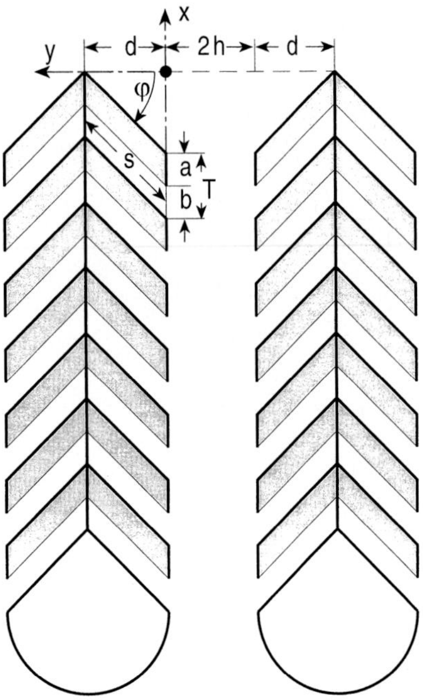

FROMMHOLD derived a correction formula to be applied to the reflection factor r for normal incidence to obtain approximately the reflection factor r_{eff} for the incident mode (claimed to be applicable for f> 70 [Hz]) :

$$r_{eff} = r\cdot\left(\sqrt{\cos\varphi} + \frac{1-\sqrt{\cos\varphi}}{1-(f[\mathrm{Hz}]/70)^3}\right) \qquad (40)$$

J.18 Single step of duct height and/or duct lining

Ref.: Mechel, [J.8]

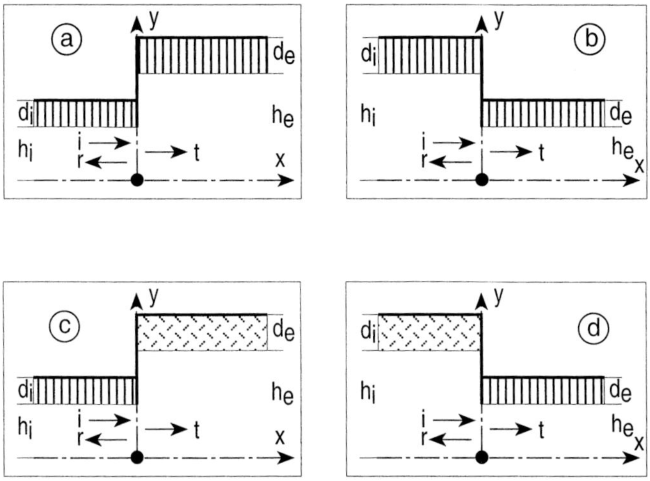

The steps are isolated, i.e. the ducts on both sides are infinitely long, in principle; in practice it will be sufficient if the next step is far enough to neglect reflections from there at the original step.

One must distinguish :

- expanding or contracting ducts (as seen in the direction of sound);
- locally or bulk reacting linings;
- same or different type (local or bulk) on both sides of the step;
- if the lining in the narrow duct is bulk reacting, whether its head is open or covered with a hard sheet.

There are more combinations than the few examples shown above. Vertical hatching will be used in the sketches below to indicate a locally reacting lining; crossed hatching indicates bulk reacting absorbers.

An index ß= i,e will indicate the duct (and its parameters) on the side of the incident sound, and on the exit side. The incident sound p_i may be a sum of modes of the entrance duct or, as a special case, one mode of order μ . Both the reflected wave p_r and the transmitted wave p_t are formulated as sums of modes in their ducts.

Fundamental relations in a duct with a locally reacting lining :

Field formulations :

$$\begin{aligned} p_i(x,y) &= \sum_m P_{im} \cdot \cos(\varepsilon_{im} y) \cdot e^{-\gamma_{im} x} \\ v_{ix}(x,y) &= \frac{1}{jk_0 Z_0} \sum_m P_{im} \cdot \gamma_{im} \cos(\varepsilon_{im} y) \cdot e^{-\gamma_{im} x} \end{aligned} \tag{1}$$

In the special case of a single incident mode : $P_{im} \to \delta_{m,\mu} \cdot P_{i\mu}$. (2)

$$\begin{aligned} p_r(x,y) &= \sum_m P_{rm} \cdot \cos(\varepsilon_{im} y) \cdot e^{+\gamma_{im} x} \\ v_{rx}(x,y) &= \frac{-1}{jk_0 Z_0} \sum_m P_{rm} \cdot \gamma_{im} \cos(\varepsilon_{im} y) \cdot e^{+\gamma_{im} x} \end{aligned} \tag{3}$$

$$\begin{aligned} p_t(x,y) &= \sum_n P_{tn} \cdot \cos(\varepsilon_{en} y) \cdot e^{-\gamma_{en} x} \\ v_{tx}(x,y) &= \frac{1}{jk_0 Z_0} \sum_n P_{tn} \cdot \gamma_{en} \cos(\varepsilon_{en} y) \cdot e^{-\gamma_{en} x} \end{aligned} \tag{4}$$

Secular equation : $\varepsilon_{\beta k}^2 = k_0^2 + \gamma_{\beta k}^2 \quad ; \quad \beta = i,e \quad ; \quad k = \mu, m, n$ (5)

if there is a higher mode in z direction : $\varepsilon_{\beta k}^2 = k_0^2 + \gamma_{\beta k}^2 - \varepsilon_z^2 \quad ; \quad \beta = i,e \quad ; \quad k = \mu, m, n$ (6)

Characteristic equation for the modes :

$$(\varepsilon_{\beta k} h_\beta) \cdot \tan(\varepsilon_{\beta k} h_\beta) = j k_0 h_\beta \cdot Z_0 G_\beta \tag{7}$$

and if $G_\beta = 0$: $\varepsilon_{\beta k} = (k-1)\pi$; $k = 0,1,2,\ldots$ (8)

Mode norms $N_{\beta k}$:

$$\begin{aligned} &\frac{1}{h_\beta} \int_0^{h_\beta} \cos(\varepsilon_{\beta k} y) \cdot \cos(\varepsilon_{\beta k'} y)\, dy = \delta_{k,k'} \cdot N_{\beta k} \\ &N_{\beta k} := \frac{1}{h_\beta} \int_0^{h_\beta} \cos^2(\varepsilon_{\beta k} y)\, dy = \frac{1}{2}\left[1 + \frac{\sin(2\varepsilon_{\beta k} h_\beta)}{2\varepsilon_{\beta k} h_\beta}\right] \end{aligned} \tag{9}$$

Mode coupling coefficients of modes of both ducts : (10)

$$R_{m,n}(h,\beta,\beta') := \frac{1}{h} \int_0^h \cos(\varepsilon_{\beta m} y) \cdot \cos(\varepsilon_{\beta' n} y)\, dy = \frac{1}{2}\left[\frac{\sin(\varepsilon_{\beta m} - \varepsilon_{\beta' n})h}{(\varepsilon_{\beta m} - \varepsilon_{\beta' n})h} + \frac{\sin(\varepsilon_{\beta m} + \varepsilon_{\beta' n})h}{(\varepsilon_{\beta m} + \varepsilon_{\beta' n})h}\right]$$

Evidently $R_{m,n}(h,\beta,\beta') = R_{n,m}(h,\beta',\beta)$. Other coupling coefficients over the range Δh of the height difference :

$$S_{m,n}(h,ß,ß') := \frac{1}{h_ß}\int_h^{h_ß} \cos(\varepsilon_{ßm}y)\cdot\cos(\varepsilon_{ß'n}y)\,dy = R_{m,n}(h_ß,ß,ß') - \frac{h}{h_ß}R_{m,n}(h,ß,ß') \quad (11)$$

$$\xrightarrow[h\to h_ß]{} 0$$

In ducts with a locally reacting lining : $S_{m,n}(h,ß,ß) = \delta_{m,n}\cdot N_{ßm} - \frac{h}{h_ß}R_{m,n}(h,ß,ß)$ (12)

Fundamental relations in a duct with a bulk reacting lining :

In order to write the mode field in a single line, a "switch function" is introduced by :

$$s_y(a,b) := \begin{cases} 1 \ ; \ a \le y < b \\ 0 \ ; \ \text{else} \end{cases} \quad (13)$$

Field formulations : (14)

$$p_i(x,y) =$$

$$\sum_m P_{im}\cdot e^{-\gamma_{im}x}\left[s_y(0,h_i)\frac{\cos(\varepsilon_{im}y)}{\cos(\varepsilon_{im}h_i)} + s_y(h_i,h_i+d_i)\frac{\cos(\kappa_{im}(y-h_i-d_i))}{\cos(\kappa_{im}d_i)}\right]$$

$$v_{ix}(x,y) =$$

$$\frac{1}{jk_0Z_0}\sum_m P_{im}\cdot\gamma_{im}\,e^{-\gamma_{im}x}\left[s_y(0,h_i)\frac{\cos(\varepsilon_{im}y)}{\cos(\varepsilon_{im}h_i)} - j\frac{s_y(h_i,h_i+d_i)}{\Gamma_{in}Z_{in}}\frac{\cos(\kappa_{im}(y-h_i-d_i))}{\cos(\kappa_{im}d_i)}\right]$$

$$p_r(x,y) =$$

$$\sum_m P_{rm}\cdot e^{+\gamma_{im}x}\left[s_y(0,h_i)\frac{\cos(\varepsilon_{im}y)}{\cos(\varepsilon_{im}h_i)} + s_y(h_i,h_i+d_i)\frac{\cos(\kappa_{im}(y-h_i-d_i))}{\cos(\kappa_{im}d_i)}\right]$$

$$v_{rx}(x,y) =$$

$$\frac{-1}{jk_0Z_0}\sum_m P_{rm}\cdot\gamma_{im}\,e^{+\gamma_{im}x}\left[s_y(0,h_i)\frac{\cos(\varepsilon_{im}y)}{\cos(\varepsilon_{im}h_i)} - j\frac{s_y(h_i,h_i+d_i)}{\Gamma_{in}Z_{in}}\frac{\cos(\kappa_{im}(y-h_i-d_i))}{\cos(\kappa_{im}d_i)}\right]$$

$$p_t(x,y) =$$

$$\sum_n P_{tn}\cdot e^{-\gamma_{en}x}\left[s_y(0,h_e)\frac{\cos(\varepsilon_{en}y)}{\cos(\varepsilon_{en}h_e)} + s_y(h_e,h_e+d_e)\frac{\cos(\kappa_{en}(y-h_e-d_e))}{\cos(\kappa_{en}d_e)}\right]$$

$$v_{tx}(x,y) =$$

$$\frac{1}{jk_0Z_0}\sum_n P_{tn}\cdot\gamma_{en}\,e^{-\gamma_{en}x}\left[s_y(0,h_e)\frac{\cos(\varepsilon_{en}y)}{\cos(\varepsilon_{en}h_e)} - j\frac{s_y(h_e,h_e+d_e)}{\Gamma_{en}Z_{en}}\frac{\cos(\kappa_{en}(y-h_e-d_e))}{\cos(\kappa_{en}d_e)}\right]$$

The matching of the modal sound pressures on both sides of the absorber surface is done by introducing the denominators.

Secular equation :

$$\varepsilon_{ßk}^2 = k_0^2 + \gamma_{ßk}^2 \quad ; \quad \kappa_{ßk}^2 = \gamma_{ßk}^2 - \Gamma_ß^2 = \varepsilon_{ßk}^2 - k_0^2 - \Gamma_ß^2 \tag{15}$$

Characteristic equation in the case of a simple porous layer :

$$\varepsilon_{ßk} h_ß \cdot \tan(\varepsilon_{ßk} h_ß) = -j k_0 h_ß \cdot \frac{\kappa_{ßk}}{\Gamma_ß Z_{ßn}} \cdot \tan(\kappa_{ßk} d_ß) \tag{16}$$

Relation of mode orthogonality :

$$\frac{1}{h_ß}\left[\int_0^{h_ß} \frac{\cos(\varepsilon_{ßm} y)}{\cos(\varepsilon_{ßm} h_ß)} \frac{\cos(\varepsilon_{ßn} y)}{\cos(\varepsilon_{ßn} h_ß)} dy - \right.$$

$$\left. - \frac{j}{\Gamma_{ßn} Z_{ßn}} \int_{h_ß}^{h_ß+d_ß} \frac{\cos\big(\kappa_{ßm}(y - h_ß - d_ß)\big)}{\cos(\kappa_{ßm} d_ß)} \frac{\cos\big(\kappa_{ßn}(y - h_ß - d_ß)\big)}{\cos(\kappa_{ßn} d_ß)} dy \right] = \delta_{m,n} \cdot M_{ßm} \tag{17}$$

Mode norms : (18)

$$M_{ßk} = \frac{1}{2} \frac{1}{\cos^2(\varepsilon_{ßk} h_ß)} \left[1 + \frac{\sin(2\varepsilon_{ßk} h_ß)}{2\varepsilon_{ßk} h_ß}\right] - \frac{1}{2} \frac{d_ß}{h_ß} \frac{j}{\Gamma_{ßk} Z_{ßk}} \frac{1}{\cos^2(\kappa_{ßk} d_ß)} \left[1 + \frac{\sin(2\kappa_{ßk} d_ß)}{2\kappa_{ßk} d_ß}\right]$$

Mode coupling coefficients : (19)

$$Q'_{m,n}(h, ß, ß') := \frac{1}{h} \int_0^h \frac{\cos(\varepsilon_{ßm} y)}{\cos(\varepsilon_{ßm} h_ß)} \frac{\cos(\varepsilon_{ß'n} y)}{\cos(\varepsilon_{ß'n} h_{ß'})} dy = \frac{R_{m,n}(h, ß, ß')}{\cos(\varepsilon_{ßm} h_ß) \cdot \cos(\varepsilon_{ß'n} h_{ß'})}$$

$$Q''_{m,n}(h_ß, ß, ß') := \frac{1}{h_ß} \int_{h_ß}^{h_ß+d_ß} \frac{\cos\big(\kappa_{ßm}(y - h_ß - d_ß)\big)}{\cos(\kappa_{ßm} d_ß)} \frac{\cos(\varepsilon_{ß'n} y)}{\cos(\varepsilon_{ß'n} h_{ß'})} dy$$

$$= \frac{1}{\cos(\varepsilon_{ß'n} h_{ß'}) \cos(\kappa_{ßm} d_ß)} \cdot$$

$$\cdot \left[\frac{\varepsilon_{ß'n} h_{ß'}}{(\varepsilon_{ß'n}^2 - \kappa_{ßm}^2) h_ß h_{ß'}} \sin\big(\varepsilon_{ß'n}(h_ß + d_ß)\big) - \frac{\sin(\varepsilon_{ß'n} h_ß - \kappa_{ßm} d_ß)}{2(\varepsilon_{ß'n} + \kappa_{ßm}) h_ß} - \frac{\sin(\varepsilon_{ß'n} h_ß + \kappa_{ßm} d_ß)}{2(\varepsilon_{ß'n} - \kappa_{ßm}) h_ß}\right]$$

with combination :

$$Q_{m,n}(h, ß, ß') := Q'_{m,n}(h, ß, ß') - \frac{j}{\Gamma_{ßn} Z_{ßn}} Q''_{m,n}(h_ß, ß, ß') \tag{20}$$

Other coupling coefficients needed are :

$$T'_{m,n}(h, ß, ß') = \frac{1}{h_ß} \int_h^{h_ß} \frac{\cos(\varepsilon_{ßm} y)}{\cos(\varepsilon_{ßm} h_ß)} \frac{\cos(\varepsilon_{ß'n} y)}{\cos(\varepsilon_{ß'n} h_{ß'})} dy = \frac{S_{m,n}(h, ß, ß')}{\cos(\varepsilon_{ßm} h_ß) \cos(\varepsilon_{ß'n} h_{ß'})} \tag{21}$$

$$T''_{m,n}(h_ß,ß,ß)=\frac{1}{h_ß}\int_{h_ß}^{h_ß+d_ß}\frac{\cos(\kappa_{ßm}(y-h_ß-d_ß))}{\cos(\kappa_{ßm}d_ß)}\frac{\cos(\kappa_{ßn}(y-h_ß-d_ß))}{\cos(\kappa_{ßn}d_ß)}dy$$
$$=\frac{d_ß/h_ß}{2\cos(\kappa_{ßm}d_ß)\cos(\kappa_{ßn}d_ß)}\left[\frac{\sin(\kappa_{ßm}-\kappa_{bn})d_ß}{(\kappa_{ßm}-\kappa_{bn})d_ß}+\frac{\sin(\kappa_{ßm}+\kappa_{bn})d_ß}{(\kappa_{ßm}+\kappa_{bn})d_ß}\right] \tag{22}$$

with combination :

$$T_{m,n}(h,ß,ß)=T'_{m,n}(h,ß,ß)-\frac{j}{\Gamma_{ßn}Z_{ßn}}T''_{m,n}(h_ß,ß,ß) \tag{23}$$

Expanding, local → local :

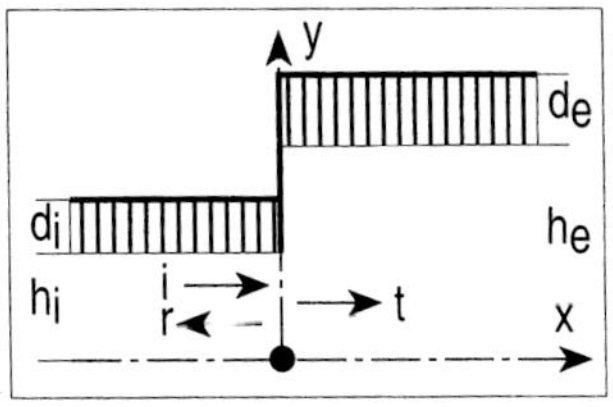

$$P_{rm}=-\delta_{\mu,m}\cdot P_{i\mu}+\frac{1}{N_{im}}\sum_n P_{tn}\cdot R_{m,n}(h_i,i,e) \tag{24}$$

$$\sum_n P_{tn}\left[\delta_{n,k}\cdot\gamma_{ek}h_e N_{ek}+\sum_m\gamma_{im}h_i\frac{R_{m,k}(h_i,i,e)\cdot R_{m,n}(h_i,i,e)}{N_{im}}\right]$$
$$=2\sum_m P_{im}\gamma_{im}h_i R_{m,k}(i,e) \tag{25}$$

Alternatively (with reduced precision, however):

$$P_{tn}=\frac{1}{\gamma_{en}h_e N_{en}}\sum_m(P_{im}-P_{rm})\cdot\gamma_{im}h_i R_{m,n}(h_i,i,e)$$

$$\sum_m P_{rm}\left[\delta_{m,k}N_{ik}+\gamma_{im}h_i\sum_n\frac{R_{k,n}(h_i,i,e)\cdot R_{m,n}(h_i,i,e)}{\gamma_{en}h_e N_{en}}\right]=$$
$$=\sum_m P_{im}\left[-\delta_{m,k}N_{ik}+\gamma_{im}h_i\sum_n\frac{R_{k,n}(h_i,i,e)\cdot R_{m,n}(h_i,i,e)}{\gamma_{en}h_e N_{en}}\right] \tag{26}$$

Contracting, local → local :

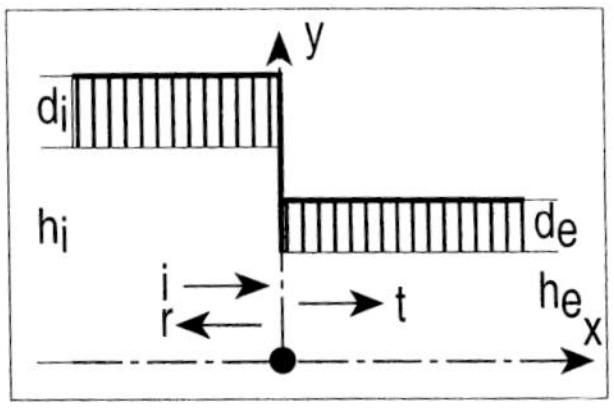

$$P_{rm}=P_{im}-\frac{1}{\gamma_{im}h_i N_{im}}\sum_n P_{tn}\gamma_{en}h_e R_{n,m}(h_e,e,i) \tag{27}$$

$$\sum_n P_{tn}\left[\delta_{k,n}N_{ek}+\gamma_{en}h_e\sum_m\frac{R_{k,m}(h_e\,e,i)\cdot R_{n,m}(h_e\,e,i)}{\gamma_{im}h_i N_{im}}\right]=$$
$$=2\sum_m P_{im}R_{k,m}(h_e,e,i) \tag{28}$$

Alternatively :

$$P_{tn} = \frac{1}{N_{en}} \sum_m (P_{im} + P_{rm}) R_{n,m}(h_e, e, i) \tag{29}$$

$$\sum_m P_{rm} \left[\delta_{m,k} \gamma_{im} h_i N_{im} + \sum_n \gamma_{en} h_e \frac{R_{n,k}(h_e,e,i) \cdot R_{n,m}(h_e,e,i)}{N_{en}} \right] = \\ = \sum_m P_{im} \left[\delta_{m,k} \gamma_{im} h_i N_{im} - \sum_n \gamma_{en} h_e \frac{R_{n,k}(h_e,e,i) \cdot R_{n,m}(h_e,e,i)}{N_{en}} \right] \tag{30}$$

Expanding, lateral → lateral, open head:

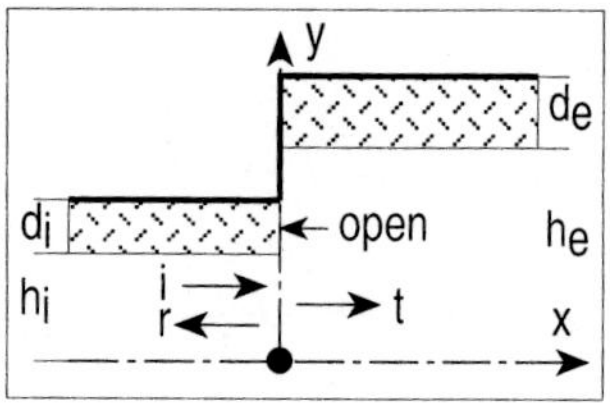

$$P_{rm} = -P_{im} + \frac{1}{M_{im}} \sum_n P_{tn} \cdot Q_{m,n}(h_i, i, e) \tag{31}$$

$$\sum_n P_{tn} \left[\delta_{n,k} \gamma_{ek} h_e M_{ek} + \sum_m \gamma_{im} h_i \frac{Q_{m,k}(h_i,i,e) \cdot Q_{m,n}(h_i,i,e)}{M_{im}} \right] \\ = \sum_m P_{im} \gamma_{im} h_i Q_{m,k}(h_i,i,e)\left(1 + \delta_{m,k}\right) \tag{32}$$

Alternatively :

$$P_{tn} = \frac{1}{\gamma_{en} h_e M_{en}} \sum_m (P_{im} - P_{rm}) \cdot \gamma_{im} h_i Q_{m,n}(h_i, i, e) \tag{33}$$

$$\sum_m P_{rm} \left[\delta_{m,k} M_{ik} - \gamma_{im} h_i \sum_n \frac{Q_{k,n}(h_i,i,e) \cdot Q_{m,n}(h_i,i,e)}{\gamma_{en} h_e M_{en}} \right] = \\ = \sum_m P_{im} \left[-\delta_{m,k} M_{ik} + \gamma_{im} h_i \sum_n \frac{Q_{k,n}(i,e) \cdot Q_{m,n}(i,e)}{\gamma_{en} h_e M_{en}} \right] \tag{34}$$

Contracting, lateral → lateral, open head:

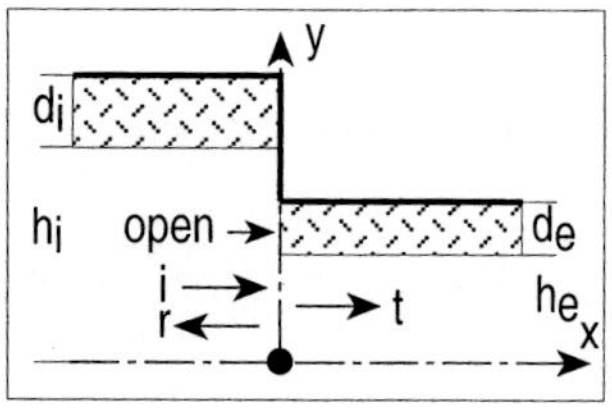

$$P_{rm} = P_{im} - \frac{1}{\gamma_{im} h_i M_{im}} \sum_n P_{tn} \gamma_{en} h_e Q_{n,m}(h_e, e, i) \tag{35}$$

$$\sum_n P_{tn} \left[\delta_{n,k} M_{ek} + \gamma_{en} h_e \sum_m \frac{Q_{k,m}(h_e,e,i) \cdot Q_{n,m}(h_e,e,i)}{\gamma_{im} h_i M_{im}} \right] = \\ = \sum_m P_{im} Q_{k,m}(h_e,e,i)\left(1 + \delta_{m,k}\right) \tag{36}$$

Alternatively :

$$P_{tn} = \frac{1}{M_{en}} \sum_m (P_{im} + P_{rm}) Q_{n,m}(h_e,e,i) \tag{37}$$

$$\begin{aligned} &\sum_m P_{rm}\left[\delta_{m,k}\,\gamma_{ik} h_i\,M_{ik} + \sum_n \gamma_{en} h_e \frac{Q_{n,k}(h_e,e,i)\cdot Q_{n,m}(h_e,e,i)}{M_{en}}\right] = \\ &= \sum_m P_{im}\left[\delta_{m,k}\,\gamma_{ik} h_i\,M_{ik} - \sum_n \gamma_{en} h_e \frac{Q_{n,k}(h_e,e,i)\cdot Q_{n,m}(h_e,e,i)}{M_{en}}\right] \end{aligned} \tag{38}$$

Expanding, lateral → lateral, covered head:

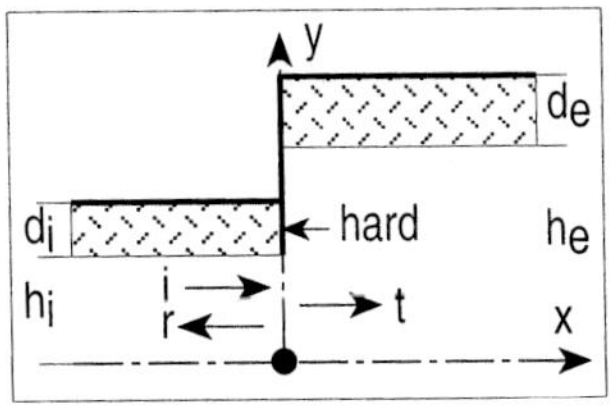

Combined system of equations for the $\{P_{rn}, P_{tn}\}$:

$$\begin{aligned} &\sum_n P_{tn}\,Q'_{k,n}(h_i,i,e) - \sum_m P_{rm}\left[\delta_{m,k}\,M_{ik} + \frac{j}{\Gamma_{in} Z_{in}} T''_{k,m}(h_i,i,i)\right] = \\ &= \sum_m P_{im}\left[\delta_{m,k}\,M_{ik} - \frac{j}{\Gamma_{in} Z_{in}} T''_{k,m}(h_i,i,i)\right] \end{aligned} \tag{39}$$

$$\begin{aligned} &\sum_m P_{rm}\left[\delta_{k,m}\,\gamma_{ik} h_i\,M_{ik} - \frac{h_i}{h_e}\gamma_{im} h_i \sum_n \frac{Q'_{k,n}(h_i,i,e)\cdot Q'_{m,n}(h_i,i,e)}{M_{en}}\right] = \\ &= \sum_m P_{im}\left[\delta_{k,m}\gamma_{ik} h_i\,M_{ik} - \frac{h_i}{h_e}\gamma_{im} h_i \sum_n \frac{Q'_{k,n}(h_i,i,e)\cdot Q'_{m,n}(h_i,i,e)}{M_{en}}\right] \end{aligned} \tag{40}$$

Contracting, lateral → lateral, covered head:

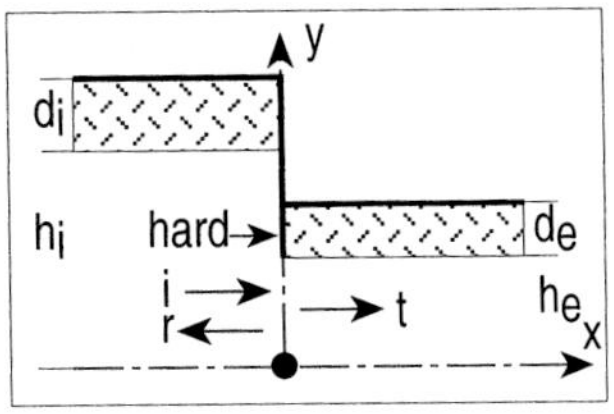

Combined system of equations for the $\{P_{rn}, P_{tn}\}$:

$$\begin{aligned} &\sum_n P_{tn}\left[\delta_{n,k}\,M_{ek} + \frac{j}{\Gamma_{en} Z_{en}} T''_{k,n}(h_e,e,e)\right] - \\ &\sum_m P_{rm}\,Q'_{k,m}(h_e,e,i) = \sum_m P_{im}\,Q'_{k,m}(h_e,e,i) \end{aligned} \tag{41}$$

$$\begin{aligned} &\sum_m P_{rm}\left[\delta_{m,k}\,\gamma_{ik} h_i\,M_{ik} - \gamma_{im} h_i \frac{h_e}{h_i} \sum_n \frac{Q'_{n,k}(h_e,e,i)\cdot Q'_{n,m}(h_e,e,i)}{M_{en}}\right] = \\ &= \sum_m P_{im}\left[\delta_{m,k}\,\gamma_{ik} h_i\,M_{ik} - \gamma_{im} h_i \frac{h_e}{h_i} \sum_n \frac{Q'_{n,k}(h_e,e,i)\cdot Q'_{n,m}(h_e,e,i)}{M_{en}}\right] \end{aligned} \tag{42}$$

Expanding, local → lateral :

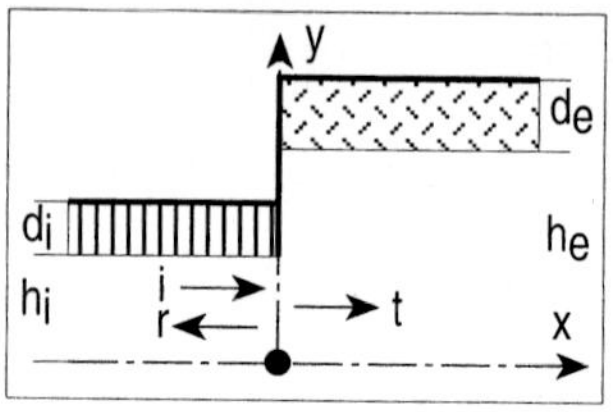

$$P_{tn} = \frac{1}{\gamma_{en} h_e M_{en} \cos(\varepsilon_{en} h_e)} \sum_m (P_{im} - P_{rm}) \gamma_{im} h_i R_{m,n}(h_i,i,e) \tag{43}$$

$$\sum_m P_{rm} \left[\delta_{m,k} N_{ik} + \gamma_{im} h_i \sum_n \frac{R_{k,n}(h_i,i,e) \cdot R_{m,n}(h_i,i,e)}{\gamma_{en} h_e M_{en} \cos^2(\varepsilon_{en} h_e)} \right] = \\ = \sum_m P_{im} \left[-\delta_{m,k} N_{im} + \gamma_{im} h_i \sum_n \frac{R_{k,n}(h_i,i,e) \cdot R_{m,n}(h_i,i,e)}{\gamma_{en} h_e M_{en} \cos^2(\varepsilon_{en} h_e)} \right] \tag{44}$$

Alternatively :

$$P_{rm} = -P_{im} + \frac{1}{N_{im}} \sum_n P_{tn} \frac{R_{m,n}(i,e)}{\cos(\varepsilon_{en} h_e)} \tag{45}$$

$$\sum_n P_{tn} \left[\delta_{n,k} \gamma_{ek} h_e M_{ek} + \frac{1}{\cos(\varepsilon_{ek} h_e) \cos(\varepsilon_{en} h_e)} \sum_m \gamma_{im} h_i \frac{R_{m,k}(h_i,i,e) \cdot R_{m,n}(h_i,i,e)}{N_{im}} \right] = \\ = \sum_m P_{im} \frac{\gamma_{im} h_i R_{m,k}(h_i,i,e)}{\cos(\varepsilon_{ek} h_e)} \left(1 + \delta_{m,k}\right) \tag{46}$$

Contracting, lateral → local :

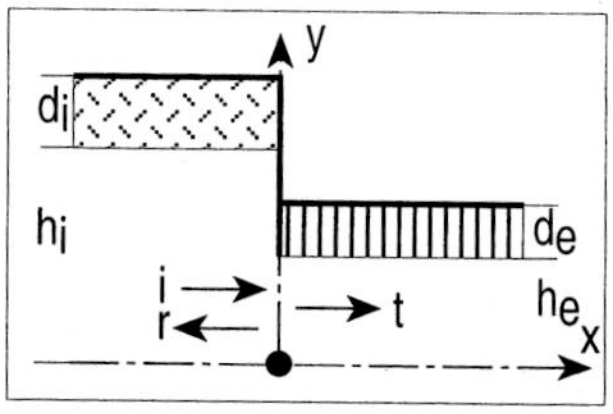

$$P_{rm} = P_{im} - \frac{1}{\gamma_{im} h_i M_{im} \cos(\varepsilon_{im} h_i)} \sum_n P_{tn} \gamma_{en} h_e R_{n,m}(h_e,e,i) \tag{47}$$

$$\sum_n P_{tn} \left[\delta_{n,k} N_{ek} + \gamma_{en} h_e \sum_m \frac{R_{k,m}(h_e,e,i) \cdot R_{n,m}(h_e,e,i)}{\gamma_{im} h_i M_{im} \cos^2(\varepsilon_{im} h_i)} \right] = \\ = 2 \sum_m P_{im} \frac{R_{k,m}(h_e,e,i)}{\cos(\varepsilon_{im} h_i)} \tag{48}$$

Alternatively :

$$P_{tn} = \frac{1}{N_{en}} \left[\sum_m (P_{im} + P_{rm}) \frac{R_{n,m}(h_e,e,i)}{\cos(\varepsilon_{im} h_i)} \right] \tag{49}$$

$$\sum_m P_{rm}\left[\delta_{k,m}\gamma_{ik}h_i M_{ik}+\frac{1}{\cos(\varepsilon_{ik}h_i)\cos(\varepsilon_{im}h_i)}\sum_n \gamma_{en}h_e\frac{R_{n,k}(h_e,e,i)\cdot R_{n,m}(h_e,e,i)}{N_{en}}\right]=$$
$$=\sum_m P_{im}\left[\delta_{k,m}\gamma_{ik}h_i M_{ik}-\frac{1}{\cos(\varepsilon_{ik}h_i)\cos(\varepsilon_{im}h_i)}\sum_n \gamma_{en}h_e\frac{R_{n,k}(h_e,e,i)\cdot R_{n,m}(h_e,e,i)}{N_{en}}\right] \quad (50)$$

Expanding, lateral → local, open head :

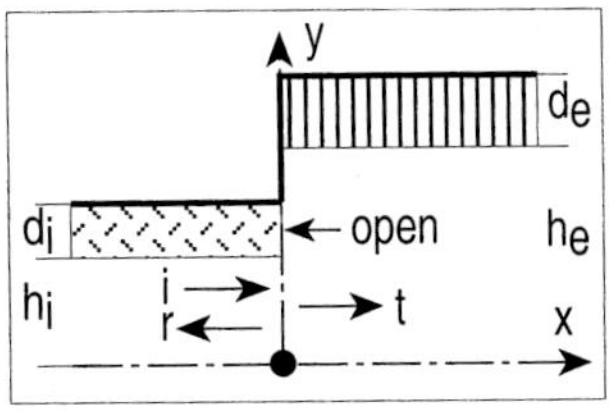

$$P_{tn}=\frac{\cos(\varepsilon_{en}h_e)}{\gamma_{en}h_e N_{en}}\sum_m (P_{im}-P_{rm})\gamma_{im}h_i Q_{m,n}(h_i,i,e) \quad (51)$$

$$\sum_m P_{rm}\left[\delta_{m,k}M_{ik}+\gamma_{im}h_i\sum_n \cos^2(\varepsilon_{en}h_e)\frac{Q_{k,n}(h_i,i,e)\cdot Q_{m,n}(h_i,i,e)}{\gamma_{en}h_e N_{en}}\right]=$$
$$=\sum_m P_{im}\left[-\delta_{m,k}M_{im}+\gamma_{im}h_i\sum_n \cos^2(\varepsilon_{en}h_e)\frac{Q_{k,n}(h_i,i,e)\cdot Q_{m,n}(h_i,i,e)}{\gamma_{en}h_e N_{en}}\right] \quad (52)$$

Alternatively :

$$P_{rm}=-P_{im}+\frac{1}{M_{im}}\sum_n P_{tn}Q_{m,n}(h_i,i,e)\cos(\varepsilon_{en}h_e) \quad (53)$$

$$\sum_n P_{tn}\left[\delta_{n,k}\gamma_{ek}h_e N_{ek}+\cos(\varepsilon_{ek}h_e)\cos(\varepsilon_{en}h_e)\sum_m \gamma_{im}h_i\frac{Q_{m,k}(h_i,i,e)\cdot Q_{m,n}(h_i,i,e)}{M_{im}}\right]=$$
$$=\sum_m P_{im}\gamma_{im}h_i\cos(\varepsilon_{ek}h_e)Q_{m,k}(h_i,i,e)\left(1+\delta_{m,k}\right) \quad (54)$$

Contracting, local → lateral, open head :

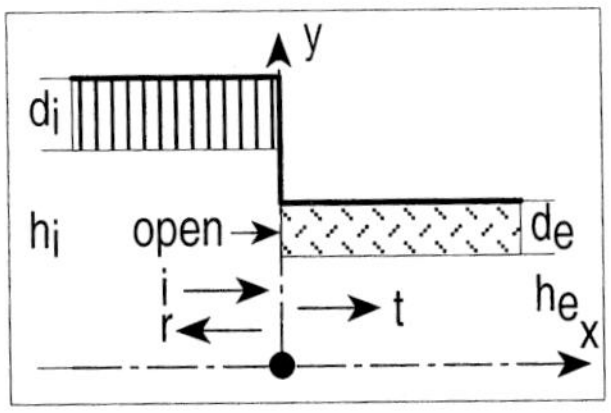

$$P_{rm}=P_{im}-\frac{1}{\gamma_{im}h_i N_{im}}\sum_n P_{tn}\gamma_{en}h_e\left[\frac{R_{n,m}(h_e,e,i)}{\cos(\varepsilon_{en}h_e)}+\right.$$
$$\left.+Q''_{n,m}(h_e,e,i)\cos(\varepsilon_{im}h_i)\right] \quad (55)$$

$$\sum_n P_{tn}\Bigg[\ \delta_{n,k}M_{ek}+\gamma_{en}h_e\sum_m\frac{1}{\gamma_{im}h_i\,N_{im}}\cdot\left(\frac{R_{k,m}(h_e,e,i)}{\cos(\varepsilon_{ek}h_e)}-\frac{j}{\Gamma_{en}Z_{en}}Q''_{k,m}(h_e,e,i)\cos(\varepsilon_{im}h_i)\right)\cdot$$
$$\cdot\left(\frac{R_{n,m}(h_e,e,i)}{\cos(\varepsilon_{en}h_e)}+Q''_{n,m}(h_e,e,i)\cos(\varepsilon_{im}h_i)\right)\Bigg]= \tag{56}$$
$$=\sum_m P_{im}\left[\frac{R_{k,m}(h_e,e,i)}{\cos(\varepsilon_{ek}h_e)}-\frac{j}{\Gamma_{en}Z_{en}}Q''_{k,m}(h_e,e,i)\cos(\varepsilon_{im}h_i)\right]\left(1+\delta_{m,k}\right)$$

Expanding, lateral → local, covered head :

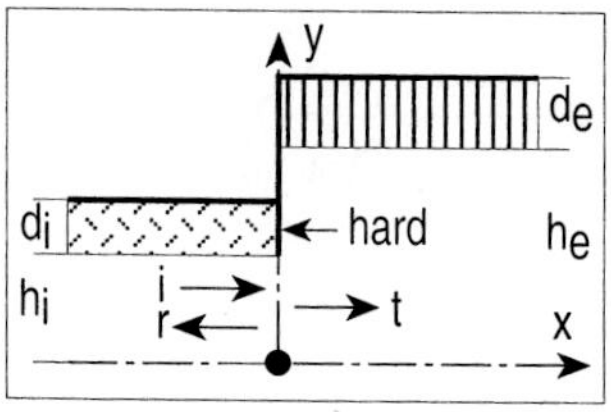

Two combined systems of equations for the P_{rn}, P_{tn} :

$$\sum_m P_{rm}\Bigg[\ \delta_{m,k}\,\gamma_{ik}h_i\,N_{ik}-$$
$$-\gamma_{im}h_i\frac{h_i}{h_e}\sum_n\cos^2(\varepsilon_{en}h_e)\frac{Q'_{k,n}(h_i,i,e)\cdot Q'_{m,n}(h_i,i,e)}{N_{ek}}\Bigg]=$$
$$=\sum_m P_{im}\left[\delta_{m,k}\,\gamma_{ik}h_i\,N_{ik}-\gamma_{im}h_i\frac{h_i}{h_e}\sum_n\cos^2(\varepsilon_{en}h_e)\frac{Q'_{k,n}(h_i,i,e)\cdot Q'_{m,n}(h_i,i,e)}{N_{ek}}\right] \tag{57}$$

$$\sum_n P_{tn}\,Q'_{k,n}(h_i,i,e)\cos(\varepsilon_{en}h_e)-\sum_m P_{rm}\left[\delta_{m,k}M_{ik}+\frac{j}{\Gamma_{in}Z_{in}}T''_{k,m}(h_i,i,i)\right]=$$
$$=\sum_m P_{im}\left[\delta_{m,k}M_{ik}+\frac{j}{\Gamma_{in}Z_{in}}T''_{k,m}(h_i,i,i)\right] \tag{58}$$

Contracting, local → lateral, covered head :

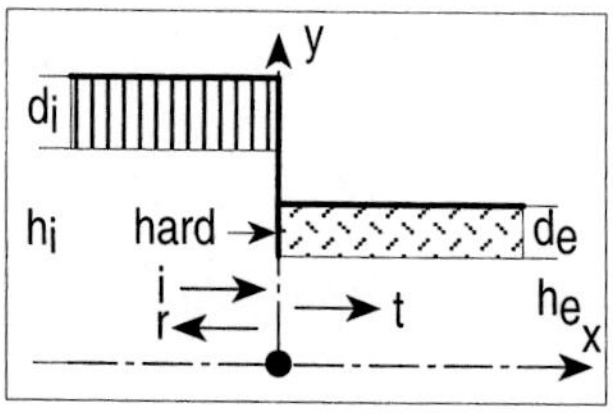

Two combined systems of equations for the P_{rn}, P_{tn} :

$$\sum_m P_{rm}\left[\delta_{m,k}\,\gamma_{ik}h_i\,N_{ik}-\cos(\varepsilon_{ik}h_i)\cos(\varepsilon_{im}h_i)\sum_n\gamma_{en}h_e\frac{Q'_{n,k}(h_e,e,i)\cdot Q'_{n,m}(h_e,e,i)}{M_{en}}\right]=$$
$$=\sum_m P_{im}\left[\delta_{m,k}\,\gamma_{ik}h_i\,N_{ik}-\cos(\varepsilon_{ik}h_i)\cos(\varepsilon_{im}h_i)\sum_n\gamma_{en}h_e\frac{Q'_{n,k}(h_e,e,i)\cdot Q'_{n,m}(h_e,e,i)}{M_{en}}\right] \tag{59}$$

$$\sum_m P_{rm}\, Q'_{k,m}(h_e,e,i)\cos(\varepsilon_{im}h_e) - \sum_n P_{tn}\left[\delta_{n,k}\, M_{ek} + \frac{j}{\Gamma_{en}Z_{en}} T''_{k,n}(h_e,e,e)\right] = \\ = -\sum_m P_{im}\, Q'_{k,m}(h_e,e,i)\cos(\varepsilon_{im}h_e) \tag{60}$$

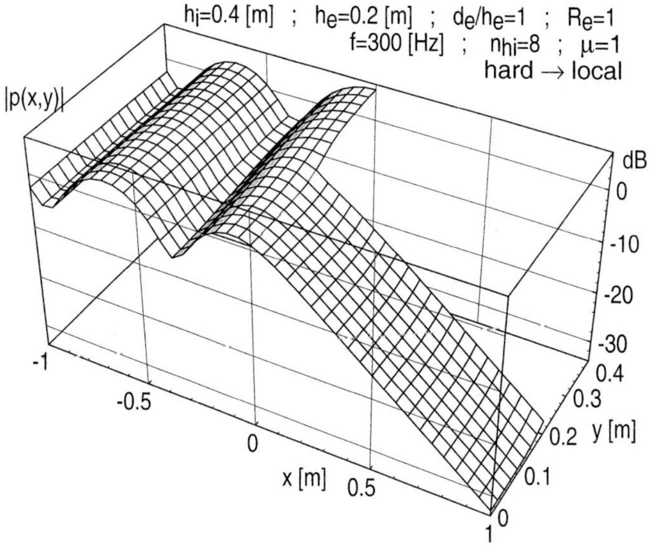

Sound pressure magnitude at a contracting duct step from hard to locally absorbing duct for the lowest duct mode incident.

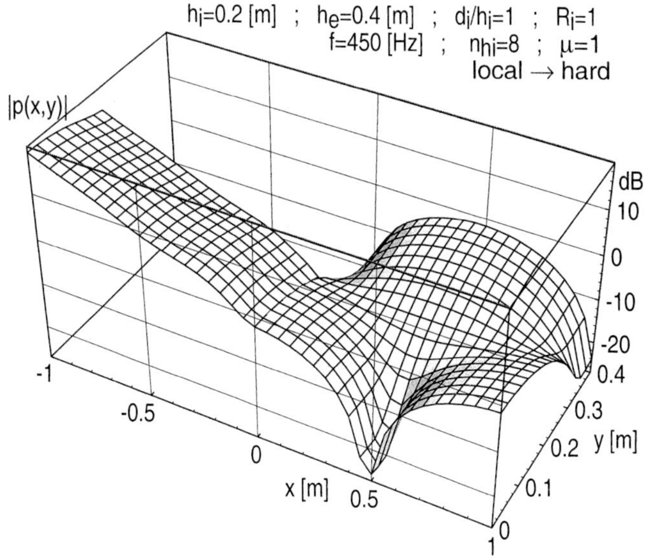

Sound pressure magnitude at an expanding duct step from a locally absorbing duct to a hard duct with the least attenuated mode as incident wave. The 1st higher mode in the hard duct is a cut-on mode.

J.19 Sections and cascades of silencers without feedback

Ref.: Mechel, [J.9]

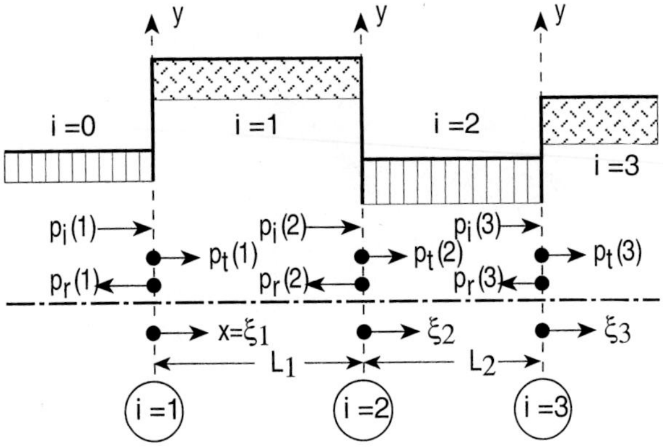

Lined duct sections $i= 1,2,3,\ldots$ have finite lengths L_i (except the entrance duct $i=0$ and the last duct). The ducts may have bulk or locally reacting linings, or are hard. The (half) duct heights are h_i .

The condition is that there *is no feedback*, i.e. at a duct step i the reflected waves from the steps $i-1$ and/or $i+1$ can be neglected.

Axial section co-ordinates are ξ_i with :
$$x = \xi_i + \begin{cases} 0 & ; \; i=1 \\ \sum_{k=1}^{i-1} L_k & ; \; i>1 \end{cases} \tag{1}$$

The incident wave is a sum of modes of the duct $i=0$ with a list of amplitudes $\{P_{im}(1)\}$ or a single mode with the amplitude $P_{i_\mu}(1)$ (note that the index i in field components and amplitudes counts the duct steps).

Field formulations of the incident wave $p_i(i)$, reflected wave $p_r(i)$, transmitted wave $p_t(i)$ at the i-th duct step :

$$\begin{aligned}
p_i(i) &= p_i(i,\xi_i,y) = \sum_m p_{im}(i,\xi_i,y) = \sum_m P_{im}(i)\cdot q_m(i-1,y)\cdot e^{-\gamma_m(i-1)\cdot\xi_i} \\
p_i(1) &= p_i(1,\xi_1,y) = \sum_m p_{im}(1,\xi_1,y) = \sum_m P_{im}(1)\cdot q_m(0,y)\cdot e^{-\gamma_m(0)\cdot\xi_1} \\
p_r(i) &= p_r(i,\xi_i,y) = \sum_m p_{rm}(i,\xi_i,y) = \sum_m P_{rm}(i)\cdot q_m(i-1,y)\cdot e^{+\gamma_m(i-1)\cdot\xi_i} \\
p_t(i) &= p_t(i,\xi_i,y) = \sum_n p_{tn}(i,\xi_i,y) = \sum_n P_{tn}(i)\cdot q_n(i,y)\cdot e^{-\gamma_n(i)\cdot\xi_i}
\end{aligned} \tag{2}$$

The $p_{\beta k}(i,\xi_i,y)$; $\beta= i,r,t$; are the mode components; $q_m(i,y)$ are their lateral profiles; $P_{\beta k}(i)$ are their amplitudes. The secular equations between lateral mode wave number $\varepsilon_k(i)$ and axial propagation constant $\gamma_k(i)$ are :

$$\gamma_k^2(i) = \varepsilon_k^2(i) - k_0^2 \qquad \text{if the field is constant in he z direction,} \tag{3}$$

$$\gamma_k^2(i) = \varepsilon_k^2(i) - k_0^2 - \varepsilon_z^2(i) \qquad \text{if there is a mode in the z direction.} \tag{4}$$

The lateral wave numbers are solutions of the characteristic equation in the duct section i.
The lateral mode pressure profiles are : (5)

$$q_n(i,y) = \begin{cases} \cos(\varepsilon_n(i)y) \;\; ; \;\; \text{locally reacting} \\ s_y(0,h_i)\dfrac{\cos(\varepsilon_n(i)y)}{\cos(\varepsilon_n(i)h_i)} + s_y(h_i,h_i+d_i)\dfrac{\cos(\kappa_n(i)(y-h_i-d_i))}{\cos(\kappa_n(i)d_i)} \;\; ; \;\; \text{bulk reacting} \end{cases}$$

the lateral profiles of the axial particle velocity are : (6)

$$q_{vn}(i,y) = \begin{cases} \cos(\varepsilon_n(i)y) \;\; ; \;\; \text{locally reacting} \\ s_y(0,h_i)\dfrac{\cos(\varepsilon_n(i)y)}{\cos(\varepsilon_n(i)h_i)} - \dfrac{j s_y(h_i,h_i+d_i)}{\Gamma_{in} Z_{in}}\dfrac{\cos(\kappa_n(i)(y-h_i-d_i))}{\cos(\kappa_n(i)d_i)} \;\; ; \;\; \text{bulk reacting} \end{cases}$$

with the "switch function" : $s_y(a,b) := \begin{cases} 1 \; ; \; a \le y < b \\ 0 \; ; \; \text{else} \end{cases}$; the (half) duct height h_i and the layer thickness d_i for a bulk reacting absorber consisting of a simple porous layer with normalised characteristic values Γ_{in}, Z_{in} .

The identity $\qquad p_{im}(i+1,\xi_{i+1},y) = p_{tm}(i,\xi_i,y) \;\; ; \;\; i = 1,2,\ldots,I-1$

leads to : $\qquad P_{im}(i+1) = P_{tm}(i) \cdot e^{-\gamma_m(i) \cdot L_i} \;\; ; \;\; i = 1,2,\ldots,I-1 \qquad (7)$

where the $P_{im}(1)$ are given.

The field evaluation is an iterative procedure, which uses the equations of the previous section J.18 . Begin with the given $P_{im}(1)$ and evaluate with those equations the transmitted mode amplitudes $P_{tm}(1)$ at the 1st duct step. With the last relation from above, they give the incident mode amplitudes $P_{im}(i+1)$ at the next step. Repeat the evaluation until the last step.

J.20 A section with feedback between sections without feedback

Ref.: Mechel, [J.9]

A duct section is said to have feedback if the reflected waves from its exit influence the boundary conditions at its entrance. Neglecting feedback (because the section is long and/or its attenuation is high) simplifies the field evaluation in cascades to a straightforward computation. Feedback in all sections, on the other hand, leads to chains of systems of equations. The amount of computational work is still reasonably low, if only one section is assumed to have feedback between sections without feedback.

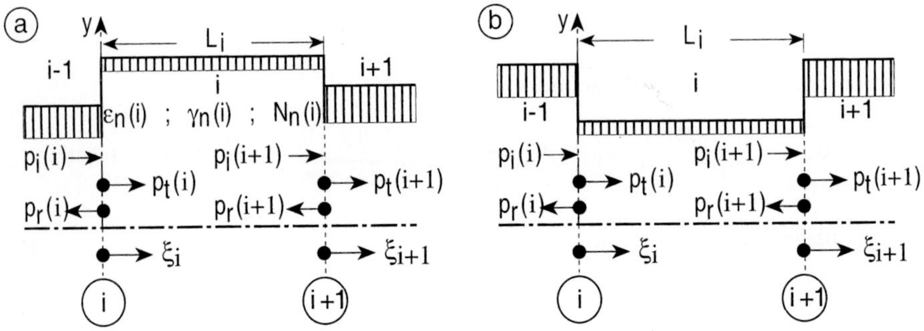

The sketches show two examples of a duct section i with feedback between duct sections i–1 , i+1 without feedback. $\varepsilon_n(i)$, $\gamma_n(i)$, $N_n(i)$ be, respectively, the lateral mode wave numbers, axial mode propagation constants, and mode norms in the section i . The (half) duct height are h_i ; ξ_i are the axial co-ordinates of the sections (see section J.18).

Sound pressure and axial particle velocity conditions at the entrance $\xi_i = 0$ of the i-th section :

$$\sum_m \left(P_{im}(i) + P_{rm}(i)\right) \cdot \cos\left(\varepsilon_n(i-1)y\right) = \sum_n \left(P_{tn}(i) + P_{rn}(i+1) \cdot e^{-\gamma_n(i) \cdot L_i}\right) \cdot \cos\left(\varepsilon_n(i)y\right) \tag{1}$$

$$\begin{aligned} &\sum_m \left(P_{im}(i) - P_{rm}(i)\right)\gamma_n(i-1) \cdot \cos\left(\varepsilon_n(i-1)y\right) = \\ &\quad = \sum_n \left(P_{tn}(i) - P_{rn}(i+1) \cdot e^{-\gamma_n(i) \cdot L_i}\right)\gamma_n(i) \cdot \cos\left(\varepsilon_n(i)y\right) \end{aligned} \tag{2}$$

Sound pressure and axial particle velocity conditions at the exit $\xi_i = L_i$ of the i-th section :

$$\sum_n \left(P_{tn}(i) \cdot e^{-\gamma_n(i) \cdot L_i} + P_{rn}(i+1)\right) \cdot \cos\left(\varepsilon_n(i)y\right) = \sum_n P_{tn}(i+1) \cdot \cos\left(\varepsilon_n(i+1)y\right) \tag{3}$$

$$\begin{aligned} &\sum_n \left(P_{tn}(i) \cdot e^{-\gamma_n(i) \cdot L_i} - P_{rn}(i+1)\right)\gamma_n(i) \cdot \cos\left(\varepsilon_n(i)y\right) = \\ &\quad = \sum_n P_{tn}(i+1)\gamma_n(i+1) \cdot \cos\left(\varepsilon_n(i+1)y\right) \end{aligned} \tag{4}$$

Special case of a wide section i with locally reacting lining between narrow sections i–1, i+1 with locally reacting linings (see sketch (a) above):

There exist two coupled linear systems of equations for the double vector of amplitudes $\{P_{tn}(i), P_{rn}(i+1)\}$ (the mode coupling coefficients $R_{m,n}(h,\beta,\beta')$ are defined in section J.18) :

$$\sum_n P_{tn}(i)\left[\delta_{k,n}\gamma_k(i)h_i N_k(i)+\sum_m \gamma_m(i-1)h_{i-1}\frac{R_{m,k}(h_{i-1},i-1,i)\cdot R_{m,n}(h_{i-1},i-1,i)}{N_m(i-1)}\right]-$$
$$\sum_n P_{rn}(i+1)e^{-\gamma_n(i)\cdot L_i}\,.$$
$$\cdot\left[\delta_{k,n}\gamma_k(i)h_i N_k(i)-\sum_m \gamma_m(i-1)h_{i-1}\frac{R_{m,k}(h_{i-1},i-1,i)\cdot R_{m,n}(h_{i-1},i-1,i)}{N_m(i-1)}\right]= \tag{5}$$
$$=2\sum_m P_{im}(i)\gamma_m(i-1)h_{i-1}R_{m,k}(h_{i-1},i-1,i)$$

and : (6)

$$\sum_n P_{tn}(i)\,e^{-\gamma_n(i)\cdot L_i}\,.$$
$$\cdot\left[\delta_{k,n}\gamma_k(i)h_i N_k(i)-\sum_m \gamma_m(i+1)h_{i+1}\frac{R_{m,k}(h_{i+1},i+1,i)\cdot R_{m,n}(h_{i+1},i+1,i)}{N_m(i+1)}\right]-$$
$$\sum_n P_{rn}(i+1)\left[\delta_{k,n}\gamma_k(i)h_i N_k(i)+\sum_m \gamma_m(i+1)h_{i+1}\frac{R_{m,k}(h_{i+1},i+1,i)\cdot R_{m,n}(h_{i+1},i+1,i)}{N_m(i+1)}\right]=0$$

With the solutions evaluate :

$$P_{tm}(i+1)=\frac{1}{N_m(i+1)}\sum_n\left(P_{tn}(i)\cdot e^{-\gamma_n(i)\cdot L_i}+P_{rn}(i+1)\right)\cdot R_{m,n}(h_{i+1},i+1,i) \tag{7}$$

$$P_{rm}(i)=-P_{im}(i)+\frac{1}{N_m(i-1)}\sum_n\left(P_{tn}(i)+P_{rn}(i+1)e^{-\gamma_n(i)\cdot L_i}\right)\cdot R_{m,n}(h_{i-1},i-1,i) \tag{8}$$

Special case of a narrow section i between wider sections i–1, i+1 , all with locally reacting lining (see sketch (b) above) :

There exist two coupled linear systems of equations for the double vector of amplitudes $\{P_{tn}(i), P_{rn}(i+1)\}$:

$$\sum_n P_{tn}(i)\left[\delta_{k,n}N_k(i)+\gamma_n(i)h_i\sum_m\frac{R_{k,m}(h_i,i,i-1)\cdot R_{n,m}(h_i,i,i-1)}{\gamma_m(i-1)h_{i-1}N_m(i-1)}\right]+$$
$$\sum_n P_{rn}(i+1)e^{-\gamma_n(i)\cdot L_i}\left[\delta_{k,n}N_k(i)-\gamma_n(i)h_i\sum_m\frac{R_{k,m}(h_i,i,i-1)\cdot R_{n,m}(h_i,i,i-1)}{\gamma_m(i-1)h_{i-1}N_m(i-1)}\right]= \tag{9}$$
$$=2\sum_m P_{im}(i)R_{k,m}(h_i,i,i-1)$$

and :

$$\sum_n P_{tn}(i) e^{-\gamma_n(i)\cdot L_i} \left[\delta_{k,n} N_k(i) - \gamma_n(i) h_i \sum_m \frac{R_{k,m}(h_i,i,i+1) \cdot R_{n,m}(h_i,i,i+1)}{\gamma_m(i+1) h_{i+1} N_m(i+1)} \right] +$$
$$\sum_n P_{rn}(i+1) \left[\delta_{k,n} N_k(i) + \gamma_n(i) h_i \sum_m \frac{R_{k,m}(h_i,i,i+1) \cdot R_{n,m}(h_i,i,i+1)}{\gamma_m(i+1) h_{i+1} N_m(i+1)} \right] = 0 \tag{10}$$

With the solutions, evaluate : (11)

$$P_{tm}(i+1) = \\ = \frac{1}{\gamma_m(i+1) h_{i+1} N_m(i+1)} \sum_n \left(P_{tn}(i) e^{-\gamma_n(i)\cdot L_i} - P_{rn}(i+1) \right) \gamma_n(i) h_i \, R_{n,m}(h_i,i,i+1)$$

$$P_{rm}(i) = \\ = P_{im}(i) - \frac{1}{\gamma_m(i-1) h_{i-1} N_m(i-1)} \sum_n \left(P_{tn}(i) - P_{rn}(i+1) e^{-\gamma_n(i)\cdot L_i} \right) \gamma_n(i) h_i \, R_{n,m}(h_i,i,i-1)$$

Other step configurations, as in section J.18 , can be treated as follows. The equations at the exit of section i do not change with feedback. The equations at the entrance are modified by :

$$\begin{aligned} P_{tn}(i) &\rightarrow \left(P_{tn}(i) + P_{rn}(i+1) e^{-\gamma_n(i)\cdot L_i} \right) \quad &; \text{ pressure} \\ P_{tn}(i)\gamma_n(i) &\rightarrow \left(P_{tn}(i) - P_{rn}(i+1) e^{-\gamma_n(i)\cdot L_i} \right) \gamma_n(i) \quad &; \text{ axial velocity} \end{aligned} \tag{12}$$

The systems of equations from the boundary conditions at the entrance will have the form :

$$\sum_n P_{tn}(i) \left[\delta_{k,n} \cdot A_k + B_{k,n} \right] \mp \sum_n P_{rn}(i+1) e^{-\gamma_n(i)\cdot L_i} \left[\delta_{k,n} \cdot A_k - B_{k,n} \right] = \\ = \sum_m P_{im}(i) \cdot C_{k,m} \quad ; \quad \begin{cases} \text{expanding} \\ \text{contracting} \end{cases} \tag{13}$$

The A_k, $B_{k,n}$, $C_{k,m}$ can be taken from the corresponding systems for sections without feedback.

J.21 Concentrated absorber in an otherwise homogeneous lining

Ref.: Mechel, [J.10]

Sometimes the attenuation produced by a homogeneous lining (say locally reacting with an admittance G) can be improved around some (preferably low) frequency. The idea is to place some isolated resonators into the lining; the resonator orifice is the "concentrated absorber".

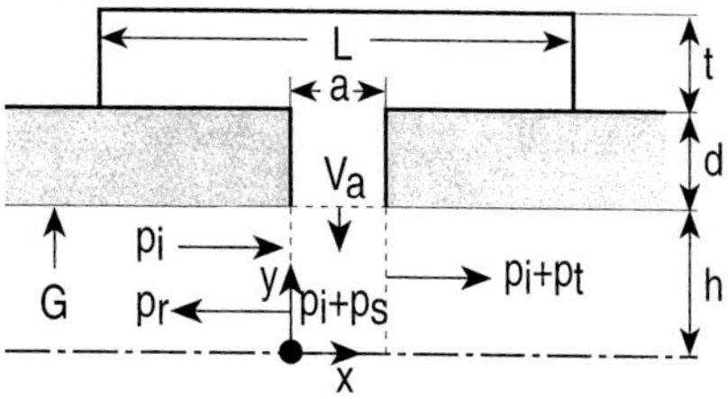

The task is treated with the method of a fictitious volume source in the orifice, having the particle velocity V_a . The sound field $p = p_r + p_s + p_t$ which the source produces in the duct satisfies the inhomogeneous wave equation :

$$\left(\Delta + k_0^2\right) p(x,y) = -j k_0 Z_0 \cdot V_a(x_0) dx_0 \cdot \delta(x - x_0) \cdot \delta(y - y_0) \tag{1}$$

with $y_0 = h$; $0 \le x_0 \le a$. The Dirac delta function $\delta(y - y_0)$ is synthesised with modes of the homogeneously lined duct having the lateral profiles $q_n(y)$ and axial propagation constants γ_n , and the mode norms N_{hn} :

$$\delta(y - y_0) = \sum_n c_n \cdot q_n(y) \quad ; \quad c_n = \frac{q_n(y_0)}{y_0 \cdot N_{hn}} \quad ; \quad N_{hn} = \frac{1}{h}\int_0^h q_n^2(y)\,dy \tag{2}$$

The sound pressure contribution $dp(x,y)$ of the elementary source $V_a(x_0) \cdot dx_0$ is :

$$dp(x,y) = \frac{j k_0 Z_0 V_a(x_0) dx_0}{2h} \sum_n \frac{q_n(h) \cdot q_n(y)}{\gamma_n N_{hn}} e^{-\gamma_n |x - x_0|} \tag{3}$$

The source contributions p_r, p_s, p_t ahead of, in front of, and down the orifice are :

$$\begin{aligned}
p_r(x,y) &= \frac{j k_0 a}{2} \sum_n \frac{q_n(h)}{\gamma_n h N_{hn}} \cdot q_n(y) e^{+\gamma_n x} \cdot Ir_n(0) \quad ; \quad x < 0 \\
p_s(x,y) &= \frac{j k_0 a}{2} \sum_n \frac{q_n(h)}{\gamma_n h N_{hn}} \cdot q_n(y) \left[e^{-\gamma_n x} \cdot It_n(x) + e^{+\gamma_n x} \cdot Ir_n(x) \right] \quad ; \quad 0 < x < a \\
p_t(x,y) &= \frac{j k_0 a}{2} \sum_n \frac{q_n(h)}{\gamma_n h N_{hn}} \cdot q_n(y) e^{-\gamma_n x} \cdot It_n(a) \quad ; \quad x > a
\end{aligned} \tag{4}$$

with the integrals :

$$Ir_n(x) := \frac{1}{a}\int_x^a Z_0 V_a(x_0)\cdot e^{-\gamma_n x_0}\,dx_0 \quad ; \quad It_n(x) := \frac{1}{a}\int_0^x Z_0 V_a(x_0)\cdot e^{+\gamma_n x_0}\,dx_0 \tag{5}$$

If the source occupies a resonator neck $0 \le x \le a$ with hard walls, its velocity profile can be synthesised :

$$V_a(x_0) = \sum_m A_m \cos(k_{xm} x_0) \tag{6}$$

which leads to :

$$It_n(x) = \sum_m A_m\, It_{n,m}(x) \quad ; \quad Ir_n(x) = \sum_m A_m\, Ir_{n,m}(x) \tag{7}$$

with :

$$\begin{aligned} It_{n,m}(x) &:= \frac{1}{a}\int_0^x \cos(k_{xm} x_0)\cdot e^{+\gamma_n x_0}\,dx_0 \\ &= \gamma_n a \frac{-1 + e^{+\gamma_n x}\big(\cos(m\pi x/a) + (m\pi/\gamma_n a)\sin(m\pi x/a)\big)}{(\gamma_n a)^2 + (m\pi)^2} \\ &\xrightarrow[x=a]{} \gamma_n a \frac{-1 + (-1)^m e^{+\gamma_n a}}{(\gamma_n a)^2 + (m\pi)^2} \\ &\xrightarrow[m=0]{} \frac{-1 + e^{+\gamma_n x}}{\gamma_n a} \xrightarrow[x=a]{} \frac{-1 + e^{+\gamma_n a}}{\gamma_n a} \end{aligned} \tag{8}$$

$$\begin{aligned} Ir_{n,m}(x) &:= \frac{1}{a}\int_x^a \cos(k_{xm} x_0)\cdot e^{-\gamma_n x_0}\,dx_0 \\ &= \gamma_n a \frac{(-1)^{m+1} e^{-\gamma_n a} + e^{-\gamma_n x}\big(\cos(m\pi x/a) - (m\pi/\gamma_n a)\sin(m\pi x/a)\big)}{(\gamma_n a)^2 + (m\pi)^2} \\ &\xrightarrow[x=0]{} \gamma_n a \frac{1 - (-1)^m e^{-\gamma_n a}}{(\gamma_n a)^2 + (m\pi)^2} \\ &\xrightarrow[m=0]{} \frac{e^{-\gamma_n x} - e^{+\gamma_n a}}{\gamma_n a} \xrightarrow[x=0]{} \frac{1 - e^{-\gamma_n a}}{\gamma_n a} \end{aligned} \tag{9}$$

Special case:
- The μ-th mode of the homogeneous duct is the incident wave p_i ;
- only a plane wave in the resonator neck.

$$p_i(x,y) = P_{i\mu}\cdot\cos(\varepsilon_\mu y)\cdot e^{-\gamma_\mu x} \quad ; \quad -\infty < x < +\infty \tag{10}$$

Field formulations with $q_n(y) = \cos(\varepsilon_n y)$:

$$p_r(x,y)=\sum_n P_{rn}\cdot\cos(\varepsilon_n y)\cdot e^{+\gamma_n x} \quad ; \quad -\infty<x<0$$
$$p_t(x,y)=\sum_n P_{tn}\cdot\cos(\varepsilon_n y)\cdot e^{-\gamma_n x} \quad ; \quad a<x<\infty \tag{11}$$
$$p_s(x,y)=\sum_n P_{sn}\cdot\cos(\varepsilon_n y)\cdot\left[a_n(x)+b_n e^{+\gamma_n x}+c_n e^{-\gamma_n x}\right] \quad ; \quad 0<x<a$$

The lateral mode wave numbers $\varepsilon_n h$ are solutions of the characteristic equation :

$$\varepsilon_n h\cdot\tan(\varepsilon_n h)=j\cdot k_0 h Z_0 G =: j\cdot U \quad ; \quad n=1,2,\ldots \tag{12}$$

and have the axial wave numbers and mode norms :

$$\gamma_n h=\sqrt{(\varepsilon_n h)^2-(k_0 h)^2}\ \xrightarrow[G=0]{}\ \sqrt{((n-1)\pi)^2-(k_0 h)^2}$$
$$N_{hn}=\frac{1}{2}\left(1+\frac{\sin(2\varepsilon_n h)}{2\varepsilon_n h}\right)\ \xrightarrow[G=0]{}\ \frac{1}{\delta_{n-1}} \quad ; \quad \delta_i=\begin{cases}1\ ;\ i=0\\ 2\ ;\ i>0\end{cases} \tag{13}$$

The particle velocity profile $V_a(x_0)$ of the fictitious source is :

$$-V_a(x_0)=v_{ay}(x_0)-G\cdot p_i(x_0,h) \tag{14}$$

If, on the other hand, the orifice input admittance G_a is known :

$$\begin{aligned}-V_a(x_0)&\approx G_a\cdot\langle p_i(x_0,h)+p_s(x_0,h)\rangle_a-G\cdot\langle p_i(x_0,h)\rangle_a \quad ; \quad 0<x_0<a\\ &=\langle p_i(x_0,h)\rangle_a\cdot(G_a-G)-\langle p_s(x_0,h)\rangle_a\cdot G_a\end{aligned} \tag{15}$$

(with $\langle\ldots\rangle_a$ the average over the orifice) and $V_a(x_0)\to\langle V_a(x_0)\rangle\to A_0$.

This gives :

$$p_r(x,y)=\frac{jk_0 h}{2}Z_0 A_0\sum_n\frac{1-e^{-\gamma_n a}}{(\gamma_n h)^2 N_{hn}}q_n(h)\cdot q_n(y)e^{+\gamma_n x} \quad ; \quad x<0$$
$$p_s(x,y)=\frac{jk_0 h}{2}Z_0 A_0\sum_n\frac{1}{(\gamma_n h)^2 N_{hn}}q_n(h)\cdot q_n(y)\left[2-e^{-\gamma_n x}+e^{+\gamma_n(x-a)}\right] \quad ; \quad 0<x<a \tag{16}$$
$$p_t(x,y)=\frac{jk_0 h}{2}Z_0 A_0\sum_n\frac{-1+e^{+\gamma_n a}}{(\gamma_n h)^2 N_{hn}}q_n(h)\cdot q_n(y)e^{-\gamma_n x} \quad ; \quad x>a$$

with the required average values :

$$\langle p_i(x,h)\rangle_a=P_{i\mu}q_\mu(h)\frac{1-e^{-\gamma_\mu a}}{\gamma_\mu a}$$
$$\langle p_s(x,h)\rangle_a=jZ_0 A_0\,k_0 h\sum_n\frac{q_n^2(h)}{(\gamma_n h)^2 N_{hn}}\left(1-\frac{1-e^{-\gamma_n a}}{\gamma_n a}\right) \tag{17}$$

The final equation for A_0 is : (18)

$$Z_0A_0 = \left[P_{i\mu}q_\mu(h)\frac{1-e^{-\gamma_\mu a}}{\gamma_\mu a}(Z_0G_a - Z_0G)\right]\cdot\left[1 + jZ_0G_a\,k_0h\sum_n\frac{q_n^2(h)}{(\gamma_n h)^2 N_{hn}}\left(1-\frac{1-e^{-\gamma_n a}}{\gamma_n a}\right)\right]^{-1}$$

The numerical example shows the attenuation $D_h = -\Delta L/h$ as the sound pressure level decrease per travel distance h in a duct with a (locally reacting) porous layer of thickness d and a normalised flow resistance $R = \Xi\cdot d/Z_0$ having T-shaped Helmholtz resonators (see initial sketch) at distances L with an orifice width a and a normalised flow resistance R_f in the orifice (e.g. by a wire mesh). The lowest duct mode $\mu=1$ is incident.

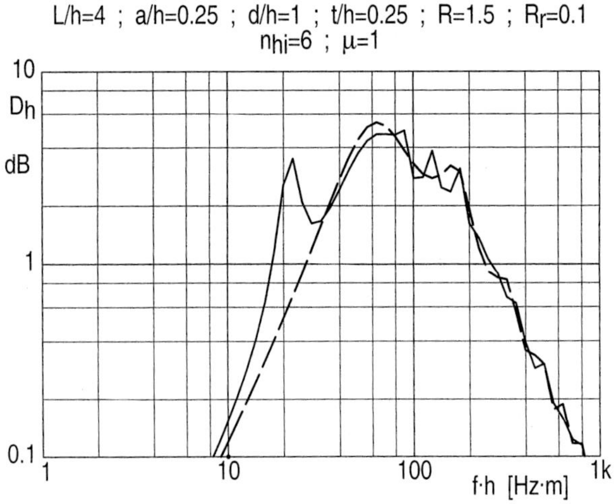

Attenuation D_h in a duct with homogeneous lining (dashed) and additional T-shaped Helmholtz resonators (full line) .

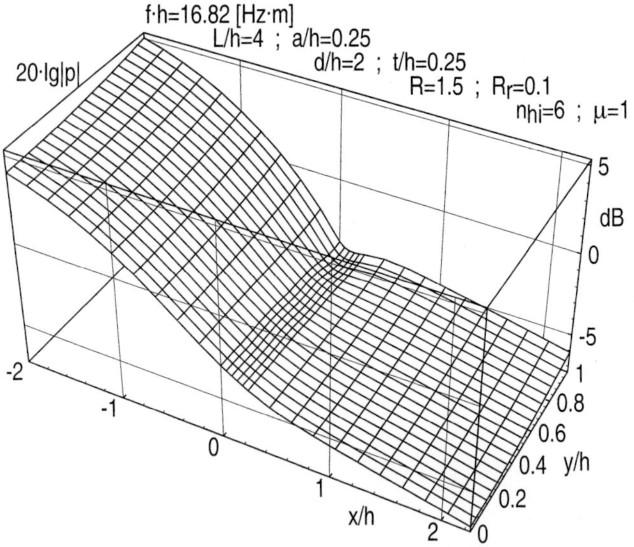

Sound pressure magnitude in a lined duct with single T-shaped Helmholtz resonators, at the resonance frequency.

J.22 Wide splitter type silencer with locally reacting splitters

Ref.: Mechel, [J.11]

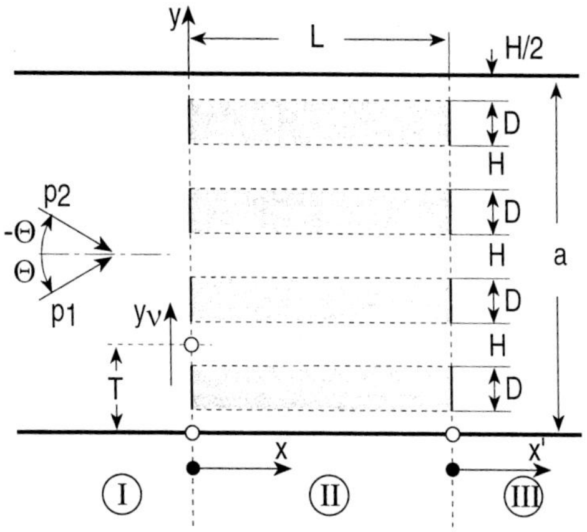

Splitters (or "baffles") with thickness D= 2d and length L are at mutual distances H= 2h . They form a periodic structure with period length T= D+H . The heads of the splitters are hard.

This section treats an arrangement with many splitters, i.e. with a large lateral extent a . A single plane wave p_1 is assumed as the incident wave. The next section will treat splitters in a hard duct; then an additional mirror-reflected wave p_2 will be incident, and the angle Θ of incidence is determined so that p_1+p_2 make a mode of the hard duct.

Incident plane wave :

$$p_1(x,y) = P_1 \cdot e^{-jk_x x} \cdot e^{-jk_y y} \quad ; \quad v_{1x}(x,y) = \frac{k_x}{k_0 Z_0} \cdot p_1(x,y)$$
$$k_x = k_0 \cos\Theta \quad ; \quad k_y = k_0 \sin\Theta \tag{1}$$

Backscattered wave p_s as sum of spatial harmonics :

$$p_s(x,y) = \sum_{n=-\infty}^{+\infty} A_n \cdot e^{j\kappa_n x} \cdot e^{-j\beta_n y}$$
$$\beta_n = \beta_0 + \frac{2\pi n}{T} \quad ; \quad n = 0, \pm 1, \pm 2, \ldots \quad ; \quad \beta_0 = k_y = k_0 \sin\Theta \tag{2}$$
$$\kappa_0 = k_x = k_0 \cos\Theta \quad ; \quad \kappa_n = \sqrt{k_0^2 - \beta_n^2} = k_0\sqrt{1-(\sin\Theta + n\lambda_0/T)^2}$$

Field in Zone I $p_I = p_1 + p_s$:

$$p_I(x,y) = P_I\, e^{-jk_x x}\, e^{-jk_y y} + \sum_{n=-\infty}^{+\infty} A_n \cdot e^{j\kappa_n x} \cdot e^{-j\beta_n y}$$

$$v_{Ix}(x,y) = \frac{j}{k_0 Z_0}\frac{\partial p_{I1}}{\partial x} = \frac{k_x}{k_0 Z_0} P_I\, e^{-jk_x x}\, e^{-jk_y y} - \frac{1}{Z_0}\sum_{n=-\infty}^{+\infty} \frac{\kappa_n}{k_0} A_n \cdot e^{j\kappa_n x} \cdot e^{-j\beta_n y} \tag{3}$$

Field in Zone III :

$$p_{III}(x',y) = \sum_{n=-\infty}^{+\infty} D_n\, e^{-j\kappa_n x'}\, e^{-j\beta_n y} = e^{-j\beta_0 y} \sum_{n=-\infty}^{+\infty} D_n\, e^{-j\kappa_n x'}\, e^{-jn2\pi y/T}$$

$$v_{IIIx}(x',y) = \frac{1}{Z_0} e^{-j\beta_0 y} \sum_{n=-\infty}^{+\infty} D_n \frac{\kappa_n}{k_0} e^{-j\kappa_n x'}\, e^{-jn2\pi y/T} \tag{4}$$

Field in the ν-th splitter duct in Zone II as sum of silencer modes :

$$p_{II}(x,y_\nu) = e^{-j\beta_0 \nu T} \sum_{m=0}^{\infty} [B_m\, e^{-\gamma_m x} + C_m\, e^{+\gamma_m x}] \cdot q_m(y_\nu) \tag{5}$$

with lateral mode profiles :

$$q_m(y_\nu) = \begin{cases} \cos(\varepsilon_m y_\nu) & ; \quad m = 0,2,4,\ldots \\ \sin(\varepsilon_m y_\nu) & ; \quad m = 1,3,5,\ldots \end{cases} \quad ; \quad \gamma_m = \sqrt{\varepsilon_m^2 - k_0^2} \quad ; \quad \mathrm{Re}\{\gamma_m\} \ge 0 \tag{6}$$

The lateral mode wave numbers $\varepsilon_m h$ are solutions of the equations (the splitter surfaces are locally reacting with admittance G):

$$jk_0 h \cdot Z_0 G = \begin{cases} \varepsilon_m h \cdot \tan(\varepsilon_m h) & ; \quad m = 0,2,4,\ldots \\ -\varepsilon_m h \cdot \cot(\varepsilon_m h) & ; \quad m = 1,3,5,\ldots \end{cases} \tag{7}$$

Mode norms :

$$\frac{1}{2h}\int_{-h}^{+h} q_m(\varepsilon_m y_\nu) \cdot q_{m'}(\varepsilon_{m'} y_\nu)\, dy_\nu = \begin{cases} 0 & ; m \ne m' \\ N_m & ; m = m' \end{cases}$$

$$N_m = \begin{cases} \frac{1}{2}\left(1 + \frac{\sin(2\varepsilon_m h)}{(2\varepsilon_m h)}\right) & ; \quad m = 0,2,4,\ldots \\ \frac{1}{2}\left(1 - \frac{\sin(2\varepsilon_m h)}{(2\varepsilon_m h)}\right) & ; \quad m = 1,3,5,\ldots \end{cases} \tag{8}$$

Auxiliary amplitudes :

$$X_m =: B_m - C_m \quad ; \quad Y_m =: B_m\, e^{-\gamma_m L} - C_m\, e^{+\gamma_m L} \tag{9}$$

The field matching at the zone limits gives for them two coupled, linear, homogeneous systems of equations :

$$\sum_{m'=0}^{\infty} X_{m'}\left[-j\frac{\gamma_{m'}}{k_0}\frac{h}{T}\sum_n \frac{k_0}{\kappa_n} S_{m,n}S_{m',n} + 2\delta_{m',m}(-1)^m N_m \frac{1+e^{-2\gamma_m L}}{1-e^{-2\gamma_m L}}\right] = \tag{10}$$

$$= 2P_1 S_{m,0} + 4(-1)^m N_m \frac{e^{-\gamma_m L}}{1-e^{-2\gamma_m L}}\cdot Y_m$$

$$\sum_{m'=0}^{\infty} Y_{m'}\left[-j\frac{\gamma_{m'}}{k_0}\frac{h}{T}\sum_n \frac{k_0}{\kappa_n} S_{m,n}S_{m',n} + 2\delta_{m',m}(-1)^m N_m \frac{1+e^{-2\gamma_m L}}{1-e^{-2\gamma_m L}}\right] = \tag{11}$$

$$= 4(-1)^m N_m \frac{e^{-\gamma_m L}}{1-e^{-2\gamma_m L}}\cdot X_m$$

i.e. a system of equations of the form:

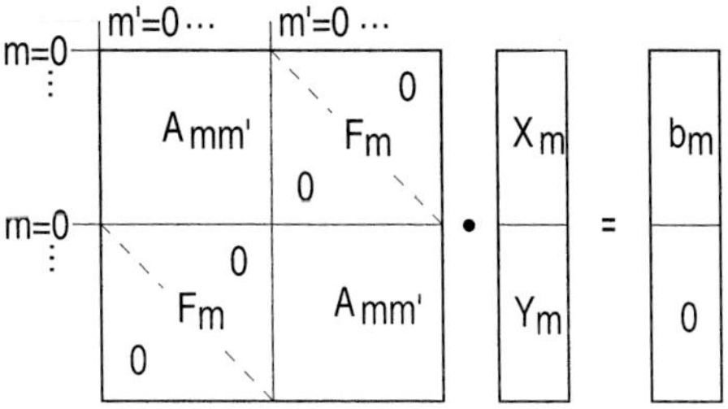

with the coefficients :

$$F_m = -4(-1)^m N_m \frac{e^{-\gamma_m L}}{1-e^{-2\gamma_m L}} \tag{12}$$

$$b_m = 2P_1\, S_{m,0}$$

$$A_{m,m'} = -j\frac{\gamma_{m'}}{k_0}\frac{h}{T}\sum_n \frac{k_0}{\kappa_n} S_{m,n}S_{m',n} + 2\delta_{m',m}(-1)^m N_m \frac{1+e^{-2\gamma_m L}}{1-e^{-2\gamma_m L}} \tag{13}$$

Therein $S_{m,n}$ are coupling coefficients between spatial harmonics and silencer modes :

$$S_{m,n} =: \frac{1}{h}\int_{-h}^{+h} e^{j\beta_n y}\, q_m(y)\,dy =$$

$$= \begin{cases} 2\dfrac{(\varepsilon_m h)\sin(\varepsilon_m h)\cos(\beta_n h) - (\beta_n h)\cos(\varepsilon_m h)\sin(\beta_n h)}{\left(\varepsilon_m^2 - \beta_n^2\right)h^2} & ; \quad m = 0,2,4,\ldots \\ 2j\dfrac{(\beta_n h)\sin(\varepsilon_m h)\cos(\beta_n h) - (\varepsilon_m h)\cos(\varepsilon_m h)\sin(\beta_n h)}{\left(\varepsilon_m^2 - \beta_n^2\right)h^2} & ; \quad m = 1,3,5,\ldots \end{cases} \tag{14}$$

$$\xrightarrow[\beta_n \to -\beta_n]{} (-1)^m S_{m,n}$$

$$\xrightarrow[\beta_n \to \varepsilon_m]{} \begin{cases} 2N_m & ; \quad m = 0,2,4,\ldots \\ 2jN_m & ; \quad m = 1,3,5,\ldots \end{cases}$$

With solutions for X_m, Y_m follow the amplitudes in the field formulations :

$$A_n = \frac{k_0}{\kappa_n}\left[\delta_{0,n}P_1\cos\Theta + j\frac{h}{T}\sum_{m=0}^{\infty} X_m \frac{\gamma_m}{k_0} S_{m,n}\right] \quad ; \quad D_n = -j\frac{k_0}{\kappa_n}\frac{h}{T}\sum_{m=0}^{\infty} Y_m \frac{\gamma_m}{k_0} S_{m,n} \tag{15}$$

$$B_m = \frac{X_m - Y_m\, e^{-\gamma_m L}}{1-e^{-2\gamma_m L}} \quad ; \quad C_m = \frac{X_m\, e^{-\gamma_m L} - Y_m}{1-e^{-2\gamma_m L}}\, e^{-\gamma_m L}$$

Sound transmission coefficient $\tau = \Pi_i/\Pi_t$, with incident effective sound power Π_i (on one baffle unit) :

$$\Pi_i = T\frac{|P_1|^2}{2Z_0}\cos\Theta \tag{16}$$

and transmitted effective sound power :

$$\Pi_t = \frac{1}{2}\mathrm{Re}\int_0^T p_{III}(0,y)\cdot \overset{*}{v}_{III1}(0,y)\,dy = \frac{T}{2Z_0}\sum_{n_s}|D_{n_s}|^2\sqrt{1-(\sin\Theta + n_s\frac{\lambda_0}{T})^2} \tag{17}$$

where the summation index n_s extends over the range of "radiating spatial harmonics" :

$$-(1+\sin\Theta)\cdot T/\lambda_0 < n_s < (1-\sin\Theta)\cdot T/\lambda_0 \tag{18}$$

(harmonics with orders outside this range only contribute to near fields).

The spatial harmonics in Zone III are plane waves with effective intensity :

$$I_{n_s} = \frac{1}{2Z_0}|D_{n_s}|^2\sqrt{1-(\sin\Theta + n_s\frac{\lambda_0}{T})^2} \tag{19}$$

and angle of radiation (relative to the x-axis) :

$$\vartheta_{n_s} = \arctan\frac{\beta_{n_s}}{\kappa_{n_s}} = \arctan\frac{\sin\Theta + n_s\dfrac{\lambda_0}{T}}{\sqrt{1-(\sin\Theta + n_s\dfrac{\lambda_0}{T})^2}} \tag{20}$$

Therewith the radiation directivity of the transmitted sound can be evaluated.

J.23 Splitter type silencer with locally reacting splitters in a hard duct :

Ref.: Mechel, [J.11]

See the sketch in section J.22 . There are K splitter elements in the main duct; the splitter ducts are counted with $\nu = 0,1,2,\ldots,K$. Let the incident wave be the μ-th mode of the hard duct ahead of the splitters. It is made up of two plane waves p_1, p_2 incident at angles $\pm\Theta_\mu$ (relative to the x axis) :

$$p_1(x,y) = P_1 \cdot e^{-jk_{x\mu}x} \cdot e^{-jk_{y\mu}y} \quad ; \quad p_2(x,y) = P_1 \cdot e^{-jk_{x\mu}x} \cdot e^{+jk_{y\mu}y}$$

$$k_{y\mu} = \frac{\mu\pi}{a} = k_0 \sin\Theta_\mu \quad ; \quad k_{x\mu} = k_0 \cos\Theta_\mu = \sqrt{k_0^2 - (\frac{\mu\pi}{a})^2} \quad ; \quad \mu = 0,1,2,\ldots \tag{1}$$

$$\sin\Theta_\mu = \frac{k_{y\mu}}{k_0} = \frac{\mu\lambda_0}{2a}$$

Field formulations :

$$p_{I\mu}(x,y) = 2P_1\, e^{-jk_0x\cos\Theta_\mu} \cdot \cos(k_0 y \sin\Theta_\mu) + 2\sum_{n=-\infty}^{+\infty} A_n\, e^{j\kappa_n x} \cdot \cos(\beta_n y)$$

$$v_{I\mu x}(x,y) = \frac{2}{Z_0}\left[P_1 \cos\Theta_\mu \cdot e^{-jk_0x\cos\Theta_\mu} \cdot \cos(k_0 y \sin\Theta_\mu) - \sum_{n=-\infty}^{+\infty} A_n \frac{\kappa_n}{k_0} e^{j\kappa_n x} \cdot \cos(\beta_n y) \right] \tag{2}$$

$$p_{III\mu}(x',y) = 2\sum_{n=-\infty}^{+\infty} D_n\, e^{-j\kappa_n x'} \cdot \cos(\beta_n y) \tag{3}$$

$$p_{II\mu}(x,y_\nu) = 2\cos(\nu\beta_0 T)\sum_{m=0}^{\infty} [B_m\, e^{-\gamma_m x} + C_m\, e^{+\gamma_m x}] \cdot q_m(y_\nu) \tag{4}$$

With these formulations the amplitudes A_n, D_n, B_m, C_m are evaluated as in the previous section J.22 .

The range of the index n_s for radiating spatial harmonics now can be formulated as :

$$-(\frac{T}{\lambda_0} + \frac{\mu}{2K}) < n_s < \frac{T}{\lambda_0} - \frac{\mu}{2K} \qquad \text{or:} \qquad -(\frac{2a}{\lambda_0} + \mu) < 2n_s K < \frac{2a}{\lambda_0} - \mu \tag{5}$$

The incident effective sound power is :

$$\Pi_{i\mu} = \frac{2a}{\delta_\mu Z_0} \cos\Theta_\mu\, |P_1|^2 \quad ; \quad \delta_\mu = \begin{cases} 1\,;\, \mu = 0 \\ 2\,;\, \mu > 0 \end{cases} \tag{6}$$

The transmitted effective sound power is :

$$\Pi_{t\mu} = \frac{2a}{\delta_\mu Z_0} \sum_{n_s} |D_{n_s}|^2 \sqrt{1 - (\frac{\lambda_0}{2a})^2 (\mu + 2n_s K)^2} \tag{7}$$

The effective sound power reflected at the front side of the splitters is :

$$\Pi_{r\mu} = \frac{2a}{\delta_\mu Z_0} \left[|A_0|^2 \cos\Theta_\mu + \sum_{n_s \neq 0} |A_{n_s}|^2 \sqrt{1 - (\frac{\lambda_0}{2a})^2 (\mu + 2n_s K)^2} \right] \tag{8}$$

Special cases :

The transmission loss of the entrance plane of the splitters can be studied with $L \to \infty$. Then $C_m \to 0$, $D_m \to 0$; $X_m \to B_m$, $Y_m \to 0$. In the system of equations of section J.22 go $F_m \to 0$, and :

$$A_{m,m'} = -j\frac{\gamma_{m'}}{k_0}\frac{h}{T}\sum_n \frac{k_0}{\kappa_n} S_{m,n}S_{m',n} \quad ; \quad b_m = 2P_1\, S_{m,0} \tag{9}$$

Special case of incident plane wave μ=0 : Consequences :

$$\mu = 0 \quad ; \quad \Theta = \Theta_\mu = 0$$

$$\beta_0 = 0 \quad ; \quad \beta_n = \frac{2\pi n}{T}$$

$$\kappa_0 = k_0 \quad ; \quad \kappa_n = k_0\sqrt{1-(\frac{2\pi n}{k_0 T})^2} \tag{10}$$

$$\beta_{-n} = -\beta_n \quad ; \quad \kappa_{-n} = \kappa_n$$

$$S_{m,n} = \begin{cases} 2\dfrac{(jk_0 h Z_0 G)\cos(2n\pi h/T) - (2n\pi h/T)\sin(2n\pi h/T)}{(\varepsilon_m h)^2 - (2n\pi h/T)^2}\cos(\varepsilon_m h) \quad ; \\ m = 0,2,4,\ldots \\ \\ 2\dfrac{j(2n\pi h/T)\cos(2n\pi h/T) - (k_0 h Z_0 G)\sin(2n\pi h/T)}{(\varepsilon_m h)^2 - (2n\pi h/T)^2}\sin(\varepsilon_m h) \quad ; \\ m = 1,3,5,\ldots \end{cases} \tag{11}$$

$$S_{m,0} = \begin{cases} 2jk_0 h Z_0 G \cdot \dfrac{\cos(\varepsilon_m h)}{(\varepsilon_m h)} & ; \quad m = 0,2,4,\ldots \\ 0 & ; \quad m = 1,3,5,\ldots \end{cases} \tag{12}$$

$$S_{m,-n} = (-1)^m \cdot S_{m,n}$$

The sums $\sum_{n=-\infty}^{\infty} \frac{k_0}{\kappa_n} S_{m,n}S_{m',n}$ disappear in the coefficients $A_{m,m'}$; therefore anti-symmetrical waves play no role. The system of equations for the auxiliary amplitudes X_m , Y_m has the coefficients :

$$A_{m,m'} = -j\frac{\gamma_{m'}}{k_0}\frac{h}{T}\sum_{n\ge 0} \delta_n \frac{k_0}{\kappa_n} S_{m,n}S_{m',n} + 2\delta_{m',m}(-1)^m N_m \frac{1+e^{-2\gamma_m L}}{1-e^{-2\gamma_m L}}$$

$$F_m = 4(-1)^m N_m \frac{e^{-\gamma_m L}}{1-e^{-2\gamma_m L}} \tag{13}$$

$$b_m = 2P_1\, S_{m,0}$$

With its solution the other amplitudes are evaluated from :

$$A_{-n} = A_n = \delta_{0,n}\, P_I + \frac{j}{\kappa_n T}\sum_m X_m \cdot \gamma_m h \cdot S_{m,n}$$

$$D_{-n} = D_n = \frac{-j}{\kappa_n T}\sum_m Y_m \cdot \gamma_m h \cdot S_{m,n} \tag{14}$$

$$B_m = \frac{X_m - Y_m\, e^{-\gamma_m L}}{1-e^{-2\gamma_m L}} \quad ; \quad C_m = \frac{X_m\, e^{-\gamma_m L} - Y_m}{1-e^{-2\gamma_m L}}\, e^{-\gamma_m L}$$

The sound fields in the zones then follow from :

$$p_I(x,y) = 2P_I e^{-jk_0 x} + 2\sum_{n=0}^{\infty} \delta_n \cdot A_n e^{jk_0 x\sqrt{1-(n\lambda_0/T)^2}} \cdot \cos(\frac{2\pi n}{T} y)$$

$$p_{III}(x',y) = 2\sum_{n=0}^{\infty} \delta_n \cdot D_n e^{-jk_0 x'\sqrt{1-(n\lambda_0/T)^2}} \cdot \cos(\frac{2\pi n}{T} y) \tag{15}$$

$$p_{II}(x,y_\nu) = 2\sum_m [B_m e^{-\gamma_m x} + C_m e^{+\gamma_m x}] \cdot \cos(\varepsilon_m y_\nu)$$

(The number K of the splitters in the main duct has disappeared, as expected.) The range of radiating spatial harmonics is :

$$-\frac{T}{\lambda_0} < n_s < \frac{T}{\lambda_0} \tag{16}$$

The expressions for the effective sound powers simplify to :

$$\Pi_i = 2a\,|P_I^2|/Z_0$$

$$\Pi_r = \frac{2a}{Z_0}\sum_{n_s} \delta_{n_s} \cdot |A_{n_s}|^2 \cdot \sqrt{1-(n_s\lambda_0/T)^2} \tag{17}$$

$$\Pi_t = \frac{2a}{Z_0}\sum_{n_s} \delta_{n_s} \cdot |D_{n_s}|^2 \cdot \sqrt{1-(n_s\lambda_0/L)^2}$$

Approximations :

Neglect the reflections at the splitter duct exit,
i.e. set everywhere $C_m=0$. The consequence is
$X_m = B_m$; $Y_m = B_m \cdot e^{-\gamma_m L} = X_m \cdot e^{-\gamma_m L}$. (18)
This simplifies the system of equations with the coefficients :

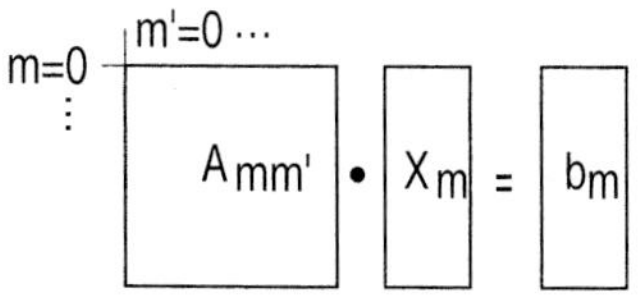

$$A_{m,m'} = -j\frac{\gamma_{m'}}{k_0}\frac{h}{T}\sum_n \frac{k_0}{\kappa_n} S_{m,n} S_{m',n} + 2\delta_{m,m'}(-1)^m N_m \tag{19}$$

$$b_m = 2P_1\, S_{m,0}$$

The amplitudes A_n are evaluated as before, the D_n follow from :

$$D_n = -j\frac{k_0}{\kappa_n}\frac{h}{T}\sum_{m=0}^{\infty} X_m\, e^{-\gamma_m L}\frac{\gamma_m}{k_0} S_{m,n} \tag{20}$$

Assume only a single mode in the splitter duct (usually the least attenuated mode) :
The X_m, Y_m follow from the two equations :

$$\begin{aligned} A_{m,m}\cdot X_m + F_m \cdot Y_m &= b_m \\ F_m \cdot X_m + A_{m,m}\cdot Y_m &= 0 \end{aligned} \tag{21}$$

with the coefficients :

$$A_{m,m} = -j\frac{\gamma_m}{k_0}\frac{h}{T}\sum_n \frac{k_0}{\kappa_n} S_{m,n}^2 + 2(-1)^m N_m \coth\gamma_m L$$

$$F_m = 2(-1)^{m+1} N_m \frac{1}{\sinh\gamma_m L} \tag{22}$$

$$b_m = 2P_1 S_{m,0}$$

or as explicit solutions :

$$X_m = \frac{A_{m,m} b_m}{A_{m,m}^2 - F_m^2} \quad ; \quad Y_m = -\frac{F_m b_m}{A_{m,m}^2 - F_m^2} \tag{23}$$

and the amplitudes in the main duct from :

$$A_n = \delta_{0,n} P_1 + j\frac{h}{T}\frac{\gamma_m}{\kappa_n} S_{m,n}\cdot X_m \quad ; \quad D_n = -j\frac{h}{T}\frac{\gamma_m}{\kappa_n} S_{m,n}\cdot Y_m \tag{24}$$

Assume only a single mode in the splitter duct and neglect its reflection at the splitter duct exit :
Then with the coefficients from above :

$$X_m = B_m = \frac{b_m}{A_{mm} + F_m\, e^{-\gamma_m L}} \tag{25}$$

$$A_n = \delta_{0,n} P_1 + j\frac{h}{T}\frac{\gamma_m}{\kappa_n} S_{m,n}\cdot X_m \quad ; \quad D_n = -j\frac{h}{T}\frac{\gamma_m}{\kappa_n} S_{m,n}\cdot X_m \cdot e^{-\gamma_m L} \tag{26}$$

Assume an incident plane wave (μ=0)*, and neglect reflections at the splitter duct exit* : The simplified system of equations for the X_m from above has the coefficients :

$$A_{m,m'} = 2\delta_{m,m'}(-1)^m N_m - j\frac{\gamma_{m'}}{k_0}\frac{h}{T}\sum_{n=0}^{\infty}\delta_n \frac{k_0}{\kappa_n} S_{m,n}S_{m',n} \quad ; \quad b_m = 2P_1\, S_{m,0} \tag{27}$$

and the amplitudes in the main duct are :

$$\begin{aligned} A_{-n} &= A_n = \delta_{0,n}\, P_1 + \frac{j}{\kappa_n T}\sum_m X_m \cdot \gamma_m h \cdot S_{m,n} \\ D_{-n} &= D_n = -j\frac{k_0}{\kappa_n}\frac{h}{T}\sum_m X_m\, e^{-\gamma_m L}\frac{\gamma_m}{k_0} S_{m,n} \end{aligned} \tag{28}$$

The splitters in the numerical example shown consist of layers of mineral fibres with a flow resistivity $\Xi = 11$ [kPa·s/m^2], covered with a porous foil of surface mass density $m'' = 0.2$ [kg/m^2] and a flow resistance $Z_{ser} = 1 \cdot Z_0$. The points are measured transmission loss values for plane wave incidence; the full curve is evaluated as explained above; the dash-dotted curve represents the propagation loss of the least attenuated mode of the splitter duct; the dashed curve shows the loss by reflection at the entrance of the splitter ducts.

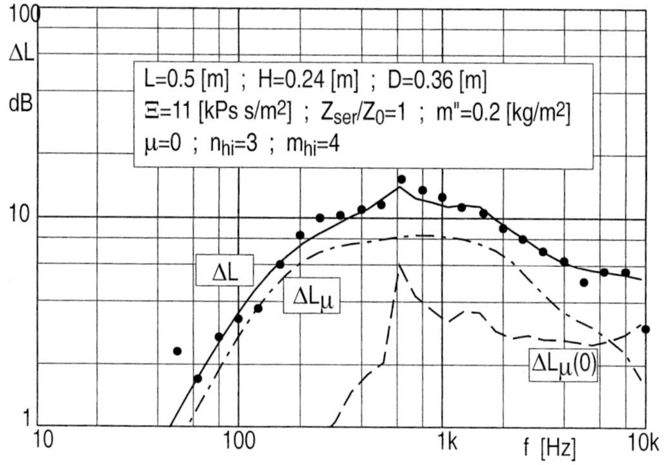

Transmission loss of a splitter silencer;
points: measured;
full: theory;
dash-dotted: propagation loss of the least attenuated mode;
dashed: loss by reflection at the splitter duct entrance.

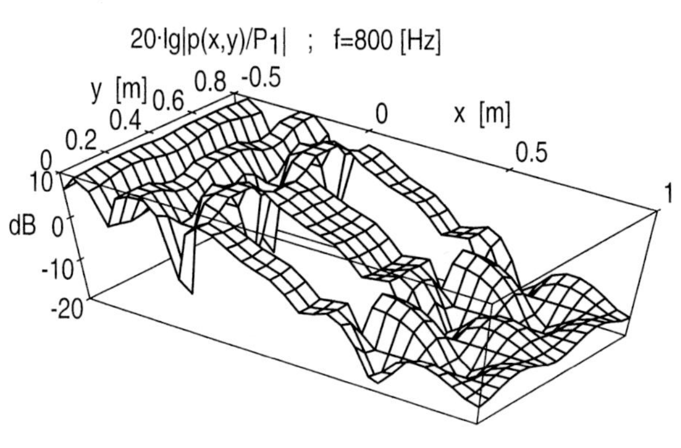

Sound pressure level ahead, in, and behind a splitter silencer with two splitters in a main duct.

J.24 Splitter type silencer with simple porous layers as bulk reacting splitters

Ref.: Mechel, [J.12]

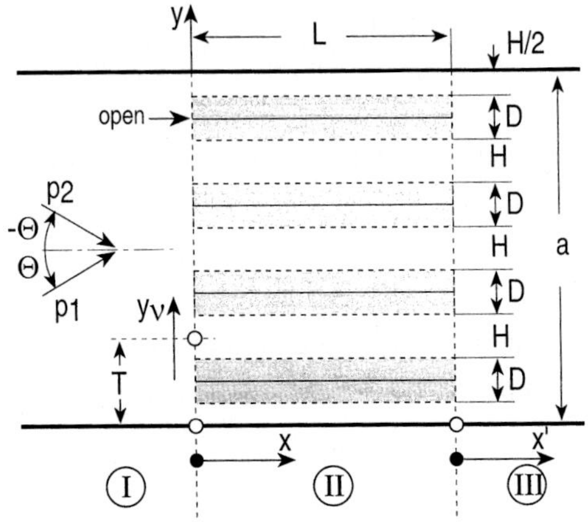

The splitters consist of simple porous layers with characteristic values of the material Γ_a, Z_a , or in normalised form $\Gamma_{an}= \Gamma_a/k_0$, $Z_{an}= Z_a/Z_0$. The heads of the splitters are open. The splitters are not sound transmissive from one splitter duct to the neighbouring splitter duct, either due to a sufficiently high flow resistance of the splitter or due to a central partition (the condition is not necessary for plane wave incidence parallel to the x axis).
H= 2h ; D= 2d .

The incident wave is a mode of the hard main duct; it is composed of two plane waves p_1+p_2 with mirror-reflected incidence under the modal angle $\Theta \to \Theta_\mu$:

$$p_1(x,y) = P_1 \cdot e^{-jk_x x} \cdot e^{-jk_y y} \quad ; \quad p_2(x,y) = P_1 \cdot e^{-jk_x x} \cdot e^{+jk_y y}$$

$$k_y \to k_{y\mu} = k_0 \sin\Theta_\mu \;\; ; \;\; k_x \to k_{x\mu} = k_0 \cos\Theta_\mu = \sqrt{k_0^2 - (\frac{\mu\pi}{a})^2} \tag{1}$$

$$\sin\Theta_\mu = \frac{k_{y\mu}}{k_0} = \frac{\mu\lambda_0}{2a}$$

The backscattered wave p_s in Zone I and the transmitted wave p_t in Zone III are sums of spatial harmonics; thus the fields in these zones are formulated as :

$$p_{I\mu}(x,y) = 2P_1 e^{-jk_0 x\cos\Theta_\mu} \cdot \cos(k_0 y \sin\Theta_\mu) + 2\sum_{n=-\infty}^{+\infty} A_n\, e^{j\kappa_n x} \cdot \cos(\beta_n y) \tag{2}$$

$$p_{III\mu}(x',y) = 2\sum_{n=-\infty}^{+\infty} D_n\, e^{-j\kappa_n x'} \cdot \cos(\beta_n y) \tag{3}$$

with :

$$\beta_n = \beta_0 + \frac{2\pi n}{T} \quad ; \quad n = 0, \pm1, \pm2, \ldots \quad ; \quad \beta_0 = k_y = k_0 \sin\Theta_\mu \tag{4}$$

$$\kappa_0 = k_x = k_0 \cos\Theta_\mu \quad ; \quad \kappa_n = \sqrt{k_0^2 - \beta_n^2} = k_0\sqrt{1-(\sin\Theta_\mu + n\lambda_0/T)^2} \tag{5}$$

The field in the ν-th splitter duct is a sum of splitter duct modes :

$$p_{II\,\mu}(x,y_\nu) = 2\cos(\nu\beta_0 T)\sum_{m=0}^{\infty}[B_m\, e^{-\gamma_m x} + C_m\, e^{+\gamma_m x}]\cdot q_m(y_\nu) \tag{6}$$

with lateral mode profiles :

$$q_m(y) = \begin{cases} s_{|y|}(0,h)\cdot\dfrac{\cos(\varepsilon_m y)}{\cos(\varepsilon_m h)} + s_{|y|}(h,h+d)\cdot\dfrac{\cos(\alpha_m(y-h-d))}{\cos(\alpha_m d)} & ;\ m=0,2,4\ldots \\ s_{|y|}(0,h)\cdot\dfrac{\sin(\varepsilon_m y)}{\sin(\varepsilon_m h)} + s_{|y|}(h,h+d)\cdot\dfrac{y}{|y|}\cdot\dfrac{\cos(\alpha_m(y-h-d))}{\cos(\alpha_m d)} & ;\ m=1,3,5\ldots \end{cases} \tag{7}$$

which use the "switch function" :

$$s_y(a,b) = \begin{cases} 1 & ;\ a \le y < b \\ 0 & ;\ \text{else} \end{cases} \tag{8}$$

The wave equations in the splitter duct and in the absorber material imply :

$$\varepsilon_m^2 = k_0^2 + \gamma_m^2 \quad ; \quad \alpha_m^2 = \gamma_m^2 - \Gamma_a^2 = \varepsilon_m^2 - k_0^2 - \Gamma_a^2 \tag{9}$$

and the lateral mode wave numbers are solutions of :

$$\begin{aligned} \varepsilon_m h\cdot\tan(\varepsilon_m h) &= -j\frac{h}{d}\frac{\alpha_m d}{\Gamma_{an} Z_{an}}\tan(\alpha_m d) \quad ;\ m=0,2,4,\ldots \\ \varepsilon_m h\cdot\cot(\varepsilon_m h) &= +j\frac{h}{d}\frac{\alpha_m d}{\Gamma_{an} Z_{an}}\tan(\alpha_m d) \quad ;\ m=1,3,5,\ldots \end{aligned} \tag{10}$$

The modes of all orders are orthogonal to each other over (–T/2,+T/2) with the mode norms M_m :

$$\frac{1}{h}\left[\int_0^h q_m(y)\cdot q_{m'}(y)\,dy - \frac{j}{\Gamma_{an} Z_{an}}\int_h^{h+d} q_m(y)\cdot q_{m'}(y)\,dy\right] = \delta_{m,m'}\cdot M_m \tag{11}$$

$$M_m = $$

$$= \begin{cases} \dfrac{1}{2}\dfrac{1}{\cos^2(\varepsilon_m h)}\left[1+\dfrac{\sin(2\varepsilon_m h)}{2\varepsilon_m h}\right] - \dfrac{j}{2}\dfrac{d/h}{\Gamma_{an} Z_{an}}\dfrac{1}{\cos^2(\alpha_m d)}\left[1+\dfrac{\sin(2\alpha_m d)}{2\alpha_m d}\right] & ;\ m=0,2,4\ldots \\ \dfrac{1}{2}\dfrac{1}{\sin^2(\varepsilon_m h)}\left[1-\dfrac{\sin(2\varepsilon_m h)}{2\varepsilon_m h}\right] - \dfrac{j}{2}\dfrac{d/h}{\Gamma_{an} Z_{an}}\dfrac{1}{\cos^2(\alpha_m d)}\left[1+\dfrac{\sin(2\alpha_m d)}{2\alpha_m d}\right] & ;\ m=1,3,5\ldots \end{cases}$$

Coupling coefficients between modes in the hard main duct and in the splitter duct, respectively, will be needed :

$$S_{m,n} = \begin{cases} \dfrac{1}{d+h}\displaystyle\int_0^h \cos(\beta_n y)\cdot q_m(|y|\le h)\,dy = \dfrac{1}{d+h}\int_0^h \cos(\beta_n y)\dfrac{\cos(\varepsilon_m y)}{\cos(\varepsilon_m h)}\,dy \;; \\ m = 0,2,\ldots \\ \dfrac{j}{d+h}\displaystyle\int_0^h \sin(\beta_n y)\cdot q_m(|y|\le h)\,dy = \dfrac{j}{d+h}\int_0^h \sin(\beta_n y)\dfrac{\sin(\varepsilon_m y)}{\sin(\varepsilon_m h)}\,dy \;; \\ m = 1,3,\ldots \end{cases} \tag{12}$$

$$R_{m,n} = \begin{cases} \dfrac{1}{d+h}\displaystyle\int_h^{h+d} \cos(\beta_n y)\cdot q_m(|y|\ge h)\,dy = \dfrac{1}{d+h}\int_h^{h+d} \cos(\beta_n y)\dfrac{\cos(\alpha_m(y-h-d))}{\cos(\alpha_m d)}\,dy \;; \\ m = 0,2,\ldots \\ \dfrac{j}{d+h}\displaystyle\int_h^{h+d} \sin(\beta_n y)\cdot q_m(|y|\ge h)\,dy = \dfrac{j}{d+h}\int_h^{h+d} \sin(\beta_n y)\dfrac{\cos(\alpha_m(y-h-d))}{\cos(\alpha_m d)}\,dy \;; \\ m = 1,3,\ldots \end{cases} \tag{13}$$

Their values are :

$$S_{m,n} = \begin{cases} \dfrac{1}{2(1+d/h)\cos(\varepsilon_m h)}\left(\dfrac{\sin(\beta_n-\varepsilon_m)h}{(\beta_n-\varepsilon_m)h}+\dfrac{\sin(\beta_n+\varepsilon_m)h}{(\beta_n+\varepsilon_m)h}\right) \;;\; m=0,2,\ldots \\ \dfrac{j}{2(1+d/h)\sin(\varepsilon_m h)}\left(\dfrac{\sin(\beta_n-\varepsilon_m)h}{(\beta_n-\varepsilon_m)h}-\dfrac{\sin(\beta_n+\varepsilon_m)h}{(\beta_n+\varepsilon_m)h}\right) \;;\; m=1,3,\ldots \end{cases} \tag{14}$$

$$R_{m,n} = \begin{cases} \dfrac{1}{2(1+d/h)\cos(\alpha_m d)}\cdot \;;\; m=0,2,\ldots \\ \left(\dfrac{2\beta_n h}{(\beta_n^2-\alpha_m^2)h^2}\sin(\beta_n h(1+d/h)) - \dfrac{\sin(\beta_n h+\alpha_m d)}{(\beta_n-\alpha_m)h} - \dfrac{\sin(\beta_n h-\alpha_m d)}{(\beta_n+\alpha_m)h}\right) \\ \dfrac{j}{2(1+d/h)\cos(\alpha_m d)}\cdot \;;\; m=1,3,\ldots \\ \left(\dfrac{-2\beta_n h}{(\beta_n^2-\alpha_m^2)h^2}\cos(\beta_n h(1+d/h)) + \dfrac{\cos(\beta_n h+\alpha_m d)}{(\beta_n-\alpha_m)h} + \dfrac{\cos(\beta_n h-\alpha_m d)}{(\beta_n+\alpha_m)h}\right) \end{cases} \tag{15}$$

Introduce the auxiliary amplitudes :

$$X_m =: B_m - C_m \quad ; \quad Y_m =: B_m\, e^{-\gamma_m L} - C_m\, e^{+\gamma_m L} \tag{16}$$

The boundary conditions of field matching at x=0 and x=L give for them the coupled systems of linear equations :

$$\sum_{m'} X_{m'}\left[-j\frac{\gamma_{m'}}{k_0}\sum_n \frac{k_0}{\kappa_n}\left(S_{m,n}-\frac{j}{\Gamma_{an}Z_{an}}R_{m,n}\right)\left(S_{m',n}+\frac{j}{\Gamma_{an}Z_{an}}R_{m',n}\right)+\right.$$
$$\left.+\,\delta_{m',m}(-1)^m\frac{H}{T}M_m\frac{1+e^{-2\gamma_m L}}{1-e^{-2\gamma_m L}}\right]-2(-1)^m\frac{H}{T}M_m\frac{e^{-\gamma_m L}}{1-e^{-2\gamma_m L}}\cdot Y_m= \tag{17}$$
$$=2P_1\left(S_{m,0}-\frac{j}{\Gamma_{an}Z_{an}}R_{m,0}\right)\quad;\quad\begin{cases}m,m'=0,1,2,\ldots\\ n=0,\pm1,\pm2\ldots\end{cases}$$

$$\sum_{m'} Y_{m'}\left[-j\frac{\gamma_{m'}}{k_0}\sum_n \frac{k_0}{\kappa_n}\left(S_{m,n}-\frac{j}{\Gamma_{an}Z_{an}}R_{m,n}\right)\left(S_{m',n}+\frac{j}{\Gamma_{an}Z_{an}}R_{m',n}\right)+\right.$$
$$\left.+\,\delta_{m',m}(-1)^m\frac{H}{T}M_m\frac{1+e^{-2\gamma_m L}}{1-e^{-2\gamma_m L}}\right]-2(-1)^m\frac{H}{T}M_m\frac{e^{-\gamma_m L}}{1-e^{-2\gamma_m L}}\cdot X_m=0 \tag{18}$$

With their solutions the amplitudes are evaluated as follows :

$$A_n=\delta_{n,0}P_1+j\frac{k_0}{\kappa_n}\sum_m X_m\frac{\gamma_m}{k_0}\left(S_{m,n}+\frac{j}{\Gamma_{an}Z_{an}}R_{m,n}\right)$$
$$D_n=-j\frac{k_0}{\kappa_n}\sum_m Y_m\frac{\gamma_m}{k_0}\left(S_{m,n}+\frac{j}{\Gamma_{an}Z_{an}}R_{m,n}\right) \tag{19}$$
$$B_m=\frac{X_m-Y_m\,e^{-\gamma_m L}}{1-e^{-2\gamma_m L}}\quad;\quad C_m=\frac{X_m\,e^{-\gamma_m L}-Y_m}{1-e^{-2\gamma_m L}}\,e^{-\gamma_m L}$$

The effective incident and transmitted sound powers are evaluated as in section J.23 , and the transmission coefficient follows from there.

J.25 Splitter type silencer with splitters of porous layers covered with a foil

Ref.: Mechel, [J.12]

The arrangement and sound incidence are as in the previous section J.24 , but the splitters are covered with a (poro-elastic) foil having a partition impedance Z_s .

The field formulations in the Zones I and III remain as in section J.24 . The field in a splitter duct and in the porous layer is formulated as :

$$p_{II\mu}(x,y_\nu) = 2\cos(\nu\beta_0 T)\sum_{m=0}^{\infty}[B_m\, e^{-\gamma_m x} + C_m\, e^{+\gamma_m (x-L)}]\cdot q_m(y_\nu) \tag{1}$$

with the mode profiles :

$$q_m(y) = \\ = \begin{cases} s_{|y|}(0,h)\cdot\cos(\varepsilon_m y) + s_{|y|}(h,h+d)\cdot b_m\cos(\alpha_m(y-h-d)) & ;\ m=0,2,4\ldots \\ s_{|y|}(0,h)\cdot\sin(\varepsilon_m y) + s_{|y|}(h,h+d)\cdot\dfrac{y}{|y|}\cdot b_m\cos(\alpha_m(y-h-d)) & ;\ m=1,3,5\ldots \end{cases} \tag{2}$$

($s_y(a,b)$ the switch function as in J.24). The wave equations in the splitter duct and in the porous layer imply :

$$(\alpha_m h)^2 = (\varepsilon_m h)^2 - (k_0 h)^2\left(1+\Gamma_{an}^2\right) \tag{3}$$

and either the $\varepsilon_m h$ or the $\alpha_m h$ are solutions of :

$$\begin{vmatrix} \varepsilon_m h\cdot\sin(\varepsilon_m h) & \dfrac{j\alpha_m h\cdot\sin(\alpha_m d)}{\Gamma_{an}Z_{an}} \\ \left(k_0 h\cdot\cos(\varepsilon_m h) + \dfrac{jZ_s}{Z_0}\varepsilon_m h\cdot\sin(\varepsilon_m h)\right) & -k_0 h\cdot\cos(\alpha_m d) \end{vmatrix} = 0\ ;\ m=0,2,4,\ldots \tag{4}$$

$$\begin{vmatrix} \varepsilon_m h\cdot\cos(\varepsilon_m h) & \dfrac{-j\alpha_m h\cdot\sin(\alpha_m d)}{\Gamma_{an}Z_{an}} \\ \left(k_0 h\cdot\sin(\varepsilon_m h) - \dfrac{jZ_s}{Z_0}\varepsilon_m h\cdot\cos(\varepsilon_m h)\right) & -k_0 h\cdot\cos(\alpha_m d) \end{vmatrix} = 0\ ;\ m=1,3,\ldots \tag{5}$$

The amplitudes B_m, C_m are solutions of the coupled systems of linear equations :

$$\sum_m B_m\left[\left(T_{m,n}+Q_{m,n}\right) - j\frac{\gamma_m}{\kappa_n}\left(T_{m,n}+\frac{j}{\Gamma_{an}Z_{an}}Q_{m,n}\right)\right] + \\ + C_m\, e^{-\gamma_m L}\left[\left(T_{m,n}+Q_{m,n}\right) + j\frac{\gamma_m}{\kappa_n}\left(T_{m,n}+\frac{j}{\Gamma_{an}Z_{an}}Q_{m,n}\right)\right] = 2\delta_{0,n}\cdot P_l \tag{6}$$

$$\sum_m B_m\, e^{-\gamma_m L}\left[\left(T_{m,n}+Q_{m,n}\right) + j\frac{\gamma_m}{\kappa_n}\left(T_{m,n}+\frac{j}{\Gamma_{an}Z_{an}}Q_{m,n}\right)\right] + \\ + C_m\left[\left(T_{m,n}+Q_{m,n}\right) - j\frac{\gamma_m}{\kappa_n}\left(T_{m,n}+\frac{j}{\Gamma_{an}Z_{an}}Q_{m,n}\right)\right] = 0 \tag{7}$$

with mode coupling coefficients :

$$T_{m,n} = \begin{cases} S_{m,n}\cos(\varepsilon_m h) & ;\ m=0,2,4\ldots \\ S_{m,n}\sin(\varepsilon_m h) & ;\ m=1,3,5\ldots \end{cases} \tag{8}$$

$$Q_{m,n} = R_{m,n} b_m \cos(\alpha_m d) \quad ; \quad m = 0,1,2\ldots \tag{9}$$

where :

$$b_m = \begin{cases} j\Gamma_{an} Z_{an} \dfrac{\varepsilon_m \sin(\varepsilon_m h)}{\alpha_m \sin(\alpha_m d)} & ; \quad m = 0,2,4,\ldots \\ -j\Gamma_{an} Z_{an} \dfrac{\varepsilon_m \cos(\varepsilon_m h)}{\alpha_m \sin(\alpha_m d)} & ; \quad m = 1,3,5,\ldots \end{cases} \tag{10}$$

With the solutions B_m, C_m the other amplitudes follow as :

$$\begin{aligned} A_n &= -\delta_{0,n} \cdot P_1 + \sum_m \left(B_m + C_m\, e^{-\gamma_m L}\right)\left(T_{m,n} + Q_{m,n}\right) \\ D_n &= \sum_m \left(B_m e^{-\gamma_m L} + C_m\right)\left(T_{m,n} + Q_{m,n}\right) \end{aligned} \tag{11}$$

J.26 Lined duct corners and junctions

Ref.: Mechel, [J.13]

See also Section J.41 about TV-splitters.

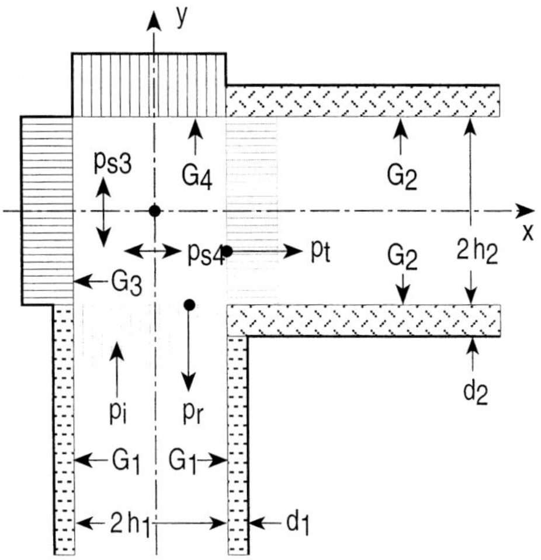

Two lined ducts i= 1,2 form a corner. The corner walls i= 3,4 opposite the ducts are lined too. All linings are supposed to be locally reacting (for ease of formulation, mainly) with surface admittances G_i .

Let the incident wave p_i be the μ-th mode of the duct i=1 . Each of the corner linings with G_3, G_4 , when mirror-reflected at the y axis and x axis, respectively, will form a fictitious lined duct i= 3,4 .

The reflected wave p_r in the duct i=1 is formulated as a mode sum of that duct, and the transmitted wave p_t in the duct i=2 is formulated as mode sum of the duct i=2 . The scattered waves p_{s3} , p_{s4} in the corner area are written as mode sums of the fictitious ducts i= 3,4 .

$$A_k = \frac{1}{N1_k}\Big[-\delta_{\mu,k}\cdot P_i\cdot N1_\mu + + \sum_\alpha B_\alpha(1+R_\alpha)\cdot S_{\alpha,k} + \sum_ß C_ß\cdot q4_ß(-h_2)\Big(e^{-\gamma 4_ß h_1}\cdot Ia_{ß,k} + R_ß\, e^{+\gamma 4_ß h_1}\cdot Ib_{ß,k}\Big)\Big] \quad (13)$$

$$D_k = \frac{1}{N2_k}\sum_\alpha B_\alpha q3_\alpha(h_1)\Big(e^{-\gamma 3_\alpha h_2}\cdot IB_{\alpha,k} + R_\alpha\, e^{+\gamma 3_\alpha h_2}\cdot IA_{\alpha,k}\Big) + \sum_ß C_ß\cdot(1+R_ß)\cdot T_{ß,k} \quad (14)$$

The linings in the numerical examples shown are simple glass fibre layers of thickness d_i with flow resistivity Ξ_i, made locally absorbing (if a duct or corner wall is hard, then d_i=0 and no Ξ value given).

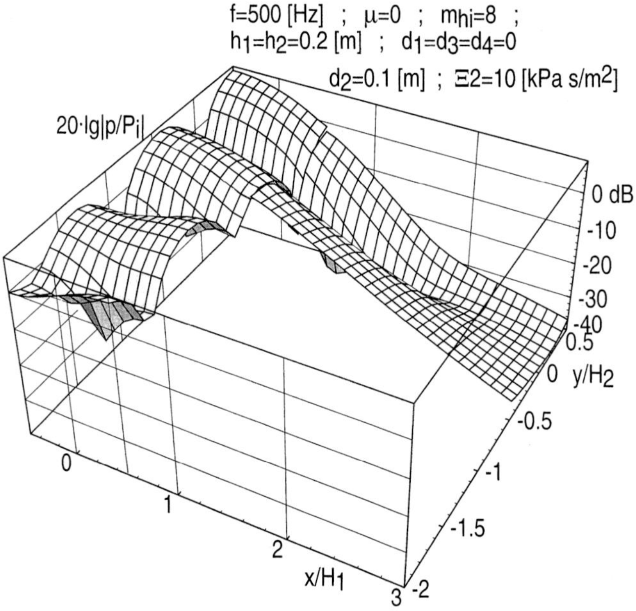

Sound pressure level in two ducts and their corner. The entrance duct i=1 and the corner walls are hard, only the exit duct i=2 is lined. Because the standing wave pattern in the corner agrees well with the 1st higher mode pattern in the exit duct, this higher mode is predominantly excited, and a high extra corner transmission loss is produced (as compared with the least attenuated mode propagation loss).

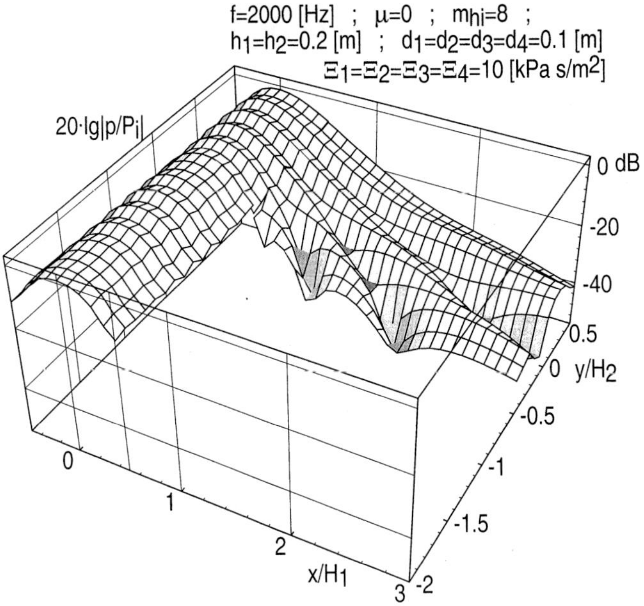

Sound pressure level in two lined ducts and their corner. Both ducts and the corner walls are equally lined.

T-joints and cross-joints of ducts can be approximately evaluated with the present method, if one or both corner wall linings are attributed a lining admittance $G_i = 1/Z_0$. See the Reference for a more precise method.

J.27 Sound radiation from a lined duct orifice

Ref.: Mechel, [J.14, J.15]

A two-dimensional, flat duct of width $2h$ with locally reacting lining of surface admittance G has its orifice in a hard baffle wall. The μ-th duct mode p_μ is incident on the orifice.

Cartesian co-ordinates x,y are used inside the duct, outside the duct an elliptic-hyperbolic system of co-ordinates ρ,ϑ is applied. The orifice is in the plane $x=0$ and $\rho=0$.

The reflected wave p_r inside the duct is composed of duct modes; the radiated field p_s is formulated as a sum of azimuthal and radial Mathieu functions $ce_m(\vartheta), Hc_m^{(2)}(\rho)$.

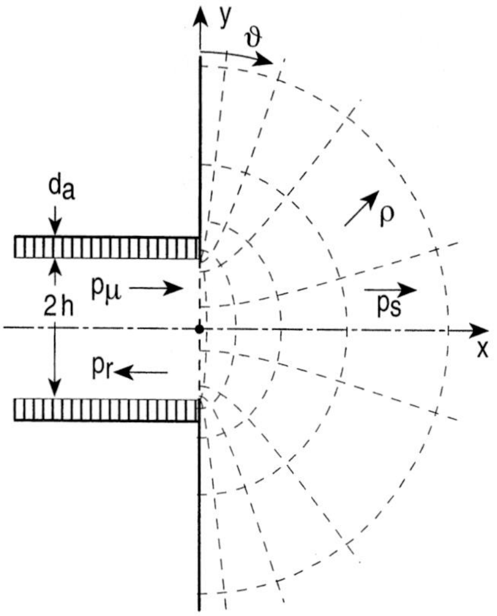

Field formulations :

$$p_\mu(x,y) = P_\mu \cdot \cos(\varepsilon_\mu y) \cdot e^{-\gamma_\mu x} \quad ; \quad v_{\mu x} = -P_\mu \frac{j\gamma_\mu}{k_0 Z_0} \cos(\varepsilon_\mu y) \cdot e^{-\gamma_\mu x}$$
$$\gamma_\mu^2 = \varepsilon_\mu^2 - k_0^2 \qquad (1)$$

$$p_r(x,y) = \sum_n A_n \cdot \cos(\varepsilon_n y) \cdot e^{+\gamma_n x} \quad ; \quad v_{nx}(0,y) = \frac{j\gamma_n h}{k_0 h Z_0} A_n \cdot \cos(\varepsilon_n y)$$
$$p_s(\rho,\vartheta) = \sum_m D_m \cdot Hc_m^{(2)}(\rho) \cdot ce_m(\vartheta) \qquad (2)$$

$$v_{m\rho}(0,\vartheta) = v_{mx}(0,\vartheta) = \frac{j}{k_0 Z_0 \cdot h \sin\vartheta} \left. \frac{\partial p_{s,m}}{\partial \rho} \right|_{\rho=0} = \frac{j}{k_0 Z_0 \cdot h \sin\vartheta} D_m \cdot Hc_m'^{(2)}(0) \cdot ce_m(\vartheta) \qquad (3)$$

Both the duct modes and the azimuthal Mathieu functions are orthogonal with norms :

$$N_n := \frac{1}{h} \int_{-h}^{h} q_n^2(y)\, dy = 1 + \frac{\sin(2\varepsilon_n h)}{2\varepsilon_n h}$$
$$\int_0^\pi ce_m^2(\vartheta)\, d\vartheta = \frac{\pi}{2} \qquad (4)$$

Coupling coefficients :

$$R_{m,n} := \int_0^\pi ce_m(\vartheta)\cdot\cos(\varepsilon_n h\cdot\cos\vartheta)\,d\vartheta = \pi\sum_{s\ge 0} A_{2s}\cdot(-1)^s J_{2s}(\varepsilon_n h) \tag{5}$$

where $J_k(z)$= Bessel function, and A_{2s}= Fourier coefficients for the representation of $ce_m(\vartheta)$ (see [J.15]).

The field matching in the orifice gives a system of linear equations for the A_n :

$$\begin{aligned}\sum_n A_n\cdot\left[\delta_{n,k}\cdot\gamma_k h\, N_k - \frac{2}{\pi}\sum_m \frac{Hc'^{(2)}_m(0)}{Hc^{(2)}_m(0)} R_{m,k}\cdot R_{m,n}\right] &= \\ = P_\mu\cdot\left[\delta_{\mu,k}\cdot\gamma_\mu h\, N_\mu + \frac{2}{\pi}\sum_m \frac{Hc'^{(2)}_m(0)}{Hc^{(2)}_m(0)} R_{m,k}\cdot R_{m,\mu}\right]\end{aligned} \tag{6}$$

(a prime indicates the derivative). With its solutions the D_m can be determined from :

$$D_m = \frac{2}{\pi Hc^{(2)}_m(0)}\sum_n \left(\delta_{\mu,n}P_\mu + A_n\right)\cdot R_{m,n} \tag{7}$$

A "radiation loss" can be defined by $\Delta L = -10\cdot\lg(\tau_\mu)$ with the transmission coefficient τ_μ being the ratio of the radiated effective power Π'_s to the effective incident power Π'_μ of the μ-th duct mode :

$$\begin{aligned}\Pi_\mu &= \frac{1}{2}\int_{-h}^{h} p_\mu(0,y)\cdot v^*_{\mu x}(0,y)\,dy = \frac{1}{2}\frac{j\gamma^*_\mu}{k_0 Z_0}|P_\mu|^2\int_{-h}^{h}|\cos(\varepsilon_\mu y)|^2\,dy \\ &= \frac{1}{2}\frac{j\gamma^*_\mu h}{k_0 Z_0}|P_\mu|^2\left(\frac{\sin(2\varepsilon'_\mu h)}{2\varepsilon'_\mu h} + \frac{\sinh(2\varepsilon''_\mu h)}{2\varepsilon''_\mu h}\right)\end{aligned} \tag{8}$$

$$\begin{aligned}\Pi_s &= \frac{1}{2}\int_{-h}^{h} p_s(0,y)\cdot v^*_{sx}(0,y)\,dy \\ &= \frac{h}{2}\int_0^\pi p_s(0,\vartheta)\cdot v^*_{s\rho}(0,\vartheta)\,\sin\vartheta\,d\vartheta \\ &= \frac{\pi}{4k_0 Z_0}\sum_m |D_m|^2\,Yc'_m(0)\cdot\left(Jc_m(0) - j\,Yc_m(0)\right)\end{aligned} \tag{9}$$

where $Jc_m(z)$, $Yc_m(z)$ are Mathieu-Bessel and Mathieu-Neumann functions associated with $ce_m(\vartheta)$. The transmission coefficient is thus (writing $\varepsilon_n = \varepsilon'_n + j\varepsilon''_n$; $\gamma_n = \gamma'_n + j\gamma''_n$; $\gamma^*_n = \gamma'_n - j\gamma''$):

$$\tau_\mu = \frac{\Pi'_s}{\Pi'_\mu} = \frac{\pi}{2\gamma''_\mu h} \frac{\sum\limits_m \frac{|D_m|^2}{|P_\mu|^2} Yc'_m(0)\cdot Jc_m(0)}{\left(\frac{\sin(2\varepsilon'_\mu h)}{2\varepsilon'_\mu h} + \frac{\sinh(2\varepsilon''_\mu h)}{2\varepsilon''_\mu h}\right)} \tag{10}$$

In the special case of a hard duct :

$$\tau_\mu = \frac{\pi\delta_\mu}{4\sqrt{(k_0h)^2-(\mu\pi)^2}} \sum_m \frac{|D_m|^2}{|P_\mu|^2} Yc'_m(0)\cdot Jc_m(0) \quad ; \quad \delta_\mu = \begin{cases} 1 \; ; \; \mu=0 \\ 2 \; ; \; \mu>0 \end{cases} \tag{11}$$

(with $\gamma_\mu h = \sqrt{(\mu\pi)^2-(k_0h)^2}$).

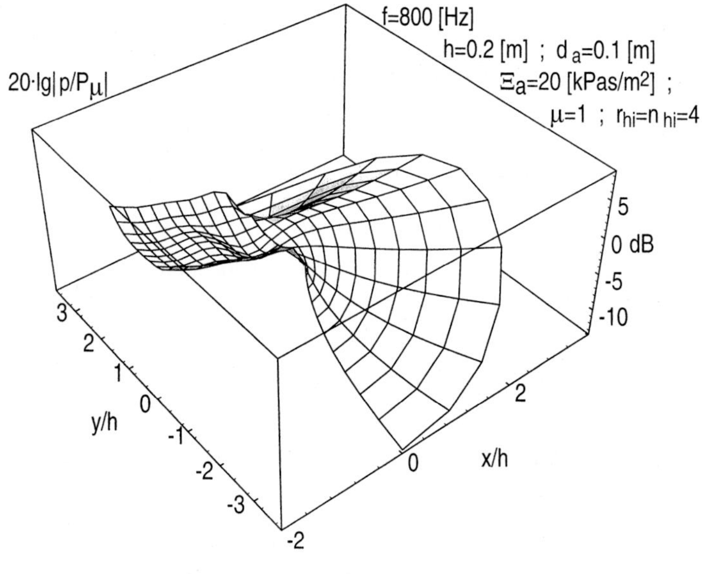

3D-plot of sound pressure level inside and behind a lined duct.

Approximate determination of the radiation loss ΔL :

The radiation loss ΔL of the orifice (in a baffle wall) of a lined duct can be evaluated approximately by :

$$\Delta L = -10\cdot\lg(1-|R|^2) \quad ; \quad R = \frac{Z_r - Z_\mu}{Z_r + Z_\mu}$$
$$Z_\mu = jk_0Z_0/\gamma_\mu \tag{12}$$

Therein Z_μ is the axial wave impedance of the incident μ-th mode, and Z_r is the radiation impedance either of a piston radiator with an area equal to the orifice area (if μ belongs to the lowest or least attenuated mode), or of a cylindrical radiator with radius $a = 2h/\pi$ (see sections F.4 , F.7 for radiation impedances).

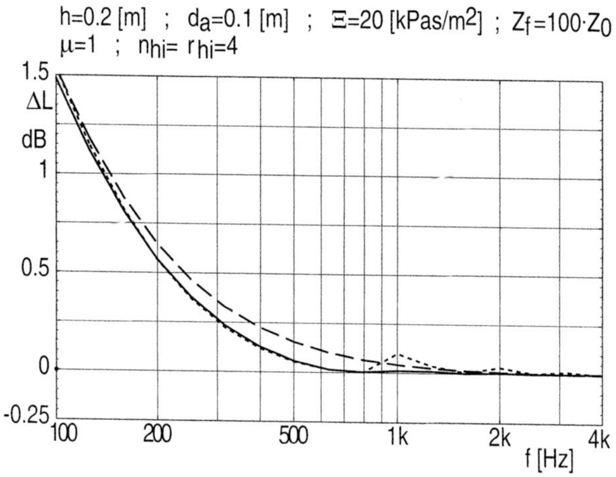

Radiation loss ΔL of the orifice of a duct, lined with a layer of glass fibres, covered with a resistive foil having a flow resistance Z_f.

Full: with a piston radiator;

long dash: with a cylindrical radiator;

short dash: with elliptic co-ordinates.

The radiation loss depends on the volume angle Ω into which the orifice radiates ($\Omega = 4\pi$: free space; $\Omega = 2\pi$: orifice in a baffle wall; $\Omega = \pi$: orifice in the corner of two walls; $\Omega = \pi/2$: orifice in the corner of three walls).

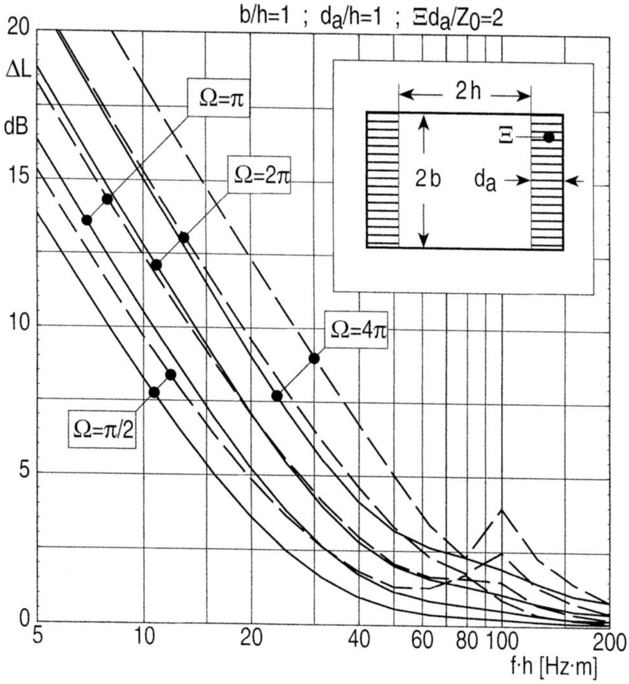

Radiation loss ΔL of duct orifices of a hard duct (full) and of a lined duct (dashed) for different radiation angles Ω. Evaluated with the radiation impedance Z_r of a hemi-spherical radiator with radius $a = \sqrt{S/\Omega}$ (S= orifice area).

J.28 Conical duct transitions; special case: hard walls

Ref.: Mechel, [J.16]

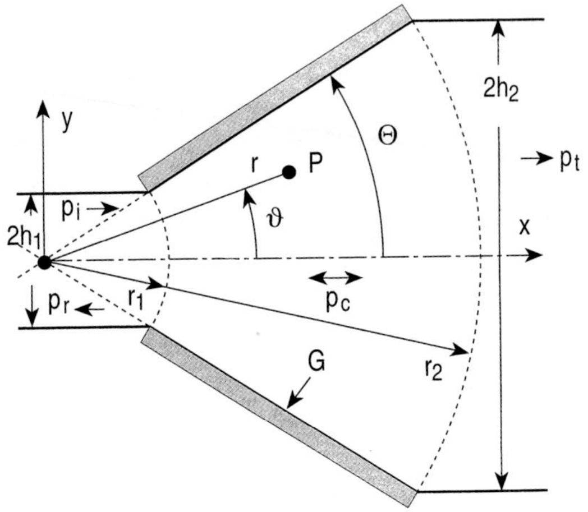

It seems to appropriate to describe the sound field in conical (wedge-shaped) duct transitions in cylindrical co-ordinates r,ϑ , because the flanks would then be on co-ordinate surfaces, and to separate modes in them as $p(r,\vartheta)= T(\vartheta)\cdot R(r)$ with azimuthal profiles : (1)

$$T(\vartheta)=\begin{cases}\cos(\eta\vartheta) & ; \text{ symmetrical}\\ \sin(\eta\vartheta) & ; \text{ antisymmetrical}\end{cases}$$

for symmetrical or anti-symmetrical distributions relative to the x axis. Such modes would be orthogonal over ϑ and, therefore, would be suitable for modal analysis of the fields. If the lining of the cone is locally reacting with a surface admittance G , the azimuthal wave numbers η have to be solutions of the equation :

$$j\,k_0 r\cdot\Theta\, Z_0 G=\begin{cases}(\eta\Theta)\cdot\tan(\eta\Theta) & ; \text{ symmetrical}\\ -(\eta\Theta)\cdot\cot(\eta\Theta) & ; \text{ anti}-\text{symmetrical}\end{cases} \qquad (2)$$

In general $\eta=\eta(r)$, and this prevents a separation of the mode into factors depending on only one co-ordinate.

Special cases with separation are :

- $G=0$ (i.e. hard flank) : $$\eta\Theta=\begin{cases}m\pi & ; \text{ symmetrical}\\ (m+1/2)\pi & ; \text{ anti}-\text{symmetrical}\end{cases} ;\quad m=0,1,2,\ldots \qquad (3)$$

- $G=\infty$ (i.e. soft flank) $$\eta\Theta=\begin{cases}(m+1/2)\pi & ; \text{ symmetrical}\\ m\pi & ; \text{ anti}-\text{symmetrical}\end{cases} ;\quad m=0,1,2,\ldots \qquad (4)$$

- $G=G(r)\sim 1/r$ $\quad \eta\Theta=\text{const}(r) \qquad (5)$

- Parallel walls (in distance $2h$) with characteristic equation :

$$j\,k_0 h\cdot Z_0 G=\begin{cases}(\eta h)\cdot\tan(\eta h) & ; \text{ symmetrical}\\ -(\eta h)\cdot\cot(\eta h) & ; \text{ anti}-\text{symmetrical}\end{cases} \qquad (6)$$

In these cases the radial part of the wave equation becomes the Bessel differential equation :

$$\left[\frac{d^2}{d(k_0r)^2}+\frac{1}{k_0r}\frac{d}{d(k_0r)}+\left(\kappa^2-\frac{\eta^2}{(k_0r)^2}\right)\right]R(k_0r)=0 \quad ; \quad \kappa^2=1-(k_z/k_0)^2 \tag{7}$$

where $k_z \neq 0$ if in the z-direction a field variation with $\cos(k_z z)$, $\sin(k_z z)$, $e^{\pm jk_z z}$ or a linear combination thereof exists. Thus, if $\kappa=1$, the radial factors $R(k_0r)$ are Bessel, Neumann, and Hankel functions of order η.

This section further deals with the *1st special case of hard flanks*; the next sections will present methods for the evaluation of sound fields in lined cones.

Both the entrance duct with height $2h_1$ and the exit duct with height $2h_2$ are assumed to be hard also (if they are lined, mainly the lateral mode wave numbers in the ducts $\varepsilon_{i,n}$ have to be solutions of the characteristic equations in these ducts). The terminating ducts are infinite; the μ-th mode of the entrance duct is the incident wave p_i; it is assumed to be a symmetrical mode. The reflected wave p_r in the entrance duct, the transmitted wave p_t in the exit duct, and the field p_c in the cone are formulated as mode sums. The fields are matched with respect to their pressures and radial particle velocities $v_r = v_x\cdot\cos\vartheta + v_y\cdot\sin\vartheta$ at the arcs r_i; $i=1,2$; with $h_i = r_i\cdot\sin\Theta$. On these arcs is : $x = r_i\cdot\cos\vartheta$; $y = r_i\cdot\sin\vartheta$.

Field formulations :

$$\begin{aligned}
p_i(x,y) &= P_i\cdot\cos(\varepsilon_{1\mu}y)\cdot e^{-\gamma_{1\mu}x} \\
p_r(x,y) &= \sum_{n\geq 0} A_n\cdot\cos(\varepsilon_{1n}y)\cdot e^{+\gamma_{1n}x} \quad ; \quad \varepsilon_{in}h_i = n\cdot\pi \quad ; \quad \gamma_{in}^2=\varepsilon_{in}^2-k_0^2 \\
p_t(x,y) &= \sum_{n\geq 0} D_n\cdot\cos(\varepsilon_{2n}y)\cdot e^{-\gamma_{2n}x} \\
p_c(x,y) &= \sum_{m\geq 0}\cos(\eta_m\vartheta)\cdot\left(B_m\cdot J_{\eta_m}(k_0r)+C_m\cdot Y_{\eta_m}(k_0r)\right) \quad ; \quad \eta_m\Theta = m\cdot\pi
\end{aligned} \tag{8}$$

$$Nc_m := \frac{1}{2\Theta}\int_{-\Theta}^{+\Theta}\cos^2(\eta_m\vartheta)\,d\vartheta = \frac{1}{2}\left(1+\frac{\sin(2\eta_m\Theta)}{2\eta_m\Theta}\right)=\frac{1}{\delta_m} \quad ; \quad \delta_m=\begin{cases}1\,;m=0\\2\,;m>0\end{cases} \tag{9}$$

Introduce the integrals (i=1,2) :

$$\begin{aligned}
Ii_{n,k}^{(\pm)}(r_i) &:= \frac{1}{2\Theta}\int_{-\Theta}^{+\Theta}\cos(\varepsilon_{in}r_i\sin\vartheta)\cdot e^{\pm\gamma_{in}r_i\cos\vartheta}\cdot\cos(\eta_k\vartheta)\,d\vartheta \\
Ji_{n,k}^{(\pm)}(r_i) &:= \frac{1}{2\Theta}\int_{-\Theta}^{+\Theta}\cos(\varepsilon_{in}r_i\sin\vartheta)\cdot e^{\pm\gamma_{in}r_i\cos\vartheta}\cdot\cos(\eta_k\vartheta)\cdot\cos\vartheta\,d\vartheta \\
Ki_{n,k}^{(\pm)}(r_i) &:= \frac{1}{2\Theta}\int_{-\Theta}^{+\Theta}\sin(\varepsilon_{in}r_i\sin\vartheta)\cdot e^{\pm\gamma_{in}r_i\cos\vartheta}\cdot\cos(\eta_k\vartheta)\cdot\sin\vartheta\,d\vartheta
\end{aligned} \tag{10}$$

The integrals must be evaluated by numerical integration.

(because the integrands are even in ϑ, they can be evaluated as $\frac{1}{2\Theta}\int_{-\Theta}^{+\Theta}\ldots d\vartheta = \frac{1}{\Theta}\int_{0}^{+\Theta}\ldots d\vartheta$)

Application of the operator $$\frac{1}{2\Theta}\int_{-\Theta}^{+\Theta}\ldots\cdot\cos(\eta_k\vartheta)\,d\vartheta \tag{11}$$

on the boundary condition for the sound pressure $p_i(r_1,\vartheta)+p_r(r_1,\vartheta)=p_c(r_1,\vartheta)$ on the arc with r_1 gives the system of equations :

$$P_i\cdot I1^{(-)}_{\mu,k}(r_1)+\sum_{n\ge 0}A_n\cdot I1^{(+)}_{n,k}(r_1)=Nc_k\left[B_k\cdot J_{\eta_k}(k_0r_1)+C_k\cdot Y_{\eta_k}(k_0r_1)\right] \tag{12}$$

The same operator applied to the boundary condition for the radial particle velocity at r_1 leads to (the prime indicates the derivative) :

$$\begin{aligned}&P_i\cdot\left[\frac{-\gamma_{1\mu}}{k_0}J1^{(-)}_{\mu,k}(r_1)-\frac{\varepsilon_{1\mu}}{k_0}K1^{(-)}_{\mu,k}(r_1)\right]+\sum_{n\ge 0}A_n\cdot\left[\frac{\gamma_{1n}}{k_0}J1^{(+)}_{n,k}(r_1)-\frac{-\varepsilon_{1n}}{k_0}K1^{(+)}_{n,k}(r_1)\right]:=\\ &P_i\cdot b_{\mu,k}+\sum_{n\ge 0}A_n\cdot a_{n,k}=Nc_k\left[B_k\cdot J'_{\eta_k}(k_0r_1)+C_k\cdot Y'_{\eta_k}(k_0r_1)\right]\end{aligned} \tag{13}$$

On the arc with r_2 drop terms with P_i as factor; substitute $r_1\to r_2$; $A_n\to D_n$; $\varepsilon_{1n}\to\varepsilon_{2n}$; $\gamma_{1n}\to-\gamma_{2n}$; and substitute the integrals correspondingly; this gives :

$$\sum_{n\ge 0}D_n\cdot I2^{(-)}_{n,k}(r_2)=Nc_k\left[B_k\cdot J_{\eta_k}(k_0r_2)+C_k\cdot Y_{\eta_k}(k_0r_2)\right] \tag{14}$$

$$\begin{aligned}&\sum_{n\ge 0}D_n\cdot\left[\frac{-\gamma_{2n}}{k_0}J2^{(-)}_{n,k}(r_2)-\frac{-\varepsilon_{2n}}{k_0}K2^{(-)}_{n,k}(r_2)\right]:=\\ &\sum_{n\ge 0}D_n\cdot d_{n,k}=Nc_k\left[B_k\cdot J'_{\eta_k}(k_0r_2)+C_k\cdot Y'_{\eta_k}(k_0r_2)\right]\end{aligned} \tag{15}$$

Solve these two equations at r_2 (with fixed but arbitrary integer $k\ge 0$) for B_k, C_k :

$$\begin{aligned}B_k&=\frac{\pi k_0r_2}{2Nc_k}\sum_{n\ge 0}D_n\cdot\left(I2^{(-)}_{n,k}(r_2)\cdot Y'_{\eta_k}(k_0r_2)-d_{n,k}\cdot Y_{\eta_k}(k_0r_2)\right)\\ C_k&=-\frac{\pi k_0r_2}{2Nc_k}\sum_{n\ge 0}D_n\cdot\left(I2^{(-)}_{n,k}(r_2)\cdot J'_{\eta_k}(k_0r_2)-d_{n,k}\cdot J_{\eta_k}(k_0r_2)\right)\end{aligned} \tag{16}$$

This inserted into the equations at r_1 leads to two coupled systems of linear equations for the sets A_n, D_n of amplitudes :

$$\sum_{n\geq 0} A_n \cdot I1_{n,k}^{(+)}(r_1) + \frac{\pi k_0 r_2}{2} \sum_{n\geq 0} D_n \cdot \Big[d_{n,k} \cdot \big(J_{\eta_k}(k_0 r_1) \cdot Y_{\eta_k}(k_0 r_2) - J_{\eta_k}(k_0 r_2) \cdot Y_{\eta_k}(k_0 r_1) \big) + \\ + I2_{n,k}^{(-)}(r_2) \cdot \big(J'_{\eta_k}(k_0 r_2) \cdot Y_{\eta_k}(k_0 r_1) - J_{\eta_k}(k_0 r_1) \cdot Y'_{\eta_k}(k_0 r_2) \big) \Big] = -P_i \cdot I1_{\mu,k}^{(-)}(r_1) \tag{17}$$

$$\sum_{n\geq 0} A_n \cdot a_{n,k} + \frac{\pi k_0 r_2}{2} \sum_{n\geq 0} D_n \cdot \Big[d_{n,k} \cdot \big(J'_{\eta_k}(k_0 r_1) \cdot Y_{\eta_k}(k_0 r_2) - J_{\eta_k}(k_0 r_2) \cdot Y'_{\eta_k}(k_0 r_1) \big) + \\ + I2_{n,k}^{(-)}(r_2) \cdot \big(J'_{\eta_k}(k_0 r_2) \cdot Y'_{\eta_k}(k_0 r_1) - J'_{\eta_k}(k_0 r_1) \cdot Y'_{\eta_k}(k_0 r_2) \big) \Big] = -P_i \cdot b_{\mu,k} \tag{18}$$

The B_k, C_k can be evaluated with the solutions D_n. The sound field is determined.

J.29 Lined conical duct transition, evaluated with stepping duct sections

Ref.: Mechel, [J.17]

This section makes use of the last special case in section J.28 , i.e. it composes a lined (locally reacting) duct cone with stepping duct sections having parallel walls.

All duct sections may have equal linings (for ease of representation) with surface admittance G .

The duct section i=0 is the entrance duct (infinitely long); it sends a mode mix to the stepping sections. The last section i=I is terminated with an admittance G_t .

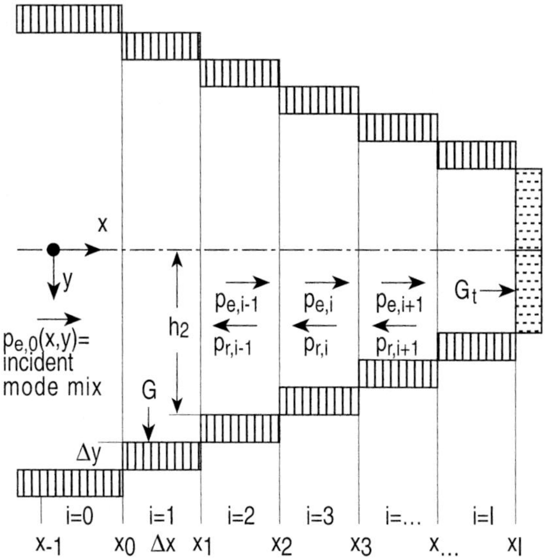

The heads of the steps are assumed to be hard. In each section the forward wave $p_{e,i}$ and the backward wave $p_{r,i}$ are written as sums of modes of that section. The fields are matched at the section limits with their sound pressure and axial particle velocity.

Field formulation :

$$p_{e,i}(x,y) = \sum_{m\geq 0} A_{i,m} \cdot q_{i,m}(y) \cdot e^{-\gamma_{i,m}(x-x_{i-1})}$$
$$p_{r,i}(x,y) = \sum_{m\geq 0} B_{i,m} \cdot q_{i,m}(y) \cdot e^{+\gamma_{i,m}(x-x_{i})} \qquad (1)$$

with lateral mode profiles :

$$q_{i,m}(y) = \begin{cases} \cos(\varepsilon_{i,m} y) & ; \quad \text{symmetrical} \\ \sin(\varepsilon_{i,m} y) & ; \quad \text{anti-symmetrical} \end{cases} \qquad (2)$$

The axial propagation constants $\gamma_{i,m}$ are obtained by $\gamma_{i,m}^2 = \varepsilon_{i,m}^2 - k_0^2 \quad ; \quad \mathrm{Re}\{\gamma_{i,m}\} \geq 0$ (3)
from the lateral wave numbers, and these in turn are solutions of the characteristic equations :

$$(\varepsilon_{i,m} h_i) \cdot \tan(\varepsilon_{i,m} h_i) = j k_0 h_i \cdot G \quad ; \quad \text{symmetrical}$$
$$(\varepsilon_{i,m} h_i) \cdot \cot(\varepsilon_{i,m} h_i) = -j k_0 h_i \cdot G \quad ; \quad \text{anti-symmetrical} \qquad (4)$$

Mode norms are :

$$\frac{1}{h_i}\int_0^{h_i} q_{i,m}(y) \cdot q_{i,n}(y)\, dy = \delta_{m,n} \cdot N_{i,m} \quad ; \quad N_{i,m} = \frac{1}{2}\left(1 \pm \frac{\sin(2\varepsilon_{i,m} h_i)}{2\varepsilon_{i,m} h_i}\right) \qquad (5)$$

and mode coupling coefficients C(i,m; k,n) of the mode of order m in the section i with the mode of order n in the section k :

$$\frac{1}{h_i}\int_0^{h_i} q_{i,m}(y) \cdot q_{k,n}(y)\, dy = C(i,m;k,n)$$
$$C(i,m;k,n) = \frac{1}{2}\left(\frac{\sin(\varepsilon_{i,m} - \varepsilon_{k,n}) h_i}{(\varepsilon_{i,m} - \varepsilon_{k,n}) h_i} \pm \frac{\sin(\varepsilon_{i,m} + \varepsilon_{k,n}) h_i}{(\varepsilon_{i,m} + \varepsilon_{k,n}) h_i}\right) \qquad (6)$$

(± for symmetrical or anti-symmetrical modes, respectively; cross-coupling coefficients between both types of symmetry are zero).

By the special termination of the wedge with an admittance G_t one has:

$$\{B_{I,m}\} = \{M_t\} \circ \{A_{I,m}\} = \left\{r_m \cdot e^{-\gamma_{I,m}\Delta x}\right\} \circ \{A_{I,m}\} \qquad (7)$$

where $\{M_t\}$ is a general coupling matrix (∘ is the symbol for matrix multiplication), which in the present case is a diagonal matrix with the values $r_m \cdot e^{-\gamma_{I,m}\Delta x}$ on the main diagonal, r_m being the modal reflection factors at the exit of the last section i=I :

$$r_m = \frac{g_m - G_t}{g_m + G_t} = \frac{j\gamma_{I,m}/k_0 + G_t}{j\gamma_{I,m}/k_0 - G_t} \qquad (8)$$

and g_m the normalised axial modal admittances of the modes of $p_{e,I}(x,y)$:

$$g_m = Z_0 \frac{vx_{eI,m}(x_I)}{p_{eI,m}(x_I)} = -j\frac{\gamma_{I,m}}{k_0} \tag{9}$$

Converging cone :

The boundary condition for the sound pressure at the entrance $x=x_{i-1}$ of the i-th section $(i \geq 1)$ is :

$$\begin{gathered} p_{e,i-1}(x_{i-1},y)+p_{r,i-1}(x_{i-1},y) \stackrel{!}{=} p_{e,i}(x_{i-1},y)+p_{r,i}(x_{i-1},y) \quad ; \quad 0 \leq y \leq h_i \\ \sum_m \left(A_{i-1,m} \cdot e^{-\gamma_{i-1,m}\Delta x} + B_{i-1,m}\right) \cdot q_{i-1,m}(y) \stackrel{!}{=} \sum_m \left(A_{i,m} + B_{i,m} \cdot e^{-\gamma_{i,m}\Delta x}\right) \cdot q_{i,m}(y) \end{gathered} \tag{10}$$

Application of the operation $\frac{1}{h_i}\int_0^{h_i} \ldots \cdot q_{i,m}(y)\, dy$ on both sides gives the system of equations:

$$\left(A_{i,m} + B_{i,m} \cdot e^{-\gamma_{i,m}\Delta x}\right) \cdot N_{i,m} = \sum_n \left(A_{i-1,n} \cdot e^{-\gamma_{i-1,n}\Delta x} + B_{i-1,n}\right) \cdot C(i,m;i-1,n) \tag{11}$$

It is an upward iteration scheme.

The boundary condition for the axial particle velocity at $x = x_{i-1}$ is :

$$v_{xe,i-1}(x_{i-1},y)+v_{xr,i-1}(x_{i-1},y) \stackrel{!}{=} \begin{cases} 0 & ; \quad h_i \leq y \leq h_{i-1} \\ v_{xe,i}(x_{i-1},y)+v_{xr,i}(x_{i-1},y) & ; \quad 0 \leq y \leq h_i \end{cases}$$

$$\begin{gathered} \sum_m \left(A_{i-1,m} \cdot e^{-\gamma_{i-1,m}\Delta x} - B_{i-1,m}\right) \cdot \gamma_{i-1,m} \cdot q_{i-1,m}(y) \stackrel{!}{=} \\ \begin{cases} 0 \quad ; \quad h_i \leq y \leq h_{i-1} \\ \sum_m \left(A_{i,m} - B_{i,m} \cdot e^{-\gamma_{i,m}\Delta x}\right) \cdot \gamma_{i,m} \cdot q_{i,m}(y) \quad ; \quad 0 \leq y \leq h_i \end{cases} \end{gathered} \tag{12}$$

Application of the operators $\frac{1}{h_i}\int_0^{h_i} \ldots \cdot q_{i,m}(y)\, dy$ right ; $\frac{1}{h_i}\int_0^{h_{i-1}} \ldots \cdot q_{i,m}(y)\, dy$ left

produces the system of equations :

$$\begin{gathered} \left(A_{i,m} - B_{i,m} \cdot e^{-\gamma_{i,m}\Delta x}\right) \cdot \gamma_{i,m} \cdot N_{i,m} = \\ = \frac{h_{i-1}}{h_i} \sum_n \left(A_{i-1,n} \cdot e^{-\gamma_{i-1,n}\Delta x} - B_{i-1,n}\right) \cdot \gamma_{i-1,n} \cdot C(i-1,n;i,m) \end{gathered} \tag{13}$$

One gets, by combination of both systems of equations :

$$A_{i,m}=\frac{1}{2N_{i,m}}\sum_n A_{i-1,n}\cdot e^{-\gamma_{i-1,n}\Delta x}\cdot\left(C(i,m;i-1,n)+C(i-1,n;i,m)\frac{\gamma_{i-1,n}h_{i-1}}{\gamma_{i,m}h_i}\right)+$$
$$+\;B_{i-1,n}\cdot\left(C(i,m;i-1,n)-C(i-1,n;i,m)\frac{\gamma_{i-1,n}h_{i-1}}{\gamma_{i,m}h_i}\right) \tag{14}$$

$$B_{i,m}=\frac{1}{2N_{i,m}\cdot e^{-\gamma_{i,m}\Delta x}}\sum_n A_{i-1,n}\cdot e^{-\gamma_{i-1,n}\Delta x}\cdot\left(C(i,m;i-1,n)-C(i-1,n;i,m)\frac{\gamma_{i-1,n}h_{i-1}}{\gamma_{i,m}h_i}\right)+$$
$$+\;B_{i-1,n}\cdot\left(C(i,m;i-1,n)+C(i-1,n;i,m)\frac{\gamma_{i-1,n}h_{i-1}}{\gamma_{i,m}h_i}\right)$$

The upward iterations begin at $i=1$, where $A_{i-1,m}=A_{0,m}$ have given numerical values, and $B_{0,m}$ are unknown symbols. At any step i of the iteration one will have systems of equations of the form:

$$A_{i,m}=\sum_n a_{i,n}+b_{i,n}\cdot B_{0,n}\quad;\quad B_{i,m}=\sum_n \alpha_{i,n}+\beta_{i,n}\cdot B_{0,n} \tag{15}$$

with numerical $a_{i,n}, b_{i,n}, \alpha_{i,n}, \beta_{i,n}$. The iteration ends with $i=I$ where on the left-hand sides of the iterative equations stand $\{A_{I,m}\}$ and $\{B_{I,m}\}$ which, with the above relation of reflection, reduces to only the $\{A_{I,m}\}$ as yet unknown amplitudes. Thus the equations for $i=I$ are two linear systems of equations in the two sets of amplitudes $\{A_{I,m}\}$, $\{B_{0,n}\}$, and they are inhomogeneous systems of equations because of the numerical terms $a_{I,n}, \alpha_{I,n}$. After they are solved for $\{B_{0,n}\}$ all amplitudes $\{A_{i,m}\}$, $\{B_{i,m}\}$ can be evaluated by insertion. The described iteration with mixed numerical and symbolic expressions can easily be performed with *Mathematica* or other computer programs for both numerical and symbolic mathematics.

Diverging cone :

One gets in a similar way the two downward iterative systems of equations :

$$A_{i-1,m}=\frac{1}{2N_{i-1,m}\cdot e^{-\gamma_{i-1,m}\Delta x}}\cdot\sum_n A_{i,n}\cdot\left(C(i-1,m;i,n)+\frac{\gamma_{i,n}h_i}{\gamma_{i-1,m}h_{i-1}}\cdot C(i,n;i-1,m)\right)+$$
$$+\;B_{i,n}\cdot e^{-\gamma_{i,n}\Delta x}\cdot\left(C(i-1,m;i,n)-\frac{\gamma_{i,n}h_i}{\gamma_{i-1,m}h_{i-1}}\cdot C(i,n;i-1,m)\right)$$

$$B_{i-1,m}=\frac{1}{2N_{i-1,m}}\cdot\sum_n A_{i,n}\cdot\left(C(i-1,m;i,n)-\frac{\gamma_{i,n}h_i}{\gamma_{i-1,m}h_{i-1}}\cdot C(i,n;i-1,m)\right)+ \tag{16}$$
$$+\;B_{i,n}\cdot e^{-\gamma_{i,n}\Delta x}\cdot\left(C(i-1,m;i,n)+\frac{\gamma_{i,n}h_i}{\gamma_{i-1,m}h_{i-1}}\cdot C(i,n;i-1,m)\right)$$

If one begins the iteration with $i=I$, the equations have the form:

$$A_{I-1,m} = \sum_n b_{I,n} \cdot A_{I,n} \quad ; \quad B_{I-1,m} = \sum_n ß_{I,n} \cdot A_{I,n} \tag{17}$$

with still unknown amplitudes $\{A_{I,n}\}$, and in the general step i :

$$A_{i-1,m} = \sum_n b_{i,n} \cdot A_{I,n} \quad ; \quad B_{i-1,m} = \sum_n ß_{i,n} \cdot A_{I,n} \tag{18}$$

with numerical values of the $b_{i,n}, ß_{i,n}$. At the end with $i=1$ one has the known amplitudes $\{A_{0,m}\}$ of the incident modes on the left-hand side of the 1st equation. Thus it can be solved for the $\{A_{I,n}\}$ and with these all other amplitudes $\{A_{i,m}\},\{B_{i,m}\}$ are computed by insertion.

The numerical examples show 3D-plots of the sound pressure level; the spatial co-ordinates are k_0x, k_0y .

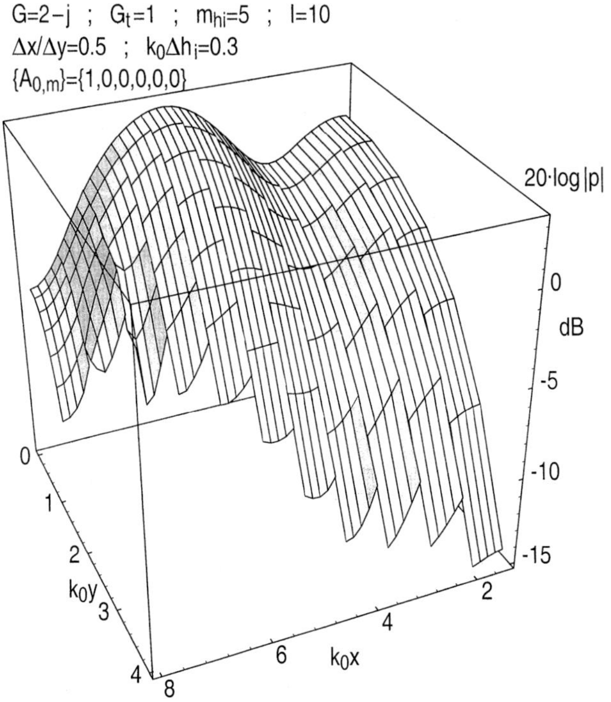

Sound pressure level in a converging cone, with the fundamental duct mode incident.

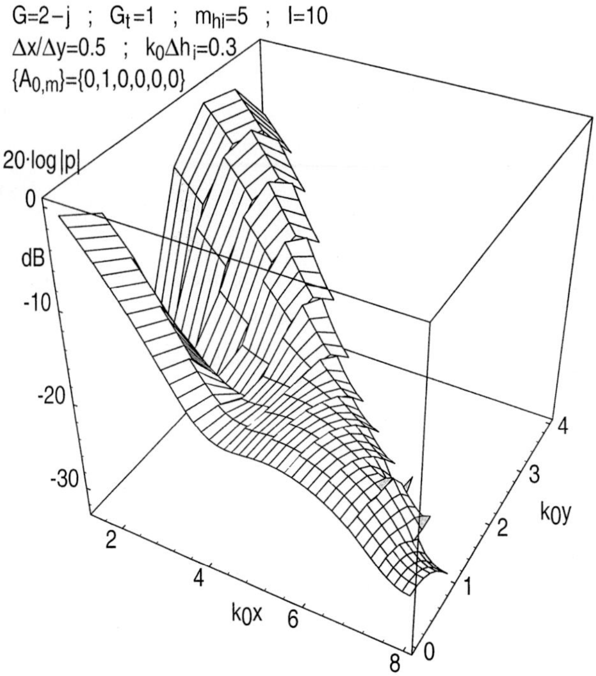

As above, but with the 1st higher duct mode incident.

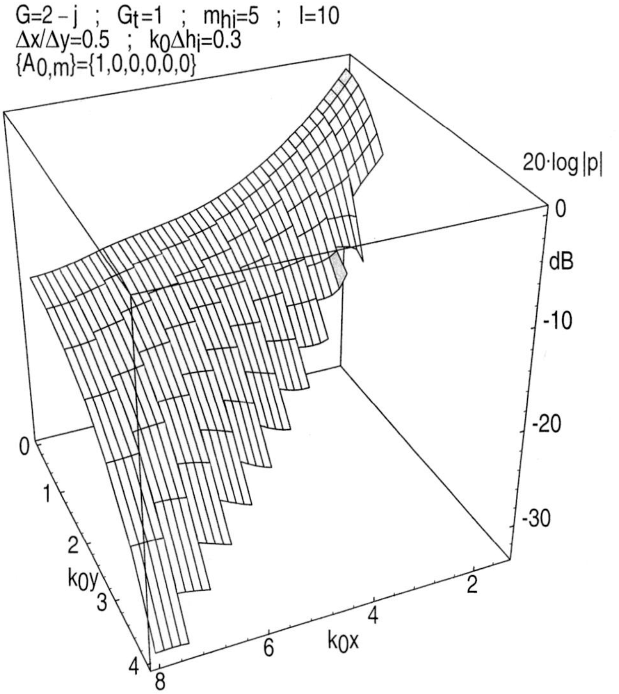

Sound pressure level in a diverging cone, with the fundamental duct mode incident.

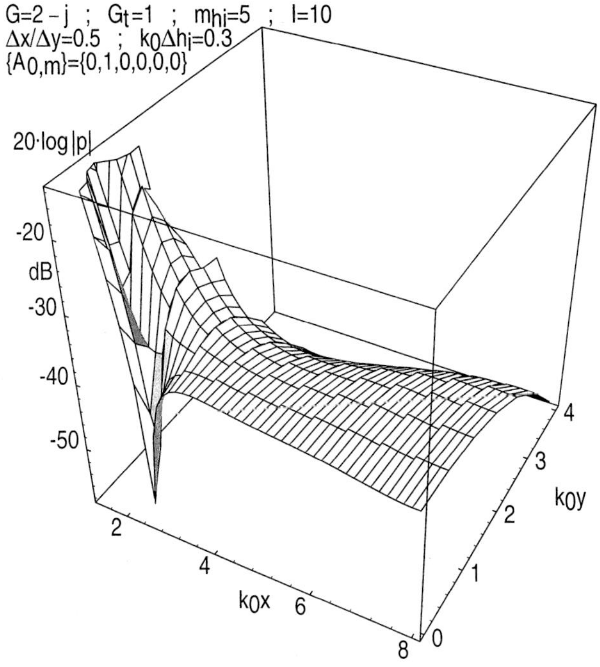

As in the previous graph, but now with the 1st higher mode incident.

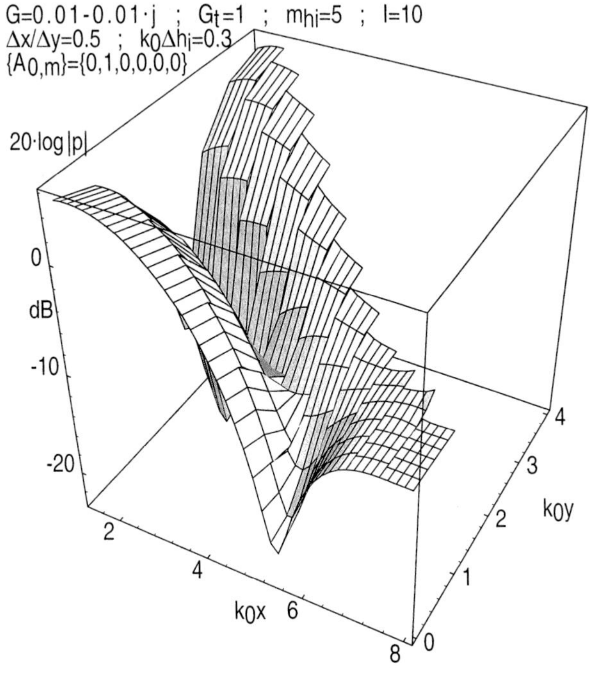

Sound pressure level in a converging, nearly hard cone, with the 1st higher mode incident. It becomes cut-off inside the cone.

J.30 Lined conical duct transition, evaluated with stepping admittance sections

Ref.: Mechel, [J.18]

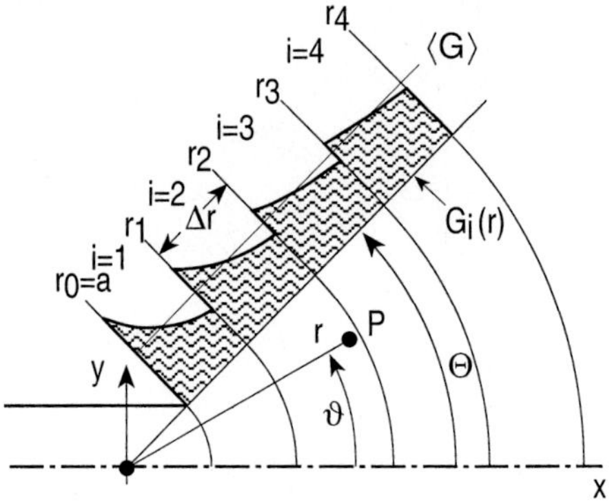

This section applies the 3rd special case of section J.28 : $\eta\Theta$= const(r) if the lining admittance G= G(r)~ 1/r . This condition is not satisfied all over the radial range, but in radial sections i= 1,2,... , such that $G_i(r)$~ 1/r , and the average admittance in the sections equals a given value G :

$$\langle G_i \rangle = G \tag{1}$$

The computational sectional admittances are:

$$G_i(r) = \frac{G}{r}\,\frac{r_i - r_{i-1}}{\ln\left(r_i/r_{i-1}\right)} \tag{2}$$

If the section width Δr is small compared to the wavelength, and if the variation of $G_i(r)$ is not too strong, the sectored lining will approximately produce a sound field like that for a homogeneous lining with admittance G (the conditions mentioned exclude a lining reaching to the origin r =0).

The sound fields in the sections can be written as sums of modes, which are orthogonal over $0 \le \vartheta \le \Theta$:

$$p_i(r,\vartheta,z) = \sum_\eta R_\eta(kr)\cdot T(\eta\vartheta)\cdot Z(k_z z) \tag{3}$$

with $k^2 = k_0^2 - k_z^2$ if the variation $Z(k_z z)$ in the z direction is like $\cos(k_z z)$, $\sin(k_z z)$, $e^{\pm j k_z z}$ or a linear combination thereof, with

$$T(\eta\vartheta) = \begin{cases} \cos(\eta\vartheta) & ;\ \text{symmetrical modes} \\ \sin(\eta\vartheta) & ;\ \text{anti-symmetrical modes} \end{cases} \tag{4}$$

the radial functions $R_\eta(kr)$ being Bessel, Neumann, or Hankel functions of order η , and with $\eta\Theta$ solutions of :

$$\begin{aligned} (\eta\Theta_0)\cdot\tan(\eta\Theta_0) &= jkr\cdot\Theta_0 G \quad ;\ \text{symmetrical modes} \\ (\eta\Theta_0)\cdot\cot(\eta\Theta_0) &= -jkr\cdot\Theta_0 G \quad ;\ \text{anti-symmetrical modes} \end{aligned} \tag{5}$$

or more definitely (with the mode counter n in the i-th section) :

$$(\eta_{i,n}\Theta)\cdot\begin{Bmatrix}\tan\\ \cot\end{Bmatrix}(\eta_{i,n}\Theta)=\pm j\Theta\frac{k_0\Delta r}{\ln(r_i/r_{i-1})}G \qquad \begin{Bmatrix}\text{rigid}\\ \text{soft}\end{Bmatrix}\ \text{flank at}\ \vartheta=0 \tag{6}$$

The mode norms are :

$$N_{i,n}=\frac{1}{\Theta}\int_0^{\Theta}\begin{matrix}\cos^2\\ \sin^2\end{matrix}(\eta_{i,n}\vartheta)\,d\vartheta=\frac{1}{2}\left(1\pm\frac{\sin(2\eta_{i,n}\Theta)}{2\eta_{i,n}\Theta}\right) \tag{7}$$

The sound field in the zone i is formulated as (henceforth only symmetrical modes are assumed; a possible variation in the z direction with $Z(k_z z)$ will be dropped):

$$\begin{aligned} p_i(r,\vartheta)&=\sum_{m\geq 0}\left[A_{i,m}\cdot H^{(1)}_{\eta_{i,m}}(kr)+B_{i,m}\cdot H^{(2)}_{\eta_{i,m}}(kr)\right]\cdot\cos(\eta_{i,m}\vartheta)\\ Z_0\cdot v_{r,i}&=\frac{jk}{k_0}\sum_{m\geq 0}\left[A_{i,m}\cdot H'^{(1)}_{\eta_{i,m}}(kr)+B_{i,m}\cdot H'^{(2)}_{\eta_{i,m}}(kr)\right]\cdot\cos(\eta_{i,m}\vartheta)\end{aligned} \tag{8}$$

(a prime indicates the derivative). The amplitudes $A_{i,m}$, $B_{i,m}$ are determined by field matching at the section limits.

In a simple example of application of the method assume that:

1st: a given radial particle velocity distribution $V_0(\vartheta)$ on the arc r=a ;
(some incident duct mode in the duct in front of r=a would lead to a method as described in the previous section J.29);

2nd: the cone is infinitely long ;
(in practice it is sufficient if it is so long, that the admittance step at the outer zone limit becomes small, so that reflections at it can be neglected, and the cone has an anechoic termination; other terminations are handled as in the section J.29 , see also below).

One needs coupling coefficients between modes of adjacent zones given by the integrals ($T_{i,m}(\vartheta)$ are the azimuthal mode functions):

$$\begin{aligned} X^{(i)}_{m,n}&=\frac{1}{\Theta}\int_0^{\Theta}T_{i,m}(\vartheta)\cdot T_{i+1,n}(\vartheta)\,d\vartheta\\ Y^{(i)}_{m,n}&=\frac{1}{\Theta}\int_0^{\Theta}T_{i,m}(\vartheta)\cdot T_{i-1,n}(\vartheta)\,d\vartheta=X^{(i-1)}_{n,m}\end{aligned} \tag{9}$$

They assume the values if the flank at $\vartheta=0$ is rigid:

$$X^{(i)}_{m,n}=\frac{1}{2}\left[\frac{\sin\left((\eta_{i,m}-\eta_{i+1,n})\Theta\right)}{(\eta_{i,m}-\eta_{i+1,n})\Theta}+\frac{\sin\left((\eta_{i,m}+\eta_{i+1,n})\Theta\right)}{(\eta_{i,m}+\eta_{i+1,n})\Theta}\right] \tag{10}$$

$$Y_{m,n}^{(i)} = \frac{1}{2}\left[\frac{\sin((\eta_{i,m}-\eta_{i-1,n})\Theta)}{(\eta_{i,m}-\eta_{i-1,n})\Theta} + \frac{\sin((\eta_{i,m}+\eta_{i-1,n})\Theta)}{(\eta_{i,m}+\eta_{i-1,n})\Theta}\right] \tag{11}$$

and if that flank at $\vartheta=0$ is soft:

$$X_{m,n}^{(i)} = \frac{1}{2}\left[\frac{\sin((\eta_{i,m}-\eta_{i+1,n})\Theta)}{(\eta_{i,m}-\eta_{i+1,n})\Theta} - \frac{\sin((\eta_{i,m}+\eta_{i+1,n})\Theta)}{(\eta_{i,m}+\eta_{i+1,n})\Theta}\right]$$
$$Y_{m,n}^{(i)} = \frac{1}{2}\left[\frac{\sin((\eta_{i,m}-\eta_{i-1,n})\Theta)}{(\eta_{i,m}-\eta_{i-1,n})\Theta} - \frac{\sin((\eta_{i,m}+\eta_{i-1,n})\Theta)}{(\eta_{i,m}+\eta_{i-1,n})\Theta}\right] \tag{12}$$

The boundary condition (source condition) at r=a is :

$$Z_0 \cdot v_{r,1} = \frac{jk}{k_0}\sum_{m\ge 0}\left[A_{1,m}\cdot H'^{(1)}_{\eta_{1,m}}(ka) + B_{1,m}\cdot H'^{(2)}_{\eta_{1,m}}(ka)\right]\cdot\cos(\eta_{1,m}\vartheta) \overset{!}{=} Z_0 V_0(\vartheta) \tag{13}$$

leading to :

$$\frac{jk\,N_{1,m}}{k_0}\left[A_{1,m}\cdot H'^{(1)}_{\eta_{1,m}}(ka) + B_{1,m}\cdot H'^{(2)}_{\eta_{1,m}}(ka)\right] = \frac{1}{\Theta}\int_0^{\Theta} Z_0 V_0(\vartheta)\cdot\cos(\eta_{1,m}\vartheta)\,d\vartheta \tag{14}$$

with known Fourier coefficients on the right-hand side.

Applying the integral operation $\frac{1}{\Theta}\int_0^{\Theta}\ldots\cdot\cos(\eta_{i+1,m}\vartheta)\,d\vartheta$ on both sides of the boundary condition for the sound pressure at the zone limit r_i between two zones i and i+1 gives :

$$\left[A_{i+1,m}\cdot H^{(1)}_{\eta_{i+1,m}}(kr_i) + B_{i+1,m}\cdot H^{(2)}_{\eta_{i+1,m}}(kr_i)\right]\cdot N_{i+1,m} =$$
$$= \sum_{n\ge 0}\left[A_{i,n}\cdot H^{(1)}_{\eta_{i,n}}(kr_i) + B_{i,n}\cdot H^{(2)}_{\eta_{i,n}}(kr_i)\right]\cdot X_{n,m}^{(i)} \tag{15}$$

and for the radial particle velocity :

$$\left[A_{i+1,m}\cdot H'^{(1)}_{\eta_{i+1,m}}(kr_i) + B_{i+1,m}\cdot H'^{(2)}_{\eta_{i+1,m}}(kr_i)\right]\cdot N_{i+1,m} =$$
$$= \sum_{n\ge 0}\left[A_{i,n}\cdot H'^{(1)}_{\eta_{i,n}}(kr_i) + B_{i,n}\cdot H'^{(2)}_{\eta_{i,n}}(kr_i)\right]\cdot X_{n,m}^{(i)} \tag{16}$$

Elimination of the $B_{i+1,m}$ and use of the Wronski determinant for Hankel functions returns the upward iterative systems of equations:

$$A_{i+1,m} = j\frac{\pi\, kr_i}{4N_{i+1,m}} \sum_{n\geq 0}\Big[A_{i,n}\cdot\Big(H^{(1)}_{\eta_{i,n}}(kr_i)\cdot H'^{(2)}_{\eta_{i+1,m}}(kr_i) - H'^{(1)}_{\eta_{i,n}}(kr_i)\cdot H^{(2)}_{\eta_{i+1,m}}(kr_i)\Big) + \\ + B_{i,n}\cdot\Big(H^{(2)}_{\eta_{i,n}}(kr_i)\cdot H'^{(2)}_{\eta_{i+1,m}}(kr_i) - H'^{(2)}_{\eta_{i,n}}(kr_i)\cdot H^{(2)}_{\eta_{i+1,m}}(kr_i)\Big)\Big]\cdot X^{(i)}_{n,m} \tag{17}$$

$$B_{i+1,m} = -j\frac{\pi\, kr_i}{4N_{i+1,m}} \sum_{n\geq 0}\Big[A_{i,n}\cdot\Big(H^{(1)}_{\eta_{i,n}}(kr_i)\cdot H'^{(1)}_{\eta_{i+1,m}}(kr_i) - H^{(1)}_{\eta_{i+1,m}}(kr_i)\cdot H'^{(1)}_{\eta_{i,n}}(kr_i)\Big) + \\ + B_{i,n}\cdot\Big(H^{(2)}_{\eta_{i,n}}(kr_i)\cdot H'^{(1)}_{\eta_{i+1,m}}(kr_i) - H^{(1)}_{\eta_{i+1,m}}(kr_i)\cdot H'^{(2)}_{\eta_{i,n}}(kr_i)\Big)\Big]\cdot X^{(i)}_{n,m}$$

If we begin with $i=1$, the $B_{1,n}$ on the right-hand sides can be expressed by the (numerical) Fourier coefficients of the particle velocity distribution $Z_0V_0(\vartheta)$ at $r_0=a$ and the symbolic $A_{1,n}$. The equations will have the general form during the iteration :

$$A_{i+1,m} = j\frac{\pi\, kr_i}{4N_{i+1,m}} \sum_{n\geq 0}\left(a_{i,n}\cdot A_{1,n} + b_{i,n}\right) \quad ; \quad B_{i+1,m} = -j\frac{\pi\, kr_i}{4N_{i+1,m}} \sum_{n\geq 0}\left(\alpha_{i,n}\cdot A_{1,n} + \beta_{i,n}\right) \tag{18}$$

where $a_{i,n}$, $b_{i,n}$, $\alpha_{i,n}$, $\beta_{i,n}$ are numerical quantities. When the iteration has proceeded up to a value $i=I$, for which the admittance step at $r=r_I$ is small enough to neglect the inward reflection, i.e. $B_{i\geq I,m}=0$, the equations will have the form :

$$A_{I+1,m} = j\frac{\pi\, kr_I}{4N_{I+1,m}} \sum_{n\geq 0}\left(a_{I,n}\cdot A_{1,n} + b_{I,n}\right) \quad ; \quad 0 = -j\frac{\pi\, kr_I}{4N_{I+1,m}} \sum_{n\geq 0}\left(\alpha_{I,n}\cdot A_{1,n} + \beta_{I,n}\right) \tag{19}$$

Then they are a coupled system of equations for the amplitude sets $A_{1,n}$, $A_{I+1,n}$. After its solution, all other amplitudes follow by insertion.

If the cone is not infinitely long, but ends with some termination at $r=r_I$, this termination will give a prescription of how to express the $B_{I+1,m}$ by the $A_{I+1,m}$, and the procedure remains the same, in principle.

In general, the upper limit n_{hi} of the required mode orders will not be high, except if $V_0(\vartheta)$ has many details.

J.31 Mode mixtures

Ref.: Mechel, [J.19]

Modes are elementary solutions of the wave equation and of the boundary conditions. For some kinds of boundaries they are orthogonal over the duct cross-section, and therefore suited for a synthesis of sound fields in the duct. Like the "science fiction" of a diffuse sound field in room acoustics, it may be useful to define in duct acoustics mode mixtures in which the modes obey some rules of mixing, but may have random phases.

Consider a rectangular hard duct with the duct axis in the z direction and the origin of the transversal co-ordinates x,y in a duct corner. A sound wave propagating in the z direction may be described by :

$$p(x,y,z)=\sum_{m,n} p_{m,n}(x,y,z)=\sum_{m,n} A_{m,n}\cdot q_{m,n}(x,y)\cdot e^{-jk_{m,n}z} \tag{1}$$

with mode profiles (containing both symmetrical and anti-symmetrical modes with respect to the duct central axis) :

$$q_{m,n}(x,y)=\cos(\varepsilon_m x)\cdot\cos(\eta_n y)\quad;\quad \varepsilon_m a=m\cdot\pi\quad;\quad \eta_n b=n\cdot\pi\quad;\quad m,n=0,1,2,\ldots \tag{2}$$

and :

$$k_0^2=\varepsilon_m^2+\eta_n^2+k_{m,n}^2\quad;\quad \kappa_{m,n}^2=\varepsilon_m^2+\eta_n^2=(m\pi/a)^2+(n\pi/b)^2 \tag{3}$$

with mode norms $N_{m,n}$ (S= duct cross-section area) :

$$\iint_S q_{m,n}\cdot q_{\mu,\nu}\,dx\,dy=\begin{cases}0 & ;\ m,n\neq\mu,\nu\\ S\cdot N_{m,n} & ;\ m,n=\mu,\nu\end{cases}\quad;\quad N_{m,n}=\frac{1}{\delta_m\cdot\delta_n}\quad;\quad \delta_k=\begin{cases}1\ ;\ k=0\\ 2\ ;\ k>0\end{cases} \tag{4}$$

The modal angles (relative to the duct axis) are :

$$\Phi_{m,n}=\arccos\frac{k_{m,n}}{k_0}=\arcsin\frac{\kappa_{m,n}}{k_0}\quad;\quad \cos\Phi_{m,n}=\sqrt{1-(m\pi/k_0a)^2-(n\pi/k_0b)^2} \tag{5}$$

Thus the modes have the form:

$$\begin{aligned}p_{m,n}(x,y,z)&=A_{m,n}\cdot\cos\frac{m\pi}{a}x\cdot\cos\frac{n\pi}{b}y\cdot e^{-jk_{m,n}z}\\ v_{zm,n}(x,y,z)&=\frac{k_{m,n}}{k_0Z_0}p_{m,n}(x,y,z)=G_{zm,n}\cdot p_{m,n}(x,y,z)\end{aligned} \tag{6}$$

where $G_{zm,n}$ are the modal axial field admittances :

$$G_{zm,n}=\frac{k_{m,n}}{k_0Z_0}=\frac{1}{Z_0}\sqrt{1-(\kappa_{m,n}/k_0)^2}=\frac{1}{Z_0}\sqrt{1-(m\pi/k_0a)^2-(n\pi/k_0b)^2}=\frac{\cos\Phi_{m,n}}{Z_0} \tag{7}$$

The modal axial effective intensity at a point $\mathbf{x}$ is :

$$I_{zm,n}(\mathbf{x})=\frac{1}{2}\mathrm{Re}\left\{p_{m,n}(\mathbf{x})\cdot v_{zm,n}^{*}(\mathbf{x})\right\}=\frac{1}{2}|p_{m,n}(\mathbf{x})|^2\cdot\mathrm{Re}\{G_{zm,n}\} \tag{8}$$

The axial effective intensity of the sound wave (a mode mixture) is :

$$I_z(\mathbf{x}) = \frac{1}{2}\mathrm{Re}\{\sum_{m,n} p_{m,n}(\mathbf{x}) \cdot v^*_{zm,n}(\mathbf{x})\} = \frac{1}{2}\sum_{m,n} | p_{m,n}(\mathbf{x}) |^2 \cdot \mathrm{Re}\{G_{zm,n}\} \tag{9}$$

The effective sound power through a duct cross-section is :

$$\begin{aligned} \Pi &= \frac{1}{2}\iint_S \mathrm{Re}\{p \cdot v_z^*\} dS = \frac{1}{2} S \cdot \mathrm{Re}\left\{ \sum_{m,n} \frac{k_{m,n}}{k_0 Z_0} \cdot | A_{m,n} |^2 \cdot N_{m,n} \right\} \\ &= \frac{1}{2} S \sum_{m,n} \frac{1}{\delta_m \delta_n} \cdot | A_{m,n} |^2 \cdot \mathrm{Re}\{G_{zm,n}\} = \sum_{m,n} \Pi_{m,n} \end{aligned} \tag{10}$$

where $\Pi_{m,n}$ are the modal effective powers.

A mode can transport effective power only if it is cut-on (propagating); the condition for cut-on is (the summations in Π extend up to these limits) :

$$(\kappa_{m,n} / k_0)^2 < 1 \quad \text{or:} \quad (m/a)^2 + (n/b)^2 < (k_0/\pi)^2 = 4/\lambda_0^2 = (2f/c_0)^2 \tag{11}$$

Below, the sound power Π sometimes will be referred to the sound power Π_0 of a plane wave with sound pressure p_0 such that $\Pi_0 = \dfrac{S\, p_0^2}{2\, Z_0} = 1$. (12)

The condition $\Pi/\Pi_0 = 1$ is equivalent with :

$$\sum_{m,n} \frac{1}{\delta_m \delta_n} \left| \frac{A_{m,n}}{p_0} \right|^2 \cdot \mathrm{Re}\{k_{m,n}/k_0\} = \sum_{m,n} \frac{1}{\delta_m \delta_n} \left| \frac{A_{m,n}}{p_0} \right|^2 \cdot \mathrm{Re}\{Z_0 G_{zm,n}\} = 1 \tag{13}$$

Mode mixture with equal modal amplitudes $A_{m,n}$:

$$\left| \frac{A_{m,n}}{p_0} \right|^2 \stackrel{!}{=} \left[\sum_{m,n} \frac{\mathrm{Re}\{Z_0 G_{zm,n}\}}{\delta_m \delta_n} \right]^{-1} = \mathrm{const}(m,n) \tag{14}$$

Mode mixture with equal modal sound powers (or intensities) :

$$\sum_{m,n} \frac{\Pi_{m,n}}{\Pi_0} = 1 \quad ; \quad \frac{\Pi_{m,n}}{\Pi_0} = \mathrm{const}(m,n) = \frac{1}{N} \tag{15}$$

where N is the total number of cut-on modes. This leads to the mode amplitudes :

$$\left| \frac{A_{m,n}}{p_0} \right|^2 \stackrel{!}{=} \frac{\delta_m \delta_n}{N} \frac{1}{\mathrm{Re}\{Z_0 G_{zm,n}\}} = \frac{1}{N} \frac{\delta_m \delta_n}{\cos \Phi_{m,n}} \tag{16}$$

If a mode approaches cut-off, $\cos\Phi_{m,n} \to 0$; i.e. this mode mixing model would require large mode amplitudes near cut-off, and also for large mode orders, because then $\Phi_{m,n} \to \pi/2$.

Mode mixture with equal mode energy density $E_{m,n}$:

The mode energy density averaged over the duct cross-section follows from the mode power:

$\Pi_{m,n} = c_{gm,n} \cdot S \cdot E_{m,n}$ with the modal group velocity : (17)

$$c_g = \frac{1}{dk/d\omega} = \frac{c_{ph}}{1 - \dfrac{\omega}{c_{ph}} \dfrac{\partial c_{ph}}{\partial \omega}}$$

$$k_{m,n} = \sqrt{(\omega/c_0)^2 - \kappa_{m,n}^2} \quad ; \quad \frac{dk_{m,n}}{d\omega} = \frac{1}{c_0} \frac{1}{\sqrt{1-(\kappa_{m,n}/k_0)^2}} \tag{18}$$

$$c_{gm,n} = c_0 \sqrt{1-(\kappa_{m,n}/k_0)^2} = c_0 \cdot Z_0 G_{zm,n}$$

Therefore the averaged modal energy density is :

$$E_{m,n} = \frac{\Pi_{m,n}}{S \cdot c_{gm,n}} = \frac{1}{2} \frac{1}{\delta_m \delta_n} |A_{m,n}|^2 \frac{\mathrm{Re}\{G_{zm,n}\}}{c_0 Z_0 G_{zm,n}} \tag{19}$$

The model of equal modal energy density demands (with restriction to propagating modes, for which $\mathrm{Re}\{G_{zm,n}\} = G_{zm,n}$) :

$$E_{m,n} = \frac{1}{2} \frac{1}{c_0 Z_0} \frac{|A_{m,n}|^2}{\delta_m \delta_n} = \mathrm{const}(m,n) \tag{20}$$

or, with the above power normalisation :

$$\left| \frac{A_{m,n}}{p_0} \right|^2 = \frac{\delta_m \delta_n}{\sum\limits_{m,n} \mathrm{Re}\{Z_0 G_{zm,n}\}} = \frac{\delta_m \delta_n}{\sum\limits_{m,n} \cos\Phi_{m,n}} \tag{21}$$

This is the most plausible model for the simulation of random sound fields.

J.32 Mode excitation coefficients

Ref.: Mechel, [J.19]

Sometimes one is interested in exciting predominantly higher modes (because they are easier to attenuate than lower modes); sometimes one would like to avoid the excitation of higher modes (e.g. in experiments with low modes). It is reasonable to introduce a coefficient which describes the excitation probability of a mode under some standard conditions.

Consider a flat lined duct extending over $-h \le y \le +h$ (other duct geometries are treated similarly) with a sound field formulated as a mode sum, with mode norms N_n :

$$p(x,y) = \sum_n A_n \cdot q_n(y) \cdot e^{-\gamma_n x} \quad ; \quad N_n = \frac{1}{2h} \int_{-h}^{h} q_n^2(y)\, dy \tag{1}$$

If the excitation is performed by a given sound pressure profile $p_i(0,y)$ in the plane $x=0$, the mode amplitudes are :

$$A_n = \frac{1}{N_n} \cdot \frac{1}{2h} \int_{-h}^{h} p_i(0,y) \cdot q_n(y)\, dy \tag{2}$$

If the excitation is done by a given axial particle velocity $v_{ix}(0,y)$, the mode amplitudes are :

$$A_n = \frac{j}{N_n \cdot \gamma_n / k_0} \cdot \frac{1}{2h} \int_{-h}^{h} Z_0 v_{ix}(0,y) \cdot\cdot q_n(y)\, dy \tag{3}$$

One plausible standard excitation is the excitation by a plane wave pressure profile $p_i(0,y)=1$, and to introduce mode excitation coefficients F_n for that excitation :

$$F_n = \frac{1}{N_n} \cdot \frac{1}{2h} \int_{-h}^{h} q_n(y)\, dy \tag{4}$$

For modes with symmetrical (relative to $y=0$) profiles $q_n(y)=\cos(\varepsilon_n y)$: (5)

$$F_n = 2 \frac{\sin(\varepsilon_n h)/(\varepsilon_n h)}{1 + \sin(2\varepsilon_n h)/(2\varepsilon_n h)} \tag{6}$$

(for a mode with $\varepsilon_n h=0$ becomes $F_n=1$). For anti-symmetrical modes with $q_n(y)=\sin(\varepsilon_n y)$:

$$F_n = 2 \frac{(1 - \cos(\varepsilon_n h))/(\varepsilon_n h)}{1 - \sin(2\varepsilon_n h)/(2\varepsilon_n h)} \tag{7}$$

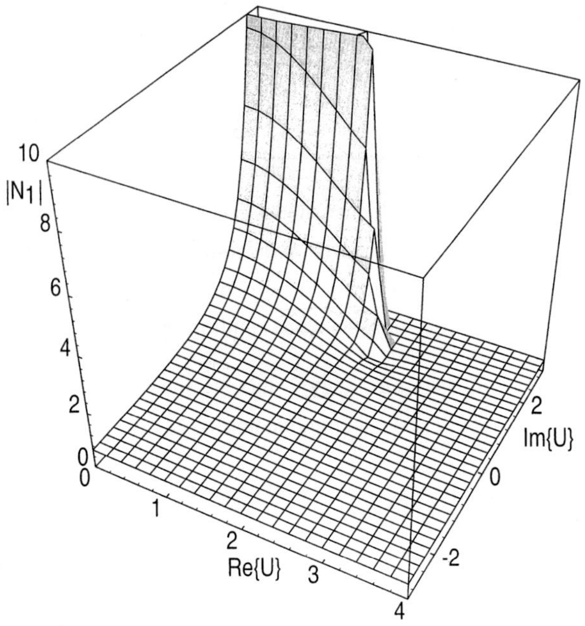

Magnitude of the mode norm in the range of the 1st mode in a flat, lined (locally reacting) duct over the plane $U = k_0 h \cdot Z_0 G$.

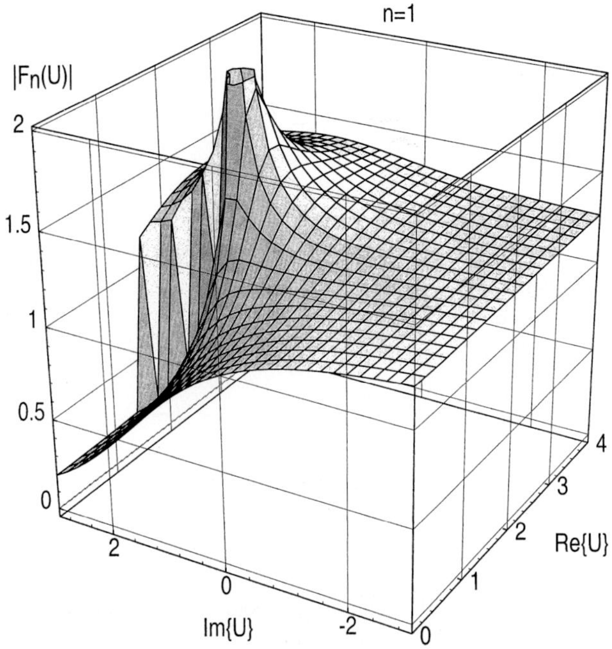

Magnitude of the mode excitation coefficient $F_n(U)$ for n=1 in a flat lined (locally reacting) duct over the plane $U = k_0 h \cdot Z_0 G$. The low values are in the range of the surface wave mode; the peak maximum is at the branch point between the 1st and 2nd modes.

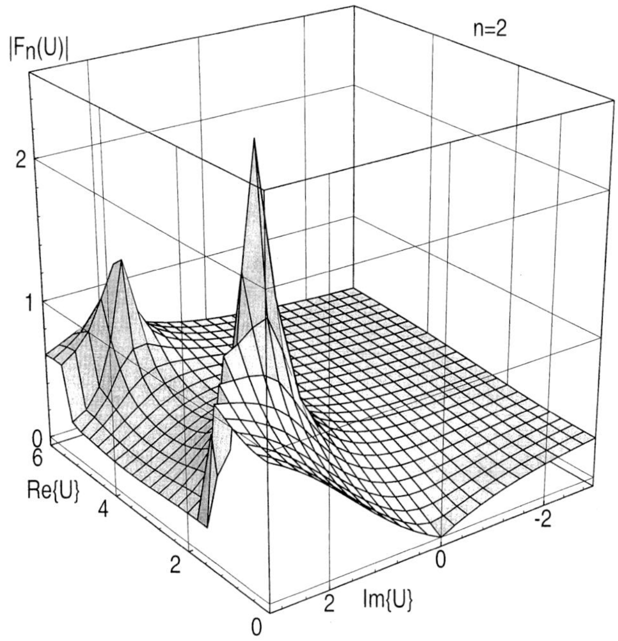

Magnitude of the mode excitation coefficient $F_n(U)$ for n=2 in a flat lined (locally reacting) duct over the plane $U = k_0 h \cdot Z_0 G$. Low values are in the range of the surface wave mode; the peak maxima are at the branch points between the 1st and 2nd and the 2nd and 3rd modes.

J.33 CREMER's admittance

Ref.: Cremer, [J.20]; Mechel, [J.21]

(The author doubted whether he should include this section, because its topic needs more wording than formulas; but the use of CREMER's admittance is a modern design of silencers, if powerful computing programs for sound absorbers are available. Duct linings are assumed to be locally reacting.)

CREMER'*s question* :
Under what condition will the least attenuated mode in a lined duct have its maximum attenuation ?

Answer (for a flat duct):

If
$$U := k_0 h \cdot Z_0 G \overset{!}{=} U_{b,1} = 2.05998 + j \cdot 1.65061 \tag{1}$$
where G is the lining admittance, and $U_{b,1}$ is the value of U in the 1st branch point of symmetrical modes.

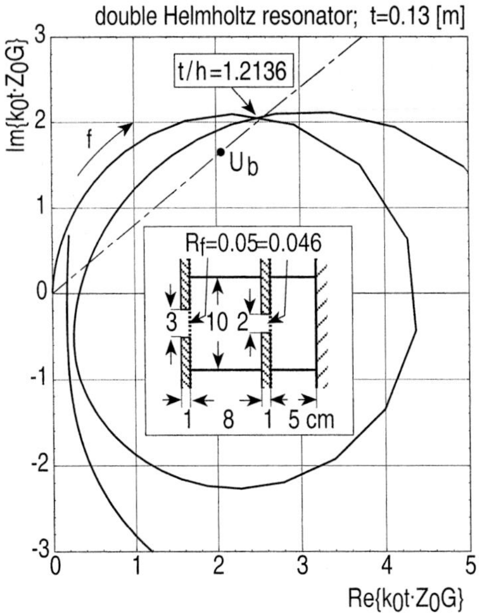

Function U_t in the complex plane for two Helmholtz resonators in series.

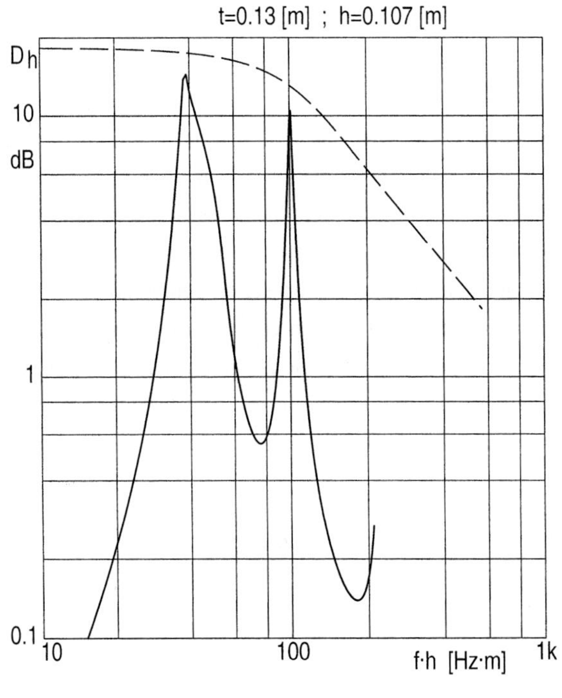

Attenuation D_h for the above absorber arrangement with a design ratio t/h= 1.2136 . The two maxima belong to the two crossings of the line (0, U_b) at about the same design point.

The next example, for a triple Helmholtz resonator with a front side porous layer, shows the influence of the selection of the design point on the attenuation curve.

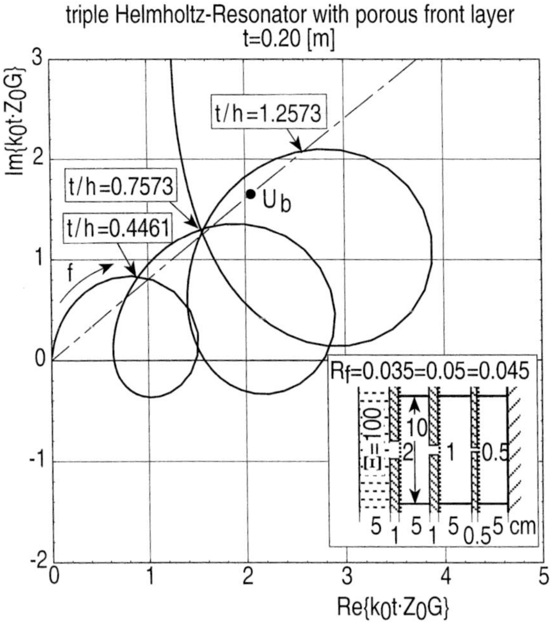

Function U_t in the complex plane for three Helmholtz resonators in series with a porous front layer. Three design points can be selected.

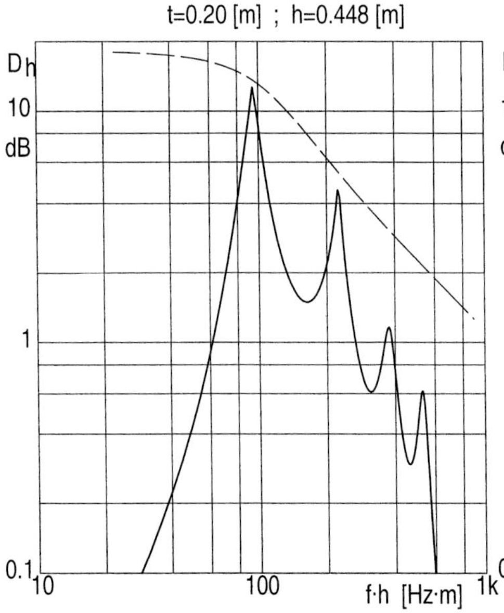

t/h= 0,4461

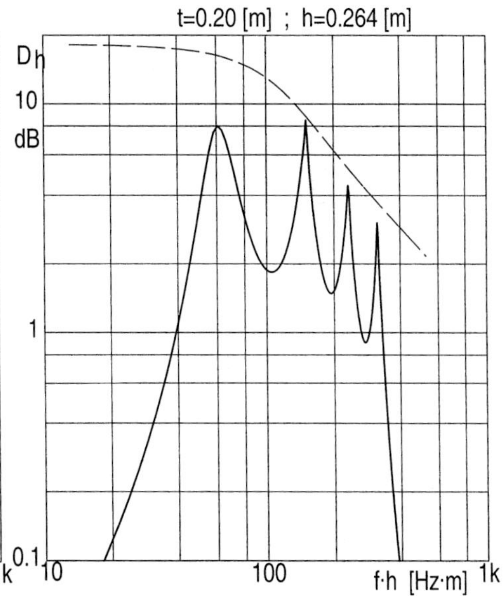

t/h= 0.7573

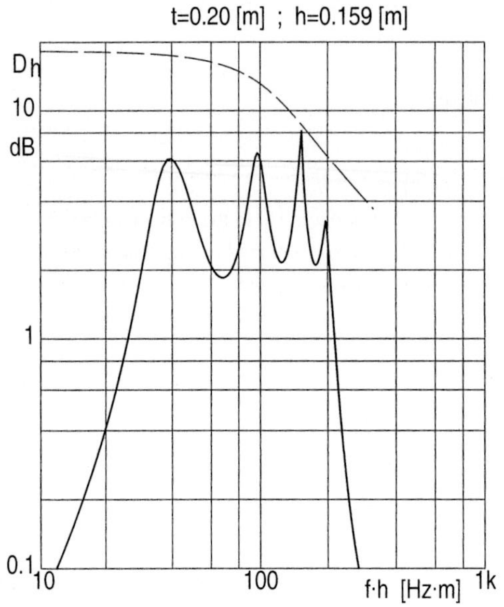

t/h= 1.2573

It is possible to construct design points with multiple cross-overs with resonators in series, but the individual crossings are separated by full resonance circles and, therefore, in the D_h curves by wide frequency steps. This can be avoided with resonators in a parallel arrangement.

J.34 CREMER's admittance with parallel resonators :

Ref.: Mechel, [J.21]; Press et al., [J.22]

The lining consists of repeated couples of absorbers. If the dimensions of the absorbers (in the direction of the duct axis) are small compared to the wavelength, the weighted (with the absorber surface areas) average of their admittances will determine the attenuation. One of the absorbers will be named the *primary* absorber (with index p), the other the *adjoint* absorber (with index a). Let F_p, F_a be the surface areas of the absorbers, G_p, G_a their surface admittances, the surface ratio $ß= F_a/F_p$, and t a common characteristic length of both absorbers. Then the extended principle of CREMER's admittance (see previous section J.33) demands that:

$$\langle U_t \rangle = \frac{F_p U_{t,p} + F_a U_{t,a}}{F_p + F_a} = \frac{U_{t,p} + ßU_{t,a}}{1+ß} \overset{!}{=} \frac{t}{h} U_b \quad ; \quad U_{t,\alpha} = k_0 t \cdot Z_0 G_\alpha \quad ; \quad \alpha = p,a \qquad (1)$$

This conditional equation defines the U function $\hat{U}_{t,a}$ of a *fictitious adjoint absorber* :

$$\hat{U}_{t,a} \overset{!}{=} \frac{1}{ß}\left[(1+ß)\cdot t/h\cdot U_b - U_{t,p}\right] \tag{2}$$

It is named "fictitious", because it is not sure whether and how it can be realised. If, for example, the real part of the brackets is negative the associate admittance $\hat{G}_a$ should have a negative real part, which cannot be realised with passive absorber elements. The conditional equation gives the "rule of construction" for $\hat{U}_{t,a}$ in the U plane :

$$ß\cdot\left[t/h\cdot U_b - \hat{U}_{t,a}\right] \overset{!}{=} -\left[t/h\cdot U_b - U_{t,p}\right] \tag{3}$$

According to this condition the point $\hat{U}_{t,a}$ belonging to a value $U_{t,p}$ is on the straight line through $U_{t,p}$ and the design point $t/h\cdot U_b$ on the opposite side (with respect to $t/h\cdot U_b$) at a distance ß times the distance between $U_{t,p}$ and $t/h\cdot U_b$. So, if $U_{t,p}$ is above the line (0, U_b), the point for $\hat{U}_{t,a}$ is below that line; if $U_{t,p}$ is above (0, U_b) and right-turning (which is normal), $\hat{U}_{t,a}$ is below (0, U_b) and right-turning also. All curves for functions U of passive absorbers begin at sufficiently low frequencies near the origin of the U plane; thus $\hat{U}_{t,a}$ begins near the line (0, U_b) beyond U_b , which physically is not possible. Therefore no realisation of $\hat{U}_{t,a}$ is possible at very low frequencies.

First the steps of the procedure for finding a lining with parallel absorbers with an effective CREMER admittance will be described (with a concrete example), then an algorithm for finding a suitable adjoint absorber will be derived. The first example simply consists of two porous absorbers with different flow resistivity values Ξ and thicknesses t_p, t_a , arranged side by side. The characteristic length is $t = t_p$.

1. Find a suitable primary absorber (i.e. an absorber with the U function on an arc above the line (0, U_b) , possibly crossing that line beyond U_b).

2. Conceive an adjoint absorber the function $U_{t,a}$ of which approximates $\hat{U}_{t,a}$ in some frequency interval of interest ($\hat{U}_{t,a}$ can be drawn with the above rule of construction).

Thus $U_{t,p}$, $U_{t,a}$ are known as functions of frequency, and the average function $\langle U_t\rangle$ of them. Plot these curves in the U plane.

The diagram below shows these curves together with the line (0, U_b) and the curve $\langle U_t\rangle$ for three surface ratios $ß = F_a/F_p$. A design point at $t/h = 0.2694$ is marked.

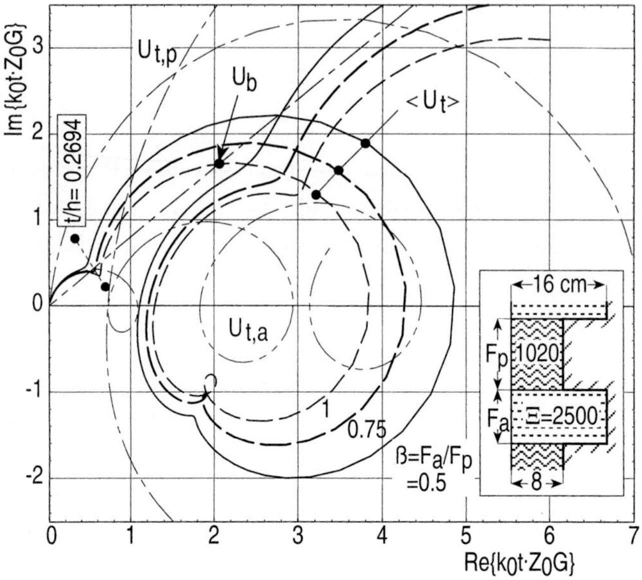

Branch point U_b, curves $U_{t,p}$, $U_{t,a}$ for the component absorbers, and average $\langle U_t \rangle$ in the U plane.
A design point t/h= 0.2694 is marked, at which two resonances have contracted to a dent.

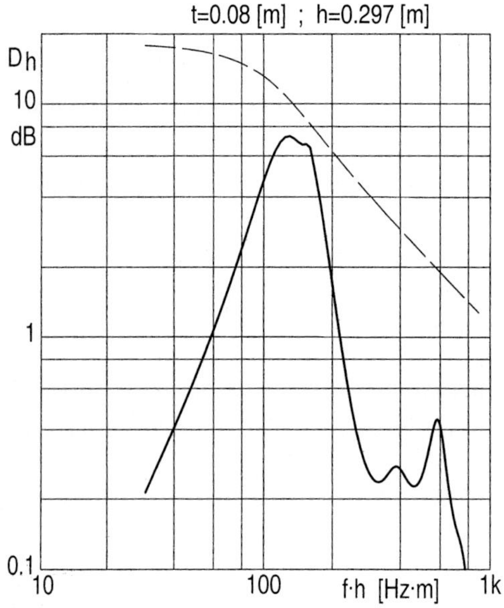

Maximum possible attenuation (dashed) and attenuation curve of the least attenuated mode in a duct with the lining from above at a surface ratio $\beta = F_a/F_p = 0.2694$.

The next example is for a similar arrangement, but now with equal thicknesses $t = t_p = t_a = 8$ [cm]. The adjoint absorber is covered with a tight, limp foil with surface mass density $m_f =$

0.06 [kg/m^2]. Two design points are of interest, one at t/h= 0.5971 , the other at t/h= 0.7330

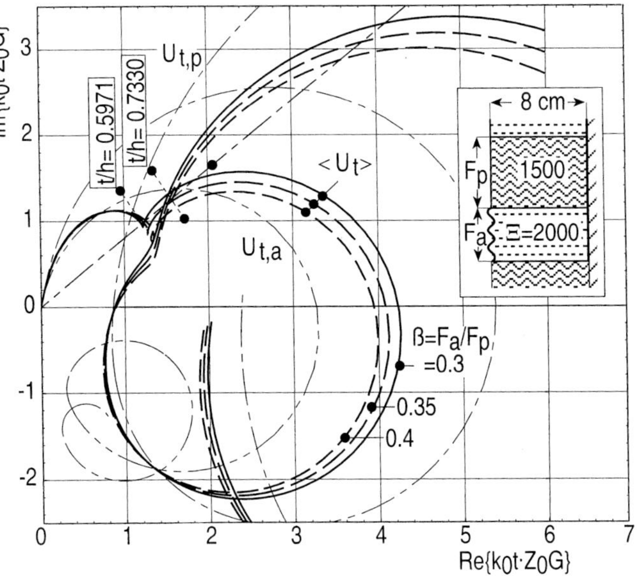

A surface area ratio $ß= F_a/F_p= 0.3$ is selected for the attenuation curves at the two design points.

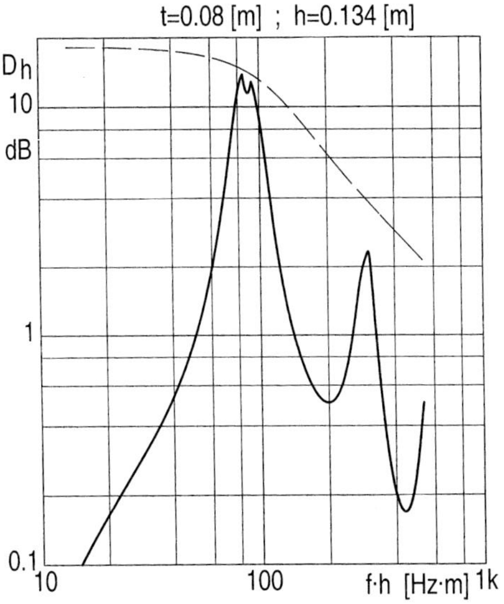

t/h= 0.5972 ; ß= 0.3

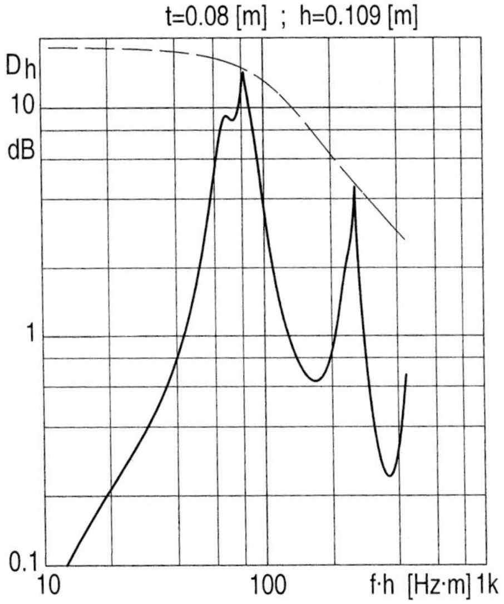

t/h= 0.7330 ; ß= 0.3

The weight function was w(x)=1 . The (computed) design point is indicated as a point on the straight line $(0, U_b)$.

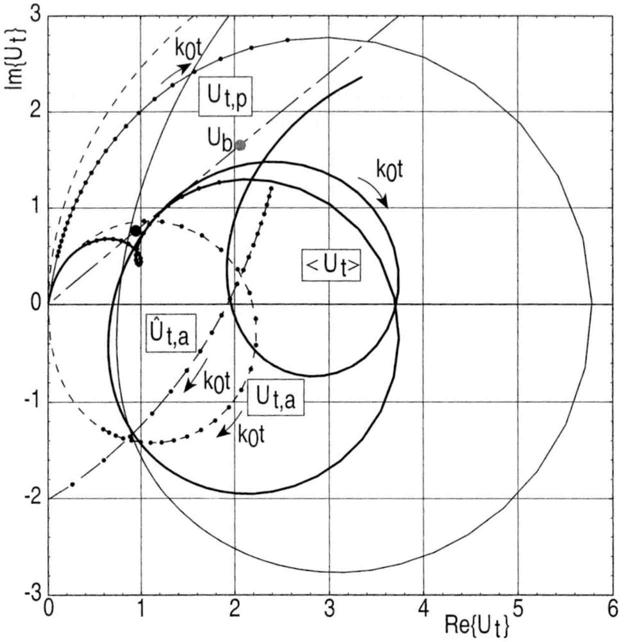

The design point is on the straight line $(0, U_b)$ near the small loop in the curve of $\langle U_t \rangle$.

t=0.10 [m] ; h=0.218 [m] ; t/h=0.4585 ; ß=0.583

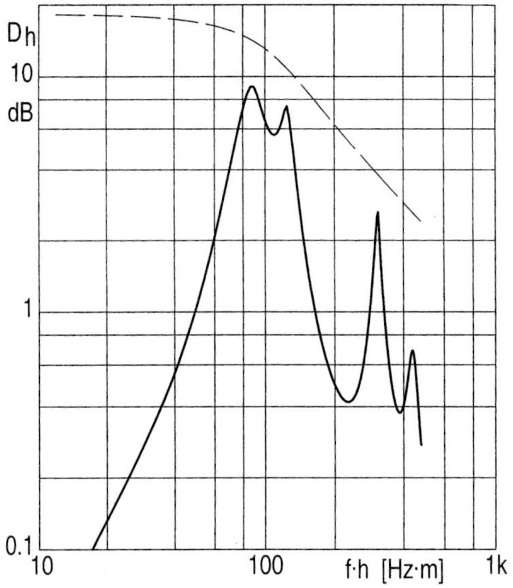

Attenuation curve D_h for the combination of a porous layer primary absorber with a Helmholtz resonator as adjoint absorber with the optimised values of the free parameters listed above.

J.35 Influence of flow on attenuation

Ref.: Mechel, [J.23, J.24]

Consider a duct with axial co-ordinate x and transversal co-ordinate y . A stationary flow with velocity profile V(y) is in the +x direction, if V>0 , and in the –x direction, if V<0 . Sound waves are assumed to propagate in the +x direction. In this section the simplifying assumption V(y)=const is made; for more details see the chapter "Flow Acoustics".

The presence of flow will modify the fundamental equations mainly by the replacement of the partial time derivative $\partial/\partial t$ by the "substantial derivative" D/Dt :

$$\rho_0 \operatorname{div} \mathbf{v} + \frac{\partial \rho}{\partial t} = 0 \xrightarrow[V \neq 0]{} \rho_0 \operatorname{div} \mathbf{v} + \frac{D\rho}{Dt} = 0$$

$$\rho_0 \frac{\partial \mathbf{v}}{\partial t} + \mathbf{grad}\, p = 0 \xrightarrow[V \neq 0]{} \rho_0 \frac{D\mathbf{v}}{Dt} + \mathbf{grad}\, p = 0 \qquad (1)$$

$$(\Delta - \frac{1}{c_0^2}\frac{\partial^2}{\partial t^2})p = 0 \xrightarrow[V \neq 0]{} (\Delta - \frac{1}{c_0^2}\frac{D^2}{Dt^2})p = 0$$

The substitution for the time derivative can be written for an assumed time factor $e^{j\omega t}$ and a sound wave of the form $p(x,y) = P_0 \cdot q(y) \cdot e^{-\Gamma x}$ as : (2)

$$\frac{\partial}{\partial t} = j\omega = jc_0 k_0 \xrightarrow[V \neq 0]{} \frac{D}{Dt} = \frac{\partial}{\partial t} + V\frac{\partial}{\partial x} = c_0[jk_0 + M\frac{\partial}{\partial x}]$$
$$= jc_0 k_0 [1 + jM\frac{\Gamma}{k_0}] \qquad (3)$$

(with $M = V/c_0$ the Mach number). So this effect of the flow can be taken into account by the substitution :

$$k_0 \xrightarrow[V \neq 0]{} k_0[1 - j\frac{M}{k_0}\frac{\partial}{\partial x}] = k_0[1 + jM\frac{\Gamma}{k_0}] \qquad (4)$$

With the abbreviation w one can write :

$$w = [1 - j\frac{M}{k_0}\frac{\partial}{\partial x}] = [1 + jM\frac{\Gamma}{k_0}]$$
$$k_0 \xrightarrow[V \neq 0]{} k_0 \cdot w \qquad (5)$$

The wave equation, for example, becomes : $\left(\Delta + k_0^2 w^2\right)p(x,y) = 0$ (6)

If the lateral sound wave profile q(y) is (for example) $q(y) = \cos(\varepsilon y)$

the characteristic equation for the determination of εh in a duct of (half) width h and with a locally reacting lining with surface admittance G changes :

$$\varepsilon h\cdot\tan(\varepsilon h)=j\,k_0h\,Z_0G\xrightarrow[V\neq 0]{}\varepsilon h\cdot\tan(\varepsilon h)=j\,k_0h\,Z_0G\cdot w \tag{7}$$

This form assumes that the boundary conditions at the lining surface are the continuity of sound pressure and normal particle velocity v_y . Some authors claim that not the particle velocity should be continuous, but the elongation e_y with $v_y=\partial e_y/\partial t$. This time derivative introduces a new factor w wherever G appears :

$$\varepsilon h\cdot\tan(\varepsilon h)=j\,k_0h\,Z_0G\xrightarrow[V\neq 0]{}\varepsilon h\cdot\tan(\varepsilon h)=j\,k_0h\,Z_0G\cdot w^2 \tag{8}$$

One can combine both theories of the boundary condition to give :

$$\varepsilon h\cdot\tan(\varepsilon h)=j\,k_0h\,Z_0G\xrightarrow[V\neq 0]{}\varepsilon h\cdot\tan(\varepsilon h)=j\,k_0h\,Z_0G\cdot w^{\alpha}\quad;\quad\alpha=1,2 \tag{9}$$

The secular equation, which follows from the wave equation, changes to :

$$(\Gamma/k_0)^2+1-(\varepsilon/k_0)^2=0\xrightarrow[V\neq 0]{}(\Gamma/k_0)^2\left(1-M^2\right)+2jM\cdot\Gamma/k_0+1-(\varepsilon/k_0)^2=0 \tag{10}$$

Because the solution shall be without flow :

$$\frac{\Gamma}{k_0}\xrightarrow[M\to 0]{}\frac{1}{k_0h}\sqrt{(\varepsilon h)^2-(k_0h)^2} \tag{11}$$

the solution with flow is :

$$\frac{\Gamma}{k_0}=\frac{-j}{1-M^2}\left[M+\frac{j}{k_0h}\sqrt{(1-M^2)(\varepsilon h)^2-(k_0h)^2}\right] \tag{12}$$

Thus w has the form :

$$w=\frac{1}{1-M^2}\left[1+j\frac{M}{k_0h}\sqrt{(\varepsilon h)^2(1-M^2)-(k_0h)^2}\right] \tag{13}$$

The appearance of εh in w modifies the characteristic equation significantly, and also the method for its numerical solution. In general, one will solve the characteristic equation first for $M=0$, i.e. $w=1$, and then will increase M iteratively to its final value, taking solutions for the earlier M as start solutions zs_i in the numerical procedure:

$$zs_1=\varepsilon h(M_{k-3})\quad,\quad zs_2=\varepsilon h(M_{k-2})\quad,\quad zs_3=\varepsilon h(M_{k-1}) \tag{14}$$

One must take very small steps ΔM , especially at the beginning of the iteration through M . A better choice of start solutions is at the beginning of the iteration :

$$zs_1=\varepsilon h(0)\quad,\quad zs_2=(zs_1+zs_3)/2\quad,\quad zs_3=\varepsilon h(0)+\Delta M\cdot d(\varepsilon h)/dM\,|_{M=0} \tag{15}$$

and at later steps :

$$zs_1 = \varepsilon h(M_{k-2}) \;,\; zs_2 = \varepsilon h(M_{k-1}) \;,\; zs_3 = \varepsilon h(M_{k-1}) + \Delta M \cdot d(\varepsilon h)/dM\,|_{M(k-1)} \tag{16}$$

The required derivatives $d(\varepsilon h)/dM$ are for symmetrical modes, for both exponents $\alpha = 1,2$, with the abbreviation : $qw = \sqrt{(\varepsilon h)^2(1-M^2)-(k_0h)^2}$ (17)

<u>α=1</u>:

$$\frac{d(\varepsilon h)}{dM} = Z_0G\frac{-(\varepsilon h)^2(1-M^2)+(k_0h)^2(1+M^2)+2\,j\,k_0h\cdot M\cdot qw}{qw\cdot(1-M^2)\left[M\dfrac{\varepsilon h}{qw}Z_0G+\dfrac{\varepsilon h}{\cos^2(\varepsilon h)}+\tan(\varepsilon h)\right]} \tag{18}$$

<u>α=2</u>: (19)

$$\frac{d(\varepsilon h)}{dM} = Z_0G\frac{2(k_0h+jM\cdot qw)\left[-(\varepsilon h)^2(1-M^2)+(k_0h)^2(1+M^2)+2\,j\,k_0h\cdot M\cdot qw\right]}{qw\cdot(1-M^2)^3\left[2j\dfrac{M^2}{1-M^2}\dfrac{\varepsilon h}{k_0h}Z_0G+2\dfrac{M}{1-M^2}\dfrac{\varepsilon h}{qw}Z_0G+\dfrac{\varepsilon h}{\cos^2(\varepsilon h)}+\tan(\varepsilon h)\right]}$$

For anti-symmetrical modes replace $1/\cos^2(\varepsilon h) \to -1/\sin^2(\varepsilon h)$, $\tan(\varepsilon h) \to \cot(\varepsilon h)$ and $Z_0G \to -Z_0G$.

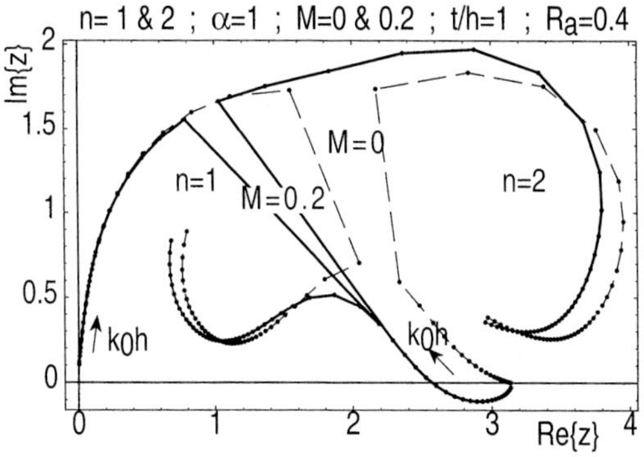

The first two mode solutions $z=\varepsilon h$ in a flat duct with a locally reacting lining, consisting of a simple layer of glass fibres with thickness t and normalised flow resistance $R_a = \Xi t/Z_0$.
For Mach numbers M=0 and M=0.2 with the boundary condition form $\alpha = 1$.

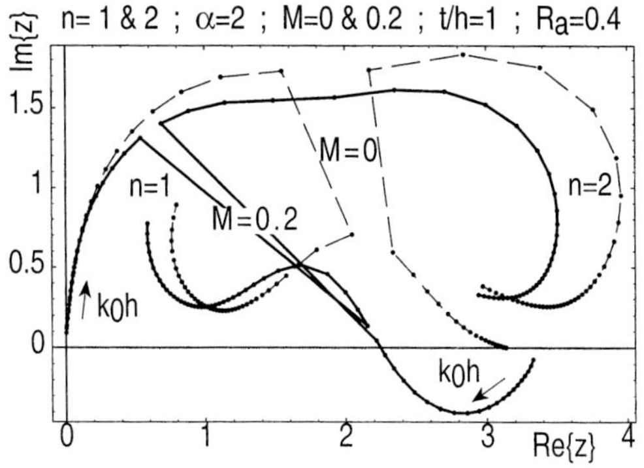

Same as above, but for the boundary condition form $\alpha = 2$.

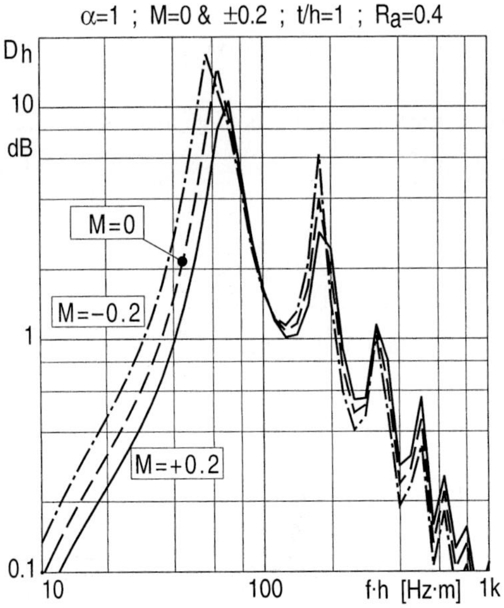

Attenuation curves D_h for the least attenuated mode in a duct as above, for Mach numbers M=0 and M= ±0.2 with the boundary condition form $\alpha = 1$.

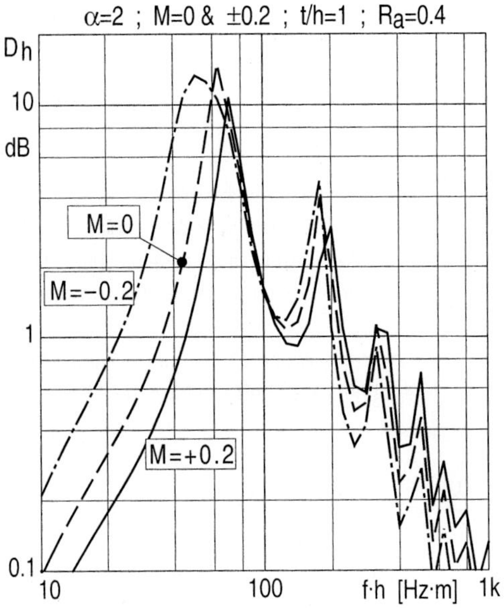

As above, but for the boundary condition form $\alpha = 2$.

One needs, in a number of applications, the branch points of the complex transformation which is induced by the characteristic equation. We write this equation for symmetrical modes in a flat duct of (half) width h with a locally reacting lining having a surface admittance G in the form :

$$f(z;M) = -j\frac{z\cdot\tan z}{w^{\alpha}} \overset{!}{=} U \quad ; \quad \alpha = 1,2 \quad ; \quad z = \varepsilon h \quad ; \quad U = k_0 h\cdot Z_0 G \tag{20}$$

The branch points $z_{b,n}(M)$ are determined as solutions of the equation :

$$\frac{f'(z;0)}{f(z;0)} \overset{!}{=} \alpha\cdot\frac{w'(z;M)}{w(z;M)} \tag{21}$$

with :

$$\frac{f'(z;0)}{f(z;0)} = \frac{1}{z} + \frac{1}{\sin z\cdot\cos z}$$

$$\frac{w'(z;M)}{w(z;M)} = j\frac{M(1-M^2)}{k_0 h}\frac{z}{\sqrt{z^2(1-M^2)-(k_0h)^2}\left[1+j\frac{M}{k_0h}\sqrt{z^2(1-M^2)-(k_0h)^2}\right]} \tag{22}$$

For small Mach numbers the branch points can be approximated by :

$$z_{b,n}(M) \approx z_{b,n}(0) + M\cdot\left.\frac{dz_b}{dM}\right|_{M=0} \tag{23}$$

with the derivative : (24)

$$\begin{aligned}\frac{dz_b}{dM} = &\Big\{\left(2jM(1-M^2)+k_0h(1-M^2)/qw\right)\cdot z_b^2 + k_0h\cdot qw + jM\cdot qw^2 + \\ &+\left[k_0h\cdot qw + jM\left(-(k_0h)^2+(1-\alpha)(1-M^2)\cdot z_b^2\right)\right]\cdot\cos^2(z_b)+\\ &+\left[2j(1-\alpha)(1-M^2)M\cdot z_b + k_0h(1-M^2)\cdot z_b/qw\right]\cdot\cos(z_b)\cdot\sin(z_b)-\\ &-\left[k_0h\cdot qw + jM\left(-(k_0h)^2+(1-\alpha)(1-M^2)\cdot z_b^2\right)\right]\cdot\sin^2(z_b)\Big\}\cdot\\ &\Big\{\left(2jM^2+k_0hM/qw\right)\cdot z_b^3 - jz_b\cdot qw^2 + \\ &+\left[\left(2jM^2(1-\alpha)+k_0hM/qw\right)\cdot z_b^2 - j\left(-(k_0h)^2+(1-\alpha)(1-M^2)\cdot z_b^2\right)\right]\cdot\cos(z_b)\cdot\sin(z_b)\Big\}^{-1}\end{aligned}$$

which for M=0 becomes (with the abbreviation $z_b = z_{b,n}(0)$) :

$$\left.\frac{dz_b}{dM}\right|_{M=0} = j\,k_0h\frac{4z_b^2-2(k_0h)^2+2(z_b^2-(k_0h)^2)\cos(2z_b)+z_b\sin(2z_b)}{\sqrt{z_b^2-(k_0h)^2}\left[2z_b(z_b^2-(k_0h)^2)+\left((1-\alpha)z_b^2-(k_0h)^2\right)\sin(2z_b)\right]} \tag{25}$$

One gets for the 1st branch point n=1 with $z_b = 2.1062 + 1.12536\cdot j$:

$$\begin{aligned}&\underline{\text{for } k_0h=2;\ \alpha=1:} \quad z_{b,1}(M) \approx z_{b,1}(0) + (-1.33228 + j\cdot 0.403591)M\\ &\underline{\text{for } k_0h=2;\ \alpha=2:} \quad z_{b,1}(M) \approx z_{b,1}(0) + (-0.66614 + j\cdot 0.201793)M\\ &\underline{\text{for } k_0h=1;\ \alpha=1:} \quad z_{b,1}(M) \approx z_{b,1}(0) + (-0.599752 + j\cdot 0.406326)M\end{aligned} \tag{26}$$

for $k_0h=1;\ \alpha=2:\quad z_{b,1}(M)\approx z_{b,1}(0)+(-0.299878+j\cdot 0.203163)M$ (27)

The images $U_{b,n}(M)$ of the branch points $z_{b,n}(M)$ can be approximated by :

$$U_{b,n}(M)\approx U_{b,n}(0)+M\cdot\left.\frac{dU_b}{dM}\right|_{M=0} \tag{28}$$

with the derivative :

$$\left.\frac{dU_b}{dM}\right|_{M=0}= \\ =j\frac{(k_0h)^2}{2}\frac{(2z_b+\sin(2z_b))\left(4z_b^2-2(k_0h)^2+2(z_b^2-(k_0h)^2)\cos(2z_b)+z_b\sin(2z_b)\right)}{2z_b(z_b^2-(k_0h)^2)+\left((1-\alpha)z_b^2-(k_0h)^2\right)\sin(2z_b)} \tag{29}$$

One gets for the 1st branch point n=1 with $U_b=2.05998+j\cdot 1.65061$:

$$\begin{aligned}
&\text{for } k_0h=2;\ \alpha=1: && U_{b,1}(M)\approx U_{b,1}(0)+(-7.206913\cdot 10^{-6}+j\cdot 0.000152866)\,M\\
&\text{for } k_0h=2;\ \alpha=2: && U_{b,1}(M)\approx U_{b,1}(0)+(-3.603788\cdot 10^{-6}+j\cdot 0.0000764332)M\\
&\text{for } k_0h=1;\ \alpha=1: && U_{b,1}(M)\approx U_{b,1}(0)+(0.0000218572+j\cdot 0.0000352125)M\\
&\text{for } k_0h=1;\ \alpha=2: && U_{b,1}(M)\approx U_{b,1}(0)+(0.0000109286+j\cdot 0.0000176063)M
\end{aligned} \tag{30}$$

Influence of the flow on the attenuation for a lining with $U=U_{b,1}(0)$:

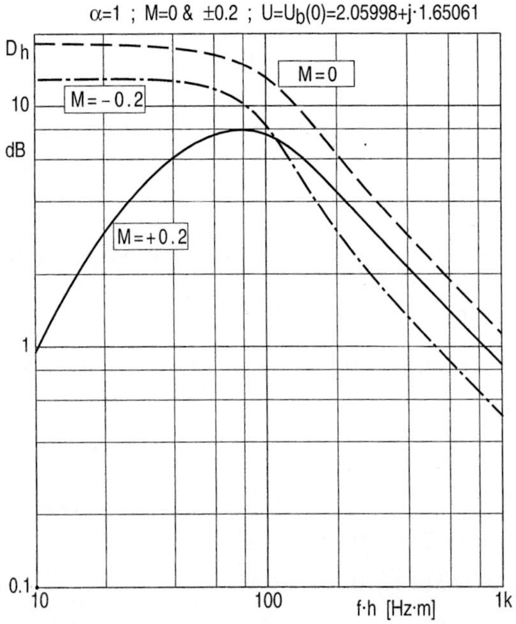

Influence of flow on the attenuation D_h of the least attenuated mode in a duct the lining of which has a U function $U=U_{b,1}(0)$ which for M=0 is in the 1st branch point.

If, however, the lining has a U function $U=U_{b,1}(M)$, the attenuation remains high.

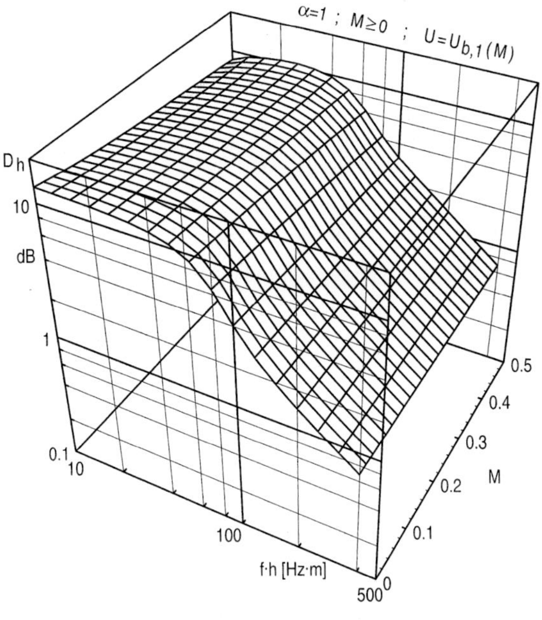

Attenuation D_h of the least attenuated mode in a flat duct having a locally reacting lining with $U = U_{b,1}(M)$, for $M \geq 0$ and boundary condition form $\alpha=1$.

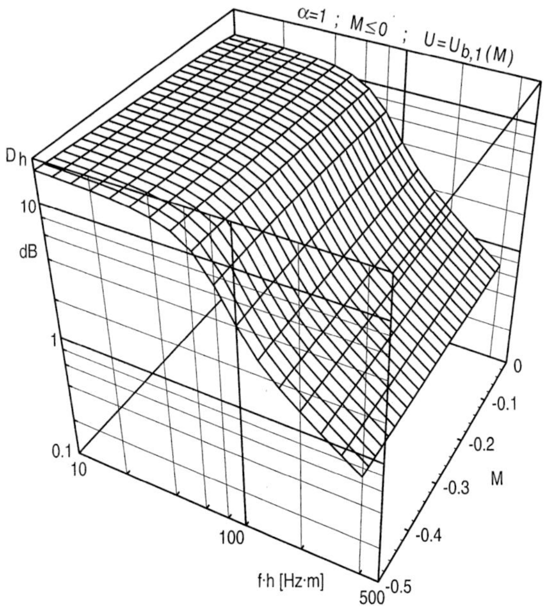

As above, i.e. for $\alpha=1$, but with $M \leq 0$.

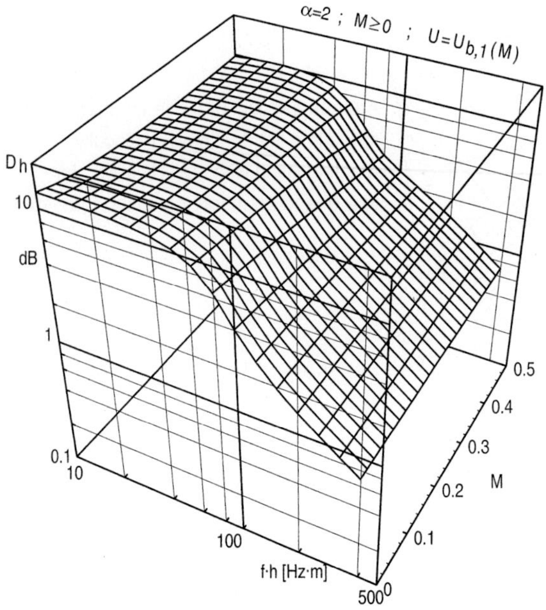

As above, but for the boundary condition form $\alpha=2$ and with $M \geq 0$.

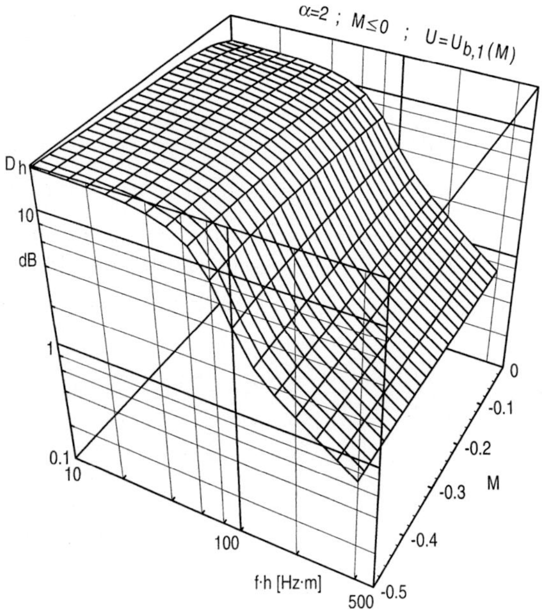

As above, i.e. for $\alpha=2$, but with $M \leq 0$.

J.36 Influence of temperature on attenuation

Ref.: Mechel, [J.24]

Silencers are used in gas flows with a wide variety of temperatures. The question is how to take the operation temperature into consideration in the design of a silencer.

If the attenuation is evaluated as a non-dimensional quantity (like $D_h = 8.6858 \cdot \mathrm{Re}\{\Gamma h\}$) using only non-dimensional parameters, and if it is plotted over a non-dimensional variable, then the result is valid for all fluids like air, and also for air at different temperatures. Some different influences of the temperature on the attenuation may be distinguished. Below, T is the operation temperature (in Kelvin), T_0 is the standard temperature.

1st: *Influence of representation* :
In a plot of D_h over $f \cdot h$ the abscissa comes from $f \cdot h = c_0/(2\pi) \cdot k_0 h$. The (linear) abscissa should be multiplied with $c_0(T)/c_0(T_0)$.

2nd: *Temperature dependent input parameters* :
Some parameters, like frequency f, geometrical dimensions, porosities, shape factors etc. are not changed by temperature. Other parameters, like bulk densities of porous materials, surface mass densities m of foils and plates remain virtually unchanged. If, however, a non-dimensional parameter $M = m/(\rho_0 d)$ is used, with the air density ρ_0 and some thickness d , M becomes $M(T) = \rho_0(T_0)/\rho(T) \cdot M_0$. Often impedances or admittances are made non-dimensional (normalised) with the free field wave impedance $Z_0 = \rho_0 c_0$. This reference impedance changes as $Z_0(T) = \rho_0(T) c_0(T)/(\rho_0 c_0) \cdot Z_0(T_0)$. If the impedance which is normalised with Z_0 is a mass reactance Z_m of a solid element (e.g. foil or plate), it is not modified by the temperature; thus the variation in Z_m/Z_0 comes from Z_0 . Resistances often used are the flow resistance $\Xi \cdot d$ of a porous layer (d its thickness; Ξ the material flow resistivity) or the flow resistance Z_f of a porous foil or plate. One can always write $\Xi \cdot a^2/\eta = f(d)$, where a, d are characteristic lengths (e.g. a= fibre radius, d= fibre distance) and η is the dynamic viscosity of air. Thus $\Xi(T) = \eta(T)/\eta(T_0) \cdot \Xi_0$. This transformation holds for all other resistances based on the friction of air.

3rd: *Temperature dependent non-dimensional material data of air* :
Theories for the characteristic propagation constant Γ_a and wave impedance Z_a of porous materials contain not only the flow resistivity Ξ , but also material data of air, such as the adiabatic exponent κ and the Prandtl number Pr . The best procedure is to evaluate Γ_a and Z_a with a physical model theory and to use material data of air at the operation temperature.

An important parameter is the product $f_{cr}\cdot d$ of the critical frequency f_{cr} and thickness d of an elastic plate. From the relation

$$\frac{f_{cr}d}{f\,d}=\left(\frac{k_b}{k_0}\right)^2=c_0\sqrt{\frac{m}{B}} \tag{1}$$

and with the assumption that the surface mass density m and the bending modulus B do not (or only slightly) change with temperature, the parameter $f_{cr}\cdot d$ changes as $c_0(T)$.

The section B.2 contains material data for air and relations for their temperature dependence. For some approximations it may be sufficient to use the ideal gas relations :

$$\begin{aligned}
&\rho_0(T)=\rho_0(T_0)\cdot T_0/T \\
&c_0(T)=\sqrt{\kappa(T)\rho_0(T_0)/\kappa(T_0)\rho_0(T)}\cdot c_0(T_0)\approx c_0(T_0)\cdot\sqrt{T/T_0} \\
&\kappa(T)\approx\kappa(T_0) \qquad Pr(T)\approx Pr(T_0) \\
&Z_0(T)=Z_0(T_0)/\sqrt{T/T_0} \qquad k_0(T)=k_0(T_0)/\sqrt{T/T_0} \\
&\eta(T)=\eta(T_0)\cdot\sqrt{T/T_0} \qquad \Xi(T)=\Xi(T_0)\cdot\sqrt{T/T_0} \\
&R(T)=R(T_0)\cdot T/T_0 \qquad E(T)=E(T_0)\cdot(T/T_0)^{-3/2}
\end{aligned} \tag{2}$$

R is the gas constant; $E=\rho_0 f/\Xi$ is a non-dimensional input parameter for some porous material model theories.

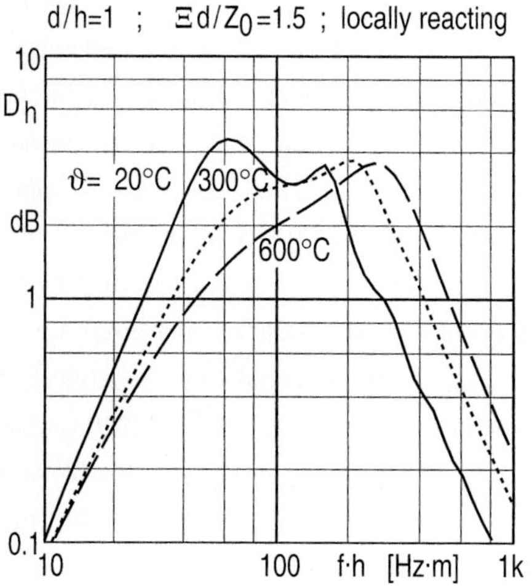

Attenuation D_h of the least attenuated mode in a flat duct with a locally reacting glass fibre layer as lining, for three operation temperatures.

J.37 Stationary flow resistance of splitter silencers

Ref.: Mechel, [J.24]

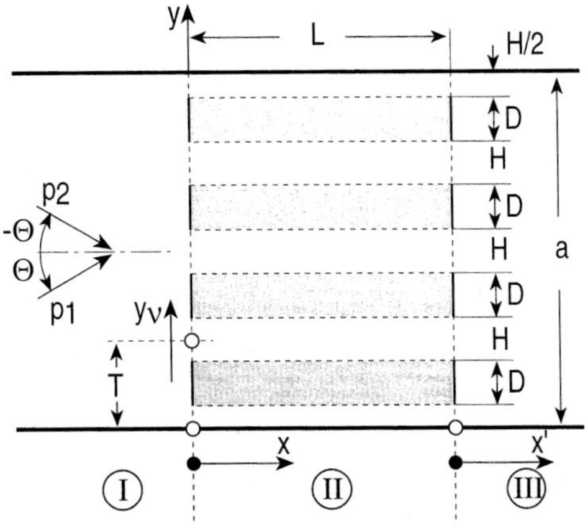

The acoustic design of silencers is often in conflict with the static pressure loss of the stationary flow, especially in splitter silencers.

The stationary flow resistance usually is described by the ζ value of the silencer :

$$\zeta = \frac{\Delta P_{with} - \Delta P_{no}}{\rho_0 \langle V \rangle^2 / 2} \tag{1}$$

where ΔP_{with} is the static pressure drop over the silencer, ΔP_{no} is the static pressure drop in the empty duct over the same distance and $\langle V \rangle$ is the average flow velocity in the duct in front of the splitters. If, alternatively, the average flow velocity $\langle V_s \rangle$ is determined in a splitter duct, the corresponding ζ value is :

$$\zeta_s = \zeta / (1 + D/H)^2 \tag{2}$$

A great number of measurements with splitter silencers (the splitters having rectangular corners) can be summarised by :

$$\zeta_s = 0.53 + 0.66 \cdot \lg \frac{D}{H} + \left(0.027 - \frac{0.004}{D/H} \right) \cdot \frac{L}{H} \tag{3}$$

Rounding the splitter heads reduces ζ by about $\Delta\zeta \approx 0.5 - 1.5$.

J.38 Nonlinearities by amplitude and/or flow

Ref.: Mechel, [J.25]; Ronneberger, [J.26]; Cummings, [J.27]

High sound amplitudes and stationary flow produce nonlinearities in some absorber components, especially in fences and perforated sheets. The references combine their own measurements with a survey of the literature.

1st: *Amplitude nonlinearity of fences* :

Let Δp_S be the sound pressure drop across the fence, v_S the particle velocity in the fence orifice (both averaged over the orifice), and σ the porosity of the fence, then the nonlinear contribution to the normalised partition impedance $Z_S = (\Delta p_S / v_S)/Z_0$ of the fence opening can be written as :

$$\Delta Z_s = \Delta\left(\frac{1}{Z_0}\frac{p_s}{v_s}\right) = \sigma^2 K(\sigma)\frac{v_s}{c_0} \tag{1}$$

The following diagram gives values of the factor $K(\sigma)$ over the porosity σ .

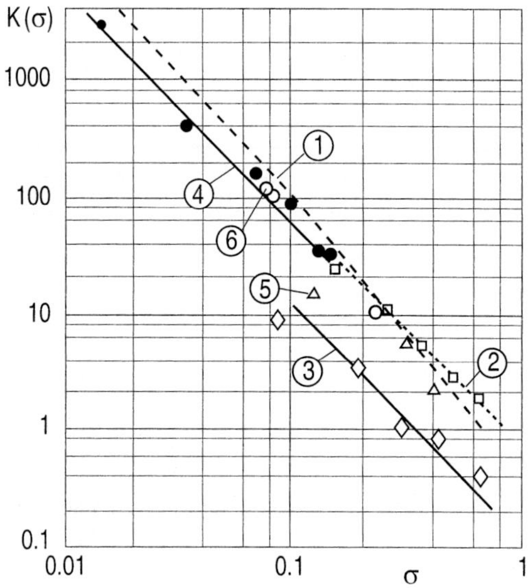

Values of the factor $K(\sigma)$ in ΔZ_S .

(1): With the stationary flow resistance coefficient ζ defined by $\Delta P = \zeta \cdot \rho_0/2 \cdot U^2$, where ΔP= static pressure drop, and U/σ= average flow velocity through the fence opening, the relation is: $K(\sigma) = 0.42 \cdot \zeta$ (2)

(2): Slit-shaped orifice with sharp corners : $K(\sigma) = 0.675/\sigma^2$ (3)

(3): Slit-shaped orifice with rounded corners: $K(\sigma) = 0.119/\sigma^2$ (4)

(4): Thin perforated sheet: $K(\sigma) = 0.58/\sigma^2$ (5)

(5),(6): Some other values for perforated sheets are taken from the literature.

2nd: *Nonlinearity by flow over orifices* :

A flow with velocity U is past the orifice with diameter d=2a of a neck in the duct wall. Experimental results by CUMMINGS for the real part Z' of the orifice input impedance and for the orifice end correction $\Delta\ell$ can be represented by the relations (f= frequency; ℓ= neck length):

$$\frac{Z'}{\rho_0 f d} = [12.52\cdot(\ell/d)^{-0.32} - 2.44]\cdot(U^*/fd) - 3.2$$

$$\frac{\Delta\ell}{\Delta\ell_0} = \begin{cases} 1 \quad ; \quad U^*/f\ell \le 0.12\, d/\ell \\ (1+0.6\,\ell/d)\cdot e^{-(U^*/f\ell+0.12\,d/\ell)/(0.25+\ell/d)} - 0.6\,\ell/d \quad ; \quad U^*/f\ell > 0.12\,d/\ell \end{cases} \tag{6}$$

Therein $\Delta\ell_0$ is the orifice end correction without flow; U^* is the flow shear velocity. It is evaluated from :

$$U^* = \sqrt{\lambda/8}\cdot\langle U\rangle \quad ; \quad \lambda = 0.306\cdot Re^{-1/4} \tag{7}$$

where $\langle U\rangle$ is the average velocity in the duct; λ is the coefficient of flow resistance by viscous shear; Re is the Reynolds number of the flow, using the duct diameter (the relation between λ and Re is for square ducts; the corresponding relation in circular ducts with diameter 2R is $\lambda = 0.316\cdot Re^{-1/4}$).

3rd: *Nonlinearity by flow through an orifice* :

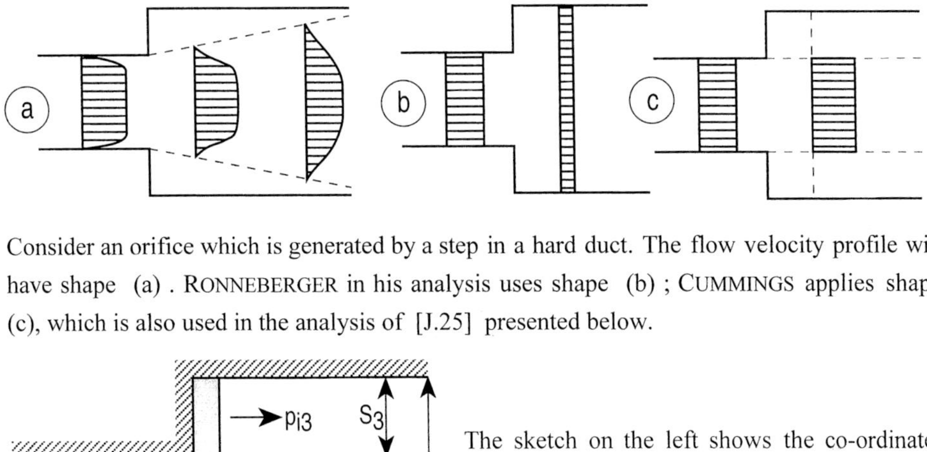

Consider an orifice which is generated by a step in a hard duct. The flow velocity profile will have shape (a) . RONNEBERGER in his analysis uses shape (b) ; CUMMINGS applies shape (c), which is also used in the analysis of [J.25] presented below.

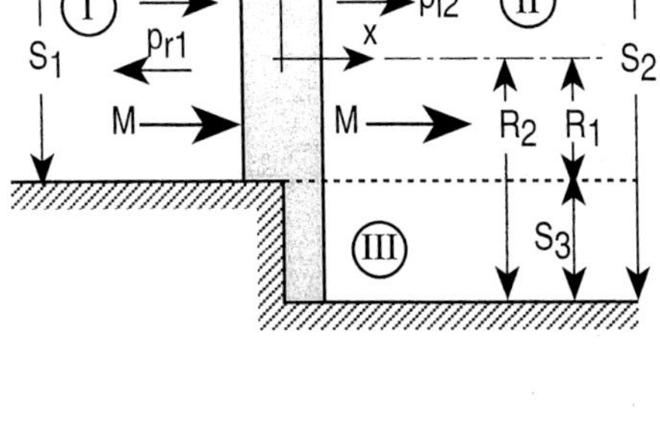

The sketch on the left shows the co-ordinates x,r , the duct areas S_i , the field zones I, II, III and the component sound fields. M is the Mach number with the average flow velocity. The grey area with limits near to x=0 may contain near fields, which will not be considered in detail.

Integrals of conservation of the mass, the impulse and the energy can be assumed to exist in this transition volume :

$$\int_A \rho\mathbf{v}\cdot d\mathbf{A}=0 \quad ; \quad \int_A (\rho\mathbf{v})\mathbf{v}\cdot d\mathbf{A}+\int_A p\cdot d\mathbf{A}=0 \quad ; \quad \int_A \rho H\mathbf{v}\cdot d\mathbf{A}=0 \tag{8}$$

where **A** is the surface of the volume; ρ, p, **v** are density, pressure and velocity, respectively; H is the stagnation enthalpy. The approximation $p_{i3}=p_{i1}+p_{r1}$ will be used at the step. The density variations are $\rho=(p+\delta)/c_0^2$, where δ is the pressure produced by variations of the enthalpy S :

$$S=\frac{-\delta}{\rho_0 T_0(\kappa-1)} \tag{9}$$

(T_0= stationary temperature; κ= adiabatic exponent). The stagnation enthalpy is :

$$H=T\cdot S+\frac{p}{\rho}+\frac{1}{2}|\mathbf{v}|^2 \tag{10}$$

Let the x factors of the sound fields p_{i2}, p_{i3} be e^{-jKk_0x} with a correction factor K for the free field wave number k_0. The axial particle velocity in zone II is :

$$v_{i2}=\frac{K\,p_{i2}}{\rho_0 c_0(1-MK)} \tag{11}$$

At the limit between zones II and III let $p_{i3}=p_{i2}$; let the fields within a zone be (approximately) constant in radial direction. Then, with the porosity $\sigma=S_1/S_2$, the integrals give :

$$\begin{aligned}
(1+M)p_{i1}-(1-M)p_{r1} &= \left(\frac{KS_3}{S_1}+M+\frac{K}{1-MK}\right)p_{i2}+M\delta \\
\left(\frac{1}{\sigma}+2M+M^2\right)p_{i1}+\left(\frac{1}{\sigma}-2M+M^2\right)p_{r1} &= \left(\frac{1}{\sigma}+M^2+\frac{2MK}{1-MK}\right)p_{i2}+M^2\delta \\
(1+M)p_{i1}+(1-M)p_{r1} &= \left(1+\frac{MK}{1-MK}\right)p_{i2}-\frac{\delta}{\kappa-1}
\end{aligned} \tag{12}$$

with the Bessel function $J_0(z)$ and the Neumann function $Y_0(z)$; the amplitude A follows from the condition of zero radial particle velocity at the outer radius of zone III .
The boundary conditions are $p_{i2}(R_1)=p_{i3}(R_1)$ and :

$$\frac{\partial p_{i3}(R_1)}{\partial r}=\frac{1}{(1-MK)^2}\frac{\partial p_{i2}(R_1)}{\partial r} \tag{13}$$

They lead to a characteristic equation for K :

$$\frac{J_1(k_0R_1\sqrt{1-K^2})\cdot Y_1(k_0R_2\sqrt{1-K^2})-J_1(k_0R_2\sqrt{1-K^2})\cdot Y_1(k_0R_1\sqrt{1-K^2})}{J_0(k_0R_1\sqrt{1-K^2})\cdot Y_1(k_0R_2\sqrt{1-K^2})-J_1(k_0R_2\sqrt{1-K^2})\cdot Y_0(k_0R_1\sqrt{1-K^2})}- \tag{14}$$

$$\frac{\sqrt{(1-MK)^2-K^2}\cdot J_1(k_0R_1\sqrt{(1-MK)^2-K^2})}{(1-MK)^2\sqrt{1-K^2}\cdot J_0(k_0R_1\sqrt{(1-MK)^2-K^2})}=0$$

A start value for its numerical solution is $K\approx 1/(1+\sigma M)$.

The following diagrams show the magnitude of the reflection factor $r_M = p_{r1}(x=0)/p_{i1}(x=0)$.

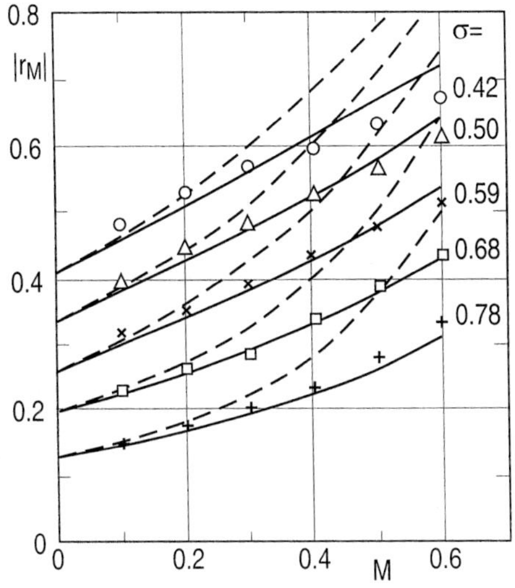

Magnitude of the reflection factor r_M over the Mach number M for different porosities $\sigma = S_1/S_2$.
Points: measured by RONNEBERGER;
dashed: computed by RONNEBERGER;
full: present computation, as in [J.27].

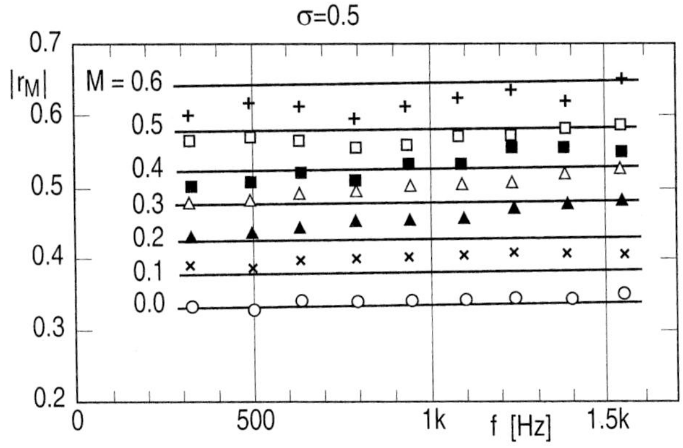

Magnitude of the reflection factor r_M over the frequency for porosity $\sigma = 0.5$ and different Mach numbers M .
Points: measured;
curves: present evaluation.

4th: *Non-linearity by flow along mineral fibre absorbers* :

The flow resistivity $\Xi(U)$ of fibrous absorbers with flow along their surface from measurements with flow velocities up to U= 80 [m/s] can be represented by :

$$\frac{\Xi(U)}{\Xi(0)}=(1-A_f U)^{-4} \quad ; \quad A_{f\,[s/m]}\approx\frac{0.085}{\sqrt{f_{[Hz]}}} \tag{15}$$

5th: *Non-linearity by flow through porous absorbers* :

The following representation of the characteristic propagation constant Γ_a and wave impedance Z_a with flow through the porous material does not include the possibility that the material is compressed by the flow !

$$\Gamma_a=\sqrt{j\omega\sigma C_{eff}\,[F_\eta\Xi+2\xi_t\cdot|U|+j\omega\rho_0\frac{\chi}{\sigma}]}$$

$$Z_a=\sqrt{\frac{\rho_0\chi/\sigma^2+\frac{1}{j\omega\sigma}(F_\eta\Xi+2\xi_t|U|)}{C_{eff}}} \tag{16}$$

with : (17)

$$C_{eff}=\frac{1}{\rho_0 c_0^2}[1+(\kappa-1)\frac{\tan(k_{\alpha0}h)}{k_{\alpha0}h}]$$

$$F_\eta=\frac{1}{3}\frac{(\frac{\Theta_\rho}{\Theta_\alpha}-1)k_v h\cdot\tan(k_v h)}{1-\frac{1}{\sqrt{\kappa\,Pr}}\frac{\Theta_\rho}{\Theta_\alpha}\tan(k_{\alpha0}h)+(\frac{\Theta_\rho}{\Theta_\alpha}-1)\tan(k_v h)}$$

f=	frequency;
ω=	angular frequency;
ρ_0=	air density;
c_0=	sound velocity in air;
U=	stationary flow velocity;
κ=	adiabatic exponent;
Pr=	Prandtl number;
σ=	material porosity;
χ=	1.362= structure factor;
Ξ=	measured flow resistivity;
ξ_t=	quadratic term of flow resistivity;
k_v=	viscosity wave number;
$k_{\alpha0}$=	thermal wave number;
Θ_ρ, Θ_α	see section B.1

$$k_{\alpha0}h=\sqrt{\kappa\,Pr}\cdot k_v h \quad ; \quad k_v h=\sqrt{-j6\pi E} \quad ; \quad E=\rho_0 f/\Xi \tag{18}$$

and ξ_t from a ΔP-U record of stationary flow with velocity U through a material layer with thickness Δz according to :

$$\frac{-\Delta P}{\Delta z\cdot U}=\Xi+\xi_t\cdot U \tag{19}$$

(make a quadratic regression through measured ΔP-U values; the coefficient of the linear term in U gives Ξ; the coefficient of the quadratic term gives ξ_t).

The following diagrams show measured points and computed curves for the characteristic values Γ_a, Z_a in a polyurethane foam for velocities U= 0; 0.82; 1.96 [m/s] .

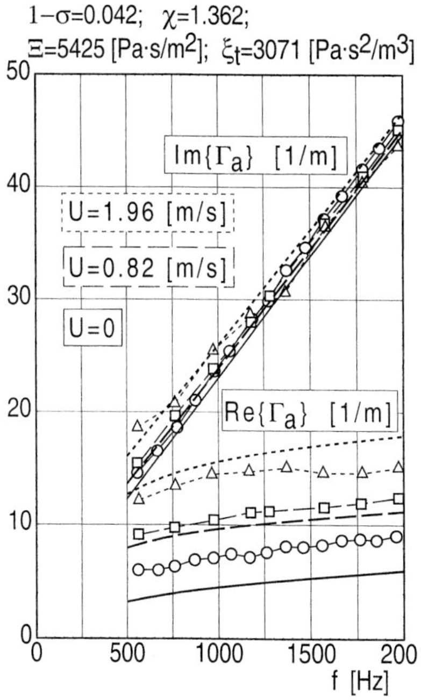

Real and imaginary parts of the propagation constant Γ_a in a PU foam with three flow velocities.

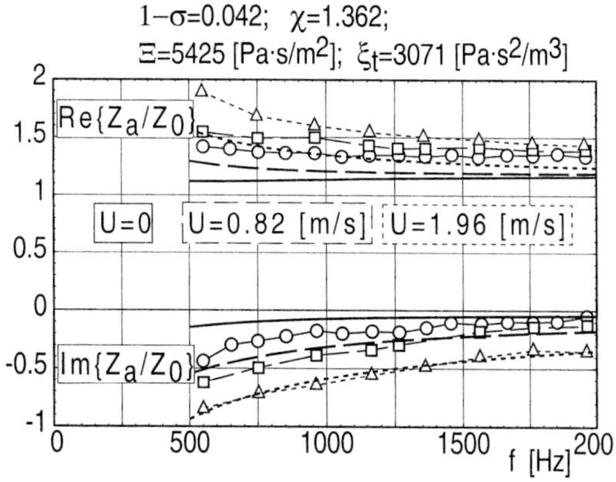

Real and imaginary parts of the normalised wave impedance Z_a/Z_0 in PU foam, with three flow velocities.

J.39 Flow-induced nonlinearity of perforated sheets

Ref.: Mechel, [J.25]; Coelho, [J.28]

The following table gives partition impedances $Z_m = Z'_m + j \cdot Z''_m$ for perforated sheets, which are mostly based on experimental data. The perforations are circular with radius a at mutual distances b ; the porosity is $\sigma = \pi a^2/b^2$; the sheet thickness is t . The Mach number M_0 for high sound levels is $M_0 = v/c_0$, with v the particle velocity in the exit orifice; in the Mach number $M_\infty = U_\infty/c_0$ the flow velocity U_∞ belongs to the undisturbed flow parallel to the sheet. Ranges of the sound pressure level L_p (relative to 20 [μPa]) are given by L_{0l}, L_{0h} without flow, and L_{Ul}, L_{Uh} with flow; ν is the kinematic viscosity. Other symbols are explained in the table below.

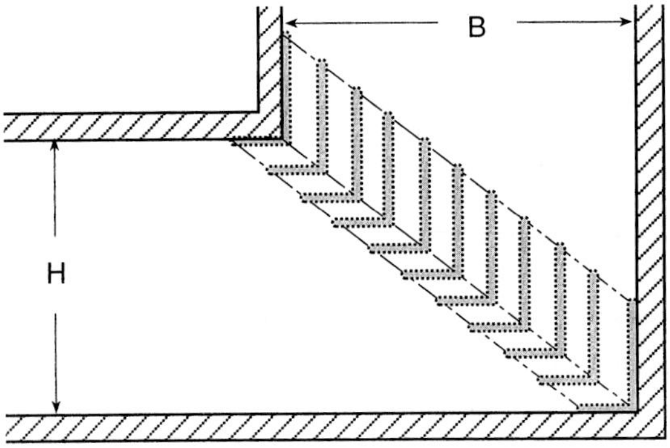

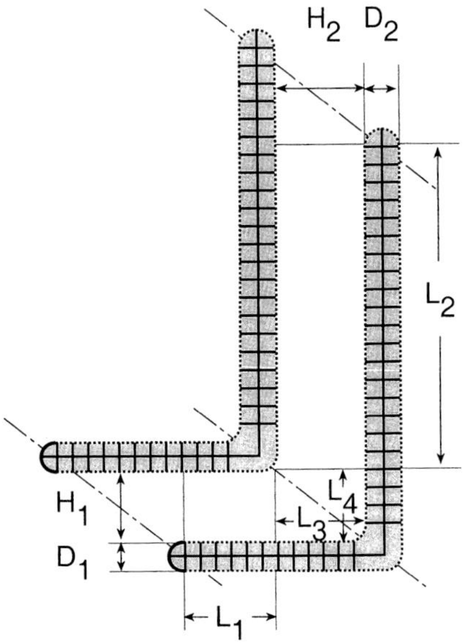

The details of the splitters are shown in the graph: The splitters are assumed to be locally reacting in the numerical example given below, which is indicated by internal partitions; it is also assumed that the splitters have a hard central sheet.

The splitters are treated as a combination of two lined ducts, one with a length L_1 and width H_1, the other with L_2 and H_2, plus a lined duct corner with dimensions $L_3 = H_2$ and $L_4 = H_1$. (the possible additional transmission loss due to the additional lengths L_3 and L_4 is neglected below).

Such objects are treated in the Section J.26, "Lined duct corners and junctions".

The insertion loss of the turning-vane splitter silencer is the sum of the transmission losses of the two straight ducts plus the insertion loss of the corner which is evaluated as in Section J.26.

The dimensions of the ducts in the example shown are H= 10 m, and B= 17 m (in a gas turbine test cell, with a meanflow Mach number in the empty stack of M= 0.0424, and in the silencer ducts of M= 0.106). In the parameter list of the example is $H_1 = 2\cdot h1$; $D_1 = 2\cdot d1$; $H_2 = 2\cdot h2$; $D_2 = 2\cdot d2$. The spliiter branches consist of layers of glass fibre (made locally reacting) with flow resistivity values $\Xi 1$, $\Xi 2$, covered with porous steel foils of thickness df1, df2 with normalised flow resistances Rf1, Rf2 , respectively. It is assumed that a plane wave is incident.

The lengths L_i ; i=1,2; should be $L_i \geq 2H_i$.

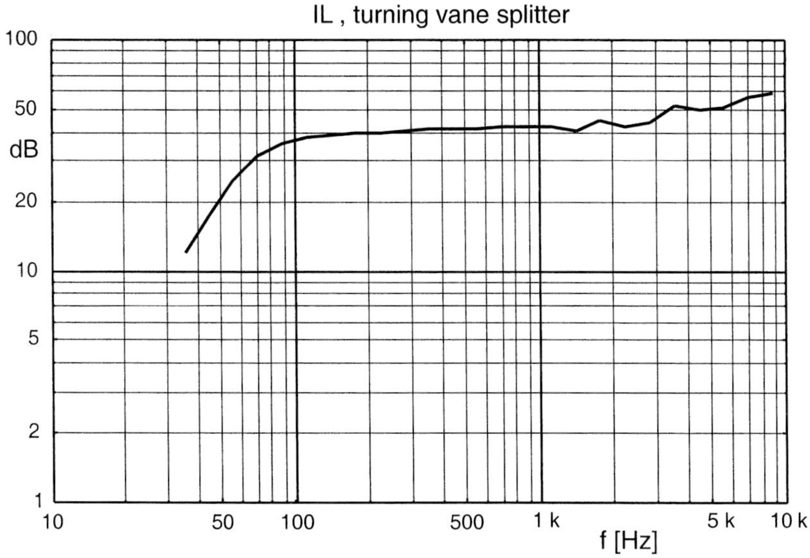

Parameters:

IL with flow, turning vane splitter silencer
locally lined duct corner
M(B)=0.0424 empty, mode index limit mhi=8

Containing ducts:
B[m]=17., H[m]=10.

Duct(1):
L1[m]=3.
h1[m]=0.2 , d1[m]=0.3
Ξ1[Pa s/m^2]=2000.
rof1[kg/m^3]=7800. , df1[m]=0.0005 , Rf1=3.

Duct(2):
L2[m]=3.
h2[m]=0.34 , d2[m]=0.51
Ξ2[Pa s/m^2]=2000.
rof2[kg/m^3]=7800. , df2[m]=0.0005 , Rf2=1.

Corner(3):
h3[m]=0.2 , d3[m]=0.3
Ξ3[Pa s/m^2]=2000.
rof3[kg/m^3]=7800. , df3[m]=0.0005 , Rf3=3.

Corner(4):
h4[m]=0.34, d4[m]=0.51
Ξ4[Pa s/m^2]=2000.
rof4[kg/m^3]=7800. , df4[m]=0.0005 , Rf4=1.

[J.21] MECHEL, F.P.
"Schallabsorber", Vol. III, Chapter 41:
"CREMER admittance and hybrid absorbers"
S.Hirzel Verlag, Stuttgart, 1998

[J.22] PRESS, W.H., FLANNERY, B.P., TEUKOLSKY, S.A., VETTERLING, W.T.
"Numerical Recipes",
Cambridge Univerity Press, 1989, N.Y.

[J.23] MECHEL, F.P.
"Schallabsorber", Vol. III, Chapter 25.4:
"Superposition of flow"
S.Hirzel Verlag, Stuttgart, 1998

[J.24] MECHEL, F.P.
"Schallabsorber", Vol. III, Chapter 42:
"Influence of flow and temperature on attenuation"
S.Hirzel Verlag, Stuttgart, 1998

[J.25] MECHEL, F.P.
"Schallabsorber", Vol. II, Chapter 28:
"Nonlinearities by amplitude and flow"
S.Hirzel Verlag, Stuttgart, 1995

[J.26] RONNEBERGER, D. Acustica 19 (1967/68) 222-235
"Experimentelle Untersuchungen zum akustischen Reflexionsfaktor von unstetigen Querschnittsänderungen in einem luftdurchströmten Rohr"

[J.27] CUMMINGS, A. J.Sound Vibr. 38 (1975) 149-155
"Sound Transmission at Sudden Area Expansions in Circular Ducts with Superimposed Mean Flow"

[J.28] COELHO, J.L.B.
"Acoustic Characteristics of Perforate Liners in Expansion Chambers"
Thesis, Fac. Engineer., Inst. Sound Vibr., Southampton, 1983

[J.29] Young-Chung CHO J.Acoust. Soc.Amer. 67 (1980) 1421-1426
"Reciprocity principle in duct acoustics"

[J.30] MECHEL, F.P.
"Schallabsorber", Vol. III, Chapter 33:
"Duct with steps"
S.Hirzel Verlag, Stuttgart, 1998

K
Muffler Acoustics
by M.L. Munjal

The performance of acoustic mufflers relies heavily on reflections at duct discontinuities, such as steps of the cross-section or of the duct lining, following each other in short distances. Such transitions, being mostly neglected in the chapter J about long, homogeneous silencers (often called "industrial silencers"), together with the always necessary consideration of mean flow and often of high temperatures, give the Muffler Acoustics a special character.

K.0 Conventions in the present chapter

Muffler acoustics is preferably formulated with the field quantities pressure p and volume flow velocity u in the duct. Therefore mostly the volume flow impedance (or flow impedance), defined by the ratio $p/u = p/(v \cdot S)$ is used, where v is the particle velocity (as usual in this book) and S is the duct cross section. The flow impedance is indicated by underlining:

$$\underline{Z}_x = \frac{p_x}{u_x} = \frac{p_x}{v_x S_x}$$

where p_x, u_x, v_x, S_x respectively are the sound pressure, volume flow, particle velocity, and duct cross-section at a position x of the duct. An exception to this rule may be the symbol $\underline{Z}_0 = \rho_0 c_0/S$ if the cross section S is unspecified, and $\underline{Z}_r$ which stands for a radiation impedance in the dimensions of a flow impedance.

It is convenient to formulate expressions for sound fields in a steady flow with *convected quantities* (see section K.1). Such convected quantities will be indicated with an index c .

Matrix formulations play an important role in muffler acoustics. The conventions for writing a vector are {v}= {p, u}, and [M] for a matrix in the running text. A matrix equation may be written either as {u}= [M]·{v} or with the elements as

$$\begin{bmatrix} u_1 \\ u_2 \end{bmatrix} = \begin{bmatrix} m_{11} & m_{12} \\ m_{21} & m_{22} \end{bmatrix} \cdot \begin{bmatrix} v_1 \\ v_2 \end{bmatrix}$$

Most graphs in this chapter will indicate by points • u and • d duct cross-sections just above the upsound and just below the downsound cross-sections, respectively, between which a transformation matrix will be developed.

$\underline{Z}_r(M)$ differs from the corresponding stationary medium $\underline{Z}_r(0)$ because of not only the convective effect, but also the interaction of the outgoing (radiated) wave with an unstable cylindrical vortex layer of the meanflow jet, [K.7]. In fact, the total acoustic power in the farfield Π_F is less than the acoustic power Π_T transmitted out, [K.8], particularly at low frequencies:

$$\Pi_T \approx \Pi_F \frac{(k_0 r_0)^2}{2M + (k_0 r_0)^2} \tag{6}$$

K.3 Transfer matrix representation

Transfer matrix representation is ideally suited for the analysis of cascaded one-dimensional systems such as acoustic filters or mufflers. The performance of a muffler may be obtained readily in terms of the four-pole parameters or transfer matrix of the entire system, which in turn may be computed by means of successive multiplication of the transfer matrices of the constituent elements.

Transfer matrices of different elements constituting commercial mufflers are given in the subsequent sections of this chapter. Transformation from classical state invariables p and u to the convective ones p_c and u_c may be obtained as follows.

Let $\{S\} = \{p, u\}^T$ and $\{S_c\} = \{p_c, u_c\}^T$. Below, vectors are denoted by braces { } and matrices by brackets [ ].

Subscripts u and d denote the upstream end and downstream end respectively of a muffler element, and

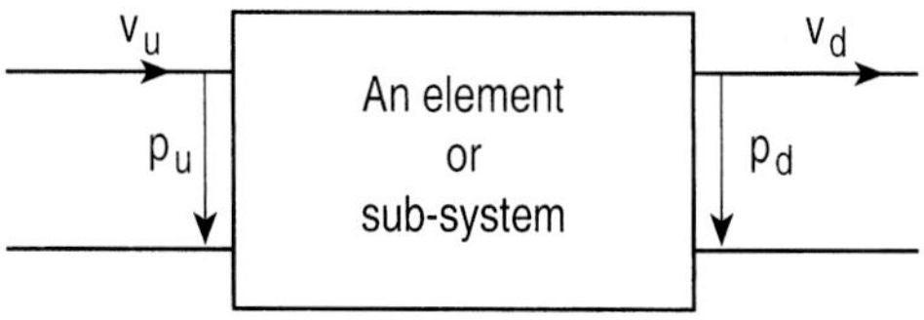

$$\{S\}_u = [T]\{S\}_d$$
$$\{S_c\}_u = [T_c]\{S_c\}_d \tag{1}$$

where [T], [T_c] are transfer matrices, then :

$$[T_c] = [C]_u [T] [C]_d^{-1} \tag{2}$$

Conversely :

$$[T] = [C]_u^{-1} [T_c] [C]_d \tag{3}$$

where [C] is the transformation matrix:

$$[C] = \begin{bmatrix} 1 & M\underline{Z}_0 \\ M/\underline{Z}_0 & 1 \end{bmatrix} \tag{4}$$

Thus, one can work with classical state variables or convective state variables as per personal preference, and skip from one system to the other at the end as necessary.

K.4 Muffler performance parameters

The performance of a muffler is measured in terms of one of the following parameters:

- Insertion loss, IL ;
- Transmission loss, TL ;
- Level difference, LD, or noise reduction, NR .

Insertion loss, IL is defined as the difference between the acoustic power radiated without any muffler and that with the muffler (inserted, as it were, between the source and the radiation load impedance), as shown in the figure.

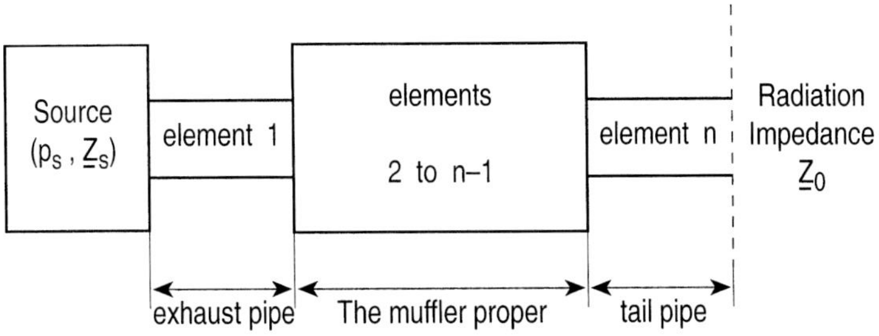

Writing $\Pi(M)$ in the form which defines $\underline{R}_e$, [K.9] :

$$\Pi(M)=\frac{|u|^2}{2}\left(\underline{R}_r+M\underline{Z}_0+M|\underline{Z}_r|^2/\underline{Z}_0\right)=\frac{|u|^2}{2}\underline{R}_e \tag{1}$$

the *insertion loss* may be expressed in terms of the product transfer matrix parameters as :

$$IL=20\log\left[\left(\frac{R_{e1}}{R_{en}}\right)^{1/2}\left|\frac{Z_{rn}T_{11}+T_{12}+Z_{rn}Z_sT_{21}+Z_sT_{22}}{Z_{r1}+Z_s}\right|\right] \tag{2}$$

with :

$$\begin{aligned}\underline{R}_{e1}&=\underline{R}_{r1}+M_1\underline{Z}_1+M_1|Z_{r1}|^2/\underline{Z}_1\\ \underline{R}_{en}&=\underline{R}_{rn}+M_n\underline{Z}_n+M_n|Z_{rn}|^2/\underline{Z}_n\end{aligned} \tag{3}$$

where $\underline{Z}_{rn}=\underline{R}_{rn}+j\cdot\underline{X}_{rn}$ is the (flow) radiation impedance of the n-th element, especially $\underline{Z}_{r1}$ that of the exhaust pipe without any muffler, $\underline{Z}_{rn}$ that of the tail pipe, and $\underline{Z}_n$ is the characteristic flow impedance in the n-th element. Further, $\underline{Z}_s$ the source (flow) impedance is defined with respect to classical field variables. T_{11}, T_{12}, T_{21} and T_{22} are the four-pole parameters of the product transfer matrix of the entire muffler, obtained by successive multiplication of the transfer matrices of exhaust pipe (element number.1, next to the source), elements of the muffler proper, ending with the tail pipe (element n), as shown in the figure.

K.6 Sudden area changes

Subscripts u and d indicate points just upstream and just downstream of the sudden area discontinuity. Typically, the meanflow Mach number M in the smaller diameter tube is M< 0.2 .

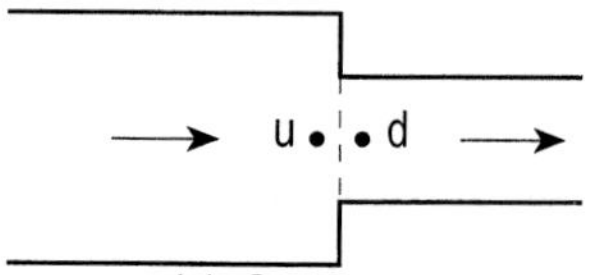

(a) Sudden contraction

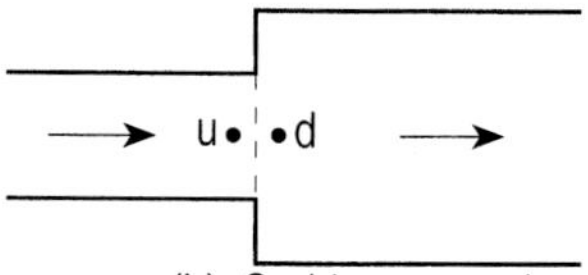

(b) Sudden expansion

For sudden expansion and sudden contraction, the equations of mass continuity, momentum balance, stagnation pressure drop and entropy fluctuations, [K.11, K.12], for plane waves and incompressible mean flow yield the transfer matrix relation, [K.3] :

$$\begin{bmatrix} p_{c,u} \\ u_{c,u} \end{bmatrix} = \begin{bmatrix} 1-\dfrac{K_d M_d^2}{1-M_d^2} & K_d M_d \underline{Z}_d \\ \dfrac{(\kappa-1)K_d M^3{}_d}{(1-M_d^2)\underline{Z}_d} & 1-\dfrac{(\kappa-1)K_d M_d^2}{1-M_d^2} \end{bmatrix} \begin{bmatrix} p_{c,d} \\ u_{c,d} \end{bmatrix} \tag{1}$$

K_d is the K-factor indicating the drop in stagnation pressure of the incompressible mean flow in terms of the dynamic head $\rho_0 V_d^2/2$:

$$\begin{aligned} K_d &= (1-S_d/S_u)/2 \quad \text{for sudden contration} \\ K_d &= \left[(S_d/S_u)-1\right]^2 \quad \text{for sudden expansion} \end{aligned} \tag{2}$$

If M<0.2 in the smaller diameter tube, the foregoing transfer matrix may be approximated as, [K.3] :

$$\begin{bmatrix} p_{c,u} \\ u_{c,u} \end{bmatrix} = \begin{bmatrix} 1 & K_d M_d \underline{Z}_d \\ 0 & 1 \end{bmatrix} \begin{bmatrix} p_{c,d} \\ u_{c,d} \end{bmatrix} \tag{3}$$

This relation, on transformation to the classical state variables, and on incorporating the low Mach number simplification, becomes :

$$\begin{bmatrix} p_u \\ u_u \end{bmatrix} = \begin{bmatrix} 1 & (1+K_d)M_d \underline{Z}_d - M_u \underline{Z}_u \\ 0 & 1 \end{bmatrix} \begin{bmatrix} p_d \\ u_d \end{bmatrix} \tag{4}$$

In the simplified, approximate, form, the end-correction effect or evanescent higher-order modes effect can be incorporated readily as an inline lumped inertance, [K.13] :

$$\begin{bmatrix} p_u \\ u_u \end{bmatrix} = \begin{bmatrix} 1 & (1+K_d)M_d\underline{Z}_d - M_u\underline{Z}_u + j\omega\, H(\alpha)\cdot 0.85\, r_p / S_p \\ 0 & 1 \end{bmatrix} \begin{bmatrix} p_d \\ u_d \end{bmatrix} \tag{5}$$

where r_p is the radius of the narrower pipe;
$\Delta\ell = 0.85\, r_p$ is the end correction of the orifice in a baffle wall;
$H(\alpha) = 1-1.25\, r_p/r_{ch}$ for co-axial tubes;
r_{ch} is radius of the chamber (or the larger diameter tube).

If the junction of the pipe (the smaller diameter tube) and chamber (the larger diameter tube) is not co-axial, then the factor $H(\alpha)$ is given by the polynomial, [K.12] :

$$\begin{aligned} H(\alpha) = {} & 1.442+3.516\,\delta - 5.403\alpha - 0.068\, kr - 11.067\,\delta^2 + 10.462\,\delta^2 - 0.099\,(kr)^2 + \\ & + 2.517\,\delta\alpha - 0.197\,\delta.kr + 1.024\,\alpha.kr + 7.774\,\delta^3 - 8.15\,\alpha^3 - 0.05\,(kr)^3 - \\ & - 0.841\,\delta^2\alpha + 0.131\,\delta^2.kr - 3.378\,\delta\alpha^2 - 1.311\,\alpha^2.kr + \\ & + 0.141(kr)^2.\,\delta - 0.067\,(kr)^2\alpha - 0.031\,\delta.\alpha.kr \end{aligned} \tag{6}$$

where $kr = k_0 r_{ch}$; $\alpha = r_p / r_{ch}$; $\delta = \text{offset distance} / (r_{ch} - r_p)$.

For the case of a stationary medium, the transfer matrix would reduce to :

$$\begin{bmatrix} 1 & j\omega H(\alpha)\, 0.85\, r_p / S_p \\ 0 & 1 \end{bmatrix} \tag{7}$$

The description of sudden area changes would become simple if one neglected the evanescent higher-order mode effects, because then the transfer matrix would reduce to a unit matrix, implying $p_u = p_d$ and $u_u = u_d$.

Decomposing the standing wave on the entrance side into the forward moving and reflected progressive waves, for anechoic termination, one gets, [K.3] :

Reflection factor :
$$R = \frac{\underline{Z}_d - \underline{Z}_u}{\underline{Z}_d + \underline{Z}_u} \approx \frac{S_u - S_d}{S_u + S_d} \tag{8}$$

Transmission loss :
$$TL = 10\lg\left[\frac{(\underline{Z}_u + \underline{Z}_d)^2}{4\underline{Z}_u\underline{Z}_d}\right] \approx 10\cdot\log\left[\frac{(S_d + S_u)^2}{4S_dS_u}\right] \tag{9}$$

These relationships represent the principle of Impedance Mismatch which is the underlying principle of reflective (or reactive, or non-dissipative) mufflers. When the characteristic impedance undergoes a sudden change or jump, a significant portion of the incident acoustic power is reflected back to the source. This impedance jump may be obtained in several ingeneous ways, one of which is sudden area changes. The symmetry of the expression for TL shows that what matters is a sudden jump or change in characteristic impedance, not whether it increases (as in sudden contraction) or decreases (as in sudden expansion).

K.8 Conical tube

With tube length ℓ and tube radii r_u, r_d, the following relations are used:

$$z_1 = \frac{r_u}{r_d - r_u}\ell \quad ; \quad z_2 = z_1 + \ell$$

$$\underline{Z}_d = \frac{\rho_0 c_0}{\pi r_d^2}$$

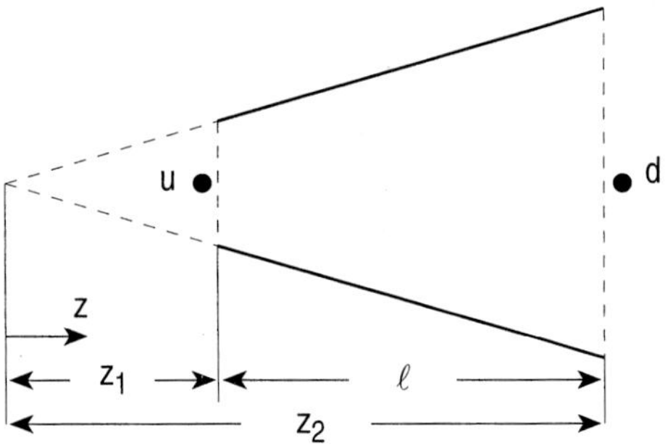

(a) a conical tube

For stationary medium, the transfer matrix relationship for a conical tube is given by, [K.3] :

$$\begin{bmatrix} p_u \\ u_u \end{bmatrix} = \begin{bmatrix} \frac{z_2}{z_1}\cos(k_0\ell) - \frac{\sin(k_0\ell)}{k_0 z_1} & j\underline{Z}_d \frac{z_2}{z_1}\sin(k_0\ell) \\ \frac{j}{\underline{Z}_d}\left\{\frac{z_1}{z_2}\left(1+\frac{1}{k_0^2 z_1 z_2}\right)\sin(k_0\ell) - \left(1-\frac{z_1}{z_2}\right)\frac{\cos(k_0\ell)}{k_0 z_2}\right\} & \frac{\sin(k_0\ell)}{k_0 z_2} + \frac{z_1}{z_2}\cos(k_0\ell) \end{bmatrix} \begin{bmatrix} p_d \\ u_d \end{bmatrix} \qquad (1)$$

These expressions hold for a convergent tube as well as a divergent tube.

For a moving medium, assuming that the flare is small enough to avoid separation of boundary layer, Easwaran and Munjal, [K.15], have solved the wave equation with variable coefficients analytically to obtain the transfer matrix parameters for a conical tube. Dokumaci has obtained these parameters numerically by means of matrizants, [K.16]. The resultant expressions are rather complicated. Fortunately, however, the convective effect of incompressible mean flow (for $M<0.2$) is negligible in the case of conical tubes as well as uniform tubes.

K.9 Exponential horn

The medium is assumed to be stationary. The horn (or tube section) is characterised by the relations (with $r(z)$ the radius at position z ; m is the flare constant) :

$$r(z) = r(0)\, e^{mz} \quad ; \quad S(z) = S(0)\, e^{2mz}$$

$$\underline{Z}_u = \frac{\rho_0 c_0}{\pi r_u^2} \quad ; \quad r_u = r(0) \qquad (1)$$

$$\begin{bmatrix} p_u \\ u_u \end{bmatrix} = \begin{bmatrix} e^{m\ell}\left(\cos k'\ell - \frac{m}{k'}\sin k'\ell\right) & je^{-m\ell}\frac{k_0}{k'}\underline{Z}_u \sin k'\ell \\ \frac{j}{\underline{Z}_u} e^{m\ell}\frac{k_0}{k'}\sin k'\ell & e^{-m\ell}\left(\cos k'\ell + \frac{m}{k'}\sin k'\ell\right) \end{bmatrix} \begin{bmatrix} p_d \\ u_d \end{bmatrix} \tag{2}$$

where $k' = (k_0^2 - m^2)^{1/2}$; $k_0 = \omega / c_0$ (3)

These expressions hold for a convergent horn as well as divergent horn.

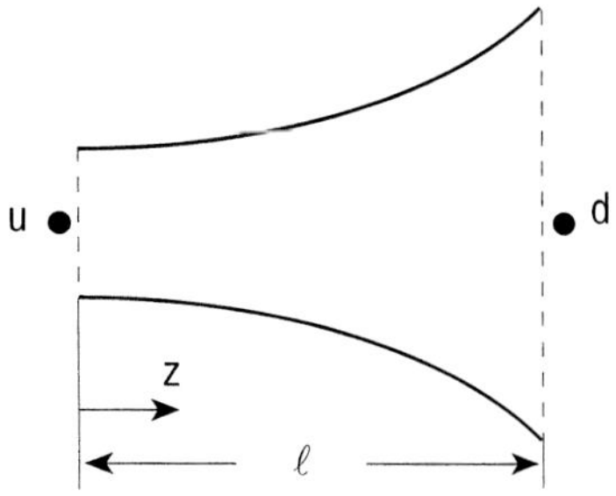

(b) a hyperbolic tube

For a moving medium, assuming that the flare is small enough to avoid separation of boundary layer, Easwaran and Munjal, [K.17], have solved the wave equation with variable coefficients analytically to obtain the transfer matrix parameters for an exponential tube. Dokumaci has obtained the same numerically by means of matrizants, [K.16]. The resultant expressions are rather too complicated. Fortunately, however, the convective effect of incompressible mean flow (for M< 0.2) is negligible in the case of exponential tubes as well as conical tubes.

K.10 Hose

A hose is a uniform tube of interior radius r_i with compliant walls. Incorporating the local wall impedance in the mass continuity equation, and the losses due to visco-thermal friction and turbulent eddies in the momentum equation, and neglecting entropy fluctuations for the linear plane waves, yields the following transfer matrix relationship [K.17] :

$$\begin{bmatrix} p_u \\ u_u \end{bmatrix} = \frac{e^{j(k^+-k^-)\ell}}{\underline{Z}^+ + \underline{Z}^-} \begin{bmatrix} (\underline{Z}^- e^{-jk^+\ell} + \underline{Z}^+ e^{jk^-\ell}) & \underline{Z}^+\underline{Z}^-(e^{+jk^-\ell} - e^{-jk^+\ell}) \\ e^{jk^-\ell} - e^{-jk^+\ell} & \underline{Z}^+ e^{-jk^+\ell} + \underline{Z}^- e^{jk^-\ell} \end{bmatrix} \begin{bmatrix} p_d \\ u_d \end{bmatrix} \tag{1}$$

Here, the wave numbers are given by :

$$k^{\pm} = \frac{k_i}{1-M^2}\left[\left\{\left(1 - j\,\frac{\alpha_M r_i + G_i}{k_i r_i}\right)^2 + (1-M^2)\left(\frac{\alpha_M r_i + G_i}{k_i r_i}\right)^2\right\}^{1/2} \mp M\left(1 - j\,\frac{\alpha_M r_i + G_i}{k_i r_i}\right)\right] \tag{2}$$

$$\begin{bmatrix} p_1(0) \\ p_2(0) \\ \underline{Z}_1\, u_1(0) \\ \underline{Z}_2\, u_2(0) \end{bmatrix} = \begin{bmatrix} T_{11} & T_{12} & T_{13} & T_{14} \\ T_{21} & T_{22} & T_{23} & T_{24} \\ T_{31} & T_{32} & T_{33} & T_{34} \\ T_{41} & T_{42} & T_{43} & T_{44} \end{bmatrix} \begin{bmatrix} p_1(\ell) \\ p_2(\ell) \\ \underline{Z}_1\, u_1(\ell) \\ \underline{Z}_2\, u_2(\ell) \end{bmatrix} \tag{1}$$

where $[T]=[A(0)]\,[A(1)]^{-1}$. Elements of the matrix $[A(z)]$ are given by $i=1,2,3,4$:

$$A_{1,i} = \psi_{3,i}\, e^{\beta_i z} \quad ; \quad A_{2i} = \psi_{4\,i}\, e^{\beta_i z}$$

$$A_{3,i} = -\frac{e^{\beta_i z}}{j k_0 + M_1 \beta_i} \quad ; \quad A_{4i} = -\frac{\psi_{2,i}\, e^{\beta_i z}}{j k_0 + M_2 \beta_i} \tag{2}$$

$[\psi]$ and $\{\beta\}$ are respectively the eigenmatrix (or modal matrix) and eigenvector of the matrix :

$$\begin{bmatrix} -\alpha_1 & -\alpha_3 & -\alpha_2 & -\alpha_4 \\ -\alpha_5 & -\alpha_7 & -\alpha_6 & -\alpha_8 \\ 1 & 0 & 0 & 0 \\ 0 & 1 & 0 & 0 \end{bmatrix}$$

where :

$$\alpha_1 = -\frac{jM_1}{1-M_1^2}\left(\frac{k_a^2+k_0^2}{k_0}\right) \quad ; \quad \alpha_2 = \frac{k_a^2}{1-M_1^2}$$

$$\alpha_3 = \frac{jM_1}{1-M_1^2}\left(\frac{k_a^2-k_0^2}{k_0}\right) \quad ; \quad \alpha_4 = -\left(\frac{k_a^2-k_0^2}{1-M_1^2}\right)$$

$$\alpha_5 = \frac{jM_2}{1-M_2^2}\left(\frac{k_b^2-k_0^2}{k_0}\right) \quad ; \quad \alpha_6 = -\left(\frac{k_b^2-k_0^2}{1-M_2^2}\right) \tag{3}$$

$$\alpha_7 = -\frac{jM_2}{1-M_2^2}\left(\frac{k_b^2+k_0^2}{k_o}\right) \quad ; \quad \alpha_8 = \frac{k_b^2}{1-M_2^2}$$

with :

$$k_0 = \omega / c_0 \; ; \quad M_1 = V_1 / c_0 \quad ; \quad M_2 = V_2 / c_0$$

$$k_a^2 = k_0^2 - \frac{4 j k_0}{d_i \zeta} \quad ; \quad k_b^2 = k_0^2 - \frac{4 j k_0 d_1}{\left(d_2^2 - d_1^2\right)\zeta} \tag{4}$$

ζ is the normalised partition impedance of the perforate. For different flow conditions, ζ is given by the following empirical expressions :

Perforates with-cross flow, [K.19] :

$$\zeta = \frac{p}{\rho_0 c_0 v} = \left[0.514\frac{d_1 M}{\ell\sigma} + j0.95 k_0 (t + 0.75 d_h\right]/\,\sigma \tag{5}$$

where
- d_1 is the diameter of the perforated tube;
- M is the mean-flow Mach number in the tube;
- ℓ is the length of perforate;
- σ is the porosity;
- f is the frequency;
- t is the thickness of the perforated tubc;
- d_h is the hole diameter;
- v is the (average) radial particle velocity at the perforate.

For perforates in stationary media, [K.20] :

$$\zeta = [0.006 + jk_0(t + 0.75\, d_h] / \sigma \tag{6}$$

Perforates with grazing flow, [K.21] :

$$\zeta = [\, 7.337 \text{ x } 10^{-3}(1 + 72.23\ M) + j\, 2.2245 \text{ x } 10^{-5}(1 + 51\ t)\,(1 + 204\ d_h)\, f\,] / \sigma \tag{7}$$

where wall thickness t and hole diameter d_h are in meters.

The desired 2x2 transfer matrix for a particular two-duct element may be obtained from the 4x4 matrix [T] by making use of the appropriate upstream and downstream variables and two boundary conditions characteristic of the element. The final results, [K.3], are given below for various two-duct elements shown in figures.

(a) Concentric-tube resonator :

$$\begin{bmatrix} p_1(0) \\ \underline{Z}_1\, u_1(0) \end{bmatrix} = \begin{bmatrix} T_a & T_b \\ T_c & T_d \end{bmatrix} \begin{bmatrix} p_2(\ell) \\ \underline{Z}_1 u_1(\ell) \end{bmatrix}$$

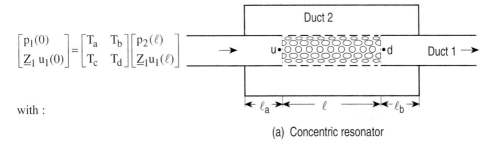

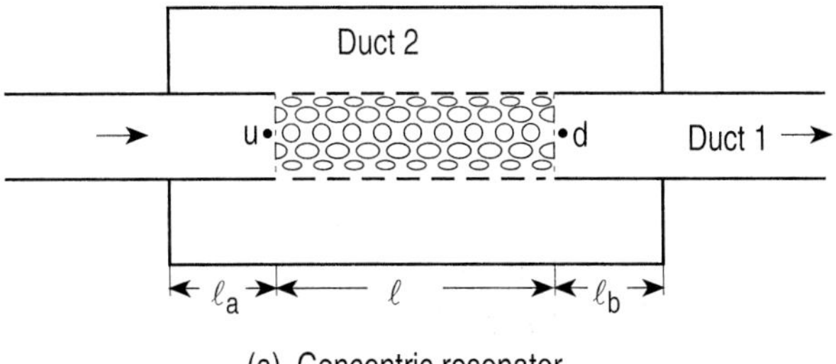

(a) Concentric resonator

with :

$$T_a = T_{11} + A_1A_2 \quad ; \quad T_b = T_{13} + B_1A_2$$
$$T_c = T_{31} + A_1B_2 \quad ; \quad T_d = T_{33} + B_1B_2 \tag{9}$$

$$A_1 = (X_1T_{21} - T_{41}) / F_1 \quad ; \quad B_1 = (X_1T_{23} - T_{43}) / F_1$$
$$A_2 = T_{12} + X_2T_{14} \quad ; \quad B_2 = T_{32} + X_2T_{34} \tag{10}$$

$$F_1 = T_{42} + X_2T_{44} - X_1(T_{22} + X_2T_{24})$$
$$X_1 = -j\tan(k_0\ell_a) \quad ; \quad X_2 = +j\tan(k_0\ell_b) \tag{11}$$

(b) Cross-flow expansion element :

$$\begin{bmatrix} p_1(0) \\ \underline{Z}_1 u_1(0) \end{bmatrix} = \begin{bmatrix} T_a & T_b \\ T_c & T_d \end{bmatrix} \begin{bmatrix} p_2(\ell) \\ \underline{Z}_2 u_2(\ell) \end{bmatrix}$$

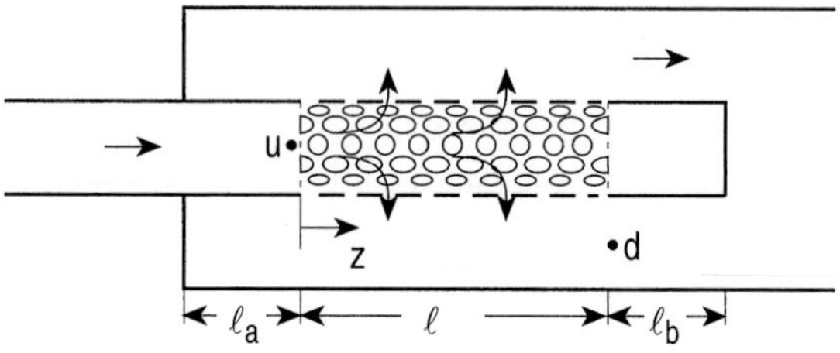

(b) Cross-flow expansion element

with :

$$T_a = T_{12} + A_1 A_2 \quad ; \quad T_b = T_{14} + B_1 A_2$$
$$T_c = T_{32} + A_1 B_2 \quad ; \quad T_d = T_{34} + B_1 B_2 \tag{13}$$

$$A_1 = (X_1 T_{22} - T_{42}) / F_1 \quad ; \quad B_1 = (X_1 T_{24} - T_{44}) / F_1$$
$$A_2 = T_{11} + X_2 T_{13} \quad ; \quad B_2 = T_{31} + X_2 T_{33} \tag{14}$$

$$F_1 = T_{41} + X_2 T_{43} - X_1 (T_{21} + X_2 T_{23})$$
$$X_1 = -j \tan(k_0 \ell_a) \quad ; \quad X_2 = j \tan(k_0 \ell_b) \tag{15}$$

(c) Cross-flow contraction element :

$$\begin{bmatrix} p_2(0) \\ \underline{Z}_2 u_2(0) \end{bmatrix} = \begin{bmatrix} T_a & T_b \\ T_c & T_d \end{bmatrix} \begin{bmatrix} p_1(\ell) \\ \underline{Z}_1 u_1(\ell) \end{bmatrix}$$

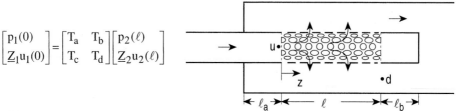

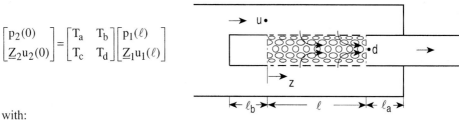

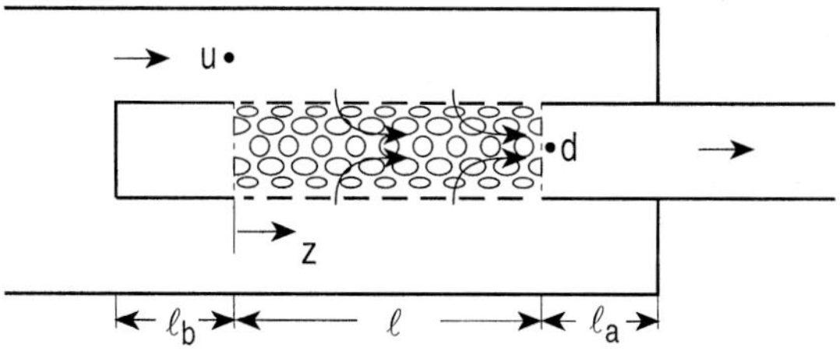

(c) Cross-flow contraction element

with:

$$T_a = T_{21} + A_1 A_2 \quad ; \quad T_b = T_{23} + B_1 A_2$$
$$T_c = T_{41} + A_1 B_2 \quad ; \quad T_d = T_{43} + B_1 B_2 \tag{17}$$

$$A_1 = (X_1 T_{11} - T_{31}) / F_1 \quad ; \quad B_1 = (X_1 T_{13} - T_{33}) / F_1$$
$$A_2 = T_{22} + X_2 T_{24} \quad ; \quad B_2 = T_{42} + X_2 T_{44} \tag{18}$$

$$F_1 = T_{32} + X_2 T_{34} - X_1 (T_{12} + X_2 T_{14})$$
$$X_1 = -j \tan(k_0 \ell_b) \quad ; \quad X_2 = j \tan(k_0 \ell_b) \tag{19}$$

(d) Reverse-flow expansion element :

$$\begin{bmatrix} p_1(0) \\ \underline{Z}_1 u_1(0) \end{bmatrix} = \begin{bmatrix} T_a & -T_b \\ T_c & -T_d \end{bmatrix} \begin{bmatrix} p_2(0) \\ \underline{Z}_2 u_2(0) \end{bmatrix} \quad (20)$$

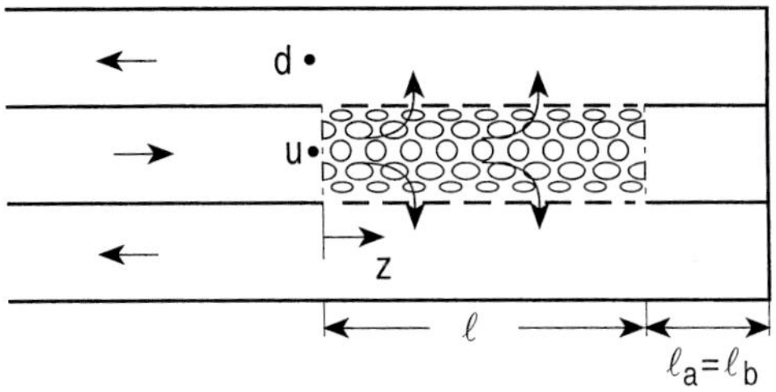

(d) Reverse-flow expansion element

The minus sign with T_b and T_d is due to reversal in the direction of $u_2(0)$, which is needed to make the foregoing relation adaptable to similar relations for other downstream elements.

Therein :

$$\begin{bmatrix} T_a & T_b \\ T_c & T_d \end{bmatrix} = \begin{bmatrix} A_1 & A_2 \\ A_3 & A_4 \end{bmatrix} \begin{bmatrix} B_1 & B_2 \\ B_3 & B_4 \end{bmatrix}^{-1} \quad (21)$$

$$A_1 = T_{11} + X_2 T_{13} \quad ; \quad A_2 = T_{12} + X_2 T_{14}$$
$$A_3 = T_{31} + X_2 T_{33} \quad ; \quad A_4 = T_{32} + X_2 T_{34} \quad (22)$$

$$B_1 = T_{21} + X_2 T_{23} \quad ; \quad B_2 = T_{22} + X_2 T_{24}$$
$$B_3 = T_{41} + X_2 T_{43} \quad ; \quad B_4 = T_{42} + X_2 T_{44} \quad (23)$$

$$X_2 = j \tan(k_0 \ell_b) \quad (24)$$

(e) Reverse-flow contraction element :

$$\begin{bmatrix} p_2(0) \\ \underline{Z}_2 u_2(0) \end{bmatrix} = \begin{bmatrix} -T_a & -T_b \\ T_c & T_d \end{bmatrix} \begin{bmatrix} p_1(0) \\ \underline{Z}_1 u_1(0) \end{bmatrix} \quad (25)$$

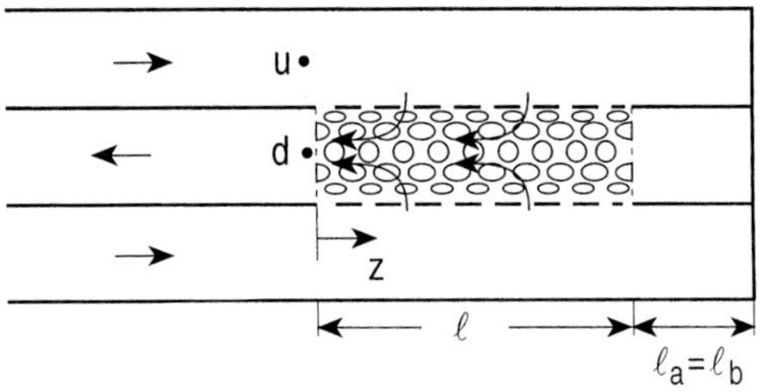

(e) Reverse-flow contraction element

with

$$\begin{bmatrix} T_a & T_b \\ T_c & T_d \end{bmatrix} = \begin{bmatrix} A_1 & A_2 \\ A_3 & A_4 \end{bmatrix}^{-1} \begin{bmatrix} B_1 & B_2 \\ B_3 & B_4 \end{bmatrix} \quad (26)$$

$$A_1 = T_{32} + X_a T_{12} \quad ; \quad A_1 = T_{34} + X_a T_{14}$$
$$A_3 = T_{42} + X_a T_{22} \quad ; \quad A_4 = T_{44} + X_a T_{24} \quad (27)$$

$$B_1 = T_{31} + X_a\, T_{11} \quad ; \quad B_2 = T_{33} + X_a\, T_{21}$$
$$B_3 = T_{41} + X_a\, T_{21} \quad ; \quad B_4 = T_{43} + X_a\, T_{23} \tag{28}$$

$$X_a = j\, \tan(k_0 \ell_a) \tag{29}$$

(f) Reversal-expansion, two-duct, open-end, perforated element :

$$\begin{bmatrix} p_u \\ \underline{Z}_u u_u \end{bmatrix} = \begin{bmatrix} T_a & -T_b \\ T_c & -T_d \end{bmatrix} \begin{bmatrix} p_d \\ \underline{Z}_d u_d \end{bmatrix} \tag{30}$$

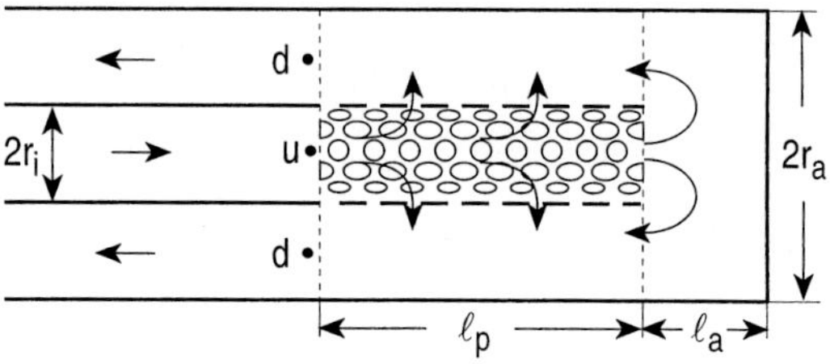

(f) Reversal-expansion, 2-duct, open-end perforated element

with :

$$\begin{bmatrix} T_a & T_b \\ T_c & T_d \end{bmatrix} = \begin{bmatrix} A_{11} & A_{12} \\ A_{21} & A_{22} \end{bmatrix} \begin{bmatrix} B_{11} & B_{12} \\ B_{21} & B_{22} \end{bmatrix}^{-1} \tag{31}$$

$$\begin{aligned} A_{11} &= T_{11}\, F_{11} + T_{12} + T_{13}\; F_{21} \\ A_{12} &= T_{11}\, F_{12} + T_{14} + T_{13}\; F_{22} \\ A_{21} &= T_{31}\, F_{11} + T_{32} + T_{33}\; F_{21} \\ A_{22} &= T_{31}\, F_{12} + T_{34} + T_{33}\; F_{22} \end{aligned} \tag{32}$$

$$\begin{aligned} B_{11} &= T_{21}\, F_{11} + T_{22} + T_{23}\; F_{21} \\ B_{12} &= T_{21}\, F_{12} + T_{24} + T_{23}\; F_{22} \\ B_{21} &= T_{41}\, F_{11} + T_{42} + T_{43}\; F_{21} \\ B_{22} &= T_{41}\, F_{12} + T_{44} + T_{43}\; F_{22} \end{aligned} \tag{33}$$

$$F_{11} = E_{11} \quad ; \quad F_{12} = -E_{12} / \underline{Z}_u$$
$$F_{21} = E_{21} \underline{Z}_d \quad ; \quad F_{22} = -E_{22}\, \underline{Z}_d / \underline{Z}_u \tag{34}$$

The matrix [E] is the 2x2 transfer matrix of the reversal expansion element of section K.7(c). [T] is the 4x4 matrix for the perforated section of the two interacting ducts derived earlier, (see equation (K.11.1)).

(g) Reversal-contraction, two-duct, open-end perforated element :

$$\begin{bmatrix} p_u \\ \underline{Z}_u u_u \end{bmatrix} = \begin{bmatrix} T_a & T_b \\ -T_c & -T_d \end{bmatrix} \begin{bmatrix} p_d \\ \underline{Z}_d u_d \end{bmatrix} \quad (35)$$

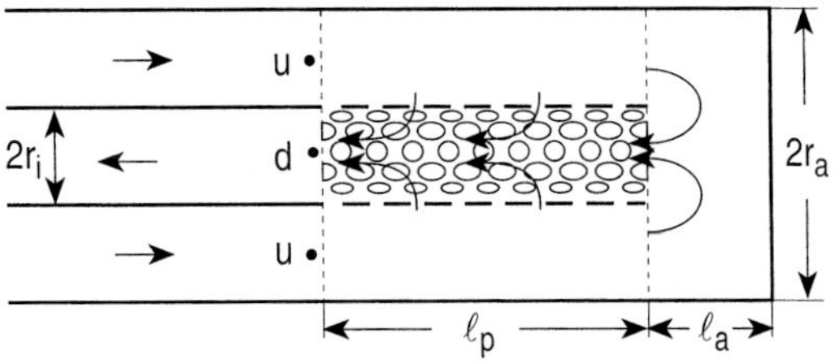

(g) Reversal-contraction, 2-duct, open-end perforated element

with :

$$\begin{bmatrix} T_a & T_b \\ T_c & T_d \end{bmatrix} = \begin{bmatrix} C_{11} & C_{12} \\ C_{21} & C_{22} \end{bmatrix} \begin{bmatrix} B_{11} & B_{12} \\ B_{21} & B_{22} \end{bmatrix}^{-1} \quad (36)$$

$$[B] = [P] + [Q][F]$$
$$[C] = [R] + [U][F] \quad (37)$$

$$\begin{aligned} P_{11} &= A_{11} \;;\; P_{12} = A_{13} \;;\; P_{21} = A_{31} \;;\; P_{22} = A_{33} \\ Q_{11} &= A_{12} \;;\; Q_{12} = A_{14} \;;\; Q_{21} = A_{32} \;;\; Q_{22} = A_{34} \\ R_{11} &= A_{12} \;;\; R_{12} = A_{23} \;;\; R_{21} = A_{41} \;;\; R_{22} = A_{43} \\ U_{11} &= A_{22} \;;\; U_{12} = A_{24} \;;\; U_{21} = A_{42} \;;\; U_{22} = A_{44} \end{aligned} \quad (38)$$

$$[A] = [T]^{-1} \quad (39)$$

[T] is the 4x4 matrix for the perforated section of the two interacting ducts, derived above.

$$F_{11} = D_{11} \;;\; F_{12} = D_{12} / \underline{Z}_d$$
$$F_{21} = D_{21}\underline{Z}_u \;;\; F_{22} = -D_{22}\underline{Z}_u / \underline{Z}_d \quad (40)$$

[D] is the 2x2 transfer matrix of the reversal contraction element of section K7(d).

(h) Perforated extended outlet :

$$\begin{bmatrix} p_u \\ u_u \end{bmatrix} = \begin{bmatrix} C_{11} & C_{12}\underline{Z}_d \\ C_{21} / \underline{Z}_u & C_{22}\underline{Z}_d / \underline{Z}_u \end{bmatrix} \begin{bmatrix} p_d \\ u_d \end{bmatrix}$$

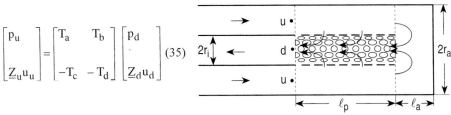

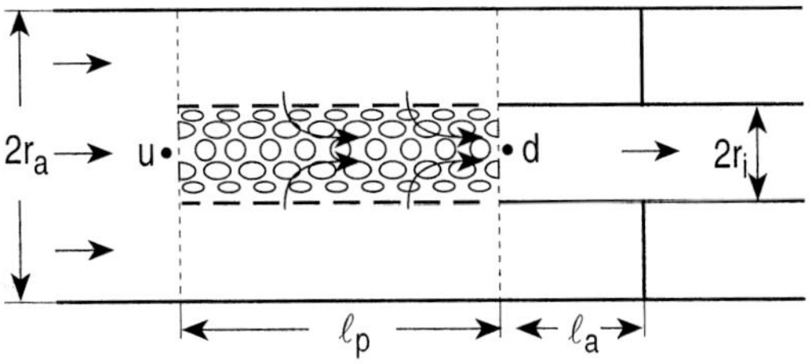

(h) Perforated extended outlet

with :

$$[C]=[A]^{-1}[B] \tag{42}$$

$$A_{11}=1-f_3\ \underline{Z}_{2u} \quad ; \quad A_{12}=M_u(1-2f_3\underline{Z}_{2u})$$
$$A_{21}=M_u-f_4\underline{Z}_{2u} \quad ; \quad A_{22}=1-2f_4\underline{Z}_{2u}M_u \tag{43}$$

$$B_{11}=T_{11}+M_dK_{p1}T_{31}-f_3\left[T_{21}+\underline{Z}_{21}\{T_{11}+2M_uT_{31}\}\right]$$
$$B_{12}=T_{13}+M_dK_{p1}T_{33}-f_3\left[T_{23}+\underline{Z}_{21}\{T_{13}+2M_uT_{33}\}\right]$$
$$B_{21}=\underline{Z}_{u2}\ T_{u1}+\underline{Z}_{u1}\{T_{31}+M_dT_{11}\}-\text{num1}\cdot f_4 \tag{44}$$
$$B_{22}=\underline{Z}_{u2}\ T_{43}+\underline{Z}_{u1}\{T_{33}+M_dT_{13}\}-\text{num2}\cdot f_4$$

$$f_3=\text{num3}/\text{den} \quad ; \quad f_4=\text{num4}/\text{den} \tag{45}$$

$$\text{num1}=T_{21}+\underline{Z}_{21}\{T_{11}+2M_uT_{31}\}$$
$$\text{num2}=T_{23}+\underline{Z}_{21}\{T_{13}+2M_uT_{33}\}$$
$$\text{num3}=T_{12}+X_aT_{14}+M_dK_{p1}\{T_{32}+X_aT_{34}\} \tag{46}$$
$$\text{num4}=\underline{Z}_{u2}\{T_{42}+X_aT_{44}\}+\underline{Z}_{u1}\left[T_{32}+M_dT_{12}+X_a\{T_{34}+M_dT_{14}\}\right]$$

$$\text{den}=T_{22}+X_aT_{24}+\underline{Z}_{21}\left[T_{12}+2M_uT_{32}+X_a\{T_{14}+2M_uT_{34}\}\right] \tag{47}$$

$$\underline{Z}_{21}=\underline{Z}_2/\underline{Z}_1 \quad ; \quad \underline{Z}_{2u}=\underline{Z}_2/\underline{Z}_u$$
$$\underline{Z}_{u2}=\underline{Z}_u/\underline{Z}_2 \quad ; \quad \underline{Z}_{u1}=\underline{Z}_u/\underline{Z}_1$$
$$\underline{Z}_d=\rho_0c_0/S_d \quad ; \quad \underline{Z}_u=\rho_0c_0/S_u \tag{48}$$
$$\underline{Z}_2=\rho_0c_0/S_a \quad ; \quad \underline{Z}_1=\underline{Z}_d$$

$$S_2=S_u-S_d \quad ; \quad X_a=j\tan(k\ell_a)$$
$$K_c=\frac{1}{2}\left\{1-\left(\frac{r_i}{r_{sh}}\right)^2\right\} \quad ; \quad K_{p1}=K_c+1 \tag{49}$$

[T] is the 4x4 matrix for the perforated section of the two interacting ducts, derived above.

(i) Perforated extended inlet :

$$\begin{bmatrix} p_u \\ u_u \end{bmatrix}=\begin{bmatrix} D_{11} & D_{12}\underline{Z}_d \\ D_{31}/\underline{Z}_u & D_{32}\underline{Z}_d/\underline{Z}_u \end{bmatrix}\begin{bmatrix} p_d \\ u_d \end{bmatrix} \tag{50}$$

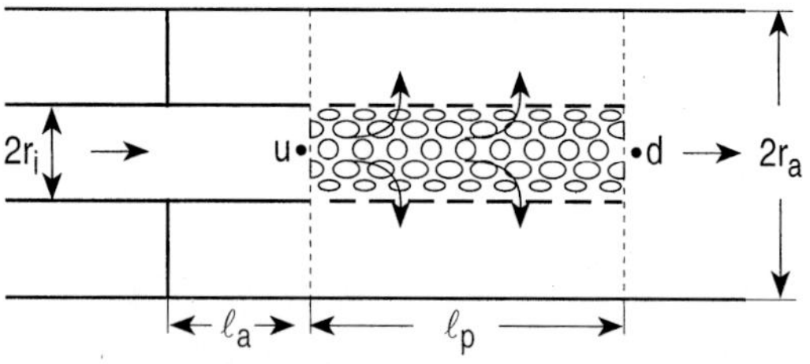

(i) Perforated extended inlet

where [D] is a 4x2 matrix given by :

$$[D]=[T]\,[A]\,[B] \tag{51}$$

(51)

[A] is a 4x4 matrix and [B] is a 4x2 matrix, elements of which are as follows :

$$\begin{aligned}
&A_{11}=1 \;;\; A_{12}=0 \;;\; A_{13}=M_u \;;\; A_{14}=0 \\
&A_{21}=M_1/\underline{Z}_w \;;\; A_{22}=0 \;;\; A_{23}=1/\underline{Z}_u \;;\; A_{24}=1/\underline{Z}_d \\
&A_{31}=1/\underline{Z}_u \;;\; A_{32}=1/\underline{Z}_d \;;\; A_{33}=2M_u/\underline{Z}_u \;;\; A_{34}=0 \\
&A_{41}=T_{41}-X_aT_{21} \;;\; A_{42}=T_{42}-X_aT_{22} \;; \\
&A_{43}=T_{43}-X_aT_{23} \;;\; A_{44}=T_{44}-X_aT_{24}
\end{aligned} \tag{52}$$

$$\begin{aligned}
&B_{11}=1 \;;\; B_{12}=M_d(1+K_e) \;;\; B_{21}=M_d/\underline{Z}_d \;;\; B_{22}=1/\underline{Z}_d \\
&B_{31}=1/\underline{Z}_d \;;\; B_{32}=2M_d/\underline{Z}_d \;;\; B_{41}=0 \;;\; B_{42}=0
\end{aligned} \tag{53}$$

$$\underline{Z}_u=\rho_0c_0/S_u \;;\; \underline{Z}_d=\rho_0c_0/S_d \;;\; \underline{Z}_2=\rho_0c_0/(S_d-S_u) \tag{54}$$

$$X_a=-j\tan(k\ell_a) \;;\; K_e=\left(1-(r_u/r_d)^2\right)^2 \tag{55}$$

[T] is the 4x4 matrix for the perforated section of the two interactive ducts, derived above.

K.12 Three-duct perforated elements

The two-duct perforated section shown in section K.11 is common to the two-duct perforated muffler elements, and it was treated in the section K.11.

Similarly, the three-duct perforated section shown on the right is common to the three-duct muffler elements which will be discussed below.

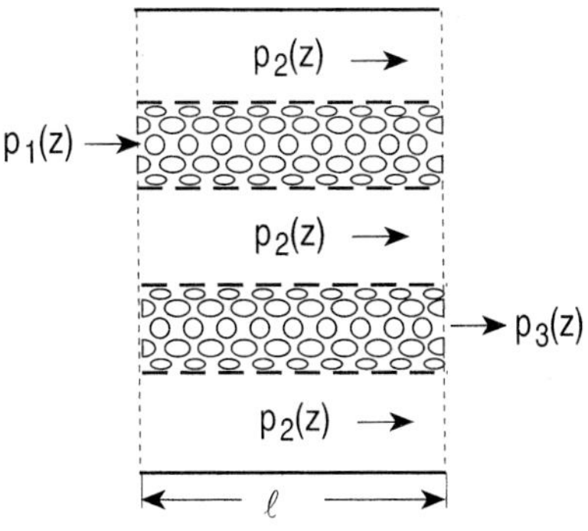

The common two-duct perforation

This common three-duct perforated section is described by the transfer matrix relation, [K.3] :

$$\begin{bmatrix} p_1(0) \\ p_2(0) \\ p_3(0) \\ \underline{Z}_1u_1(0) \\ \underline{Z}_2u_2(0) \\ \underline{Z}_3u_3(0) \end{bmatrix} = \begin{bmatrix} T_{11} & T_{12} & T_{13} & T_{14} & T_{15} & T_{16} \\ T_{21} & T_{22} & T_{23} & T_{24} & T_{25} & T_{26} \\ T_{31} & T_{32} & T_{33} & T_{34} & T_{35} & T_{36} \\ T_{41} & T_{42} & T_{43} & T_{44} & T_{45} & T_{46} \\ T_{51} & T_{52} & T_{53} & T_{54} & T_{55} & T_{56} \\ T_{61} & T_{62} & T_{63} & T_{64} & T_{65} & T_{66} \end{bmatrix} \begin{bmatrix} p_1(\ell) \\ p_2(\ell) \\ p_3(\ell) \\ \underline{Z}_1u_1(\ell) \\ \underline{Z}_2u_2(\ell) \\ \underline{Z}_3u_3(\ell) \end{bmatrix} \tag{1}$$

where :

$$[T]=[A(0)]\,[A(\ell)]^{-1} \tag{2}$$

The elements of the matrix $[A(z)]$ are given by, i=1,2,…,6 :

$$\begin{aligned}
&A_{1,i} = \psi_{4,i}\; e^{\beta_{i,z}} \;;\; A_{2,i} = \psi_{5,i}\; e^{\beta_{i,z}} \;;\; A_{3,i} = \psi_{6,i}\; e^{\beta_{i,z}} \\
&A_{4,i} = -e^{\beta_{i,z}} / (jk_0 + M_1\beta_i) \;;\; A_{5,i} = -\psi_{2,i}\; e^{\beta_{i,z}} / (jk_0 + M_2\beta_i) \\
&A_{6,i} = -\psi_{3,i} e^{\beta_{i,z}} / (jk_0 + M_3\beta_i)
\end{aligned} \tag{3}$$

$[\psi]$ and $\{\beta\}$ are the eigenmatrix (or modal matrix) and eigenvector of the following matrix, [K.18] :

$$\begin{bmatrix}
-\alpha_1 & \alpha_3 & 0 & -\alpha_2 & -\alpha_4 & 0 \\
-\alpha_5 & -\alpha_7 & -\alpha_9 & -\alpha_6 & -\alpha_8 & -\alpha_{10} \\
0 & -\alpha_{11} & -\alpha_{13} & 0 & -\alpha_{12} & -\alpha_{14} \\
1 & 0 & 0 & 0 & 0 & 0 \\
0 & 1 & 0 & 0 & 0 & 0 \\
0 & 0 & 1 & 0 & 0 & 0
\end{bmatrix} \tag{4}$$

where :

$$\begin{aligned}
&\alpha_1 = \frac{-jM_1}{1-M_1^2}\left(\frac{k_a^2 + k_0^2}{k_0}\right) \;;\; \alpha_2 = \frac{k_a^2}{1-M_1^2} \;;\; \alpha_3 = \frac{jM_1}{1-M_1^2}\left(\frac{k_a^2 - k_0^2}{k_0}\right) \\
&\alpha_4 = \frac{k_a^2 - k_0^2}{1-M_1^2} \;;\; \alpha_5 = \frac{jM_2}{1-M_2^2}\left(\frac{k_b^2 - k_0^2}{k_0}\right) \;;\; \alpha_6 = \frac{k_b^2 - k_0^2}{1-M_2^2} \\
&\alpha_7 = \frac{-jM_2}{1-M_2^2}\left(\frac{k_b^2 - k_c^2}{k_0}\right) \;;\; \alpha_8 = \frac{k_b^2 + k_c^2 - k_0^2}{1-M_2^2} \;;\; \alpha_9 = \frac{jM_2}{1-M_2^2}\,\frac{k_c^2 - k_0^2}{k_0} \\
&\alpha_{10} = \frac{k_c^2 - k_0^2}{1-M_2^2} \;;\; \alpha_{11} = \frac{jM_3}{1-M_3^2}\left(\frac{k_b^2 - k_0^2}{k_0}\right) \;;\; \alpha_{12} = \frac{k_d^2 - k_0^2}{1-M_3^2} \\
&\alpha_{13} = \frac{-jM_3}{1-M_3^2}\left(\frac{k_d^2 + k_0^2}{k_0}\right) \;;\; \alpha_{14} = \frac{k_d^2}{1-M_3^2}
\end{aligned} \tag{5}$$

and :

$$\begin{aligned}
&M_1 = V_1/c_0 \;;\; M_2 = V_2/c_0 \;;\; M_3 = V_3/c_0 \\
&k_a^2 = k_0^2 - \frac{4jk_0}{d_1\zeta_1} \;;\; k_b^2 = k_0^2 - \frac{4jk_0 d_1}{(d_2^2 - d_1^2 - d_3^2)\zeta_1} \;;\; k_c^2 = k_0^2 - \frac{4jk_0 d_3}{(d_2^2 - d_1^2 - d_3^2)\zeta_2} \\
&k_d^2 = k_0^2 - \frac{4jk_0}{d_3\zeta_2}
\end{aligned} \tag{6}$$

The desired 2x2 transfer matrix for a particular three-duct perforated element may be obtained from the 6x6 transfer matrix [T] above, making use of the appropriate upstream and downstream variables and four boundary conditions characteristic of the element. The final results are given below for different three-duct perforated elements shown in graphs.

(a) Cross-flow, three-duct, closed-end element :

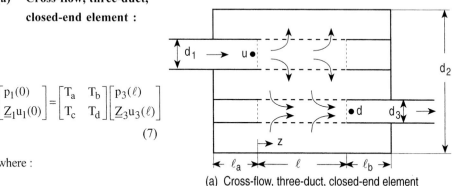

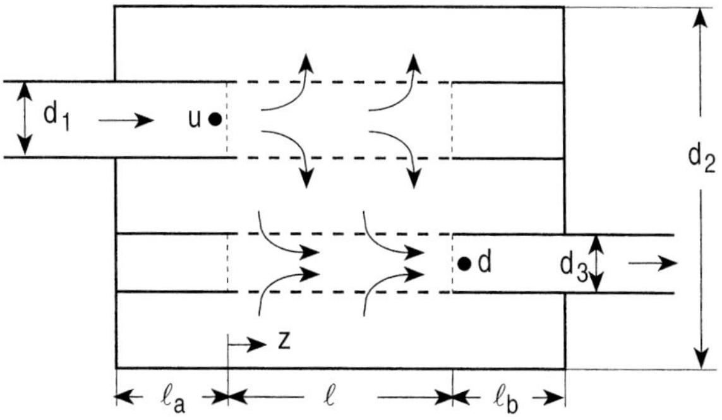

(a) Cross-flow, three-duct, closed-end element

$$\begin{bmatrix} p_1(0) \\ Z_1u_1(0) \end{bmatrix} = \begin{bmatrix} T_a & T_b \\ T_c & T_d \end{bmatrix} \begin{bmatrix} p_3(\ell) \\ Z_3u_3(\ell) \end{bmatrix} \tag{7}$$

where :

$$T_a = TT_{1,2} + A_3C_3 \quad ; \quad T_b = TT_{1,4} + B_3C_3$$
$$T_c = TT_{3,2} + A_3D_3 \quad ; \quad T_d = TT_{3,4} + B_3D_3 \tag{8}$$

$$A_3 = (TT_{2,2}X_2 - TT_{4,2})/F_2 \quad ; \quad B_3 = (TT_{2,4}X_2 - TT_{4,4})/F_2$$
$$C_3 = TT_{1,1} + X_1TT_{1,3} \quad ; \quad D_3 = TT_{3,1} + X_1TT_{3,3} \tag{9}$$

$$F_2 = TT_{4,1} + X_1TT_{4,3} - X_2(TT_{2,1} + X_1TT_{2,3}) \tag{10}$$

with :

$$TT_{1,1} = A_1A_2 + T_{1,2}\ ;\ TT_{1,2} = B_1A_2 + T_{1,3}\ ;\ TT_{1,3} = C_1A_2 + T_{1,5}\ ;\ TT_{1,4} = D_1A_2 + T_{1,6}$$
$$TT_{2,1} = A_1B_2 + T_{2,2}\ ;\ TT_{2,2} = B_1B_2 + T_{2,3}\ ;\ TT_{2,3} = C_1B_2 + T_{2,5}\ ;\ TT_{2,4} = D_1B_2 + T_{2,6}$$
$$TT_{3,1} = A_1C_2 + T_{4,2}\ ;\ TT_{3,2} = B_1C_2 + T_{4,3}\ ;\ TT_{3,3} = C_1C_2 + T_{4,5}\ ;\ TT_{3,4} = D_1C_2 + T_{4,6}$$
$$TT_{4,1} = A_1D_2 + T_{5,2}\ ;\ TT_{4,2} = B_1D_2 + T_{5,3}\ ;\ TT_{4,3} = C_1D_2 + T_{5,5}\ ;\ TT_{4,4} = D_1D_2 + T_{5,6} \tag{11}$$

$$A_1 = (T_{3,2}X_2 - T_{6,2})/F_1 \quad ; \quad B_1 = (T_{3,3}X_2 - T_{6,3})/F_1$$
$$C_1 = (T_{3,5}X_2 - T_{6,5})/F_1 \quad ; \quad D_1 = (T_{3,6}X_2 - T_{6,6})/F_1$$
$$A_2 = T_{1,1} + T_{1,4}X_1 \quad ; \quad B_2 = T_{2,1} + T_{2,4}X_1$$
$$C_2 = T_{4,1} + T_{4,4}X_1 \quad ; \quad D_2 = T_{5,1} + T_{5,4}X_1 \tag{12}$$

$$F_1 = T_{6,1} + X_1T_{6,4} - X_2(T_{3,1} + X_1T_{3,4})$$
$$X_1 = j\tan(k_0\ell_b) \quad ; \quad X_2 = -j\tan(k_0\ell_a) \tag{13}$$

$$TM_{i,4} = T_{i,4} \cdot \underline{Z}_1 \quad ; \quad TM_{i,5} = T_{i,5} \cdot \underline{Z}_2 \quad ; \quad TM_{i,6} = T_{i,6} \cdot \underline{Z}_1$$
$$TM_{4,i} = T_{4,i} / \underline{Z}_1 \quad ; \quad TM_{5,i} = T_{5,i} / \underline{Z}_2 \quad ; \quad TM_{6,i} = T_{6,i} / \underline{Z}_1 \qquad (25)$$

$\underline{Z}_1$ is the volume-flow impedance of the upstream/downstream or inlet/outlet ducts

$\underline{Z}_1 = \rho_0 c_0 / S_u$;

$\underline{Z}_2$ is the volume-flow impedance of the annular duct

$\underline{Z}_2 = \rho_0 c_0 / (S_{shell} - 2S_u)$

neglecting the duct wall thickness;

[T] is the 6x6 transfer matrix of the common perforated section of the three interacting ducts, (see equation (K.12.1)).

(d) Reverse-flow, open-end, three-duct element

This element is a combination of elements (b) and (c) inasmuch as at the left-hand end it is like the closed-end element (b) and at the right-hand end it is like the open-end of the element (c). The transfer matrix relationship between the upstream point u and the downstream point d is given by, [K.22] :

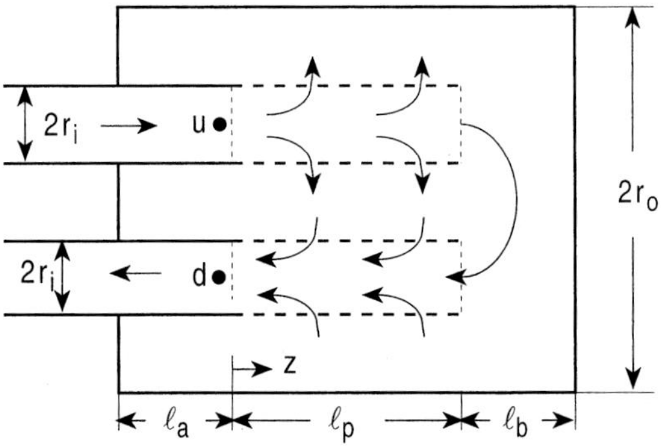

(d) Reverse-flow, three-duct, open-end element

$$\begin{bmatrix} p_u \\ u_u \end{bmatrix} = \begin{bmatrix} T_a & -T_b \underline{Z}_i \\ T_c / Z_i & -T_d \end{bmatrix} \begin{bmatrix} p_d \\ u_d \end{bmatrix} \qquad (26)$$

where :

$$\begin{aligned} T_a &= B_{11}C_{11} + (B_{12} + B_{15}X_a)\, C_{21} + B_{13}\, C_{31} + B_{14}\, C_{41} + B_{16}\, C_{51} \\ T_b &= B_{11}C_{12} + (B_{12} + B_{15}X_a)\, C_{22} + B_{13}\, C_{32} + B_{14}\, C_{42} + B_{16}\, C_{52} \\ T_c &= B_{41}C_{11} + (B_{42} + B_{45}X_a)\, C_{21} + B_{43}\, C_{31} + B_{44}\, C_{41} + B_{46}\, C_{51} \\ T_d &= B_{41}C_{12} + (B_{42} + B_{45}X_a)\, C_{22} + B_{43}\, C_{32} + B_{44}\, C_{42} + B_{46}\, C_{52} \end{aligned} \qquad (27)$$

$$X_a = j\tan(k\ell_a) \quad ; \quad X_b = -j\tan(k\ell_b) \qquad (28)$$

Therein $[C] = [A]^{-1}$, and [A] is a 5x5 matrix, the elements of which are related to the 6x6 matrix [T] as follows :

$$A_{11} = T_{31}\,;\quad A_{12} = T_{32} + X_a T_{35}\,;\quad A_{13} = T_{33}\,;\quad A_{14} = T_{34}$$
$$A_{15} = T_{36}\,;\quad A_{21} = T_{61}\,;\quad A_{22} = T_{62} + X_a T_{65}\,;\quad A_{23} = T_{63} \qquad (29)$$
$$A_{24} = T_{64}\,;\quad A_{25} = T_{66}\,;\quad A_{31} = T_{51} - X_b T_{21}\,;\quad A_{32} = T_{52} - X_b T_{22} + X_a T_{55} - X_b T_{25}$$

$$A_{33} = T_{53} - X_b T_{23}\,;\quad A_{34} = T_{54} - X_b T_{24}\,;\quad A_{35} = T_{56} - X_b T_{26}$$
$$A_{41} = 1\,;\quad A_{42} = 0\,;\quad A_{43} = -1\,;\quad A_{44} = M_i\,;\quad A_{45} = M_i(1 + K_{re} + K_{rc})$$
$$A_{51} = M_i / \underline{Z}_i\,;\quad A_{52} = X_a / \underline{Z}_6\,;\quad A_{53} = -M_i / \underline{Z}_i \qquad (29)$$
$$A_{54} = 1/\underline{Z}_i\,;\quad A_{55} = 1/\underline{Z}_i$$

K_{re} , K_{rc} are the pressure loss factors for reversal-expansion and reversal-contraction,

M_i is the mean flow Mach number in the inner pipes of radius r_i ;

$\underline{Z}_i$ is the volume flow impedance of the inner pipe, $\underline{Z}_i = \rho_0 c_0 / S_i$; $S_i = \pi\, r_i^2$;

[T] is the 6x6 transfer matrix for the common perforated section of the three interacting ducts, given above in equation (1).

K.13 Three-duct perforated elements with extended perforations

Three possible configurations in this class of elements are shown below. The derivation of the transfer matrix between the upstream point u and downstream point d calls for simultaneous solution of equations representing

(1) the common three-duct perforated section of length ℓ_2 ;

(2) two-duct extended perforated sections on either end, of lengths ℓ_1 and ℓ_3;

(3) the closed-end cavities of lengths ℓ_a and ℓ_b, and l_1 and l_3 in the closed-end configuration (b).

The algebraic equations for item (1) are in the form of a 6x6 transfer matrix, discussed earlier in Section K.12. Equations for item (2) are in the form of two 4x4 transfer matrices discussed in Section K.11; and equations for the end-cavities of item (3) are in the form of an impedance expression.

In the open-end elements (a) and (c) below, as indicated for similar elements earlier, most of the mean flow grazes the perforations; very little flows across the perforations. So, for convenience, the entire flow may be assumed to be of the grazing type for selection of the appropriate expression for the perforated impedance. In the closed-end configuration (b) below, however, the entire flow has to get across the perforations, calling for the cross-flow or through-flow expression for perforate impedance.

(b) Cross-flow, closed-end, extended perforation element :

The overall transfer matrix between the upstream point u and down-stream point d of the configuration of the figure is given by :

$$\begin{bmatrix} p_u \\ u_u \end{bmatrix} = \begin{bmatrix} T_a & T_b \underline{Z}_d \\ T_c / \underline{Z}_u & T_d \underline{Z}_d / \underline{Z}_u \end{bmatrix} \begin{bmatrix} p_d \\ u_d \end{bmatrix} \tag{12}$$

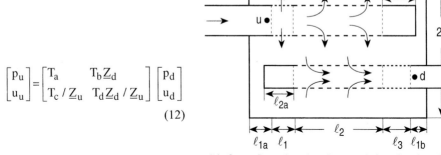

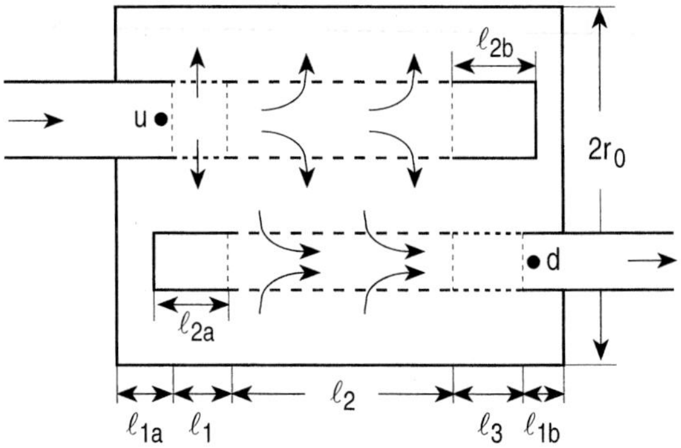

(b) Cross-flow, closed-end, extended-perforation element

where :

$$\begin{aligned} T_a &= (E_{12} + X_{1b}E_{14})\ NP/DN + E_{11} \\ T_b &= (E_{12} + X_{1b}E_{14})\ NV/DN + E_{13} \\ T_c &= (E_{32} + X_{1b}E_{34})\ NP/DN + E_{31} \\ T_d &= (E_{32} + X_{1b}E_{34})\ NV/DN + E_{33} \end{aligned} \tag{13}$$

$$\begin{aligned} NP &= X_{1a}E_{21} - E_{41} \quad ; \quad NV = X_{1a}E_{23} - E_{43} \\ DN &= E_{42} + X_{1b}\,E_{44} - (E_{22} + X_{1b}\,E_{24})X_{1a} \end{aligned} \tag{14}$$

$$\begin{aligned} X_{1a} &= -j\tan(k_0\ell_{a1}) \quad ; \quad X_{1b} = j\tan(k_0\ell_{b1}) \\ X_{2a} &= -j\tan(k_0\ell_{a2}) \quad ; \quad X_{2b} = j\tan(k_0\ell_{b2}) \end{aligned} \tag{15}$$

$$[E] = [A]\,[D]\,[C] \tag{16}$$

[A] and [C] are 4x4 transfer matrices for the extended perforated pipes of lengths ℓ_1 and ℓ_3, respectively (see figure).

$$\begin{aligned} D_{11} &= T_{11}\,G_2 + T_{13} + T_{14}\,G_2\,X_{2b} \quad ; \quad D_{12} = T_{11}\,G_1 + T_{12} + T_{14}\,G_1\,X_{2b} \\ D_{13} &= T_{11}\,G_4 + T_{16} + T_{14}\,G_4\,X_{2b} \quad ; \quad D_{14} = T_{11}\,G_3 + T_{15} + T_{14}\,G_3\,X_{2b} \\ D_{21} &= T_{21}\,G_2 + T_{23} + T_{24}\,G_2\,X_{2b} \quad ; \quad D_{22} = T_{21}\,G_1 + T_{22} + T_{24}\,G_1\,X_{2b} \\ D_{23} &= T_{21}\,G_4 + T_{26} + T_{24}\,G_4\,X_{2b} \quad ; \quad D_{24} = T_{21}\,G_3 + T_{25} + T_{24}\,G_3\,X_{2b} \\ D_{31} &= T_{41}\,G_2 + T_{43} + T_{44}\,G_2\,X_{2b} \quad ; \quad D_{32} = T_{41}\,G_1 + T_{42} + T_{44}\,G_1\,X_{2b} \\ D_{33} &= T_{41}\,G_4 + T_{46} + T_{44}\,G_4\,X_{2b} \quad ; \quad D_{34} = T_{41}\,G_3 + T_{45} + T_{44}\,G_3\,X_{2b} \\ D_{41} &= T_{51}\,G_2 + T_{53} + T_{54}\,G_2\,X_{2b} \quad ; \quad D_{42} = T_{51}\,G_1 + T_{52} + T_{54}\,G_1\,X_{2b} \\ D_{43} &= T_{51}\,G_4 + T_{56} + T_{54}\,G_4\,X_{2b} \quad ; \quad D_{44} = T_{51}\,G_3 + T_{55} + T_{54}\,G_3\,X_{2b} \end{aligned} \tag{17}$$

[T] is the 6x6 transfer matrix for the common perforated section of the three interacting ducts, given in equation (K.12.1).

$$G_1 = \left(X_{2a}T_{32} - T_{62}\right)/h \quad ; \quad G_2 = \left(X_{2a}T_{33} - T_{63}\right)/h$$
$$G_3 = \left(X_{2a}T_{35} - T_{65}\right)/h \quad ; \quad G_4 = \left(X_{2a}T_{36} - T_{66}\right)/h \tag{18}$$

$$h = T_{61} - X_{2a}T_{31} + X_{2b}\,(T_{64} - X_{2a}T_{34})$$
$$\underline{Z}_u = \rho_0 c_0 / S_u \quad ; \quad \underline{Z}_d = \rho_0 c_0 / S_d \tag{19}$$

(c) Reverse-flow, open-end, extended-perforated element :

For the configuration shown in the figure, the final transfer matrix relationship is given by, [K.24]:

$$\begin{bmatrix} p_u \\ u_u \end{bmatrix} = \begin{bmatrix} T_a & T_b \underline{Z}_d \\ T_c / \underline{Z}_u & T_d \underline{Z}_d / \underline{Z}_u \end{bmatrix} \begin{bmatrix} p_d \\ u_d \end{bmatrix} \tag{20}$$

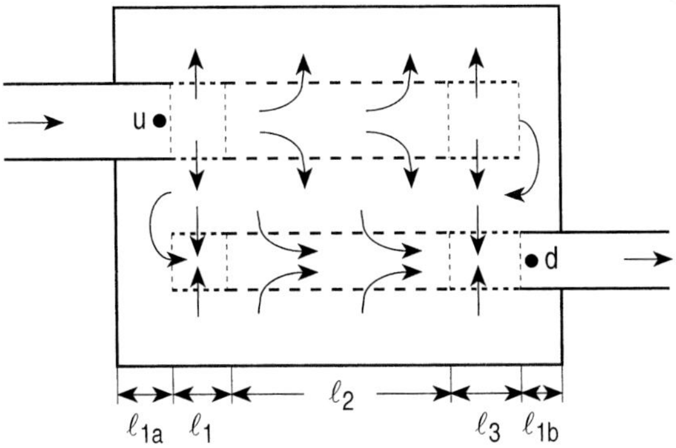

(c) Reverse-flow, open-end, extended-perforetion element

where :

$$T_a = B_{11}W_{11} + B_{12}U_{11} + B_{13} + B_{14}W_{21} + B_{15}U_{21}$$
$$T_b = B_{11}W_{12} + B_{12}U_{12} + B_{16} + B_{14}W_{22} + B_{15}U_{22}$$
$$T_c = B_{41}W_{11} + B_{42}U_{11} + B_{43} + B_{44}W_{21} + B_{45}U_{21}$$
$$T_d = B_{41}W_{12} + B_{42}U_{12} + B_{46} + B_{44}W_{22} + B_{45}U_{22} \tag{21}$$

$$[W] = [G][U] \quad ; \quad [U] = [Q][R] \quad ; \quad [Q] = [X]^{-1} \tag{22}$$

$$R_{11} = H_{11}B_{33} + H_{12}B_{63} - B_{23} \quad ; \quad R_{12} = H_{11}B_{36} + H_{12}B_{66} - B_{26}$$
$$R_{21} = H_{21}B_{33} + H_{22}B_{63} - B_{53} \quad ; \quad R_{22} = H_{21}B_{36} + H_{22}B_{66} - B_{56} \tag{23}$$

$$X_{11} = B_{21}G_{11} + B_{22} + B_{24}G_{21} - H_{11}(B_{31}G_{11} + B_{32} + B_{34}G_{21}) - H_{12}(B_{61}G_{11} + B_{62} + B_{64}G_{21})$$
$$X_{12} = B_{21}G_{12} + B_{25} + B_{24}G_{22} - H_{11}(B_{31}G_{12} + B_{35} + B_{34}G_{22}) - H_{12}(B_{61}G_{12} + B_{65} + B_{64}G_{22})$$
$$X_{21} = B_{51}G_{11} + B_{52} + B_{54}G_{21} - H_{21}(B_{31}G_{11} + B_{32} + B_{34}G_{21}) - H_{22}(B_{61}G_{11} + B_{62} + B_{64}G_{21})$$
$$X_{22} = B_{51}G_{12} + B_{55} + B_{54}G_{22} - H_{21}(B_{31}G_{12} + B_{35} + B_{34}G_{22}) - H_{22}(B_{61}G_{12} + B_{65} + B_{64}G_{22}) \tag{24}$$

$$M_1 = U_1 / c_0 \quad ; \quad M_2 = U_2 / c_0 \quad ; \quad M_3 = U_3 / c_0 \quad ; \quad k_0 = \omega / c_0 \tag{8}$$

Use of the boundary conditions of the annular duct, and rearranging yields a reduced, and more useful form, of the transfer matrix relationship :

$$\begin{bmatrix} p_1(0) \\ V_1(0) \\ p_2(0) \\ V_2(0) \\ p_3(0) \\ V_3(0) \end{bmatrix} = \begin{bmatrix} T_{11} & T_{12} & T_{13} & T_{14} & T_{15} & T_{16} \\ T_{21} & T_{22} & T_{23} & T_{24} & T_{25} & T_{26} \\ T_{31} & T_{32} & T_{33} & T_{34} & T_{35} & T_{36} \\ T_{41} & T_{42} & T_{43} & T_{44} & T_{45} & T_{46} \\ T_{51} & T_{52} & T_{53} & T_{54} & T_{55} & T_{56} \\ T_{61} & T_{62} & T_{63} & T_{64} & T_{65} & T_{66} \end{bmatrix} \begin{bmatrix} p_1(\ell) \\ V_1(\ell) \\ p_2(\ell) \\ V_2(\ell) \\ p_3(\ell) \\ V_3(\ell) \end{bmatrix} \tag{9}$$

$$T_{ij} = TM_{ij} + \frac{(TM_{i7} + X_{pb} TM_{i8})(X_{pa} TM_{7j} - TM_{8j})}{TM_{87} + X_{pb} TM_{88} - X_{pa}(TM_{77} + X_{pb} TM_{78})} \quad ; \quad i,j = 1,2,......,6 \tag{10}$$

$$X_{pa} = -j\tan(k_0 \ell_a) \quad ; \quad X_{pb} = -j\tan(k_0 \ell_b) \tag{11}$$

Partitioning the foregoing matrix equation yields :

$$\begin{bmatrix} S_1(0) \\ S_2(0) \\ S_3(0) \end{bmatrix} = \begin{bmatrix} D & E & F \\ G & H & K \\ P & Q & R \end{bmatrix} \begin{bmatrix} S_1(\ell_p) \\ S_2(\ell_p) \\ S_3(\ell_p) \end{bmatrix} \tag{12}$$

where $\{S_i\} = [p_i \;\; V_i]^T$ and D, E, F, G, H, K, P, Q and R are 2x2 sub-matrices as becomes clear from comparison of the two corresponding matrix equations.

(a) Flush-tube three-pass perforated element :

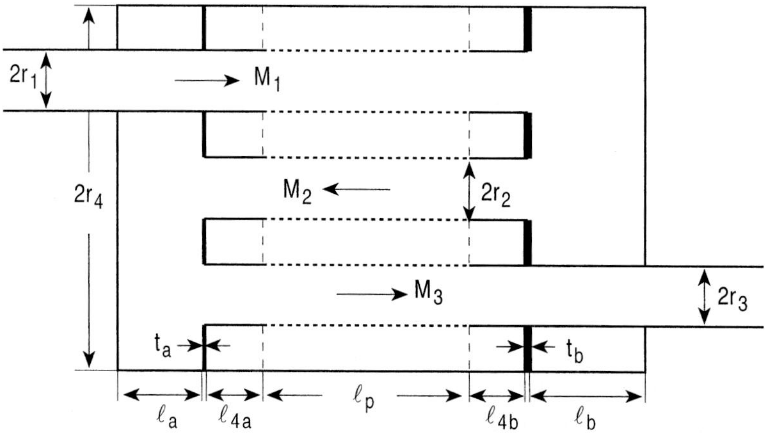

(a) Flush-tube, three-pass perforated element chamber

Now, the desired 2x2 transfer matrix relationship between the upstream point u and the downstream point d in the configuration shown is given by, [K.25] :

$$\begin{bmatrix} p_1(0) \\ u_1(0) \end{bmatrix} = \begin{bmatrix} C_{11} & C_{12} \cdot \underline{Z}_d \\ C_{21} / \underline{Z}_u & C_{22} \underline{Z}_d / \underline{Z}_u \end{bmatrix} \begin{bmatrix} p_3(\ell_p) \\ u_3(\ell_p) \end{bmatrix} \tag{13}$$

where :

$$\begin{aligned} &[C] = [D]\,[A]\,[W] + [E]\,[W] + [F] \\ &[W] = [[G][A] + [H] - [B][P][A] - [B][Q]]^{-1}\,[[B][R] - [K]] \end{aligned} \tag{14}$$

[A] is the product of the transfer matrices of

(1) a duct of length $\delta_1 + \ell_{4b} + t_b$;

(2) reversal expansion element;

(3) sudden contraction element;

(4) a duct of length $\delta_2 + t_b + \ell_{4b}$.

[B] is the product of transfer matrices of

(1) a duct of length $\delta_1 + \ell_{4a} + t_a$;

(2) reversal expansion element;

(3) sudden contraction element;

(4) a duct of length $\delta_3 + t_a + \ell_{4a}$.

Here, δ_a and δ_b are end corrections. These as well as other constituent transfer matrices are given in Sections K.5 and K.6.

(b) Extended-tube three-pass perforated element :

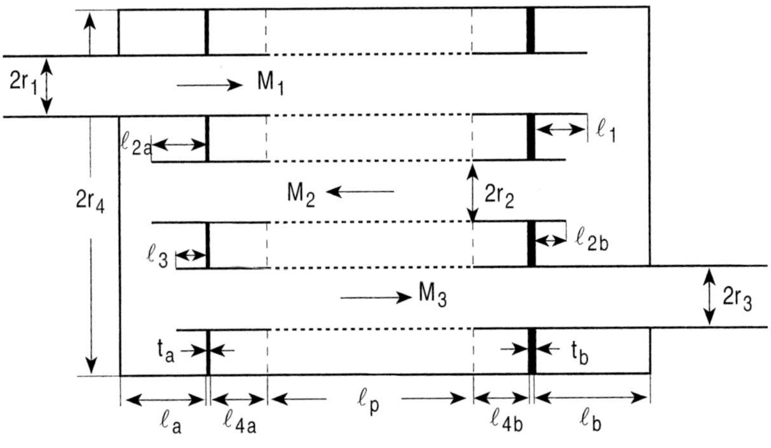

(b) Extended-tube, three-pass perforated element chamber

The foregoing transfer matrix relationship between the state vectors $[p_1(0)\ \ u_1(0)]^T$ and $[p_3(\ell_p)\ \ u_3(\ell_p)]^T$ would hold for the extended-tube three-pass perforated element shown above, with the difference that end-matrices [A] and [B] would now be different. For the extended-tube end chambers,

[A] would be the product of the transfer matrices of, [K. 26] :

(1) a duct of length $\delta_1 + \ell_{4b} + t_b + \ell_1$;
(2) a reversal expansion element;
(3) an extended outlet element;
(4) a duct of length $\delta_2 + t_b + \ell_{4b} + \ell_{2b}$.

[B] would be product of the transfer matrices of:

(1) a duct of length $\delta_2 + \ell_{4a} + t_a + \ell_{2a}$;
(2) a reversal expansion element;
(3) an extended outlet element;
(4) a duct of length $\delta_3 + \ell_3 + t_a + \ell_{4a}$.

Transfer matrices of the extended-tube elements are given in Section K.7 while the end-correcting $\delta_1, \delta_2, \delta_3$ and other transfer matrices are given in Sections K.5 and K.6.

K.15 Catalytic converter elements

Catalytic converters are used often in series with exhaust mufflers for control of air emissions by means of oxidation of unburnt carbon particles and carbon monoxide to carbon dioxide. These converters involve area changes and porous blocks of catalyst pellets, or a bank of capillary tubes as shown below.

(a) Pellet block element :

While transfer matrices of simple uniform tubes and sudden area changes have been given earlier in sections K.5 and K.6, the transfer matrix of a pellet block element is given by :

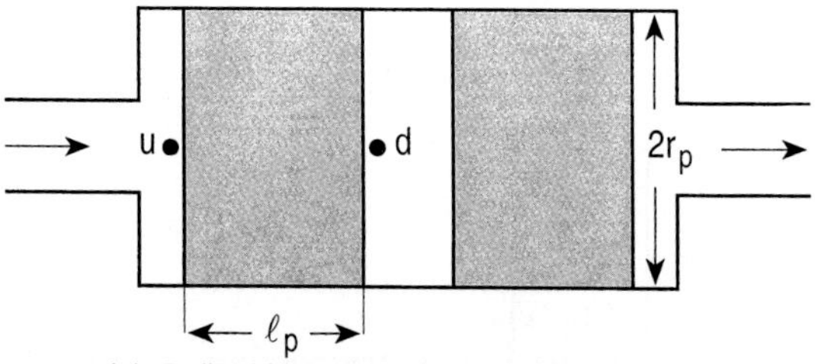

(a) Pellet-block catalytic converter element

$$\begin{bmatrix} \cos(k\ell) & j\underline{Z}\sin(k\ell) \\ \frac{j}{\underline{Z}}\sin(k\ell) & \cos(k\ell) \end{bmatrix} \qquad (1)$$

where $\underline{Z}$ and k for granular pellets are given by Attenborough's expression, [K.27] :

$$\frac{k}{k_0} = q\left(\frac{1+(\kappa-1)T(C)}{1-T(B)}\right)^{1/2} \quad ; \quad \frac{\underline{Z}}{\underline{Z}_0} = \frac{q^2}{\sigma}\frac{1}{1-T(B)}\frac{k_0}{k} \tag{2}$$

where q is a tortuosity factor, κ is the ratio of specific heats for the gaseous medium, and :

$$T(x) = \frac{2J_1(x)}{xJ_o(x)} \quad ; \quad x = B \text{ or } C$$

$$B = (-j)^{1/2}\lambda_p \quad ; \quad C = B\, N_{pr}^{1/2} \tag{3}$$

$$\lambda_p^2 = 8\rho_0 q^2 S\omega / (n^2 \sigma E)$$

q	is the steady flow shape factor;
n	is the dynamic shape factor, n=2–S ;
N_{Pr}	is the Prandtl number.

(b) Capillary tube monolith :

The transfer matrix of a monolith or bank of capillary tubes (coated with catalyst), is given by :

$$\begin{bmatrix} \cos(k_m\ell) & j\underline{Z}\sin(k_m\ell)/\Phi \\ \dfrac{j\Phi}{\underline{Z}}\sin(k_m\ell) & \cos(k_m\ell) \end{bmatrix} \tag{4}$$

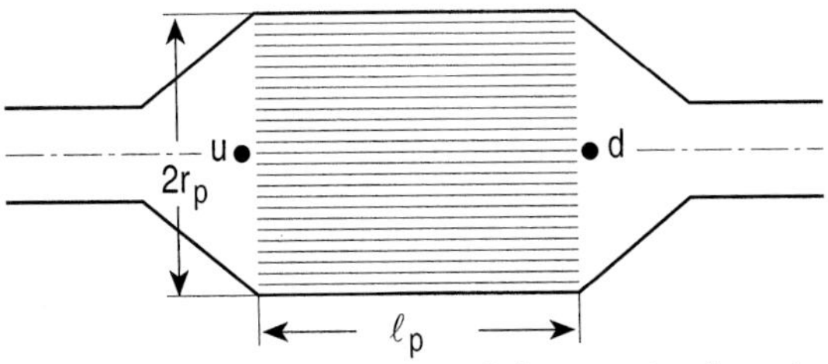

(b) Capillary-tube monolyth catalytic converter element

where Φ is the open area ratio.

$$\underline{Z} = \rho_m c_m / S \quad ; \quad k_m = k_0 c_0 / c_m$$

$$\frac{c_0}{c_m} = \left[(1+\Phi E G/D)\left(\kappa - \frac{\kappa - 1}{1+\Phi E G'/(D\,Pr)}\right)\right]^{1/2} \tag{5}$$

with :

$$D = j\omega\rho_0 \quad ; \quad G = \frac{-ab/4}{1-2b/a} \quad ; \quad G' = \frac{-a'b'/4}{1-2b'/a'}$$

$$a = s(-j)^{1/2} \quad ; \quad b = J_1(a)/J_0(a)$$

$$a' = s(-j)^{1/2}\,Pr^{1/2} \quad ; \quad b' = J_1(a')/J_0(a') \tag{6}$$

$$\rho_m = \rho_0 + E\,\Phi\,G/(j\omega)$$

$$s = \alpha\left(\frac{8\,\omega\,\rho_0}{E\,\Phi}\right)^{1/2}$$

E is the specific flow resistance for laminar flow, $E = 32\mu/(\Phi\, d^2)$. For air, $\kappa = 1.4$, $Pr = 0.7$, $\mu = 1.81\cdot 10^{-5}$ Pa·s . Typically, E $\approx$ 500 Pa·s/m^2 and α= 1.07 .

K.16 Helmholtz resonator

See Sections H.4 – H.16 for a more detailed description of Helmholtz resonators without flow, and Sections J.38 - J.39 for nonlinear effects of flow.

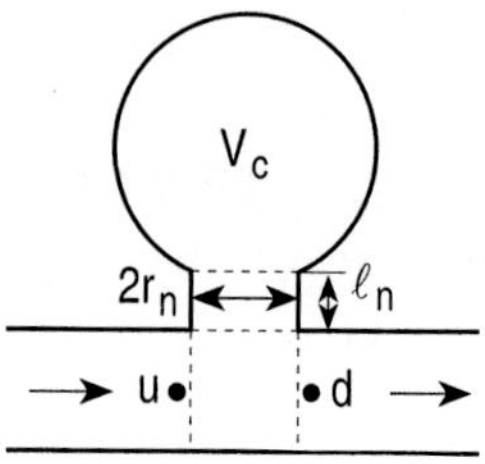

The transfer matrix of a Helmholtz resonator as shown is given by

$$\begin{bmatrix} p_u \\ u_u \end{bmatrix} = \begin{bmatrix} 1 & 0 \\ 1/\underline{Z}_r & 1 \end{bmatrix} \begin{bmatrix} p_d \\ u_d \end{bmatrix} \qquad (1)$$

where $\underline{Z}_r$ is the flow impedance of the resonator at the junction, made up of a resistance term, an inertance term and a compliance term :

$$\underline{Z}_r = \rho_0 \left[\frac{\omega^2}{\pi c_0} \left\{ 2 - \frac{r_n}{r_u} \right\} + 0.425 \frac{M c_0}{S_n} + j \left\{ \frac{\omega\, \ell_{eq}}{S_n} - \frac{c_0^2}{\omega V_c} \right\} \right]$$
$$\ell_{eq} = \ell_n + t_w + 0.85\, r_n \left(2 - \frac{r_n}{r_u} \right) \qquad (2)$$

r_u is the radius of the upstream (or downstream) duct; r_n and ℓ_n are the radius and length of the neck of the resontor; V_c is the volume of the resonator cavity; $S_n = \pi r_n^2$.

K.17 In-line cavity :

The transfer matrix of an in-line cavity as shown is given by :

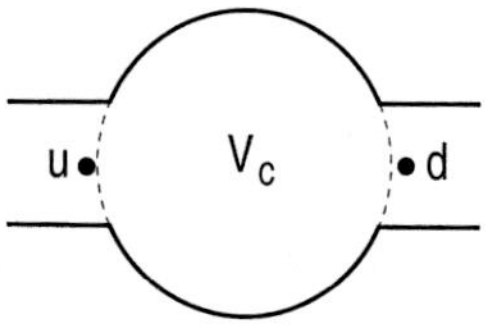

$$\begin{bmatrix} p_u \\ u_u \end{bmatrix} = \begin{bmatrix} 1 & 0 \\ 1/\underline{Z} & 1 \end{bmatrix} \begin{bmatrix} p_d \\ u_d \end{bmatrix} \qquad (1)$$

where $\underline{Z}$ is the compliance-type flow impedance, V_c the volume of the cavity :

$$\underline{Z} = \frac{\rho_0 c_0^2}{j\omega V_c} \qquad (2)$$

K.18 Bellows

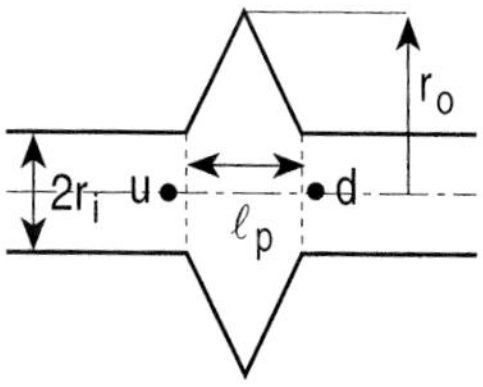

A bellow is characterized by wall compliance coupled with a gradual area change. The matrizant approach leads to the following transfer matrix, [K.32], for the divergent conical part of a single step of a bellow :

$$\begin{bmatrix} T_{11} & T_{12} \\ T_{21} & T_{22} \end{bmatrix} \tag{1}$$

where :

$$T_{11} = \frac{e^{\beta \ell/2}}{2\delta'}(2\delta' \cos \delta'\ell - \beta \sin \delta'\ell)$$

$$T_{12} = j\frac{e^{\beta \ell/2}}{2\delta'}\frac{k_0 \underline{Z}(\ell)}{\rho_0 \delta'}\sin \delta'\ell$$

$$T_{21} = j\frac{-\rho_0 e^{\beta \ell/2}}{k_0 \delta' \underline{Z}(0)}\left(\frac{\beta^2}{4} + \delta'^2\right)\sin \delta'\ell \tag{2}$$

$$T_{22} = \frac{e^{\beta \ell/2}}{2\delta'}(\beta \sin \delta'\ell + 2\delta' \cos \delta'\ell)\frac{\underline{Z}(0)}{\underline{Z}(\ell)}$$

with :

$$\delta' = j\delta \;;\; \underline{Z}(0) = \frac{\rho_0 c_0}{\pi r_i^2} \;;\; \underline{Z}(\ell) = \frac{\rho_0 c_0}{\pi r_o^2}$$

$$\delta = \frac{1}{2}\left\{\beta^2 - 4\left(k_0^2 - jk_0\alpha\right)\right\} \;;\; \beta = 2a\alpha/B \;;\; \ell = \ell_p/2 \tag{3}$$

$$a = (r_o - r_i)/\ell \;;\; \alpha = \frac{B \ln(r_o/r_i)}{a\ell} \;;\; B = 2\rho_0 c_0/Z_w$$

Z_w is the wall impedance given earlier in Section K.10 on Hoses.

By interchanging the inlet and outlet radii, we obtain the transfer matrix for the convergent, conical part of the bellows. Successive multiplication of the transfer matrices of the two halves of the step shown above yields the transfer matrix of the full stop (single bellow). Extension of this multiplication process would yield the transfer matrix of multi-step bellows.

Evaluation of the transfer matrix of flexible-wall bellows, incorporating the convective effect of mean flow is done by means of the same matrizant approach as given in Ref [K.32]. However, the more important effect of flow separation and the consequent losses have been neglected; only the convective effect of the mean flow is considered.

$$M_a = M_u - \frac{S_b}{S_a} M_b \quad ; \quad S_a = S_u \quad ; \quad M_b = M_u / (f_{11} + S_3 / S_2)$$

$$f_{11} = \left\{ \frac{d_a}{\ell_a} \left(\frac{\ell_b}{d_b} + 187.5 \right) \right\}^{1/2} \tag{7}$$

It may be noted that generally M_b will be much less than M_a because of the transverse connection and the requirement of equal pressure drop across the two parallel arms.

K.21 Annular airgap lined duct :

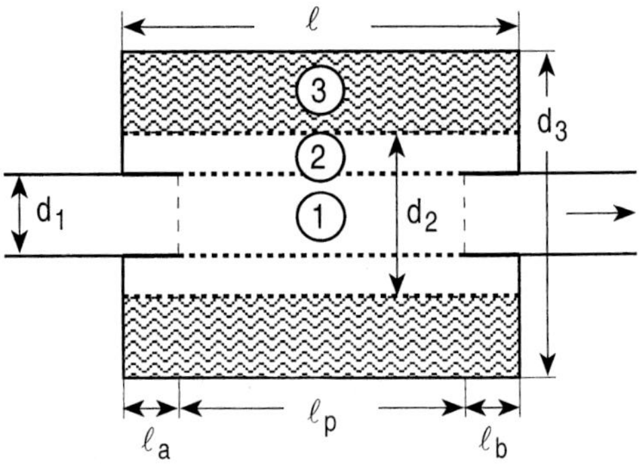

Automotive exhaust systems are characterized by hot gas flows, often containing some unburnt carbon particles. In an acoustically lined duct with high velocity grazing flow, there is a strong possibility of some fibres of a fibrous absorptive material such as glass wool, ceramic wool, or mineral/rock wool being swept away continuously, and the perforated protective plate getting progressively clogged with unburnt carbon particles and possibly lubricating oil. One of the alternatives would be to make use of another perforated cylinder between the inner flow pipe and the absorptive layer with an airgap inbetween, as shown. This element involves acoustic interaction between three ducts; viz., the inner flow duct , the annular airgap duct, and the outer acoustically lined duct. Therefore, the analysis of this element runs on the same lines as those of Section K.12, with the important difference that the medium in the outer lined duct is different; its wave number k_W and characteristic impedance $\underline{Z}_W$ are given in terms of the specific flow resistance by Mechel's expressions, [K.31]. The final transfer matrix is given by, [K.35] :

$$\begin{bmatrix} p_u \\ u_u \end{bmatrix} = \begin{bmatrix} ACP_{11} & ACV_{11} \underline{Z}_d \\ ACP_{61} / \underline{Z}_u & ACV_{61} \, \underline{Z}_d / \underline{Z}_u \end{bmatrix} \begin{bmatrix} p_d \\ u_d \end{bmatrix} \tag{1}$$

where [ACP] and [ACV] are 10x10 matrices given by :

$$[ACP] = [A]^{-1} \{CP\} \quad ; \quad [ACV] = [A]^{-1} \{CV\} \tag{2}$$

with $\underline{Z}_u = \underline{Z}_d = \rho_0 c_0 / (\pi \, d_1^2 / 4)$ and :

$$\begin{aligned}\{CP\} &= \left[T_{11}\ \ T_{21}\ T_{31}\ \ T_{41}\ \ T_{51}\ \ T_{61}\ 0\ \ 0\ \ 0\ \ 0\ \right]^T \\ \{CV\} &= \left[T_{14}\ \ T_{24}\ T_{34}\ \ T_{44}\ \ T_{54}\ \ T_{64}\ 0\ \ 0\ \ 0\ \ 0\ \right]^T\end{aligned} \tag{3}$$

$$[A] = \begin{bmatrix} 1 & 0 & 0 & -T_{12} & -T_{13} & 0 & 0 & 0 & -T_{15} & -T_{16} \\ 0 & 1 & 0 & -T_{22} & -T_{23} & 0 & 0 & 0 & -T_{25} & -T_{26} \\ 0 & 0 & 1 & -T_{32} & -T_{33} & 0 & 0 & 0 & -T_{35} & -T_{36} \\ 0 & 0 & 0 & -T_{42} & -T_{43} & 1 & 0 & 0 & -T_{45} & -T_{46} \\ 0 & 0 & 0 & -T_{52} & -T_{53} & 0 & 1 & 0 & -T_{55} & -T_{56} \\ 0 & 0 & 0 & -T_{62} & -T_{63} & 0 & 0 & 1 & -T_{65} & -T_{66} \\ 0 & -X_1 & 0 & 0 & 0 & 0 & 1 & 0 & 0 & 0 \\ 0 & 0 & 0 & -X_2 & 0 & 0 & 0 & 0 & 1 & 0 \\ 0 & 0 & 0 & o & 0 & 0 & 0 & 1 & -X_4 & 1 \\ 0 & 0 & 0 & -X_4 & 0 & 0 & 0 & 0 & 0 & 1 \end{bmatrix} \tag{4}$$

$$\begin{aligned} X_1 &= -j\tan(k_0\ell_a) \quad ; \quad X_2 = -j\tan(k_0\ell_b) \\ X_3 &= -j\tan(k_w\ell_a) \quad ; \quad X_4 = -j\tan(k_w\ell_b) \end{aligned} \tag{5}$$

[T] is the 6x6 transfer matrix for the common perforated section where all three duct sections interact. It is evaluated exactly as shown in Section K.12; only the values of α's in the coefficient matrix are different . These are as follows :

$$\begin{aligned} &\alpha_1 = -\ (0.5\ f_0 M_1 + f_1)/\text{den} \quad ; \quad \alpha_2 = k_0^2/\text{den} \quad ; \quad \alpha_3 = f_1/\text{den} \quad ; \quad \alpha_4 = f_2/\text{den} \\ &\alpha_5 = 0 \quad ; \quad \alpha_6 = f_4 \quad ; \quad \alpha_7 = 0 \quad ; \alpha_8 = -k_0^2 - f_4 - f_6 \quad ; \quad \alpha_9 = 0 \quad ; \quad \alpha_{10} = f_6 \\ &\alpha_{11} = 0 \quad ; \quad f_{12} = f_7 \quad ; \quad \alpha_{13} = 0 \quad ; \quad \alpha_{14} = k_w^2 - f_7 \end{aligned} \tag{6}$$

where : $\quad f_0 = 4jk_0 \quad ; \quad f_1 = 4M_1\xi_1/d_1 \quad ; \quad f_2 = f_0\xi_1/d_1$ (7)

Y_W and k_W are given by MECHEL's formulae given in Section K.25 ; and $\text{den} = 1 - M_1^2$.

ξ_1 is the reciprocal of the non-dimensional grazing-flow impedance of the perforate given in Section K.11, at the interface 1-2 (diameter d_1); ξ_2 is the reciprocal of the non-dimensional stationary-flow impedance of the perforate given in Section K.12, at the interface 2-3 (diameter d_2).

K.22 Micro-perforated Helmholtz panel parallel baffle muffler

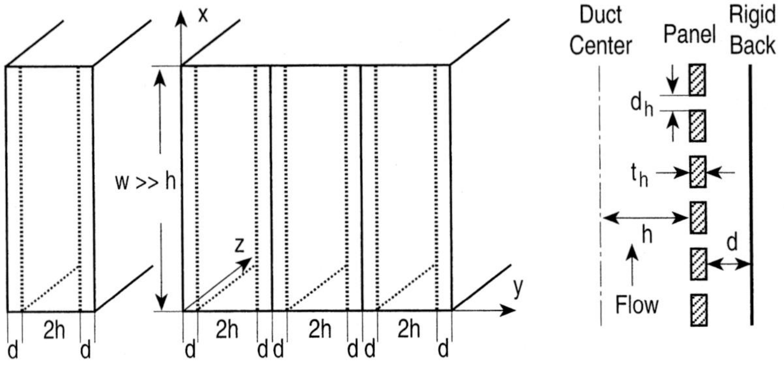

This element, in a way, is a combination of a concentric tube resonator (see Section K.11(a)) and a parallel baffle muffler (see Section K.24) without absorbing material. Following WU [K.36], the transfer matrix of this element across a length ℓ_p is given by :

$$\begin{bmatrix} \cos(k_z\ell_p) & j\,\underline{Z}\sin(k_z\ell_p) \\ j\sin(k_z\ell_p)/\underline{Z} & \cos(k_z\ell_p) \end{bmatrix} \tag{1}$$

where :

$$\underline{Z} = \underline{Z}_0\, k_0 / k_z \quad ; \quad \underline{Z}_0 = \rho_0 c_0 / \left(2n_p hW\right)$$
$$k_z = (k_0^2 - k_y^2)^{1/2} \tag{2}$$

k_y is a root of the transcendental equation :

$$k_y h \cdot \tan(k_y h) - j\, k_0 h\, \rho_0 c_0 / Z = 0 \tag{3}$$

Z is the grazing-flow impedance of the micro-perforated panel :

$$Z = (1+M)\left\{ \frac{32\nu\, \rho_0 \cdot t_h}{\sigma\, d_h^2}\left[\left(1 + x^2/32\right)^{1/2} + 0.177 \cdot x \cdot d_h/t_h\right] + \right.$$
$$\left. + \frac{j\omega \cdot t_h \cdot \rho_0}{\sigma}\left[1 + \frac{1}{(9 + x^2/2)^{1/2}} + 0.85\frac{d_h}{t_h}\right] - j\,\frac{\rho_0 c_0}{\tan(k_0 d)}\right\} \tag{4}$$

with $x = d \cdot h\left(10^5 \cdot f\right)^{1/2}$; M is the mean flow Mach number in the flow passage; d, t_h, d_h, h and W are shown in the figure; $\nu = 1.56 \cdot 10^{-5}$ Pa.s for air.

K.23 Acoustically lined circular duct :

Circular and annular ducts are extensively treated in the Chapter J . Whereas that chapter describes steps in such ducts with a multi-mode analysis, the present chapter uses representations with a fundamental-mode analysis. Thus, in order to get some completeness in the collection of muffler elements in the present chapter, this and the following Section K.26 will present the transfer matrix method for elements which are contained in earlier chapters.

A uniform circular duct is often lined on the inside for acoustical absorption or dissipation into heat, as shown. For mechanical protection against the eroding effect of the moving medium, the acoustic layer is covered on the exposed side with a very thin membrane such as Mylar, or a highly perforated thin metallic plate. This protective cover affects the absorptive properties of the acoustic lining at certain frequencies, and therefore needs to be included in the acoustic model.

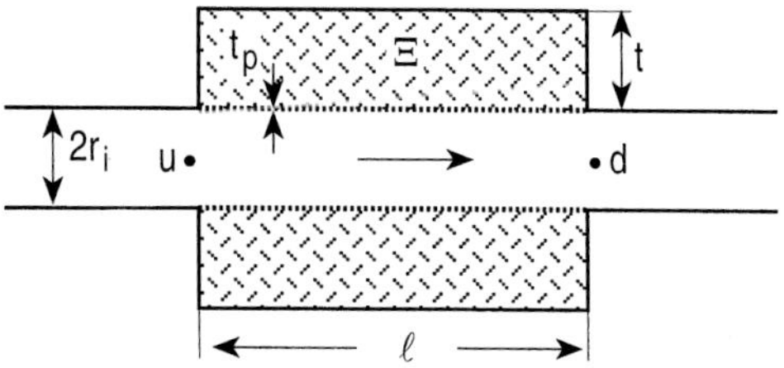

The protective layer is characterized by a radial partition impedance that would support a pressure difference between the two media. The porous/fibrous acoustic layer is characterized by a complex wave number k_W ($=-j\cdot\Gamma_a$ in earlier chapters) and a characteristic impedance Z_W ($=Z_a$ in earlier chapters). These in turn may be expressed as a function of the flow resistivity Ξ or of an "absorber variable" E , given by :

$$E=\frac{\rho_0 f}{\Xi}=\frac{\rho_0 c_0}{\lambda_0 \Xi} \tag{1}$$

with λ_0 the free field wavelength (see the Chapter G). For the fibrous materials, the characteristic constants, Z_W and k_W are approximately described by the empirical formulae of DELANY and BAZLEY, [K.30], as modified and improved by MECHEL, [K.31] :

$$\frac{Z_W}{\rho_0 c_0}=\begin{cases}1+0.0485\,E^{-0.754}-j\,0.087\,E^{-0.73} & ;\quad E>1/60\\[4pt] \dfrac{0.5/(\pi E)+j\,1.4}{(-1.466+j\,0.212/E)^{1/2}} & ;\quad E<1/60\end{cases} \tag{2}$$

$$\frac{k_W}{k_0}=\begin{cases}1-j\,0.189\,E^{-0.6185}+j\,0.0978\,E^{-0.6929} & ;\quad E>1/60\\[4pt] (-1.466+j\,0.212/E)^{1/2} & ;\quad E<1/60\end{cases} \tag{3}$$

The transfer matrix for a bulk reacting lined duct shown above is given by, [K.29] :

$$\begin{bmatrix} p_u \\ u_u \end{bmatrix} = \begin{bmatrix} \cos(k_z \ell) & j\underline{Z} \sin(k_z \ell) \\ (j/\underline{Z}) \sin(k_z \ell) & \cos(k_z \ell) \end{bmatrix} \begin{bmatrix} p_d \\ u_d \end{bmatrix} \tag{4}$$

where $\underline{Z} = \underline{Z}_0 k_0 / k_z$; $\underline{Z}_0 = \rho_0 c_0 / S$; $k_0 = \omega / c_0$.

The axial wave number for the fundamental mode is the first (lowest) root of the transcendental equation :

$$j\frac{\omega \rho_0 J_0(k_{r,0} r_i)}{k_{r,0} J_1(k_{r,0} r_i)} = j\frac{\omega \rho_w}{k_{r,w}} \frac{J_0(k_{r,w} r_i) + C \cdot Y_0(k_{r,w} r_i)}{J_o(k_{r,w} r_i) + C \cdot Y_1(k_{r,\omega} r_i)} + Z_p(\omega) \tag{5}$$

$$k_{r,0} = (k_0^2 - k_z^2)^{1/2} \; ; \; k_{r,w} = (k_w^2 - k_z^2)^{1/2} \; ; \; C = -\frac{J_1(k_w r_o)}{Y_1(k_w r_o)} \; ; \; r_o = r_i + t \tag{6}$$

Z_p, the partition impedance of a thin impermeable protective foil is given by : $Z_p = j\omega \rho_p t_p$, where t_p and ρ_p are the thickness and material density of the foil, respectively.

For a perforated plate, in the presence of a grazing mean flow, the corresponding expression is, [K.21] :

$$Z_p = \rho_0 c_0 \left[7.337 \cdot 10^{-3} (1 + 72.23\, M + j\, 2.2245 \cdot 10^{-5} f\, (1 + 51\, t_p)(1 + 204\, d_h) \right] / \sigma \tag{7}$$

where d_h is the hole diameter in m; M is the mean flow Mach number; and σ is the porosity of the perforated plate.

A locally reacting lining would not support waves inside the lining in the axial direction. Therefore, for a locally reacting lining, $k_{r,w} = k_w$, and therefore, the right-hand side of the foregoing transcendental equation would be independent of the variable k_z .

In either case (for either type of lining), the appropriate transcendental equation can be solved for the axial wave number k_z by means of a Newton-Raphson scheme, making use of the starting approximation :

$$\left(k_{r,0} r_o\right)^2 \approx \frac{96 + 36jQ \pm \left(9216 + 2304jQ - 912Q^2\right)^{1/2}}{12 + jQ} \; ; \; Q = (k_0 r_o) \frac{\rho_0 c_0}{Z_w} \tag{8}$$

and Z_w is the right-hand side of the transcendental equation above for $k_{r,w} = k_w$, i.e. for the locally reacting lining case.

K.24 Parallel baffle muffler (multi-pass lined duct)

See the introductory comment of the previous Section K.23 .

The figure shows a typical parallel baffle muffle used in general to increase the contact area and to obtain the required attenuation within a short axial length. For plane-wave propagation, each of the passes will act as a two-dimensional rectangular duct as shown in the separate figure.

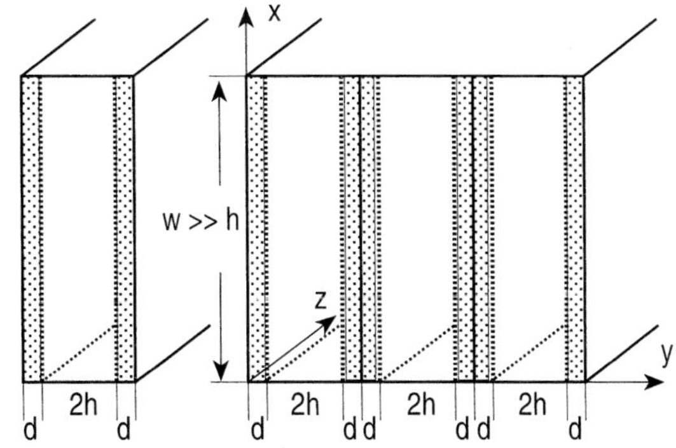

The transfer matrix for axial wave propagation for the element is the same as for a lined circular duct (see Section K.23). The difference lies in the value of the axial wave number, k_z. The eigenequation for a two-dimensional rectangular duct with bulk-reacting lining is :

$$jZ_0\frac{k_0}{k_{y,0}}\cot(k_{y,0}h)=Z_p-jZ_w\cot(k_{y,w}d) \tag{1}$$

where :

$$Z_0=\rho_0c_0 \quad ; \quad Z_w=\rho_wc_w \quad ; \quad k_{y,0}=(k_0^2-k_z^2)^{1/2} \quad ; \quad k_{y,w}=(k_w^2-k_z^2)^{1/2} \tag{2}$$

k_w, Z_w and Z_p are as given in Section K.25.

For locally reacting linings, the foregoing transcendental equation would hold with the simplification : $k_{y,w}=k_w$.

In either case, the transcendental equation or eigenequation may be solved for k_z by means of a Newton-Raphson iteration scheme, with the starting value being given by :

$$(k_{y,0}h/2)^2\approx\frac{2.47+Q+\left[(2.47+Q)^2-1.87Q\right]^{1/2}}{0.38} \quad ; \quad Q=jk_0\frac{h}{2}\frac{\rho_0c_0}{Z_w} \tag{3}$$

Incidentally, the transmission loss or attenuation of a lined duct of length ℓ may be obtained from the imaginary component of the axial wave number k_z , by means of the equation: :

$$TL=-8.68\ \ \mathrm{Im}\{k_z\ell\},\ \ \mathrm{dB} \tag{4}$$

or else from the transfer matrix of the lined duct or parallel baffle muffler, making use the expression given in Section K.4.

[K.30] DELANY, M.E., BAZLEY, B.N., Applied Acoustics **3** (1970) 106-116
"Acoustical characteristics of fibrous absorbent materials"

[K.31] MECHEL, F.P., Acustica **35** (1976) 210-213
"Extension to low frequencies of the formulae of Delany and Bazley for absorbing materials" (in German)

[K.32] SINGHAL, V., MUNJAL, M.L., Internat. J. Acoust. and Vibr. **4** (1999) 181-188
"Prediction of the acoustic performance of flexible bellows incorporating the convective effect of incompressible mean flow"

[K.33] SELAMET A., DICKEY N.S., NOVAK, J.M., J. Ac. Soc. Amer. **96** (1994) 3177-99
"The Herschel-Quincke tube: A theoretical, computational, and experimental investigation"

[K.34] VENKATESHAM, B., M.E. Dissertation, I.I.Sc., Bangalore, Jan 2001
"Aeroacoustic analysis of complex muffler elements"

[K.35] MUNJAL, M.L., VENKATESHAM, B., IUTAM International Symposium an Designing for Quietness, I.I.Sc., Bangalore, Dec. 2000.
"Analysis and Design of an annular airgap lined duct for hot exhaust systems"

[K.36] WU, M.Q., Noise Control Eng. J. **45** (1997) 69-77
"Micro-perforated panels for duct silencing"

L
Capsules and Cabins

As usual, capsules and cabins are combined in one chapter, like here, although their tasks and the analytical methods are quite different. The task of a capsule is to reduce the sound pressure level from a noise source in the environment; the task of cabins is to produce a quiet space in a noisy environment. Suppose we have a noise source with constant sound power output, whatever the sound field around the source may be, and suppose we have a capsule surrounding the source with some transmission loss of its walls, but with no sound absorption, either inside or in the walls. The sound pressure level inside the capsule will rise, until all the sound power produced will be radiated by the capsule again. Thus sound absorption in a capsule plays an important role.

L.1 The energetic approximation for the efficiency of capsules

Ref.: Mechel, [L.1]

Consider two arrangements of a source Q :

(a): the source is placed in free space and radiates the (effective) sound power Π.

(b): the source is surrounded by a capsule; the question is, what sound power Π_t will be radiated ?

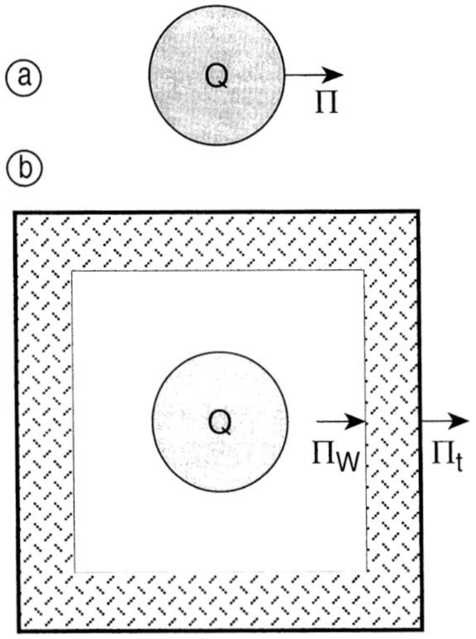

The capsule efficiency is measured by its insertion loss

$$D_e = -10 \cdot \lg(\tau_e) = -10 \cdot \lg(\Pi_t / \Pi) \text{ dB} \qquad (1)$$

A widely used proposal by GOESELE evaluates

$$D_e = R - 10 \cdot \lg(1/\alpha) \qquad (2)$$

where $R = -10 \cdot \lg(\Pi_t / \Pi_w)$ is the sound transmission loss of the capsule wall, and α is the sound absorption coefficient of the interior side of the capsule wall (measured with a hard backing of the wall). This proposal produces $D_e \to -\infty$ for $\alpha \to 0$.

By identical transformations :

$$\tau_e = \frac{\Pi_t}{\Pi} = \frac{\Pi_w}{\Pi} \cdot \frac{\Pi_t}{\Pi_w} = c_w \cdot \tau_{VS} \quad ; \quad c_w = \frac{\Pi_w}{\Pi} \quad ; \quad \tau_{VS} = \frac{\Pi_t}{\Pi_w} \tag{3}$$

where τ_{VS} is the sound transmission coefficient of the capsule wall (possibly a combination of an interior absorber layer and an outer tight wall). If r_{VS} is the symbol for the interior reflection factor of the capsule wall (including its radiation to the outside), the sound power Π_V loss inside the capsule is $\Pi_V = (1 - |r_{VS}|^2) \cdot \Pi_W$. The "energetic approximation", which was proposed by HENNIG, assumes that at the equilibrium $\Pi = \Pi_V$; thus :

$$\tau_e = \frac{\tau_{VS}}{1 - |r_{VS}|^2} = \frac{\tau_{VS}}{\alpha_{VS}} \quad ; \quad \alpha_{VS} = 1 - |r_{VS}|^2 \tag{4}$$

The difference from GOESELE's proposal is the fact that the absorption coefficient α_{VS} now also contains the radiated power. If the sound transmission factor t_{VS} through the wall of the capsule is used, with $\tau_{VS} = |t_{VS}|^2$, the insertion power ratio becomes :

$$\tau_e = \frac{|t_{VS}|^2}{1 - |r_{VS}|^2} \tag{5}$$

Simple example : *porous interior layer and outer metal sheet* :

It is assumed that the interior sound field can be described by plane waves p_e incident on the capsule wall at a polar angle ϑ. The capsule wall consists of an interior porous layer of thickness d_a and with characteristic values Γ_a, Z_a of its material plus an outer metal sheet with partition impedance Z_T.

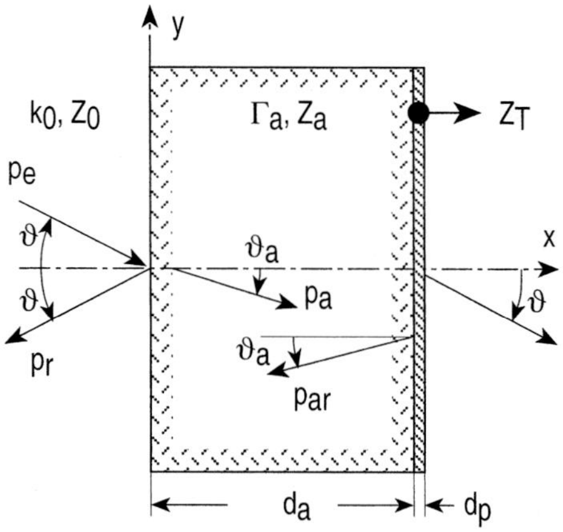

In total, it assumed that the capsule is large and has plane walls.
Field formulations :

$$\begin{aligned}
p_e(x,y) &= P_e \cdot e^{j k_0 y \sin\vartheta} \cdot e^{-j k_0 x \cos\vartheta} \\
p_r(x,y) &= P_r \cdot e^{j k_0 y \sin\vartheta} \cdot e^{+j k_0 x \cos\vartheta} \\
p_t(x,y) &= P_t \cdot e^{j k_0 d_a \cos\vartheta} \cdot e^{j k_0 y \sin\vartheta} \cdot e^{-j k_0 x \cos\vartheta}
\end{aligned} \tag{6}$$

$$p_a(x,y) = P_a \cdot e^{\Gamma_a y \sin\vartheta_a} \cdot e^{-\Gamma_a x \cos\vartheta_a}$$
$$p_{ar}(x,y) = P_{ar} \cdot e^{-\Gamma_a d_a \cos\vartheta_a} \cdot e^{\Gamma_a y \sin\vartheta_a} \cdot e^{+\Gamma_a x \cos\vartheta_a} \tag{7}$$

with interior angle in the porous layer : $\sin\vartheta_a = \dfrac{jk_0}{\Gamma_a} \cdot \sin\vartheta$ (8)

and partition impedance Z_T :

$$\frac{Z_T}{Z_0} = 2\pi Z_m F[\eta F^2 \cdot \sin^4\vartheta_a + j(1 - F^2 \cdot \sin^4\vartheta_a)] \; ; \; Z_m = \frac{f_{cr} d_p}{Z_0}\rho_p \; ; \; F = \frac{f}{f_{cr}} \tag{9}$$

where η is the bending loss factor of the sheet; ρ_p the density of its material; f_{cr} the critical frequency.

The boundary conditions give the system of equations :

$$\begin{pmatrix} 1 & -1 & -e^{-a} & 0 \\ 1 & b & -be^{-a} & 0 \\ 0 & e^{-a} & 1 & -(1 + Z_{Tn}\cos\vartheta) \\ 0 & be^{-a} & -b & -1 \end{pmatrix} \cdot \begin{pmatrix} P_r \\ P_a \\ P_{ar} \\ P_t \end{pmatrix} = \begin{pmatrix} -P_e \\ P_e \\ 0 \\ 0 \end{pmatrix} \tag{10}$$

with the abbreviations :

$$a = \Gamma_a d_a \cdot \cos\vartheta_a = k_0 d_a \sqrt{\Gamma_{an}^2 + \sin^2\vartheta} \quad ; \quad b = \frac{Z_0 \cos\vartheta_a}{Z_a \cos\vartheta} = \frac{1}{\Gamma_{an} Z_{an}} \frac{\sqrt{\Gamma_{an}^2 + \sin^2\vartheta}}{\cos\vartheta} \tag{11}$$

($\Gamma_{an} = \Gamma_a/k_0$; $Z_{an} = Z_a/Z_0$) . It can be solved with KRAMER's rule :

$$\det = -(1+b)^2 + (1-b)^2 \cdot e^{-2a} - b\left(1 + b + (1-b)e^{-2a}\right) Z_T / Z_0$$
$$\det 1 = -(1-b^2)(1-e^{-2a}) + b\left(1 - b + (1+b)e^{-2a}\right) Z_T / Z_0$$
$$\det 2 = -2\left(1 + b(1 + Z_T/Z_0)\right) \tag{12}$$
$$\det 3 = 2\left(1 - b(1 + Z_T/Z_0)\right)e^{-a}$$
$$\det 4 = -\,4b \cdot e^{-a}$$

and the desired amplitude ratios are :

$$\frac{P_r}{P_e} = r_{Vs} = \frac{\det 1}{\det} \quad ; \quad \frac{P_t}{P_e} = t_{Vs} = \frac{\det 4}{\det} \quad ; \quad \frac{P_a}{P_e} = \frac{\det 2}{\det} \quad ; \quad \frac{P_{ar}}{P_e} = \frac{\det 3}{\det} \tag{13}$$

The insertion coefficient for a single incident plane wave (with angle ϑ) becomes :

$$\tau_e(\vartheta)=\frac{\tau_{vs}}{1-|r_{vs}|^2}=\frac{|t_{vs}|^2}{1-|r_{vs}|^2}=\frac{|\det 4|^2}{|\det|^2-|\det 1|^2}$$

$$=\left|4b\cdot e^{-a}\right|^2\cdot\left[\ \left|(1+b)^2-(1-b)^2\cdot e^{-2a}+b\cdot\left(1+b+(1-b)\cdot e^{-2a}\right)\cdot Z_T/Z_0\right|^2-\right. \qquad (14)$$

$$\left.-\left|(1-b^2)(1-e^{-2a})+b\cdot\left(1-b+(1+b)\cdot e^{-2a}\right)\cdot Z_T/Z\right|^2\ \right]^{-1}$$

Special cases : (15)

$$\tau_e(\vartheta)\xrightarrow[\text{no abs.}]{}4\Big/\left[\,|2+Z_T/Z_0|^2-|Z_T/Z_0|^2\right]\xrightarrow[\eta\to0]{}1$$

$$\tau_e(\vartheta)\xrightarrow[|Z_T|\to\infty]{}\frac{|t_{vs}|^2}{1-|r_A|^2}=\frac{|\det 4|^2}{|\det|^2\left[1-|r_A|^2\right]}\quad;\quad r_{Vs}\xrightarrow[|Z_T|\to\infty]{}r_A=\frac{(1+b)e^{-2a}+1-b}{(1-b)e^{-2a}+1+b}$$

$$\tau_e(\vartheta)\xrightarrow[|a|>>1]{}\frac{|\det 4|^2}{|\det|^2_{|a|>>1}-|\det 1|^2_{|a|>>1}}=\left[\left|4b\cdot e^{-a}\right|^2\right]\Big/\left[4\,\mathrm{Re}\{b\}\cdot\left|1+b\cdot(1+Z_T/Z_0)\right|^2\right]$$

For three-dimensional diffuse sound incidence : $\tau_{3-dif}=\dfrac{2}{\sin^2\vartheta_{hi}}\displaystyle\int_0^{\vartheta_{hi}}\tau_e(\vartheta)\cos\vartheta\sin\vartheta\,d\vartheta$ (16)

For two-dimensional diffuse sound incidence: $\tau_{2-dif}=\dfrac{1}{\sin\vartheta_{hi}}\displaystyle\int_0^{\vartheta_{hi}}\tau_e(\vartheta)\cos\vartheta\,d\vartheta$ (17)

The example shows the sound transmission loss $R=-10\cdot\lg(\tau_{VS})$ of a capsule wall (thin curves) and the insertion loss of a capsule (thick curves) with these walls, for three flow resistivity values Ξ of the absorber layer (sound incidence under $\vartheta=45°$).

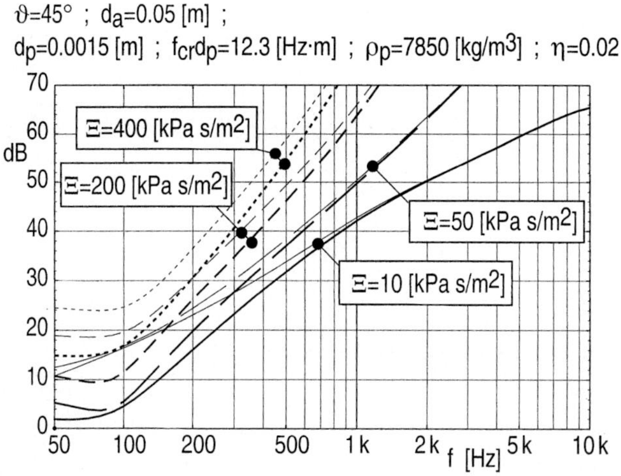

Sound transmission loss R of capsule walls and insertion loss D_e of a capsule for three flow resistivity values Ξ of the porous layer.

L.2 Absorbent sound source in a capsule

Ref.: Mechel, [L.1]

Sound absorption inside a capsule may be produced not only by an absorber layer on the capsule walls, but also by the source itself. This effect will be illustrated with a model, in which the capsule and the source are two-dimensional; the source offers at its surface an impedance Z_i to incident waves. Let Z_F be the field impedance at the source surface; then the sound pressure and particle velocity at its surface are given by :

$$p(x_{Qu}) = \frac{1}{1+Z_i/Z_F} \cdot p_{Qu} \quad ; \quad v(x_{Qu}) = \frac{1}{1+Z_F/Z_i} \cdot v_{Qu} \tag{1}$$

where p_{Qu}, v_{Qu} are generally used to characterise a source, and which belong to the special cases :

$$p(x_{Qu}) \xrightarrow[Z_F \to \infty]{} p_{Qu} \quad ; \quad v(x_{Qu}) \xrightarrow[Z_F \to 0]{} v_{Qu} \tag{2}$$

The relation exists (Helmholtz's source theorem) : $p_{Qu} = Z_i \cdot v_{Qu}$, and :

$$\frac{p(x_{Qu})}{v(x_{Qu})} = \frac{1+Z_F/Z_i}{1+Z_i/Z_F} \cdot \frac{p_{Qu}}{v_{Qu}} = \frac{Z_F}{Z_i} \cdot \frac{p_{Qu}}{v_{Qu}} \tag{3}$$

Because of the finite interior impedance Z_i of the source, the condition of the energetic approximation of the previous section L.1 , that the source power Π is constant for whatever exterior sound field, no longer holds.

The model consists of a plane sound source Qu , which radiates a plane wave towards both sides at an angle ϑ , which is given by the wave number k_{qu} along the source surface with $\sin\vartheta = k_{qu}/k_0$. The walls of the capsule are equal on both sides (for simplicity), but possibly have different distances $t_\pm$ to the source. They consist of a porous layer and a metal sheet or plate.

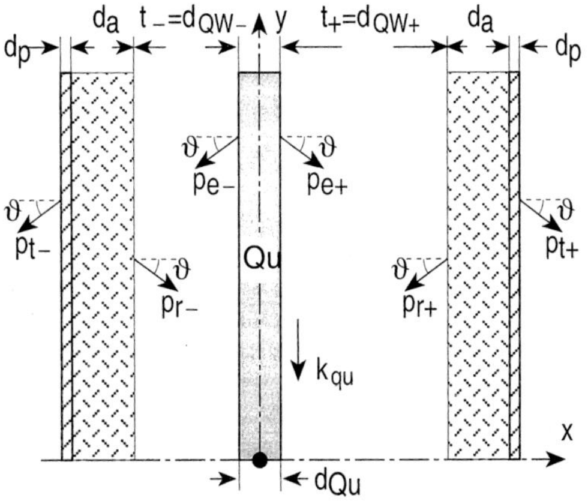

The source will not be transmissive for incident sound (such as largemachines).
Thus both sides of the source are independent of each other; sound fields will be written only for one side; the fields on the other side follow by simple substitutions. Because of this

independence, the source thickness can be taken to be $d_{Qu}=0$ (or the co-ordinate x is shifted correspondingly).

Field formulations :

$$\begin{aligned}
p_{e+}(x,y) &= P_{e+}\cdot e^{jk_0 y\sin\vartheta}\cdot e^{-jk_0(x-t_+)\cos\vartheta}\\
p_{r+}(x,y) &= P_{r+}\cdot e^{jk_0 y\sin\vartheta}\cdot e^{+jk_0(x-t_+)\cos\vartheta}\\
p_{t+}(x,y) &= P_{t+}\cdot e^{jk_0 d_a\cos\vartheta}\cdot e^{jk_0 y\sin\vartheta}\cdot e^{-jk_0(x-t_+)\cos\vartheta}\\
p_{a+}(x,y) &= P_{a+}\cdot e^{\Gamma_a y\sin\vartheta_a}\cdot e^{-\Gamma_a(x-t_+)\cos\vartheta_a}\\
p_{ar+}(x,y) &= P_{ar+}\cdot e^{-\Gamma_a d_a\cos\vartheta_a}\cdot e^{\Gamma_a y\sin\vartheta_a}\cdot e^{+\Gamma_a(x-t_+)\cos\vartheta_a}
\end{aligned} \qquad (4)$$

with interior angle in the porous layer : $\sin\vartheta_a = \dfrac{jk_0}{\Gamma_a}\cdot\sin\vartheta$ (5)

and partition impedance Z_T of the metal sheet :

$$\frac{Z_T}{Z_0} = 2\pi Z_m F[\eta F^2\cdot\sin^4\vartheta_a + j(1-F^2\cdot\sin^4\vartheta_a)] \;;\; Z_m = \frac{f_{cr}d_p}{Z_0}\rho_p \;;\; F = \frac{f}{f_{cr}} \qquad (6)$$

where η is the bending loss factor of the sheet; ρ_p the density of its material; f_{cr} the critical frequency. The amplitudes P_{e+}, P_{r+}, P_{a+} are "defined" at $x=t_+$, the amplitudes P_{ar+}, P_{t+} at $x=t_+ + d_a$. The source strength is described by its surface velocity profile :

$$v_{Qu}(y) = V_{Qu}\cdot e^{jk_q y\sin\vartheta} \qquad (7)$$

The condition at $x=x_{Qu}=0$: $p(0,y) + Z_i\cdot v(0,y) = Z_i\cdot v_{Qu}$ together with the boundary conditions leads to the system of equations :

$$\begin{pmatrix} 1 & -1 & -e^{-a} & 0 & 1\\ 1 & b & -be^{-a} & 0 & -1\\ 0 & e^{-a} & 1 & -(1+Z_T/Z_0) & 0\\ 0 & be^{-a} & -b & -1 & 0\\ d & 0 & 0 & 0 & c \end{pmatrix}\cdot\begin{pmatrix} P_{r+}\\ P_{a+}\\ P_{ar+}\\ P_{t+}\\ P_{e+}\end{pmatrix} = \begin{pmatrix} 0\\0\\0\\0\\ Z_i\cdot V_{Qu}\end{pmatrix} \qquad (8)$$

with the abbreviations ($\Gamma_{an}=\Gamma_a/k_0$; $Z_{an}=Z_a/Z_0$) :

$$a := \Gamma_a d_a\cdot\cos\vartheta_a = k_0 d_a\sqrt{\Gamma_{an}^2+\sin^2\vartheta} \;;\; b := \frac{Z_0\cos\vartheta_a}{Z_a\cos\vartheta} = \frac{1}{\Gamma_{an}Z_{an}}\frac{\sqrt{\Gamma_{an}^2+\sin^2\vartheta}}{\cos\vartheta}$$
$$c := (1+\cos\vartheta\cdot Z_i/Z_0)\cdot e^{+jk_0 t_+\cos\vartheta} \;;\; d := (1-\cos\vartheta\cdot Z_i/Z_0)\cdot e^{-jk_0 t_+\cos\vartheta} \qquad (9)$$

The transmission and reflection factors of the capsule walls are :

$$t_{Vs} = \frac{P_{t\pm}}{P_{e\pm}} \quad ; \quad r_{Vs} = \frac{P_{r\pm}}{P_{e\pm}} \tag{10}$$

The sound intensities I_+ radiated by the capsule (on one side) and I_0 by the source, if it is in the free space ($Z_F(0)=Z_0/\cos\vartheta$), are :

$$I_+ = \frac{\cos\vartheta}{2Z_0} |P_{t+}|^2$$

$$I_0 = \frac{\cos\vartheta}{2Z_0} |p_+(0)|^2 = \frac{\cos\vartheta}{2Z_0} \cdot \left|\frac{Z_i \cdot Z_F(0)}{Z_i + Z_F(0)}\right|^2 \cdot |V_{Qu}|^2 = \frac{\cos\vartheta}{2Z_0} \frac{|Z_i V_{Qu}|^2}{|Z_i/Z_0 \cdot \cos\vartheta + 1|^2} \tag{11}$$

Thus the insertion power coefficient for one side becomes :

$$\tau_{e+} = \frac{I_+}{I_0} = \left|1 + Z_i/Z_0 \cdot \cos\vartheta\right|^2 \cdot \left|\frac{P_{t+}}{Z_i V_{Qu}}\right|^2 \tag{12}$$

and for both sides together :

$$\tau_e = \frac{I_+ + I_-}{2I_0} = \frac{\left|1 + Z_i/Z_0 \cdot \cos\vartheta\right|^2}{2} \cdot \left[\left|\frac{P_{t+}}{Z_i V_{Qu}}\right|^2 + \left|\frac{P_{t-}}{Z_i V_{Qu}}\right|^2\right] \tag{13}$$

The determinants needed in KRAMER' rule for the solution of the system of equations are :

$$\begin{aligned} \det = {} & -(1+b)^2 c - (1-b^2)d + \left((1-b)^2 c - (1-b^2)d\right) \cdot e^{-2a} - \\ & - b \cdot Z_T/Z_0 \cdot \left[\left((1-b)c + (1+b)d\right) \cdot e^{-2a} + (1+b)c + (1-b)d\right] \end{aligned} \tag{14}$$

$$\begin{aligned} \det 1 &= Z_i V_{Qu}\left[-(1-b^2) - b(1-b)Z_T/Z_0 + (1+b)\left(1 - b(1+Z_T/Z_0) \cdot e^{-2a}\right)\right] \\ \det 2 &= Z_i V_{Qu}\left[-2 \cdot \left(1 + b(1+Z_T/Z_0)\right)\right] \\ \det 3 &= Z_i V_{Qu}\left[2 \cdot \left(1 - b(1+Z_T/Z_0)\right) \cdot e^{-a}\right] \\ \det 4 &= Z_i V_{Qu}\left[-4 \cdot b \cdot e^{-a}\right] \\ \det 5 &= Z_i V_{Qu}\left[-(1+b)^2 - b(1+b)Z_T/Z_0 + (1-b)\left(1 - b(1+Z_T/Z_0)\right) \cdot e^{-2a}\right] \end{aligned} \tag{15}$$

and the required ratio :

$$\begin{aligned} \frac{P_{t\pm}}{Z_i V_{Qu}} &= \frac{\det 4}{\det} = \\ &= \left[4 \cdot b \cdot e^{-a}\right] \cdot \left\{(1+b)^2 c_\pm + (1-b^2)d_\pm - \left((1-b)^2 c_\pm + (1-b^2)d_\pm\right) \cdot e^{-2a} + \right. \\ &\quad \left. + b \cdot Z_T/Z_0 \cdot \left[(1+b)c_\pm + (1-b)d_\pm + \left((1-b)c_\pm + (1+b)d_\pm\right) \cdot e^{-2a}\right]\right\}^{-1} \end{aligned} \tag{16}$$

Special case: *the source is a pressure source* , i.e. $Z_i \to 0$:

$$\tau_{e+} \to \tau_{ep+} = \left|4\,b\cdot e^{-a}\right|^2 \cdot \left|\left[(1+b)^2-(1-b)^2\cdot e^{-2a}+b\cdot Z_T/Z_0\left(1+b+(1-b)\cdot e^{-2a}\right)\right]\cdot e^{+jk_0t_+\cos\vartheta} + \right. \tag{17}$$
$$\left. +\left[(1-b^2)\cdot(1-e^{-2a})+b\cdot Z_T/Z_0\left(1-b+(1+b)\cdot e^{-2a}\right)\right]\cdot e^{-jk_0t_+\cos\vartheta}\right|^{-2}$$

Special case: *the source is a velocity source* , i.e. $Z_i \to \infty$:

$$\tau_{e+} \to \tau_{ev+} = \left|4\,b\cdot e^{-a}\right|^2 \cdot \left|\left[(1+b)^2-(1-b)^2\cdot e^{-2a}+b\cdot Z_T/Z_0\left(1+b+(1-b)\cdot e^{-2a}\right)\right]\cdot e^{+jk_0t_+\cos\vartheta} - \right. \tag{18}$$
$$\left. -\left[(1-b^2)(1-e^{-2a})+b\cdot Z_T/Z_0\left(1-b+(1+b)\cdot e^{-2a}\right)\right]\cdot e^{-jk_0t_+\cos\vartheta}\right|^{-2}$$

Special case: *no absorber layer* , i.e. $d_i \to 0$; $a \to 0$; $b \to 1$: (19)

$$\tau_{e+} \to \tau_{e0+} = 4\frac{|1+Z_i/Z_0\cdot\cos\vartheta|^2}{|2c+(c+d)\cdot Z_{Tnx}|^2} = 4\frac{|1+Z_i/Z_0\cdot\cos\vartheta|^2}{\left|2+\left(1+\dfrac{(1-Z_i/Z_0)}{(1+Z_i/Z_0)}\cdot e^{-2jk_0t_+\cos\vartheta}\right)\cdot Z_T/Z_0\right|^2}$$

If Z_i of the source is large :

$$\tau_{e0+} \to 4\frac{|1+Z_i/Z_0\cdot\cos\vartheta|^2}{\left|2+(1+e^{-2jk_0t_+\cos\vartheta})\cdot Z_T/Z_0\right|^2} \tag{20}$$

This quantity oscillates strongly with frequency and/or distance t_+ :

$$\tau_{e0+,max} \to |1+Z_i/Z_0\cdot\cos\vartheta|^2 \quad ; \quad \tau_{e0+,min} \to \frac{|1+Z_i/Z_0\cdot\cos\vartheta|^2}{|1+Z_T/Z_0|^2} \tag{21}$$

Special case of *a narrow capsule* , i.e. $t_\pm << \lambda_0$:

$$c \to (1+\cos\vartheta\cdot Z_i/Z_0)\cdot(1+jk_0d_+\cos\vartheta) \quad ; \quad d \to (1-\cos\vartheta\cdot Z_i/Z_0)\cdot(1-jk_0d_+\cos\vartheta) \tag{22}$$

$$\tau_{en+} = \left|4\,b(1+Z_i/Z_0\cdot\cos\vartheta)\cdot e^{-a}\right|^2 \cdot$$
$$\cdot\left|(1-Z_i/Z_0\cdot\cos\vartheta)\cdot(1-jk_0t_+\cos\vartheta)\left[(1-b^2)(1-e^{-2a})+b\cdot Z_T/Z_0\left(1-b+(1+b)\cdot e^{-2a}\right)\right]+\right.$$
$$\left.(1+Z_i/Z_0\cdot\cos\vartheta)\cdot(1+jk_0t_+\cos\vartheta)\left[(1+b)^2-(1-b)^2e^{-2a}+b\cdot Z_T/Z_0\left(1+b+(1-b)\cdot e^{-2a}\right)\right]\right|^{-2} \tag{23}$$

$\vartheta=0°$; $Z_i/Z_0=10$; $d_a=0.05$ [m] ; $\Xi=10$ [kPa s/m^2]
$d_p=0.0015$ [m] ; $f_{cr}d_p=12.3$ [Hz·m] ; $\rho_p=7850$ [kg/m^3] ; $\eta=0.02$

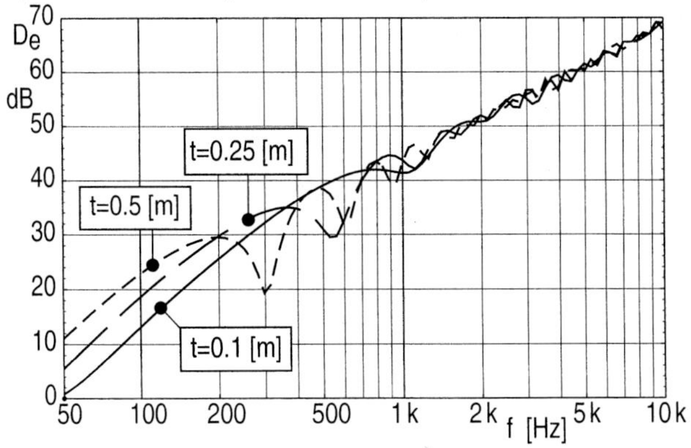

Insertion loss D_e of a capsule, for only normal incidence $\vartheta=0$, with a rather "high-ohmic" source $Z_i/Z_0= 10$, for three distances t between source and wall .

ϑ=var° ; $Z_i/Z_0=10$; $t=0.5$ [m] ; $d_a=0.05$ [m] ; $\Xi=10$ [kPa s/m^2]
$d_p=0.0015$ [m] ; $f_{cr}d_p=12.3$ [Hz·m] ; $\rho_p=7850$ [kg/m^3] ; $\eta=0.02$

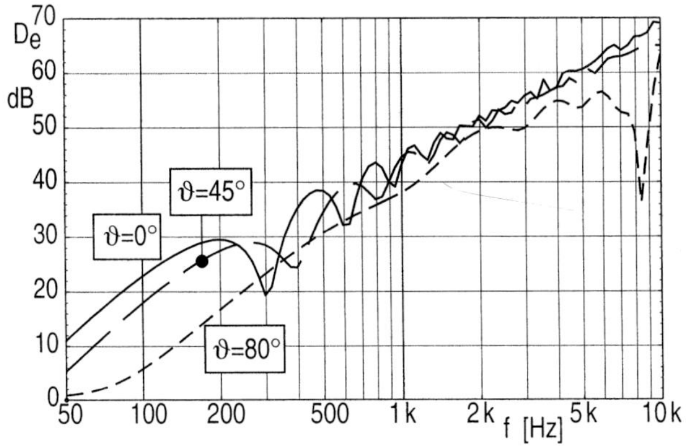

As before, but with one distance t= 0.5 [m] and three angles of incidence ϑ .

$Z_i/Z_0=0.1$

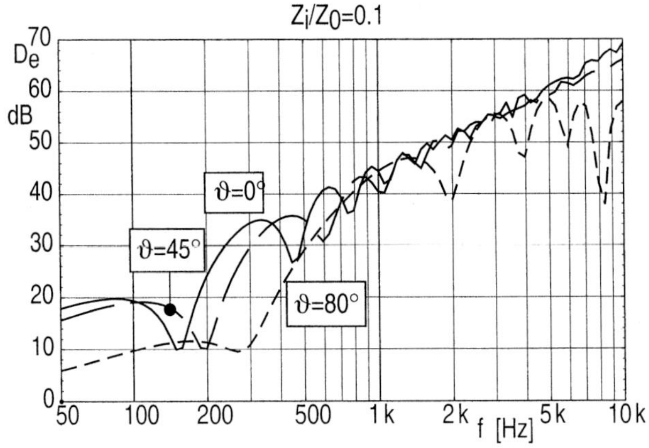

As before, but with a "low-ohmic" source.

ϑ=diff. ; Z_i/Z_0=10 ; t=var. ; d_a=0.05 [m] ; Ξ=10 [kPa s/m^2]
d_p=0.0015 [m] ; $f_{cr}d_p$=12.3 [Hz·m] ; ρ_p=7850 [kg/m^3] ; η=0.02

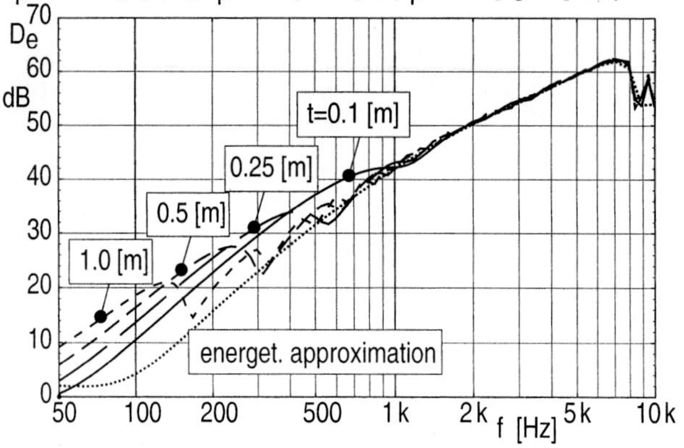

Insertion loss D_e of a capsule for diffuse sound incidence, with three distances t of source and wall. The energetic approximation is shown for comparison.

ϑ=diff. ; $Z_i/Z_0=\infty$; t=0.5 [m] ; d_a=0.05 [m] ; Ξ=var.
d_p=0.0015 [m] ; $f_{cr}d_p$=12.3 [Hz·m] ; ρ_p=7850 [kg/m^3] ; η=0.02

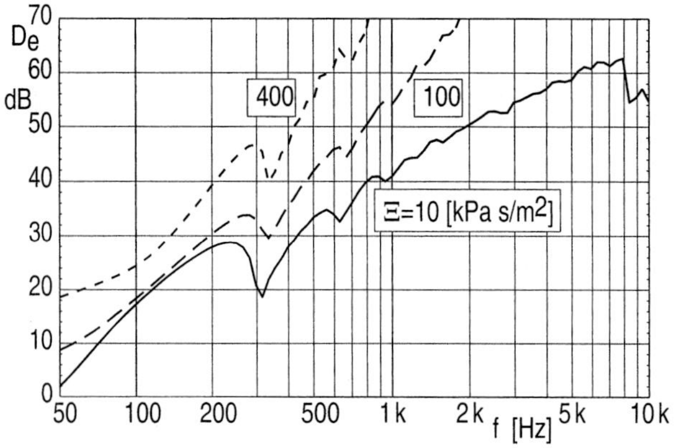

Insertion loss D_e of a capsule for diffuse sound incidence, with three flow resistivity values Ξ of the porous layer material, for a velocity source.

Z_i/Z_0=0

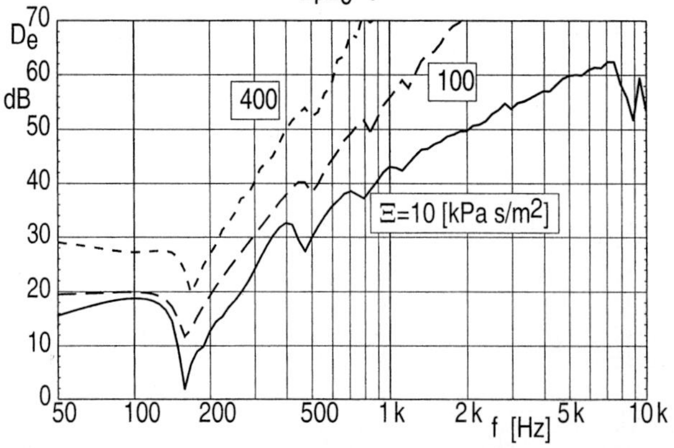

As before, but for a pressure source.

L.3 Semicylindrical source and capsule

Ref.: Mechel, [L.1]

The assumption of a two-dimensional model for a capsule may seem to be taking the abstraction of the model too far.

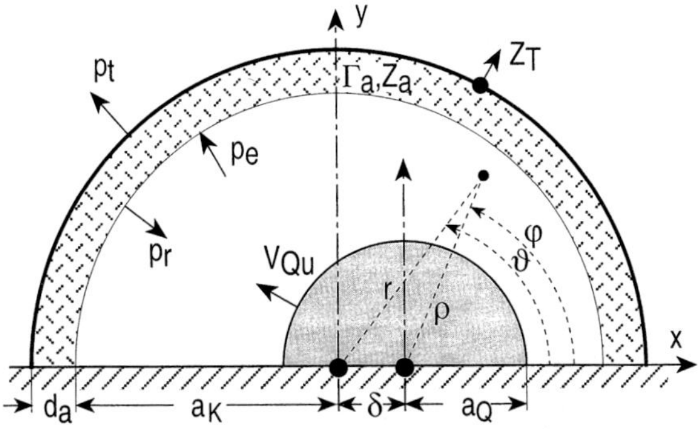

The model is now a semi-circular capsule and a similar source. In principle the source can have an eccentric position. The radiated wave can be expanded in cylindrical harmonics in the co-ordinate system (ρ,φ) of the source. These harmonics, in turn, can be expanded, by the addition theorem for Bessel functions, in cylindrical harmonics with the co-ordinates (r,ϑ). For simplicity this double expansion is avoided here; the source is concentric with the capsule.

The strength of the source is described by its radial particle velocity profile v_{Qu}, which, if necessary, is expanded as :

$$v_{Qu}(\varphi,z) = \sum_n V_n \cdot \cos(n\varphi) \cdot \cos(k_{zn} z) \tag{1}$$

The interior source impedance Z_i is assumed to be constant in z,φ. The relation holds :

$$p(\rho = a_Q,\varphi,z) + Z_i \cdot v_r(\rho = a_Q,\varphi,z) = Z_i \cdot v_{Qu}(\varphi,z) \tag{2}$$

Let the field in the interspace between source and capsule wall be $p_i = p_e + p_r$; in the absorber layer p_a ; in the outer space p_t. Let the floor of the capsule be hard.

Field formulations :

$$p_i(r,\vartheta,z) = \sum_{n\ge 0} \left[P_{e,n} \cdot \frac{H_n^{(2)}(k_r r)}{H_n^{(2)}(k_r a_Q)} + P_{r,n} \cdot \frac{H_n^{(1)}(k_r r)}{H_n^{(1)}(k_r a_K)} \right] \cdot \cos(n\vartheta) \cdot \cos(k_{zn} z)$$

$$p_a(r,\vartheta,z) = \sum_{n\ge 0} \left[P_{a,n} \cdot \frac{H_n^{(2)}(k_{ar} r)}{H_n^{(2)}(k_{ar} a_K)} + Q_{a,n} \cdot \frac{H_n^{(1)}(k_{ar} r)}{H_n^{(1)}(k_{ar}(a_K + d_a))} \right] \cdot \cos(n\vartheta) \cdot \cos(k_{zn} z) \tag{3}$$

$$p_t(r,\vartheta,z) = \sum_{n\ge 0} P_{t,n} \cdot \frac{H_n^{(2)}(k_r r)}{H_n^{(2)}(k_r (a_K + d_a))} \cdot \cos(n\vartheta) \cdot \cos(k_{zn} z) \tag{3}$$

with : $$k_r^2+k_{zn}^2=k_0^2 \quad ; \quad k_{a,r}^2+k_{zn}^2=k_a^2=-\Gamma_a^2 \tag{4}$$

This relation defines a modal angle of incidence Θ_n :

$$\left(\frac{k_r}{k_0}\right)^2+\left(\frac{k_{zn}}{k_0}\right)^2=1=\cos^2\Theta_n+\sin^2\Theta_n \tag{5}$$

which is zero for conphase excitation along the z axis. The relevant angle χ_n for the evaluation of the partition impedance Z_T of the outer shell of the capsule is given by :

$$\sin\chi_n=\frac{1}{k_0}\sqrt{k_{zn}^2+(n/a)^2}=\sqrt{\sin^2\Theta_n+\left(n/k_0(a_K+d_a)\right)^2} \tag{6}$$

The boundary conditions with a concentric source :

$$\begin{aligned}
&p_i(a_K)\overset{!}{=}p_a(a_K) \quad ; \quad v_{ir}(a_K)\overset{!}{=}v_{ar}(a_K)\\
&p_a(a_K+d_a)-p_t(a_K+d_a)\overset{!}{=}Z_T\cdot v_{tr}(a_K+d_a)\\
&v_{ar}(a_K+d_a)\overset{!}{=}v_{tr}(a_K+d_a)\\
&p_i(a_Q)+Z_i\cdot v_{ip}(a_Q)\overset{!}{=}Z_i\cdot v_{Qu}
\end{aligned} \tag{7}$$

hold term-wise and produce the system of equations :

$$\begin{pmatrix} \vdots\cdots & \cdots & \cdots\vdots \\ & \ddots \quad \iddots & \\ \vdots & a_{i,k} & \vdots \\ & \iddots \quad \ddots & \\ \vdots\cdots & \cdots & \cdots\vdots \end{pmatrix}\cdot\begin{pmatrix} P_{e,n} \\ P_{r,n} \\ P_{a,n} \\ Q_{a,n} \\ P_{t,n} \end{pmatrix}=\begin{pmatrix} 0 \\ 0 \\ 0 \\ 0 \\ Z_iV_n \end{pmatrix} \tag{8}$$

with coefficients (a prime indicates the derivative) :

$$a_{11}=\frac{H_n^{(2)}(k_ra_K)}{H_n^{(2)}(k_ra_Q)} \; ; \; a_{12}=1 \; ; \; a_{13}=-1 \; ; \; a_{14}=-\frac{H_n^{(1)}(k_{ar}a_K)}{H_n^{(1)}(k_{ar}(a_K+d_a))} \; ; \; a_{15}=0 \tag{9}$$

$$\begin{aligned}
&a_{21}=jk_ra_K\cdot\frac{H_n'^{(2)}(k_ra_K)}{H_n^{(2)}(k_ra_Q)} \; ; \; a_{22}=jk_ra_K\cdot\frac{H_n'^{(1)}(k_ra_K)}{H_n^{(1)}(k_ra_K)} \; ;\\
&a_{23}=\frac{k_{a,r}a_K}{\Gamma_{an}Z_{an}}\frac{H_n'^{(2)}(k_{ar}a_K)}{H_n^{(2)}(k_{ar}a_K)} \; ; \; a_{24}=\frac{k_{a,r}a_K}{\Gamma_{an}Z_{an}}\frac{H_n'^{(1)}(k_{ar}a_K)}{H_n^{(1)}(k_{ar}(a_K+d_a))} \; ; \; a_{25}=0
\end{aligned} \tag{10}$$

$$\begin{aligned}
&a_{31}=a_{32}=0 \; ; \; a_{33}=-\frac{H_n^{(2)}(k_{ar}(a_K+d_a))}{H_n^{(2)}(k_{ar}a_K)} \; ; \; a_{34}=-1 \; ;\\
&a_{35}=1+j\frac{k_rZ_T}{k_0Z_0}\frac{H_n'^{(2)}(k_r(a_K+d_a))}{H_n^{(2)}(k_r(a_K+d_a))}
\end{aligned} \tag{11}$$

$$a_{41} = a_{42} = 0 \;;\; a_{43} = \frac{k_{a,r}a_K}{\Gamma_{an}Z_{an}} \frac{H_n'^{(2)}(k_{ar}(a_K + d_a))}{H_n^{(2)}(k_{ar}a_K)} \;;$$

$$a_{44} = \frac{k_{a,r}a_K}{\Gamma_{an}Z_{an}} \frac{H_n'^{(1)}(k_{ar}(a_K + d_a))}{H_n^{(1)}(k_{ar}(a_K + d_a))} \;;\; a_{45} = j k_r a_K \cdot \frac{H_n'^{(2)}(k_r(a_K + d_a))}{H_n^{(2)}(k_r(a_K + d_a))} \tag{12}$$

$$a_{51} = 1 + j\frac{k_r Z_i}{k_0 Z_0} \frac{H_n'^{(2)}(k_r a_Q)}{H_n^{(2)}(k_r a_Q)} \;;\; a_{52} = \frac{H_n^{(1)}(k_r a_Q)}{H_n^{(1)}(k_r a_K)} + j\frac{k_r Z_i}{k_0 Z_0} \frac{H_n'^{(1)}(k_r a_Q)}{H_n^{(1)}(k_r a_K)} \;; \tag{13}$$

$$a_{53} = a_{54} = a_{55} = 0$$

The radiated (effective) power is the sum of the modal powers. The radiated modal power of the free source (radius $a= a_Q$) without z factors (they cancel in τ) is :

$$\Pi_n^{(0)} = \frac{\pi a_Q}{2\delta_n Z_0} \cdot \frac{\mathrm{Re}\{Z_n(a_Q)/Z_0\}}{|Z_i/Z_0 + Z_n(a_Q)/Z_0|^2} \cdot |Z_i V_n|^2 \tag{14}$$

the radiated (effective) modal power of the capsule (radius $a= a_K+d_a$) is :

$$\begin{aligned}\Pi_n &= \frac{a_K + d_a}{2} Z_0 \cdot \mathrm{Re}\{Z_n(a_K + d_a)/Z_0\} \cdot \int_0^{\pi} |v_{tr,n}(a_K + d_a, \vartheta)|^2 \, d\vartheta \\ &= \frac{\pi(a_K + d_a)}{2\delta_n} Z_0 \cdot \mathrm{Re}\{Z_n(a_K + d_a)/Z_0\} \cdot \frac{|k_r P_{t,n}|^2}{(k_0 Z_0)^2} \left| \frac{H_n'^{(2)}(k_r(a_K + d_a))}{H_n^{(2)}(k_r(a_K + d_a))} \right|^2\end{aligned} \tag{15}$$

Thus the insertion power coefficient of the capsule becomes :

$$\begin{aligned}\tau_K &= \frac{\sum_{n\ge 0} \Pi_n}{\sum_{n\ge 0} \Pi_n^{(0)}} = \\ &= \frac{a_K + d_a}{a_Q} \cdot \frac{\sum_{n\ge 0} \frac{1}{\delta_n} \mathrm{Re}\{Z_n(a_K + d_a)/Z_0\} \cdot \left| \frac{k_r}{k_0} \cdot \frac{H_n'^{(2)}(k_r(a_K + d_a))}{H_n^{(2)}(k_r(a_K + d_a))} \right|^2 \cdot |P_{t,n}|^2}{\sum_n \frac{1}{\delta_n} \frac{\mathrm{Re}\{Z_n(a_Q)/Z_0\}}{|(Z_i + Z_n(a_Q))/Z_0|^2} \cdot |Z_i V_n|^2}\end{aligned} \tag{16}$$

with $\delta_0=1$; $\delta_{n>0}=2$ and the modal radiation impedances :

$$\frac{Z_n(a)}{Z_0} = -j\frac{k_0}{k_r} \frac{H_n^{(2)}(k_r a)}{H_n'^{(2)}(k_r a)} \tag{17}$$

One needs information about the $Z_i V_n$ for further valuation, if the source pattern is not mono-modal.

$\Theta=0°$; $n=0$; $Z_i/Z_0=10$; $a_K=1$ [m] ;
$d_a=0.05$ [m] ; $\Xi=10$ [kPa s/m²]
$d_p=0.0015$ [m] ; $f_{cr}d_p=12.3$ [Hz·m] ; $\rho_p=7850$ [kg/m³] ; $\eta=0.02$

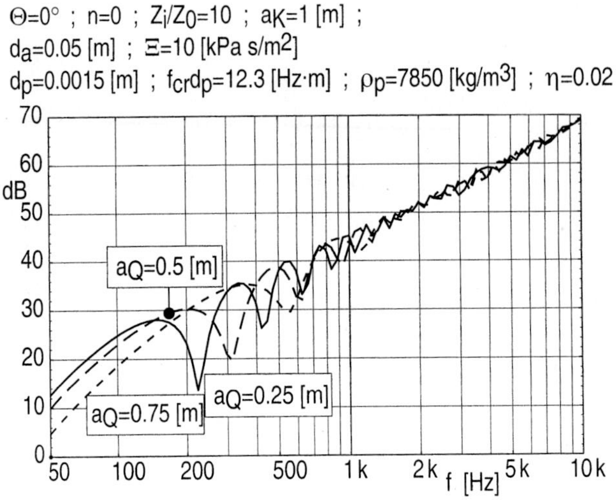

Insertion loss D_e for monomodal excitation, $n=0$, by a high-ohmic source, $Z_i/Z_0=10$, with three source radii a_Q.

$\Theta=0°$; $n=0$; $Z_i/Z_0=0.1$; $a_K=1$ [m] ;
$d_a=0.05$ [m] ; $\Xi=10$ [kPa s/m²]
$d_p=0.0015$ [m] ; $f_{cr}d_p=12.3$ [Hz·m] ; $\rho_p=7850$ [kg/m³] ; $\eta=0.02$

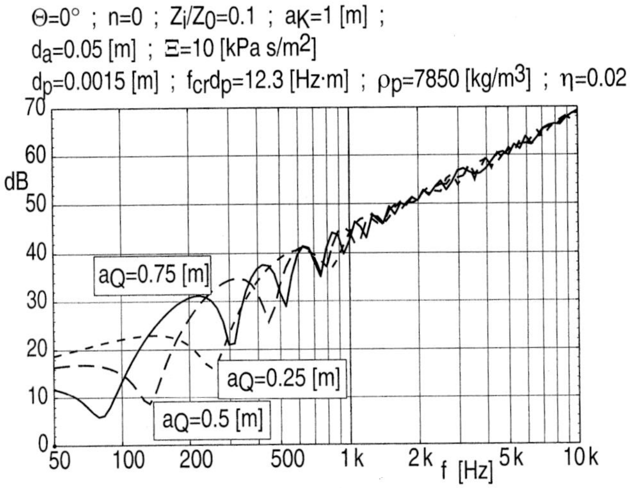

As before, but for a low-ohmic source.

Heuristic assumptions about the source mode amplitudes could be $Z_iV_n=$ const or $Z_iV_n \sim 1/(n+1)$. The following diagram shows the influence of such assumptions on D_e for a multi-modal excitation with the mode orders $n=0,...,4$ (with a low-ohmic source $Z_i/Z_0=0.1$) :

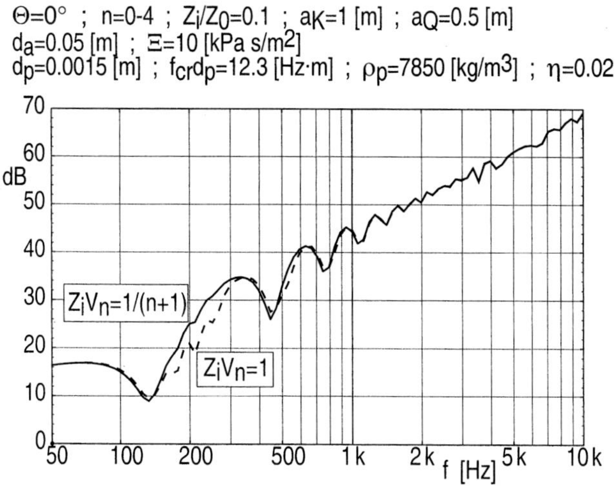

Insertion loss D_e for a multi-modal excitation, n=0,...,4, with two assumptions about the modal strength.

L.4 Hemispherical source and capsule

Ref.: Mechel, [L.1]

The object is similar to the object of the previous section L.3 , but now the source and the capsule are hemispherical. An eccentric source could again be treated with the addition theorem for Bessel functions, but, for simplicity, a concentric source will mainly be considered below. The source strength is described by a surface radial velocity profile v_{Qu} .

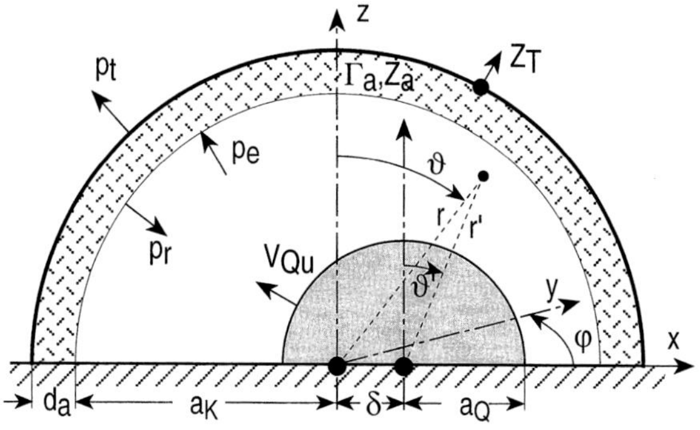

The sound field inside the capsule is $p_i = p_e + p_r$, the sound field in the interior absorber layer is p_a , the radiated sound outside the capsule is represented by p_t .

Field formulation :

$$v_{Qu}(\vartheta,\varphi) = \sum_{n\geq 0} V_n \cdot P_n^m(\cos\vartheta) \cdot \cos(m\varphi) \tag{1}$$

(for ease of writing, only one azimuthal mode m≥0 is supposed to exist; in the final equations below a sum of azimuthal modes will be considered)

$$p_i(r,\vartheta,\varphi)=\sum_{n\geq 0}\left[P_{e,n}\cdot\frac{h_n^{(2)}(k_0 r)}{h_n^{(2)}(k_0 a_Q)}+P_{r,n}\cdot\frac{h_n^{(1)}(k_0 r)}{h_n^{(1)}(k_0 a_K)}\right]\cdot P_n^m(\cos\vartheta)\cdot\cos(m\varphi)$$

$$p_a(r,\vartheta,\varphi)=\sum_{n\geq 0}\left[P_{a,n}\cdot\frac{h_n^{(2)}(k_a r)}{h_n^{(2)}(k_a a_K)}+Q_{a,n}\cdot\frac{h_n^{(1)}(k_a r)}{h_n^{(1)}(k_a(a_K+d_a))}\right]\cdot P_n^m(\cos\vartheta)\cdot\cos(m\varphi) \qquad (2)$$

$$p_t(r,\vartheta,\varphi)=\sum_{n\geq 0}P_{t,n}\cdot\frac{h_n^{(2)}(k_0 r)}{h_n^{(2)}(k_0(a_K+d_a))}\cdot P_n^m(\cos\vartheta)\cdot\cos(m\varphi)$$

$h_n^{(\alpha)}(z);\ \alpha=1,2;$ are spherical Hankel functions $\qquad h_n^{(\alpha)}(z)=\sqrt{\frac{\pi}{2z}}H_n^{(\alpha)}(z) \qquad (3)$

$P_n^m(z)$ are associate Legendre functions with special values :

$$P_n^m(z)\equiv 0\ ;\quad m>n$$

$$P_n^m(0)\xrightarrow[n+m=odd]{}0\ ;\quad \frac{dP_n^m(z)}{dz}\xrightarrow[n+m=even]{}0 \qquad (4)$$

For m=0 they go over to the Legendre polynomes $P_n(z)$:

$$P_n^m(z)\xrightarrow[m=0]{}P_n(z)$$

$$P_n^m(z)=(-1)^m\cdot(1-z^2)^{m/2}\frac{d^m P_n(z)}{dz^m}\ ;\quad z=\text{real},\ |z|\leq 1 \qquad (5)$$

The partition impedance Z_T of the outer, elastic shell follows from : the bending wave equation as :

$$Z_T:=\frac{\delta p}{v_\perp}=\frac{B}{j\omega}\frac{\left[\Delta_{\vartheta,\varphi}\Delta_{\vartheta,\varphi}-k_B^4\right]v_\perp(\vartheta,\varphi)}{v_\perp(\vartheta,\varphi)}=\frac{B}{j\omega}\frac{\left[\Delta_{\vartheta,\varphi}\Delta_{\vartheta,\varphi}-k_B^4\right]T(\vartheta)\cdot P(\varphi)}{T(\vartheta)\cdot P(\varphi)}$$
$$=\frac{B}{j\omega}\left(k_{trace}^4-k_B^4\right)=j\omega m\left[1-(k_{trace}/k_B)^4\right]=j\omega m\left[1-(f/f_{cr})^2\sin^4\Theta\right] \qquad (6)$$

where B= bending modulus; k_B= free bending wave number; f_{cr}= critical frequency of the shell (if it were a plane plate); k_{trace}= wave number of the trace of the exciting wave along the shell; Θ= polar angle of incidence on the shell with $k_{trace}=k_0\cdot\sin\Theta$. It is (a= shell radius) :

$$k_{trace}^4=\frac{n(n^2-1)(2+n)}{a^4}\ ;\quad \left(\frac{k_{trace}}{k_B}\right)^4=\frac{n(n^2-1)(2+n)}{(k_0 a)^4}\left(\frac{f}{f_{cr}}\right)^2 \qquad (7)$$

and therefore: $\qquad \sin^4\Theta=\frac{n(n^2-1)(2+n)}{(k_0 a)^4} \qquad (8)$

The radiated effective power of the source into free space is the sum of modal powers :

$$\Pi^{(0)}_{m,n}=\frac{a_Q^2 N_{m,n}}{4Z_0}\cdot\frac{\mathrm{Re}\{Z_n(a_Q)/Z_0\}}{|Z_i/Z_0+Z_n(a_Q)/Z_0|^2}\cdot|Z_iV_n|^2 \tag{9}$$

with the mode norms $N_{m,n}$:

$$\frac{a^2}{2}N_{m,n}=\int_0^{2\pi}d\varphi\int_0^{\pi/2}a^2\sin\vartheta\cdot(P_n^m(\cos\vartheta))^2\cdot\cos^2(m\varphi)\,d\vartheta=\frac{2\pi a^2}{\delta_m}\frac{1}{2n+1}\frac{(n+m)!}{(n-m)!} \tag{10}$$

($\delta_0=1$; $\delta_{n>0}=2$), and with the modal radiation impedance (a prime denotes the derivative) :

$$\frac{Z_n(a)}{Z_0}=-j\frac{h_n^{(2)}(k_0a)}{h_n'^{(2)}(k_0a)} \tag{11}$$

The modal effective power radiated by the capsule is correspondingly :

$$\Pi_{m,n}=\frac{(a_K+d_a)^2N_{m,n}}{4Z_0}\cdot\mathrm{Re}\{Z_n(a_K+d_a)/Z_0\}\cdot\left|\frac{h_n'^{(2)}(k_0(a_K+d_a))}{h_n^{(2)}(k_0(a_K+d_a))}\right|^2|P_{t,n}|^2 \tag{12}$$

The insertion power coefficient for multi-modal excitation finally becomes :

$$\tau_e=\frac{\sum\limits_{m,n\geq0}\Pi_{m,n}}{\sum\limits_{m,n\geq0}\Pi^{(0)}_{m,n}}=$$

$$=\frac{(a_K+d_a)^2}{a_Q^2}\cdot\frac{\sum\limits_{m,n\geq0}N_{m,n}\,\mathrm{Re}\{Z_n(a_K+d_a)/Z_0\}\cdot\left|\frac{h_n'^{(2)}(k_0(a_K+d_a))}{h_n^{(2)}(k_0(a_K+d_a))}\right|^2\cdot\left|P_{t,m,n}\right|^2}{\sum\limits_{m,n\geq0}N_{m,n}\frac{\mathrm{Re}\{Z_n(a_Q)/Z_0\}}{\left|(Z_i+Z_n(a_Q))/Z_0\right|^2}\cdot\left|Z_iV_{m,n}\right|^2} \tag{13}$$

The amplitudes $P_{t,m,n}$ therein follow from the system of equations of the previous section L.3 after substitution $H_n^{(\alpha)}(k_r r)\to h_n^{(\alpha)}(k_0 r)$ in the coefficients a_{ik} of that system, and $Z_iV_n\to Z_iV_{m,n}$ on the right-hand side.

The examples shown below illustrate the influence of the source interior impedance Z_i , and of the source radius a_Q (with a fixed capsule radius a_K= 1 [m]) for mono-modal and multi-modal excitations.

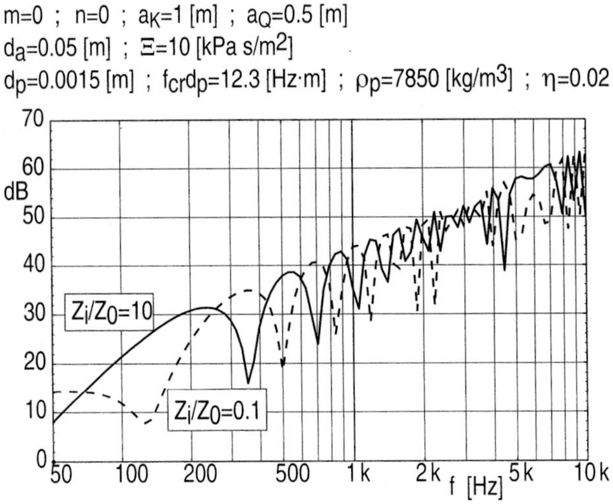

Insertion loss D_e for mono-modal excitation with two source impedances Z_i.

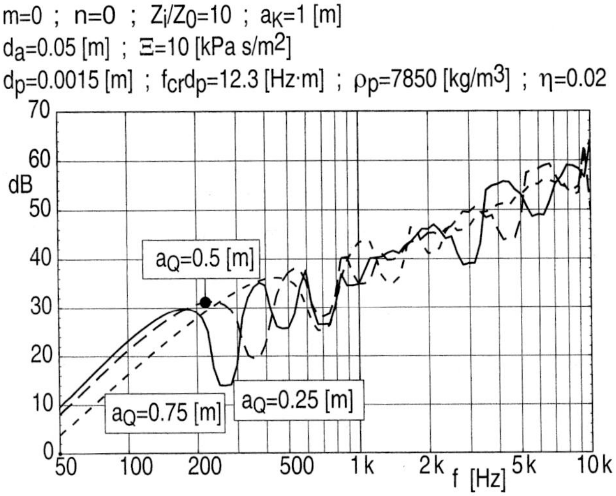

Insertion loss D_e for mono-modal excitation with three source radii a_Q.

m=0 ; n=0-5 ; Z_i/Z_0=10 ; a_K=1 [m]
d_a=0.05 [m] ; Ξ=10 [kPa s/m^2]
d_p=0.0015 [m] ; $f_{cr}d_p$=12.3 [Hz·m] ; ρ_p=7850 [kg/m^3] ; η=0.02

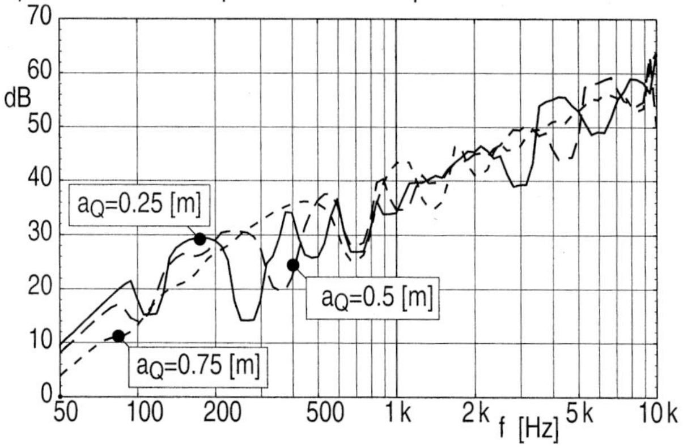

Insertion loss D_e for multi-modal excitation and three source radii a_Q.
The excitation mode amplitudes decay as Z_iV_n= 1/(n+1) .

m=0 ; n=0-5 ; Z_i/Z_0=1 ; a_K=1 [m]
d_a=0.05 [m] ; Ξ=10 [kPa s/m^2]
d_p=0.0015 [m] ; $f_{cr}d_p$=12.3 [Hz·m] ; ρ_p=7850 [kg/m^3] ; η=0.02

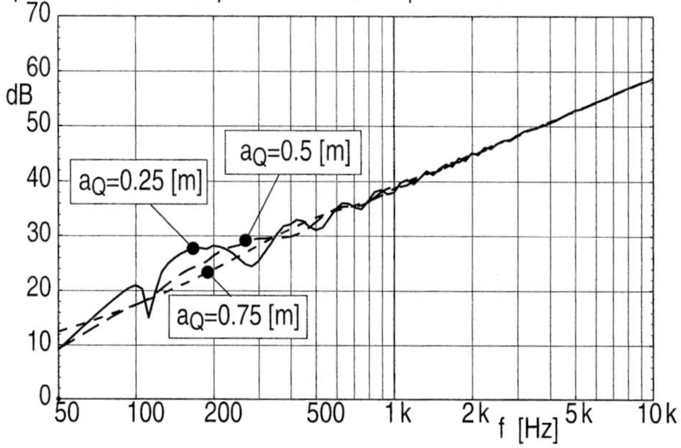

As before, but for a source which is "matched" to the exterior field, Z_i/Z_0=1 .

L.5 Cabins, semicylindrical model

Ref.: Mechel, [L.1]

Cabins are exposed to an exterior sound field p_e . A suitable quantity for the qualification of the efficiency of the cabin is the sound pressure level difference (*sound protection measure*)

$$\Delta L = -10 \cdot \lg \frac{\langle |p_i|^2 \rangle_{V_K}}{\langle |p_e|^2 \rangle_{V_K}} \quad \text{dB} \tag{1}$$

where p_i is the interior sound field, and $\langle \ldots \rangle_{V_K}$ indicates the spatial average over the volume V_K of the cabin. However, because cabins are often relatively small with a low mode density in them, the measurement of the average may be difficult. Alternatively a sound pressure level in some defined point r_0 in the cabin could be used :

$$\Delta L_0 = -10 \cdot \lg \frac{|p_i(r_0)|^2}{|p_e(r_0)|^2} \quad \text{dB} \tag{2}$$

This definition reduces the amount of numerical computation considerably.

The model to be treated below is a semicylindrical cabin on a hard floor. It consists of an outer elastic shell with partition impedance Z_T , and an interior layer of porous material with characteristic constants Γ_a, Z_a (or in normalised form $\Gamma_{an}= \Gamma_a/k_0$, $Z_{an}= Z_a/Z_0$).

A plane wave p_e^+ is assumed as incident wave (see below for diffuse sound incidence); a mirror-reflected wave p_e^- simulates the hard floor.

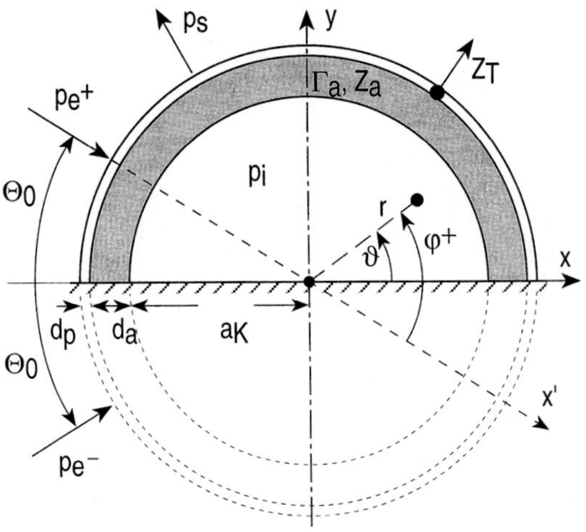

A variation of the incident wave in the z direction can be as $\cos(k_z z)$, $\sin(k_z z)$, $e^{\pm j k_z z}$ or a linear combination thereof. It would influence the radial wave number k_r by $k_z^2 + k_r^2 = k_0^2$. The field factor in the z direction can be dropped, because it is the same in all field parts.

The exciting field $p_e = p_e^+ + p_e^-$ is : (3)

$$p_e(r,\vartheta) = P_e \sum_{n\geq 0} \delta_n(-j)^n \cdot J_n(k_r r)\cdot\left[\cos\left(n(\vartheta+\Theta_0)\right)+\cos\left(n(\vartheta-\Theta_0)\right)\right]$$

$$= 2P_e \sum_{n\geq 0} \delta_n(-j)^n \cdot \cos(n\Theta_0)\cdot J_n(k_r r)\cdot\cos(n\vartheta) \qquad (4)$$

$$:= \sum_{n\geq 0} P_{en}\cdot J_n(k_r r)/J_n\left(k_r(a_K+d_a)\right)\cdot\cos(n\vartheta)$$

The last form uses the abbreviation :

$$P_{en} = 2P_e\cdot\delta_n(-j)^n\cdot\cos(n\Theta_0)\cdot J_n\left(k_r(a_K+d_a)\right) \qquad (5)$$

with $\delta_0=1$; $\delta_{n>0}=2$.

The interior field p_i , in the absorber layer p_a , and the exterior scattered field p_s are formulated as :

$$p_i(r,\vartheta) = \sum_{n\geq 0} P_{in}\cdot J_n(k_r r)/J_n(k_r a_K)\cdot\cos(n\vartheta)$$

$$p_a(r,\vartheta) = \sum_{n\geq 0}\left[P_{an}\cdot\frac{H_n^{(2)}(k_{ar}r)}{H_n^{(2)}(k_{ar}a_K)}+Q_{an}\cdot\frac{H_n^{(1)}(k_{ar}r)}{H_n^{(1)}\left(k_{ar}(a_K+d_a)\right)}\right]\cdot\cos(n\vartheta) \qquad (6)$$

$$p_s(r,\vartheta) = \sum_{n\geq 0} P_{sn}\cdot H_n^{(2)}(k_r r)/H_n^{(2)}\left(k_r(a_K+d_a)\right)\cdot\cos(n\vartheta)$$

The radial wave number k_{ar} in the absorber layer is given by $k_{ar}^2+k_z^2=k_a^2=-\Gamma_a^2$ .The unknown amplitudes P_{in}, P_{sn}, P_{an}, Q_{an} follow from the boundary conditions which hold term-wise :

$$p_{in}(a_K)\overset{!}{=}p_{an}(a_K) \quad ; \quad v_{irn}(a_K)\overset{!}{=}v_{arn}(a_K)$$

$$p_{an}(a_K+d_a)-p_{en}(a_K+d_a)-p_{sn}(a_K+d_a)\overset{!}{=}Z_{Tn}\cdot v_{arn}(a_K+d_a) \qquad (7)$$

$$v_{arn}(a_K+d_a)\overset{!}{=}v_{ern}(a_K+d_a)+v_{srn}(a_K+d_a)$$

The modal partition impedance Z_{Tn} of the shell is evaluated from :

$$\frac{Z_{Tn}}{Z_0} = 2\pi Z_m F[\eta F^2\cdot\sin^4\chi_n + j(1-F^2\cdot\sin^4\chi_n)] \quad ; \quad Z_m = \frac{f_{cr}d_p}{Z_0}\rho_p \quad ; \quad F=\frac{f}{f_{cr}}$$

$$\sin\chi_n = \frac{1}{k_0}\sqrt{k_z^2+(n/(a_K+d_a))^2} \qquad (8)$$

(d_p= shell thickness; ρ_p= shell material density; f_{cr}= critical frequency of the shell as a plane plate, η= bending loss factor of the shell).

The system of equations of the boundary conditions has the form, with the abbreviation :

$$C_n = k_r(a_K + d_a)\frac{J'_n(k_r(a_K + d_a))}{J_n(k_r(a_K + d_a))} \tag{9}$$

(a prime indicates the derivative) :

$$\begin{pmatrix} 1 & 0 & -1 & a_{1,4} \\ a_{2,1} & 0 & a_{2,3} & a_{2,4} \\ 0 & -1 & a_{3,3} & a_{3,4} \\ 0 & a_{4,2} & a_{4,3} & a_{4,4} \end{pmatrix} \cdot \begin{pmatrix} P_{in} \\ P_{sn} \\ P_{an} \\ Q_{an} \end{pmatrix} = \begin{pmatrix} 0 \\ 0 \\ P_{en} \\ C_n \cdot P_{en} \end{pmatrix} \tag{10}$$

The matrix coefficients are : (11)

$$a_{1,4} = -\frac{H'^{(1)}_n(k_r a_K)}{H^{(1)}_n(k_r(a_K + d_a))}$$

$$a_{2,1} = k_r a_K \frac{J'_n(k_r a_K)}{J_n(k_r a_K)} \quad ; \quad a_{2,3} = \frac{-j k_{ar} a_K}{\Gamma_{an} Z_{an}} \frac{H'^{(2)}_n(k_{ar} a_K)}{H^{(2)}_n(k_{ar} a_K)} \quad ;$$

$$a_{2,4} = \frac{-j k_{ar} a_K}{\Gamma_{an} Z_{an}} \frac{H'^{(1)}_n(k_{ar} a_K)}{H^{(1)}_n(k_{ar}(a_K + d_a))}$$

$$a_{3,3} = \frac{H^{(2)}_n(k_{ar}(a_K + d_a))}{H^{(2)}_n(k_{ar} a_K)} + \frac{k_{ar}/k_0 \cdot Z_{Tn}/Z_0}{\Gamma_{an} Z_{an}} \frac{H'^{(2)}_n(k_{ar}(a_K + d_a))}{H^{(2)}_n(k_{ar} a_K)} \quad ;$$

$$a_{3,4} = 1 + \frac{k_{ar}/k_0 \cdot Z_{Tn}/Z_0}{\Gamma_{an} Z_{an}} \frac{H'^{(1)}_n(k_{ar}(a_K + d_a))}{H^{(1)}_n(k_{ar}(a_K + d_a))}$$

$$a_{4,2} = -k_r(a_K + d_a)\frac{H'^{(2)}_n(k_r(a_K + d_a))}{H^{(2)}_n(k_r(a_K + d_a))} \quad ; \quad a_{4,3} = \frac{j k_{ar}(a_K + d_a)}{\Gamma_{an} Z_{an}} \frac{H'^{(2)}_n(k_{ar}(a_K + d_a))}{H^{(2)}_n(k_{ar} a_K)} \quad ;$$

$$a_{4,4} = \frac{j k_{ar}(a_K + d_a)}{\Gamma_{an} Z_{an}} \frac{H'^{(1)}_n(k_{ar}(a_K + d_a))}{H^{(1)}_n(k_{ar}(a_K + d_a))}$$

C_n and some matrix coefficients have the form :

$$z\frac{Z'_n(z)}{Z_n(z)} = -z\frac{Z_{n+1}(z)}{Z_n(z)} + n \tag{12}$$

with $Z_n(z)$ some cylinder function.

The wanted quantity P_{in} is :

$$P_{in} = -\frac{(a_{1,4} \cdot a_{2,3} + a_{2,4}) \cdot (a_{4,2} + C_n)}{(a_{1,4} \cdot a_{2,1} - a_{2,4}) \cdot (a_{3,3} \cdot a_{4,2} + a_{4,3}) + (a_{2,1} + a_{2,3}) \cdot (a_{3,4} \cdot a_{4,2} + a_{4,4})} P_{en} \tag{13}$$

With this, the field inside the cabin is known. Factors with a variation in the z direction will cancel in the ratio for ΔL after averaging; therefore the average over the area $A_K = \pi\, a_K^2$ is sufficient. The average of the exterior field is (with $k_y = k_r \cdot \sin\Theta_0$) :

$$\langle |p_e|^2 \rangle_{A_K} = \frac{2}{A_K} \int_{-a_K}^{+a_K} dx \int_0^{y(x)} |p_e|^2 \, dy = 2|P_e|^2 \left(1 + \frac{J_1(2k_y a_K)}{2k_y a_K} \right) \quad ; \quad y(x) = \sqrt{a_K^2 - x^2} \tag{14}$$

The average inside the cabin is :

$$\langle |p_i|^2 \rangle_{A_K} = \sum_{n \ge 0} \frac{1}{\delta_n} |P_{in}|^2 \left(1 - \frac{J_{n-1}(k_r a_K) \cdot J_{n+1}(k_r a_K)}{J_n^2(k_r a_K)} \right) \tag{15}$$

The sound protection measure, based on the average squared pressure magnitudes, for a single incident plane wave p_e^+ is finally :

$$\Delta L = -10 \cdot \lg \frac{\sum_{n \ge 0} \frac{1}{\delta_n} \left| \frac{P_{in}}{P_e} \right|^2 \left(1 - \frac{J_{n-1}(k_r a_K) \cdot J_{n+1}(k_r a_K)}{J_n^2(k_r a_K)} \right)}{2 \left(1 + \frac{J_1(2k_y a_K)}{2k_y a_K} \right)} \tag{16}$$

The sound protection measure, based on the level difference in the cabin centre, is :

$$\Delta L_0 = -10 \cdot \lg \left[\frac{1}{4} \left| \frac{P_{i0}}{P_e} \right|^2 \frac{1}{J_0^2(k_r a_K)} \right] \tag{17}$$

Because the angle of incidence Θ_0 is not contained, it also holds for two-dimensional diffuse incidence.

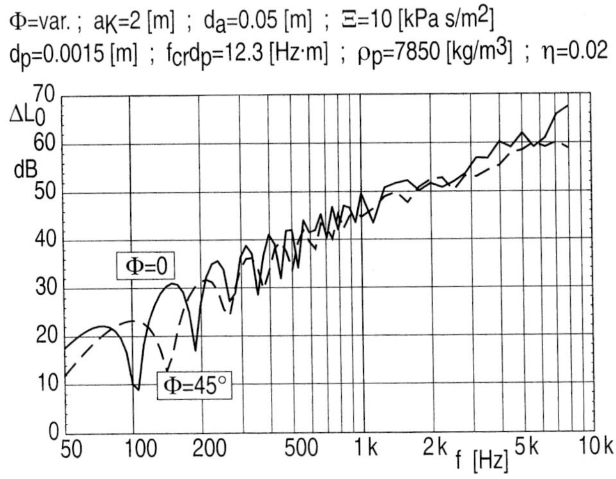

Sound protection measure DL_0 for sound incidence normal and oblique to the cabin axis.

Φ=45° ; a_K=var. ; d_a=0.1 [m] ; Ξ=10 [kPa s/m2]
d_p=0.0015 [m] ; $f_{cr}d_p$=12.3 [Hz·m] ; ρ_p=7850 [kg/m3] ; η=0.02

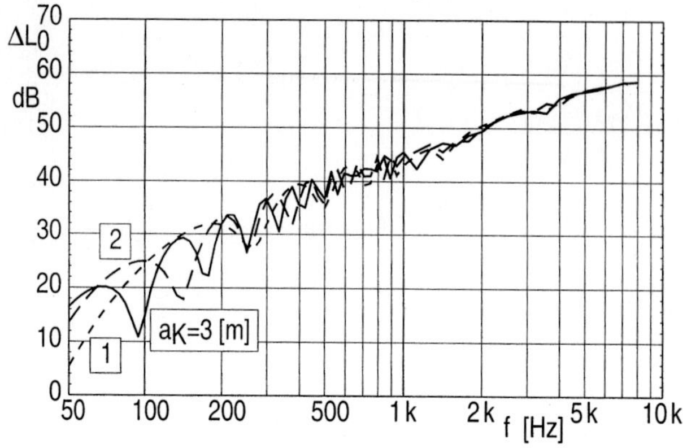

Sound protection measure DL_0 for oblique sound incidence and three cabin radii a_K.

L.6 Cabin with plane walls

Ref.: Mechel, [L.1]

The model is two-dimensional : two walls consisting of an exterior plate and an interior porous absorber layer. This model is an approximation to reality if the lateral dimension of the cabin in the x direction is large (at least compared to t), and if sound incidence comes mainly from the half space in front of one wall.

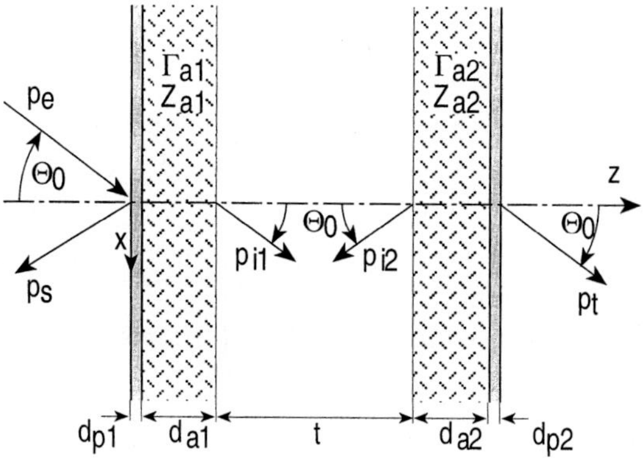

Under these conditions the cabin is just a multi-layer absorber; the desired interior sound field is the field in one layer. This view of the task permits the application of the equivalent network method (see chapter C).

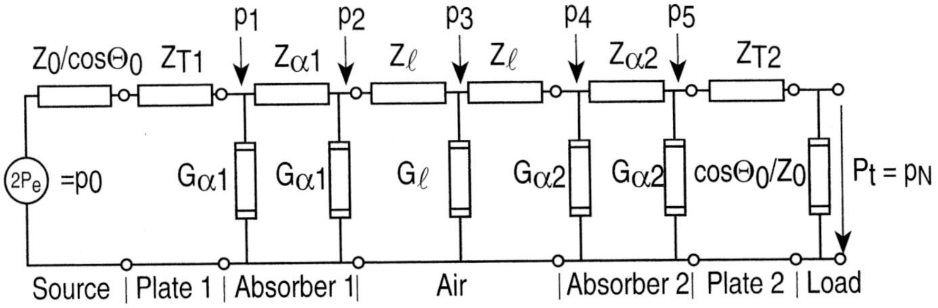

The source pressure is $p_0 = 2P_e$ with the amplitude P_e of the incident plane wave. The interior impedance of the source and the load impedance are $Z_0/\cos\Theta_0$.

The sound pressure inside the cabin, i.e. in the layer named "Air" above, is :

$$\begin{aligned} p_i(x,y) &= \left[P_{i1}\cdot e^{-jk_x(x-d_{a1})} + P_{i2}\cdot e^{+jk_x(x-d_{a1}-t)}\right]\cdot e^{-jk_y y} \\ k_x &= k_0\cos\Theta_0 \quad ; \quad k_y = k_0\sin\Theta_0 \end{aligned} \tag{1}$$

with the relation between amplitudes and circuit node pressures :

$$\begin{aligned} p_i(x=d_{a1}) &= \left[P_{i1} + P_{i2}\cdot e^{-jk_x t}\right] \stackrel{!}{=} 2P_e\cdot p_2 \\ p_i(x=d_{a1}+t) &= \left[P_{i1}\cdot e^{-jk_x t} + P_{i2}\right] \stackrel{!}{=} 2P_e\cdot p_4 \end{aligned} \tag{2}$$

or :

$$P_{i1} = \frac{p_2 - p_4\cdot e^{-jk_x t}}{1-e^{-2jk_x t}}\cdot 2P_e \quad ; \quad P_{i2} = \frac{p_4 - p_2\cdot e^{-jk_x t}}{1-e^{-2jk_x t}}\cdot 2P_e \tag{3}$$

If one includes in the exciting wave the ground-reflected wave, the factor $e^{-jk_y y}$ simply changes to $\cos(k_y y)$; it cancels anyway in the averages for ΔL and ΔL_0 .

The equivalent circuit elements are (i= 1,2) :

$$\begin{aligned} \frac{Z_{\alpha,i}}{Z_0} &= Z_{an,i}\frac{\sinh(\Gamma_{a,i}d_{ai}\cdot\cos\Theta_{\alpha,i})}{\cos\Theta_{\alpha,i}} \quad ; \quad Z_0 G_{\alpha,i} = \frac{\cos\Theta_{\alpha,i}}{Z_{an,i}}\frac{\cosh(\Gamma_{a,i}d_{ai}\cdot\cos\Theta_{\alpha,i})-1}{\sinh(\Gamma_{a,i}d_{ai}\cdot\cos\Theta_{\alpha,i})} \\ \cos\Theta_{\alpha,i} &= \frac{\sqrt{\Gamma_{an,i}^2+\sin^2\Theta_0}}{\Gamma_{an,i}} \quad ; \quad \Gamma_{a,i}d_{ai}\cdot\cos\Theta_{\alpha,i} = k_0 d_{ai}\sqrt{\Gamma_{an,i}^2+\sin^2\Theta_0} \end{aligned} \tag{4}$$

$$\frac{Z_\ell}{Z_0} = j\frac{1-\cos(k_0 t\cdot\cos\Theta_0)}{\cos\Theta_0\cdot\sin(k_0 t\cdot\cos\Theta_0)} \quad ; \quad Z_0 G_\ell = j\cos\Theta_0\cdot\sin(k_0 t\cdot\cos\Theta_0) \tag{5}$$

$$\frac{Z_{T,i}}{Z_0} = 2\pi\cdot Z_{m,i}\cdot F_i\left[\eta_i F_i^2\cdot\sin^4\Theta_0 + j\left(1-F_i^2\cdot\sin^4\Theta_0\right)\right] \tag{6}$$

with: $$Z_{m,i}=\frac{f_{cr,i}d_i}{Z_0}\rho_i \quad ; \quad F_i=\frac{f}{f_{cr,i}}=\frac{f\cdot d_i}{f_{cr,i}d_i} \qquad (7)$$

(d_i= plate thickness; ρ_i= plate material density; $f_{cr,i}$= critical frequency; η_i= bending loss factor)

With the abbreviations :

$$\begin{aligned}
&z1=1+\frac{Z_{T,2}}{Z_0}\cos\Theta_0 \quad ; \quad \alpha=\cos\Theta_0+Z_0G_{\alpha,2}\cdot z1\\
&z2=z1+\frac{Z_{a,2}}{Z_0}\cdot\alpha \quad ; \quad \beta=\alpha+Z_0G_{\alpha,2}\cdot z2\\
&z3=\frac{Z_\ell}{Z_0}\cdot\beta \quad ; \quad z4=z2+z3 \quad ; \quad \gamma=\beta+Z_0G_\ell\cdot z4\\
&z5=z4+\frac{Z_\ell}{Z_0}\cdot\gamma \quad ; \quad \delta=\gamma+Z_0G_{\alpha,1}\cdot z5 \quad ; \quad z6=z5+\frac{Z_{a,1}}{Z_0}\cdot\delta
\end{aligned} \qquad (8)$$

becomes :

$$p2=\frac{z5}{\left(z6+(\delta+Z_0G_{\alpha,1}\cdot z6)\left(\frac{Z_{T,1}}{Z_0}+\frac{1}{\cos\Theta_0}\right)\right)} \qquad (9)$$

$$\begin{aligned}
p_4=z2\cdot\Bigg\{&z5+\frac{Z_{a,1}}{Z_0}\cdot\delta+\left[\frac{Z_{T,1}}{Z_0}+\frac{1}{\cos\Theta_0}\right]\cdot\\
&\cdot\left[\delta+Z_0G_{\alpha,1}\cdot\left(z5+\frac{Z_{a,1}}{Z_0}\cdot\left(\gamma+Z_0G_{\alpha,1}\cdot\left(z2+\frac{Z_\ell}{Z_0}\cdot(\alpha+Z_0G_{\alpha,2}\cdot(z1+\frac{Z_{a,2}}{Z_0}\cdot\alpha)\right)+\right.\right.\right.\\
&\left.\left.\left.+\frac{Z_\ell}{Z_0}\cdot\left(\beta+Z_0G_\ell\cdot\left(z2+\frac{Z_\ell}{Z_0}\cdot\left(\alpha+Z_0G_{\alpha,2}\cdot\left(z1+\frac{Z_{a,1}}{Z_0}\cdot(\cos\Theta_0+Z_0G_{\alpha,2}\cdot z_1^2)\right)\right)\right)\right)\right)\right)\right]\Bigg\}^{-1}
\end{aligned}$$

Thus the square of the sound pressure magnitude in the cabin will be (k_x, k_y real) :

$$\begin{aligned}
|p_i(x,y)|^2=\frac{4|P_e|^2}{\sin^2(k_xt)}\Big[&|p_2|^2\cdot\sin^2(k_x(x-d_{a1}-t))+|p_4|^2\cdot\sin^2(k_x(x-d_{a1}))-\\
&-2\,\mathrm{Re}\{p_2p_4^*\}\cdot\sin(k_x(x-d_{a1}-t))\cdot\sin(k_x(x-d_{a1}))\Big]
\end{aligned} \qquad (10)$$

with an average :

$$\langle|p_i(x,y)|^2\rangle=$$
$$\frac{|P_e|^2}{\sin^2(k_xt)}\left[2\left(|p_2|^2+|p_4|^2\right)\left(1-\frac{\sin(2k_xt)}{2k_xt}\right)-4\,\mathrm{Re}\{p_2p_4^*\}\left(\cos(k_xt)-\frac{\sin(k_xt)}{k_xt}\right)\right] \qquad (11)$$

The sound protection measure ΔL is finally : (12)

$$\Delta L = -10\cdot \lg\left\{\frac{1}{\sin^2(k_x t)}\left[2\left(|p_2|^2+|p_4|^2\right)\left(1-\frac{\sin(2k_x t)}{2k_x t}\right)-4\,\mathrm{Re}\{p_2 p_4^*\}\left(\cos(k_x t)-\frac{\sin(k_x t)}{k_x t}\right)\right]\right\}$$

With the sound pressure in the cabin centre :

$$|p_i(x=d_{a1}+t/2,y)|^2=\frac{4|P_e|^2\sin^2(k_x t/2)}{\sin^2(k_x t)}\left[|p_2|^2+|p_4|^2+2\,\mathrm{Re}\{p_2 p_4^*\}\right] \tag{13}$$

the sound protection measure ΔL_0 becomes :

$$\Delta L_0=-10\cdot \lg\frac{2(1-\cos(k_x t))}{\sin^2(k_x t)}\left[|p_2|^2+|p_4|^2+2\,\mathrm{Re}\{p_2 p_4^*\}\right] \tag{14}$$

Until now it was tacitly assumed that the shadow field of the cabin, i.e. the field behind the cabin which is generated there by scattering of p_e, is much lower than the sound pressure of the incident wave at the front side of the cabin. A different extreme situation would be a scattered sound field behind the cabin, which would be strong enough to inhibit the radiation of the sound which has traversed the cabin at the back side of the cabin. Then the equivalent network ends in the node with p_5. One gets in this case with the abbreviations :

$$\begin{aligned}
&\alpha=1+G_{\alpha,2}\cdot Z_{\alpha,2} \quad ; \quad g1=Z_0 G_{\alpha,2}\cdot(1+\alpha)\\
&\beta=\alpha+\frac{Z_\ell}{Z_0}\cdot g1 \quad ; \quad g2=g1+Z_0 G_\ell\cdot\beta\\
&\gamma=\beta+\frac{Z_\ell}{Z_0}\cdot g2 \quad ; \quad g3=g2+Z_0 G_{\alpha,1}\cdot\gamma\\
&\delta=\gamma+\frac{Z_{\alpha,1}}{Z_0}\cdot g3
\end{aligned} \tag{15}$$

the node pressures :

$$\begin{aligned}
p_2&=\gamma\cdot\left[\delta+\left(g3+Z_0 G_{\alpha,1}\left(\gamma+Z_0 G_{\alpha,1}\cdot g3\right)\right)\left(\frac{Z_{T1}}{Z_0}+\frac{1}{\cos\Theta_0}\right)\right]^{-1}\\
p_4&=\alpha\cdot\left[\delta+\left(g3+Z_0 G_{\alpha,1}\cdot\delta\right)\left(\frac{Z_{T1}}{Z_0}+\frac{1}{\cos\Theta_0}\right)\right]^{-1}
\end{aligned} \tag{16}$$

The sound protection measures then follow as above. Numerical checks in a number of examples have shown that the results agree with the former results within the precision of graphical representation.

A further special case can be easily treated: a *coherent* sound incidence with equal strength takes place on both sides of the cabin. Then the equivalent circuit ends in the node with p_3 after substitution $t \to t/2$. The required node pressures are :

$$p_2 = \alpha \cdot \left[\beta + \left(g1 + Z_0 G_{\alpha,1} \cdot \beta \right) \left(\frac{Z_{T1}}{Z_0} + \frac{1}{\cos\Theta_0} \right) \right]^{-1}$$
$$p_4 = \left[\beta + \left(g1 + Z_0 G_{\alpha,1} \cdot \beta \right) \left(\frac{Z_{T1}}{Z_0} + \frac{1}{\cos\Theta_0} \right) \right]^{-1} \tag{17}$$

with the abbreviations :

$$\alpha = 1 + G_\ell \cdot Z_\ell \; ; \; g1 = Z_0 G_\ell + Z_0 G_{\alpha,1} \cdot \alpha \; ; \; \beta = \alpha + \frac{Z_{\alpha,1}}{Z_0} \cdot g1 \tag{18}$$

ΔL follows as before, except for an additional factor $1/2$ in the argument of the logarithm.

With *incoherent* sound incidence from both sides the contributions of each side to the interior sound pressure magnitude are evaluated separately and added.

In the examples shown below, the walls on both sides of the cabin are equal, for simplicity.

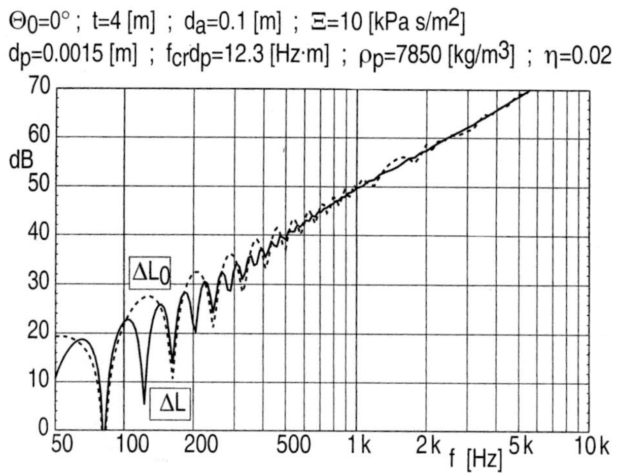

The two sound pressure measures DL and DL_0 in a 2-dimensional cabin.

$\Theta_0=0°$; $t=4$ [m] ; d_a=var. ; $\Xi=10$ [kPa s/m^2]
$d_p=0.0015$ [m] ; $f_{cr}d_p=12.3$ [Hz·m] ; $\rho_p=7850$ [kg/m^3] ; $\eta=0.02$

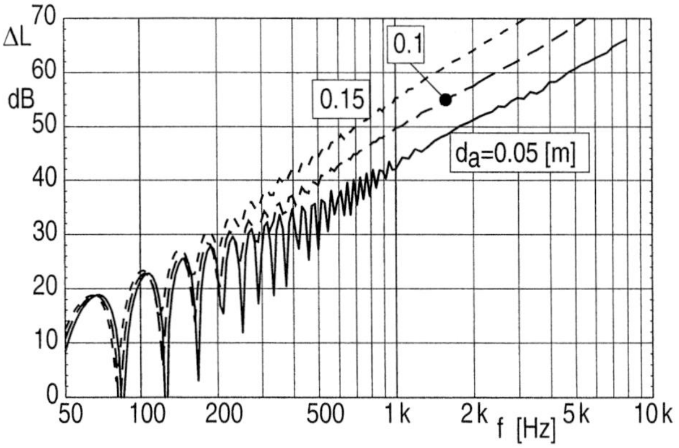

Influence of the absorber layer thickness d_a on the sound protection measure DL for normal sound incidence on the cabin wall.

Θ_0=diff. ; $t=4$ [m] ; d_a=var. ; $\Xi=10$ [kPa s/m^2]
$d_p=0.0015$ [m] ; $f_{cr}d_p=12.3$ [Hz·m] ; $\rho_p=7850$ [kg/m^3] ; $\eta=0.02$

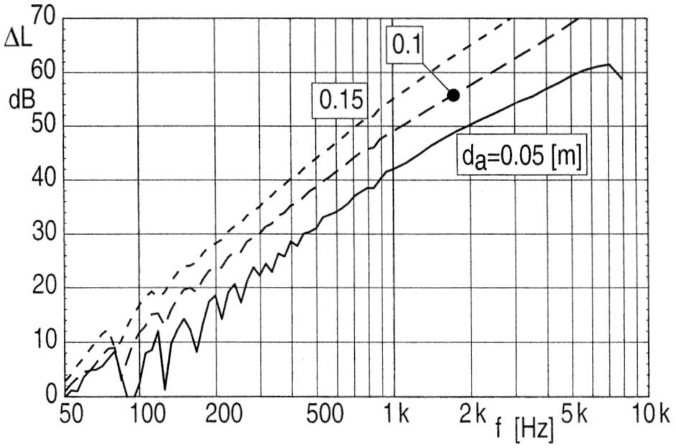

As above, but for diffuse sound incidence.

L.7 Cabin with rectangular cross section :

Ref.: Mechel, [L.1]

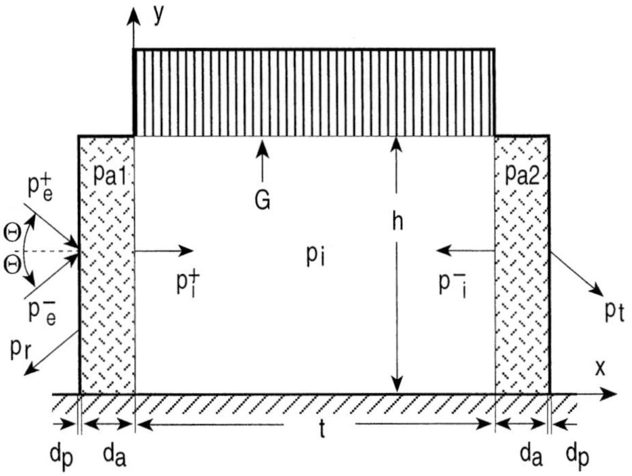

The cabin still is two-dimensional, but with a ceiling which (for simplicity) is locally reacting with a surface admittance G .

The incident wave p_e with $p_e = p_e^+ + p_e^-$ considers also the ground reflection.

The field p_i inside the cabin is formulated as a mode sum of a locally lined flat duct with the cabin ceiling as lining. The lateral wave numbers ε_m are solutions of the characteristic equation of the duct.

With the usual dimensions of cabins it makes no significant difference whether the back wall radiates a wave p_t into the free space, or if the back wall has a hard termination. Therefore this possibility of simplification will be used below.

Field formulations :

$$p_e = P_e \cdot e^{-jk_x x} \cdot \cos(k_y y) \quad ; \quad k_x = k_0 \cdot \cos\Theta \quad ; \quad k_y = k_0 \cdot \sin\Theta$$

$$p_r = P_r \cdot e^{+jk_x x} \cdot \cos(k_y y)$$

$$p_i = \sum_m \left[A_m \cdot e^{-\gamma_m x} + B_m \cdot e^{+\gamma_m x}\right] \cdot \cos(\varepsilon_m y) \quad ; \quad \gamma_m = \sqrt{\varepsilon_m^2 - k_0^2} \tag{1}$$

$$p_{a1} = \sum_m \left[P_{am} \cdot e^{-\gamma_{am} x} + Q_{am} \cdot e^{+\gamma_{am} x}\right] \cdot \cos(\varepsilon_m y) \quad ; \quad \gamma_{am} = \sqrt{\varepsilon_m^2 - \Gamma_a^2}$$

$$p_{a2} = \sum_m R_{am} \cosh\left(\gamma_{am}(x - t - d_a)\right) \cdot \cos(\varepsilon_m y)$$

The exciting and the reflected field at x=0 are synthesised with duct modes : (2)

$$p_e(0,y) = P_e \cdot \cos(k_y y) = \sum_m P_{em} \cdot \cos(\varepsilon_m y) \quad ; \quad p_r(0,y) = P_r \cdot \cos(k_y y) = \sum_m P_{rm} \cdot \cos(\varepsilon_m y)$$

with modal amplitudes :

$$P_{em} = \frac{P_e}{N_m} \cdot \frac{1}{h}\int_0^h \cos(k_y y)\cdot\cos(\varepsilon_m y)\,dy = \frac{S_m}{N_m}\cdot P_e \qquad (3)$$

$$P_{rm} = \frac{S_m}{N_m}\cdot P_r$$

using mode norms :

$$N_m = \frac{1}{h}\int_0^h \cos^2(\varepsilon_m y)\,dy = \frac{1}{2}\left(1+\frac{\sin(2\varepsilon_m h)}{(2\varepsilon_m h)}\right) \qquad (4)$$

and mode coupling coefficients :

$$S_m = \frac{1}{2}\left(\frac{\sin(\varepsilon_m - k_y)h}{(\varepsilon_m - k_y)h} + \frac{\sin(\varepsilon_m + k_y)h}{(\varepsilon_m + k_y)h}\right) \qquad (5)$$

The boundary conditions (which hold term-wise) at the front and back side walls :

$$\begin{aligned} &p_e(-d_a)+p_r(-d_a)-p_{a1}(-d_a) \overset{!}{=} Z_T\cdot v_{a1x}(-d_a) \\ &v_{ex}(-d_a)+v_{rx}(-d_a) \overset{!}{=} v_{a1x}(-d_a) \\ &p_{a1}(0) \overset{!}{=} p_i(0) \quad ; \quad v_{a1x}(0) \overset{!}{=} v_{ix}(0) \\ &p_i(t) \overset{!}{=} p_{a2}(t) \quad ; \quad v_{ix}(t) \overset{!}{=} v_{a2x}(t) \end{aligned} \qquad (6)$$

(Z_T is the partition impedance of an outer cover plate of the walls; see the previous section L.6) lead to the system of equations :

$$\begin{pmatrix} \ddots & \cdots & \cdot^{\cdot^{\cdot}} \\ \vdots & a_{i,k} & \vdots \\ \cdot^{\cdot^{\cdot}} & \cdots & \ddots \end{pmatrix} \cdot \begin{pmatrix} P_{rm} \\ A_m \\ B_m \\ P_{am} \\ Q_{am} \\ R_{am} \end{pmatrix} = e^{+jk_x d_a}\cdot P_{em}\cdot \begin{pmatrix} 1 \\ 1 \\ 0 \\ 0 \\ 0 \\ 0 \end{pmatrix} \qquad (7)$$

with the matrix coefficients : (8)

$$a_{1,1} = -e^{-jk_x d_a} \quad ; \quad a_{1,2} = a_{1,3} = a_{1,6} = 0$$

$$a_{1,4} = e^{+\gamma_{am} d_a}\left(1+\frac{\gamma_{am} d_a\, Z_T/Z_0}{k_0 d_a\cdot \Gamma_{an} Z_{an}}\right) \quad ; \quad a_{1,5} = e^{-\gamma_{am} d_a}\left(1-\frac{\gamma_{am} d_a\, Z_T/Z_0}{k_0 d_a\cdot \Gamma_{an} Z_{an}}\right)$$

$$a_{2,1}=+e^{-jk_xd_a} \; ; \; a_{2,2}=a_{2,3}=a_{2,6}=0$$

$$a_{2,4}=e^{+\gamma_{am}d_a}\frac{\gamma_{am}d_a}{k_xd_a\cdot\Gamma_{an}Z_{an}} \; ; \; a_{2,5}=-e^{-\gamma_{am}d_a}\frac{\gamma_{am}d_a}{k_xd_a\cdot\Gamma_{an}Z_{an}}$$

$$a_{3,1}=a_{3,6}=0 \; ; \; a_{3,2}=1 \; ; \; a_{3,3}=e^{-\gamma_m t} \; ; \; a_{3,4}=a_{3,5}=-1$$

$$a_{4,1}=a_{4,6}=0 \; ; \; a_{4,2}=1 \; ; \; a_{4,3}=-e^{-\gamma_m t}$$

$$a_{4,4}=-\frac{j\gamma_{am}d_a/\gamma_m t}{\Gamma_{an}Z_{an}}\frac{t}{d_a} \; ; \; a_{4,5}=+\frac{j\gamma_{am}d_a/\gamma_m t}{\Gamma_{an}Z_{an}}\frac{t}{d_a} \tag{8}$$

$$a_{5,1}=a_{5,4}=a_{5,5}=0 \; ; \; a_{5,2}=e^{-\gamma_m t} \; ; \; a_{5,3}=1$$

$$a_{5,6}=\frac{-1}{2}\left(e^{+\gamma_{am}d_a}+e^{-\gamma_{am}d_a}\right)$$

$$a_{6,1}=a_{6,4}=a_{6,5}=0 \; ; \; a_{6,2}=e^{-\gamma_m t} \; ; \; a_{6,3}=-1$$

$$a_{6,6}=\frac{-j\gamma_{am}d_a/\gamma_m t}{2\Gamma_{an}Z_{an}}\frac{t}{d_a}\left(e^{+\gamma_{am}d_a}-e^{-\gamma_{am}d_a}\right)$$

With the abbreviations :

$$X_m := j\Gamma_{an}Z_{an}\cdot\gamma_m t+\gamma_{am}t \; ; \; Y_m := j\Gamma_{an}Z_{an}\cdot\gamma_m t-\gamma_{am}t \tag{9}$$

the required amplitudes A_m, B_m are :

$$\begin{aligned}A_m=\;&4\Gamma_{an}Z_{an}\cdot k_0d_a\cdot k_xt\cdot\gamma_{am}d_a\left(X_m\cdot e^{+\gamma_{am}d_a}+Y_m\cdot e^{-\gamma_{am}d_a}\right)\cdot e^{+jk_xd_a}\cdot P_{em}\cdot\\ &\cdot\Big\{\Big[\left(e^{-2\gamma_{am}d_a}+e^{-2\gamma_m t}\right)\cdot Y_m+\left(1+e^{-2(\gamma_{am}d_a+\gamma_m t)}\right)\cdot X_m\Big]\cdot\\ &\cdot\big[\left(\Gamma_{an}Z_{an}\cdot k_0d_a\cdot k_xd_a-\gamma_{am}d_a\cdot(k_0d_a+k_xd_a\cdot Z_T/Z_0)\right)\cdot X_m-\\ &-\left(\Gamma_{an}Z_{an}\cdot k_0d_a\cdot k_xd_a+\gamma_{am}d_a\cdot(k_0d_a+k_xd_a\cdot Z_T/Z_0)\right)\cdot e^{+2\gamma_{am}d_a}\cdot Y_m\big]\Big\}^{-1}\end{aligned} \tag{10}$$

$$\begin{aligned}B_m=\;&4\Gamma_{an}Z_{an}\cdot k_0d_a\cdot k_xt\cdot\gamma_{am}d_a\left(X_m\cdot e^{-\gamma_{am}d_a}+Y_m\cdot e^{+\gamma_{am}d_a}\right)\cdot e^{-\gamma_m t}\cdot e^{+jk_xd_a}\cdot P_{em}\cdot\\ &\cdot\Big\{\Big[\left(e^{-2\gamma_{am}d_a}+e^{-2\gamma_m t}\right)\cdot Y_m+\left(1+e^{-2(\gamma_{am}d_a+\gamma_m t)}\right)\cdot X_m\Big]\cdot\\ &\cdot\big[\left(\Gamma_{an}Z_{an}\cdot k_0d_a\cdot k_xd_a-\gamma_{am}d_a\cdot(k_0d_a+k_xd_a\cdot Z_T/Z_0)\right)\cdot X_m-\\ &-\left(\Gamma_{an}Z_{an}\cdot k_0d_a\cdot k_xd_a+\gamma_{am}d_a\cdot(k_0d_a+k_xd_a\cdot Z_T/Z_0)\right)\cdot e^{+2\gamma_{am}d_a}\cdot Y_m\big]\Big\}^{-1}\end{aligned} \tag{11}$$

References to part L :

Capsules and Cabins

[C.1] MECHEL, F.P.
"Schallabsorber", Vol. III, Chapter 20:
"Capsules and Cabins"
S.Hirzel Verlag, Stuttgart, 1998

M
Room Acoustics
by M. Vorländer and F.P. Mechel

M.1 Eigenfunctions in parallelepipeds

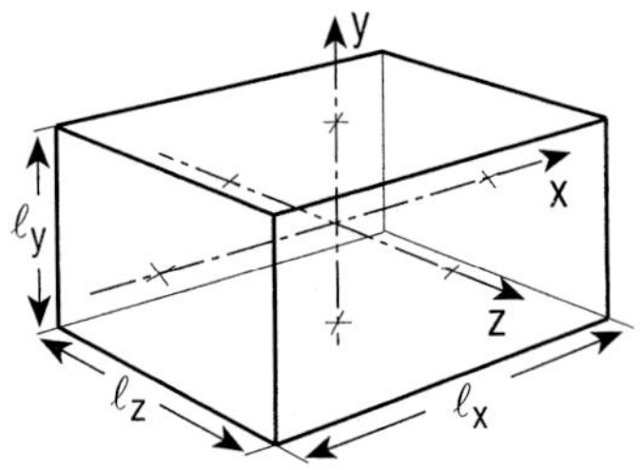

Rooms with rectangular walls belong to the few examples in room acoustics in which a modal analysis can be performed with a reasonable amount of analytical and numerical work. As such they may serve as gauge objects for conceptions and methods.

The corner lengths are ℓ_x, ℓ_y, ℓ_z, respectively. The wall surface admittances are G_x, G_y, G_z, if the walls on opposite sides are equal, otherwise G_{x1}, G_{x2}, etc.

The aim is to find elementary solutions (eigenfunctions or modes) with which sound fields for an arbitrary sound source in the room can be synthesized. They must obey the wave equation, symmetry conditions, and boundary conditions.

Alternative writing :

$x, y, z = x_1, x_2, x_3 \; ; \; \ell_x, \ell_y, \ell_z = \ell_1, \ell_2, \ell_3 \; ; \; G_x, G_y, G_z = G_1, G_2, G_3$.

Wave equation :

$$\frac{\partial^2 p}{\partial x^2}+\frac{\partial^2 p}{\partial y^2}+\frac{\partial^2 p}{\partial z^2}+k_0^2 p=0 \tag{1}$$

Fundamental solutions separate :

$$p(x_1,x_2,x_3)=q_1(k_1x_1)\cdot q_2(k_2x_2)\cdot q_3(k_3x_3) \tag{2}$$

with

$$q_i=\begin{cases}\cos(k_ix_i) & ; \text{ symmetrical rel. } x_i=0\\ \sin(k_ix_i) & ; \text{ anti-symmetrical rel. } x_i=0\end{cases} \quad ; \quad i=1,2,3 \tag{3}$$

They satisfy the wave equation if (secular equation) :

$$k_0^2 \stackrel{!}{=} k_1^2 + k_2^2 + k_3^2 \tag{4}$$

If the room is symmetrical in the direction of x_i (i.e. $G_{i1} = G_{i2} = G_i$) and the field is symmetrical (depending on the directivity and position of the source) : $q_i = \cos(k_i x_i)$; with room symmetry in the direction of x_i and anti-symmetrical field: $q_i = \sin(k_i x_i)$.
Else : $q_i = a_i \cos(k_i x_i) + b_i \sin(k_i x_i)$.

The boundary conditions at the walls lead to the characteristic equations for the k_i . Define

$U_i = (k_0 \ell_i / 2) \cdot Z_0 G_i$ for symmetrical walls

$U_{i1} = (k_0 \ell_i / 2) \cdot Z_0 G_{i1}$; $U_{i2} = (k_0 \ell_i / 2) \cdot Z_0 G_{i2}$ for unsymmetrical walls.

General case (unsymmetrical room):

With : $U_{si} = \frac{1}{2}(U_{i1} + U_{i2})$; $U_{ai} = \frac{1}{2}(U_{i1} - U_{i2})$

the characteristic equation is written as :

$$\left[(k_i \ell_i / 2) \cdot \tan(k_i \ell_i / 2) - j U_{si}\right] \cdot \left[(k_i \ell_i / 2) \cdot \cot(k_i \ell_i / 2) + j U_{si}\right] \stackrel{!}{=} U_{ai}^2 \tag{5}$$

with the amplitude ratio of the anti-symmetrical to the symmetrical part of the mode :

$$\frac{b_i}{a_i} = -\cot(k_i \ell_i / 2) \frac{(k_i \ell_i / 2) \cdot \tan(k_i \ell_i / 2) - j U_{i2}}{(k_i \ell_i / 2) \cdot \tan(k_i \ell_i / 2) + j U_{i2}} \tag{6}$$

Special case: symmetrical room ($G_{i1} = G_{i2} = G_i$), symmetrical mode; characteristic equation :

$$(k_i \ell_i / 2) \cdot \tan(k_i \ell_i / 2) = j U_i \tag{7}$$

Special case: symmetrical room ($G_{i1} = G_{i2} = G_i$), anti-symmetrical mode; characteristic equation :

$$(k_i \ell_i / 2) \cdot \cot(k_i \ell_i / 2) = -j U_i \tag{8}$$

Special case: both walls normal to x_i are hard ($G_{i1} = G_{i2} = G_i = 0$), symmetrical mode:

$$k_i \ell_i / 2 = m_i \pi \quad ; \quad m_i = 0, 1, 2, \ldots \tag{9}$$

Special case: both walls normal to x_i are hard ($G_{i1} = G_{i2} = G_i = 0$), anti-symmetrical mode:

$$k_i \ell_i / 2 = (m_i + 1/2)\pi \quad ; \quad m_i = 0, 1, 2, \ldots \tag{10}$$

In these equations $G_i \neq 0$ may represent either a locally reacting or a bulk reacting wall. In the first case G_i is independent of the k_i ; in the second case $G_i = G_i(k_1, k_2, k_3)$. In both cases the modes are orthogonal, see Sections B.6, B.7, and as such suited for field synthesis.

The three characteristic equations (i=1,2,3) and the secular equation (4) in general cannot be solved simultaneously for all frequencies. In the special case of only hard walls and symmetrical modes, one finds (with $l=m_1$, $m=m_2$, $n=m_3$) eigenfrequencies f_{lmn}:

$$k_0^2 = (2\pi f_{lmn}/c_0)^2 = \sum_i k_i^2 = \sum_i (2\pi m_i/\ell_i)^2$$
$$f_{lmn} = c_0\sqrt{(l/\ell_1)^2 + (m/\ell_2)^2 + (n/\ell_3)^2} \qquad (11)$$

If the mode is anti-symmetrical in some direction, substitute $m_i \to m_i+1/2$.

In cases with $G_i\neq 0$ the secular equation (4) with the solutions k_i of the characteristic equations must be solved numerically for eigenfrequencies. Under some restrictive conditions one can derive approximations. This will be shown for symmetrical, locally reacting walls and symmetrical modes, i.e. for equ.(7). Write that equation as

$$z_i \cdot \tan z_i = jU_i = jk_0\ell_i \cdot Z_0G_i \quad ; \quad z_i = k_i\ell_i/2 \qquad (12)$$

With the continued fraction expansion of $\tan(z_i) = \tan(z_i - m\pi)$; $m_i = 0,1,2\ldots$ with writing :

$$a_1/\left(a_2 - a_3/\left(a_4 - a_5/\left(a_6 - \ldots\right)\right)\right) = \frac{a_1}{a_2 -}\,\frac{a_3}{a_4 -}\,\frac{a_5}{a_6 -}\ldots$$

one gets (see Section J.7) :

$$z_i^2 = m_i\pi \cdot z_i + jk_0\ell_i \cdot Z_0G_i \cdot \left[1 - \frac{(z_i - m_i\pi)^2}{3 -}\,\frac{(z_i - m_i\pi)^2}{5 -}\ldots\right]$$
$$\xrightarrow[z_i \to m_i\pi]{} (m_i\pi)^2 + jk_0\ell_i \cdot Z_0G_i \xrightarrow[k_0\ell_i \to 0 \text{ and/or } G_i \to 0]{} (m_i\pi)^2 \qquad (13a)$$

Thus:

$$k_i^2 = z_i^2 (2/\ell_i)^2 \approx 4\frac{(m_i\pi)^2 + jk_0\ell_i \cdot Z_0G_i}{\ell_i^2} \qquad (13b)$$

This inserted into the secular equation gives an approximate equation for the eigenfrequencies (contained in k_0):

$$k_0^2 - 4jk_0\left(Z_0G_1/\ell_1 + Z_0G_2/\ell_2 + Z_0G_3/\ell_3\right) - 4\pi^2\left((m_1/\ell_1)^2 + (m_2/\ell_2)^2 + (m_3/\ell_3)^2\right) \stackrel{!}{=} 0 \qquad (14)$$

Similar procedures may be applied for other cases of symmetry. It should be noticed that the condition used $z_i \to m_i\pi$ implies small admittance values $|G_i|$. But even this equation cannot be discussed further without knowledge about the functions $G_i(k_0)$.

In a formal manner one can write for the solutions (with $k_{lmn} = 2\pi f_{lmn}/c_0$; f_{lmn} from (11)):

$$k_0 \to k_{lmn} = \left(\omega_{lmn} + j\delta_{lmn}\right)/c_0 \tag{15}$$

where δ_{lmn} represents a modal damping constant (which itself generally is complex).

Half-widths of modes (resonance curve):

$$(\Delta f)_{lmn} = \frac{\delta_{lmn}}{\pi} \tag{16}$$

The transfer function between two points in a room is calculated by superposition of damped modes (resonance curves):

$$p(x_1,x_2,x_3,\omega) = \sum_{l,m,n} \frac{A_{lmn}(\omega)}{\left(\omega^2 - \omega_{lmn}^2 - 2j\delta_{lmn}\omega_{lmn}\right)} \tag{17}$$

with $\delta_{lmn} << \omega_{lmn}$ and $\underline{A}_{lmn}(\omega)$ depending on the source (x_{S1}, x_{S2}, x_{S3}) and receiver positions (x_1, x_2, x_3) :

$$A_{lmn}(\omega) = \underline{p}_{lmn}(x_1,x_2,x_3)\underline{p}_{lmn}(x_{S1},x_{S2},x_{S3}) \Big/ \iiint_V \left|\underline{p}_{lmn}(x_1,x_2,x_3)\right|^2 dx_1 dx_2 dx_3 \tag{18}$$

M.2 Density of eigenfrequencies in rooms

Let N be the number of eigenfrequencies below the frequency f ; let n= dN/df be the number of eigenfrequencies in an interval of 1 Hz.

Volume with smallest corner length $>\lambda_0/2$:
(V= volume; S= room surface; L= sum of corner lengths ; f= frequency)

$$\begin{aligned} N &= \frac{4\pi V \cdot f^3}{3c_0^3} + \frac{\pi S \cdot f^2}{4c_0^2} + \frac{L \cdot f}{2c_0} \\ n &= \frac{4\pi V \cdot f^2}{c_0^3} + \frac{\pi S \cdot f}{2c_0^2} + \frac{L}{2c_0} \end{aligned} \tag{1}$$

Flat volume with smallest corner $a < \lambda_0/2$, other corners $b,c > \lambda_0/2$:

$$n = \frac{2\pi \cdot b \cdot c \cdot f}{c_0^2} \tag{2}$$

Tube of length ℓ with hard walls :

$$n = 2\ell / c_0 \tag{3}$$

The *mode overlap* m is defined as the ratio of the half value bandwidth of a room resonance and the average frequency separation between neighbouring resonances around a frequency f . The mode overlap in the diffuse reverberant field is (T= reverberation time):

$$m = 0.69\frac{V}{T}(f/1000)^2 \tag{4}$$

The mode overlap must exceed some lower limit value for the application of statistical theories in room acoustics. This defines a lower limit frequency f_s for such theories.

Limit frequency for $m \geq 10$: $\quad f_s > 4000\sqrt{T/V}$ (5)

Limit frequency for $m \leq 10$: $\quad f_s > 2000\sqrt{T/V}$ (6)

The modulus of the room transfer function, $|p(\omega)|$, can be estimated by (see also equation M.1.17)

$$|p(\omega)| \approx \frac{|A_{lmn}|}{\sqrt{(\omega^2 - \omega_{lmn}^2)^2 + 4\omega^2\delta_{lmn}^2}} \tag{7}$$

The probability density of the transfer function modulus $z = |p(\omega)|$ (Rayleigh distribution) is

$$P(z)\,dz = \frac{\pi}{2} e^{-\pi z^2/4}\, z\,dz \quad , \tag{8}$$

and the probability density of transfer function phase $\varphi = \arg(p(\omega))$:

$$P(\varphi)\,d\varphi = \frac{1}{2\pi} d\varphi \tag{9}$$

M.3 Geometrical room acoustics in parallelepipeds

Assumptions:

- the room is a parallelepiped with corner lengths ℓ_x, ℓ_y, ℓ_z;
- a small isotropic source is placed in the room centre.

Room volume: $\quad V = \ell_x \cdot \ell_y \cdot \ell_z$ (1)

Walls: $\quad S_x = \ell_y \cdot \ell_z \;;\; S_y = \ell_z \cdot \ell_x \;;\; S_z = \ell_x \cdot \ell_y$ (2)

Interior room surface: $\quad S = 2(S_x + S_y + S_z)$ (3)

Number of mirror sources up to the order n= 1,2,... : $s_n = 4n^2 + 2$ (4)

Positions of the mirror sources : $\{\pm n\ell_x , \pm n\ell_y , \pm n\ell_z\}$ (5)

Number of reflections including the order n : $\Sigma_n = \frac{2}{3}n\left(2n^2 + 3n + 4\right)$ (6)

Estimation of the temporal density of reflections, [2]:

$$\frac{\Delta n(t)}{\Delta t} \approx \frac{4\pi c_0^3}{V} t^2 \tag{7}$$

and of the total number of reflections between times 0 and t :

$$n(t) \approx \frac{4\pi c_0^3}{3V} t^3 \tag{8}$$

Mean free path length of sound :

energetic average :

$$\ell_{me}^2 = \lim_{n\to\infty} \frac{4n^2+2}{n^2} \Bigg/ \sum_{i=12}^{4n^2+2} (1/\ell_{i,n}^2) \approx \frac{3}{4}\left[\frac{1}{\ell_x^2+\ell_y^2} + \frac{1}{\ell_y^2+\ell_z^2} + \frac{1}{\ell_z^2+\ell_x^2}\right]^{-1} \tag{9}$$

geometrical average :

$$\ell_{mg} = \lim_{n\to\infty} \frac{1}{n} \sum_{i=12}^{4n^2+2} \frac{\ell_{i,n}}{4n^2+2} \approx \frac{1}{6}\left[\sqrt{\ell_x^2+\ell_y^2} + \sqrt{\ell_y^2+\ell_z^2} + \sqrt{\ell_z^2+\ell_x^2}\right] \tag{10}$$

value, often used in room acoustics, [3], :

$$\ell_{ma} \approx 4V/S \tag{11}$$

(Effective) intensity of the direct sound field :
(Π_q= effective power of source ; d= distance source to receiver)

$$I_D = \frac{\Pi_q}{4\pi d} \tag{12}$$

Intensity of the reverberant field :

$$I_R = \sum_{n=1}^{\infty} I_{Rn} = \Pi_q \sum_{n=1}^{\infty} \frac{(1-\overline{\alpha})^n (4n^2+2)}{4\pi n^2 \ell_{me}^2} \xrightarrow[\overline{\alpha}<0.1]{} \Pi_q \frac{1-\overline{\alpha}}{\pi\overline{\alpha}\,\ell_{me}^2} \tag{13}$$

with : $S \cdot \overline{\alpha} = \sum \alpha_i S_i$ (14)

Intensity of the reverberant field, including absorption of air :

$$I_R = \frac{\Pi_q}{\pi \ell_{me}^2} \frac{(1-\bar{\alpha}) \cdot e^{-\beta \ell_{me}}}{1-(1-\bar{\alpha}) \cdot e^{-\beta \ell_{me}}} \approx \frac{\Pi_q}{\pi \ell_{me}^2} \frac{1-\bar{\alpha}-\beta \ell_{me}}{\bar{\alpha}+\beta \ell_{me}} \tag{15}$$

Approximation for $\bar{\alpha} > 0.1$:

$$I_R \approx \Pi_q \frac{1-\bar{\alpha}}{\pi \ell_{me}^2}\left[\frac{2}{\bar{\alpha}}+\frac{\pi^2}{6}\right] \approx \Pi_q \frac{(1-\bar{\alpha})(6+5\bar{\alpha})}{3\pi \bar{\alpha} \ell_{me}^2} \tag{16}$$

Level steps in the reverberation plot :

$$\Delta L_1 = 10 \cdot \lg\left[1+\frac{3\bar{\alpha}\ell_{me}^2}{2d^2(1-\bar{\alpha})(6+5\bar{\alpha})}\right] \text{ dB} \tag{17}$$

if an order of reflection is missing (independent of n !) :

$$\Delta L_2 = -10 \cdot \lg(1-\bar{\alpha}) \text{ dB} \tag{18}$$

Decay rate (slope of the level decay):

$$m = -\Delta L \frac{c_0}{\ell_{me}} \tag{19}$$

Reverberation time :

$$T = -60/m \tag{20}$$

M.4 Statistical room acoustics

The diffuse sound field is a scientific artefact; it is a model for sound fields in large rooms.

A sound field is said to be diffuse if *in the average over some time interval* the effective sound intensity (as a vector) in any field point is omnidirectional with constant magnitude for all directions. An immediate conscquence is a zero effective power through a (small) reference volume around a field point. Without the permission of a finite time averaging the necessary consequence would be a zero sound field.

$Z_0 = \rho_0 c_0 =$ free field wave impedance;
$r =$ distance source to field point;
$Q =$ source directivity;
$w =$ energy density;
$A =$ total absorption area;
$V =$ room volume;
$S =$ room interior surface area;
$a, b, c =$ corner lengths of a cubic room;
$\bar{\alpha} =$ average absorption coefficient;
$S_i =$ absorber surface areas;
$\alpha_i =$ absorption coefficient of S_i ;
$\beta =$ propagation attenuation of power;
$d =$ distance of limit between direct and reverberant field;

Energy balance in steady state conditions, [4]:

$$V\frac{dw}{dt} = \Pi_q - V \cdot \bar{n} \cdot \bar{\alpha} \cdot w \tag{1}$$

with $\bar{n}$ denoting the average reflection rate, i.e. the expected number of reflections per time unit. It is calculated from the mean free path (see equ. M.3.11) by

$$\bar{n} = \frac{c_0}{\ell_{ma}} \tag{2}$$

The energy components in parallelepiped rooms can be related to reflections which depend on room shape, the reflection rate and wall absorption. The expectation value of the magnitude of the intensity of a reflection of order i is:

$$\langle I(t) \rangle = \frac{\Pi_q}{4\pi(c_0 t)^2}(1-\bar{\alpha})^i \tag{3}$$

where $\bar{\alpha}$ denotes the average absorption coefficient. Follows with the mean reflection rate $\bar{n}$

$$\langle I(t) \rangle = \frac{\Pi_q}{4\pi(c_0 t)^2}(1-\bar{\alpha})^{\bar{n}\,t} \tag{4}$$

In a diffuse field the intensity vectors are independent on the direction. Therefore the total sound field is obtained by superposition of incoherent contributions of intensity moduli. With the number of reflections per time interval dt as given in M.3 above (equ. M.3.7), the total time-differential intensity is

$$\left\langle \frac{dI(t)}{dt} \right\rangle = \frac{c_0 \cdot \Pi_q}{V}(1-\bar{\alpha})^{\bar{n}\,t} \tag{5}$$

The expectation value of the total time-differential energy density, $dw(t)$, is accordingly (differential energy density impulse response or: energy time curve)

$$\left\langle \frac{dw(t)}{dt} \right\rangle = \frac{\Pi_q}{V}(1-\bar{\alpha})^{\bar{n}\,t} \tag{6}$$

The energy density is thus

$$\langle w \rangle = \int_{t_0}^{\infty} \frac{\Pi_q}{V}(1-\bar{\alpha})^{\bar{n}\,t}\,dt \tag{7}$$

with $t_0 = 0$ and $\bar{n}$ according to equ. (M.3.11) yields

$$\langle w \rangle = \frac{4\Pi_q}{c_0 S\bar{\alpha}} \tag{8}$$

which can also be expressed in terms of the mean sound pressure ($\overline{|p|^2} = Z_0 c_0 \langle w \rangle$)

$$\overline{|p|^2} = Z_0 c_0 \langle w \rangle = \frac{4 Z_0 \Pi_q}{S\overline{\alpha}} = \frac{4 Z_0 \Pi_q}{A} \tag{9}$$

With $t_0 = 1/\overline{n}$ (see (2)) this yields

$$\overline{|p|^2} = \frac{4 Z_0 \Pi_q}{S\overline{\alpha}}(1-\overline{\alpha}) = \frac{4 Z_0 \Pi_q}{A}(1-\overline{\alpha}) \tag{10}$$

More general, including air attenuation as well as wall absorption:

$$\langle dw(t) \rangle = \frac{\Pi_q \cdot dt}{V}(1-\overline{\alpha})^{\overline{n}t} e^{-\beta_0 c_0 t}, \tag{11}$$

and again, with $t_0 = 1/\overline{n}$, :

$$\overline{|p|^2} = \frac{4 Z_0 \Pi_q}{S\overline{\alpha}} e^{-A/S} = \frac{4 Z_0 \Pi_q}{A} e^{-A/S} \tag{12}$$

Relation between the sound pressure $p(r)$ at a field point with distance r to the source of a diffuse sound field and the effective power Π_q of a (small) source:

$$\overline{|p(r)|^2} = Z_0 c_0 \Pi_q \left[\frac{Q}{4\pi r^2} + \frac{4}{A} e^{-A/S} \right] \tag{13}$$

with Q = directivity of the source.

$A = S\overline{\alpha} + 4\beta V$ is the equivalent absorption area (14)

$$S\overline{\alpha} = \sum_i S_i \alpha_i \tag{15}$$

Expectation of level decay

$$L(t) = L_0 + 4.34\, \overline{n}\, t \cdot \ln(1-\overline{\alpha}) \quad \text{dB} \tag{16}$$

Decay rate (slope of the level decay):

$$m = 4.34\, \overline{n} \cdot \ln(1-\overline{\alpha}) \tag{17}$$

Reverberation times (level decay over 60 dB)

$$T = -\frac{60}{4.34\, \overline{n}\, \ln(1-\overline{\alpha})} \tag{18}$$

according to *Eyring*

$$T_{Ey} = -\frac{24 \ln(10)}{c_0} \frac{V}{S \ln(1-\overline{\alpha})} \tag{19}$$

$$T_{Ey} = \frac{60\,V}{-1.086\,c_0\,S \cdot \ln(1-\overline{\alpha})} = -0.161\frac{V}{S \cdot \ln(1-\overline{\alpha})} \tag{20}$$

according to *Sabine* :

$$T_{Sab} = \frac{60\,V}{1.086\,c_0 \sum \alpha_i S_i} = 0.161\frac{V}{S \cdot \overline{\alpha}_{Sab} + 4ßV} \tag{21}$$

according to *Millington-Sette* :

$$T_{MS} = \frac{60\,V}{-1.086\,c_0 \sum S_i \cdot \ln(1-\alpha_i)} = -0.161\frac{V}{\sum S_i \cdot \ln(1-\alpha_i)} \tag{22}$$

according to *Pujolle* (for rectangular rooms):

$$T_{Pu} = -\frac{6\,\ell_{mg}}{c_0 \log(1-\overline{\alpha})} \approx -\frac{\sqrt{\ell_x^2+\ell_y^2}+\sqrt{\ell_y^2+\ell_z^2}+\sqrt{\ell_z^2+\ell_x^2}}{c_0 \log(1-\overline{\alpha})} \tag{23}$$

according to *Pujolle* (for rectangular rooms) including absorption of air:

$$T_{Pu} = \frac{13.8\,\ell_{mg}}{c_0\left[ß\,\ell_{mg} - \log(1-\overline{\alpha})\right]} = \frac{6\,\ell_{mg}}{c_0\left[0.43ß\,\ell_{mg} - \log(1-\overline{\alpha})\right]} \tag{24}$$

The distance, d, of limit between direct and reverberant field is determined by the equilibrium of direct and reverberant sound (equ. M.4.13):

$$d = \sqrt{\frac{Q\,A}{16\pi}e^{A/S}} = 0.1\sqrt{\frac{Q\,V}{\pi\,T}e^{A/S}} \tag{25}$$

or in approximation for low absorption (A << S):

$$d = \sqrt{\frac{Q\,A}{16\pi}} = 0.1\sqrt{\frac{Q\,V}{\pi\,T}} \tag{26}$$

M.5 The mirror source model

Ref. MECHEL, [M.7]

The task is the evaluation of the sound pressure at a receiver point P inside a room for a simple source placed at Q . Analytically the sound field must satisfy the source condition and the boundary conditions at all room walls. Analytical solutions, however, are possible only for very simple room geometries (see Section M.3). The classical tool for sound field evaluations in room acoustics is the mirror source model. It will be displayed here in some detail, going farther than the usual textbook example of parallelepipedic rooms, because the mirror source model is often unjustly said to be inapplicable in practical tasks due to the enormous number of mirror sources which it supposedly needs. One can find in the literature the number $n_W(n_W-1)^{(o-1)}$ of mirror sources "needed" for the reflection order o in a room with n_W walls. It will be shown that the number of sources really needed is much smaller.

The word "mirror source" will be abbreviated as MS , because of its frequent occurrence. The word "source" and the symbol q may stand for both the original source Q and a mirror source. Sometimes we speak of a "mother source" which creates at a wall (or "mother wall", if necessary) a "daughter source".

M.5.1 Foundation of the mirror source approximation

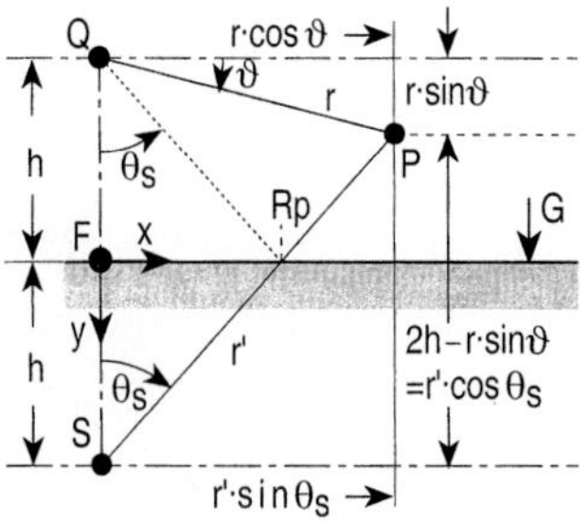

The mirror source method is exact only for a single, infinite, plane wall with ideal reflection (either hard or soft). Then the superposition of the fields of the source Q and the mirror source S satisfy the boundary condition at the wall. (See Section D.15 through D.20).

The analysis of that elementary task returns the result for the reflected field p_r :

$$p_r(r',\theta_s) \xrightarrow[k_0 r'>>1]{} R(\theta_s)\cdot\begin{cases} H_0^{(2)}(k_0 r') & ;\ \text{line source} \\ h_0^{(2)}(k_0 r') & ;\ \text{point source} \end{cases} \tag{1}$$

with: $H_0^{(2)}(z)$ zero order cylindrical Hankel function of 2nd kind;

$h_0^{(2)}(z)$ zero order spherical Hankel function of 2nd kind;

$R(\theta_S)$ is the reflection factor of a plane wave incident on the wall under the angle which the connection of the mirror point S with the field point P includes with the normal to the wall. The forms (1) do not satisfy the wave equation if $R(\theta_S) \neq const(\theta_S)$, because an azimuthal factor strictly cannot be associated with a Hankel function of zero order. This should also be kept in mind, if the original source has a directivity factor $D(\vartheta)$ (in 2D) or $D(\vartheta,\varphi)$ (in 3D).

This is the *mirror source approximation* (MS approximation). One should keep in mind:

- The MS solution is only approximate if $R(\theta_S) \neq const(\theta_S)$, i.e. for $G \neq 0$ or $|G| \neq \infty$;
- Then it violates the wave equation;
- It determines precisely the meaning of $R(\theta_S)$! ;
- With that definition (and only with that) it satisfies the wall boundary condition;
- It supposes $k_0 r' >> 1$, i.e. large distances dist(S,P);
  or more precisely: a great sum of the heights of Q and P over the wall;
- It supposes that P is not under an angle θ_S with a strong angular variation of $R(\theta_S)$;
- For grazing incidence, i.e. Q and P on the wall, the influence of higher terms $R^{(n)}(\theta_S)$ in (8) is important !
- The derivation further supposes that the wall does not guide a surface wave (but this is only rarely the case in a restricted frequency range below a high-quality resonance; but even then (8) is an approximation to the field in points not too close to the wall).

The mentioned facts have important consequences :

- on the one hand, it makes no sense to try to compute with a higher precision than the precision of the fundamental process of the MS method;
- on the other hand, the approximate character of this process does not give a justification to fantastic modifications of the MS method.

M.5.2 General criteria for mirror sources

Mirror sources are created at a wall by a source inside the wall by the steps:

- mirror-reflect the source position to behind the wall;
- multiply its source factor by $R(\theta_S)$;
- if the source has a directivity factor $D(\vartheta)$, reflect that directivity, i.e. rotate it.

The continued multiplication, for increasing order o of reflection, creates a product of reflection factors $R(\theta_S)$, which will be called the "source factor" and symbolised with ΠR .

The form of the MS and the co-ordinates used should, if possible, be such that these steps can be performed easily in the computations.

The right of a MS to exist is the satisfaction, together with its mother source, of the boundary condition at the wall at which it was created by its mother source. Nothing else !

The first criterion for the generation of a daughter MS at a wall is that the mother source irradiates the interior surface of that wall. As a consequence: if a source is outside a wall, it does not create a daughter source at that wall. In particular, a mirror source will never be mirror-reflected back to the position of its mother source. We call this rule the "inside criterion". The chain of MS-production is interrupted if this criterion is violated; otherwise the daughter source would be "illegal" (see below for other criteria of interruption).

M.5.3 Field angle of a mirror source

The field angle of a MS gives a further important criterion for the interrupt of MS-production. The field angle is explained for the case of reflection of a source Q at a plane wall which is subdivided in two sections with different reflection factors R_i in them.

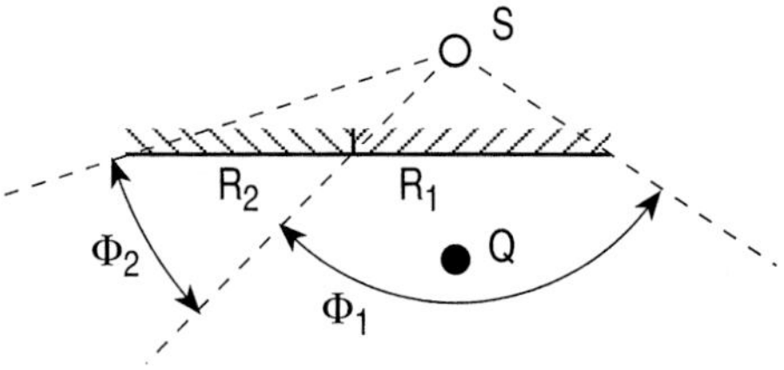

Although the source Q has only one position S for a MS, there are indeed positioned two MSs with different angular ranges Φ_1, Φ_2 of their fields because there are two different source factors R_1, R_2 in both ranges.

The field is unsteady at the common flank of the field angles. This is a consequence of the character of the MS method as an approximate solution which must be tolerated. In 3D the field angle Φ is given by a polygonal pyramid subtended by a wall W and with the source S in the apex.

The "field angle criterion" states two things:

- a MS creates in P a field contribution only if P is in its field angle (more precisely: ...and on the interior side of the creating wall); otherwise we say "the MS is ineffective";
- a MS generates a daughter MS at a wall W only, if the wall is inside the range of its field angle (again on the interior side of the creating wall); otherwise no boundary condition must be satisfied at W for MS , and therefore the daughter MS would be "illegal".

The additions in parentheses will be important for convex corners (see below). We can describe the effect of the field angle with the word "visibility". A source q sees an object only if that object is inside its field angle. If q does not see P , then q is ineffective; if q does not see a wall W, then q does not produce a daughter source with W .

The chain of MS production is continued for an ineffective MS (because a daughter MS may become effective), but the chain is interrupted at an illegal MS.

Some problems are caused by walls W which are only partially inside Φ . In a strict procedure one would have to subdivide the wall at the intersection with the flank of Φ , but such a "dynamical" definition of walls would produce much computational work. It is sufficient, within the frame of precision of room acoustical computations, to check whether the wall section inside Φ exceeds some size limit (e.g. λ_0); if not, that wall is neglected for that MS. It is a good compromise between precision and amount of computation to check whether the centre C of W is inside Φ . This check is done by a repeated test whether C is inside the walls of the polygonal pyramid with the MS at its apex, and subtended by W . The repetition can be interrupted if C is outside one of the pyramid walls.

In general, MSs with increasing order are displaced farther and farther away from the interior of the considered room. Thus their field angles become smaller and smaller; so fewer walls have to be considered for the production of further MSs.

The mentioned additional condition that either P or an other wall must be in the field angle on the interior side of the generating wall is important, as can bee seen from the next figure.

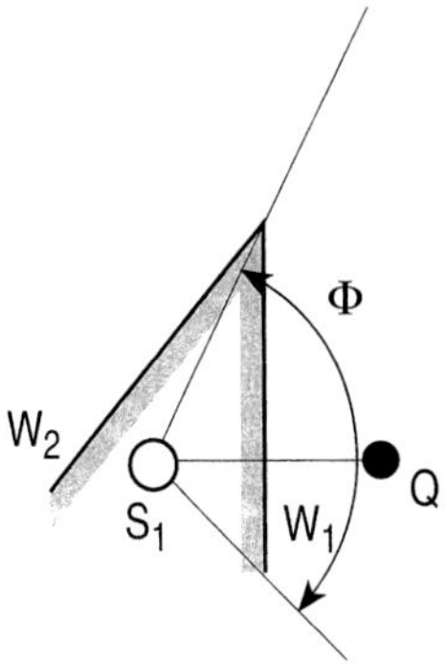

Here the source Q creates at W_1 a MS, S_1 , which in the case shown is outside both walls W_1, W_2 . The inside criterion would interrupt a further production of MSs anyhow.
If, however, Q is displaced farther away from the wall W_1 , then S_1 may fall on the interior side of W_2 . Nevertheless S_1 will not produce a MS at W_2 , because W_2 would be in Φ but not on the interior side of the generating wall W_1 .

The additional condition (W_2 inside W_1) is relevant only for convex corners.

M.5.4 Multiple covering of MS positions

In the sketch of Section M.5.3 two MSs occupied the same position; both are legal, because they are different from each other. If two walls form a space wedge, and if the wedge angle

Θ of a couple of walls is a rational multiple of π, the MS beginning with some higher order will fall upon positions of MSs of lower orders.

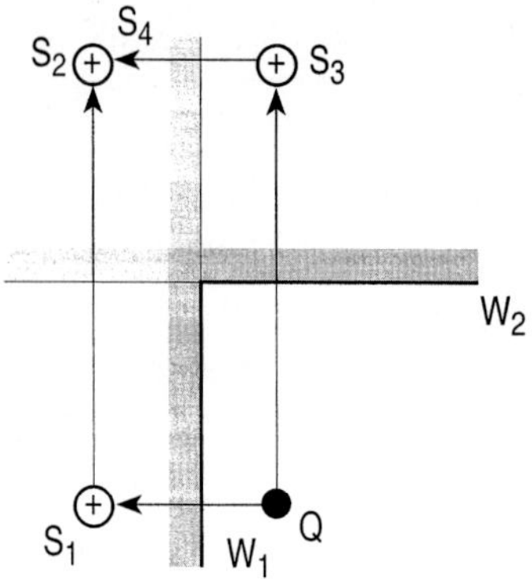

The outer sides of the (possibly extended) walls are greyed.

The chain of the MS production ends with S_2, S_4, because both MSs are outside both walls.
The source factors of both MS S_2, S_4 are equal.

Such coincident MSs with the same source factors are illegal, because they violate the source condition.

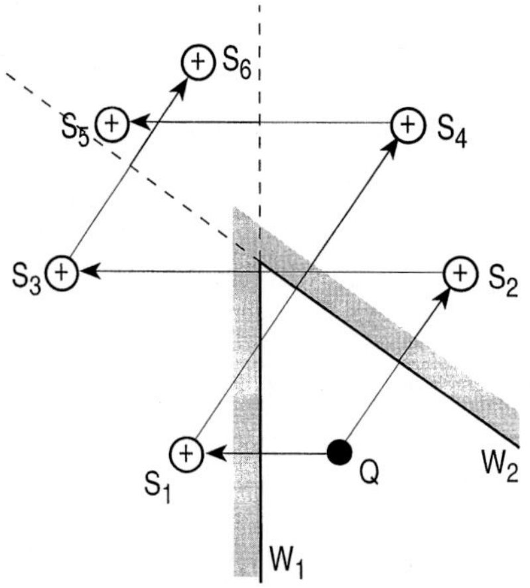

In this example the MS production ends with S_5, S_6, because both MSs are outside both walls.

Depending on the wedge angle Θ and on the position of Q , different numbers of MSs can be constructed (higher for small Θ, infinitely high for parallel walls with a part of the room between the walls, i.e. for $\Theta=0$).

M.5.5 Convex corners

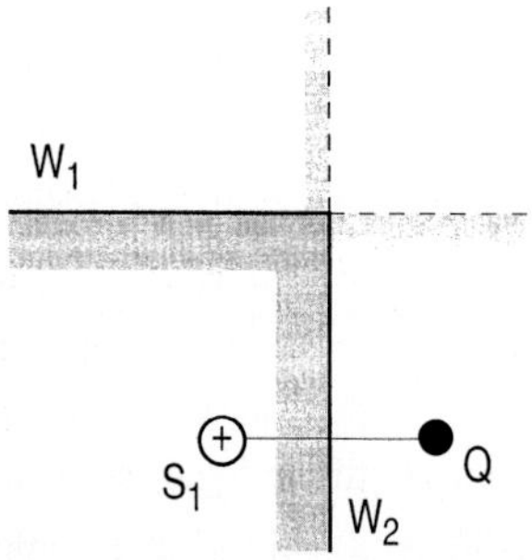

Convex corners, i.e., with wedge angles $\Theta>\pi$, cannot be treated with the "traditional" MS method. Only one MS can be constructed at this convex rectangular corner. That is evidently not enough to represent the sound field at such a corner. Consequently, the traditional MS method fails at convex corners !

M.5.6 Interrupt criteria in the MS method

The MS method is often blamed for the apparent exorbitantly high numbers of mirror sources involved. In such statements inherent interruption criteria of the MS method are ignored.

Conditions for the interrupt of the chain of MS production are :

1) The source q is outside the mirror wall W :
there exists no boundary condition for q at W ;
the MS chain can be interrupted.
2) A wall W is outside the field angle Φ of q :
q does not produce a daughter MS at W .
For walls with a common corner this is equivalent to 1).
3) A MS would fall into the interior room space :
then there would be, except for Q , a new pole position of the field;
this offends the condition of regularity of the field.
This case is met with only for convex corners.
4) If the new MS would fall on Q :
from then on the MS positions would be repeated;
this violates the source condition which demands that the volume flow through a small enclosure around Q must be the same as of the original source.
5) The product $\Pi|R|$, which is the source factor of q, would become < limit , which is a pre-set limit. This would make the field contribution negligible, also for all daughter sources of q .
6) If the sound field is the target quantity, the distance dist(q,P) of a source q from P may be limited to be <dmax . For point sources the amplitude ratio of q at P , relative to the amplitude of Q at P , is dist(Q,P)/dist(q,P) . This ratio generally becomes even smaller for daughter sources of q .
7) A limitation $\Pi|R|\cdot$dist(Q,P)/dist(q,P) < limit$\cdot$dmax would be more significant.
8) With some arbitrariness one sets an upper limit of the orders o=1,2,…,omax of the MS.

In interrupt checks using $\Pi|R|$ it may be sufficient to use approximate values for the reflection factors R (or their magnitudes $|R|$), for example the reflection factor for normal sound incidence or the reflection coefficient $|R|^2$ from the absorption coefficient for diffuse sound incidence. Then the construction of the MS becomes independent of the position of P . The reflection factors are the only quantities which introduce the frequency into the construction of the MS . If one takes for the interruption check a lower limit or an average value of $|R|$ over the frequency interval considered, the construction of the MS is independent of the frequency also. One should apply tests with the true $|\Pi R|$ in the step of evaluation of the field contributions of the sources, when the ΠR are available. But then these tests must be applied on a smaller number of MSs than in the phase of MS-construction.

M.5.7 Computational parts of the MS method

The traditional MS method consists of three computational parts :

- Find the positions of the MS (considering the inside and field angle criteria);
- Determine the source factors of the MS, i.e., of the reflection factors $R(\theta_S)$, (depending on the acoustical quality of the mirror wall and of the relative positions of the MS and P):
- Evaluate the contributions of the MS to the field at P .

Most programming is needed for the finding of the MS positions, although the single steps are elementary geometrical tasks. Less intensive in programming is the evaluation of the reflection factors of absorbent walls. This task will be delegated to subroutines for the wall surface admittance G . Most simple is the evaluation of the field contributions; only a number of Hankel functions of zero order must be evaluated; this sub-task is fast-computing for a spherical Hankel function (which are given by cos(x), sin(x)), and is fast-computing also for cylindrical Hankel functions when using the known polynomial approximations for Bessel and Neumann functions of zero order.

The traditional MS method proceeds with the order o of mirror reflections.

- At the order o=1 the MSs S(1) are determined in turn for all walls (consider the inside criterion for convex corners!);
- At the order o=2 all S(1) are potential mother sources for the generation of the MSs of 2nd order S(2) at all walls (except for the wall at which S(1) was produced), unless the inside and field angle criteria exclude S(2);
- Continue until a final interrupt criterion is met.

M.5.8 Inside checks

Checks for interruption and efficiency form the main part of the computational work in the traditional MS method. They are fundamental tasks of computational geometry. But because they are repeated very often, they should compute fast. We break down all tests to "inside checks". An inside check examines whether a point q is on the interior side of a wall plane, in the wall plane, or on the exterior side of the wall plane (the sides are defined by the rotational sense of the edges E_k of a wall $W=\{E_1, E_2, E_3,\ldots\}$). One could do the inside check with the help of direction cosines of the connecting lines between q and the E_k . The evaluation of angles, however, is slow. An inside check in 3D uses the vector triple product (scalar product of a vector and a vector product) if the wall is given by three of its edges. If the parameters of the reduced normal form of the wall equation

$$a\cdot x+b\cdot y+c\cdot z+d=0 \tag{2}$$

are known, the inside check needs three multiplications and three additions (see the Appendix) with $q=\{x, y, z\}$:

$$a\cdot x+b\cdot y+c\cdot z+d \begin{cases} >0\,; & \text{q inside the W plane} \\ =0\,; & \text{q on the W plane} \\ <0\,; & \text{q outside the W plane} \end{cases} \tag{3}$$

if W, q form a right-handed system; otherwise the signs change.

Another, often used, test examines if a point P is inside the polygonal pyramid which has the point q as apex and is subtended by a wall W . This test is done by a repetition of inside checks for P and the triangles forming the sides of the pyramid. The loop over the triangles can be aborted with a negative answer for the test if one of the inside checks fails (distinguish whether W and q are a right-handed or left-handed system).

The shading of a point P or a wall W by a convex corner, and the visibility of P or W from a point q through an aperture (formed by convex corners with a free interspace between them) are also tested with inside checks.

M.5.9 What is needed in the traditional MS method ?

One needs as input :

- the list of walls $\{W_1, W_2, W_3,\ldots\}$ which themselves are lists of edges $\{E_1, E_2, E_3,\ldots\}$, $E_i=\{x_i,y_i,z_i\}$;
- the source point $Q=\{x_Q, y_Q, z_Q\}$;
- the field point $P=\{x,y,z\}$;
- the list of wall admittances $G=\{G_1,G_2, G_3,\ldots\}$;
- the limits omax, limit, dmax for the order o , the source factors $|\Pi R|$, dist(q,P), respectively.

It is supposed that the dist(Q,P) , the wall centres C_W , and the parameters a,b,c,d of the reduced normal forms of the wall equations $a\cdot x+b\cdot y+c\cdot z+d=0$ are evaluated (see Appendix). One needs, for the evaluation of the reflection factors and of the contributions in P :

1) the position q of a source (either Q or a MS);
2) the counting index w of the wall W at which q was generated;

3) the distance dist(q,P) ;
4) the amplitude factor $\Pi R(\theta_s)$;
5) a flag which signals with flag=0 that q is an effective source, and with flag=1 that q is ineffective.

These data are collected in "source lists" {q, w, dist(q,P), $\Pi R(\theta_s)$, flag} , and the source lists for a given order o=0,1,2,...,omax are collected in tables

tab(o)= {...,{q, w, dist(q,P), $\Pi R(\theta_s)$, flag},...} .

Let the counting index of a source list within tab(o) be s . The source table for the order o=0, i.e., for the original source q=Q , has the form tab(0)={{Q, 0, dist(Q,P), 1, 0}} (for rooms with concave corners; see below for rooms with convex corners).

One can delegate the task of mirror reflection of a mother source q_m , represented by its source list {q_m, w_m, dist(q_m,P), $\Pi_m R(\theta_s)$, $flag_m$}, at a wall W_w , given by its index w , including all tests of interrupt and effectivity, into a subroutine, which should be carefully checked and economised with respect to computing time. That subroutine returns

- the source list {q, w, dist(q,P), $\Pi R(\theta_s)$, flag} of the daughter source, if no interrupt criterion is met;
- the value 0 , if an interrupt criterion is met.

Such a subroutine for 3D rooms with concave corners is a program of about 25 program lines in Mathematica® language (the geometrical sub-tasks inside the subroutine are delegated to subroutines).

The traditional MS method works in three nested loops :

1) The outer loop over the order o=1,2,...,omax :
   it produces the source table tab(o);
2) The middle loop over the counting index s=1,2,... of the sources in tab(o–1) ;
3) The innermost loop over the counting index w=1,2,... of walls ;
   it calls the above mentioned subroutine;
   if that subroutine does not return 0 , the new source list is appended to tab(o) .

This traditional MS method is attractive by its computational simplicity. A frame program for the evaluation of the tab(o) in Mathematica® typically is a program of about 12 lines, if the frame program calls a subroutine for the MS evaluation with all checks inside the subroutine. It may be of some advantage to perform the checks of legitimacy of a new MS (mother source inside the mirror wall; mirror wall in the field cone of the mother source) in the frame program (this is true especially when the MS method is applied to rooms with convex corners).

The returned tables tab(o) contain also ineffective sources (flag=1). One can select the effective sources with flag=0 and collect them in tables tabeff(o) . So one has available all data which are needed to evaluate and sum up the field contributions in P of the effective sources.

M.5.10 The object

The geometrical object is a room formed by plane walls W_w . It is clearly defined what is inside and outside of the room. The (original) source Q and the field point P are always inside. A right-handed Cartesian system of co-ordinates x,y,z is laid over the room.

The walls W_w are plane. They are described by lists of edges, $W_w = \{E_{w1}, E_{w2}, E_{w3}, \ldots\}$ which are ordered such that the sense of rotation in that order and the direction pointing to the inside of the room make a right-handed system. Because the first three edges are used for the determination of the unit normal vector of the wall, these edges should not be collinear, and they should agree with the general sense of rotation of wall edges. Edges may be cyclically interchanged in a wall list. The order w=1,2,3,... of the walls is arbitrary.

Wall couples form a room wedge; they either have a (straight) real corner if the walls are succeeding each other, or they have a virtual corner if other walls are placed between the flanking walls. The walls of a couple include a wedge angle Θ (measured inside the room). Real corners are "concave" when $0 < \Theta \le \pi$, and "convex" for $\pi < \Theta \le 2\pi$.

The acoustic qualification of a wall will use its surface admittance G_w . The MS method applies for the acoustic qualification of a wall its reflection factor $R_w(\theta)$ which is the reflection factor for a plane wave with incidence under the polar angle θ formed by the normal to the wall and the connection line of the MS with P . $R_w(\theta)$ depends on that angle, and thereby on the position of P, as :

$$R_w(\theta) = \frac{\cos\theta - Z_0 G_w}{\cos\theta + Z_0 G_w} \tag{4}$$

If the wall is bulk reacting one further has $G_w = G_w(\theta)$.

It is strongly recommended not to use too small parts of walls. The subdivision of the room envelop into too small wall sections not only produces analytical and numerical nonsense, but the computational work is increased immensely by the subsequent generation of MSs at such small faces. If one likes to consider the acoustic effect of, e.g., pillars and/or handrails , it would be easier and faster to solve the task of scattering at suitable scatterers (e.g. cylinders or spheres). As a general rule for the dimensions of walls to be considered one can neglect walls with dimensions smaller than about λ_0 .

M.5.10 A concave model room, as an example

We consider a 3D model room which could go as a simple concert hall (see figures below). It has w=1,2,…,19 walls, two of them are coplanar, and two couples have parallel walls on opposite sides of the room. The floors of the stage and of the seat area are inclined. Balconies cannot be modelled with concave rooms. The following 3D plots are computed from the input data (such plots are parts of the checks of input data). The 1st figure shows the room as a 3D-wire-plot, together with the source Q and the field point P , the 2nd figure shows an outside view of the room. The co-ordinate units are arbitrary. In the 2nd figure and in later diagrams for the mirror sources they are scaled down with a factor 1/10 , as compared to the 1st figure.

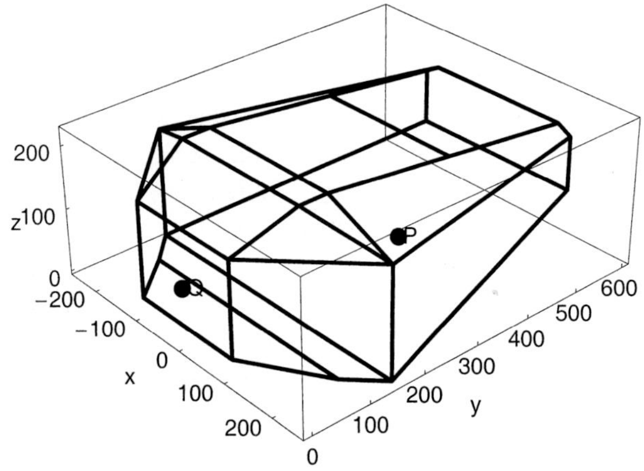

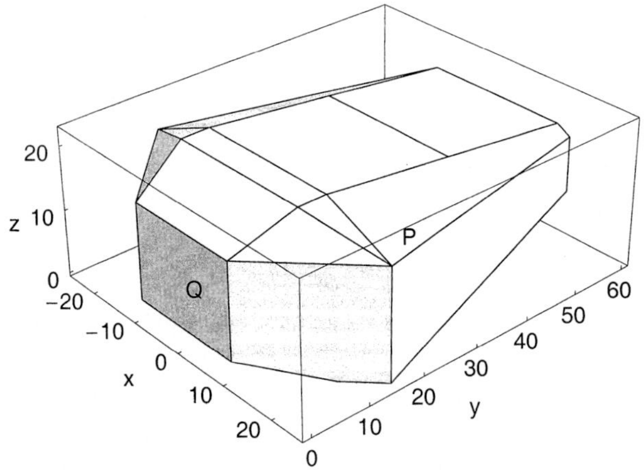

The normalised wall admittances Z_0G_W for the example shown produce absorption coefficients α_{dif} for diffuse sound incidence as given below. They are not exceptional in any sense.

$\alpha_{dif} \approx \{0.10, 0.10, 0.40, 0.71, 0.20, 0.60, 0.20, 0.40, 0.20, 0.40, 0.40, 0.20, 0.20, 0.20, 0.20, 0.20, 0.20, 0.20, 0.50\}$

The following diagrams show mirror sources (as points) for two orders o= 1, 3 , if only back-reflection (into the position of the mother source) is avoided. Such diagrams would be the basis for reported numbers of mirror sources "needed".

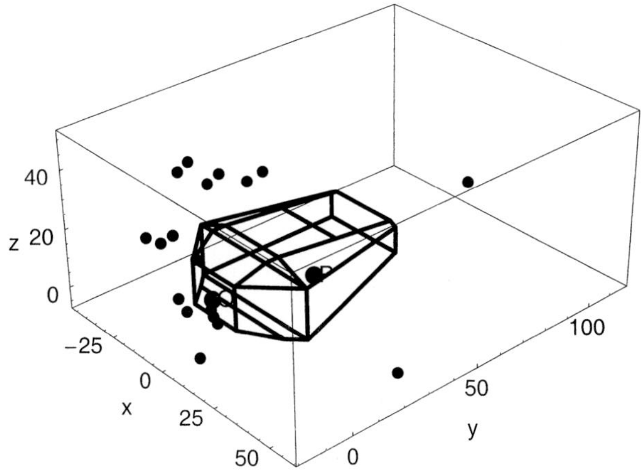

Mirror sources of order o=1 ,
with only back-reflection criterion.
Number of MSs: 19

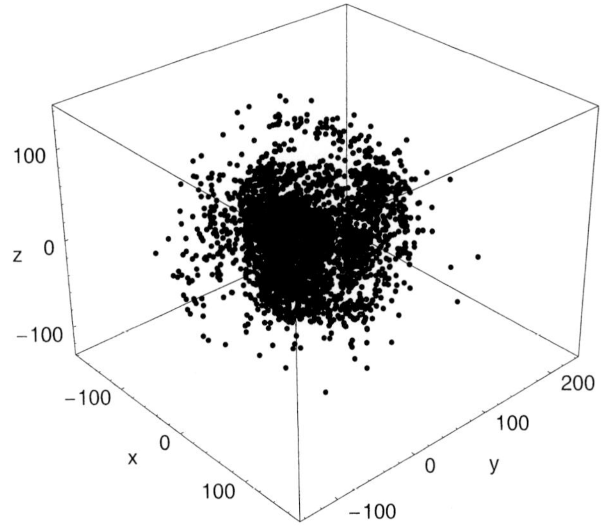

Mirror sources of order o=3 ,
with only back-reflection criterion.
Number of MSs: 6156

The next diagrams show the *effective* mirror sources with all interrupt criteria applied. As the criterion for exclusion of a wall as a mirror wall it was checked whether the wall centre is inside the field angle cone of the mother source. It should be noticed that already for the order o=1 the number of MSs is reduced from 19 to 10 (mainly by the efficiency check).

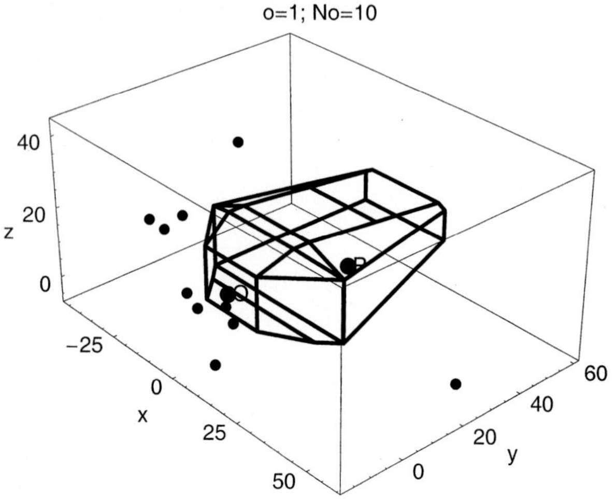

Effective mirror sources of order o=1 ,with all interrupt criteria. Number of MSs: 10

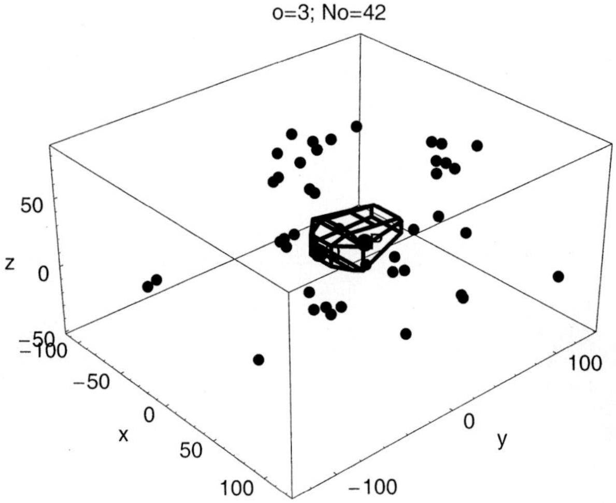

Effective mirror sources of order o=3 ,with all interrupt criteria. Number of MSs: 42

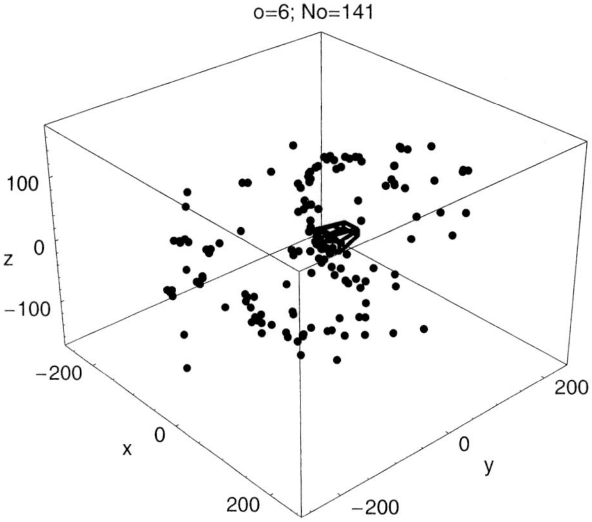

Effective mirror sources of order o=6 ,with all interrupt criteria. Number of MSs: 141

The published estimates $N \cdot (N-1)^{(o-1)}$ for the MSs of order o needed in a room with N walls would give for the order o=6 (with N=19 for our room) a number 35901792 !! The computing time for the tabeff(o) up to o=6 was 116 s (on a 400 MHz laptop computer with non-compiled Mathematica® programs).

The numbers of MSs in orders o for different applied criteria are collected in the following Table 1.

o	Back-Reflection	& Inside q and wall	& Effectivity
1	19	19	10
2	342	97	25
3	6 256	261	42
4	110 808	478	69
5	1 994 544	755	99
6	35 901 792	1 059	141

Table 1: Numbers of MSs in several orders o for different interrupt criteria applied.

The next Table 2 collects, for each order o, minimum and maximum reductions in the level of the sound pressure contribution in P due to |ΠR|, to dist(Q,P)/dist(q,P), and to their product.

o	|ΠR|		dist(Q,P)/dist(q,P)		|ΠR|·dist(Q,P)/dist(q,P)	
	min	max	min	max	min	max
1	7.34	0.566	4.69	0.122	7.46	0.948
2	21.05	1.32	10.23	0.331	25.57	4.36
3	22.61	1.69	14.2	4.36	32.29	6.93
4	22.96	3.06	15.91	4.86	36.02	8.89
5	36.21	3.58	18.86	7.34	49.29	12.86
6	37.72	4.19	19.01	8.98	52.91	14.53

Table 2: Minimum and maximum reductions in the level of field contributions in the order o by the source factor ΠR, the distance ratio dist(Q,P)/dist(q,P), and their product

The limits were set to |ΠR|<0.01 ~ –40 dB ; dist(q,P)>10·dist(Q,P) (corresponding to –20 dB of the distance ratio); so neither of the two limitations restricted the number of effective mirror sources. The table also shows that a limitation of the orders to $o \le o_{max}=6$ is reasonable, because the highest contribution of a mirror source for o=6 is –14.53 dB below the contribution of the original source Q.

As an example of application, we plot the profiles of the sound pressure level in places $X=(x,y,z_P)$ around $P=(x_P,y_P,z_P)$ as 3D-plots of $20\cdot\lg|p(X)/p(P)|$ over k_0x, k_0y for a fixed frequency. It is supposed that the distances dist(X,P) are small enough to neglect the influence of variation of X on $R(\theta_s)$. (Such patterns could not be computed with modified MS methods using $|p_s(P)|$ or rays or sound particles as field descriptors.)

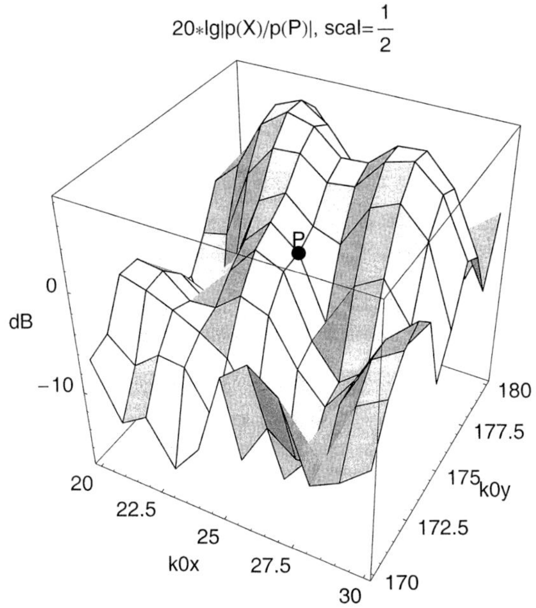

Profile of the sound pressure level around P , for a higher frequency.

M.5.11 The MS method in rooms with convex corners

As shown above, the traditional MS method fails for the description of the sound field around convex corners (we shall solve this problem below). But one can apply the traditional form of the MS-method in such rooms, if one supposes that the scattered sound field in the shadow zone behind a convex corner can be neglected in comparison with the contributions of mirror sources which radiate into the shadow zone without scattering at the convex corner.

The mentioned supposition introduces the concept of "shading" into the MS-method (see below). It is an important statement that a convex corner can be treated in the computations like a concave corner, if the source q "sees" both flanks of the corner (from inside), because then no shadow is created. This condition is easily checked by "q inside both flanks". The next figure shows a possible situation at a strongly convex corner, in which the legal daughter source S_1 of Q lies inside the room. Shading concerns a possible contribution of a source at P (efficiency of the sourcc), as well as a possible continuation of MS production, if a wall is shaded.

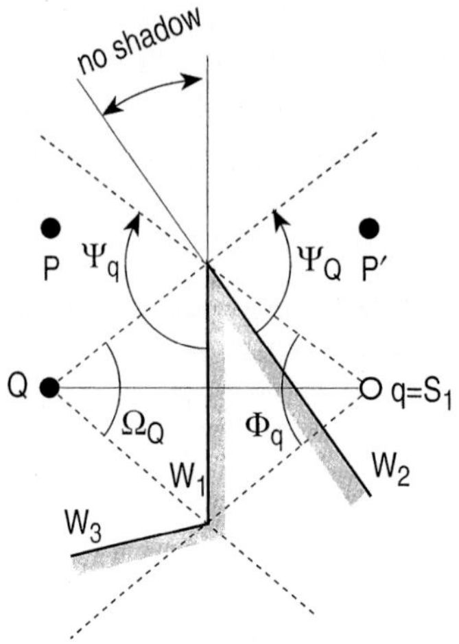

Q in the example is ineffective if the field point is at P′ ; P′ is in the shade angle Ω_Q which Q subtends with the wall W_1 .

If we consider the mirror source q , it does not produce a daughter source at W_2 although it is inside that wall. Now Φ_q is the field angle of q as defined and used in previous sections. The prescription that q shall not produce a daughter source at W_2 is covered, when we expand the condition for walls W at which a source q can produce a daughter source (W is inside Φ_q) by the additional requirement that W is inside the mother wall (here W_1) as well.

With that expanded rule, q can legally produce a daughter source at W_3 .

The convex corner in the figure is treated like a concave corner, if the source is in the range indicated with "no shadow". This "no shadow" range gets larger for "mildly convex" corners. Thus the sound field evaluated in a room having only mildly convex corners will not be much different from the field in a similar room with only concave corners, as would be expected.

The situation is more complicated if two (or more) convex corners form an "aperture" which subdivides the room.

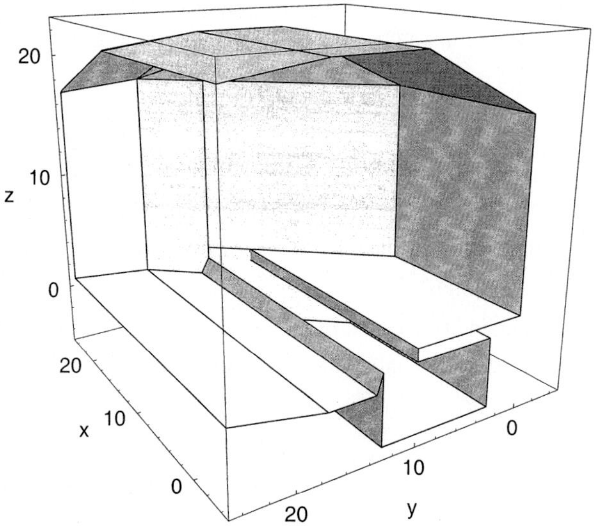

The figure shows, as an example, a 3D-view from inside the room on the stage and into an orchestra pit (this model will be used below for application of the MS method).

The head of the stage and the balustrade of the orchestra pit form an aperture A .

One must not only distinguish if the original source Q and the receiver P are on different sides of the aperture (for effectivity), but also a wall (for further MS production) must be seen by a source through the aperture, i.e., either P or a wall must be inside the aperture angle Φ_A which is the cone subtended by the aperture A and with the source q in the apex.

The "inside check" for P inside the cone(q, A) is simple, see above . In concave rooms one can similarly test if a mirror wall is inside the field angle Φ of a source by checking whether the wall centre is inside Φ . This kind of check for the visibility of a wall through an aperture would be too rough; important sound paths from one subspace (e.g. the orchestra pit) to the other subspace (e.g. the stage or the auditorium) could be missed with that kind of test.

The source q and the aperture A produce a "bright patch" F in the plane of an opposite wall W (F is the polygon formed in the plane of W by the intersection points X_i of the side corners of the cone(q, A) with the plane of W ; they can be evaluated). Three cases of visibility should be distinguished; they are shown in the figure.

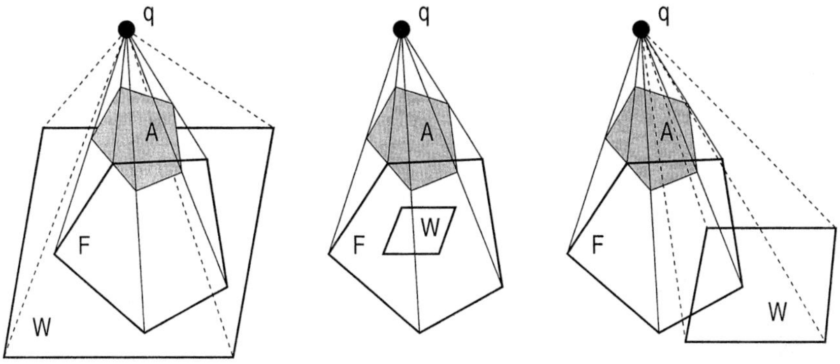

a b c

One could describe the condition of visibility by the requirement that at least one edge of F is within W or at least one edge of W is within F . The implementation of this test would need the evaluation of the intersection points X_i and the test of whether a "point is within a polygon" (which should not be confused with "a point is inside the polygon plane"). One can avoid these (computer-intensive) sub-tasks by using the cone(q, W) . Then the visibility check reduces to the tests "at least one edge of A inside the cone(q, W)" or "at least one edge of W inside the cone(q, A)".

The MS method in rooms with convex corners (in the supposed approximation which neglects corner scattering) has the same aim as in concave rooms, namely to find source lists {q, w, dist(q,P), ΠR, flag} for legal sources q . As previously the frame program operates in three nested loops over the order o , the counting index s of the sources in the order o–1, the counting index w of the walls. Because of the many decisions which must be made by the frame program, anyhow, it is advisable to write a subroutine for the mirror source evaluation, which is applicable for flanks of concave and convex corners, i.e. which internally only makes interrupt checks related to $|\Pi R|$ and to dist(q,P) and efficiency tests for the new MS, whereas the frame program performs all interrupt tests. One can summarise the modifications of the MS method in rooms with convex corners as follows:

1) Find the lists of wall couples forming convex corners and of exclusive couples.
2) Determine the aperture A (if any).
3) An efficiency check must be performed already for the order o=0 , i.e. for q=Q .
   (Q is ineffective if P is on the other side of A , but not in Φ_A of Q ; or, for a single convex corner when A is not defined, if P is in the shade angle Ω_Q (a pyramid with Q at the apex and subtended by a wall W which is a flank of the convex corner)). Therefore determine tabo(0) in the frame program.
4) The source lists of the order o=1 (i.e. with Q as mother source) must also be determined separately in the frame program (because Q has no mother wall).
5) For orders o>1 the frame program in its innermost loop over the wall indices w has to check the interrupt conditions:
   - $w=w_m$, the index of the mother wall;
   - the mother source q_m outside the wall with W(w) ;
   - for walls w, w_m on the same side of A , if the centre of the wall w is outside the cone(q_m, W(w_m)) ;
   - for walls w, w_m on different sides of A , if W(w) is not visible for q_m through A .
6) If none of the tests in 5) is positive, the frame program calls a subroutine for the evaluation of the source list of a new mirror source.

The subroutine causes an interruption (skip of w), if $|\Pi R| < \text{limit}$ or dist(q, P) > dmax. It further makes the efficiency tests. These efficiency checks should be clear from the explanations given above.

M.5.12 A model room with convex corners

The model room of this section widely corresponds to the concave model room in the previous section, but an orchestra pit is added below the stage (the 3D-view above showed a part of the room). A 3D-wire-plot of the walls (unscaled) is contained in the next figure, then a 3D-view from outside is shown (scaled with the factor 1/10 as in the subsequent diagrams with mirror sources). The original source Q is placed in the orchestra pit; the receiving point P has its former position. The number of walls of the room is n_W=29 . There are three important convex corners: the upper corner of the balustrade at the orchestra pit, and the upper and lower corners of the head of the stage floor.

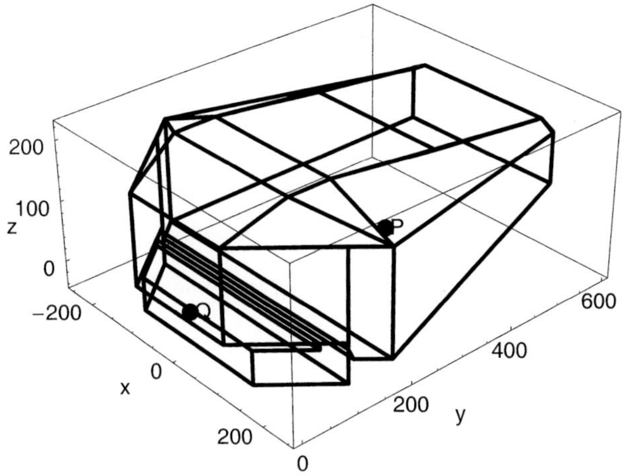

A 3D-wire plot of the room, showing the positions of the source Q and of the field point P .

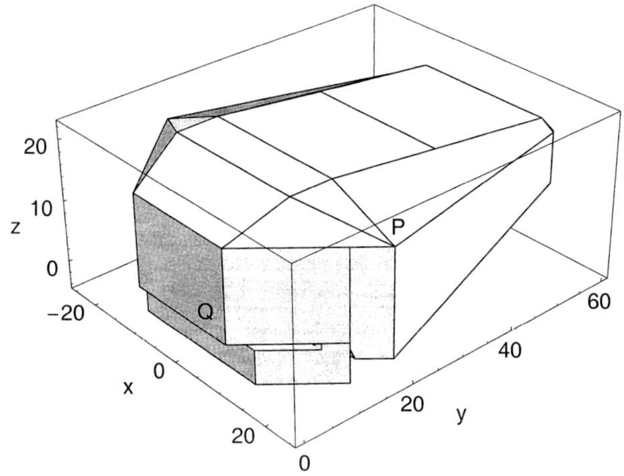

A 3D-view from outside, with scaled co-ordinates.

Table 3 gives the numbers of the legal and of the effective mirror sources in the orders $o=1,2,\ldots,o_{max}=6$.

Type :	**o=1**	**o=2**	**o=3**	**o=4**	**o=5**	**o=6**
legal	24	109	286	637	1 306	2 467
effective	0	19	44	96	186	272

Table 3: Number of legal and effective MSs of orders o=1,2,...,6 .

All MSs of the order o=1 are ineffective, like the original source Q . The total number of effective sources is 617 up to o=6 ; the predicted number would be (with $n_W=29$) :

$$1+\sum_{o=1}^{6} n_W(n_W-1)^{(o-1)} = 517\,585\,882$$

The following diagrams show legal and effective MSs in 3D-plots for some orders o (legal first, then effective, except for o=1 where no effective MS exists).

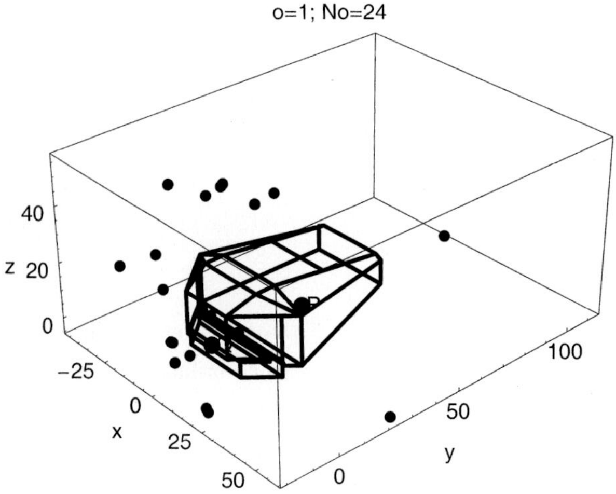

Legal MSs for o=1 .

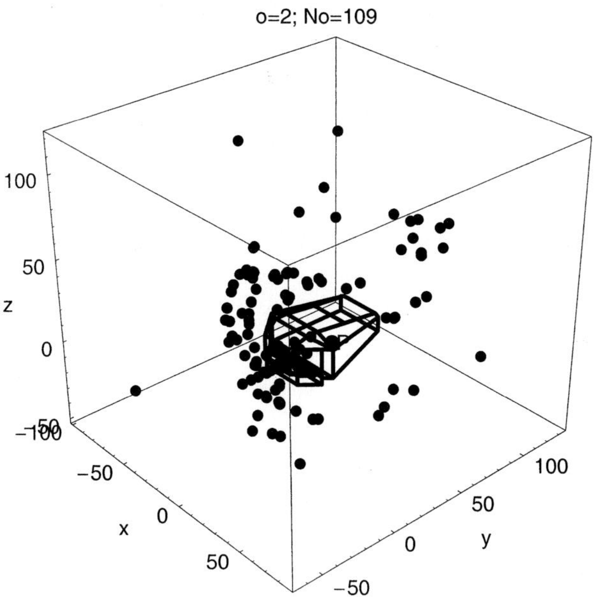

Legal MSs for o=2 .

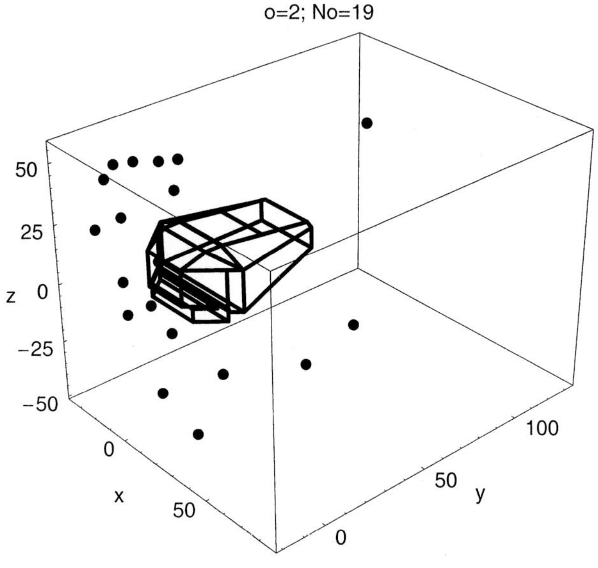

Effective MSs for o=2 .

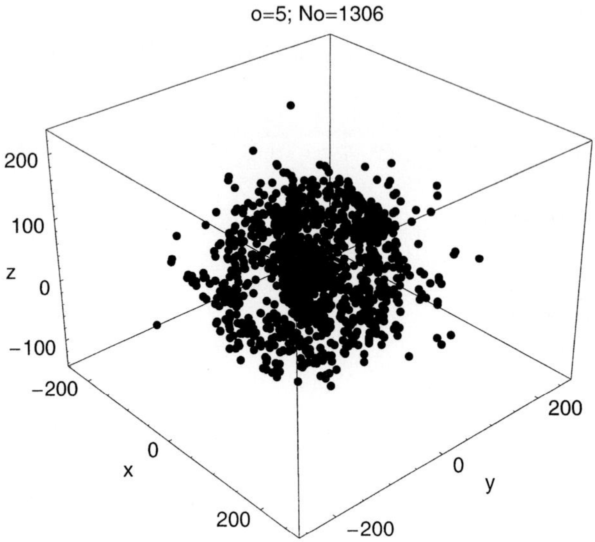

Legal MSs for o=5 .

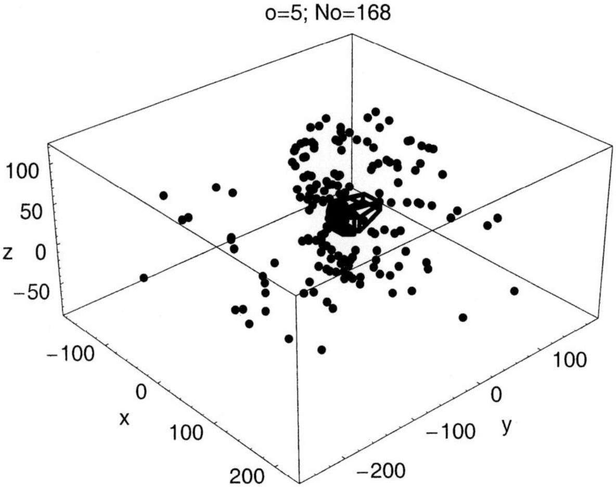

Effective MSs for o=5 .

Table 4 gives the lowest values for the source factor |ΠR| (in dB) and the lowest and highest values of |ΠR|·dist(Q,P)/dist(q/P) (in dB) within the orders. The table shows that, when using the product |ΠR|·dist(Q,P)/dist(q/P) as interrupt criterion, some MSs in the orders o=5, 6 could be dropped, but the last row shows that the order o_{max}=6 may not be high enough.

Level Change by	o=2	o=3	o=4	o=5	o=6
$\|\Pi R\|$, min	–4.38	–19.02	–23.81	–31.05	–34.59
$\|\Pi R\|$·dist(Q,P)/dist(q/P), min	–12.37	–25.17	–30.88	–37.89	–47.13
$\|\Pi R\|$·dist(Q,P)/dist(q/P), max	-3.37	-4.22	-3.38	-4.25	-6.98

Table 4: Level changes of contributions in the order o .

Again we plot the sound pressure level profile around P (see above for the assumptions made) and refer the level to the sound pressure which the free source Q would produce at P. The figure is for a higher frequency.

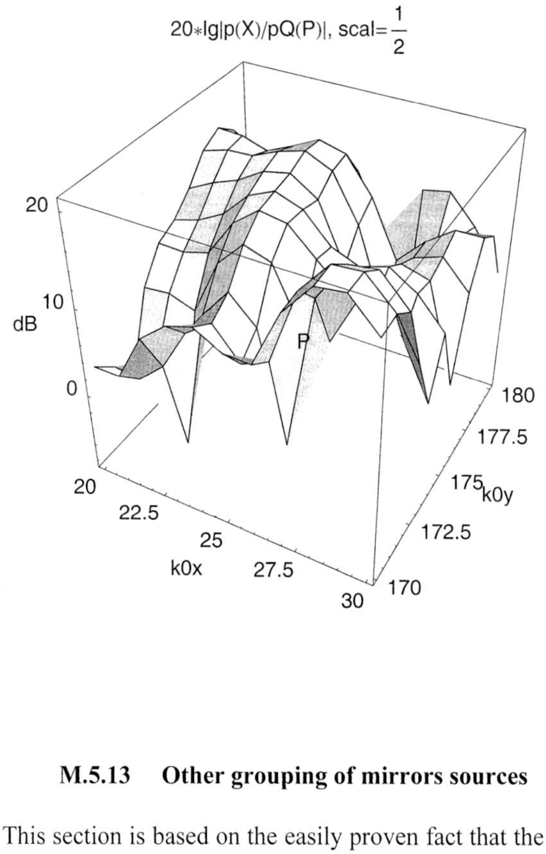

The contributions of the higher order MSs (the orders o=0, 1 are missing) lift the sound pressure level in the room significantly over the level in free space.

M.5.13 Other grouping of mirrors sources

This section is based on the easily proven fact that the original source Q and all mirror sources which are created by it and its daughter sources at a couple of walls are arranged on a circle ("MS circle") which contains Q and has its centre in the foot point Z of Q (normal projection) on the intersection line of the walls. So we are dealing with flanks of a corner; the corner is "real" if the flanks are subsequent walls of the room, or the corner is "virtual" if other walls are arranged between the flanks. The corners (either real or virtual) may be concave or convex, but for the moment we consider only concave corners. A special case form anti-parallel flanks; their (necessarily virtual) corner line is at infinity; the MS circle becomes a straight line through Q normal to the flanks.

Because the plane containing the MS-circle is normal to the flank corner, we are dealing with a 2-dimensional problem (a further advantage of grouping the mirror sources in groups of MSs at wall couples). It is suggesting to discuss the problem in a cylindrical co-ordinate system r, ϑ centred at Z and with the reference for ϑ preferably (but not necessarily) in one of the flanks. The radius of the MS circle will be symbolised with r_q and the angle for Q with ϑ_Q, whereas the angle for a MS will be called ϑ_q. The co-ordinates of the field point P are r, ϑ, ζ, where ζ is the co-ordinate along the corner line, with $\zeta=0$ for Z, Q, q. The transformation between the Cartesian co-ordinates x, y, z of the room and the cylindrical co-ordinates r, ϑ, ζ of a flank couple is described in the Appendix.

The next figure shows a couple of flanks F_1, F_2, the original source Q and its mirror sources on the MS circle.

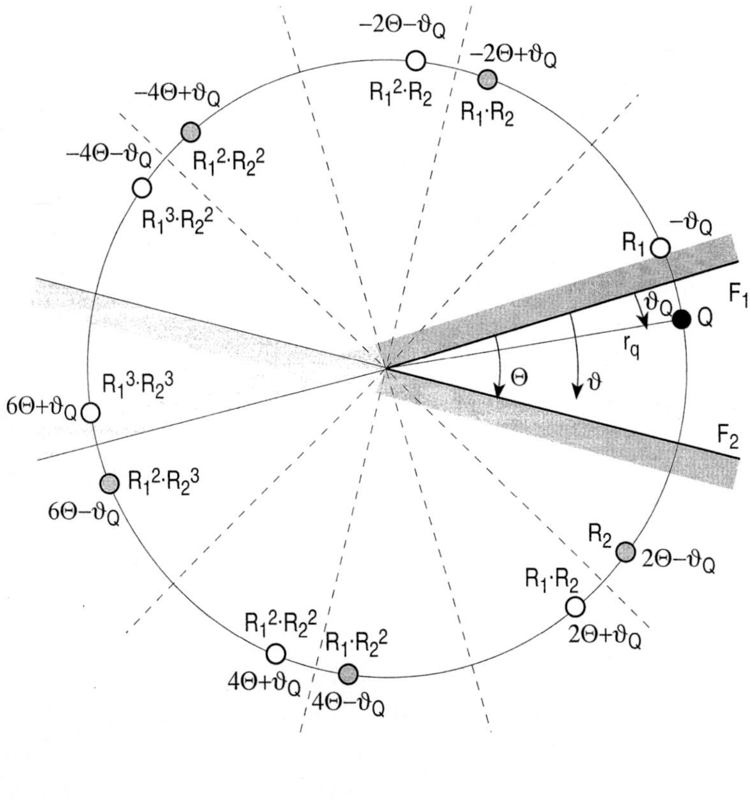

The figure shows all legal MSs (the number is rather high because the wedge angle Θ is small, by intention). The source at $6\Theta-\vartheta_Q$ formally could be mirror-reflected once again, but then the daughter source would coincide with the MS at $6\Theta+\vartheta_Q$ with the same source factor, so by the coincidence criterion this daughter source would be illegal. The indicated source factors as products of R_1, R_2 should not be interpreted too literally; the factors in the powers of these reflection factors may contain different angles θ_s.

The MSs on both circular arcs $\vartheta<0$ and $\vartheta>0$ have source angles ϑ_q within the limits :

$$\Theta-\pi \overset{!}{<} \vartheta_q = -2s\cdot\Theta \pm \vartheta_Q \overset{!}{<} -\pi \quad ; \quad s=0,\pm 1,\pm 2,\dots \tag{5}$$

The mirror sources lastly were generated at F_1 for $s<0$; they would produce a daughter source on the lower circular arc with F_2 if that is not excluded by the inside criterion, and vice versa for $s>0$.

The range (12) leads to limits for the counter s :

on the upper arc $\vartheta<0$:
$$0 \ge s' > \frac{1}{2}\left(\frac{\pi \pm \vartheta_Q}{\Theta} - 1\right) \tag{6a}$$

on the lower arc $\vartheta>0$:
$$0 < s < \frac{\pi \mp \vartheta_Q}{2\Theta} \tag{6b}$$

The sum of the counters is restricted to
$$s + |s'| < \frac{\pi}{\Theta} - \frac{1}{2} \tag{7}$$

Thus the number of MSs decreases with increasing Θ. If $\Theta=\pi$, only the MS with $n=0$ is legal. Evidently, anti-parallel wall couples with $\Theta=0$ are a special case, which will be discussed separately below.

One needs, for the implementation of the MS method, the distance between q and P, and the angle θ_s formed by the connection line (q,P) with the normal of the flank.

The distance between q and P is :

$$\mathrm{dist}(q,P) = \sqrt{\zeta^2 + r^2 + r_q^2 - 2r\,r_q\cos(\vartheta-\vartheta_q)} \tag{8}$$

The cosine $\cos(\theta_s)$ is easily evaluated in the Cartesian co-ordinates, but this would need a co-ordinate transformation for all q. It is better to use the cylindrical co-ordinates of P and to evaluate $\cos(\theta_s)$ in that system. This task will be described in the Appendix.

The advantage of the described grouping is evident: one need not find the MSs by trial-and-error, but they are evaluated in a straightforward method, if the flanks have a real corner (where exclusion by interrupt and efficiency checks play no role). Further, one is sure to have covered all MSs for a wall couple. The question is whether one possibly introduces too many MSs if the flanks have a virtual corner. See the next figure for that question.

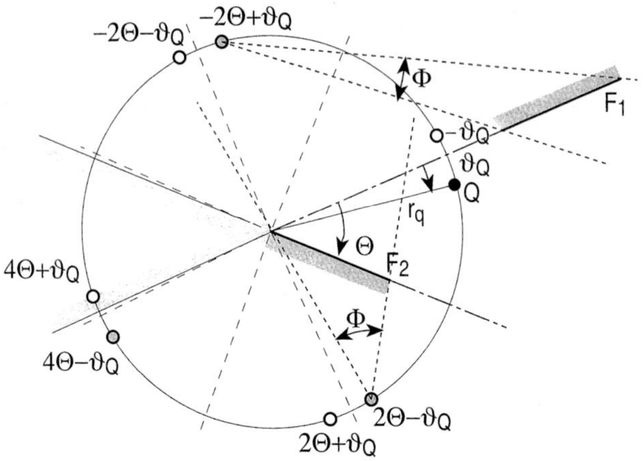

This graph illustrates the case that both flanks F_1, F_2 are on different sides of the MS circle. The MSs lastly generated at the outer flank F_1 never "see" the other flank F_2, and the MSs lastly generated at F_2 may have F_1 in their field angles Φ, or not.

Conclusion: for flanks with a virtual corner the tests "F_i in cone(q, F_j)" must be made as interrupt checks; however, there are situations that are to describe easily in which the number of these tests can be reduced. Similarly, the efficiency checks, which ask whether the projection P' of P into the plane of the MS circle is in the angle Φ, not only is simplified because this check now is a 2D-check, but conditions which make such checks unnecessary can be formulated. The tests "F_i in cone(q, F_j)" remain three-dimensional if the flanking walls have a virtual corner and no intersection with the MS circle.

The set of mirror sources so constructed (we call it "corner set") is "complete"; the sum of their contributions to $p(P)$ is a precise field description (with the principal limitation of precision of the MS method), if only the flanking walls of the couple exist. We call their sound field the "corner field". If P lies in between the flanks, it is plausible that this field represents the most important contributions of wall reflections at P. Further contributions come from other wall couples and their corner sets if P is in the field angle Ψ of those couples (see below), and by reflections of the corner set of a wall couple at other walls within Ψ. The combination of corner sets to the ensemble of mirror sources of a room ("room set"), or the completion of corner fields to the room field, will be discussed in the next section.

M.5.14 Combination of corner fields to the room field

The room may have N walls W . Find solutions for corners with couples W_i, W_j ; i,j=1,2,…,N ; i≠j ; that is N·(N–1)/2 combinations (the combination W_i,W_j is equivalent with the combination W_j,W_i).

For the preparation of the next idea we take the most simple examples of two-dimensional triangular and rectangular rooms. If the room had a 3D tetrahedral shape, for example, the main difference would be the inclination of the MS circles relative to each other (it is simpler to compute the situation in a 3D room than to present it in a graph). Below, MSs from the traditional MS method are drawn with interrupts according to the inside criterion. This will be sufficient for the ensuing argumentation.

The MSs created at a wall couple are collected on circles. In the next figure for the triangular room three of the MSs appear twice (they are marked with a grey fill); they are the MS of first order. In the subsequent figure for the rectangular room the vertical and horizontal lines through Q represent the limit cases of MS circles for the two anti-parallel wall couples. In that figure the four MSs of first order appear three times. Long arrows in the figures indicate where the MSs come from. Short arrows indicate at which wall the MSs shown should be mirror-reflected in the next step of MS production. These arrows point to opposite walls.

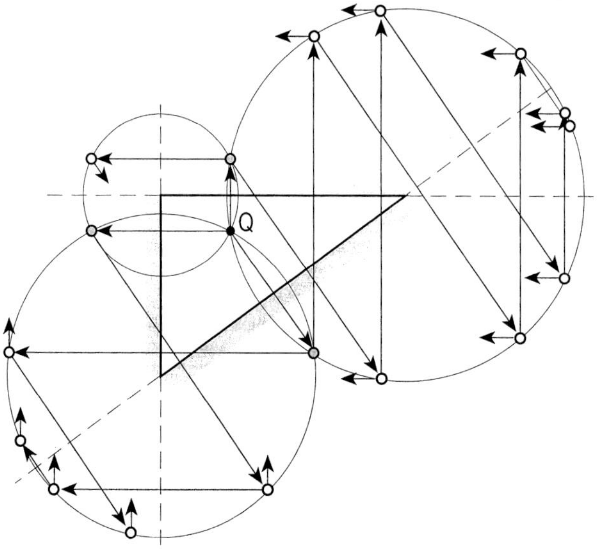

The original source Q appears three times in the corner sets, the MS of first order appear twice.

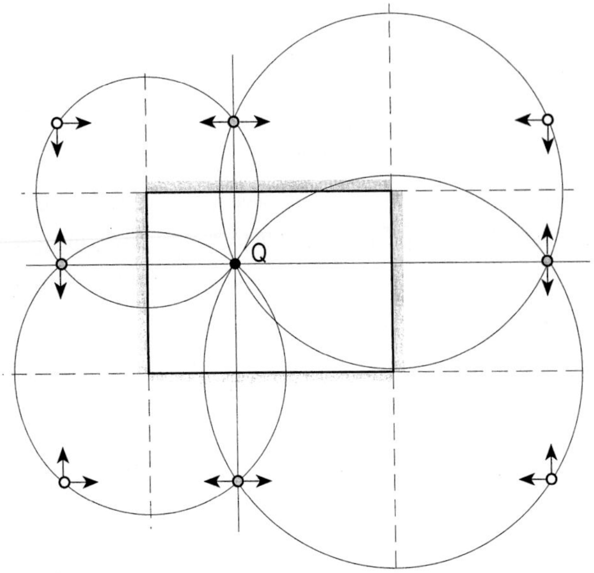

The original source Q appears four times in the corner sets, the MSs of first order appear three times.

The facts that the MSs of a wall couple are placed on a circle around the wall corner with Q on it, and that the continuation of MS production would imply opposite walls, suggest to collect the MSs of a wall couple into one equivalent source for the corner field. The following rules are evident from the above discussion:

- exclude the original source from the field of that source;
- exclude the MSs of first order from that source;
- define the field angle Ψ of the source.

We shall see in the next section that the new source is placed in the foot point Z of Q on the corner line. The field angle Ψ therefore is the angle defined by the two flanks and the corner line as apex line.

M.5.15 Collection of the MSs of a wall couple in a corner source

Up to now the graphs in the previous section indicate nothing more than an involved, but legitimate, procedure of MS production.

Now we take advantage of the fact that the MSs of a wall couple are arranged on a circle around the intersection line (normal to that line) which also contains Q . The intersection line between the walls must not really exist (see the rectangular room). But first we exclude the special case of parallel walls (it is specially treated below). Let the radius of the MS circle be designated as r_q .

The field of a MS at P is described by :

$$p_q(P) = \Pi R \cdot h_0^{(2)}\left(k_0 d_{qP}\right) = \frac{\Pi R}{k_0 d_{qP}} \cdot \left[\sin\left(k_0 d_{qP}\right) + j\cos\left(k_0 d_{qP}\right)\right]$$
$$= j\,\Pi R \frac{e^{-jk_0 d_{qP}}}{k_0 d_{qP}} \tag{9}$$

with d_{qP}=dist(q, P) . The addition theorem for spherical Bessel functions, when applied to the above expression, leads to :

$$h_0^{(2)}(k_0 d_{qP}) = \sum_{n\geq 0} (2n+1) \cdot P_n\left(\frac{r}{r_p}\cos(\vartheta_q - \vartheta)\right) \cdot \begin{cases} j_n(k_0 r_p) \cdot h_n^{(2)}(k_0 r_q) & ; \quad r_p < r_q \\ j_n(k_0 r_q) \cdot h_n^{(2)}(k_0 r_p) & ; \quad r_p > r_q \end{cases} \tag{10}$$

where $j_n(x)$ are spherical Bessel functions of order n , $h_n^{(2)}(x)$ are spherical Hankel functions of the 2nd kind, $P_n(x)$ are Legendre polynomials, and the geometrical quantities are taken from the next figure.

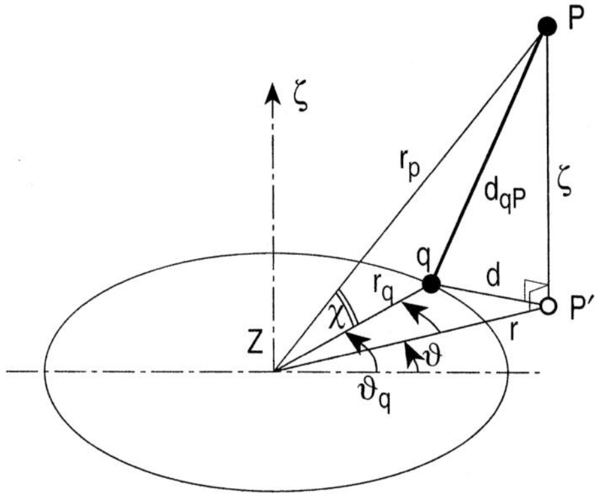

The circle is the MS circle; q is a source on it with the cylindrical co-ordinates r_q, ϑ_q, ζ=0 ; P is the field point with the cylindrical co-ordinates r, ϑ, ζ; P′ is the projection of P on the plane of the MS circle.

The following relations exist between the geometrical quantities:

$$d_{qP}^2 = r_q^2 + r_p^2 - 2r_q r_p \cos\chi$$
$$r_p^2 = r^2 + \zeta^2 \tag{11}$$

$$d_{qP}^2 = d^2 + \zeta^2$$
$$d^2 = r_q^2 + r^2 - 2r_q r\cos(\vartheta_q - \vartheta) \tag{12}$$

from which it follows that :

$$\cos\chi = \frac{r}{r_p}\cos(\vartheta_q - \vartheta) \tag{13}$$

The first line in eq.(11) was used for the addition theorem.

Summation over the sources on the MS circle gives for the field contribution at P of that sources :

$$p(P) = j\sum_s \Pi_s R(\theta_s) \sum_{n\geq 0} (2n+1)\cdot P_n\left(\frac{r}{r_p}\cos(\vartheta_{qs} - \vartheta)\right)\cdot\begin{cases} j_n(k_0 r_p)\cdot h_n^{(2)}(k_0 r_q) & ; \quad r_p < r_q \\ j_n(k_0 r_q)\cdot h_n^{(2)}(k_0 r_p) & ; \quad r_p > r_q \end{cases} \tag{14}$$

The summation index s can be taken from eq.(5), if all sources are added. As explained above, it is recommended to avoid the summation over Q and the MSs of first order (their field contributions will be added to the field in their traditional forms); and illegal or inefficient sources will also be left out of the summation over s . Equation (14) represents an important group of mirror sources in an explicit formula. The terms represent radially standing waves in r_p if $r_p<r_q$, and outward propagating waves if $r_p>r_q$. The sound field is steady at $r_p=r_q$. The sum satisfies the boundary conditions at the flanks, Sommerfeld's far field condition, the source condition and the edge condition (which requires that the volume flow through a small cylinder around a corner or a sphere around an edge does not exceed the volume flow through a similar cylinder or sphere, respectively, around the original source Q; the edge condition mostly is used for the selection of permitted radial functions, like Sommerfeld's far field condition). But equ.(14) in general does not satisfy the wave equation, because the factors $\Pi_s(\theta_s)$ in general are neither constant nor do they have the form which is required by the wave equation for angular factors to Bessel functions of the order n . But satisfaction of the wave equation could not be expected with the MS method as basis for equ.(14).

This representation, however, has a numerical problem: the convergence and the precision are critical for $r_p=r_q$, i.e., if the field point lies on the sphere which has the MS circle as equator circle. Physically the numerical problem comes from the fact that the sphere surface contains the poles of the sources q . The evaluation of eq.(14) at $r_p=r_q$ needs a careful check of the summation limit for n (a detailed discussion of the convergence check can be found in [9]). This problem in general does not appear after a mirror-reflection of equ.(14) at an opposite wall (see below). One could avoid it by using at or near $r_p=r_q$ the traditional MS method, but this would need the programming of both methods. An easier method in the case $r_p=r_q$ is the evaluation of eq.(14) on both sides of that limit, in some distance, and then to take the mean value.

The fact that a set of spherical Bessel and Hankel functions with integer orders must be evaluated causes no problem; the set can be obtained from two start values of the order by the known recursions of such functions, and also the Legendre polynomials are easily computed.

The radial arguments are constant for all sources on the MS circle and for a fixed immission point P .

Important conclusions can be drawn from eq.(14). The components of the sum over n are spherical wave terms, which are centred at the centre Z of the MS circle on the corner line. Therefore we say that eq.(14) represents the field of a "corner source". It can be introduced into a continued MS generation like any directional source. The advantages of the corner source are evident. Its position need not be found in a complicated search algorithm; it is explicitly defined by the room geometry and the position of Q . Also its field angle Ψ is immediately given; it is the angle between the flanking walls. The difference to the field angle Φ of a traditional MS is remarkable (Φ is the angle of the cone subtended by the mother wall and the MS in the apex). Efficiency checks (P in Ψ ?) and interrupt checks (an opposite wall in Ψ ?) are much easier to perform.

It is not proven, but plausible, that one can stop the field evaluation after the corner sources of all wall couples have been evaluated and mirror-reflected once at their opposite (visible) walls. All really important field contributions will be obtained with that procedure.

The mirror-reflection of the corner source is done by a simple modification of eq.(14) if one applies the reciprocity of the next section. One must not evaluate the positions of the mirror-reflected sources, but one mirror-reflects P , multiplies the $\Pi R(\theta_s)$ in the sum over s with the new reflection factor, and evaluates equ.(14) with the geometrical parameters of the new position of P (which also can be determined directly from the room input data and the position of P). This procedure avoids the reflection of the directivity of the corner source.

M.5.16 A kind of reciprocity in the MS method

Evidently, a source q , generated by a mother source q_m at a mirror wall W will produce at a receiver point P the same contribution $p(P)$ that the mother source q_m would produce at the point P' , which is the mirror-reflected point to P with respect to W , after multiplication with $R(\theta_s)$.

So one could set-up a "mirror-receiver method" instead of a "mirror-source method" by recasting all interrupt and efficiency rules. Numerically there would be no advantage as compared with the MS method for isotropic sources Q . If, however, the original source Q has a directivity $D(\vartheta,\varphi)$, the mirror-receiver method avoids the mirror-reflection of the directivity.

M.5.17 Limit case of parallel walls

Because only anti-parallel walls will be considered here, we use the abbreviation "parallel" walls. In principle the number of MSs needed for the sound field in the space between parallel walls is infinitely high (on both sides of the walls). Up to now we have only an interrupt criterion if the walls are absorbent due to the reduction of the source factor ΠR with increasing order.

In the limit case of parallel walls in equ.(14) the limits $r, r_p, r_q \to \infty$ and $\vartheta \to 0$ are assumed; the MS circle becomes a straight line normal to the walls..

First we derive with the next figure a plausible interrupt criterion for the MS production with parallel walls by consideration of allowable errors, then we sum up the MSs to a single equivalent source.

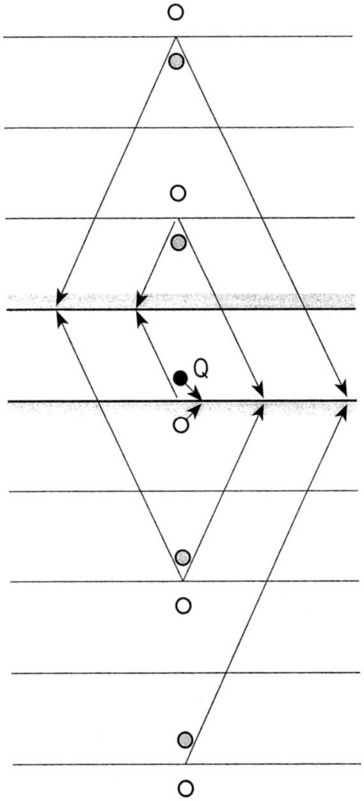

The arrows indicate which couples of sources satisfy the boundary condition at which wall.

If one interrupts one has a couple without "partner" for the boundary condition (this is the lowest couple in the graph).

An interruption makes an error in the boundary condition at a wall. The absolute error is in the order of magnitude of the field contribution of the uncompensated couple. It decreases with increasing order of reflection

- because of the increasing distance of the couple to the wall (geometrical reduction),
- because of the decrease of ΠR with absorbent walls (acoustical reduction).

The relative error, which is important, further decreases with increasing order, because the reference quantity is the sum of contributions of "complete" couples.

From experiences of field evaluations in flat ducts (which is the object at hand) one can conclude that a relative boundary value error $\Delta_{rel} \approx 1/10$ to $1/20$ is tolerable. Neglecting geometrical and acoustical reductions, this leads to a permitted interrupt at about $s_{hi} = 10$ to 20 MSs. This will be sufficient if geometrical reduction with increasing order is taken into account, and if one considers that real walls never are ideally reflecting. A reflection coefficient $|R|=0.9$ produces for an order $o=15$ a source factor of about $0.9^{15}=0.206$.

One can perform an analysis as in section M.5.15 for parallel walls also. Its principal result is a recommendation to write the sound field of the MS for parallel walls as the field of the original source multiplied with a directional factor. But, knowing this goal it is easier to derive that form of the MS field directly. The problem is characterised geometrically by the straight line, normal to the flanks F_1, F_2 , which contains the original source Q and the MS q , and the field point P . We therefore take a right-handed Cartesian co-ordinate system x′, y′, z′ with the y′ axis on the source line, and the x′ axis in one of the flanks so that P is in the x′,y′ plane (it follows from the system x, y, z of the room by a rotation and shift); see next figure.

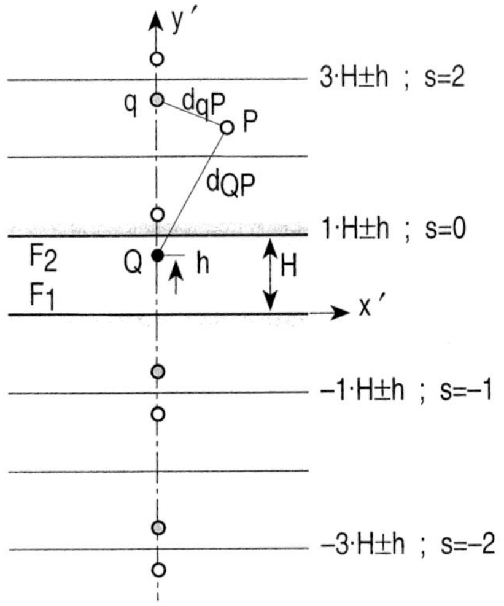

The source positions are given by $x'=0$, $z'=0$ and :

$$x'_q = (2s+1)\cdot H \pm h \quad ; \quad s = 0, \pm 1, \pm 2, \ldots \tag{15a}$$

if the co-ordinate origin is chosen on F_1 as in the figure, otherwise by :

$$x'_q = 2s\cdot H \pm h \quad ; \quad s = 0, \pm 1, \pm 2, \ldots \tag{15b}$$

The field contribution $p_q(P)$ at P of a source q can be written in terms of the contribution $p_Q(P)$ of Q as :

$$p_q(P) = \Pi_q R(\theta_s)\cdot \frac{k_0 d_{QP}}{k_0 d_{qP}} e^{-jk_0(d_{qP}-d_{QP})} \cdot p_Q(P) \tag{16}$$

with :

$$d_{qP} = \sqrt{(x'_P - x'_q)^2 + (y'_P - y'_q)^2} \quad ; \quad d_{QP} = \sqrt{(x'_P - x'_Q)^2 + (y'_P - y'_Q)^2} \tag{17}$$

The angles θ_s for the reflection factors are easily obtained in the x',y' co-ordinates. The equivalent source and its contribution representing all mirror sources are given by :

$$p(P) = p_Q(P) \sum_s \Pi_q R(\theta_s) \cdot \frac{k_0 d_{QP}}{k_0 d_{qP}} e^{-jk_0(d_{qP} - d_{QP})} \tag{18}$$

For the combination with corner sources of non-parallel wall couples it is again recommended to leave the original source and the MSs of first order out of the summation over s . For the mirror-reflection of this source at an opposite wall use the reciprocity, i.e. determine the position of the reflected field point P' in the co-ordinates x', y', z' .

If the parallel flanks F_1, F_2 are not directly opposite to each other, but with a parallel offset, interrupt checks for mirror sources are preferably performed in the co-ordinates x', y', z' (skip s for illegal sources in equ.(18)), as well as efficiency checks (skip s for inefficient sources). The total source equ.(18) is ineffective if P is not in the space between the planes of the flanks F_1, F_2 , and the mirror-reflected combined source is ineffective if P' is not in that space. So indeed one has to form two sums as in equ.(18), one for the "legal" equivalent source, which is used in its mirror-reflection, and one for the final evaluation of the field contribution. But the decision if equ.(18) must be evaluated at all can be made before any computation.

One still has the problem with convex corners and their scattered field. This problem can be solved rather easily, within the frame of a MS method, by combination of the MS method with the "second principle of superposition" (PSP) . See Section B.10 for that principle. It will be briefly repeated below, because it will take special forms in combination with the MS method.

M.5.18 The 2nd principle of superposition (PSP)

It should be stated in advance:

- the PSP , when applied to single concave corners, does not result in significant saving of computation, as compared to the traditional MS method;
- unless it is globally applied to symmetrical rooms (see below);
- but its application is necessary with convex corners;
- because it is applicable to both convex and concave corners, and some of its features are more easily explained with concave corners, these are treated here also.

The objects of the 2nd principle of superposition (PSP) are an arbitrary (also multi-modal) source Q and a scattering object which has a plane of symmetry M . When the PSP is applied to room edges, the flanking walls are supposed to extend from their line of intersection to infinity. If they are absorbent, the absorption should be the same at both flanks (symmetrical flanks; see below for unsymmetrical flanks). It is not necessary that the flanking walls of a room have a real line of intersection; they may be couples of walls, with other walls forming the connection on the apex side.

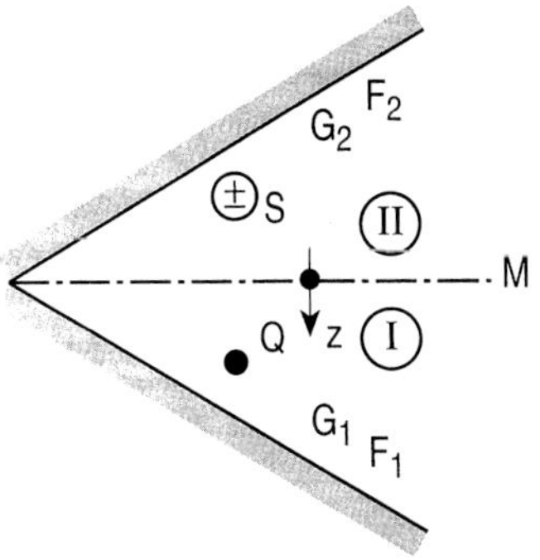

We assume equal wall admittance values G_i at both flanks F_i . We further assume a co-ordinate system with a co-ordinate z (which favourably is an azimuthal co-ordinate) normal to the median plane M . This assumption is not necessary; it just simplifies the description.

The median plane M subdivides the wedge into two halves:

- (I) on the source side of M ,
- (II) on the back side of M (as seen from Q).

With the above choice of z , mirror-reflected points in both halves are distinguished just by ±z .

The PSP composes the solution of the original scattering task by the solutions of two sub-tasks $\sigma = h, w$ of scattering in the zone (I) :

- in the 1st sub-task, σ=h, the plane of symmetry M is hard;
- in the 2nd sub-task, σ=w, the plane of symmetry M is soft.

The field of each sub-task is composed of the source field p_Q and a scattered field $p_s^{(\sigma)}$: $p^{(\sigma)} = p_Q + p_s^{(\sigma)}$. The fields of the original task in both zones (I), (II) then are :

$$\begin{aligned} p_I(x,z) &= p_Q(x,z) + \tfrac{1}{2}\left(p_s^{(h)}(x,z) + p_s^{(w)}(x,z)\right) \\ p_{II}(x,z) &= \tfrac{1}{2}\left(p_s^{(h)}(x,-z) - p_s^{(w)}(x,-z)\right) \end{aligned} \qquad (19a)$$

(x represents the co-ordinates other than z). One can also decompose the field as $p^{(\sigma)} = p_Q + p_r + p_s^{(\sigma)}$, where p_r is the same in both sub-tasks; then we have :

$$\begin{aligned} p_I(x,z) &= p_Q(x,z) + p_r(x,z) + \tfrac{1}{2}\left(p_s^{(h)}(x,z) + p_s^{(w)}(x,z)\right) \\ p_{II}(x,z) &= \tfrac{1}{2}\left(p_s^{(h)}(x,-z) - p_s^{(w)}(x,-z)\right) \end{aligned} \qquad (19b)$$

The scattered fields here are generally different from those of equs.(19a), but in our problem p_r is just a member of the scattered field terms (see below) so that these remain unchanged.

In summary: the PSP solves the scattering task on the source side (I) for the sub-tasks σ=h,w and computes with them the field on the back side (II) (just by mirror-reflection at M). One should keep in mind this "detour" of the field evaluation in (II) via the zone (I) .

The *derivation* of the PSP uses, in addition to the source Q(x,z), sources Q(x,–z) which are mirror-reflected at M ; in the 1st sub-task the MS has the same amplitude as Q; in the 2nd sub-task it is multiplied by –1 . The eqs.(19a) just describe the superposition of both sub-tasks with Q and ±MS as sources. The following features should be observed:

- The simplicity of the derivation shows the general validity of the PSP (under the mentioned condition of symmetry of the object).
- The PSP is an exact description if the scattered fields of the sub-tasks can be determined exactly.
- The PSP is suitable for combination with the MS method !
- If, in the course of the PSP, mirror sources are created at M (to satisfy the boundary conditions there), one should remember that mirror sources at ideally reflecting planes give an exact description of the field. With absorbing flanks, the errors of the MS method remain.
- The fields of the mirror sources S_i form the "scattered fields".
  If S_i is created by mirror-reflection at F_1 on the source side, the sign of the MS is the same in both sub-tasks; this will be indicated in the sketches below with (+) .
  If S_i is created by mirror-reflection at M , the sign of S_i is different in both sub-tasks; this will be marked in the sketches with (±) , i.e., (+) for σ=h , (–) for σ=w. If a MS which was created at M (i.e. marked with (±)) afterwards is reflected at F_1 , the double sign (±) remains.

The resulting fields of the PSP are:

$$\begin{aligned} p_I(x,z) &= Q+\tfrac{1}{2}\left(\sum_i S_i^{(h)}(x,z)+\sum_i S_i^{(w)}(x,z)\right) \\ p_{II}(x,z) &= \tfrac{1}{2}\left(\sum_i S_i^{(h)}(x,-z)-\sum_i S_i^{(w)}(x,-z)\right) \end{aligned} \qquad (20a)$$

The change h → w of the median plane M does not influence the position and number of the MSs; therefore the sums have the same counting and summation index i .

If the first MS is created at F_1 (on the source side), it has in both sub-tasks the same sign (it will be indicated by S_0). In $p_I(x,z)$ it can be pulled to outside the parentheses and needs no superscript (h) or (w). The remaining MSs under the sums (with superscripts) have undergone at least one mirror-reflection at M . So one can write :

$$p_I(x,z) = Q + S_0(x,z) + \tfrac{1}{2}\left(\sum_i S_i^{(h)}(x,z) + \sum_i S_i^{(w)}(x,z)\right)$$
$$p_{II}(x,z) = \tfrac{1}{2}\left(\sum_i S_i^{(h)}(x,-z) - \sum_i S_i^{(w)}(x,-z)\right) \tag{20b}$$

One must distinguish with the production of MSs in the PSP:

1.) "mirror reflected at..."

(a) the wall F_1 ; then the sign is the same as for the mother source; if F_1 is absorbent, a reflection factor R arises;

(b) at M , then the sign in the sub-task h is that of the mother source, in the sub-task w the sign is changed; i.e., the new factor in the source factor ΠR is $R - \pm 1$.;

2.) two sub-tasks

(a) σ=h : no sign change at mirror reflections;

(b) σ=w: sign change for mirror reflections at M (but not at F_1);

3) two "paths" of MS production:

(a) 1st path: begins with mirror reflection at F_1 ;
MSs on this path will be designated with even indices i=0,2,4,... ;

(b) 2nd path: begins with mirror reflection at M ;
MSs on this path will have odd indices i=1,3,5,...

Whereas the sub-tasks h,w do not change the position of a MS of some order, the positions of the MSs of both paths are generally different from each other.

As a detailed example we take the concave rectangular corner of the next figure.

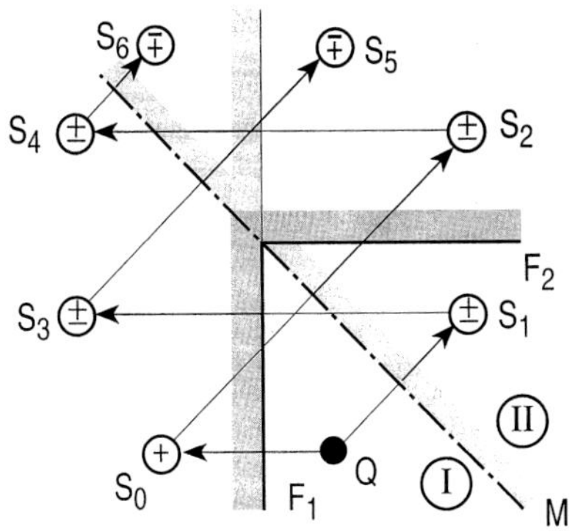

Here the MS production is continued (irrespective of the inside criterion) until the MSs begin to fall on positions of previously created MSs. The paths from then on would begin to be continued backwards (with multiple covering of the positions).

The following schemes represent the chains with multiple reflections at F_1 and M for both paths. Mirror reflection at F_1 is marked by a simple arrow $\rightarrow$ and reflection at M by a double arrow $\Rightarrow$. The cases σ=h are arranged in the upper line, and σ=w in the lower line. MSs in vertical columns of the scheme have equal positions. The flanking wall F_1 is initially assumed to be rigid.

Below the symbols of the MSs in the scheme are written the sums appearing in the PSP (for one path; recall that MSs in a column of the scheme have equal positions) :

$$\sum S = \sum_{i\geq 0} S_i^{(h)} + S_i^{(w)} \quad \text{in zone (I)}$$

$$\Delta S = \sum_{i\geq 0} S_i^{(h)} - S_i^{(w)} \quad \text{in zone (II).}$$

1st path:

$$Q \overset{h}{\underset{w}{\rightarrow}} S_0 \begin{array}{l} \overset{h}{\nearrow} +S_2 \rightarrow +S_4 \Rightarrow +S_6 \rightarrow +S_8 \Rightarrow +S_{10} \rightarrow +S_{12} \\ \underset{w}{\searrow} -S_2 \rightarrow -S_4 \Rightarrow +S_6 \rightarrow +S_8 \Rightarrow -S_{10} \rightarrow -S_{12} \end{array}$$

$$\sum S = \sum_{i\geq 0} S_i^{(h)} + S_i^{(w)} = \quad 2S_0 \quad 0 \quad 0 \quad 2S_6 \quad 2S_8 \quad 0 \quad 0$$

$$\Delta S = \sum_{i\geq 0} S_i^{(h)} - S_i^{(w)} = 0 \quad 2S_2 \quad 2S_4 \quad 0 \quad 0 \quad 2S_{10} \quad 2S_{12} \tag{21a}$$

2nd path:

$$Q \begin{array}{l} \overset{h}{\nearrow} +S_1 \rightarrow +S_3 \Rightarrow +S_5 \rightarrow +S_7 \Rightarrow +S_9 \rightarrow +S_{11} \\ \underset{w}{\searrow} -S_1 \rightarrow -S_3 \Rightarrow +S_5 \rightarrow +S_7 \Rightarrow -S_9 \rightarrow -S_{11} \end{array}$$

$$\sum S = \sum_{i\geq 0} S_i^{(h)} + S_i^{(w)} = \quad 0 \quad 0 \quad 2S_5 \quad 2S_7 \quad 0 \quad 0$$

$$\Delta S = \sum_{i\geq 0} S_i^{(h)} - S_i^{(w)} = \quad 2S_1 \quad 2S_3 \quad 0 \quad 0 \quad 2S_9 \quad 2S_{11} \tag{21b}$$

If one sums up the MSs of both paths, the PSP gives :

$$\begin{aligned} p_I(x,z) &= Q(x,z) + S_0(x,z) + \sum_{n=1,3,5\ldots} \sum_{k=1}^{4} S_{4n+k}(x,z) \\ p_{II}(x,z) &= \sum_{n=0,2,4\ldots} \sum_{k=1}^{4} S_{4n+k}(x,-z) \end{aligned} \tag{22}$$

Now we complete the scheme for the case of (symmetrical) absorption at the walls. The reflection factors R_i in an order of MS are not changed by $h \rightarrow w$, except for the sign.

1st path:

$$Q \underset{w}{\overset{h}{\rightarrow}} R_0 S_0 \begin{array}{l} \overset{h}{\nearrow} +R_0 S_2 \rightarrow +R_0 R_4 S_4 \Rightarrow +R_0 R_4 S_6 \rightarrow +R_0 R_4 R_8 S_8 \Rightarrow \cdots \\ \underset{w}{\searrow} -R_0 S_2 \rightarrow -R_0 R_4 S_4 \Rightarrow +R_0 R_4 S_6 \rightarrow +R_0 R_4 R_8 S_8 \Rightarrow \cdots \end{array}$$

$$\sum S = \sum_{i \ge 0} S_i^{(h)} + S_i^{(w)} = 2R_0 S_0 \qquad 0 \qquad 0 \qquad 2R_0 R_4 S_6 \qquad 2R_0 R_4 R_8 S_8$$

$$\Delta S = \sum_{i \ge 0} S_i^{(h)} - S_i^{(w)} = 0 \qquad 2R_0 S_2 \qquad 2R_0 R_4 S_4 \qquad 0 \qquad 0$$

(23a)

2nd path:

$$Q \begin{array}{l} \overset{h}{\nearrow} +S_1 \rightarrow +R_3 S_3 \Rightarrow +R_3 S_5 \rightarrow +R_3 R_7 S_7 \Rightarrow +R_3 R_7 S_9 \rightarrow \cdots \\ \underset{w}{\searrow} -S_1 \rightarrow -R_3 S_3 \Rightarrow +R_3 S_5 \rightarrow +R_3 R_7 S_7 \Rightarrow -R_3 R_7 S_9 \rightarrow \cdots \end{array}$$

$$\sum S = \sum_{i \ge 0} S_i^{(h)} + S_i^{(w)} = 0 \qquad 0 \qquad 2R_3 S_5 \qquad 2R_3 R_7 S_7 \qquad 0$$

$$\Delta S = \sum_{i \ge 0} S_i^{(h)} - S_i^{(w)} = 2S_1 \qquad 2R_3 S_3 \qquad 0 \qquad 0 \qquad 2R_3 R_7 S_9$$

(23b)

The most important advantage of the combination MS and PSP is the fact that with it convex corners become tractable; the traditional MS method fails there completely. Since the wedge angle of a couple of walls always is $\Theta \le 2\pi$, the angle between F and M is always $\le \pi$; the zone (I) in which the scattered field has to be determined is concave; the MS method can be applied there.

In concave corners the application of the PSP possibly produces a higher precision because the MSs have new positions. But this should be checked.

The following sketches mainly assume hard flanks (for simplicity reasons); absorbing flanks are specially mentioned.

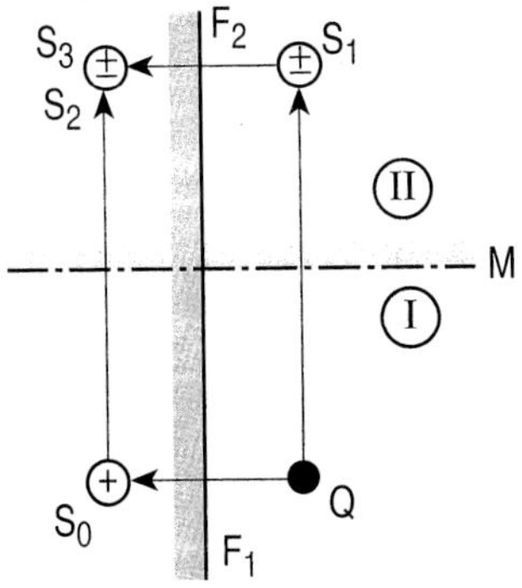

This is the simple case of a source above a plane wall.

The MSs are shown until interruption by the inside criterion. There is coincidence of S_2, S_3 (with same source factor R for absorbent F_1, F_2).

It is an important question whether such coinciding MS have to be counted a multiple of times or if the MS production is interrupted when coincidence begins.

S_1, S_2, S_3 compensate each other in the sum ΣS of $p_{(I)}$. The field in (I) is built up by Q, S_0, as expected. In the difference ΔS of $p_{(II)}$ the contributions of S_1, S_2, S_3 would sum up, if the MSs were used as drawn. As a consequence, the field would be unsteady at M.

Consequently, coincidence of MS with same source factor ΠR should be avoided ! Taking this interrupt criterion into account, the field is correctly given by the MS and PSP method. This method thus has a further interrupt criterion, as compared with the traditional MS method. This example also illustrates well the procedure in the MS and PSP method for the evaluation in zone (II) : One first evaluates with the significant MSs the scattered field in zone (I), and then mirror-reflects that field at M into zone (II).

An other instructive example is the concave, rectangular edge.

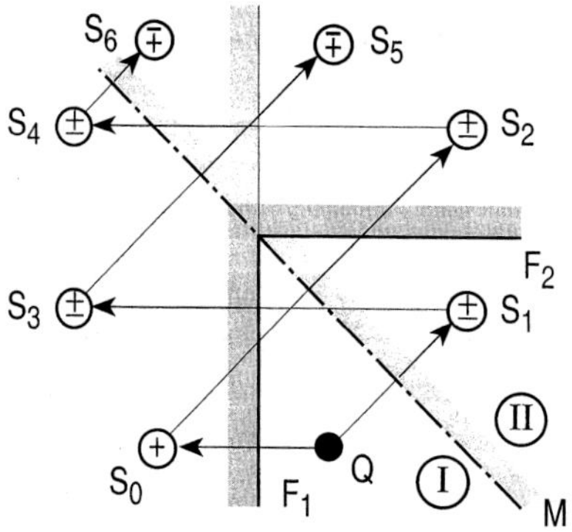

The MSs are drawn as far as it is permitted by the inside and coincidence criteria (a further reflection of S_5 at F_1, which would be permitted by the inside criterion, would produce coincidence with S_6)

According to equ.(22), the field $p_I(x,z)$ is created by the superposition of the fields of Q, S_0, S_5, S_6, and the field $p_{II}(x,-z)$ by the mirror sources S_1, S_2, S_3, S_4. The boundary conditions at the flanks F_1, F_2 are evidently satisfied. Because both groups of MS can be transformed into each other by a rotation by $\pi/4$ and a mirror-reflection, the field is also steady at M ; thus it is a solution of the task.

A special case which can easily be understood is obtained if the source Q approaches the flank F_1. With a hard flank F_1 again the case of a source above a hard wall F_2 is achieved.

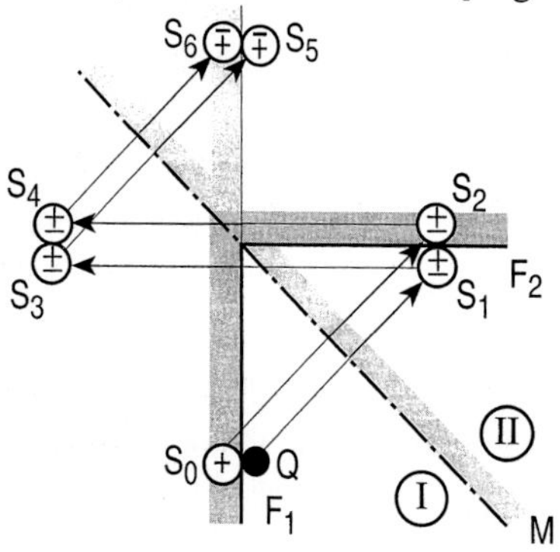

The effective source is a double source $Q+S_0$.

As expected, the field is symmetrical relative to F_1 and F_2. It is also steady at M.

The field is completely and precisely represented.

In a further limiting case Q approaches the median plane M ; the field again is correctly represented as can be seen from the sketch below.

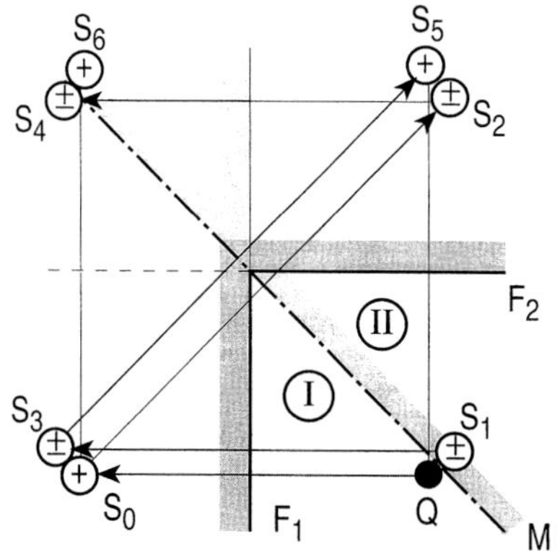

The MSs are again drawn according to the inside and coincidence criteria.

The source Q and the MSs marked with (+) determine the field in zone (I) ; the MSs marked with (±) compose (first in zone (I)) to the field in zone (II).

The boundary conditions at F_1, F_2 and condition of steadiness at M are satisfied.

The examples presented again illustrate the importance of the determination of p_{II} via the detour over zone (I) . The examples contain MSs with (±) in (II) . Without the detour, this would mean poles of the field in zone (II) . That must be excluded according to the condition of regularity of the scattered field.

M.5.19 The PSP for unsymmetrical absorption

The condition of symmetrical absorption for the application of the PSP to sound fields between couples of walls is a sensible restriction in applications. It will be attempted, therefore, to resolve that restriction, if necessary by an approximation, which, however, should not introduce errors exceeding the errors of the traditional MS method with absorbing walls.

If the flanks are symmetrical, the field in the sub-task σ=h is symmetrical with respect to M (i.e. M is hard); in the sub-task σ=w the field is anti-symmetrical (i.e. M is soft). Symmetry and anti-symmetry of the fields are solely determined by the source Q and the auxiliary sources of the PSP . If the walls are different, an anti-symmetrical part of the field will arise due to the asymmetry of the walls.

It is always possible to compose an unsymmetrical field p(x,z) from two *geometrically* equal halves (I) and (II) of a wedge with symmetrical and anti-symmetrical field components :

$$p(x,z) = p^{(sy)}(x,z) + p^{(as)}(x,z)$$
$$p^{(sy)}(x,-z) = p^{(sy)}(x,z) \quad ; \quad p^{(as)}(x,-z) = -p^{(as)}(x,z) \tag{24}$$

From this it follows that :

$$p^{(sy)}(x,z)=\tfrac{1}{2}\big(p(x,z)+p(x,-z)\big)$$
$$p^{(as)}(x,z)=\tfrac{1}{2}\big(p(x,z)-p(x,-z)\big) \qquad (25)$$

M by definition is hard for the part $p^{(sy)}$ and soft for the part $p^{(as)}$. Therefore these parts have a common characteristic with $p_s^{(h)}, p_s^{(w)}$ of the PSP with symmetrical flanks.

Further, the field p(x,z) shall satisfy the boundary conditions at the flanks F_1, F_2 :

$$p(F_1)=p^{(sy)}(F_1)+p^{(as)}(F_1)\stackrel{!}{=}v_\perp(F_1)/G_1$$
$$p(F_2)=p^{(sy)}(F_1)-p^{(as)}(F_1)\stackrel{!}{=}v_\perp(F_2)/G_2 \qquad (26)$$

In these boundary conditions the parts $p^{(sy)}, p^{(as)}$ appear in the same combination as $p_s^{(h)}, p_s^{(w)}$ in the PSP. Thus, from formal considerations one arrives at an approximation for the PSP when applied to walls with different absorption :

Complete the PSP for acoustically different walls as follows :

- Apply in the PSP for the evaluation of the field $p_{(I)}$ in zone (I) the sum

$$p_{(I)}(x,z)=Q+\tfrac{1}{2}\left(\sum_i p_{s,i}^{(h)}(x,z)+\sum_i p_{s,i}^{(w)}(x,z)\right)$$

using for the reflection factors the admittance of the flank F_1.

- Apply in PSP for the evaluation of the field $p_{(II)}$ in zone (II) the difference

$$p_{(II)}(x,z)=\tfrac{1}{2}\left(\sum_i p_{s,i}^{(h)}(x,-z)-\sum_i p_{s,i}^{(w)}(x,-z)\right)$$

using for the reflection factors the admittance of the flank F_2.

This rule creates a field which
- satisfies the wave equation in the approximation of the MS method for absorbent walls,
- satisfies the source condition,
- satisfies the boundary conditions at F_1, F_2 ,
- but is not steady at M for unsymmetrical absorption.

It is possible to derive a better approximation which is steady also at M , but it is more complicated to handle and therefore will not be presented here. The MS method also has similar errors.

M.5.20 A global application of the PSP

Most auditoria have a constructional plane of symmetry M . So the room as a whole satisfies the condition for the application of the PSP.

That means: one solves the task of field evaluation twice in the half of the room containing the source Q ; once M is assumed to be hard, and once M is assumed to be soft.

The computational advantage can be easily quantified in 2D (similar relations hold in 3D). The room is supposed to have N walls. In the sub-tasks of the PSP will appear the following numbers of walls (M included) :

- N/2+2 walls, if M ends on both sides on walls,
- (N-1)/2+2 walls, if M ends on one side with a wall, and on the other side in a corner
- N/2 walls, if M ends on both sidcs in corners.

With this remark ends that part which is concerned with the evaluation of the stationary sound field in rooms. It should be noted that the methods described yield complex sound pressures in P . The computations will be somewhat simplified if one is satisfied with the magnitudes $|p_q|$ of the contributions of the effective sources (MS and corner sources). This is mostly done in room acoustical papers, although it is impossible to conclude from $|p_q|$ to $|p(P)|$ (the magnitude of a sum mostly is different from the sum of magnitudes...). We shall be confronted with that difference in the next section which deals with the determination of the room reverberation using the results of the field evaluation. Although reverberation is a non-stationary process, it should be possible to evaluate the most important room acoustical qualifier, the reverberation time, from the results of a computational field model. In doing that, one will be confronted with some lack of definition of reverberation in usual descriptions.

M.5.21 Reverberation time with results of the MS method

The method described above delivers the stationary, monochromatic field of a stationary, harmonic source in a room. It returns the complex sound pressure $p(P)$ at a point P, i.e. with magnitude and phase.

The most important room acoustical qualifier is the reverberation time; it is described in the literature (more or less) by:

> "The reverberation time is the elapsed time for a decay of the sound pressure (or sound pressure level) of 60 dB after termination of a stationary sound excitation."

It is tacitly understood that "sound pressure" means the magnitude of the sound pressure, and in most experimental determinations of the reverberation time band noise is used with an average over the band-width. It is not mentioned (but it is important as we shall see below) that the rectification of the received signal (for the magnitude and band average) implies averaging over time intervals.

The reverberation process is non-stationary; our field evaluation is for stationary fields. Therefore one needs a "switch-off model" for the evaluations. It should be recalled in that context that the sound field in the room is created in the MS-method (as well as in the modified MS and PSP method) by equivalent sources, which means that, after the equivalent sources have been installed in the right places and with the right amplitudes, the walls of the room can be taken away. The equivalent sources radiate into the free space.

One can imagine the distributed sources as a network of loudspeakers driven by the same signal generator. The network lines contain attenuators which model the source factors ΠR. If the signal generator is switched off, all loudspeakers are switched off instantly, but the sound waves radiated before switch-off still propagate. This model has the advantage that the boundary conditions at the walls are satisfied everytime, because as long as a sound wave from a source hits a wall, the sound wave from its daughter source with that wall will be present also.

The end of the contribution of a source q arrives at P at a time t after switch-off

$$t=\frac{\mathrm{dist}(q,P)}{c_0}=\frac{k_0\cdot\mathrm{dist}(q,P)}{\omega}\quad;\quad \omega=2\pi/T_p$$

$$\frac{t}{T_p}=\frac{k_0\cdot\mathrm{dist}(q,P)}{2\pi} \qquad (27)$$

Therein T_p is the time period of the sound wave. It is reasonable to measure t in units of periods T_p.

Imagine all evaluated MSs (and their field contributions $p_s(P)$) sorted with increasing $k_0 \cdot dist(q,P)$ and indexed in this order with a number s ($s=0$ may represent the original source Q). After some elapsed time t/T_p , those contributions that are still travelling from their source to P will be summed at P. The decay curve $L(P)$ at P expressed as sound pressure level is therefore :

$$L\left(t/T_p\right) = 10 \cdot \lg \left| \sum_{k_0 \cdot dist(q(s),P) \geq 2\pi \cdot t/T_p} p_s(P) \right|^2 \qquad (28)$$

With increasing t/T_p the summation is performed over smaller and smaller remainders of the set of effective MS. This evaluation will therefore produce a steeper slope of $L(t/T_p)$ at the end of a time interval of observation when this end approaches fewer and fewer remaining MS. This increase in slope must not be confused with the slope produced by the decreasing amplitudes of MS with increasing distance (due to geometrical and/or acoustical reduction), but is a consequence of the finite size of the set of MS .

Equ.(28) is a direct transcription of the reverberation process defined above verbally. Formation of the magnitude and square is applied to the sum of contributions. Instead of proceeding on the t/T_p axis in unit steps of s , one can proceed in steps $\Delta t/T_p$. Contributions within the interval $\Delta t/T_p$ are summed up (linearly!). Below we shall see that the decay curves thus evaluated have only little similarity to an expected reverberation curve. A clearer definition describes the reverberation as the "decay of the average energy density". The (effective) energy density implies the square of the sound pressure; averaging is performed over directions of the sound intensity and time intervals which are short compared with the reverberation time. Because tacitly the contributions of different sources are also assumed to be incoherent, the contributions of sources to the energy density can be added and they are proportional to the magnitude squared of their contributions $p_q(P)$ to the sound pressure $p(P)$:

$$E_q(P) = \frac{1}{2\rho_0 c_0^2} \, | \, p_q(P) \, |^2 \qquad (29)$$

This definition leads to a reverberation curve

$$L\left(t/T_p\right) = 10 \cdot \lg \left[\frac{1}{2\rho_0 c_0^2} \sum_{n > \frac{t/T_p}{\Delta t/T_p}} \frac{1}{\Delta s} \sum_{k_0 \cdot dist(q(s),P) \subset 2\pi \cdot \Delta t/T_p} | \, p(s) \, |^2 \right] \qquad (30)$$

The outer sum indicates summation in steps of time intervals $\Delta t/T_p$ in which Δs sources are found and that this outer summation includes a decreasing number n of such steps. The inner sum forms a squared average (with the factor $1/\Delta s$) of the contributions in a time interval (this summation is skipped if $\Delta s=0$).

M.5.22 A room with concave edges as an example

We take the concave room of the previuous section as a model room, with the same positions of the source Q and the receiver P as there. The evaluation of the mirror sources is performed for orders up to $o_{max}=8$. The lower limit of $|\Pi R|$ was set to limit= 0.01 ; the upper limit for the distances was with dmax=100·dist(Q,P) set so high that it did not exclude a legal source. This produces 837 effective sources; the computing time for the determination of source positions and source factors was 320 s.

The next diagram shows over $k_0 r$, with r=dist(q,P) , the sound pressure levels $p_q(P)$ of the contributions at P , relative to the contribution $p_Q(P)$ of the original source Q . The cloud of points has a typical triangular shape: the upper border has an approximately constant slope after a somewhat steeper slope for smaller $k_0 r$. The lower border lines are not so well defined, because there the points are disperse. The upwards going lower border line towards the right is predominantly determined by the termination with o_{max} . One can expect that the upper border line has some similarity with the reverberation curve.

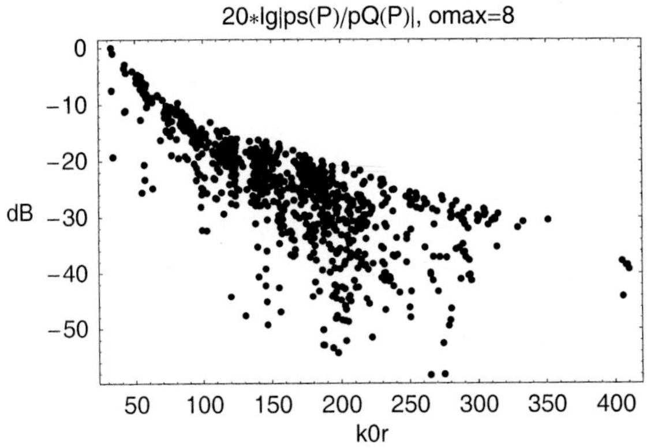

Levels of sound pressure contributions at P of the original and mirror sources up to order $o_{max} = 8$.

The evaluation of equ.(28) returns the reverberation curves of the next figures which differ from each other only in the value used for the time interval $\Delta t/T_p$. They do not resemble reverberation curves with which one is acquainted.

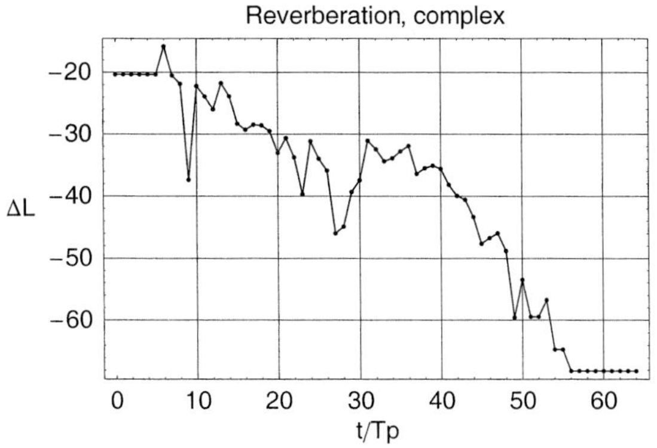

Summation of complex contributions, with time interval $\Delta t/T_p = 1$.

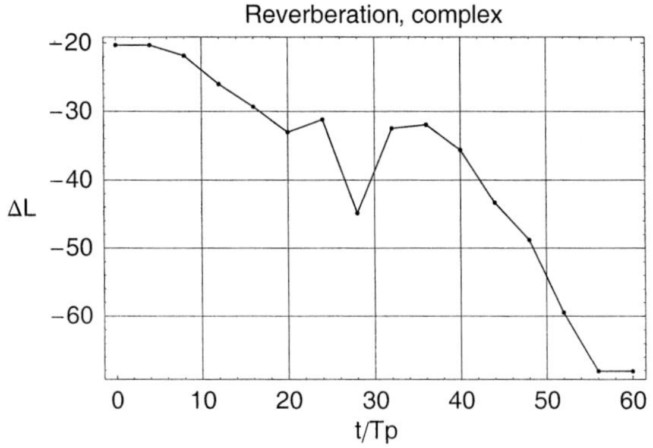

Summation of complex contributions, with time interval $\Delta t/T_p = 4$.

The next diagram combines results of the evaluation of equ.(30) with squared averaging in time intervals, again for different time intervals $\Delta t/T_p$ (within which now an averaging of squared magnitudes takes place); the values of the curves from high to low are $\Delta t/T_p = 1, 2, 4$ The points represent centres of the time intervals. The constant factor $1/(\rho_0 c_0^2)$ is omitted.

Except for the steep ends of the curves, which come from the termination of the MS evaluation with o_{max} , as explained above, the curves now represent usual reverberation curves. They have a steeper "early reverberation" and not so steep "late reverberation". The choice of $\Delta t/T_p$ may influence to some degree the value of the reverberation time obtained from such curves, for example as the coefficient of a linear regression through the points within a given interval of observation for t/T_p .

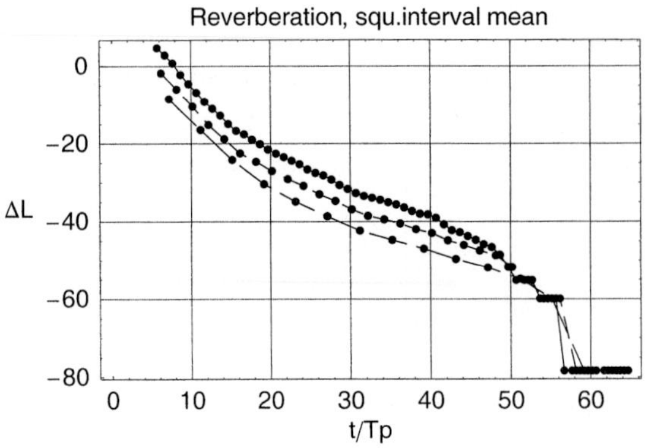

Reverberation curves with equ. (30) , for different values $\Delta t/T_p = 1, 2, 4$ (from high to low).

The last diagram combines the points of the reverberation curve for $\Delta t/T_p = 1$ with the linear regression within the interval $20 \le t/T_p \le 50$. The reverberation time T_r in units of T_p is $T_r/T_p = 67.46$.

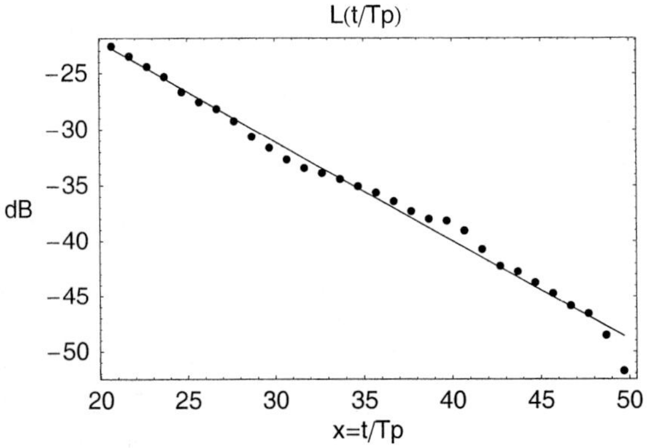

M.5 Appendix: Geometrical sub-tasks

A three-dimensional, right-handed Cartesian co-ordinate system x,y,z is assumed. Points, lines and planes will be considered in 3D. Corresponding relations in 2D are obtained either by setting one co-ordinate to a zero value, identically, or by easy direct derivations. Walls are defined by a list of subsequent edge points; the sequence of the edges in the list is such that they define a rotation which with the direction towards the interior of the room forms a right-handed system. Below, mostly not the polygon of a wall is considered, but the plane which contains the wall.

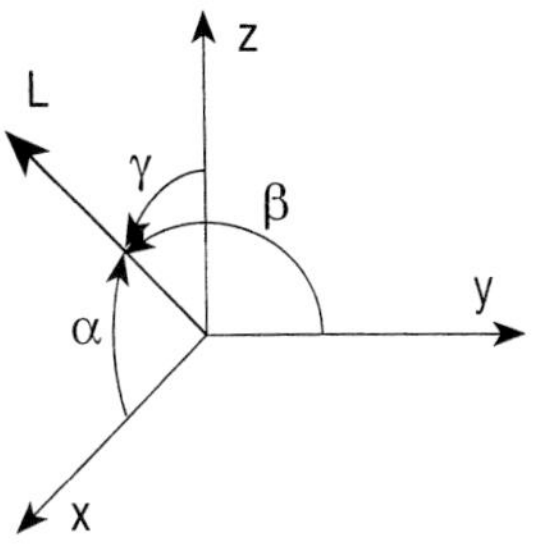

The direction angles α,β,γ of an oriented line L are the angles between the axes and the line.

1. *Distance* d *between two points* $P_1(x_1,y_1,z_1)$, $P_2(x_2,y_2,z_2)$:

$$dPP = \sqrt{(x_2 - x_1)^2 + (y_2 - y_1)^2 + (z_2 - z_1)^2} \tag{A.1}$$

2. *Direction cosines of the line from* $P_1(x_1,y_1,z_1)$ to $P_2(x_2,y_2,z_2)$:

$$\cos\alpha = \frac{x_2 - x_1}{dPP} \quad ; \quad \cos\beta = \frac{y_2 - y_1}{dPP} \quad ; \quad \cos\gamma = \frac{z_2 - z_1}{dPP} \tag{A.2}$$

3. *Cosine of the angle* φ *between two lines* :

The directions of the lines given by their direction angles α_i , β_i , γ_i

$$\cos\varphi = \cos\alpha_1 \cos\alpha_2 + \cos\beta_1 \cos\beta_2 + \cos\gamma_1 \cos\gamma_2 \tag{A.3}$$

4. *Normal form* $A\cdot x+B\cdot y+C\cdot z+D=0$ *of a plane* :

The plane is given by three points P_1, P_2, P_3 on it. A possible form of the plane equation (coming of a zero value of the vector triple product of the vectors (P,P_1), (P_2,P_1), (P_3,P)) is :

$$\begin{vmatrix} x - x_1 & y - y_1 & z - z_1 \\ x_2 - x_1 & y_2 - y_1 & z_2 - z_1 \\ x_3 - x_1 & y_3 - y_1 & z_3 - z_1 \end{vmatrix} = 0 \tag{A.4}$$

whence follow the parameters A,B,C,D :

$$\begin{aligned} A &= y_1\,(z_2 - z_3) + y_2\,(z_3 - z_1) + y_3\,(z_1 - z_2) \\ B &= -x_1\,(z_2 - z_3) - x_2\,(z_3 - z_1) - x_3\,(z_1 - z_2) \\ C &= x_1\,(y_2 - y_3) + x_2\,(y_3 - y_1) + x_3\,(y_1 - y_2) \\ D &= x_1\,(y_3 z_2 - y_2 z_3) + y_1\,(x_2 z_3 - x_3 z_2) + z_1\,(x_3 y_2 - x_2 y_3) \end{aligned} \tag{A.5}$$

5. *Reduced normal form* $a\cdot x+b\cdot y+c\cdot z+d=0$ *of a plane* :

$$a = \frac{A}{\sqrt{A^2 + B^2 + C^2}} \quad ; \quad b = \frac{B}{\sqrt{A^2 + B^2 + C^2}} \quad ; \quad c = \frac{C}{\sqrt{A^2 + B^2 + C^2}} \tag{A.6}$$

$$d = \frac{D}{\sqrt{A^2 + B^2 + C^2}} \tag{A.6}$$

This reduced normal form should not be confused with Hesse's normal form :

$$a' \cdot x + b' \cdot y + c' \cdot z - p = 0$$

$$a' = \frac{A}{\pm\sqrt{A^2 + B^2 + C^2}} \quad ; \quad b' = \frac{B}{\pm\sqrt{A^2 + B^2 + C^2}} \quad ; \quad c' = \frac{C}{\pm\sqrt{A^2 + B^2 + C^2}} \tag{A.7}$$

$$p = \frac{D}{\pm\sqrt{A^2 + B^2 + C^2}} \geq 0$$

where the sign of the root is taken so that p is positive. This additional convention in Hesse's normal form makes it unsuited for inside checks.

The parameters a, b, c of the reduced normal form are the direction cosines $\cos\alpha, \cos\beta, \cos\gamma$ of the normal vector on the plane (pointing to the interior side).

6. *Foot point* $P=(x,y,z)$ *of a point* $P_1=(x_1,y_1,z_1)$ *on a plane* :

The "foot point" P is the projection of P_1 on the plane. Let the plane be given by the parameters of its reduced normal form.

$$\begin{aligned} x &= (b^2 + c^2)x_1 - a(d + b y_1 + c z_1) \\ y &= (a^2 + c^2)y_1 - b(d + a x_1 + c z_1) \\ z &= (a^2 + b^2)z_1 - c(d + a x_1 + b y_1) \end{aligned} \tag{A.8}$$

7. *Mirror point* $P=(x,y,z)$ *of a point* $P_1=(x_1,y_1,z_1)$ *at a plane* :

Let P_F be the foot point of P_1 on the plane. Then $P = 2 \cdot P_F - P_1$. (A.9)

8. *Direction cosines of intersection line of two planes* :

The planes W_i be given by the parameters a_i, b_i, c_i, d_i of their reduced normal form.

$$\cos\alpha = \frac{\Delta_1}{\Delta} \quad ; \quad \cos\beta = \frac{\Delta_2}{\Delta} \quad ; \quad \cos\gamma = \frac{\Delta_3}{\Delta} \quad ;$$

$$\Delta_1 = \begin{vmatrix} b_1 & c_1 \\ b_2 & c_2 \end{vmatrix} \quad ; \quad \Delta_2 = \begin{vmatrix} c_1 & a_1 \\ c_2 & a_2 \end{vmatrix} \quad ; \quad \Delta_3 = \begin{vmatrix} a_1 & b_1 \\ a_2 & b_2 \end{vmatrix} \quad ; \tag{A.10}$$

$$\Delta = \sqrt{\Delta_1^2 + \Delta_2^2 + \Delta_2^2}$$

The rotation $W_1 \to W_2$ and the direction of the intersection line form a right-handed system.

9. *Point of intersection* $X=(x, y, z)$ *of a line through two points* $P_i=(x_i, y_i, z_i)$ *with a plane* :

Let the plane W be given by the parameters a, b, c, d of its reduced normal form.

$$\begin{aligned} x &= -d\,(x_1 - x_2) + b\,(x_2\, y_1 - x_1\, y_2) + c\,(x_2\, z_1 - x_1\, z_2) / xx \\ y &= -d\,(y_1 - y_2) + a\,(x_1\, y_2 - x_2\, y_1) + c\,(y_2\, z_1 - y_1\, z_2) / xx \\ z &= -d\,(z_1 - z_2) + a\,(x_1\, z_2 - x_2\, z_1) + b\,(y_1\, z_2 - y_2\, z_1) / xx \\ xx &= a\,(x_1 - x_2) + b\,(y_1 - y_2) + c\,(z_1 - z_2) \end{aligned} \qquad (A.11)$$

10. *Foot point* $P=(x, y, z)$ *of a point* $P_1=(x_1, y_1, z_1)$ *on the intersection line of two planes* W_1, W_2 :

Let the planes W_i be given by the parameters a_i, b_i, c_i, d_i of their reduced normal form.

$$\begin{aligned} x &= \big(\Delta_3\,(b_1\, d_2 - b_2\, d_1) + \Delta_2\,(c_2\, d_1 - c_1\, d_2) + \Delta_1 \cdot (\Delta_1\, x_1 + \Delta_2\, y_1 + \Delta_3\, z_1)\big) / \Delta^2 \\ y &= \big(\Delta_3\,(a_2\, d_1 - a_1\, d_2) + \Delta_1\,(c_1\, d_2 - c_2\, d_1) + \Delta_2 \cdot (\Delta_1\, x_1 + \Delta_2\, y_1 + \Delta_3\, z_1)\big) / \Delta^2 \\ z &= \big(\Delta_2\,(a_1\, d_2 - a_2\, d_1) + \Delta_1\,(b_2\, d_1 - b_1\, d_2) + \Delta_3 \cdot (\Delta_1\, x_1 + \Delta_2\, y_1 + \Delta_3\, z_1)\big) / \Delta^2 \end{aligned} \qquad (A.12)$$

with Δ and Δ_i from (A.10) .

11. *Bisectrice plane between two intersecting planes* W_1, W_2 :

Let the planes W_i be given by the parameters a_i, b_i, c_i, d_i of their reduced normal form. The parameters of the bisectrice plane (containing the intersection line) are :

$$a = (a_1 + \lambda a_2) \;;\quad b = (b_1 + \lambda b_2) \;;\quad c = (c_1 + \lambda c_2) \;;\quad d = (d_1 + \lambda d_2) \qquad (A.13)$$

with $\lambda = \pm 1$.

12. *Two planes parallel or anti-parallel* :

Parallel: the three parameters a, b, c of the reduced normal form are pairwise equal;

Anti-parallel: two of the parameters are pairwise equal, the other differs in sign.

13. *Distance between two anti-parallel planes* :

Let the planes W_i be given by the parameters a_i, b_i, c_i, d_i of their reduced normal form. Distance δ :

$$\delta = |d_1 - d_2| \qquad (A.14)$$

14. *Inside check of a point* $P=(x, y, z)$ *relative to a plane given by three points* $P_i=(x_i, y_i, z_i)$:

The check is performed with and returns:

$$\operatorname{sign}\begin{vmatrix} x-x_1 & y-y_1 & z-z_1 \\ x-x_2 & y-y_2 & z-z_2 \\ x-x_3 & y-y_3 & z-z_3 \end{vmatrix} = \begin{cases} +1 & \text{if P is inside W} \\ 0 & \text{if P is on W} \\ -1 & \text{if P is outside W} \end{cases} \tag{A.15}$$

where $|\ldots|$ indicates a determinant, and sign(x) checks the sign of x .

15. *Inside check of a point* P=(x, y, z) *relative to a plane given by its reduced normal form parameters* :

The check is performed with and returns:

$$\operatorname{sign}(a\cdot x+b\cdot y+c\cdot z+d) = \begin{cases} +1 & \text{if P is inside W} \\ 0 & \text{if P is on W} \\ -1 & \text{if P is outside W} \end{cases} \tag{A.16}$$

16. *Co-ordinate transformation* :

The system x, y, z is rotated and shifted as shown in the graph below. The new axis $z'=\zeta$ in the applications is the intersection line of two planes F_1, F_2 ; the new origin Z is the foot point of the original source Q on the intersection line.

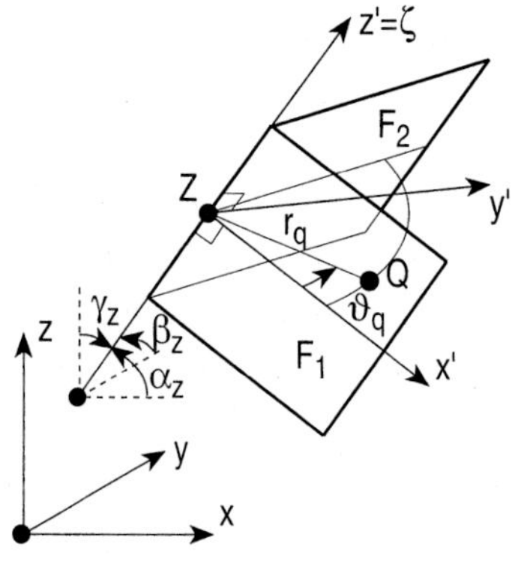

x′, y′, z′ is a right-handed system, like x, y, z . The rotation $F_1 \rightarrow F_2$ forms with z′ a right-handed system.

The transformation $x, y, z \rightarrow x', y', z'$ is done by :

$$\begin{pmatrix} x' \\ y' \\ z' \end{pmatrix} = \begin{pmatrix} \cos\alpha_x & \cos\beta_x & \cos\gamma_x \\ \cos\alpha_y & \cos\beta_y & \cos\gamma_y \\ \cos\alpha_z & \cos\beta_z & \cos\gamma_z \end{pmatrix} \cdot \begin{pmatrix} x-x_Z \\ y-y_Z \\ z-z_Z \end{pmatrix} \tag{A.17}$$

The inverse transformation $x', y', z' \rightarrow x, y, z$ is done (with the transposed matrix) by :

$$\begin{pmatrix} x \\ y \\ z \end{pmatrix} = \begin{pmatrix} \cos\alpha_x & \cos\alpha_y & \cos\alpha_z \\ \cos\beta_x & \cos\beta_y & \cos\beta_z \\ \cos\gamma_x & \cos\gamma_y & \cos\gamma_z \end{pmatrix} \cdot \begin{pmatrix} x' \\ y' \\ z' \end{pmatrix} + \begin{pmatrix} x_Z \\ y_Z \\ z_Z \end{pmatrix} \tag{A.18}$$

M.6 Ray tracing models

Ray tracing models are based on geometrical acoustics. Accordingly the wavelengths are considered small compared with characteristic dimensions of the room. Sound is regarded as energetic process. Calculations of quantities proportional to sound pressure are estimations with the assumption of:

δ:	random-incidence scattering coefficient;
e:	ray energy;
N:	number of rays;
r_d:	detector radius;
Δt:	sampling interval;
W:	total sound energy in a volume V
z, z_1, z_2:	random numbers;

a) Broadband stationary or transient signals
b) Superposition of energy or other quadratic field quantities.

These assumptions are both related to a consideration of sound energy in terms of a particle rather than a wave. Accordingly, ray tracing cannot be used for simulation of interference effects like standing waves, modes or diffraction.

Scattering, however, is often accounted for by a statistical approach by means of modelling the statistical case. The statistical case of scattering is described by Lambert's law. The direction of scattered sound is independent of the direction of incidence. Furthermore, the directions of scattering are distributed according to a cosine law, thus resulting in a constant emission of energy into all spatial angles. (In the equivalent phenomenon in optics, a surface with sound scattering according to Lambert's law would be observed with constant light intensity from all directions.) The crucial parameter for surface scattering is the size and the shape of the surface corrugation. Sound wavelengths are distributed from being "large" to "small", compared with the surface corrugations. This fact can be related to assumption of two ideal cases: (a) geometrical reflection and (b) random scattering. Therefore low frequencies are best treated with specular reflections, intermediate frequencies with random scattering, high frequencies with small-scale geometrical reflections.

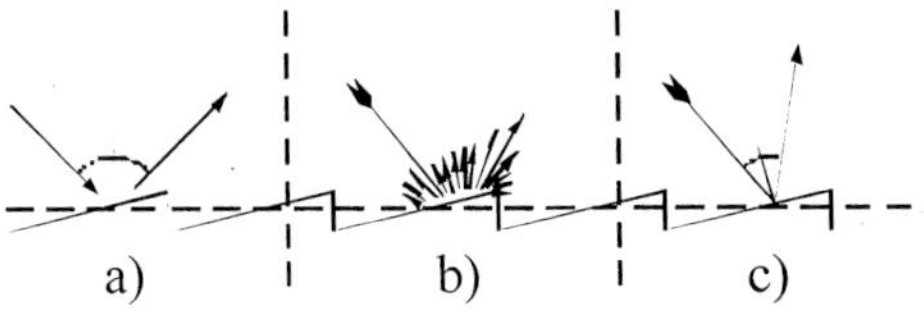

Reflection from rough surfaces, at low (a), mid (b) and high (c) frequencies.

Since reflections in room acoustics and particularly the reverberation process are built up by numerous reflections, the average behaviour is much more important than each individual scattering characteristic. It is thus sufficient to consider a mixed model of geometrical and diffuse reflections with a "switching" parameter and a random process with appropriate probability distribution. The parameter which decides which model is used is the random-incidence scattering coefficient δ :

$$\delta = 1 - \frac{E_{spec}}{E_{total}} \tag{1}$$

Monte Carlo method:

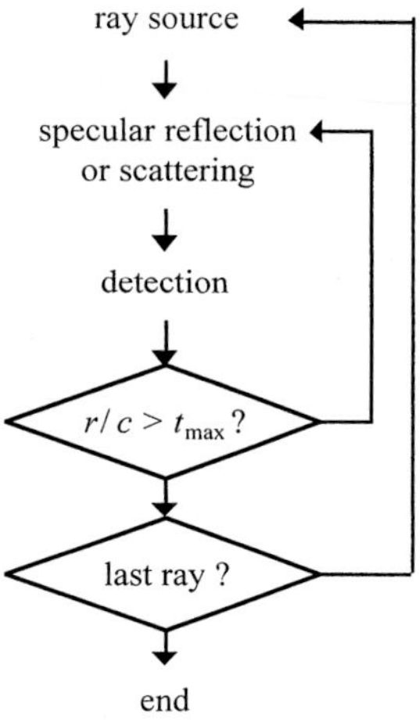

N rays are radiated from a source point. Typically N is larger than 10000. Each ray is carrying a portion of sound energy e_0. In this method the ray detection is provided by counting the rays hitting a detector (for instance, a sphere with radius r_d) and sampling the counts in time intervals Δt .

In a diffuse sound field the expectation value of the energy decay (energy time curve) is

$$\langle e(t) \rangle = e_0 N (1 - \overline{\alpha})^{\overline{n}t} \frac{\pi r_d^2 c_0 \Delta t}{V} \tag{2}$$

with room volume V and $\overline{n}$ the mean reflection rate, and $\overline{\alpha}$ the average absorption coefficient of the room (see M.4).

Wall absorption can be modelled by energy reduction according to multiplication of the ray energy by $(1 - \alpha)$, α being the random-incidence absorption coefficient of the wall.

Alternatively, a random number $z \in [0;1]$ can be chosen and the ray can be absorbed if $z < \alpha$. The probability density that the ray is absorbed at the next wall reflection is

$$w(\overline{n}t) = (1 - \overline{\alpha})^{\overline{n}t - 1} \overline{\alpha} \tag{3}$$

Whether Lambert scattering or specular reflection is used is decided by a random number $z \in [0;1]$. The ray is scattered if $z < \delta$.

Uncertainty of the Monte Carlo method:

Relative standard deviation of the statistical counts in the energy decay curve:

$$\frac{\sigma_e}{\langle e \rangle} = \sqrt{\frac{V}{N \pi r_d^2 c_0 \Delta t}} \quad \text{(energy absorption by multiplication), or} \tag{4}$$

$$\frac{\sigma_e}{\langle e \rangle} = \sqrt{\frac{V}{N (1 - \overline{\alpha})^{\overline{n}t} \pi r_d^2 c_0 \Delta t}} \quad \text{(energy absorption by random absorption)} \tag{5}$$

Relative standard deviation of the total energy $\langle W\rangle$ in the energy decay curve (energy integral):

$$\frac{\sigma_W}{\langle W\rangle}=\frac{\sigma_{\overline{|p|^2}}}{\left\langle\overline{|p|^2}\right\rangle}=\sqrt{\frac{A}{8\pi N r_d^2}} \qquad \text{(energy absorption by multiplication)} \qquad (6)$$

which is related to a sound level variation of

$$\sigma_L=4.34\sqrt{\frac{A}{8\pi N r_d^2}} \qquad (7)$$

and

$$\frac{\sigma_W}{\langle W\rangle}=\frac{\sigma_{\overline{|p|^2}}}{\left\langle\overline{|p|^2}\right\rangle}=\sqrt{\frac{A}{4\pi N r_d^2}} \qquad \text{(energy absorption by random absorption)} \qquad (8)$$

which is related to a sound level variation of

$$\sigma_L=4.34\sqrt{\frac{A}{4\pi N r_d^2}} \qquad (9)$$

The cone, beam or pyramid approach :

In contrast to the Monte Carlo approach of ray tracing, deterministic models of ray tracing are known. Here, the energy time curves are not calculated by counting rays, but by determination of ray paths and corresponding geometrical energy reductions. There are several ways of finding physically correct paths by associating the rays with cones or beams with constant solid angle and, thus, increasing spatial spread. In a diffuse sound field the expectation value of each ray energy is calculated by

$$\langle e(t)\rangle=e_0\frac{(1-\overline{\alpha})^{\overline{n}t}}{r^2} \qquad (10)$$

with r denoting the distance between the ray source and the receiving point.

Another important difference to the Monte Carlo approach is that the ray energy can be recorded in an arbitrarily high time resolution. In the case of a time resolution that is sufficient for audio processing, impulse responses can be composed from a set of reflections (or image sources). Each contribution contains a frequency function H_j which is based on the Fourier transform of the reflection pulse (a Dirac pulse) and multiplied by various frequency functions corresponding on the filter effects on the ray path:

$$H_j = \frac{e^{-jk_0 r_j}}{r_j} H_S H_R H_a \prod_{i=1}^{n_j} R_i \qquad (11)$$

with r_j denoting the distance between image source and receiver, H_S the (directional) spectrum of the source, H_R the directional head-related transfer function (HRTF, right or left ear) of the receiver person (in the case of binaural processing), H_a the spectrum of air attenuation, and R_i the reflection factors of the walls involved in the ray path (or the mirror source). The total binaural impulse response (r, l = right, left ear) is then obtained by inverse Fourier transformation:

$$p_{r,l}(t) = \mathsf{F}^{-1}\left\{ \sum_{j=1}^{N} H_{j,r,l} \right\} \qquad (12)$$

Ray sources (deterministic, random) :

Consider an omnidirectional source, with angles of ray direction in relation to source-related spherical co-ordinates φ, ϑ. The angles are

$$\varphi = 2\pi z_1 \quad ; \quad \vartheta = \arccos z_2 \qquad (13), (14)$$

with z_1 = random number of the interval [0;1], and z_2 = random number of the interval [0;1] independent of z_1.

Algorithm of reflections (scouting of reflection points) :

Last point of ray history P (vector $\vec{p}$), calculation of next wall hit at point S (vector $\vec{s}$):

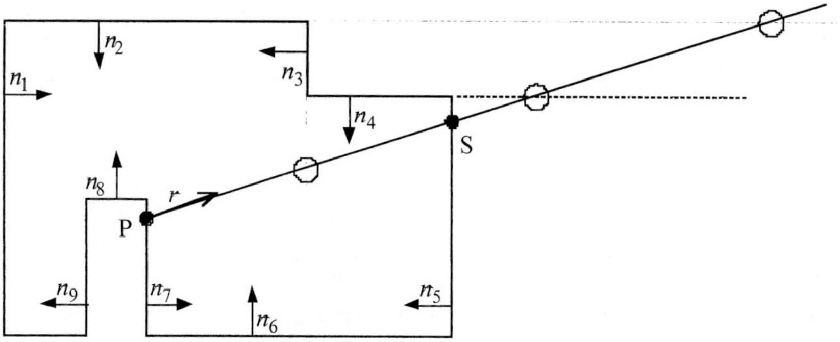

$\vec{n}_i$: normal vector of wall i, (all wall normal vectors direction towards inside the room),

$\vec{r}$: vector of actual flight direction

step 1 : ray must hit the wall plane facing the room (inside) $\vec{n}_i \cdot \vec{r} < 0$ (15)

step 2 : all walls passing step 1, ray must hit the next wall in positive flight direction: Calculation of $(\vec{p} - \vec{s}) \cdot \vec{r} < 0$ (16)

step 3 : after sorting of distances $\overline{PS_i}$ (i = 3, 5, 4, 2 in the figure above), starting with the nearest wall intersection point, test of “intersection point within polygon” (see M.5). In the case of failure (wall 3 in the figure above), the next point S_i is to be checked. If a special wall type is defined, "non-shadowing wall", this test can be skipped for each wall that is non-shadowing. Shadowing walls can block the free line of sight between two arbitrary observer points in the room; non-shadowing walls never block any two arbitrary lines within the room (in the figure above, i = 3, 4, 7, 8, 9 are shadowing walls, i = 1, 2, 5, 6 are non-shadowing walls). It is worthwhile to divide the wall polygons into the two categories.

step 4 : if a non-shadowing wall is reached or if S_i lies within the wall polygon, the ray is reflected (or absorbed) at wall i.

Wall scattering (Lambert's law). Independent on the angle of incidence, the reflection angle $\Omega(\varphi, \vartheta)$ is randomly chosen according to:

$$w(\vartheta)d\Omega = \frac{1}{\pi}\cos\vartheta d\Omega \tag{17}$$

with $w(\vartheta)$= probability density of the reflection polar angle ϑ. The azimuth angles are equally distributed in $[0, 2\pi]$. This is obtained by two independent random numbers $z_{1,2} \in [0;1]$ in:

$$\vartheta = \arccos\sqrt{z_1} \quad ; \quad \varphi = 2\pi z_2 \tag{18}$$

M.7 Room impulse responses, decay curves, and reverberation times

$r^2(t)$:	decay curve;
T30, Tx:	reverberation time;
EDT:	Early Decay Time

The results of mirror source or ray tracing algorithms are discrete energy density impulse responses.

$$w(t)\Delta t \approx w(t)dt \propto p^2(t)dt \tag{1}$$

Decay curve (expectation value for interrupted noise decay):

$$r^2(t) = \int_0^\infty p^2(\tau)\,d\tau - \int_0^t p^2(\tau)\,d\tau = \int_t^\infty p^2(\tau)\,d\tau = \int_\infty^t p^2(\tau)\,d(-\tau) \qquad (2)$$

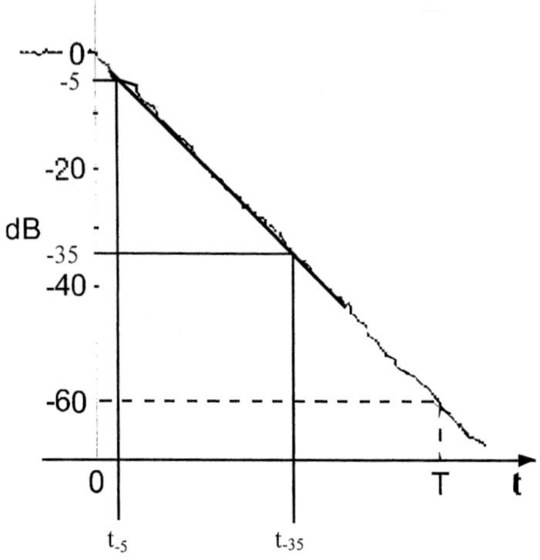

Reverberation time T30 :

Linear regression from –5 dB to –35 dB of $10\log[r^2(t)]$ for determination of t_{-5} and t_{-35}, extrapolated to –60 dB by $T30 = 2\,(t_{-35} - t_{-5})$.

General reverberation time Tx:

Linear regression from –5 dB to –(x+5) dB of $10\log[r^2(t)]$ for determination of t_{-5} and $t_{-(x+5)}$, extrapolated to –60 dB by $Tx = 60/x\,(t_{-(5+x)} - t_{-5})$.

Early decay time EDT
(characterising the subjectively perceived reverberance of a room):

Linear regression from 0.1 dB to –10.1 dB of $10\log[r^2(t)]$ for determination of $t_{-0.1}$ and $t_{-10.1}$, extrapolated to –60 dB by $EDT = 6\,(t_{-10.1} - t_{-0.1})$.

M.8 Other room acoustical parameters

Impulse responses can be evaluated for calculation of quantities with specific correlation to subjective impressions. All quantities are based on integrals of the squared impulse responses.

The *sound strength* G is the relative level between the sound field in the room and the level in a free sound field at 10 m distance, with the power output of the source and the

p(t):	sound pressure impulse response;
$p_{10}(t)$:	sound pressure impulse response in a reference source-to-receiver distance of 10 m;
$p_r(t)$:	sound pressure impulse response obtained for the right ear of a test subject or a dummy head (see equ. M.6.12);
$p_l(t)$:	sound pressure impulse response obtained for a test subject or a dummy head (see equ. M.6.12);

direction of the axis of source and receiver remaining the same.

$$G = 10\log\left(\int_0^\infty |p(t)|^2\, dt \Big/ \int_0^\infty |p_{10}(t)|^2\, dt\right) \text{ dB} \tag{3}$$

$p_L(t)$: lateral sound pressure impulse response (figure-of-eight directionality) with its directional null pointed towards the source.

Definition D or *early-to-late energy ratio* (characterising the speech intelligibility):

$$D = \int_0^{50ms} |p(t)|^2 dt \Big/ \int_0^\infty |p(t)|^2 dt \tag{4}$$

Clarity C_X, early-to-late energy ratio for music (x = 80 ms) and for speech (x = 50 ms) (characterising the subjective transparency or speech intelligibility, respectively):

$$C_{80} = 10\log \int_0^{80ms} |p(t)|^2 dt \Big/ \int_{80ms}^\infty |p(t)|^2 dt \text{ dB} \tag{5}$$

$$C_{50} = 10\log \int_0^{50ms} |p(t)|^2 dt \Big/ \int_{50ms}^\infty |p(t)|^2 dt \text{ dB} \tag{6}$$

Relation between D and C_{50}:

$$C_{50} = 10\log\left(\frac{D}{1-D}\right) \tag{7}$$

Centre time :

(1st moment of the impulse response, characterising the reverberance and speech intelligibility)

$$TS = \int_0^\infty t|p(t)|^2 dt \Big/ \int_0^\infty |p(t)|^2 dt \tag{8}$$

Lateral energy fraction :

Early lateral sound ratio (characterising the subjective spatial impression "apparent source width")

$$LF = \int_{5ms}^{80ms} |p_L(t)|^2 dt \Big/ \int_0^{80ms} |p(t)|^2 dt \tag{9}$$

with $p_L(t)$ denoting the sound pressure weighted with a figure-of-eight directionality, with its directional null pointed towards the source.

Interaural cross-correlation function :

$$\mathrm{IACF}_{t_1,t_2}(\tau) = \int_{t_1}^{t_2} p_l(t)\cdot p_r(t+\tau)\,dt \Bigg/ \sqrt{\int_{t_1}^{t_2} p_l^2(t)dt \int_{t_1}^{t_2} p_r^2(t)\,dt} \tag{10}$$

Interaural cross-correlation coefficient
(characterising the subjective spatial impression):

$$\mathrm{IACC} = \max\left[\mathrm{IACF}_{-1ms,1ms}(\tau)\right] \tag{11}$$

Late lateral sound level :
(characterising the subjective spatial impression "listener envelopment"), [5]

$$LG_{80}^{\infty} = \int_{80ms}^{\infty} |p_L(t)|^2\,dt \Bigg/ \int_0^{\infty} |p_{10}(t)|^2\,dt \tag{12}$$

Estimates of the room acoustical parameters by using just the reverberation time can be achieved by consideration of a purely exponential decay, [6]:

Energy integral from time t to infinity :

$$i_t = (31200\,T/V)\cdot e^{-13.82t/T} \tag{13}$$

with r denoting the source-to-receiver distance in m, T the reverberation time in s, and V the room volume in m^3. Contribution of direct, early and late (limit 80 ms) sound energy at a source-to-receiver distance r :

$$e_d = 100/d^2 \tag{14}$$

$$e_e = (31200\,T/V)\cdot e^{-0.04d/T}(1-e^{-1.11/T}) \tag{15}$$

$$e_l = (31200\,T/V)\cdot e^{-0.04d/T}\cdot e^{-1.11/T} \tag{16}$$

and accordingly :

$$G = 10\cdot\log(e_d + e_e + e_l) = 10\cdot\log(100/d^2 + 31200\,T/V) \tag{17}$$

and

$$C_{80} = 10\cdot\log[(e_d + e_e)/e_l] \tag{18}$$

References to part M :

Room Acoustics

[1a] PUJOLLE, Revue d'Acoustique 25 (1973) 121-123

[1b] PUJOLLE, Rev.Techn.Radio Télevis. 25 (1972) 25

[2] CREMER, L.
"Die wissenschaftlichen Grundlagen der Raumakustik",
Band I : Geometrische Raumakustik,
Hirzel Verlag Stuttgart 1948

[3] KOSTEN, C.W. ACUSTICA 10 (1960) 245
"The mean free path in room acoustics."

[4] KUTTRUFF, H.
"Room Acoustics", 4th edition
E&FN SPON, London 2000

[5] BRADLEY, J., SOULODRE, G.A. J. Acoust. Soc. Am 98 (1995) 2590
"Objective measures of listener envelopment."

[6] BARRON, M., LEE, L.-J. J. Acoust. Soc. Am. 84 (1988) 618
"Energy relations in concert auditoriums."

[7] MECHEL, F.P. submitted to J. Sound Vibr.
"Improved Mirror Source Method in Room Acoustics"

N
Flow Acoustics
by P. Költzsch

Sound propagation in a flowing medium is treated also in the chapters "J. Duct Acoustics" and "K. Acoustic Mufflers", but there mostly with simplifying assumptions.

This chapter uses throughout the "double subscript summation rule", i.e., terms in an expression in which a subscript (for example i) appears twice represent a sum over that term with the multiple subscript as summation index. Terms containing x_i^2 , for example, are also cases of the summation rule.

The general convention to symbolise the density of air with ρ_0 and sound velocity with c_0 must be suspended in this chapter, because these quantities may be used in other than standard conditions. These conditions will always be defined in the context.

N.1 Concepts and notations in fluid mechanics, in connection with the field of aeroacoustics

Ref.: Morfey, [N.1]; Lauchle, [N.2]; Douglas, [N.3]; Roger, [N.4];

N.1.1 Types of fluids

Ideal fluids: $\mu = 0,\ \lambda = 0$ — μ: dynamic viscosity; λ: thermal conductivity

Newtonian fluid: $\mu = \text{constant}$

Non-Newtonian fluid: $\mu \neq \text{constant}$

The relationship between shear stress τ and velocity gradient $\partial v/\partial n$ is non-linear.

N.1.2 Properties of fluids

Density: ρ, mass per volume, $[\rho] = \dfrac{\text{kg}}{\text{m}^3}$

Pressure — p, normal force pushing against a plane area divided by the area

$$[p] = \frac{N}{m^2} = Pa$$

Viscosity:

dynamic viscosity $\mu, \; [\mu] = \frac{N \cdot s}{m^2} = Pa \cdot s$

kinematic viscosity $\nu, \; [\nu] = \frac{m^2}{s}$

Gas constant $R, \; [R] = \frac{J}{kg \cdot K}$

Specific heats

at constant volume
$$c_V = T\left(\frac{\partial s}{\partial T}\right)_\rho = \left(\frac{\partial u}{\partial T}\right)_\rho \tag{1}$$

at constant pressure
$$c_p = T\left(\frac{\partial s}{\partial T}\right)_p = \left(\frac{\partial h}{\partial T}\right)_p \tag{2}$$

$$\left[c_p, c_V\right] = \frac{J}{kg \cdot K}$$

with:

s	specific entropy
u	specific internal energy
h	specific enthalpy .

Specific heat ratio $\kappa = c_p / c_V$, ratio of the specific heat at constant pressure to that at constant volume. (3)

Speed of sound $c, \; [c] = \frac{m}{s}$

Bulk modulus — K, expresses the compressibility of a fluid, $[K] = Pa$,

adiabatic or isentropic bulk modulus:
$$K_S = \rho\left(\frac{\partial p}{\partial \rho}\right)_S \tag{4}$$

isothermal bulk modulus:
$$K_T = \rho\left(\frac{\partial p}{\partial \rho}\right)_T \tag{5}$$

The reciprocal $1/K_S$ or $1/K_T$ is the adiabatic or isothermal compressibility.

Coefficient of expansion
$$\beta = -\frac{1}{\rho}\left(\frac{\partial \rho}{\partial T}\right)_p \qquad [\beta] = \frac{1}{K} \tag{6}$$

Thermal conductivity $\lambda, \; [\lambda] = \frac{W}{m \cdot K}$

Shear stress $\tau, \; [\tau] = Pa$

N.1.3 Models of fluid flows

Real flow: flow without any assumptions;

ideal flow: flow without viscosity and thermal conductivity .

Inviscid flow: flow without viscosity;

viscous flow: $\mu \neq 0$.

Incompressible flow: $\rho = \text{constant}$;

compressible flow: $\rho \neq \text{constant}$.

Adiabatic flow: flow without heat transfer;

isentropic flow: $\frac{Ds}{Dt} = 0$, the specific entropy of each fluid particle along its path is constant, but may vary from a particle to another (ROGER); inviscid and non heat-conducting gas flow, also frictionless adiabatic flow

homentropic flow: $s = \text{constant}$ throughout the flow, uniform specific entropy;

isothermal flow: $T = \text{constant}$.

Steady flow: no time dependence for v, p, ρ, T, …; thus $\frac{\partial \ldots}{\partial t} = 0$;

stationary flow: $\frac{\partial \overline{A}}{\partial t} = 0$, with $\overline{A} = \overline{v}, \overline{p}, \overline{\rho}, \overline{T}, \ldots$ and $\overline{A} = \lim_{T\to\infty} \frac{1}{T}\int_0^T A\,dt$; (7)

unsteady flow: $\frac{\partial A}{\partial t} \neq 0$, possibly also $\frac{\partial \overline{A}}{\partial t} \neq 0$.

Uniform flow: $\frac{\partial \vec{v}}{\partial s} = 0$;

non-uniform flow: $\frac{\partial \vec{v}}{\partial s} \neq 0$.

Rotational flow:

$$\vec{\omega} = \text{rot}\vec{v} = \text{curl}\vec{v} = \nabla \times \vec{v} = \begin{vmatrix} \vec{e}_x & \vec{e}_y & \vec{e}_z \\ \frac{\partial}{\partial x} & \frac{\partial}{\partial y} & \frac{\partial}{\partial z} \\ u & v & w \end{vmatrix} \neq 0 \qquad (8)$$

with: $v_x = u,\; v_y = v,\; v_z = w$.

Vorticity: $\vec{\omega} = \text{rot}\vec{v} = \nabla \times \vec{v}$ is a measure of local fluid rotation .

Irrotational flow: $\vec{\omega} = \text{rot}\vec{v} = \nabla \times \vec{v} = 0$. (9)

Comment:

From Crocco's form of the momentum equations it follows that (stationary flow with constant stagnation enthalpy) $\vec{\omega} \times \vec{v} = T \cdot \text{grad}\, s$.

Consequences:

a rotational flow cannot exist with uniform entropy;

a homentropic flowmust be irrotational (except when vorticity field and velocity field are parallel).

Laminar flow: viscous or streamline flow, without turbulence; the particles of the fluid moving in an orderly manner and retaining the same relative positions in successive flow cross-sections.

Turbulent flow, Turbulence: a random, non-deterministic motion of eddying fluid flow; characterized by (MORFEY):

- three-dimensional velocity fluctuations field;
- unsteady flow;
- viscous flow;
- rotational flow;
- flow with viscous dissipation of energy;
- viscous dissipation takes place at the smallest length scales of eddies, far removed from the larger scales eddies contain most of the kinetic energy; the smallest scales >> molecular scales;
- fluctuations cover a wide frequency range and a wide range of eddy sizes or length scales;
- occurring at high Reynolds numbers .

Turbulence level: based on the averaging of the specific kinetic energy

$$\frac{1}{2}\overline{v_i^2} = \frac{1}{2}\overline{(\bar{v}_i + v_i')(\bar{v}_i + v_i')} = \frac{1}{2}\bar{v}_i^2 + \frac{1}{2}\overline{v_i'^2} \tag{10}$$

three-dimensional

$$\frac{1}{2}\overline{v_i^2} = \frac{1}{2}\left(\bar{u}^2 + \bar{v}^2 + \bar{w}^2\right) + \frac{1}{2}\left(\overline{u'^2} + \overline{v'^2} + \overline{w'^2}\right) \tag{11}$$

turbulence level :

$$\mathrm{Tu} = \sqrt{\frac{\frac{1}{3}\left(\overline{u'^2} + \overline{v'^2} + \overline{w'^2}\right)}{\left(\bar{u}^2 + \bar{v}^2 + \bar{w}^2\right)}} \tag{12}$$

in the special case of isotropic turbulence and unidirectional flow $\bar{v}_i = \{\bar{v}_x = U;0;0\}$:

$$\mathrm{Tu} = \frac{\sqrt{\overline{u'^2}}}{U} = \frac{u'_{rms}}{U} \tag{13}$$

Transition: the fluid flow change from laminar to turbulent flow .

Boundary layer flow: • in the mean flow sense (MORFEY):

flow next to a solid surfaces within which the mean flow $\overline{u}(y)$ varies with distance y from the wall, from zero at the wall (at $y = 0$) to 99 % of its free-stream value at $y = \delta$, δ is the boundary layer thickness ;

• in the acoustic sense (MORFEY):

a thin region produced by a sound field next to a solid boundary, within which the oscillatory velocity parallel to the wall drops to zero as the wall is approached, as a result of viscosity. The acoustic boundary layer thickness is

$$\delta = \sqrt{\frac{2\nu}{\omega}} << \lambda \tag{14}$$

Reynolds stress: • $\rho v_i v_j$ in unsteady fluid flow;

v_i, v_j are fluid velocity components in any of the three orthogonal Cartesian coordinate directions;

$\rho v_i v_j$ represents the transfer rate of j-components fluid momentum, per unit area;

the double divergence of $\rho v_i v_j$ represents a source term in Lighthill's inhomogeneous wave equation (acoustic analogy for aerodynamic sound generation);

• in turbulent flows:

the time-average Reynolds stress $\overline{\rho v_i' v_j'}$ is a term in the time-averaged momentum equation, as the negative of an effective stress;

$\overline{\rho v_i' v_j'}$ represents the mean momentum flux due to turbulent eddies;

the Reynolds stress tensor is $\tau_{ij} = \rho \overline{v_i v_j}$ with normal stress if $i = j$, and shear stress if $i \neq j$ (MORFEY)

N.2 Some tools in fluid mechanics and aeroacoustics

Ref.: Telionis, [N.5]; Johnson, [N.6]; Schlichting, [N.7]; Lauchle, [N.2]; Liu, [N.8]; Hussain, [N.9]; Reynolds [N.10]

N.2.1 Averaging

General quantity: $f(\vec{x}, t)$

Spatial average: $$\bar{f}_{spatial} = \frac{1}{V}\int_V f(\vec{x},t)\,dV = \bar{f}^{\,s}(t) \qquad (1)$$

Time average: $$\bar{f}_{time} = \lim_{T\to\infty}\frac{1}{T}\int_0^T f(\vec{x},t)\,dt = \bar{f}^{\,t}(\vec{x}) = \bar{f} \qquad (2)$$

with: abbreviation: $\bar{f}$.

Root mean square: $$f_{rms} = \sqrt{\overline{f^2}} = \sqrt{\lim_{T\to\infty}\frac{1}{T}\int_0^\infty f^2(\vec{x},t)\,dt} = \bar{f}^{\,rms}(\vec{x}) \qquad (3)$$

the square root of the mean square value .

Ensemble average: over N repeated experiments

$$\bar{f}_{ensemble} = \langle f(\vec{x},t)\rangle = \lim_{N\to\infty}\frac{1}{N}\sum_{\alpha=1}^{N} f^{(\alpha)}(\vec{x},t) = \bar{f}^{\,en}(\vec{x},t) \qquad (4)$$

Phase average: for periodic flows

$$\bar{f}_{phase} = \langle f(\vec{x},t)\rangle = \lim_{N\to\infty}\frac{1}{N}\sum_{n=0}^{N} f(\vec{x},t+n\tau) = \bar{f}^{\,ph}(\vec{x},t) \qquad (5)$$

with: τ period of an externally imposed fluctuation ;

$$\bar{f}_{phase} = \langle f(\vec{x},t)\rangle = \lim_{N\to\infty}\frac{1}{N}\sum_{n=0}^{N} f(\vec{x},\varphi_0+n\omega\tau) = \bar{f}^{\,ph}(\vec{x},\varphi_0) \qquad (6)$$

with: $0<\varphi_0<2\pi$; $\varphi=\varphi_0+n\omega\tau$ phase of the periodic flow.

Reynolds averaging: decomposition of a general quantity in the flow in the following form:

$$f = \bar{f} + f'$$

with: $\bar{f} = f_0 = \lim_{T\to\infty}\frac{1}{T}\int_0^T f\,dt$ mean quantity; (7)

$\overline{f'} = \lim_{T\to\infty}\frac{1}{T}\int_0^T f'\,dt = 0$ fluctuating quantity. (8)

Mass-weighted or Favre averaging:

decomposition of a general quantity in the flow in the following form:

$$f = \tilde{f}^{f} + f''^{f}$$

with: $\tilde{f}^{f} = \dfrac{\overline{\rho f}}{\bar{\rho}}$ filtered part of f ;

f''^{f} unresolved or subgrid part of f

$\widetilde{f''^{f}} = 0$.

N.2.2 Decomposition (in general)

Decomposition a general flow quantity in three (or four) parts:

$$f(\vec{x},t)=\bar{f}(\vec{x})+\tilde{f}(\vec{x},t)+f'(\vec{x},t) \tag{9}$$

with: $\bar{f}(\vec{x})$ time-averaged or mean component,

obtained by Reynolds averaging : $\bar{f}(\vec{x})=\lim\limits_{T\to\infty}\frac{1}{T}\int_0^T f(\vec{x},t)\,dt$;

$\bar{\tilde{f}}=0$

$\overline{f'}=0$;

$\tilde{f}(\vec{x},t)$ organized fluctuation: periodicities are in time,

periodic mean component of low can be split into the odd modes $\tilde{f}^{odd}$ and in the even modes $\tilde{f}^{even}$ (see LIU) :

$$\tilde{f}(\vec{x},t)=\tilde{f}^{odd}(\vec{x},t)+\tilde{f}^{even}(\vec{x},t)\ ; \tag{10}$$

$f'(\vec{x},t)$ random fluctuations, incoherent fluctuating flow quantities,

e. g. small-scale stochastic fluctuations of the fine-grained turbulence ;

$$\bar{f}_{phase}=\langle f(\vec{x},t)\rangle=\lim_{N\to\infty}\frac{1}{N}\sum_{n=0}^{N}f(\vec{x},t+n\tau)=\bar{f}+\tilde{f} \tag{11}$$

with: $\langle f'(\vec{x},t)\rangle=0$.

$$\tilde{f}=\langle f(\vec{x},t)\rangle-\bar{f} \tag{12}$$

Phase averaging with period 2τ is denoted by $\langle\langle\ \rangle\rangle$ so that $\langle\langle\tilde{f}^{odd}\rangle\rangle=0$.

Therefore the even modes are obtained from

$$\langle\langle\tilde{f}^{odd}+\tilde{f}^{even}\rangle\rangle=\tilde{f}^{even} \tag{13}$$

and the odd modes from subtracting

$$\tilde{f}^{odd}=\tilde{f}-\langle\langle\tilde{f}^{odd}+\tilde{f}^{even}\rangle\rangle \tag{14}$$

N.2.3 Decomposition of the physical quantities in the basic equations

Decomposition: $\rho=\bar{\rho}+\rho'$

$p=\bar{p}+p'$

$v_i=\bar{v}_i+v_i'$.

Continuity equation :

$$\frac{\partial\rho}{\partial t}+\frac{\partial(\rho v_i)}{\partial x_i}=0 \tag{15}$$

$$\frac{\partial\rho'}{\partial t}+\frac{\partial}{\partial x_i}\left[\bar{\rho}\bar{v}_i+\bar{\rho}v_i'+\bar{v}_i\rho'+v_i'\rho'\right]=0 \tag{16}$$

with assumptions: $\rho', v_i' << \overline{\rho}, \overline{v}_i$ and $\overline{\rho} \neq f(\vec{x})$

Continuity equation in the case of mean flow:
$$\frac{\partial \overline{v}_i}{\partial x_i} = 0 \tag{17}$$

and in the case of fluctuating flow
$$\frac{\partial \rho'}{\partial t} + \overline{\rho}\frac{\partial v_i'}{\partial x_i} + \overline{v}_i \frac{\partial \rho'}{\partial x_i} = 0 \tag{18}$$

and with the equation of state $\rho' = \frac{p'}{c_0^2}$:

$$\frac{\partial p'}{\partial t} + \overline{v}_i \frac{\partial p'}{\partial x_i} + \overline{\rho} c_0^2 \frac{\partial v_i'}{\partial x_i} = 0 \tag{19}$$

$$\frac{\overline{D} p'}{Dt} + \overline{\rho} c_0^2 \frac{\partial v_i'}{\partial x_i} = 0 \tag{20}$$

with:
$$\frac{\overline{D}}{Dt} = \frac{\partial}{\partial t} + \overline{v}_i \frac{\partial}{\partial x_i} \tag{21}$$

<u>Momentum equation</u> (without viscosity)

$$\frac{\partial}{\partial t}(\rho v_i) + \frac{\partial}{\partial x_j}\left(\rho v_i v_j + p\delta_{ij}\right) = 0 \tag{22}$$

mean flow
$$\overline{\rho}\,\overline{v}_j \frac{\partial \overline{v}_i}{\partial x_j} + \frac{\partial \overline{p}}{\partial x_i} = 0 \tag{23}$$

fluctuating flow (with linearisation)
$$\overline{\rho}\frac{\partial v_i'}{\partial t} + \overline{\rho}\,\overline{v}_j \frac{\partial v_i'}{\partial x_j} + \rho' \overline{v}_j \frac{\partial \overline{v}_i}{\partial x_j} + \overline{\rho} v_j' \frac{\partial \overline{v}_i}{\partial x_j} + \frac{\partial p'}{\partial x_i} = 0 \tag{24}$$

with constant mean flow
(assumption: $\overline{v}_i$ is uniform)
$$\overline{\rho}\frac{\partial v_i'}{\partial t} + \overline{\rho}\,\overline{v}_j \frac{\partial v_i'}{\partial x_j} + \frac{\partial p'}{\partial x_i} = 0 \tag{25}$$

<u>Wave equation</u>

following from the continuity equation
$$\frac{\partial p'}{\partial t} + \overline{v}_i \frac{\partial p'}{\partial x_i} + \overline{\rho} c_0^2 \frac{\partial v_i'}{\partial x_i} = 0 \tag{26}$$

and the momentum equations
$$\overline{\rho}\frac{\partial v_i'}{\partial t} + \overline{\rho}\,\overline{v}_j \frac{\partial v_i'}{\partial x_j} + \frac{\partial p'}{\partial x_i} = 0 \tag{27}$$

Result: convective wave equation :
$$\frac{\partial^2 p'}{\partial x_i^2} - \frac{1}{c_0^2}\left[\frac{\partial}{\partial t} + \overline{v}_i \frac{\partial}{\partial x_i}\right]^2 p' = 0 \tag{28}$$

with :
$$\left[\frac{\partial}{\partial t} + \overline{v}_i \frac{\partial}{\partial x_i}\right]^2 = \left(\frac{\partial^2}{\partial t^2} + 2\overline{v}_i \frac{\partial^2}{\partial x_i \partial t} + \overline{v}_i \overline{v}_j \frac{\partial^2}{\partial x_i \partial x_j}\right) \tag{29}$$

Navier-Stokes equation :

Double decomposition of quantities and time averaging: Reynolds averaging.

Assumptions: incompressible flow

for the mean flow : $$\overline{v}_j \frac{\partial \overline{v}_i}{\partial x_j} = -\frac{1}{\rho}\frac{\partial \overline{p}}{\partial x_i} + \nu \frac{\partial^2 \overline{v}_i}{\partial x_j^2} - \frac{\partial\left(\overline{v_i' v_j'}\right)}{\partial x_j} \quad \text{Reynolds equation} \tag{30}$$

with stress tensor of fluctuating flow $$\begin{pmatrix} \sigma_x' & \tau_{xy}' & \tau_{xz}' \\ \tau_{xy}' & \sigma_y' & \tau_{yz}' \\ \tau_{xz}' & \tau_{yz}' & \sigma_z' \end{pmatrix} = -\begin{pmatrix} \rho\overline{u'^2} & \rho\overline{u'v'} & \rho\overline{u'w'} \\ \rho\overline{u'v'} & \rho\overline{v'^2} & \rho\overline{v'w'} \\ \rho\overline{u'w'} & \rho\overline{v'w'} & \rho\overline{w'^2} \end{pmatrix} \tag{31}$$

Triple decomposition of quantities and time averaging (TELIONIS) :

$$f(\vec{x},t) = \overline{f}(\vec{x}) + \tilde{f}(\vec{x},t) + f'(\vec{x},t)$$

for the mean flow : $$\overline{v}_j \frac{\partial \overline{v}_i}{\partial x_j} = -\frac{1}{\rho}\frac{\partial \overline{p}}{\partial x_i} + \nu \frac{\partial^2 \overline{v}_i}{\partial x_j^2} - \frac{\partial\left(\overline{v_i' v_j'}\right)}{\partial x_j} - \frac{\partial\left(\overline{\tilde{v}_i \tilde{v}_j}\right)}{\partial x_j} \tag{32}$$

with two Reynolds stress terms on the right-hand side of the equation: non-linear contributions due to the random fluctuations and due to organized fluctuations.

for the organized fluctuations :

$$\frac{\partial \tilde{v}_i}{\partial t} + \overline{v}_j \frac{\partial \tilde{v}_i}{\partial x_j} + \tilde{v}_j \frac{\partial \overline{v}_i}{\partial x_j} + \tilde{v}_j \frac{\partial \tilde{v}_i}{\partial x_j} = -\frac{1}{\rho}\frac{\partial \tilde{p}}{\partial x_i} + \nu \frac{\partial^2 \tilde{v}_i}{\partial x_j^2} + \frac{\partial\left(\overline{\tilde{v}_i \tilde{v}_j}\right)}{\partial x_j} + \frac{\partial\left(\overline{v_i' v_j'}\right)}{\partial x_j} - \frac{\partial\left(\left\langle v_i' v_j' \right\rangle\right)}{\partial x_j} \tag{33}$$

N.2.4 Correlations

$$R_p(\tau) = \lim_{T\to\infty} \frac{1}{T}\int_0^T p(t)p(t+\tau)dt \quad \text{autocorrelation function} \tag{34}$$

$$R_p(0) = \overline{p^2(t)} \tag{35}$$

$$R_p(\xi) = \frac{1}{L}\int_0^L p(x)p(x+\xi)dx \tag{36}$$

$$R = \frac{\overline{v_1' v_2'}}{\sqrt{\overline{v_1'^2}\,\overline{v_2'^2}}} \quad \text{correlation function, fluctuations of velocity} \tag{37}$$

N.2.5 Scales

$$\tau_c = \frac{1}{R_p(0)} \int_0^\infty R_p(\tau)\,d\tau$$ integral time scale (38)

$$\Lambda = \frac{1}{R_p(0)} \int_0^\infty R_p(\xi)\,d\xi$$ integral length scale (39)

$$\Lambda_x = \frac{E_{11}(k_1)_{k_1\to 0}}{4 v'^2_{rms}}$$ integral length scale, limiting value of the power spectrum as k_1 approaches zero. (40)

$$E(k) = \alpha \frac{\overline{v'^2_{rms}}}{k_e} \frac{(k/k_e)^4}{\left(1+(k/k_e)^2\right)^{17/6}} e^{-2(k/k_v)^2}$$ power spectral density for isotropic turbulence (von Kármán spectrum) (41)

with (LONGATTE) :

$\alpha \approx 1.453$

$k_e \approx 0.747 / \Lambda$

$\Lambda = (\overline{v^2})^{3/2} / \varepsilon$

ε= dissipation rate of turbulent kinetic energy

$k_v = (\varepsilon / v^3)^{1/4}$

$$l_\tau^2 = \frac{\overline{p^2(t)}}{\overline{\left(\frac{\partial p}{\partial t}\right)^2}}$$ differential time scale (42)

$$l_\tau^2 = -\frac{R_p(0)}{\dfrac{\partial^2 R_p(0)}{\partial \tau^2}}$$ (43)

N.3 The basic equations of fluid motion

Ref.: Bangalore/Morris, [N.11]; Bailly/Lafon/Candel, [N.12] ;

N.3.1 Continuity equation, momentum equation, energy equation :

Continuity equation :

$$\frac{\partial\rho}{\partial t}+\frac{\partial(\rho v_i)}{\partial x_i}=0 \qquad \text{respectively} \qquad \frac{\partial\rho}{\partial t}+\mathrm{div}(\rho\vec{v})=0 \tag{1}$$

$$\frac{\partial\rho}{\partial t}+\rho\frac{\partial v_i}{\partial x_i}+v_j\frac{\partial\rho}{\partial x_j}=0 \tag{2}$$

$$\frac{D\rho}{Dt}+\rho\frac{\partial v_i}{\partial x_i}=0 \qquad \text{with} \qquad \frac{D}{Dt}=\frac{\partial}{\partial t}+v_j\frac{\partial}{\partial x_j} \tag{3}$$

Momentum equation :

Euler equations (viscous terms are neglected) :

$$\frac{\partial v_i}{\partial t}+v_j\frac{\partial v_i}{\partial x_j}+\frac{1}{\rho}\frac{\partial p}{\partial x_i}=0 \qquad \text{respectively} \qquad \frac{\partial\vec{v}}{\partial t}+(\vec{v}\cdot\nabla)\vec{v}+\frac{1}{\rho}\mathrm{grad}p=0 \tag{4}$$

$$\frac{Dv_i}{Dt}+\frac{1}{\rho}\frac{\partial p}{\partial x_i}=0 \tag{5}$$

Reformulation with help of continuity equation :

$$\frac{\partial(\rho v_i)}{\partial t}+\frac{\partial(\rho v_i v_j)}{\partial x_j}+\frac{\partial p}{\partial x_i}=0 \tag{6}$$

Energy equation :

$$\frac{\partial e}{\partial t}+\frac{\partial}{\partial x_i}\left[v_i(e+p)\right]=0 \qquad \text{respectively} \qquad \frac{\partial e}{\partial t}+\mathrm{div}\left[\vec{v}(e+p)\right]=0 \tag{7}$$

with:

$$e=\rho u+\frac{1}{2}\rho|v_i|^2 \qquad \text{fluid energy density (per unit volume)} \tag{8}$$

$$u=\frac{p}{\rho(\kappa-1)} \qquad \text{specific internal energy} \tag{9}$$

with:

$$u=c_V T=c_V\frac{p}{\rho R}=c_V\frac{p}{\rho}\frac{1}{c_V(\kappa-1)}=\frac{p}{\rho(\kappa-1)} \tag{10}$$

$$e=\frac{p}{\kappa-1}+\frac{1}{2}\rho|v_i|^2 \tag{11}$$

Reformulation with help of continuity and momentum equation :

$$\rho\frac{Du}{Dt}+p\frac{\partial v_i}{\partial x_i}=0 \qquad \text{respectively} \qquad \frac{Dp}{Dt}+\kappa p\frac{\partial v_i}{\partial x_i}=0 \tag{12}$$

N.3.1 Thermodynamic relationships

The law of energy conservation $dq = du + pdv - dr$ (13)

with:

dq	supplied heat
du	internal energy
pdv	mechanical work
dr	friction loss

Internal energy :

u specific internal energy per unit volume

$$du = Tds - pd\left(\frac{1}{\rho}\right) = c_v dT \tag{14}$$

$$u = \frac{p}{\rho(\kappa-1)} \tag{15}$$

Fluid energy :

e fluid energy density per unit volume

$$e = \rho u + \frac{1}{2}\rho|\vec{v}|^2 = \frac{p}{\kappa-1}+\frac{1}{2}\rho|\vec{v}|^2 \tag{16}$$

Enthalpy :

$B = h + \frac{1}{2}v^2 (= h_s)$ total enthalpy, stagnation enthalpy, per unit volume (17)

$h = u + \frac{p}{\rho}$ specific enthalpy per unit volume (18)

$$dh = Tds + \frac{dp}{\rho} = c_p dT \tag{19}$$

Entropy :

s specific entropy per unit volume

$$ds = \frac{dq+dr}{T} \tag{20}$$

Further relationships :

$p = \rho RT$	equation of state of an ideal gas	(21)
$R = c_p - c_v$	gas constant, equal to the difference in the specific heats	(22)
$\kappa = \dfrac{c_p}{c_v}$	ratio of the specific heats	(23)
$c^2 = \kappa \dfrac{p_0}{\rho_0}$	isentropic speed of sound	(24)

The pressure-density relation :

$$d\rho = \left(\frac{\partial \rho}{\partial p}\right)_s dp + \left(\frac{\partial \rho}{\partial s}\right)_p ds = \frac{dp}{c^2} + \left(\frac{\partial \rho}{\partial T}\right)_p \left(\frac{\partial T}{\partial s}\right)_p ds = \frac{dp}{c^2} - \frac{\beta \rho T}{c_p} ds = \frac{dp}{c^2} - \frac{\rho}{c_p} ds \tag{25}$$

with:

$\left(\dfrac{\partial \rho}{\partial p}\right)_s = \dfrac{1}{c^2}$	isentropic sound speed	(26)
$c_p = T\left(\dfrac{\partial s}{\partial T}\right)_p$	specific heat at constant pressure	(27)
$\beta = -\dfrac{1}{\rho}\left(\dfrac{\partial \rho}{\partial T}\right)_p = \dfrac{1}{T}$	coefficient of expansion	(28)

N.3.2 Non-linear perturbation equations, non-linear Euler equations

General form of the three-dimensional fluid flow equations :

$$\frac{\partial E}{\partial t} + \frac{\partial F}{\partial x} + \frac{\partial G}{\partial y} + \frac{\partial H}{\partial z} = 0 \tag{29}$$

The general flow quantity A is split into a mean quantity A_0 and a perturbation quantity A', so that the nonlinear disturbance equations follow :

$$\frac{\partial E'}{\partial t} + \frac{\partial F'}{\partial x} + \frac{\partial G'}{\partial y} + \frac{\partial H'}{\partial z} + \frac{\partial F_n'}{\partial x} + \frac{\partial G_n'}{\partial y} + \frac{\partial H_n'}{\partial z} = Q = -\left(\frac{\partial F_0}{\partial x} + \frac{\partial G_0}{\partial y} + \frac{\partial H_0}{\partial z}\right) \tag{30}$$

with:

F', G', H'	linear perturbation terms
F_n', G_n', H_n'	non-linear perturbation terms
F_0, G_0, H_0	Q: sum of the divergence of mean convective fluxes

The abbreviations are calculated from the following three-dimensional equations:

Continuity equation :

$$\frac{\partial \rho}{\partial t}+\frac{\partial(\rho u)}{\partial x}+\frac{\partial(\rho v)}{\partial y}+\frac{\partial(\rho w)}{\partial z}=0 \tag{31}$$

Momentum equations :

$$\begin{aligned}
&\frac{\partial(\rho u)}{\partial t}+\frac{\partial(\rho uu)}{\partial x}+\frac{\partial(\rho uv)}{\partial y}+\frac{\partial(\rho uw)}{\partial z}+\frac{\partial p}{\partial x}=0\\
&\frac{\partial(\rho v)}{\partial t}+\frac{\partial(\rho uv)}{\partial x}+\frac{\partial(\rho vv)}{\partial y}+\frac{\partial(\rho wv)}{\partial z}+\frac{\partial p}{\partial y}=0\\
&\frac{\partial(\rho w)}{\partial t}+\frac{\partial(\rho uw)}{\partial x}+\frac{\partial(\rho vw)}{\partial y}+\frac{\partial(\rho ww)}{\partial z}+\frac{\partial p}{\partial z}=0
\end{aligned} \tag{32}$$

Energy equation :

$$\frac{\partial e}{\partial t}+\frac{\partial[u(e+p)]}{\partial x}+\frac{\partial[v(e+p)]}{\partial y}+\frac{\partial[w(e+p)]}{\partial z}=0 \tag{33}$$

$$\text{with:}\quad e=\frac{p}{\kappa-1}+\frac{1}{2}\rho\left(u^2+v^2+w^2\right) \tag{34}$$

The abbreviations in the general form of the nonlinear disturbance equations mean in the sequence of the equations below :

- Continuity equation
- Euler equation in x direction
- Euler equation in y direction
- Euler equation in z direction
- Energy equation

$$E'\begin{cases}\rho'\\ \rho_0 u'+\rho' u_0+\rho' u'\\ \rho_0 v'+\rho' v_0+\rho' v'\\ \rho_0 w'+\rho' w_0+\rho' w'\\ e'\end{cases}\qquad F'\begin{cases}\rho_0 u'+u_0\rho'\\ u_0^2\rho'+2\rho_0 u_0 u'+p'\\ \rho_0 u_0 v'+\rho_0 v_0 u'+u_0 v_0\rho'\\ \rho_0 u_0 w'+\rho_0 w_0 u'+u_0 w_0\rho'\\ u'(e_0+p_0)+u_0(e'_{lin}+p')\end{cases} \tag{35, 36}$$

$$G'\begin{cases}\rho_0 v'+v_0\rho'\\ \rho_0 u_0 u'+\rho_0 u_0 v'+u_0 v_0\rho'\\ v_0^2\rho'+2\rho_0 v_0 v'+p'\\ \rho_0 v_0 w'+\rho_0 w_0 v'+v_0 w_0\rho'\\ v'(e_0+p_0)+v_0(e'_{lin}+p')\end{cases}\qquad H'\begin{cases}\rho_0 w'+w_0\rho'\\ \rho_0 w_0 u'+\rho_0 u_0 w'+u_0 w_0\rho'\\ \rho_0 w_0 v'+\rho_0 v_0 w'+v_0 w_0\rho'\\ w_0^2\rho'+2\rho_0 w_0 w'+p'\\ w'(e_0+p_0)+w_0(e'_{lin}+p')\end{cases} \tag{37, 38}$$

$$F'_n \begin{cases} \rho' u' \\ 2u_0\rho' u' + \rho_0 u'^2 + \rho' u'^2 \\ \rho_0 u'v' + u_0\rho' v' + v_0\rho' u' + \rho' u'v' \\ \rho_0 u'w' + u_0\rho' w' + w_0\rho' u' + \rho' u'w' \\ u'(e' + p') + u_0 e'_{nonlin} \end{cases} \qquad G'_n \begin{cases} \rho' v' \\ \rho_0 u'v' + u_0\rho' v' + v_0\rho' u' + \rho' u'v' \\ 2v_0\rho' v' + \rho_0 v'^2 + \rho' v'^2 \\ \rho_0 v'w' + v_0\rho' w' + w_0\rho' v' + \rho' v'w' \\ v'(e' + p') + v_0 e'_{nonlin} \end{cases} \tag{39, 40}$$

$$H'_n \begin{cases} \rho' w' \\ \rho_0 u'w' + u_0\rho' w' + w_0\rho' u' + \rho' u'w' \\ \rho_0 v'w' + v_0\rho' w' + w_0\rho' v' + \rho' v'w' \\ 2w_0\rho' w' + \rho_0 w'^2 + \rho' w'^2 \\ w'(e' + p') + w_0 e'_{nonlin} \end{cases} \tag{41}$$

$$F_0 \begin{cases} \rho_0 u_0 \\ \rho_0 u_0^2 + p_0 \\ \rho_0 u_0 v_0 \\ \rho_0 u_0 w_0 \\ u_0(e_0 + p_0) \end{cases} \qquad G_0 \begin{cases} \rho_0 v_0 \\ \rho_0 u_0 v_0 \\ \rho_0 v_0^2 + p_0 \\ \rho_0 v_0 w_0 \\ v_0(e_0 + p_0) \end{cases} \qquad H_0 \begin{cases} \rho_0 w_0 \\ \rho_0 u_0 w_0 \\ \rho_0 v_0 w_0 \\ \rho_0 w_0^2 + p_0 \\ w_0(e_0 + p_0) \end{cases} \tag{42, 43, 44}$$

Fluctuating part of the energy density :

$$e' = \frac{p'}{\kappa - 1} + \frac{1}{2}\rho'\left(u_0^2 + v_0^2 + w_0^2\right) + \frac{1}{2}(\rho_0 + \rho')\left(u'^2 + v'^2 + w'^2\right) + + (\rho_0 + \rho')(u'u_0 + v'v_0 + w'w_0) \tag{45}$$

subdivided in two terms:

• linear term

$$e'_{lin} = \frac{p'}{\kappa - 1} + \frac{1}{2}\rho'\left(u_0^2 + v_0^2 + w_0^2\right) + \rho_0(u'u_0 + v'v_0 + w'w_0) \tag{46}$$

• non-linear term

$$e'_{nonlin} = \frac{1}{2}(\rho_0 + \rho')\left(u'^2 + v'^2 + w'^2\right) \tag{47}$$

Stationary part of the energy density :

$$e_0 = \frac{p_0}{\kappa - 1} + \frac{1}{2}\rho_0\left(u_0^2 + v_0^2 + w_0^2\right) \tag{48}$$

N.3.3 Formulation of the Euler equations to use in the computational aeroacoustics (CAA)

Basic equations :

$$\frac{D\rho}{Dt}+\rho\frac{\partial v_i}{\partial x_i}=0 \tag{49}$$

$$\frac{Dv_i}{Dt}+\frac{1}{\rho}\frac{\partial p}{\partial x_i}=0 \tag{50}$$

Decomposition : $$\vec{v}(\vec{x},t)=\vec{v}_0(\vec{x},t)+\vec{v}'(\vec{x},t) \tag{51}$$

with: $$\vec{v}_0(\vec{x},t)=\frac{1}{T}\int_t^{t+T}\vec{v}(\vec{x},t)\,dt \tag{52}$$

$$T_1 \ll T \underset{\approx}{\overset{\ll}{-}} T_2 \tag{53}$$

T_1 time scale of the turbulent fluctuations

T_2 time scale of the large variations of the mean flow

$$\frac{1}{T}\int_t^{t+T}\vec{v}'(\vec{x},t)\,dt=\overline{\vec{v}'}=0 \tag{54}$$

The short forms $v=v_0+v'$; $p=p_0+p'$; $\rho=\rho_0+\rho'$ are introduced : (55)

- in the continuity equation :

$$\frac{\partial\rho'}{\partial t}+v_{0j}\frac{\partial\rho'}{\partial x_j}+v'_j\frac{\partial\rho_0}{\partial x_j}+\rho_0\frac{\partial v'_i}{\partial x_i}+\rho'\frac{\partial v_{0i}}{\partial x_i}=-\frac{\partial\rho_0}{\partial t}-v_{0j}\frac{\partial\rho_0}{\partial x_j}-v'_j\frac{\partial\rho'}{\partial x_j}-\rho_0\frac{\partial v_{0i}}{\partial x_i}-\rho'\frac{\partial v'_i}{\partial x_i} \tag{56}$$

with condition of the incompressibility: $\rho_0=\text{const.}$ and $\frac{\partial v_{0i}}{\partial x_i}=0$:

$$\frac{\partial\rho'}{\partial t}+v_{0j}\frac{\partial\rho'}{\partial x_j}+\rho_0\frac{\partial v'_i}{\partial x_i}=-v'_j\frac{\partial\rho'}{\partial x_j}-\rho'\frac{\partial v'_i}{\partial x_i} \tag{57}$$

- in the momentum equation :

$$\frac{\partial v'_i}{\partial t}+v_{0j}\frac{\partial v'_i}{\partial x_j}+v'_j\frac{\partial v_{0i}}{\partial x_j}-\frac{\rho'}{\rho_0^2}\frac{\partial p_0}{\partial x_i}+\frac{1}{\rho_0}\frac{\partial p'}{\partial x_i}=-\frac{\partial v_{0i}}{\partial t}-v_{0j}\frac{\partial v_{0i}}{\partial x_j}-v'_j\frac{\partial v'_i}{\partial x_j}+\frac{\rho'}{\rho_0^2}\frac{\partial p'}{\partial x_i}-\frac{1}{\rho_0}\frac{\partial p_0}{\partial x_i}$$

with: $$\frac{1}{\rho}=\frac{1}{\rho_0+\rho'}=\frac{\frac{1}{\rho_0}}{1+\frac{\rho'}{\rho_0}}\approx\frac{1}{\rho_0}\left(1-\frac{\rho'}{\rho_0}\right)=\frac{1}{\rho_0}-\frac{\rho'}{\rho_0^2} \tag{59}$$

$(\rho' \ll \rho_0)$

Averaging over the time T_1 :

$$\frac{\partial v_{0i}}{\partial t}+v_{0j}\frac{\partial v_{0i}}{\partial x_j}+\frac{1}{\rho_0}\frac{\partial p_0}{\partial x_i}=-\overline{v'_j\frac{\partial v'_i}{\partial x_j}}+\overline{\frac{\rho'}{\rho_0^2}\frac{\partial p'}{\partial x_i}} \tag{60}$$

$$\frac{\partial v'_i}{\partial t}+v_{0j}\frac{\partial v'_i}{\partial x_j}+v'_j\frac{\partial v_{0i}}{\partial x_j}-\frac{\rho'}{\rho_0^2}\frac{\partial p_0}{\partial x_i}+\frac{1}{\rho_0}\frac{\partial p'}{\partial x_i}=-v'_j\frac{\partial v'_i}{\partial x_j}+\frac{\rho'}{\rho_0^2}\frac{\partial p'}{\partial x_i}+\overline{v'_j\frac{\partial v'_i}{\partial x_j}}-\overline{\frac{\rho'}{\rho_0^2}\frac{\partial p'}{\partial x_i}} \tag{61}$$

Introducing the decomposition in turbulent and acoustic fluctuating quantities:

$$\rho'=\rho_t+\rho_a \quad ; \quad \vec{v}'=\vec{v}_t+\vec{v}_a \quad ; \quad p'=p_t+p_a \tag{62}$$

furthermore introducing the approximate condition of incompressibility of turbulent flow: $\rho_t\approx 0$ and $\partial v_{ti}/\partial x_i=0$ results :

- for the continuity equation :

$$\frac{\partial \rho_a}{\partial t}+v_{0j}\frac{\partial \rho_a}{\partial x_j}+\rho_0\frac{\partial v_{ai}}{\partial x_i}=Q_{at}^{\text{cont.eq.}}+Q_{aa}^{\text{cont.eq.}} \tag{63}$$

with:

$$Q_{at}^{\text{cont.eq.}}=-v_{tj}\frac{\partial \rho_a}{\partial x_j}-\rho'\frac{\partial v'_i}{\partial x_i} \tag{64}$$

$$Q_{aa}^{\text{cont.eq.}}=-v_{aj}\frac{\partial \rho_a}{\partial x_j}-\rho_a\frac{\partial v_{ai}}{\partial x_i} \tag{65}$$

- for the momentum equation :

$$\frac{\partial v_{ai}}{\partial t}+v_{0j}\frac{\partial v_{ai}}{\partial x_j}+\frac{1}{\rho_0}\frac{\partial p_a}{\partial x_i}+v_{aj}\frac{\partial v_{0i}}{\partial x_j}-\frac{\rho_a}{\rho_0^2}\frac{\partial p_0}{\partial x_i}=Q_{ot}^{\text{mom.eq.}}+Q_{tt}^{\text{mom.eq.}}+Q_{at}^{\text{mom.eq.}}+Q_{aa}^{\text{mom.eq.}}$$

with:

$$Q_{0t}^{\text{mom.eq.}}=-\frac{\partial v_{ti}}{\partial t}-v_{0j}\frac{\partial v_{ti}}{\partial x_j}-v_{tj}\frac{\partial v_{0i}}{\partial x_j}-\frac{1}{\rho_0}\frac{\partial p_t}{\partial x_i} \tag{67}$$

sound source due to the interaction between the mean flow and turbulent flow: partly shear noise;

$$Q_{tt}^{\text{mom.eq.}}=-v_{tj}\frac{\partial v_{ti}}{\partial x_j}+\overline{v_{tj}\frac{\partial v_{ti}}{\partial x_j}} \tag{68}$$

sound generated by turbulent interaction: self-noise;

$$Q_{at}^{\text{mom.eq.}}=-v_{tj}\frac{\partial v_{ai}}{\partial x_j}-v_{aj}\frac{\partial v_{ti}}{\partial x_j}+\frac{\rho_a}{\rho_0^2}\frac{\partial p_t}{\partial x_i}+\overline{v_{tj}\frac{\partial v_{ai}}{\partial x_j}}+\overline{v_{aj}\frac{\partial v_{ti}}{\partial x_j}}-\overline{\frac{\rho_a}{\rho_0^2}\frac{\partial p_t}{\partial x_i}} \tag{69}$$

sound generated by interaction between turbulence and sound;

$$Q_{aa}^{mom.eq.} = -v_{aj}\frac{\partial v_{ai}}{\partial x_j} + \frac{\rho_a}{\rho_0^2}\frac{\partial p_a}{\partial x_i} + \overline{v_{aj}\frac{\partial v_{ai}}{\partial x_j}} - \overline{\frac{\rho_a}{\rho_0^2}\frac{\partial p_a}{\partial x_i}} \tag{70}$$

sound generated by sound, e. g. scattering of sound.

N.4 The equations of linear acoustics

The basic equations of fluid mechanics to use in the acoustics are :

• equation of continuity, the law of conservation of mass :

$$\frac{\partial \rho}{\partial t} + \frac{\partial(\rho v_i)}{\partial x_i} = \dot{M} \quad \text{(in tensor notation, double suffix summation convention)} \tag{1}$$

with: $\dot{M}$ mass flux per unit volume;

• equation of motion, the law of conservation of momentum, Newton's law of motion, momentum equation, Navier-Stokes equations (for a viscous fluid) :

$$\rho\frac{\partial v_i}{\partial t} + \rho v_j \frac{\partial v_i}{\partial x_j} = \rho F_i - \frac{\partial}{\partial x_j}\left(p\delta_{ij} - \tau_{ij}\right) \tag{2}$$

and in the form due to Reynolds :

$$\frac{\partial}{\partial t}(\rho v_i) + \frac{\partial}{\partial x_j}\left(\rho v_i v_j + p_{ij}\right) = \rho F_i + \dot{M} v_i \tag{3}$$

with: $p_{ij} = p\delta_{ij} - \tau_{ij}$ (4)

• equation of state (the pressure-density relation) :

$$d\rho = \frac{dp}{c^2} - \frac{\rho}{c_p} ds \tag{5}$$

and with the isentropic condition $ds = 0$:

$$d\rho = \frac{dp}{c^2} \tag{6}$$

Premises of linear acoustics :

Assumptions, applied to the basic equations of fluidmechanics :

• without mass sources: $\dot{M} = 0$;

• without external forces: $F_i = 0$;

- inviscid flow: $p_{ij} = p \quad (\tau_{ij} = 0)$;

- decomposition of all physical quantities in mean values and fluctuating components:

$$p = \overline{p} + p' \qquad \rho = \overline{\rho} + \rho' \qquad v_i = \overline{v}_i + v'_i \ ;$$

with the following assumptions and definitions:

$\overline{p}, \overline{\rho}$ constant in time and space;

$p' << \overline{p}$ (with: p' sound pressure);

$\rho' << \overline{\rho}$ (with: ρ' acoustic fluctuation of mass density);

$\overline{v}_i = 0$ without mean flow;

v'_i acoustic part of fluid velocity, particle velocity;

- linearisation of equations of fluid mechanics;

- irrotational flow: $\vec{\omega} = \mathrm{rot}\vec{v} = \nabla \times \vec{v} = 0$.

The basic equations of linear acoustics :

$$\frac{\partial \rho'}{\partial t} + \overline{\rho}\frac{\partial v'_i}{\partial x_i} = 0$$ linearised continuity equation; (7)

$$\overline{\rho}\frac{\partial v'_i}{\partial t} + \frac{\partial p'}{\partial x_i} = 0$$ linearised Euler equation (momentum equation); (8)

$$\frac{p'}{\rho'} = c^2$$ linearised equation of state (9)

with: p', ρ', v'_i sound pressure, acoustic density fluctuation, particle velocity;

The homogeneous wave equation of linear acoustics;
there are the following homogeneous wave equations:

- pressure fluctuations (sound pressure): $$\frac{\partial^2 p}{\partial x_i^2} - \frac{1}{c_0^2}\frac{\partial^2 p}{\partial t^2} = 0 \qquad (10)$$

- density fluctuations : $$\frac{\partial^2 \rho}{\partial x_i^2} - \frac{1}{c_0^2}\frac{\partial^2 \rho}{\partial t^2} = 0 \qquad (11)$$

- velocity fluctuations : $$\frac{\partial^2 v_i}{\partial x_j^2} - \frac{1}{c_0^2}\frac{\partial^2 v_i}{\partial t^2} = 0 \quad \text{resp.} \quad \Delta\vec{v} - \frac{1}{c_0^2}\frac{\partial^2 \vec{v}}{\partial t^2} = 0 \qquad (12)$$

with: $$\Delta = \frac{\partial^2}{\partial x_i^2} = \frac{\partial^2}{\partial x^2} + \frac{\partial^2}{\partial y^2} + \frac{\partial^2}{\partial z^2}$$ Laplace operator (13)

- velocity potential : $$\frac{\partial^2 \Phi}{\partial x_i^2} - \frac{1}{c_0^2}\frac{\partial^2 \Phi}{\partial t^2} = 0 \quad \text{resp.} \quad \Delta\Phi - \frac{1}{c_0^2}\frac{\partial^2 \Phi}{\partial t^2} = 0 \qquad (14)$$

The wave equation for uniform flow :

$$\Delta\Phi - \left(\frac{1}{c_0}\frac{\partial}{\partial t} + \frac{U_\infty}{c_0}\frac{\partial}{\partial x}\right)^2 \Phi = 0 \qquad \text{convective wave equation} \tag{15}$$

$$\Delta\Phi - \frac{1}{c_0^2}\frac{D^2\Phi}{Dt^2} = 0 \tag{16}$$

with: $$\frac{D^2}{Dt^2} = \left(\frac{\partial}{\partial t} + U_\infty\frac{\partial}{\partial x}\right)^2 = \frac{\partial^2}{\partial t^2} + 2U_\infty\frac{\partial^2}{\partial x\partial t} + U_\infty^2\frac{\partial^2}{\partial x^2} \tag{17}$$

U_∞ uniform time-averaged velocity in the x direction;

$$\left(1 - \frac{U_\infty^2}{c_0^2}\right)\frac{\partial^2\Phi}{\partial x^2} + \frac{\partial^2\Phi}{\partial y^2} + \frac{\partial^2\Phi}{\partial z^2} - 2\frac{U_\infty}{c_0^2}\frac{\partial^2\Phi}{\partial x\partial t} - \frac{1}{c_0^2}\frac{\partial^2\Phi}{\partial t^2} = 0 \tag{18}$$

$$\left[\Delta - \left(M\frac{\partial}{\partial x} + \frac{1}{c_0}\frac{\partial}{\partial t}\right)^2\right]\Phi = 0 \tag{19}$$

with: $M = \dfrac{U_\infty}{c_0}$ Mach number;

With harmonic components, separating the time factor $e^{j\omega t}$:

$$\left\{\Delta - M^2\frac{\partial^2}{\partial x^2} - 2jkM\frac{\partial}{\partial x} + k^2\right\}\Phi = 0 \tag{20}$$

with: Φ complex amplitude of the velocity potential,

$k = \omega/c$ wave number.

Cylindrical co-ordinate system:
(with Laplace operator in cylindrical co-ordinates)

$$\Delta\Phi - \frac{1}{c_0^2}\frac{D^2\Phi}{Dt^2} = 0 \qquad \text{convective wave equation of the velocity potential} \tag{21}$$

becomes :

$$\left(\frac{\partial^2}{\partial x^2} + \frac{\partial^2}{\partial r^2} + \frac{1}{r}\frac{\partial}{\partial r} + \frac{1}{r^2}\frac{\partial^2}{\partial \Theta^2} - \frac{1}{c_0^2}\frac{D^2}{Dt^2}\right)\Phi = 0 \tag{22}$$

Spherical co-ordinate system:
(with Laplace operator in spherical co-ordinates)

$$\left\{\left[\frac{1}{r^2}\frac{\partial}{\partial r}\left(r^2\frac{\partial}{\partial r}\right)\right] + \frac{1}{r^2\sin^2\vartheta}\frac{\partial^2}{\partial\varphi^2} + \frac{1}{r^2\sin\vartheta}\frac{\partial}{\partial\vartheta}\left(\sin\vartheta\frac{\partial}{\partial\vartheta}\right) - \frac{1}{c_0^2}\frac{D^2}{Dt^2}\right\}\Phi = 0 \tag{23}$$

N.5 The inhomogeneous wave equation, Lighthill's acoustic analogy

Ref.: Lighthill, [N.13]; Howe [N.14]; Crighton et al., [N.15]; Goldstein, [N.16]; Curle, [N.17]

N.5.1 Lighthill's inhomogeneous wave equation

Equation of continuity, the law of conservation of mass :

$$\frac{\partial \rho}{\partial t} + \frac{\partial (\rho v_i)}{\partial x_i} = 0 \tag{1}$$

Equation of motion, in fact in the form due to Reynolds :

$$\frac{\partial}{\partial t}(\rho v_i) + \frac{\partial}{\partial x_j}\left(\rho v_i v_j + p_{ij}\right) = 0 \tag{2}$$

with: $p_{ij} = p\delta_{ij} - \tau_{ij}$ compressive stress tensor (3)

$\tau_{ij} = \mu\left(\frac{\partial v_i}{\partial x_j} + \frac{\partial v_j}{\partial x_i}\right) - \frac{2}{3}\mu\frac{\partial v_k}{\partial x_k}\delta_{ij}$ viscous stresses (4)

LIGHTHILL's equation is obtained by taking the time derivative of the continuity equation and subtracting the divergence of the momentum equation :

$$\frac{\partial}{\partial t}(\text{continuity equation}) - \frac{\partial}{\partial x_i}(\text{momentum equation})$$

eliminating the term ρv_i , that is the mass density flux in the continuity equation but the momentum density in the momentum equation :

$$\frac{\partial^2 \rho}{\partial t^2} = \frac{\partial^2}{\partial x_i \partial x_j}\left(\rho v_i v_j + p_{ij}\right) \tag{5}$$

addition the term $-c_0^2\frac{\partial^2 \rho}{\partial x_i^2}$ (with c_0 characteristic speed of sound in the medium surrounding the flow region), gives :

$$\frac{\partial^2 \rho}{\partial t^2} - c_0^2\frac{\partial^2 \rho}{\partial x_i^2} = \frac{\partial^2}{\partial x_i \partial x_j}\left(\rho v_i v_j + p_{ij} - c_0^2\rho\delta_{ij}\right) = q \tag{6}$$

LIGHTHILL's inhomogeneous wave equation (6)

with: $T_{ij} = \rho v_i v_j + p_{ij} - c_0^2\rho\delta_{ij} = \rho v_i v_j + \left(p - c_0^2\rho\right) - \tau_{ij}$ (7)

T_{ij} Lighthill's stress tensor ,

q source term.

The fluid mechanical problem of calculating the aerodynamic sound is formally equivalent to solving this equation for the radiation into a stationary ideal fluid produced by a distribution of quadrupole sources whose strength per unit volume is the Lighthill stress tensor T_{ij} (HOWE).

Incompressible approximation: $T_{ij} \approx \rho_0 v_i v_j$ (8)

with assumptions:

- low Mach number, velocity fluctuations are of order $\rho_0 Ma^2$,
- isentropic flow,
- high Reynolds number, viscous effects are much smaller than inertial effects, the viscous stress tensor is neglected compared with the Reynolds stresses $\rho v_i v_j$,
- furthermore: viscous terms in T_{ij} are

$$\tau_{ij} = \mu \frac{\partial v_i}{\partial x_j}, \text{ so that } \frac{\partial^2 T_{ij}}{\partial x_i \partial x_j} = \mu \frac{\partial^3 v_i}{\partial x_j \partial x_i \partial x_j}, \qquad (9)$$

corresponding to a octupole source (a very ineffective sound radiator).

$T_{ij} \approx \rho_0 v_i v_j$ can be used as a source term, generating the acoustic field.

Lighthill's development can be expanded as an inhomogeneous wave equation in general form, starting from :

- equation of continuity with external sources :

$$\frac{\partial \rho}{\partial t} + \frac{\partial(\rho v_i)}{\partial x_i} = \dot{M} \qquad (10)$$

with: $\dot{M}$ external source flux of mass (per unit volume);

- equation of motion with external forces :

$$\rho \frac{\partial v_i}{\partial t} + \rho v_j \frac{\partial v_i}{\partial x_j} = F_i - \frac{\partial}{\partial x_j}\left(p\delta_{ij} - \tau_{ij}\right) \qquad (11)$$

with: F_i external forces (per unit volume),

or in the form due to Reynolds:

$$\frac{\partial}{\partial t}(\rho v_i) + \frac{\partial}{\partial x_j}\left(\rho v_i v_j + p_{ij}\right) = F_i + \dot{M} v_i \qquad (12)$$

Inhomogeneous wave equation in general form :

$$\frac{\partial^2\rho}{\partial t^2} - c_0^2\frac{\partial^2\rho}{\partial x_i^2} = \frac{\partial\dot{M}}{\partial t} - \frac{\partial}{\partial x_i}\left(F_i + \dot{M}v_i\right) + \frac{\partial^2}{\partial x_i \partial x_j}\left(\rho v_i v_j + p_{ij} - c_0^2\rho\delta_{ij}\right) = q \tag{13}$$

Source terms :

$$q = \frac{\partial\dot{M}}{\partial t} - \frac{\partial}{\partial x_i}\left(\dot{M}v_i + F_i\right) + \frac{\partial^2}{\partial x_i \partial x_j}T_{ij} \tag{14}$$

with:

$\frac{\partial\dot{M}}{\partial t}$ monopole source (15)

$-\frac{\partial}{\partial x_i}\left(F_i + \dot{M}v_i\right)$ dipole source (16)

$\frac{\partial^2 T_{ij}}{\partial x_i \partial x_j}$ quadrupole source (17)

N.5.2 Solutions of inhomogeneous wave equation

Using the generalisation of Kirchhoff's equation :

$$p(x_i,t) = \frac{1}{4\pi}\int_S\left(\frac{1}{c_0 r}\frac{\partial r}{\partial n}\frac{\partial p}{\partial t} + \frac{1}{r^2}\frac{\partial r}{\partial n}p + \frac{1}{r}\frac{\partial p}{\partial n}\right)_\tau dS + \frac{1}{4\pi}\int_V\left(\frac{q}{r}\right)_\tau dV \tag{18}$$

a formal solution of the inhomogeneous wave equation follows :

$$\begin{aligned} p(x_i,t) = {} & \int_V \frac{1}{4\pi r}\left(\frac{\partial\dot{M}}{\partial t}\right)_\tau dV - \int_S \frac{1}{4\pi r}\left(\frac{\partial(\rho v_i)}{\partial t}\right)_\tau n_i dS \\ & - \frac{\partial}{\partial x_i}\int_V \frac{1}{4\pi r}\left(F_i + \dot{M}v_i\right)_\tau dV + \frac{\partial}{\partial x_i}\int_S \frac{1}{4\pi r}\left(\rho v_i v_j + p_{ij}\right)_\tau n_j dS \\ & + \frac{\partial^2}{\partial x_i \partial x_j}\int_V \frac{1}{4\pi r}\left(T_{ij}\right)_\tau dV \end{aligned} \tag{19}$$

with:

$(....)_\tau$ retarded source strength,

$\tau = t - \frac{r}{c_0}$ retardation time,

$r = |x_i - y_i|$ distance between source point and observer point,

x_i vector from origin of co-ordinates to the observer point,

y_i vector from origin of co-ordinates to the source point.

Free-space solution :

$$p(x_i,t) = \int_V \frac{1}{4\pi r}\left(\frac{\partial\dot{M}}{\partial t}\right)_\tau dV - \frac{\partial}{\partial x_i}\int_V \frac{1}{4\pi r}\left(F_i + \dot{M}v_i\right)_\tau dV + \frac{\partial^2}{\partial x_i \partial x_j}\int_V \frac{1}{4\pi r}\left(T_{ij}\right)_\tau dV \tag{20}$$

Decomposition into the near-field and the far-field solution : (21)

$$p(x_i,t)=\int_V \frac{1}{4\pi r}\left(\frac{\partial \dot{M}}{\partial t}\right)_\tau dV+$$

$$+\frac{1}{4\pi}\int_V \frac{1}{r^2}\left(F_i+\dot{M}v_i\right)_\tau \frac{(x_i-y_i)}{r}dV+\frac{1}{4\pi c_0}\int_V \frac{1}{r}\frac{\partial}{\partial t}\left(F_i+\dot{M}v_i\right)_\tau \frac{(x_i-y_i)}{r}dV+$$

$$+\frac{1}{4\pi}\int_V \frac{1}{r^2}\left(T_{ij}\right)_\tau\left[\frac{3}{r^2}(x_i-y_i)(x_j-y_j)-\delta_{ij}\right]dV+\frac{1}{4\pi c_0^2}\int_V \frac{1}{r}\frac{\partial^2\left(T_{ij}\right)_\tau}{\partial t^2}\left[\frac{1}{r^2}(x_i-y_i)(x_j-y_j)\right]dV$$

with the far-field solution :

$$p(x_i,t)=\int_V \frac{1}{4\pi r}\left(\frac{\partial \dot{M}}{\partial t}\right)_\tau dV+\frac{1}{4\pi c_0}\int_V \frac{1}{r}\frac{\partial}{\partial t}\left(F_i+\dot{M}v_i\right)_\tau \frac{(x_i-y_i)}{r}dV+$$

$$+\frac{1}{4\pi c_0^2}\int_V \frac{1}{r}\frac{\partial^2\left(T_{ij}\right)_\tau}{\partial t^2}\left[\frac{1}{r^2}(x_i-y_i)(x_j-y_j)\right]dV \qquad (22)$$

Solution with solid boundaries in flow (CURLE) :

$$p(x_i,t)=\int_V \frac{1}{4\pi r}\left(\frac{\partial \dot{M}}{\partial t}\right)_\tau dV-\int_S \frac{1}{4\pi r}\left(\frac{\partial(\rho v_i)}{\partial t}\right)_\tau n_i dS$$

$$-\frac{\partial}{\partial x_i}\int_V \frac{1}{4\pi r}\left(F_i+\dot{M}v_i\right)_\tau dV+\frac{\partial}{\partial x_i}\int_S \frac{1}{4\pi r}\left(\rho v_i v_j+p_{ij}\right)_\tau n_j dS \qquad (23)$$

$$+\frac{\partial^2}{\partial x_i \partial x_j}\int_V \frac{1}{4\pi r}\left(T_{ij}\right)_\tau dV$$

with: surface S is stationary,
the body can have mass injection or suction: $v_i n_i = v_n \neq 0$,
body vibrations: $\dfrac{\partial v_n}{\partial t}\neq 0$.

Solution when the boundaries S are rigid and not permeable :

$$p(x_i,t)=\int_V \frac{1}{4\pi r}\left(\frac{\partial \dot{M}}{\partial t}\right)_\tau dV$$

$$-\frac{\partial}{\partial x_i}\int_V \frac{1}{4\pi r}\left(F_i+\dot{M}v_i\right)_\tau dV+\frac{\partial}{\partial x_i}\int_S \frac{1}{4\pi r}\left(p_{ij}\right)_\tau n_j dS \qquad (24)$$

$$+\frac{\partial^2}{\partial x_i \partial x_j}\int_V \frac{1}{4\pi r}\left(T_{ij}\right)_\tau dV$$

Sound from free turbulence :

$$p(x_i,t)=\frac{1}{4\pi}\int_V \frac{1}{r}\left(\frac{\partial^2 T_{ij}}{\partial y_i \partial y_j}\right)_\tau dV \tag{25}$$

$$p(x_i,t)=\frac{1}{4\pi}\frac{\partial^2}{\partial x_i \partial x_j}\int_V \frac{1}{r}\left(T_{ij}\right)_\tau dV \tag{26}$$

$$p(x_i,t)=\frac{1}{4\pi c_0^2}\int_V \frac{1}{r}\left\{\frac{\partial^2 \left(T_{ij}\right)}{\partial t^2}\right\}_\tau \left[\frac{1}{r^2}(x_i-y_i)(x_j-y_j)\right] dV \tag{27}$$

in the far-field $x_i >> y_i$:

$$p(x_i,t)=\frac{1}{4\pi r}\int_V \left(\frac{\partial^2 T_{ij}}{\partial y_i \partial y_j}\right)_\tau dV \tag{28}$$

$$p(x_i,t)=\frac{1}{4\pi r}\frac{\partial^2}{\partial x_i \partial x_j}\int_V \left(T_{ij}\right)_\tau dV \tag{29}$$

$$p(x_i,t)=\frac{1}{4\pi c_0^2 r}\frac{x_i x_j}{r^2}\int_V \left[\frac{\partial^2 \left(T_{ij}\right)}{\partial t^2}\right]_\tau dV \tag{30}$$

$$p(x_i,t)=\frac{1}{4\pi c_0^2 r}\int_V \left[\frac{\partial^2 \left(T_{rr}\right)}{\partial t^2}\right]_\tau dV \tag{31}$$

with: $T_{rr}=\frac{x_i x_j}{r^2}T_{ij}$.

Solutions using a Fourier transformation :

$$p(x_i,\omega)=-\frac{\omega^2}{4\pi c_0^2}\frac{e^{-j\omega x/c_0}}{r}\int_0^T \int_V e^{j\vec{k}\cdot\vec{y}}e^{-j\omega t}T_{rr}(y_i,t)\,dV\,dt \tag{32}$$

Solutions neglecting viscous stresses: $T_{ij}=\rho v_i v_j$:

$$p(x_i,t)=\frac{x_i x_j}{4\pi c_0^2 r^3}\int_V \frac{\partial^2 \left(\rho v_i v_j\right)_\tau}{\partial t^2} dV \tag{33}$$

N.6 Acoustic analogy with source terms using the pressure

Ref.: Ribner, [N.18]; Meecham, [N.19], [N.20];

N.6.1 Lighthill's representation of the source term with use of pressure

Reformulation of the source strength with use the equations of continuity and momentum to introduce the pressure as a source (see Lighthill) :

$$\frac{\partial}{\partial t}\left(\rho v_i v_j\right) = p_{ik}\frac{\partial v_j}{\partial x_k} + p_{jk}\frac{\partial v_i}{\partial x_k} - \frac{\partial}{\partial x_k}\left(\rho v_i v_j v_k + p_{ik} v_j + p_{jk} v_i\right) \tag{1}$$

neglecting the viscous stresses and the octupole source :

$$\frac{\partial}{\partial t}\left(\rho v_i v_j\right) \approx p\left(\frac{\partial v_i}{\partial x_j} + \frac{\partial v_j}{\partial x_i}\right) \tag{2}$$

calculating the source term :

$$\frac{\partial^2}{\partial t^2}T_{ij} \approx \frac{\partial^2}{\partial t^2}\left(\rho v_i v_j\right) \approx \frac{\partial}{\partial t}\left[p\left(\frac{\partial v_i}{\partial x_j} + \frac{\partial v_j}{\partial x_i}\right)\right] = \frac{\partial p}{\partial t}\left(\frac{\partial \overline{v}_i}{\partial x_j} + \frac{\partial \overline{v}_j}{\partial x_i}\right) + \frac{\partial}{\partial t}\left[p\left(\frac{\partial v_i'}{\partial x_j} + \frac{\partial v_j'}{\partial x_i}\right)\right] \tag{3}$$

first term: shear-noise,
second term: self-noise.

N.6.2 Pressure-source theory (Ribner)

Decomposition of the pressure fluctuations inside the flow $\quad p - p_0 = p' + p_a$

with:
- $p - p_0$ pressure fluctuations,
- p_0 constant pressure,
- p' pseudo-sound: pressure fluctuations in a nearly incompressible flow, that is, inside the flow pressure fluctuations are dominated by inertial effects rather than compressibility,
- p_a acoustic pressure.

Pseudo-sound pressure is a solution of the Poisson's equation:
use of divergence operator to the momentum equation

$$\frac{\partial}{\partial x_i}\left[\frac{\partial}{\partial t}(\rho v_i)\right]+\frac{\partial^2}{\partial x_i \partial x_j}\left(\rho v_i v_j + p_{ij}\right)=0 \tag{4}$$

Incompressible flow $\frac{\partial v_i}{\partial x_i}=0$, (without viscosity) leads to the Poisson equations :

$$\frac{\partial^2 p}{\partial x_i^2}=-\frac{\partial^2\left(\rho v_i v_j\right)}{\partial x_i \partial x_j} \tag{5}$$

$$\frac{\partial^2 p'}{\partial x_i^2}=-\frac{\partial^2\left(\rho v_i v_j\right)}{\partial x_i \partial x_j} \tag{6}$$

$$\frac{\partial^2 p'}{\partial x_i^2}=-\rho_0\frac{\partial^2\left(v_i v_j\right)}{\partial x_i \partial x_j}=-\rho_0\frac{\partial v_i}{\partial x_j}\frac{\partial v_j}{\partial x_i} \tag{7}$$

Inhomogeneous wave equation :

$$\frac{\partial^2 p}{\partial x_i^2}-\frac{1}{c_0^2}\frac{\partial^2 p}{\partial t^2}=\frac{\partial^2 p'}{\partial x_i^2} \tag{8}$$

and with $p-p_0=p'+p_a$:

$$\frac{\partial^2 p_a}{\partial x_i^2}-\frac{1}{c_0^2}\frac{\partial^2 p_a}{\partial t^2}=\frac{1}{c_0^2}\frac{\partial^2 p'}{\partial t^2} \qquad \text{Ribner's inhomogeneous wave equation} \tag{9}$$

The far-field solution :

$$p(x_i,t)=-\frac{1}{4\pi c_0^2}\int_V \frac{1}{r}\left(\frac{\partial^2 p'}{\partial t^2}\right)_\tau dV \tag{10}$$

in comparison with Lighthill :

$$p(x_i,t)=\frac{1}{4\pi c_0^2}\int_V \frac{1}{r}\frac{\partial^2\left(T_{ij}\right)_\tau}{\partial t^2}\left[\frac{1}{r^2}(x_i-y_i)(x_j-y_j)\right]dV \tag{11}$$

(In far-field both equations are identical!)

N.6.3 Pressure-source theory (Meecham)

Expanding the field about an incompressible flow, low Mach number fluctuating flow

$$p=p_0+p_a \tag{12}$$

$$\rho=\rho_0+\rho_1+\rho_a \tag{13}$$

$$v = v_0 + v_a \tag{14}$$

with: subscript 0 for incompressible flow quantities,
p_a, ρ_a, v_a acoustic field.

Definition of ρ_1 :
The change in density ρ_1 is caused by the nearly incompressible pressure change p_0 :

$$\frac{\overline{D}\rho_1}{Dt} = \frac{1}{c^2}\frac{\overline{D}p_0}{Dt} \tag{15}$$

with: $\frac{\overline{D}}{Dt} = \frac{\partial}{\partial t} + \vec{v}_0 \cdot \nabla$ substantial derivative, following the incompressible flow v_0

Wave equation :

$$\frac{1}{c_0^2}\frac{\partial^2 p_a}{\partial t^2} - \nabla^2 p_a = -\frac{\partial^2 \rho_1}{\partial t^2} + \nabla \cdot \left(\rho_1 \frac{\partial \vec{v}_0}{\partial t}\right) \tag{16}$$

The second source term on the right side may be neglected.

Reformulation with
$$\frac{\partial p_0}{\partial t} = c_0^2 \frac{\partial \rho_1}{\partial t} \tag{17}$$

leads to :

$$\frac{1}{c_0^2}\frac{\partial^2 p_a}{\partial t^2} - \nabla^2 p_a = -\frac{1}{c_0^2}\frac{\partial^2 p_0}{\partial t^2} \qquad \text{Meecham's inhomogeneous wave equation.} \tag{18}$$

Solution :

$$p(x_i, t) = -\frac{1}{4\pi c_0^2}\int_V \frac{1}{r}\left(\frac{\partial^2 p_0}{\partial t^2}\right)_\tau dV \tag{19}$$

(see Ribner's mentioned above).

The incompressible pressure fluctuations are the solution of the Poisson's differential equation :

$$\frac{\partial^2 p_0}{\partial x_i^2} = -\rho_0 \frac{\partial^2 v_i v_j}{\partial x_i \partial x_j} \tag{20}$$

N.7 Acoustic analogy with mean flow effects, in the form of convective inhomogeneous wave equation

Ref.: Phillips, [N.21]; Pao, [N.22]; Lilley, [N.23], [N.24], [N.35]; Legendre, [N.25]; Morfey, [N.26]; Goldstein/Howes, [N.27]; Ribner, [N.28]; Albring/Detsch/Dittmar, [N.29], [N.30], [N.31]

N.7.1 Phillips's convective inhomogeneous wave equation

$$\frac{D^2\Pi}{Dt^2}-\frac{\partial}{\partial x_i}\left[c^2\frac{\partial\Pi}{\partial x_i}\right]=\kappa\frac{\partial v_i}{\partial x_j}\frac{\partial v_j}{\partial x_i}+\frac{D}{Dt}\left(\frac{\kappa}{c_p}\frac{Ds}{Dt}\right)-\frac{\partial}{\partial x_i}\left\{\frac{\kappa}{\rho}\frac{\partial\tau_{ij}}{\partial x_j}\right\}+\frac{D}{Dt}\left(\frac{\kappa}{\rho}\dot{M}\right)-\frac{\partial}{\partial x_i}\left\{\frac{\kappa}{\rho}F_i\right\} \quad (1)$$

with: $\Pi=\ln\frac{p}{p_0}$ definition of a dimensionless logarithmic pressure ratio,

p_0 constant reference pressure,

$$\tau_{ij}=\mu\left(\frac{\partial v_i}{\partial x_j}+\frac{\partial v_j}{\partial x_i}\right)-\frac{2}{3}\mu\frac{\partial v_k}{\partial x_k}\delta_{ij} \quad \text{viscous stresses,} \quad (2)$$

$$\frac{D}{Dt}=\frac{\partial}{\partial t}+v_i\frac{\partial}{\partial x_i} \quad \text{substantial derivative.}$$

left-hand side: corresponding with a wave equation in a moving medium with variable speed of sound (PHILLIPS);

right-hand side: contains propagation terms (in the first member) and source terms, generation of pressure fluctuations by velocity fluctuations in the fluid, by effects of entropy fluctuations, of fluid viscosity and by external mass and force sources (GOLDSTEIN).

Neglecting the effects of viscosity and thermal conductivity, furthermore the external mass and force sources :

$$\frac{D^2\Pi}{Dt^2}-\frac{\partial}{\partial x_i}\left[c^2\frac{\partial\Pi}{\partial x_i}\right]=\kappa\frac{\partial v_i}{\partial x_j}\frac{\partial v_j}{\partial x_i} \quad (2)$$

in comparison with Lighthill's equation :

$$\frac{\partial^2\rho}{\partial t^2}-c_0^2\frac{\partial^2\rho}{\partial x_i^2}=\frac{\partial^2(\rho v_i v_j)}{\partial x_i\partial x_j} \quad (3)$$

(Both equations are identical with assumptions: low Mach number and constant speed of sound.)

For example: shear flow $\quad v_i = \overline{v}_i + v_i' \qquad \overline{v}_i = \{\overline{v}_x(y);0;0\}$

$$\frac{\overline{D}^2\Pi}{Dt^2} - \frac{\partial}{\partial x_i}\left[\overline{c^2}\frac{\partial\Pi}{\partial x_i}\right] = 2\kappa\frac{\partial\overline{v}_x}{\partial y}\frac{\partial v_y'}{\partial x} + \kappa\frac{\partial v_j'}{\partial x_i}\frac{\partial v_i'}{\partial x_j} \tag{4}$$

with: $\dfrac{\overline{D}}{Dt} = \dfrac{\partial}{\partial t} + \overline{v}_i\dfrac{\partial}{\partial x_i}$,

$\overline{c^2}$ mean value the square of the speed of sound ;

Right-hand side: first term: shear noise,

second term: self noise.

In comparison with the equation of sound propagation in non-uniform flow :

$$\frac{\overline{D}^2\Pi'}{Dt^2} - \frac{\partial}{\partial x_i}\left[c_0^2\frac{\partial\Pi'}{\partial x_i}\right] = 2\kappa\frac{\partial\overline{v}_x}{\partial y}\frac{\partial v_y'}{\partial x} + \kappa\left[\frac{\partial v_j'}{\partial x_i}\frac{\partial v_i'}{\partial x_j} - \overline{\frac{\partial v_j'}{\partial x_i}\frac{\partial v_i'}{\partial x_j}}\right] \tag{5}$$

with: $\Pi' = \ln\dfrac{p'}{p_0}$; $\qquad p' = p_a$.

N.7.2 Lilley's convective inhomogeneous wave equation

Continuity equation :
$$\frac{D\Pi}{Dt} + \kappa\,\mathrm{div}\,\vec{v} = \frac{\kappa}{c_p}\frac{Ds}{Dt} \tag{6}$$

with: $\Pi = \ln\dfrac{p}{p_0}$,

following from continuity equation in the form $\dfrac{1}{\rho}\dfrac{D\rho}{Dt} + \mathrm{div}\,\vec{v} = 0$,

with $\dfrac{d\rho}{\rho} = \dfrac{1}{\kappa}\dfrac{dp}{p} - \dfrac{ds}{c_p}$.

Momentum equation :
$$\kappa\frac{Dv_i}{Dt} = -c^2\frac{\partial\Pi}{\partial x_i} + \frac{\kappa}{\rho}\frac{\partial\tau_{ij}}{\partial x_j} \tag{7}$$

Lilley's wave equation :

$$\frac{D}{Dt}\left(\frac{D^2\Pi}{Dt^2} - \frac{\partial}{\partial x_i}\left[c^2\frac{\partial\Pi}{\partial x_i}\right]\right) + 2\frac{\partial v_j}{\partial x_i}\frac{\partial}{\partial x_j}\left[c^2\frac{\partial\Pi}{\partial x_i}\right] = -2\kappa\frac{\partial v_j}{\partial x_i}\frac{\partial v_k}{\partial x_j}\frac{\partial v_i}{\partial x_k} + $$
$$+2\frac{\partial v_j}{\partial x_i}\frac{\partial}{\partial x_j}\left[\frac{\kappa}{\rho}\frac{\partial\tau_{ij}}{\partial x_j}\right] - \frac{D}{D\tau}\left\{\frac{\partial}{\partial x_i}\left[\frac{\kappa}{\rho}\frac{\partial\tau_{ij}}{\partial x_j}\right]\right\} + \frac{D^2}{Dt^2}\left[\frac{\kappa}{c_p}\frac{Ds}{Dt}\right] \tag{8}$$

interpretation: third-order equation,
left-hand side contains all propagation effects,
right-hand side includes all source terms .

Neglecting the effects of viscosity and thermal conductivity, furthermore introducing the mean values in the propagations terms on the left-hand sides,
replacing v_i by $\overline{v}_i$ and c^2 by $\overline{c^2}$:

$$\frac{\overline{D}}{Dt}\left(\frac{\overline{D}^2\Pi}{Dt^2}-\frac{\partial}{\partial x_i}\left[\overline{c^2}\frac{\partial\Pi}{\partial x_i}\right]\right)+2\frac{\partial\overline{v}_j}{\partial x_i}\frac{\partial}{\partial x_j}\left[\overline{c^2}\frac{\partial\Pi}{\partial x_i}\right]=-2\kappa\frac{\partial v_j}{\partial x_i}\frac{\partial v_k}{\partial x_j}\frac{\partial v_i}{\partial x_k} \tag{9}$$

<u>For example: shear flow</u> : $v_i=\overline{v}_i+v_i'$ $\quad\overline{v}_i=\{\overline{v}_x(y);0;0\}$

$$\frac{\overline{D}}{Dt}\left(\frac{\overline{D}^2\Pi}{Dt^2}-\frac{\partial}{\partial x_i}\left[c^2\frac{\partial\Pi}{\partial x_i}\right]\right)+2\frac{\partial\overline{v}_x}{\partial y}\frac{\partial}{\partial x}\left[c^2\frac{\partial\Pi}{\partial y}\right]=q \tag{10}$$

with: q different source terms but no terms which are linear in fluctuating velocities.

In comparison with equation applicable to the sound propagation in a shear flow :

$$\frac{\overline{D}}{Dt}\left(\frac{1}{c_0^2}\frac{\overline{D}^2\Pi'}{Dt^2}-\frac{\partial^2\Pi'}{\partial x_i^2}\right)+2\frac{\partial\overline{v}_x}{\partial y}\frac{\partial^2\Pi'}{\partial x\partial y}=0 \tag{11}$$

$$\frac{\overline{D}}{Dt}\left(\frac{1}{c_0^2}\frac{\overline{D}^2p_a}{Dt^2}-\frac{\partial^2p_a}{\partial x_i^2}\right)+2\frac{\partial\overline{v}_x}{\partial y}\frac{\partial^2p_a}{\partial x\partial y}=0 \tag{12}$$

(This equation follows as a result of the Lilley's equation with assumptions:
$q=0$ $\quad\Pi=\ln\frac{p}{p_0}=\ln\frac{p_0+p_a}{p_0}\approx\frac{p_a}{p_0}=\Pi'$ with $p_a\ll p_0$ $\quad c^2=c_0^2$)

N.7.3 Lilley's wave equation with a new Lighthill stress tensor

Definition of a new Lighthill stress tensor :

$$T_{ij}'=\rho v_i'v_j'-\tau_{ij}+\left(p-c_\infty^2\rho\right)\delta_{ij} \tag{13}$$

with: v' velocity fluctuations; that is: the Lighthill tensor involves only quadratic fluctuations of the velocity field.

Assuming a uniform mean flow: $\overline{v}_i=\{\overline{v}_x(y);0;0\}$;
decomposition: $v_i=\overline{v}_i+v_i'$;

convective operator for the mean flow $\quad \frac{\overline{D}}{Dt}=\frac{\partial}{\partial t}+\overline{v}_x(y)\frac{\partial}{\partial x}$;

this leads to the generalised linear convective wave equation, a third-order equation :

$$\frac{\overline{D}}{Dt}\left(\frac{\overline{D}^2}{Dt^2}-c_0^2\Delta\right)\rho+2c_0^2\frac{d\overline{v}_x}{dy}\frac{\partial^2\rho}{\partial x\partial y}=\frac{\overline{D}}{Dt}\frac{\partial^2 T'_{ij}}{\partial x_i\partial x_j}-2\frac{d\overline{v}_x}{dy}\frac{\partial^2 T'_{yi}}{\partial x\partial x_j} \tag{14}$$

on the left-side hand: all linear fluctuating terms;

on the right-hand side: generation terms, all quadratic in the fluctuations.

N.7.4 Convected wave equation for the dilatation (Legendre)

$$\frac{D}{Dt}\left(\frac{1}{c^2}\frac{D\Theta}{Dt}\right)-\Delta\Theta=-\frac{D}{Dt}\left(\frac{1}{c^2}\frac{\partial v_i}{\partial x_j}\frac{\partial v_j}{\partial x_i}-\frac{\nabla \ln h}{c^2}\frac{Dv_i}{Dt}\right)-\frac{\Delta v_i}{c^2}\frac{Dv_i}{Dt}-2\frac{\partial v_j}{\partial x_i}\frac{\partial}{\partial x_j}\left(\frac{1}{c^2}\frac{Dv_i}{Dt}\right) \tag{15}$$

with: $\Theta=\nabla\cdot\vec{v}=-\frac{1}{\rho}\frac{D\rho}{Dt}=-\frac{D}{Dt}\left(\ln\frac{\rho}{\rho_0}\right)$ dilatation

h specific enthalpy .

N.7.5 Goldstein's third-order inhomogeneous wave equation

Assumptions: parallel shear flow in the x direction, mean velocity U, density $\overline{\rho}$, sound speed $\overline{c}$: all independent of x and t:

$$\frac{\overline{D}}{Dt}\left[\frac{1}{\overline{c}^2}\frac{\overline{D}^2 p}{Dt^2}-\overline{\rho}\nabla\cdot\left(\frac{1}{\overline{\rho}}\nabla p\right)\right]+2\frac{\partial}{\partial x}(\nabla U\cdot\nabla p)=\overline{\rho}\left\{\frac{\overline{D}^2 q}{Dt^2}-\frac{\overline{D}}{Dt}\left(\nabla\cdot\vec{f}\right)+2\frac{\partial}{\partial x}\left(\vec{f}\cdot\nabla U\right)\right\} \tag{16}$$

with: $\frac{\overline{D}}{Dt}=\frac{\partial}{\partial t}+U\frac{\partial}{\partial x}$ material derivative,

$\vec{f}$ force per unit mass,

q mass flux per unit mass,

$\nabla\cdot\vec{v}=q-\frac{1}{\rho}\frac{D\rho}{Dt}$ source term in the continuity equation.

N.7.6 Goldstein-Howes inhomogeneous wave equation

Assumptions: high Reynolds number, no entropy fluctuations;
wave operator is linearised by assuming that only the first order interaction between the mean flow and the fluctuating field is retained;
the mean flow is decomposed into a mean part and a fluctuating part: $v_i=U(y)\delta_{xi}+v'_i$.

Introducing the convective derivative : $\frac{D}{Dt}=\frac{\partial}{\partial t}+U\frac{\partial}{\partial x}$

use of Phillips wave equation : $\frac{D^2\Pi}{Dt^2}-c_0^2\nabla^2\Pi-2\kappa\frac{\partial U}{\partial y}\frac{\partial v_y'}{\partial x}=\kappa\frac{\partial v_j'}{\partial x_i}\frac{\partial v_i'}{\partial x_j}$

convective derivative and introduction the momentum equation (no viscosity) leads to :

$$\frac{D}{Dt}\left\{\frac{D^2\Pi}{Dt^2}-c_0^2\nabla^2\Pi\right\}+2c_0^2\frac{dU}{dy}\frac{\partial^2\Pi}{\partial x\partial y}=\kappa\frac{D}{Dt}\left(\frac{\partial v_i'}{\partial x_j}\frac{\partial v_j'}{\partial x_i}\right)-2\kappa\frac{\partial U}{\partial y}\frac{\partial}{\partial x}\left(v_i'\frac{\partial v_y'}{\partial x_i}\right) \tag{17}$$

This is an acoustic analogy, like Lighthill's equation.

Further assumptions: acoustic fluctuations are negligible compared to the turbulent fluctuations in the source volume, turbulent velocity is incompressible, these lead to another formulation of the wave equation :

$$\begin{aligned}\frac{D}{Dt}\left\{\frac{D^2\Pi}{Dt^2}-c_0^2\nabla^2\Pi\right\}+2c_0^2\frac{dU}{dy}\frac{\partial^2\Pi}{\partial x\partial y}&=\\=\kappa\frac{\partial^2}{\partial x_i\partial x_j}\left(\frac{Dv_i'v_j'}{Dt}\right)-4\kappa\frac{\partial U}{\partial y}\frac{\partial}{\partial x}\left(v_i'\frac{\partial v_y'}{\partial x_i}\right)-\kappa\frac{\partial}{\partial x}\left(\frac{d^2U}{dy^2}v_y'^2\right)\end{aligned} \tag{18}$$

A new form of the inhomogeneous convective wave equation :

$$\begin{aligned}\frac{D^2\Gamma}{Dt^2}-c_0^2\nabla^2\Gamma=\kappa\frac{\partial^2}{\partial x_i\partial x_j}\left(\frac{Dv_i'v_j'}{Dt}\right)-4\kappa\frac{\partial}{\partial x}\left[\frac{\partial U}{\partial y}\frac{\partial}{\partial x_i}\left(v_i'v_y'+\frac{c_0^2}{\kappa}\pi\delta_{yi}\right)\right]\\-\kappa\frac{\partial}{\partial x}\left[\frac{d^2U}{dy^2}\left(v_y'^2+\frac{c_0^2}{\kappa}\pi\right)\right]\end{aligned} \tag{19}$$

with the new variable: $$\Gamma=\frac{D\Pi}{Dt}=\frac{\partial\Pi}{\partial t}+U\frac{\partial\Pi}{\partial x} \tag{20}$$

alternatively :

$$\frac{D^2\Gamma}{Dt^2}-c_0^2\nabla^2\Gamma=\kappa\frac{\partial^2}{\partial x_i\partial x_j}\left(\frac{Dv_i'v_j'}{Dt}\right)+4\kappa\frac{\partial}{\partial x}\left[\frac{\partial U}{\partial y}\frac{Dv_y'}{Dt}\right]-\kappa\frac{\partial}{\partial x}\left[\frac{d^2U}{dy^2}\left(v_y'^2+\frac{c_0^2}{\kappa}\pi\right)\right] \tag{21}$$

Simplification of the source term:

In the mixing layer of a jet: gradient of the mean velocity is constant $\frac{d}{dy}\left(\frac{dU}{dy}\right)=0$

from which follows the source term q (right-hand side of wave equation) :

$$q=\kappa\frac{\partial^2}{\partial x_i\partial x_j}\left(\frac{Dv_i'v_j'}{Dt}\right)+4\kappa\frac{\partial}{\partial x}\left[\frac{\partial U}{\partial y}\frac{Dv_y'}{Dt}\right] \tag{22}$$

N.7.7 Ribner's recent reformulation of Lighthill's source term

Lighthill's wave equation:

$$\frac{1}{c_0^2}\frac{\partial^2 p}{\partial t^2}-\Delta p=\frac{\partial^2 \rho v_i v_j}{\partial x_i \partial x_j}+\frac{1}{c_0^2}\frac{\partial^2 p}{\partial t^2}-\frac{\partial^2 \rho}{\partial t^2} \tag{23}$$

Assumptions :

- instantaneous local velocity: $v_i=\overline{v}_i+v_i'$,
- unidirectional, transversely sheared, mean flow: $\overline{v}_i=\left[U(y),0,0\right]$.

$$\frac{1}{c_0^2}\frac{\partial^2 p}{\partial t^2}-\Delta p=\frac{\partial^2 \rho v_i' v_j'}{\partial x_i \partial x_j}+2\frac{\partial U}{\partial y}\frac{\partial \rho v_y'}{\partial x}+\frac{1}{c_0^2}\frac{\partial^2 p}{\partial t^2}-\frac{\overline{D}^2\rho}{Dt^2} \tag{24}$$

with: $\frac{\overline{D}}{Dt}=\frac{\partial}{\partial t}+U\frac{\partial}{\partial x}$ convective derivative following the mean flow.

With the approximation: $\frac{\overline{D}^2\rho}{Dt^2}=\frac{1}{\overline{c}^2}\frac{\overline{D}^2 p}{Dt^2}$ $(\overline{c}=\overline{c}(\vec{x})$ local time-averaged sound speed) :

$$\frac{1}{\overline{c}^2}\frac{\overline{D}^2 p}{Dt^2}-\Delta p=\frac{\partial^2 \rho v_i' v_j'}{\partial x_i \partial x_j}+2\frac{\partial U}{\partial y}\frac{\partial \rho v_y'}{\partial x} \tag{25}$$

In the case of an exact wave equation with the following notations:

$v_i=\overline{v}_i+v_i'$, $\langle v_i\rangle_{av}=\overline{v}_i(\vec{x})$, $\rho=\overline{\rho}(\vec{x})+\rho'$, $\langle\rho\rangle_{av}=\overline{\rho}(\vec{x})$,

there follows the wave equation :

$$\begin{aligned}\frac{1}{c_0^2}\frac{\partial^2 p}{\partial t^2}-\Delta p=\;&2\rho\frac{\partial \overline{v}_i}{\partial x_j}\frac{\partial v_j'}{\partial x_i}+\frac{1}{c_0^2}\frac{\partial^2 p}{\partial t^2}-\frac{\overline{D}^2\rho}{Dt^2}+\frac{\partial^2 v_i' v_j'}{\partial x_i \partial x_j}+2v_j'\frac{\partial \overline{v}_i}{\partial x_j}\frac{\partial \rho}{\partial x_i}+\\&+2\frac{\partial v_i' v_j'}{\partial x_j}\frac{\partial \rho}{\partial x_i}+v_i' v_j'\frac{\partial^2 \rho}{\partial x_i \partial x_j}+2\frac{\partial}{\partial x_j}\left(\rho v_j'\frac{\partial \overline{v}_i}{\partial x_i}\right)+\rho\left(\frac{\partial \overline{v}_j}{\partial x_i}\frac{\partial \overline{v}_i}{\partial x_j}+\frac{\partial \overline{v}_i}{\partial x_i}\frac{\partial \overline{v}_j}{\partial x_j}\right)\end{aligned} \tag{26}$$

$$\text{with:}\quad \frac{\overline{D}^2}{Dt^2}=\left(\frac{\partial}{\partial t}+\overline{v}_i\frac{\partial}{\partial x_i}\right)^2=\frac{\partial^2}{\partial t^2}+2\overline{v}_i\frac{\partial^2}{\partial x_i \partial t}+\overline{v}_i\overline{v}_j\frac{\partial^2}{\partial x_i \partial x_j} \tag{27}$$

N.7.8 Inhomogeneous wave equation including the stream function (Albring/Detsch)

Inhomogeneous wave equation in the approximated form of Lighthill

$$\frac{1}{c_0^2}\frac{\partial^2 p}{\partial t^2}-\frac{\partial^2 p}{\partial x_i^2}=\rho_0\frac{\partial^2\left(v_i v_j\right)}{\partial x_i \partial x_j}=q \tag{28}$$

Introducing the stream function in a two-dimensional flow

$$v_x = \frac{\partial \Psi}{\partial y} \qquad v_y = -\frac{\partial \Psi}{\partial x}$$

leads to :

$$q = 2\rho_0 \left[\left(\frac{\partial^2 \Psi}{\partial x \partial y} \right)^2 - \frac{\partial^2 \Psi}{\partial x^2} \frac{\partial^2 \Psi}{\partial y^2} \right] \tag{29}$$

Decomposition $\Psi = \overline{\Psi} + \Psi'$ gives for the source term :

$$q = 2\rho_0 \left\{ 2\left(\frac{\partial^2 \overline{\Psi}}{\partial x \partial y} \frac{\partial^2 \Psi'}{\partial x \partial y} \right) + \left(\frac{\partial^2 \Psi'}{\partial x \partial y} \right)^2 - \frac{\partial^2 \overline{\Psi}}{\partial x^2} \frac{\partial^2 \Psi'}{\partial y^2} - \frac{\partial^2 \overline{\Psi}}{\partial y^2} \frac{\partial^2 \Psi'}{\partial x^2} - \frac{\partial^2 \Psi'}{\partial x^2} \frac{\partial^2 \Psi'}{\partial y^2} \right\} \tag{30}$$

with: term number 1, 3, 4: shear noise;
term number 2, 5: self-noise .

N.8 Acoustic analogy in terms of vorticity, wave operators for enthalpy

Ref.: Powell, [N.32], [N.33]; Howe, [N.14], [N.34]; Crighton et al., [N.15]; Möhring, [N.36], [N.37]; Möhring/Obermeier, [N.38]; Doak, [N.39], [N.40];

N.8.1 Powell's theory of vortex sound

Assumption: fluid motion is isentropic.

Use of vector identities :

$$\nabla\left(\frac{1}{2}v^2\right) = (\vec{v}\cdot\nabla)\vec{v} - (\nabla\times\vec{v})\times\vec{v} = (\vec{v}\cdot\nabla)\vec{v} - \omega\times\vec{v} \tag{1}$$

with: $\vec{\omega} = \nabla\times\vec{v}$ vorticity vector ,

or :

$$\nabla\left(\frac{1}{2}v^2\right) = v_j \frac{\partial v_i}{\partial x_j} - \left[\left(\frac{\partial v_k}{\partial x_i} - \frac{\partial v_i}{\partial x_k} \right) v_k - \left(\frac{\partial v_j}{\partial x_i} - \frac{\partial v_i}{\partial x_j} \right) v_j \right] \tag{2}$$

of the continuity equation :

$$\frac{\partial \rho}{\partial t} + (\vec{v} \cdot \nabla)\rho + \rho \nabla \cdot \vec{v} = 0 \tag{3}$$

and of the momentum equation (neglecting the viscous stress tensor τ_{ij})

$$\rho \frac{\partial \vec{v}}{\partial t} + \rho(\vec{\omega} \times \vec{v}) + \rho \nabla\left(\frac{1}{2}|\vec{v}|^2\right) = -\nabla p \tag{4}$$

leads to an inhomogeneous wave equation :

$$\frac{1}{c_0^2}\frac{\partial^2 p_a}{\partial t^2} - \Delta p_a = \nabla\left\{\rho(\vec{\omega} \times \vec{v}) + \nabla\left(\rho \frac{v^2}{2}\right) - \frac{v^2}{2}\nabla\rho - \vec{v}\frac{\partial \rho}{\partial t} + \nabla\left(p - c_0^2 \rho\right)\right\}$$

With the assumption: $|\vec{v}| << c_0$ Powell's equation follows in the theory of vortex sound :

$$\frac{1}{c_0^2}\frac{\partial^2 p_a}{\partial t^2} - \Delta p_a = \nabla\left\{\rho(\vec{\omega} \times \vec{v}) + \nabla\left(\rho \frac{v^2}{2}\right)\right\} \tag{5}$$

Solution in free space :

$$p(\vec{x},t) = \frac{\partial}{\partial x_i}\int_V \frac{1}{4\pi r}\left[\rho(\vec{\omega} \times \vec{v})\right]_\tau dV + \frac{\partial^2}{\partial x_i^2}\int_V \frac{1}{4\pi r}\left[\frac{1}{2}\rho|\vec{v}|^2\right]_\tau dV \tag{6}$$

$$p(\vec{x},t) = \frac{\rho_0 x_i}{4\pi c_0 r^2}\frac{\partial}{\partial t}\int_V \left[(\vec{\omega} \times \vec{v})_i\right]_\tau dV + \frac{\rho_0}{4\pi c_0^2 r}\frac{\partial^2}{\partial t^2}\int_V \left[\frac{1}{2}|\vec{v}|^2\right]_\tau dV \tag{7}$$

Source terms :

first term: dipole source, vorticity distribution, incompressible approximation of the fluctuating flow, rate of change of vortex stretching by the fluid flow
$\rightarrow$ principal source of sound at low Mach number ;

second term: quadrupole source, isotropic quadrupole (three longitudinal quadrupoles with undirected radiation characteristic), rate of change of the kinetic energy of the source flow.

N.8.2 Howe's formulation of the acoustic analogy equation for the total enthalpy

Momentum equation (Crocco's formulation, neglecting the effects of the external forces) :

$$\frac{\partial v_i}{\partial t} + \frac{\partial B}{\partial x_i} = -(\vec{\omega} \times \vec{v}) + T\frac{\partial s}{\partial x_i} + \frac{1}{\rho}\frac{\partial \tau_{ij}}{\partial x_j} \tag{8}$$

with: thermodynamic relations:

$B = h + \frac{1}{2}v^2$ total enthalpy, stagnation enthalpy, per unit volume,

$h = u + \frac{p}{\rho}$ specific enthalpy (per unit volume),

u specific internal energy (per unit volume);

vector relations :

$\vec{\omega} = \mathrm{rot}\vec{v} = \mathrm{curl}\vec{v} = \nabla \times \vec{v}$ vorticity,

$\frac{D\vec{v}}{Dt} = \frac{\partial \vec{v}}{\partial t} + (\vec{v} \cdot \nabla)\vec{v} = \frac{\partial \vec{v}}{\partial t} + \vec{\omega} \times \vec{v} + \nabla\left(\frac{1}{2}v^2\right)$ vector identity,

$\vec{L} = (\vec{\omega} \times \vec{v})$ Lamb vector, unsteady vortical lifting force (per mass unit)

or $\vec{L} = (\vec{v} \cdot \vec{\nabla})\vec{v} = \vec{\omega} \times \vec{v} + \mathrm{grad}\frac{1}{2}v^2$ (9)

Continuity equation :

$$\frac{1}{\rho}\frac{D\rho}{Dt} + \mathrm{div}\,\vec{v} = 0 \quad \text{or} \quad \frac{1}{\rho c^2}\frac{Dp}{Dt} + \mathrm{div}\,\vec{v} = \frac{\beta T}{c_p}\frac{Ds}{Dt} \tag{10}$$

with: elimination of $\frac{D\rho}{Dt}$ by means of the relations :

$$d\rho = \frac{dp}{c^2} + \left(\frac{\partial \rho}{\partial s}\right)_p ds = \frac{dp}{c^2} + \left(\frac{\partial \rho}{\partial T}\right)_p \left(\frac{\partial T}{\partial s}\right)_p ds = \frac{dp}{c^2} - \frac{\beta \rho T}{c_p} ds \tag{11}$$

Subtraction the divergence of the momentum equation from the time derivative of the continuity equation gives :

$$\frac{\partial}{\partial t}\left(\frac{1}{\rho c^2}\frac{Dp}{Dt}\right) - \Delta B = \mathrm{div}\left(\vec{\omega} \times \vec{v} - T\nabla s - \vec{\sigma}\right) + \frac{\partial}{\partial t}\left(\frac{\beta T}{c_p}\frac{Ds}{Dt}\right) \tag{12}$$

with: $\sigma_i = \frac{1}{\rho}\frac{\partial \tau_{ij}}{\partial x_j}$,

τ_{ij} viscous stress tensor .

Howe's inhomogeneous wave equation in terms of the total enthalpy :

$$\left\{\frac{D}{Dt}\left(\frac{1}{c^2}\frac{D}{Dt}\right) - \frac{\nabla p \cdot \nabla}{\rho c^2} - \Delta\right\}B = \left(\mathrm{div} + \frac{\nabla p}{\rho c^2}\right) \cdot \left(\vec{\omega} \times \vec{v} - T\nabla s - \vec{\sigma}\right) + \frac{\partial}{\partial t}\left(\frac{\beta T}{c_p}\frac{Ds}{Dt}\right) + \\ + \frac{D}{Dt}\left[\frac{1}{c^2}\left(T\frac{Ds}{Dt} + \vec{v} \cdot \vec{\sigma}\right)\right] \tag{13}$$

Special cases :

• Absence of viscous dissipation and heat transfer, with momentum equation $\rho\frac{D\vec{v}}{Dt} = -\nabla p$

$$\left\{\frac{D}{Dt}\left(\frac{1}{c^2}\frac{D}{Dt}\right) + \frac{1}{c^2}\frac{D\vec{v}}{Dt}\cdot\nabla - \Delta\right\}B = \left[\operatorname{div} - \frac{1}{c^2}\frac{D\vec{v}}{Dt}\right]\left(\vec{\omega}\times\vec{v} - T\nabla s\right) \qquad (14)$$

• High Reynolds number, homentropic flow:
(dissipation and heat transfer are neglected, $s = \text{const.}$)

$$\left\{\frac{D}{Dt}\left(\frac{1}{c^2}\frac{D}{Dt}\right) - \frac{1}{\rho}\nabla\cdot(\rho\nabla)\right\}B = \frac{1}{\rho}\operatorname{div}\left(\rho\vec{\omega}\times\vec{v}\right) \qquad (15)$$

• Low Mach number, $\rho = \rho_0$ and $c = c_0$, neglecting non-linear effects of propagation and the scattering of sound by the vorticity :

$$\left(\frac{1}{c_0^2}\frac{\partial^2}{\partial t^2} - \Delta\right)B = \operatorname{div}\left(\vec{\omega}\times\vec{v}\right) \qquad (16)$$

• Nonhomentropic source flow, fluid is temporarily incompressible, dissipation is ignored, mean flow is irrotational, mean velocity $\vec{v}(\vec{x})$, density $\rho(\vec{x})$, sound velocity $c(\vec{x})$:

$$\left\{\left(\frac{\partial}{\partial t} + \vec{v}\cdot\nabla\right)\left[\frac{1}{c^2}\left(\frac{\partial}{\partial t} + \vec{v}\cdot\nabla\right)\right] - \frac{1}{\rho}\nabla\cdot(\rho\nabla)\right\}B = \operatorname{div}\left(\vec{\omega}\times\vec{v} - T\nabla s\right) + \frac{\partial}{\partial t}\left(\frac{\beta T}{c_p}\frac{Ds}{Dt}\right) \qquad (17)$$

• At very small Mach number :

$$\left\{\frac{1}{c_0^2}\frac{\partial^2}{\partial t^2} - \Delta\right\}B = \operatorname{div}\left(\vec{\omega}\times\vec{v} - T\nabla s\right) + \frac{\partial}{\partial t}\left(\frac{\beta T}{c_p}\frac{Ds}{Dt}\right) \qquad (18)$$

right-hand side:
sources of noise are vorticity, entropy gradients and unsteady heating of the fluid; the last source is equivalent to a volume monopole of strength $q(x_i,t) = \frac{\beta T}{c_p}\frac{Ds}{Dt}$.

• At low Mach number, mean density, entropy and sound velocity are constant, isentropic flow ;

$$\left\{\frac{1}{c_0^2}\left(\frac{\partial}{\partial t} + \vec{v}\cdot\nabla\right)^2 - \Delta\right\}B = \operatorname{div}\left(\vec{\omega}\times\vec{v}\right) \qquad (19)$$

The acoustic pressure p can be calculated from the fluctuations in total enthalpy by

$$\frac{p}{\rho_0} = B \qquad (20)$$

This is a linearised relation of acoustic pressure p in the far field, to the first order in Mach number, furthermore considering a very low mean flow Mach number.

Formulation the wave equation in terms of the pressure :

$$\left\{\frac{1}{c_0^2}\frac{\partial^2}{\partial t^2}-\Delta\right\}p=\rho_0\,\mathrm{div}(\vec{\omega}\times\vec{v}) \tag{21}$$

The right-hand side represents an aerodynamic source for incompressible flow, the generated sound is dipole in nature, $\vec{\omega}\times\vec{v}$ is a force distribution ;
Lamb vector: $\vec{L}=(\vec{\omega}\times\vec{v})$, quantity ρL vortex force (per unit volume);
$\vec{v}$ can be calculated directly from $\vec{\omega}$:

$$\vec{v}(\vec{x},t)=\mathrm{curl}\left[\int\frac{\vec{\omega}(\vec{y},t)}{4\pi|\vec{x}-\vec{y}|}dV\right] \qquad \text{(Biot-Savart law)} \tag{22}$$

Therefore the source term depends only on the vorticity. See also the Helmholtz vorticity equation :

$$\frac{\partial\vec{\omega}}{\partial t}+\mathrm{curl}(\vec{\omega}\times\vec{v})=0 \tag{23}$$

It is a direct relationship between the non-linear term $\vec{\omega}\times\vec{v}$ and the linear term in the vorticity.

Solution :

$$p(x_i,t)=-\rho_0\int\frac{1}{r}\frac{\partial}{\partial y_i}(\vec{\omega}\times\vec{v})_i\left(y_i,t-\frac{r}{c_0}\right)dV \tag{24}$$

or: $$p(x_i,t)=-\rho_0\frac{\partial}{\partial x_i}\int\frac{1}{r}(\vec{\omega}\times\vec{v})_i\left(y_i,t-\frac{r}{c_0}\right)dV \tag{25}$$

In the far field (with the approximations of low Mach number flow, compact turbulent eddies):

$$p(x_i,t)=-\frac{\rho_0 x_i}{4\pi c_0 x^2}\frac{\partial}{\partial t}\int(\vec{\omega}\times\vec{v})_i\left(y_i,t-\frac{r}{c_0}\right)dV \tag{26}$$

$$p(x_i,t)=-\frac{\rho_0 x_i}{4\pi c_0^2 x^3}\frac{\partial^2}{\partial t^2}\int(\vec{x}\cdot\vec{y})\vec{x}\cdot(\vec{\omega}\times\vec{v})\left(y_i,t-\frac{r}{c_0}\right)dV \tag{27}$$

N.8.3 Möhring's equation with source term linear dependent on vorticity field

Definition of a vector Green's function: $\operatorname{curl}\vec{G} = \operatorname{grad} G$ (with: G a scalar Green's function).
Solution of the wave equation :

$$\left(\frac{1}{c_0^2}\frac{\partial^2}{\partial t^2} - \Delta\right)B = \operatorname{div}(\vec{\omega}\times\vec{v}) \tag{28}$$

Introduce a Green's function $G(\vec{y},\tau|\vec{x},t)$ which satisfies :

$$\left\{\frac{1}{c^2}\frac{\partial^2}{\partial\tau^2} - \frac{\partial^2}{\partial y_i^2}\right\}G = \delta(\vec{x}-\vec{y}, t-\tau) \tag{29}$$

Use the combination $$B(\vec{x},t) = \int_V \operatorname{div}(\vec{\omega}\times\vec{v})G\,dV d\tau \tag{30}$$

in the far field $$p_a(\vec{x},t) = -\rho_0 \int_V (\vec{\omega}\times\vec{v})_i \frac{\partial G}{\partial y_i} dV d\tau \tag{31}$$

with the vector Green's function $$p_a(\vec{x},t) = -\rho_0 \int_V (\vec{\omega}\times\vec{v}) \operatorname{curl}\vec{G}\, dV d\tau \tag{32}$$

with Helmholtz vorticity equation $$p_a(\vec{x},t) = \rho_0 \int_V \frac{\partial\vec{\omega}}{\partial\tau}\vec{G} dV d\tau \tag{33}$$

to obtain : $$p_a(\vec{x},t) = -\rho_0 \int \vec{\omega}\frac{\partial\vec{G}}{\partial\tau} dV d\tau = \rho_0 \frac{\partial}{\partial t}\int \vec{\omega}\vec{G} dV d\tau \tag{34}$$

This equation does not contain the flow velocity. It depends linearly on the vorticity field, that is, the contributions from several vortices add linearly.

Vorticity sound in a low Mach number flow in unbounded space (see Dowling):

$$p(\vec{x},t) = \frac{\rho_0}{12\pi c_0^2 r^3}\frac{\partial^3}{\partial t^3}\int (\vec{x}\cdot\vec{y})\vec{y}\cdot[\vec{\omega}\times\vec{x}]_{t-\frac{x}{c_0}} dV \tag{35}$$

right-hand side:
Only components of vorticity perpendicular to the observer's position vector $\vec{x}$ contribute to the sound far field.

N.8.4 Convected wave operators for total enthalpy in comparison

Möhring:

$$\boldsymbol{L}_{Möhring}B = \nabla\cdot\left(\rho\nabla B - \frac{\rho\vec{v}}{c^2}\frac{DB}{Dt}\right) - \frac{\partial}{\partial t}\frac{\rho}{c^2}\frac{DB}{Dt} = -\mathrm{div}\rho\vec{L} + \frac{\partial}{\partial t}\frac{\partial\rho}{\partial s}\frac{\partial s}{\partial t} + \mathrm{div}\frac{\partial\rho}{\partial s}\vec{v}\frac{\partial s}{\partial t} \tag{36}$$

with: $\vec{L} = \vec{\omega}\times\vec{v} - T\nabla s$

Möhring/Obermeier:

$$\boldsymbol{L}_{Möhring/Obermeier}B = \nabla\cdot(\rho\nabla B) - \rho\frac{D}{Dt}\left(\frac{1}{c^2}\frac{DB}{Dt}\right) = -\mathrm{div}\rho\vec{L} - \frac{\partial\vec{v}}{\partial t}\frac{\partial\rho}{\partial s}\nabla s \tag{37}$$

with: $\vec{L} = \vec{\omega}\times\vec{v} = \mathrm{curl}\vec{v}\times\vec{v}$

Howe:

$$\boldsymbol{L}_{Howe}B = \Delta B - \frac{1}{c^2}\frac{D\vec{v}}{Dt}\cdot\nabla B - \frac{D}{Dt}\left(\frac{1}{c^2}\frac{DB}{Dt}\right) \tag{38}$$

Doak:

$$\boldsymbol{L}_{Doak}B = \Delta B - \frac{1}{c^2}\left[\frac{\partial^2 B}{\partial t^2} + \left(2\vec{v}\frac{\partial}{\partial t} + \vec{\omega}\times\vec{v} + T\nabla s - 2\nabla h\right)\nabla B + \vec{v}\vec{v}\cdot\nabla\cdot\nabla B\right] \tag{39}$$

Those terms of these three convected wave operators that contain second derivatives of B agree. This means, that they agree in the high frequency limit of geometrical acoustics (MÖHRING/OBERMEIER).

N.8.5 Doak's theory of aerodynamic sound including the fluctuating total enthalpy as a basic generalized acoustic field for a fluid

Continuity equation :

$$\frac{\partial\rho}{\partial t} + \frac{\partial(\rho v_i)}{\partial x_i} = \dot{M} \tag{40}$$

with: $\dot{M}$ rate of mass creation per unit volume.

Momentum equation :

$$\frac{\partial(\rho v_i)}{\partial t} + \frac{\partial(\rho v_i v_j)}{\partial x_j} + \frac{\partial p_{ij}}{\partial x_j} = \dot{M}v_i + \dot{I}_i + F_i \tag{41}$$

with: $\dot{I}_i$ rate of production of momentum per unit volume by internal processes such as chemical reactions,

F_i rate of production of momentum per unit volume due to external forces such as gravitational and electromagnetic fields.

Energy equation :

$$\frac{\partial\left(\rho u+\frac{1}{2}\rho v_i^2\right)}{\partial t}+\frac{\partial\left(\rho u v_j+\frac{1}{2}\rho v_i^2 v_j+p_{ij}v_i-\lambda\frac{\partial T}{\partial x_j}\right)}{\partial x_j}=\dot{M}v_i^2+\dot{I}_i v_i+F_i v_i+\rho q+\rho Q \qquad (42)$$

with: u internal energy per unit volume,
λ coefficient of thermal conductivity,
Q rate of external heat addition per unit mass,
ρq rate of energy production per unit volume due to internal processes,
ρQ rate of energy production per unit volume due to external processes.

Inhomogeneous convected scalar wave equation for the fluctuating total enthalpy :

$$\frac{\partial^2 B'}{\partial x_i^2}-\left\{\frac{1}{c^2}\left[\frac{\partial^2 B'}{\partial t^2}+\left(2v_i\frac{\partial}{\partial t}-(\vec{v}\times\vec{\omega})_i+\boldsymbol{V}_i-2\frac{\partial h}{\partial x_i}\right)\frac{\partial B'}{\partial x_i}+v_i v_j\frac{\partial^2 B'}{\partial x_i\partial x_j}\right]\right\}'=$$

$$=\frac{\partial}{\partial x_i}\left[(\vec{v}\times\vec{\omega})_j'+\boldsymbol{V}_j'\right]-\left\{\frac{1}{c^2}\left[-(\vec{v}\times\vec{\omega})_i+\boldsymbol{V}_i-2\frac{\partial h}{\partial x_i}+v_i v_j\frac{\partial}{\partial x_j}\right]\left[\left((\vec{v}\times\vec{\omega})_j'+\boldsymbol{V}_j'\right)+\left[\frac{\partial v_i'}{\partial t}\right]\right]\right\}'$$

$$+\left[\frac{\partial}{\partial t}\left(\frac{1}{c^2}\right)'\frac{Dh}{Dt}\right]'-\left[\frac{\partial}{\partial t}\left(\frac{1}{R}\frac{Ds}{Dt}+\frac{\dot{M}}{\rho}\right)'\right]'-\left[\frac{1}{c^2}v_i\frac{\partial \boldsymbol{V}_i'}{\partial t}\right]'$$

with: fluctuating part of the quantity is denoted by the prime,
all quantities without primes are the sum of the respective mean and fluctuating parts;

$\frac{D}{Dt}=\frac{\partial}{\partial t}+v_i\frac{\partial}{\partial x_i}$ material derivative;

B' fluctuating total enthalpy, basic generalized acoustic field for a fluid (Doak);

$\boldsymbol{V}_i=T\frac{\partial s}{\partial x_i}+\frac{1}{\rho}\frac{\partial\tau_{ij}}{\partial x_i}+\frac{\dot{I}_i}{\rho}+\frac{F_i}{\rho}$ sum of accelerations (“forces” per unit mass)

with: first term vanishes if the motion is homentropic,
second term vanishes if the fluid is inviscid,
third term vanishes if there is no production of momentum by

internal processes such as chemical processes,
fourth term vanishes if there are no external forces.

Rewriting in a more compact form :

$$\frac{\partial^2 B'}{\partial x_i^2}-\left\{\frac{1}{c^2}\left[\frac{\partial^2}{\partial t^2}+\left(2v_i\frac{\partial}{\partial t}-(\vec{v}\times\vec{\omega})_i+\boldsymbol{V}_i-2\frac{\partial h}{\partial x_i}\right)\frac{\partial}{\partial x_i}+v_iv_j\frac{\partial^2}{\partial x_i\partial x_j}\right]B'\right\}'=$$

$$=\left\{\left[\frac{\partial}{\partial x_i}-\frac{1}{c^2}\left(-(\vec{v}\times\vec{\omega})_i+\boldsymbol{V}_i-2\frac{\partial h}{\partial x_i}+v_iv_j\frac{\partial}{\partial x_i}\right)\right]\left(\left[(\vec{v}\times\vec{\omega})_j'+\boldsymbol{V}_j'\right]+\left[\frac{\partial v_i'}{\partial t}\right]\right)\right\}' \quad (44)$$

$$+\left[\frac{\partial}{\partial t}\left(\frac{1}{c^2}\right)'\frac{Dh}{Dt}\right]'-\left[\frac{\partial}{\partial t}\left(\frac{1}{R}\frac{Ds}{Dt}+\frac{\dot{M}}{\rho}\right)'\right]'-\left[\frac{1}{c^2}v_i\frac{\partial\boldsymbol{V}_i'}{\partial t}\right]'$$

Reformulation:

The idealized case of the homentropic flow of the lossless fluid, under no external force, mass creation or heat addition :

$$\frac{\partial^2 B'}{\partial x_i^2}-\left\{\frac{1}{c^2}\left[\frac{\partial^2}{\partial t^2}+\left(2v_i\frac{\partial}{\partial t}-(\vec{v}\times\vec{\omega})_i-2\frac{\partial h}{\partial x_i}\right)\frac{\partial}{\partial x_i}+v_iv_j\frac{\partial^2}{\partial x_i\partial x_j}\right]B'\right\}'=$$

$$=\left\{\left[\frac{\partial}{\partial x_i}-\frac{1}{c^2}\left(-(\vec{v}\times\vec{\omega})_i-2\frac{\partial h}{\partial x_i}+v_iv_j\frac{\partial}{\partial x_i}\right)\right]\left((\vec{v}\times\vec{\omega})_j'+\left[\frac{\partial v_i'}{\partial t}\right]\right)\right\}'+\left[\frac{\partial}{\partial t}\left(\frac{1}{c^2}\right)'\frac{Dh}{Dt}\right]'$$

If the flow is also irrotational :

$$\frac{\partial^2 B'}{\partial x_i^2}-\left\{\frac{1}{c^2}\left[\frac{\partial^2}{\partial t^2}+\left(2v_i\frac{\partial}{\partial t}-2\frac{\partial h}{\partial x_i}\right)\frac{\partial}{\partial x_i}+v_iv_j\frac{\partial^2}{\partial x_i\partial x_j}\right]B'\right\}'=$$

$$=\left\{\left[\frac{\partial}{\partial x_i}-\frac{1}{c^2}\left(-2\frac{\partial h}{\partial x_i}+v_iv_j\frac{\partial}{\partial x_i}\right)\right]\left(\left[\frac{\partial v_i'}{\partial t}\right]\right)\right\}'+\left[\frac{\partial}{\partial t}\left(\frac{1}{c^2}\right)'\frac{Dh}{Dt}\right]' \quad (46)$$

If the flow is also time-stationary :

$$\frac{\partial^2 B'}{\partial x_i^2}-\left\{\frac{1}{c^2}\left[\frac{\partial^2}{\partial t^2}+\left(2v_i\frac{\partial}{\partial t}-2\frac{\partial h}{\partial x_i}\right)\frac{\partial}{\partial x_i}+v_iv_j\frac{\partial^2}{\partial x_i\partial x_j}\right]B'\right\}'=\left[\frac{\partial}{\partial t}\left(\frac{1}{c^2}\right)'\frac{Dh}{Dt}\right]' \quad (47)$$

For flows in which temperature variations are rather negligible :

$$\frac{\partial^2 B'}{\partial x_i^2}-\left\{\frac{1}{c^2}\left[\frac{\partial^2}{\partial t^2}+\left(2v_i\frac{\partial}{\partial t}-(\vec{v}\times\vec{\omega})_i\right)\frac{\partial}{\partial x_i}+v_iv_j\frac{\partial^2}{\partial x_i\partial x_j}\right]B'\right\}' =$$
$$=\left\{\left[\frac{\partial}{\partial x_i}-\frac{1}{c^2}\left(-(\vec{v}\times\vec{\omega})_i+v_iv_j\frac{\partial}{\partial x_i}\right)\right]\left((\vec{v}\times\vec{\omega})_j'+\left[\frac{\partial v_i'}{\partial t}\right]\right)\right\}' \tag{48}$$

In the field of aeroacoustics:

- external forces and heat addition are of relatively less significance,
- like the effects of temperature and entropy fluctuations, mean entropy gradients and viscous diffusion on propagation,
- also the dependence of the inhomogeneous source terms on these quantities is of less significance;
- therefore one can suggest neglecting them, and linearise the equation mentioned above in the fluctuations :

$$\frac{\partial^2 B'}{\partial x_i^2}-\frac{1}{\bar{c}^2}\left[\frac{\partial^2}{\partial t^2}+\left(2\bar{v}_i\frac{\partial}{\partial t}-\left(\bar{\vec{v}}\times\bar{\vec{\omega}}\right)_i-\frac{2}{\kappa-1}\frac{\partial\bar{c}^2}{\partial x_i}\right)\frac{\partial}{\partial x_i}+\bar{v}_i\bar{v}_j\frac{\partial^2}{\partial x_i\partial x_j}\right]B' =$$
$$=\left\{\frac{\partial}{\partial x_i}-\frac{1}{\bar{c}^2}\left[-\left(\bar{\vec{v}}\times\bar{\vec{\omega}}\right)_i-\frac{2}{\kappa-1}\frac{\partial\bar{c}^2}{\partial x_i}+\bar{v}_i\bar{v}_j\frac{\partial}{\partial x_j}\right]\left[\left(\vec{v}'\times\bar{\vec{\omega}}\right)_i+\left(\bar{\vec{v}}\times\vec{\omega}'\right)_i\right]\right\}$$
$$-\frac{1}{\bar{c}^2}\left[-\left(\vec{v}'\times\bar{\vec{\omega}}\right)_i-\left(\bar{\vec{v}}\times\vec{\omega}'\right)_i-\frac{2}{\kappa-1}\frac{\partial c^{2\prime}}{\partial x_i}+\left(v_i'\bar{v}_j+\bar{v}_iv_j'\right)\frac{\partial}{\partial x_j}\right]\left[\frac{\partial v_i'}{\partial t}\right] \tag{49}$$

with the approximation $\bar{h}=\dfrac{\bar{c}^2}{\kappa-1}\left(=\dfrac{\kappa R\bar{T}}{\kappa-1}\right)$ (50)

If the flow is time-stationary, then the last line of the last equation is zero.

N.9 Acoustic analogy with effects of solid boundaries

Ref.: Ffowcs Williams/Hawkings, [N.41]; Howe, [N.14]; Crighton et al., [N.15]; Farassat, [N.42]; Prieur/Rahier, [N.43]; Long/Watts, [N.44]; Pilon/Lyrintzis, [N.45]; Farassat/Myers, [N.46]; Brentner/Farassat, [N.47]; Brentner, [N.48]; Farassat, [N.49]

N.9.1 Ffowcs Williams-Hawkings (FW-H) inhomogeneous wave equation, FW-H equation in differential und integral form

The aeroacoustic analogy in the representation of Ffowcs Williams-Hawkings is a very general Lighthill's acoustic analogy. Primarily it is used for problems of sound generation by flow with moving boundaries and by moving sources interacting with such boundaries.

Notations in connection with moving surface S :

$f = f(x_i, t) = 0$,

$u_i(x_i, t)$ velocity of the surface,

$\vec{n} = \nabla f$; subscript n indicates the projection of a vector quantity in the surface normal direction,

$u_n = u_i n_i$, $v_n = v_i n_i$,

$|\nabla f| = 1$ on the surface,

$$\frac{Df}{Dt} = \frac{\partial f}{\partial t} + u_i \frac{\partial f}{\partial x_i} = 0 \ .$$

Heaviside function $H(f)$:

$H(f) = 1$ for $f(x_i, t) > 0$, that is in the fluid region, exterior of S ,

$H(f) = 0$ for $f(x_i, t) < 0$, that is in the volume enclosed by the surface S interior of S ,

$$\frac{\partial H(f)}{\partial t} = \frac{\partial H(f)}{\partial f}\frac{\partial f}{\partial t} = \delta(f)\frac{\partial f}{\partial t}$$
$$\frac{\partial H(f)}{\partial x_i} = \frac{\partial H(f)}{\partial f}\frac{\partial f}{\partial x_i} = \delta(f)\frac{\partial f}{\partial x_i} = \delta(f) n_i \qquad (1)$$

In what follows the function $(\rho - \rho_0)H(f)$ is defined in the continuity and momentum equations. That is, the derived equations are valid for all space. The term $(\rho - \rho_0)H(f)$ is determined only in the region of space that is of interest, i.e., space occupied by fluid.

Continuity equation :

$$\frac{\partial}{\partial t}(\rho - \rho_0) + \frac{\partial}{\partial x_i}(\rho v_i) = 0 \qquad (2)$$

$$\frac{\partial}{\partial t}\left[(\rho-\rho_0)H(f)\right]+\frac{\partial}{\partial x_i}\left[\rho v_i H(f)\right]=\left[\rho_0 u_i+\rho(v_i-u_i)\right]\delta(f)\frac{\partial f}{\partial x_i}$$
$$=\left[\rho_0 u_i+\rho(v_i-u_i)\right]\frac{\partial H(f)}{\partial x_i}=Q\delta(f) \tag{3}$$

with:
$$Q=\left[\rho_0 u_i+\rho(v_i-u_i)\right]\frac{\partial f}{\partial x_i}=\left[\rho_0 u_n+\rho(v_n-u_n)\right]=$$
$$=\rho_0\left[\left(1-\frac{\rho}{\rho_0}\right)u_n+\frac{\rho}{\rho_0}v_n\right]=\rho_0 U_n \tag{4}$$

Q mass flux per unit volume,
ρ_0 density of undisturbed medium,
u_n local normal velocity of the surface.

Momentum equation :

$$\frac{\partial}{\partial t}(\rho v_i)+\frac{\partial}{\partial x_j}\left(\rho v_i v_j+p_{ij}\right)=0 \tag{5}$$

with: $p_{ij}=p\delta_{ij}-\tau_{ij}$.

$$\frac{\partial}{\partial t}\left[\rho v_i H(f)\right]+\frac{\partial}{\partial x_j}\left[\left(\rho v_i v_j+p_{ij}\right)H(f)\right]=F_i\delta(f) \tag{6}$$

with: $$F_i=\left[p_{ij}+\rho v_i\left(v_j-u_j\right)\right]\frac{\partial f}{\partial x_j}=\left[p_{ij}n_j+\rho v_i\left(v_n-u_n\right)\right] \tag{7}$$

F_i force per unit area acting on the medium.

Eliminate $\rho v_i H(f)$ by cross differentiation :

$$\frac{\partial^2}{\partial t^2}\left[(\rho-\rho_0)H(f)\right]-c_0^2\frac{\partial^2\left[(\rho-\rho_0)H(f)\right]}{\partial x_i^2}=\frac{\partial}{\partial t}\left[Q\delta(f)\right]-\frac{\partial}{\partial x_i}\left[F_i\delta(f)\right]+$$
$$+\frac{\partial^2}{\partial x_i\partial x_j}\left[\left(\rho v_i v_j+p_{ij}\right)H(f)\right]-c_0^2\frac{\partial^2\left[(\rho-\rho_0)H(f)\right]}{\partial x_i^2} \tag{8}$$

and with Lighthill tensor $T_{ij}=\rho v_i v_j+p_{ij}-c_0^2(\rho-\rho_0)\delta_{ij}$ (9)

folloows :

<u>Ffowcs Williams-Hawkings equation in differential form</u>

$$\left\{\frac{\partial^2}{\partial t^2}-c_0^2\frac{\partial^2}{\partial x_i^2}\right\}\left[(\rho-\rho_0)H(f)\right]=\frac{\partial}{\partial t}\left[Q\delta(f)\right]-\frac{\partial}{\partial x_i}\left[F_i\delta(f)\right]+\frac{\partial^2}{\partial x_i\partial x_j}\left[T_{ij}H(f)\right] \tag{10}$$

This equation is valid throughout the whole of space. Inside the fluid, $H=1$, are Lighthill's quadrupole sources. In addition to these volume sources there exist surface sources in the form of dipoles and monopoles with :

dipole strength density:	surface stress,
monopole strength density:	rate at which mass is transmitted across unit area of surface.

Source contributions caused by the permeability of the surface:

- monopole source: $\frac{\partial}{\partial t}\left[\rho(v_n - u_n)\delta(f)\right]$
- dipole source: $-\frac{\partial}{\partial x_i}\left[\rho v_i(v_n - u_n)\delta(f)\right]$

<u>Ffowcs Williams-Hawkings equation in integral form</u> :

$$H c_0^2(\rho - \rho_0) = \frac{\partial^2}{\partial x_i \partial x_j}\int_{V(\tau)}\left[T_{ij}\right]_\tau \frac{dV}{4\pi|x_i - y_i|} - \frac{\partial}{\partial x_i}\int_{S(\tau)}\left[\rho v_i(v_j - u_j) + p_{ij}\right]_\tau \frac{dS_j(y_i)}{4\pi|x_i - y_i|}$$
$$+\frac{\partial}{\partial t}\int_{S(\tau)}\left[\rho(v_j - u_j) + \rho_0 u_j\right]_\tau \frac{dS_j(y_i)}{4\pi|x_i - y_i|} \quad (11)$$

with: $\tau = t - \frac{|x_i - y_i|}{c_0} = t - \frac{|\vec{x} - \vec{y}|}{c_0}$ retarded time,

surface integrals over the retarded surface $S(\tau)$ defined by $f(y_i, t) = 0$,
surface element dS_i directed into the region $V(\tau)$ where $f > 0$.

The control surface is a non-porous surface (impenetrable): $\rho(v_i - u_i) = 0$

Monopole term: $$Q = \rho_0 u_i \frac{\partial f}{\partial x_i} = \rho_0 u_n \quad (12)$$

Dipole term: $$F_i = p_{ij}\frac{\partial f}{\partial x_j} = p_{ij} n_j \quad (13)$$

with: $p_{ij} = p\delta_{ij} - \tau_{ij}$,

$F_i = p\frac{\partial f}{\partial x_i} = p n_i$, if viscous stresses are neglected;

p	local surface pressure,
n_i	local unit outward normal to the surface.,

Sources can be interpreted

- as a volume distribution of quadrupoles $\frac{\partial^2 T_{ij}}{\partial x_i \partial x_j}$ in the outer region of the surfaces, due to the turbulent flow,
- as a surface distribution of dipoles $-\frac{\partial}{\partial x_i}\left[F_i\delta(f)\right]$, due to the interaction of the flow with the moving bodies, especially due to surface pressure and stress fluctuations on

the bodies in the flow,

- as a surface distribution of monopoles $\frac{\partial}{\partial t}[Q\delta(f)]$, due to the kinematics of the bodies, especially from normal accelerations of the body surfaces.

Detailed representation of the Ffowcs Williams-Hawkings equation in differential form :

$$\Box^2[\phi H(f)] = \frac{1}{c_0^2}\frac{\partial^2}{\partial t^2}[\phi H(f)] - \frac{\partial^2}{\partial x_i^2}[\phi H(f)] =$$

$$= \frac{\partial^2}{\partial x_i \partial x_j}[T_{ij}H(f)] - \frac{\partial}{\partial x_i}\{[(p-p_0)\delta_{ij} - \tau_{ij}]n_j\delta(f)\} - \frac{\partial}{\partial x_i}[\rho v_i(v_n - u_n)\delta(f)] + \qquad (14)$$

$$+ \frac{\partial}{\partial t}[\rho_0 u_n \delta(f)] + \frac{\partial}{\partial t}[\rho(v_n - u_n)\delta(f)]$$

with: $\phi = c_0^2(\rho - \rho_0)$ wave variable,
in linear acoustics corresponding to sound pressure.

Reformulation (Pilon/Lyrintzis) :

$$\Box^2[\phi H(f)] = -\left(\frac{\partial\phi}{\partial n} + \frac{M_n}{c_0}\frac{\partial\phi}{\partial t_x}\right)\delta(f) - \frac{1}{c_0}\frac{\partial}{\partial t}[M_n\phi\delta(f)] - \frac{\partial}{\partial x_i}[\phi n_i\delta(f)] + \frac{\partial^2 T_{ij}}{\partial x_i \partial x_j}H(f) \qquad (15)$$

with: subscript x in the time derivative denoting differentiation with respect to time, holding the observer co-ordinates fixed .

In comparison to the generalized wave equation, which is the governing equation for the Kirchhoff formulation:

- the domain is considered in terms of wave propagation;
- the generalised pressure perturbation: $p' = \begin{cases} p' & f > 0 \\ 0 & f < 0 \end{cases}$

the generalised wave equation :

$$\Box^2 p' = \frac{1}{c_0^2}\frac{\partial^2 p'}{\partial t^2} - \frac{\partial^2 p'}{\partial x_i^2} = -\left(\frac{\partial p'}{\partial t}\frac{M_n}{c} + \frac{\partial p'}{\partial n}\right)\delta(f) - \frac{\partial}{\partial t}\left[p'\frac{M_n}{c}\delta(f)\right] - \frac{\partial}{\partial x_i}[p' n_i \delta(f)] \qquad (16)$$

with: $M_n = \frac{u_n}{c}$ local normal Mach number on S

$$\Box^2 p_a = q_{KIRCHHOFF} \qquad (17)$$

Manipulation the FW-H source terms into the form of Kirchhoff source terms (inviscid fluid):

$$\Box^2 p' = q_{FW-H}$$

$$= q_{KIRCHHOFF} + \frac{\partial^2}{\partial x_i \partial x_j}\left[T_{ij} H(f)\right] - \frac{\partial}{\partial x_j}\left[\rho v_i v_j\right] n_i \delta(f) - \frac{\partial}{\partial x_j}\left[\rho v_i v_n \delta(f)\right] + \qquad (18)$$

$$+ \frac{\partial}{\partial t}\left[p' - c^2 \rho'\right] \frac{M_n}{c} \delta(f) + \frac{\partial}{\partial t}\left[\left(p' - c^2 \rho'\right) \frac{M_n}{c} \delta(f)\right]$$

Extra source terms are of second order in perturbation quantities:

- $q_{FW-H} \approx q_{KIRCHHOFF}$ equivalent in linear region $p' \approx c_0^2 \rho'$; $v_i << 1$;
- $q_{FW-H} \neq q_{KIRCHHOFF}$ not equivalent in non-linear flow region .

There are four types of source terms in the inhomogeneous wave equation in the form of the FW-H-equation and Kirchhoff equation:

$$\Box^2 p_a = \frac{\partial^2}{\partial x_i \partial x_j}\left[T_{ij} H(f)\right] \qquad \Box^2 p_a = Q(x_i, t)\delta(f) \qquad (19, 20)$$

$$\Box^2 p_a = \frac{\partial}{\partial t}\left[Q \delta(f)\right] \qquad \Box^2 p_a = \frac{\partial}{\partial x_i}\left[Q_i \delta(f)\right] \qquad (21, 22)$$

N.9.2 Curle’s equation

Curle's equation is a special case of the FW-H equation in integral form, with control surface stationary and rigid: $u_i = 0$:

$$H c_0^2 (\rho - \rho_0) = \frac{\partial^2}{\partial x_i \partial x_j} \int_{V(\tau)} \left[T_{ij}\right]_\tau \frac{dV}{4\pi |x_i - y_i|} - \frac{\partial}{\partial x_i} \int_{S(\tau)} \left[p_{ij}\right]_\tau \frac{dS_j(y_i)}{4\pi |x_i - y_i|} \qquad (23)$$

N.10 Acoustic analogy in terms of entropy, heat sources as sound sources, sound generation by turbulent two-phase flow

Ref.: Howe, [N.14]; Crighton/Dowling et al., [N.15]; Morfey, [N.50]; Strahle, [N.51], [N.52], [N.53]; Boineau/Gervais, [N.54]; Perrey-Debain et al., [N.55]; Crighton/Ffowcs Williams, [N.56];

N.10.1 Acoustic analogy in terms of entropy, sound generation by fluctuating heat sources (Dowling, Howe)

Lighthill's inhomogeneous wave equation :

$$\frac{\partial^2 \rho}{\partial t^2} - c_0^2 \Delta\rho = \frac{\partial^2}{\partial x_i \partial x_j}\left\{\rho v_i v_j + \left[(p-p_0) - c_0^2(\rho-\rho_0)\right]\delta_{ij} - \tau_{ij}\right\} \tag{1}$$

with: τ_{ij} viscous stress tensor ,

p_0, ρ_0, c_0 mean value in acoustic field .

Reformulation by Dowling (1992) :

$$\frac{1}{c_0^2}\frac{\partial^2 p}{\partial t^2} - \Delta p = \frac{\partial^2}{\partial x_i \partial x_j}\left\{\rho v_i v_j - \tau_{ij}\right\} - \frac{\partial^2 \rho_e}{\partial t^2} \tag{2}$$

with: $\rho_e = \rho - \rho_0 - \dfrac{(p-p_0)}{c_0^2}$

„excess" density ρ_e vanishing in the far field but is nonzero in regions where the entropy is different from ambient.

$$\begin{aligned}\frac{\partial \rho_e}{\partial t} = \frac{\alpha\rho_0}{c_p\rho}\left[\sum_{n=1}^{N}\left.\frac{\partial h}{\partial Y_n}\right|_{\rho,p,Y_m} \rho\frac{DY_n}{DT} + \nabla\cdot q_i - \frac{\partial v_i}{\partial x_j}\tau_{ij}\right] - \nabla\cdot(v_i\rho_e) \\ - \frac{1}{c_0^2}\left[\left(1-\frac{\rho_0 c_0^2}{\rho c^2}\right)\frac{Dp}{Dt} - \frac{(p-p_0)}{\rho}\frac{D\rho}{Dt}\right]\end{aligned} \tag{3}$$

α volumetric expansion coefficient, for an ideal gas equal to $ß=T^{-1}$;

h enthalpy;

Y_n mass fraction of the n-th species;

N number of N (possible reacting) species;

$\vec{q}$ heat flux;

$$\rho\frac{DY_n}{Dt} = w_n - \nabla\cdot J_{n,i}\ ;$$

w_n production rate per unit volume of species n by reaction ;
$J_{n,i}$ flux of species n by diffusion.

Inhomogeneous wave equation :

$$\frac{1}{c_0^2}\frac{\partial^2 p}{\partial t^2} - \Delta p = -\frac{\partial}{\partial t}\left\{\frac{\alpha\rho_0}{c_p\rho}\left[\sum_{n=1}^{N}\left.\frac{\partial h}{\partial Y_n}\right|_{\rho,p,Y_m}\rho\frac{DY_n}{DT} + \nabla\cdot q_i - \frac{\partial v_i}{\partial x_j}\tau_{ij}\right]\right\} + \frac{\partial^2}{\partial x_i\partial x_j}\left(\rho v_i v_j - \tau_{ij}\right)$$
$$+\frac{1}{c_0^2}\frac{\partial}{\partial t}\left[\left(1-\frac{\rho_0 c_0^2}{\rho c^2}\right)\frac{Dp}{Dt} - \frac{(p-p_0)}{\rho}\frac{D\rho}{Dt}\right] + \frac{\partial^2}{\partial x_i\partial t}\left(v_i\rho_e\right) \tag{4}$$

Solution, with the help of the Green's function in unbounded space and with the far field approximations :

$$(p-p_0)(x_i,t) = -\frac{1}{4\pi r}\frac{\partial}{\partial t}\int\left\{\frac{\alpha\rho_0}{c_p\rho}\left[\sum_{n=1}^{N}\left.\frac{\partial h}{\partial Y_n}\right|_{\rho,p,Y_m}\rho\frac{DY_n}{DT} + \nabla\cdot q_i - \frac{\partial v_i}{\partial x_j}\tau_{ij}\right]\right\}_\tau dV$$
$$+\frac{x_i x_j}{4\pi r^3 c_0^2}\frac{\partial^2}{\partial t^2}\int\left(\rho v_i v_j - \tau_{ij}\right)_\tau dV$$
$$+\frac{1}{4\pi r c_0^2}\frac{\partial}{\partial t}\int\left[\left(1-\frac{\rho_0 c_0^2}{\rho c^2}\right)\frac{Dp}{Dt} - \frac{(p-p_0)}{\rho}\frac{D\rho}{Dt}\right]_\tau dV \tag{5}$$
$$-\frac{x_i}{4\pi r^2 c_0}\frac{\partial^2}{\partial t^2}\int\left(v_i\rho_e\right)_\tau dV$$

Right-hand side: the thermoacoustic source mechanisms

- first term: sound generated by irreversible flow processes, including the diffusion of mass and heat ; → monopole sources;
- second term: sound generated by momentum flux density fluctuations and by fluctuations of viscous stress (Lighthill's jet noise theory) : → quadrupole and octupole sources ;
- third term: sound generated by unsteady flow regions with different mean density and sound speed from the ambient fluid ; → dipole sources ;
- fourth term: sound generated by effects of momentum changes of density inhomogeneities ; → dipole sources .

Other formulation by Howe (1998) :

$$\frac{1}{c_0^2}\frac{\partial^2 p}{\partial t^2} - \Delta p = \frac{\partial}{\partial t}\left(\frac{\rho_0}{c_p}\frac{Ds}{Dt}\right) + \frac{\partial^2}{\partial x_i \partial x_j}\left(\rho_0 v_i v_j\right)$$
$$+\frac{1}{c_0^2}\frac{\partial}{\partial t}\left[\left(1-\frac{\rho_0 c_0^2}{\rho c^2}\right)\frac{\partial p}{\partial t}\right] - \rho_0 \operatorname{div}\left[\left(\frac{1}{\rho}-\frac{1}{\rho_0}\right)\nabla p\right] \tag{6}$$

Solution :

$$p(x_i,t) = \frac{\rho_0}{4\pi r}\frac{\partial}{\partial t}\int\left[\frac{q(y_i,t)}{c_p \rho T}\right]_\tau dV + \frac{x_i x_j}{4\pi r^3 c_0^2}\frac{\partial^2}{\partial t^2}\int\left(\rho_0 v_i v_j\right)_\tau$$
$$+\frac{\rho_0 x_j}{4\pi r^2 c_0}\frac{\partial}{\partial t}\int\left[\left(\frac{1}{\rho_0}-\frac{1}{\rho}\right)\frac{\partial p}{\partial y_j}\right]_\tau dV + \frac{\rho_0}{4\pi r}\frac{\partial^2}{\partial t^2}\int\left[\left(\frac{1}{\rho_0 c_0^2}-\frac{1}{\rho c^2}\right)(p-p_0)\right]_\tau dV \tag{7}$$

Right-hand side: the thermoacoustic source mechanisms

- first term: direct combustion noise → monopole sources;
- second term: jet noise (Lighthill) → quadrupole sources;
- third term: indirect combustion noise, "entropy" noise → dipole sources;

  dipole source strength is proportional to $-\left[\frac{1}{\rho}-\frac{1}{\rho_0}\right]\nabla p$,

  that is, to the difference between the acceleration of fluid of density ρ in the jet and which fluid of ambient mean density ρ_0 would experience in the same pressure gradient;

  other interpretation:

  $$\rho_0\left(\frac{1}{\rho_0}-\frac{1}{\rho}\right) = \frac{\rho-\rho_0}{\rho} \approx \frac{\Delta T}{T}$$

  that is, the dipole source strength is proportional to the fractional temperature difference between the hot spot and its environment
- fourth term: indirect combustion noise → monopole sources ;

  monopole source strength is proportional to the difference between the adiabatic compressibility $(1/\rho c^2)$ in the hot source region and in the ambient medium.

To the third and fourth terms :

The "hot spots" or "entropy inhomogeneities" behave as scattering centres at which dynamic pressure fluctuations are converted directly into sound.

N.10.2 Acoustic analogy in terms of heat release, turbulent density fluctuations and turbulent velocity fluctuations on the outer flame surface (Strahle)

Solution of the inhomogeneous wave equation :

$$p(\vec{x},t) = \frac{(\kappa-1)}{4\pi r c_0^2}\int_V \left[\frac{\partial q}{\partial t}\right]_\tau dV \tag{8}$$

with: q rate of heat release per unit volume.

$$\rho(\vec{x},t) = \frac{1}{4\pi r c_0^2}\frac{\partial^2}{\partial t^2}\int_V [\rho_t]_\tau dV \tag{9}$$

with: ρ_t turbulent fluctuating components of density, is considered negligible outside of the reacting region of the flame, even if the non-reacting portions of the flow field are turbulent, if the Mach number is low;

ρ acoustic fluctuating components of density.

$$\rho(\vec{x},t) = \frac{\overline{\rho}_1}{4\pi r c_0}\frac{\partial}{\partial t}\int_{S_r} v_{i,t} n_i dS \tag{10}$$

with: $\overline{\rho}_1$ mean density behind flame,

S_r surface of the direct combustion region, of the turbulent reaction zone,

$v_{i,t}$ turbulent velocity fluctuations on the outer flame surface, produced by interior density fluctuations .

N.10.3 Sound power radiated by a turbulent flame

Far-field acoustic pressure :

$$p(x_i,t) = \frac{(\kappa-1)}{4\pi r c_0^2}\frac{\partial}{\partial t}\int_V [Q]_\tau dV \tag{11}$$

with: q heat release rate per unit volume ,

$q(y_i,t) = \frac{\partial}{\partial t}S(y_i,t) = \frac{\partial}{\partial t}\left[\overline{\rho}(y_i)\overline{c}_p(y_i)T'(y_i,t)\right]$ acoustic source strength ,

$T'(y_i,t)$ temperature fluctuations in the combustion zone.

Sound power spectrum radiated by a turbulent flame :

$$W(\omega)=\frac{(\kappa-1)^2}{4\pi\rho_0 c_0}k^4\int_{V_y}\int_{V_\eta} S_{rms}(y_i')S_{rms}(y_i'')\Gamma(y_i',\eta_i,\omega)dy_i'd\eta_i \tag{12}$$

with: k acoustic wave number ,

Γ Fourier transform of acoustic sources, space-time correlation coefficient ,

η_i separation vector between two acoustic sources ,

S_{rms} rms amplitude of acoustic sources .

Respectively in the case of axisymmetrical flow :

$$W(\omega)=\frac{(\kappa-1)^2}{4\pi\rho_0 c_0}k^4\int_{V_{y'}}\int_{V_{y''}} S_{rms}(y_i')S_{rms}(y_i'')\Gamma(\xi_z,\xi_r,\omega)\times$$

$$\times\sqrt{P(y_i',\omega)}\sqrt{P(y_i'',\omega)}J_0^2\left(\frac{k_{ac}r_i}{2}\right)J_0^2\left(\frac{k_{ac}r_j}{2}\right)dy_i'dy_i'' \tag{13}$$

with: $$\Gamma(\xi_z,\xi_r,\omega)=\exp\left(\frac{-\pi\xi_z^2}{L_{cz}^2}\right)\exp\left(\frac{-\pi\xi_r^2}{L_{cr}^2}\right)\frac{\sqrt{\pi}}{\omega_t}\exp\left(\frac{-\omega^2}{4\omega_t^2}\right) \tag{14}$$

$\Gamma(\xi_z,\xi_r,\omega)$ spatial and frequency coherence function ;

ξ_z,ξ_r longitudinal and radial separation distances respectively between acoustic sources ;

L_{cz},L_{cr} longitudinal and radial coherence scale respectively ;

$\omega_t=2\pi C_\omega\dfrac{\varepsilon}{k}$

ω_t characteristic angular frequency of turbulence ;

$C_\omega=1{,}5$

$P(\omega)=\dfrac{2a}{\pi\left(1+a^2\omega^2\right)}$

$P(\omega)$ normalised temperature fluctuation spectrum, containing the characteristic time for temperature fluctuations, which is a function of the turbulence level and the characteristic turbulence angular frequency ;

$a=\dfrac{k}{C_S\varepsilon}$ characteristic time of temperature fluctuations ;

k turbulent kinetic energy ;

ε dissipation rate of turbulent kinetic energy ;

$C_S=6{,}4$ see Sanders, J. P. H. and P. G. G. Lammers ("Combustion and Flame", 1994).

N.10.4 Sound generation by turbulent two-phase flow

Mass continuity equation for the α-phase :

$$\frac{\partial}{\partial t}(1-\beta)\rho^{\alpha}+\frac{\partial}{\partial x_j}(1-\beta)\rho^{\alpha}v_j^{\alpha}=0 \tag{15}$$

with:
- α — α-phase, e. g. water ,
- β — β-phase, e. g. gas bubbles,
- ρ^{α}, ρ^{β} — mass of α-phase (resp. β-phase) per volume occupied by α-phase (resp. β-phase),
- $(1-\beta)\rho^{\alpha}$ — mass of α-phase in unit volume of mixture;
- $(1-\beta)\rho^{\alpha}+\beta\rho^{\beta}$ — total mass per unit volume.

Reformulation :

$$\frac{\partial}{\partial t}\rho^{\alpha}+\frac{\partial}{\partial x_j}\rho^{\alpha}v_j^{\alpha}=Q \tag{16}$$

with: $$Q=-\rho^{\alpha}\left(\frac{\partial}{\partial t}+v_j^{\alpha}\frac{\partial}{\partial x_j}\right)\ln(1-\beta) \tag{17}$$

Momentum equation for α-phase :

$$\frac{\partial}{\partial t}(1-\beta)\rho^{\alpha}v_i^{\alpha}+\frac{\partial}{\partial x_j}\left[(1-\beta)\rho^{\alpha}v_i^{\alpha}v_j^{\alpha}+p_{ij}\right]=F_i \tag{18}$$

with:
- p_{ij} — stress tensor,
- F_i — interphase force.

Reformulation :

$$\frac{\partial}{\partial t}\rho^{\alpha}v_i^{\alpha}+\frac{\partial}{\partial x_j}\left[(1-\beta)\rho^{\alpha}v_i^{\alpha}v_j^{\alpha}+p_{ij}\right]=G_i \tag{19}$$

with: $$G_i=F_i+G_i'=F_i+\frac{\partial}{\partial t}\left(\beta\rho^{\alpha}v_i^{\alpha}\right) \tag{20}$$

Inhomogeneous wave equation :

$$\left(\frac{\partial^2}{\partial t^2}-c_{\alpha}^2\nabla^2\right)\rho^{\alpha}=\frac{\partial Q}{\partial t}-\frac{\partial G_i}{\partial x_i}+\frac{\partial^2 T_{ij}}{\partial x_i\partial x_j} \tag{21}$$

with: $T_{ij}=(1-\beta)\rho^{\alpha}v_i^{\alpha}v_j^{\alpha}+p_{ij}-c_{\alpha}^2\rho^{\alpha}\delta_{ij}$,

c_{α} — sound speed in pure water .

Sound is generated in turbulent two-phase flow by three physical mechanisms:

- first term: distribution of monopoles, of strength Q, equal to the rate of mass injection into the α-phase;

- second term: distribution of dipoles, of strength G_i, equal to the effective force on the α-fluid, composed in part of the interphase force F_i, and in part of the term G_i', which represents the momentum defect arising from the fact that a fraction β of the total volume is not occupied by α-phase;
- third term: distribution of quadrupoles, of strength T_{ij} (LIGHTHILL), is dominated by the Reynolds stress (viscous contributions are neglected).

N.11 Acoustics of moving sources

Ref.: Lowson, [N.57]; Tanna/Morfey, [N.58], [N.59], [N.60]; Roger, [N.4]; Ianniello, [N.61]; Lyrintzis, [N.62];

N.11.1 Sound field of moving point sources

Monopole point source:

Starting point :

$$p(x_i,t)=\frac{1}{4\pi}\int\limits_V\left[\frac{1}{r}\frac{\partial(\dot{M}\delta)}{\partial t}\right]_\tau dV \tag{1}$$

with:

$\dot{M}$	mass flux, point source;
$\delta=\delta[y_i-y_{qi}(t)]$	three-dimensional delta function ;
y_i	coordinates of the source point ;
$y_{qi}(t)$	time-variable place of the mass flux source ;
$v_{qi}=\dfrac{\partial y_{qi}}{\partial t}=c_0 M_{qi}$	velocity of the source point ;
$M_{qi}=\dfrac{v_{qi}}{c_0}$	Mach number, based on the source velocity ;
$\tau=t-\dfrac{r_\tau}{c_0}$	retarded time, time of radiation;
t	time of observation;
r_τ	observer – source separation, dependent on τ.

Sound pressure (general) :

$$p(x_i,t)=\frac{1}{4\pi(1-M_{qr})_\tau}\left\{\frac{1}{r}\frac{\partial\dot{M}}{\partial t}+\frac{\partial}{\partial y_i}\left[\frac{\dot{M}c_0M_{qi}}{r(1-M_{qr})}\right]+\frac{(x_i-y_i)}{r^2}\frac{\partial}{\partial t}\left[\frac{\dot{M}M_{qi}}{(1-M_{qr})}\right]\right\}_\tau \tag{2}$$

Sound pressure in the far field :

$$p(x_i,t) = \left\{ \frac{1}{4\pi r\left(1-M_{qr}\right)^2} \left[\frac{\partial \dot{M}}{\partial t} + \frac{\dot{M}}{\left(1-M_{qr}\right)} \frac{\partial M_{qr}}{\partial t} \right] \right\}_\tau \tag{3}$$

with: $M_{qr} = \frac{(x_i - y_i)}{|x_i - y_i|} M_{qi}$ projection of the Mach number vector M_{qi} in the direction of sound radiation $\vec{r} = \vec{x} - \vec{y}$;

$\dot{M}$ mass flux .

Other formulation :

$$p(x_i,t) = \left\{ \frac{1}{4\pi r\left(1-M_{qr}\right)} \frac{\partial \dot{M}}{\partial t} + \frac{(x_i - y_i)}{4\pi r^2\left(1-M_{qr}\right)^2} \left[\frac{\partial\left(\dot{M} M_{qi}\right)}{\partial t} + \frac{\dot{M} M_{qi}}{\left(1-M_{qr}\right)} \frac{\partial M_{qr}}{\partial t} \right] \right\}_\tau \tag{4}$$

Proportionalities :

- first term: temporal change of the mass flux

$$p \sim \rho_0 \frac{L}{r} \frac{1}{\left(1-M_{qr}\right)} U^2 \quad \cong \quad \text{monopole source ;} \tag{5}$$

- second term: temporal change of the momentum flux

$$p \sim \frac{\rho_0}{c_0} \frac{L}{r} \frac{1}{\left(1-M_{qr}\right)^2} U^3 \quad \cong \quad \text{dipole source ;} \tag{6}$$

- third term: accelerated motion of the momentum flux

$$p \sim \frac{\rho_0}{c_0^2} \frac{L}{r} \frac{1}{\left(1-M_{qr}\right)^3} U^4 \quad \cong \quad \text{quadrupole source.} \tag{7}$$

Sound pressure in the far field, uniform motion of the monopole point source :

$$p(x_i,t) = \left\{ \frac{1}{4\pi r\left(1-M_{qr}\right)} \frac{\partial \dot{M}}{\partial t} \right\}_\tau \tag{8}$$

Dipole point source:

Starting point :

$$p(x_i,t) = -\frac{1}{4\pi} \int_V \left[\frac{1}{r} \frac{\partial (F_i \delta)}{\partial y_i} \right]_\tau dV \tag{9}$$

with: F_i force, point source .

Sound pressure (general) :

$$p(x_i,t) = \frac{1}{4\pi\left(1-M_{qr}\right)_\tau}\left\{F_{i\tau}\frac{\partial\left(\frac{1}{r}\right)_\tau}{\partial y_i} + \left[\frac{(x_i-y_i)}{c_0 r^2}\frac{\partial F_i}{\partial t}\right]_\tau + \left[\frac{\partial}{\partial y_j}\left(\frac{(x_i-y_i)}{r^2}\frac{F_i M_{qj}}{\left(1-M_{qr}\right)}\right)\right]_\tau + \right. \tag{10}$$

$$\left. + \left[\frac{(x_i-y_i)(x_j-y_j)}{c_0 r^3}\frac{\partial}{\partial t}\left(\frac{F_i M_{qj}}{\left(1-M_{qr}\right)}\right)\right]_\tau\right\}$$

Sound pressure in the far field :

$$p(x_i,t) = \left\{\frac{(x_i-y_i)}{4\pi r^2 c_0\left(1-M_{qr}\right)^2}\left[\frac{\partial F_i}{\partial t} + \frac{F_i}{\left(1-M_{qr}\right)}\frac{\partial M_{qr}}{\partial t}\right]\right\}_\tau \tag{11}$$

Proportionalities :

- first term: temporal change of the force

$$p \sim \frac{\rho_0}{c_0}\frac{L}{r}\frac{1}{\left(1-M_{qr}\right)^2}U^3 \quad \cong \quad \text{dipole source ;} \tag{12}$$

- second term: accelerated motion of the force

$$p \sim \frac{\rho_0}{c_0^2}\frac{L}{r}\frac{1}{\left(1-M_{qr}\right)^3}U^4 \quad \cong \quad \text{quadrupole source.} \tag{13}$$

Other formulation (LIGHTHILL) :

$$p(x_i,t) = \left\{\frac{(x_i-y_i)}{r c_0\left(1-M_{qr}\right)}\frac{\partial}{\partial t}\left[\frac{F_i}{4\pi r\left(1-M_{qr}\right)}\right]\right\}_\tau \tag{14}$$

Sound pressure in the far field, uniform motion of the dipole point source :

$$p(x_i,t) = \left\{\frac{(x_i-y_i)}{4\pi r^2 c_0\left(1-M_{qr}\right)^2}\frac{\partial F_i}{\partial t}\right\}_\tau \tag{15}$$

Quadrupole point source:

Starting point :

$$p(x_i,t) = \frac{1}{4\pi}\int_V\left[\frac{1}{r}\frac{\partial^2\left(T_{ij}\delta\right)}{\partial y_i\partial y_j}\right]_\tau dV \tag{16}$$

with: T_{ij} pressure-stress tensor, Lighthill tensor, point source.

Sound pressure :

$$p(x_i,t)=\left\{\frac{(x_i-y_i)(x_j-y_j)}{4\pi r^3 c_0^2(1-M_{qr})^3}\left[\frac{\partial^2 T_{ij}}{\partial t^2}+\frac{3}{(1-M_{qr})}\frac{\partial T_{ij}}{\partial t}\frac{\partial M_{qr}}{\partial t}+\frac{T_{ij}}{(1-M_{qr})}\frac{\partial^2 M_{qr}}{\partial t^2}+\right.\right.$$
$$\left.\left.+\frac{3T_{ij}}{(1-M_{qr})^2}\left(\frac{\partial M_{qr}}{\partial t}\right)^2\right]\right\}_\tau \qquad (17)$$

Proportionalities :

- first term: second time derivative of the momentum flux density
  $p\sim\frac{\rho_0}{c_0^2}\frac{L}{r}\frac{1}{(1-M_{qr})^3}U^4 \quad \cong \quad$ quadrupole source; (18)
- second/third term: accelerated motion of the momentum flux density, respectively, and the time derivative of this quantity
  $p\sim\frac{\rho_0}{c_0^3}\frac{L}{r}\frac{1}{(1-M_{qr})^4}U^5 \quad \cong \quad$ octupole source (19)
- fourth term: strong accelerated motion of the momentum flux density
  $p\sim\frac{\rho_0}{c_0^4}\frac{L}{r}\frac{1}{(1-M_{qr})^5}U^6 \quad \cong \quad$ sexdecupole source (20)

Other formulation (LIGHTHILL) :

$$p(x_i,t)=\left\{\frac{(x_i-y_i)}{rc_0(1-M_{qr})}\frac{\partial}{\partial t}\left[\frac{(x_j-y_j)}{rc_0(1-M_{qr})}\frac{\partial}{\partial t}\left(\frac{T_{ij}}{4\pi r(1-M_{qr})}\right)\right]\right\}_\tau \qquad (21)$$

Sound pressure in the far field, uniform motion of the quadrupole point source :

$$p(x_i,t)=\left\{\frac{(x_i-y_i)(x_j-y_j)}{4\pi r^3 c_0^2(1-M_{qr})^3}\frac{\partial^2 T_{ij}}{\partial t^2}\right\}_\tau \qquad (22)$$

N.11.2 Formulation the equation of sound sources in motion based on the Ffowcs Williams-Hawkings equation

Far-field solution :

$$p(x_i,t)=\frac{\partial^2}{\partial x_i\partial x_j}\int_V\left[\frac{T_{ij}}{4\pi r(1-M_r)}\right]_\tau dV-\frac{\partial}{\partial x_i}\int_S\left[\frac{p_{ij}n_j}{4\pi r(1-M_r)}\right]_\tau dS+\frac{\partial}{\partial t}\int_S\left[\frac{\rho_0 u_n}{4\pi r(1-M_r)}\right]_\tau dS \qquad (23)$$

with: $p_{ij} n_j dS$ force on the fluid from each surface element;

u_n normal velocity field on surfaces;

$M_r = \dfrac{x_i u_i}{r c_0}$ Mach number of the sources;

$1 - M_r$ Doppler amplification factor related to the projected motion on the line between the source and the observer point .

Quadrupole source :

$$p(\vec{x},t) = \frac{\partial^2}{\partial x_i \partial x_j} \int_V \left[\frac{T_{ij}}{4\pi r |1 - M_r|} \right]_\tau dV \tag{24}$$

respectively :

$$p(\vec{x},t) = \frac{1}{c_0^2} \frac{\partial^2}{\partial t^2} \int_V \left[\frac{T_{rr}}{4\pi r |1 - M_r|} \right]_\tau dV + \frac{1}{c_0} \frac{\partial}{\partial t} \int_V \left[\frac{3T_{rr} - T_{ii}}{4\pi r^2 |1 - M_r|} \right]_\tau dV + \int_V \left[\frac{3T_{rr} - T_{ii}}{4\pi r^3 |1 - M_r|} \right]_\tau dV \tag{25}$$

with: M_r projection of the rotational Mach number in the source-observer direction ;

$T_{rr} = T_{ij} \hat{r}_i \hat{r}_j$ Lighthill stress tensor in the radiation direction;

$\hat{\vec{r}}$ unit vector in the radiation direction, with components $\hat{r}_i$.

N.11.3 Moving Kirchhoff surfaces

Kirchhoff's surface S :

- encloses all the non-linear effects and sound sources;
- outside Kirchhoff's surface the acoustic field is linear.

Wave equation :

$$\frac{1}{c_0^2} \frac{\partial^2 \Phi}{\partial t^2} - \Delta \Phi = 0 \tag{26}$$

with: Φ acoustic quantity satisfying the wave equation in the exterior of surface S .

Classical Kirchhoff formulation for a stationary control surface :

$$4\pi \Phi(\vec{x},t) = \int_S \left[\frac{1}{r^2} \Phi \frac{\partial r}{\partial n} - \frac{1}{r} \frac{\partial \Phi}{\partial n} + \frac{1}{r c_0} \frac{\partial r}{\partial n} \frac{\partial \Phi}{\partial \tau} \right]_\tau dS \tag{27}$$

with: τ retarded (emission) time;

r distance between observer and source;

$$\frac{\partial r}{\partial n} = \cos\Theta$$

Θ angle between the normal vector $\vec{n}$ on the surface and the radiation direction $\vec{r} = \vec{x} - \vec{y}$.

Interpretation :

- integral representation of Φ at points exterior to S in terms of quantities on the control surface S,
- computation of noise at an arbitrary point, if the solution is known on surface S ,
- first term: is not significant in the far field.

Uniformly moving surface :

Convective wave equation :

$$\Delta\Phi - \frac{1}{c_0^2}\left(\frac{\partial}{\partial t} + U_\infty \frac{\partial}{\partial x}\right)^2 \Phi = 0 \tag{28}$$

with: U_∞ uniform velocity of the control surface S .

Distance between the observer and the surface point :

$$r_0 = \sqrt{(x - x')^2 + \beta^2\left[(y - y')^2 + (z - z')^2\right]} \tag{29}$$

with:

$\vec{x} = (x, y, z, t)$	observer location;
$\vec{y} = (x', y', z', \tau')$	source location;
$\tau' = t - \tau = t - r/c_0$	source time, retarded (emission) time;
$\tau = \left[r_0 - M_\infty(x - x')\right]/(c\beta^2)$	time delay between emission and detection;
$x_0 = x,\ y_0 = \beta y,\ z_0 = \beta z$	Prandtl-Glauert transformation;
$\beta = \sqrt{1 - M_\infty^2}$.	

Solution for the case of subsonically moving surface :

$$4\pi\Phi(\vec{x}, t) = \int_{S_0} \left[\frac{1}{r_0^2}\Phi\frac{\partial r_0}{\partial n_0} - \frac{1}{r_0}\frac{\partial \Phi}{\partial n_0} + \frac{1}{r_0 c_0 \beta^2}\frac{\partial \Phi}{\partial \tau}\left(\frac{\partial r_0}{\partial n_0} - M_\infty \frac{\partial x_0}{\partial n_0}\right)\right]_\tau dS_0 \tag{30}$$

with: subscript 0: transformed variable, e. g. $\vec{n}_0$ is the outward pointing vector normal to the surface S_0 .

Solution for the case of supersonically moving surface :

$$4\pi\Phi(\vec{x}, t) = \int_{S_0} \left[\frac{1}{r_0^2}\Phi\frac{\partial r_0}{\partial n_0} - \frac{1}{r_0}\frac{\partial \Phi}{\partial n_0} + \frac{1}{r_0 c_0 \beta^{*2}}\frac{\partial \Phi}{\partial \tau}\left(\pm\frac{\partial r_0}{\partial n_0} - M_\infty \frac{\partial x_0}{\partial n_0}\right)\right]_{\tau^\pm} dS_0 \tag{31}$$

with: $\beta^* = \sqrt{M_\infty^2 - 1}$;

$\tau^{\pm} = \dfrac{\left[\pm r_0 - M_\infty (x - x')\right]}{c\beta^{*2}}$ time delay;

$x_0 = x,\ y_0 = \beta^* y,\ z_0 = \beta^* z$ Prandtl-Glauert transformation;

$r_0 = \sqrt{(x - x')^2 + \beta^{*2}\left[(y - y')^2 + (z - z')^2\right]}$

distance between the observer and the surface point in Prandtl-Glauert co-ordinates;

sign $\pm$ evaluation at both retarded times τ^+ and τ^- .

Arbitrarily moving surface (subsonically-moving, rigid surface) :

$$4\pi\Phi(\vec{x},t) = \int\limits_S \left\{ \frac{\dfrac{\partial r}{\partial n}}{r^2(1 - M_r)}\Phi - \frac{1}{r(1 - M_r)}\left[\frac{\partial \Phi}{\partial n} + \frac{M_n}{c_0}\frac{\partial \Phi}{\partial \tau}\right] + \right.$$
$$\left. + \frac{1}{c_0(1 - M_r)}\frac{\partial}{\partial \tau}\left[\frac{\dfrac{\partial r}{\partial n} - M_n}{r(1 - M_r)}\Phi\right]\right\}_\tau dS \qquad (32)$$

with: $\vec{y}, \vec{r}$ functions of time: $\vec{y}(\tau'), \vec{r}(\tau')$;

observer point is stationary;

all the quantities are evaluated for the retarded (emission) time, which is the root of the equation $\tau - t + r(\tau)/c_0 = 0$;

$M_r = \vec{u} \cdot \vec{r}/(rc_0)$ Mach number in the direction of wave propagation from the source to the observer;

$\vec{u} = \partial \vec{y}/\partial \tau$ local source-surface velocity;

$M_n = u_n/c_0$ local normal Mach number;

u_n local normal velocity of S with respect to the undisturbed medium;

$\dfrac{\partial \Phi}{\partial \tau}, \dfrac{\partial \Phi}{\partial n}$ are taken with respect to source co-ordinates and time.

N.12 Aerodynamic sound sources in practice

Ref.: Bailly, [N.63]; Goldstein, [N.16]; Fuchs/Michalke, [N.64]; Ribner, [N.65]; Roger, [N.4]; Goldstein/Howes, [N.66]; Béchara et al., [N.67]; Goldstein/Rosenbaum, [N.68]; Heckl, [N.69]; Ffowcs Williams/Hawkings, [N.70]; Lowson, [N.71]; Farassat/Brentner, [N.72]; Singer/Brentner, [N.73]; Brentner, [N.48]; Ianiello, [N.61]; Költzsch, [N.75], [N.76], [N.77]

N.12.1 Jet noise

The acoustic intensity and power radiated by a jet are evaluated.

Starting point: Lighthill's inhomogeneous wave equation :

$$\frac{\partial^2\rho}{\partial t^2}-c_0^2\frac{\partial^2\rho}{\partial x_i^2}=\frac{\partial\dot{M}}{\partial t}-\frac{\partial}{\partial x_i}\left(F_i+\dot{M}\nu_i\right)+\frac{\partial^2 T_{ij}}{\partial x_i\partial x_j}=q \qquad (1)$$

with: $T_{ij}=\rho v_i v_j+p_{ij}-c_0^2\rho\delta_{ij}$ Lighthill tensor .

Solution :

$$p(x_i,t)=\frac{1}{4\pi}\int_V\frac{1}{r}\left(\frac{\partial^2 T_{ij}}{\partial y_i\partial y_j}\right)_\tau dV \qquad (2)$$

$$p(x_i,t)=\frac{1}{4\pi c_0^2}\int_V\frac{1}{r}\left\{\frac{\partial^2(T_{ij})}{\partial t^2}\right\}_\tau\left[\frac{1}{r^2}(x_i-y_i)(x_j-y_j)\right]dV \qquad (3)$$

in the far field :

$$p(x_i,t)=\frac{1}{4\pi c_0^2 r}\frac{x_i x_j}{r^2}\int_V\left[\frac{\partial^2(T_{ij})}{\partial t^2}\right]_\tau dV \qquad (4)$$

Calculation the acoustic intensity I with the help of the two-point correlation function R_a of the acoustic pressure fluctuations :

$$R_a(x_i,\tau)=R_{pp}(x_i,\tau)=\frac{\overline{p(x_i,t)p(x_i,t+\tau)}}{\rho_0 c_0}=\lim_{T\to\infty}\frac{1}{T}\int_0^T\frac{p(x_i,t)p(x_i,t+\tau)}{\rho_0 c_0}dt \qquad (5)$$

$I(x_i)=R_a(x_i,\tau=0)$ overall intensity at the observation point.

Power spectral density of the acoustic pressure at the observer point :

$$S_a(x_i,\omega) = S_{pp}(x_i,\omega) = \frac{1}{2\pi}\int_{-\infty}^{+\infty} R_a(x_i,\tau)e^{i\omega\tau}d\tau \tag{6}$$

Introducing Lighthill's solution :

$$R_a(x_i,\tau) = \frac{1}{16\pi^2 c_0^5 \rho_0 x^2}\frac{x_i x_j x_k x_l}{x^4}\int_{V'}\int_{V''}\overline{\frac{\partial^2 T_{ij}}{\partial t^2}(y_i',t')\frac{\partial^2 T_{kl}}{\partial t^2}(y_i'',t'')}dV'\,dV'' \tag{7}$$

with: the retarded times

$$t' = t - \frac{|x_i - y_i'|}{c_0} \quad \text{and} \quad t'' = t + \tau - \frac{|x_i - y_i''|}{c_0}.$$

Formulation with coherent source regions :

volume correlation :

$$\int_{V'}\int_{V''}\overline{\left[\frac{\partial^2 T_{ij}}{\partial t^2}(y_i',t')\right]\left[\frac{\partial^2 T_{kl}}{\partial t^2}(y_i'',t'')\right]}dV'\,dV'' = \int_{V'}\overline{\left[\frac{\partial^2 T_{ij}}{\partial t^2}(y_i',t')\right]^2}V_{corr}dV' \tag{8}$$

$$\text{with:}\quad V_{corr} = \int_{V''}\frac{\overline{\left[\frac{\partial^2 T_{ij}}{\partial t^2}(y_i',t')\right]\left[\frac{\partial^2 T_{kl}}{\partial t^2}(y_i'',t'')\right]}}{\overline{\left[\frac{\partial^2 T_{ij}}{\partial t^2}(y_i',t')\right]^2}}dV'' \tag{9}$$

Assumption of stationary properties of turbulence :

$$R_a(x_i,\tau) = \frac{1}{16\pi^2\rho_0 c_0^5}\frac{x_i x_j x_k x_l}{x^6}\frac{\partial^4}{\partial\tau^4}\int_{V'}\int_{V''}\overline{T_{ij}(y_i',t')T_{kl}(y_i'',t'')}dV'\,dV'' \tag{10}$$

Far-field approximation :

$$R_a(x_i,\tau) = \frac{1}{16\pi^2\rho_0 c_0^5}\frac{x_i x_j x_k x_l}{x^6}\frac{\partial^4}{\partial\tau^4}\int_{V'}\int_{V''}R_{ijkl}\left(y_i',\eta_i,\tau+\frac{x_i\cdot\eta_i}{xc_0}\right)dV'\,dV_\eta \tag{11}$$

$$\text{with:}\quad R_{ijkl}(y_i',\eta_i,\tau) = \overline{T_{ij}(y_i',t)T_{kl}(y_i'+\eta_i,t+\tau)} \tag{12}$$

R_{ijkl} two-point fourth order correlation tensor of the turbulent velocity;

$\eta_i = y_i'' - y_i'$ distance vector between the two points y_i' and y_i'' in the source volume;

$\eta << x$ far-field assumption.

$$S_a(x_i,\omega) = S_{pp}(x_i,\omega) = \frac{\omega^4}{32\pi^3 \rho_0 c_0^5} \frac{x_i x_j x_k x_l}{x^6} \int_{-\infty}^{+\infty} \iint e^{i\omega\left(\tau - \frac{\vec{x}}{x}\frac{\vec{\eta}}{c_0}\right)} R_{ijkl}(\vec{y}',\vec{\eta},\tau)\,dV'dV_\eta d\tau \qquad (13)$$

Reformulation with following assumptions :

- isotropic turbulence in a frame moving with the mean convection velocity U_c;
- introducing the new coordinate (Ffowcs Williams) $\vec{\xi} = \vec{\eta} - U_c \tau \vec{y}_1$ ($\vec{y}_1$ direction of mean flow) and the variable $\lambda = \alpha U_c \tau$, with $u' = \alpha U_c$ (typical turbulence velocity, α small parameter, corresponding the turbulence level);
- definition the new correlation tensor with the change of variables $(\vec{\eta},\tau) \to (\vec{\xi},\lambda)$:

$$R^*_{ijkl}(\vec{y},\vec{\eta},\tau) = R_{ijkl}\left(\vec{y},\vec{\xi} = \vec{\eta} - U_c \tau \vec{y}_1,\lambda\right) \qquad (14)$$

This leads to an expression for the far field correlation for noise, generated by convected isotropic turbulence :

$$R_a(\vec{x},\tau) = \frac{1}{16\pi^2 \rho_0 c_0^5} \frac{x_i x_j x_k x_l}{x^6} \iint_V \frac{1}{C^5} \frac{\partial^4}{\partial t^4} R_{ijkl}\left(\vec{y}',\vec{\xi},t = \frac{\tau}{C}\right) dV'dV_\xi \qquad (15)$$

with: $C = \sqrt{(1 - M_c \cos\Theta)^2 + \alpha^2 M_c^2}$ convection factor, Doppler factor;

$M_c = U_c / c_0$ convection Mach number;

Θ angle between the mean flow direction $\vec{y}_1$ and the observer point $\vec{x}$;

$\cos\Theta = \dfrac{\vec{x}\cdot\vec{y}_1}{x}$;

$\alpha^2 = \dfrac{\omega^2 L^2}{\pi c_0^2}$ (usually $\alpha = \text{const.}$, e. g. $\alpha \approx 0{,}55$ (Ribner), $\alpha \approx 0{,}3$ (Davies), often approximately $\alpha \approx 0$, i.e., using the simplified convection factor $C = 1 - M_c \cos\Theta$).

Acoustic intensity :

$$I(\vec{x}) = \frac{1}{16\pi^2 \rho_0 c_0^5} \frac{x_i x_j x_k x_l}{x^6} \iint_V \frac{1}{C^5} \frac{\partial^4}{\partial t^4} R_{ijkl}\left(\vec{y}',\vec{\xi},0\right) dV'dV_\xi \qquad (16)$$

Dimensional development leads to a power law for the acoustic intensity (Lighthill) :

$$I(\vec{x}) \sim \frac{1}{\rho_0 c_0^5 x^2} \frac{1}{C^5} \left(\frac{U}{D}\right)^4 \left(\rho^2 U^4\right) D^6 \qquad (17)$$

$$I(\vec{x}) \sim \frac{D^2}{x^2} \frac{\rho^2}{\rho_0} \frac{U_c^3 M_c^5}{\left[(1 - M_c \cos\Theta)^2 + \alpha^2 M_c^2\right]^{5/2}} \qquad (18)$$

with: D typical length of the source volume;
ρ mean flow density;
U typical velocity;
$U_c \approx \frac{2}{3} U$ for a round jet.

$$I(\vec{x}) \sim U^8 \quad \text{(Lighthill)} \tag{19}$$

Acoustic power :
may be obtained by integrating the acoustic intensity over the sphere of radius r :

$$P = \int_0^{\pi} I(r,\Theta) 2\pi r^2 \sin\Theta \, d\Theta \tag{20}$$

Integration with a simplified convection factor $C = 1 - M_c \cos\Theta$ for the subsonic region $M_c \le 1$:

$$P \sim \frac{\rho^2}{\rho_0} D^2 \frac{U^8}{c_0^5} \frac{1+M_c^2}{\left(1-M_c^2\right)^4} \tag{21}$$

(for practical applications: with the proportionality factor $K \approx 5 \cdot 10^{-5}$).

Acoustic efficiency η :

$$\eta = \frac{P}{P_{mech}} \sim \frac{\rho^2}{\rho_0} \frac{U^5}{c_0^5} \frac{1+M_c^2}{\left(1-M_c^2\right)^4} \tag{22}$$

$$\text{with:} \quad P_{mech} = \frac{\pi}{4} D^2 \rho U^3 \, . \tag{23}$$

The convection amplifies the radiated acoustic power by the factor $\dfrac{1+M_c^2}{\left(1-M_c^2\right)^4}$.

For supersonic jet flow $M_c > 1$:

$$P \sim \frac{\rho^2}{\rho_0} D^2 U^3 \tag{24}$$

$$\eta \sim \frac{\rho}{\rho_0} \tag{25}$$

Power spectral density of the far-field noise :
(statistical source models in jet noise of Ribner 1969).

Notations:

Density autocorrelation function :

$$C_{\rho\rho}(\vec{x},\tau) = \frac{1}{\rho_0 c_0^{-3}} \overline{\left[\rho(\vec{x},t+\tau)-\rho_0\right]\left[\rho(\vec{x},t)-\rho_0\right]} \tag{26}$$

$$C_{\rho\rho}(\vec{x},\tau) = \frac{\rho_0}{16\pi^2 c_0^5}\frac{x_i x_j x_k x_l}{x^6}\iint_V \overline{\frac{\partial^2}{\partial t^2} v_i' v_j'(\vec{y}',t')\frac{\partial^2}{\partial t^2} v_k'' v_l''(\vec{y}'',t'')}\,dV'dV'' \tag{27}$$

with: $\vec{y}',\ \vec{y}''$ two running points in the source domain;

$$t' = t - \frac{|\vec{x}-\vec{y}'|}{c_0}, \quad t'' = t+\tau-\frac{|\vec{x}-\vec{y}''|}{c_0}\,.$$

$$C_{\rho\rho}(\vec{x},\tau) = \frac{\rho_0}{16\pi^2 c_0^5}\frac{x_i x_j x_k x_l}{x^6}\iint_V \frac{\partial^4}{\partial\tau^4} R_{ijkl}\left(\vec{y}',\vec{\eta},\tau+\frac{\vec{\eta}\cdot\vec{x}}{c_0 x}\right)dV'dV_\eta \tag{28}$$

with: $R_{ijkl}(\vec{y}',\vec{\eta},\tau) = \overline{v_i' v_j'(\vec{y}',t)v_k'' v_l''(\vec{y}'',t+\tau)}$

two-point time-delayed fourth order correlation tensor .

Directional acoustical intensity spectrum,
which is the temporal Fourier transform of the density autocorrelation $C_{\rho\rho}(\vec{x},\tau)$:

$$I_\omega(\vec{x}) = \frac{1}{2\pi}\int_{-\infty}^{+\infty} C_{\rho\rho}(\vec{x},\tau)e^{j\omega\tau}d\tau \tag{29}$$

Acoustical power spectrum (emitted from a unit volume located at $\vec{y}$) :

$$P_\omega(\vec{y}) = 2\pi r^2\int_0^\pi I_\omega(\vec{x},\Theta|\vec{y})\sin\Theta\, d\Theta \tag{30}$$

Total acoustic power :

$$P = \int_V\int_{-\infty}^{+\infty} P_\omega(\vec{y})\,dV\,d\omega \tag{31}$$

Far-field noise radiated: acoustic intensity from an elementary volume of jet :

$$dI_{\text{noise}}^{\text{self}}(\vec{x}) \sim \frac{\rho_0 \overline{u'^2}^2 L^3\omega_t^4}{c_0^5 r^2 C^5} \tag{32}$$

$$dI_{\text{noise}}^{\text{shear}}(\vec{x}) \sim \frac{\rho_0 \overline{u'^2} L^5\omega_t^4}{c_0^5 r^2 C^5}\left(\frac{\partial U}{\partial y_2}\right)^2 D_\Theta \tag{33}$$

respectively the intensity spectrum :

$$\overset{\text{self}}{dI_{\omega}^{\text{noise}}}(\vec{x},\Theta|\vec{y}) = \frac{\rho_0 \overline{u'^2}^2 L^3}{128\pi^{5/2} c_0^5 r^2} \frac{\omega^4}{\omega_t} \exp\left(-\frac{\omega^2 C^2}{8\omega_t^2}\right) \tag{34}$$

$$\overset{\text{shear}}{dI_{\omega}^{\text{noise}}}(\vec{x},\Theta|\vec{y}) = \frac{\rho_0 \overline{u'^2} L^5}{24\pi^{7/2} c_0^5 r^2} \left(\frac{\partial U}{\partial y_2}\right)^2 \frac{\omega^4}{\omega_t} \exp\left(-\frac{\omega^2 C^2}{4\omega_t^2}\right) D_\Theta \tag{35}$$

with: $D_\Theta = \frac{1}{2}\left(\cos^2\Theta + \cos^4\Theta\right)$ directivity of the shear-noise component

(isotropic directivity of self noise is a necessary consequence of the isotropy of the turbulence);

C — convection factor ;
(at Goldstein/Howes: C^{-3} instead of C^{-5})

$\omega_t = 2\pi \frac{\varepsilon}{k}$;

ω_t — local characteristic frequency of turbulence, related to the eddy life time;

$L \approx \frac{k^{3/2}}{\varepsilon}$;

L — integral length scale of turbulence;

u' — turbulent velocity;

$U(y_2)$ — mean velocity in the direction of y_1, dependent on the co-ordinate y_2 .

Other developments (Goldstein/Rosenbaum) :

Acoustic intensity per unit source volume :

$$\overset{\text{self}}{dI^{\text{noise}}}(\vec{x}) \sim \frac{\rho_0 \overline{u'^2}^2 L_1 L_2^2 \omega_t^4}{c_0^5 r^2 C^5} D_1 \tag{36}$$

$$\overset{\text{shear}}{dI^{\text{noise}}}(\vec{x}) \sim \frac{\rho_0 \overline{u_1'^2} L_1 L_2^4 \omega_t^4}{c_0^5 r^2 C^5} \left(\frac{\partial U_1}{\partial y_2}\right)^2 D_2 \tag{37}$$

respectively :

$$\overset{\text{self}}{dI_{\omega}^{\text{noise}}}(\vec{x},\Theta|\vec{y}) = \frac{\rho_0 \overline{u_1'^2}^2 L_1 L_2^2}{40\sqrt{2}\pi^{3/2} c_0^5 r^2} \frac{\omega^4}{\omega_t} D_1 \exp\left(-\frac{\omega^2 C^2}{8\omega_t^2}\right) \tag{38}$$

$$\overset{\text{shear}}{dI_{\omega}^{\text{noise}}}(\vec{x},\Theta|\vec{y}) = \frac{\rho_0 \overline{u_1'^2} L_1 L_2^4}{\pi^{3/2} c_0^5 r^2} \frac{\omega^4}{\omega_t} \left(\frac{\partial U_1}{\partial y_2}\right)^2 D_2 \exp\left(-\frac{\omega^2 C^2}{4\omega_t^2}\right) \tag{39}$$

with :

$$D_1 = 1 + 2\left(\frac{M}{9} - N\right)\cos^2\Theta\sin^2\Theta + \frac{1}{3}\left[\frac{M^2}{7} + M - \frac{3N}{2}\left(3 - 3N + \frac{3}{2\Delta^2} - \frac{\Delta^2}{2}\right)\right]\sin^4\Theta \qquad (40)$$

$$D_2 = \cos^2\Theta\left[\cos^2\Theta + \frac{1}{2}\left(\frac{1}{\Delta^2} - 2N\right)\sin^2\Theta\right] \qquad (41)$$

with: anisotropic structure of the turbulence ;

$$\Delta = \frac{L_2}{L_1}$$

$$M = \left[\frac{3}{2}\left(\Delta - \frac{1}{\Delta}\right)\right]^2 \qquad (42)$$

$$N = 1 - \frac{\overline{u_2'^2}}{\overline{u_1'^2}} \qquad (43)$$

$\overline{u_1'^2}, \overline{u_2'^2}$ axial and transversal turbulent kinetic energy;

U_1 axial mean flow velocity;

L_1, L_2 integral length scale in the direction of flow and in the transverse direction;

$$L_2 \approx \frac{1}{3} L_1 \qquad (44)$$

$$L_1 \approx \frac{(2k/3)^{3/2}}{\varepsilon} \qquad (45)$$

$\omega_t = 2\pi\dfrac{\varepsilon}{k}$ angular frequency of turbulence;

for isotropic turbulence: $\overline{u_1'^2} = \dfrac{2}{3}k$; (46)

for anisotropic turbulence: $\overline{u_1'^2} = \dfrac{2}{3}k - \nu_t \dfrac{\partial \overline{U}_1}{\partial x_1}$; (47)

$$\overline{u_2'^2} = \frac{2}{3}k - 2\nu_t \frac{\partial \overline{U}_2}{\partial x_2}, \qquad (48)$$

with: $\nu_t = 0{,}09\dfrac{k^2}{\varepsilon}$ (49)

kinematic turbulent viscosity.

N.12.2 Rotor noise

Computation of rotor noise, based on the Ffowcs Williams-Hawkings equation :

$$\Box^2 p_a(\vec{x}, t) = \frac{\partial}{\partial t}[Q\delta(f)] - \frac{\partial}{\partial x_i}[L_i\delta(f)] + \frac{\partial^2}{\partial x_i \partial x_j}[T_{ij}H(f)] \qquad (50)$$

Assumptions :

- (at first) neglecting the quadrupole sources,
- moving surface is non-porous.

Blade thickness noise :

$$\Box^2 p_T = \frac{\partial}{\partial t}\left[\rho_0 u_n \delta(f)\right] \tag{51}$$

Solution :

$$p_T(x_i,t) = \frac{\partial}{\partial t}\int_S \left[\frac{\rho_0 u_n}{4\pi r(1-M_r)}\right]_\tau dS = \int_S \left[\frac{1}{1-M_r}\frac{\partial}{\partial \tau}\frac{\rho_0 u_n}{4\pi r(1-M_r)}\right]_\tau dS \tag{52}$$

for open, rotating blades with a subsonic tip Mach number :

$$p_T(\vec{x},t) = \int_{f=0} \left[\frac{\rho_0(\dot{u}_n + u_{\dot{n}})}{4\pi r(1-M_r)^2}\right]_\tau dS + \int_{f=0} \left\{\frac{\rho_0 u_n\left[r\dot{M}_r + c_0\left(M_r - M^2\right)\right]}{4\pi r^2(1-M_r)^3}\right\}_\tau dS \tag{53}$$

with:
- u_n local velocity of the blade surface in the directional normal to $f=0$;
- $\dot{u}_n = \dot{u}_i n_i$;
- $u_{\dot{n}} = u_i \dot{n}_i = u_i \dfrac{\partial n_i}{\partial \tau}$;
- $\vec{n}$ unit outward normal vector to surface, with components n_i;
- $M = |\vec{M}|$;
- $\vec{M}$ local Mach number vector of source, with components M_i ;
- M_r component of velocity in the radiation direction normalised by c_0
- $f=0$ function that describes the rotor blade surface;
- c_0 sound speed in quiescent medium;

The dot over a symbol implies source-time differentiation of that symbol, e. g. $\dot{M}_r = \left(\dfrac{\partial M_i}{\partial \tau}\right) r_i$.

Loading noise :

$$\Box^2 p_L = -\frac{\partial}{\partial x_i}\left[F_i \delta(f)\right] \tag{54}$$

Solution :

$$\begin{aligned} p_L(x_i,t) &= -\frac{\partial}{\partial x_i}\int_{f=0}\left[\frac{F_i}{4\pi r(1-M_r)}\right]_\tau dS \\ &= \frac{1}{c}\frac{\partial}{\partial t}\int_{f=0}\left[\frac{F_r}{4\pi r(1-M_r)}\right]_\tau dS + \int_{f=0}\left[\frac{F_r}{4\pi r^2(1-M_r)}\right]_\tau dS \end{aligned} \tag{55}$$

for open, rotating blades with a subsonic tip Mach number:

$$p_L(\vec{x},t) = \frac{1}{c_0} \int\limits_{f=0} \left[\frac{\dot{L}_r}{4\pi r(1-M_r)^2} \right]_\tau dS + \int\limits_{f=0} \left[\frac{L_r - L_M}{4\pi r^2(1-M_r)^2} \right]_\tau dS + \frac{1}{c_0} \int\limits_{f=0} \left\{ \frac{L_r\left[r\dot{M}_r + c_0\left(M_r - M^2\right)\right]}{4\pi r^2(1-M_r)^3} \right\}_\tau dS \tag{56}$$

with: L_i components of local force that acts on the fluid (L_i is identical with L_i used above);

$L_i = \left[p_{ij}n_j + \rho v_i(v_n - u_n)\right]$;

$L_r = L_i r_i$;

L_r component of the local force that acts on the fluid (due to the body), in the radiation direction;

$L_M = L_i M_i$;

M_i velocity of the surface $f = 0$ normalised to the ambient sound speed;

M_r component of velocity in the radiation direction normalised to c_0 ;

r distance from a source point on the surface to the observer;

The dot over a symbol implies source-time differentiation of that symbol, e. g. $\dot{L}_r = \left(\frac{\partial L_i}{\partial \tau}\right) r_i$.

Other formulation: **Thickness and loading noise together**

Integral representation of solution (of FW-H equation):

$$p_a(\vec{x},t) = \int\limits_{f=0} \left[\frac{\dot{Q} + \dot{L}_r/c_0}{4\pi r(1-M_r)^2} \right]_\tau dS + \int\limits_{f=0} \left[\frac{L_r - L_M}{4\pi r^2(1-M_r)^2} \right]_\tau dS + \int\limits_{f=0} \left\{ \frac{(Q + L_r/c_0)\left[r\dot{M}_r + c_0\left(M_r - M^2\right)\right]}{4\pi r^2(1-M_r)^3} \right\}_\tau dS \tag{57}$$

Quadrupole noise :

$$\Box^2 p_a(\vec{x},t) = \frac{\partial^2}{\partial x_i \partial x_j}\left[T_{ij}H(f)\right] \tag{58}$$

Solution :

$$p_T(x_i,t) = \frac{\partial}{\partial t}\int_S \left[\frac{\rho_0 u_n}{4\pi r(1-M_r)}\right]_\tau dS = \int_S \left[\frac{1}{1-M_r}\frac{\partial}{\partial \tau}\frac{\rho_0 u_n}{4\pi r(1-M_r)}\right]_\tau dS \tag{59}$$

$$p_Q(\vec{x},t) = \frac{\partial^2}{\partial x_i \partial x_j}\int_V \left[\frac{T_{ij}}{4\pi r|1-M_r|}\right]_\tau dV \tag{60}$$

respectively : (61)

$$p_Q(\vec{x},t) = \frac{1}{c_0^2}\frac{\partial^2}{\partial t^2}\int_V \left[\frac{T_{rr}}{4\pi r|1-M_r|}\right]_\tau dV + \frac{1}{c_0}\frac{\partial}{\partial t}\int_V \left[\frac{3T_{rr}-T_{ii}}{4\pi r^2|1-M_r|}\right]_\tau dV + \int_V \left[\frac{3T_{rr}-T_{ii}}{4\pi r^3|1-M_r|}\right]_\tau dV$$

with:

M_r	projection of the rotational Mach number in the source-observer direction;
$T_{rr} = T_{ij}\hat{r}_i\hat{r}_j$	Lighthill stress tensor in the radiation direction;
$\hat{\vec{r}}$	unit vector in the radiation direction, with components $\hat{r}_i$.

Rotor noise in practice :

sound far field in terms of the source strength spectrum
(rotor and stator noise, propeller noise, helicopter rotor noise etc.)

Monopole :

$$p(x_i,\omega) = \frac{e^{-2\pi j\omega\frac{r}{c_0}}}{4\pi r}\sum_{n=-\infty}^{+\infty} m(\phi,\omega-n\omega_0)e^{-jn\left(\frac{\pi}{2}-\phi\right)}J_n\left(-\frac{2\pi\omega R}{c_0}\sin\Theta\right) \tag{62}$$

with:

$m(\phi,\omega-n\omega_0)$	point monopole, spectrum of the source strength;
ϕ	angular position of the point source at time $t=0$;
ω_0	rotational angular velocity ;
R	position vector of the point source .

Dipole :

$$p(x_i,\omega) = -\frac{e^{-2\pi j\omega\frac{r}{c_0}}}{4\pi r}\frac{2\pi j\omega}{c_0}\hat{r}_i\sum_{n=-\infty}^{+\infty} f_i(\phi,\omega-n\omega_0)e^{-jn\left(\frac{\pi}{2}-\phi\right)}J_n\left(-\frac{2\pi\omega R}{c_0}\sin\Theta\right) \tag{63}$$

with:

$f_i(\phi,\omega-n\omega_0)$	point dipole, spectrum of the source strength;
$\hat{r}_i = \frac{\vec{r}}{\lvert\vec{r}\rvert}$	component of the unit vector in the direction i

Quadrupole :

$$p(x_i,\omega) = -\frac{e^{-2\pi j\omega\frac{r}{c_0}}}{4\pi r}\left(2\pi\frac{\omega}{c_0}\right)^2 \hat{r}_i\hat{r}_j \sum_{n=-\infty}^{+\infty} t_{ij}(\phi,\omega-n\omega_0)e^{-jn\left(\frac{\pi}{2}-\phi\right)}J_n\left(-\frac{2\pi\omega R}{c_0}\sin\Theta\right) \qquad (64)$$

with: $t_{ij}(\phi,\omega-n\omega_0)$ point quadrupole, spectrum of the source strength;

$\hat{r}_i, \hat{r}_j$ components of the unit vector in the direction i, j .

Rotor monopole sound :

$$P = \frac{\dot{M}_B^2\omega_0^2}{8\pi\rho_0 c_0}\sum_{m=1}^{\infty}(mB)^2\int_0^{\pi} J_{mB}^2(mBM_q\sin\Theta)\sin\Theta\, d\Theta \qquad (65)$$

with: B number of rotor blades;

$\dot{M}_B = B\dot{M}$ overall fluid mass displaced per unit time by all rotor blades;

ω_0 rotor angular frequency;

m integer;

J_{mB} Bessel function (first kind) of the order mB;

$M_q = \omega_0 R/c_0$ Mach number ;

R radius of rotor, equivalent radius from the hub point on the rotor blade, in which the source strength is concentrated;

Θ angle between the rotor axis and the vector from the rotor center to the observer point.

Rotor dipole sound due to stationary rotating forces :

$$p_m = \frac{mB\omega_0}{2\pi c_0 r_0}(-j)^{mB+1}e^{-jmB\frac{\omega_0 r_0}{c_0}}\left(F_T\cos\Theta - \frac{F_D}{M_q}\right)J_{mB}(mBM_q\sin\Theta) \qquad (66)$$

$$P = \frac{\omega_0^2}{4\pi\rho_0 c_0^3}\sum_{m=1}^{\infty}\int_0^{\pi}\left(F_T\cos\Theta - \frac{F_D}{M_q}\right)^2(mB)^2 J_{mB}^2(mBM_q\sin\Theta)\sin\Theta\, d\Theta \qquad (67)$$

with: p_m m-th sound pressure harmonic;

F_T, F_D thrust and drag respectively which act on the air and in the site direction, effect of all rotor blades together.

Rotor dipole sound due to rotating periodic time-variable forces :

• sound radiation of the rotor:

$$p_{m\mu} = \frac{jmB\omega_0}{2\pi c_0 r_0}(-j)^{mB-\mu}\left(F_{T\mu}\cos\Theta - \frac{mB-\mu}{mB}\frac{F_{D\mu}}{M_q}\right)J_{mB-\mu}(mBM_q\sin\Theta) \qquad (68)$$

$$P_{m\mu} = \frac{\omega_0^2}{4\pi\rho_0 c_0^3} \int_0^{\pi} \left(F_{T\mu} \cos\Theta - \frac{mB-\mu}{mB} \frac{F_{D\mu}}{M_q} \right)^2 (mB)^2 J_{mB-\mu}^2 \left(mBM_q \sin\Theta \right) \sin\Theta \, d\Theta \tag{69}$$

with: $p_{m\mu}$ / $P_{m\mu}$ m-th harmonic of the sound pressure / sound power radiated by the μ-th harmonic of the fluctuating forces (blade loading harmonic) of the rotor blades;

$F_{T\mu}, F_{D\mu}$ Fourier components of the periodic time-variable blade forces (thrust and drag);

in the case of B rotor blades and V stator vanes: $\mu = kV$.

• sound radiation of the stator :

$$p_{mk} = -\frac{jmBV\omega_0}{2\pi c_0 r_0} j^{mB-kV} \left[F_{Tm} \cos\Theta - \left(\frac{mB-kV}{mB} \right) \frac{F_{Dm}}{M_q} \right] J_{mB-kV}\left(mBM_q \sin\Theta \right) \tag{70}$$

with: p_{mk} m-th harmonic of the sound pressure radiated by the k-th harmonic of the fluctuating forces on the stator vanes owing to rotor-stator interaction.

Rotor dipole sound by rotating random blade forces :

$$p_\omega(x_i,\omega) = -\frac{\omega}{2c_0 r_0} e^{-j\omega \frac{r_0}{c_0}} \sum_{n=-\infty}^{+\infty} \left(F_{T\Omega} \cos\Theta - \frac{n\omega_0}{\omega} \frac{F_{D\Omega}}{M_q} \right) j^{n+1} J_n\left(\frac{\omega}{\omega_0} M_q \sin\Theta \right) \tag{71}$$

$$P_\omega = \frac{z_R \omega^2}{4\pi\rho_0 c_0^3} \int_0^{\pi} \sum_{n=-\infty}^{+\infty} \left(F_{T\Omega} \cos\Theta - \frac{n\omega_0}{\omega} \frac{F_{D\Omega}}{M_q} \right)^2 J_n^2\left(\frac{\omega}{\omega_0} M_q \sin\Theta \right) \sin\Theta \, d\Theta \tag{72}$$

with: $F_{T\Omega}, F_{D\Omega}$ Fourier transforms of the stochastic time-variable blade forces (thrust and drag);

$\omega = \Omega + n\omega_0$;

Ω frequency variable of the force spectrum .

N.13 Power law of the aerodynamic sound sources

Ref.: Ffowcs Williams, [N.74]; Költzsch, [N.76], [N.78];

The power law of aerodynamic sound sources is presented based on developments with help of dimensional analysis, generalised for multipoles of arbitrary order and of variable number of space dimensions, for compact and noncompact aerodynamic multipoles.

Order of multipoles N:

Monopole source $\rightarrow N = 0$;

Dipole source $\rightarrow N = 1$;

Quadrupole source $\rightarrow N = 2$.

Power law of the compact aerodynamic multipoles:

$$(\rho - \rho_0)(\vec{x}) \sim \bar{\rho}_q \left(\frac{L}{r}\right)^{\frac{n-1}{2}} M_t^{N+\frac{n+1}{2}} \tag{1}$$

with: $\rho - \rho_0$ acoustic density fluctuation in the far field;

ρ_0 density of the ambient fluid;

$\bar{\rho}_q$ mean density inside the flow;

L scale of coherent regions in the flow,characteristic length scale;

$M_t = v'/c_0$ Mach number, ratio of characteristic turbulence velocity v' to the speed of sound in the uniform environment, measure of compactness;

N order of multipoles;

n number of space dimensions in which the wave field spreads.

Measure of compactness $\rightarrow M_t > kL$; $\frac{L}{\lambda} < \frac{1}{2\pi} M_t$

in general: M_t is much less than 2π, that is $\frac{L}{\lambda} \ll 1$;

acoustic wavelength is much greater than the length scale characteristic of the turbulent source flow;

it follows: such sources are acoustically compact.

The following overview presents (in the three-dimensional case):

- power laws for acoustic density fluctuations,
- the sound power P ,
- the acoustic-aerodynamically efficiency η (ratio of sound power to the flow power):

Assumptions: $M_t \sim M = \dfrac{U}{c_0}$ with: U a characteristic flow velocity;

$\eta_{ac} = \dfrac{P_{ac}}{P_{mech}}$ acoustic-aerodynamically efficiency

with: P_{ac} sound power;

P_{mech} flow mechanical power: $P_{mech} \sim \overline{\rho}_q L^2 U^3$.

General: $$(\rho - \rho_0)(\vec{x}) \sim \overline{\rho}_q \left(\frac{L}{r}\right) M^{2+N} \sim \overline{\rho}_q \left(\frac{L}{r}\right) U^2 M^N \tag{2}$$

$$P \sim \frac{\overline{\rho}_q}{\rho_0} L^2 U^{4+2N} \sim \frac{\overline{\rho}_q}{\rho_0} L^2 U^3 M^{1+2N} \tag{3}$$

$$\eta_{ac} \sim \frac{\overline{\rho}_q}{\rho_0} M^{1+2N} \tag{4}$$

Monopole: $$(\rho - \rho_0)(\vec{x}) \sim \overline{\rho}_q \left(\frac{L}{r}\right) M^2 \sim \overline{\rho}_q \left(\frac{L}{r}\right) U^2 \tag{5}$$

$$P \sim \frac{\overline{\rho}_q}{\rho_0} L^2 U^4 \sim \frac{\overline{\rho}_q}{\rho_0} L^2 U^3 M \tag{6}$$

$$\eta_M \sim \frac{\overline{\rho}_q}{\rho_0} M \tag{7}$$

Dipole: $$(\rho - \rho_0)(\vec{x}) \sim \overline{\rho}_q \left(\frac{L}{r}\right) M^3 \sim \overline{\rho}_q \left(\frac{L}{r}\right) U^2 M \tag{8}$$

$$P \sim \frac{\overline{\rho}_q}{\rho_0} L^2 U^6 \sim \frac{\overline{\rho}_q}{\rho_0} L^2 U^3 M^3 \tag{9}$$

$$\eta_{ac} \sim \frac{\overline{\rho}_q}{\rho_0} M^3 \tag{10}$$

Quadrupole: $$(\rho - \rho_0)(\vec{x}) \sim \overline{\rho}_q \left(\frac{L}{r}\right) M^4 \sim \overline{\rho}_q \left(\frac{L}{r}\right) U^2 M^2 \tag{11}$$

$$P \sim \frac{\overline{\rho}_q}{\rho_0} L^2 U^8 \sim \frac{\overline{\rho}_q}{\rho_0} L^2 U^3 M^5 \tag{12}$$

$$\eta_{ac} \sim \frac{\overline{\rho}_q}{\rho_0} M^5 \tag{13}$$

Power law of the noncompact aerodynamic multipoles:

measure of compactness → $M_t < kL$; $\dfrac{L}{\lambda} > \dfrac{1}{2\pi} M_t$; acoustic wavelength λ is smaller than $2\pi L / M_t$; → such sources are acoustically noncompact.

Power law: $$(\rho-\rho_0)(\vec{x}) \sim \bar{\rho}_q \left(\frac{L}{r}\right)^{\frac{n-1}{2}} M_t \tag{14}$$

Power law of the moving aerodynamic sources :

$$(\rho-\rho_0)(\vec{x}) \sim \bar{\rho}_q \left(\frac{L}{r}\right) \frac{M^{2+N}}{\left|1-M_{qr}\right|^{1+N}} \tag{15}$$

with: $M_{qr} = \frac{(x_i - y_i)}{|x_i - y_i|} M_{qi}$ projection of the Mach number vector M_{qi} in the direction of sound radiation $\vec{r} = \vec{x} - \vec{y}$.

References to part N.

Flow Acoustics

[N.1] MORFEY, C. L.
Dictionary of acoustics
Academic Press, San Diego etc., 2001

[N.2] LAUCHLE, G. C.
Fundamentals of flow-induced noise
Graduate program in acoustics, Penn State University, 1996

[N.3] DOUGLAS, J. F. et al.
Fluid Mechanics
Longman Scientific & Technical, Harlow, Essex, England, 1986

[N.4] ROGER, M.
Applied aero-acoustics: prediction methods
Lecture Series 1996-04, von Kármán Institute for Fluid Dynamics, Belgium, 1996

[N.5] TELIONIS, D. P.
Unsteady Viscous Flows
Springer-Verlag, New York etc., 1981

[N.6] JOHNSON, R. W. (ED.)
The Handbook of Fluid Dynamics
CRC Press, Boca Raton, FL, Springer-Verlag, Heidelberg, 1998

[N.7] SCHLICHTING, H. und K. GERSTEN
Grenzschicht-Theorie
Springer-Verlag, Berlin etc., 1997

[N.8] LIU, J. T. C.
Contributions to the understanding of large-scale coherent structures in developing free turbulent shear flows.
Advances in Applied Mechanics (ed. by J. W. HUTCHINSON AND T. Y. WU) Vol. 26, Boston etc.: Academic Press, Inc. 1988, pp. 183 - 309

[N.9] HUSSAIN, A. K. M. F. and W. C. REYNOLDS
The mechanics of an organized wave in turbulent shear flow.
J. Fluid Mech. 41 (1970) 2, 241 – 258

[N.10] REYNOLDS, W. C. and A. K. M. F. HUSSAIN
The mechanics of an organized wave in turbulent shear flow.
J. Fluid Mech. 54 (1972) 2, 263 – 288

[N.11] BANGALORE, A.; P. J. MORRIS and L. N. LONG
A parallel three-dimensional computational aeroacoustics method using non-linear disturbance equations.
AIAA 96-1728, 2nd AIAA/CEAS Aeroacoustics Conference State College, PA, 1996

[N.12] BAILLY, C.; P. LAFON and S. CANDEL
Computation of noise generation and propagation for free and confined turbulent flows.
AIAA 96-1732, 2nd AIAA/CEAS Aeroacoustics Conference, State College, PA, 1996

[N.13] LIGHTHILL, M. J.
On sound generated aerodynamically.
Proc. Roy. Soc., London (A), Part I: 211 (1952) 564 – 587; Part II: 222 (1954) 1 - 31

[N.14] HOWE, M. S.
Acoustics of fluid-structure interactions.
University Press, Cambridge, 1998

[N.15] CRIGHTON, D. G.; A. P. DOWLING, J. E. FFOWCS WILLIAMS, M. HECKL and F. G. LEPPINGTON
Modern methods in analytical acoustics.
Springer-Verlag, Berlin etc., 1992

[N.16] GOLDSTEIN, M. E.
Aeroacoustics.
McGraw-Hill International Book Company, New York etc., 1976

[N.17] CURLE, N.
The influence of solid boundaries upon aerodynamic sound.
Proc. Roy. Soc., London (A), 231 (1955) 505 – 514

[N.18] RIBNER, H. S.
Aerodynamic sound from fluid dilatations.
UTIA-Report No. 86, Toronto, Canada, 1958;
J. Acoust. Soc. Am. 31 (1959) 245 - 246

[N.19] MEECHAM, W. C. and G. W. FORD
Acoustic radiation from isotropic turbulence.
J. Acoust. Soc. Am. 30 (1958) 318 – 322

[N.20] MEECHAM, W. C.
Discussion of the pressure-source aerosonic theory and of Doak's criticism.
J. Acoust. Soc. Am. 69 (1981) 3, 643 – 646

[N.21] PHILLIPS, O. M.
On the generation of sound by supersonic turbulent shear layers.
J. Fluid. Mech. 9 (1960) 1 - 28

[N.22] PAO, S. P.
Developments of a generalized theory of jet noise.
AIAA Journal 10 (1972) 5, S. 596 - 602

[N.23] LILLEY, G. M.
On the noise from air jets.
ronautical Research Council ARC 20376, U. K. 1958

[N.24] LILLEY, G. M.
On the refraction of aerodynamic noise.
6th Internat. Congr. on Sound and Vibration, Copenhagen 1999, S. 3581 – 3588

[N.25] LEGENDRE, R.
Bruits émis par la turbulence.
ONERA Publ. 1981-3, 1981

[N.26] MORFEY, C. L.
Fundamental problems in aeroacoustics.
7th Internat. Congr. on Sound and Vibration, Garmisch 2000, pp. 59 – 74

[N.27] GOLDSTEIN, M. E. and W. L. HOWES
New aspects of subsonic aerodynamic noise theory.
NASA TN D-7158

[N.28] RIBNER, H. S.
Effects of jet flow on jet noise via an extension to Lighthill model.
J. Fluid Mech. 321 (1996) 1 – 24

[N.29] ALBRING, W.
Elementarvorgänge fluider Wirbelbewegungen.
Akademie-Verlag, Berlin, 1981

[N.30] DETSCH, F. und F. E. DETSCH
Über die Schallerzeugung in Wirbelfeldern.
Dissertation TU Dresden 1976

[N.31] DITTMAR, R.
Zum Zusammenhang zwischen Turbulenz- und Schallspektrum.
Dissertation TU Dresden 1983

[T32] POWELL, A.
Theory of vortex sound.
J. Acoust. Soc. Am. 36 (1964) 1, 177 - 195

[T33] POWELL, A.
Mechanisms of aerodynamic sound production.
AGARD-Report No. 466, 1963

[N.34] HOWE, M. S.
Contributions to the theory of aerodynamic sound, with application to excess jet noise and the theory of the flute.
J. Fluid Mech. 71 (1975) 625 - 673

[N.35] LILLEY, G. M.
On the noise radiated from a turbulent high speed jet.
In: HARDIN, J. C. and M. Y. HUSSAINI (Editors): Computational Aeroacoustics. Springer-Verlag, New York etc., 1993, pp. 85 - 115

[N.36] MÖHRING, W.
On vortex sound at low Mach number.
J. Fluid Mech. 85 (1978) 685 - 691

[N.37] MÖHRING, W.
A well posed acoustic analogy based on a moving acoustic medium.
Proceedings 1st Aeroacoustic Workshop (in connection with the German research project SWING) 1999, Dresden

[N.38] MÖHRING, W. and F. OBERMEIER
Vorticity – the voice of flows.
Proceed. 6th Internat. Congr. on Sound and Vibration, Copenhagen, pp. 3617 - 3626

[N.39] DOAK, P. E.
Fluctuating total enthalpy as a generalized acoustic field.
Acoustical Physics 41 (1995) 5, pp. 677 – 685

[N.40] DOAK, P. E.
Fluctuating total enthalpy as the basic generalized acoustic field.
Theoret. Comput. Fluid Dynamics (1998) 10, pp. 115 – 133

[N.41] FFOWCS WILLIAMS, J. E. and D. L. HAWKINGS
Sound generation by turbulence and surfaces in arbitrary motion.
Phil. Trans. of the Roy. Soc. London 264 (1969) pp. 321 –342

[N.42] FARASSAT, F.
The Ffowcs Williams-Hawkings equation – Fifteen years of research.
IUTAM Symposium Lyon 1985. Berlin etc.: Springer-Verlag 1986

[N.43] PRIEUR, J. AND G. RAHIER
Comparison of Ffowcs Williams-Hawkings and Kirchhoff rotor noise calculations.
AIAA 98-2376, pp. 984 – 994

[N.44] LONG, L. N. and G. A. WATTS
Arbitrary motion aerodynamics using an aeroacoustic approach.
AIAA Journal 25 (1987) 11, pp. 1442 – 1448

[N.45] PILON, A. R. and A. S. LYRINTZIS
Development of an improved Kirchhoff method for jet aeroacoustics.
AIAA Journal 36 (1998) 5, pp. 783 – 790

[N.46] FARASSAT, F. and M. K. MYERS
Extension of Kirchhoff's formula to radiation from moving surfaces.
J. Sound Vibr. 123 (1988) 3, pp. 451 – 460

[N.47] BRENTNER, K. S. and F. FARASSAT
Analytical comparison of the acoustic analogy and Kirchhoff formulation for moving surfaces.
AIAA Journal 36 (1998) 8, pp. 1379 - 1386

[N.48] BRENTNER, K. S.
A superior Kirchhoff method for aeroacoustic noise prediction:
The Ffowcs Williams-Hawkings equation.
134th Meeting of the ASA, San Diego, CA 1997

[N.49] FARASSAT, F.
Acoustic radiation from rotating blades – the Kirchhoff method in aeroacoustics.
J. Sound Vibr. 239 (2001) 4, pp. 785 – 800

[N.50] MORFEY, C. L.
Amplification of aerodynamic noise by convected flow inhomogeneities.
J. Sound Vibr. 31 (1973) 4, 391 - 397

[N.51] STRAHLE, W. C.
Some results in combustion generated noise.
J. Sound Vibr. 23 (1972) 1, 113 - 125

[N.52] STRAHLE, W. C.
On combustion generated noise.
J. Fluid Mech. 49 (1971) 2, 399 - 414

[N.53] STRAHLE, W. C.
Convergence of theory and experiment in direct combustion-generated noise.
AIAA-Paper 75-522, 2nd Aeroacoustic Conference, Hampton, Va., 1975

[N.54] BOINEAU, PH. ; Y. GERVAIS and M. TOQUARD
Application of combustion noise calculation model to several burners.
AIAA-Paper 98-2271, 4th Aeroacoustic Conference, Toulouse, France, 1998

[N.55] PERREY-DEBAIN, E. ; P. BOINEAU and Y. GERVAIS
A numerical study of refraction effects in combustion-generated noise.
Proceed. 6th Internat. Congr. on Sound and Vibration, Copenhagen, pp. 3361 - 3368

[N.56] CRIGHTON, D. G. and J. E. FFOWCS WILLIAMS
Sound generation by turbulent two-phase flow.
J. Fluid Mech. 36 (1969) 3, pp. 585 – 603

[N.57] LOWSON, M. V.
The sound field for singularities in motion.
Proc. Roy. Soc., London (A) 286 (1965) pp. 559 – 572

[N.58] TANNA, H. K. and C. L. MORFEY
Sound radiation from point sources in circular motion.
J. Sound Vibr. 16 (1971) 3, pp. 337 - 348

[N.59] MORFEY, C. L. and H. K. TANNA
Sound radiation from a point force in circular motion.
J. Sound Vibr. 15 (1971) 3, pp. 325 – 351

[N.60] TANNA, H. K.
Sound radiation from point acoustic stresses in circular motion.
J. Sound Vibr. 16 (1971) 3, pp. 349 – 363

[N.61] IANNIELLO, S.
Quadrupole noise predictions through the Ffowcs Williams-Hawkings equation.
AIAA Journal 37 (1999) 9, pp. 1048 – 1054

[N.62] LYRINTZIS, A. S.
Modelling of turbulent mixing noise. Application to subsonic and supersonic jet noise.
Lecture Series 1997-07, von Kármán Institute for Fluid Dynamics, Belgium, 1997

[N.64] FUCHS, H. V. and A. MICHALKE
Introduction to aerodynamic noise theory.
In: Progress in Aerospace Sciences, Vol. 14 (1973) pp. 227 - 297

[N.65] RIBNER, H. S.
Quadrupole correlations governing the pattern of jet noise.
J. Fluid Mech. 38 (1969) 1, 1-24

[N.66] GOLDSTEIN, M. E. and W. L. HOWES
New aspects of subsonic aerodynamic noise theory.
NASA TN D-7158, 1973

[N.67] BÉCHARA, W.; P. LAFON and C. BAILLY
Application of a $k-\varepsilon$ –turbulence model to the prediction of noise for simple and coaxial free jets.
J. Acoust. Soc. Am. 97 (1995) 6, 3518 - 3531

[N.68] GOLDSTEIN, M. E. and B. ROSENBAUM
Effect of anisotropic turbulence on aerodynamic noise.
J. Acoust. Soc. Amer. 54 (1973) 3, 630 - 645

[N.69] HECKL, M.
Strömungsgeräusche.
Fortschr.-Ber. VDI-Z. Reihe 7, Nr. 20. Düsseldorf: VDI-Verlag 1969

[N.70] FFOWCS WILLIAMS, J. E. and D. L. HAWKINGS
Theory relating to the noise of rotating machinery.
J. Sound Vibr. 10 (1969) 1, 10 – 21

[N.71] LOWSON, M. V.
Theoretical analysis of compressor noise.
J. Acoust. Soc. Amer. 47 (1970) 1(2), 371 – 385

[N.72] FARASSAT, F. and K. S. BRENTNER
The acoustic analogy and the prediction of the noise of rotating blades.
Theoret. Comput. Fluid Dynamics (1998) 10, pp. 155 – 170

[N.73] SINGER, B. A.; K. S. BRENTNER, D. P. LOCKARD and G. M. LILLEY
Simulation of acoustic scattering from a trailing edge.
AIAA 99-0231, 37th Aerospace Sciences Meeting & Exhibit Reno 1999

[N.74] FFOWCS WILLIAMS, J. E.
Hydrodynamic noise.
Annual Review of Fluid Mechanics 1 (1969) 197 – 222

[N.75] KÖLTZSCH, P.
Beitrag zur Berechnung des Wirbellärms von Axialventilatoren.
In: Ventilatoren (Herausgeber: L. Bommes, J. Fricke, K. Klaes),
Vulkan-Verlag, Essen, 1994, S. 434 – 453

[N.76] KÖLTZSCH, P.
Strömungsmechanisch erzeugter Lärm.
Dissertation B (Habilitationsschrift), Technische Universität Dresden, 1974

[N.77] KÖLTZSCH, P.
Berechnung der Schallleistung von axialen Strömungsmaschinen.
(Calculation of sound power of axial flow machines.)
Freiberger Forschungshefte A 721, Deutscher Verlag für Grundstoffindustrie,
Leipzig, 1986

[N.78] KÖLTZSCH, P.
Wozu werden Ähnlichkeitskennzahlen in der Akustik verwendet?
Preprint ET-ITA-01-1998, Technische Universität Dresden, Dresden 1998

O
Analytical and Numerical Methods in Acoustics

by M. Ochmann and F.P. Mechel

Numerous analytical and numerical methods are displayed in this book together with the solutions for special tasks. This chapter contains analytical and numerical methods to be applied in acoustics, going beyond the scope of single examples. The description of a method unavoidably needs more textual explanations than the representation of just the resulting formulas. Section O.1 describes a procedure for optimisation of the parameters of a sound absorber; the Section O.2 outlines a method for the evaluation of many concatenated transfer matrices. The Section O.3 will present five standard problems of numerical acoustics which frequently occur in practical applications. In Sects. O.4 – O.6 three important methods for the numerical solution of these problems will be described. The source simulation technique and the boundary element method are mainly used for exterior problems such as the radiation or the scattering problem (see Sects. O.4 and O.5). The finite element method is especially suited for computing sound fields in interior spaces (see Sect. O.6). The fluid-structure interaction problem can be treated by a combined finite element and boundary element approach, for example with the method of Sect. O.6. The transmission problem can be formulated in terms of boundary integral equations (see Sect. O.5). Analytical field solutions for benchmark models are given in Sects. O.7 , O.8 .

O.1 Computational optimisation of sound absorbers :

See section J.34 for a similar task with duct lining absorbers.

The situation:
There exist precise and fast computing algorithms for the evaluation of a variety of absorbers, even those with complicated structures (e.g. multi-layer absorbers with foils and/or resonator neck plates in front of and/or between the layers of the absorber; see chapters D., G., H.). If one intends to design an absorber with a good performance in some aspect (e.g. sound absorption), one has to optimise by trial and error. This is an optimisation in the space of the absorber parameters which often becomes a 10- to 20-dimensional space, if all parameters are to be optimised in one run.

The task:
Write a computer program which performs this optimisation of the absorber parameters.

What does "optimisation" mean ?

First, one has to fix the absorber quantity which will be improved by variation, e.g. the sound absorption coefficient α .

Second, optimisation is generally understood with respect to a frequency response curve of that quantity. This introduces a frequency interval (f_{lo},f_{hi}) for optimisation, and a stepping through the interval, with either linear or logarithmic steps. Logarithmic stepping may be preferable, because it accentuates lower frequencies, which is often wanted.

Third, optimisation cannot be understood as a general optimisation of all parameters, because then the result can be found in most cases without any computation. For example, the general optimisation with respect to sound absorption of a multi-layer absorber would produce a porous layer with a huge thickness and a tiny flow resistivity of the material, with no surface cover. Therefore a reasonable optimisation supposes some given *structure* of the absorber (e.g. multi-layer absorber with locally reacting layers and possibly layer covers, or multiple-layer absorbers with bulk reacting layers, or locally reacting and bulk reacting layers in a mixed sequence, or resonators of different structures, etc.).

The following description of the method uses as a special example (just for illustration) a multi-layer absorber with locally reacting layers (in fact the layers must not be partitioned if the flow resistivity values of porous materials come out with sufficiently high values). Next, a "parallel" absorber will be optimised.

Fundamentals:

The more acoustical evaluation begins with the computation of the normalised input admittance Z_0G of the arrangement. This evaluation needs as input most of the absorber parameters. The next step is the evaluation of the reflection coefficient $|r|^2$ from which the absorption coefficient follows as $\alpha = 1-|r|^2$.

$$\alpha(\Theta) = 1-|\,r(\Theta)\,|^2 \quad ; \quad r(\Theta) = \frac{\cos\Theta - Z_0G}{\cos\Theta + Z_0G} \tag{1}$$

The absorption coefficient α_{dif} for diffuse sound incidence on locally reacting absorbers is obtained with $Z_0G = g' + j \cdot g''$ by :

$$\alpha_{dif} = 8g'\left[1+\frac{g'^2-g''^2}{g''}\arctan\frac{g''}{g'+g'^2+g''^2} - g'\ln\left(1+\frac{1+2g'}{g'^2+g''^2}\right)\right]$$
$$\xrightarrow[g''\to 0]{} 8g'\left[1+\frac{g'}{1+g'} - g'\ln\left(1+\frac{1+2g'}{g'^2}\right)\right] \tag{2}$$

If the absorber is bulk reacting, or has mixed bulk and locally reacting layers, α_{dif} is obtained from $\alpha(\Theta)$ by evaluation for a set of incidence angles Θ and numerical list integration of the intermediate results. We further write α and $|r|^2$ commonly for the different possible kinds of incidence.

It is important that one knows an "ideal value" for α, i.e. $\alpha = 1$ (this value can be used as the goal, even knowing that for locally reacting absorbers the highest possible absorption coefficient for diffuse incidence is $\alpha \approx 0.95$). Then in the relation $\alpha = 1-|r|^2$ the reflection coefficient $|r|^2$ can be interpreted as a squared "error" of the actual value α compared with the ideal value $\alpha = 1$. The algorithm for optimisation minimises the averaged square error

$$\langle |r|^2\rangle = q(a_1,a_2,\ldots;b_1,b_2,\ldots) = \sum_n w_n\cdot|r(f_n;a_1,a_2,\ldots;b_1,b_2,\ldots)|^2 \Big/ \sum_n w_n \overset{!}{=} \text{Min} \tag{3}$$

where f_n are sampling frequencies over (f_{lo},f_{hi}) ; $w(f)$ is a weight function; $w_n = w(f_n)$; the $\{a_i\}$ are the variable absorber parameters ; the $\{b_k\}$ are fixed parameters. The minimisation is performed by variation of the $\{a_i\}$.

The use of a weight function $w(f)$ introduces the possibility to perform either a broad-banded optimisation over (f_{lo},f_{hi}), e.g. with $w(f)=1$, or a centred optimisation, if $w(f)$ has a central maximum in (f_{lo},f_{hi}) (this form of optimisation can be used, for example, for the down-tuning of resonators, together with an improvement).

There exist algorithms in the literature for finding a minimum of a real function in the multi-dimensional space (e.g. W.H. PRESS et al. "Numerical Recipes", Cambridge University Press). A principal distinction must be made in such algorithms whether the partial derivatives $\partial q / \partial a_i$ can be evaluated or not. This generally is not possible in the present task. Some start values $\{a_i\}_{start}$ must be given for the start of the minimum search. Most algorithms described for minimisation extend the search over the whole (real) space of the $\{a_i\}$, i.e. also to negative values. In our case, however, the parameters a_i represent geometrical lengths or material data which should be positive; some of them, like the porosity σ, have to respect lower and higher limits $(0<\sigma<1)$. Thus range limits for the a_i must be transmitted to the minimisation algorithm.

The procedure for optimisation of the absorption coefficient α (and similarly for other target quantities) works as follows :

(1) Make the decision for the target quantity (e.g. the sound absorption coefficient α for diffuse incidence).

(2) Make the decision for a structure of the absorber (e.g. a multi-layer absorber with locally reacting layers with (possibly) porous cover foils and/or perforated plates on the front sides of the layers). This decision determines the theoretical description of the absorber.

(3) Fix the frequency interval (f_{lo},f_{hi}), the kind of frequency steps (linear Δf or logarithmic $\Delta lg(f)$,which often is preferable), the step width (it should not be too large, otherwise $\langle|r|^2\rangle = q(a_1,a_2,\ldots;b_1,b_2,\ldots)$ may not be steady enough for finding a minimum); fix the type of the weight function $w(f)$ (see above).

(4) Conceive a "start configuration" of the absorber, i.e. select values for all required absorber parameters (they define what will be called here the "original absorber").

It is supposed that at this stage the data input is structured as follows (just for giving an example for a two-layer absorber; entries (**text* *) indicate comment text):

```
(*Input frequency*)
flo= 50. ; fhi= 2000. ; lgfstep= 0.1 ;
(*Input layers*)
gn=  2 ;                           (*number N of layers*)
matlist= {1 , 1} ;                 (*types of porous material*)
xilist=  {10000., 10000.};         (*flow resistivities Ξ *)
tlist=   {0.05 , 0.05 };           (*layer thickness t in meter*)
(*Input foils*)
roflist=  {2700., 2700.};          (*foil material density ρf in kg/m³*)
dflist=   {0.0002, 0.0002};        (*foil thickness df in meter *)
rflist=   {1000. , 2. };           (*normalised foil flow resistance Rf *)
(*Input perforates*)
shapelist= {1 , 1};            (*perforation shape; 1= hole; 2= slit *)
siglist=  {0. , 0. };          (*porosity σ *)
dialist=  {0. , 0. };          (*diameter of perforation in meter*)
dplist=   {0. , 0.};           (*plate thickness dp in meter *)
```

(Values d_f=0 or d_p=0 indicate that there is no foil or plate in that position; the other foil or plate parameter entries in that position are then neglected).

(5) Determine which absorber parameters should be varied for optimisation. Some parameters evidently cannot be varied, such as the number N of layers, the layer material type (in matlist), the shape of the perforations (in shapelist); others should be kept constant, because a variation could lead to values which cannot be realised (such as the foil material density ρ_f in roflist).

The variable parameters could be signalled by lists of flags like

```
(*Input layers*)
xiflag=  {1 , 1 };          (*flow resistivities Ξ *)
tflag=   {0 , 0 };          (*layer thickness t in meter*)
(*Input foils*)
dfflag=    {0 , 0 };        (*foil thickness df in meter *)
rfflag=    {1 , 1 };        (*normalised foil flow resistance Rf *)
(*Input perforates*)
sigflag=   {0 , 0 };        (*porosity σ *)
diaflag=   {0 , 0 };        (*diameter of perforation in meter*)
dpflag=    {0 , 0 };        (*plate thickness dp in meter *)
```

A value 1 indicates that the corresponding parameter belongs to the $\{a_i\}$, a value 0 signals that the parameter belongs to the $\{b_k\}$. The number of unit values determines the dimension of the space of variables for the minimisation.

(6) Some algorithms for minimisation need more than one set of start parameters, for example to ir

```
(*Input layers*)
xilo=     {xx , xx };       (*flow resistivities Ξ *)
tlo= {0 , 0 };              (*layer thickness t in meter*)
(*Input foils*)
dflo=       {0 , 0 };       (*foil thickness df in meter *)
rflo=       {xx, xx };      (*normalised foil flow resistance Rf *)
(*Input perforates*)
siglo=      {0 , 0 };       (*porosity σ *)
dialo=      {0 , 0 };       (*diameter of perforation in meter*)
dplo=       {0 , 0 };       (*plate thickness dp in meter *)
```

and

```
(*Input layers*)
xihi=     {xx , xx };       (*flow resistivities Ξ *)
thi= {0 , 0 };              (*layer thickness t in meter*)
(*Input foils*)
dfhi=       {0 , 0 };       (*foil thickness df in meter *)
rfhi=       {xx, xx };      (*normalised foil flow resistance Rf *)
(*Input perforates*)
sighi=      {0 , 0 };       (*porosity σ *)
diahi=      {0 , 0 };       (*diameter of perforation in meter*)
dphi=       {0 , 0 };       (*plate thickness dp in meter *)
```

(7) If one writes a general algorithm for minimisation of a function $q(\{a_i\},\{b_k\})$, which should be usable for a changing number and composition of the $\{a_i\},\{b_k\}$, then one should write a subroutine which transmits the variable absorber parameters to the variables $\{a_i\}$

(8) One needs correspondingly a subroutine which transmits the $\{a_i\}$ back to the right positions of the absorber parameters (the information comes from the lists of flags).

This subroutine, at the same time, takes care of the range limits for the parameters. A lower zero limit for example will be considered, if the subroutine transmits $|a_i|$. A lower or a higher non-zero limit can be introduced by the replacement of a_i by the limit values ...lo or ...hi whenever a_i exceeds the limits; this procedure replaces the variable function $q(a_i;b_k)$ outside the limits by a constant value (the minimisation program consequently will avoid ranges outside the limits). With this back transmission of the a_i to the absorber parameters one must not be worried when the minimisation program returns negative values for the a_i .

Examples of 2-layer absorbers:

Some examples will illustrate the procedure. The absorber is a two-layer absorber with locally reacting layers, possibly with foils and/or perforated plates (resonator neck plates) at the front sides of the layers. Diffuse sound incidence is applied.

First example:

This example uses the data input for the "original absorber" as given in the above lists; the variable parameters are `xilist[[1]]`, `xilist[[2]]`, `grflist[[1]]`, `grflist[[2]]` ([[n]] indicates the position n of an element in a list). The diagram shows α_{dif} for both the original absorber (dashed curve) and the optimised absorber (full line); the print-out of the input data indicates the optimised parameters with bold printing.

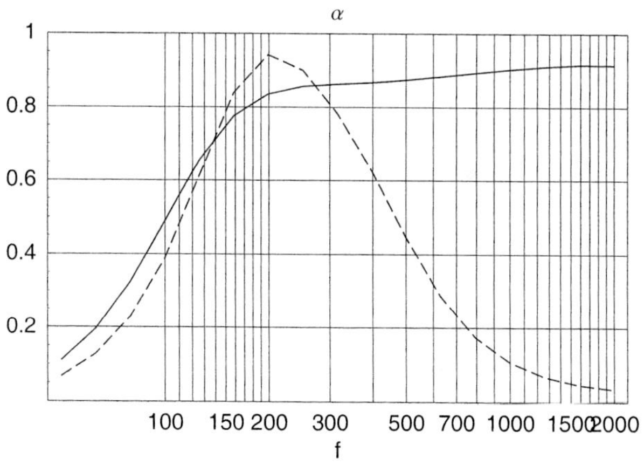

```
Input & optimised parameters
(*Input frequency*)
flo=50., fhi=2000., Dlg(f)=0.1
(*Flags*)
theta=0., Reflection=diffuse, Weight=no
(*Number of layers & dimensions*)
Nlayer=2, dim=4
(*Layer parameters*)
mat={1,1}
Xi={13468.5,4435.31}
t={0.05,0.05}

(*Foil parameters*)
rof={2700.,2700.}
df={0.0002,0.0002}
Rf={1.28767,1.67335}
(*Perforate parameters*)
shapes={1,1}
sig={0.,0.}
dia={0.,0.}
d={0.,0.}
(*Minimum value of <|r|^2>*)
<|r|^2>min=0.2783
(*Varied parameters*)
{xilist[[1]],xilist[[2]],grflist[[1]],grflist[[2]]}
```

The reduction of absorption in the resonance maximum could be reduced by choosing a smaller frequency interval (f_{lo},f_{hi}) and a weight function w(f) with a central weighting (e.g. in the form of a cosine arc).

Second example:

The second example uses a resonator neck plate between the layers; its porosity is constant with $\sigma = 0.15$. The second layer makes up the resonator volumes; it consists of air (mat=0). The Helmholtz resonance is marked by a rather low maximum of α of the original absorber at higher frequencies. Varied parameters are `xilist[[1]]`, `tlist[[2]]`, `grflist[[1]]`, `dialist[[2]]`, `dplist[[2]]` .

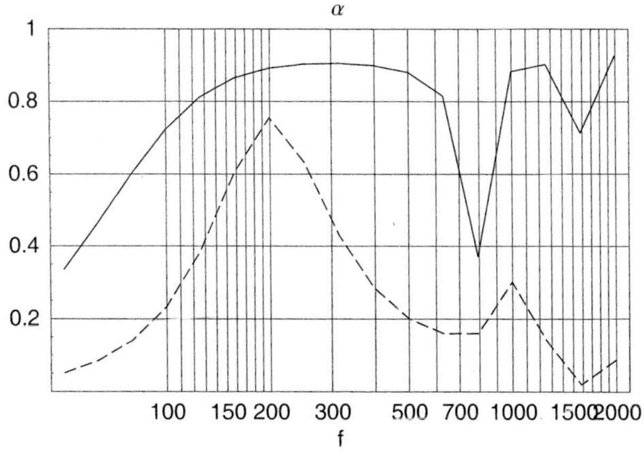


```
Input & optimised parameters
(*Input frequency*)
flo=50., fhi=2000., Dlg(f)=0.1
(*Flags*)
theta=0., Reflection=diffuse, Weight=no
(*Number of layers & dimensions*)
Nlayer=2, dim=5
(*Layer parameters*)
mat={1,0}
Xi={32183.4,0.}
t={0.02,0.185033}
(*Foil parameters*)
rof={2700.,0.}
df={0.0002,0.}
Rf={0.903345,0.}
(*Perforate parameters*)
shapes={0,1}
sig={0.,0.15}
dia={0.,0.000174437}
d={0.,0.00628449}
(*Minimum value of <|r|^2>*)
<|r|^2>min=0.233324
(*Varied parameters*)
{xilist[[1]],tlist[[2]],grflist[[1]],dialist[[2]],dlist[[2]]}
```

The third example is more of a theoretical than a practical interest. Thickness and flow resistance of the cover foil and of the first layer are varied. The absorption coefficients α for diffuse incidence of both the original absorber (curve) and the optimised absorber (upper horizontal line) are full.

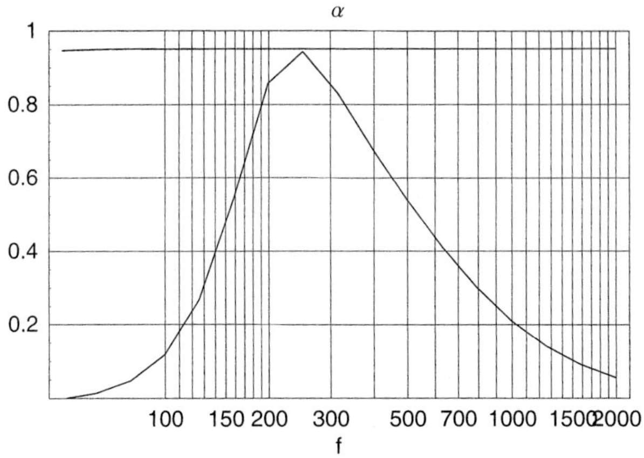


```
Input & optimised parameters
(*Input frequency*)
flo=50., fhi=2000., Dlg(f)=0.1
(*Flags*)
theta=0., Reflection=diffuse, Weight=no
(*Number of layers & dimensions*)
Nlayer=2, dim=4
(*Layer parameters*)
mat={1,1}
Xi={328.465,10000.}
t={3.26549,0.05}
(*Foil parameters*)
rof={2700.,2700.}
df={0.000328666,0.0002}
Rf={0.513109,2.}
(*Perforate parameters*)
shapes={1,1}
sig={0.,0.}
dia={0.,0.}
d={0.,0.}
(*Minimum value of <|r|^2>*)
<|r|^2>min=0.0492101
(*Varied parameters*)
{xilist[[1]],tlist[[1]],dflist[[1]],grflist[[1]]}
```

As one could have anticipated with this choice of variable parameters, the front layer has an enormous thickness and a very small flow resistivity, and the cover foil is thin with a low flow resistance. It is of some interest that the average reflection coefficient $\langle|r|^2\rangle$ obtained by optimisation is very precisely the theoretical minimum for diffuse sound incidence on locally reacting absorbers.

Example of a "parallel" absorber :

The examples of multi-layer absorbers, shown above, can be named "series" absorbers. A "parallel" absorber, in contrast, is composed of different elementary absorbers which are placed side by side. As long as the lateral dimensions of the component absorbers and the composition are small compared to the wavelength, only the average admittance $\langle G\rangle$ counts.

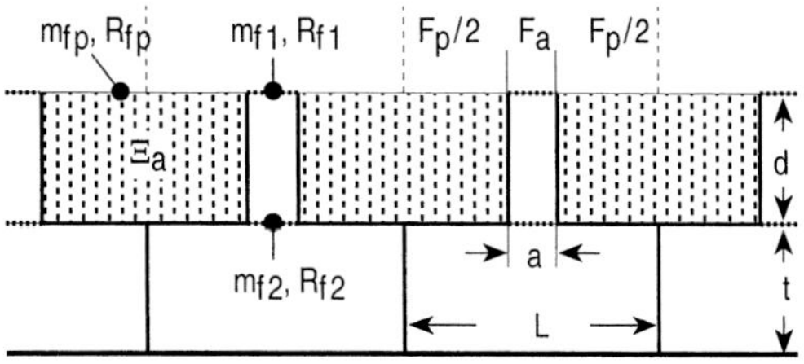

Boxes of width L–a and depth d contain a porous absorber material (e.g. glass fibres) with flow resistivity Ξ . Between the boxes are gaps of width a which form the necks of Helmholtz resonators. The boxes and/or one or both neck orifices may be covered with foils having a surface mass density $m_f = \rho_f \cdot d_f$ and a normalised flow resistance R_f. With G_p for the input admittance of the boxes, and G_a for the input admittance of the resonator necks, the average admittance is :

$$\langle G\rangle = \frac{G_p + ß\cdot G_a}{1+ß} \quad ; \quad ß = \frac{F_a}{F_p} = \frac{a/L}{1-a/L} \tag{4}$$

Possibly varied parameters are :

```
pars= {L, a/L, d/L, t/L, Ξ, dfp, Rfp, df1, Rf1, df2, Rf2}
```

i.e. up to 11 variables in the search for a minimum.

The dashed curve of the "original" absorber in the following diagram for the sound absorption coefficient with diffuse sound incidence was obtained after "manual" optimisation. The full curve was obtained after application of the optimisation algorithm on the printed input data.

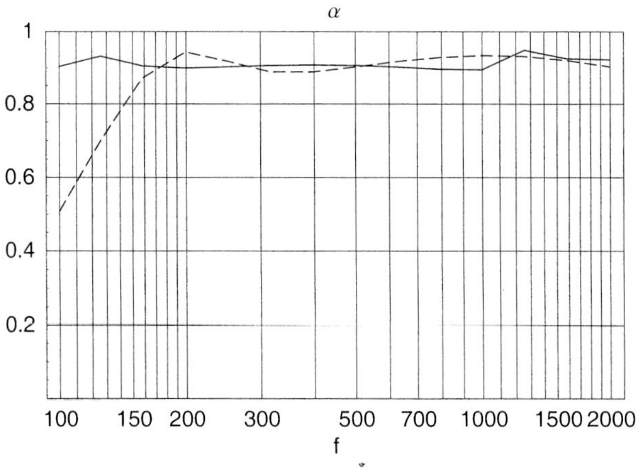

```
Input & optimised parameters
(*Input frequency*)
flo=100. [Hz], fhi=2000. [Hz], Dlg(f)=0.1
(*Flags*)
Θ=0., Reflection=diffuse, Weight=no
(*Dimensions*)
L=0.2 [m], a/L=0.1, d/L=0.3, t/L=0.5
(*Layer parameters*)
Mat=1
Ξ=10000. [Pa·s/m^2]
(*Foil parameters*)
ρfp=2700. [kg/m^3]
dfp=0.0005 [m]
Rfp=1.
Foil Position=1
rof1=2700. [kg/m^3]
df1=0.0002 [m]
Rf1=0.2
rof2=0.
df2=0.
Rf2=0.
(*Minimum value of <|r|^2>*)
<|r|^2>min=0.0886165
(*Varied parameters*)
{a/L,d/L,df1,Rf1}={0.0862694,0.554809,0.0000810588,0.35581}
```

Post-processing:

One should notice that the algorithms for finding a minimum generally search for local minima, a number of which may exist in the range of the variable parameters, especially if the dimension of the parameter space is large.

(9) You may try to find other minima either by starting with a different "original absorber" and/or by using other parameter ranges.

(10) You may modify the optimum found for example by an exchange of a large resonator volume depth with a smaller neck plate porosity. Then put the other optimised parameter values among the $\{b_k\}$ and append a run of optimisation with the parameters between which a "trading" shall be tried.

(11) It is a good practice to evaluate the target quantity with some "manual" modifications of the optimised parameter values, in order to see whether the absorber is sensitive with

respect to small changes of the parameter values. In a similar procedure it can be tested, whether parameter values which are obtained by optimisation, but which are not available, can be replaced by nearby values of available absorber components.

(12) Do not forget that the optimum found may depend also on the frequency range used $\{f_{lo}, f_{hi}\}$.

A final remark should be made. The program for finding the minimum may make wide excursions in the parameter space. Therefore the evaluations for the input admittance Z_0G and the reflection coefficient $|r|^2$ should apply formulas with analytical foundations, in order to avoid false or even nonsense results. This implies that the evaluation of characteristic data of porous materials should not use formulas which stem of regressions through experimental data (like the Delany-Bazley approximation), because the range of the flow resistivity, for example, in which the data were measured may be exceeded. Evaluations are recommended which are based either on analytical models of porous materials or on analytical models which are fitted to experimental data. Further, the diameter of resonator necks may become very small within the search for a minimum. It is recommended to use the propagation constant and wave impedance in capillaries, which include viscous and thermal losses and which go to $j \cdot k_0$, Z_0 for wider necks.

O.2 Computing with mixed numeric-symbolic expressions, illustrated with silencer cascades

See sections J.19, J.20 dealing with silencer cascades, where the problem of this section is avoided by neglecting acoustical feedback in the sections, and see section J.29 with a more simple application, due to the monotonic variation of cross-sections.

There are many tasks in acoustics which lead to iterative linear systems of equations. Such tasks are, for example, duct cascades, mufflers, conical ducts, wedges, a medium with spatial variation, etc. In principle one could consider the system of systems of linear equations as a large system for the combined vector of variables of those systems. However, this procedure mostly fails in its numerical realisation.

On the other hand, there exist mathematical programs which support both numerical and symbolic computations. This feature can be used to design straightforward solutions for the mentioned tasks, avoiding large systems of equations and many inversions of matrices. The method will be explained and illustrated with the example of a cascade of sections of lined ducts with different cross-sections and/or linings.

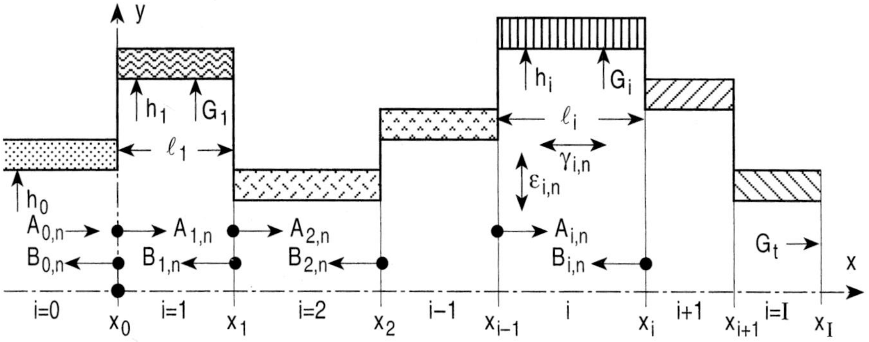

A sequence i=1,2,…,I of two-dimensional, flat duct sections follow each other, with half heights h_i, lengths ℓ_i, and lined with locally reacting absorbers with input admittances G_i. The fields in the sections are formulated as mode sums of forward propagating modes with amplitudes $A_{i,n}$ and backward propagating modes with amplitudes $B_{i,n}$. The lateral wave numbers $\varepsilon_{i,n}$ follow from the characteristic equations of the sections, the axial propagating constants $\gamma_{i,n}$ from the secular equation. Some kinds of excitation at the entrance $x=x_0$ will be considered, and some kinds of terminations at $x=x_I$. The "heads" of duct steps are assumed to be hard.

Field formulations :

$$p_i(x,y)=\sum_n\left[A_{i,n}\cdot e^{-\gamma_{i,n}(x-x_{i-1})}+B_{i,n}\cdot e^{+\gamma_{i,n}(x-x_i)}\right]\cdot q_{i,n}(y)$$

$$Z_0 v_{i,x}(x,y)=-j\sum_n\frac{\gamma_{i,n}}{k_0}\left[A_{i,n}\cdot e^{-\gamma_{i,n}(x-x_{i-1})}-B_{i,n}\cdot e^{+\gamma_{i,n}(x-x_i)}\right]\cdot q_{i,n}(y) \tag{1}$$

with the mode profiles:

$$q_{i,n}(y)=\cos(\varepsilon_{i,n}y) \tag{2}$$

and the lateral wave numbers $\varepsilon_{i,n}$ solutions of the characteristic equations :

$$(\varepsilon_{i,n}h_i)\cdot\tan(\varepsilon_{i,n}h_i)=jk_0h_i\cdot Z_0G_i \tag{3}$$

the axial propagation constants from the secular equations:

$$\gamma_{i,n}h_i=\sqrt{(\varepsilon_{i,n}h_i)^2-(k_0h_i)^2}\quad;\quad \mathrm{Re}\{\gamma_{i,n}h_i\}\ge 0 \tag{4}$$

Mode norms $N_{i,m}$:

$$\frac{1}{h_i}\int_0^{h_i} q_{i,m}(y)\cdot q_{i,n}(y)\,dy=\delta_{m,n}\cdot N_{i,m}\quad;\quad N_{i,m}=\frac{1}{2}\left(1+\frac{\sin(2\varepsilon_{i,m}h_i)}{2\varepsilon_{i,m}h_i}\right) \tag{5}$$

Mode coupling coefficients $C(i,m;k,n)$:

$$C(i,m;k,n)=\frac{1}{h_i}\int_0^{h_i} q_{i,m}(y)\cdot q_{k,n}(y)\,dy\quad;\quad k=\begin{cases}i-1\\ i+1\end{cases}$$

$$C(i,m;k,n)=\frac{1}{2}\left(\frac{\sin(\varepsilon_{i,m}-\varepsilon_{k,n})h_i}{(\varepsilon_{i,m}-\varepsilon_{k,n})h_i}+\frac{\sin(\varepsilon_{i,m}+\varepsilon_{k,n})h_i}{(\varepsilon_{i,m}+\varepsilon_{k,n})h_i}\right) \tag{6}$$

Boundary conditions at $x=x_i$ for sound pressure and axial particle velocity :

expanding duct :
($h_{i+1}\ge h_i$) (7)

$$p_i(x_i,y)\overset{!}{=}p_{i+1}(x_i,y)\quad \text{in } y=(0,h_i)$$

$$Z_0v_{i+1,x}(x_i,y)\overset{!}{=}\begin{cases}Z_0v_{i,x}(x_i,y) & \text{in } y=(0,h_i)\\ 0 & \text{in } y=(h_i,h_{i+1})\end{cases}$$

contracting duct :
($h_{i+1}<h_i$) (8)

$$p_i(x_i,y)\overset{!}{=}p_{i+1}(x_i,y)\quad \text{in } y=(0,h_{i+1})$$

$$Z_0v_{i,x}(x_i,y)\overset{!}{=}\begin{cases}Z_0v_{i+1,x}(x_i,y) & \text{in } y=(0,h_{i+1})\\ 0 & \text{in } y=(h_{i+1},h_i)\end{cases}$$

Notations (for ease of writing) :

$$\gamma_{i,n}h_{i+1}=\gamma h_{i,n}\quad;\quad \gamma_{i,n}\ell_i=\gamma\ell_{i,n}\quad;\quad A_{i,n}\cdot N_{i,n}=\overline{A}_{i,n}\quad;\quad B_{i,n}\cdot N_{i,n}=\overline{B}_{i,n} \tag{9}$$

By use of the orthogonality, with m= any mode order :

expanding duct : $(h_{i+1} \ge h_i)$

$$\left[\overline{A}_{i,m}\cdot e^{-\gamma\ell_{i,m}}+\overline{B}_{i,m}\right]=\sum_n\left[\overline{A}_{i+1,n}+\overline{B}_{i+1,n}\cdot e^{-\gamma\ell_{i+1,n}}\right]\cdot\frac{C(i,m;i+1,n)}{N_{i+1,n}}$$

$$\gamma h_{i+1,m}\cdot\left[\overline{A}_{i+1,m}-\overline{B}_{i+1,m}\cdot e^{-\gamma\ell_{i+1,m}}\right]=\sum_n\gamma h_{i,n}\cdot\left[\overline{A}_{i,n}\cdot e^{-\gamma\ell_{i,n}}-\overline{B}_{i,n}\right]\cdot\frac{C(i,n;i+1,m)}{N_{i,n}} \tag{10}$$

contracting duct : $(h_{i+1} < h_i)$

$$\left[\overline{A}_{i+1,m}+\overline{B}_{i+1,m}\cdot e^{-\gamma\ell_{i+1,m}}\right]=\sum_n\left[\overline{A}_{i,n}\cdot e^{-\gamma\ell_{i,n}}+\overline{B}_{i,n}\right]\cdot\frac{C(i+1,m;i,n)}{N_{i,n}}$$

$$\gamma h_{i,m}\cdot\left[\overline{A}_{i,m}\cdot e^{-\gamma\ell_{i,m}}-\overline{B}_{i,m}\right]=\sum_n\gamma h_{i+1,n}\cdot\left[\overline{A}_{i+1,n}-\overline{B}_{i+1,n}\cdot e^{-\gamma\ell_{i+1,n}}\right]\cdot\frac{C(i+1,n;i,m)}{N_{i+1,n}} \tag{11}$$

The ratios of mode coupling coefficients and mode norms on the right-hand sides form matrices $\{\text{matrix}\}$; we symbolise with $\{\text{matrix}\}^{-1}$ the inverse of a matrix. Then the above systems (of couples of systems) of linear equations for the mode amplitudes lead to the iterative systems:

expanding duct : $(h_{i+1} \ge h_i)$

$$\overline{A}_{i+1,m}=\frac{1}{2}\sum_n\overline{A}_{i,n}\cdot e^{-\gamma\ell_{i,n}}\left(\left\{\frac{C(i,n;i+1,m)}{N_{i+1,m}}\right\}^{-1}+\frac{\gamma h_{i,n}}{\gamma h_{i+1,m}}\frac{C(i,n;i+1,m)}{N_{i,n}}\right)+ \overline{B}_{i,n}\left(\left\{\frac{C(i,n;i+1,m)}{N_{i+1,m}}\right\}^{-1}-\frac{\gamma h_{i,n}}{\gamma h_{i+1,m}}\frac{C(i,n;i+1,m)}{N_{i,n}}\right) \tag{12}$$

$$\overline{B}_{i+1,m}\cdot e^{-\gamma\ell_{i+1,m}}=\frac{1}{2}\sum_n\overline{A}_{i,n}\cdot e^{-\gamma\ell_{i,n}}\left(\left\{\frac{C(i,n;i+1,m)}{N_{i+1,m}}\right\}^{-1}-\frac{\gamma h_{i,n}}{\gamma h_{i+1,m}}\frac{C(i,n;i+1,m)}{N_{i,n}}\right)+ \overline{B}_{i,n}\left(\left\{\frac{C(i,n;i+1,m)}{N_{i+1,m}}\right\}^{-1}+\frac{\gamma h_{i,n}}{\gamma h_{i+1,m}}\frac{C(i,n;i+1,m)}{N_{i,n}}\right) \tag{13}$$

contracting duct : $(h_{i+1} < h_i)$

$$\overline{A}_{i+1,m}=\frac{1}{2}\sum_n\overline{A}_{i,n}\cdot e^{-\gamma\ell_{i,n}}\left(\frac{C(i+1,m;i,n)}{N_{i,n}}+\frac{\gamma h_{i,n}}{\gamma h_{i+1,m}}\left\{\frac{C(i+1,m;i,n)}{N_{i+1,m}}\right\}^{-1}\right)+ \overline{B}_{i,n}\left(\frac{C(i+1,m;i,n)}{N_{i,n}}-\frac{\gamma h_{i,n}}{\gamma h_{i+1,m}}\left\{\frac{C(i+1,m;i,n)}{N_{i+1,m}}\right\}^{-1}\right) \tag{14}$$

$$\overline{B}_{i+1,m}\cdot e^{-\gamma\ell_{i+1,m}}=\frac{1}{2}\sum_n\overline{A}_{i,n}\cdot e^{-\gamma\ell_{i,n}}\left(\frac{C(i+1,m;i,n)}{N_{i,n}}-\frac{\gamma h_{i,n}}{\gamma h_{i+1,m}}\left\{\frac{C(i+1,m;i,n)}{N_{i+1,m}}\right\}^{-1}\right)+$$
$$\overline{B}_{i,n}\left(\frac{C(i+1,m;i,n)}{N_{i,n}}+\frac{\gamma h_{i,n}}{\gamma h_{i+1,m}}\left\{\frac{C(i+1,m;i,n)}{N_{i+1,m}}\right\}^{-1}\right) \qquad (15)$$

There are more unknown mode amplitudes $A_{i,n}$, $B_{i,n}$ than equations at the duct section limits. One needs *source conditions* and *termination conditions*.

Alternative *source conditions* :

Sound pressure source :
At the entrance $x=x_0$ a sound pressure profile $P(y)$ is given (over the full height h_1, else the task would be ill posed). Expand the pressure profile in modes of the section $i=1$:

$$P(y)=\sum_n a_n\cdot\cos(\varepsilon_{1,n}y)\quad;\quad \bar{a}_n=N_{1,n}\cdot a_n=\frac{1}{h_1}\int_0^{h_1}P(y)\cdot\cos(\varepsilon_{1,n}y)\,dy \qquad (16)$$

The boundary condition for the sound pressure at $x=x_0$ gives :

$$\overline{A}_{1,n}+\overline{B}_{1,n}\cdot e^{-\gamma\ell_{1,n}}=\bar{a}_n \qquad (17)$$

One of the amplitudes $\overline{A}_{1,n}$, $\overline{B}_{1,n}$ of the first section can be expressed by the other amplitude and a known number $\bar{a}_n$.

Particle velocity source :
At the entrance $x=x_0$ a particle velocity profile $V(y)$ is given; the source may cover a height $h_0\le h_1$, assuming the other part (h_0,h_1) of the entrance plane is hard. Expand in modes of the section $i=1$:

$$Z_0v_{1,x}(x_0,y)=\begin{cases}Z_0V(y) & \text{in } y=(0,h_0)\\ 0 & \text{in } y=(h_0,h_1)\end{cases}=\sum_n a_n\cdot\cos(\varepsilon_{1,n}y)$$
$$\bar{a}_n=N_{1,n}\cdot a_n=\frac{1}{h_1}\int_0^{h_1}\left\{\begin{matrix}Z_0V(y)\\0\end{matrix}\right\}\cdot\cos(\varepsilon_{1,n}y)\,dy \qquad (18)$$

The boundary condition at $x=x_0$ for the particle velocity gives :

$$\overline{A}_{1,n}-\overline{B}_{1,n}\cdot e^{-\gamma\ell_{1,n}}=\frac{jk_0h_1}{\gamma h_{1,n}}\bar{a}_n \qquad (19)$$

Again one of the amplitudes $\overline{A}_{1,n}$, $\overline{B}_{1,n}$ of the first section can be expressed by the other amplitude and a known number containing $\bar{a}_n$.

Incident wave from the entrance duct i=0 :

A sound wave with given mode amplitudes $A_{0,n}$ is incident from the entrance duct i=0 which is supposed to be anechoic or infinite for $x \to -\infty$.

Setting formally $\ell_0 = 0$, i.e. defining the amplitudes $A_{0.n}$ in the plane $x= x_0$, the above equations can also be used for that cross-section. The right-hand sides contain only the amplitudes $B_{0,n}$ as unknown quantities.

Depending on the selected source condition, the above systems of equations will have the general forms (imagine the iteration to be performed up to i):

for pressure or velocity source :

$$\overline{A}_{i+1,m} = \frac{1}{2}\sum_n \overline{B}_{I,n} \cdot \alpha_{i,n} + \beta_{i,n}$$
$$\overline{B}_{i+1,m} = \frac{1}{2}\sum_n \overline{B}_{I,n} \cdot \alpha'_{i,n} + \beta'_{i,n} \tag{20}$$

with numerical values $\alpha_{i,n}$, $\beta_{i,n}$, $\alpha'_{i,n}$, $\beta'_{i,n}$ and symbolic $\overline{B}_{I,n}$.

for incident wave :

$$\overline{A}_{i+1,m} = \frac{1}{2}\sum_n \overline{B}_{0,n} \cdot \alpha_{i,n} + \beta_{i,n}$$
$$\overline{B}_{i+1,m} = \frac{1}{2}\sum_n \overline{B}_{0,n} \cdot \alpha'_{i,n} + \beta'_{i,n} \tag{21}$$

with numerical values $\alpha_{i,n}$, $\beta_{i,n}$, $\alpha'_{i,n}$, $\beta'_{i,n}$ and symbolic $\overline{B}_{0,n}$.

Alternative *termination conditions* :

Section i+1= I anechoic : i.e. $\overline{B}_{I,m} = 0$

Then the last of the above equations is a system of linear inhomogeneous equations for either $\overline{B}_{I,n}$ or $\overline{B}_{0,n}$. After its solution, all other mode amplitudes follow by insertion in the former systems of equations.

Section i+1= I terminated with an admittance G_t:

This termination condition gives the relations between $\overline{A}_{I,n}, \overline{B}_{I,n}$:

$$\gamma h_{I,n}\left[\overline{A}_{I,n} \cdot e^{-\gamma\ell_{I,n}} - \overline{B}_{I,n}\right] = jk_0h_I \cdot Z_0G_t \cdot \left[\overline{A}_{I,n} \cdot e^{-\gamma\ell_{I,n}} + \overline{B}_{I,n}\right] \tag{22}$$

so one of the amplitudes $\overline{A}_{I,n}, \overline{B}_{I,n}$ can be expressed by the other (e.g. $\overline{A}_{I,n}$ by $\overline{B}_{I,n}$). Thus the last of the above iterative systems of equations will give two coupled systems of equations for the amplitudes for the amplitudes $\overline{B}_{I,n}, \overline{B}_{0,n}$ or $\overline{B}_{I,n}, \overline{B}_{1,n}$ (depending on the source condition). After the solution, all other amplitudes follow by insertion.

General termination :

With other terminations, e.g. a radiating duct end of the section $i=I$, one always gets a relation of the form (the termination with an admittance G_t is a special case thereof with a diagonal matrix):

$$\{\overline{A}_{I,m}\} = \{\{\text{matrix}\}\} \bullet \{\overline{B}_{I,n}\} \tag{23}$$

so one will end again in the two coupled systems of equations mentioned above.

The general problem with cascades of ducts and layers comes from the fact that one gets enough systems of equations only after the source condition has been concatenated with the termination condition. And one gets solvable systems of equations after the amplitudes $\overline{A}_{i,n}, \overline{B}_{i,n}$ for the intermediate sections are eliminated. This elimination can be done in an analytical manner only for a low number I of sections. A solution which is suited for numerical evaluation is obtained if the systems of equations from the boundary conditions are transformed to iterative systems, and a mixed numerical-symbolic evaluation ("hybrid" evaluation) is applied to that system.

The hybrid evaluation makes use of the ability of mathematical programs (like *Mathematica*® or *Maple*®) to handle hybrid expressions. They automatically simplify numerical terms in the expressions and give them canonical forms, so that the numerical coefficients or the symbolic factors can be extracted. Thus the final system(s) of equations for the numerical evaluation can easily be obtained. And if the right-hand sides of the intermediate equations are saved, one gets the numerical values of the left-hand sides (after solution of the final system(s)) just by calling them, because the key solutions ($\overline{B}_{I,n}, \overline{B}_{0,n}, \overline{B}_{1,n}$) then are automatically inserted.

O. 3 Five standard problems of numerical acoustics

Five standard problems of numerical acoustics are presented below which frequently occur in practical applications.

O.3.1. The radiation problem

A vibrating structure radiates sound into the surrounding space. The radiated sound field is characterized by the sound pressure p, particle velocity $\vec{v}$, and derived quantities such as the sound intensity $\vec{I}$, the radiated sound power Π, the radiation efficiency σ etc., which will be calculated by numerical methods.

As shown in the figure, the bounded volume of the radiating structure in three-dimensional space is denoted by B (like **B**ody). The interior of B is called B_i and the exterior B_e. The surface normal n should be directed into the exterior B_e.

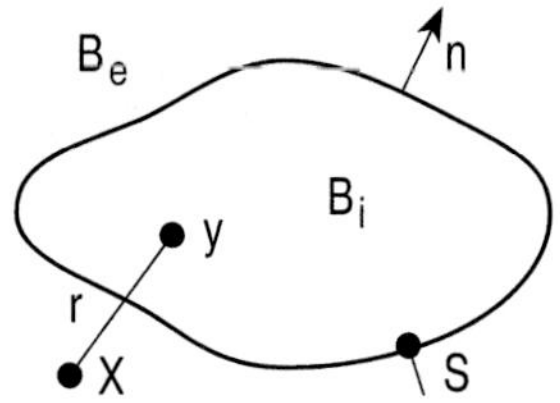

The complex sound pressure p, radiated into the free, three-dimensional space has to satisfy the Helmholtz equation

$$\Delta p + k_0^2 p = 0 \quad \text{in } B_e , \tag{1}$$

where $k_0 = \omega / c_0$ is the wave number, ω is the angular frequency, c_0 the speed of sound, and Δ is the Laplace operator. All time-varying quantities should obey the time dependence $\exp(j\omega t)$ with $j = \sqrt{-1}$. The fluid in the outer space is assumed to be loss-free, homogeneous, and at rest. For a complete description of the problem, boundary conditions on the surface of the radiator and at infinity are needed. If the body has edges corner conditions must be applied also, see Section B.16. The most important *Neumann* boundary value problem describes a body, which vibrates with a given normal velocity v. Therefore, the pressure gradient

$$\frac{\partial p}{\partial n} = -j\omega\rho_0 v \quad \text{on } S \tag{2a}$$

is prescribed on S. Here, ρ_0 is the fluid density and $\partial / \partial n$ is the derivative in the direction of the outward normal n. Sometimes, the sound pressure

$$p = p_0 \qquad \text{on} \quad S \tag{2b}$$

is given, which is called the *Dirichlet* problem. The most general form of a local boundary condition is the *Robin* or impedance problem with

$$R(p) := \frac{\partial p}{\partial n} + j\omega\rho_0 G p = f(S) \quad \text{on } S\,, \tag{2c}$$

where f(S) is a function defined on S. R(p) is a linear boundary operator. G is the field admittance in the direction of the normal n . For $G \to 0$, we obtain the Neumann problem, for $G \to \infty$ the Dirichlet problem is obtained. If $Z := 1/G$ is a nonzero but finite quantity, the general impedance problem results, which describes locally reacting absorbing surfaces.

In addition, the physical requirement that all radiated waves are outgoing leads to the Sommerfeld radiation condition

$$\lim_{R\to\infty} R\left[\frac{\partial p}{\partial R} + jk_0 p\right] = 0\,, \tag{3}$$

which can be interpreted as a boundary condition at infinity. Here, $R = |x| = \sqrt{x_1^2 + x_2^2 + x_3^2}$ denotes the distance from x to the origin, where points in space are denoted by simple letters such as $x = (x_1, x_2, x_3)$. The Sommerfeld condition generally leads to the choice of functions among different mathematically possible alternatives (like the corner condition does) ; therefore it decides about the sign of exponents and of roots, and if a field function has branch points, it gives rules about which branch of multi-valued functions should be selected.

Eqs. (1), (2a, b, or c), and (3) describe the radiation problem for the radiated pressure p. With the knowledge of p and $\vec{v}$ the effective sound intensity

$$\vec{I} = \frac{1}{2}\mathrm{Re}\{p\vec{v}^*\}\,, \tag{4}$$

the effective sound power

$$\Pi(p,v) := \frac{1}{2}\iint_S \mathrm{Re}\{pv^*\}ds\,, \tag{5}$$

and the radiation efficiency

$$\sigma = \Pi(p,v) / \Pi(\rho_0 c_0 v, v) \tag{6}$$

can be easily calculated. Here, $\Pi(\rho_0 c_0 v, v)$ is the power, of a plane wave with $p = \rho_0 c_0 v$ (7) on S, [1]. The asterisk denotes the complex conjugate and Re{...} the real part of the quantity in brackets.

Due to definition (6), the quantity $\Pi(\rho_0 c_0 v, v)$ can be considered as a sound power with radiation efficiency σ=1. Hence, the numerical computation of the radiated sound field allows a numerical sound intensity method to be used. Such a method can be used for the localisation of acoustical sources on the surface of the vibrating machine structure or for detecting cracks that cause changes of the acoustical surface intensity, [2].

O.3. 2. The scattering problem

See also chapter E for scattering.

For the related scattering problem, an incident wave p_{in} is impinging on the body B and causes a scattered wave p_S. The scattering problem for the scattered pressure p_S is again described by the Helmholtz equation

$$\Delta p_s + k_0^2 p_s = 0 \quad \text{in } B_e \tag{8}$$

and the corresponding radiation condition

$$\lim_{R\to\infty} R\left[\frac{\partial p_s}{\partial R} + jk_0 p_s\right] = 0 \tag{9}$$

Now, the boundary condition is homogeneous, since the body is at rest, and it must be formulated for the total pressure $p_T = p_{in} + p_S$. Similar to the radiation problem, the Neumann problem for an acoustically rigid scatterer is given by $\partial p_T/\partial n = 0$ on S. (10a)

A perfectly (acoustically) soft scatterer leads to the Dirichlet condition $p_T = 0$ on S. (10b)

The general form of the impedance boundary condition is described by

$$R(p_T) := \frac{\partial p_T}{\partial n} + j\omega\rho_0 G\, p_T = 0 \tag{10c}$$

where again G is the field admittance at the surface S.

The scattering problem can be formulated as an equivalent radiation problem by the following procedure: Considering the hard scatterer, the normal velocity v_{in} of the incident pressure wave p_{in} will be evaluated at the surface S where the scatterer is assumed (for the moment) to be sound transparent. If B is now vibrating with the negative normal velocity $(-v_{in})$, the radiated sound pressure is identical to the pressure p_S scattered from B due to the incident wave p_{in}. Hence, instead equ.(10a), we simply have

$$\frac{\partial p_S}{\partial n} = -j\omega\rho_0(-v_{in}) \tag{11a}$$

for the scattering problem, which again is an inhomogeneous boundary condition like eq.(2a). Analogously, the impedance boundary condition (10c) can be written as $R(p_s) = f$ (11b) where f is the known function $f = -R(p_{in})$. Equs. (8), (9), (11a/11b) describe the scattering problem as an equivalent "radiation problem" for the scattered pressure p_S. In conclusion, the radiation and the scattering problem can be treated by numerical methods in a uniform way.

Having calculated the main quantities p_s and $\vec{v}_s$, the scattered effective intensity (or intensity of return) is given by

$$\vec{I}_s = \frac{1}{2}\mathrm{Re}\{p_s \vec{v}_s{}^*\} . \tag{12}$$

Assuming that the incident wave is a plane wave with incident intensity I_{in} (in the direction of incidence), the target strength TS is defined by

$$TS = 10 \lg \frac{I_s(r = 1\,m)}{I_{in}} , \ dB \tag{13}$$

where $I_s(r = 1\text{ m})$ is the effective intensity of the sound returned by the scatterer at a distance of 1 m from its acoustical center in some specified direction. In equ.(13), lg denotes the logarithm to the base 10. The target strength is often used in the context of underwater sound (see Urick, [3], for more details).

O.3. 3. The sound field in interior spaces

If a sound field in an interior space B_i is considered, the Helmholtz equation (1) must be satisfied in B_i. This is a classical problem of room acoustics. At the boundary S of the enclosure, there may exist a rigid part S_1 with $\partial p / \partial n = 0$, an acoustically soft part S_2 with p=0, and an absorbing surface S_3 with $\partial p / \partial n + j\omega\rho_0 G\, p = 0$, such that $S = S_1 \cup S_2 \cup S_3$. In contrast to exterior radiation or scattering problems, a finite volume is considered, and no radiation condition is necessary.

If no acoustical sources such as vibrating walls or bodies are inside the room, the Helmholtz equation together with the boundary conditions represents an eigenvalue problem for the determination of the eigenfrequencies and eigenmodes of the enclosed fluid.

On the other hand, sources such as vibrating parts of the boundary, point sources, etc. lead to forced vibrations of the fluid.

Sometimes, interior and exterior acoustic problems can be coupled. Such a situation occurs when an interior acoustic space is connected to an exterior one through openings, [4]. A simple example is a duct with an open end rising into the surrounding infinite space.

The problem of a half space belongs to the class of exterior problems rather than to interior problems (see Section O.5.5).

Numerical methods based on the solution of the wave equation or Helmholtz equation (reduced wave equation) are only useful for the treatment of small enclosures at low frequencies, i.e. for small $k_0 a$ numbers, where a is a characteristic dimension of the room (for example the diameter or one side of the room). For problems with high $k_0 a$ numbers, it is more convenient to use methods of geometrical acoustics (see the Chapter N).

O.3. 4. The coupled fluid–elastic structure interaction problem

If the radiating or scattering structure B consists of an elastic material, the interaction between the body and the surrounding fluid must be taken into account. In addition, the structure can be coupled with an internal acoustic cavity. The problem is significantly simplified if the acoustic loading of sound inside the cavity can be neglected.

Following Soize, [5], the boundary value problem can be described as follows: B is the bounded domain occupied by the linearly elastic structure. The boundary S is divided into three parts: $S = S_0 \cup S_1 \cup S_2$. On S_0 the boundary is fixed. This means that u=0 on S_0 where $u = (u_1, u_2, u_3)$ denotes the displacement field. On $S_1 \cup S_2$, it is free (see the figure).

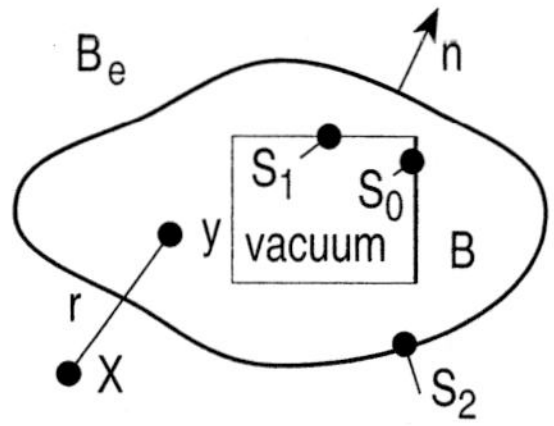

For simplicity, we only consider the coupling between a structure and an external fluid. Systems coupled with internal acoustic cavities are described by Soize in [6]. The structure is subject to a body force $f = (f_1, f_2, f_3)$ and a surface force $f_{S_2} = (f_{s_2,1}, f_{s_2,2}, f_{s_2,3})$. The steady state response of the linearly elastic structure is given by

$$-\omega^2 \rho_s u_i - \sigma_{ij,j} = f_i \quad \text{in } B \tag{14a}$$

$$\sigma_{ij} n_j = -p n_i \quad \text{on } S_2 \tag{14b}$$

$$\sigma_{ij} n_j = f_{S_1,i} \quad \text{on } S_1 \tag{15}$$

$$u_i = 0 \quad \text{on } S_0 \tag{16}$$

where the summation convention over repeated indices is used. ρ_s is the mass density of the structure and

$$\sigma_{ij,j} = \sum_{j=1}^{3} \frac{\partial \sigma_{ij}}{\partial x_j}$$

For a linear viscoelastic material, the stress tensor is given by

$$\sigma_{ij} = a_{ijkh}(x,\omega)\, \varepsilon_{kh}(u) + b_{ijkh}(x,\omega)\, \varepsilon_{kh}(j\omega u)$$

where summation over indices k and h must be performed, and the linearized strain tensor is

$$\varepsilon_{kh}(u) = \left(\frac{\partial u_k}{\partial x_h} + \frac{\partial u_h}{\partial x_k} \right) / 2$$

The coefficients $a_{ijkh}(x,\omega)$ and $b_{ijkh}(x,\omega)$ are real and depend on the properties of the elastic medium.

The coupling term between the external fluid and the structure is the pressure p that acts like a fluid load. Introducing the velocity potential ψ by $\operatorname{grad}\psi = \vec{v}$, (17)
the external fluid is described by the Helmholtz equation for the potential

$$\Delta\psi + k_0^2\psi = 0 \; . \tag{18}$$

According to Newton's law the external pressure is given by $p = -j\omega\rho_0\psi$. (19)

At the outer boundary S_2 the normal velocity of the structure $\partial u_n/\partial t = j\omega\, u_n$
and of the fluid $v_n = \partial\psi/\partial n$
must be equal $\partial\psi/\partial n = j\omega\, u_n$. (20)

In addition, the potential ψ has to fulfill the Sommerfeld radiation condition (3). For solving the coupled fluid-structure problem, equs. 14, 15, 16, 18, 20 have to be solved. Clearly, the vibrating structure can be coupled to an external fluid and to an internal acoustic cavity leading to very similar equations.

O.3. 5. The transmission problem

See also the Chapter I on sound transmission.

The transmission problem is characterized by the fact that the incident sound wave p_{in} can penetrate into the body B, which is assumed to have acoustic constants (sound speed c_i and density ρ_i) different from those of the surrounding medium c_0 and ρ_0.

The total pressure $p = p_{in} + p_s$ in B_e and the interior pressure p_i in B_i have to satisfy the Helmholtz equations

$$\Delta p + k_0^2\, p = 0 \quad \text{in } B_e , \qquad p_i + k_i^2 p_i = 0 \quad \text{in } B_i \tag{21}$$

with $k_i = \omega/c_i$, the radiation condition

$$\lim_{R\to\infty} R\left[\frac{\partial p}{\partial R} + jk_0 p \right] = 0 \tag{22}$$

and the transmission conditions for the pressure and normal velocities

$$p - p_i = f \qquad \text{on } S \tag{23}$$

$$v - v_i = g \Leftrightarrow \frac{j}{\omega\rho_0}\frac{\partial p}{\partial n} - \frac{j}{\omega\rho_i}\frac{\partial p_i}{\partial n} = g \qquad \text{on } S \tag{24}$$

where f and g are given continuous functions on S. For f=g=0 pressure and velocity are continuous at the boundary. It can be shown (see [7], p. 101) that the transmission problem has a unique solution.

O. 4 The source simulation technique (SST)

The source simulation technique is a general tool for calculating the sound radiation or scattering from complex-shaped structures into the three-dimensional space, [1-6]. Hence, it can be used for the numerical solution of the first and second standard problem (see Sect. O.3, problem 1 and problem 2). It should be noted that many names are in use for the same or similar methods such as *source simulation method*, *multipole method*, *superposition method*, *spherical wave synthesis* , etc. [1-21]. The basic idea of the method consists in replacing the structure by a system of acoustical sources placed in the interior of the structure. By definition, these source functions have to satisfy the Helmholtz equation and the radiation condition. For solving the radiation or scattering problem completely, the source system also has to fulfil the boundary conditions on the surface of the body or, equivalently, a certain boundary equation. The better the system of sources satisfies the boundary condition on the surface of the structure, the closer is the agreement between the original and the simulated sound field. Spherical wave functions are often used as sources, since they can easily be calculated. Bodies of arbitrary shape are treated by taking into account spherical wave functions with different source locations. To solve the boundary equation, which minimizes the boundary error, the method of weighted residuals is applied. Depending on the choice of the weighting functions, different variants of the source simulation technique are obtained, for example, the *null-field equations* and the *full-field equations*. The full-field equations often lead to better conditioned sets of equations than the null-field equations. The null-field and the full-field equations can also be derived from the interior or exterior Helmholtz integral equation, respectively (see [2, 4]). This illustrates the close relationship between the source simulation technique and the boundary element method (see Sect. O.5). This presentation follows the lines given in the review article [2], where more details and examples of calculations for several radiation and scattering problems can be found. In the list of references many different aspects of the SST can be found: for instance, the calculation of sound fields in the interior of enclosures containing scattering objects, [17], the treatment of scattering and radiation from bodies of revolution, [22], a formulation in the time domain, [23], or the use of special surface sources weighted according to a Gaussian distribution, [24].

In Section O.3.2 it is shown that the scattering problem can be considered as an equivalent radiation problem. Thus, it is sufficient to treat only the radiation problem in the following. Exceptionally, only in this chapter, the time convention $e^{-j\omega t}$ is used, in order to be in agreement with most of the related literature about the SST.

O.4. 1 General description of the source simulation technique

The basic idea of the SST consists of replacing the vibrating body by a system of sources placed in the interior of the body. The sources are denoted by $q(x,y)$ where x is an arbitrary point in space and y is the position of the singularity, i.e., the location of the source point. Now, for the SST in its most general form, it is assumed that the pressure can be presented in the form

$$p(x) = \iiint_Q c(y)\, q(x,y)\, dy \tag{1}$$

where Q is a region which is fully contained in B_i and embodies all sources; $c(y)$ is the yet unknown source density, which gives every source a certain source strength. Every single source function (also called trial function) $q(x,y)$ itself can consist of a finite or infinite sum of elementary sources such as monopoles, dipoles, etc.: $q(x,y) = d_0 q_0(x,y) + d_1 q_1(x,y) + \ldots$ The volume integral in eq. (1) reduces to a surface integral or a contour integral if the region Q is a surface or a line, respectively. The integral turns into a finite sum if isolated point sources are used or if the integral must be discretized for numerical reasons, since we cannot work with infinitely many sources. The system of functions $c(y)\, q(y)$ with $y \in Q$ will be called the source system. All functions of the source system have by definition to satisfy eqs. (O.3.1) and (O.3.3) (with respect to x). The source system also has to satisfy the boundary condition on the surface S. As a consequence of the present time convention $e^{-j\omega t}$, the boundary conditions (O.3.2a) and (O.3.2c) must be written with a different sign

$$\frac{\partial p}{\partial n} = j\omega\rho_0 v\,, \qquad R(p) := \frac{\partial p}{\partial n} - j\omega\rho_0 G\, p = f \quad \text{on } S \tag{2a, 2b}$$

Hence, by substituting eq. (1) into eq. (2a) or into the general radiation boundary condition (2b), one gets for the rigid radiator

$$\iiint_Q c(y)\frac{\partial}{\partial n} q(x,y)\,dy - j\omega\rho_0 v = 0 \quad \text{on } S \tag{2c}$$

or, in the general case

$$R\left[\iiint_Q c(y)\, q(x,y)\,dy\right] - f = 0 \quad \text{on } S \tag{2d}$$

These equations "live" on the boundary, and they are called the boundary equations of the SST. Thus by using sources as trial functions, the original domain problem can be transformed into a boundary problem. This illustrates that the SST can be considered as a counterpart to the finite element method (FEM, see Section O.6), which solves the original domain problem. If the boundary equations can be solved exactly, the coefficients $c(y)$ are determined such that the sound field generated by the vibrating structure is identical to the field produced by the source system. This follows from the unique solvability of the exterior problems described in Section O.3 (see [25]). Such a source system is called an *equivalent source system*. Consequently, the exact solution of the radiation problem can be found if it is possible to construct an equivalent source system. However, such exact solutions are only known for special geometries in standard coordinate systems (e.g. in spherical coordinates). For nearly all relevant practical problems the surface S has a complicated shape. Therefore, we are only looking for approximate solutions of the boundary equations by minimizing the so-called *boundary error* or *residual*

$$\varepsilon(x) = \iiint_Q c(y)\frac{\partial}{\partial n(y)} q(x,y)\,dy - j\omega\rho_0 v \quad \text{on } S \tag{3a}$$

or

$$\varepsilon(x) = R\left[\iiint_Q c(y)\, q(x,y)\,dy\right] - f \quad \text{on } S \tag{3b}$$

The minimization process can be performed by means of the method of weighted residuals, which is a very general approach (see [5]). It consists in choosing a complete family of weighting functions w_n; $n = 1, 2, 3, \ldots$ and demanding that

$$\iint_S \varepsilon(x)\, w_\ell(x)\,ds = \iint_S \left\{ R\left[\iiint_Q c(y)\, q(x,y)\,dy\right] - f \right\} w_\ell(x)\,ds = 0 \;\; ; \;\; \ell = 1,2,3\ldots \tag{4}$$

The completeness ensures that the residual will go to zero if the number of weighting functions tends to infinity. Eqs. (4) are called the *weighted residual equations* of the SST for the determination of the source density c(y). It can be seen that different variants of the SST stem from different choices of sources and weighting functions. For example, the kind, number, and locations of source and corresponding weighting functions are important parameters of the method. Important source functions are the spherical wave functions, which will be introduced in the next section together with corresponding symmetry relations.

O.4. 2 Spherical wave functions and symmetry relations

By definition, sources must be radiating wave functions. However, analytical solutions of the Helmholtz equation can only be constructed explicitly in separable coordinate systems, [26, p. 494]. In three-dimensional space, the spherical wave functions present the simplest form of such solutions. Hence they are the type of sources most often used, and they are given by

$$\psi_{nm}^{c,s}(x) = \Gamma_{nm}\, h_n^{(1)}(k_0 r)\, P_n^m(\cos\vartheta) \begin{cases} \cos(m\varphi) \\ \sin(m\varphi) \end{cases} \tag{5}$$

where the $P_n^m(\cos\vartheta)$ are the associated Legendre polynomials, [27]. Here, spherical co-ordinates $\{r,\vartheta,\varphi\}$ are introduced by $x = \{r\sin\vartheta\cos\varphi,\ r\sin\vartheta\sin\varphi,\ r\cos\vartheta\}$. The superscript c (or s) indicates that the cosine (or sine) is used. The cylindrical functions $h_n^{(1)}(z)$ are the spherical Hankel functions of the first kind, [27]. The normalizing factors

$$\Gamma_{nm} = \left[\frac{\varepsilon_m}{4\pi}(2n+1)\frac{(n-m)!}{(n+m)!}\right]^{1/2} \quad ; \quad \varepsilon_m = \begin{cases} 1; & m = 0 \\ 2; & m > 0 \end{cases} \tag{6}$$

are chosen in such a way that the spherical harmonics

$$y_{nm}^{c,s}(x) = \Gamma_{nm}\, P_n^m(\cos\vartheta) \cdot \begin{cases} \cos(m\varphi) \\ \sin(m\varphi) \end{cases} \tag{7}$$

are orthonormal with respect to the integration over the unit sphere:

$$\int_0^{2\pi}\int_0^{\pi} y_{mn}^{\alpha} \cdot y_{\mu\nu}^{\beta} \sin\vartheta \; d\vartheta\, d\varphi = \begin{cases} 1; & \text{if } \alpha = \beta,\ \mu = m\,,\ \text{and}\ \nu = n \\ 0; & \text{else} \end{cases} \tag{8}$$

where α and β stand for c or s, respectively. Taking into account that there are only wave functions of cosine type for m=0, the number of different spherical wave functions up to an index n_0 is given by

$$\sum_{j=0}^{n_0} (2j+1) = (n_0+1)^2 \tag{9}$$

For simplicity, we denote the $\psi_{nm}^{c,s}$ and $y_{nm}^{c,s}$ by ψ_ℓ and y_ℓ, respectively, where the index $\ell = 0,1,2,\ldots$ runs through all combinations of m and n for c and s. The regular wave functions $\chi_l = \mathrm{Re}\{\psi_l\}$ (10)

present standing waves, where Re{} denotes the real part of the quantity in brackets. The regular wave functions contain spherical Bessel functions $j_n(z)$ instead of Hankel functions like the radiating wave functions, since $h_n^{(1)}(z) = j_n(z) + j\,n_n(z)$, with $n_n(z)$ spherical Neumann functions.

For deriving the null-field and the full-field equations from the weighted residual equations (4), the following symmetry relations are very important

$$\iint_S \left(\psi_\ell \frac{\partial \psi_m}{\partial n} - \psi_m \frac{\partial \psi_\ell}{\partial n} \right) ds = 0 \tag{12a}$$

$$\iint_S \left(\chi_\ell \frac{\partial \psi_m}{\partial n} - \psi_m \frac{\partial \chi_\ell}{\partial n} \right) ds = \delta_{\ell m} \frac{j}{k_0} \tag{12b}$$

$$\iint_S \left(\psi_\ell^* \frac{\partial \psi_m}{\partial n} - \psi_m \frac{\partial \psi_\ell^*}{\partial n} \right) ds = \delta_{\ell m} \frac{2j}{k_0} \tag{12c}$$

where the asterisk denotes the complex conjugate, and $\delta_{\ell m}$ is the Kronecker Delta. Eq. (12a) is valid for all radiating wave functions as shown in [5]. The proofs for eqs. (12a)-(12c) can be found in [4] together with [1]. Since the bilinear form

$$[u,v] := \iint_S u\,v\,dy \tag{13}$$

is symmetric, i.e. $[u,v]=[v,u]$, all three symmetry relations (12) for the operator $D = \partial/\partial n$ can be extended to the more general boundary operator R (see eq. (2b)):

$$[\psi_\ell, R\psi_m] - [R\psi_\ell, \psi_m] = 0 \tag{14a}$$

$$[\chi_\ell, R\psi_m] - [R\chi_\ell, \psi_m] = \delta_{\ell m} \frac{j}{k_0} \tag{14b}$$

$$[\psi_\ell^*, R\psi_m] - [R\psi_\ell^*, \psi_m] = \delta_{\ell m} \frac{2j}{k_0} \tag{14c}$$

It is well-known that the spherical wave functions can be interpreted as multipoles. Hence in related works the name *multipole method* or *multipole radiator synthesis* is used, especially in the case when a sum of multipoles is located at a few isolated points in the interior B_i. By combining the results of Sections 2 and 3, the SST with spherical wave functions is obtained.

O.4. 3 Variants of the SST with spherical wave functions

In the following, spherical wave functions are used as equivalent sources. Only sparsely scattered attempts can be found in the literature, in which other types of source functions are used. For example, functions in prolate spheroidal coordinates were applied in [28], and surface sources weighted according to a Gaussian distribution were employed in [24]. The reason is that it is more difficult to deal with such functions than with spherical wave functions. Fortunately, working with spherical functions is adequate, since the use of several source locations distributed over the interior B_i enables us to treat complex non-spherical geometries. This is shown in Sections 5 and 6. In the present section, only spherical wave functions with respect to one source location (placed in the origin) as defined in eq.(5) will be considered. Hence, the source area is simply $Q=\{0\}$, and the source function $q(x,0)$ (see eq.(1)) consists of a series of spherical wave functions with increasing order in the most general case. Hence, in accordance with eqs.(1) and (5) it can be written

$$p(x)=c(0)q(x,0)=\sum_{m=0}^{\infty} c_m\psi_m \; . \tag{15}$$

By introducing expansion (15) into the weighted residuals (4) one gets the weighted residual equations with spherical wave functions as sources:

$$\sum_{m=0}^{\infty} c_m \iint_S R[\psi_m(y)]w_\ell(y)\,ds=\iint_S f(y)\,w_\ell(y)\,ds \quad ; \quad \ell = 0,\,1,\,2,\,3\,\ldots\,. \tag{16}$$

These equations are called *spherical weighted residual equations* . Three important variants of the SST result from specifying the weighting functions in eq.(16). All other parameters are fixed.

O.4. 3.1 Null-field-like equations

As mentioned in Section 3, the family of weighting functions has to form a complete system in the Hilbert space $H=L_2(S)$ of square integrable functions over the surface S equipped with the usual scalar product

$$(u,v)=\iint_S u\,v^*ds \quad \text{for} \quad u,\,v\in H\,. \tag{17}$$

Recall that a system of functions is complete if the only function orthogonal to all other functions is the null function (see for example [29], p. 15). In [30] Vekua has shown that the spherical wave functions are complete in the Hilbert space H for every wave number k if S is a Lyapunov surface as assumed in Section 2.1 (see Theorem 1 of [1]). Hence, the spherical wave functions are also admissible as weighting functions: $w_\ell=\psi_\ell$. (18)

Such an approach is known as the *Galerkin method* where trial and weighting functions are identical. By introducing these weighting functions into the spherical weighted residual equations (16), we obtain

$$\sum_{m=0}^{\infty} c_m \iint_S R[\psi_m(y)]\, \psi_\ell(y)\, ds = \iint_S f(y)\, \psi_\ell(y)\, ds\,; \quad \ell = 0,\ 1,\ 2,\ \ldots\,. \tag{19}$$

The symmetry relation (14a) yields

$$\iint_S \sum_{m=0}^{\infty} c_m \psi_m(y) R[\psi_\ell(y)]\, ds = \iint_S f(y)\, \psi_\ell(y)\, ds\,; \quad \ell = 0,\ 1,\ 2,\ \ldots \tag{20}$$

or equivalently

$$\iint_S p(y) R[\psi_\ell(y)]\, ds = \iint_S f(y)\, \psi_\ell(y)\, ds\,; \quad \ell = 0,\ 1,\ 2,\ \ldots \tag{21}$$

by taking into account expansion (15). Choosing $R = \partial / \partial n$ and $f = j\omega\rho_0 v$ yields exactly the null-field equations for Neumann data (see [31], Eq. 3.3). Hence, the following result is obtained: the null-field equations for Neumann data use the spherical wave functions as weighting functions. If in addition the pressure p is expanded into spherical wave functions, then the null-field equations and the spherical SST are identical. Similarly, the null-field equations for Dirichlet data (see [31], eq. (3.2)) are obtained if the weighting functions $w_l = \partial\psi_n / \partial n$ are used. The null-field equations were described in many papers (see for example [25], [31], [32], [33]). Theorem 6 of [1] shows that the the null-field equations minimise the amplitude of the reactive power of the error field

$$\Delta p = \sum_{m=0}^{N} c_m \psi_m - p\,, \tag{22}$$

but not the sound power radiated by this error field (the error disappears if $N \to \infty$, see eq.(15)).

For most radiation problems it would be more desirable to minimise the sound power of the radiated error wave directly. Unfortunately, to the author's knowledge, such a variational principle has not yet been found (see [1]).

In section 6 of [2] the null-field equations were derived from the Helmholtz integral equation for the interior field, which is an ill-posed integral equation of the first kind. This shows that the null-field equations and the related T-matrix approach of Waterman, [34], can lead to unstable systems of equations. Hence, alternative formulations of the SST can be found by choosing other sets of weighting functions.

O.4. 3.2 The full-field equations

If the regular wave functions $\chi_\ell = \mathrm{Re}\{\psi_\ell\}$ are selected as weighting functions and introduced into the spherical weighted residual equations (16) one obtains

$$\sum_{m=0}^{\infty} c_m \iint_S R[\psi_m(y)]\,\chi_\ell(y)\,ds = \iint_S f(y)\chi_\ell(y)\,ds\,; \quad \ell = 0,1,2,\dots\,. \tag{23}$$

By employing the symmetry relation (14b) we find

$$\frac{j}{k_0} c_\ell + \sum_{m=0}^{\infty} c_m \iint_S \psi_m(y) R[\chi_\ell(y)]\,ds = \iint_S f(y)\chi_\ell(y)\,ds\,; \quad \ell = 0,1,2,\dots\,. \tag{24}$$

These equations are called *full-field equations of the first kind* (**FFE1**). They are derived and discussed in [4]. In [2,4], the FFE1 are also derived from the exterior Helmholtz integral equation for the "full" outer space, which clarifies the term "full-field equations" in contrast to the "null-field equations". It can be shown that the diagonal elements $\mathrm{diag1} = j/k_0$ on the left-hand side of eq.(24) have a stabilizing effect on the resulting system of equations.

However, the FFE1 has the following disadvantage: The family of weighting functions $\{\chi_\ell\}$ is not complete and hence does not constitute a basis for the Hilbert space $L_2(S)$ whenever k_0 is an eigenvalue of the interior Dirichlet problem (see [34], Appendix A). This problem is closely related to the appearance of critical frequencies in the boundary element method (see [35] and Section O.5.4). However, the completeness of weighting functions is necessary for inverting the system of equations (16) (see [31], Appendix 2). To avoid such difficulties at certain critical frequencies, it is recommended to use the complex conjugate spherical wave functions (see [4])

$$w_\ell = \psi_\ell^* \tag{25}$$

as weighting functions. Since the ψ_ℓ are complete, it follows that the ψ_ℓ^* are also complete and constitute a basis for $L_2(S)$ at every wave number k_0. On the other hand, the symmetry relation (14c) also contains a diagonal element on the right-hand side similar to eq.(14b). Hence, by inserting the functions (25) into eq.(16) and applying eq.(14c) one finds the equations

$$\frac{2j}{k_0} c_\ell + \sum_{m=0}^{\infty} c_m \iint_S \psi_m R[\psi_\ell^*]\,ds = \iint_S f\psi_\ell^* ds\,; \quad \ell = 0,1,2,\dots \tag{26}$$

As suggested in [4], eqs.(26) were called *full-field equations of the second kind* (**FFE2**). Again, the diagonal elements

$$\mathrm{diag2} = 2\,\mathrm{diag1} = 2\frac{j}{k_0} \tag{27}$$

arising from the application of eq.(14c), have a stabilizing effect. This was demonstrated in section 10.3 of [2]. The idea of using the complex conjugate functions for constructing energy expressions goes back to Cremer and Wang [14].

O.4. 3.3 The least squares minimization technique

Finally, we consider the weighting functions $$w_\ell = \frac{\partial \psi_\ell^*}{\partial n} \tag{28}$$

It is known that these functions are complete in $L_2(S)$ (see [1], p. 517). By introducing these functions into the spherical weighted residual equations (16) with Neumann data $(R = \partial / \partial n,\ f = j\omega\rho_0 v)$ one gets

$$\sum_{m=0}^{\infty} c_m \iint_S \frac{\partial \psi_m}{\partial n} \frac{\partial \psi_\ell^*}{\partial n} ds = \iint_S f(y) \frac{\partial \psi_\ell^*}{\partial n} ds\ ; \quad \ell = 0,1,2,\ldots \tag{29}$$

Eqs.(29) have the advantage that the corresponding matrix is Hermitian since

$$a_{m\ell} := \frac{\partial \psi_m}{\partial n} \frac{\partial \psi_\ell^*}{\partial n} = a_{\ell m}^* \tag{30}$$

It can be shown that eqs. (29) minimise the surface velocity error $E/j\omega\rho_0$ with

$$E = \iint_S \left| \sum_{m=0}^{\infty} c_m \frac{\partial \psi_m}{\partial n} - j\omega\rho v \right|^2 ds \tag{31}$$

in the mean square sense (see [1], chap. 6.2). Hence, eqs.(31) can be interpreted as the normal equations of the *least squares method*. It is also possible to consider the least squares method as an orthogonalization method and to employ the Gram-Schmidt technique as described briefly in [1], p. 519. An approach which deals directly with orthonormalized functions can be found in [36].

For extending the least squares method to the mixed boundary conditions (2b) we suggest using the weighting functions

$$w_\ell = (R\psi_\ell)^* = \frac{\partial \psi_\ell^*}{\partial n} - G^* \psi_\ell^* . \tag{32}$$

According to Millar ([37], see also [25], p. 97), these functions are complete in the space $L_2(S)$ if the assumption $$\mathrm{Im}\{G\} \le 0 \tag{33}$$
is satisfied. Then, the weighting functions (32) are admissible and again lead to a set of equations with a corresponding Hermitian matrix

$$\sum_{m=0}^{\infty} c_m \iint_S R\psi_m (R\psi_\ell)^* ds = \iint_S f (R\psi_\ell)^* ds ; \quad \ell = 0,1,2,\ldots \tag{34}$$

As before, eqs.(34) are the normal equations of the least squares method which now minimize the error functional

$$E = \iint_S \left| \sum_{m=0}^{\infty} c_m R\psi_m - f \right|^2 ds \tag{35}$$

O.4. 3.4 Summary: weighting functions and corresponding methods

Table 1 provides an overview of all the variants of the spherical SST described, depending on the weighting functions chosen. In all cases only spherical wave functions with respect to one source location are used as source functions (cf. eq.(15)).

Table 1: Methods and weighting functions

Method	Weighting functions
Null-field equations for Neumann data	ψ_ℓ
Null-field equations for Dirichlet data	$\partial\psi_\ell / \partial n$
Full-field equations of the first kind	$\chi_\ell = \mathrm{Re}\{\psi_\ell\}$
Full-field equations of the second kind	ψ_ℓ^*
Least squares method	$(\partial\psi_\ell / \partial n)^*$ or $(R\psi_\ell)^*$

O.4. 4 Extension of the SST to bodies of arbitrary geometry

Up to now only spherical wave functions with a singularity at the origin of the coordinate system as defined in eq.(5) are used. This kind of SST is called *one-point multipole method*, [1], and is well suited for sphere-like radiators. But the more the shape of the body deviates from a sphere the slower the rate of convergence of the one-point multipole expansion (15) will become. For this reason a more flexible source system can be constructed by using spherical wave functions with several source locations x_q located in the interior B_i as source functions (multi-point multipole method, [1]):

$$\psi_\ell^q(x) = \psi_\ell(x - x_q) \quad ; \quad q = 1,\dots,Q\,; \ \ell = 0,\dots,N\ . \tag{36}$$

Such source systems allow the boundary conditions to be satisfied on complex-shaped structures. For example, by choosing $N = 0$ the source system consists of Q monopoles distributed over an interior auxiliary surface. Such an approach can be considered as derived from an acoustic single layer potential (see O.5.1 and [3]). Also, it can be found in American [7], [8], French [24], and Russian papers [16], [19] under the names *superposition method* or *equivalent source method*. Now, instead of eq.(15) the pressure is expanded into the double series

$$p(x) = \sum_{q=1}^{Q}\sum_{m=0}^{\infty} c_m^q \psi_m^q(x) \tag{37}$$

First, the null-field equations are extended by using the ψ_ℓ^q as source and weighting functions. Proceeding along the same lines as in Section 4.1, we obtain the generalized null-field equations

$$\iint_S \sum_{q=1}^{Q}\sum_{m=0}^{\infty} c_m^q \psi_m^q(x) R\left[\psi_\ell^s(y)\right] ds = \iint_S f(y)\psi_\ell^s(y)\, ds \quad ; \quad s = 1,\dots,Q\,; \ \ell = 0,\ 1,\ \dots \tag{38}$$

instead of the null-field equations (20). Here we have taken into account the symmetry relation (14a), which is valid for all kinds of radiating wave functions.

However, in deriving generalized full-field equations some care is needed since the symmetry relations (14b, 14c) are only valid for wave functions with the same source location. For example, instead of (14b) one gets

$$\iint_S \left(\chi_\ell^s R\psi_m^q - \psi_m^q R\chi_\ell^s\right) ds = \begin{cases} \dfrac{j}{k_0} \ \text{ for } \ l = m \ \text{ and } \ s = q & \text{(39a)} \\ 0 \ \text{ for } \ l \neq m \ \text{ and } \ s = q & \text{(39b)} \\ ? \ \text{ for } \ s \neq q & \text{(39c)} \end{cases}$$

Eqs. (39a, 39b) follow directly from eq.(14b), which does not depend on the special choice of the source location $x_q \in B_i$. For wave functions with different source locations, the present author has not found analogous symmetry relations. This does not matter since the symmetry relations have only to be applied to the diagonal $\ell = m$ and $s = q$ which yields the generalized full-field equations of the first kind

$$\frac{j}{k_0} c_\ell^s + \sum_{q=1}^{Q}\sum_{m=0}^{\infty} c_m^q \iint_S g_{\ell,m}^{s,q} ds = \iint_S f\chi_\ell^s ds \quad ; \quad s = 1,\dots,Q\,; \ \ell = 0,\ 1,\ \dots \tag{40a}$$

with

$$g_{\ell,m}^{s,q} = \begin{cases} \psi_m^q R\left[\chi_\ell^s\right], & \text{for } \ell = m \text{ and } s = q \\ R\left[\psi_m^q\right]\chi_\ell^s, & \text{elsewhere} \end{cases} \tag{40b}$$

Analogously, the weighting functions $\left(\psi_l^s\right)^*$ lead to the generalized generalized full-field equations of the second kind

$$\frac{2j}{k_0}c_\ell^s + \sum_{q=1}^{Q}\sum_{m=0}^{\infty} c_m^q \iint_S h_{\ell,m}^{s,q}\,ds = \iint_S f\left(\psi_\ell^s\right)^* ds \quad ; \quad s = 1,\dots,Q\,;\ \ell = 0,\ 1,\ \dots \tag{41a}$$

$$h_{\ell,m}^{s,q} = \begin{cases} \psi_m^q R\left(\psi_l^s\right)^* & ;\ \text{for } \ell = m \text{ and } s = q \\ \left(\psi_\ell^s\right)^* R\left[\psi_m^q\right] & ;\ \text{elsewhere} \end{cases} \tag{41b}$$

Eqs.(41) for Neumann data were derived in [4]. They possess the following advantages: the diagonal terms diag2 ensure an improved stability, no critical frequencies occur, and arbitrary surface geometries can be treated if the source locations are chosen in an appropriate manner. This will be considered in the next section.

The analogous derivation of the generalized least squares method is straightforward and hence will be omitted.

O.4. 5 Position of sources and their optimal choice

In practical applications the important question arises of how to find optimal source locations in the interior B_i of the structure. Clearly, the main requirement is that the source locations have to be chosen such that the boundary error, for example the surface velocity error (31) for Neumann data, becomes minimal.

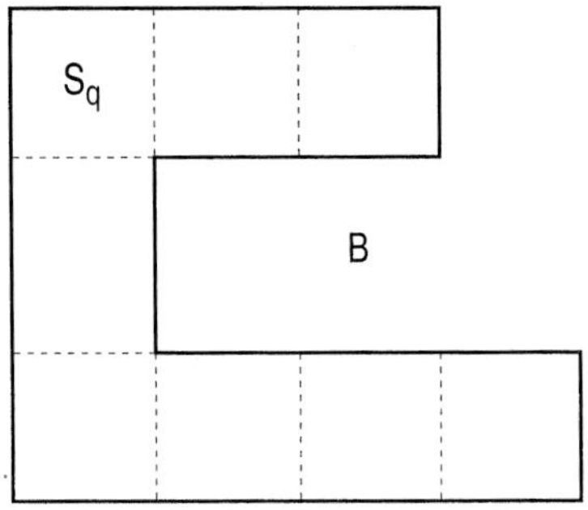

No general rule exists for achieving this, but a rule of thumb was given in [5]: The structure B should be divided into Q substructures as indicated by broken lines in the figure. Then, one multipole has to be placed at the centroid of each substructure. The shape of every substructure S_q (q = 1, ...,Q) should be as sphere-like as possible. In addition, it has to be star-like with respect to its centroid, which means that every part of the surface can be seen from the position of the centroid. Such a choice seems to make sense from an intuitive point of view since the "influence area" of each multipole represents a sphere-like substructure. Many numerical calculations confirm that such source systems may lead to smaller boundary errors than other source configurations. However, no strong mathematical proof has been given up to now.

In [38] an attempt was made to optimize numerically the source locations. For this purpose a nonlinear least squares problem has to be solved, which can be done iteratively by applying the Levenberg-Marquardt algorithm. The investigation of idealized radiating structures like cubes and cylinders showed that the additional optimization of the multipole locations had a strong effect on the boundary error and did improve the accuracy of the sound field approximation remarkably, [38]. On the other hand, such an automatic optimisation needs a greater amount of computer time, which may be justified if the manual choice as described above does not lead to satisfactory results.

After having presented the theory of the SST, numerical aspects and one example of the method will be considered.

O.4. 6 Numerical aspects

O.4. 6.1 Numerical implementation

In practical calculations, only a finite number $N_W=(N+1)Q$ of source functions ψ_m^q with $m = 0,\ldots, N$ and $q = 1,\ldots,Q$ can be used. For example, if only monopoles $(N=0)$ and two source locations $(Q=2)$ are used, the full-field equations of the second kind (41) for Neumann data (2a) lead to a system of equations $A\vec{c}=\vec{f}$ with

$$A=\begin{pmatrix} 2j/k_0+\iint\limits_S \psi_0^1 \dfrac{\partial\left(\psi_0^1\right)^*}{\partial n}ds & \iint\limits_S \dfrac{\partial\psi_0^2}{\partial n}\left(\psi_0^1\right)^* ds \\ \iint\limits_S \dfrac{\partial\psi_0^1}{\partial n}\left(\psi_0^2\right)^* ds & 2j/k_0+\iint\limits_S \psi_0^2 \dfrac{\partial\left(\psi_0^2\right)^*}{\partial n}ds \end{pmatrix}, \tag{42a}$$

$$\vec{c}=\begin{pmatrix} c_0^1 \\ c_0^2 \end{pmatrix} \quad \text{and} \quad \vec{f}=j\omega\rho_0\iint\limits_S \begin{pmatrix} v\left(\psi_0^1\right)^* \\ v\left(\psi_0^2\right)^* \end{pmatrix} ds\ . \tag{42b}$$

The solution of eqs.(42) gives the unknown vector of coefficients $\vec{c}$ which determines the source strength of both monopoles. For performing the surface integration, the surface is divided into M boundary elements. The simplest integration scheme is obtained if one assumes that pressure and normal velocity are constant over a single surface element. Such constant elements are often used, but, as in BEM, linear, quadratic, or more sophisticated elements may lead to better numerical results (see Section O.5.2). One main advantage of the SST can be seen from eqs.(42): In general, the number of equations will be much smaller than the number of boundary elements, i.e. $N_W \ll M$, especially if fine-meshed grids are used for the purpose of high-frequency calculations. Hence, the SST may lead to faster numerical algorithms than boundary element techniques, which work with $N_W = M$ (if constant elements are used). On the other hand, the SST will give approximate solutions for $N_W < M$, whereas the BEM normally provides exact solutions limited only by the discretization error (see [39]). An example may illustrate this situation: let us consider a body that vibrates like a pulsating sphere. Then only one equation has to be solved for finding the exact solution if the equivalent monopole is placed at the right position. In contrast, the BEM has to solve as many equations as the body has boundary elements. However, a complex shaped body with a complicated vibration pattern may require that nearly $N_W \approx M$ sources should be taken into account.

The spherical wave functions involve the spherical Bessel and Hankel functions. Depending on order and argument, these functions may assume large numerical values. Hence, we recommend working with normalized functions $\psi_\ell^q(x)/K_\ell^q$ instead of using the spherical wave functions directly. We have chosen the Hankel functions $h_n^{(1)}(k_0 a)$ as normalizing constants in most calculations with a being a typical dimension of the structure.

O.4. 6.2 Stability and condition number

The variants of the SST may be distinguished by their different stability behaviour, which is especially valid for the null-field equations and the full-field equations. The stability of a system of equations $A \cdot x = b$ can be investigated by considering the condition number, [40], [41]

$$\kappa = \mathrm{cond}(A) = \|A\| \left\|A^{-1}\right\| \tag{43}$$

where $\|\ldots\|$ is a matrix norm and κ is a real number greater than or equal to one. The condition number measures the sensitivity of the solution x with respect to perturbations in A or in b. An unstable or badly conditioned system has a large condition number, [40], [41]. The condition number depends on the chosen matrix norm. The Euclidean (or spectral) condition number κ_{spec} is defined as the ratio of the largest and the smallest singular value of A. As suggested by Tobocman, [42], the matrix norm

$$\kappa_F = \frac{1}{n} \|A\|_F \left\|A^{-1}\right\|_F \tag{44a}$$

can also be used which is based on the Frobenius norm

$$\|A\|_F = \sqrt{\sum_{i=1}^{N_w} \sum_{j=1}^{N_w} |a_{ij}|^2} \,. \tag{44b}$$

Here again, $A = (a_{ij})$ is the above mentioned $N_w \times N_w$ matrix. However, the calculation of κ_{spec} or κ_F may be very time consuming due to the calculation of the singular values or the inverse matrix, respectively. For this reason, it is advisable to use approximations for the condition number as given in [41].

Numerical examples show that the null-field equations may lead to large condition numbers, especially if the number of sources is large, [2,4]. This drawback can be removed by applying the singular-value decomposition or by using the full-field equations instead of the null-field equations.

O.4. 6.3 Calculation of field quantities and sound power

After having determined the coefficients c_i , i.e. the source strengths of the equivalent sources by inverting the matrix A , $\vec{c} = A^{-1}\vec{f}$, all field quantities in the exterior space B_e can easily be computed by a simple source superposition as indicated in eqs.(1), (15), or (37). For example, the acoustic pressure p is obtained from eqs.(1), (15), or (37) directly. Then, the velocity vector $\vec{v} = (v_x, v_y, v_z)$ can be obtained from the well-known formula

$$\vec{v} = (j\omega\rho_0)^{-1} \operatorname{grad} p \tag{45}$$

From the knowledge of p and v the sound intensity and the sound power can be calculated (see Sect. O.3.1). If the one-point multipole method is used, the radiated effective sound power Π is proportional to the sum of the source coefficients squared

$$\Pi = \Pi(p, v) = \frac{1}{2} \mathrm{Re}\left\{ \iint_{\Omega_{R_a}} p(x) v^*(x) \, dx \right\} = \frac{1}{2\rho_0 c_0 k_0^2} \sum_{m=0}^{N} |c_m|^2 \tag{46}$$

where Ω_{R_a} is a sphere surrounding the radiating body. Formula (46) can be derived from eqs.(5), (8), and (15). If the multi-point multipole method is used, the sound power cannot be expressed in such a simple manner, since the source functions ψ_m^q are not orthogonal over an exterior sphere. Therefore, the integration over a closed surface has to be performed numerically.

Some more useful definitions will be given for the Neumann problem where the normal velocity v is prescribed on the surface S. Clearly, the normal velocity $w = v_{sim}$ simulated by the source system may differ from the real v, since the SST will only give approximate solutions (see Section 2). Hence, the relative surface velocity error

$$F_{rel} = \iint_S |v - w|^2 ds \,/\, \iint_S |v|^2 ds \tag{47}$$

is a measure for the accuracy of the solution depending on the number and on the locations of the sources. For every calculation such a boundary error (eq.(47) valid for Neumann data or a corresponding error for other boundary conditions) can be computed even if the solution is not known a priori. For Neumann data we can also differentiate between the sound power $\Pi(p, v)$, which is calculated, using the prescribed velocity v and the sound power $\Pi(p, v_{sim})$ for the simulated velocity (see eq.(O.3.5)). In both cases the pressure p is the simulated pressure since the true pressure is unknown.

O.4. 7 A numerical example: Sound scattering from a non-convex cat's-eye structure

To illustrate the theory of the SST numerical results for one specially chosen scattering problem will be presented. More results for additional radiation and scattering problems concerning idealized structures, a propeller, and a cylinder can be found in [2] and in [43]-[47].

In the example shown the SST is used for treating the scattering from complex-shaped, non-convex structures. In [4], [39], [48] such a structure was studied which consisted of a sphere where the positive octant, i.e. the part corresponding to $x > 0,\ y > 0,\ z > 0$, was cut out. The region of the missing octant is called "cat's-eye", since it acts like a three-dimensional reflector. As shown in Fig. 1a, the finite element model of the cat's-eye structure consists of 7911 boundary elements to achieve six elements per wavelength at $k_0 a = 20.9$ (a is the radius of the corresponding sphere). The incidence direction n_i of the single frequency, plane wave is along the negative bisector of the angle between the x and y axes. Hence the incident wave illuminates the reflecting area of the cat's-eye and leads to multiple reflections. According to the "rule of thumb" of Section 6 the source system was constructed as follows: multipoles of order N were placed in the middle of each of the seven octants.

Only a few results are presented here. More details and results can be found in [4]. In Fig. 1b the directivity pattern of the target strength TS (see section O.3.2, eq.(13)) is shown in the xy-plane. (The target strength was calculated in the far-field and then projected back to the distance of 1 m from the scatterer.) The surface is assumed to be rigid. The total number of sources was 700, since multipoles of order up to $N = 9$ at each of the seven source locations were taken into account. For the incidence direction n_i the maximum of the forward and backward scattering can be seen in the xy-plane. Obviously, the data of the null-field equations, full-field equations of 2nd kind, and the plane wave approximation (see Section O.3.1) agree well at forward scattering. However, the backscattering maximum is only predicted by the full-field equations of 2nd kind ($\kappa_{spec} = 4.6 \cdot 10^4,\ F_{rel} = 23\%$). The null-field equations produce too large results since we have $\kappa_{spec} = 1.1 \cdot 10^8,\ F_{rel} = 700\%$. The plane wave approximation could not find any backscattering since it principally neglects multiple reflections.

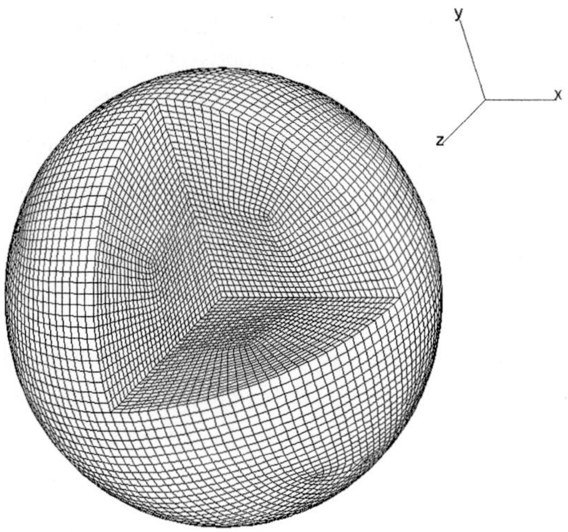

Fig. 1 a :
Finite element model for the cat's-eye structure (at $k_0a = 20.9$)

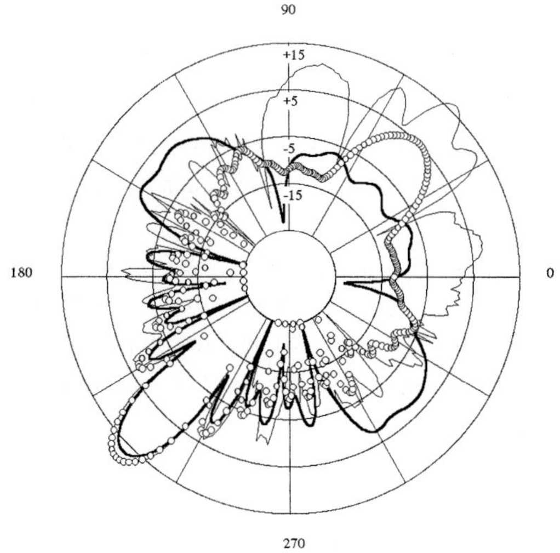

Fig. 1 b :
Directivity pattern of the target strength TS in the x,y plane for a rigid surface, direction of incidence along the negative bisector of the angle between the x and y axes; 700 sources were used;
thick curve: plane wave approximation,
thin curve: null-field equations,
circles: full-field equations of the 2nd kind.

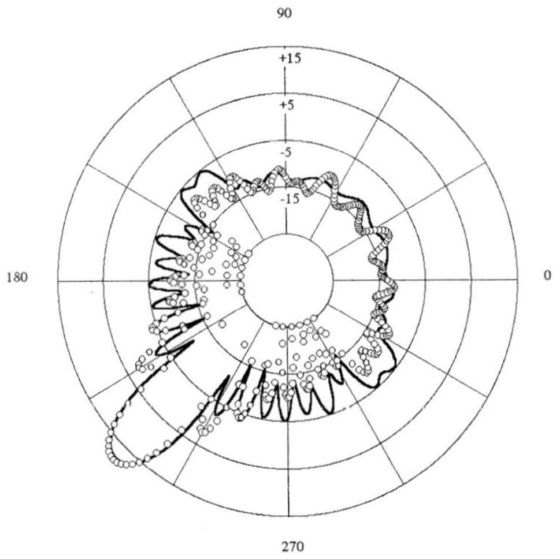

Fig. 1 c :
Directivity pattern of the target strength TS in the x,y plane for an absorbing impedance, direction of incidence along the negative bisector of the angle between the x and y axes;
567 sources were used,
thick curve: plane wave approximation,
circles: least squares method

O.4. 8 Concluding remarks

A general assumption of the SST is that all sources must be located in the interior B_i of the structure. If the sources are placed on the boundary itself, the corresponding BEM is obtained (see Section R.5). In addition, the SST as well as the BEM can be considered as methods of weighted residuals, [5]. Advantages and drawbacks of both methods are presented and compared in [1] with the following result: the application of the BEM is easier and more automatic than the SST, since no source system must be constructed explicitly. However, a BEM computation can become extremely time-consuming for the treatment of complex structures involving a large number of elements. Consequently, the SST should be applied if the surface model of the structure is very finely discretized and the structural vibration shows a not too complicated pattern, such as a pulsating or oscillating body. This implies that the number of sources needed will be much smaller than the number of boundary elements, and a smaller system of equations has to be solved. Also, complex structures which vibrate in a complicated manner, can be treated by the SST in a very efficient way if one looks for an approximate solution with explicitly determined boundary errors.

The main topic of the presented overview is the application of the SST to the calculation of sound radiation or scattering into the unbounded, three-dimensional space. In addition, the SST can be used for the treatment of several other acoustical problems. For example, two-dimensional sound fields, problems with axisymmetric or cyclic symmetry, sound fields in interior spaces or in half spaces, or scattering and radiation from elastic structures can be investigated by means of the SST. Moreover, Leviatan and his co-workers have analyzed various scattering problems by means of a source-model technique in a series of papers (see, e.g. [49] where more references can be found).

O. 5 The boundary element method (BEM)

The BEM is mainly used for solving the radiation and scattering problem (see Section O.3, standard problems 1 and 2). The interior problem (standard problem 3) can also be treated with the help of the BEM. The fluid-structure interaction problem can be solved by combining the BEM with the finite element method (see Section O.6.3).

O.5. 1 Boundary integral equations

The most frequently used integral equation formulation in acoustics is the well-known Helmholtz integral equation for exterior field problems. The Helmholtz integral equation is obtained by applying Green's second theorem to the Helmholtz equation (O.3.1) (see for example [1] or [2]). Depending on the location of the field point x , the Helmholtz integral equation takes the form

$$\iint_S \left[p(y)\frac{\partial g(x,y)}{\partial n(y)} - \frac{\partial p(y)}{\partial n(y)} g(x,y) \right] ds = \begin{cases} p(x)\ , & x \in B_e & \text{(1a)} \\ \tfrac{1}{2} p(x), & x \in S & \text{(1b)} \\ 0\ , & x \in B_i & \text{(1c)} \end{cases}$$

where

$$g(x,y) = \frac{1}{4\pi\tilde{r}} e^{-jk_0\tilde{r}} \quad \text{with} \quad \tilde{r} = \tilde{r}(x,y) = \|x - y\| \tag{2}$$

is the free-space Green's function, and y is a spatial point on the structural surface S. The geometrical notations are chosen as described in Section O.3.1. Eqs. (1a), (1b), and (1c) are called exterior Helmholtz integral equation, the surface Helmholtz integral equation, and the interior Helmholtz integral equation, respectively. There also exists an analogous Helmholtz integral formulation for interior field problems, which will be considered in Section O.5.5. The interior Helmholtz integral equation (1c) considered here gives a null-field in B_i which obviously does not represent a physical solution, and should not be confused with the Helmholtz integral formulation for interior problems. Solutions of the Helmholtz integral equation automatically satisfy the radiation condition (O.3.3). Many boundary element formulations in acoustics use the surface Helmholtz integral equation as a starting point since it is a second kind Fredholm integral equation for the familiar Neumann boundary condition with satisfactory numerical stability, [1]. Fast numerical solvers for the discretized version of the surface Helmholtz integral equation can be obtained by using, for example, iterative algorithms, [3], or multigrid methods, [4].

The Helmholtz formula (1) is valid if the surface S is assumed to be closed and sufficiently smooth, i.e. there is a unique tangent to S at every $x \in S$. For the general case, where no unique tangent plane exists at $x \in S$ (for example, when x is lying on a corner or an edge), the surface Helmholtz integral equation has to be modified slightly, [5]

$$\iint_S \left[p(y)\frac{\partial g(x,y)}{\partial n(y)} - \frac{\partial p(y)}{\partial n(y)} g(x,y) \right] ds = \frac{C(x)}{4\pi} p(x) \qquad (3a)$$

where

$$C(x) = 4\pi + \iint_S \frac{\partial}{\partial n(y)} \left(\frac{1}{\tilde{r}(x,y)} \right) ds(y) \qquad (3b)$$

is the solid angle seen from x, [6]. For a smooth surface $C(x) = 2\pi$ is in agreement with eq.(1b).

For calculating the quantities of the sound field, two steps are necessary. First, the surface Helmholtz integral equation (1b) has to be solved which gives pressure and normal velocity on the surface of the structure. This procedure requires the main effort, since a complex, fully populated, and unsymmetrical system of linear equations has to be solved. Second, the sound field in the whole outer space can be calculated with the help of the exterior Helmholtz integral equation by a simple integration over the surface S.

The numerical treatment of the surface Helmholtz integral equation involves two characteristic difficulties. The equation possesses a weakly singular kernel, and it has no unique solution at the so-called critical frequencies. The interior Helmholtz integral equation does not suffer from these disadvantages. However, as an integral equation of the first kind, it provides a less satisfactory basis for numerical calculations, [1]. Its numerical treatment needs extreme care, and regularisation methods should be applied, [2,7].

Another formulation of acoustical boundary integral methods is the potential-layer approach. By representing the pressure p as a single-layer potential (i.e. a layer of monopoles)

$$p(x) = \iint_S \sigma(y) g(x,y)\, ds(y) \quad ; \quad x \in B_e \ \text{or} \ x \in B_i \qquad (4a)$$

one obtains the boundary integral equation

$$\frac{\sigma(x)}{2} - \iint_S \sigma(y) \frac{\partial g(x,y)}{\partial n(x)} ds(y) = j\omega\rho_0 v \quad ; \quad x \in S \qquad (4b)$$

for the determination of the density σ, if the exterior Neumann problem with

$$\partial p / \partial n = -j\omega\rho_0 v \qquad (5)$$

is considered (see Theorem 3.16 of [2]).

For the interior Neumann problem, one gets

$$\frac{\sigma(x)}{2}+\iint_S \sigma(y)\frac{\partial g(x,y)}{\partial n(x)}ds(y)=-j\omega\rho_0 v \quad ; \quad x\in S \tag{6}$$

The double-layer potential (i.e. a layer of dipoles)

$$p(x)=\iint_S \psi(y)\frac{g(x,y)}{\partial n(y)}ds(y) \quad ; \quad x\in B_e \text{ or } x\in B_i \tag{7a}$$

leads to the integral equation

$$\frac{\psi(x)}{2}+\iint_S \psi(y)\frac{\partial g(x,y)}{\partial n(y)}ds(y)=p_0(x) \quad ; \quad x\in S \tag{7b}$$

for the exterior Dirichlet problem with $\quad p=p_0 \;$ on $\; S$. (8)

It leads to

$$\frac{\psi(x)}{2}-\iint_S \psi(y)\frac{\partial g(x,y)}{\partial n(y)}ds(y)=-p_0(x) \quad ; \quad x\in S \tag{9}$$

for the corresponding interior Dirichlet problem (Theorem 3.15 of [2]).

Whereas the single- or double-layer approach can be interpreted as a layer of monopoles or dipoles, respectively, the Helmholtz integral equation contains both layers of monopoles and dipoles.

All the above presented layer approaches lead to boundary equations of the second kind. It is also possible to derive equations of the first kind. Following Colton and Kress, [2], the single-layer potential (4a) solves the interior and exterior Dirichlet problem with boundary conditions (8) if the density σ is a solution of the integral equation (Theorem 3.28 of [2])

$$\iint_S \sigma(y)g(x,y)\,ds(y)=p_0 \quad ; \quad x\in S \tag{10}$$

The double-layer potential (7a) solves the interior and exterior Neumann problem with boundary condition (5), provided the density ψ is a solution of the singular integral equation (Theorem 3.31 of [2])

$$\frac{\partial}{\partial n(x)}\iint_S \psi(y)\frac{\partial g(x,y)}{\partial n(y)}ds(y)=-j\omega\rho_0 v \quad ; \quad x\in S \tag{11}$$

Such integral equations of the first kind are improperly posed. The ill-posed nature of these equations is described in [2, p. 90]. However, there exist several attempts to deal with equations of the first kind too, and advances have been made in their numerical analysis. References can be found in [2,7].

In addition, eq.(11) contains the normal derivative of the double-layer potential, which in general does not exist on the boundary. Even if it exists, eq.(11) becomes strongly singular and a regularization is required.

Table 1 provides an overview about the exterior integral equations considered above, where the Neumann boundary conditions, [5], are taken into account. Only slight modifications are necessary if other boundary data, such as Dirichlet or Robin data, are prescribed (see [8,9]). Table 1 shows the surface equation, which has to be solved first. The corresponding equation for the pressure in the exterior space is given in the right column.

Table 1: Boundary integral formulations for the exterior domain;
HIE= Helmholtz integral equation;
SLP= Single-layer potential;
DLP= Double-layer potential

	Surface equation on S	Exterior equation for p in B_e
HIE	$p = Kp + R(j\omega\rho_0 v)$	$p = \frac{1}{2}[Kp + R(j\omega\rho_0 v)]$
SLP	$\sigma - K'\sigma = 2j\omega\rho_0 v$	$p = \frac{1}{2}R\sigma$
DLP	$T\psi = -2j\omega\rho_0 v$	$p = \frac{1}{2}K\psi$

Here, the following abbreviations for integral operators (after [2]) are used

$$(K\varphi)(x) := 2\iint_S \varphi(y)\frac{\partial g(x,y)}{\partial n(y)}\,ds(y) \tag{12}$$

$$(K'\varphi)(x) := 2\iint_S \varphi(y)\frac{\partial g(x,y)}{\partial n(x)}\,ds(y) \tag{13}$$

$$(R\varphi)(x) := 2\iint_S \varphi(y)g(x,y)\,ds(y) \tag{14}$$

$$(T\varphi)(x) := 2\frac{\partial}{\partial n(x)}\iint_S \varphi(y)\frac{\partial g(x,y)}{\partial n(y)}\,ds(y) \tag{15}$$

O.5. 2 Discretization of the boundary integral equation

In the following, only the exterior radiation problem with Neumann data is considered as the model problem. For the numerical solution of one of the above three surface integral equations, the continuous equation is discretised and transformed into a system of linear equations. For this reason, the surface of the radiator is approximated by a finite element model consisting of N finite surface elements $F_k\ (k=1,\ldots,N)$. The generation of such finite element grids for complex machine structures, which often consists of several thousand elements, requires an immense effort, even if special finite element model generators ("preprocessors") will be used.

The easiest approach to performing the surface integration in the Helmholtz integral equation (1b) is to consider pressure and normal velocity as constant over each single element. This approach is called *collocation method*, since it means that a certain value of pressure and normal velocity is assigned to the centroid of each element. These simple and often used approaches are described by Rao and Raju, [10], under the name "method of moments".

The collocation method transforms the Helmholtz integral equation into the following system of $N \times N$ linear equations

$$DP + MV = P/2 \tag{16}$$

where the matrix $D = (d_{ik})$ consists of the dipole terms

$$d_{ik} = \iint_{F_k} \frac{\partial g(x_i, y)}{\partial n(y)}\, ds(y) \tag{17a}$$

and the matrix $M = (m_{ik})$ consists of the monopole terms

$$m_{ik} = j\omega\rho_0 \iint_{F_k} g(x_i, y)\, ds(y) \tag{17b}$$

P is the N-dimensional vector of pressure values in the centroids of the N elements, and V is the corresponding normal velocity vector.

For practical purposes, often plane triangular or rectangular elements are used. Hence, the discretised surface will be not smooth, but it will be equipped with several edges and corners. Nevertheless, it is not necessary to use modification (3), since the pressure is only evaluated in the element centroids. In addition, corners and edges can be considered as slightly rounded, which will not influence the sound radiation remarkably.

Higher order elements such as linear, quadratic or cubic elements are used, too. A quadratic isoparametric element formulation with six nodes for the triangular and eight nodes for the quadrilateral curvilinear element was suggested in [5], in which both the surface elements and the acoustic variables are represented by second-order shape functions.

Often, complex structures consisting of a large number of plane triangular and rectangular elements occur in practical industrial problems. For such finite element models it is recommended to perform the integrations appearing in eqs.(17) as follows: choose variables that are constant over a single element and transform each element F_k on to the unit triangular or rectangular element (see [11,12]). Then use Gaussian integration rules of the desired order of accuracy.

The average size of the elements of a boundary grid determines the highest frequency that can be treated. A famous rule of thumb is the "six elements per wavelength rule". It means that at least six elements per wavelength should be taken into account if constant or linear elements are used. This is approximately valid for $k_0 d \le 1$ where d is a typical dimension of the element. A detailed discussion about this rule can be found in [13]. Fig. (O.4.1a) shows a boundary element mesh consisting of 7869 rectangular and 42 triangular elements, which can be used up to $k_0 a \le 21$, where a is the radius of the corresponding sphere.

O.5. 3 Solution of the linear system of equations

One of the main problems of the BEM is that the matrix A of the system (16) written in the form

$$A\,P = F \tag{18}$$

with $A = D - 0.5I$ (I = unity matrix) and $F = -MV$ (V = prescribed normal velocity vector) is fully occupied, complex and unsymmetrical. This is an essential disadvantage of the BEM in comparison with the FEM, which leads to symmetrical and weakly populated matrices with small bandwidth (see Section O.6.2). However, many technical structures consist of several thousands of elements leading to huge systems of equations. The numerical solution of such systems needs much computer time. Often, it is not possible to store the whole matrix A on the disk. In addition, if complete spectra should be calculated, a full system of equations has to be solved for every single frequency. For this reason, direct solvers such as the Gaussian elimination or the LU decomposition with backsubstitution, [14], should only be used for systems up to a few hundreds of equations, [15, p. 147], since the numerical effort is of order N^3.

For vibrating structures with symmetry, simplified boundary integral equations with a reduced number of unknowns can be derived leading to reduced computer costs. For example, the sound radiation from axisymmetric bodies is treated in [16,17]. Under the condition that the boundary conditions are axisymmetric too, the introduction of elliptic integrals leads to further simplifications, [17].

For larger systems without special symmetry properties, iterative solvers should be preferred, since they lead to a numerical cost of order N^2 approximately. The shape of the system (16) suggests the iteration scheme (called Picard iteration)

$$\tfrac{1}{2}P^{(i+1)} = DP^{(i)} + MV \tag{19}$$

for a prescribed normal velocity vector V. Starting with an initial guess for the pressure vector $P^{(0)}$, a sequence of iteration vectors $P^{(i)}$ is obtained. Such an iteration scheme corresponds to the basic iterative method of Jacobi, [15, p. 148]. Unfortunately, the Jacobi iteration (19) does not converge generally. Convergence only takes place if the powers of the matrices D converge to a matrix of zeros, [15, p. 148], or if the eigenvalues of the corresponding integral operators do not lie inside of the unit circle, [18]. For example, this is not the case for the spheroid investigated by Chertock, [18]. However, by introducing a suitable chosen relaxation parameter β into eq.(19) and using

$$\tfrac{1}{2}P^{(i+1)} = \tfrac{1}{2}\beta P^{(i)} + (1-\beta)\left[DP^{(i)} + MV\right] \tag{20}$$

Chertock achieved convergence in some cases. The iteration scheme (20) corresponds to the *Successive Overrelaxation* (SOR) method. But, Kleinman and Wendland, [19], showed that a convergent series of successive approximations for the Helmholtz integral equation is only obtained by (20) if the wave number is sufficiently small. Hence, method (20) is not a satisfactory basis for practical applications.

In [20] Kleinman and Roach presented an iterative method, which is convergent for all wave numbers k_0. The key point of the method is a self-adjoint formulation of the Helmholtz integral equation which is obtained by multiplying eq.(18) with the complex conjugate matrix A*

$$A^*AP = A^*F \ . \tag{21}$$

Then again, a successive overrelaxation method is applied to eq.(21) leading to

$$P^{(i+1)} = \left(P^{(i)} - \alpha A^*AP^{(i)}\right) + \alpha\, A^*F \tag{22}$$

with relaxation parameter α for the radiation problem considered in this chapter. Makarov and Ochmann, [3], applied a variant of this method to the high-frequency acoustic scattering from complex bodies consisting of up to 60.000 surface elements. One major problem arising from method (22) is to find the optimal parameter α. Hence, it is more convenient to use parameter-free iterative methods such as the Generalized Minimum Residual Method , which was successfully applied to acoustic and medium-to-high electromagnetic scattering in [21-23].

Another promising iterative method for the solution of the Helmholtz integral equation is the multigrid method, [4]. It consists mainly of two steps, the smoothing step and the coarse grid correction. Starting with an initial value P_0, one Picard iteration (19) is performed for determining the part $P_{1/2}$ of the solution vector P that is highly oscillating with respect to the spatial coordinate on the surface S. This is the so-called smoothing step. It forces the solution of the system of equations for the error $P - P_{1/2}$ to be smoother, so that it can be determined on a coarser finite element grid without significant loss of accuracy. For obtaining a coarser grid, two neighbouring grid points are condensed into one. The calculated error is projected by linear interpolation from the coarser grid to the finer one. If only two grids are used, the approach is called the *two-grid method.* The key idea of the multigrid method is as follows: The system for the error is not solved directly. Instead, a two-grid method is used again for obtaining an approximate solution, where one has to go over to an even coarser grid. This procedure is repeated until one arrives at a grid that contains only a small number of elements. Thus, the corresponding system of equation is small enough too, and can be solved with little effort by a direct solver. A detailed description of the multigrid method and corresponding calculations for radiating structures can be found in [4].

O.5. 4 Critical frequencies and other singularities

The integral equation formulation for the exterior problem involves a characteristic problem: At the so-called critical wave numbers k_c the integral equation is not solvable or not uniquely solvable. This phenomenon has been known for a long time (see Kupradze, [24], or Smirnow, [25]). Copley, [26], considered it in connection with the acoustical radiation problem and pointed out that the integral equation for the single-layer potential (4b) does not possess a solution if the wave number k_c is an eigenvalue of the interior Dirichlet problem

$$\Delta p + k_c^2 = 0 \quad \text{in } B_i; \quad p=0 \quad \text{on } S \tag{23, 24}$$

Physically, such a behaviour can be interpreted as a resonance phenomenon of the interior space, [26]. The surface Helmholtz integral equation, however, possesses solutions at the critical wave numbers, but these solutions are not uniquely determined. This was shown by Schenk, [1], in detail, based on the general theory of Fredholm integral equations. He suggested to use the combined integral equation formulation.

O.5. 4.1 Combined integral equation formulation (CHIEF)

The idea is as follows. At wave numbers k_c the surface Helmholtz integral equation (1b) has infinitely many solutions (i.e. a nontrivial null-space). However, it can be shown that the physically relevant solution of (1b) is the only one that satisfies the interior Helmholtz integral equation (1c) simultaneously. Hence, the idea of CHIEF is to solve the surface Helmholtz integral equation on S and the interior Helmholtz integral equation at certain selected interior points x_i (so-called CHIEF points) simultaneously:

$$\iint_S \left[p(y)\frac{\partial g(x,y)}{\partial n(y)} - \frac{\partial p(y)}{\partial n(y)} g(x,y) \right] ds = \tfrac{1}{2} p(x) \quad ; \quad x \in S \tag{25a}$$

$$\iint_S \left[p(y)\frac{\partial g(x_i,y)}{\partial n(y)} - \frac{\partial p(y)}{\partial n(y)} g(x_i,y) \right] ds = 0 \quad ; \quad x_i \in B_i \tag{25b}$$

The discretization of both integral equations leads to an overdetermined system of equations. If N surface elements and M CHIEF points are used, then a $(N+M)\times N$ system results. Such a system can be solved by a least-squares orthonormalising procedure. An alternative approach is that of Rosen et al. ,[27], who suggested to create a square matrix by introducing a vector λ of Lagrange multipliers:

$$\begin{pmatrix} A & B^* \\ B & I \end{pmatrix} \begin{pmatrix} P \\ \lambda \end{pmatrix} = \begin{pmatrix} F \\ F_C \end{pmatrix} \tag{26}$$

A (see eq.(18)) and B are the matrices of coefficients resulting from the discretisation of eqs.(25a) and (25b), respectively. F and F_C are the corresponding right-hand sides, B^* is the $N\times M$ conjugate transpose of B, and I is the $M\times M$ identity matrix. The Lagrange multipliers

$$\lambda = F_c - BP \tag{27}$$

are the residuals in satisfying the M CHIEF constraint equations.

The CHIEF method suffers from the fact that the interior points x_i are not allowed to lie on the node surfaces of the interior standing wave field. However, these nodes are not known for general radiator geometries, and their number increases with increasing frequency. On the other hand, the higher the frequency the more CHIEF points are needed, and no rules are known how to choose the number and locations of these points optimally. For avoiding this problem, the following similar method was proposed.

O.5. 4.2 Combination with the null-field equation

Stupfel et al., [28], suggested the following method, originally based on an idea of Jones, [29]. The surface pressure p has simultaneously to satisfy the surface Helmholtz integral equation and additional M null-field equations of the form

$$\iint_S p(y)\frac{\partial \psi_\ell(y)}{\partial n(y)}\, ds(y) = \iint_S \frac{\partial p(y)}{\partial n(y)} \psi_\ell(y)\, ds(y) \quad ; \quad \ell = 1,2,\ldots \tag{28}$$

where the spherical wave functions ψ_ℓ are defined in Section O.4.3. In this way, all critical wave numbers $k \le k_M$ are suppressed, where the eigenvalues of the interior Dirichlet problem (23), (24) are ordered so that $k_1 \le k_2 \le k_3 \le \ldots$.

A detailed description of the null-field equations can be found in Section O.4.4.1. Eqs.(28) are identical with the generalized null-field equations (O.4.21) for Neumann data.

The surface Helmholtz integral equation as an integral equation of the second kind is well-posed, and the null-field equations do not suffer from the non-uniqueness problem, [30,31]. Hence, the combination of both types of equations seems to be a promising approach – also for numerical calculations in the high-frequency regime.

O.5. 4.3 Combination with the differentiated integral equation

According to Table 1 the surface Helmholtz integral equation for the exterior Neumann problem in operator notation can be written as $p = K\,p + R(j\omega\rho_0 v)$ (29a)

A second relationship can be obtained by formally differentiating the surface Helmholtz integral equation in the direction of the normal. This results in the boundary integral equation

$$-j\omega\rho_0 v = \frac{\partial p}{\partial n} = T\,p + K'(j\omega\rho_0 v) \tag{29b}$$

of the first kind for the unknown boundary value p on S, where the integral operators T and K' are defined below Table 1 above. Combining both equations leads to the equation

$$p - K\,p + j\eta T\,p = R(j\omega\rho_0 v) - j\eta\left(j\omega\rho_0 v + K'(j\omega\rho_0 v)\right) \tag{30}$$

of the second kind. Burton and Miller showed in reference [32] that the linear combination (30) has a unique solution for all real values of the wave number k, if the real coupling parameter η is not zero. This approach is called the *Burton and Miller method* or the *Composite Outward Normal Derivative Overlap Relation* (with acronym CONDOR). A drawback of the method is that eq.(30) includes the derivative of the double-layer Helmholtz potential

$$(Tp)(x) := 2\frac{\partial}{\partial n(x)}\iint_S p(y)\frac{\partial g(x,y)}{\partial n(y)}\,ds(y)$$

which is a hypersingular operator. This operator must be transformed to reduce the strength of the singularity, and two ways of regularisation are discussed in [32].

In a similar way, the solution of the exterior Neumann problem can be sought in the form of a combined single- and double-layer potential

$$p(x)=\iint_S \left\{ g(x,y)+j\eta\frac{\partial g(x,y)}{\partial n(y)} \right\} \sigma(y)\, ds(y) \tag{31}$$

leading to the hypersingular integral equation (see Table 1 and [2, p. 92])

$$\sigma - K'\sigma - j\eta T\sigma = 2j\omega\rho_0 v \tag{32}$$

for the unknown density σ, which is the adjoint of the combined Green's formula integral equation (30). Again, it can be shown that the combined single- and double-layer integral equation (32) is uniquely solvable for all wave numbers if $\eta \neq 0$, [2, Theorem 3.34]. Kress and Spassow, [33], had analyzed how to choose the coupling parameter η appropriately in order to minimise the condition number of the integral operators appearing in eq.(32).

O.5. 4.4 Modified Green's functions

Another approach leading to uniquely solvable integral equations for exterior boundary value problems was developed by Jones, [29], Ursell, [34, 35], and Kleinman and Roach, [20, 36, 37]. They suggested to add a series of radiating wave functions ψ_ℓ (all definitions are given in Section O.4.3) to the free field Green's function $g(x,y)$ resulting in a modified Green's function

$$g_m(x,y) := g(x,y) + \sum_{\ell=0}^{\infty} c_\ell \psi_\ell \ . \tag{33}$$

Now, all surface integral equations from Table 1 can be modified by using the Green's function $g_m(x,y)$ instead of $g(x,y)$ in the definition of the operators (12)-(15). For example, the modified double-layer potential is of the form

$$p(x)=\iint_S \Psi(y)\frac{g_m(x,y)}{\partial n(y)}\, ds(y) \quad ; \quad x \notin B_e \text{ or } x \notin B_i \tag{34}$$

with density ψ .

It can be shown that the corresponding surface integral equations are uniquely solvable for all wave numbers provided that the coefficients c_l appearing in (33) satisfy certain inequalities (see [2, Theorem (3.35)] and [20]). In addition, it was shown by Jones, [29], (see also [36]) that the integral equations will still be uniquely solvable for $k_0 \leq k_N$, where k_N is a certain critical wave number, if the c_k are chosen to vanish for $\ell > N$. Hence, the Green's function has only to be modified with a finite number of terms. However, this shows a drawback of the method: the higher the frequency the more terms in the series must be included, which may lead to numerical problems if high-frequency radiation or scattering is considered.

For a numerical analysis, it can be of advantage to get an estimate about the locations of the critical frequencies appearing in a certain frequency range. In [28, p. 928] it was noted that the inequalities

$$\frac{K_n}{A} \le k_n \le \frac{K_n}{a} \tag{35}$$

are valid, where K_n is the n-th eigenvalue of the interior Dirichlet problem (23), (24) for the unit sphere. A and a are the radii of spheres lying completely in the exterior B_e and the interior B_i, respectively.

O.5. 4.4 Treatment of singularities

The integral equations considered involve singularities, since the Green's function $g(x,y)$ and its normal derivations become singular for $x \to y$. First, the surface Helmholtz integral equation contains the weekly singular monopole terms m_{ii}, eq.(17b). This singularity is of order $\tilde{r}^{-1} = \|x - y\|^{-1}$ and can be removed by introducing polar coordinates. Following Everstine, [38], it is assumed that v is constant over a small circular area of radius b_i with centroid x_i so that the monopole terms (17b) can be written as

$$m_{ii} = j\omega\rho_0 \iint_{F_i} g(x_i,y)\, ds(y) = j\omega\rho_0 \int_0^{2\pi} \int_0^{b_i} \frac{e^{-jk_0 r}}{4\pi r}\, r\, dr\, d\varphi \tag{36}$$

where b_i is such that $\pi b_i^2 = F_i$ gives the total area of the element F_i assigned to the point x_i. Hence, the following result is obtained

$$m_{ii} = j\omega\rho_0 F_i / (2\pi b_i) \tag{37}$$

with $b_i = \sqrt{F_i / \pi}$.

Second, the dipole self terms d_{ii} (17a) must be evaluated. It can be shown, [39], that for plane elements $d_{ii} = 0$, [39], since $\tilde{r} \perp n$ and hence $\partial\tilde{r} / \partial n = 0$. If the curvature c_i of the radiating surface element is taken into account, it can be shown that approximately, [38],

$$d_{ii} = -(1 + j k_0 b_i)(c_i F_i) / (4\pi b_i) \; . \tag{38}$$

O.5. 5 The interior problem: sound fields in rooms and half-spaces

The numerical calculation of a sound field in an interior space B_i is the third standard problem of Section O.3.3. If a problem without acoustical sources in the interior B_i is considered, the application of Green's second theorem to the Helmholtz equation (O.3.1) (see for example [2], [40], or [41]) leads to the interior Helmholtz integral formula

$$\iint_S \left[p(y)\frac{\partial g(x,y)}{\partial n(y)} - \frac{\partial p(y)}{\partial n(y)} g(x,y) \right] ds = \begin{cases} 0, & x \in B_e & (39a) \\ -\frac{1}{2}p(x), & x \in S & (39b) \\ -p, & x \in B_i & (39c) \end{cases}$$

for the pressure p. Again, the Green's function g(x,y) is given by eq.(2). The interior formula is very similar to the exterior Helmholtz integral formula (1). Only the sign and the role of the interior and exterior space have changed. Now, the BEM can be applied to the solution of the interior problem in just the same way as described for the radiation problem (e.g. [41]). Boundary conditions for the pressure or the pressure gradient have to be inserted into the surface integral equation (39b). Clearly, a radiation condition is not necessary. Discretisation of eq.(39b) leads to a system of equations for the determination of the second acoustic surface variable as described in Section 2 and 3. Fortunately, critical frequencies cannot occur, since the adjoint problem now is the exterior problem which does not possess any discrete eigenvalues. Hence, the regularisation procedures described in section 3 are not needed. However, fully interior problems are mainly the domain of the finite element method, which can deal with very general interior fluid-structure interactions problems (see Section O.6).

The BEM is well suited for the calculation of sound radiation into a half-space, [42], too. In many applications the radiator is situated on a locally reacting plane S_{plane}. Such an infinite plane can be taken into account by using a half-space Green's function

$$g_H(x,y) = \frac{1}{4\pi\tilde{r}}e^{-jk_0\tilde{r}} + R\frac{1}{4\pi\tilde{r}'}e^{-jk_0\tilde{r}'} \quad \text{with} \quad \tilde{r} = \|x-y\| \quad \text{and} \quad \tilde{r}' = \|x'-y\| \qquad (40)$$

instead of the free-space Green's function g(x,y) in the surface Helmholtz integral equation (1b).

Here, x' is the image point of x behind the plane and R is the reflection coefficient of the plane. For a rigid plane R=1 and for a free surface R=–1 . For other values of R, eq.(40) is only an approximation. The exact representation of the Green's function over an absorbing impedance plane is more complicated, and a detailed discussion of various formulas can be found in [43]; for exact solutions and approximations see Sections D.14 - D.20. By using the Green's function (40) one only has to extend the integration in eq.(1b) over the surface of the radiator S. The infinite plane S_{plane} does not have to be taken into account !

If the radiating body is in contact with the plane, a slightly modified version of the surface Helmholtz integral equation (3a) has to used, where the coefficient C(x) is now given by

$$C(x) = (1+R)\left[2\pi - \iint\limits_{S_0 \cup S_c} \frac{\partial}{\partial n(y)}\left(\frac{1}{\tilde{r}}\right) ds(y)\right] . \tag{41}$$

Here $S = S_0 \cup S_c$, where S_c is the part of the radiator surface S that is in contact with S_{plane} . The derivation of $C(x)$ and more details can be found in [44]. Eq.(41) is valid for a normal pointing into the body.

The method of mirror sources for constructing the Green's function as a kernel for an appropriate integral equation can be generalized to regard additional plane boundaries around the radiating body. For example, if the vibrating structure is situated in a rectangular room, the resulting sound field can be computed by using two different representations of the Green's function for a rectangular enclosure with dimensions $\ell_x \times \ell_y \times \ell_z$, [45]. The first formula for the Green' function can be constructed by mirror sources. Hence it consists of an infinite series of exponentials, and converges for small distances from the radiator.

For larger distances, the approximate Green's function

$$G_E(x,y) = \sum_{n=0}^{\infty} \frac{\Phi_n(x)\,\Phi_n(y)}{V\,\Gamma_n\left(k_n^2 - k_0^2 - j\tau_n\right)} \tag{42}$$

which is combined of the known eigenmodes of a rectangular room, [45, 46], shows a better convergence behaviour.

The eigenfunctions are given by

$$\Phi_n(x) = \cos(k_{nx}x)\cos(k_{ny}y)\cos(k_{nz}z)$$

with

$$k_{nx} = \frac{\pi n_x}{\ell_x},\; k_{ny} = \frac{\pi n_y}{\ell_y},\; k_{nz} = \frac{\pi n_z}{\ell_z}$$

The damping factor of the n-th mode is given by

$$\tau_n = k_0\left[\varepsilon_{nx}\left(\frac{\beta_{xo} + \beta_{xl}}{\ell_x}\right) + \varepsilon_{ny}\left(\frac{\beta_{yo} + \beta_{yl}}{\ell_y}\right) + \varepsilon_{nz}\left(\frac{\beta_{zo} + \beta_{zl}}{\ell_z}\right)\right] ,$$

where β_{xl} (for example) is the specific acoustic admittance at the wall at $x = \ell_x$. Γ_n is given by $\Gamma_n = 1/\left(\varepsilon_{nx}\varepsilon_{ny}\varepsilon_{nz}\right)$,

where the Neumann symbol ε_n is defined by

$$\varepsilon_n = \begin{cases} 1, & \text{for } n=0 \\ 2, & \text{otherwise} \end{cases} ;$$ V is the volume of the room.

This method seems to be superior to the FEM for complex structures in large rectangular rooms, since it only requires to divide the surface of the structure into elements, whereas the FEM has to discretise the whole interior volume of the room into finite elements.

The solution of coupled interior–exterior acoustics problems with the help of the BEM was investigated by Seybert et al. [47].

O.5. 6 The scattering and the transmission problem

As explained in Section O.3.2, the scattering problem can be considered as an equivalent radiation problem with respect to the scattered pressure p_S. However, sometimes it is more convenient to have an explicit boundary integral equation for the total pressure $p = p_T = p_s + p_{in}$ as he starting point for a numerical calculation. The scattered wave p_s has to fulfil the exterior Helmholtz formula (1). The incident pressure p_{in} is assumed to have no singularities in B_i, and hence it must satisfy the interior Helmholtz formula (39). By adding both eqs.(1) and (39), one gets the Helmholtz formula

$$p_{in} + \iint_S \left[p(y)\frac{\partial g(x,y)}{\partial n(y)} - \frac{\partial p(y)}{\partial n(y)} g(x,y) \right] ds = \begin{cases} p(x), & x \in B_e & (43a) \\ \frac{1}{2} p(x), & x \in S & (43b) \\ 0, & x \in B_i & (43c) \end{cases}$$

for the total pressure p. Assuming that the surface of the scatterer is rigid, the pressure gradient $\partial p / \partial n = -j\omega\rho_0 v$ on the surface is zero. Hence, for a rigid scatterer the boundary integral equation (43b) can be written as

$$p(x) = 2\iint_S p(y)\frac{\partial g(x,y)}{\partial n(y)}\, ds(y) + 2p_{in} \qquad (44)$$

For an arbitrary surface velocity distribution v , eq.(43b), takes the form

$$p(x) = 2\iint_S p(y)\frac{\partial g(x,y)}{\partial n(y)}\, ds(y) + 2\iint_S j\omega\rho_0 v(y)\, g(x,y)\, ds(y) + 2p_i \qquad (45)$$

For the general impedance boundary value problem, the normal impedance Z=p/v is introduced at each point on the surface S of the scatterer. Substitution of the normalized impedance $\overline{Z} = Z / (\rho_0 c_0)$ into eq.(45) gives the boundary integral equation

$$p(x) = 2\iint_S p(y)\frac{\partial g(x,y)}{\partial n(y)}\, ds(y) + 2\iint_S \frac{jk_0}{\overline{Z}(y)} p(y)\, g(x,y)\, ds(y) + 2p_{in} \qquad (46)$$

for an impedance scatterer. It should be emphasised that the local impedance $\overline{Z}(y)$ can vary with the surface point y. Calculations for structures with varying normal surface impedance based on an iterative BE solver can be found in [3].

The transmission problem is described in Section O.3.5. The total pressure $p_1 := p$ in the surrounding medium with constants c_1 and ρ_1 has to satisfy eq.(43b) in the form

$$p_1(x) = 2\iint_S p_1(y)\frac{\partial g_1(x,y)}{\partial n(y)}\,ds(y) - 2\iint_S \frac{\partial p_1(y)}{\partial n(y)} g_1(x,y)\,ds(y) + 2p_{in} \tag{47}$$

with

$$g_1(x,y) = \frac{1}{4\pi\tilde{r}} e^{-jk_1\tilde{r}} \quad \text{with} \quad \tilde{r} = \|x-y\| \quad \text{and} \quad k_1 = \frac{\omega}{c_1} \tag{48}$$

The pressure $p_2 := p_i$ inside of the scattering body with constants c_2 and ρ_2 must satisfy eq.(39b)

$$p_2(x) = -2\iint_S p_2(y)\frac{\partial g_2(x,y)}{\partial n(y)}\,ds(y) + 2\iint_S \frac{\partial p_2(y)}{\partial n(y)} g_2(x,y)\,ds(y) \tag{49}$$

with

$$g_2(x,y) = \frac{1}{4\pi\tilde{r}} e^{-jk_2\tilde{r}} \quad \text{with} \quad \tilde{r} = \|x-y\| \quad \text{and} \quad k_2 = \frac{\omega}{c_2} \tag{50}$$

For f=g=0 , the transmission conditions (O.3.23) and (O.3.24) take the form

$$p_1 = p_2 \quad ; \quad \frac{\partial p_1}{\partial n} = \frac{1}{\varsigma}\frac{\partial p_2}{\partial n} \quad ; \quad \varsigma = \rho_2 / \rho_1 \tag{51, 52}$$

By introducing eqs.(51) and (52) into eqs.(47) and (49), two coupled integral equations are obtained

$$p_1(x) = 2\iint_S p_1(y)\frac{\partial g_1(x,y)}{\partial n(y)}\,ds(y) - 2\iint_S \frac{\partial p_1(y)}{\partial n(y)} g_1(x,y)\,ds(y) + 2p_{in} \tag{53}$$

$$p_1(x) = -2\iint_S p_1(y)\frac{\partial g_2(x,y)}{\partial n(y)}\,ds(y) + 2\varsigma\iint_S \frac{\partial p_1(y)}{\partial n(y)} g_2(x,y)\,ds(y) \tag{54}$$

for the determination of the surface variables p_1 and $\partial p_1 / \partial n$.

For equal sound velocities $c_1 = c_2$, the Green's functions also are equal $g_1(x,y) = g_2(x,y)$, and one single integral equation can be derived by combining eqs.(53) and (54) in a suitable way

$$p_1(x) = \frac{2(\varsigma - 1)}{\varsigma + 1} \iint_S p_1(y) \frac{\partial g_1(x,y)}{\partial n(y)} \, ds(y) + 2\frac{\varsigma}{\varsigma + 1} p_{in} \quad . \tag{55}$$

The as yet unpublished boundary integral equation (55) was derived by S. Makarov (Worcester Polytechnic Institute, MA, USA) and communicated to the author.

For $\varsigma \to \infty$, i.e. $\rho_2 >> \rho_1$, the integral equation (44) for the rigid scatterer is obtained as expected for physical reasons.

Additional recent formulations and numerical implementations of the BEM can be found in the book [48].

O.6 The finite element method (FEM)

O.6. 1 Introduction

The finite element method (FEM) is especially suited for the numerical calculation of sound fields in irregularly formed inner spaces, since such spaces are of finite dimensions. Originally, the FEM was developed for predicting the static or dynamical response of structures under certain loads in mechanical engineering. Several high-developed FEM packages exist which are commercially available. Some of these programs contain modules for acoustical computations. In principle, acoustical calculations can be directly performed with programs which were originally developed for structural computations, since a mechanical analogy for fluid motion can be used, [1]. Consider the acoustical wave equation for the pressure with spatial varying fluid density ρ in Cartesian coordinates

$$\frac{\partial}{\partial x}\left(\frac{1}{\rho}\frac{\partial p}{\partial x}\right)+\frac{\partial}{\partial y}\left(\frac{1}{\rho}\frac{\partial p}{\partial y}\right)+\frac{\partial}{\partial z}\left(\frac{1}{\rho}\frac{\partial p}{\partial z}\right)=\frac{1}{\chi}\frac{\partial^2 p}{\partial t^2} , \qquad (1)$$

where $\chi=\rho c^2$ is the bulk modulus, and the equation for the equilibrium of stresses in a particular fixed direction x

$$\frac{\partial\sigma_{xx}}{\partial x}+\frac{\partial\tau_{xy}}{\partial y}+\frac{\partial\tau_{xz}}{\partial z}=\rho_s\frac{\partial^2 u_x}{\partial t^2} , \qquad (2)$$

where u_x is the structural displacement in the x direction, σ_{xx}, τ_{xy}, τ_{xz} are stress components, and ρ_s is the structural mass density. The acoustic-structural analogy can be established by comparing eqs.(1) and (2) and taking

$$u_x = p \ ; \ \rho_s = 1/\chi$$

$$\sigma_{xx}=\frac{1}{\rho}\frac{\partial p}{\partial x}=-\frac{\partial^2 w_x}{\partial t^2}$$

$$\tau_{xy}=\frac{1}{\rho}\frac{\partial p}{\partial y}=-\frac{\partial^2 w_y}{\partial t^2}$$

$$\tau_{xz}=\frac{1}{\rho}\frac{\partial p}{\partial z}=-\frac{\partial^2 w_z}{\partial t^2}$$

Here, the acoustic equilibrium equation

$$\nabla p+\rho\frac{\partial^2 \vec{w}}{\partial t^2}=0 \qquad (3)$$

was taken into account, where $\vec{w}$ is the particle displacement within the fluid. For completing the analogy, the structural displacement components u_y and u_z must be set equal to zero and the general stress-strain relationship must be modified in a suitable way (details can be found in [1]). Hence, all the tools of classical FEM programs such as the variety of element types, solution methods etc. are available for the acoustical analysis, too.

Typical areas of application of the acoustical FEM are sound fields in small rooms at low frequencies as mentioned in Section O.3.3 where aspects of wave propagation play an essential role. Also, all problems involving fluid-structure interaction are treated by the FEM with preference.

At first, the principle of FEM will be explained for the simple example of an air-filled enclosure with rigid boundaries. Afterwards, more advanced applications will be shortly discussed.

O.6. 2 The sound field in irregular shaped cavities with rigid walls

This problem belongs to the standard problems described in Section O.3.3, where the notation is explained. The Helmholtz equation (O.3.1) has to be satisfied in the interior B_i of the cavity. The whole boundary S is assumed to be rigid with $\partial p / \partial n = 0$. For simplicity, only the two-dimensional case is considered. Since no acoustical sources are specified, the eigenmodes and eigenfrequencies of the cavity are quantities which should be calculated. Following the presentation given in [2], the starting point of the FEM calculation is a variational principle for the Helmholtz equation, which must be minimized. In the present case, the functional, [3-5],

$$L = \frac{1}{2} \iint_{B} \left[(\text{grad } p)^2 - k_0^2 p^2 \right] dv \tag{4}$$

must be minimised, where the integral has to be extended over the volume (or area in two dimensions) B, and the gradient is denoted by grad. The functional L can be interpreted as a Lagrange function, i.e. as the difference between kinetic and potential energy of the vibrating fluid. Hence, the minimisation of L corresponds to Hamilton's principle. Therefore, the Helmholtz equation is the Euler-Lagrange equation of the functional (4). The rigid boundary conditions are so-called natural boundary conditions and will be automatically satisfied.

The second step consists in dividing the cavity B into simple elements. In the following, only triangular elements are used. It is important to note that the process of triangularization should be adapted to the particular problem. This means, for example, that parts of the area, in which the solution changes more rapidly, should be modelled with more and smaller elements than other parts, where the change of the solution is slower. In addition, the triangular elements should not have too acute angles. Such requirements will be checked automatically in most commercial FE programs. If B is divided into N triangular elements E_n, the discretized functional is given by

$$L_D = \frac{1}{2} \sum_{n=1}^{N} \iint_{E_n} \left[(\mathrm{grad}\,p)^2 - k_0^2 p^2 \right] dv \tag{5}$$

The third step is to choose approximate functions for the sound pressure at each single element. Frequently, polynomials are used satisfying certain continuity conditions between adjacent elements. For fulfilling such continuity conditions in selected points of the element, named nodal points or nodes, the approximate function $p^{(e)}$ for the e-th element is represented by

$$p^{(e)}(x,y) = \sum_{k=1}^{K_e} p_k^{(e)} N_k^{(e)}(x,y) = P^{(e)^T} N^{(e)}(x,y) \tag{6}$$

where the so-called nodal variables $p_k^{(e)}$ are the sound pressure values in the nodes (for example, in the three corner points (K_e=3) of the triangular element), and the $N_k^{(e)}$ are the shape functions. Moreover, the vectors

$$P^{(e)^T} = \left(P_1^{(e)}, P_2^{(e)}, \ldots, P_{K_e}^{(e)} \right) \tag{7a}$$

$$N^{(e)^T}(x,y) = \left(N_1^{(e)}, N_2^{(e)}, \ldots, N_{K_e}^{(e)} \right) \tag{7b}$$

were introduced, where the superscript T denotes transposition. The shape functions have to satisfy the interpolation property

$$N_i^{(e)}\left(x_k^{(e)}, y_k^{(e)}\right) = \begin{cases} 1 & \text{for } i = k \\ 0 & \text{for } i \neq k \end{cases} \tag{8}$$

in the nodal points $Q_i^{(e)} = \left(x_i^{(e)}, y_i^{(e)}\right)$ of the e-th element. The global representation of pressure p in the whole area B is composed of all element pressures $p^{(e)}$. By numbering all nodes lying in B from 1 to K successively, one obtains

$$p(x,y) = \sum_{k=1}^{K} p_k N_k(x,y) \tag{9}$$

where the so-called global shape functions $N_k(x,y)$ consist of the union of all element shape functions $N_k^{(e)}(x,y)$ which possess the value 1 at the nodal point Q_k. It follows from (8) that the global shape functions are different from zero only in a small part of B, i.e. they are functions with local support, which is a key property of the FEM.

For example, if the linear substitution $p(x,y) = c_1 + c_2 x + c_3 y$ (10)

is used in the unit triangle, the shape functions

$$N_1(x,y) = 1 - x - y \; ; \; N_2(x,y) = x \; ; \; N_3(x,y) = y \tag{11}$$

are obtained (see [2, p. 90]).

The fourth step is to introduce the approximation (9) into the discretized functional (5). The integrals have to performed element by element and can be solved analytically for polynomial shape functions. For a given triangle T_n with nodes $Q_k(x_k, y_k)$, k=1,2,3 and the linear substitution (10) the following results are obtained (see [2, pp. 71])

$$\iint\limits_{T_n} \left[\left(\frac{\partial p}{\partial x} \right)^2 + \left(\frac{\partial p}{\partial y} \right)^2 \right] dxdy = P^{(e)^T} S_e P^{(e)} \tag{12}$$

$$\iint\limits_{T_n} p^2 dxdy = P^{(e)^T} M_e P^{(e)} \tag{13}$$

where the stiffness matrix S_e and the mass matrix M_e are composed of the four basis matrices

$$S_1 = \frac{1}{2} \begin{pmatrix} 1 & -1 & 0 \\ -1 & 1 & 0 \\ 0 & 0 & 0 \end{pmatrix} \; ; \; S_2 = \frac{1}{2} \begin{pmatrix} 2 & -1 & -1 \\ -1 & 0 & 1 \\ -1 & 1 & 0 \end{pmatrix}$$
$$S_3 = \frac{1}{2} \begin{pmatrix} 1 & 0 & -1 \\ 0 & 0 & 0 \\ -1 & 0 & 1 \end{pmatrix} \; ; \; S_4 = \frac{1}{24} \begin{pmatrix} 2 & 1 & 1 \\ 1 & 2 & 1 \\ 1 & 1 & 2 \end{pmatrix} \tag{14}$$

in the following way

$$S_e = aS_1 + bS_2 + cS_3 \quad \text{and} \quad M_e = J\, S_4 \tag{15}$$

where the constants

$$a = \left[(x_3 - x_1)^2 + (y_3 - y_1)^2 \right] / J \tag{16a}$$

$$b = -\left[(x_3 - x_1)(x_2 - x_1) + (y_3 - y_1)(y_2 - y_1) \right] / J \tag{16b}$$

$$c = \left[(x_2 - x_1)^2 + (y_2 - y_1)^2 \right] / J \tag{16c}$$

$$J = (x_2 - x_1)(y_3 - y_1) - (x_3 - x_1)(y_2 - y_1) \tag{16d}$$

only depend on the geometry of the triangle considered.

Hence, the contribution of a single triangular element to the whole functional is given by

$$L_D^{(e)} = \tfrac{1}{2} P^{(e)^T} S_e P^{(e)} - \tfrac{1}{2} k_0^2 P^{(e)^T} M_e P^{(e)} \tag{17}$$

Summing up all contributions element by element leads to the discretized global functional of quadratic form

$$L_D = \tfrac{1}{2} P^T S P - \tfrac{1}{2} k_0^2 P^T M P \tag{18}$$

where P is the vector of all nodal pressure variables. S and M are the global stiffness and mass matrix, respectively. The requirement that L_D takes a minimum leads to the equation

$$SP = k_0^2 MP \tag{19}$$

which is a generalised eigenvalue problem for the eigenvalue parameter $\lambda = k_0^2$. Such a problem was solved in [2] with three different numerical methods for an idealised automobile passenger compartment. For this specific example, it was shown that the simultaneous inverse vector iteration was the most efficient method. Details about numerical methods for solving the generalised eigenvalue problem can be found in [2].

A great advantage of the FEM is the fact that the global matrices S and M are symmetric and weakly populated, in contrast to the BEM. By using optimal numbering procedures such as the algorithms of Rosen or Cuthill-McKee (see [2]) for the nodal variables, the bandwidth of such sparse matrices can be minimised which leads to a large reduction of computer time and storage capacity.

O.6. 3 Supplementary aspects and fluid-structure coupling

The simple model problem of a rigid two-dimensional cavity, which is discretised with triangular elements, can be extended into different directions. For example, elements or shape functions of higher orders can be used. Also, the generalization to the three-dimensional case can be easily done. Especially, the sound field in enclosures and automobile passenger compartment was investigated by several authors: Craggs, [6], used tetrahedral and cuboid finite elements with a linear variation of the pressure between the nodes in order to find the eigenfrequencies and modes of a three-dimensional enclosure. Shuku and Ishihara, [7], used triangular elements with cubic polynomial functions for the pressure. Petyt et al., [8], developed a twenty-node, isoparametric acoustic finite element for analysing the acoustics modes of irregular shaped cavities. Richards and Jha, [9], preferred quadratic triangular elements with six nodes.

If only the part S_0 of the surface S is rigid, whereas the normal velocity v is prescribed on S_1 and an absorbing material with impedance Z is specified on S_2, the following boundary conditions are obtained

$$\frac{\partial p}{\partial n}=0 \quad \text{on } S_0 \;;\quad \frac{\partial p}{\partial n}=-j\omega\rho_0 v \quad \text{on } S_1 \;;\quad \frac{\partial p}{\partial n}=-j\omega\rho_0\frac{p}{Z} \quad \text{on } S_2 \tag{20}$$

Instead of eq.(4) the corresponding variational principle is now given by [3]

$$L=\tfrac{1}{2}\iiint_B\left[(\text{grad}p)^2-k_0^2p^2\right]dv+\iint_{S_1} j\omega\rho_0 pv\,ds+\tfrac{1}{2}\iint_{S_2} j\omega\rho_0\frac{p^2}{Z}\,ds \tag{21}$$

which is minimised according to the same rules as described in section 2. The resulting system of equations can be used for the investigation of passenger compartments or silencers which are partially lined with absorbing material (see [3] and [10], where more references can be found). In eq.(21), the three-dimensional case is considered, and hence the first integral is extended about the volume B. In [11], wave propagation within the porous absorber is taken into account.

The coupled fluid-structure interaction problem is described in Section O.3.4 as the fourth standard problem. A mathematically rigorous description of the corresponding finite dimension approximation, which can be associated with a finite element mesh, is given by Soize, [12-14]. Instead, the method of Everstine and Henderson, [15], will be described here, which is based on a coupled FE/BE approach. A very similar formulation was given by Smith, Hunt, and Barach, [16]. The structure is assumed to be modeled with finite elements. This leads to a matrix equation of motion for the structural degrees of freedom

$$Z_sV_s=F_s-GA_sP \tag{22}$$

where Z_s is the structural impedance matrix, V_s is the global velocity vector, F_s is the vector of mechanical forces applied to the structure, G is a transformation matrix in order to transform a vector of outward normal forces at the so-called wet points (which are in contact with the fluid) to a vector of forces at all points in the selected coordinate system, A_s is the diagonal matrix of areas for the wet surface, and P is the vector of total acoustic pressures. Eq.(22) can be derived in a similar way as described in the last section for the acoustical case. The structural impedance matrix Z_s is given by

$$Z_s=\frac{1}{j\omega}\left(-\omega^2M_s+j\omega D_s+K_s\right) \tag{23}$$

where M_s, D_s and K_s are the structural mass, damping and stiffness matrices, respectively. The integral equation of the scattering problem for an incident pressure wave p_i was derived in Section O.5.6 and is given by (see eq.(O.5.45))

$$p(x) = 2\iint_S p(y)\frac{\partial g(x,y)}{\partial n(y)}\,ds(y) + 2\iint_S j\omega\rho_0 v(y) g(x,y)\,ds(y) + 2p_i \tag{24}$$

Corresponding to the BE approach described by eqs.(O.5.16-O.5.18), the discretised version of the integral equation (24) is

$$-AP = MV + P_i \quad \text{with} \quad A := D - 0.5I \ , \tag{25}$$

where P_i is the vector of incident pressures, V is the vector of normal velocities, and D and M are the dipole and monopole matrices, respectively, defined in eqs.(O.5.17). According to Everstine and Henderson, [15], the vector of normal velocities V is transformed into the vector V_S of total structural velocities by applying the transposed matrix G^T to V_S:

$$V = G^T V_s \tag{26}$$

Now, by combining eqs.(22), (25), and (26), the velocity vectors V and V_S can be eliminated, which leads to the coupled fluid-structure equation

$$HP = Q + P_i \tag{27}$$

with

$$H := -A + MG^T Z_s^{-1} GA_s \quad \text{and} \quad Q = MG^T Z_s^{-1} F_s \tag{28, 29}$$

Having solved system (27) for the pressure P, the vector V_S of structural velocities is obtained from eq.(22) by solving the equation

$$V_s = Z_s^{-1} F_s - Z_s^{-1} GA_s P \tag{30}$$

With the knowledge of the fluid surface variables P and V all acoustics quantities in the exterior field can be calculated by evaluating the exterior Helmholtz formula (O.5.43a).

In [17-19], a combined finite element and spectral approach is proposed: the FEM is used for calculating the vibration of the elastic structure and the acoustic field inside of a finite fluid sphere which totally surrounds the vibrating structure. At the boundary of the sphere, a perfect absorption condition is given explicitly in terms of the spherical wave functions (O.4.5). This method was generalised and implemented into the structural analysis code NASTRAN resulting in a commercially available FEM package. Some recent results of this technique can be found in [20].

Another interesting idea is to use absorbing boundary condition operators, [21], on the artificial boundary of the finite domain instead of coupling with analytical wave functions. The perfect absorbing boundary condition is modeled by the so-called BGT operator named after Bayliss, Gunzburger and Turkel, [22]. The first-order operator B_1 is defined by

$$B_1 p := \left(\frac{\partial}{\partial R} + jk_0 + \frac{1}{2R}\right) p \tag{31}$$

where p is the radiated sound pressure. R and k_0 are defined as in the Sommerfeld radiation condition (O.3.3) which is approximated with increasing accuracy for increasing order of the BGT operator. The second operator B_2 is given by

$$B_2 p := \left(\frac{\partial}{\partial R} + jk_0 + \frac{5}{2R} \right) \left(\frac{\partial}{\partial R} + jk_0 + \frac{1}{2R} \right) p \tag{32}$$

and, in general, the m-th order operator can be introduced by, [21,22]

$$B_m := \prod_{\ell=1}^{m} \left(\frac{\partial}{\partial R} + \frac{2\ell - \frac{3}{2}}{R} + jk_0 \right) \tag{33}$$

It can be shown, [21, 22], that the operator B_m annihilates terms in $1/r$ of the asymptotic far field solution p_s

$$p_s = \frac{e^{-jk_0 R}}{\sqrt{R}} \left(a_0 + \frac{a_1(\varphi)}{R} + \frac{a_2(\varphi)}{R^2} + \ldots \right) \tag{34}$$

such that the accuracy in approximating the radiation condition is

$$B_m p_s = O\left(\frac{1}{r^{2m+1+1/2}} \right) \tag{35}$$

Here, a_0 is a constant, and $a_1(\varphi), a_2(\varphi), \ldots$ are functions of the polar angle φ. These formulas are given for the two-dimensional case. The incorporation into a finite element model is described in [21].

Some more special examples from the huge number of papers dealing with the application of the FEM to acoustical problems should be mentioned: The formulation of the FEM for cavities with fluid-structure interaction can be found in [14, 23], for example. The sound transmission between enclosures using plate and acoustic finite elements is studied in [24]. A comparison between FEM and BEM for the calculation of sound fields is performed in [25].

O.7 The Cat's Eye Model

Many variations and improvements of numerical methods are studied in the literature. It is helpful to have benchmark models available for which analytical solutions are known, and which are not over-simplified, so that comparisons between different variants of numerical methods can be performed. One favourite benchmark model is the Cat's Eye model for which the analytical solution of the sound field will be described in this Section. An other model, with more simple numerics of the analytical solution is the "Orange" model which will be described in Section O.8 . Both sections treat the radiation problem for the special case of a constant surface velocity of the sphere (monopole radiation problem). It will be explained at the end of this Section, how multipole radiation problems and scattering problems can be treated.

O.7.1 Cat's Eye model and general fundamental solutions :

The Cat's Eye model is a sphere with one octant of the sphere taken away.

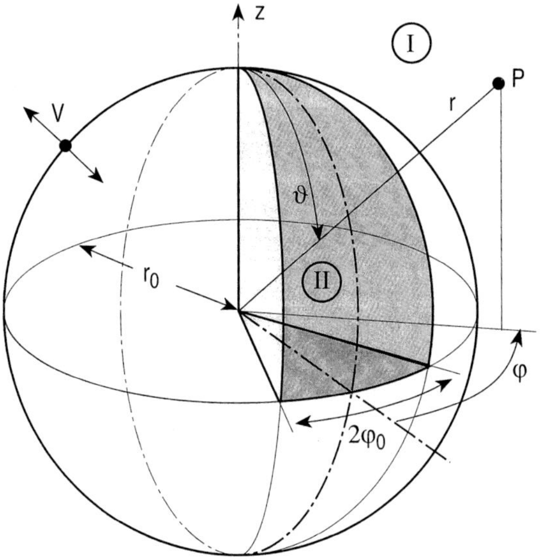

An octant with cuts at $\varphi = \pm\varphi_0$; $\vartheta = \pi/2$ is taken away of a sphere with radius r_0 . The walls of the cuts are hard.

The remainder of the spherical surface oscillates with a constant velocity $V \neq 0$ (monopole radiation problem).

Target quantities are :

- the sound field in the outer zone (I);
- the sound field in the inner zone (II).

The time factor $e^{j\omega t}$ is dropped.

The problem is symmetrical with respect to $\varphi=0$ (*viz.* the axis $\varphi=0$ is selected so that this holds).

Particle velocity components in the spherical co-ordinates are :

$$\mathbf{v} = \frac{j}{k_0 Z_0}\,\mathrm{grad}\,p$$
$$v_r = \frac{j}{k_0 Z_0}\frac{\partial p}{\partial r} \;;\; v_\vartheta = \frac{j}{k_0 Z_0}\frac{\partial p}{r\cdot\partial\vartheta} \;;\; v_\varphi = \frac{j}{k_0 Z_0}\frac{\partial p}{r\sin\vartheta\cdot\partial\varphi} \xrightarrow[\vartheta\to\pi/2]{} \frac{j}{k_0 Z_0}\frac{\partial p}{r\cdot\partial\varphi} \tag{1}$$

General form of a fundamental solution $p(r,\vartheta,\varphi)$ in spherical co-ordinates :

$$\begin{aligned} &p(r,\vartheta,\varphi) = R(k_0 r)\cdot\Theta(\vartheta)\cdot\Phi(\varphi) \\ &R(k_0 r) = A\cdot\mathfrak{R}_\nu^{(1)}(k_0 r) + B\cdot\mathfrak{R}_\nu^{(2)}(k_0 r) \\ &\Theta(\vartheta) = a\cdot P_\nu^\mu(\cos\vartheta) + b\cdot Q_\nu^\mu(\cos\vartheta) \\ &\Phi(\varphi) = \alpha\cdot\sin(\mu\varphi) + \beta\cdot\cos(\mu\varphi) \end{aligned} \tag{2}$$

with: $\mathfrak{R}_\nu^{(i)}(k_0 r)$ 2 linear independent spherical Bessel functions;

$P_\nu^\mu(z)\,;\,Q_\nu^\mu(z)$ associated Legendre functions of 1st and 2nd kind, respectively;

From the symmetry condition in φ follows: $\Phi(\varphi) = \cos(\mu\varphi)$ (3)

Outer field in (I) :

From Sommerfeld's condition follows : $\mathfrak{R}_\nu(k_0 r) = h_\nu^{(2)}(k_0 r)$ (4)
(spherical Hankel function of 2nd kind).

The $Q_\nu^\mu(z)$ have logarithmic singularities at $z=\pm 1$, i.e., at $\vartheta=0$, $\vartheta=\pi$; the sound field should be regular there, consequently the $Q_\nu^\mu(z)$ are dropped.

The outer field is periodic in ϑ, φ with a period 2π ; consequently the μ, ν are integer.

The outer field is composed as the sum of the exciting field p_e , which is a monopole field in the present case, plus the scattered field p_s :

$$\begin{aligned} p^{(I)}(r,\vartheta,\varphi) &= p_e(r) + p_s(r,\vartheta,\varphi) = \\ &= P_e\cdot h_0^{(2)}(k_0 r) + \sum_{n,m\ge 0} a_{n,m}\cdot h_n^{(2)}(k_0 r)\cdot P_n^m(\cos\vartheta)\cdot\cos(m\varphi) \end{aligned} \tag{5a}$$

$$\begin{aligned} Z_0 v_r^{(I)}(r,\vartheta,\varphi) &= \\ &= jP_e\cdot h_0'^{(2)}(k_0 r) + j\sum_{n,m\ge 0} a_{n,m}\cdot h_n'^{(2)}(k_0 r)\cdot P_n^m(\cos\vartheta)\cdot\cos(m\varphi) \end{aligned} \tag{5b}$$

(a prime indicates the derivative with respect to the argument).

To be noticed: $P_n^m(\cos\vartheta) = 0$ für $m > n$.

which follows from : $$P_n^m(x) = (-1)^m (1-x^2)^{m/2}\frac{d^m P_n(x)}{dx^m} \tag{6}$$

for m,n = integers, with Legendre polynomials of n-th degree $P_n(x)$.

The relation between the reference pressure P_e and V :

$$v_{er}^{(I)}(r_0,\vartheta,\varphi) = V = \frac{jP_e}{Z_0} h_0'^{(2)}(k_0 r) \tag{7}$$

Interior field in (II) :

From regularity at r = 0 : $\Re_\nu(k_0 r) = j_\nu(k_0 r)$ (8)

with spherical Bessel functions.

From symmetry in φ : $\Phi(\varphi) = \cos(\mu\varphi)$ (9)

Regularity at ϑ=0 again excludes the $Q_\nu^\mu(z)$.

The condition $v_\varphi \xrightarrow[\varphi\to\pm\varphi_0]{} 0$ has the consequence :

$$\sin(\mu\varphi_0) \overset{!}{=} 0 \;\Rightarrow\; \mu\varphi_0 = m\pi \;;\; m = 0,1,2,\ldots \;\Rightarrow$$
$$\mu = \frac{m\pi}{\varphi_0} \xrightarrow[\varphi_0\to\pi/4]{} 4m \tag{10}$$

The angle φ_0 is restricted henceforth to an integer part of π , $\varphi_0 = \pi/N$; N = 2,3,4,…, from which condition follows $\mu = m \cdot N$ (11)

The condition $v_\vartheta \xrightarrow[\vartheta\to\pi/2]{} 0$ implies : $\Theta'(\vartheta)\big|_{\vartheta=\pi/2} \sim P_\nu'^\mu(0) \overset{!}{=} 0$ (12)

from which with

$$P_\nu'^\mu(0) = \frac{2^{\mu+1}}{\sqrt{\pi}} \sin\big((\nu+\mu)\pi/2\big) \frac{\Gamma(\nu/2+\mu/2+1)}{\Gamma(\nu/2-\mu/2+1/2)} \tag{13}$$

follows : $\sin\big((\nu+\mu)\pi/2\big) \overset{!}{=} 0$ (14)

This is satisfied for integer μ with $\nu = \text{integer} \;;\; \mu+\nu \overset{!}{=} \text{even}$ (15)

The condition $v_\vartheta \xrightarrow[\vartheta\to 0]{} 0$ leads to $\sin\vartheta \cdot P_\nu'^\mu(\cos\vartheta)\big|_{\vartheta=0} \overset{!}{=} 0$

which is satisfied because $P_\nu'^\mu(1)$ is regular.

Finally, the wall of the cut at $\vartheta=\pi/2$, i.e. at $x = \cos\vartheta = 0$, being hard, the field in (II) must be symmetrical relative to this wall. Using :

$$P_\nu^\mu(-z) = e^{\mp j\nu\pi} P_\nu^\mu(z) - \frac{2}{\pi} e^{-j\mu\pi} \cdot \sin\big((\nu+\mu)\pi\big) \cdot Q_\nu^\mu(z)$$
$$\xrightarrow[\nu+\mu=\text{even}]{} e^{\mp j\nu\pi} P_\nu^\mu(z) \tag{16}$$

it follows : $\nu = \text{even} \rightarrow \mu = \text{even}\ .$ (17)

Consequently, the field formulation in (II) is :

$$p^{(II)}(r,\varphi,\vartheta) = \sum_{\nu,\mu} b_{\nu,\mu} \cdot j_\nu(k_0 r) \cdot P_\nu^\mu(\cos\vartheta) \cdot \cos(\mu\varphi) \quad ; \quad \mu = m \cdot N \stackrel{!}{=} \text{even} \ ; \nu \stackrel{!}{=} \text{even} \tag{18a}$$

$$Z_0 v_r^{(II)}(r,\varphi,\vartheta) = j \sum_{\nu,\mu} b_{\nu,\mu} \cdot j'_\nu(k_0 r) \cdot P_\nu^\mu(\cos\vartheta) \cdot \cos(\mu\varphi) \tag{18b}$$

O.7.2 Orthogonality :

Because of symmetry in φ only the range $0 \le \varphi \le \pi$ must be considered.

In φ *direction*:

In the outer zone (I) :

$$\frac{1}{\pi}\int_0^\pi \cos(m\varphi) \cdot \cos(m'\varphi)\, d\varphi = \frac{1}{2}\left[\frac{\sin((m+m')\pi)}{(m+m')\pi} + \frac{\sin((m-m')\pi)}{(m-m')\pi}\right] =$$
$$= \begin{cases} 1 \ ; & m = m' = 0 \\ 1/2 \ ; & m = m' \ne 0 \\ 0 \ ; & m \ne m' \end{cases} = \frac{\delta_{m,m'}}{\delta_m} \tag{19a}$$

with Kronecker's symbol $\delta_{n,m} = 0 \, ; n \ne m \ ; \ \delta_{n,m} = 1 \, ; n = m$ and Heaviside's symbol $\delta_{n=0} = 1 \ ; \ \delta_{n \ne 0} = 2$.

In the inner zone (II) , with $\mu\varphi_0 = m\pi \, ; \mu'\varphi_0 = m'\pi$ holds:

$$\frac{1}{\varphi_0}\int_0^{\varphi_0} \cos(\mu\varphi) \cdot \cos(\mu'\varphi)\, d\varphi = \frac{1}{2}\left[\frac{\sin((\mu+\mu')\varphi_0)}{(\mu+\mu')\varphi_0} + \frac{\sin((\mu-\mu')\varphi_0)}{(\mu-\mu')\varphi_0}\right]$$
$$= \frac{1}{2}\left[\frac{\sin((m+m')\pi)}{(m+m')\pi} + \frac{\sin((m-m')\pi)}{(m-m')\pi}\right] = \frac{\delta_{m,m'}}{\delta_m} \tag{19b}$$

In ϑ *direction*:

With $x = \cos\vartheta$:

$$\int_{-1}^{1} P_\nu^\mu(x) \cdot P_{\nu'}^\mu(x)\, dx = \begin{cases} 0 \ ; & \nu \ne \nu' \\ N_\nu^\mu \ ; & \nu = \nu' \end{cases} = \delta_{\nu,\nu'} N_\nu^\mu \quad ; \nu, \nu', \mu = \text{integer} \tag{20}$$

$$\int_{-1}^{1} P_\nu^\mu(x)\cdot P_\nu^{\mu'}(x)\frac{dx}{1-x^2}=\begin{cases}0 \; ; & \mu\neq\mu' \\ M_\nu^\mu \; ; & \mu=\mu'\end{cases} = \delta_{\mu,\mu'} M_\nu^\mu \quad ; \nu,\mu',\mu = \text{integer} \tag{21}$$

with the norms N_ν^μ , M_ν^μ .

Because of the symmetry in (II) relative to $\vartheta=\pi/2$, ortogonality holds also in the range ϑ = $(0, \pi/2)$ (with halved norms).

O.7.3 Remaining boundary conditions :

There are still two boundary conditions of matching the sound pressure and the radial particle velocity at the zone limits :

$$p^{(II)}(r_0,\vartheta,\varphi) \stackrel{!}{=} p^{(I)}(r_0,\vartheta,\varphi) \quad \text{in} \quad \vartheta=(0,\pi/2)\ \&\ \varphi=(-\varphi_0/2,+\varphi_0/2) \tag{22}$$

$$v_r^{(I)}(r_0,\vartheta,\varphi) \stackrel{!}{=} \begin{cases} v_r^{(II)}(r_0,\vartheta,\varphi) & \text{in} \quad \vartheta=(0,\pi/2)\ \&\ |\varphi|\le\varphi_0/2 \\ V & \text{on remainder of the sphere}\end{cases} \tag{23a}$$

or after subtraction of V on both sides :

$$v_{sr}^{(I)}(r_0,\vartheta,\varphi) \stackrel{!}{=} \begin{cases} v_r^{(II)}(r_0,\vartheta,\varphi)-V & \text{in} \quad \vartheta=(0,\pi/2)\ \&\ |\varphi|\le\varphi_0/2 \\ 0 & \text{on remainder of the sphere}\end{cases} \tag{23b}$$

Boundary condition for p :

Apply on both sides of eq.(22) the integral (with ι,κ= even, from the value range of μ,ν) :

$$\frac{1}{\varphi_0}\int_0^{\varphi_0} d\varphi \int_0^1 \ldots \cdot P_\kappa^\iota(x)\cdot\cos(\iota\varphi)\,dx \tag{24}$$

On the left-hand side :

$$\sum_{\nu,\mu}\frac{\delta_{\mu,\iota}}{\delta_\iota}\cdot\delta_{\nu,\kappa}\cdot\frac{N_\kappa^\mu}{2}\cdot b_{\nu,\mu}\cdot j_\nu(k_0 r_0) = \frac{N_\kappa^\iota}{2\delta_\iota}\cdot b_{\kappa,\iota}\cdot j_\kappa(k_0 r_0) \tag{25a}$$

on the right-hand side:

$$\sum_{n,m\ge 0}\left(\delta_{0,m}\delta_{0,n}\cdot P_e + a_{n,m}\right)\cdot h_n^{(2)}(k_0 r_0)\cdot K_{n,\kappa}^{m,\iota}\cdot I_{m,\iota} \tag{25b}$$

with the mode coupling integrals :

$$I_{m,\iota} = \frac{1}{\varphi_0} \int_0^{\varphi_0} \cos(m\varphi) \cdot \cos(\iota\varphi)\, d\varphi \tag{26}$$

$$K_{n,\kappa}^{m,\iota} = \int_0^1 P_n^m(x) \cdot P_\kappa^\iota(x)\, dx \tag{27}$$

Setting (25a) = (25b) yields a linear, inhomogeneous system of equations, in which the vector of solutions $\{b_{\kappa,\iota}, a_{n,m}\}$ are matrices, as well as the coefficient matrix of the system is composed of sub-matrices, as are the right-hand sides of the equations.

Boundary condition for v_r *:*

Apply on both sides of (23b) the integral

$$\frac{1}{\pi}\int_0^{\pi} d\varphi \int_{-1}^{1} \ldots \cdot P_k^i(x) \cdot \cos(i\varphi)\, dx \tag{28a}$$

which on the right-hand side reduces to :

$$\frac{1}{\pi}\int_0^{\varphi_0} d\varphi \int_0^1 \ldots \cdot P_k^i(x) \cdot \cos(i\varphi)\, dx \tag{28b}$$

One gets on the left-hand side :

$$j \sum_{n,m\geq 0} \frac{\delta_{m,i}}{\delta_i} \cdot \delta_{n,k} N_n^m \cdot h_n'^{(2)}(k_0 r_0) \cdot a_{n,m} = j \frac{N_k^i}{\delta_i} \cdot h_k'^{(2)}(k_0 r_0) \cdot a_{k,i} \tag{29a}$$

and on the right-hand side :

$$j \sum_{\nu,\mu} b_{\nu,\mu} \cdot j_\nu'(k_0 r_0) \cdot L_{\nu,k}^{\mu,i} \cdot J_{\mu,i} - Z_0 V \cdot Q_{i,k} \tag{29b}$$

with the integrals :

$$J_{\mu,i} = \frac{1}{\pi} \int_0^{\varphi_0} \cos(\mu\varphi) \cdot \cos(i\varphi)\, d\varphi = \frac{\varphi_0}{\pi} I_{i,\mu} \tag{30}$$

$$L_{\nu,k}^{\mu,i} = \int_0^1 P_\nu^\mu(x) \cdot P_k^i(x)\, dx = K_{k,\nu}^{i,\mu} \tag{31}$$

$$Q_{i,k} = \frac{1}{\pi} \int_0^{\varphi_0} d\varphi \int_0^1 P_k^i(x) \cdot \cos(i\varphi)\, dx \tag{32}$$

where $\mu = m \cdot N$=even , ν= ven, and i,k= 0,1,2,….

(29a) = (29b) is a second inhomogeneous, linear system system of equations for the $\{a_{k,i}$, $b_{\nu,\mu}\}$. Because the left-hand sides of both systems are diagonal, they can be reduced to one system of equations (see below).

O.7.4 Coupling integrals :

In principle, the coupling integrals can be evaluated numerically, because the integration interval is finite. But because the integrands oscillate, analytical solutions are preferable.

$$J_{\mu,i} = \frac{1}{\pi}\int_0^{\varphi_0} \cos(\mu\varphi)\cdot\cos(i\varphi)\,d\varphi = \frac{\varphi_0}{\pi} I_{i,\mu} \tag{30}$$

$$J_{\mu,i} = \frac{1}{\pi}\int_0^{\varphi_0} \cos(\mu\varphi)\cdot\cos(i\varphi)\,d\varphi = \frac{\varphi_0}{2\pi}\left[\frac{\sin((i+mN)\varphi_0)}{(i+mN)\varphi_0} + \frac{\sin((i-mN)\varphi_0)}{(i-mN)\varphi_0}\right]$$

$$\xrightarrow[i=mN=0]{} \frac{\varphi_0}{\pi} \quad ; \quad \xrightarrow[i=mN\neq 0]{} \frac{\varphi_0}{2\pi}\left[1+\frac{\sin(2mN\varphi_0)}{2mN\varphi_0}\right] = \frac{\varphi_0}{2\pi}\left[1+\frac{\sin(2m\pi)}{2m\pi}\right] = \frac{\varphi_0}{2\pi} \tag{33}$$

$$L^{\mu,i}_{\nu,k} = \int_0^1 P^{\mu}_{\nu}(x)\cdot P^{i}_{k}(x)\,dx = K^{i,\mu}_{k,\nu} \tag{31}$$

Because of symmetry :

$$L^{\mu,i}_{\nu,k} = \int_0^1 P^{\mu}_{\nu}(x)\cdot P^{i}_{k}(x)\,dx = \begin{cases} 0 & ; \quad i+k=\text{odd} \\ \dfrac{1}{2}\displaystyle\int_{-1}^{1} P^{\mu}_{\nu}(x)\cdot P^{i}_{k}(x)\,dx & ; \quad i+k=\text{even} \end{cases} \tag{34}$$

with :

$$\int_0^{\pi} P^{m}_{n}(\cos\vartheta)\cdot P^{i}_{k}(\cos\vartheta)\cdot\sin\vartheta\,d\vartheta \xrightarrow[\substack{\cos\vartheta\to x \\ -\sin\vartheta\,d\vartheta\to dx}]{} \int_{-1}^{1} P^{m}_{n}(x)\cdot P^{i}_{k}(x)\,dx = I(i,k,m,n) \tag{35a}$$

and the values for I(i,k,m,n) as given by :

$$Q_{i,k} = \frac{1}{\pi}\int_0^{\varphi_0} d\varphi \int_0^1 P^{i}_{k}(x)\cdot\cos(i\varphi)\,dx = \frac{\varphi_0}{\pi}\frac{\sin(i\varphi_0)}{i\varphi_0}\int_0^1 P^{i}_{k}(x)\,dx \xrightarrow[i=0]{} \frac{\varphi_0}{\pi}\int_0^1 P^{i}_{k}(x)\,dx \tag{36a}$$

If i+k = even :

$$\int_0^1 P_k^i(x)\,dx = \frac{1}{2}\int_{-1}^{1} P_k^i(x)\,dx = \begin{cases} \dfrac{i\cdot((k/2)!)^2(i+k)!}{k\cdot((k-i)/2)!\,((k+i)/2)!\,(k+1)!} & ; \; i=0,2,4,\ldots ; k=2,4,\ldots \\ 1 & ; \; i=k=0 \\ \dfrac{i\cdot\pi\cdot(i+k)!\,(k+1)!}{2^{2k+2}\cdot k\cdot(((k+1)/2)!)^2\,((k-i)/2)!\,((k+i)/2)!} & ; \; i,k=1,3,5,\ldots \end{cases} \tag{36b}$$

No analytical solutions were found for $i+k = \text{odd}$.

Solutions for $I(i,k,m,n)$ are :

$$I(i,k,m,n) = \begin{cases} 0 & ; \; m>n \text{ oder } i>k \\ 0 & ; \; \begin{cases} i=m\,;\,k\neq n \\ i=m+2\,; \begin{cases} n>k \\ k-m=\text{odd} \end{cases} \end{cases} \\ \dfrac{2(n+m)!}{(2n+1)(n-m)!} & ; \; i=m\,;\,k=n \\ -\dfrac{2(n+m)!}{(2n+1)(n-m-2)!} & ; \; i=m+2\,;k=n \\ \dfrac{4(m+1)(n+m)!}{(n-m)!} & ; \; i=m+2\,;k=n+2 \\ (-1)^{(3m+i)/2}\dfrac{2(n+m)!}{(2n+1)(n-i)!} & ; \; k=n\,;m+i=\text{even}\,;i\geq m \end{cases} \tag{37}$$

O.7.5 Reduction of the system of equations :

Formally one can eliminate from (25a,b) the $b_{\kappa,\iota}$ and insert them into the system (29a)=(29b). But a better convergence of the system of equations is obtained by elimination of the $a_{k,i}$ from (29a,b) and insertion into (25a)=(25b) .

$$a_{k,i} = \frac{\delta_i}{N_k^i\cdot h_k'^{(2)}(k_0r_0)}\left[\frac{\varphi_0}{\pi}\sum_{\nu,\mu} b_{\nu,\mu}\cdot j_\nu'(k_0r_0)\cdot K_{k,\nu}^{i,\mu}\cdot I_{i,\mu} + jZ_0V\cdot Q_{i,k}\right] \tag{38}$$

After insertion and rearrangement of the sums, one gets :

$$\sum_{\nu,\mu} b_{\nu,\mu}\cdot\left\{\delta_{\kappa,\nu}\delta_{\iota,\mu}\frac{N_\kappa^\iota}{2\delta_\iota}\cdot j_\kappa(k_0r_0) - \right.$$

$$\left. -\frac{\varphi_0}{\pi}j_\nu'(k_0r_0)\sum_{n,m\geq 0}\frac{\delta_m\cdot h_n^{(2)}(k_0r_0)\cdot K_{n,\kappa}^{m,\iota}\cdot K_{n,\nu}^{m,\mu}\cdot I_{m,\iota}\cdot I_{m,\mu}}{N_n^m\cdot h_n'^{(2)}(k_0r_0)}\right\} = \tag{39a}$$

$$= P_e\cdot h_0^{(2)}(k_0r_0)\cdot K_{0,\kappa}^{0,\iota}\cdot I_{0,\iota} + jZ_0V\sum_{n,m\geq 0}\frac{\delta_m\cdot h_n^{(2)}(k_0r_0)\cdot K_{n,\kappa}^{m,\iota}\cdot I_{m,\iota}}{N_n^m\cdot h_n'^{(2)}(k_0r_0)}\cdot Q_{m,n}$$

or with a unified reference amplitude V (see (7)) on the right-hand side :

$$\sum_{\nu,\mu} b_{\nu,\mu}\cdot\left\{\delta_{\kappa,\nu}\delta_{\iota,\mu}\frac{N_\kappa^\iota}{2\delta_\iota}\cdot j_\kappa(k_0r_0)-\right.$$

$$\left.-\frac{\varphi_0}{\pi}j'_\nu(k_0r_0)\sum_{n,m\ge 0}\frac{\delta_m\cdot h_n^{(2)}(k_0r_0)\cdot K_{n,\kappa}^{m,\iota}\cdot K_{n,\nu}^{m,\mu}\cdot I_{m,\iota}\cdot I_{m,\mu}}{N_n^m\cdot h_n'^{(2)}(k_0r_0)}\right\}= \qquad (39b)$$

$$=jZ_0V\left\{-\frac{h_0^{(2)}(k_0r_0)}{h_0'^{(2)}(k_0r)}\cdot K_{0,\kappa}^{0,\iota}\cdot I_{0,\iota}+\sum_{n,m\ge 0}\frac{\delta_m\cdot h_n^{(2)}(k_0r_0)\cdot K_{n,\kappa}^{m,\iota}\cdot I_{m,\iota}}{N_n^m\cdot h_n'^{(2)}(k_0r_0)}\cdot Q_{m,n}\right\}$$

After solution of this system of equations (ι,κ from the range of μ,ν) for the $b_{\nu,\mu}$, the $a_{k,i}$ are obtained by insertion in (38) . Thus the sound field in and around the Cat's Eye is known.

The formulas above are derived for the radiation problem in the simple case of a monopole excitation (m,n=0). For a multipole excitation substitute $V(\vartheta,\varphi)$=const with the velocity distribution of a multipole source. For any given distribution $V(\vartheta,\varphi)$ first expand that distribution as sum of multipole sources. Then solve the multipole radiation task for each sum term and add the results.

In a scattering task, e.g. for an incident plane wave, first expand the plane wave in spherical harmonics :

$$e^{-j\mathbf{k}\cdot\mathbf{r}}=\sum_{n=0}^{\infty}(-j)^n(2n+1)\sum_{m=0}^{\infty}\delta_m\frac{(n-m)!}{(n+m)!}\cos\big(m(\varphi-v)\big)\,P_n^m(\cos u)\,P_n^m(\cos\vartheta)\,j_n(k_0r) \qquad (40)$$

where the vector **r** shows from the origin to the field point $P=(r,\varphi,\vartheta)$, and the wave number vector **k** has the length k_0 and has the spherical angles ϑ=u , φ=v . The particle velocities of the terms at the surface $r=r_0$ are considered as given velocities of a multipole radiation task; these are solved in turn, and the fields are added.

0.8 The Orange model

For tests of some numerical methods the much simpler Orange model may be sufficient also. This model consists of a sphere with "orange slices" taken away.

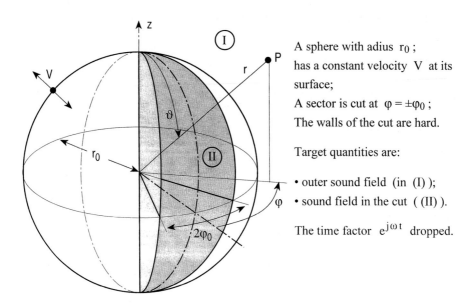

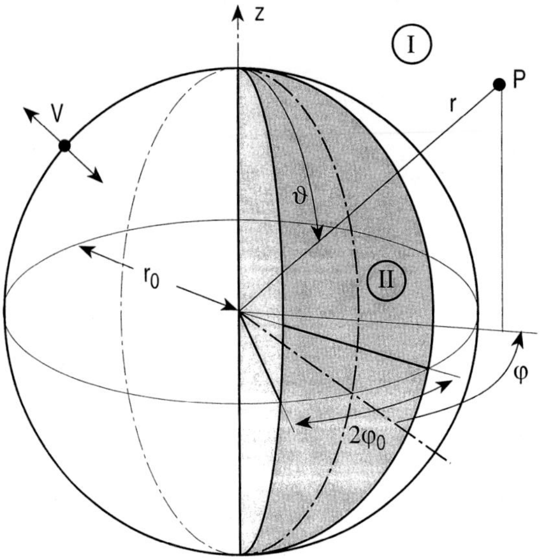

A sphere with adius r_0 ;
has a constant velocity V at its surface;
A sector is cut at $\varphi = \pm\varphi_0$;
The walls of the cut are hard.

Target quantities are:

- outer sound field (in (I));
- sound field in the cut ((II)).

The time factor $e^{j\omega t}$ dropped.

The problem is symmetrical in φ relative to $\varphi=0$, and in ϑ relative to $\vartheta=\pi/2$. Because neither the object nor the excitation (in the present case) has a variation along ϑ (except the "natural" variation in spherical co-ordinates), sectoral spherical harmonics are sufficient for the field synthesis. They are characterised by n=m in the associate Legendre functions $P_n^m(x)$.

O.8.1 Elementary solutions and field formulations :

A general elementary solution in spherical co-ordinates is :

$$\begin{aligned} p(r,\vartheta,\varphi) &= R(k_0r)\cdot\Theta(\vartheta)\cdot\Phi(\varphi) \\ R(k_0r) &= A\cdot\Re_\nu^{(1)}(k_0r)+B\cdot\Re_\nu^{(2)}(k_0r) \\ \Theta(\vartheta) &= a\cdot P_\nu^\mu(\cos\vartheta)+b\cdot Q_\nu^\mu(\cos\vartheta) \\ \Phi(\varphi) &= \alpha\cdot\sin(\mu\varphi)+\beta\cdot\cos(\mu\varphi) \end{aligned} \tag{1}$$

with: $\Re_\nu^{(i)}(k_0r)$ 2 linearly independent spherical Bessel functions;

$P_\nu^\mu(z)\,;\,Q_\nu^\mu(z)$ associated Legendre functions of 1st and 2nd kind, respectively;

From symmetry in φ follows: $\Phi(\varphi) = \cos(\mu\varphi)$ (2)

Fields in both zones (I) and (II) are regular at $\vartheta=0$ and $\vartheta=\pi$; from which follows that the $Q_\nu^\mu(z)$ are dropped.

Because of the sectoral character of the object and the excitation, $\mu=\nu$ in $P_\nu^\mu(z)$.

Outer field in (I) :

From Sommerfeld's condition follows: $\mathfrak{R}_\nu(k_0r) = h_\nu^{(2)}(k_0r)$ (3)

with spherical Hankel function of the 2nd kind.

The outer field is periodic in φ, ϑ with period 2π ; therefore $m,\nu = m,n$ are integers. The outer field is composed as the sum of the exciting field p_e , which is a monopole field in the present case, plus the scattered field p_s :

$$p^{(I)}(r,\vartheta,\varphi) = p_e(r) + p_s(r,\vartheta,\varphi) = \\ = P_e \cdot h_0^{(2)}(k_0r) + \sum_{m\geq 0} a_m \cdot h_m^{(2)}(k_0r) \cdot P_m^m(\cos\vartheta) \cdot \cos(m\varphi) \tag{4a}$$

$$Z_0 v_r^{(I)}(r,\vartheta,\varphi) = \\ = jP_e \cdot h_0^{\prime(2)}(k_0r) + j\sum_{m\geq 0} a_m \cdot h_m^{\prime(2)}(k_0r) \cdot P_m^m(\cos\vartheta) \cdot \cos(m\varphi) \tag{4b}$$

The symmetry at $\vartheta=\pi/2$ is given by formulation. The relation between the reference pressure P_e and V is :

$$v_{er}^{(I)}(r_0,\vartheta,\varphi) = V = \frac{jP_e}{Z_0} h_0^{\prime(2)}(k_0r) \tag{5}$$

Interior field in (II) :

From regularity at $r=0$ follows: $\mathfrak{R}_\nu(k_0r) = j_\nu(k_0r)$ (6)

with spherical Bessel functions.

From symmetry in φ follows: $\Phi(\varphi) = \cos(\mu\varphi)$ (7)

Regularity at $\vartheta=0$ again excludes the $Q_\nu^\mu(z)$.

The condition $v_\varphi \xrightarrow[\varphi\to\pm\varphi_0]{} 0$ has the consequence :

$$\sin(\mu\varphi_0) \stackrel{!}{=} 0 \;\Rightarrow\; \mu\varphi_0 = m\pi \;\; ; \;\; m = 0,1,2,\ldots \;\Rightarrow\; \mu = \frac{m\pi}{\varphi_0} \xrightarrow[\varphi_0\to\pi/4]{} 4m \tag{8}$$

The angle φ_0 is restricted henceforth to an integer part of π , $\varphi_0 = \pi/N$; $N = 2,3,4,\ldots$, from which condition $\mu = m \cdot N$ (9)

The use of sectoral harmonics with $\mu=\nu$ in $P_\nu^\mu(z)$ satifies the condition of symmetry at $\vartheta=0$.

Thus the interior field can be formulated as :

$$p^{(II)}(r,\varphi,\vartheta)=\sum_\mu b_\mu\cdot j_\mu(k_0 r)\cdot P_\mu^\mu(\cos\vartheta)\cdot\cos(\mu\varphi) \tag{10a}$$

$$Z_0 v_r^{(II)}(r,\varphi,\vartheta)=j\sum_\mu b_\mu\cdot j'_\mu(k_0 r)\cdot P_\mu^\mu(\cos\vartheta)\cdot\cos(\mu\varphi) \tag{10b}$$

O.8.2 Orthogonality :

Because of symmetry in φ only the range $0\le\varphi\le\pi$ must be considered. Also the range $0\le\vartheta\le\pi/2$ would be sufficient in the cinsiderations; but one finds more integrals in the literature for the range $0\le\vartheta\le\pi$.

In φ direction :

In the outer zone (I) :

$$\frac{1}{\pi}\int_0^\pi\cos(m\varphi)\cdot\cos(m'\varphi)\,d\varphi=\frac{1}{2}\left[\frac{\sin((m+m')\pi)}{(m+m')\pi}+\frac{\sin((m-m')\pi)}{(m-m')\pi}\right]=\frac{\delta_{m,m'}}{\delta_m} \tag{11a}$$

with Kronecker's symbol $\delta_{n,m}=0\,;\,n\ne m$; $\delta_{n,m}=1\,;\,n=m$ and Heaviside's symbol $\delta_{n=0}=1$; $\delta_{n\ne0}=2$.

In the inner zone (II) , with $\mu\varphi_0=m\pi\,;\,\mu'\varphi_0=m'\pi$:

$$\frac{1}{\varphi_0}\int_0^{\varphi_0}\cos(\mu\varphi)\cdot\cos(\mu'\varphi)\,d\varphi=\frac{1}{2}\left[\frac{\sin((\mu+\mu')\varphi_0)}{(\mu+\mu')\varphi_0}+\frac{\sin((\mu-\mu')\varphi_0)}{(\mu-\mu')\varphi_0}\right]=\frac{\delta_{m,m'}}{\delta_m} \tag{11b}$$

In ϑ direction : With $x=\cos\vartheta$:

$$\int_{-1}^{1}P_n^n(x)\cdot P_{n'}^{n'}(x)\,dx=\begin{cases}0\;; & n\ne n'\\ N_n\;; & n=n'\end{cases}=\delta_{n,n'}N_n\quad;n,n'=\text{integer}$$

$$N_n=\frac{2(2n)!}{2n+1} \tag{12}$$

O.8.3 Boundary conditions at the zone limits :

There are still two boundary conditions of matching the sound pressure and the radial particle velocity at the zone limits :

$$p^{(II)}(r_0,\vartheta,\varphi)\overset{!}{=}p^{(I)}(r_0,\vartheta,\varphi)\quad\text{in}\quad\vartheta=(0,\pi)\;\&\;\varphi=(-\varphi_0/2,+\varphi_0/2) \tag{13}$$

$$v_r^{(I)}(r_0,\vartheta,\varphi) \stackrel{!}{=} \begin{cases} v_r^{(II)}(r_0,\vartheta,\varphi) & \text{in } \vartheta=(0,\pi)\ \&\ |\varphi|\le\varphi_0/2 \\ V & \text{in remainder of the sphere} \end{cases} \tag{14a}$$

or, after subtraction of V on both sides :

$$v_{sr}^{(I)}(r_0,\vartheta,\varphi) \stackrel{!}{=} \begin{cases} v_r^{(II)}(r_0,\vartheta,\varphi)-V & \text{in } \vartheta=(0,\pi)\ \&\ |\varphi|\le\varphi_0/2 \\ 0 & \text{in remainder of the sphere} \end{cases} \tag{14b}$$

In (13) one applies the orthogonality in (II) , and in (14b) the orthogonality in (I) .

Boundary condition for p :

Apply on both pressure distributions

$$p^{(II)}(r,\varphi,\vartheta)=\sum_\mu b_\mu\cdot j_\mu(k_0r)\cdot P_\mu^\mu(\cos\vartheta)\cdot\cos(\mu\varphi)$$

$$\begin{aligned} p^{(I)}(r,\vartheta,\varphi) &= p_e(r)+p_s(r,\vartheta,\varphi)= \\ &= P_e\cdot h_0^{(2)}(k_0r)+\sum_{m\ge0} a_m\cdot h_m^{(2)}(k_0r)\cdot P_m^m(\cos\vartheta)\cdot\cos(m\varphi) \\ &= \sum_{m\ge0}\left(\delta_{0,m}\cdot P_e+a_m\right)\cdot h_m^{(2)}(k_0r)\cdot P_m^m(\cos\vartheta)\cdot\cos(m\varphi) \end{aligned}$$

(with $P_0^0(x)=1$) the integral (with ι in the value range of μ) :

$$\frac{1}{\varphi_0}\int_0^{\varphi_0} d\varphi\int_{-1}^{1}\ldots\cdot P_\iota^\iota(x)\cdot\cos(\iota\varphi)\,dx \tag{15}$$

giving on the left-hand side of the condition (13) :

$$\sum_\mu\frac{\delta_{\mu,\iota}}{\delta_\iota}\cdot\frac{N_\iota}{2}\cdot b_\mu\cdot j_\mu(k_0r_0)=\frac{N_\iota}{2\delta_\iota}\cdot b_\iota\cdot j_\iota(k_0r_0) \tag{16a}$$

and on the right-hand side :

$$\sum_{m\ge0}\left(\delta_{0,m}\cdot P_e+a_m\right)\cdot h_m^{(2)}(k_0r_0)\cdot K_{m,\iota}\cdot I_{m,\iota} \tag{16b}$$

with the mode coupling integrals:

$$I_{m,\iota}=\frac{1}{\varphi_0}\int_0^{\varphi_0}\cos(m\varphi)\cdot\cos(\iota\varphi)\,d\varphi \tag{17}$$

$$K_{m,\iota}=\int_{-1}^{1}P_m^m(x)\cdot P_\iota^\iota(x)\,dx \tag{18}$$

(16a) = (16b) is a linear, inhomogeneous system of equations for the vector of solutions $\{b_\iota, a_m\}$.

Boundary condition for v_r :

Apply on both particle velocity distributions

$$Z_0 v_{sr}^{(I)}(r,\vartheta,\varphi) = j\sum_{m\geq 0} a_m \cdot h_m'^{(2)}(k_0 r)\cdot P_m^m(\cos\vartheta)\cdot\cos(m\varphi)$$

$$Z_0 v_r^{(II)}(r,\varphi,\vartheta) = j\sum_{\mu} b_\mu \cdot j_\mu'(k_0 r)\cdot P_\mu^\mu(\cos\vartheta)\cdot\cos(\mu\varphi)$$

the integral (with i= integer, out of the value range of m) :

$$\frac{1}{\pi}\int_0^\pi d\varphi\int_{-1}^{1}\ldots\cdot P_i^i(x)\cdot\cos(i\varphi)\,dx \tag{19a}$$

which on the right-hand side of (14b) becomes the integral

$$\frac{1}{\pi}\int_0^{\varphi_0} d\varphi\int_{-1}^{1}\ldots\cdot P_i^i(x)\cdot\cos(i\varphi)\,dx \tag{19b}$$

and get on the left-hand side :

$$j\sum_{m\geq 0}\frac{\delta_{m,i}}{\delta_i}N_m\cdot h_m'^{(2)}(k_0 r_0)\cdot a_m = j\frac{N_i}{\delta_i}\cdot h_i'^{(2)}(k_0 r_0)\cdot a_i \tag{20a}$$

and on the right-hand side :

$$j\sum_{\mu} b_\mu\cdot j_\mu'(k_0 r_0)\cdot L_{\mu,i}\cdot J_{\mu,i} - Z_0 V\cdot Q_i \tag{20b}$$

with the integrals :

$$J_{\mu,i} = \frac{1}{\pi}\int_0^{\varphi_0}\cos(\mu\varphi)\cdot\cos(i\varphi)\,d\varphi = \frac{\varphi_0}{\pi}I_{i,\mu} \tag{21}$$

$$L_{\mu,i} = \int_{-1}^{1} P_\mu^\mu(x)\cdot P_i^i(x)\,dx = K_{i,\mu} \tag{22}$$

$$Q_i = \frac{1}{\pi}\int_0^{\varphi_0} d\varphi\int_{-1}^{1} P_i^i(x)\cdot\cos(i\varphi)\,dx \tag{23}$$

(20a) = (20b) is a second inhomogeneous, linear system of equations for $\{a_i , b_\mu\}$.

O.8.4 Integrals :

$$J_{\mu,i} = \frac{1}{\pi}\int_0^{\varphi_0} \cos(\mu\varphi)\cdot\cos(i\varphi)\, d\varphi = \frac{\varphi_0}{\pi} I_{i,\mu} \tag{24}$$

$$J_{\mu,i} = \frac{1}{\pi}\int_0^{\varphi_0} \cos(\mu\varphi)\cdot\cos(i\varphi)\, d\varphi = \frac{\varphi_0}{2\pi}\left[\frac{\sin((i+mN)\varphi_0)}{(i+mN)\varphi_0} + \frac{\sin((i-mN)\varphi_0)}{(i-mN)\varphi_0}\right]$$
$$\xrightarrow[i=mN=0]{} \frac{\varphi_0}{\pi} \quad ; \quad \xrightarrow[i=mN\neq 0]{} \frac{\varphi_0}{2\pi}\left[1+\frac{\sin(2mN\varphi_0)}{2mN\varphi_0}\right] = \frac{\varphi_0}{2\pi}\left[1+\frac{\sin(2m\pi)}{2m\pi}\right] = \frac{\varphi_0}{2\pi} \tag{25}$$

$$L_{\mu,i} = \int_{-1}^{1} P_\mu^\mu(x)\cdot P_i^i(x)\, dx = K_{i,\mu} \tag{26}$$

$$L_{\mu,i} = K_{i,\mu} = \int_{-1}^{1} P_\mu^\mu(x)\cdot P_i^i(x)\, dx = \begin{cases} 0 \;;\; i \neq \mu \\ \dfrac{2(2i)!}{2i+1} \;;\; i=\mu \end{cases} = \delta_{i,\mu} N_i \tag{27}$$

$$Q_i = \frac{1}{\pi}\int_0^{\varphi_0} d\varphi \int_{-1}^{1} P_i^i(x)\cdot\cos(i\varphi)\, dx = \frac{\varphi_0}{\pi}\frac{\sin(i\varphi_0)}{i\varphi_0}\int_{-1}^{1} P_i^i(x)\, dx \xrightarrow[i=0]{} \frac{\varphi_0}{\pi}\int_{-1}^{1} P_i^i(x)\, dx \tag{28}$$

$$\int_{-1}^{1} P_i^i(x)\, dx = \begin{cases} \dfrac{2i\cdot((i/2)!)^2(2i)!}{i\cdot i!\cdot(i+1)!} \;;\; i=0,2,4,\ldots \\ 2 \;;\; i=0 \\ \dfrac{i\cdot\pi\cdot(2i)!\,(i+1)!}{2^{2i+1}\cdot i\cdot i!\cdot(((i+1)/2)!)^2} \;;\; i=1,3,5,\ldots \end{cases} \tag{29}$$

For integer m,n :

$$P_n^m(x) = (-1)^m (1-x^2)^{m/2}\frac{d^m P_n(x)}{dx^m} \tag{30a}$$

and therefore

$$P_m^m(x) = (-1)^m (1-x^2)^{m/2}\frac{d^m P_m(x)}{dx^m} \tag{30b}$$

The Legendre polynomials $P_m(x)$ are polynomes of m-th degree; thus the derivative in (30b) is a constant.

$$P_m(x)=\sum_{n=0}^{[m/2]} ß_n\cdot x^{m-2n}=\frac{1}{2^m}\sum_{n=0}^{[m/2]}\frac{(-1)^n(2m-2n)!}{n!\cdot(m-n)!\cdot(m-2n)!}x^{m-2n} \tag{31}$$

with $[m/2]$ = highest integer $\leq m/2$.

$$\text{Whence : }\quad \frac{d^m P_m(x)}{dx^m}=ß_0\frac{d^m x^m}{dx^m}=ß_0\cdot m!=\frac{1}{2^m}\frac{(2m)!}{(m)!} \tag{32}$$

$$\text{and : }\quad P_m^m(x)=\frac{(-1)^m(2m)!}{2^m(m)!}(1-x^2)^{m/2}\xrightarrow[x=\cos\vartheta]{}\frac{(-1)^m(2m)!}{2^m(m)!}\sin^m\vartheta \tag{33}$$

Above integrals can be readily evaluated with these expressions :

$$\int_{-1}^{1} P_i^i(x)\,dx=\sqrt{\pi}\frac{(-1)^i\cdot(2i)!\cdot\Gamma(1+i/2)}{2^i\cdot i!\cdot\Gamma((3+i)/2)} \tag{34}$$

$$L_{\mu,i}=K_{i,\mu}=\int_{-1}^{1} P_m^m(x)\cdot P_i^i(x)\,dx=\sqrt{\pi}\frac{(-1)^{i+m}\cdot(2i)!\cdot(2m)!\cdot\Gamma((2+i+m)/2)}{2^{i+m}\cdot i!\cdot m!\cdot\Gamma((3+i+m)/2)} \tag{35}$$

O.8.5 Reduction of the system of equations :

We eliminate from (20a)=(20b) the a_i :

$$a_i=\frac{\varphi_0}{\pi}\frac{\delta_i}{N_i\cdot h_i'^{(2)}(k_0r_0)}\sum_\mu b_\mu\cdot j_\mu'(k_0r_0)\cdot K_{i,\mu}\cdot I_{i,\mu}+jZ_0V\cdot\frac{\delta_i}{N_i\cdot h_i'^{(2)}(k_0r_0)}Q_i \tag{36}$$

and insert them in (16a)=(16b), getting after rearrangement :

$$\sum_\mu b_\mu\cdot\left\{\frac{N_\iota\delta_{\iota,\mu}}{2\delta_\iota}\cdot j_\iota(k_0r_0)-\right.$$
$$\left.-\frac{\varphi_0}{\pi}j_\mu'(k_0r_0)\sum_{m\geq 0}\frac{\delta_m\cdot h_m^{(2)}(k_0r_0)\cdot K_{m,\iota}\cdot K_{m,\mu}\cdot I_{m,\iota}\cdot I_{m,\mu}}{N_m\cdot h_m'^{(2)}(k_0r_0)}\right\}= \tag{37}$$
$$=P_e\cdot h_0^{(2)}(k_0r_0)\cdot K_{0,\iota}\cdot I_{0,\iota}+jZ_0V\cdot\sum_{m\geq 0}\frac{\delta_m\cdot h_m^{(2)}(k_0r_0)Q_m\cdot K_{m,\iota}\cdot I_{m,\iota}}{N_m\cdot h_m'^{(2)}(k_0r_0)}$$

The right-hand side of this system of equations for the b_μ can be normalised either to P_e or to V with (5) . After solution for the b_μ , insertion of them in (36) delivers the a_i ; and the sound field is known.

References to part

O. Analytical and Numerical Methods in Acoustics

Section O.3 :

[1] JUNGER, M. C., FEIT, D.
"Sound, Structures, and their Interaction"
The Massachusetts Institute of Technology Press, Cambridge, 1972.

[2] KOOPMANN, G. H., PERRAUD, J. C.
"Crack location by means of surface acoustic intensity measurements",
ASME publication for presentation at the Winter Annual Meeting, Washington, 1981.

[3] URICK, R. J.
"Principles of Underwater Sound"
McGraw-Hill, New York, 3rd ed., 1983.

[4] SEYBERT, A. F., CHENG, C. Y., WU, T. W. J.Acoust.Soc.Amer. 88 (1990) 1612-1618
"The solution of coupled interior/exterior acoustic problems using the boundary element method"

[5] SOIZE, C. J.Acoust.Soc.Amer. 103 (1998) 3393-3406
"Reduced models in the medium-frequency range for general external structural-acoustic systems"

[6] SOIZE, C. J.Acoust.Soc.Amer. 106 (1999) 3362-3374
"Reduced models for structures in the medium-frequency range coupled with internal acoustic cavities"

[7] COLTON, D., KRESS, R.
"Integral Equations in Scattering Theory"
Wiley-Interscience Publication, New York, 1983.

Section O.4 :

[1] OCHMANN, M. Acustica 81 (1995) 512-527
"The source simulation technique for acoustic radiation problems"

[2] OCHMANN, M.
"Source simulation techniques for solving boundary element problems",
in: Boundary Elements in Acoustics,Advances and Applications in Boundary Element, ed. by Otto von Estorff, Chapter 7,
WIT Press, Computat. Mechan. Publ., Southampton, Boston 2000.

[3] KRESS, R. AND MOHSEN, A. Math.Meth.in Appl.Sci.., 8 (1986) 585-597
"On the simulation source technique for exterior problems in acoustics"

[4] OCHMANN, M. J.Acoust.Soc.Am. 105 (1999) 2574-2584
"The full-field equations for acoustic radiation and scattering"

[5] OCHMANN, M. Acustica 72 (1990) 233-246
"Multipole radiator synthesis - an effective method for calculating the radiated sound field of vibrating structures of arbitrary surface configuration", in German.

[6] OCHMANN, M., WELLNER, F. Rev.Franc.Mécanique, Numéro spécial 1991, pp. 457–471
"Calculation of sound radiation from complex machine structures using the multipole radiator synthesis and the boundary element multigrid method"

[7] KOOPMANN, G., SONG, L., FAHNLINE, J. J.Acoust.Soc.Amer. 86 (1989) 2433-2438
"A method for computing acoustic fields based on the principle of wave superposition"

[8] FAHNLINE, J. B., KOOPMANN, G. H. J.Acoust.Soc.Amer. 90 (1991) 2808-2819
"A numerical solution for the general radiation problem based on the combined methods of superposition and singular-value decomposition"

[9] JEANS, R., MATHEWS, I. C. J.Acoust.Soc.Amer. 92 (1992) 1156-1166
"The wave superposition method as a robust technique for computing acoustic fields"

[10] OCHMANN, M.
"The multipole method: Contributions of Manfred Heckl and new developments for high-frequency acoustic scattering"
Proc. 16th ICA and 135th meeting of the ASA, 1998, pp. 1223-1224.

[11] OCHMANN, M., HOMM, A.
"Calculation of the acoustic radiation from a propeller using the multipole method", Proc. 3rd Int. Congress on Air- and Structure-Borne Sound and Vibration, Vol. 2, International Scientific Pub., Auburn, pp. 1215-1220, 1994.

[12] ATTALA, N., WINCKELMANS, G., SGARD, F. Acust / Acta Acust 85 (1999) 47-53
"A multiple multipole expansion approach for predicting the sound power of vibrating structures"

[13] HECKL, M. Acustica 68 (1989) 251-257
"Remarks on the calculation of sound radiation using the method of spherical wave synthesis", in German.

[14] CREMER, L., WANG, M. Acustica 65 (1988) 53-74
"Synthesis of spherical wave fields to generate the sound radiated from bodies of arbitrary shape, its realisation by calculation and experiment", in German.

[15] MASSON, P., REDON, E., PRIOU, J.-P., GERVAIS, Y.
"The application of the Trefftz Method to acoustics"
Proc. 3rd Int. Congr. on Air- and Structure-Borne Sound and Vibration, Vol. 3, International Scientific Pub., Auburn, pp. 1809-1816, 1994.

[16] BOBROVNITSKII, YU. I., TOMILINA, T. M. Acoust. Physics 41 (1995) 649-660
"General properties and fundamental errors of the method of equivalent sources"

[17] JOHNSON, M. E., ELLIOTT, S. J., BAEK, K-H., GARCIA-BONITO, J.
"An equivalent source technique for calculating the sound field inside an enclosure containing scattering objects" J.Acoust.Soc.Amer. 104 (1998) 1221-1231

[18] KARAGEORGHIS, A., FAIRWEATHER, G. J.Acoust.Soc.Amer. 104 (1998) 3212-3218
"The method of fundamental solutions for axisymmetric acoustic scattering and radiation problems"

[19] BOBROVNITSKII, YU. I., TOMILINA, T. M. Sov.Phys.Acoust. 36 (1990) 334-338
"Calculation of radiation from finite elastic bodies by the method of auxiliary sources"

[20] HWANG, J., CHANG, S. J.Acoust.Soc.Amer. 90 (1991) 1167-1179
"A retracted boundary integral equation for exterior acoustic problem with unique solution for all wave numbers"

[21] CUNEFARE, A. K., KOOPMANN, G. H., BROD, K. J.Acoust.Soc.Amer. 85 (1989) 39-48
"A boundary element method for acoustic radiation valid for all wavenumbers"

[22] STEPANISHEN, P. R. J.Acoust.Soc.Am. 101 (1997) 3270-3277
"A generalized internal source density method for the forward and backward projection of harmonic pressure fields from complex bodies"

[23] KROPP, W., SVENSSON, U. Acustica 81 (1995) 528-543
"Application of the time domain formulation of the method of equivalent sources to radiation and scattering problems"

[24] GUYADER, J.L.
"Methods to reduce computing time in structural acoustic prediction"
Proc. 3rd Int. Congr. on Air- and Structure-Borne Sound and Vibration, Vol. 1, International Scientific Pub., Auburn, pp. 5-20, 1994.

[25] COLTON, D., KRESS, R.
"Integral Equations in Scattering Theory"
Wiley-Interscience Publication, New York, 1983.

[26] MORSE, P.M., FESHBACH, H.
"Methods of Theoretical Physics", Part I
McGraw-Hill, New York, 1953.

[27] ABRAMOWITZ, M., STEGUN, I. A.
"Handbook of Mathematical Functions", 9th Printing
Dover Publications, Inc., New York, 1972.

[28] HACKMAN, R. H. J.Acoust.Soc.Amer. 75 (1984) 35-45
"The transition matrix for acoustic and elastic wave scattering in prolate spheroidal coordinates"

[29] HIGGINS, J. R.
"Completeness and Basis Properties of Sets of Special Functions"
Cambridge University Press, Cambridge, 1977.

[30] VEKUA, N. P. Dokl.Akad.Nauk SSSR 90 (1953) 715-718
"On the completeness of the system of metaharmonic functions" (in Russian).

[31] KLEINMAN, R. E., ROACH, G. F.,STRÖM, S. E. G.
Proc.Roy.Soc.Lond., A 394 (1984) 121-136
"The null field method and modified Green functions"

[32] MARTIN, P. A. Wave Motion 4 (1982) 391-408
"Acoustic scattering and radiation problems, and the null-field method"

[33] STUPFEL, B., LAVIE, A.,DECARPIGNY, J. N. J.Acoust.Soc.Amer. 83 (1988) 927-941
"Combined integral formulation and null-field method for the exterior acoustic problem"

[34] WATERMAN, P. C. J.Acoust.Soc.Amer. 45 (1969) 1417-1429
"New formulation of acoustic scattering"

[35] SCHENCK, H. A. J.Acoust.Soc.Amer. 44 (1968) 41-58
"Improved integral formulation for acoustic radiation problems"

[36] WU, S. F., YU, J. J.Acoust.Soc.Amer. 104 (1998) 2054-2060
"Reconstructing interior acoustic pressure fields via Helmholtz equation least-squares method"

[37] MILLAR, R. F. IMA J.Appl.Math. 30 (1983) 27-37
"On the completeness of sets of solutions to the Helmholtz equation"

[38] OCHMANN, M.
"Calculation of sound radiation from complex structures using the multipole radiator synthesis with optimized source locations"
Proc. 2nd Int. Congr. on Recent Developm. in Air- and Struct.-Borne Sound and Vibration, International Scientific Pub., Auburn, pp. 1887-1194, 1992.

[39] MAKAROV, A. N., OCHMANN, M. J.Acoust.Soc.Am. 103 (1998) 742-750
"An iterative solver of the Helmholtz integral equation for high-frequency acoustic scattering"

[40] PRESS, W. H., FLANNERY, B. P., TEUKOLSKY, S. A., VETTERLING, W. T.
"Numerical Recipes, the Art of Scientific Computing"
Cambridge University Press, Cambridge, 2nd ed., 1990.

[41] PÄRT-ENANDER, E., SJÖBERG, A., MELIN, B., ISAKSSON, P.
"The MATLAB Handbook", Addison-Wesley, Harlow, 1996.

[42] TOBOCMAN, W. J.Acoust.Soc.Amer. 77 (1985) 369-374
"Comparison of the T-matrix and Helmholtz integral equation methods for wave scattering calculations"

[43] HOMM, A., SCHNEIDER, H-G.
"Application of BE and coupled FE/BE in underwater acoustics",
in: Boundary Elements in Acoustics,Advances and Applications in Boundary Element (ed. by Otto von Estorff), Chapter 13, WIT Press, Computational Mechanics Publications, Southampton, Boston 2000.

[44] OCHMANN, M. VDI-Berichte No. 816 (1990) 801-810
"Calculation of sound radiation from complex machine structures with application to gearboxes", in German.

[45] Ochmann, M., Heckl, M.
"Numerical methods in technical acoustics", Chapter 3,
"Taschenbuch der Technischen Akustik", eds. M. Heckl and H.A. Müller, 2nd Printing, Springer, Berlin, Heidelberg, New York, 1994, in German.

[46] OCHMANN, M., HOMM, A. Acust / Acta Acust 82 (1996, suppl.1) S159
"The source simulation technique for acoustic scattering"

[47] HOMM, A., OCHMANN, M.
"Sound scattering of a rigid test cylinder using the source simulation technique for numerical calculations"
Proc. 4th Int. Congress on Sound and Vibration,
International Scientific Pub., Auburn, pp. 133-138, 1996.

[48] OCHMANN, M., HOMM, A.
"Calculation of sound scattering from bodies with variable surface impedance and multiple reflections"
Proc. 23rd German Acoustics DAGA Conference, Fortschritte der Akustik, DEGA, Oldenburg, pp. 167-168, 1997, in German.

[49] EREZ, E., LEVIATAN, Y. J.Acoust.Soc.Amer. 93 (1993) 3027-3031
"Analysis of scattering from structures containing a variety of length scales using a source-model technique"

Section O.5 :

[1] SCHENCK, H. A. J.Acoust.Soc.Amer. 44 (1968) 41-58
"Improved integral formulation for acoustic radiation problems"

[2] COLTON, D., KRESS, R.
"Integral Equations in Scattering Theory"
Wiley-Interscience Publication, New York, 1983.

[3] MAKAROV, A. N., OCHMANN, M. J.Acoust.Soc.Amer. 103 (1998) 742-750
"An iterative solver of the Helmholtz integral equation for high-frequency acoustic scattering"

[4] OCHMANN, M., WELLNER, F. Rev.Franc.Mécan., num.spec. (1991) 457–471
"Calculation of sound radiation from complex machine structures using the multipole radiator synthesis and the boundary element multigrid method"

[5] SEYBERT, A. F., SOENARKO B., RIZZO, F. J., SHIPPY, D. J.
J.Acoust.Soc.Amer. 77 (1985) 362-368
"An advanced computational method for radiation and scattering of acoustic waves in three dimensions"

[6] PETER, J.
"Iterative solution of the direct collocation BEM equations"
Proc. 7th Int. Congr. Sound and Vibration, Garmisch-Partenkirchen, pp. 2077-2084, July 2000.

[7] COLTON, D., KRESS, R.
"Inverse Acoustic and Electromagnetic Scattering Theory"
Springer, Berlin, Heidelberg, New York, 1992.

[8] ANGELL, T. S., KLEINMAN, R. E. Math.Meth.in Appl.Sci. 4 (1982) 164-193
"Boundary integral equations for the Helmholtz equation: The third boundary value problem"

[9] KLEINMAN, R. E. , ROACH, G. F.
"Boundary integral equations for the three-dimensional Helmholtz equation"
Siam Review, 16, pp. 214-236, 1974.

[10] RAO, S. M., RAJU, P. K. J.Acoust.Soc.Amer. 86 (1989) 1143-1148
"Application of the method of moments to acoustic scattering from multiple bodies of arbitrary shape"

[11] CHEN, L. H., SCHWEIKERT, D. G. J.Acoust.Soc.Amer. 35 (1963) 1626-1632
"Sound radiation from an arbitrary body"

[12] SCHWARZ, H. R.
"Methode der finiten Elemente"
Teubner, Stuttgart, 1980.

[13] MARBURG, S. Forthcoming in Journ.Computation Acoust
"Six elements per wavelength – is that enough?"

[14] PRESS, W. H., FLANNERY, B. P., TEUKOLSKY, S. A., VETTERLING, W. T.
"Numerical recipes, the art of scientific computing"
Cambridge University Press, Cambridge, 1990.

[15] STUMMEL, F., HAINER, K.
"Praktische Mathematik"
Teubner, Stuttgart, 1971.

[16] AKYOL, T. P. Acustica 61 (1986) 200-212
"Schallabstrahlung von Rotationskörpern"

[17] SEYBERT, A. F., SOENARKO B., RIZZO, F. J., SHIPPY, D. J. J.Acoust.Soc.Amer. 80 (1986) 1241-1247
"A special integral equation formulation for acoustic radiation and scattering for axisymmetric bodies and boundary conditions"

[18] CHERTOCK, G. Quart.Appl.Math. 26 (1968) 268-272
"Convergence of iterative solutions to integral equations for sound radiation"

[19] KLEINMAN, R. E., WENDLAND, W. L. J.Math.Anal.Appl. 57 (1977) 170-202
"On Neumann's method for the exterior Neumann problem for the Helmholtz equation"

[20] KLEINMAN, R. E., ROACH, G. F. Proc.Roy.Soc.Lond., A 417 (1988) 45-57
"Iterative solutions of boundary integral equations in acoustics"

[21] OCHMANN, M., HOMM, A., SEMENOV, S., MAKAROV, S.
"An iterative GMRES-based solver for acoustic scattering from cylinder-like structures" ; will appear in: Proc. 8th Int. Congress on Sound and Vibration (ICSV8), Hong Kong, China, July 2001.

[22] MAKAROV, S., OCHMANN, M., LUDWIG, R.
"Comparison of GMRES and CG iterations on the normal form of magnetic field integral equation", submitted.

[23] MAKAROV, S., OCHMANN, M., LUDWIG, R.
"An iterative solution for magnetic field integral equation", submitted.

[24] KUPRADSE, W. D.
"Randwertaufgaben der Schwingungstheorie und Integralgleichungen"
VEB Deutscher Verlag der Wissenschaften, Berlin, 1956.

[25] SMIRNOV, W. I.
"Lehrgang der Höheren Mathematik", Teil IV,
VEB Deutscher Verlag der Wissenschaften, Berlin, 1977.

[26] COPLEY, L. G. J.Acoust.Soc.Amer. 44 (1968) 28-32
"Fundamental results concerning integral representations in acoustic radiation"

[27] ROSEN, E. M., CANNING, X. C., COUCHMAN, L. S. J.Acoust.Soc.Amer. 98 (1995) 599-610
"A sparse integral equation method for acoustic scattering"

[28] STUPFEL, B., LAVIE, A., DECARPIGNY, J,. N. J.Acoust.Soc.Amer. 83 (1988) 927-941
"Combined integral equation formulation and null-field method for the exterior acoustic problem"

[29] JONES, D. S. Q.J. Mech.Appl.Math. 27 (1974) 129-142
"Integral equation for the exterior acoustic problem"

[30] MARTIN, P. A. Q. J. Mech.Appl.Math. 33 (1980) 385-396
"On the null field equations for the exterior problems of acoustics"

[31] COLTON, D., KRESS, R. Q. J. Mech.Appl.Math. 36 (1983) 87-95
"The unique solvability of the null field equations of acoustics"

[32] BURTON, A. J., MILLER, G. F. Proc.Roy.Soc.Lond., A 323 (1971) 201-210
"The application of integral equation methods to the numerical solution of some exterior boundary-value problems"

[33] KRESS, R., SPASSOW, W. Numer. Math., 42 (1983) 77-95
"On the condition number of boundary integral operators for the exterior Dirichlet problem for the Helmholtz equation"

[34] URSELL, F. Proc.Camb.Phil.Soc., 74 (1973) 117-125
"On the exterior problems of acoustics"

[35] URSELL, F. Proc.Camb.Phil.Soc., 84 (1978) 545-548
"On the exterior problems of acoustics: II"

[36] KLEINMAN, R. E., ROACH, G. F., SCHUETZ, L. S., SHIRRON, J.
J.Acoust.Soc.Amer. 84 (1988) 385-391
"An iterative solution to acoustic scattering by rigid objects"

[37] KLEINMAN, R. E., ROACH, G. F. Proc.Roy.Soc.Lond., A 383 (1982) 313-323
"On modified Green functions in exterior problems for the Helmholtz equation"

[38] EVERSTINE, G. C., HENDERSON, F. M. J.Acoust.Soc.Amer. 87 (1990) 1938-1947
"Coupled finite element/boundary element approach for fluid-structure interaction"

[39] KOOPMANN, G. H., BENNER, H. J.Acoust.Soc.Amer. 71 (1982) 78-89
"Method for computing the sound power of machines based on the Helmholtz integral"

[40] SKUDRZYK, E.
"The foundations of acoustics"
Springer, Wien, New York, 1971.

[41] SEYBERT, A. F., CHENG C. Y. R. J. Vib. Ac. Stress Rel. Design, 109, pp. 15-21, 1987
"Applications of the boundary element method to acoustic cavity response and muffler analysis"

[42] SEYBERT, A. F., SOENARKO, B.
"Radiation and scattering of acoustic waves from bodies of arbitrary shape in a three-dimensional half space", ASME Trans. J. Vib. Acoust. Stress Reliab. Design., 110, pp. 112-117, 1988.

[43] MECHEL, F. P.
"Schallabsorber, Band I: Äußere Schallfelder - Wechselwirkungen"
Hirzel, Stuttgart, 1989.

[44] SEYBERT, A. F., WU, T. W. J.Acoust.Soc.Amer. 85 (1989) 19-23
"Modified Helmholtz integral equation for bodies sitting on an infinite plane"

[45] LAM, Y. W., HODGSON, D. C. J.Acoust.Soc.Amer. 88 (1990) 1993-2000
"The prediction of the sound field due to an arbitrary vibrating body in a rectangular enclosure"

[46] MORSE, P. M., INGARD, K. U.
"Theoretical acoustics"
McGraw-Hill, New York, 1968, chap. 9, pp. 579-580.

[47] SEYBERT, A. F., CHENG, C. Y. R., WU, T. W.
"The solution of coupled interior/exterior acoustic problems using the boundary element method" J. Acoust. Soc. Amer., 88, pp. 1612-1618, 1990.

[48] ESTORFF, O. v. (ed.)
"Boundary Elements in Acoustics, Advances and Applications, (Advances in Boundary Element Series)"
WIT Press, Computational Mechanics Publications, Southampton, Boston, 2000.

Section O.6 :

[1] GOCKEL, M. A. (ed.)
"Handbook for dynamic analysis", Chap. 7.3, (MSC/NASTRAN)
Los Angeles: The Mac Neal-Schwendler Corporation, 1983.

[2] SCHWARZ, H. R.
„Methode der finiten Elemente"
Stuttgart, Teubner, 1980.

[3] PETYT, M.
"Finite element techniques for acoustics, Chap. 2 in: Theoretical acoustics and numerical techniques", (ed. P. Filippi)
Wien, New York, Springer, 1983.

[4] PETYT, M.
"Finite element techniques for structural vibration and acoustics, Chap. 15, 16 in: Noise and vibration", (ed. R. G. White, J. G. Walker)
New York, John Wiley, 1982.

[5] GLADWELL, G. M. L., ZIMMERMANN, G. J.Sound Vib. 3 (1966) 233-241
"On energy and complementary energy formulations of acoustic and structural vibration problems"

[6] CRAGGS, A. J.Sound Vib. 23 (1972) 331-339
"The use of simple three-dimensional acoustic finite elements for determining the natural modes and frequencies of complex shaped enclosures"

[7] SHUKU, T., ISHIHARA, K. J.Sound Vib. 29 (1973) 67-76, .
"The analysis of the acoustics field in irregularly shaped rooms by the finite element method"

[8] PETYT, M., LEA, J., KOOPMANN, G. H. J.Sound Vib. 45 (1976) 495-502
"A finite element method for determining the acoustic modes of irregular shaped cavities"

[9] RICHARDS, T. L., JHA, S. K. J.Sound Vib. 63 (1979) 61-72
"A simplified finite element method for studying acoustic characteristics inside a car cavity"

[10] MUNJAL, M. L.
"Acoustics of ducts and mufflers"
New York, John Wiley, 1987.

[11] SCHULZE HOBBELING, H. Acustica, 67 (1989) 275-283
„Berechnung komplexer Absorptions-/ Reflexions-Schalldämpfer mit Hilfe der Finite-Element-Methode"

[12] SOIZE C. Eur.J. Mech. A/Solids, 17 (1998) 657-685
"Reduced models in the medium frequency range for general dissipative structural-dynamics systems"

[13] SOIZE C. J.Acoust.Soc.Amer. 103 (1998) 3393-3406
"Reduced models in the medium-frequency range for general external structural-acoustic systems"

[14] SOIZE C. J.Acoust.Soc.Amer., 106 (1999) 3362-3374
"Reduced models for structures in the medium-frequency range coupled with internal acoustic cavities"

[15] EVERSTINE, G. C., HENDERSON, F. M. J.Acoust.Soc.Amer. 87 (1990) 1938-1947
"Coupled finite element/boundary element approach for fluid-structure interaction"

[16] SMITH, R. R., HUNT, J. T., BARACH, T. J.Acoust.Soc.Amer. 54 (1973) 1277-1288
"Finite element analysis of acoustically radiating structures with applications to sonar transducers"

[17] HUNT, T. J., KNITTEL, M. R., BARACH, D. J.Acoust.Soc.Amer. 55 (1974) 269-280
"Finite element approach to acoustic radiation from elastic structures"

[18] MASMOUDI, M. Numer. Math., 51 (1987) 87-101
"Numerical solution for exterior problems"

[19] KIRSCH, A., MONK, P. IMA J.Numer.Anal., 9 (1990) 425-447
"Convergence analysis of a coupled finite element and spectral method in acoustic scattering"

[20] ZIMMER, H., OCHMANN, M., HOLZHEUER, C.
"A finite element approach combined with analytical wave functions for acoustic radiation from elastic structures"
Proc. 17th Int. Congress on Acoustics (ICA), Rome, Italy, 2001, to appear.

[21] GAN, H., LEVIN, P. L., LUDWIG, R. J.Acoust.Soc.Amer. 94 (1993) 1651-1662
"Finite element formulation of acoustic scattering phenomena with absorbing boundary condition in the frequency domain"

[22] BAYLISS, A., GUNZBURGER, M., TURKEL, E. SIAM J.Appl.Math. 42 (1982) 430-451
"Boundary conditions for the numerical solution of elliptic equations in exterior regions"

[23] NEFSKE, D. J., WOLF, J. A. JR, HOWELL, L. J. J.SoundVib. 80 (1982) 247-266
"Structural-acoustic finite element analysis of the automobile passenger compartment: a review of current practice"

[24] CRAGGS, A., STEAD, G. Acustica 35 (1976) 89-98
"Sound transmission between enclosures – a study using plate and acoustic finite elements"

[25] BECKER, P., WALLER, H. Acustica 60 (1986) 21-33
„Vergleich der Methoden der Finiten Elemente und der Boundary-Elemente bei der numerischen Berechnung von Schallfeldern"

P
Variational Principles in Acoustics
by A. Cummings

In this chapter, some applications of *variational principles* in acoustics are discussed and several examples are given, mainly in duct acoustics. This subject area is not necessarily restrictive, since the reader will be able to see how, by extension, the ideas may be applied to other types of problems. In Section B.11 of this book, *Hamilton's Principle* is described. In its application to particles, the time average of the *Lagrangian* of a system, $L = E_{kin} - E_{pot}$, is minimised and so its first variation is equated to zero, *viz.* $\delta\langle L\rangle = 0$ if there is no external work input. In spatially distributed systems, it is the space-time average of the *Lagrange density* that is minimised. Hamilton's Principle gives rise to *Lagrange's equations*, otherwise known as the *Lagrange-Euler* equations or the *Euler* equations (see, for example, chapter 3 of the book by Morse and Feshbach [P.1]). One may apply Hamilton's Principle to sound waves in the absence of dissipation. Here, the Lagrange density is given by

$$\Lambda = (\rho_0 / 2)|\mathbf{u}|^2 - (\kappa / 2)p^2,$$

where ρ_0 is fluid density, $\mathbf{u}$ is acoustic particle velocity, p is sound pressure and κ is isentropic fluid compressibility, $1/\rho_0 c_0^2$, where c_0 is the sound speed. To proceed on the basis of the classical Lagrange equations, one must express both p and $\mathbf{u}$ in terms of the velocity potential ϕ, where $p = \rho_0 \partial\phi / \partial t$, $\mathbf{u} = -\nabla\phi$ (see chapter 6 of the book by Morse and Ingård [P.2]). Substitution of Λ into the appropriate Lagrange equation then yields the acoustic wave equation

$$\nabla^2\phi - (1/c_0^2)\partial^2\phi / \partial t^2 = 0$$ or, in terms of sound pressure,

$$\nabla^2 p - (1/c_0^2)\partial^2 p / \partial t^2 = 0 \,.$$

Hamilton's Principle is not in fact restricted to conservative systems, and may also include a potential related to dissipative forces (see, for example, the book by Achenbach [P.3]).

Hamilton's Principle and Lagrange's equations will yield the governing differential equation(s) of a physical system, but not their solution. Variational methods can be employed to find approximate solutions, by the use of *trial functions*. A trial function is one that contains a number of arbitrary coefficients, which can be altered to change the shape of the function. Values of these coefficients can be found by substituting the trial function into a *functional* (such as the space-time average of the Lagrange density), and then finding a stationary value of this with respect to each of these coefficients in turn. One can employ variational methods to solve eigenvalue problems too. Such methods will be illustrated in the various examples given in this chapter. Rather than using Lagrange's equations as the starting point of the analysis, an alternative technique is to find a variational principle such that the Euler equations are the governing differential equations of the problem and as many as possible of the prevailing physical boundary conditions. This approach is adopted here in the case of simple harmonic time dependence. The comments of Morse and Feshbach [P.1] (see p1107), on variational methods *vis-à-vis* perturbation techniques are worth repeating, "the variational method....permits the exploitation of any information bearing on the problem such as might be available from purely intuitional considerations". There is considerable versatility in the application of variational techniques in engineering acoustics particularly and, indeed, intuition can provide valuable information in the solution of problems, as will be seen here.

It should be noted that the trial functions employed in each of the various examples described here could, in principle, be extended to a full finite element discretization. This would give greater numerical accuracy, but at the expense of considerably increased computational effort. In the present context, it is preferable to maintain a level of relative simplicity in the analysis.

P.1 Eigenfrequencies of a rigid-walled cavity and modal cut-on frequencies of a uniform flat-oval duct with zero mean fluid flow

The problem of finding the eigenfrequencies of a rigid-walled cavity will first be considered, together with the related problem of modal cut-on frequencies in a rigid-walled "flat-oval" duct, a problem of some importance in the acoustics of air-conditioning ducts. In [P.1] (see p.1112) it is shown that, for the scalar Helmholtz equation, $\nabla^2\psi + k^2\psi = 0$ (ψ being a scalar field variable in a volume Ω satisfying homogeneous Dirichlet or Neumann boundary conditions on the bounding surface, and k being the wavenumber), a variational principle exists such that

$$\delta k^2 = \delta[-\iiint_\Omega \psi\nabla^2\psi d\Omega / \iiint_\Omega \psi^2 d\Omega] = 0 \tag{1}$$

By the use of Green's theorem, it is shown [P.1] that

$$k^2 = \iiint_\Omega (\nabla\psi)^2 d\Omega / \iiint_\Omega \psi^2 d\Omega \tag{2}$$

This expression may be applied to the acoustic eigenmodes of a rigid-walled cavity, in which case p replaces ψ and k_0^2 ($k_0=\omega / c_0$ being the acoustic wavenumber, where ω is the radian frequency) replaces k^2. If the cavity is of irregular shape, then a trial function for the sound pressure amplitude distribution, $\tilde{p}(\mathbf{x}) = p(\mathbf{x}) + \varepsilon\eta(\mathbf{x})$ ($\mathbf{x}$ being a position vector, ε being a small parameter and η being an arbitrary function), may be used to replace the (presumably unknown) exact form of sound pressure amplitude $p(\mathbf{x})$ in this expression. One then has

$$\tilde{k}_0^2 = k_0^2 + O(\varepsilon^2) = \iiint_\Omega \nabla\tilde{p} \cdot \nabla\tilde{p} d\Omega \Big/ \iiint_\Omega \tilde{p}^2 d\Omega \tag{3}$$

giving the approximate value of the acoustic wavenumber of a particular eigenmode. If a suitable function for $\tilde{p}$ can be found for a particular mode, this expression may be used to find the approximate eigenfrequency.

It is also possible to apply this formula to a uniform rigid-walled duct of flat-oval cross-section, see [P.4].

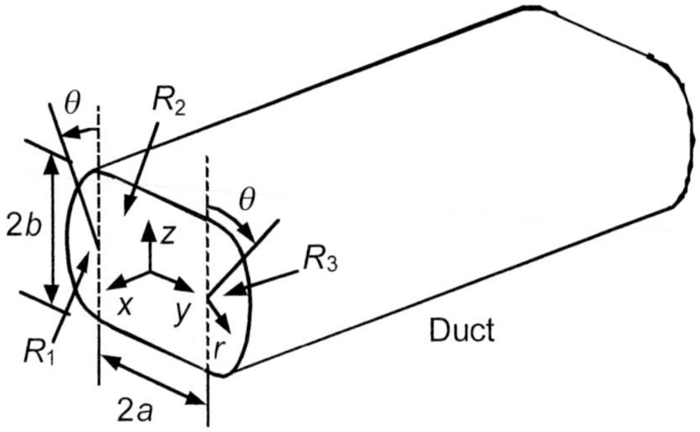

This has two opposite flat sides (width 2a) and two opposite semi-circular sides (diameter 2b) as depicted.

Only the (1,0), (0,1) and (1,1) modes will be discussed here, since higher modes than these present problems in the choice of trial functions. If the mode functions are represented as $p_i = X_i(x)P_i(y,z)\exp(j\omega t)$, then the cut-on frequency for the i-th mode can be found from the acoustic wavenumber $\tilde{k}_i$ corresponding to the trial function $\tilde{P}_i$ for the cross-sectional sound pressure pattern, given by

$$\tilde{k}_i^2 = \iint_R \nabla_t\tilde{P}_i \cdot \nabla_t\tilde{P}_i dR \Big/ \iint_R \tilde{P}_i^2 dR \tag{4}$$

where $R = R_1 + R_2 + R_3$ and the subscript "t" on ∇ signifies a gradient in two dimensions, on the duct cross-section. Suitably continuous trial functions for these modal pressure patterns, satisfying the rigid-wall boundary condition, were found [P.5] from intuition and experience (it was assumed that the nodal lines were straight, and this assumption was verified by experiment). These trial functions are as follows:

$$\tilde{P}_{10} = \begin{cases} -A & \text{on } R_1 \\ A\sin(\pi y/2a) & \text{on } R_2 \\ A & \text{on } R_3 \end{cases} \tag{5a}$$

$$\tilde{P}_{01} = \begin{cases} A\left(2r/b - r^2/b^2\right)\cos\theta & \text{on } R_{1,3} \\ A\left(2z/b - \mathrm{sgn}(z)\cdot z^2/b^2\right) & \text{on } R_2 \end{cases} \tag{5b}$$

$$\tilde{P}_{11} = \begin{cases} -A\left(2r/b - r^2/b^2\right)\cos\theta & \text{on } R_1 \\ A\left(2z/b - \mathrm{sgn}(z)\cdot z^2/b^2\right)\sin(\pi y/2a) & \text{on } R_2 \\ A\left(2r/b - r^2/b^2\right)\cos\theta & \text{on } R_3 \end{cases} \tag{5c}$$

The acoustic wavenumbers at cut-on for these three modes are determined by inserting these trial functions into equation (4) and evaluating the two integrals. The cut-on frequencies ($\tilde{f}_{mn} = \tilde{k}_{mn}c_0/2\pi$) are then given as follows:

$$\tilde{f}_{10} = \frac{c_0}{4a}\sqrt{\frac{1}{1+\pi b/2a}} \;;\; \tilde{f}_{01} = \frac{c_0}{2\pi}\sqrt{\frac{5\pi/4+16a/3b}{11\pi b^2/30+32ab/15}}$$

$$\tilde{f}_{11} = \frac{c_0}{2\pi}\sqrt{\frac{5\pi/4+4\pi^2 b/15a+8a/3b}{11\pi b^2/30+16ab/15}} \tag{6a–c}$$

Experimental data were taken on a flat-oval cavity, to find the cut-on frequencies for a duct with $a = b = 50$mm, and comparisons were also made with predictions from a finite difference numerical scheme [P.5]. The comparisons between experiment and numerical prediction are summarised in the table below (the % figure referring to the numerical prediction accuracy as compared to the measured cut-on frequency).

Mode numbers (m,n)	Measured $\tilde{f}_{mn}$ (Hz)	$\tilde{f}_{mn}$ from FD method (Hz)	$\tilde{f}_{mn}$ from equs. (6a-c) (Hz)
(1,0)	959.9	970 (+1%)	1084 (+13%)
(0,1)	1829.9	1845 (+0.8%)	1859 (+1.5%)
(1,1)	2206	2232 (+1.2%)	2257 (+2.3%)

It can be seen that the prediction accuracy of the variational formulae is only modest for the (1,0) mode, but is much better for the (0,1) and (1,1) modes. In the case of the (1,0) mode, the assumed constant sound pressure distributions in the half-cylindrical parts of the duct would not be a particularly accurate representation of the actual pattern. As one would expect, the predicted frequencies are always too high as is the case, for example, in the use of Rayleigh's method in vibration analysis.

P.2 Sound propagation in a uniform narrow tube of arbitrary cross-section with zero mean fluid flow

See also Sections J.1, J.2, J.3 for sound propagation in flat or circular capillaries.

This second example is not, strictly speaking, an acoustics problem at all (since the thermodynamic processes involved are not isentropic) but relates to wave propagation in narrow tubes, a topic connected with the acoustics of porous media. The procedure here will be to write a functional that has the correct Euler equations and then to proceed to find suitable trial functions (see [P.6]).

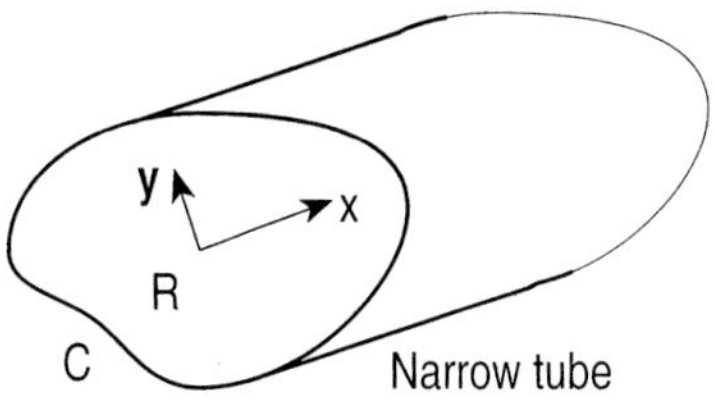

The geometry of the problem is as depicted; x is the axial co-ordinate and **y** is a position vector in the transverse plane.

The tube is assumed to have a uniform cross-section and rigid heat-conducting walls, which are at a constant temperature. Subject to the usual boundary-layer approximations, appropriate forms of the linearised Navier-Stokes and thermal energy equations may be written, for simple-harmonic time dependence, as

$$(\nabla_t^2 - j\omega / \nu)u = (-jk_x / \mu)p \quad ; \quad (\nabla_t^2 - j\omega\rho_0 C_p / K)T = (-j\omega / K)p \qquad (1)$$

Here, ν, μ, C_p and K are the fluid kinematic viscosity, dynamic viscosity, specific heat at constant pressure and thermal conductivity respectively, k_x is the axial wavenumber of the fluid wave, T is the temperature perturbation and p is the sound pressure. It will be noted that the above two equations are isomorphic and also have identical boundary conditions, namely zero axial velocity and temperature perturbations at the duct wall. Stinson [P.7] wrote both these equations in the form :

$$(\nabla_t^2 - j\omega/\eta)\psi = -j\omega/\eta \tag{2}$$

where $\psi \equiv (\omega\rho_0 / k_x p)u$, $\eta \equiv \nu$ in the velocity equation and $\psi \equiv (\rho_0 C_p / p)T$, $\eta \equiv \nu / Pr$ in the temperature equation, Pr being the fluid Prandtl number. The axial wavenumber can be found [P.7] as :

$$k_x = (\omega / c_0)\{[\gamma - (\gamma-1)F(\nu/Pr)]/F(\nu)\}^{1/2} \tag{3}$$

where $F(\eta)$ is defined as $\langle\psi\rangle$, *viz.* the average of ψ over R , the cross-section of the tube, and γ is the ratio of principal specific heats. The problem now is to find an approximate solution to this equation that satisfies the Dirichlet boundary condition $\psi = 0$ on C.

A variational principle for this problem may be obtained as follows, if ψ is expressed as $\psi = \Psi(\mathbf{y})\exp[j(\omega t - k_x x)]$. A functional may be defined [P.6],

$$\Phi = \tfrac{1}{2}\iint_R [\nabla_t\Psi\cdot\nabla_t\Psi + (j\omega/\eta)\Psi^2 - (j2\omega/\eta)\Psi]\,dR \tag{4}$$

By putting $\delta\Phi = 0$ and using Green's formula, it is easily shown that the Euler equations for Φ are the governing differential equation (2) and the "natural" Neumann boundary condition $\nabla\Psi\cdot\mathbf{n} = 0$, where $\mathbf{n}$ is the outward unit normal to the tube surface. This natural boundary condition is not, in fact, the aforementioned physical boundary condition $\Psi = 0$ on C , and this latter condition must be imposed as a "forced" boundary condition with the proviso that $\delta\Psi = 0$ on C . This is done by choosing a trial function $\tilde{\Psi}$ such that $\tilde{\Psi} = 0$ and $\delta\tilde{\Psi} = 0$ on C . For the sake of simplicity, two trial functions will be used here, the first intended to apply in the low frequency limit where the velocity and temperature perturbations are quasi-steady, and the second at high frequencies where both viscous and thermal boundary layers are thin.

Low frequencies :

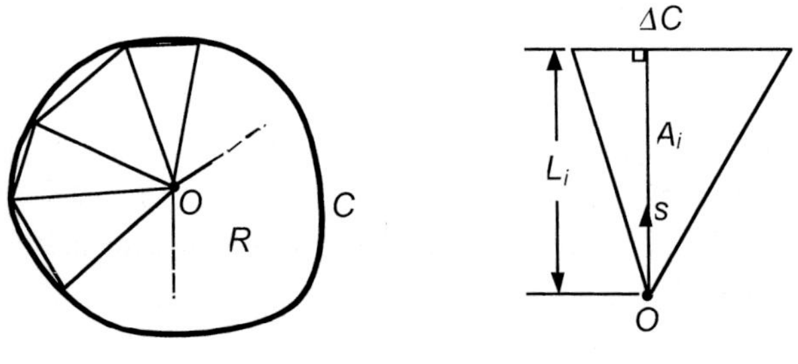

The tube cross-section is divided, approximately, into triangles as shown above. At point O, both velocity and temperature perturbations are assumed to have their maximum value. At sufficiently low frequencies in a narrow circular tube, both temperature and velocity fields exhibit a parabolic radial distribution of the field variable (as can be seen by solving equation (2)), and so an obvious choice for a low frequency trial function is :

$$\tilde{\Psi} = \Psi_0(1-\zeta^2) \tag{5}$$

where $\zeta = s / L_i$. It can be noted that $\delta\tilde{\Psi} = \partial\tilde{\Psi} / \partial\Psi_0 = 1-\zeta^2$, which is zero everywhere on C, thus satisfying the aforementioned requirement for the forced boundary condition. The location of O is obvious in certain cases (e.g. tubes with circular or square cross-section) but may be less so in other cases. It is suggested in [P.6] that, in general, O be located at the centre of the *largest possible* inscribed circle within R. For a single triangular area element, the contribution to Φ is :

$$\Phi_i = \tfrac{1}{2}[\Psi_0^2 \Delta C / L_i + (j\omega / \eta)\Psi_0^2 A_i / 3 - (j2\omega / \eta)\Psi_0 A_i / 2] \tag{6}$$

and therefore

$$\Phi = \sum_i \Phi_i = \frac{1}{2}\left(\Psi_0^2 \oint_C dC / L + j\omega\Psi_0^2 R / 3\eta - j\omega\Psi_0 R / \eta \right) \tag{7}$$

Taking $\delta\Phi = 0$ is equivalent to putting $\partial\Phi / \partial\Psi_0 = 0$, and this gives :

$$\Psi_0 = (j\omega R / 2\eta) \Big/ \left(\oint_C dC / L + j\omega R / 3\eta\right) \tag{8}$$

and $$F(\eta) = \Psi_0 / 2 = (j\omega R / 4\eta) \Big/ \left(\oint_C dC / L + j\omega R / 3\eta\right) \tag{9a}$$

The integral over C can readily be evaluated in most cases of interest and an expression for k_x found from equation (3). For a regular polygon

$$F(\eta) = (j\omega r_h^2 / 4\eta) \big/ (2 + j\omega r_h^2 / 3\eta) \tag{9b}$$

with r_h = hydraulic radius (= 2 x area/perimeter).

It can be assumed that the low frequency approximation is valid at frequencies where both viscous and thermal boundary layer thicknesses are greater than the largest value of L (L_{max}, say). By using expressions for these boundary layer thicknesses (see [P.2], p.286), respectively $\delta_v = \sqrt{2\nu / \omega}, \delta_t = \sqrt{2\nu / \omega \Pr}$, one may then express the upper limiting frequency for the low frequency model as :

$$f_l = \min(\nu / \pi L_{max}^2, \nu / \pi L_{max}^2 \Pr) \tag{10}$$

High frequencies :

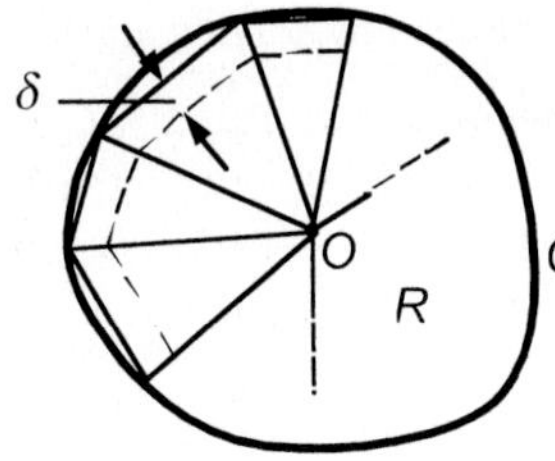

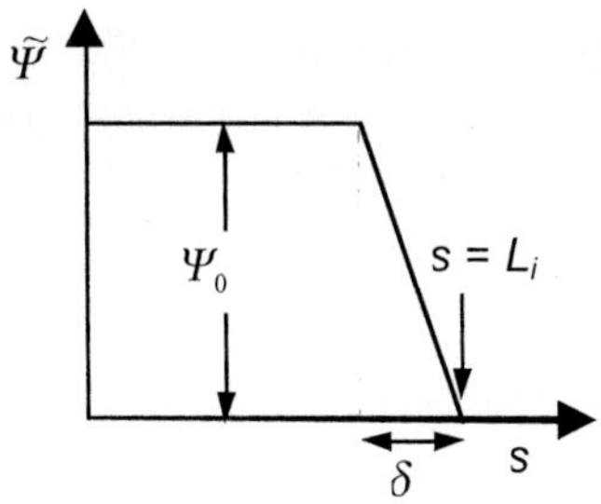

At high frequencies, one may assume that the field variables have an approximately constant value in the central part of the tube, and that a boundary layer exists between this region and the wall. The most basic approximation for the boundary layer profile is linear, as shown above. Here, δ is the boundary layer thickness (viscous or thermal, as the case may be). Now the contribution to Φ from a single element is :

$$\begin{aligned}\Phi_i = \tfrac{1}{2}[\Psi_0^2 A_i(2\varepsilon_i - \varepsilon_i^2)/\delta^2 + (j\omega/\eta)\Psi_0^2 A_i(1 - 4\varepsilon_i/3 + \varepsilon_i^2/2) - \\ - (j2\omega/\eta)\Psi_0 A_i(1 - \varepsilon_i + \varepsilon_i^2/3)]\end{aligned} \tag{11}$$

where $\varepsilon_i = \delta / L_i$, and so

$$\Phi = \tfrac{1}{2}[\Psi_0^2 I_1/\delta^2 + (j\omega/\eta)\Psi_0^2 I_2 - (j2\omega/\eta)\Psi_0 I_3] \tag{12}$$

where

$$\begin{aligned} I_1 &= \oint_C (\delta - \delta^2/2L)\,dC \quad ; \quad I_2 = \oint_C (L/2 - 2\delta/3 + \delta^2/4L)\,dC \\ I_3 &= \oint_C (L/2 - \delta/2 + \delta^2/6L)\,dC. \end{aligned} \tag{13}$$

Putting $\delta\Phi = 0$ now gives :

$$\Psi_0 = (j\omega/\eta)I_3/[I_1/\delta^2 + (j\omega/\eta)I_2] \quad ; \quad F(\eta) = (j\omega/\eta R)I_3^2[I_1/\delta^2 + (j\omega/\eta)I_2] \tag{14}$$

In the case of a regular polygon, one may define $\varepsilon = \delta / L = \delta / r_h$ and utilise the fact that $\delta = \sqrt{2\eta/\omega}$ to write : (15)

$$F(\eta) = 2j(1 - \varepsilon + \varepsilon^2/3)^2 / [(2\varepsilon - \varepsilon^2) + 2j(1 - 4\varepsilon/3 + \varepsilon^2/2)] \tag{16}$$

This expression should be valid at frequencies above a limit

$$f_2 = \max(\nu / \pi L_{min}^2, \nu / \pi L_{min}^2 \, Pr) \tag{17}$$

A circular section tube : (see also Section J.3)

A circle is the limiting case of a regular polygon with an infinite number of sides. Accordingly, at low frequencies, the axial wavenumber here is immediately found by putting $r_h = a$, the tube radius, in (9) and utilising (3). At high frequencies, k_x is found by putting $\varepsilon = \delta / a$ in (16) (with δ expressed in terms of η from (15)) and using (3). The predictions for both the real and imaginary parts of k_x are in close agreement with the exact solution, given in [P.7 or Section J.3]. The region $f_1 < f < f_2$, in which neither the low frequency nor high frequency approximation is valid, is fairly narrow in most cases of practical interest, since $f_2 / f_1 = Pr$. It is easy to connect the two curves graphically.

A parallel slit : (see also Section J.1)

Neither of the trial functions above is appropriate in the case of the slit, and equivalent trial functions in one dimension only – with a parabolic profile and a linear boundary layer profile respectively – may be employed in this case. A process analogous to that above yields :

$$F(\eta) = (j\omega a^2 / 12\eta) / (1 + j\omega a^2 / 10\eta) \tag{18}$$

at low frequencies (where a is the *width* of the slit) and

$$F(\eta) = j(1-\varepsilon)^2 / [\varepsilon + j(1 - 4\varepsilon / 3)] \tag{19}$$

(with $\varepsilon = \delta / a$) at high frequencies. The low frequency formula is in excellent agreement with the exact solution [P.7 or Section J.1] and the high frequency formula yields good accuracy in this comparison, though this degenerates slightly as f_2 is approached from above.

Other geometries :

In [P.6], tubes with equilateral triangular, square, rectangular, hexagonal and semi-circular sections are examined and in all cases, the method described above yields predictions that are in at least reasonable agreement with other reported exact or numerical solutions. The detailed formulae for each case can readily by obtained from the above expressions.

P.3 Sound propagation in a uniform, rigid-walled, duct of arbitrary cross-section with a bulk-reacting lining and no mean fluid flow: low frequency approximation

See also Chapter J for circular and rectangular ducts.

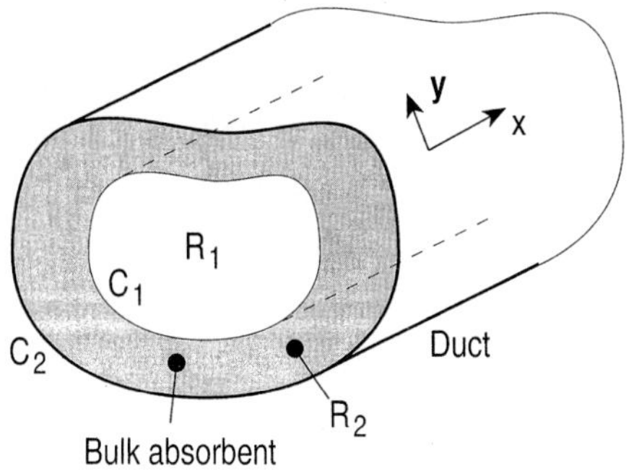

Consider the above problem, in which a rigid-walled duct of uniform but arbitrary cross-sectional geometry has an internal lining of "bulk-reacting" sound absorbent (initially considered to be isotropic), describable as an equivalent fluid, and characterised by a complex characteristic impedance and acoustic wavenumber. The duct contains no mean fluid flow. It is required to find an approximate expression for the axial wavenumber of the fundamental "coupled mode" in this duct, such that the axial wavenumbers for the sound fields in R_1 and R_2 (the central passage and lining respectively) are identical. Astley [P.8] has (in a more general formulation) derived a variational principle with a functional

$$\Phi = \tfrac{1}{2}\iint_{R_1} (\nabla_t P \cdot \nabla_t P - \kappa^2 P^2)\, dR_1 + (\rho_0 / 2\rho_a)\iint_{R_2} (\nabla_t P \cdot \nabla_t P - \kappa_a^2 P^2)\, dR_2 \tag{1}$$

where the sound pressure in the fundamental mode in R_1 and R_2 is expressed in the form $p = P(\mathbf{y})\exp(j[\omega t - k_x x])$, and $\kappa^2 = k_0^2 - k_x^2$, $\kappa_a^2 = k_a^2 - k_x^2$, k_a being the (complex) effective acoustic wavenumber in the absorbent. The Euler equations of this functional are the Helmholtz equations in R_1 and R_2 (*viz*., $(\nabla_t^2 + \kappa^2)P = 0$ and $(\nabla_t^2 + \kappa_a^2)P = 0$), together with the conditions of continuity of normal particle displacement on C_1 and zero particle displacement on C_2. Note that continuity of sound pressure on C_1 is not a natural boundary condition, and must therefore be satisfied by the trial function; the imposition of a forced boundary condition is not necessary here. The simplest trial function for the sound pressure is just a constant, representing a plane wave, *viz*. $\tilde{P} = P_0$ (this satisfies the requirement of continuity of sound pressure on C_1). This function would normally be a reasonable approximation to reality at low frequencies. Inserting this expression into (1) and taking $\delta\Phi = 0$ (which simply involves taking $\partial\Phi / \partial P_0 = 0$) then yields an explicit dispersion relationship

$$k_x = k_a \sqrt{\frac{1 + \phi\rho_a k_0^2 / \rho_0 k_a^2}{1 + \phi\rho_a / \rho_0}} \tag{2}$$

where ρ_a is the (complex) effective density of the absorbent and $\phi = R_1 / R_2$. Clearly, this formula gives the correct limiting behaviour as $\phi \to 0$ and $\phi \to \infty$ (namely $k_x \to k_a$ and $k_x \to k_0$ respectively). Cummings [P.9] has reported a version of this formula that is valid for an anisotropic absorbent, with acoustic wavenumber k_{ax} in the axial direction and k_{ay} in any transverse direction (independent of direction in the transverse plane), and effective density ρ_{ay} in any transverse direction :

$$k_x = k_{ax} \sqrt{\frac{1 + \phi\rho_{ay} k_0^2 / \rho_0 k_{ay}^2}{1 + \phi\rho_{ay} k_{ax}^2 / \rho_0 k_{ay}^2}} \tag{3}$$

which, again, can be seen to give the correct limiting behaviour as $\phi \to 0$ and $\phi \to \infty$.

P.4 Sound propagation in a uniform, rigid-walled, rectangular flow duct containing an anisotropic bulk-reacting wall lining or baffles

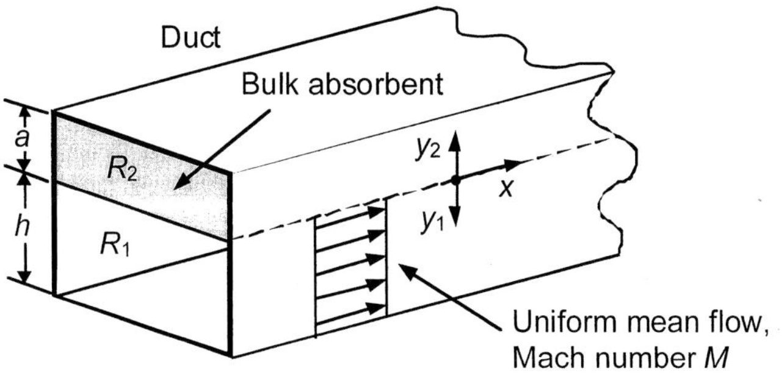

The duct is depicted above. In its simplest form, the lining geometry involves one layer of bulk absorbent (treated as an equivalent fluid), placed against one wall of the duct. The bulk acoustic properties of the absorbent are assumed to be different in the x and y directions. A uniform mean gas flow (Mach number M) is assumed to be present in the remaining part of the duct cross-section. To treat a baffle silencer, one would assume the baffle width to be 2a and the "airway" width to be 2h . The following analysis would then be representative of the fundamental coupled acoustic mode in this arrangement. For the sake of simplicity, the sound field will be assumed to be two-dimensional. The extension to three dimensions is trivial. Cummings [P.10] has reported a variational statement of this problem. If it is assumed that the sound fields in R_1 and R_2 can be expressed in the form $p = P(y_{1,2})\exp(j(\omega t - k_x x))$, a variational functional may be defined :

$$\Phi = \frac{1}{2(1-MK)^2}\iint_{R_1}\{\nabla_t P\cdot\nabla_t P - k_0^2[(1-MK)^2 - K^2]P^2\}dR_1 + \frac{\rho_0}{2\rho_{ay}}\iint_{R_2}[\nabla_t P\cdot\nabla_t P - k_0^2(1-K^2/\gamma_{ax}^2)\gamma_{ay}^2 P^2]dR_2, \tag{1}$$

where $K = k_x / k_0$, $\gamma_{ax} = k_{ax} / k_0$ and $\gamma_{ax} = k_{ay} / k_0$. The Euler equations of this functional are: the convected wave equation in R_1, the Helmholtz equation in the lining, the rigid-wall boundary condition on the duct walls and continuity of normal particle displacement on the interface between the lining and the airway. Suitable trial functions for the sound pressure may be written separately for regions R_1 and R_2 :

$$\tilde{P}_1 = A + \sum_{m=1}^{M} B_m \sin[(2m-1)\pi y_1 / 2h] \quad ; \quad \tilde{P}_2 = A + \sum_{n=1}^{N} C_n \sin[(2n-1)\pi y_2 / 2a] \tag{2 a,b}$$

It is desirable to use separate functions here for $\tilde{P}_1$ and $\tilde{P}_2$ since the normal gradient of the sound pressure is discontinuous at $y_1 = y_2 = 0$, and this discontinuity cannot be represented by a finite number of terms. Only odd integers are included in the arguments of the sine functions because it is necessary for the trial functions to satisfy the rigid-wall boundary condition at $y_1 = h, y_2 = a$. Note that the boundary condition of continuity of sound pressure at $y_1 = 0$, $y_2 = 0$ (not one of the natural boundary conditions) is satisfied by these trial functions. By truncating these summations at appropriate values of M and N, solutions of the desired accuracy may be achieved. An example with $M = 3$, $N = 3$ (i.e. a seven degree-of-freedom trial function) will be discussed in detail here. Equations (2a,b) are inserted into (1), with $\nabla_t\tilde{P}\cdot\nabla_t\tilde{P} \equiv (d\tilde{P}/dy_{1,2})^2$, and the appropriate integrations are carried out. Next, putting $\delta\Phi = 0$ involves taking $\partial\Phi/\partial A = 0, \partial\Phi/\partial B_i = 0, \partial\Phi/\partial C_i = 0$ (with $i = 1,....3$). This gives rise to a homogeneous system of simultaneous linear equations in A, B_i, C_i :

$$A\{[k_0h/(1-MK)^2][(1-MK)^2-K^2]+k_0a(\rho_0/\rho_{ay})[(1-K^2/\gamma_{ax}^2)\gamma_{ay}^2]\}+$$
$$B_1\{[k_0h/(1-MK)^2](2/\pi)[(1-MK)^2-K^2]\}+B_2\{[k_0h/(1-MK)^2](2/3\pi)[(1-MK)^2-$$
$$K^2]\}+B_3\{[k_0h/(1-MK)^2](2/5\pi)[(1-MK)^2-K^2]\}+$$
$$C_1\{k_0a(\rho_0/\rho_{ay})(2/\pi)(1-K^2/\gamma_{ax}^2)\gamma_{ay}^2\}+C_2\{k_0a(\rho_0/\rho_{ay})(2/3\pi)(1-K^2/\gamma_{ax}^2)\gamma_{ay}^2\}+$$
$$C_3\{k_0a(\rho_0/\rho_{ay})(2/5\pi)(1-K^2/\gamma_{ax}^2)\gamma_{ay}^2\}=0, \tag{3a}$$

$$A\{-(4/\pi)[(1-MK)^2-K^2]\}+B_1\{(\pi/2k_0h)^2-[(1-MK)^2-K^2]\}+B_2\{0\}+B_3\{0\}+$$
$$C_1\{0\}+C_2\{0\}+C_3\{0\}=0, \tag{3b}$$

$$A\{-(4/3\pi)[(1-MK)^2-K^2]\}+B_1\{0\}+B_2\{9(\pi/2k_0h)^2-[(1-MK)^2-K^2]\}+B_3\{0\}+$$
$$C_1\{0\}+C_2\{0\}+C_3\{0\}=0, \tag{3c}$$

$$A\{-(4/5\pi)[(1-MK)^2-K^2]\}+B_1\{0\}+B_2\{0\}+B_3\{25(\pi/2k_0h)^2-[(1-MK)^2-K^2]\}+$$
$$C_1\{0\}+C_2\{0\}+C_3\{0\}=0, \tag{3d}$$

$$A\{-(4/\pi)(1-K^2/\gamma_{ax}^2)\gamma_{ay}^2\}+B_1\{0\}+B_2\{0\}+B_3\{0\}+C_1\{(\pi/2k_0a)^2-(1-K^2/\gamma_{ax}^2)\gamma_{ay}^2\}+$$
$$C_2\{0\}+C_3\{0\}=0, \tag{3e}$$

$$A\{-(4/3\pi)(1-K^2/\gamma_{ax}^2)\gamma_{ay}^2\}+B_1\{0\}+B_2\{0\}+B_3\{0\}+C_1\{0\}+C_2\{9(\pi/2k_0a)^2-$$
$$(1-K^2/\gamma_{ax}^2)\gamma_{ay}^2\}+C_3\{0\}=0, \tag{3f}$$

$$A\{-(4/5\pi)(1-K^2/\gamma_{ax}^2)\gamma_{ay}^2\}+B_1\{0\}+B_2\{0\}+B_3\{0\}+C_1\{0\}+C_2\{0\}+$$
$$C_3\{25(\pi/2k_0a)^2-(1-K^2/\gamma_{ax}^2)\gamma_{ay}^2\}=0. \tag{3g}$$

Equations (3a-g) may be written in the form :

$$[A_{ij}](A \quad B_1 \quad B_2 \quad B_3 \quad C_1 \quad C_2 \quad C_3)^T=\{0\} \tag{4a}$$

and a dispersion relation follows as :

$$\left|[A_{ij}]\right|=0 \tag{4b}$$

The elements of the (7x7) square matrix in (4a,b) are the coefficients of $A, B_1, \dots C_3$ in equations (3a-g). The solution of (4b) for k_x is readily accomplished by standard numerical techniques such as Newton's method or Muller's method, given a suitable starting value for k_x. This can be done for the fundamental mode or higher modes of propagation, though the most accurate results will be obtained for the fundamental mode. The seven degree-of-freedom trial function yields accurate results for the fundamental mode [P.10], even up to frequencies of several kHz. Better accuracy could be obtained by taking more terms in the summations in (2a,b). Of course, there is an exact dispersion relationship for this problem, involving circular functions (see, e.g., [P.11]), and therefore the Rayleigh-Ritz method described here is to be regarded as an alternative method rather than one to be used out of necessity in this particular case. The mode shape (e.g., from $B_1 / A, B_2 / A, \dots$ and equations (2a,b)) may readily be found from equations (3a-g), once k_x has been determined.

P.5 Sound propagation in a uniform, rigid-walled, flow duct of arbitrary cross-section, with an inhomogeneous, anisotropic bulk lining

This is a more difficult problem than that described in section P.4, and is one which does not, in general, give rise to an exact dispersion relationship.

General formulation :

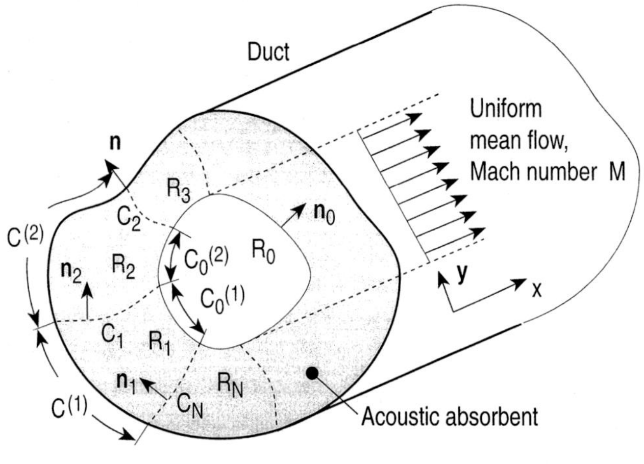

The duct geometry is shown in the diagram. The duct is of uniform but arbitrary cross-section, as is the lining, which is assumed to behave as an equivalent fluid. The bulk acoustic properties of the lining are assumed different in the x direction and in any transverse direction, independent of direction, in the transverse **y** plane.

The lining is divided into sections 1, 2,….N. Within each of these sections, the properties are assumed uniform, but may vary between sections. The outer duct wall is rigid and the central passage carries a uniform mean gas flow.

As in sections P.3 and P.4, coupled mode solutions will be sought for the sound field, which will be taken to have the form $p = P(\mathbf{y})\exp(j(\omega t - k_x x))$, where the acoustic pressure amplitude is defined piecewise in the various parts of the cross-section. It will prove convenient to find a variational functional that will have, as its Euler equations, not only the governing wave equations in R_0 and R_i together with the rigid-wall boundary condition on $C^{(i)}$ and continuity of normal particle displacement on $C_0^{(i)}$, but also continuity of normal particle displacement and sound pressure on C_i, since the last two boundary conditions are not readily satisfied by the trial function in what follows. Such a functional is (see [P.12]) :

$$\Phi = \frac{1}{2(1-MK)^2} \iint_{R_0} \{\nabla_t P \cdot \nabla_t P - k_0^2[(1-MK)^2 - K^2]P^2\}\, dR_0 + \sum_{i=1}^{N} \frac{\rho_0}{2\rho_{ayi}} \iint_{R_i} [\nabla_t P_i \cdot \nabla_t P_i - k_0^2(1-K^2/\gamma_{axi}^2)\gamma_{ayi}^2 P_i^2]\, dR_i + \rho_0\omega^2 \sum_{i=1}^{N} \int_{C_i} \xi_i (P_{i+1} - P_i)\, dC_i \tag{1}$$

where (again) $K = k_x / k_0$ and the summations are over the N sub-regions R_i of absorbent, separated by the boundaries C_i. The last summation ensures that the aforementioned two boundary conditions on C_i are natural boundary conditions.

Application to a duct with two cross-sectional lines of symmetry :

One important application of this formulation is to dissipative vehicle exhaust silencers of oval cross-section. These almost invariably have a central circular-section gas flow passage and a cross-section with two lines of symmetry at right angles. The lining material can be not only anisotropic, but also inhomogeneous, in its bulk properties.

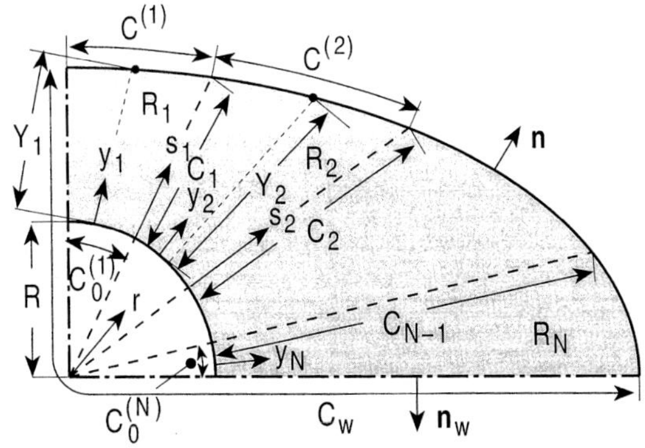

For the fundamental acoustic mode, it is clear that only one-quarter of the duct cross-section need be considered, and the appropriate geometry is shown in the figure.

The portion C_w of the boundary is effectively rigid, from considerations of symmetry. At fairly low frequencies, it should be reasonable to assume a purely radial variation is acoustic pressure in both R_0 and in each segment of lining R_i (though without any oscillatory behaviour), together with a circumferential variation between the lining segments. Accordingly, the lining is segmented radially, as shown. The functional in (1) still applies, except that the second summation is taken to N–1 rather than N, and a lower limit of 2 is thus imposed on N. The Euler equations now include the rigid-wall boundary condition on C_w. A composite trial function for this geometry is appropriate, having the following form :

$$\tilde{P}_0 = a + b\cos(\pi r / 2R) \qquad \text{in } R_0;$$

$$\tilde{P}_i = a + c_i \sin(\pi y_i / 2Y_i) \qquad \text{in } R_i \ (i = 1,\ldots N), \tag{2 a-c}$$

$$\tilde{\xi}_i = d_i \sin(\pi s_i / 2C_i) \qquad \text{on } C_i \ (i = 1,\ldots N-1).$$

Here, a, b, c_i and d_i are (complex) arbitrary constants, r, y_i and s_i are co-ordinates (as shown) and R, Y_i and C_i are the lengths shown in the diagram. Continuity of pressure on $C_0^{(i)}$ (not a natural boundary condition) is satisfied by (2a,b). It is convenient to divide the quarter of the absorbent into N *equal segments*.

The next step is to insert the trial functions (2a-c) into (1) and carry out the integrations. The area integrals over the sub-regions of absorbent as depicted above are not easily found, and so it will be assumed that these radial segments are equivalent to segments of circular annuli, having an inner radius R and an outer radius $R + Y_i$, with the radial sound pressure distribution as specified in (2b). This greatly simplifies these area integrals. The integrals over C_i are readily found exactly for the geometry above. The variational functional may now be written :

$$\Phi = a^2 f_1 + b^2 f_2 + abf_3 + \sum_{i=1}^{N} c_i^2 g_i + a\sum_{i=1}^{N} c_i h_i + \sum_{i=1}^{N-1} d_i c_{i+1} u_i - \sum_{i=1}^{N-1} d_i c_i v_i \tag{3}$$

where

$$f_1 = -(\pi / 8)[k_0 R / (1 - MK)]^2 [(1 - MK)^2 - K^2] - (\pi / 4N)\sum_{i=1}^{N} (\rho_0 / \rho_{ayi})(1 - K^2 / \gamma_{axi}^2)\gamma_{ayi}^2 [k_0^2 R Y_i + (k_0 Y_i)^2 / 2] \tag{4a}$$

$$f_2 = [\pi / 4(1 - MK)^2]\{(\pi^2 / 4)(1/4 + 1/\pi^2) - (k_0 R)^2 [(1 - MK)^2 - K^2](1/4 - 1/\pi^2)\} \tag{4b}$$

$$f_3 = -[k_0 R / (1 - MK)]^2 [(1 - MK)^2 - K^2](1 - 2/\pi) \tag{4c}$$

$$g_i = \{(\pi^2/4)(R/2Y_i + 1/4 - 1/\pi^2) - (1 - K^2/\gamma_{axi}^2)\gamma_{ayi}^2[k_0^2 R Y_i/2 + (k_0 Y_i)^2(1/4 + 1/\pi^2)]\}(\pi\rho_0/4N\rho_{ayi}) \tag{4d}$$

$$h_i = -(1 - K^2/\gamma_{axi}^2)\gamma_{ayi}^2(k_0 Y_i/\pi)(2k_0 Y_i/\pi + k_0 R)(\pi\rho_0/N\rho_{ayi}) \tag{4e}$$

$$u_i = \begin{cases} [2\rho_0\omega^2 Y_{i+1}/\pi(Y_{i+1}^2/C_i^2 - 1)]\cos(\pi C_i/2Y_{i+1}), & Y_{i+1} \neq C_i \\ \rho_0\omega^2 C_i/2, & Y_{i+1} = C_i \end{cases} \tag{4f}$$

$$v_i = \begin{cases} [2\rho_0\omega^2 Y_i/\pi(Y_i^2/C_i^2 - 1)]\cos(\pi C_i/2Y_i), & Y_i \neq C_i \\ \rho_0\omega^2 C_i/2, & Y_i = C_i \end{cases} \tag{4g}$$

Now $\delta\Phi = 0$ is equivalent to taking

$$\partial\Phi/\partial a = 0, \partial\Phi/\partial b = 0 \quad ; \quad \partial\Phi/\partial c_i = 0\ (i = 1,\ldots N) \quad ; \quad \partial\Phi/\partial d_i = 0\ (1 = 1\ldots N-1)$$

This gives rise to a system of linear equations :

$$2af_1 + bf_3 + \sum_{i=1}^{N} c_i h_i = 0 \quad ; \quad 2bf_2 + af_3 = 0 \quad ; \quad 2c_1 g_1 + ah_1 - d_1 v_1 = 0 \,;$$

$$2c_i g_i + ah_i + d_{i-1}u_{i-1} - d_i v_i = 0 \quad (i = 2,\ldots N-1) \quad ; \quad 2c_N g_N + ah_N + d_{N-1}u_{N-1} = 0 \,; \tag{5 a-f}$$

$$c_{i+1}u_i - c_i v_i = 0 \quad (i = 1,\ldots N-1)$$

There are $2N+1$ of these equations in the $2N+1$ unknowns, a, b,…., and the system of equations may be written :

$$[A_{ij}](a \quad b \quad c_1\ldots \quad d_1\ldots)^T = \{0\} \tag{6a}$$

where A_{ij} are the coefficients in (5a-f) and, as in Section P.4, a dispersion relation may be written :

$$\left|[A_{ij}]\right| = 0 \tag{6b}$$

This may be solved numerically for k_x, as in the example in Section P.4.

From equations (5a-f), expressions for the ratios of coefficients, $b/a, c_i/a$ and d_i/a, may easily be found, from which the mode shape may be determined via equations (2a-c) :

$$b/a = -f_3/2f_2 \quad ; \quad c_1/a = (f_3^2/2f_2 - 2f_1)\Big/\left\{h_1 + \sum_{i=1}^{N-1}\left[\prod_{j=1}^{i}(v_j/u_j)\right]h_{i+1}\right\} \tag{7 a-b}$$

$$c_i/a = (v_{i-1}/u_{i-1})c_{i-1}/a, \quad (i = 2,\ldots N) \quad ; \quad d_1/a = (2g_1 c_1/a + h_1)/v_1 \tag{7 c-d}$$

$$d_i / a = (2g_i c_i / a + h_i + u_{i-1} d_{i-1} / a) / v_i \,, \quad (i = 2, \ldots N-2) \quad ;$$
$$d_{N-1} / a = -(2g_N c_N / a + h_N) / u_{N-1} \qquad (7\text{ e-f})$$

Clearly, some degree of substitution is required in implementing these formulae, e.g. c_1 / a in (7d) has first to be found from (7b).

This method has been shown by Cummings [P.12] to yield predictions that are in good agreement with measured data. Where mean flow is present, it may be necessary to account for the effects of mean gas flow, within the absorbent itself, on the bulk properties of the absorbent (see [P.12]). This aspect of the problem is, however, beyond the scope of this chapter and in any case does not affect the details of the analysis. The effects of a perforated tube, separating the gas flow passage from the absorbent, may – if desired – also be incorporated in the formulation provided a suitable model for the perforate impedance is available. This feature is, however, omitted from the above formulation for the sake of simplicity.

In the case of a circular duct with a uniform (isotropic or anisotropic) lining of constant thickness and a circular gas flow passage, the above formulation may be used, with $N = 2$. Otherwise it may be more convenient to treat this case separately. The method of Section P.4 may be applied here, with the same functional and a similar trial function. The area integrals have a different form, of course, but the formulation is simple and straightforward. This approach has been shown [P.10] to yield excellent results, and it has the possible advantage over the exact solution that Bessel functions do not have to be computed.

P.6 Sound propagation in a uniform duct of arbitrary cross-section with one or more plane flexible walls, an isotropic bulk lining and a uniform mean gas flow

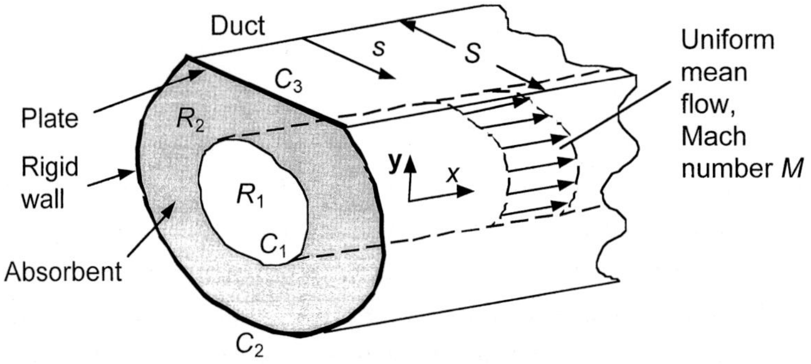

The duct is depicted above. It is shown as having one flexible wall, consisting of a flat elastic plate, though the analysis that follows is equally valid for an arbitrary number of such walls.

For simplicity, it will be assumed that there is just one flexible wall; the extension of the treatment to the case of multiple flexible walls is obvious. The other parts of the duct wall are rigid. There is a uniform gas flow in the central passage.

In this problem, there is wave motion not only in the fluid in the duct and the lining, but also in the plate forming the flexible wall. Coupled mode solutions are sought, and accordingly the axial wavenumbers in the central gas flow passage (R_1), the lining (R_2) and the plate (C_3) are all identical.

General formulation :

It will be assumed that the sound pressure can be represented as $p = P(\mathbf{y})\exp(j(\omega t - k_x x))$ and the outward plate displacement as $u = U(s)\exp(j(\omega t - k_x x))$. Then the three governing differential equations in R_1, R_2 and the plate are, respectively :

$$\nabla_t^2 P + k_0^2[(1-MK)^2 - K^2]P = 0 \quad ; \quad \nabla_t^2 P + (k_a^2 - k_x^2)P = 0 \quad ;$$
$$g[(d^2/ds^2 - k_x^2)^2 U - k_p^4 U] = P_p(s) \tag{1 a-c}$$

where $P_p(s)$ is the transverse factor in the acoustic pressure difference (inside-outside) across the plate, forcing its motion (it is assumed the axial dependence of this pressure difference is the same as that of p and u), g is the flexural rigidity of the plate and the plate wavenumber is $k_p = (m\omega^2/g)^{1/4}$ (m being the mass/unit area of the plate). Other notation here is as in section P.3 .

A variational statement of this problem has been made by Astley [P.8], and a functional defined :

$$\Phi = \rho_0\omega^2 \int_{C_3} \{(g/2)[(d^2U/ds^2)^2 + 2k_x^2(dU/ds)^2 + (k_x^4 - k_p^4)U^2] - UP_p\}ds +$$
$$\frac{1}{2(1-MK)^2}\iint_{R_1} \{\nabla_t P\cdot\nabla_t P - k_0^2[(1-MK)^2 - K^2]P^2\}dR_1 + \tag{2}$$
$$\frac{\rho_0}{2\rho_a}\iint_{R_2} [\nabla_t P\cdot\nabla_t P - (k_a^2 - k_x^2)P^2]dR_2 .$$

The Euler equations of this functional are obtained by taking variations of Φ with respect to P and U: (1a,b) (provided P is continuous on C_1), (1c) (with thc constraint of zero displacement at the edges of the flexible wall), the rigid-wall boundary condition on C_2 , equality of the normal plate displacement and the normal acoustic particle displacement in the internal sound field on C_3 (provided that the normal gradient of sound pressure is allowed to vary freely at the outer surface of the flexible wall) and continuity of normal particle displacement on C_1 (provided that the normal gradient of sound pressure is allowed to vary independently on C_1 in R_1 and R_2).

Low frequency approximation :

As in section P.3, the trial function for the internal acoustic field embodies a uniform transverse sound pressure distribution in R_1 and R_2, i.e. $P(\mathbf{y}) = P \equiv$ const., representative of the fundamental mode. Furthermore, one may write $U(s) = P_u U^*(s,k_x)$, where $U^*(s,k_x)$ is the *solution* of (1c), with a unit pressure on the right-hand side, subject to the prevailing boundary conditions, and P_u is defined by the foregoing equation. With this trial function, the functional may be written (after some manipulation) :

$$\Phi = j k_0 \langle\beta(k_x)\rangle S(PP_u - \tfrac{1}{2}P^2) + \tfrac{1}{2}k_0^2 P^2 \left\{ R_1\left[\left(\frac{K}{1-MK}\right)^2 - 1\right] + R_2\left(\frac{\rho_0}{\rho_a}\right)[K^2 - (k_a/k_0)^2]\right\} \quad (3)$$

where $\langle\beta(k_x)\rangle$ is the space-average (over C_3) of the dimensionless admittance of the flexible wall, $j\omega\rho_0 c_0 U^*(s,k_x)$, and S is the width of the flexible duct wall. This expression is now minimised with respect to P and P_u , and P is equated to P_u on the basis that the external radiation load on the flexible duct wall is negligibly small, to give the dispersion equation :

$$j k_0 \langle\beta(k_x)\rangle S + k_0^2 \left\{ R_1\left[\left(\frac{K}{1-MK}\right)^2 - 1\right] + R_2\left(\frac{\rho_0}{\rho_a}\right)[K^2 - (k_a/k_0)^2]\right\} = 0 \quad (4)$$

This can be solved by an appropriate standard root-finding method. Certain special cases are of interest [P.8], as follows.

(i) A duct with rigid walls and mean gas flow :

In this case, $\langle\beta(k_x)\rangle = 0$ and equation (4) becomes :

$$R_1\left[\left(\frac{K}{1-MK}\right)^2 - 1\right] + R_2\left(\frac{\rho_0}{\rho_a}\right)[K^2 - (k_a/k_0)^2] = 0. \quad (5)$$

(ii) A duct with rigid walls and no mean gas flow :

The result here is identical to that in section P.3.

(iii) A central region of gas flow in a rigid-walled duct, surrounded by a stagnant region :

An example of this is an enclosed gas jet in a duct, such as that which is formed downstream of an abrupt area expansion in a flow duct. Here, $k_a = k_0, \rho_a = \rho_0$ and $\langle\beta(k_x)\rangle = 0$. The dispersion equation now becomes :

$$R_1\left[\left(\frac{K}{1-MK}\right)^2 - 1\right] + R_2(K^2 - 1) = 0. \tag{6}$$

(iv) <u>A duct with a flexible wall, no mean flow and no lining</u> :

In this case, $M = 0$ and $R_2 = 0$, and equation (4) yields :

$$k_x = k_0\sqrt{1 - jS\langle\beta(k_x)\rangle / k_0 R_1} \tag{7}$$

which is identical to the expression obtained by Cummings [P.13].

If the flexible wall is clamped along both edges, an exact solution of equation (1c) exists :

$$\begin{aligned}\langle\beta(k_x)\rangle = j\omega\rho_0 c_0 \Big\{ &\frac{A_1}{\alpha_1 S}\sin(\alpha_1 S) - \frac{A_2}{\alpha_1 S}[\cos(\alpha_1 S) - 1] + \frac{A_3}{\alpha_2 S}\sinh(\alpha_2 S) + \\ &\frac{A_4}{\alpha_2 S}[\cosh(\alpha_2 S) - 1] + \frac{1}{g(k_x^4 - k_p^4)}\Big\}\end{aligned} \tag{8}$$

where $\alpha_1 = \sqrt{k_p^2 - k_x^2}$, $\alpha_2 = \sqrt{k_p^2 + k_x^2}$ and

$$\begin{aligned}A_1 = \{&\alpha_1[1 + \cos(\alpha_1 S) - \cosh(\alpha_2 S) - \cos(\alpha_1 S)\cosh(\alpha_2 S)] + \alpha_2\sin(\alpha_1 S)\sinh(\alpha_2 S)\} / \\ &g(k_x^4 - k_p^4)[2\alpha_1\cos(\alpha_1 S)\cosh(\alpha_2 S) - 2\alpha_1 + (\alpha_1^2/\alpha_2 - \alpha_2)\sin(\alpha_1 S)\sinh(\alpha_2 S)]\end{aligned}$$

$$\begin{aligned}A_2 = \{&\alpha_1\sin(\alpha_1 S)[1 - \cosh(\alpha_2 S)] + \alpha_2\sinh(\alpha_2 S)[1 - \cos(\alpha_1 S)]\} / \\ &g(k_x^4 - k_p^4)[2\alpha_1\cos(\alpha_1 S)\cosh(\alpha_2 S) - 2\alpha_1 + (\alpha_1^2/\alpha_2 - \alpha_2)\sin(\alpha_1 S)\sinh(\alpha_2 S)]\end{aligned} \tag{9 a-d}$$

$$A_3 = -A_1 - 1/g(k_x^4 - k_p^4) \quad ; \quad A_4 = -A_2\alpha_1/\alpha_2$$

Equation (7) has been shown [P.13] to give predictions of axial phase speed for the fundamental coupled structural/acoustic mode that are in excellent agreement with measured data, for a duct of square cross-section (and, therefore, effectively having four flexible walls, clamped along their edges, with no rigid walls). Transverse structural resonance effects in the wall are very prominent in the wall admittance expression (8) and, of course, in the axial wavenumber in the duct.

P.7 Sound propagation in a rectangular section duct with four flexible walls, an anisotropic bulk lining and no mean gas flow

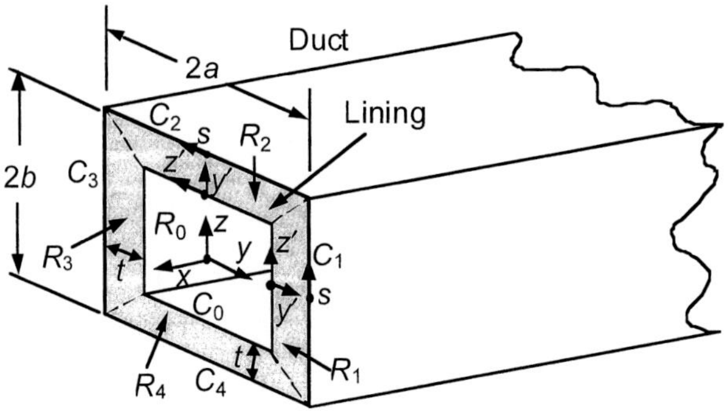

The duct geometry and co-ordinate systems are shown in the diagram. The open central channel R_0 is surrounded by layers of bulk absorbent, all of thickness t , placed against the four flexible walls. These layers are denoted $R_1,\ldots R_4$. The perimetral co-ordinates s are local to each of the four walls of the duct, $C_1,\ldots C_4$, and C_0 is the interface between the central channel and the lining.

A global co-ordinate system x,y,z is centred on the duct axis, and local co-ordinate systems $xy'z'$ are used to define position in the four layers of absorbent. The bulk acoustic properties of the absorbent are different in the y' and x,z' directions, and the properties of all layers are identical. Coupled eigenmodes are sought, for the sound pressure in the duct and the outward wall displacement, having the form :

$$p(x,y,z;t) = P(y,z)\cdot \exp\big(j(\omega t - k_x x)\big) \quad ; \quad u(x,s;t) = U(s)\cdot \exp\big(j(\omega t - k_x x)\big) \qquad (1\text{ a-b})$$

The acoustic wave equations in R_0 and in the absorbent $(R_1,\ldots R_4)$ yield Helmholtz equations, respectively having the forms :

$$\frac{\partial^2 P}{\partial y^2}+\frac{\partial^2 P}{\partial z^2}+(k_0^2-k_x^2)P=0 \quad ; \quad \frac{\partial^2 P}{\partial y'^2}+\frac{k_{ay}^2}{k_{ax}^2}\frac{\partial^2 P}{\partial z'^2}+(k_{ay}^2-k_x^2\frac{k_{ay}^2}{k_{ax}^2})P=0 \qquad (2\text{ a-b})$$

where k_{ax} is the acoustic wavenumber of the absorbent in the x,z' plane (independent of direction) and k_{ay} is the acoustic wavenumber in the y' direction. The equation of motion in the duct walls may be written in the form :

$$g[\frac{d^4U}{ds^4}-2k_x^2\frac{d^2U}{ds^2}+(k_x^4-k_p^4)U]=P \qquad \text{on } C_1,\ldots,C_4 \qquad (2c)$$

where k_p is defined in section P.6.

A variational statement for this problem is reported by Cummings and Astley [P.14], with a functional :

$$\Phi = \frac{1}{2}\left(\iint_{R_0} [(\partial P/\partial y)^2 + (\partial P/\partial z)^2 + (k_x^2 - k_0^2)P^2]\, dR_0 + \right.$$

$$\sum_{i=1}^{4} (\rho_0/\rho_{ay}) \iint_{R_i} [(\partial P/\partial y')^2 + (k_{ay}^2/k_{ax}^2)(\partial P/\partial z')^2 + (k_x^2 k_{ay}^2/k_{ax}^2 - k_{ay}^2)P^2]\, dR_i + \qquad (3)$$

$$\left. \sum_{i=1}^{4} \rho_0 \omega^2 \int_{C_i} \{g[(d^2U/ds^2)^2 + 2k_x^2 (dU/ds)^2 + (k_x^4 - k_p^4)U^2] - 2UP\}\, ds \right)$$

The notation here is essentially that of sections P.4 and P.6. The Euler equations of this functional are the same as those outlined in section P.6 (except for the rigid-wall boundary condition), with the additional feature of anisotropy in the absorbent.

A Rayleigh-Ritz approximation for the coupled eigenproblem may be found by the use of trial functions, having (for example) polynomial form, for the sound pressures in the central passage and lining, and for the wall displacement [P.14]. These trial functions – intended to represent the fundamental coupled mode – are, respectively :

$$\tilde{P} = f_1(y,z)P_1 + f_2(y,z)P_2 + f_3(y,z)P_3 \qquad \text{in } R_0 \qquad (4a)$$

$$\tilde{P} = g_1(y',z')P_1 + g_2(y',z')P_2 + g_3(y',z')P_3 \qquad \text{in } R_1,\ldots,R_4 \qquad (4b)$$

(where $f_1 = [1-(y/a_1)^2][1-(z/b_1)^2]$, $a_1 = a - t$, $b_1 = b - t$,

$f_2 = 1 - [1-(y/a_1)^2][1-(z/b_1)^2]$, $f_3 = 0$, $g_1 = 0$, $g_2 = 1-(y'/t)$, $g_3 = y'/t$),

$$\tilde{U} = h_1(s)U_1 + h_2(s)U_2 + h_3(s)\theta \qquad (4c)$$

(where $h_1 = 1 - 3(s/b)^2 + 2(s/b)^3$, $h_2 = 0$, $h_3 = -s[(s/b) - (s/b)^2]$ for $0 \le s \le b$ on the right-hand vertical side of the duct depicted above, with equivalent expressions for the other parts of the walls).

In these equations, the constants P_1, P_2, P_3 have the dimensions of sound pressure and are identified as follows: P_1 is the sound pressure amplitude on the axis of the central passage, P_2 is the (constant) sound pressure amplitude at the liner/air interface and P_3 is the (constant) sound pressure amplitude on the inner surface of the duct wall, according to the assumed form of the trial function for the sound field. In the vibration field, constants U_1, U_2 are the displacement amplitudes at the mid-points of two adjacent sides of the duct wall and constant θ is the amplitude of the angle of rotation of the corner (assumed to remain right-angled) between these sides.

The insertion of equations (4a-c) into (3) yields (after some manipulation) an expression for the functional which may conveniently be written in matrix form, [P.14],

$$\Phi = \tfrac{1}{2}\mathbf{U}^T[\mathbf{A} + k_x^2\mathbf{B} + k_x^4\mathbf{C}]\mathbf{U} - \mathbf{P}^T\mathbf{T}\mathbf{U} + \tfrac{1}{2}\mathbf{P}^T[\mathbf{E} + k_x^2\mathbf{G}]\mathbf{P} \tag{5}$$

where **P** and **U** are column vectors containing the acoustic coefficients P_1, P_2, P_3 and the structural coefficients $U_1,\ U_2,\ \theta$ respectively, and **A, B, C, T, E, G** are 3×3 matrices, the elements of which are given by :

$$A_{jk} = \sum_{i=1}^{4} \rho_0\omega^2 g \int_{C_i} [(d^2h_j/ds^2)(d^2h_k/ds^2) - k_p^4 h_j h_k]\, ds \tag{6a}$$

$$B_{jk} = \sum_{i=1}^{4} \rho_0\omega^2 g \int_{C_i} 2(dh_j/ds)(dh_k/ds)\, ds \tag{6b}$$

$$C_{jk} = \sum_{i=1}^{4} \rho_0\omega^2 g \int_{C_i} h_j h_k\, ds \quad ; \quad T_{jk} = \sum_{i=1}^{4} \rho_0\omega^2 \int_{C_i} g_j h_k\, ds \tag{6 c-d}$$

$$\begin{aligned} E_{jk} = {} & \iint_{R_0} [(df_j/dy)(df_k/dy) + (df_j/dz)(df_k/dz) - k_0^2 f_j f_k]\, dydz + \\ & \sum_{i=1}^{4} (\rho_0/\rho_{ay}) \iint_{R_i} [(\partial g_j/\partial y')(\partial g_k/\partial y') + \\ & (k_{ay}^2/k_{ax}^2)(\partial g_j/\partial z')(\partial g_k/\partial z') - k_{ay}^2 g_j g_k]\, dy'dz' \end{aligned} \tag{6e}$$

$$G_{jk} = \iint_{R_0} f_j f_k\, dydz + \sum_{i=1}^{4} (\rho_0/\rho_{ay}) \iint_{R_i} (k_{ay}^2/k_{ax}^2) g_j g_k\, dy'dz' \tag{6f}$$

The integrals in these equations are readily evaluated analytically for the trial functions chosen, but could otherwise be evaluated numerically, for example by Gaussian quadrature.

As before, putting $\delta\Phi = 0$ involves minimising Φ with respect to the acoustic variables P_1, P_2, P_3 and the structural variables $U_1,\ U_2,\ \theta$. The former process gives the relationship :

$$[\mathbf{E} + k_x^2\mathbf{G}]\mathbf{P} - \mathbf{T}\mathbf{U} = \mathbf{0} \tag{7a}$$

and the latter yields :

$$[\mathbf{A} + k_x^2\mathbf{B} + k_x^4\mathbf{C}]\mathbf{U} - \mathbf{T}^T\mathbf{P} = \mathbf{0} \tag{7b}$$

These two equations constitute a coupled eigenvalue problem in k_x, the coupling occurring *via* the matrix **T**. It is worth noting that, if **T** is removed from equations (7a,b), two uncoupled eigenvalue problems result :

$$[\mathbf{E} + k_x^2 \mathbf{G}]\mathbf{P} = \mathbf{0} \quad ; \quad [\mathbf{A} + k_x^2 \mathbf{B} + k_x^4 \mathbf{C}]\mathbf{U} = \mathbf{0} \qquad (8\ a,b)$$

The first of these relates to acoustic modes in a lined duct with rigid walls, and the second to structural modes in an elastic-walled duct in which the acoustic loading on the walls is neglected.

Solution of the eigenproblem posed in equations (7a,b) is not entirely straightforward, particularly since the coupled mode types are – broadly – divided into "acoustic" type modes, in which the power flow is predominantly in the fluid, and "structural" type modes, in which most of the power flow is in the elastic walls. Astley, who was responsible for the Rayleigh-Ritz formulation reported here, described, [P.14], a robust iterative method of solution for this problem, based on the foregoing arguments. Although a detailed description of this is not appropriate here, the interested reader is referred to [P.14].

The predictive accuracy of the method described here is surprisingly good (see [P.14]), considering the relative crudity of the acoustic and structural trial functions. Better accuracy could be obtained by the use of trial functions containing more degrees of freedom, but one should then also consider a full finite element discretization as an alternative.

References to part

P. Variational Principles in Acoustics

[P.1] MORSE, P.M. and FESHBACH, H.
"Methods of Theoretical Physics",
McGraw-Hill, N.Y., 1953

[P.2] MORSE, P.M. and INGÅRD, K.U.
"Theoretical Acoustics",
McGraw-Hill, N.Y., 1968

[P.3] ACHENBACH, J.D.
"Wave Propagation in Elastic Solids",
North Holland, Amsterdam, 1973

[P.4] CUMMINGS, A. and CHANG, I.-J. J.Sound Vibr. 106 (1986) 17-33
"Noise Breakout from Flat-Oval Ducts"

[P.5] CUMMINGS, A. and CHANG, I.-J. J.Sound Vibr. 106 (1986) 35-43
"Sound Propagation in a Flat-Oval Waveguide"

[P.6] CUMMINGS, A. J. Sound Vibr. 162 (1993) 27-42
"Sound Propagation in Narrow Tubes of Arbitrary Cross-Section"

[P.7] STINSON, M.R. J. Acoust. Soc. Amer. 89 (1991) 550-558
"The Propagation of Plane Sound Waves in Narrow and Wide Circular Tubes, and Generalization to Uniform Tubes of Arbitrary Cross-Sectional Shape"

[P.8] ASTLEY, R.J. Proc. Inter-Noise 90 (1990) 575-578
"Acoustical Modes in Lined Ducts with Flexible Walls: a Variational Approach"

[P.9] CUMMINGS, A. J. Sound Vibr. 151 (1991) 63-75
"Impedance Tube Measurements on Porous Media: the Effects of Air-Gaps around the Sample"

[P.10] CUMMINGS, A.
Proc. 2nd Internatl. Conference on Recent Developments in Air- and Structure-Borne Sound and Vibration, Auburn University, USA (1992) 689-696.
"Sound Absorbing Ducts"

[P.11] CUMMINGS, A. J. Sound Vibr. 49 (1976) 9-35
"Sound Attenuation in Ducts Lined on Two Opposite Sides with Porous Material, with Some Applications to Splitters"

[P.12] CUMMINGS, A. J. Sound Vibr. 187 (1995) 23-37
"A Segmented Rayleigh-Ritz Method for Predicting Sound Transmission in a Dissipative Exhaust Silencer of Arbitrary Cross-Section"

[P.13] CUMMINGS, A. J. Sound Vibr. 61 (1978) 327-345
"Low Frequency Acoustic Transmission through the Walls of Rectangular Ducts"

[P.14] CUMMINGS, A. and ASTLEY, R.J. J. Sound Vibr. 179 (1995) 617-646
"The Effects of Flanking Transmission on Sound Attenuation in Lined Ducts"

Q
Elasto-Acoustics
by W. Maysenhölder and F.P. Mechel

Some fundamental relations, and relations concerning sound transmission through plates may be found also in the Chapter "I. Sound Transmission".

Q.1 Fundamental equations of motion

Ref.: Achenbach, [Q.1]; Maysenhölder, [Q.2]

Used notation (including some quantities of later sections):

x_i (i = 1, 2, 3) or x, y, z	Cartesian co-ordinates of position
u_i	displacement
$v_i = \partial u_i/\partial t$	velocity
c	phase velocity
C	group velocity
λ_w	wavelength (various subscripts)
k	wavenumber
k_i	wavevector
ε_{ij}	strain
σ_{ij}	stress
p	sound pressure
ρ	mass density
λ , μ	Lamé's constants (μ ≡ shear modulus)
K	compression (bulk) modulus
E	Young's modulus
σ	Poisson's ratio for isotropic media (no subscript)
ν_{ij}	Poisson's ratios for anisotropic media
C_{ijkl}	elasticity tensor
e_{kin}	kinetic energy density
e_{pot}	potential energy density
L	Lagrangian density
I_i	intensity
δ_{ij}	Kronecker's delta (identity matrix)

Sometimes vectors will be written with arrows, e. g. $\vec{x}$ for x_i .

The summation rule is applied:

$$C_{ijkl}\,\varepsilon_{kl} \equiv \sum_{k=1}^{3}\sum_{l=1}^{3} C_{ijkl}\,\varepsilon_{kl} \qquad \varepsilon_{kk} \equiv \sum_{k=1}^{3} \varepsilon_{kk} \qquad \text{(summation over repeated subscripts)}$$

Unlike electrodynamics the theory of elasticity is genuinely non-linear, since the exact relation between strain field ε_{ij} and displacement field u_i is non-linear. This chapter, however, is confined to the linearised theory, i. e. with

$$\varepsilon_{ij} = \frac{1}{2}\left(\frac{\partial u_i}{\partial x_j} + \frac{\partial u_j}{\partial x_i}\right) \tag{1}$$

and the generalised version of Hooke's law

$$\sigma_{ij} = C_{ijkl}\,\varepsilon_{kl} \; . \tag{2}$$

The decomposition of the strain tensor,

$$\varepsilon_{ij} = \frac{\Delta V}{3V}\,\delta_{ij} + \tilde{\varepsilon}_{ij} \; , \tag{3}$$

is invariant with respect to co-ordinate transformations and therefore physically essential. The first term involving the trace $(\varepsilon_{kk} = \Delta V/V)$ represents a change of volume without change of shape, whereas the strain deviator $\tilde{\varepsilon}_{ij}$ with zero trace describes pure shear deformations (change of shape without change of volume). The corresponding decomposition for the stress tensor is $\sigma_{ij} = -p\,\delta_{ij} + \tilde{\sigma}_{ij}$ with the pressure p $(\sigma_{kk} = -3p)$ and the stress deviator $\tilde{\sigma}_{ij}$ with zero trace for the shear stresses. In isotropic media Hooke's law decomposes into the part for (isotropic) compression, $p = -K\cdot\Delta V/V$ with compression (bulk) modulus K, and the shear part $\tilde{\sigma}_{ij} = 2\mu\,\tilde{\varepsilon}_{ij}$ with shear modulus μ. Both parts are combined in the convenient form $\sigma_{ij} = \lambda\,\varepsilon_{kk}\,\delta_{ij} + 2\mu\,\varepsilon_{ij}$ with the Lamé constant $\lambda = K - \frac{2}{3}\mu$.

If dissipation effects are ignored, the Lagrangian density is given by $L = e_{kin} - e_{pot}$ with

$$e_{kin} = \frac{1}{2}\rho\,v_i\,v_i \; , \qquad e_{pot} = \frac{1}{2}\sigma_{ij}\,\varepsilon_{ij} = \frac{1}{2}\varepsilon_{ij}\,C_{ijkl}\,\varepsilon_{kl} = \frac{1}{2}\frac{\partial u_i}{\partial x_j}C_{ijkl}\frac{\partial u_k}{\partial x_l} \; , \tag{4}$$

where the last equal sign is justified because of the symmetries of the elastic tensor,

$$C_{ijkl} = C_{jikl}, \qquad C_{ijkl} = C_{klij} \; .$$

The equation of motion (for time-independent material properties) may be obtained from Hamilton's principle,

$$\delta \int L\,dt\,dx_1\,dx_2\,dx_3 = 0 \tag{5}$$

leading to the Lagrange-Euler equations

$$\frac{d}{dt}\left(\frac{\partial L}{\partial v_i}\right) + \frac{d}{dx_j}\left(\frac{\partial L}{\partial\left(\frac{\partial u_i}{\partial x_j}\right)}\right) = 0 \tag{6}$$

which finally result in the partial differential equations :

$$\rho\frac{\partial^2 u_i}{\partial t^2} = C_{ijkl}\frac{\partial^2 u_k}{\partial x_j \partial x_l} + \frac{\partial C_{ijkl}}{\partial x_j}\frac{\partial u_k}{\partial x_l} \tag{7}$$

The last term vanishes for homogeneous media with position-independent material properties.

Alternative derivation:

The right-hand side of (7) equals the divergence $\partial\sigma_{ij}/\partial x_j$ of the stress tensor, which is zero for local elastostatic equilibrium in the absence of external forces. According to D'Alembert's principle, the combination of this balance of forces with the inertia term (left-hand side of (7)) again yields the above equation of motion. In the case of external body forces, like gravity, a volume density of forces [N/m^3] must be added.

In the special case of locally isotropic media one obtains with

$$C_{ijkl} = \lambda\delta_{ij}\delta_{kl} + \mu\left(\delta_{ik}\delta_{jl} + \delta_{il}\delta_{jk}\right) \tag{8}$$

the simplified equation of motion

$$\rho\frac{\partial^2 u_i}{\partial t^2} = (\lambda + \mu)\frac{\partial^2 u_j}{\partial x_i \partial x_j} + \mu\frac{\partial^2 u_i}{\partial x_j \partial x_j} + \frac{\partial\lambda}{\partial x_i}\frac{\partial u_j}{\partial x_j} + \frac{\partial\mu}{\partial x_j}\frac{\partial u_i}{\partial x_j} + \frac{\partial\mu}{\partial x_j}\frac{\partial u_j}{\partial x_i}, \tag{9}$$

where the last three terms vanish in homogeneous media.

Q.2 Anisotropy and isotropy

Ref.: Helbig, [Q.3, p. 68-92]; Jones, [Q.4, p. 56-70]; Lai et al. [Q.5, p. 293-314]

For an explicit description of anisotropic elasticity the fourth-rank tensor C_{ijkl} is often transformed to the symmetric 6x6-matrix c_{IJ} with subscript relations

$ij \to I$: $11 \to 1, \; 22 \to 2, \; 33 \to 3, \; 23 \to 4, \; 31 \to 5, \; 12 \to 6$.

(contracted or Voigt's notation). The most general (triclinic) anisotropy is described by a fully occupied matrix (21 independent elastic constants), which is shown here with four-subscript entries in order to illustrate the above subscript relations:

$$\{c_{IJ}\} = \begin{pmatrix} C_{1111} & C_{1122} & C_{1133} & C_{1123} & C_{1113} & C_{1112} \\ & C_{2222} & C_{2233} & C_{2223} & C_{1322} & C_{1222} \\ & & C_{3333} & C_{2333} & C_{1333} & C_{1233} \\ & & & C_{2323} & C_{1323} & C_{1223} \\ & \text{sym} & & & C_{1313} & C_{1213} \\ & & & & & C_{1212} \end{pmatrix} \tag{1}$$

triclinic

"sym" means symmetric completion. Since c_{IJ} is not a tensor, transformation of Hooke's law (see Q.1) is not trivial:

$$\sigma_I = c_{IJ}\,\varepsilon_J \text{ with } \quad \begin{array}{l} \{\sigma_I\} = (\sigma_{11} \;\; \sigma_{22} \;\; \sigma_{33} \;\; \sigma_{23} \;\; \sigma_{13} \;\; \sigma_{12}) \\ \{\varepsilon_I\} = (\varepsilon_{11} \;\; \varepsilon_{22} \;\; \varepsilon_{33} \;\; 2\varepsilon_{23} \;\; 2\varepsilon_{13} \;\; 2\varepsilon_{12}) \end{array} \tag{2}$$

Similarly, with the compliance tensor S_{ijkl}, which is the inverse of C_{ijkl}, the contracted form is :

$$\varepsilon_J = s_{IJ}\,\sigma_I \; ; \qquad s_{IJ} = c_{IJ}^{-1} \, . \tag{3}$$

The most general anisotropy admissible in thin-plate theory, which implies the middle plane of the plate to be a plane of symmetry, needs 13 independent elastic constants (monoclinic), whereas orthotropic anisotropy, characterised by three mutually perpendicular planes of symmetry, requires nine:

$$\begin{pmatrix} c_{11} & c_{12} & c_{13} & 0 & 0 & c_{16} \\ & c_{22} & c_{23} & 0 & 0 & c_{26} \\ & & c_{33} & 0 & 0 & c_{36} \\ & & & c_{44} & c_{45} & 0 \\ & \text{sym} & & & c_{55} & 0 \\ & & & & & c_{66} \end{pmatrix} \qquad \begin{pmatrix} c_{11} & c_{12} & c_{13} & 0 & 0 & 0 \\ & c_{22} & c_{23} & 0 & 0 & 0 \\ & & c_{33} & 0 & 0 & 0 \\ & & & c_{44} & 0 & 0 \\ & \text{sym} & & & c_{55} & 0 \\ & & & & & c_{66} \end{pmatrix} \tag{4}$$

monoclinic — plane of symmetry: x_3 = const; orthotropic (≡ orthorhombic)

In "engineering notation" orthotropic anisotropy is expressed by three Young's moduli E_i, six Poisson numbers ν_{ij}, and three shear moduli G_{ij}. Their physical meaning may be deduced from the compliance representation :

$$\{s_{IJ}\} = \begin{pmatrix} \frac{1}{E_1} & -\frac{\nu_{21}}{E_2} & -\frac{\nu_{31}}{E_3} & 0 & 0 & 0 \\ -\frac{\nu_{12}}{E_1} & \frac{1}{E_2} & -\frac{\nu_{32}}{E_3} & 0 & 0 & 0 \\ -\frac{\nu_{13}}{E_1} & -\frac{\nu_{23}}{E_2} & \frac{1}{E_3} & 0 & 0 & 0 \\ 0 & 0 & 0 & \frac{1}{G_{23}} & 0 & 0 \\ 0 & 0 & 0 & 0 & \frac{1}{G_{31}} & 0 \\ 0 & 0 & 0 & 0 & 0 & \frac{1}{G_{12}} \end{pmatrix}, \tag{5}$$

which also provides for the three relations between the six Poisson numbers due to $s_{IJ} = s_{JI}$. In terms of the c_{IJ} ,[Q.6]:

$$E_1 = \frac{N}{c_{22}\,c_{33} - c_{23}^2}, \qquad E_2 = \frac{N}{c_{11}\,c_{33} - c_{13}^2}, \qquad E_3 = \frac{N}{c_{11}\,c_{22} - c_{12}^2}, \tag{6}$$

$$N = c_{11}\,c_{22}\,c_{33} - c_{11}\,c_{23}^2 - c_{22}\,c_{13}^2 - c_{33}\,c_{12}^2 + 2c_{12}\,c_{13}\,c_{23}, \tag{7}$$

$$\nu_{21} = \frac{c_{12}\,c_{33} - c_{13}\,c_{23}}{c_{11}\,c_{33} - c_{13}^2}, \qquad \nu_{31} = \frac{c_{13}\,c_{22} - c_{12}\,c_{23}}{c_{11}\,c_{22} - c_{12}^2}, \qquad \nu_{32} = \frac{c_{23}\,c_{11} - c_{12}\,c_{13}}{c_{11}\,c_{22} - c_{12}^2},$$

$$\nu_{12} = \frac{c_{12}\,c_{33} - c_{13}\,c_{23}}{c_{22}\,c_{33} - c_{23}^2}, \qquad \nu_{13} = \frac{c_{13}\,c_{22} - c_{12}\,c_{23}}{c_{22}\,c_{33} - c_{23}^2}, \qquad \nu_{23} = \frac{c_{23}\,c_{11} - c_{12}\,c_{13}}{c_{11}\,c_{33} - c_{13}^2}. \tag{8}$$

Backward transformation:

$$c_{11} = \frac{1 - \nu_{23}\,\nu_{32}}{\Delta} E_1, \qquad c_{22} = \frac{1 - \nu_{13}\,\nu_{31}}{\Delta} E_2, \qquad c_{33} = \frac{1 - \nu_{12}\,\nu_{21}}{\Delta} E_3,$$

$$c_{12} = \frac{\nu_{21} + \nu_{31}\,\nu_{23}}{\Delta} E_1, \qquad c_{23} = \frac{\nu_{32} + \nu_{12}\,\nu_{31}}{\Delta} E_2, \qquad c_{13} = \frac{\nu_{13} + \nu_{23}\,\nu_{12}}{\Delta} E_3, \tag{9}$$

$$c_{44} = G_{23}, \qquad c_{55} = G_{31}, \qquad c_{66} = G_{12},$$

$$\Delta = 1 - \nu_{12}\,\nu_{21} - \nu_{23}\,\nu_{32} - \nu_{31}\,\nu_{13} - 2\nu_{21}\,\nu_{32}\,\nu_{13}.$$

Elastic stability requires all diagonal elements of both c_{IJ} and s_{IJ} as well as N and Δ to be positive. In addition, $\nu_{IJ}^2 < E_I/E_J$ for $I \neq J$. For further constraints on Poisson's ratios see [Q.4, p. 69].

Further frequent cases with even lower anisotropies (i. e. higher symmetries) are transversely isotropic (equivalent to hexagonal) and cubic with five and three independent elastic constants, respectively :

$$\begin{pmatrix} c_{11} & c_{12} & c_{13} & 0 & 0 & 0 \\ & c_{11} & c_{13} & 0 & 0 & 0 \\ & & c_{33} & 0 & 0 & 0 \\ & & & c_{44} & 0 & 0 \\ & \text{sym} & & & c_{44} & 0 \\ & & & & & c_{66} \end{pmatrix} \qquad \begin{pmatrix} c_{11} & c_{12} & c_{12} & 0 & 0 & 0 \\ & c_{11} & c_{12} & 0 & 0 & 0 \\ & & c_{11} & 0 & 0 & 0 \\ & & & c_{44} & 0 & 0 \\ & \text{sym} & & & c_{44} & 0 \\ & & & & & c_{44} \end{pmatrix} \tag{10}$$

with $c_{66} = \frac{1}{2}(c_{11} - c_{12})$ — cubic

transverse isotropy (hexagonal)

The engineering notation is also used for transversely isotropic materials ($E_1 = E_2$, $G_{31} = G_{23}$, $\nu_{32} = \nu_{31}$, $G_{12} = E_1/[2(1+\nu_{21})]$; further $\nu_{12} = \nu_{21}$, $\nu_{13}/E_1 = \nu_{31}/E_3$, $\nu_{23}/E_1 = \nu_{32}/E_3$; $\nu_{13} = \nu_{23}$; therefore $E_1, \nu_{21}, E_3, \nu_{31}$ and G_{23} provide a complete description. For cubic materials E_1, ν_{21} and G_{23} suffice (however, $G_{23} \neq E_1/[2(1+\nu_{21})]$, if not isotropic!).

The elastic properties of cubic materials are conveniently expressed by one modulus of compression K (bulk modulus) for changes of volume with constant shape and two shear moduli μ and μ' for changes of shape with constant volume :

$$K = \tfrac{1}{3}(c_{11} + 2c_{12}), \qquad \mu' = \tfrac{1}{2}(c_{11} - c_{12}), \qquad \mu = c_{44}. \tag{11}$$

This anisotropy of shear deformation may be characterised by the dimensionless measure :

$$a = \frac{\mu' - \mu}{\mu' + \mu} \quad \text{with} \quad -1 < a < 1. \tag{12}$$

With $a = 0$ one proceeds from cubic anisotropy to isotropy, which is determined by two independent parameters. Several pairs are in common use, e. g. compression (bulk) modulus K and shear modulus μ, Lamé's constants λ and μ, finally Young's modulus E and Poisson's ratio σ. The dependencies are summarised in the table below. See also Table 4 in Section I.8.

Elastic stability requires $K > 0$ and $\mu > 0$, leading to

$$\lambda > -\tfrac{2}{3}\mu, \quad E > 0, \quad -1 < \sigma < \tfrac{1}{2}, \quad c_{11} > 0, \quad -\tfrac{1}{2}c_{11} < c_{12} < c_{11}. \tag{13}$$

To: \ From:	K, μ	λ, μ	E, σ	$c_{11}, c_{12}\ (c_{44})$
$K =$	K	$\lambda + \frac{2}{3}\mu$	$\dfrac{E}{3(1-2\sigma)}$	$\frac{1}{3}(c_{11} + 2c_{12})$
$\mu =$	μ	μ	$\dfrac{E}{2(1+\sigma)}$	$\frac{1}{2}(c_{11} - c_{12}) = c_{44}$
$\lambda = c_{12} =$	$K - \frac{2}{3}\mu$	λ	$\dfrac{\sigma E}{(1+\sigma)(1-2\sigma)}$	c_{12}
$E =$	$\dfrac{9K\mu}{3K+\mu}$	$\dfrac{(3\lambda+2\mu)\mu}{\lambda+\mu}$	E	$\dfrac{(c_{11}+2c_{12})(c_{11}-c_{12})}{c_{11}+c_{12}}$
$\sigma =$	$\dfrac{3K-2\mu}{6K+2\mu}$	$\dfrac{\lambda}{2(\lambda+\mu)}$	σ	$\dfrac{c_{12}}{c_{11}+c_{12}}$
$c_{11} =$	$K + \frac{4}{3}\mu$	$\lambda + 2\mu$	$\dfrac{E(1-\sigma)}{(1+\sigma)(1-2\sigma)}$	c_{11}

Table 1: Interrelations between isotropy parameters. For a more extensive table including eight additional pair combinations see [Q.7, p. 74].

Moduli associated with elementary deformations and corresponding waves :

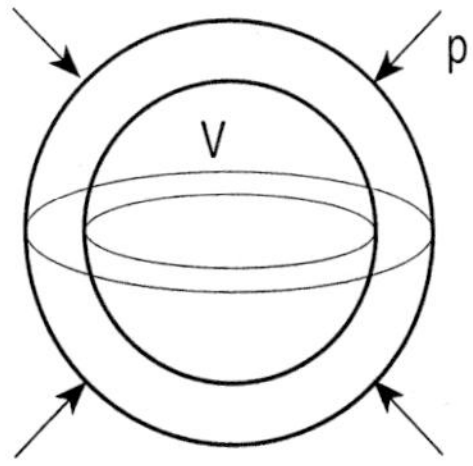

Compression modulus (bulk modulus) K:

$$p = -K\frac{\Delta V}{V} \tag{14}$$

K^{-1} = compressibility; no associated wave.

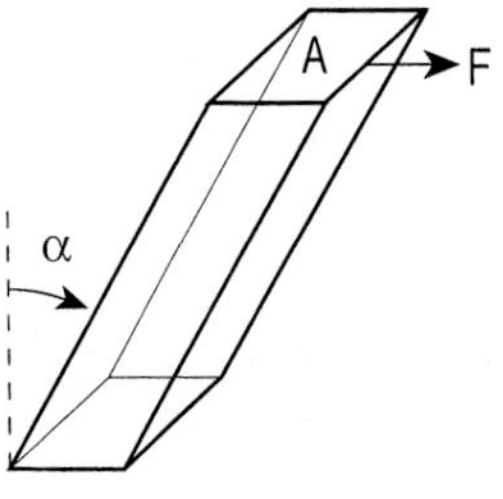

Shear modulus μ:

$$\frac{F}{A} = \mu\alpha \qquad (\alpha\text{: engineering shear strain}) \tag{15}$$

Velocity of transversal waves (shear waves and torsional waves):

$$c_T = \sqrt{\mu/\rho}\,. \tag{16}$$

Young's modulus E *and Poisson's ratio (lateral contraction)* σ:

$$\frac{F}{A} = E\frac{\Delta L}{L} \quad \text{(x-direction along L)} \tag{17}$$

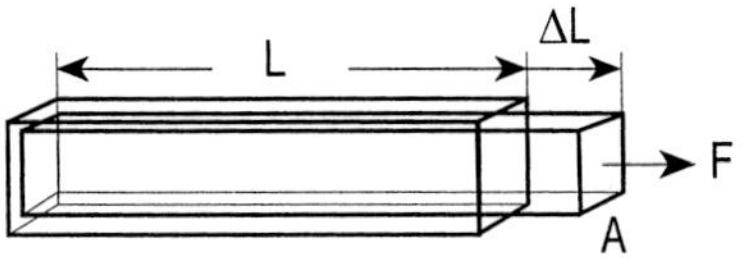

$$\sigma = -\frac{\varepsilon_{yy}}{\varepsilon_{xx}} \tag{18}$$

Velocity of (quasi-) longitudinal waves in bars: $c_{QL} = \sqrt{E/\rho}$. (19)

Modulus for longitudinal waves D:

$$-p = \sigma_{xx} = D\varepsilon_{xx} = c_{11}\varepsilon_{xx} \tag{20}$$

Velocity of (pure) longitudinal waves (with compression/dilatation and shear):

$$c_L = \sqrt{D/\rho} \; . \tag{21}$$

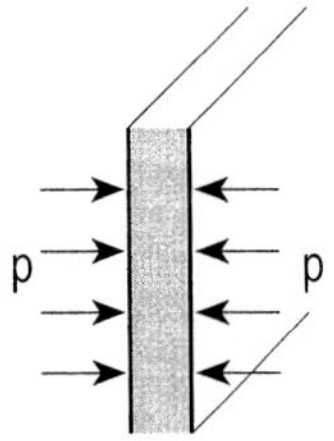

Q.3 Interface conditions, reflection and refraction of plane waves

Ref.: Auld, [Q.8, p. 1-62]

At the interface (boundary) between two elastic media I and II the requirements for tight contact are continuity of displacements, $u_i^I = u_i^{II}$, and balance of forces

$$\left(\sigma_{ij}^{I} - \sigma_{ij}^{II}\right) n_j = 0 \tag{1}$$

with unit normal vector n_i on the interface. If the medium II is a fluid with sound pressure p^{II} at the interface, then $\sigma_{ij}^{I} n_j = -p^{II} n_i$, since $\sigma_{ij}^{II} = -p^{II}\delta_{ij}$. With the usual slip assumption only the displacement component normal to the interface, $n_i u_i$, must be continuous.

A plane elastic wave incident on a plane interface between two homogeneous elastic solids with different material properties (tight contact) generates up to three reflected waves and up to three refracted (transmitted) waves, since there are up to three different wave velocities in an anisotropic medium. The propagation directions of these scattered waves are given by Snell's law, the relations for their amplitudes are called Fresnel equations. The boundary conditions require that the wavevector component tangent to the boundary is the same for all waves. This immediately leads to Snell's law :

$$\frac{\sin\Theta_i}{c_i} = \frac{\sin\Theta_0}{c_0} = \frac{1}{c_{trace}}, \tag{2}$$

with phase velocities c and angles Θ $(0 \le \Theta < \pi)$ between wavevector $\vec{k}$ and normal $\vec{n}$. Subscript i : reflected and refracted waves; subscript 0: incident wave. For $\sin\Theta_i > 1$, the i-th wave becomes evanescent (exponentially decaying perpendicular to the interface) and propagates with trace velocity c_{trace} along the interface (e. g. for total reflection).

Since in anisotropic media the phase velocities are no longer independent from the propagation direction (see Q.9.1), the evaluation of Snell's law is more involved. A helpful geometrical technique makes use of slowness surfaces, which provide the magnitude of $\vec{k}/\omega$ as a function of the direction of $\vec{k}$ [Q.3, p. 32-38; Q.9, p. 393-413].

In anisotropic media, where the direction of energy propagation (see Q.5) may be different from the wavevector direction, the incident wave should be characterised by the former rather than by the latter. The energy directions need not lie in the plane defined by the wavevectors! [Q.10; Q.11].

For Fresnel equations see e. g. [Q.1, p. 168-187] and [Q.8, p. 21-43].

Q.4 Material damping

Ref.: Gaul, [Q.12]

The conventional viscoelastic generalisation of the relation $\tilde{\sigma}_{ij}(t) = 2\mu\,\tilde{\varepsilon}_{ij}(t)$ between stress and strain deviators for pure shear deformation in an isotropic solid (μ: shear modulus; see Q.1) is given by :

$$\sum_{k=0}^{M} p_k \frac{d^k}{dt^k}\tilde{\sigma}_{ij}(t) = \sum_{k=0}^{N} q_k \frac{d^k}{dt^k}\tilde{\varepsilon}_{ij}(t) \tag{1}$$

with integer k and real coefficients p_k and q_k. Classical viscoelastic models are the Zener model :

$$\tilde{\sigma}_{ij}(t) + p_1\frac{d\tilde{\sigma}_{ij}(t)}{dt} = q_0\,\tilde{\varepsilon}_{ij}(t) + q_1\frac{d\tilde{\varepsilon}_{ij}(t)}{dt} \tag{2}$$

and – as its descendants – the Kelvin-Voigt model ($p_1 = 0$) and the Maxwell model ($q_0 = 0$). In the following, however, the further generalisation to time derivatives of fractional order will be considered :

$$\sum_{k=0}^{M} p_k \frac{d^{\alpha_k}}{dt^{\alpha_k}} \tilde{\sigma}_{ij}(t) = \sum_{k=0}^{N} q_k \frac{d^{\beta_k}}{dt^{\beta_k}} \tilde{\varepsilon}_{ij}(t), \tag{3}$$

which normally assures causality and leads in many cases to improved curve-fitting of measured data with less parameters ($\alpha_0 = \beta_0 = 0;\ 0 < \alpha_k, \beta_k < 1$). An extension to fractional orders beyond one is possible. An alternative formulation in terms of relaxation functions reads :

$$\tilde{\sigma}_{ij}(t) = 2\int_{-\infty}^{t} G(t-\tau)\frac{d\tilde{\varepsilon}_{ij}(\tau)}{d\tau}\,d\tau, \qquad \tilde{\varepsilon}_{ij}(t) = 2\int_{-\infty}^{t} J(t-\tau)\frac{d\tilde{\sigma}_{ij}(\tau)}{d\tau}\,d\tau \tag{4}$$

with $\displaystyle \frac{d}{dt}\int_0^t 2G(\tau)J(t-\tau)\,d\tau = 1.$

The relaxation modulus $G(t)$ and the creep compliance $J(t)$ describe the fading memory of the (linear viscoelastic) material with respect to the loading history. Any elastic modulus M may be generalised in the same manner.

For time-harmonic fields, $\tilde{\varepsilon}_{ij}(t) = \mathrm{Re}\left\{\tilde{\varepsilon}_{ij}(\omega)\exp[j\omega t]\right\}$,

one obtains :

$$\tilde{\sigma}_{ij}(\omega) = 2\mu(\omega)\,\tilde{\varepsilon}_{ij}(\omega) \quad \text{with} \quad 2\mu(\omega) = \sum_{k=0}^{N} q_k (j\omega)^{\beta_k} \Big/ \sum_{k=0}^{M} p_k (j\omega)^{\alpha_k} \tag{5}$$

The decomposition $\mu(\omega) = \mu'(\omega) + j\mu''(\omega) = \mu'(\omega)[1 + j\eta(\omega)]$ of the complex shear modulus introduces the storage modulus $\mu'(\omega)$ (real part), the loss modulus $\mu''(\omega)$ (imaginary part) and the loss factor $\eta(\omega)$, which may be interpreted as the ratio of the dissipated energy $D(\omega)$ per cycle to the 2π-fold of the stored energy $U(\omega)$:

$$\eta(\omega) = \frac{D(\omega)}{2\pi U(\omega)} = \frac{\mu''(\omega)}{\mu'(\omega)} = \frac{1}{Q(\omega)} = \frac{\Lambda}{\pi} = \tan\delta(\omega) \tag{6}$$

with quality factor $Q(\omega)$, logarithmic decrement $\Lambda(\omega)$ and loss tangent $\tan\delta(\omega)$. The relationship between the complex modulus $\mu(\omega)$ and the relaxation modulus $G(t)$ amounts to

$$\mu'(\omega) = \mu_0 + \omega\int_0^{\infty} \hat{G}(\tau)\sin(\omega\tau)\,d\tau, \qquad \mu''(\omega) = \omega\int_0^{\infty} \hat{G}(\tau)\cos(\omega\tau)\,d\tau \tag{7}$$

with real static or equilibrium modulus $\mu_0 = \mu(\omega = 0) = G\,(t = \infty)$ and $\hat{G}(t) = G(t) - \mu_0$, hence $\hat{G}(\infty) = 0$.

Because of causality and linearity the real and imaginary parts of any modulus $M = M' + j\,M''$ are connected via the Kramers-Kronig relations :

$$M'(\omega) = \frac{2}{\pi} P \int_0^\infty \frac{\Omega\, M''(\Omega)}{\Omega^2 - \omega^2}\, d\Omega \,, \qquad M''(\omega) = -\frac{2}{\pi} P \int_0^\infty \frac{\Omega M'(\Omega)}{\Omega^2 - \omega^2}\, d\Omega \,, \tag{8}$$

where P denotes Cauchy's principal value [Q.13, p. 20]. Regarding practical applications of these relations see, e. g. [Q.14]. The often utilised (non-viscoelastic) model with M' and M'' independent of frequency ('constant hysteresis damping model', 'hysteretic model' or 'structural damping model') violates causality.

The elastic-viscoelastic correspondence principle states that the solution of a viscoelastic problem may be obtained from the solution of the corresponding elastic problem. In the case of time-harmonic problems this means a straightforward substitution of the elastic moduli by the corresponding frequency-dependent and complex viscoelastic moduli. Otherwise, Fourier transformations have to be performed before and after this substitution. (An alternative realisation uses Laplace transforms and impact response functions.)

***Example: Five-parameter fractional-derivative model** (**generalised Zener model**)*

The complex shear modulus for a highly damped polymer,

$$\mu(\omega) = \mu_0 \frac{1 + b(j\omega)^\beta}{1 + a(j\omega)^\alpha} \tag{9}$$

with $\mu_0 = 87$ kPa, $a = 0.039$, $\alpha = 0.39$, $b = 0.38$, $\beta = 0.64$, yields a good fit of experimental data of $\mu'(\omega)$ and $\eta(\omega)$ from 1 Hz to almost 10 kHz.

Special case: Four-parameter fractional-derivative model [Q.15]

Thermodynamic constraints like non-negative rate of dissipated energy and the requirement of a finite viscoelastic wave speed impose the conditions

$$\mu > 0, \quad b > a > 0, \quad 0 < \alpha = \beta < 2 \tag{10}$$

on the generalised Zener model, thus leading to the simplified version with four parameters. For some modulus M it may be written as :

$$M(\omega) = \frac{M_0 + (j\omega\tau_r)^{\alpha} M_{\infty}}{1 + (j\omega\tau_r)^{\alpha}} \qquad \text{or} \qquad \frac{M(\omega) - M_{\infty}}{M_0 - M_{\infty}} = \frac{1}{1 + (j\omega\tau_r)^{\alpha}} \tag{11}$$

(Cole-Cole equation) with the static modulus M_0, the high-frequency limit M_{∞} and a relaxation time τ_r. With normalised frequency $\nu = \omega\tau_r$, $\Psi = \alpha\pi/2$ and ratio $\gamma = M_{\infty}/M_0$:

$$\frac{M'}{M_0} = \frac{1 + (\gamma + 1)\nu^{\alpha}\cos\Psi + \gamma\,\nu^{2\alpha}}{1 + 2\nu^{\alpha}\cos\Psi + \nu^{2\alpha}}, \qquad \frac{M''}{M_0} = \frac{(\gamma - 1)\nu^{\alpha}\sin\Psi}{1 + 2\nu^{\alpha}\cos\Psi + \nu^{2\alpha}}, \tag{12}$$

$$\eta = \frac{(\gamma - 1)\nu^{\alpha}\sin\Psi}{1 + (\gamma + 1)\nu^{\alpha}\cos\Psi + \gamma\,\nu^{2\alpha}}. \tag{13}$$

M' is monotonically increasing with frequency, while M'' and η have maxima :

$$M''_{max} = \frac{(\Delta M/2)\sin\Psi}{1 + \cos\Psi} \qquad \text{at} \qquad \omega = 1/\tau_r, \tag{14}$$

$$\eta_{max} = \frac{(\gamma - 1)\sin\Psi}{2\sqrt{\gamma} + (\gamma + 1)\cos\Psi} \qquad \text{at} \qquad \omega = \frac{1}{\tau_r\,\sqrt[2\alpha]{\gamma}} \tag{15}$$

($\Delta M = M_{\infty} - M_0$). The half-value bandwidths are for M'' and for η :

$$\left(\frac{\omega_2}{\omega_1}\right)_{M''} = \alpha\sqrt{\frac{A + \sqrt{A^2 - 1}}{A - \sqrt{A^2 - 1}}} \qquad \text{with} \qquad A = 2 + \cos\Psi, \tag{16}$$

$$\left(\frac{\omega_2}{\omega_1}\right)_{\eta} = \alpha\sqrt{\frac{B + \sqrt{B^2 - 4\gamma}}{B - \sqrt{B^2 - 4\gamma}}} \qquad \text{with} \qquad B = 4\sqrt{\gamma} + (\gamma + 1)\cos\Psi. \tag{17}$$

The model is causal and applicable for materials with one symmetric loss peak.

Example: Young's modulus E of dense PVC foam (446 kg/m³)

$E_0 = 1.82$ MPa, $E_{\infty} = 1.14$ GPa, $\alpha = 0.335$ and $\tau_r = 21.3$ ns have been determined from measurements between about 100 Hz and 10 kHz. Hence the loss factor maximum is $\eta_{max} = 0.53$ at 500 Hz.

Q.5 Energy

Q.5.1 General relations [Q.9, p. 144-145]

With the density of power $P_{ext} = F_i v_i$ supplied by an external density of forces F_i and the density of dissipated power P_{diss} the energy balance is expressed by :

$$\frac{d}{dt}\left(e_{kin} + e_{pot}\right) + \frac{\partial S_i}{\partial x_i} + P_{diss} = P_{ext} \tag{1}$$

with the energy flux density $S_i = -\sigma_{ij} v_j$, also called acoustic Poynting vector or Kirchhoff vector.

Q.5.2 Surface intensity

At the free surface $z = 0$ of an elastic solid the stress components σ_{xz} , σ_{yz} and σ_{zz} vanish, hence, with the in-surface velocity components v_x and v_y , the Kirchhoff vector becomes :

$$\vec{S} = -\begin{pmatrix} \sigma_{xx} v_x + \sigma_{xy} v_y \\ \sigma_{xy} v_x + \sigma_{yy} v_y \\ 0 \end{pmatrix} . \tag{2}$$

This quantity or its time-average, the surface intensity, can be experimentally determined from the five measurable values of ε_{xx} , ε_{xy} , ε_{yy} , v_x and v_y , if the elastic properties at the surface are known. In the isotropic case, [Q.16]

$$S_x = -2\mu \left\{ \left(\frac{\varepsilon_{xx} + \sigma\, \varepsilon_{yy}}{1-\sigma} \right) v_x + \varepsilon_{xy} v_y \right\} , \tag{3}$$

$$S_y = -2\mu \left\{ \varepsilon_{xy} v_x + \left(\frac{\sigma\, \varepsilon_{xx} + \varepsilon_{yy}}{1-\sigma} \right) v_y \right\} , \tag{4}$$

whereas in the anisotropic case the three needed stress components (including three non-measurable strain components) have to be evaluated from the system of six linear equations :

$$\begin{pmatrix} \sigma_{xx} \\ \sigma_{yy} \\ 0 \\ 0 \\ 0 \\ \sigma_{xy} \end{pmatrix} = \begin{pmatrix} c_{11} & c_{12} & c_{13} & c_{14} & c_{15} & c_{16} \\ & c_{22} & c_{23} & c_{24} & c_{25} & c_{26} \\ & & c_{33} & c_{34} & c_{35} & c_{36} \\ & & & c_{44} & c_{45} & c_{46} \\ & \text{sym} & & & c_{55} & c_{56} \\ & & & & & c_{66} \end{pmatrix} \cdot \begin{pmatrix} \varepsilon_{xx} \\ \varepsilon_{yy} \\ \varepsilon_{zz} \\ 2\varepsilon_{yz} \\ 2\varepsilon_{zx} \\ 2\varepsilon_{xy} \end{pmatrix} . \tag{5}$$

Q.5.3 Time-harmonic wavefields [Q.2, p. 10-28]; [Q.9, p. 154]

For harmonic time dependence the field quantities are advantageously written in complex notation, e. g. $\sigma_{ij}(x_i, t) = \sigma_{ij}(x_i) \exp(j\omega t)$, with the local amplitude $\sigma_{ij}(x_i)$ being complex as well. The physical quantity is understood to be the real (or imaginary) part of the complex representation. However, for energy quantities which are products of two field quantities, one has to be careful. Here the complex representation conveniently yields the time average, e. g., of thc Kirchhoff vector,

$$\langle S_i(x_i,t)\rangle_t = \left\langle \mathrm{Re}\{-\sigma_{ij}(x_i,t)\}\,\mathrm{Re}\{v_j(x_i,t)\}\right\rangle_t = \tfrac{1}{2}\mathrm{Re}\{-\sigma_{ij}(x_i)v_j^*(x_i)\} = I_i(x_i)$$

which is called intensity. Sometimes, a complex intensity is defined (spatial dependencies will be dropped from now on):

$$I_i^C = -\tfrac{1}{2}\sigma_{ij}v_j^* = I_i + jQ_i \qquad \text{with} \qquad Q_i = \tfrac{1}{2}\mathrm{Im}\{-\sigma_{ij}v_j^*\} \tag{7}$$

The sign of the reactive intensity Q_i depends on the convention of the time factor (here: exp(jωt)), whereas the sign of the (active or effective) intensity I_i does not. With the complex external power density $P_{ext}^C = F_i v_i^*$ the divergence relations

$$\frac{\partial I_i}{\partial x_i} = -\langle P_{diss}\rangle_t + \mathrm{Re}\left\{\langle P_{ext}^C\rangle_t\right\} \quad \text{and} \quad \frac{\partial Q_i}{\partial x_i} = -2\omega\langle L\rangle_t + \mathrm{Im}\left\{\langle P_{ext}^C\rangle_t\right\} \tag{8}$$

$(L = e_{kin} - e_{pot})$

hold for general time-harmonic fields in solids (P_{diss} is real). (Unlike with airborne sound fields, in elasto-acoustic fields Q_i is not in general proportional to the gradient of the time-averaged potential energy density, and the curl of Q_i is not generally zero.)

Q.5.4 Rayleigh's principle [Q.17]; [Q.18]

Rayleigh's principle states equipartition of potential and kinetic energy for time-harmonic sound fields without dissipation under certain circumstances, e. g.,

- after space and time averaging in freely vibrating finite solids.

For solids of infinite extent in at least one dimension the rule is called *Rayleigh's principle for propagating waves* and holds in homogeneous media, e. g.,

- for a plane wave in an unbounded isotropic or anisotropic solid (no averaging required);
- for a plate wave in a thick isotropic or anisotropic plate after averaging over time and thickness;

- for a beam wave in a thick isotropic or anisotropic straight beam after averaging over time and cross-section;
- for a bending wave in a thin isotropic plate or beam [Q.19, p. 103] (no averaging required).

and in periodically inhomogeneous media :

- for a Bloch wave after averaging over time and unit cell.

Q.5.5 Energy velocity and group velocity [Q.20], [Q.18]

The velocity of energy transport, defined by $\dfrac{I_i}{\left\langle e_{kin} + e_{pot} \right\rangle_t}$,

and the group velocity $C_i = \dfrac{\partial \omega}{\partial k_i}$

of a plane wave with wavevector k_i are equal in homogeneous anisotropic solids.

Therefore, $I_i = \left\langle e_{kin} + e_{pot} \right\rangle_t C_i$.

The same is true for Bloch waves in periodically inhomogeneous media after spatial averaging over a unit cell and for the fundamental waves in homogeneous plates and straight beams after averaging over the thickness or the cross-section, respectively.

Q.6 Random media

Ref.: Sornette, [Q.21]

The random inhomogeneities in a three-dimensional elastic medium may be characterised by an elastic mean free path $\ell_e \approx (nq_s)^{-1}$ with the density n of (non-absorptive) scatterers with scattering cross section q_s. The elasto-acoustic energy propagation over a distance d obeys different laws in different regimes :

I. Wavelike propagation: Weak disorder; $d < \ell_e$:

$$I_i = \left\langle e_{kin} + e_{pot} \right\rangle_t C_i^{eff} \tag{1}$$

Propagation takes place like in homogeneous media with an effective velocity C_i^{eff}. Homogenisation techniques (effective medium theories) are applicable. Even without dissipation there is an attenuation of the wave due to scattering into other directions.

II. Diffusion: Moderate disorder; $d > \ell_e$ (multiple scattering); wavelength $\lambda_w < \ell_e$:

$$I_i = -D(\omega)\frac{\partial\langle e_{kin} + e_{pot}\rangle_t}{\partial x_i} \tag{2}$$

with frequency-dependent diffusion coefficient

$$D(\omega) \approx D_0\left(1 - \gamma\left(\frac{\lambda_w}{\ell_e}\right)^2\right) \quad ; \quad D_0 = \ell_e c/3 \tag{3}$$

(c: velocity between scatterers) and γ a numerical factor of order unity. The transmission factor of a plate of thickness h is proportional to D/h. If $(\lambda_w/\ell_e)^2$ cannot be neglected in $D(\omega)$, one enters the regime of weak localisation (contributions of coherent interference effects; transition to III).

III Anderson localisation: Strong disorder; $d > \ell_e$; $\lambda_w > \ell_e$ (Ioffe-Regel criterion).

Energy propagation is slower than diffusive (D=0) due to coherent interference effects. Vibrations decay exponentially with localisation length ξ. The transmission factor of a plate is proportional to $\exp(-h/\xi)$.

Q.7 Periodic media

Ref.: Maysenhölder, [Q.2, Chapt. 5]

A periodically inhomogeneous medium may be constructed by infinite repetition of a unit cell defined by linearly independent basis vectors $\vec{a}_1, \vec{a}_2, \vec{a}_3$. Its volume is $V_0 = \vec{a}_1 \cdot \vec{a}_2 \times \vec{a}_3$. A position $\vec{r}$ in a unit cell and an equivalent position in another unit cell are connected by a lattice vector $\vec{g}_n = n_1\vec{a}_1 + n_2\vec{a}_2 + n_3\vec{a}_3$ (n_i: integer). Any function f with the periodicity of the lattice, $f(\vec{r} + \vec{g}_n) = f(\vec{r})$, may be Fourier expanded :

$$f(\vec{r}) = \sum_M f^M \exp\left[j\vec{G}^M \cdot \vec{r}\right] \tag{1}$$

with Fourier coefficients :

$$f^M = \frac{1}{V_0}\int f(\vec{r})\exp\left[-j\vec{G}^M \cdot \vec{r}\right] d^3r \tag{2}$$

The sum is over all reciprocal lattice points, the integral over one unit cell.

$$\vec{G}^M = M_1\vec{A}_1 + M_2\vec{A}_2 + M_3\vec{A}_3 \tag{3}$$

(M_i: integer) is a vector of the reciprocal lattice with the basis vectors :

$$\vec{A}_1 = \frac{2\pi}{V_0}(\vec{a}_2 \times \vec{a}_3) \;;\quad \vec{A}_2 = \frac{2\pi}{V_0}(\vec{a}_3 \times \vec{a}_1) \;;\quad \vec{A}_3 = \frac{2\pi}{V_0}(\vec{a}_1 \times \vec{a}_2) \;; \qquad (4)$$
$$\vec{a}_i \cdot \vec{A}_j = 2\pi\delta_{ij} \,.$$

Analogous to plane waves in homogeneous media the fundamental wave solutions for periodic media are Bloch waves

$$\vec{u}(\vec{r},t) = \vec{p}(\vec{r})\, e^{j(\omega t - \vec{k}\cdot\vec{r})} \qquad (5)$$

with the periodic function $\vec{p}(\vec{r} + \vec{g}_n) = \vec{p}(\vec{r})$. With the Fourier coefficients of the reciprocal density, $\tilde{\rho} = \rho^{-1}$, the Fourier transform of the equation of motion (see Q.1) for Bloch waves attains the usual form of an eigenvalue problem :

$$\omega^2 p_i^N = \sum_{L,M} \tilde{\rho}^{\,N-M} C_{ijkl}^{M-L}\left(k_l + G_l^L\right)\left(k_j + G_j^M\right) p_k^L \qquad (6)$$

Capital Latin superscripts again run over all points of the reciprocal lattice. Local isotropy in terms of the Lamé constants λ and μ leads to [Q.22] :

$$\omega^2 p_i^N = \sum_{L,M} \tilde{\rho}^{\,N-M}\Big\{\Big[\lambda^{M-L}\left(k_j + G_j^L\right)\left(k_i + G_i^M\right) + \mu^{M-L}\left(k_i + G_i^L\right)\left(k_j + G_j^M\right)\Big] p_j^L + $$
$$+ \mu^{M-L}\left(k_j + G_j^L\right)\left(k_j + G_j^M\right) p_i^L\Big\} \qquad (7)$$

Analytical expressions for energy density and intensity of Bloch waves and their averages over a unit cell including low-frequency limits are given in [Q.2, Chapt. 5].

Special case: Locally isotropic medium with one-dimensional periodicity along the x-direction (density ρ , stiffness $\zeta = \lambda + 2\mu$, spatial period h). Consider longitudinal waves in x-direction (one-dimensional problem).

a) *Example with analytical solution* :

Unit cell: two homogeneous layers (n = 1, 2) of thicknesses h_n with ρ_n and ζ_n. For given frequency ω the Bloch-wavenumber k follows from [Q.13, p. 216-219] :

$$\cos(kh) = \cos(\omega t_1)\cos(\omega t_2) - \Delta\sin(\omega t_1)\sin(\omega t_2) \qquad (8)$$

with travel times $t_n = h_n/c_n$, phase velocities $c_n = \sqrt{\zeta_n / \rho_n}$ and

$$\Delta = \frac{\rho_1 \zeta_1 + \rho_2 \zeta_2}{2\sqrt{\rho_1 \rho_2 \zeta_1 \zeta_2}} . \tag{9}$$

$|\cos(kh)| \le 1:$ pass bands; $|\cos(kh)| > 1:$ stop bands.

b) Low-frequency limit (homogenisation) :

Real formulation for $\zeta(-x) = \zeta(x)$ with spatial averages $\langle \zeta(x) \rangle = \zeta_0$ and $\langle \rho(x) \rangle = \rho_0$. For $k \ll 2\pi/h$ the displacement field of a Bloch wave may be approximated by a 'phase-modulated' wave :

$$u(x,t) \propto \cos(kx - \omega t) - k q(x) \sin(kx - \omega t) \approx \cos[k(x + q(x)) - \omega t], \tag{10}$$

with anti-symmetric periodic 'modulation function'

$$q(x) = -x + \zeta_{eff} \int_0^x \frac{dx'}{\zeta(x')}, \qquad q(0) = q(\tfrac{h}{2}) = 0 . \tag{11}$$

The effective properties are :

$$\rho_{eff} = \rho_0 \qquad \zeta_{eff} = \left\langle \frac{1}{\zeta(x)} \right\rangle^{-1} \qquad c_{eff} = \sqrt{\frac{\zeta_{eff}}{\rho_0}} \tag{12}$$

Explicitly for the example a):

$$\zeta_{eff} = \frac{h}{\dfrac{h_1}{\zeta_1} + \dfrac{h_2}{\zeta_2}}, \tag{13}$$

$$q(x) = \begin{cases} \left(\dfrac{\zeta_{eff}}{\zeta_1} - 1\right) x & \quad 0 \le x \le \dfrac{h_1}{2} \\ & \text{for} \\ \left(\dfrac{\zeta_{eff}}{\zeta_2} - 1\right) x + \dfrac{h_1 \zeta_{eff}}{2}\left(\dfrac{1}{\zeta_1} - \dfrac{1}{\zeta_2}\right) & \quad \dfrac{h_1}{2} < x \le \dfrac{h}{2} \end{cases} \tag{14}$$

For continuous variation of the stiffness, e. g. $\zeta(x) = \zeta_0 + \zeta_s \cos(2\pi x/h) > 0$, one obtains :

$$\zeta_{eff} = \sqrt{\zeta_0^2 - \zeta_s^2}, \qquad q(x) = -x + \frac{h}{\pi} \arctan\left[\sqrt{\frac{1 - \zeta_s/\zeta_0}{1 + \zeta_s/\zeta_0}} \tan\frac{\pi x}{h}\right]. \tag{15}$$

Q.8 Homogenisation

Ref.: Beltzer, [Q.13, p. 187-202]

Homogenisation substitutes an inhomogeneous medium by an 'equivalent' homogeneous medium with effective material properties in the limit of low frequencies. The definition of effective elastic constants is by spatial averages

$$\left\langle \sigma_{ij} \right\rangle = C_{ijkl}^{eff} \left\langle \varepsilon_{kl} \right\rangle ; \qquad \text{in general} \qquad C_{ijkl}^{eff} \neq \left\langle C_{ijkl} \right\rangle . \tag{1}$$

In general, the effective density may become complex [Q.23] or even a tensor [Q.3, p. 298, 322-323], however, in all cases listed below $\rho_{eff} = \langle \rho \rangle$.

Q.8.1 Bounds on effective moduli :

For an inhomogeneous elastic medium consisting of N phases with volume fractions ϕ_n and elastic constants $C_{ijkl}^{(n)}$ the *Voigt and Reuss averages,*

$$C_{ijkl}^{Voigt} = \sum_{n=1}^{N} \phi_n C_{ijkl}^{(n)} , \qquad C_{ijkl}^{Reuss} = \left[\sum_{n=1}^{N} \frac{\phi_n}{C_{ijkl}^{(n)}} \right]^{-1} , \qquad C_{ijkl}^{Reuss} \le C_{ijkl}^{eff} \le C_{ijkl}^{Voigt} , \tag{2}$$

are rigorous bounds for the effective elastic constants, which hold for arbitrary structures. These bounds are useful for small ϕ_n and small differences between the $C_{ijkl}^{(n)}$.

Narrower bounds are available for special cases, e. g. the *Hashin-Shtrikman bounds* for $N = 2$ statistically distributed isotropic phases (K_n, μ_n: compression and shear moduli; $\phi_2 \equiv \phi$) :

$$\frac{\left(K_1 + \tilde{K}\right)\phi}{K_1 + \tilde{K} + (K_2 - K_1)(1-\phi)} \le \frac{K_{eff} - K_1}{K_2 - K_1} \le \frac{\left(K_1 + \tilde{\tilde{K}}\right)\phi}{K_1 + \tilde{\tilde{K}} + (K_2 - K_1)(1-\phi)} , \tag{3}$$

$$\frac{\left(\mu_1 + \tilde{\mu}\right)\phi}{\mu_1 + \tilde{\mu} + (\mu_2 - \mu_1)(1-\phi)} \le \frac{\mu_{eff} - \mu_1}{\mu_2 - \mu_1} \le \frac{\left(\mu_1 + \tilde{\tilde{\mu}}\right)\phi}{\mu_1 + \tilde{\tilde{\mu}} + (\mu_2 - \mu_1)(1-\phi)} , \tag{4}$$

where, if $(\mu_2 - \mu_1)(K_2 - K_1) \ge 0$, then

$$\tilde{K} = \frac{4\mu_1}{3} , \quad \tilde{\tilde{K}} = \frac{4\mu_2}{3} , \quad \tilde{\mu} = \frac{3}{2}\left(\frac{1}{\mu_1} + \frac{10}{9K_1 + 8\mu_1} \right)^{-1} , \quad \tilde{\tilde{\mu}} = \frac{3}{2}\left(\frac{1}{\mu_2} + \frac{10}{9K_2 + 8\mu_2} \right)^{-1} , \tag{5}$$

while, if $(\mu_2 - \mu_1)(K_2 - K_1) \le 0$, then

$$\tilde{K} = \frac{4\mu_2}{3} , \quad \tilde{\tilde{K}} = \frac{4\mu_1}{3} , \quad \tilde{\mu} = \frac{3}{2}\left(\frac{1}{\mu_1} + \frac{10}{9K_2 + 8\mu_1} \right)^{-1} , \quad \tilde{\tilde{\mu}} = \frac{3}{2}\left(\frac{1}{\mu_2} + \frac{10}{9K_1 + 8\mu_2} \right)^{-1} . \tag{6}$$

For $\mu_1 = \mu_2$ the bounds for K_{eff} coincide and yield the exact K_{eff}.

Q.8.2 Effective moduli for particular structures :

a) Voigt and Reuss averages for polycrystals with statistical orientation of the grains (all of the same anisotropic material) are approximations for compression and shear moduli of an effective isotropic material:

$$K_{Voigt} = \tfrac{1}{9} C_{iikk} , \quad \mu_{Voigt} = \tfrac{1}{10}\left(C_{ikik} + \tfrac{1}{3} C_{iikk} \right). \tag{7}$$

Reuss averages are obtained accordingly with the compliance tensor (inverse of C_{ijkl}).

For a material with cubic grains:

$$K_{Voigt} = \tfrac{1}{3}(c_{11} + 2c_{12}) = K , \quad \mu_{Voigt} = \tfrac{1}{5}(c_{11} - c_{12} + 3c_{44}) = \tfrac{1}{5}(2\mu' + 3\mu). \tag{8}$$

In this case $K_{Reuss} = K_{Voigt} = K_{eff}$.

b) Spherical inclusions (K_s, μ_s , radius r) randomly dispersed in a homogeneous matrix with K_m and μ_m ; wavelengths much larger than r. Neglecting multiple scattering for small volume fractions ϕ of the spheres leads to the approximation :

$$K_{eff} = K_m + \frac{(3K_m + 4\mu_m)(K_s - K_m)\,\phi}{(3K_s + 4\mu_m) - 3(K_s - K_m)\,\phi} , \tag{9}$$

$$\mu_{eff} = \mu_m + \frac{\mu_m(15K_m + 20\mu_m)(\mu_s - \mu_m)\,\phi}{6\mu_s(K_m + 2\mu_m) + \mu_m(9K_m + 8\mu_m) - 6(K_m + 2\mu_m)(\mu_s - \mu_m)\,\phi} . \tag{10}$$

c) Composite sphere assembly: A sphere composed of two isotropic materials (K_s, μ_s up to inner radius a; K_m, μ_m from a to outer radius b; $\phi =$ a/b) has an effective compression modulus K_{eff}, which is equal to the above expression for spherical inclusions. An assembly of such composite spheres with different sizes, but common ratio ϕ, possesses the same K_{eff} in the long-wavelength limit (exact results).

d) Periodically spaced fibres parallel to the x-axis in isotropic matrix; ϕ = volume fraction of the fibres. Both the fibre material and the effective medium are transversely isotropic. Approximation for the effective moduli in 'engineering notation' according to Chamis [Q.24, p. 151]:

$$E_{xx} = \phi E_{xx}^f + (1-\phi)E_m , \qquad E_{yy} = E_{zz} = \frac{E_m}{1-(1-E_m/E_{yy}^f)\sqrt{\phi}} , \tag{11}$$

$$G_{xy} = G_{xz} = \frac{G_m}{1-(1-G_m/G_{xy}^f)\sqrt{\phi}} , \qquad G_{yz} = \frac{G_m}{1-(1-G_m/G_{yz}^f)\sqrt{\phi}} , \tag{12}$$

$$\nu_{xy} = \nu_{xz} = \phi \nu_{xy}^f + (1-\phi)\nu_m , \qquad \left(\nu_{yz} = -1 + \frac{E_{yy}}{2G_{yz}} \right) . \tag{13}$$

(E_m, ν_m, $G_m = E_m/[2(1+\nu_m)]$: Young's modulus, Poisson's ratio and shear modulus of isotropic matrix; superscript f denotes fibre properties.) An alternative for isotropic fibre materials (with E_f, ν_f and $G_f = E_f/[2(1+\nu_f)]$) are the Halpin-Tsai equations [Q.4, p. 151-158] :

$$E_{xx} = \phi E_f + (1-\phi)E_m , \qquad \nu_{xy} = \nu_{xz} = \phi \nu_f + (1-\phi)\nu_m , \tag{14}$$

$$\frac{M}{M_m} = \frac{1+\xi\eta\phi}{1-\eta\phi} \quad \text{with} \quad \eta = \frac{(M_f/M_m)-1}{(M_f/M_m)+\xi} , \tag{15}$$

where M is one of the quantities E_{yy}, G_{xy} or ν_{yz} of the composite and M_m or M_f are the corresponding quantities E , G or ν for the matrix and the fibre materials. The three ξ's are adjustable parameters depending on fibre and packing geometry and can range from 0 to ∞. Example: Circular fibres in a square array: Use $\xi = 2$ for E_{yy} and $\xi = 1$ for G_{xy}.

e) Symmetric stack of layers with unidirectional fibre reinforcement (for properties of transversely isotropic fibres see above). Fractional layer thickness h_n , total thickness h ; the fibre directions relative to the global x-axis may be different for different layers, but lie all in the x-y-plane of the layers. By symmetric is meant that the top half of the stack is a mirror image of the bottom half. After transformation to global co-ordinates the elastic constants (now 'monoclinic-like') of the n-th layer are denoted by c_{nIJ} , the constants of the monoclinic effective medium by c_{IJ} (no summation convention).

$$c_{33} = \left[\sum_{n=1}^{N} \frac{h_n}{c_{n33}} \right]^{-1} , \qquad c_{13} = \tfrac{1}{2} \sum_{n=1}^{N} h_n c_{n13} \left(1 + \frac{c_{33}}{c_{n33}} \right) , \tag{16}$$

$$c_{23} = \tfrac{1}{2} \sum_{n=1}^{N} h_n c_{n23} \left(1 + \frac{c_{33}}{c_{n33}} \right) , \qquad c_{36} = \tfrac{1}{2} \sum_{n=1}^{N} h_n c_{n36} \left(1 + \frac{c_{33}}{c_{n33}} \right) , \tag{17}$$

$$c_{IJ} = \sum_{n=1}^{N} h_n c_{nIJ} \quad \{I,J = 1,2,6\} , \qquad c_{IJ} = \sum_{n=1}^{N} \frac{h_n c_{nIJ}}{\Delta\, \Delta_n} \quad \{I,J = 4,5\} , \tag{18}$$

$$\Delta = \left[\sum_{n=1}^{N} \frac{h_n c_{n44}}{\Delta_n} \right] \left[\sum_{n=1}^{N} \frac{h_n c_{n55}}{\Delta_n} \right] - \left[\sum_{n=1}^{N} \frac{h_n c_{n45}}{\Delta_n} \right]^2 , \qquad \Delta_n = c_{n44} c_{n55} - (c_{n45})^2 . \tag{19}$$

The formulae are valid for wavelengths greater than the total thickness of the stack [Q.24, p. 159-160].

f) Periodic stack of transversely isotropic layers. Notation as in the previous case (h = period). Note that in both *e)* and *f)* the constituent layers are transversely isotropic, however, in *e)* the axes of symmetry lie – possibly in different directions – in the x-y-plane, whereas now the axis of symmetry is the z-axis for all layers.

$$c_{33} = \left[\sum_{n=1}^{N} \frac{h_n}{c_{n33}} \right]^{-1} , \qquad c_{13} = c_{33} \sum_{n=1}^{N} h_n \frac{c_{n13}}{c_{n33}} , \tag{20}$$

$$c_{11} = \frac{(c_{13})^2}{c_{33}} + \sum_{n=1}^{N} h_n \left(c_{n11} - \frac{(c_{n13})^2}{c_{n33}} \right), \quad c_{55} = \left[\sum_{n=1}^{N} \frac{h_n}{c_{n55}} \right]^{-1} , \quad c_{66} = \sum_{n=1}^{N} h_n \, c_{n66} \, . \tag{21}$$

The formulae are valid for wavelengths greater than the total thickness of the period [Q.3, p. 313]. If all constituent layers are isotropic, then c_{55} cannot be greater than c_{66} (because of the Cauchy-Schwartz-Kolmogorov inequality). For the inverse problem, the determination of the c_{nIJ} from the c_{IJ}, with $N = 2$ or 3 see [Q.3, p. 324-336].

For further homogenisation results see I.13 (sandwich panels) and Q.7 (periodic media).

Q.9 Plane waves in unbounded homogeneous media :

Ref.: Maysenhölder, [Q.2, p. 50-51, 91-100]; Beltzer, [Q.13, p. 95-109]

Q.9.1 Anisotropic media :

The equation of motion for homogeneous anisotropic media (see Q.1) can be satisfied by the plane-wave ansatz $u_i = A_i \exp[(j(\omega t - k_m x_m)]$ with real amplitude A_i defining the polarisation, wavenumber k and wavevector direction e_i, leading to Christoffel's equation :

$$\left(\Gamma_{il} - \rho c^2 \right) A_l = 0 \qquad \text{with} \qquad \Gamma_{il} = C_{ijkl}\, e_j\, e_k = \Gamma_{li} \tag{1}$$

(Γ_{il}: Christoffel's tensor). For a given direction e_i this is a cubic eigenvalue problem for eigenvalues ρc^2 and eigenvectors A_i ($c = \omega/k$: magnitude of phase velocity). Since without dissipative effects the symmetric Γ_{il} is also real, the polarisation are mutually orthogonal or can be chosen as such in case of degenerate eigenvalues. 'Pure modes' are – according to one definition – waves with purely longitudinal or purely transversal polarisation. Another definition requires the group velocity or intensity to be parallel to the wavevector. A wave which is not a pure mode may be termed quasi-longitudinal or quasi-transversal, if its polarisation is close to one of the pure polarisation.

The group velocity C_i is equal to the velocity of energy transport, which is in general not parallel to e_i :

$$C_j = \frac{\partial c}{\partial e_j} = \frac{A_i\,C_{ijkl}\,A_l\,e_k}{\rho A_m^2\,c}\;; \qquad C_i\,e_i = c. \tag{2}$$

Time average of energy density : $$\left\langle e_{kin} + e_{pot} \right\rangle_t = 2\left\langle e_{kin} \right\rangle_t = \tfrac{1}{2}\rho\omega^2 A_i^2 \tag{3}$$

Intensity : $$I_j = \left\langle e_{kin} + e_{pot} \right\rangle_t C_j = \frac{\omega^2}{2c} A_i\,C_{ijkl}\,A_l\,e_k \tag{4}$$

The slowness vector ℓ_m, which is useful for reflection and refraction problems (see Q.3) is defined by $\ell_m = \frac{e_m}{c} = \frac{k_m}{\omega}$:

$$\exp[(j(\omega t - k_m x_m)] = \exp[(j\omega(t - \ell_m x_m)]\,.$$

Two-dimensional example with cubic (i. e. quadratic) anisotropy:

Wave vector direction: $$e_i = \begin{pmatrix} \cos\varphi \\ \sin\varphi \end{pmatrix}. \tag{5}$$

Polarisation angle: $$\psi = \arctan\frac{A_2}{A_1} \tag{6}$$

Eigenvalues:

$$\rho c^2 = \tfrac{1}{2}\left\{ c_{11} + c_{44} \pm \sqrt{\tfrac{1}{2}(c_{11} - c_{44})^2(1 + \cos 4\varphi) + \tfrac{1}{2}(c_{12} + c_{44})^2(1 - \cos 4\varphi)} \right\} \tag{7}$$

Polarisations: $$\frac{A_2}{A_1} = \frac{\rho c^2 - c_{11}\cos^2\varphi - c_{44}\sin^2\varphi}{(c_{12} + c_{44})\cos\varphi\,\sin\varphi} \tag{8}$$

The figures below are drawn for compression modulus $K = 1$, shear modulus $\mu' = 9$ and shear modulus $\mu = 1$ (fictitious material, arbitrary units; anisotropy $a = 0.8$). The two eigen-solutions (modes) are numbered according to increasing phase velocity.

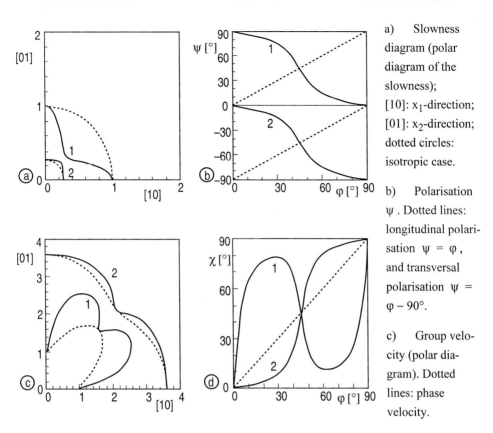

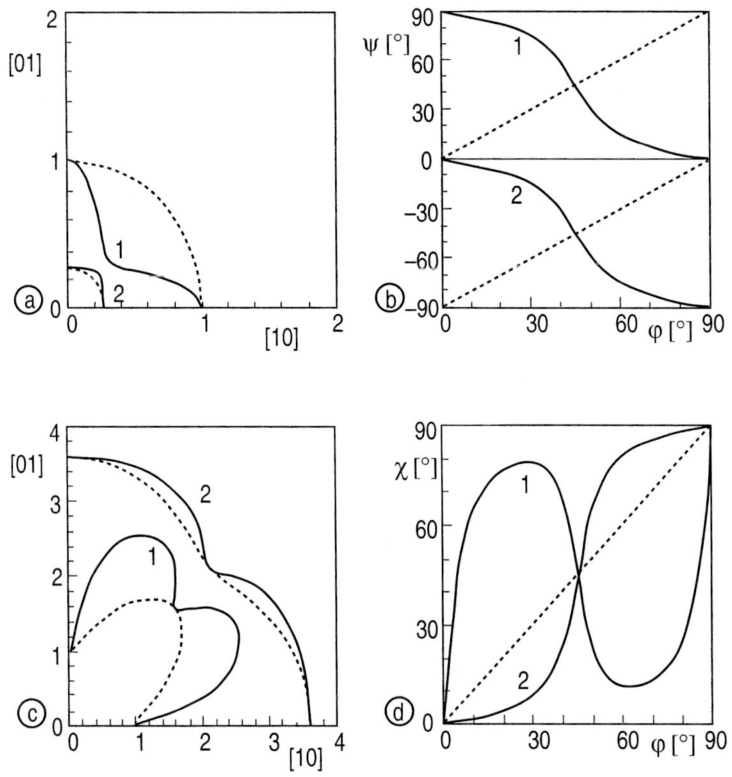

a) Slowness diagram (polar diagram of the slowness); [10]: x_1-direction; [01]: x_2-direction; dotted circles: isotropic case.

b) Polarisation ψ . Dotted lines: longitudinal polarisation $\psi = \varphi$, and transversal polarisation $\psi = \varphi - 90°$.

c) Group velocity (polar diagram). Dotted lines: phase velocity.

d) Intensity direction χ . Dotted line: propagation direction φ .

Q.9.2 Isotropic media :

In isotropic materials Christoffel's equation leads to two types of solutions: longitudinal and transversal plane waves with wave speeds

$$c_L = \sqrt{\frac{\lambda + 2\mu}{\rho}} \qquad \text{and} \qquad c_T = \sqrt{\frac{\mu}{\rho}} \tag{9}$$

The displacement field of longitudinal waves is curl free (changes of shape and volume without rotations), that of transversal waves is divergence free (change of shape without volume change). The strain and stress amplitudes for waves propagating in the x-direction with velocity amplitude v_0 are for longitudinal polarisation :

$$-\frac{v_0}{c_L}\begin{pmatrix}1&0&0\\0&0&0\\0&0&0\end{pmatrix}, \qquad -\rho v_0 c_L\begin{pmatrix}1&0&0\\0&\frac{\sigma}{1-\sigma}&0\\0&0&\frac{\sigma}{1-\sigma}\end{pmatrix} \tag{10}$$

and for transversal polarisation along the y-direction :

$$-\frac{v_0}{2c_T}\begin{pmatrix}0&1&0\\1&0&0\\0&0&0\end{pmatrix}, \qquad -\tfrac{1}{2}\rho v_0 c_T\begin{pmatrix}0&1&0\\1&0&0\\0&0&0\end{pmatrix}. \tag{11}$$

The formulas for energetic quantities are of the same form for both wave types :

$$e_{kin} = e_{pot} = \tfrac{1}{2}\rho|v_0|^2\cos^2(kx-\omega t), \qquad S_x = \rho|v_0|^2 c\cos^2(kx-\omega t), \tag{12}$$

$$\langle e_{kin}\rangle_t = \langle e_{pot}\rangle_t = \tfrac{1}{4}\rho|v_0|^2, \qquad I_x = \tfrac{1}{2}\rho|v_0|^2 c = 2\langle e_{kin}\rangle_t c. \tag{13}$$

(Add subscript $_L$ or $_T$ to the wave number k and the wave speed c as required.).
All y- and z-components and the reactive intensity vanish.

Q.10 Waves in bounded media

Q.10.1 Plate waves [Q.25]

Plate waves may be classified into two groups: shear waves and Lamb waves. Each group is further subdivided into a symmetric family (mirror symmetry of the displacement vector with respect to the plane $z = 0$) and an anti-symmetric family (sign change of the displacement vector after the mirror operation). The *shear waves*, sometimes called SH ('Shear-Horizontal') waves, possess non-vanishing displacements in the y direction only:

$$u_y = A\cos\left(\frac{n\pi z}{h}\right), \qquad n = 0, 2, 4, \ldots \qquad \text{(symmetric family);} \tag{1}$$

$$u_y = A\sin\left(\frac{n\pi z}{h}\right), \qquad n = 1, 3, 5, \ldots \qquad \text{(anti symmetric family).} \tag{2}$$

(A: Amplitude; phase factor $\exp[(j(\omega t - kx)]$ omitted.) The expressions for phase velocities c and group velocities C are valid for both families ($c_T = \sqrt{\mu/\rho}$):

$$c^2 = c_T^2\left[1 + \left(\frac{n\pi}{kh}\right)^2\right] = \frac{c_T^2}{1 - \left(\frac{n c_T}{2fh}\right)^2}, \qquad C \cdot c = c_T^2. \tag{3}$$

Conversely, $u_y \equiv 0$ for *Lamb waves*. With the abbreviations

$$\alpha_1 = \sqrt{1 - \left(\frac{c}{c_L}\right)^2}, \qquad \alpha_2 = \sqrt{1 - \left(\frac{c}{c_T}\right)^2}, \qquad \alpha_x = \frac{2\alpha_1\alpha_2}{1 + \alpha_2^2}, \tag{4}$$

$$\alpha_z = \frac{2}{1 + \alpha_2^2}, \qquad R_s = \frac{\sinh(\alpha_1 kh/2)}{\sinh(\alpha_2 kh/2)}, \qquad R_a = \frac{\cosh(\alpha_1 kh/2)}{\cosh(\alpha_2 kh/2)} \tag{5}$$

the x- and z-components of the displacement fields are for the symmetric family :

$$u_x = jA\left[\cosh(\alpha_1 kz) - \alpha_x R_s \cosh(\alpha_2 kz)\right],$$

$$u_z = \alpha_1 A\left[\sinh(\alpha_1 kz) - \alpha_z R_s \sinh(\alpha_2 kz)\right] \tag{6}$$

and for the anti symmetric family :

$$u_x = jA\left[\sinh(\alpha_1 kz) - \alpha_x R_a \sinh(\alpha_2 kz)\right],$$

$$u_z = \alpha_1 A\left[\cosh(\alpha_1 kz) - \alpha_z R_a \cosh(\alpha_2 kz)\right]. \tag{7}$$

Since measurements are usually confined to the surfaces of the plate, the displacement ratios u_x/u_z at $z = h/2$ are of particular interest [Q.26] :

$$\left.\frac{u_x}{u_z}\right|_{z=h/2} = \frac{\left(1+\alpha_2^2\right)\coth(\alpha_1 kh/2) - 2\alpha_1\alpha_2\coth(\alpha_2 kh/2)}{-j\alpha_1\left(\alpha_2^2-1\right)} \qquad \text{(symmetric family)} \qquad (8)$$

$$\left.\frac{u_x}{u_z}\right|_{z=h/2} = \frac{\left(1+\alpha_2^2\right)\tanh(\alpha_1 kh/2) - 2\alpha_1\alpha_2\tanh(\alpha_2 kh/2)}{-j\alpha_1\left(\alpha_2^2-1\right)} \qquad \text{(antisymmetric family)} \qquad (9)$$

Usually, the phase velocities have to be determined numerically from the transcendental Rayleigh-Lamb frequency equations :

$$\frac{4\alpha_1\alpha_2}{\left(1+\alpha_2^2\right)^2} = \left[\frac{\tanh(\pi\alpha_2 f h/c)}{\tanh(\pi\alpha_1 f h/c)}\right]^{\pm 1}, \qquad (10)$$

(f: frequency) with the plus sign for the symmetric family and the minus sign for the antisymmetric family. For dispersion diagrams see e. g. [Q.8, p. 76-87; Q.19, p.143]. The orthogonality relation for two Rayleigh-Lamb modes (1) and (2) with common frequency, but different wavenumbers,

$$\int_{-h/2}^{h/2} \left(u_x^{(1)}\sigma_{xx}^{(2)} - u_z^{(2)}\sigma_{zx}^{(1)}\right) dz = 0, \qquad (11)$$

holds even for the corresponding modes of a layered plate with z-dependent Lamé constants [Q.27].

If c is known, the corresponding group velocity C may be obtained analytically [Q.28] (with the same meaning of $\pm$ as above) :

$$C = \frac{c}{1 - \dfrac{f}{c}\dfrac{dc}{df}} \qquad \text{with} \qquad \frac{f}{c}\frac{dc}{df} = \frac{\pm Y}{X_\pm \pm Z}, \qquad (12)$$

$$X_+ = T_1^2 X, \quad X_- = T_2^2 X, \quad T_1 = \tanh(\pi\alpha_1 f h/c), \quad T_2 = \tanh(\pi\alpha_2 f h/c),$$

$$X = \frac{4c^3}{\pi c_T^2 f h N^2}\left[\frac{4\alpha_1^2\alpha_2^2}{N} - \alpha_1^2 - \left(\frac{\alpha_2 c_T}{c_L}\right)^2\right], \qquad N = 1 + \alpha_2^2 = 2 - \left(\frac{c}{c_T}\right)^2,$$

$$Y = \alpha_1 T_1 \alpha_2^2 K_2 - \alpha_2 T_2 \alpha_1^2 K_1, \qquad Z = \alpha_1 T_1 K_2 - \alpha_2 T_2 K_1,$$

$$K_1 = \cosh^{-2}(\pi\alpha_1 f h/c), \qquad K_2 = \cosh^{-2}(\pi\alpha_2 f h/c).$$

The group velocity C together with Rayleigh's principle for propagating waves (see Q.5.4) may be used for the calculation of the average intensity of a plate wave: $\langle I_x \rangle = 2\langle w_{kin}\rangle C$, with the time average w_{kin} of the kinetic energy density and the average over the plate thickness denoted by $\langle ... \rangle$. For both families of shear waves :

$$\langle w_{kin}\rangle = \frac{\pi^2}{2}|A|^2 \rho f^2 . \tag{13}$$

With $S_m = \dfrac{\sinh(\alpha_m kh)}{\alpha_m kh}$ (m = 1, 2) for symmetric Lamb waves : (14)

$$\begin{aligned}\langle w_{kin}\rangle = \frac{\pi^2}{2}|A|^2 \rho f^2 \Big\{ & (S_1+1) + |\alpha_x|^2 |R_s|^2 (S_2+1) + \\ & + |\alpha_1|^2 \Big[|S_1-1| + |\alpha_z|^2 |R_s|^2 |S_2-1| \Big] - 4\alpha_1^2 \alpha_z S_1 \Big\}\end{aligned} \tag{15}$$

and for anti-symmetric Lamb waves :

$$\begin{aligned}\langle w_{kin}\rangle = \frac{\pi^2}{2}|A|^2 \rho f^2 \Big\{ & |S_1-1| + |\alpha_x|^2 |R_a|^2 |S_2-1| + \\ & + |\alpha_1|^2 \Big[(S_1+1) + |\alpha_z|^2 |R_a|^2 (S_2+1) \Big] - 4|\alpha_1|^2 \alpha_z S_1 \Big\}\end{aligned} \tag{16}$$

(Note the little difference in the last term. It is essential for imaginary α_1 , i. e. for $c > c_L$.) The y- and z-components of the intensity are everywhere zero for all wave families.

Special case: *Quasi-longitudinal mode* :

This is the fundamental symmetric Lamb wave, which exhibits predominantly longitudinal character at low frequencies.

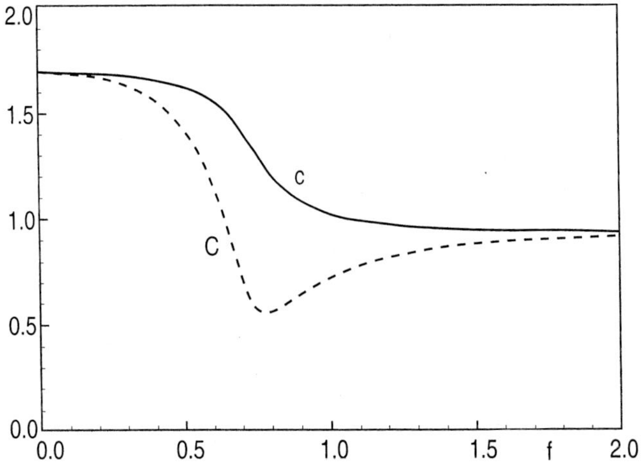

Phase velocity c and group velocity C of the quasi-longitudinal mode for Poisson's ratio $\sigma = 0.3$. Units: c_T for velocities, c_T/h for frequency.

At $f = c_T/\sqrt{2}h$ phase and group velocity are independent of Poisson's ratio σ and analytically known (Lamé wave): $c = c_T\sqrt{2}$, $C = c_T/\sqrt{2}$. This implies $kh = \pi$ (wavelength = 2h) and

$$u_x = -A\cos\left(\frac{\pi z}{h}\right), \qquad u_z = jA\sin\left(\frac{\pi z}{h}\right), \tag{17}$$

$$\langle w_{kin}\rangle = \frac{\pi^2}{2}\left|\frac{A}{h}\right|^2 \mu, \qquad \langle I_x\rangle = \frac{\pi^2}{\sqrt{2}}\left|\frac{A}{h}\right|^2 \mu c_T. \tag{18}$$

The strain field of this wave is pure shear and $u_x \equiv 0$ at the plate surfaces !

Low-frequency approximation:

$$c_{QL} = C_{QL} = c_T\sqrt{\frac{2}{1-\sigma}}, \qquad \frac{u_z}{u_x} = \frac{-j\sigma}{1-\sigma}kz, \qquad \langle w_{kin}\rangle = \pi^2|A|^2\rho f^2. \tag{19}$$

For Poisson's ratio $\sigma = 0.3$ the error of $\langle I_x\rangle = 2\langle w_{kin}\rangle C$ in this approximation is smaller than 20% (10%; 5%) for $fh/c_T < 0.4$ (0.3; 0.1).

Low-frequency expansions up to the third non-vanishing order [Q.26] :

$$\begin{aligned} c^2 &= c_{QL}^2\left[1 - \frac{\sigma^2}{12(1-\sigma)^2}(hk)^2 - \frac{\sigma^2\left(6-10\sigma-7\sigma^2\right)}{720(1-\sigma)^4}(hk)^4\right] \\ &= c_{QL}^2\left[1 - \frac{\pi^2\sigma^2}{3(1-\sigma)^2}\left(\frac{hf}{c_{QL}}\right)^2 - \frac{\pi^4\sigma^2\left(6-10\sigma-2\sigma^2\right)}{45(1-\sigma)^4}\left(\frac{hf}{c_{QL}}\right)^4\right] \end{aligned} \tag{20}$$

High-frequency approximation ($\approx$ a Rayleigh wave on each plate surface) :

$$c = C = c_R, \qquad \frac{u_x}{u_z} = \frac{j\left[\exp(\alpha_1 kz) - \alpha_x\exp(\alpha_2 kz)\right]}{\alpha_1\left[\exp(\alpha_1 kz) - \alpha_z\exp(\alpha_2 kz)\right]}, \tag{21}$$

$$\langle w_{kin}\rangle = \tfrac{\pi}{8}|A|^2\rho\,\frac{f c_R}{h}\,e^{\alpha_1 kh}\left\{\frac{1}{\alpha_1} + \alpha_1(1-4\alpha_z) + (\alpha_1\alpha_z)^2\left(\frac{1}{\alpha_2} + \alpha_2\right)\right\}. \tag{22}$$

For Poisson's ratio $\sigma = 0.3$ the error of $\langle I_x\rangle = 2\langle w_{kin}\rangle C$ in this approximation is smaller than 20% (10%; 5%) for $fh/c_T > 1.5$ (1.9; 3.0).

Special case: *Bending mode* :

This is the fundamental anti-symmetric Lamb wave, which coincides with the bending wave in thin plates at low frequencies.

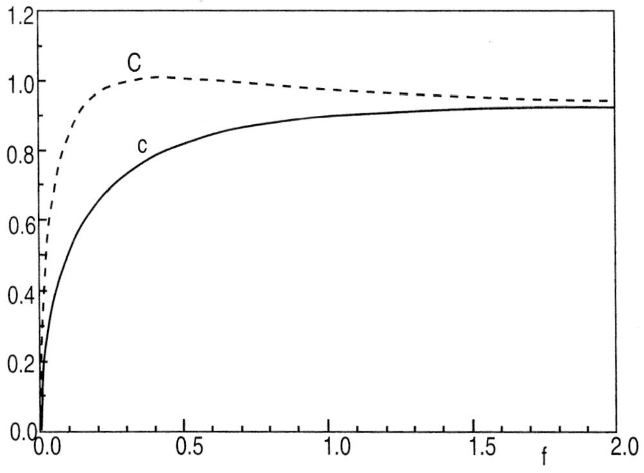

Phase velocity c and group velocity C of the bending mode for Poisson's ratio $\sigma = 0.3$. Units: c_T for velocities, c_T/h for frequency.

Low-frequency approximation (lowest order; corresponding to thin plate theory) :

$$c_B = \frac{\sqrt{2\pi f h c_T}}{\sqrt[4]{6(1-\sigma)}}\,, \qquad C_B = 2c_B\,, \qquad k_B = \sqrt[4]{6(1-\sigma)}\sqrt{\frac{2\pi f}{h c_T}}\,, \tag{23}$$

$$\lambda_B = \frac{\sqrt{2\pi h c_T}}{\sqrt[4]{6(1-\sigma)}\sqrt{f}}\,, \qquad \frac{u_x}{u_z} = -jkz\,, \qquad \langle w_{kin}\rangle = \pi^2 |A|^2 \rho f^2\,. \tag{24}$$

For Poisson's ratio $\sigma = 0.3$ the error of $\langle I_x\rangle = 2\langle w_{kin}\rangle C$ in this approximation is smaller than 20% (10%; 5%) for $fh/c_T < 0.13\ (0.065;\ 0.007)$. (For formulas with bending stiffness B see Q.10.3)

Low-frequency expansions up to the third non-vanishing order [Q.26] :

$$\begin{aligned} c^2 &= \frac{c_{QL}^2}{12}(hk)^2\left[1 - \frac{17-7\sigma}{60(1-\sigma)}(hk)^2 + \frac{489-418\sigma+62\sigma^2}{5040(1-\sigma)^2}(hk)^4\right] \\ &= \frac{\pi c_{QL}}{\sqrt{3}}(hf)\left[1 - \frac{\pi\sqrt{3}\,(17-7\sigma)}{30(1-\sigma)}\left(\frac{hf}{c_{QL}}\right) + \frac{\pi^2\left(3711-3362\sigma+211\sigma^2\right)}{4200(1-\sigma)^2}\left(\frac{hf}{c_{QL}}\right)^2\right] \end{aligned} \tag{25}$$

The high-frequency approximation (≈ a Rayleigh wave on each plate surface) is the same as with the quasi-longitudinal mode. For Poisson's ratio $\sigma = 0.3$ the error of $\langle I_x\rangle = 2\langle w_{kin}\rangle C$ in this approximation is smaller than 20% (10%; 5%) for $fh/c_T > 0.67\ (0.77;\ 3.0)$.

For additional formulas and diagrams including z-dependence of displacements, energy densities and intensities see [Q.25].

Q.10.2 Rayleigh waves :

The wave speed c_R of Rayleigh waves on a force-free surface of an isotropic half-space is obtained from the positive solution of the equation for γ^2 :

$$(2-\gamma^2)^2 - 4\sqrt{(1-\zeta^2\gamma^2)(1-\gamma^2)} = 0\,, \qquad \gamma = \frac{c_R}{c_T}\,, \qquad \zeta^2 = \left(\frac{c_T}{c_L}\right)^2 = \frac{1-2\sigma}{2-2\sigma}\,. \tag{26}$$

There is exact one such solution within the bounds $0 < c_R < c_T$. Eliminating the square root leads to the more familiar form $\gamma^6 - 8\gamma^4 + 8(3-2\zeta^2)\gamma^2 - 16(1-\zeta^2) = 0$, which, however, has additional extraneous solutions [Q.1, p. 189-191].

The non-vanishing displacement components of Rayleigh waves propagating along the x-direction on a half-space $z \le 0$ may be obtained by superposition of the two fundamental symmetric and anti-symmetric Lamb modes (for the α's with $c = c_R$ see Q.10.1):

$$u_x = j\tilde{A}\left[\exp(\alpha_1 kz) - \alpha_x \exp(\alpha_2 kz)\right], \qquad u_z = \alpha_1 \tilde{A}\left[\exp(\alpha_1 kz) - \alpha_z \exp(\alpha_2 kz)\right]. \tag{27}$$

This describes elliptical trajectories at arbitrary depth. The sense of rotation changes at a depth of about 0.2 wavelengths, where $u_x = 0$.

Displacement ratio at the surface:

$$\left.\frac{u_x}{u_z}\right|_{z=0} = \frac{j(1-\alpha_x)}{\alpha_1(1-\alpha_z)}\,. \tag{28}$$

Time average of kinetic energy per unit width in y-direction :

$$W_{kin} = \int_{-\infty}^{0} w_{kin}\, dz = \frac{\pi}{4}\rho c_R f\left|\tilde{A}\right|^2 \left\{\frac{1}{\alpha_1} + \alpha_1(1-4\alpha_z) + (\alpha_1\alpha_z)^2\left(\frac{1}{\alpha_2} + \alpha_2\right)\right\} \tag{29}$$

Time average of energy flow per unit width in y-direction:

$$\int_{-\infty}^{0} I\, dz = 2 W_{kin} c_R\,. \tag{30}$$

There are various approximations to the velocity of Rayleigh waves for the range $0 < \sigma < \frac{1}{2}$, some of which are discussed in [Q.29]. The Bergmann-Viktorov equation :

$$\gamma = \frac{c_R}{c_T} = \frac{0.87 + 1.12\sigma}{1+\sigma} \tag{31}$$

is accurate to within 0.5%. For surface waves on anisotropic half-spaces see also, [Q.30].

Q.10.3 Waves in thin plates :

Wave equations for thin plates with thickness h (thickness direction = z-direction) and mass density ρ; excitation terms like force densities, moment densities or pressure differences are omitted. Solutions: Phase velocities $c = \omega/k$ of waves propagating along the x-direction (u , w : displacements in x- and z-direction) with phase factor $\exp[j(\omega t - kx)]$.

Nota bene: The symbols c_{QL} and B have different meanings for plates and beams !
The intensity I has dimension Wm^{-2} ; the mean energy flow per unit width in a plate is $I \cdot h$.

Quasi-longitudinal waves (in-plane waves) , [Q.19, p. 86] :

For a homogeneous isotropic plate:

$$-\frac{\partial^2 u}{\partial t^2} + c_{QL}^2 \frac{\partial^2 u}{\partial x^2} = 0 \, , \qquad c_{QL} = \sqrt{\frac{E}{\rho(1-\sigma^2)}} \tag{32}$$

Transversal contraction:

$$\frac{\hat{w}}{\hat{u}} = \left(\frac{\sigma}{1-\sigma}\right)\frac{\pi h}{\lambda} \, . \qquad I_{QL} = 2\pi^2 |u_0|^2 \rho \, f^2 c_{QL} \tag{33}$$

($\hat{w}$, $\hat{u}$: maximum transversal and longitudinal displacements; u_0: amplitude of u). For range of validity see Q.10.1.

Bending waves

Inhomogeneous, locally monoclinic plate ($x \equiv x_1$, $y \equiv x_2$) , [Q.31] :

$$\rho h \frac{\partial^2 w}{\partial t^2} + \sum_{\alpha,\beta,\gamma,\delta=1}^{2} \frac{\partial^2}{\partial x_\alpha \partial x_\beta}\left(B_{\alpha\beta\gamma\delta} \frac{\partial^2 w}{\partial x_\gamma \partial x_\delta} \right) = 0 \tag{34}$$

with generalised bending stiffnesses :

$$B_{\alpha\beta\gamma\delta} = \frac{h^3}{12}\left(C_{\alpha\beta\gamma\delta} - \frac{C_{\alpha\beta 33}\, C_{\gamma\delta 33}}{C_{3333}} \right) \tag{35}$$

C_{ijkl}: elasticity tensor. ρ , h and $B_{\alpha\beta\gamma\delta}$ may depend on x_1 and x_2 .

Inhomogeneous, locally isotropic plate , [Q.32] :

$$\begin{aligned} \rho h \frac{\partial^2 w}{\partial t^2} &+ \frac{\partial^2}{\partial x^2}\left(\frac{B\,\partial^2 w}{\partial x^2}\right) + 2\frac{\partial^2}{\partial x \partial y}\left(\frac{B\,\partial^2 w}{\partial x \partial y}\right) + \frac{\partial^2}{\partial y^2}\left(\frac{B\,\partial^2 w}{\partial y^2}\right) \\ &+ \frac{\partial^2}{\partial x^2}\left(\frac{\sigma B\,\partial^2 w}{\partial y^2}\right) - 2\frac{\partial^2}{\partial x \partial y}\left(\frac{\sigma B\,\partial^2 w}{\partial x \partial y}\right) + \frac{\partial^2}{\partial y^2}\left(\frac{\sigma B\,\partial^2 w}{\partial x^2}\right) = 0 \end{aligned} \tag{36}$$

with the usual bending stiffness :

$$B = \frac{Eh^3}{12(1-\sigma^2)} = \frac{\mu h^3}{6(1-\sigma)} = \frac{\mu(\lambda+\mu)h^3}{3(\lambda+2\mu)} . \qquad (37)$$

Homogeneous orthotropic plate , [Q.33] :

$$\rho h \frac{\partial^2 w}{\partial t^2} + \left(B_x \frac{\partial^4}{\partial x^4} + 2B_{xy} \frac{\partial^4}{\partial x^2 \partial y^2} + B_y \frac{\partial^4}{\partial y^2} \right) w = 0 \qquad (38)$$

The bending stiffnesses are, [Q.34], in terms of the Voigt constants :

$$B_x = \frac{h^3}{12}\left[c_{11} - \frac{c_{13}^2}{c_{33}} \right], \quad B_{xy} = \frac{h^3}{12}\left[c_{12} + 2c_{66} - \frac{c_{13}c_{23}}{c_{33}} \right], \quad B_y = \frac{h^3}{12}\left[c_{22} - \frac{c_{23}^2}{c_{33}} \right] \qquad (39)$$

and with engineering constants (see Section Q.2) :

$$B_x = \frac{h^3}{12}\left[\frac{E_x}{1-\nu_{xy}\nu_{yx}} \right], \quad B_{xy} = \frac{h^3}{12}\left[2G_{xy} + \frac{E_x \nu_{yx}}{1-\nu_{xy}\nu_{yx}} \right], \quad B_y = \frac{h^3}{12}\left[\frac{E_y}{1-\nu_{xy}\nu_{yx}} \right]. \qquad (40)$$

A plane wave propagating at an angle Φ with the x-direction experiences a bending stiffness :
$B(\Phi) = B_x(\cos\Phi)^4 + 2B_{xy}(\cos\Phi \sin\Phi)^2 + B_y(\sin\Phi)^4$. (41)
($B(45°) = \frac{1}{4}\left(B_x + 2B_{xy} + B_y \right)$; phase velocity as for an isotropic plate with $B = B(\Phi)$).

The extremal values of $B(\Phi)$ from 0° to 90° are $B(0°) = B_x$, $B(90°) = B_y$ and

$$B(\Phi_e) = \frac{B_x B_y - B_{xy}^2}{B_x - 2B_{xy} + B_y} \quad \text{with} \quad \Phi_e = \tfrac{1}{2}\arccos\frac{-B_x + B_y}{B_x - 2B_{xy} + B_y} \quad (\text{for real } \Phi_e).(42)$$

In order that $B(\Phi) > 0$ for all Φ : $B_x > 0$, $B_y > 0$, $B_{xy} > -\sqrt{B_x B_y}$. (43)

Transition to isotropic case with $B_x = B_{xy} = B_y = B = \dfrac{Eh^3}{12(1-\sigma^2)}$. (44)

Homogeneous isotropic plate – classical theory :

$$\rho h \frac{\partial^2 w}{\partial t^2} + B\left(\frac{\partial^2}{\partial x^2} + \frac{\partial^2}{\partial y^2} \right)^2 w = 0 , \qquad c_B = \sqrt[4]{\frac{B}{\rho h}}\,\sqrt{\omega} \qquad (45)$$

with $B = \dfrac{Eh^3}{12(1-\sigma^2)} = \dfrac{\mu h^3}{6(1-\sigma)} = \dfrac{\mu(\lambda+\mu)h^3}{3(\lambda+2\mu)}$:

$$c_B = \sqrt[4]{\frac{E}{12\rho(1-\sigma^2)}}\,\sqrt{\omega h} = \frac{\sqrt{\omega h c_T}}{\sqrt[4]{6(1-\sigma)}} , \qquad \lambda_B = \sqrt[4]{\frac{E}{3\rho(1-\sigma^2)}}\,\sqrt{\frac{\pi h}{f}} = \frac{\sqrt{2\pi c_T h / f}}{\sqrt[4]{6(1-\sigma)}} .(46)$$

The accuracy of c_B is better than 10%, if the bending wavelength $\lambda_B > 6h$ (Cremer-Heckl limit [Q.19, p. 162]). The non-propagating solutions with imaginary speed $\pm jc_B$ and imaginary wavenumber $\pm jk_B$ are called nearfields.

Amplitude ratio : $$\frac{\hat{u}}{\hat{w}} = \frac{\pi h}{\lambda_B}\,. \tag{47}$$

Bending wave intensity : $$I_B = 2\pi^2 |w_0|^2 \rho f^2 C_B \tag{48}$$

Group velocity : $$C_B = 2c_B\,. \tag{49}$$

($\hat{u}$, $\hat{w}$: maximum longitudinal and transversal displacements; w_0: amplitude of w)

Measurement of the intensity of bending waves
along the x-direction (without nearfields) by two accelerometers according to :

$$I_B = -\frac{\sqrt{B\rho h}}{\omega h} \mathrm{Re}\left\{ a \int \frac{\partial a^*}{\partial x} dt \right\}, \qquad a = \frac{\partial^2 w}{\partial t^2}\,; \tag{50}$$

$\partial a^*/\partial x$ being approximated by a finite difference. The error of I_B relative to the exact result $\langle I_x \rangle$ (see Section Q.10.1) :

$$\delta = \frac{I_B - \langle I_x \rangle}{\langle I_x \rangle}\,,$$

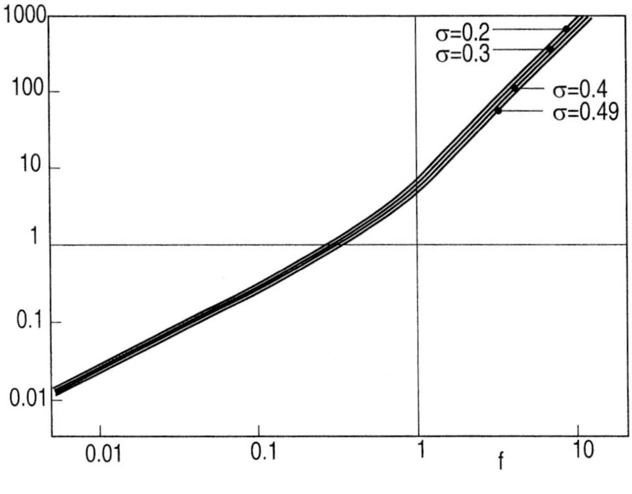

is shown in the diagram for various Poisson ratios (frequency in units of c_T/h) [Q.25].

Critical frequency f_{cr},
where $c_B = c_0$ (phase velocity in an ambient fluid) :

$$f_{cr} = \frac{c_0^2}{2\pi}\sqrt{\frac{\rho h}{B}} = \frac{c_0^2}{\pi h}\sqrt{\frac{3\rho(1-\sigma^2)}{E}} \tag{51}$$

Coincidence frequency $$f_c = \frac{f_{cr}}{\sin^2\Theta}\,, \tag{52}$$

where the trace velocity $c_0/\sin\Theta$ of a plane wave incident on the plate with polar angle Θ equals the wave speed c_B of free bending waves.

Wavenumber relations: $$k_B = \frac{2\pi}{\lambda_B} = \frac{\omega}{c_B} = \sqrt[4]{6(1-\sigma)}\sqrt{\frac{\omega}{h c_T}} = k_0\sqrt{\frac{f_{cr}}{f}}\,. \tag{53}$$

Homogeneous isotropic plate – Timoshenko-Mindlin model , [Q.13, p. 160] :
(including rotatory inertia and transverse shear effects)

$$\rho h\frac{\partial^2 w}{\partial t^2} + \left(B\nabla^2 - \frac{\rho h^3}{12}\frac{\partial^2}{\partial t^2}\right)\left(\nabla^2 - \frac{\rho}{\kappa^2\mu}\frac{\partial^2}{\partial t^2}\right)w = 0\,, \qquad \nabla^2 = \frac{\partial^2}{\partial x^2} + \frac{\partial^2}{\partial y^2} \tag{54}$$

with μ: shear modulus; κ: factor near unity. For a wave propagating in x-direction :

$$\frac{\partial^2 w}{\partial t^2} + \frac{B}{\rho h}\frac{\partial^4 w}{\partial x^4} - \left(\frac{B}{\kappa^2\mu h} + \frac{h^2}{12}\right)\frac{\partial^4 w}{\partial x^2\partial t^2} + \frac{\rho h^2}{12\kappa^2\mu}\frac{\partial^4 w}{\partial t^4} = 0\,. \tag{55}$$

The compact form of the dispersion relation :

$$\left(1 - \frac{c^2}{c_\kappa^2}\right)\left(\frac{c_{QL}^2}{c^2} - 1\right) = \frac{12}{(kh)^2} \tag{56}$$

with $c_\kappa^2 = \kappa^2 c_T^2 = \kappa^2\mu/\rho$ may be transformed to quadratic equations for $\tilde{c}^2 = c^2/c_{QL}^2$:

$$\frac{\tilde{c}^4}{\tilde{c}_\kappa^2} - \left(\frac{1}{\tilde{c}_\kappa^2} + 1 + \frac{12}{(kh)^2}\right)\tilde{c}^2 + 1 = 0, \qquad \left(\frac{1}{\tilde{c}_\kappa^2} - \frac{12 c_{QL}^2}{(\omega h)^2}\right)\tilde{c}^4 - \left(\frac{1}{\tilde{c}_\kappa^2} + 1\right)\tilde{c}^2 + 1 = 0. \tag{57}$$

with $\tilde{c}_\kappa = c_\kappa/c_{QL}$. The smaller root c^2 of the dispersion relation is the desired solution. The choice $c_\kappa = c_R$ (Rayleigh velocity) assures the correct value in the limit of high frequencies ($\kappa = 0.925$ for steel).

Q.10.4 Waves in thin beams :

Wave equations for thin, straight, isotropic, homogeneous beams with cross-sectional area A. Young's modulus E, Poisson's ratio σ and mass density ρ; excitation terms like forces or moments are omitted. Solutions: Phase velocities $c = \omega/k$ of waves propagating along the x-direction (= beam axis) with wavelengths λ_{QL} or λ_B and phase factor $\exp[j(\omega t - kx)]$.

Nota bene: The symbols c_{QL} and B have different meanings for plates and beams! The intensity I has dimension Wm^{-2} ; the mean total energy flow in a beam is I·A.

Quasi-longitudinal waves , [Q.19, p. 82-85] :

$$-\frac{\partial^2 u}{\partial t^2}+c_{QL}^2\frac{\partial^2 u}{\partial x^2}=0\,,\qquad c_{QL}=\sqrt{\frac{E}{\rho}}\,,\qquad I_{QL}=2\pi^2|u_0|^2\rho f^2 c_{QL} \tag{58}$$

u: longitudinal displacement; u_0: amplitude of u. Valid for arbitrary cross-section, if greatest thickness $d \ll \lambda_{QL}$. (For $d \gg \lambda_{QL}$, $c = c_L$.)

Lateral contraction for quadratic or circular cross-section :

$$\frac{\hat{w}}{\hat{u}}=\sigma\frac{\pi d}{\lambda_{QL}} \tag{59}$$

($\hat{w}$, $\hat{u}$: maximum lateral and longitudinal displacements).

Torsional waves , [Q.19, p. 90-94] :

$$-\frac{\partial^2 \Phi}{\partial t^2}+c_{\Phi}^2\frac{\partial^2 \Phi}{\partial x^2}=0\,,\qquad c_{\Phi}=\sqrt{\frac{T}{\Theta}}\,,\qquad I_{\Phi}=2\pi^2|\Phi_0|^2\Theta f^2 c_{\Phi} \tag{60}$$

Φ: angle of rotation of a cross-section; Φ_0: amplitude of Φ; T: torsional stiffness (rigidity) defined by $M_x = T\cdot\partial\Phi/\partial x$ for arbitrary cross-section (M_x: torsional moment), Θ: moment of inertia per unit length about the x-axis. For a hollow circular cylinder with inner radius r_i and outer radius r_o :

$$T=\frac{\pi}{2}\mu\left(r_o^4-r_i^4\right),\qquad \Theta=\frac{\pi}{2}\rho\left(r_o^4-r_i^4\right),\qquad c_{\Phi}=\sqrt{\frac{\mu}{\rho}}=c_T \tag{61}$$

(μ: shear modulus). This is the exact solution for arbitrary frequency (lowest torsional mode) including the solid cylinder ($r_i = 0$) , [Q.35].

Cross-sections without rotational symmetry do not remain plane ('warping'). Therefore the above torsional wave solution is only approximate; it always yields $c_{\Phi} < c_T$.

Results for rectangular cross-sections with dimensions $a \geq b$:

$$\Theta=\rho\frac{ab^3+a^3b}{12}=\frac{\rho A^2}{12}\left(\frac{a}{b}+\frac{b}{a}\right),\qquad T=s\mu ab^3=s\mu A^2\frac{b}{a} \tag{62}$$

a/b	1	1.5	2	3	6	10	∞
s	0.141	0.196	0.229	0.263	0.298	0.312	1/3
c_Φ	$0.920\, c_T$	$2\frac{b}{a}c_T\sqrt{\frac{3s}{1+(a/b)^{-2}}}$					$2\frac{b}{a}c_T$

Hence $c_\Phi \approx 2\frac{b}{a}c_T$ within 7% for $a/b \geq 6$.

Bending waves – Bernoulli-Euler model, [Q.19, p. 95-99] :

$$\frac{\partial^2 w}{\partial t^2}+\frac{B}{\rho A}\frac{\partial^4 w}{\partial x^4}=0\,, \qquad c_B=\sqrt[4]{\frac{B}{\rho A}}\sqrt{\omega}\,, \qquad I_B=2\pi^2|w_0|^2\rho f^2 C_B \tag{63}$$

($C_B = 2c_B$).

w: displacement along z-direction (displacement vector is in x-z-plane);

w_0: amplitude of w;

B = E J: bending stiffness;

J: second moment of cross-sectional area A about the neutral axis:

$$J=\int_A z^2\,dy\,dz \qquad \text{with} \qquad \int_A z\,dy\,dz=0$$

(Select z = 0 accordingly for the neutral axis.).

Hollow circular cylinder :

$$J=\frac{\pi}{4}\left(r_o^4-r_i^4\right)=\frac{A}{4}\left(r_o^2+r_i^2\right)\,, \qquad c_B=\sqrt[4]{\frac{E(r_i^2+r_o^2)}{4\rho}}\sqrt{\omega}\,. \tag{64}$$

Rectangular cross-section :

$$J=\frac{h^3 b}{12}\,, \qquad c_B=\sqrt[4]{\frac{E h^2}{12\rho}}\sqrt{\omega}\,. \tag{65}$$

(h: thickness in the z-direction, b: thickness in the y-direction). For a solid cylinder ($r_i = 0$) the Bernoulli-Euler model agrees with the exact solution only for $r_o/\lambda_B < 0.1$ [Q.13, p. 155-156].

Bending waves – Timoshenko model, [Q.36, p. 201-205] :
(including rotatory inertia and transverse shear effects)

$$\frac{\partial^2 w}{\partial t^2} + \frac{B}{\rho A}\frac{\partial^4 w}{\partial x^4} - \left(\frac{B}{\kappa^2 \mu A} + \frac{J}{A}\right)\frac{\partial^4 w}{\partial x^2 \partial t^2} + \frac{\rho J}{\kappa^2 \mu A}\frac{\partial^4 w}{\partial t^4} = 0\,, \qquad (66)$$

κ: factor near unity depending on the shape of the cross-section.

The smaller root c^2 of the dispersion relation :

$$\left(\frac{1}{c_\kappa^2} - \frac{A}{J\omega^2}\right)c^4 - \left(\frac{c_{QL}^2}{c_\kappa^2} + 1\right)c^2 + c_{QL}^2 = 0 \quad \text{or} \quad \frac{c^4}{c_\kappa^2} - \left(\frac{c_{QL}^2}{c_\kappa^2} + 1 - \frac{A}{Jk^2}\right)c^2 + c_{QL}^2 = 0 \qquad (67)$$

with $c_\kappa = \kappa c_T$, is the desired solution. The Rayleigh velocity c_R is a convenient choice for c_κ, which is the high frequency limit of c ($\kappa = 0.925$ for steel). For rectangular cross-section ($A/J = 12/h^2$) the dispersion relation is the same as for Timoshenko-Mindlin plates, however, with the different definition of c_{QL} for plates (see Section Q.10.3).

Q.11 Moduli of isotropic materials and related quantities

Notice:
Some notations in this and the following sections (written by F.P. Mechel) are different from corresponding notations in the previous sections of this chapter (by W. Maysenhölder).

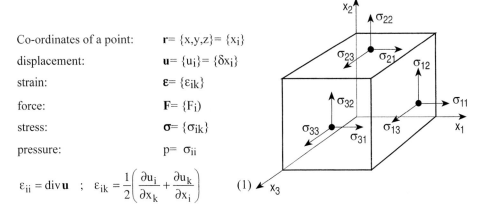

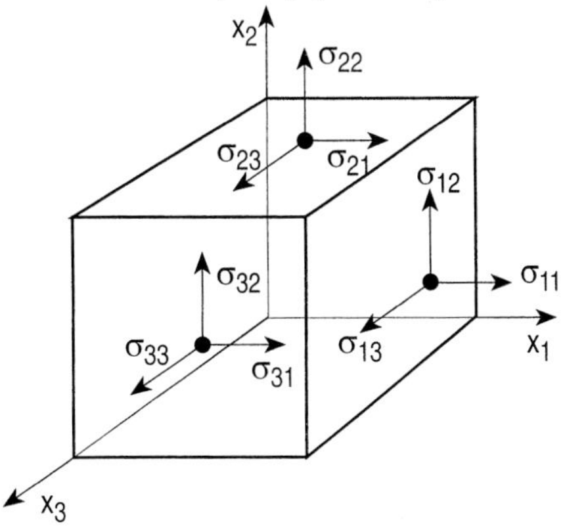

Co-ordinates of a point: $\mathbf{r} = \{x,y,z\} = \{x_i\}$

displacement: $\mathbf{u} = \{u_i\} = \{\delta x_i\}$

strain: $\boldsymbol{\varepsilon} = \{\varepsilon_{ik}\}$

force: $\mathbf{F} = \{F_i)$

stress: $\boldsymbol{\sigma} = \{\sigma_{ik}\}$

pressure: $p = \sigma_{ii}$

$$\varepsilon_{ii} = \operatorname{div}\mathbf{u} \quad ; \quad \varepsilon_{ik} = \frac{1}{2}\left(\frac{\partial u_i}{\partial x_k} + \frac{\partial u_k}{\partial x_i}\right) \qquad (1)$$

Lamé constants λ, μ :

$$\sigma_{ik} = \mu\left(\frac{\partial u_i}{\partial x_k} + \frac{\partial u_k}{\partial x_i}\right) = 2\mu\cdot\varepsilon_{ik} \quad ; \quad i \neq k \qquad (2)$$

$$\sigma_{ii} = \lambda\cdot\operatorname{div}\mathbf{u} + 2\mu\cdot\varepsilon_{ii} \qquad (3)$$

With losses: $\lambda \to \lambda(1+j\eta)$; $\mu \to \mu(1+j\eta)$

Shear modulus S :

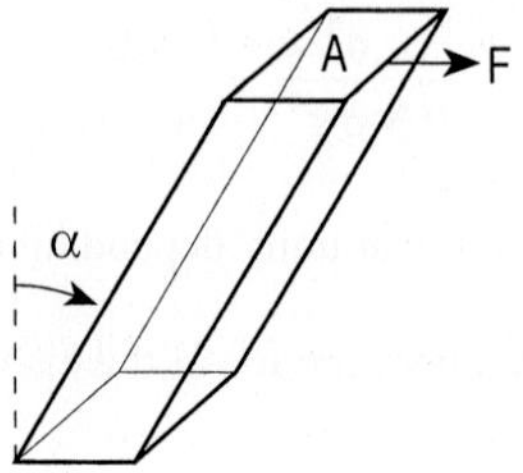

$\sigma = F/A$; $S = \sigma/\alpha$ (4)

$$S = \mu = \frac{E}{2(1+\sigma)} \qquad (5)$$

Free shear wave velocity (ρ= material density) :

$$c_S = \sqrt{S/\rho} \qquad (6)$$

Young's modulus E :

$-\sigma = F/A = E \cdot \Delta L/L = E \cdot s_{xx}$ (7)

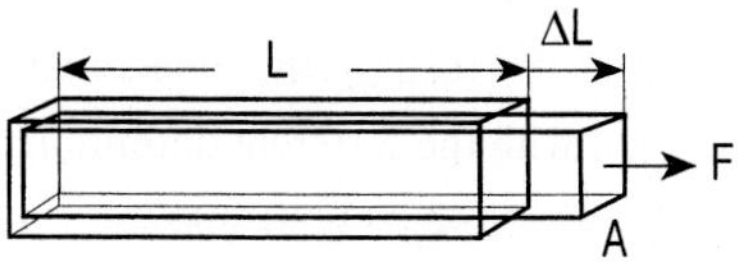

Lateral contraction:

$\sigma = -\varepsilon_{xx}/\varepsilon_{yy}$ (8)

$$E = \frac{\mu(3\lambda+2\mu)}{\lambda+\mu} = 2\mu(1+\sigma) = 2S(1+\sigma) \qquad (9)$$

Free bar longitudinal wave velocity: $c_E = \sqrt{E/\rho}$ (10)

Poisson's lateral contraction σ (Poisson ratio):

$$\sigma = -\frac{\varepsilon_{yy}}{\varepsilon_{xx}} = \frac{\lambda}{2(\lambda+\mu)} = \frac{E}{2S} - 1 \quad ; \quad -1 < \sigma < 0.5$$
$$\frac{\mu}{\lambda} = \frac{1-2\sigma}{2\sigma} \qquad (11)$$

Compression modulus K :

$$K = -\frac{p}{dV/V} \qquad (12)$$

$$K = \lambda + \frac{2}{3}\mu = S\left[\frac{2}{3} + \frac{2\sigma}{1-2\sigma}\right] = \frac{E}{3(1-2\sigma)} = \frac{1}{C} \qquad (13)$$

C= compressibility

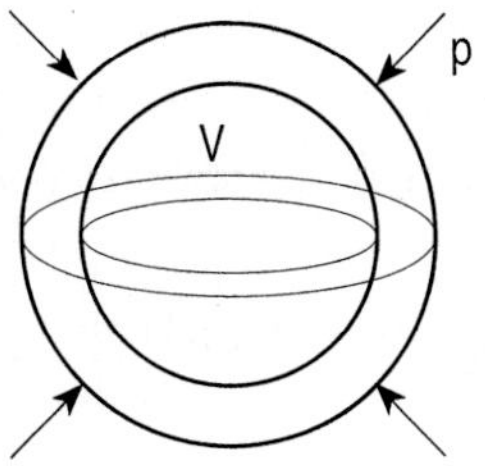

Dilatation modulus D : (1-dimensional deformation)

$p = -D \cdot \varepsilon_{xx}$ (14)

$$D = \lambda + 2\mu = 2S\frac{1-\sigma}{1-2\sigma} = E\frac{1-\sigma}{(1+\sigma)(1-2\sigma)} = \frac{1}{3}\frac{1-\sigma}{1+\sigma}K \qquad (15)$$

Free dilatational wave velocity: $c_D = \sqrt{D/\rho}$ (16)

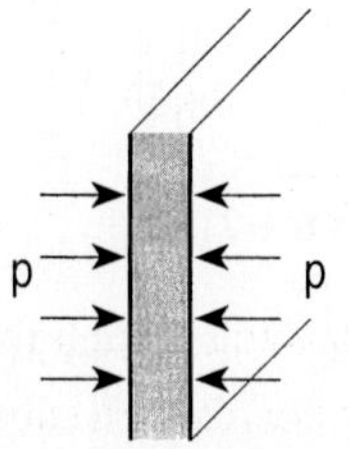

Bar bending modulus B_{St} :

$$B_{St} = E\frac{b \cdot h^3}{12} \tag{17}$$

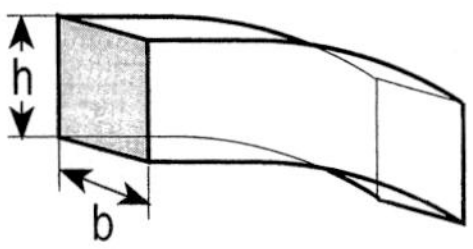

Free bar bending wave velocity: $c_{B_{St}} = \sqrt[4]{\omega^2 B_{St}/m'}$ (18)

(m'= mass per bar length)

Plate bending modulus B :

$$B = E\frac{h^3}{12(1-\sigma^2)} = S\frac{h^3}{6(1-\sigma)} = \frac{h^3}{3}\frac{\mu(\lambda+\mu)}{\lambda+2\mu} \tag{19}$$

Free plate bending wave velocity: $c_B = \sqrt[4]{\omega^2 B/m''}$ (20)

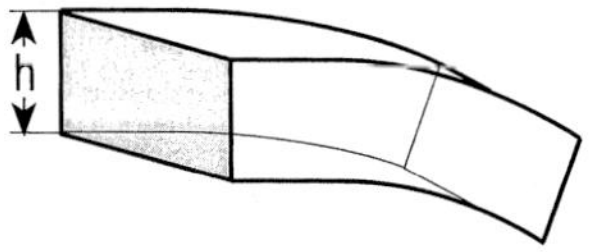

Relations for free plate bending velocity c_B :

$$c_B = \sqrt[4]{\omega^2 B/m''} = 1.347\sqrt{f\,h}\,\sqrt[4]{\frac{E}{\rho(1-\sigma^2)}} = 1.391\sqrt{f\,c_E\,h} = \frac{\sqrt{2\pi f\,c_E\,h}}{\sqrt[4]{12(1-\sigma^2)}} \tag{21}$$

Relations for free plate bending wave number $k_B = \omega/c_B$ (for $h \le \lambda_B/6$):

$$k_B = \frac{\omega}{c_B} = \frac{2\pi}{\lambda_B} = \sqrt{\omega}\,\sqrt[4]{m''/B} = \sqrt{\frac{\omega}{c_E h}}\,\sqrt[4]{12(1-\sigma^2)} = \frac{1}{h}\sqrt{k_E h}\,\sqrt[4]{12(1-\sigma^2)}$$
$$= k_0\sqrt{f_{cr}/f} \xrightarrow[\sigma=0.35]{} 4.515\sqrt{\frac{f}{c_E h}} \tag{22}$$

(f_{cr}= critical frequency)

Relations at coincidence :
The free plate bending wave speed c_B agrees with the trace speed $c_0/\sin\vartheta$ of a plane wave incident on the plate with a polar angle ϑ .

Coincidence frequency : $$f_c = f_{cr}/\sin^2\vartheta = \frac{c_0^2}{2\pi h \sin^2\vartheta}\sqrt{\frac{12\rho(1-\sigma^2)}{E}} \tag{23}$$

Critical frequency (at $\vartheta = \pi/2$; $c_0 = c_B$)

$$f_{cr} = \frac{c_0^2}{2\pi}\sqrt{\frac{m''}{B}} = \frac{c_0^2}{2\pi h}\sqrt{\frac{12\rho(1-\sigma^2)}{E}} = \frac{c_0^2}{2\pi h c_E}\sqrt{12(1-\sigma^2)} \quad ; \quad \lambda_{cr} = c_0/f_{cr} \tag{24}$$

$$k_B = k_0\sqrt{f_{cr}/f} \tag{25}$$

Effective bending moduli for sandwich panels :
(sheets and boards with subscripts 1,2 ; adhesive layers without subscript)

Table 1: Sandwich panels

No.	Sandwich	Connection
1	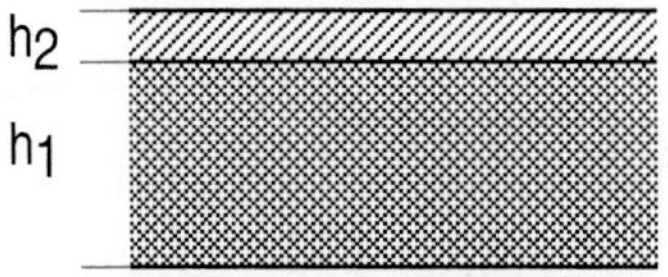	fix connection
2	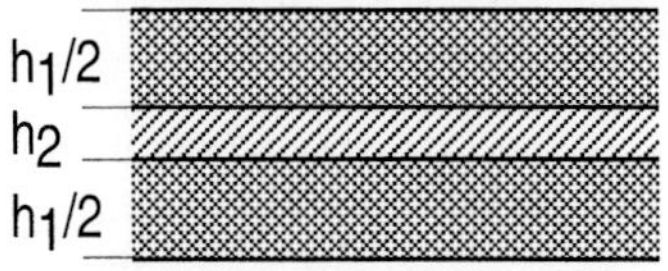	fix connection
3	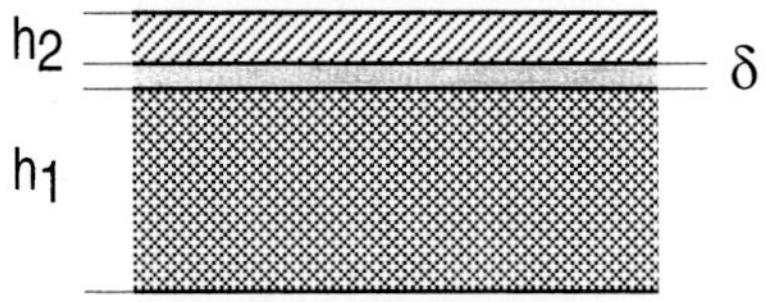	connection with shear $\delta \geq \frac{3.5\cdot 10^{-3}}{h_1}\frac{G}{E_1} - 1.3\cdot 10^{-12}$ [m] $E_2 h_2 \geq 2\cdot 10^7$ [Pa m]
4	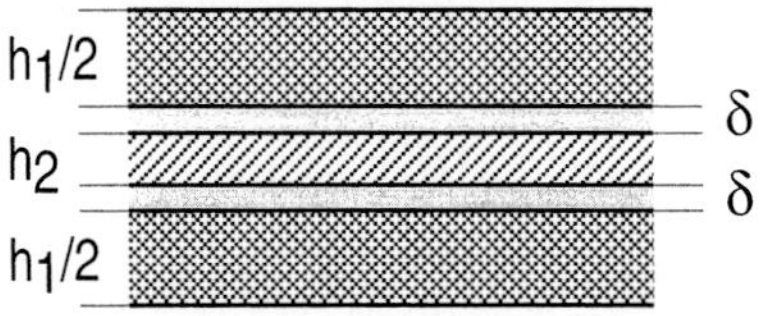	connection with shear $\delta \geq \frac{0.25\cdot 10^{-3}}{h_1}\frac{G}{E_1}$ [m] h_i in [m]; G, E_i in [Pa]

Table 2: Effective bending muduli B for sandwiches from Table 1 ..

No.	Effective bending modulus	Remark
1	$B = B_2 \dfrac{1+2\frac{E_1}{E_2}\left[2\,(\frac{h_1}{h_2})+3(\frac{h_1}{h_2})^2+2(\frac{h_1}{h_2})^3\right]+(\frac{h_1}{h_2})^2(\frac{E_1}{E_2})^4}{1+\frac{h_1}{h_2}\frac{E_1}{E_2}}$	
2	$B \approx B_1\left(1+\frac{h_2}{h_1}\right)^3 + B_2$	
3	$B \approx B_1 + B_2 + 3G\delta\frac{h_1^2}{4} + \frac{E_1 h_1 E_2 h_2 (h_1/2 + h_2/2)^2 \cdot g}{E_1 h_1 + g\cdot(E_1 h_1 + E_2 h_2)} -$ $-3G\delta\frac{h_1+h_2}{4}\frac{E_1 h_1^2 + 2gE_2 h_1 h_2}{E_1 h_1 + g\cdot(E_1 h_1 + E_2 h_2)}$	$\delta \ll h_1$ $g = \frac{G}{\delta E_1 h_1 \omega}\sqrt{B/m}$
4	$B = B_2 + 2E_1\frac{(h_1/2)^3}{12\,(1-\sigma_1^2)}$	

Q.12 Modes of rectangular plates

Ref.: Mechel, [Q.38]; Gorman, [Q.39]

A rectangular plate in the x,y-plane has the dimensions a in x-direction and b in y-direction; its shape factor is $\beta = b/a$. The non-dimensional co-ordinates are $\xi = x/a$, $\eta = y/b$. Two opposite borders are supposed to be simply supported, the other borders may have different supports. Distinguish between the wave number k_B of the "free bending wave" (in an infinite plate) and k_b the bending wave number in the finite, supported plate.

Used abbreviations (with integer m):

$$\mu_m = m\pi \quad ; \quad \gamma_m = \beta\sqrt{\left|(m\pi)^2 - (k_b a)^2\right|} \quad ; \quad \delta_m = \beta\sqrt{(m\pi)^2 + (k_b a)^2} \tag{1}$$

Boundary conditions for “classical supports” :

Fixation	**Condition**	**Symbol**
Simply supported S	$u(\xi,1) = \dfrac{\partial^2 u(\xi,1)}{\partial \eta^2} = 0$	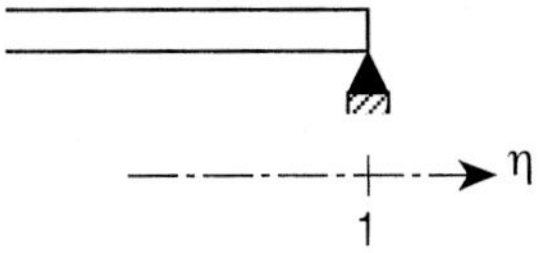
Clamped C	$u(\xi,1) = \dfrac{\partial u(\xi,1)}{\partial \eta} = 0$	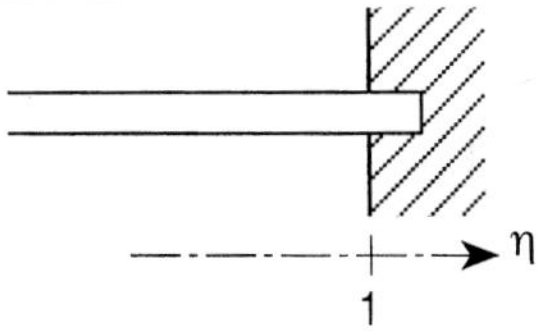
Free F	$\dfrac{\partial^2 u}{\partial \eta^2} + \sigma\beta^2 \dfrac{\partial^2 u}{\partial \xi^2} = 0$ $\dfrac{\partial^3 u}{\partial \eta^3} + (2-\sigma)\beta^2 \dfrac{\partial^3 u}{\partial \eta \partial \xi^2} = 0$	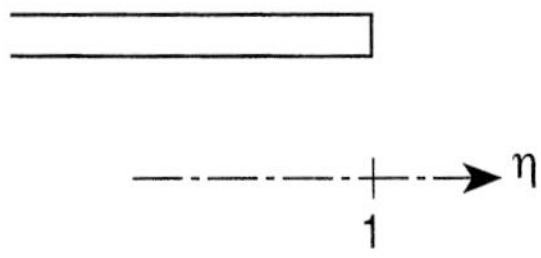 $\beta = b/a$

Table 1: Classical boundary conditions for plates.

Bending wave equation for the plate displacement $u(\xi,\eta)$:

$$\left[\frac{\partial^4}{\partial \eta^4} + 2\beta \frac{\partial^4}{\partial \eta^2 \partial \xi^2} + \beta^4 \frac{\partial^4}{\partial \xi^4} - \beta^4 (k_b a)^4\right] u(\xi,\eta) = 0 \tag{2}$$

Supposed the plate is simply supported at $\xi = 0$ and $\xi = 1$, the plate displacement field can be formulated as :

$$u(\xi,\eta) = \sum_{m=1}^{\infty} Y_m(\eta)\cdot\sin(m\pi\xi) \tag{3}$$

with the wave equation after insertion :

$$\left[\frac{d^4}{d\eta^4} - 2\beta^2\mu_m^2\frac{d^2}{d\eta^2} + \beta^4(\mu_m^4 - (k_b a)^4)\right]Y_m(\eta) = 0 \tag{4}$$

General solutions (with yet undetermined A_m, D_m) :

$\mu_m^2 > (k_b a)^2$: i.e. $m > a/(\lambda_b/2)$

$$Y_m(\eta) = A_m\cdot\cosh(\beta\eta\sqrt{\mu_m^2 + (k_b a)^2}) + B_m\cdot\sinh(\beta\eta\sqrt{\mu_m^2 + (k_b a)^2}) + C_m\cdot\cosh(\beta\eta\sqrt{\mu_m^2 - (k_b a)^2}) + D_m\cdot\sinh(\beta\eta\sqrt{\mu_m^2 - (k_b a)^2}) \tag{5}$$

$\mu_m^2 < (k_b a)^2$: i.e. $m < a/(\lambda_b/2)$

$$Y_m(\eta) = A_m\cdot\cosh(\beta\eta\sqrt{\mu_m^2 + (k_b a)^2}) + B_m\cdot\sinh(\beta\eta\sqrt{\mu_m^2 + (k_b a)^2}) + C_m\cdot\cos(\beta\eta\sqrt{(k_b a)^2 - \mu_m^2}) + D_m\cdot\sin(\beta\eta\sqrt{(k_b a)^2 - \mu_m^2}) \tag{6}$$

Eigenvalues must be found for $k_b a$; they follow from the boundary conditions. These are :

simply supported (S) : $$Y_m(\eta) = \frac{d^2Y_m(\eta)}{d\eta^2} = 0 \tag{7}$$

clamped (C) : $$Y_m(\eta) = \frac{dY_m(\eta)}{d\eta} = 0 \tag{8}$$

free (F) : $$\frac{\partial^2 u}{\partial\eta^2} + \sigma\beta^2\frac{\partial^2 u}{\partial\xi^2} = 0 \ ; \quad \frac{\partial^3 u}{\partial\eta^3} + (2-\sigma)\beta^2\frac{\partial^3 u}{\partial\eta\,\partial\xi^2} = 0 \tag{9}$$

The boundary conditions give a system of homogeneous equations for the amplitudes; for a non-trivial solution the determinant must vanish; this is the equation for k_b .

Cases :

(SSSS):

$$Y_m(\eta) = A_{m,n}\cdot\sin(n\pi\eta) \tag{10}$$

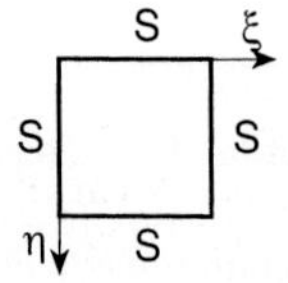

Determinant equation :

$$(\gamma_m^2 + \delta_m^2)\cdot\cosh\frac{\delta_m}{2}\cdot\cos\frac{\gamma_m}{2} = 0 \tag{11}$$

Solutions :

$$(k_b a)^2 = (m\pi)^2 + \frac{(n\pi)^2}{ß^2} \quad ; \quad m, n = 1, 2, \ldots \tag{12}$$

(SCSS) :

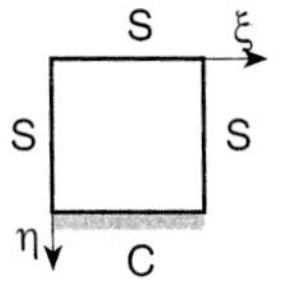

$$Y_m(\eta) = -\frac{\sin\gamma_m}{\sinh\delta_m} \cdot \sin(\gamma_m \eta) \cdot \sinh(\delta_m \eta) \tag{13}$$

Determinant equation :

$$\gamma_m \cdot \sinh\delta_m \cdot \cos\gamma_m - ß\cosh\delta_m \cdot \sin\gamma_m = 0 \tag{14}$$

The equation must be solved numerically for $k_b a$.

(SCSC) :

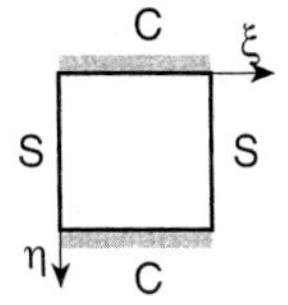

Symmetrical modes (m= 1,3,5,...):

$$Y_m(\eta) = -\frac{\cos(\gamma_m/2)}{\cosh(\delta_m/2)} \cdot \cos(\gamma_m \eta) \cdot \cosh(\delta_m \eta) \tag{15}$$

Determinant equation :

$$\gamma_m \cdot \cosh(\delta_m/2) \cdot \sin(\gamma_m/2) + \delta_m \sinh(\delta_m/2) \cdot \cos(\gamma_m/2) = 0 \tag{16}$$

Anti-symmetrical modes (m= 2,4,6,...):

$$Y_m(\eta) = -\frac{\sin(\gamma_m/2)}{\sinh(\delta_m/2)} \cdot \sin(\gamma_m \eta) \cdot \sinh(\delta_m \eta) \tag{17}$$

Determinant equation :

$$\gamma_m \cdot \sinh(\delta_m/2) \cdot \cos(\gamma_m/2) - \delta_m \cosh(\delta_m/2) \cdot \sin(\gamma_m/2) = 0 \tag{18}$$

Both equations must be solved numerically for $k_b a$.

(SFSS) :

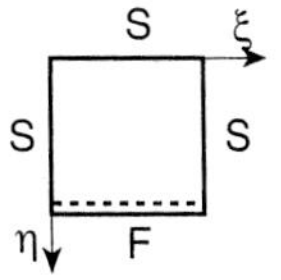

$\underline{(k_b a)^2 > (m\pi)^2}$:

$$Y_m(\eta) = \sin(\gamma_m \eta) + \frac{(\gamma_m^2 + \sigma ß^2 m^2 \pi^2)\sin\gamma_m}{(\delta_m^2 - \sigma ß^2 m^2 \pi^2)\sinh\delta_m} \cdot \sinh(\delta_m \eta) \tag{19}$$

Determinant equation (σ= Poisson ratio):

$$\begin{aligned}&\gamma_m(\gamma_m^2+(2-\sigma)\beta^2 m^2\pi^2)(\delta_m^2-\sigma\beta^2 m^2\pi^2)\cdot\sinh\delta_m\cdot\cos\gamma_m-\\&\quad\delta_m(\delta_m^2-(2-\sigma)\beta^2 m^2\pi^2)(\gamma_m^2+\sigma\beta^2 m^2\pi^2)\cdot\cosh\delta_m\cdot\sin\gamma_m=0\end{aligned} \tag{20}$$

$(k_b a)^2<(m\pi)^2$:

$$Y_m(\eta)=\sinh(\gamma_m\eta)-\frac{(\gamma_m^2-\sigma\beta^2 m^2\pi^2)\sinh\gamma_m}{(\delta_m^2-\sigma\beta^2 m^2\pi^2)\sinh\delta_m}\cdot\sinh(\delta_m\eta) \tag{21}$$

Determinant equation :

$$\begin{aligned}&\gamma_m(\gamma_m^2-(2-\sigma)\beta^2 m^2\pi^2)(\delta_m^2-\sigma\beta^2 m^2\pi^2)\cdot\sinh\delta_m\cdot\cosh\gamma_m-\\&\quad\delta_m(\delta_m^2-(2-\sigma)\beta^2 m^2\pi^2)(\gamma_m^2-\sigma\beta^2 m^2\pi^2)\cdot\cosh\delta_m\cdot\sinh\gamma_m=0\end{aligned} \tag{22}$$

More combinations of boundary conditions in [Q.39] .

Q.13 Partition impedance of plates

Ref.: Mechel, [Q.37]

The partition impedance Z_T is a useful quantity in boundary value problems. It displays its full usefulness if the plate is homogeneous, i.e., has no ribs etc., and is either infinite or at least so large that border effects can be neglected in the given task. Then the sound fields on both sides can be supposed to have the same distribution along the plate.

Suppose a Cartesian co-ordinate system x,y,z with the plate in the plane x,y at the position $z=\zeta$, and the z axis directed from the front side to the back side. The partition impedance for a plate is defined by :

$$Z_T=\frac{p_{front}(x,y,\zeta)-p_{back}(x,y,\zeta)}{v_{plate}(x,y)} \tag{1}$$

Be $p_{front}=p_e+p_r$ the sum of an incident wave p_e and a reflected wave p_r , and $p_{back}=p_t$ the transmitted wave. All waves α= e,r,t have the distributions

$$p_\alpha(x,y,z)=P_\alpha\cdot X(x)\cdot Y(y)\cdot Z_\alpha(z) \tag{2}$$

and also the plate velocity has the distribution $v_p(x,y)=V_p\cdot X(x)\cdot Y(y)$. Thus the profile $X(x)\cdot Y(y)$ cancels in Z_T .

It is supposed that the waves p_α satisfy the wave equation, Sommerfeld's far field condition, the source condition (if a source exists) and, possibly, boundary conditions at other boundaries than the plate. There exist three boundary conditions at the plate :

$$\begin{aligned} &(p_e + p_r - p_t)_{z=\zeta} \overset{!}{=} Z_T \cdot v_p \\ &\left(v_{e,z} + v_{r,z}\right)_{z=\zeta} \overset{!}{=} v_p \overset{!}{=} \left(v_{t,z}\right)_{z=\zeta} \end{aligned} \tag{3}$$

wherein $v_{\alpha,z}$, if the plate is in contact with air, follows from

$$v_{\alpha,z}(x,y,\zeta) = \frac{j}{k_0 Z_0} \cdot \operatorname{grad}_z\big(p_\alpha(x,y,\zeta)\big) \tag{4}$$

and if the plate is in contact with a porous absorber, from :

$$v_{\alpha,z}(x,y,\zeta) = \frac{-1}{\Gamma_a Z_a} \cdot \operatorname{grad}_z\big(p_\alpha(x,y,\zeta)\big) . \tag{5}$$

The plate has to satisfy the bending wave equation :

$$\left[\Delta_{x,y}\Delta_{x,y} - k_B^4\right] v_p = \frac{j\omega}{B} \cdot \delta p \tag{6}$$

in which Δ is the Laplace operator in the indicated co-ordinates, k_B is the wave number of the free bending wave on the plate, B is the bending stiffness, and $\delta p = p_{front} - p_{back}$ is the driving sound pressure difference. With the relations

$$k_B^4 = \omega^2 \frac{m''}{B} \quad ; \quad \frac{k_0}{k_B} = \sqrt{\frac{f}{f_{cr}}} \tag{7}$$

in which $\omega = 2\pi f$ is the circular frequency, m'' the surface mass density of the plate, f_{cr} the critical (coincidence) frequency, one immediately gets:

$$\frac{Z_T}{Z_0} = j k_0 \frac{m''}{\rho_0} \cdot \left[1 - \left(\frac{f}{f_{cr}}\right)^2 \frac{1}{k_0^4} \cdot \frac{\Delta_{x,y}\Delta_{x,y} v_p}{v_p}\right] \tag{8}$$

or alternatively :

$$\frac{Z_T}{Z_0} = j k_0 \frac{m''}{\rho_0} \cdot \left[1 - \left(\frac{f}{f_{cr}}\right)^2 \sin^4 \chi\right] \tag{9}$$

where the last two fractions in the brackets are replaced by the sine function of an effective angle of sound incidence χ of the incident wave p_e (defined below).

Energy dissipation in the plate can be taken into account by a loss factor η introducing a complex modulus $B \rightarrow B\cdot(1+j\eta)$. This leads to:

$$\frac{Z_T}{Z_0}=Z_{m''}F\cdot\left[\eta F^2\sin^4\chi+j\left(1-F^2\sin^4\chi\right)\right]=Z_{m''}F\cdot\left[\eta\left(\frac{f}{f_c}\right)^2+j\left(1-\left(\frac{f}{f_c}\right)^2\right)\right] \qquad (10)$$

with $$Z_{m''}=\frac{\omega_{cr}m''}{Z_0}\quad;\quad F=\frac{f}{f_{cr}} \qquad (11)$$

where $Z_{m''}$ is the normalised inertial impedance of the plate at the critical frequency ($\omega_{cr}=2\pi f_{cr}$), and f_c is the coincidence frequency at the incidence angle χ, with $f_{cr}=f_c\cdot\sin^2\chi$

It remains to determine :

$$\sin^4\chi=\frac{1}{k_0^4}\cdot\frac{\Delta_{x,y}\Delta_{x,y}v_p}{v_p}=\frac{1}{k_0^4}\cdot\frac{\Delta_{x,y}\Delta_{x,y}v_{z\alpha}(x,y,\zeta)}{v_{z\alpha}(x,y,\zeta)} \qquad (12)$$

where the last form makes use of the boundary condition that the pattern of the waves p_α at the plate agrees with that of v_p. After this determination all wave equations and all boundary conditions are satisfied, therefore the waves p_α make up a solution of the task.

Set $x=x_1$, $y=x_2$, $X(x)=X_1(x_1)$, $Y(y)=X_2(x_2)$ and suppose the wave factors $X_i(x_i)$ to have one of the forms :

$$X_i(x_i)=\begin{cases}e^{\pm jk_{xi}x_i}\\ \cos(k_{x_i}x_i)\\ \sin(k_{x_i}x_i)\end{cases} \qquad (13)$$

or a linear combination thereof. Then

$$\sin^4\chi=\frac{1}{k_0^4}\cdot\frac{\Delta_{x,y}\Delta_{x,y}v_{z\alpha}(x,y,\zeta)}{v_{z\alpha}(x,y,\zeta)}=\frac{\left(k_x^2+k_y^2\right)^2}{k_0^4} \qquad (14)$$

This corresponds to the possibility, which always exists, to transform the secular equation $k_0^2=k_x^2+k_y^2+k_z^2$ to a form :

$$1=\left((k_x/k_0)^2+(k_y/k_0)^2\right)+(k_z/k_0)^2=\sin^2\chi+\cos^2\chi \qquad (15)$$

where χ evidently is the polar angle of incidence of p_e on the plate. This holds also, if some or all of the k_x, k_y, k_z are complex.

The plates must not be thin in the sense of "thin plate theory"; it is important, that the compressibility of the plate normal to its surface is negligible. For a Timoshenko-Mindlin plate (with shear stress and rotational inertia) the result for Z_T is :

$$Z_T = j\omega m'' \frac{k_B^4 - (k_s^2 - k_L^2)(k_s^2 - k_R^2)}{k_B^4 - k_R^2(k_s^2 - k_L^2)} \quad ; \quad k_R^2 = \frac{12}{\pi^2} k_S^2 \quad ; \quad k_s^2 = k_x^2 + k_y^2 \tag{16}$$

with the characteristic wave numbers k_B , k_L , k_S of the free bending wave with the bending stiffness B, of the longitudinal wave with the plate dilatational stiffness D, and of the shear wave with the shear stiffness S . An equivalent form is:

$$Z_T = j\omega m'' \frac{1 - \dfrac{(k_x^2 + k_y^2)^2}{k_B^4} + \dfrac{h^2}{12}\left[(k_x^2 + k_y^2)\left(1 + \dfrac{c_L^2}{c_T^2}\right) - \dfrac{\omega^2}{c_T^2}\right]}{1 + \dfrac{h^2}{12}\left[(k_x^2 + k_y^2)\dfrac{c_L^2}{c_T^2} - \dfrac{\omega^2}{c_T^2}\right]} \tag{17}$$

with the plate thickness h and the speeds $c_L = \sqrt{E/\rho}$ of the longitudinal wave and $c_T = \sqrt{S/\rho}$ of the shear wave. An approximation has the form :

$$Z_T = j\omega m''\left[1 - \frac{(k_x^2 + k_y^2)^2}{\hat{k}_B^4}\right]$$
$$\hat{k}_B^2 = \frac{4.43}{24}\frac{\omega^2 m'' h^2}{B} + \sqrt{\left(\frac{4.43}{24}\frac{\omega^2 m'' h^2}{B}\right)^2 - \frac{m''\omega^2}{B}\left(\frac{0.26\omega^2 m'' h}{E} - 1\right)} \tag{18}$$

Q.14 Partition impedance of shells

Ref.: Mechel, [Q.37]

For fundamental considerations about the partition impedance Z_T see the previous section Q.13 .

Circular cylindrical shell :

In cylindrical co-ordinates r,ϑ,z the sound field near the shell and the vibration velocity of the shell with radius r=a (if necessary after expanding the incident wave in cylindrical waves) be :

$$p(r,\vartheta,z) = R(r)\cdot T(\vartheta)\cdot U(z) \quad ; \quad v_p(a,\vartheta,z) = A\cdot T(\vartheta)\cdot U(z) \tag{1}$$

The form of U(z) may be one of (or a linear combination) :

$$U(z) = \begin{cases} e^{\pm j k_z z} \\ \cos(k_z z) \\ \sin(k_z z) \end{cases} \tag{2}$$

The factor R(r) may be one of the cylinder functions and $T(\vartheta)$ a trigonometric function (or a linear combination) :

$$Z_m^{(i)}(k_r r) = \begin{cases} J_m(k_r r) & ; \ i=1 \\ Y_m(k_r r) & ; \ i=2 \\ H_m^{(1)}(k_r r) & ; \ i=3 \\ H_m^{(2)}(k_r r) & ; \ i=4 \end{cases} \quad ; \quad T(\vartheta) = \begin{cases} \cos(m\vartheta) \\ \sin(m\vartheta) \end{cases} \tag{3}$$

The Laplace operators in cylindrical co-ordinates are :

$$\Delta = \frac{\partial^2}{\partial r^2} + \frac{1}{r}\frac{\partial}{\partial r} + \frac{1}{r^2}\frac{\partial^2}{\partial \vartheta^2} + \frac{\partial^2}{\partial z^2} \quad ; \quad \Delta_{\vartheta,z} = \frac{1}{a^2}\frac{\partial^2}{\partial \vartheta^2} + \frac{\partial^2}{\partial z^2} \tag{4}$$

The wave equation is satisfied by the above field factors, when the secular equation

$$k_0^2 = k_z^2 + k_r^2 \quad ; \quad 1 = (k_z/k_0)^2 + (k_r/k_0)^2 = \sin^2\Theta + \cos^2\Theta \tag{5}$$

holds. The angle Θ is between the wave vector and the radius. The two-dimensional Laplace operator together with the Bessel differential equation for the $Z_m^{(i)}(k_r r)$ gives :

$$\Delta_{\vartheta,z}\, p(a,\vartheta,z) = -\left(\frac{m^2}{a^2} + k_z^2\right) \cdot p(a,\vartheta,z) \tag{6}$$

Therefore :

$$\sin^4\chi = \frac{1}{k_0^4}\frac{\Delta_{\vartheta,z}\Delta_{\vartheta,z} v_p}{v_p} = \frac{1}{k_0^4}\left(\frac{m^2}{a^2} + k_z^2\right)^2 = \left(\frac{m^2}{(k_0 a)^2} + \sin^2\Theta\right)^2 \tag{7}$$

With this quantity the partition impedance Z_T can be evaluated from the previous section Q.13. Because $T(\vartheta)$ is orthogonal over $0 \le \vartheta \le 2\pi$ for different values of m , the boundary conditions at the shell hold term-wise, if $p(r,\vartheta,z)$ is a sum of multi-pole terms.

Spherical shell :

Suppose spherical co-ordinates r, ϑ, φ and a shell with radius $r=a$. The field near the shell and the shell vibration velocity have the forms (if necessary after expanding the incident wave in spherical waves) :

$$p(r,\vartheta,\varphi) = R(r) \cdot T(\vartheta) \cdot P(\varphi) \quad ; \quad v_p(a,\vartheta,\varphi) = A \cdot T(\vartheta) \cdot P(\varphi) \tag{8}$$

with spherical Bessel functions for R(r) and associated Legendre functions for $T(\vartheta)$ (or linear combinations thereof) :

$$R(r)=\begin{cases} j_m(k_0r) \\ y_m(k_0r) \\ h_m^{(1)}(k_0r) \\ h_m^{(2)}(k_0r) \end{cases} \quad ; \quad T(\vartheta)=\begin{cases} P_m^n(\cos\vartheta) \\ Q_m^n(\cos\vartheta) \end{cases} \quad ; \quad P(\varphi)=\begin{cases} \cos(n\varphi) \\ \sin(n\varphi) \end{cases} \tag{9}$$

The Laplace operators are:

$$\begin{aligned} \Delta &= \frac{\partial^2}{\partial r^2}+\frac{2}{r}\frac{\partial}{\partial r}+\frac{1}{r^2}\frac{\partial^2}{\partial\vartheta^2}+\frac{1}{r^2\tan\vartheta}\frac{\partial}{\partial\vartheta}+\frac{1}{r^2\sin^2\vartheta}\frac{\partial^2}{\partial\varphi^2} \\ \Delta_{\vartheta,\varphi} &= \frac{1}{a^2}\frac{\partial^2}{\partial\vartheta^2}+\frac{1}{a^2\tan\vartheta}\frac{\partial}{\partial\vartheta}+\frac{1}{a^2\sin^2\vartheta}\frac{\partial^2}{\partial\varphi^2} \end{aligned} \tag{10}$$

and, with the above separation :

$$\Delta_{\vartheta,\varphi}\, p(a,\vartheta,\varphi)=-\frac{m(m+1)}{a^2}\cdot p(a,\vartheta,z) \tag{11}$$

Therefore :

$$\sin^4\chi=\frac{1}{k_0^4}\frac{\Delta_{\vartheta,z}\Delta_{\vartheta,z}v_p}{v_p}=\left(\frac{m(m+1)}{(k_0a)^2}\right)^2 \tag{12}$$

With this quantity the partition impedance Z_T can be evaluated from the previous Section Q.13. Because $T(\vartheta)$, $P(\varphi)$ are orthogonal over $(0,\ 2\pi)$ for different values of m or n, respectively, the boundary conditions at the shell hold term-wise, if $p(r,\vartheta,\varphi)$ is a sum of multipole terms.

Because of $P_m^n(\cos\vartheta)\equiv 0\ ;\ n>m$, the angle of incidence is $\chi=0$ for the "breathing sphere" m=n=0, which is plausible.

Q.15 Density of eigenfrequencies in plates, bars, strings, membranes

Be n the number of eigen frequencies in an interval of 1 Hertz.

String of length ℓ : $$n = 2\ell\sqrt{m / T} \tag{1}$$
(m= mass per string length ; T= string tension)

Longitudinal wave on a bar of length ℓ : $$n = 2\ell\sqrt{\rho / E} \tag{2}$$
(ρ= material density; E= Young's modulus)

Bending wave on a bar of length ℓ : $$n = \ell\sqrt[4]{m / (\omega^2 B)} \tag{3}$$
(m= mass per bar length : B= bar bending modulus)

Plate, simply supported : $$n = \tfrac{1}{2} S\sqrt{m/B} \tag{4}$$
(S= plate area; m= surface mass density; B= plate bending stiffness)

Circular membrane : $$n = \pi\frac{S\rho d\cdot f}{T} \tag{5}$$
(S= membrane area ; d= membrane thickness ; ρ= material density ; T= tension per unit length of circumference ; f= frequency)

Tube of length ℓ with outer diameter 2a and wall thickness d :
(simply supported at the ends; ρ= material density ; E= Young's modulus)

$$n = \begin{cases} \dfrac{5\pi}{2}\sqrt[4]{\dfrac{\rho^3}{E^3}}\sqrt{\omega a^3}\,\dfrac{\ell}{d} & ; \quad \omega < \dfrac{\sqrt{E/\rho}}{a} \\ 2\pi\sqrt{\dfrac{3\rho}{E}\dfrac{\ell a}{d}} & ; \quad \omega > \dfrac{\sqrt{E/\rho}}{a} \end{cases} \tag{6}$$

Q.16 Foot point impedances of forces

Ref: Fahy, [Q.40]; Cremer/Heckl, [Q.41]

Foot point impedances Z of structures for external forces are defined as ratios of

- force of a point source to structure velocity at the point of attack,
- force per length of a line source to average structure velocity at the line of attack,
- force per area of an area source to average structure velocity in the area.

It is advantageous to introduce the *foot point admittance* G= 1/Z . Be F the force of a point source or the constant force of a line source, be v(0,0) the structure velocity at the foot point

(0,0) of a point source, and v(0) the structure velocity at the line of excitation. The real part of the admittance can always be written as :

$$\mathrm{Re}\{G\} = \mathrm{Re}\{F_0/v(0,0)\} = \frac{1}{\omega\rho V_q} \qquad \text{point force} \qquad (1)$$

$$\mathrm{Re}\{G_L\} = \mathrm{Re}\{F/v(0)\} = \frac{1}{\omega\rho S_q} \qquad \text{line force} \qquad (2)$$

The quantity V_q with the dimension of a volume is called *source volume* , the quantity S_q with the dimension of an area is called *source area* , see [R.5].

The Table 1 collects valucs of the foot point admittance G and the source volume V_q for a point force on several objects. The arrows indicate the direction of the force.

h= thickness of plate or membrane;
A= cross section of bar;
ρ= material density;
m'= mass per unit length of bar;
m''= mass per unit area of plate;
c_L= longitudinal wave velocity;
c_B= bending wave velocity;
c_T= torsional wave velocity;
λ_L= longitudinal wave length;
λ_B= bending wave length;
λ_T= torsional wave length;
ρ_F= fluid density;
c_F= fluid sound speed;
T'= tension of a membrane;
B= plate bending stiffness;
σ= Poisson ratio;

Object		**Re{G}**	**Im{G}**	$\mathbf{V_q}$
Bar		$1/(\rho c_L A)$	0	$A\lambda_L/(2\pi)$
Bar, thin	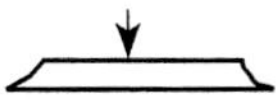	$1/(4m'c_B)$	$-1/(4m'c_B)$	$4\,A\lambda_B/(2\pi)$
Bar, thin		$1/(m'c_B)$	$-1/(m'c_B)$	$A\lambda_B/(2\pi)$
Bar, Timoshenko	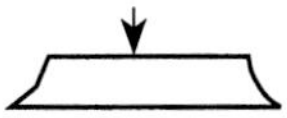	eq. (3)	eq. (3)	–

Object		**Re{G}**	**Im{G}**	$\mathbf{V_q}$
Membrane		$\omega/(4T')$	∞	$h\lambda^2/\pi^2$
Plate, thin		$1/(8\sqrt{Bm''})$	0	$2h\lambda_B^2/\pi^2$
Plate, thin		$1/(3.5\sqrt{Bm''})$	0	$0.9\,h\lambda_B^2/\pi^2$
Plate with shear stiffness	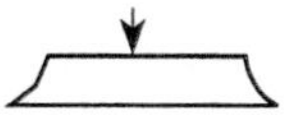	eq. (5)	eq. (5)	–
Plate with tangential force		–	$\approx 2\omega/(Eh)$	$h\lambda_T^2/(3\pi^2)$
Plate, orthotropic		eq. (7)	0	–
Plate on elastic bed	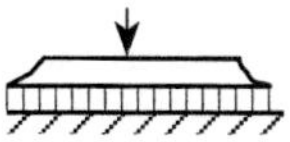	eq. (10)	eq. (10)	–
Strip of plate	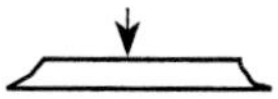	eq. (11)	eq. (11)	–
Tube	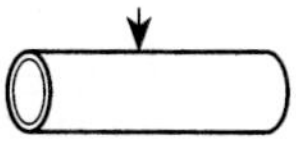	eqs. (12,13)	eqs. (12,13)	–
Elastic half space	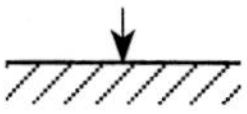	eqs. (14,15)	eqs. (14,15)	$\approx (\lambda_T / \pi)^3$
Fluid half space	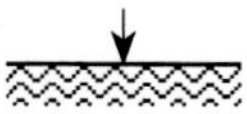	$\mathrm{Re}\{G\} = \dfrac{\omega^2}{6\pi\rho_F c_F^3}$	∞	$3\lambda_F^3 / (4\pi^2)$
Thick plate	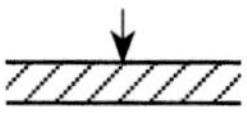	eqs. (16, 17)	eqs. (16, 17)	–

Table 1: Foot point admittances

Timoshenko bar with central excitation:

$$G = \frac{1}{2\omega m'} \frac{k_T^2 + k_I k_{II}}{k_I + k_{II}} \tag{3}$$

with:

$$k_I^2 = \frac{k_T^2 + k_L^2}{2} + \sqrt{\left(\frac{k_T^2 + k_L^2}{2}\right)^2 + \alpha k_B^2}$$
$$k_{II}^2 = \frac{k_T^2 + k_L^2}{2} - \sqrt{\left(\frac{k_T^2 + k_L^2}{2}\right)^2 + \alpha k_B^2} \tag{4}$$
$$\alpha = 1 - k_T^2 h^2 / 12$$

Plate with shear stiffness:

$$G = \frac{k_B^4}{8\omega m''\left[k_I^2 + \frac{1}{2}(k_T^2 + k_L^2\right]}(A_R + jA_I) \tag{5}$$

(see eq.(4) for k_I, k_{II}, α), and

with:

$$A_R = \begin{cases} \alpha + k_T^2\left(2k_I^2 - k_L^2 - k_T^2\right)/k_B^4 & ; \quad \alpha > 0 \\ k_T^2\left(2k_I^2 - k_L^2 - k_T^2\right)/k_B^4 & ; \quad \alpha < 0 \end{cases} \tag{6}$$

$$A_I = -\frac{2}{\pi}\left[\alpha \ln\left(\frac{k_B^2}{k_I^2}\sqrt{|\alpha|}\right) + \frac{k_T^2}{k_B^2}\left(k_I^2 \ln(k_I r) - k_{II}^2 \ln(k_{II} r)\right)\right]$$

with: r the distance between force point and measuring point.

Foot point admittance of an orthotropic plate with bending stiffnesses B_x, B_z in the x,z directions, B_σ bending stiffness induced by contraction, B_G bending stiffness induced by the shear modulus G :

$$G = \frac{K(ß)}{4\pi\left(m''^2 B_x B_z\right)^{1/4}} \approx \frac{1}{8\left(m''^2 B_x B_z\right)^{1/4}} \quad ; \quad ß = \frac{1}{2}\left(1 - \frac{B_\sigma + 2B_G}{\sqrt{B_x B_z}}\right) \tag{7}$$

where K(ß) is the complete elliptic integral of 1st kind.

Examples of orthotropic plates (with E= Young's modulus; S= shear modulus; σ= Poisson ratio):

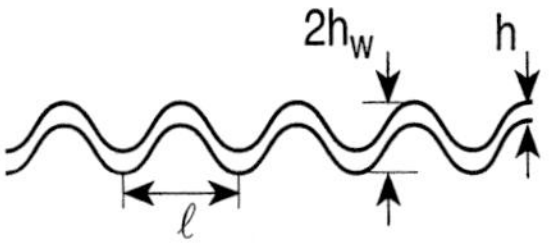

ℓ much smaller than bending wave length

$$B_x = E\cdot I \;;\; B_z = \frac{\ell}{s}\frac{Eh^3}{12(1-\sigma^2)} \;;\; B_\sigma \approx 0 \;;\; 2S \approx \frac{s}{\ell}\frac{Eh^3}{12(1-\sigma^2)}$$

$$s = \ell\left(\frac{\pi h_w}{2\ell}\right)^2 \;;\; I = \frac{h_w^2 h}{2}\left(1-\frac{0.81}{1+2.5(h_w/2\ell)^2}\right) \tag{8}$$

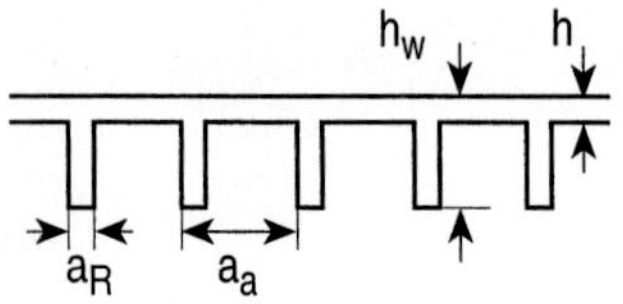

a_a much smaller than bending wave length

$$B_x = E\cdot I \;;\; B_z = \frac{Eh^3}{12}\frac{a_a}{a_a - a_R\left(1-h^3/h_w^3\right)} \;;\; B_\sigma \approx 0 \;;$$

$$S \approx \frac{E}{6(1+\sigma)}\left(h^3 + h_w^3\, a_R/a_a\right) \;;\; I = \frac{a_a}{3}\left(s_1^2 - s_2^2\right) + \frac{a_R}{3}\left(s_2^2 + s_3^2\right) \tag{9}$$

$$s_1 = \frac{1}{2}\frac{a_a h_w^2 + (a_a - a_R)h^2}{a_a h_w + (a_a - a_R)h} \;;\; s_2 = s_1 - h \;;\; s_3 = h_w - s_1$$

Foot point admittance of a thin plate on an elastic bed; the bed with a spring stiffness s (per unit area) is tuned with the surface mass density of the plate m'' to a resonance with $\omega_0^2 = s/m''$:

$$G = \frac{1}{8\sqrt{Bm''}}\begin{cases}\sqrt{1-\omega_0^2/\omega^2} & ;\; \omega_0 < \omega \\ \sqrt{\omega_0^2/\omega^2 - 1} & ;\; \omega_0 > \omega\end{cases} \tag{10}$$

Foot point admittance of a strip of plate with width l_s and thickness h for a point force at $z=0,\ x=x_0$:

$$G = \frac{1}{2\rho h l_s c_B}\left[\frac{1-j}{2}\alpha + \sum_{n=1}^{\infty}\left(\frac{1}{\sqrt{1-\kappa_n^2}} - \frac{j}{\sqrt{1+\kappa_n^2}}\right)\varphi_n^2(x_0)\right]$$

$$\kappa_n = n\pi/(k_B l_s) \;;\; \alpha = (\varphi_0(x_0))^2 \tag{11}$$

$$\varphi_n(x) = \cos\left(n\pi x/l_s\right) \;;\; n = 0,1,2,\ldots$$

Foot point admittance of a tube with outer radius a and wall thickness h :

$$G \approx (1-j)\left[2\pi a\rho h\sqrt{\omega c_L a/\sqrt{2}}\right]^{-1} \;;\; \nu = \omega a/c_L < 0.77\, h/a \tag{12}$$

$$\mathrm{Re}\{G\} \approx \begin{cases} \dfrac{0.66}{2.3c_L\rho h^2}\sqrt{\omega a/c_L} & ; \quad 0.77h/a < \nu < 0.6 \\ \left(2.3c_L\rho h^2\right)^{-1} & ; \quad \nu > 2 \end{cases} \tag{13}$$

Foot point admittance of a force acting in a small circle with radius a on an elastic half space :

$$G \approx \frac{\omega k_T}{S}(1-\sigma)\left(0.19 + j0.3/(k_T a)\right) \tag{14}$$

Foot point admittance of a force acting in a strip of width b on an elastic half space :

$$G \approx \frac{\omega}{S}(1-\sigma)\left\{0.463 + j1.5\ \ln\left[\left(1.9 - 15(\sigma-0.25)^2\right)/k_T b\right]\right\} \tag{15}$$

Foot point admittance of a force acting in a small circle with radius a on a thick plate :

$$G \approx \frac{\omega k_T}{S}\left\{\frac{0.063}{H^2} + \frac{1}{8}\left(\frac{H}{1.6+H}\right)^2 + j\left(0.06 + \frac{0.001}{H^{1.3}}\right)\frac{\lambda_T}{2a}\right\} \quad ; \quad H = k_T h/2 \tag{16}$$

Foot point admittance of a force acting in a strip of width b on a thick plate :

$$G \approx \frac{\omega}{S}\left\{\frac{1}{8H^{1.5}} + 0.31\left(\frac{H}{1.6+H}\right)^2 + j\left(\frac{-1}{8H^{1.5}} + 0.16\ln(\lambda_T/b)\right)\right\} \quad ; \quad H = k_T h/2 \tag{17}$$

Foot point admittance of a point force acting on an isotropic, thin plate with bending stiffness B and membrane stress T :

$$G = \frac{1}{8\sqrt{1+\beta^2}\sqrt{Bm''}}\left[1 + \frac{j}{\pi}\ln\left(\frac{\sqrt{1+\beta^2}+\beta}{\sqrt{1+\beta^2}-\beta}\right)\right] \quad ; \quad \beta = \frac{T}{2\omega\sqrt{Bm''}} \tag{18}$$

Foot point admittance of a point force acting in the centre of a bar of length ℓ , width w, thickness h , simply supported at both ends :

$$G = \frac{j\omega}{4EIk_B^3}\left(\tan k_B\ell - \tanh k_B\ell\right) \quad ; \quad I = wh^3/12 \tag{19}$$

Q.17 Transmission loss at steps, joints, corners

Ref. Cremer/Heckl, [Q.41]

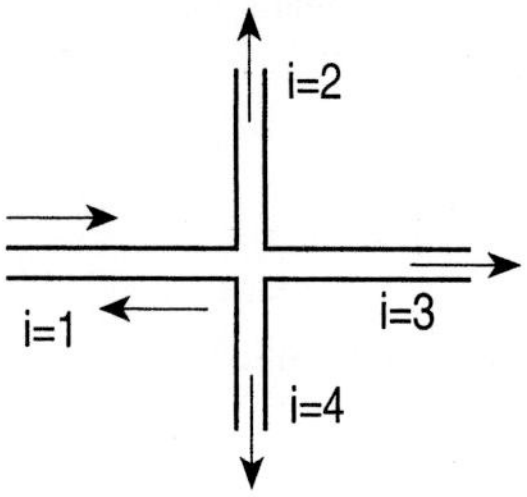

Bars or plates i=1,2,... are joined to each other at steps (of material and/or cross section), joints, corners. The branch i= 1 is the side of excitation, either by longitudinal or bending waves. The other branches are anechoic.

h_i= thickness;
A_i= cross section (of bars);
ρ_i= mass density;
m'_i= mass per length of bars;
E_i= Young's modulus;
c_i= group velocity;
K_i= radius of gyration;
c= compressional stiffness of interlayer;
m= blocking mass;
K= radius of gyration of blocking mass;
S_F= shear modulus of interlayer;
l_F= thickness of interlayer;
$s_{i,k}= h_i/h_k$ for plates; $s= s_{12}$
$s_{i,k}= A_i/A_k$ for bars; $s= s_{12}$

The transmission coefficient

$$\tau_{1i} = \frac{\Pi_i}{\Pi_1}$$

is the ratio of the effective powers Π_i, with Π_1 the incident power.

$$\kappa = \left(\frac{\rho_2 E_1 K_1^2}{\rho_1 E_2 K_2^2}\right)^{1/4} = \left(\frac{m'_2 B_1}{m'_1 B_2}\right)^{1/4}$$

$$\psi = \frac{K_2 A_2}{K_1 A_1}\sqrt{\frac{\rho_2 E_2}{\rho_1 E_1}} = \sqrt{\frac{m'_2 B_2}{m'_1 B_1}} \qquad (1)$$

$K_1 = h_1 / \sqrt{12}$ for plates

Object & Wave		**Transmission coefficient**
Longitudinal wave :		
Cross section change	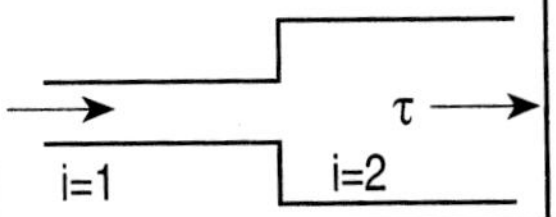	$\tau = 4\left[s_{12}^{1/2} + s_{12}^{-1/2}\right]^{-2}$
Material change	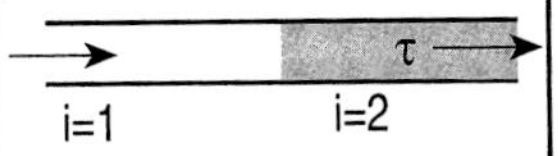	$\tau = 4\left[\left(\frac{E_1\rho_1}{E_2\rho_2}\right)^{1/4} + \left(\frac{E_1\rho_1}{E_2\rho_2}\right)^{-1/4}\right]^{-2}$
Elastic interlayer	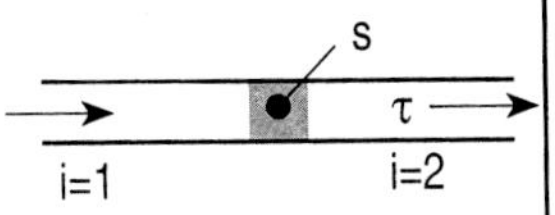	$\tau = \left(1 + (f/f_u)^2\right)^{-1}$; $f_u = \frac{c}{\pi A_1\sqrt{E_1\rho_1}}$
Blocking mass	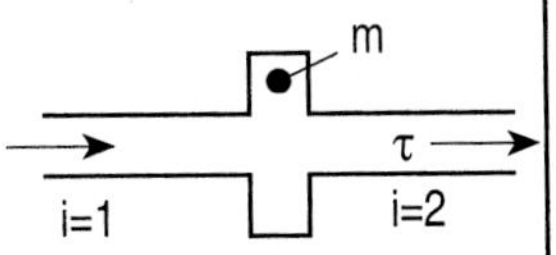	$\tau = \left(1 + (f/f_u)^2\right)^{-1}$; $f_u = \frac{A_1\sqrt{E_1\rho_1}}{\pi m}$
Bending wave :		
Cross-section change	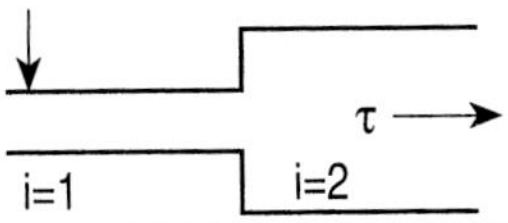	$\tau = \left[\frac{s^{-5/4} + s^{-3/4} + s^{3/4} + s^{5/4}}{s^{-2}/2 + s^{-1/2} + 1 + s^{1/2} + s^2/2}\right]^2$
Material change	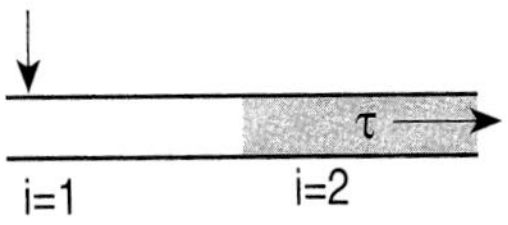	$\tau = \left[\frac{2\sqrt{\kappa\psi}(1+\kappa)(1+\psi)}{\kappa(1+\psi)^2 + 2\psi(1+\kappa^2)}\right]^2$
Corner	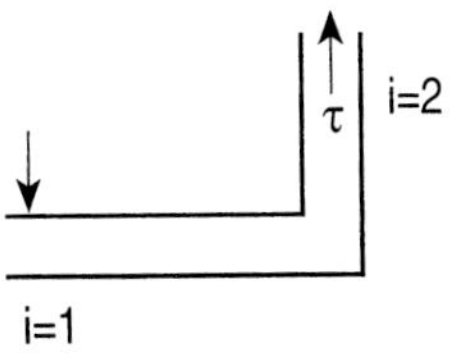	$\tau = 2\left[s^{-5/4} + s^{5/4}\right]^{-2}$
Cross	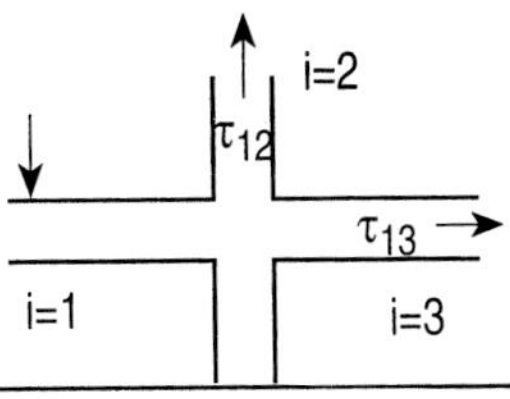	$\tau_{12} = \frac{1}{2}\left[s^{-5/4} + s^{5/4}\right]^{-2}$ $\tau_{13} = \frac{1}{2}\left[1 + 2s^{5/2} + s^5\right]^{-1}$

Object & Wave		**Transmission coefficient**
Branching	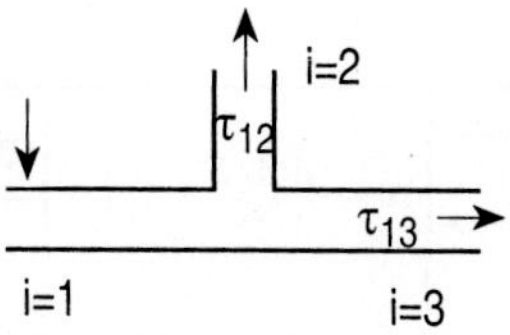	$\tau_{12}=\left[\sqrt{2}s^{-5/4}+s^{5/4}/\sqrt{2}\right]^{-2}$ $\tau_{13}=\left[2+2s^{5/2}+s^{5}/2\right]^{-1}$
Elastic interlayer	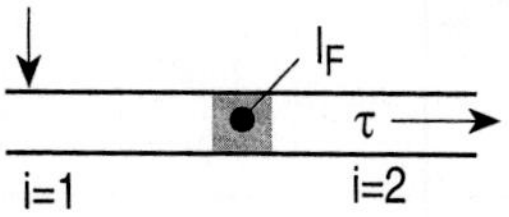	$\tau=\left(1+(f/f_u)^3\right)^{-1}$ $f_u=\left(\dfrac{G_F^2}{1.8\pi^2\rho_1\sqrt{E_1\rho_1}\,h_1 l_F^2}\right)^{1/3}$
Blocking masses	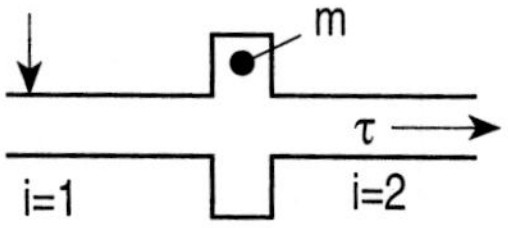	$\tau=1 \quad ; \quad f<0.5f_s$ $\tau=\left[1+f/f_u\right]^{-1} \quad ; \quad f>2f_s$ $f_s=\dfrac{K_1}{2\pi K^2}\sqrt{\dfrac{E_1}{\rho_1}}\;;\; f_u=\dfrac{2\rho_1 A_1^2 K_1\sqrt{E_1\rho_1}}{\pi m^2}$

Q.18 Cylindrical shell

Ref: Dym, [Q.42]

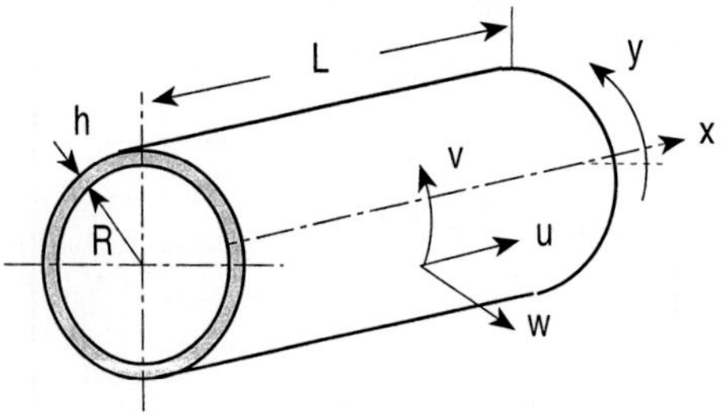

A cylindrical shell is simply supported at its ends. Used co-ordinates are :

$0 \le x \le L$

$0 \le y \le 2\pi R$.

Abbreviations :

$$a=\tfrac{1}{2}(1-\sigma) \quad ; \quad H=\frac{1}{12}(h/R)^2$$

$$\Lambda=m\pi R/L \quad ; \quad K^2=\frac{1-\sigma^2}{E}\rho R^2\omega^2 \qquad (1)$$

ρ= material density;
E= Young's modulus;
S= shear modulus;
σ= Poisson ratio;
u, v, w= elongations;

Mode shapes : with m,n= 0,1,2,...

$$u(x,y)=A\cdot\cos\frac{m\pi x}{L}\cdot\cos\frac{ny}{R}$$
$$v(x,y)=B\cdot\sin\frac{m\pi x}{L}\cdot\sin\frac{ny}{R} \qquad (2)$$
$$w(x,y)=C\cdot\sin\frac{m\pi x}{L}\cdot\cos\frac{ny}{R}$$

Special cases:	n=0	axial symmetry;
	n=1	no deformation of cross section;
	m=0	axial shear motion;
	m=1	fundamental bending motion.

General eigen value equation (for eigen values K):

$$K^6-(Q_3+Q_4)K^4+(Q_1+Q_2)K^2-Q_0=0 \qquad (3)$$

with coefficients:

$$Q_0=a\left\{(1-\sigma^2)\Lambda^4+H\left[(\Lambda^2+n^2)^4+\tfrac{9}{4}(1-\sigma^2)\Lambda^4+4\Lambda^2n^2+n^4+6\Lambda^4n^2-8\Lambda^2n^4-2n^6\right]+\right.$$
$$+H^2\left[\tfrac{1}{4}n^2-\tfrac{3}{2}\sigma^2\Lambda^4n^2-\tfrac{1}{2}n^6+\tfrac{9}{4}\Lambda^8+4\Lambda^6n^2+\tfrac{3}{2}\Lambda^4n^4+\tfrac{1}{4}n^8\right]+$$
$$\left.+H^3\left[a(1-a)\Lambda^4n^4\right]\right\}$$

$$Q_1=a\left\{(5-4a)\Lambda^2+n^2+H\left[\tfrac{9}{4}\Lambda^2+\left(\frac{1}{a}+\frac{1}{4}\right)n^2-\left(\frac{2}{a}+4a\right)\Lambda^2n^2-\frac{2}{a}n^4+\frac{1+a}{a}\left(\Lambda^2+n^2\right)^3\right]+\right.$$
$$\left.+H^2\left[\tfrac{9}{4}\Lambda^6+\left(\tfrac{11}{4}-a\right)\Lambda^4n^2+\left(\tfrac{3}{4}-a\right)\Lambda^2n^4+\tfrac{1}{4}n^6\right]\right\}$$

$$Q_2=a\left\{\left(\Lambda^2+n^2\right)^2+H\left[\tfrac{9}{4}\Lambda^4+\left(\tfrac{3}{2}+a+1/a\right)\Lambda^2n^2+\tfrac{5}{4}n^4\right]+\tfrac{1}{4}H^2n^4\right\} \qquad (4)$$

$$Q_3=1+H\left(\Lambda^2+n^2\right)$$

$$Q_4=(1+a)\left(\Lambda^2+n^2\right)+H\left[\tfrac{9}{4}\Lambda^2+(1+a/4)n^2\right]$$

Eigenfrequencies, from solutions, with eq.(1) :

$$\omega=\frac{K}{R}\sqrt{\frac{E}{\rho(1-\sigma^2)}} \qquad (5)$$

Linear approximation to K (at low frequencies):

$$K^2\approx\frac{Q_0}{Q_1+Q_2} \qquad (6)$$

Approximation with quadratic correction:

$$K^2 \approx \frac{Q_0}{Q_1+Q_2} + \frac{Q_0^2(Q_3+Q_4)}{(Q_1+Q_2)^3} \tag{7}$$

Amplitude ratios :

$$\begin{aligned}\frac{A}{C} &= \frac{1}{\Delta}\Big\{\Big[K^2 - a\big(1+\tfrac{9}{4}H\big)\Lambda^2 - (1+H)n^2\Big]\cdot\Lambda(aHn^2-\sigma) - \\ &\quad -\big(a+\sigma-\tfrac{3}{4}aH\big)\Big[\Lambda n^2 + (1+a)H\Lambda^3 n^2 + H\Lambda n^4\Big]\Big\} \\ \frac{B}{C} &= \frac{1}{\Delta}\Big\{\Big[K^2 - \Lambda^2 - a\big(1+\tfrac{1}{4}H\big)n^2\Big]\cdot\Big[n + (1+a)H\Lambda^2 n + Hn^3\Big] - \\ &\quad -\big(aHn^2-\sigma\big)\big(a+\sigma-\tfrac{3}{4}aH\big)\Lambda^2 n\Big\} \\ \Delta &= K^4 - K^2\Big[\big(1+a+\tfrac{9}{4}aH\big)\Lambda^2 + \big(1+a+\tfrac{1}{4}aH\big)n^2\Big] + \\ &\quad + a\big(1+\tfrac{9}{4}H\big)\Lambda^4 + a\big(1+\tfrac{5}{4}H+\tfrac{1}{2}H^2\big)n^4 + \big(2a+H+a^2H+\tfrac{3}{2}aH\big)\Lambda^2 n^2\end{aligned} \tag{8}$$

Special case : torsion with axial symmetry $u=w=0$

Differential equation:
$$\frac{1-\sigma}{2}\big(1+\tfrac{9}{4}H\big)\frac{\partial^2 v}{\partial x^2} = \frac{1-\sigma}{E}\rho\frac{\partial^2 v}{\partial t^2} \tag{9}$$

Modes:
$$v = \begin{cases} \sin\dfrac{m\pi x}{L} & ;\ \text{simply supported} \\ \cos\dfrac{m\pi x}{L} & ;\ \text{free} \end{cases} \tag{10}$$

Eigenvalues:
$$\begin{aligned} K^2 &= \frac{1-\sigma}{2}\big(1+\tfrac{1}{4}H\big)\Lambda^2 \\ \omega^2 &= \frac{E}{2(1+\sigma)\rho}\big(1+\tfrac{1}{4}H\big)(m\pi/L)^2 \end{aligned} \tag{11}$$

Special case : longitudinal vibration with axial symmetry $v=w=0$

Differential equation:
$$\frac{\partial^2 u}{\partial x^2} = \frac{1-\sigma^2}{E}\rho\frac{\partial^2 u}{\partial t^2} \tag{12}$$

Eigenvalues:
$$\begin{aligned} K^2 &= \Lambda^2 = \left(\frac{m\pi R}{L}\right)^2 \\ \omega^2 &= \frac{E}{(1-\sigma^2)\rho}\left(\frac{m\pi}{L}\right)^2 \end{aligned} \tag{13}$$

Special case : radial vibration with axial symmetry $u=v=0$

Differential equation: $$\frac{w}{R^2}+HR^2\frac{\partial^4 w}{\partial x^4}=\frac{w}{R^2}+\frac{h^2}{12}\frac{\partial^4 w}{\partial x^4}=-\frac{1-\sigma^2}{E}\rho\frac{\partial^2 w}{\partial t^2} \tag{14}$$

Modes: $$w\sim\sin\frac{m\pi x}{L}\quad\text{or}\quad\sim\cos\frac{m\pi x}{L} \tag{15}$$

Eigenvalues: $$K^2=1+H\Lambda^4$$ $$\omega^2=\frac{E}{\rho(1-\sigma^2)}\left[\frac{1}{R^2}+\tfrac{1}{12}h^2\left(\frac{m\pi}{L}\right)^4\right] \tag{16}$$

Special case : ring-shaped vibration $u–0$

Differential equations: $$(1+H)\frac{\partial^2 v}{\partial y^2}+\frac{1}{R}\frac{\partial w}{\partial y}-4R\frac{\partial^3 w}{\partial y^3}=\frac{1-\sigma^2}{E}\rho\frac{\partial^2 v}{\partial t^2}$$ $$\frac{1}{R}\frac{\partial v}{\partial y}-HR\frac{\partial^3 v}{\partial y^3}+\frac{w}{R^2}+HR^2\frac{\partial^4 w}{\partial y^4}=\frac{1-\sigma^2}{E}\rho\frac{\partial^2 v}{\partial t^2} \tag{17}$$

Modes: $$v=V\cdot\sin\frac{ny}{R}\quad;\quad w=W\cdot\cos\frac{ny}{R}$$ $$\frac{V}{W}=\frac{K^2-(1+H)n^2}{n+Hn^3} \tag{18}$$

Eigenvalue equation: $$K^4-K^2\left(1+n^2\right)\left(1+Hn^2\right)+Hn^2\left(n^2-1\right)^2=0 \tag{19}$$

Approximation for thin shells : $Hn^2 \ll 1$:

$$K_1^2=H\frac{n^2(n^2-1)^2}{n^2+1}\quad;\quad K_2^2=H(n^2+1)$$ $$\omega_1^2=\frac{E}{\rho(1-\sigma^2)}\frac{1}{R^2}\frac{n^2(n^2-1)^2}{n^2+1}\quad;\quad\omega_2^2=\frac{E}{\rho(1-\sigma^2)}\frac{1}{R^2}(n^2+1) \tag{20}$$

Special case : ring-shaped vibration without dilatation $nV+W=0$

Eigenvalues: $$K^2=H\,n^2(n^2-1)^2$$ $$\omega^2=\frac{E}{\rho(1-\sigma^2)}\frac{1}{R^2}n^2(n^2-1)^2 \tag{21}$$

Compare with plate in $0\le x\le L$; $0\le y\le\pi R$; simply supported:

$$K^2=H(n^2+\Lambda^2)^2$$ $$\omega^2=\frac{Eh^2}{12\rho(1-\sigma^2)}\left[\left(\frac{m\pi}{L}\right)^2+\left(\frac{n\pi}{\pi R}\right)^2\right] \tag{22}$$

Special case : ring-shaped bar, simply supported

with $A = 2\pi R h$ cross section

$I = \pi R^3 h (1 + 3H)$ moment of inertia (23)

Eigenvalues: $$\omega^2 = \frac{E}{2\rho}\left(R^2 + \tfrac{1}{4}h^2\right)\left(\frac{m\pi}{L}\right)^2 \tag{24}$$

Q.19 Similarity relations for spherical shells

Ref: Soedel, [Q.43]

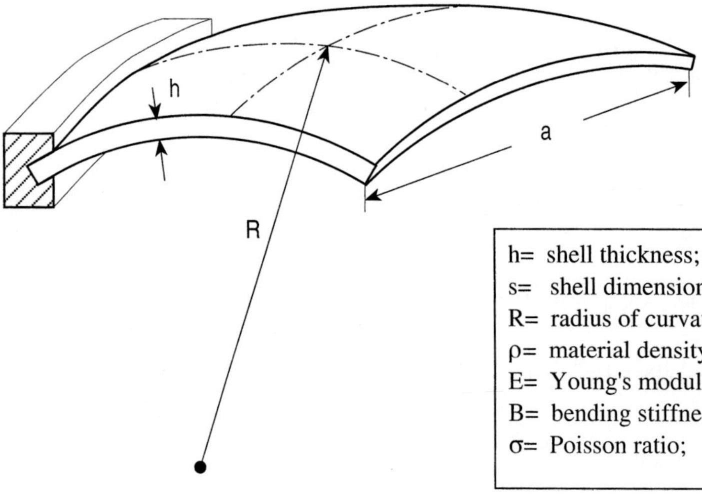

h= shell thickness;
s= shell dimension;
R= radius of curvature;
ρ= material density;
E= Young's modulus;
B= bending stiffness;
σ= Poisson ratio;

The shell be supported anyhow. The bending stiffness is defined as :

$$B = \frac{E h^3}{12(1-\sigma^2)} \tag{1}$$

Free bending wave number $k_{B\cap}$ in the shell in relation to the free bending wave number $k_{B||}$ in a plate :

$$k_{B\cap} = \left[\frac{1}{B}\left(\rho h \omega^2 - Eh/R^2\right)\right]^{1/4} = k_{B||}\sqrt{1 - \frac{E}{\rho\omega^2 R^2}} \tag{2}$$

Plates and shells with geometrically similar contours and equal supports have same mode solutions w_n/a if they agree in :

- $\frac{a^3}{B}\left(\rho h\omega_n^3 - Eh/R^2\right)$,
- σ .

Relation between the eigenfrequencies ω_{n1}, ω_{n2} of two spherical shells, i=1,2, which have :

- similar contours,
- equal supports

$$\omega_{n2} = \sqrt{\left(\frac{a_1}{a_2}\right)^3 \frac{E_2}{E_1}\left(\frac{h_2}{h_1}\right)^2 \frac{1-\sigma_1^2}{1-\sigma_2^2}\left(\frac{\rho_1}{\rho_2}\omega_{n1}^2 - \frac{E_1}{\rho_2 R_1^2}\right) + \frac{E_2}{\rho_2 R_2^2}} \tag{3}$$

Relation between the eigenfrequencies ω_{n2} of a spherical shell and ω_{n1} of a plane plate having

- similar contours,
- equal supports

$$\omega_{n2} = \sqrt{\left(\frac{a_1}{a_2}\right)^3 \frac{E_2}{E_1}\left(\frac{h_2}{h_1}\right)^2 \frac{1-\sigma_1^2}{1-\sigma_2^2}\frac{\rho_1}{\rho_2}\omega_{n1}^2 + \frac{E_2}{\rho_2 R_2^2}} \tag{4}$$

If material, contour, dimensions and support are equal :

$$\omega_{n2} = \sqrt{\omega_{n1}^2 + \frac{E}{\rho R^2}} \tag{5}$$

Q.20 Sound radiation from plates :

Ref. Cremer/Heckl, [Q.41]; Heckl, [Q.44]; Maidanik, [Q.45]

The radiation efficiency Σ , presented below, is the real part of the normalised radiation impedance. See Chapter "F. Radiation of Sound" for radiation impedances.

A plate, infinite if not otherwise stated, be excited by a point or line force, or by a sound field. If the plate has a finite area A it is supposed to be mounted in an infinite hard baffle wall.
The effective sound power Π radiated to one side will be given.

ρ_0=	density of surrounding medium;
c_0=	sound velocity in medium;
k_0=	ω/c_0 ;
Z_0=	$\rho_0 c_0$;
h=	plate thickness;
m''=	surface mass density of plate;
F_{eff}=	root mean square of force;
A=	plate area;
ρ_p=	plate material density;
E=	Young's modulus;
σ=	Poisson ratio;
η=	plate bending loss factor;
k_B=	free plate bending wave number;
c_L=	longitudinal wave speed in plate;

Plate excited by a point force :

$$\Pi = \frac{\rho_0}{2\pi c_0 m''^2} F_{eff}^2 \left[1 - \frac{\rho_0 c_0}{\omega m''} \operatorname{arctg} \frac{\omega m''}{\rho_0 c_0}\right]$$

$$\xrightarrow[k_0 m''/\rho_0 >> 1]{} \frac{\rho_0}{2\pi c_0} \frac{F_{eff}^2}{m''^2} \tag{1}$$

$$\xrightarrow[k_0 m''/\rho_0 << 1]{} \frac{k_0^2}{6\pi \rho_0 c_0} F_{eff}^2$$

Radius a of an equivalent piston radiator (i.e. piston radiator with radiation efficiency Σ=1 and same effective velocity):

$$a = \sqrt{8/\pi^3} \cdot \lambda_c = 0.286 \cdot \lambda_c \tag{2}$$

with $\lambda_c = c_0 / f_c$ and f_c= coincidence frequency.

Plate excited by a line source (Π and F_{eff} per unit length):

$$\Pi = \frac{\rho_0}{2\omega m''^2} F_{eff}^2 \left[1 - \frac{\rho_0 c_0}{\omega m''} \left(1 + (\rho_0 c_0/\omega m'')^2\right)^{-1/2}\right] \tag{3}$$

Space- and time-averaged squared velocity for excitation of a finite plate by a point source:

$$\langle v^2 \rangle_{s,t} = \frac{k_B^2}{8\eta A \omega^2 m''^2} F_{eff}^2 \tag{4}$$

Radiated power of a finite plate, excited by a point force,

follows from the definition of radiation efficiency Σ :

$$\Pi = A \cdot \rho_0 c_0 \cdot \Sigma \cdot \langle v^2 \rangle_{s,t} \tag{5}$$

$$\Pi = \frac{\rho_0 c_0 \cdot k_B^2 \cdot \Sigma}{8\eta \omega^2 m''^2} F_{eff}^2 \tag{6}$$

Velocity of a plate when excited by a diffuse sound field (p= effective sound pressure):

$$v^2 = \frac{\pi c_0^2 \cdot k_B^2 \cdot \Sigma}{2\eta m''^2} p \qquad \text{above coincidence frequency} \tag{7}$$

$$v^2 = \frac{1}{\omega^2 m''^2} \left(2 + \frac{\pi}{2} \frac{k_B^2 \cdot \Sigma}{k_0^2 \eta}\right) p \qquad \text{below coincidence frequency} \tag{8}$$

Ratio of radiated sound power by a point-excited plate to sound field excitation:

- excitation by a point force $\Pi = \alpha \cdot F_{eff}^2$,

• excitation by a diffuse sound field $v^2 = ß \cdot p^2$

follows :
$$\frac{\alpha}{ß} = \frac{\rho_0 c_0 k_0^2}{4\pi} = \frac{\rho_0 \omega^2}{4\pi c_0} \tag{9}$$

Similar relations for a line force (and 2-dimensional sound field):

$$\begin{aligned} \Pi_L &= \alpha_L \cdot F_{Leff}^2 \\ v_L^2 &= ß_L \cdot p_L^2 \\ \frac{\alpha_L}{ß_L} &= \frac{\rho_0 c_0 k_0}{4} = \frac{\rho_0 \omega}{4} \end{aligned} \tag{10}$$

Sound power Π_m fed into a plate by a diffuse sound field:

$$\Pi_m = \frac{\pi A k_B^2 \cdot \Sigma}{2 k_0^2 \omega m''} p^2 \tag{11}$$

Radiated sound power of a finite plate, with area A, periphery U, driven by a point force (approximation):

$$\Pi = \frac{\rho_0}{2\pi c_0 m''^2} F_{eff}^2 \left(1 + \frac{U}{2 A k_B \eta} \right) \quad ; \quad f << f_c \tag{12}$$

More general, if radiated power is small compared with internally lost power, i.e. $2\rho_0 c_0 \cdot \Sigma << \omega m'' \cdot \eta$:

$$\Pi = \frac{\rho_0}{2\pi c_0 m''^2} F_{eff}^2 \left(1 + \frac{2\pi c_0^2 \cdot \Sigma}{2.3 c_L h \omega \eta} \right) \tag{13}$$

Approximations for radiation efficiency of point-excited, weakly damped, finite plates:

$$\Sigma \approx \begin{cases} U\lambda_c/(\pi^2 A) \cdot \sqrt{f/f_c} & ; \quad f << f_c \\ 0.45\sqrt{U/\lambda_c} & ; \quad f = f_c \quad ; \quad \lambda_c = c_0/f_c \\ 1 & ; \quad f >> f_c \end{cases} \tag{14}$$

More precise approximation for $f<f_c$ (MAIDANIK) :

$$\Sigma = \frac{4}{\pi^2} \frac{\lambda_0 \lambda_c}{A} \frac{1-2\alpha^2}{\alpha\sqrt{1-\alpha^2}} + \frac{U\lambda_c}{4\pi^2 A} \frac{(1-\alpha^2)\ln\frac{1+\alpha}{1-\alpha} + 2\alpha}{(1-\alpha^2)^{3/2}} \quad ; \quad \alpha = \sqrt{f/f_c} \tag{15}$$

with λ_0= wave length of air-borne sound at frequency f.

Sound pressure far field of a plate excited by a point force $F = \sqrt{2} \cdot F_{eff}$
(R= radius from point of excitation; Θ= polar angle)

loss-free plate :

$$p(R,\Theta) = \frac{jk_0}{2\pi} F \frac{e^{-jk_0R}}{R} \frac{\cos\Theta}{1+\frac{jk_0m''}{\rho_0}\cos\Theta\cdot\left(1-(f/f_c)^2\sin^4\Theta\right)} \quad ; \quad f < 0.7\cdot f_c \tag{16}$$

$$p(R,\Theta) = \frac{jk_0}{2\pi} F \frac{e^{-jk_0R}}{R} \frac{(1+\varphi(\Theta))\cdot\cos\Theta}{1+\varphi(\Theta)+\frac{jk_0m''}{\rho_0}\left\{1+\left[1-\frac{1-\sigma}{24}\frac{(\pi c_L\sin\Theta)^2}{c_0^2}\right]\cdot\varphi(\Theta)\right\}} \quad ; \quad f > 0.7\cdot f_c$$

with f_c= coincidence frequency;

$$\varphi(\Theta) = \frac{2(k_0h)^2}{\pi^2(1-\sigma)}\left(\sin^2\Theta-(c_0/c_L)^2\right) \tag{17}$$

In the direction $\Theta=0$ normal to the plate:

$$p(R,0) = \frac{jk_0}{2\pi} F \frac{e^{-jk_0R}}{R} \frac{\cos\Theta}{1+\frac{jk_0m''}{\rho_0}} \tag{18}$$

In the direction of the angle of coincidence $\Theta_c = \sin^{-1}\sqrt{f_c/f}$:

$$p(R,0) = \frac{jk_0}{2\pi} F \frac{e^{-jk_0R}}{R} \sqrt{1-(f_c/f)^2} \tag{19}$$

Plate with bending loss factor $\eta<<1$:

Losses have negligible influence for $f< f_c$.

Define complex coincidence frequency: $$\omega_c = 2\pi f_c = \sqrt{12}\frac{c_0^2}{c_L h}(1+j\eta/2) \tag{20}$$

$$p(R,\Theta_c) = \frac{jk_0}{2\pi} F \frac{e^{-jk_0R}}{R} \frac{\sqrt{1-(f_c/f)^2}}{1+\eta\frac{k_0m''}{\rho_0}\sqrt{1-f_c/f}} \tag{21}$$

References to part Q

Elasto-Acoustics :

[Q.1] ACHENBACH, J. D.
"Wave propagation in elastic solids"
North-Holland , Amsterdam, 1975

[Q.2] MAYSENHÖLDER, W.
"Körperschallenergie"
S. Hirzel Verlag, Stuttgart, 1994

[Q.3] HELBIG, K.
"Foundations of anisotropy for exploration seismics"
Pergamon/Elsevier, Oxford, 1994

[Q.4] JONES, R. M.
"Mechanics of composite materials"
Taylor & Francis, Philadelphia, 1999

[Q.5] LAI, W. M.; RUBIN, D.; KREMPL; E.
"Introduction to continuum mechanics"
Pergamon Press, Oxford, 1993

[Q.6] MAYSENHÖLDER, W.
"Low-frequency sound transmission through periodically inhomogeneous plates with arbitrary local anisotropy and arbitrary global symmetry "
ACUSTICA • acta acustica 82 (1996) 628-635

[Q.7] THURSTON, R. N.
"Wave propagation in fluids and normal solids"
In: Physical Acoustics, Vol. I (Methods and devices, Part A),
W. P. Mason (ed.), Academic Press, New York, 1964, 1-110

[Q.8] AULD, B. A.
"Acoustic fields and waves in solids"
Vol. II, Krieger Publishing Company, Malabar, Florida 1990

[Q.9] AULD, B. A.
"Acoustic fields and waves in solids"
Vol. I, Krieger Publishing Company, Malabar, Florida 1990

[Q.10] ROKHLIN, S. I.; BOLLAND, T. K.; ADLER, L.
"Reflection and refraction of elastic waves on a plane interface between two generally anisotropic media" ; J. Acoust. Soc. Am. 79 (1986) 906-918

[Q.11] LANCELEUR, P.; RIBEIRO, H.; DE BELLEVAL, J.-F.
"The use of inhomogeneous waves in the reflection-transmission problem at a plane interface between two anisotropic media" ;
J. Acoust. Soc. Am. 93 (1993) 1882-1892

[Q.12] GAUL, L.
"The influence of damping on waves and vibrations"
Mechanical Systems and Signal Processing 13 (1999) 1-30

[Q.13] BELTZER, A. I.
"Acoustics of solids"
Springer-Verlag, Berlin, 1988

[Q.14] MOBLEY, J., et al.
"Kramers-Kronig relations applied to finite bandwidth data from suspensions of encapsulated microbubbles" ; J. Acoust. Soc. Am. 108 (2000) 2091-2106

[Q.15] PRITZ, T.
"Analysis of four-parameter fractional derivative model of real solid materials"
J. Sound Vib. 195 (1996) 103-115

[Q.16] PAVIC, G.
"Structural surface intensity: An alternative approach in vibration analysis and diagnosis" ; J. Sound Vib. 115 (1987) 405-422

[Q.17] PIERCE, A. D.
"The natural reference wavenumber for parabolic approximations in ocean acoustics"
Comp. & Maths. with Appls. 11 (1985) 831-841

[Q.18] MAYSENHÖLDER, W.
"Proof of two theorems related to the energy of acoustic Bloch waves in periodically inhomogeneous media" ; Acustica 78 (1993) 246-249

[Q.19] CREMER, L.; HECKL, M.
"Körperschall"
Springer-Verlag, Berlin, 1996

[Q.20] LIGHTHILL, M. J.
"Group velocity" ; J. Inst. Maths. Appls. 1 (1965) 1-28

[Q.21] SORNETTE, D.
"Acoustic waves in random media. I. Weak disorder regime.
II. Coherent effects and strong disorder regime. III. Experimental situations."
Acustica 67 (1989) 199-215; 251-265; 68 (1989) 15-25

[Q.22] SIGALAS, M. M.; ECONOMOU, E. N.
"Elastic and acoustic wave band structure" ; J. Sound Vib. 158 (1992) 377-382

[Q.23] VIKTOROVA, R. N.; TYUTEKIN, V. V.
"Physical foundations for synthesis of sound absorbers using complex-density composites" ; Acoust. Phys. 44 (1998) 275-280

[Q.24] SKELTON, E. A.; JAMES, J. H.
"Theoretical acoustics of underwater structures"
Imperial College Press, London, 1997

[Q.25] MAYSENHÖLDER, W.
"Rigorous computation of plate-wave intensity" ; Acustica 72 (1990) 166-179

[Q.26] MAYSENHÖLDER, W.
"Some didactical and some practical remarks on free plate waves"
J. Sound Vib. 118 (1987) 531-538

[Q.27] MURPHY, J. E.; LI, G.; CHIN-BING, S. A.
"Orthogonality relation for Rayleigh-Lamb modes of vibration of an arbitrarily layered elastic plate with and without fluid loading"
J. Acoust. Soc. Am. 96 (1994) 2313-2317

[Q.28] MAYSENHÖLDER, W.
"Analytical determination of the group velocity of an arbitrary Lamb wave from its phase velocity" ; Acustica 77 (1992) 208

[Q.29] MOZHAEV, V. G.
"Approximate analytical expressions for the velocity of Rayleigh waves in isotropic media and on the basal plane in high-symmetry crystals"
Sov. Phys. Acoust. 37 (1991) 186-189

[Q.30] TING, T. C. T.; BARNETT, D. M.
"Classifications of surface waves in anisotropic elastic materials"
Wave Motion 26 (1997) 207-218

[Q.31] MAYSENHÖLDER, W.
"Sound transmission through periodically inhomogeneous anisotropic plates: Generalizations of Cremer's thin plate theory" ;
Acustica 84 (1998) 668-680

[Q.32] PIERCE, A. D.
"Variational formulations in acoustic radiation and scattering"
In: Physical Acoustics, Vol. XXII (Underwater Scattering and Radiation), A. D. Pierce, R. N. Thurston (eds.), Academic Press, Boston, 1993, 195-371

[Q.33] HECKL, M.
"Untersuchungen an orthotropen Platten" ; Acustica 10 (1960) 109-115

[Q.34] HABERKERN, R.
Personal communication

[Q.35] GAZIS, D. C.
"Three-dimensional investigation of the propagation of waves in hollow circular cylinders. I. Analytical formulation" ; J. Acoust. Soc. Am. 31 (1959) 568-578

[Q.36] JUNGER, M. C.; FEIT, D.
"Sound, structures and their interactions"
MIT Press, Cambridge MA, 1986

[Q.37] MECHEL, F.P. Acta Acustica, submitted 1999
"About the Partition Impedance of Plates, Shells, and Membranes"

[Q.38] MECHEL, F.P.
"Schallabsorber", Vol. II, Chapter 27:
"Plate and Membrane Absorbers"
S.Hirzel Verlag, Stuttgart, 1995

[Q.39] GORMAN, D.J.
"Free Vibration Analysis of Rectangular Plates"
Elesevier/North Holland Inc., 1982, N.Y.

[Q.40] FAHY, F.
"Sound and Structural Vibration",
Academic Press, London, 1985

[Q.41] CREMER, L., HECKL, M.
"Körperschall"
Springer, Berlin, 1996

[Q.42] DYM, C.L. J.Sound and Vibr. 29 (1973) 189-205
"Some new results for the vibration of circular cylinders."

[Q.43] SOEDEL, J.Sound and Vibr. 29 (1973) 457-461
"A natural frequency analogy between spherically curved panels and flat plates."

[Q.44] HECKL, M. Frequenz 18 (1964) 299-304
"Einige Anwendungen des Reziprozitätsprinzips in der Akustik"

[Q.45] MAIDANIK J.Acoust.Soc.Amer. 34 (1962) 809-826
"Response of ribbed panels to reverberant acoustic fields"

R
Ultrasound Absorption in Solids

by W. Arnold

See list of symbols at the end of this Chapter, on p. 1139.

R.1 Generation of ultrasound

Surface excitation :

In most cases ultrasound is generated by piezoelectric transducers. The principle can be easily seen by a one-dimensional consideration.

Piezoelectric equations:

$$\sigma_1 = c_{11}\varepsilon_1 - e_{11}E_1 \tag{1}$$

$$P_1 = e_{11}\varepsilon_1 + \eta_{11}E_1 \tag{2}$$

E is the electric field, ε the strain, e the piezoelectric constant, η the dielectric suszeptibility, σ the stress and P the electric polarisation. Applying an electrical field $E_1(t) = E_0 e^{j\omega t}$ across the surface of a piezoelectric crystal yields for the wave equation:

$$\frac{\partial^2\xi}{dx^2} - \frac{1}{v^2}\frac{\partial^2\xi}{\partial t^2} = d_{11}\frac{\partial E_1}{\partial x} \tag{3}$$

with $d_{11} = e_{11}/c_{11}$ and the boundary condition that the surface is stress free :

$$\partial\xi / \partial x = \varepsilon_1 = d_{11}E_1(t) \tag{4}$$

one obtains:

$$\xi = j\left(d_{11}E_0 / k\right)e^{j\omega t}e^{-j(x-x_0)} \tag{5}$$

Eq.(3) shows that the gradient of the E-field is the source of ultrasound. This holds for all piezoelectric transducers.

The radiated ultrasonic energy is, see figure below :

$$S = P_{in}\left[c_{11}d_{11}^2 vAQ / \omega\varepsilon_r\varepsilon_0\left(\varepsilon_r V_g + V_r\right)\right] \tag{6}$$

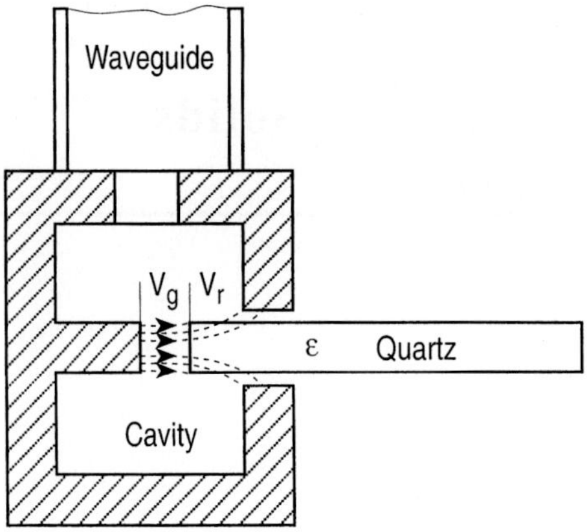

Figure: Principle of surface generation of ultrasonic waves due to piezoelectricity.

A is the cross section of the rod, Q is the quality factor of the resonator, P_{in} is the input electrical intensity. Eq.(6) is also valid correspondingly for placing the transducer into a capacitor.

Thin films and piezoelectric discs :

In the case there is a thin film or a piezoelectric disc as a transducer, waves are generated at both surfaces of the transducer.

$$S = \left(\frac{Z'}{Z}\right) \frac{A v_l \kappa K^2 E_0^2 (1-\cos kd)^2}{2\left(\sin^2 kd + \left(Z'/Z\right)^2 \cos^2 kd\right)} \tag{7}$$

Here, d is the film thickness, Z' and Z are the impedances of the transducer and material, respectively, K^2 is the piezoelectric coupling factor, κ is the dielectric permittivity, and A is the area of the transducer, see figure below.

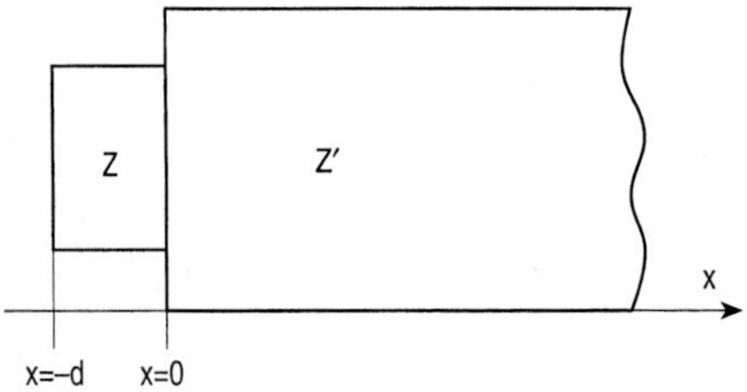

Principle of generation of ultrasonic waves due to piezoelectricity in a thin film.

R.2 Ultrasonic attenuation

General considerations :

Strain wave:

$$\sigma(x,t) = \sigma_0 e^{j(\omega t - kx)} \tag{1}$$

The relation between k-vector k and angular frequency ω is $k^2 v^2 = \omega^2$ with phase velocity v. If there is attenuation, either the velocity v, the k-vector or the frequency is complex:

$$v = v_1 + j v_2 \;; \qquad k = k_1 - j\alpha \tag{2}$$

Hence:

$$\sigma(x,t) = \sigma_0 e^{-\alpha x} e^{j(\omega t - k_1 x)} \tag{3}$$

α is the damping coefficient and v_p is the sound velocity as a real quantity; $v_p = \omega / k_1$.

With (2):

$$\alpha = \frac{\omega v_2}{\left(v_1^2 + v_2^2\right)} = \frac{\omega v_2}{|v|^2} \qquad k_1 = \frac{\omega v_1}{\left(v_1^2 + v_2^2\right)} = \frac{\omega v_1}{|v|^2} \tag{4a,b}$$

Likewise one may start with a complex elastic modulus: $c = c_1 + j c_2$ (5)

and obtains :

$$\alpha = \left(\frac{\omega v_2}{v_\rho v_1}\right) = \left(\frac{\omega (c_2)^{1/2}}{v\rho (c_1)^{1/2}}\right) \tag{6}$$

Less common is the assumption that the frequency is complex and k real, details see [R.2.1].

Definition of attenuation:

$$\sigma(x) = \sigma_0 e^{-\alpha x} \qquad \alpha = \frac{1}{x_2 - x_1} \log_e\left(\frac{\sigma(x_1)}{\sigma(x_2)}\right) \tag{7a,b}$$

then

$$\alpha = \left(\frac{1}{x_2 - x_1}\right) 20 \log_{10}\left(\frac{\sigma(x_1)}{\sigma(x_2)}\right) \quad \text{[dB/unit length]} \tag{8a}$$

$$\alpha = \left(\frac{1}{x_2 - x_1}\right) \log_e\left(\frac{\sigma(x_1)}{\sigma(x_2)}\right) \quad \text{[nepers/unit length]} \tag{8b}$$

Conversion factors :

$$\alpha\,[\text{dB / unit length}] = 8.686\,\alpha \quad \text{[nepers/unit length]} \tag{9a}$$

$$\alpha\,[\text{dB/unit time}] = \alpha\,[\text{dB/unit length}] \times \text{sound velocity} \tag{9b}$$

Logarithmic decrement:

$$\delta = \log_e\left(\frac{\sigma_n}{\sigma_{n+1}}\right) \tag{10}$$

where two consecutive oscillations are considered. Hence :

$$\delta\ [\text{nepers}] = \alpha\ [\text{nepers/cm}]\,\lambda\ [\text{cm}] \qquad \delta = \frac{\alpha\ [\text{nepers/cm}]}{\upsilon\ [1/\text{sec}]}\,v\ [\text{cm/sec}] \tag{11a,b}$$

Therefore :

$$\alpha\,[\text{dB}/\mu\text{sec}] = 8.68\times 10^{-6}\,v\,[\text{cm/sec}]\ \alpha\ [\text{nepers/cm}]$$

$$\alpha\,[\text{dB}/\mu\text{sec}] = 8.68\times 10^{-6}\,\upsilon\,[\text{sec}^{-1}]\ \delta\ [\text{nepers}] \tag{12}$$

$$\alpha\,[\text{dB}/\mu\text{sec}] = \alpha\,[\text{dB/cm}]\times 10^{-6}\,v\ [\text{cm/sec}]$$

Definition of Q-value from the bandwidth $\Delta\nu$ and resonance ν_r: $Q = \nu_r / \Delta\upsilon$ (13)
equivalent to:

$$Q = \omega_r \frac{\text{energy in the system}}{\text{energy dissipated per second}} \tag{14a}$$

or $$Q = \omega_1 \frac{W}{dW/dt} \tag{14b}$$

where W is the energy stored dissipating with dW/dt per second, hence:

$$W = W_o e^{-(\omega_1/Q)t} \tag{15}$$

There are close similarities to the behaviour of oscillators.

Geometrical losses :

Diffraction losses play an important role, particularly at low frequencies:

$$\alpha_g = \frac{1.8}{1.05a^2/\lambda}[\text{dB/cm}] \quad \text{or} \qquad \alpha_d = 1.7\frac{\upsilon}{a^2 v}[\text{dB/cm}] \tag{16}$$

a= radius of transducer, λ= wavelength, υ= frequency, and v= sound velocity.

Non-parallelism is another source of geometrical attenuation:

$$\alpha \cong 8.7\text{x}10^{-5}\,va\theta \tag{17}$$

θ is the angle between non-parallel surfaces of the sample.

Scattering losses :

Scattering at single spheres:

Scattering losses occur if there is a change in local mechanical impedance in polycrystalline and two- or multiphase materials. In case of cavities, diameter a, in elastic materials (denoted 1), this leads to a scattering cross-section for $da << 1$(Rayleigh approximation):

$$\gamma_N = \frac{4}{9} g_c (k_1 a)^4 \tag{18a}$$

$$g_c = \frac{4}{3} + 40\frac{2+3(\kappa_1/k_1)^5}{\left(4-9(\kappa_1/k_1)^2\right)^2} - \frac{3}{2}\left(\frac{\kappa_1}{k_1}\right)^2 + \frac{2}{3}\left(\frac{\kappa_1}{k_1}\right)^3 + \frac{9}{16}\left(\frac{\kappa_1}{k_1}\right)^4 \tag{18b}$$

Here (and below): $$k = \omega\left(\frac{\rho}{\lambda+2\mu}\right)^{1/2}, \kappa = \omega/(\mu/\rho)^{1/2} \tag{19}$$

with k= longitudinal wavenumber, κ= transversal wavenumber,
ρ= density, λ, μ= Lamé constants.

In case of elastic spheres (denoted 2) in matrix 1, incident longitudinal waves:

$$\gamma_N = \frac{4}{9} g_{el} (k_1 a)^4 \tag{20a}$$

$$\begin{aligned} g_{el} = &\left[\frac{3(\kappa_1/k_1)^2}{\left(3(\kappa_2/k_2)^2-4\right)(\mu_2/\mu_1)+4} - 1\right]^2 + \frac{1}{3}\left[1+2\left(\frac{\kappa_1}{k_1}\right)^3\right]\left[\left(\frac{\kappa_2}{\kappa_1}\right)^2\frac{\mu_2}{\mu_1} - 1\right]^2 + \\ &+40\left[2+3\left(\frac{\kappa_1}{k_1}\right)^5\right]\left[\frac{(\mu_2/\mu_1)-1}{2\left(3(\kappa_1/k_1)^2+2\right)(\mu_2/\mu_1)+9(\kappa_1/k_1)^2-4}\right]^2 \end{aligned} \tag{20b}$$

In case of elastic spheres (denoted 2) in matrix 1, incident transverse waves:

$$\gamma_N = \frac{4}{9} g_{el,t} (\kappa_1 a)^4 \tag{21a}$$

$$g_{el,t} = \frac{8}{3}\left[1+\frac{1}{2}\frac{k_1^3}{\kappa_1^3}\right]\left\{\frac{\left[3\frac{\kappa_1^2}{\kappa_2^2} - 3\frac{\kappa_2^2}{\kappa_1^2} - 4\frac{k_2^2\kappa_1^2}{\kappa_2^4} + 10\frac{k_2^2}{\kappa_2^2} - 6\frac{k_2^2}{\kappa_1^2}\right]}{\left[1-10\frac{\kappa_1}{k_1} + 6\frac{k_2}{k_1} - 6\frac{k_2}{k_1}\frac{\kappa_2^2}{\kappa_1^2} + 9\frac{\kappa_2^2}{\kappa_1^2}\right]^2}\right\} \tag{21b}$$

Scattering (neglecting multiple scattering) leads to an attenuation coefficient :

$$\alpha_s = \frac{1}{2} n_0 \gamma \tag{22}$$

with n_0= density of scatterers.

Scattering in polycrystalline materials :

Here, scattering at the polycrystals boundaries arises due the anisotropy of the crystals described by the anisotropy factor:

$$\alpha_L = \frac{8\pi^3}{375} \cdot \frac{Vf^4A^2}{\rho_0{}^2 v_L{}^8}\left[2+3\left(\frac{v_L}{V_T}\right)^5\right] \qquad \alpha_T = \frac{6\pi^3}{375} \cdot \frac{Vf^4A^2}{\rho_0{}^2 v_T{}^8}\left[3+2\left(\frac{v_T}{v_L}\right)^5\right] \tag{23a,b}$$

Attenuation can be written as :

$$\alpha_{L,T} = S_{L,T} V f^4 \approx S_{L,T} d^3 f^4 \tag{24}$$

where : $S_{L,T}$ is the scattering parameter for longitudinal and transverse waves, respectively. The volume of the scatterers is $V \approx d^3$, d grain size. The anisotropy factor is shown in the figures below.

For cubic crystals: $A^2 = (c_{11} - c_{12} - 2c_{44})^2$

Because $(v_L / v_T)^5 \approx 32$ transverse wave scattering is much stronger.

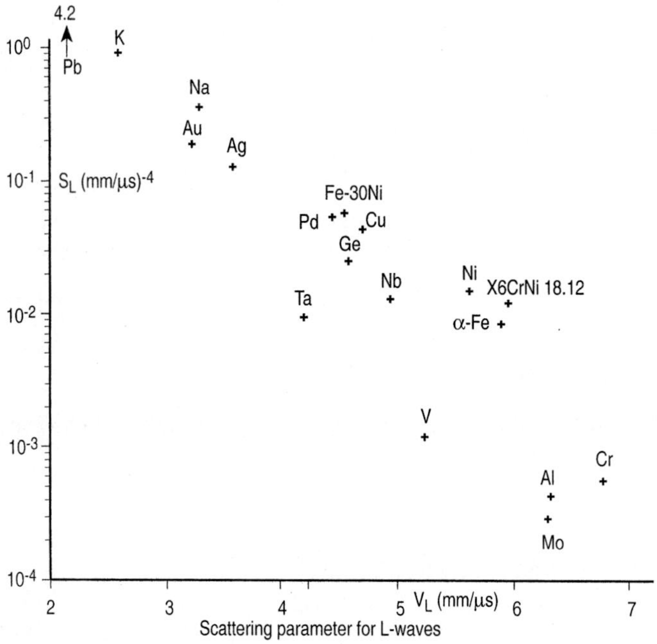

Scattering parameter for longitudinal waves.

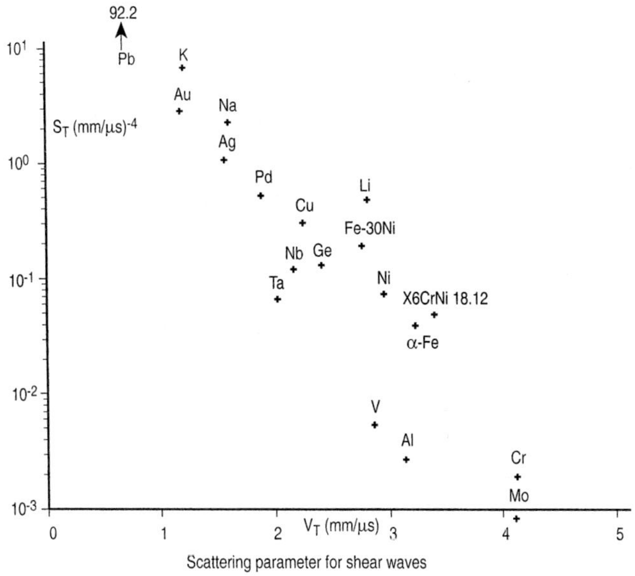

Scattering parameter for transverse waves.

Ultrasonic backscattering :

Ultrasonic backscattering may be used to characterise microstructures which are influenced by damage of various kinds, see Ref. [R.2.2, R.2.3, R.2.4]. Neglecting multiple scattering, the backscattered signal is:

Intensity: $$I_S(x) = I_0 \cdot 2\alpha_{L,T} \Delta x \exp(-2\alpha x) \tag{25a}$$

Amplitude: $$A_S(x) = A_0 (\alpha_S \Delta x)^i \cdot \exp(-\alpha x) \tag{25b}$$

with $\alpha = \alpha_{L,T} + \alpha_A$ the total attenuation coefficient, and α_A is the absorption coefficient which describes internal friction, see the following Sections.

R.3 Absorption and dispersion in solids due to dislocations :

Equation of motion:

$$\frac{\partial^2 \sigma_{ij}}{\partial x_j^2} = \rho \frac{\partial^2}{\partial t^2} \varepsilon_{ij} \tag{1}$$

Strain in the solid:

Elastic strain is determined by Hooke's law. The contribution of the dislocation to the strain is cast into ε_{dis} :

$$\varepsilon = \varepsilon_{el} + \varepsilon_{dis} \qquad \sigma = G\varepsilon_{el} \tag{2a,b}$$

where G is the shear modulus. The dislocations can undergo oscillations much like a string:

$$A\frac{\partial^2\xi}{\partial t^2} + B\frac{\partial\xi}{\partial t} - C\frac{\partial^2\xi}{\partial y^2} = b\sigma \tag{3}$$

with $A = \pi\rho b^2$ and $C = 2\,Gb^2/(\pi(1-\upsilon))$.

Here, ξ is the amplitude of the oscillating dislocation, A is their effective mass per unit length, B is the damping constant and is determined by viscous drag in the phonon and electron bath, C is the tension in the bowed-out dislocation, b is the Burgers vector, σ is given above, and ν is the Poisson ratio. Combining the above equations leads to:

$$\frac{\partial^2\sigma_{xx}}{\partial x^2} - \frac{\rho}{G}\frac{\partial^2\sigma_{xx}}{\partial t^2} = \frac{\Lambda\rho b}{l}\frac{\partial^2}{\partial t^2}\int_0^l \xi(y)dy \tag{4}$$

The integral in (4) represents the average amplitude of the dislocation oscillating in y-direction with l being their length. Solving (4) leads for the absorption α and dispersion $v(\omega)$ to:

$$\alpha(\omega) = \frac{1}{v}\left(\frac{4Gb^2}{\pi^4 C}\right)\omega_0^2\Lambda L^2\frac{\omega^2 d}{\left(\omega_0^2-\omega^2\right)^2+(\omega d)^2} \tag{5}$$

$$v(\omega) = v_0\left[1-\left(\frac{4Gb^2}{\pi^4 C}\right)\omega_0^2\Lambda L^2\frac{\omega_0^2-\omega^2}{\left(\omega_0^2-\omega^2\right)^2+(\omega d)^2}\right] \tag{6}$$

where

$$v_0 = \sqrt{G/\rho}, \qquad \omega_0 = \left(\pi / L\sqrt{C/A}\right), \qquad d = B/A \tag{7}$$

L is the loop length of the dislocation, Λ their density per area, and ω_0 is the resonance frequency of the dislocation. Two limits arise: resonance $(a \equiv d/\omega_0 << 1)$ and relaxation $(a >> 1)$. In case of relaxation eq. (5) reduces to:

$$\alpha = \left[8.68\times10^{-6}\left(\frac{4Gb^2}{\pi^4 C}\right)\Lambda L^2\omega_m\right]\left[\frac{(\omega/\omega_m)^2}{1+(\omega/\omega_m)^2}\right]\ [\mathrm{db}/\mu\mathrm{s})] \tag{8}$$

For convenience eq. (8) is written to give units dB/µs. The constant B is of the order $B \approx 10^{-4}$ [s/cm]. There are many other absorption mechanisms possible, based on dislocations dynamics, see [R.3.1-R.3.3] and references contained in [R.3.2, R.3.3].

Causes of losses and associated velocity changes

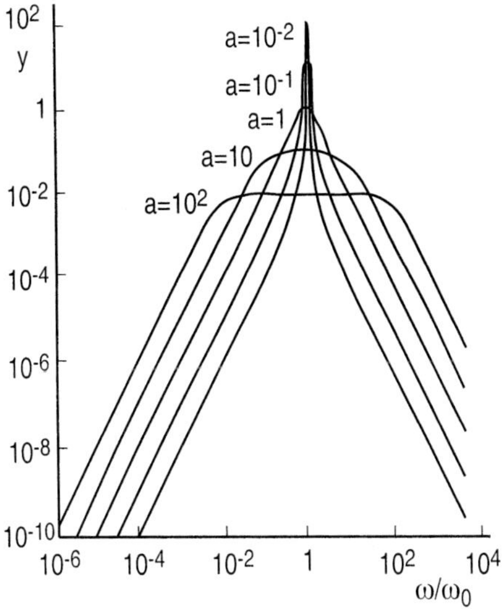

Dependence of $y = \dfrac{\alpha(\omega)}{(1/v)(4Gb^2/\pi^4 C)\,\Lambda L^2 \omega_0}$ on ω/ω_0 for various values of $a = d/\omega_0$

Absorption due to dislocations with transition from resonant ($a \equiv d/\omega_0 \ll 1$) to relaxation absorption ($a \gg 1$).

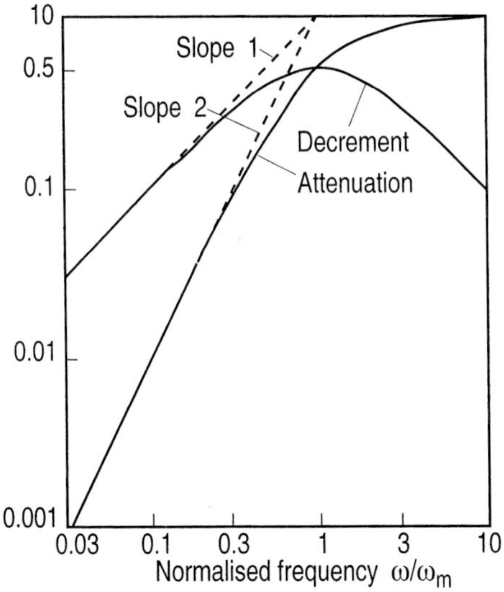

Normalised attenuation and decrement as functions of ω/ω_m for the case of large damping

Relaxation absorption from dislocations for $a \gg 1$.

R.4 Absorption due to the thermoelastic effects, phonon scattering and related effects

Thermoelastic effect :

In all solids an absorption mechanism arises because a propagating ultrasonic wave entails a temperature modulation for longitudinal waves due to thermoelasticity. This temperature modulation tends to return to equilibrium by thermal conductivity.

Phenomenological description for the stress-strain relation:

$$\tau^{-1}\sigma + \dot{\sigma} = (M_1 / \tau)\varepsilon + M_0\dot{\varepsilon} \tag{1}$$

τ is the relaxation time, σ the stress and ε the strain. The modulus defect $\Delta M/M$ describes the microscopic coupling and has to be determined in each individual case:

$$\frac{\Delta M}{M} = \frac{M_0 - M_1}{M_0} \tag{2}$$

General solution:

$$\alpha' \, v = \alpha \approx \left[\frac{\Delta M}{2M_0}\left(1 - \frac{\omega^2\tau}{1+\omega^2\tau^2}\right)\right] \tag{3}$$

$$v = \frac{\omega}{k} = \left[\frac{M_0}{\rho}\left(1 - \frac{\Delta M / M_o}{1+\omega^2\tau^2}\right)\right]^{1/2} \equiv \left(\frac{M\omega}{\rho}\right)^{1/2} \tag{4}$$

$$v_g = \left(\frac{M_0}{\rho}\right)^{1/2}\left\{1 - \frac{\Delta M}{M_0}\frac{\left(1-\omega^2\tau^2\right)}{\left(1+\omega^2\tau^2\right)^2}\right\} \tag{5}$$

Eqs.(4) and (5) indicate that there is dispersion, i.e., that the phase velocity v and group velocities v_g are not the same.

For the thermoelastic effect:

$$\frac{\Delta M}{M} = \frac{M_0 - M_1}{M_0} = \frac{E_{ad} - E_T}{E_T} = \frac{\beta^2 T}{\rho c_p}E_{ad} \approx \frac{\beta^2 T}{\rho c_p}E_T \tag{6}$$

where T is the temperature, β is the thermal expansion, E_{ad} and E_T are the adiabatic Young's modulus and at constant temperature, respectively, and c_P is the specific heat at constant pressure. The relaxation time τ_{th} is given by the thermal diffusivity D and the Debye average for the sound velocity v :

$$\tau_{th} \equiv \frac{L^2}{D} = \frac{L^2 \rho c_p}{\gamma} = \frac{D}{v^2} = \frac{\gamma}{\rho c_p v^2} \tag{7}$$

L is a length comparable to the wavelength, and γ the heat conductivity and c_p the specific heat. Over the length $L \cong \lambda$ the temperature difference generated by the ultrasonic wave due to the thermoelastic effect is equalised via thermal conductivity.

In polycrystals, heat may flow from one grain to the next because they heat up differently due to their anisotropy. This leads to an absorption, important for all technical materials:

$$\alpha = \frac{C_p - C_v R}{Cv} \frac{R}{2v} \frac{\omega^2 \tau}{1 + \omega^2 \tau^2} \tag{8}$$

with τ again the relaxation time. L is the mean grain diameter. R depends on the anisotropy for the strain energy and extends from $R \approx 10^{-6}$ for tungsten to $R = 6.5\ 10^{-2}$ for lead [R.4.1].

Phonon interactions :

Ultrasonic waves also modulate *locally* the thermal phonon distribution [R.4.1]. This holds both for longitudinal and shear waves and hence differs from the thermoelastic effect. For $\omega\tau_{th} << 1$, the absorption for longitudinal waves is :

$$\alpha_l = \frac{\Delta c_{jj}}{2\rho v_{long}^3} \frac{\omega^2 \tau_{th}}{1 + \omega^2 \tau_{th}^2} = \frac{1}{2\rho v_{long}^3} \left[\frac{3U_0}{N} \sum_i \left(\gamma_{j(i)}\right)^2 - \gamma^2 C_v \Theta \right] \frac{\omega^2 \tau_{th}}{1 + \omega^2 \tau_{th}^2} \tag{9}$$

with the modulus defect :

$$\Delta c_{jj} = \frac{3U_0}{N} \sum_i \left(\gamma_{j(i)}\right)^2 - \gamma^2 C_v \Theta \tag{10}$$

For transverse wave :

$$\Delta c_{jj} = \frac{3U_0}{N} \sum_i \left(\gamma_{j(i)}\right)^2 \tag{11}$$

$$\alpha_l = \frac{\Delta c_{jj}}{2\rho v_{long}^3} \frac{\omega^2 \tau_{th}}{1 + \omega^2 \tau_{th}^2} = \frac{1}{2\rho v_{long}^3} \left[\frac{3U_0}{N} \sum_i \left(\gamma_{j(i)}\right)^2 \right] \frac{\omega^2 \tau_{th}}{1 + \omega^2 \tau_{th}^2} \tag{12}$$

$$\tau_{th} = \frac{3k}{C_p v_0^2} \tag{13}$$

Here, $\sum_i \left(\gamma_{j(i)}\right)^2$ is an average Grüneisen constant for transverse and longitudinal phonons and C_V and C_p are the corresponding heat capacities per volume. U_0 is the total thermal energy. Note that this effect occurs in addition to the one described by the thermoelastic effect.

Eqs. (1), (12) can be simplified for practical purposes:

$$\text{eq.(1):} \quad \alpha_l = \frac{\Theta\kappa\gamma^2\omega^2}{2\rho v_{long}^5} \qquad \text{and eq. (12):} \quad \alpha_t = \frac{\Theta\kappa\gamma^2\omega^2}{2\rho v_t^5} \tag{14a,b}$$

If $\omega\tau_{th} >> 1$, a different view-point must be considered. The ultrasonic wave no longer modulates the phonon distribution, but ultrasonic phonon is scattered by the thermal phonons and corresponding conservation of momentums must be taken into account. This mechanism plays a role for low temperatures $(< 40\ \text{K})$, i.e. much below the Debye temperature Θ [R.1.1, R.4.1, R.4.2].

$$\alpha_t = \frac{\pi^3}{60} \frac{k_B{}^4 \overline{F}_l{}^2}{\rho^3 v_0^{10} \hbar^3} \omega_l T^4 \tag{15}$$

F_l is an average of second and third order elastic constants. Note the strong dependence on sound velocity which was verified experimentally. This mechanism also holds for longitudinal and transverse waves in various crystallographic orientations in crystals and the $\omega_l T^4$ is retained, however, the pre-factor is different. The pre-factor also depends on the crystal class.

R.5 Interaction of ultrasound with electrons in metals

The ultrasonic wave leads to a spatial separation of the ions from the free electrons at the Fermi surface. This causes an electrical field which eventually leads to a redistribution of the electrons and hence to an absorption coefficient α [R.1.1]:

$$\alpha = \frac{nm}{2\rho v_l \tau} \left[\frac{\frac{1}{3} k^2 l_e{}^2 \tan^{-1}(k l_e)}{k l_e - \tan^{-1}(k l_e)} - 1 \right] \tag{1}$$

Here, n is the number of conduction electrons per volume, m is the electronic mass, l_e is the mean free path for electrons, v_F is the Fermi velocity, k the wave vector of the ultrasonic wave, and v_l the longitudinal sound velocity.

In the limit of $kl_e \le 1$ this leads to :

$$\alpha = \frac{2}{15}\frac{nmv_F{}^2\tau}{\rho v_l{}^3}(1 - 9/35(k_e l)^2 + ...) \tag{2}$$

In the limit $kl_e >> 1$:

$$\alpha = \frac{\pi nmv_F}{12\rho v_l^2}\omega \tag{3}$$

The collision time τ is determined by the electrical conductivity $\sigma = ne^2\tau / m$.

Transverse waves also cause absorption because they generate indirectly an internal electrical field via an internal magnetic field. Final results are similar to the equations above.

The absorption is magnetic field dependent because the electrons follow curved trajectories in a magnetic field and are bound to the Fermi surface. For transverse waves and a magnetic field perpendicular to the polarisation vector and k-vector:

$$\frac{\alpha(H)}{\alpha(0)} = \frac{1}{1+(2\omega_c\tau)^2} \tag{4}$$

where ωc is the cyclotron resonance frequency $\omega_c = (eH/mc)$ (5)

similarly for H parallel to he polarisation vector and perpendicular to the k-vector:

$$\frac{\alpha(H)}{\alpha(0)} = \frac{1}{1+(\omega_c\tau)^2} \tag{6}$$

Various resonance phenomena may occur if the electrons can complete an orbit on the Fermi surface whose size is equal to an integer number of the wavelength. If H is perpendicular to k, geometrical resonance occurs. When the ultrasonic frequency $\omega = n\omega_c$, n integer, and k parallel to H, temporal resonance may be induced. Finally quantum oscillations may be excited. Here the effect is due to the quantisation of the electron energy in a strong magnetic field applied in the z-direction. In all three cases the ultrasonic absorption becomes periodically field dependent with the periodicity $(1/H)$:

$$\Delta(1/H) = e\lambda / \hbar k_y v_l) \tag{7}$$

k_y is the size of the orbit in k-space equivalent to the cross section of the Fermi surface.

$$\Delta(1/H) = e / \omega m v_l \tag{8}$$

$$\Delta\left(\frac{1}{H}\right) = \frac{e\hbar}{mv_l}\left[\frac{\hbar^2 k_f{}^2}{2m} - \frac{\hbar^2 k_z{}^2}{2m}\right]^{-1} = \frac{2\pi e}{\hbar c_l}\frac{1}{A(E_f, k_z)} \tag{9}$$

m is the effective mass of the electron and A is the cross section of the Fermi surface at E_f and k_z.

The latter case is analogue to the so-called de Haas van Alphen effect. Exploitation of eqs. (7)-(9) allow to measure the shape of the Fermi surface in metals and its anisotropy.

If the metal becomes superconducting, the absorption drops. This can be exploited to measure the gap function $\Delta(T)$ in the superconductor:

$$\frac{\alpha_s}{\alpha_n} = \frac{2}{\exp(\Delta / kT) + 1} \tag{10}$$

R.6 Wave propagation in piezoelectric semiconducting solids

The generation of a stress field in a piezoelectric solids leads to an accompanying electric field which accelerates electrons and leads to absorption and dispersion because the electrons undergo inelastic scattering [R.6.1]:

$$v = v_0 \left[1 + \frac{e^2}{2cp} \frac{1 + (\omega_C / \omega_D) + (\omega / \omega_D)^2}{1 + 2(\omega_C / \omega_D) + (\omega / \omega_D)^2 + (\omega_C / \omega)^2} \right] \tag{1}$$

$$\alpha = \frac{\omega}{v_0} \frac{e^2}{2cp} \left[\frac{\omega_C / \omega}{1 + 2(\omega_C / \omega_D) + (\omega / \omega_D)^2 + (\omega_C / \omega)^2} \right] \tag{2}$$

Here, ω_C is the so-called conductivity frequency and ω_D the diffusion frequency; $\omega_C = \sigma/\varepsilon_d$ (ε_d dielectric constant), $\omega_D \approx (ev^2/\mu kT)$ (μ: mobility of the electrons, k: Boltzmann constant, T: temperature).

R.7 Absorption in Amorphous Solids and Glasses

Disordered materials exhibit additional absorption mechanism, mostly due to relaxing units in the molecular or atomic structure.

Similar to eq. (R.4.3), the absorption is described as :

$$\alpha \approx \left[\frac{\Delta M}{2M_0} \left(1 - \frac{\omega^2 \tau}{1 + \omega^2 \tau^2} \right) \right] \qquad (1)$$

however, with the modulus defect or relaxation strength $\Delta M/2M_0$, orders of magnitude larger than in the thermoelastic regime, see figure below. The relaxation frequency $1/\tau$ is determined by an Arrhenius process : $1/\tau = kT/he^{-\Delta E/kT}$
with ΔE the activation energy, k the Boltzmann constant and T the temperature.

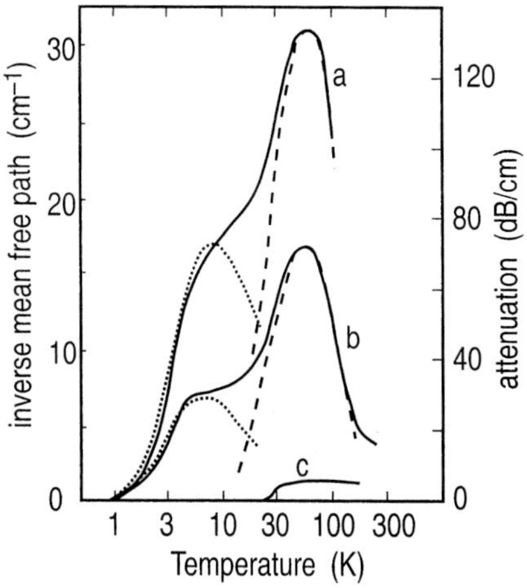

For low temperatures, amorphous solids behave in a much different way than crystalline solids. Some of the structural units can tunnel between different local spatial co-ordinates leading to a resonant absorption which depends on the ultrasonic intensity S, S_c is a critical intensity. M is the coupling coefficient ($M \approx 0.5$ eV) and n is the density of the tunnel units per energy and volume ($n_0 \approx 10^{33}$ erg^{-1}cm^{-3}).

$$\alpha_{res} = -\frac{\pi n_0 \omega M^2}{\rho v^3} \frac{\tanh(\hbar\omega / kT)}{\sqrt{1 + S/S_c}} \qquad (2)$$

The corresponding change of sound velocity is for $S \ll S_c$:

$$\Delta v/v_0 = \frac{n_0 M^2}{\rho v^2} \ln(T/T_0) \qquad (3)$$

where T_0 is a reference temperature. There is also a relaxation absorption due to the coupling to the phonon bath of the tunnelling units. An ultrasonic wave modulates the thermal occupation leading to :

$$\alpha_{rel} = \frac{3n_0 D^2 T^3}{\pi\rho^2 \hbar^4 v^3}\left(\frac{M_l^2}{v_l^5} + \frac{2M_{tl}^2}{v_t^5}\right) \tag{4}$$

Also this absorption mechanism leads to dispersion, see Ref. [R.7.1] and [R.7.2].

If crystals exhibit a certain disorder or are irradiated by electrons or neutrons, similar phenomena are observed.

R.8 Relation of Ultrasonic Absorption to Internal Friction

Internal friction discusses the absorption mechanism of mechanical waves and oscillations at low frequencies usually below 20 kHz, i.e., in the audible range. The mechanism are mostly relaxation phenomena which are described by equations like (R.7.1), where the relaxation strength is adapted to the corresponding situation. Overviews can be found in [R.8.1] - [R.8.3].

R.9 Gases and Liquids

Treatments of the ultrasonic absorption due to relaxation phenomena in gases and liquids can be found in [R.9.1] .

R.10 Kramers-Kroning Relation

Kramers-Kroning relation describe the interdependence between absorption and dispersion:

$$K_1 = \frac{2}{\pi} P \int_0^\infty \frac{\omega' K_2(\omega')}{\omega'^2 - \omega^2} d\omega' \tag{1}$$

$$K_2 = \frac{2}{\pi} P \int_0^\infty \frac{\omega' K_1(\omega')}{\omega'^2 - \omega^2} d\omega' \tag{2}$$

Here, K_1 is the real part and K_2 the imaginary part of the compressibility. Analogue expressions hold for the k-vector k and the absorption coefficient.

Practical simplification of eqs. (1) and (2) are:

$$\frac{dK_1(\omega)}{d\omega} = -\frac{2}{\rho_0 v^3(\omega)}\frac{dv(\omega)}{d\omega} \quad (3)$$

$$\frac{dv(\omega)}{d^{\omega}} = 2v^2(\omega)\alpha(\omega)/\pi\omega^2 \quad (4)$$

$$\frac{dv(\omega)}{v^2(\omega)} = \frac{2\alpha(\omega)}{\omega^2}d\omega \quad (5)$$

$$\frac{1}{v_0} - \frac{1}{v(\omega)} = \frac{2}{\pi}\int_{\omega_0}^{\omega}\frac{\alpha(\omega')}{\omega'^2}d\omega' \quad (6)$$

$$\alpha(\omega) = \frac{\pi\omega^2}{2v_0^2}\frac{dv(\omega)}{d\omega} \quad (7)$$

$$\Delta v = v(\omega) - v_0 = \frac{2v_0^2}{\pi}\int_0^{\omega}\frac{\alpha(\omega')}{\omega'^2}d\omega' \quad (8)$$

Here v_0 is the sound velocity at the frequency ω_0.

List of symbols used in this Chapter:

a	transducer radius
A	area, amplitude
B	Burger's vector
c	elastic constant
c_p	specific heat at constant pressure
d	thickness, grain size
e	piezoelectric constant
f	frequency
D	deformation potential
E	electric field, Young's modulus
G	shear modulus
H	magnetic field
k	k-vector
k_B	Boltzmann's constant
K	piezoelectric coupling factor, compressibility
l_e	mean free path for electrons
M	elastic moduli, deformation potential
n	volume density of scatterers
P	polarisation, power
Q	Q-value
S	ultrasonic energy
$S_{L,T}$	scattering parameter
T	temperature
v	sound velocity
v_F	Fermi velocity
V	volume
W	energy
x	path length
Z	acoustic impedance
α	attenuation, absorption
β	thermal expansion
ε	strain
γ	cross-section, heat conduction, Grüneisen constant
κ	dielectric permittivity, k-vector
λ	wavelength
λ, μ	Lamé constants
η	polarisability
ξ	displacement
ρ	density
τ	relaxation time
Θ	angle, Debye temperature
σ	stress, electrical conductivity
υ	frequency
ω	angular frequency

References to part R

Ultrasound Absorption in Solids :

The references are numbered [R.x.n] , where x denotes the section, and n is a running number.

[R.1.1] J.W. TUCKER, V.W. RAMPTON,
"Microwave Ultrasonic Methods", North-Holland, (1972)

[R.1.2] G. S. KINO,
"Acoustic Waves, Devices, Imaging, and Signal Processing",
Prentice-Hall, Inc. (1987)

[R.2.1] R.TRUELL, C. ELBAUM, B. B. CHICK,
"Ultrasonic Methods" in Solid State Physics, Academic Press, 1969

[R.2.2] K. GOEBBELS,
"Structure Analysis by Scattered Ultrasonic Radiation",
in *Research Techniques in ND,* Ed. R.S. Sharpe, Academic Press, London,
IV, (1980) 87-150

[R.2.3] S. HIRSEKORN, P.W. ANDEL, U. NETZELMANN,
"Ultrasonic Methods to Detect and Evaluate Damage in Steel",
Nondestr. Test. and Evaluation, 15 (1980) 373-393

[R.2.4] B. BOYD, C.P. CHIOU, B. THOMPSON, J. OLIVER,
"Development of Geometrical Models of Hard-Alpha Inclusions for Ultrasonic Analysis in Titanium Alloys",
Review of Progress in Quantitative Nondestructive Evaluation,
Eds. D.O. Thompson, D.E. Chimenti, Plenum, New York, XVIII (1998) 823-830

[R.3.1] R.TRUELL, C. ELBAUM, B. B. CHICK,
"Ultrasonic Methods" in Solid State Physics, Academic Press, 1969

[R.3.2] G. GREMAUD, S. KUSTOV,
"Theory of dislocation-solute atom interaction in solid solutions and related nonlinear anelasticity", Phys. Rev. B, 60 (1999) 9353-9364

[R.3.3] G. GREMAUD,
"Dislocation-point defect interaction" in Mechanical Spectroscopy Q-1 2001,
edited by R. Schaller, G. Fantozzi and G. Gremaud,
Chapter 3.3, Materials Science Forum 366-368 (2001) 178-247,
Trans Tech Publications, Switzerland

[R.4.1] R.T. BEYER, S.V. LETCHER,
"Physical Ultrasonics", Academic Press 1969

[R.4.2] K. DRANSFELD
J. de Physique C1 28 (1967) 157-162

[R.6.1] A. R. HUTSON, D.L. WHITE,
J. Appl. Phys. 33, 40-47 (1962)

[R.7.1] S. HUNKLINGER, W. ARNOLD,
Phys. Acoustics, Eds. W.P. Mason and R.N. Thurston, XII (1976) 156-215

[R.7.2] C. ENSS, S. HUNKLINGER,
"Tieftemperaturphysik", Springer Berlin, 2000

[R.8.1] R. DE BATIST,
"Internal Friction of Structural Defects in Solids",
North-Holland Publishing Company, Amsterdam (1972)

[R.8.2] A. S. NOWICK, B.S. BERRY,
"Anelastic Relaxation in Crystalline Solids",
Academic Press, (1972)

[R.8.3] R. SCHALLER, G. FANTOZZI, G. GREMAUD (Eds.)
"Mechanical Spectroscopy Q-1 2001",
Materials Science Forum 366-368, Trans Tech Publications, Switzerland (2001)

[R.9.1] A.B. BHATIA,
"Ultrasonic Absorption",
Clarendon Press, Oxford (1967)

[R.10.1] M. O'DONELL, E.T. JAYNES,G. MILLER,
J. Acosut. Soc. 69, 696-701 (1969)

S
Nonlinear Acoustics
by O.V. Rudenko

S.1 General formulas

See Chapter "B. General Linear Fluid Acoustics", and especially Section B.1, for linear acoustic relations.

The strength of an acoustic field can be characterised by an acoustic Mach number:

$$M = \frac{v}{c_0} = \frac{\rho_1}{\rho_0} = \frac{p_1}{p_0} \tag{1}$$

Parameters of equilibrium state of the medium are: c_0= adiabatic sound speed, ρ_0= density, and $p_0 = c_0^2\rho_0$= internal static pressure. The acoustic wave leads to variations of density $\rho_1 = \rho - \rho_0$ and pressure $p_1 = p - p_0$, as well as to the appearance of nonzero particle velocity v. The limit $M \to 0$ corresponds to linear acoustics. Nonlinear acoustics deals with small, but finite values of M.

Nonlinear properties of the medium are described by the nonlinear parameter ε. For liquid and gas it equals to

$$\varepsilon = 1 + \beta/(2A) \tag{2}$$

where A and B are determined by a series expansion of the equation of state $p = p(\rho)$ in powers of M :

$$p_1 = \left(\frac{\partial p}{\partial \rho}\right)_0 \rho_1 + \frac{1}{2}\left(\frac{\partial^2 p}{\partial \rho^2}\right)_0 \rho_1^2 + \ldots = A\left(\frac{\rho_1}{\rho_0}\right) + \frac{1}{2}B\left(\frac{\rho_1}{\rho_0}\right)^2 \ldots \tag{3}$$

The partial derivatives in eq.(3) are evaluated at the unperturbed state (p_0, ρ_0), and

$$A = \rho_0\left(\frac{\partial p}{\partial \rho}\right)_0 = c_0^2\rho_0 \quad ; \quad B = \rho_0^2\left(\frac{\partial^2 p}{\partial \rho^2}\right)_0 \tag{4}$$

The coefficients A and B can be measured experimentally or calculated theoretically. For an isentropic equation of state of a perfect gas :

$$\frac{p_1}{p_0} = \left(1 + \frac{\rho_1}{\rho_0}\right)^{\kappa} - 1 \quad ; \quad \varepsilon = \frac{\kappa + 1}{2} \tag{5}$$

For longitudinal waves propagating in isotropic solids :

$$\varepsilon = -\frac{3}{2} - \frac{1}{c_0^2 \rho_0}\left(A_L + 3B_L + C_L\right) \tag{6}$$

Here A_L, B_L, C_L are elastic moduli of the third order (or nonlinear elastic moduli) introduced by Landau to the expansion of internal energy E in powers of the deformation tensor U_{ik} :

$$E = \mu u_{ik}^2 + \left(\frac{K}{2} - \frac{\mu}{3}\right)u_{ll}^2 + \frac{1}{3}A u_{ik} u_{il} u_{kl} + B u_{ik}^2 u_{ll} + \frac{1}{3} C u_{ll}^3 \tag{7}$$

where μ is the linear shear elasticity and K is the linear volume elasticity.

The fundamental role of the numbers M and ε can be illustrated by the simplest problem of nonlinear wave propagation. Let at the boundary x=0 of a nonlinear medium the input signal be harmonic in time: $v(x = 0, t) = v_0 \sin \omega t$. The second harmonic appears in the medium because of its nonlinear properties. At small distances x the wave contains both the 1st and 2nd harmonics:

$$\frac{v}{c_0} \approx \frac{v_0}{c_0}\sin\omega\left(t - \frac{x}{c_0}\right) + \frac{\varepsilon\omega}{2c_0^3} v_0^2 \, x \sin 2\omega\left(t - \frac{x}{c_0}\right) \tag{8}$$

The 2nd harmonic increases in its amplitude with increase in both nonlinearity ε and Mach number $M = v_0/c_0$. In addition, it increases with increase in number of wavelengths x/λ traversed by the travelling wave. Such tendency in dependence of nonlinear phenomena on ε and M are typical for more complicated problems as well.

Values of ε measured experimentally are given in Table 1.

Table 1: Nonlinear parameter ε for different media.

Substance	ε	T, °C	Substance	ε	T, °C
Monatomic gas	7/6	20	Liquid Sodium	2.4	110
Diatomic gas	6/5	20	Liquid Nitrogen	5.5	-183
Distilled water	3.1	0	Beef liver	4.5	30
	3.5	20	Dog's kidney	4.6	30
	4.1	80	Human fat	6.0	30
Sea water (3.5 %)	3.6	20	Nickel steel	2.4	20
Alcohol	6.3	20	Armco iron	5.7	20
Acetone	5.6	20	Aluminum	7.2	20
Benzene	5.5	20	Sandstone	800	20
Water+air bubbles	up to $5\cdot10^3$	20	Concrete	$(0.6–1)\cdot10^5$	20

S.2 Riemann waves

The implicit function of particle velocity v :

$$v(x,t) = \Phi\left(t - \frac{x}{c_0 + \varepsilon v}\right) \tag{1}$$

is an exact solution of the one-dimensional system of hydrodynamic equations :

$$\frac{\partial \rho}{\partial t} + \frac{\partial}{\partial x}\rho v = 0 \quad ; \quad \frac{\partial v}{\partial t} + v\frac{\partial v}{\partial x} + \frac{1}{\rho}\frac{\partial p}{\partial x} = 0 \quad ; \quad p = p_0\left(\frac{\rho}{\rho_0}\right)^{\kappa} \tag{2}$$

describing the plane nonlinear wave propagating through the ideal medium. Here Φ is the arbitrary function describing the initial (at x=0) temporal profile of the incident wave: $v(x=0,t)= \Phi(t)$. The pressure and density fields corresponding to the particle velocity wave (1) are calculated by the formulas :

$$\frac{p}{p_0} = \left(1 + \frac{\kappa - 1}{2}\frac{v}{c_0}\right)^{\frac{2\kappa}{\kappa-1}} \quad ; \quad \frac{\rho}{\rho_0} = \left(1 + \frac{\kappa - 1}{2}\frac{v}{c_0}\right)^{\frac{2}{\kappa-1}} \tag{3}$$

At $M \ll 1$ the wave (1) is described by the simpler formula:

$$v = \Phi\left(\tau + \frac{\varepsilon}{c_0^2} v x\right) \tag{4}$$

where $\tau = t - x/c_0$ is the retarded time or time measured in an accompanying coordinate system which is moving with the sound speed c_0. The simplified solution (4) satisfies to the 1st order the differential evolution equation:

$$\frac{\partial v}{\partial x} - \frac{\varepsilon}{c_0^2} v \frac{\partial v}{\partial \tau} = 0 \tag{5}$$

For a harmonic initial wave the input signal takes a form $v(x=0,t)= v_0 \sin(\omega t)$. During the propagation, with increase in co-ordinate x, the nonlinear distortion of both temporal profile and spectrum go on. The solution (4) is written as:

$$\frac{v}{v_0} = \sin\omega\left(\tau + \frac{\varepsilon}{c_0^2} v x\right) = \sin\left(\omega\tau + z\frac{v}{v_0}\right) \tag{6}$$

Here the normalised distance z equal to

$$z = \frac{\varepsilon}{c_0^2}\omega v_0 x = \frac{x}{x_s} \quad ; \quad x_s = \frac{c_0^2}{\varepsilon \omega v_0} = \frac{\lambda}{2\pi\varepsilon M} \tag{7}$$

is measured in units of shock formation distance x_s . The solution (6) is valid for distances $0 < x < x_s$ or $0 < z < 1$. The series expansion for eq.(6):

$$\frac{v}{v_0} = \sum_{n=1}^{\infty} B_n(z)\sin(n\omega\tau) \quad ; \quad B_n = 2J_n(nz)/(nz) \tag{8}$$

($J_n(z)$= Bessel function) is named Bessel-Fubini solution. It describes higher harmonics generation having frequencies $n\omega$; n=2,3,... ; of the initially harmonic wave.

For the general case (1) the spectrum of a periodic Riemann wave is written as:

$$\frac{v}{v_0} = \sum_{n=-\infty}^{\infty} C_n(z)\exp(jn\omega\tau) = \sum_{n=1}^{\infty}\left(A_n(z)\cos(n\omega\tau) + B_n(z)\sin(n\omega\tau)\right) \tag{9}$$

where:

$$C_n = -\frac{j}{2\pi nz}\int_{-\pi}^{\pi}\left[e^{inz\Phi(\xi)} - 1\right]e^{-in\xi}\,d\xi \quad ; \quad A_n = C_n + C_n^* \quad ; \quad B_n = j\left(C_n - C_n^*\right) \tag{10}$$

For a non-periodic initial signal $\Phi(t)$ the Fourier spectrum of a Riemann wave is:

$$C(x,\omega) = \frac{1}{2\pi}\int_{-\infty}^{\infty}\Phi\left(\tau + \frac{\varepsilon}{c_0^2}vx\right)e^{-j\omega\tau}\,d\tau = \left(2\pi j\frac{\varepsilon}{c_0^2}\omega x\right)^{-1}\int_{-\infty}^{\infty}\left[\exp\left(j\frac{\varepsilon}{c_0^2}\omega x\Phi(\xi)\right) - 1\right]e^{-j\omega\xi}\,d\xi$$

S.3 Plane nonlinear waves in a dissipative medium

See Sections B.1, J.1–J.3 for linear acoustics in a dissipative medium.

Plane nonlinear wave propagation in a viscous and heat-conducting medium is governed by Burger's equation:

$$\frac{\partial v}{\partial x} - \frac{\varepsilon}{c_0^2}v\frac{\partial v}{\partial \tau} = \frac{b}{2c_0^3\rho_0}\frac{\partial^2 v}{\partial \tau^2} \tag{1}$$

As distinct from eq.(S.2.5), above equation (1) contains a dissipative term in its right-hand side, which is proportional to the 2nd derivative of the variable v and to the coefficient

$$b = \zeta + \frac{4}{3}\eta + \chi\left(\frac{1}{c_v} - \frac{1}{c_p}\right) \tag{2}$$

Here ζ and η are bulk and shear viscosities, and χ is thermal conductivity. At $\varepsilon = 0$ the linear solution to eq.(1) corresponding to the initial signal $\Phi(\tau)$ has a form:

$$v = \int_{-\infty}^{\infty} G(x, \tau - \tau')\, \Phi(\tau')\, d\tau' \tag{3}$$

where G is Green's function:

$$G(x,\tau) = \frac{1}{\sqrt{4\pi\delta x}} \exp\left(-\frac{\tau^2}{4\delta x}\right) \quad ; \quad \delta = \frac{b}{2c_0^3\rho_0} \tag{4}$$

For periodic $\Phi(\tau)$ the solution (3), (4) describes the decaying wave:

$$v = v_0 \exp\left(-\delta\omega^2 x\right) \sin(\omega\tau) \tag{5}$$

It is convenient to rewrite Burger's eq.(1) in normalised form:

$$\frac{\partial V}{\partial z} - V\frac{\partial V}{\partial \theta} = \Gamma \frac{\partial^2 V}{\partial \theta^2} \tag{6}$$

using non-dimensional variables:

$$V = \frac{v}{v_0} \quad ; \quad \theta = \omega\tau \quad ; \quad z = \frac{\varepsilon}{c_0^2}\omega v_0 x = \frac{x}{x_s} \tag{7}$$

Here Γ is the so-called "inverse acoustic Reynolds number" or "Goldberg's number":

$$\Gamma = \frac{b\omega}{2\varepsilon c_0\rho_0 v_0} = \delta\omega^2 x_s \tag{8}$$

At small magnitudes, $\Gamma \ll 1$, the nonlinear phenomena are well defined, and at $\Gamma \gg 1$ the dissipative phenomena predominate over weak nonlinearity.

The exact solution to eq.(6) is:

$$V = 2\Gamma \frac{\partial}{\partial\theta} \ln \int_{-\infty}^{\infty} \exp\left(-\frac{(\theta - \theta')^2}{4\Gamma z} + \frac{1}{2\Gamma}\int^{\theta'} V(z = 0, \theta'')\, d\theta''\right) d\theta' \tag{9}$$

The solution (9) for a harmonic initial signal is shown in the figure below. Temporal profiles of one period $-\pi \le \theta \le \pi$ of wave are constructed for $\Gamma = 0.1$ at different distances, $0 \le z \le 30$.

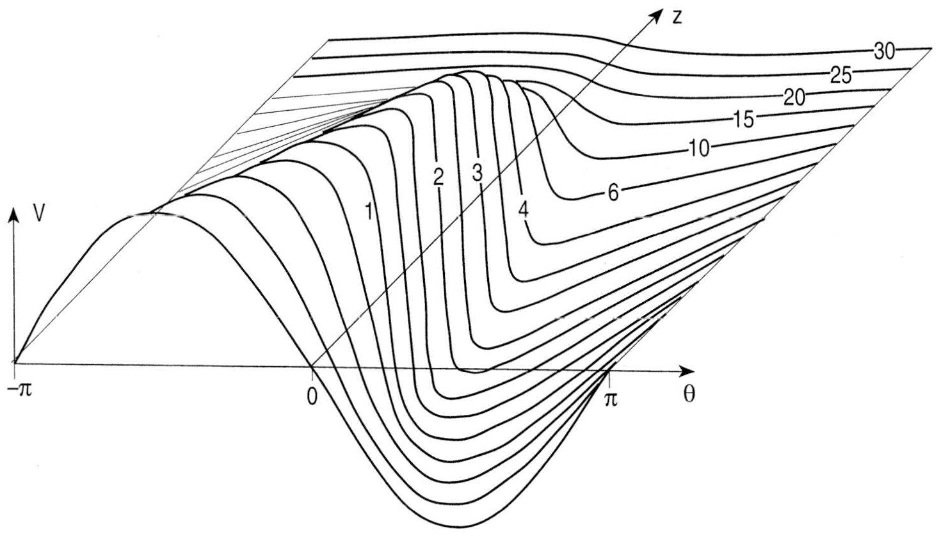

At the first stage of wave propagation, $0 \le z \le 1$, the steeping of the leading front goes on up to the shock formation distance $z=1$; here the wave profile can be described approximately (at $\Gamma \ll 1$) as a Riemann wave eq.(S.2.6). At the second stage, $0 \le z < 1/\Gamma = 10$, the wave has the sawtooth-like form described by Khokhlov's solution:

$$V = \frac{1}{1+z}\left[-\theta + \pi \tanh \frac{\pi\theta}{2\Gamma(1+z)}\right] \quad ; \quad -\pi < \theta < \pi \tag{10}$$

and its series expansion

$$V = \sum_{n=1}^{\infty} \frac{2\Gamma}{\sinh(n\Gamma(1+z))} \sin\theta \tag{11}$$

known as Fay solution. At the final stage of propagation, $z > 1/\Gamma$, the wave is weak and has the harmonic form again.

In the figures below the dependence of the harmonics $B_n(z)$; n=1,2,3; of eq.(S.2.9) on distance z is shown for different values of Γ .

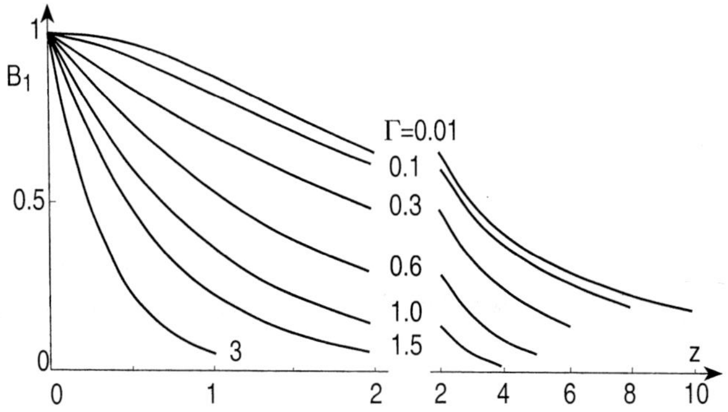

1st harmonic amplitude B_1 over distance z .

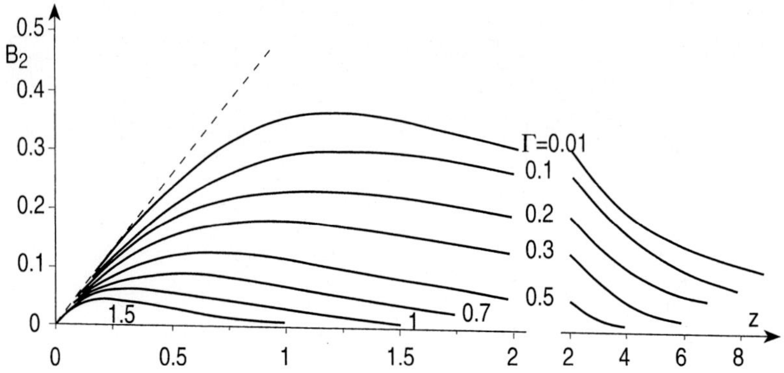

2nd harmonic amplitude B_2 over distance z

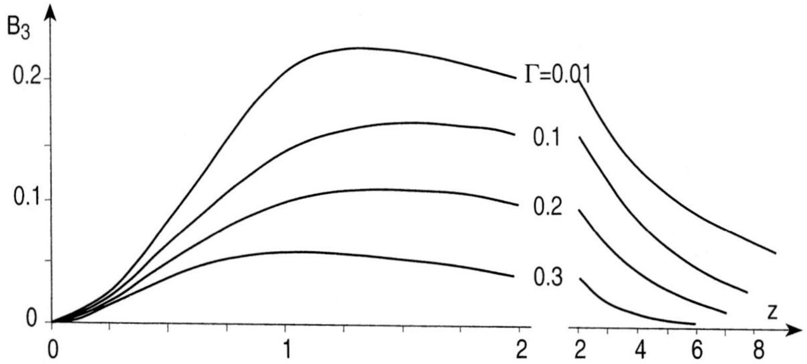

3rd harmonic amplitude B_3 over distance z

In the figure below the intensity E(z) is shown which is a sum of intensities of all harmonics:

$$E(z) = \sum_{n=1}^{\infty} B_n(z) \tag{12}$$

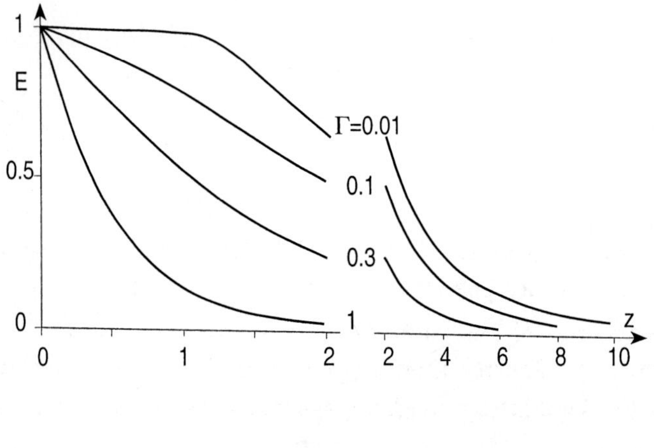

Intensity E(z) over distance z, for different values of Γ.

For most interesting cases $\Gamma = 0.01$ can be used, when nonlinear phenomena predominate over dissipative ones, the values of B_1, B_2, B_3 and E are given in Table 2.

Table 2

Amplitudes of harmonics $B_n(z)$; n=1,2,3; and intensity of signal E(z) at Γ=0.01

z	B_1	B_2	B_3	E
0.00	1.000	0.000	0.000	1.000
0.25	0.991	0.122	0.023	0.998
0.50	0.968	0.229	0.081	0.996
0.75	0.929	0.308	0.150	0.994
1.00	0.878	0.351	0.204	0.989
1.25	0.817	0.359	0.225	0.949
1.50	0.755	0.348	0.224	0.854
1.75	0.698	0.330	0.215	0.751
2.00	0.646	0.310	0.204	0.655
2.50	0.561	0.274	0.181	0.502
3.00	0.494	0.243	0.162	0.392
4.00	0.397	0.197	0.131	0.255
6.00	0.285	0.142	0.095	0.131
10.00	0.182	0.091	0.060	0.053
15.00	0.125	0.062	0.041	0.025
20.00	0.095	0.047	0.032	0.014
25.00	0.077	0.038	0.025	0.009
30.00	0.064	0.032	0.021	0.006

In the figures below, the transformation process of pulse signals is shown, both for monopolar and bipolar pulses. The initial monopolar pulse has a Gaussian form $V(z=0)=\exp(-\theta^2)$. Its transformation is calculated for Γ=0.1 at distances z= 0; 0.4; 0.7; 1; 2; 5; 10; 15; 20 (curves n=1-9, correspondingly). Nonlinearity transforms the profile from Gaussian to triangular form (see curves 5 and 6). Thereafter, the broadening of the pulse and its leading front go on.

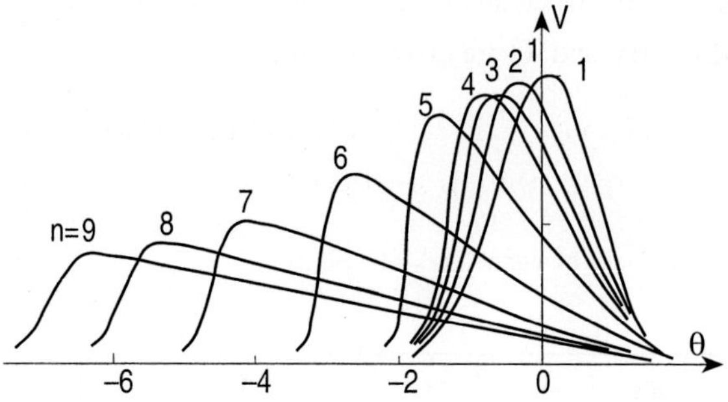

Development of monopolar pulse with distance values z_n (see text for values).

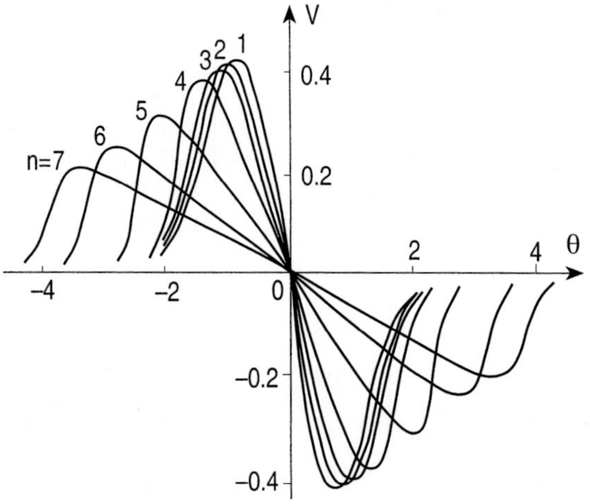

Development of bipolar pulse with distance values z_n (see text for values).

The initial bipolar pulse has a form $V(z=0,\theta)=-\theta\exp(-\theta^2)$. Its evolution is calculated for $\Gamma = 0.05$ at distances z= 0.4; 0.7; 1; 2; 5; 1; 10; 15 (curves 1-7, correspondingly). The nonlinearity leads to the transformation of this pulse to N- waves which is the typical form of pulses generated by supersonic aircraft or by explosion.

S.4 One-dimensional nonlinear waves in a dissipative medium

The general form of the differential equation describing nonlinear propagation through viscous and heat-conducting media is:

$$\frac{\partial v}{\partial x}+\frac{v}{2}\frac{d}{dx}\ln S(x)-\frac{\varepsilon}{c_0^2}v\frac{\partial v}{\partial \tau}=\frac{b}{2c_0^3\rho_0}\frac{\partial^2 v}{\partial \tau^2} \tag{1}$$

Here $S(x)$ is the cross-section of a tube, concentrator or horn varying with the distance x. In addition, all the parameters of a medium (c_0, ρ_0, ε and b) can depend on x.

Using the dimensionless variables:

$$V=\frac{v}{v_0}\sqrt{\frac{S(x)}{S(0)}} \quad ; \quad \theta=\omega\tau \quad ; \quad z=\omega\, v_0\int_0^x \frac{\varepsilon}{c_0^3\rho_0}\sqrt{\frac{S(0)}{S(s)}}\,ds \tag{2}$$

one can reduce eq.(1) to the normalized form:

$$\frac{\partial V}{\partial z}-V\frac{\partial V}{\partial \theta}=\Gamma(z)\frac{\partial^2 V}{\partial \theta^2} \tag{3}$$

Here:

$$\Gamma(z)=\left[\frac{b\omega}{2\varepsilon c_0\rho_0 v_0}\sqrt{\frac{S(x)}{S(0)}}\right]_{x=x(z)} \tag{4}$$

is Goldberg's number depending on distance z.

For an exponential concentrator: $$S(x)=S(0)\exp(-2x/x_0) \tag{5}$$

the corresponding form of generalized Burgers' eq.(3):

$$\frac{\partial V}{\partial z}-V\frac{\partial V}{\partial \theta}=\frac{\Gamma_0}{1+z/z_0}\frac{\partial^2 V}{\partial \theta^2} \quad ; \quad \Gamma_0=\frac{b\omega}{2\varepsilon c_0\rho_0 v_0} \quad ; \quad z_0=\frac{\varepsilon}{c_0^2}\omega v_0 x_0 \tag{6}$$

describes the steady-state nonlinear wave:

$$V(z,\theta)=\frac{f(\theta)}{1+z/z_0} \tag{7}$$

The form of the temporal profile $f(\theta)$ is fixed.

The eqs.(1), (3), among other processes, can describe the propagation of spherical and cylindrical waves for the cross section $$S(x)=S(0)(1\pm x/x_0)^n \tag{8}$$
where n=2 corresponds to a spherical wave and n=1 to a cylindrical wave. The plus sign must be chosen for the diverging wave, $0<x<\infty$, and the minus sign for the wave converging to the center or to the axis, $0<x<x_0$.

For example, the spherically converging wave is governed by the equation:

$$\frac{\partial V}{\partial z} - V\frac{\partial V}{\partial \theta} = \Gamma_0 \exp(-z/z_0)\frac{\partial^2 V}{\partial \theta^2} \quad ; \quad 0 \le z < \infty \tag{9}$$

and the cylindrically diverging wave by the equation:

$$\frac{\partial V}{\partial z} - V\frac{\partial V}{\partial \theta} = \Gamma_0 (1 + z/z_0)\frac{\partial^2 V}{\partial \theta^2} \quad ; \quad 0 \le z < \infty \tag{10}$$

The transformation of half-period of initial harmonic wave propagating as a spherically-diverging one is shown below for $\Gamma_0 = 0.1$ and $z_0 = 10$. The phenomena of twofold shock formation is illustrated here. As the distance increases, the steeping of the leading front goes on (see curves $x/x_s=1$ and 1.8). Then nonlinear absorption and dissipation lead to the decrease of peak and smoothing of the front. Thereafter, at the convergence to the focus (see curves $x/x_s=$ 7; 8; 8.65), the peak increases and the steep front forms again.

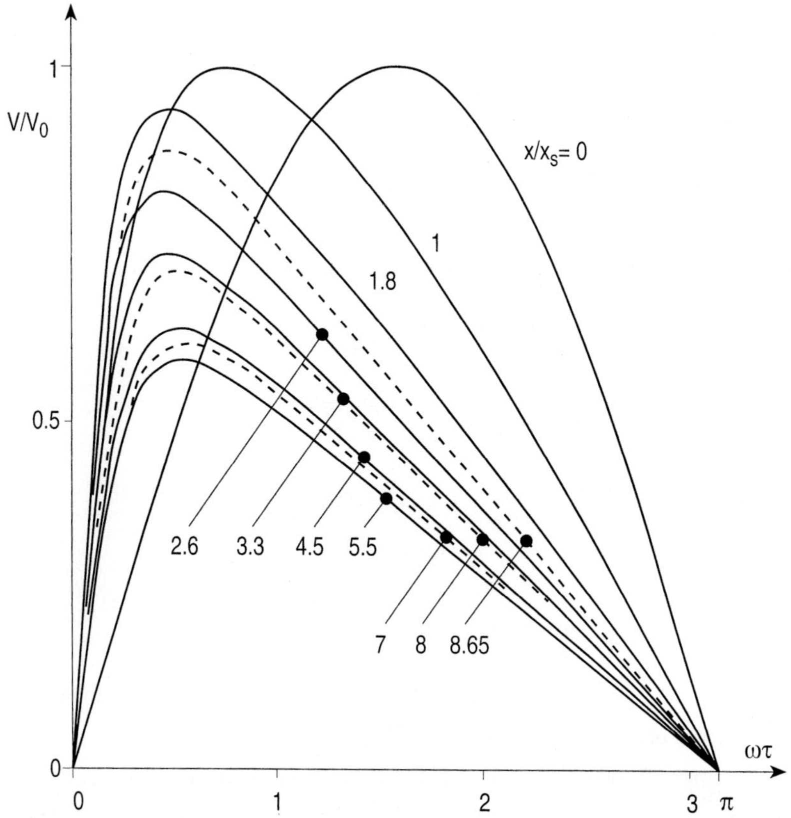

Index

The material of the Index is arranged according to the Chapters, because it is supposed that the reader is searching within some context. Capitalised entries (if not names) indicate Sections.

B General Linear Fluid Acoustics :

C Equivalent Networks :

D Reflection of Sound :

E Scattering of Sound :

F Radiation of Sound :

G Porous Absorbers :

H Compound Absorbers :

I Sound Transmission :

J Duct Acoustics :

K Muffler Acoustics :

I

L

M

N

P

Q

R

S

T

U

V

L Capsules and Cabins :

M Room Acoustics :

N Flow Acoustics :

O Analytical and Numerical Methods in Acoustics :

P Variational Principles in Acoustics :

Q Elasto-Acoustics :

R Ultrasound Absorption in Solids :

S Nonlinear Acoustics :